Quantum Mechanics

Second Edition

Contents

Library of Congress Cataloging-in-Publication Data

Zettili, Nouredine.
Quantum Mechanics: concepts and applications / Nouredine Zettili. – 2nd ed.
        p. cm.
Includes bibliographical references and index.
ISBN 978-0-470-02678-6 (cloth: alk. paper) – ISBN 978-0-470-02679-3 (pbk.: alk. paper)
1. Quantum theory. I. Title
QC174.12.Z47 2009
530.12 – dc22

2008045022

A catalogue record for this book is available from the British Library

Produced from LaTeX files supplied by the author

ISBN: 978-0-470-02678-6 (H/B)
        978-0-470-02679-3 (P/B)

Quantum Mechanics Concepts and Applications

Second Edition

Nouredine Zettili
Jacksonville State University, Jacksonville, USA

A John Wiley and Sons, Ltd., Publication

Preface

Preface to the Second Edition

It has been eight years now since the appearance of the first edition of this book in 2001. During this time, many courteous users—professors who have been adopting the book, researchers, and students—have taken the time and care to provide me with valuable feedback about the book. In preparing the second edition, I have taken into consideration the generous feedback I have received from these users. To them, and from the very outset, I want to express my deep sense of gratitude and appreciation.

The underlying focus of the book has remained the same: to provide a well-structured and self-contained, yet concise, text that is backed by a rich collection of fully solved examples and problems illustrating various aspects of nonrelativistic quantum mechanics. The book is intended to achieve a double aim: on the one hand, to provide instructors with a pedagogically suitable teaching tool and, on the other, to help students not only master the underpinnings of the theory but also become effective practitioners of quantum mechanics.

Although the overall structure and contents of the book have remained the same upon the insistence of numerous users, I have carried out a number of streamlining, surgical type changes in the second edition. These changes were aimed at fixing the weaknesses (such as typos) detected in the first edition while reinforcing and improving on its strengths. I have introduced a number of sections, new examples and problems, and new material; these are spread throughout the text. Additionally, I have operated substantive revisions of the exercises at the end of the chapters; I have added a number of new exercises, jettisoned some, and streamlined the rest. I may underscore the fact that the collection of end-of-chapter exercises has been thoroughly classroom tested for a number of years now.

The book has now a collection of almost six hundred examples, problems, and exercises. Every chapter contains: (a) a number of solved examples each of which is designed to illustrate a specific concept pertaining to a particular section within the chapter, (b) plenty of fully solved problems (which come at the end of every chapter) that are generally comprehensive and, hence, cover several concepts at once, and (c) an abundance of unsolved exercises intended for home-work assignments. Through this rich collection of examples, problems, and exercises, I want to empower the student to become an independent learner and an adept practitioner of quantum mechanics. Being able to solve problems is an unfailing evidence of a real understanding of the subject.

The second edition is backed by useful resources designed for instructors adopting the book (please contact the author or Wiley to receive these free resources).

The material in this book is suitable for three semesters—a two-semester undergraduate course and a one-semester graduate course. A pertinent question arises: How to actually use the book in an undergraduate or graduate course(s)? There is no simple answer to this question as this depends on the background of the students and on the nature of the course(s) at hand. First, I want to underscore this important observation: As the book offers an abundance of information, every instructor should certainly select the topics that will be most relevant to her/his students; going systematically over all the sections of a particular chapter (notably Chapter 2), one might run the risk of getting bogged down and, hence, ending up spending too much time on technical topics. Instead, one should be highly selective. For instance, for a one-semester course where the students have not taken modern physics before, I would recommend to cover these topics: Sections 1.1–1.6; 2.2.2, 2.2.4, 2.3, 2.4.1–2.4.8, 2.5.1, 2.5.3, 2.6.1–2.6.2, 2.7; 3.2–3.6; 4.3–4.8; 5.2–5.4, 5.6–5.7; and 6.2–6.4. However, if the students have taken modern physics before, I would skip Chapter 1 altogether and would deal with these sections: 2.2.2, 2.2.4, 2.3, 2.4.1–2.4.8, 2.5.1, 2.5.3, 2.6.1–2.6.2, 2.7; 3.2–3.6; 4.3–4.8; 5.2–5.4, 5.6–5.7; 6.2–6.4; 9.2.1–9.2.2, 9.3, and 9.4. For a two-semester course, I think the instructor has plenty of time and flexibility to maneuver and select the topics that would be most suitable for her/his students; in this case, I would certainly include some topics from Chapters 7–11 as well (but not all sections of these chapters as this would be unrealistically time demanding). On the other hand, for a one-semester graduate course, I would cover topics such as Sections 1.7–1.8; 2.4.9, 2.6.3–2.6.5; 3.7–3.8; 4.9; and most topics of Chapters 7–11.

Acknowledgments

I have received very useful feedback from many users of the first edition; I am deeply grateful and thankful to everyone of them. I would like to thank in particular Richard Lebed (Arizona State University) who has worked selflessly and tirelessly to provide me with valuable comments, corrections, and suggestions. I want also to thank Jearl Walker (Cleveland State University)—the author of *The Flying Circus of Physics* and of the Halliday–Resnick–Walker classics, *Fundamentals of Physics*—for having read the manuscript and for his wise suggestions; Milton Cha (University of Hawaii System) for having proofread the entire book; Felix Chen (Powerwave Technologies, Santa Ana) for his reading of the first 6 chapters. My special thanks are also due to the following courteous users/readers who have provided me with lists of typos/errors they have detected in the first edition: Thomas Sayetta (East Carolina University), Moritz Braun (University of South Africa, Pretoria), David Berkowitz (California State University at Northridge), John Douglas Hey (University of KwaZulu-Natal, Durban, South Africa), Richard Arthur Dudley (University of Calgary, Canada), Andrea Durlo (founder of the A.I.F. (Italian Association for Physics Teaching), Ferrara, Italy), and Rick Miranda (Netherlands). My deep sense of gratitude goes to M. Bulut (University of Alabama at Birmingham) and to Heiner Mueller-Krumbhaar (Forschungszentrum Juelich, Germany) and his Ph.D. student C. Gugenberger for having written and tested the C++ code listed in Appendix C, which is designed to solve the Schrödinger equation for a one-dimensional harmonic oscillator and for an infinite square-well potential.

Finally, I want to thank my editors, Dr. Andy Slade, Celia Carden, and Alexandra Carrick, for their consistent hard work and friendly support throughout the course of this project.

N. Zettili
Jacksonville State University, USA
January 2009

Preface to the First Edition

Books on quantum mechanics can be grouped into two main categories: textbooks, where the focus is on the formalism, and purely problem-solving books, where the emphasis is on applications. While many fine textbooks on quantum mechanics exist, problem-solving books are far fewer. It is not my intention to merely add a text to either of these two lists. My intention is to combine the two formats into a single text which includes the ingredients of both a textbook and a problem-solving book. Books in this format are practically nonexistent. I have found this idea particularly useful, for it gives the student easy and quick access not only to the essential elements of the theory but also to its practical aspects in a unified setting.

During many years of teaching quantum mechanics, I have noticed that students generally find it easier to learn its underlying ideas than to handle the practical aspects of the formalism. Not knowing how to calculate and extract numbers out of the formalism, one misses the full power and utility of the theory. Mastering the techniques of problem-solving is an essential part of learning physics. To address this issue, the problems solved in this text are designed to teach the student how to calculate. No real mastery of quantum mechanics can be achieved without learning how to derive and calculate quantities.

In this book I want to achieve a double aim: to give a self-contained, yet concise, presentation of most issues of nonrelativistic quantum mechanics, and to offer a rich collection of fully solved examples and problems. This unified format is not without cost. Size! Judicious care has been exercised to achieve conciseness without compromising coherence and completeness.

This book is an outgrowth of undergraduate and graduate lecture notes I have been supplying to my students for about one decade; the problems included have been culled from a large collection of homework and exam exercises I have been assigning to the students. It is intended for senior undergraduate and first-year graduate students. The material in this book could be covered in three semesters: Chapters 1 to 5 (excluding Section 3.7) in a one-semester undergraduate course; Chapter 6, Section 7.3, Chapter 8, Section 9.2 (excluding fine structure and the anomalous Zeeman effect), and Sections 11.1 to 11.3 in the second semester; and the rest of the book in a one-semester graduate course.

The book begins with the experimental basis of quantum mechanics, where we look at those atomic and subatomic phenomena which confirm the failure of classical physics at the microscopic scale and establish the need for a new approach. Then come the mathematical tools of quantum mechanics such as linear spaces, operator algebra, matrix mechanics, and eigenvalue problems; all these are treated by means of Dirac's bra-ket notation. After that we discuss the formal foundations of quantum mechanics and then deal with the exact solutions of the Schrödinger equation when applied to one-dimensional and three-dimensional problems. We then look at the stationary and the time-dependent approximation methods and, finally, present the theory of scattering.

I would like to thank Professors Ismail Zahed (University of New York at Stony Brook) and Gerry O. Sullivan (University College Dublin, Ireland) for their meticulous reading and comments on an early draft of the manuscript. I am grateful to the four anonymous reviewers who provided insightful comments and suggestions. Special thanks go to my editor, Dr Andy Slade, for his constant support, encouragement, and efficient supervision of this project.

I want to acknowledge the hospitality of the Center for Theoretical Physics of MIT, Cambridge, for the two years I spent there as a visitor. I would like to thank in particular Professors Alan Guth, Robert Jaffee, and John Negele for their support.

Note to the student

We are what we repeatedly do. Excellence, then, is not an act, but a habit.

Aristotle

No one expects to learn swimming without getting wet. Nor does anyone expect to learn it by merely reading books or by watching others swim. Swimming cannot be learned without practice. There is absolutely no substitute for throwing yourself into water and training for weeks, or even months, till the exercise becomes a smooth reflex.

Similarly, physics *cannot be learned passively*. Without tackling various challenging problems, the student has no other way of testing the quality of his or her understanding of the subject. Here is where the student gains the sense of satisfaction and involvement produced by a genuine understanding of the underlying principles. *The ability to solve problems is the best proof of mastering the subject.* As in swimming, the more you solve problems, the more you sharpen and fine-tune your problem-solving skills.

To derive full benefit from the examples and problems solved in the text, avoid consulting the solution too early. If you cannot solve the problem after your first attempt, try again! If you look up the solution only after several attempts, it will remain etched in your mind for a long time. But if you manage to solve the problem on your own, you should still compare your solution with the book's solution. You might find a shorter or more elegant approach.

One important observation: as the book is laden with a rich collection of fully solved examples and problems, one should absolutely avoid the temptation of memorizing the various techniques and solutions; instead, one should focus on understanding the concepts and the underpinnings of the formalism involved. It is not my intention in this book to teach the student a number of tricks or techniques for acquiring good grades in quantum mechanics classes without genuine understanding or mastery of the subject; that is, I didn't mean to teach the student how to pass quantum mechanics exams without a deep and lasting understanding. However, the student who focuses on understanding the underlying foundations of the subject and on reinforcing that by solving numerous problems and thoroughly understanding them will doubtlessly achieve a double aim: reaping good grades as well as obtaining a sound and long-lasting education.

N. Zettili

Chapter 1

Origins of Quantum Physics

In this chapter we are going to review the main physical ideas and experimental facts that defied classical physics and led to the birth of quantum mechanics. The introduction of quantum mechanics was prompted by the failure of classical physics in explaining a number of microphysical phenomena that were observed at the end of the nineteenth and early twentieth centuries.

1.1 Historical Note

At the end of the nineteenth century, physics consisted essentially of classical mechanics, the theory of electromagnetism[1], and thermodynamics. Classical mechanics was used to predict the dynamics of *material bodies*, and Maxwell's electromagnetism provided the proper framework to study *radiation*; *matter* and *radiation* were described in terms of *particles* and *waves*, respectively. As for the interactions between matter and radiation, they were well explained by the Lorentz force or by thermodynamics. The overwhelming success of classical physics— classical mechanics, classical theory of electromagnetism, and thermodynamics—made people believe that the ultimate description of nature had been achieved. It seemed that all known physical phenomena could be explained within the framework of the general theories of matter and radiation.

At the turn of the twentieth century, however, classical physics, which had been quite unassailable, was seriously challenged on two major fronts:

- **Relativistic domain:** Einstein's 1905 theory of relativity showed that the validity of Newtonian mechanics ceases at very high speeds (i.e., at speeds comparable to that of light).

- **Microscopic domain:** As soon as new experimental techniques were developed to the point of probing atomic and subatomic structures, it turned out that classical physics fails miserably in providing the proper explanation for several newly discovered phenomena. It thus became evident that the validity of classical physics ceases at the microscopic level and that *new* concepts had to be invoked to describe, for instance, the structure of atoms and molecules and how light interacts with them.

---

[1] Maxwell's theory of electromagnetism had unified the, then ostensibly different, three branches of physics: electricity, magnetism, and optics.

The failure of classical physics to explain several microscopic phenomena—such as black-body radiation, the photoelectric effect, atomic stability, and atomic spectroscopy—had cleared the way for seeking new ideas outside its purview.

The first real breakthrough came in 1900 when Max Planck introduced the concept of the *quantum* of energy. In his efforts to explain the phenomenon of blackbody radiation, he succeeded in reproducing the experimental results only after postulating that the energy exchange between *radiation* and its surroundings takes place in *discrete*, or *quantized*, amounts. He argued that the energy exchange between an *electromagnetic wave* of frequency ν and matter occurs *only in integer multiples* of $h\nu$, which he called the energy of a *quantum*, where h is a fundamental constant called *Planck's constant*. The quantization of electromagnetic radiation turned out to be an idea with far-reaching consequences.

Planck's idea, which gave an accurate explanation of blackbody radiation, prompted new thinking and triggered an avalanche of new discoveries that yielded solutions to the most outstanding problems of the time.

In 1905 Einstein provided a powerful consolidation to Planck's quantum concept. In trying to understand the photoelectric effect, Einstein recognized that Planck's idea of the quantization of the *electromagnetic* waves must be valid for *light* as well. So, following Planck's approach, he posited that *light itself is made of discrete bits of energy (or tiny particles)*, called *photons*, each of energy $h\nu$, ν being the frequency of the light. The introduction of the photon concept enabled Einstein to give an elegantly accurate explanation to the photoelectric problem, which had been waiting for a solution ever since its first experimental observation by Hertz in 1887.

Another seminal breakthrough was due to Niels Bohr. Right after Rutherford's experimental discovery of the atomic nucleus in 1911, and combining Rutherford's atomic model, Planck's quantum concept, and Einstein's photons, Bohr introduced in 1913 his model of the hydrogen atom. In this work, he argued that atoms can be found only in *discrete states* of energy and that the interaction of atoms with radiation, i.e., the emission or absorption of radiation by atoms, takes place only in *discrete amounts* of $h\nu$ because it results from transitions of the atom between its various discrete energy states. This work provided a satisfactory explanation to several outstanding problems such as atomic stability and atomic spectroscopy.

Then in 1923 Compton made an important discovery that gave the most conclusive confirmation for the corpuscular aspect of light. By scattering X-rays with electrons, he confirmed that the X-ray photons behave like particles with momenta $h\nu/c$; ν is the frequency of the X-rays.

This series of breakthroughs—due to Planck, Einstein, Bohr, and Compton—gave both the theoretical foundations as well as the conclusive experimental confirmation for the particle aspect of waves; that is, the concept that waves exhibit particle behavior at the microscopic scale. At this scale, classical physics fails not only quantitatively but even qualitatively and conceptually.

As if things were not bad enough for classical physics, de Broglie introduced in 1923 another powerful new concept that classical physics could not reconcile: he postulated that not only does radiation exhibit particle-like behavior but, conversely, *material particles* themselves display *wave-like* behavior. This concept was confirmed experimentally in 1927 by Davisson and Germer; they showed that interference patterns, a property of waves, can be obtained with material particles such as electrons.

Although Bohr's model for the atom produced results that agree well with experimental spectroscopy, it was criticized for lacking the ingredients of a theory. Like the "quantization" scheme introduced by Planck in 1900, the postulates and assumptions adopted by Bohr in 1913 were quite arbitrary and do not follow from the first principles of a theory. It was the dissatisfaction with the arbitrary nature of Planck's idea and Bohr's postulates as well as the need to fit them within the context of a consistent theory that had prompted Heisenberg and Schrödinger to search for the theoretical foundation underlying these new ideas. By 1925 their efforts paid off: they skillfully welded the various experimental findings as well as Bohr's postulates into a refined theory: *quantum mechanics*. In addition to providing an accurate reproduction of the existing experimental data, this theory turned out to possess an astonishingly reliable prediction power which enabled it to explore and unravel many uncharted areas of the microphysical world. This new theory had put an end to twenty five years (1900–1925) of patchwork which was dominated by the ideas of Planck and Bohr and which later became known as the old quantum theory.

Historically, there were two independent formulations of quantum mechanics. The first formulation, called *matrix mechanics*, was developed by Heisenberg (1925) to describe atomic structure starting from the observed spectral lines. Inspired by Planck's quantization of waves and by Bohr's model of the hydrogen atom, Heisenberg founded his theory on the notion that the only allowed values of energy exchange between microphysical systems are those that are discrete: quanta. Expressing dynamical quantities such as energy, position, momentum and angular momentum in terms of matrices, he obtained an eigenvalue problem that describes the dynamics of microscopic systems; the diagonalization of the Hamiltonian matrix yields the energy spectrum and the state vectors of the system. Matrix mechanics was very successful in accounting for the discrete quanta of light emitted and absorbed by atoms.

The second formulation, called *wave mechanics*, was due to Schrödinger (1926); it is a generalization of the de Broglie postulate. This method, more intuitive than matrix mechanics, describes the dynamics of microscopic matter by means of a *wave equation*, called the *Schrödinger equation*; instead of the matrix eigenvalue problem of Heisenberg, Schrödinger obtained a differential equation. The solutions of this equation yield the energy spectrum and the wave function of the system under consideration. In 1927 Max Born proposed his *probabilistic* interpretation of wave mechanics: he took the square moduli of the wave functions that are solutions to the Schrödinger equation and he interpreted them as *probability densities*.

These two ostensibly different formulations—Schrödinger's *wave* formulation and Heisenberg's *matrix* approach—were shown to be equivalent. Dirac then suggested a more general formulation of quantum mechanics which deals with abstract objects such as kets (state vectors), bras, and operators. The representation of Dirac's formalism in a *continuous* basis—the position or momentum representations—gives back Schrödinger's wave mechanics. As for Heisenberg's matrix formulation, it can be obtained by representing Dirac's formalism in a *discrete* basis. In this context, the approaches of Schrödinger and Heisenberg represent, respectively, the *wave* formulation and the *matrix* formulation of the general theory of quantum mechanics.

Combining special relativity with quantum mechanics, Dirac derived in 1928 an equation which describes the motion of electrons. This equation, known as Dirac's equation, predicted the existence of an antiparticle, the positron, which has similar properties, but opposite charge, with the electron; the positron was discovered in 1932, four years after its prediction by quantum mechanics.

In summary, quantum mechanics is the theory that describes the dynamics of matter at the microscopic scale. Fine! But is it that important to learn? This is no less than an otiose question, for quantum mechanics is the *only* valid framework for describing the microphysical world. It is vital for understanding the physics of solids, lasers, semiconductor and superconductor devices, plasmas, etc. In short, quantum mechanics is the founding basis of all modern physics: solid state, molecular, atomic, nuclear, and particle physics, optics, thermodynamics, statistical mechanics, and so on. Not only that, it is also considered to be the foundation of chemistry and biology.

1.2 Particle Aspect of Radiation

According to classical physics, a particle is characterized by an energy E and a momentum $\vec{p}$, whereas a wave is characterized by an amplitude and a wave vector $\vec{k}$ ($|\vec{k}| = 2\pi/\lambda$) that specifies the direction of propagation of the wave. Particles and waves exhibit entirely different behaviors; for instance, the "particle" and "wave" properties are mutually exclusive. We should note that waves can exchange *any* (continuous) amount of energy with particles.

In this section we are going to see how these rigid concepts of classical physics led to its failure in explaining a number of microscopic phenomena such as blackbody radiation, the photoelectric effect, and the Compton effect. As it turned out, these phenomena could only be explained by abandoning the rigid concepts of classical physics and introducing a new concept: the *particle* aspect of radiation.

1.2.1 Blackbody Radiation

At issue here is how radiation interacts with matter. When heated, a solid object glows and emits thermal radiation. As the temperature increases, the object becomes red, then yellow, then white. The thermal radiation emitted by glowing solid objects consists of a *continuous* distribution of frequencies ranging from infrared to ultraviolet. The continuous pattern of the distribution spectrum is in sharp contrast to the radiation emitted by heated gases; the radiation emitted by gases has a discrete distribution spectrum: a few sharp (narrow), colored lines with no light (i.e., darkness) in between.

Understanding the continuous character of the radiation emitted by a glowing solid object constituted one of the major unsolved problems during the second half of the nineteenth century. All attempts to explain this phenomenon by means of the available theories of classical physics (statistical thermodynamics and classical electromagnetic theory) ended up in miserable failure. This problem consisted in essence of specifying the proper theory of thermodynamics that describes how energy gets exchanged between radiation and matter.

When radiation falls on an object, some of it might be absorbed and some reflected. An idealized "blackbody" is a material object that absorbs all of the radiation falling on it, and hence appears as black under reflection when illuminated from outside. When an object is heated, it radiates electromagnetic energy as a result of the thermal agitation of the electrons in its surface. The intensity of this radiation depends on its frequency and on the temperature; the light it emits ranges over the entire spectrum. An object in thermal equilibrium with its surroundings radiates as much energy as it absorbs. It thus follows that a blackbody is a perfect absorber as well as a perfect emitter of radiation.

A practical blackbody can be constructed by taking a hollow cavity whose internal walls perfectly reflect electromagnetic radiation (e.g., metallic walls) and which has a very small hole on its surface. Radiation that enters through the hole will be trapped inside the cavity and gets completely absorbed after successive reflections on the inner surfaces of the cavity. The

Figure 1.1 Spectral energy density $u(\nu, T)$ of blackbody radiation at different temperatures as a function of the frequency ν.

hole thus absorbs radiation like a black body. On the other hand, when this cavity is heated[2] to a temperature T, the radiation that leaves the hole is blackbody radiation, for the hole behaves as a perfect emitter; as the temperature increases, the hole will eventually begin to glow. To understand the radiation inside the cavity, one needs simply to analyze the spectral distribution of the radiation coming out of the hole. In what follows, the term *blackbody radiation* will then refer to the radiation leaving the hole of a heated hollow cavity; the radiation emitted by a blackbody when hot is called blackbody radiation.

By the mid-1800s, a wealth of experimental data about blackbody radiation was obtained for various objects. All these results show that, at equilibrium, the radiation emitted has a well-defined, continuous energy distribution: to each frequency there corresponds an energy density which depends neither on the chemical composition of the object nor on its shape, but only on the temperature of the cavity's walls (Figure 1.1). The energy density shows a pronounced maximum at a given frequency, which increases with temperature; that is, *the peak of the radiation spectrum occurs at a frequency that is proportional to the temperature* (1.16). This is the underlying reason behind the change in color of a heated object as its temperature increases, notably from red to yellow to white. It turned out that the explanation of the blackbody spectrum was not so easy.

A number of attempts aimed at explaining the origin of the continuous character of this radiation were carried out. The most serious among such attempts, and which made use of classical physics, were due to Wilhelm Wien in 1889 and Rayleigh in 1900. In 1879 J. Stefan found *experimentally* that the total intensity (or the total power per unit surface area) radiated by a glowing object of temperature T is given by

$$P = a\sigma T^4, \tag{1.1}$$

which is known as the Stefan–Boltzmann law, where $\sigma = 5.67 \times 10^{-8} \ \mathrm{W\,m^{-2}\,K^{-4}}$ is the

---

[2]When the walls are heated uniformly to a temperature T, they emit radiation (due to thermal agitation or vibrations of the electrons in the metallic walls).

Figure 1.2 Comparison of various spectral densities: while the Planck and experimental distributions match perfectly (solid curve), the Rayleigh–Jeans and the Wien distributions (dotted curves) agree only partially with the experimental distribution.

Stefan–Boltzmann constant, and a is a coefficient which is less than or equal to 1; in the case of a blackbody $a = 1$. Then in 1884 Boltzmann provided a *theoretical* derivation for Stefan's experimental law by combining thermodynamics and Maxwell's theory of electromagnetism.

Wien's energy density distribution
Using thermodynamic arguments, Wien took the Stefan–Boltzmann law (1.1) and in 1894 he extended it to obtain the energy density per unit frequency of the emitted blackbody radiation:

$$u(\nu, T) = A\nu^3 e^{-\beta\nu/T},\tag{1.2}$$

where A and β are empirically defined parameters (they can be adjusted to fit the experimental data). **Note:** $u(\nu, T)$ has the dimensions of an energy per unit volume per unit frequency; its SI units are $J\,m^{-3}\,Hz^{-1}$. Although Wien's formula fits the high-frequency data remarkably well, it fails badly at low frequencies (Figure 1.2).

Rayleigh's energy density distribution
In his 1900 attempt, Rayleigh focused on understanding the nature of the electromagnetic radiation inside the cavity. He considered the radiation to consist of standing waves having a temperature T with nodes at the metallic surfaces. These standing waves, he argued, are equivalent to harmonic oscillators, for they result from the harmonic oscillations of a large number of electrical charges, electrons, that are present in the walls of the cavity. When the cavity is in thermal equilibrium, the electromagnetic energy density inside the cavity is equal to the energy density of the charged particles in the walls of the cavity; the average total energy of the radiation leaving the cavity can be obtained by multiplying the average energy of the oscillators by the number of modes (standing waves) of the radiation in the frequency interval ν to $\nu + d\nu$:

$$N(\nu) = \frac{8\pi\nu^2}{c^3},\tag{1.3}$$

where $c = 3 \times 10^8$ m s^{-1} is the speed of light; the quantity $(8\pi v^2/c^3)dv$ gives the number of modes of oscillation per unit volume in the frequency range v to $v+dv$. So the electromagnetic energy density in the frequency range v to $v + dv$ is given by

$$u(v, T) = N(v)\langle E \rangle = \frac{8\pi v^2}{c^3}\langle E \rangle, \tag{1.4}$$

where $\langle E \rangle$ is the average energy of the oscillators present on the walls of the cavity (or of the electromagnetic radiation in that frequency interval); the temperature dependence of $u(v, T)$ is buried in $\langle E \rangle$.

How does one calculate $\langle E \rangle$? According to the equipartition theorem of classical thermodynamics, all oscillators in the cavity have the same mean energy, irrespective of their frequencies[3]:

$$\langle E \rangle = \frac{\int_0^\infty E e^{-E/kT} dE}{\int_0^\infty e^{-E/kT} dE} = kT, \tag{1.5}$$

where $k = 1.3807 \times 10^{-23}$ J K^{-1} is the Boltzmann constant. An insertion of (1.5) into (1.4) leads to the Rayleigh–Jeans formula:

$$u(v, T) = \frac{8\pi v^2}{c^3} kT. \tag{1.6}$$

Except for low frequencies, this law is in complete disagreement with experimental data: $u(v, T)$ as given by (1.6) *diverges* for high values of v, whereas experimentally it must be finite (Figure 1.2). Moreover, if we integrate (1.6) over all frequencies, the integral *diverges*. This implies that the cavity contains an *infinite* amount of energy. This result is absurd. Historically, this was called the *ultraviolet catastrophe*, for (1.6) diverges for *high* frequencies (i.e., in the *ultraviolet* range)—a real catastrophical failure of classical physics indeed! The origin of this failure can be traced to the derivation of the average energy (1.5). It was founded on an erroneous premise: the energy exchange between radiation and matter is *continuous*; any amount of energy can be exchanged.

Planck's energy density distribution
By devising an ingenious scheme—interpolation between Wien's rule and the Rayleigh–Jeans rule—Planck succeeded in 1900 in avoiding the ultraviolet catastrophe and proposed an accurate description of blackbody radiation. In sharp contrast to Rayleigh's assumption that a standing wave can exchange *any* amount (continuum) of energy with matter, Planck considered that the energy exchange between radiation and matter must be *discrete*. He then *postulated* that the energy of the radiation (of frequency v) emitted by the oscillating charges (from the walls of the cavity) must come *only* in *integer multiples* of hv:

$$E = nhv, \qquad n = 0, 1, 2, 3, \cdots, \tag{1.7}$$

where h is a universal constant and hv is the energy of a *"quantum"* of radiation (v represents the frequency of the oscillating charge in the cavity's walls as well as the frequency of the radiation emitted from the walls, because the frequency of the radiation emitted by an oscillating charged particle is equal to the frequency of oscillation of the particle itself). That is, the energy of an oscillator of natural frequency v (which corresponds to the energy of a charge

---

[3] Using a variable change $\beta = 1/(kT)$, we have $\langle E \rangle = -\frac{\partial}{\partial \beta}\ln\left(\int_0^\infty e^{-\beta E} dE\right) = -\frac{\partial}{\partial \beta}\ln(1/\beta) = 1/\beta \equiv kT$.

oscillating with a frequency ν) must be an *integral multiple* of $h\nu$; note that $h\nu$ is not the same for all oscillators, because it depends on the frequency of each oscillator. Classical mechanics, however, puts no restrictions whatsoever on the frequency, and hence on the energy, an oscillator can have. The energy of oscillators, such as pendulums, mass–spring systems, and electric oscillators, varies continuously in terms of the frequency. Equation (1.7) is known as *Planck's quantization rule* for energy or *Planck's postulate*.

So, assuming that the energy of an oscillator is quantized, Planck showed that the *correct* thermodynamic relation for the average energy can be obtained by merely replacing the integration of (1.5)—that corresponds to an energy continuum—by a *discrete* summation corresponding to the discreteness of the oscillators' energies[4]:

$$\langle E \rangle = \frac{\sum_{n=0}^{\infty} nh\nu e^{-nh\nu/kT}}{\sum_{n=0}^{\infty} e^{-nh\nu/kT}} = \frac{h\nu}{e^{h\nu/kT} - 1}, \tag{1.8}$$

and hence, by inserting (1.8) into (1.4), the energy density per unit frequency of the radiation emitted from the hole of a cavity is given by

$$u(\nu, T) = \frac{8\pi\nu^2}{c^3} \frac{h\nu}{e^{h\nu/kT} - 1}. \tag{1.9}$$

This is known as *Planck's distribution*. It gives an exact fit to the various experimental radiation distributions, as displayed in Figure 1.2. The numerical value of h obtained by fitting (1.9) with the experimental data is $h = 6.626 \times 10^{-34}$ J s. We should note that, as shown in (1.12), we can rewrite Planck's energy density (1.9) to obtain the energy density per unit wavelength

$$\tilde{u}(\lambda, T) = \frac{8\pi hc}{\lambda^5} \frac{1}{e^{hc/\lambda kT} - 1}. \tag{1.10}$$

Let us now look at the behavior of Planck's distribution (1.9) in the limits of both low and high frequencies, and then try to establish its connection to the relations of Rayleigh–Jeans, Stefan–Boltzmann, and Wien. First, in the case of very low frequencies $h\nu \ll kT$, we can show that (1.9) reduces to the Rayleigh–Jeans law (1.6), since $\exp(h\nu/kT) \simeq 1 + h\nu/kT$. Moreover, if we integrate Planck's distribution (1.9) over the whole spectrum (where we use a change of variable $x = h\nu/kT$ and make use of a special integral[5]), we obtain the total energy density which is expressed in terms of Stefan–Boltzmann's total power per unit surface area (1.1) as follows:

$$\int_0^\infty u(\nu, T)d\nu = \frac{8\pi h}{c^3} \int_0^\infty \frac{\nu^3}{e^{h\nu/kT} - 1} d\nu = \frac{8\pi k^4 T^4}{h^3 c^3} \int_0^\infty \frac{x^3}{e^x - 1} dx = \frac{8\pi^5 k^4}{15h^3 c^3} T^4 = \frac{4}{c}\sigma T^4, \tag{1.11}$$

where $\sigma = 2\pi^5 k^4 / 15h^3 c^2 = 5.67 \times 10^{-8}$ W m^{-2} K^{-4} is the Stefan–Boltzmann constant. In this way, Planck's relation (1.9) leads to a *finite* total energy density of the radiation emitted from a blackbody, and hence avoids the ultraviolet catastrophe. Second, in the limit of *high* frequencies, we can easily ascertain that Planck's distribution (1.9) yields Wien's rule (1.2).

In summary, the spectrum of the blackbody radiation reveals the quantization of radiation, notably the particle behavior of electromagnetic waves.

---

[4]To derive (1.8) one needs: $1/(1-x) = \sum_{n=0}^{\infty} x^n$ and $x/(1-x)^2 = \sum_{n=0}^{\infty} nx^n$ with $x = e^{-h\nu/kT}$.

[5]In integrating (1.11), we need to make use of this integral: $\int_0^{+\infty} \frac{x^3}{e^x - 1} dx = \frac{\pi^4}{15}$.

The introduction of the constant h had indeed heralded the end of classical physics and the dawn of a new era: physics of the microphysical world. Stimulated by the success of Planck's quantization of radiation, other physicists, notably Einstein, Compton, de Broglie, and Bohr, skillfully adapted it to explain a host of other outstanding problems that had been unanswered for decades.

Example 1.1 (Wien's displacement law)
(a) Show that the maximum of the Planck energy density (1.9) occurs for a wavelength of the form $\lambda_{max} = b/T$, where T is the temperature and b is a constant that needs to be estimated.
(b) Use the relation derived in (a) to estimate the surface temperature of a star if the radiation it emits has a maximum intensity at a wavelength of 446 nm. What is the intensity radiated by the star?
(c) Estimate the wavelength and the intensity of the radiation emitted by a glowing tungsten filament whose surface temperature is 3300 K.

Solution
(a) Since $v = c/\lambda$, we have $dv = |dv/(d\lambda)| \, d\lambda = (c/\lambda^2) d\lambda$; we can thus write Planck's energy density (1.9) in terms of the wavelength as follows:

$$\tilde{u}(\lambda, T) = u(v, T) \left| \frac{dv}{d\lambda} \right| = \frac{8\pi hc}{\lambda^5} \frac{1}{e^{hc/\lambda kT} - 1}. \tag{1.12}$$

The maximum of $\tilde{u}(\lambda, T)$ corresponds to $\partial \tilde{u}(\lambda, T)/\partial \lambda = 0$, which yields

$$\frac{8\pi hc}{\lambda^6} \left[-5 \left(1 - e^{-hc/\lambda kT} \right) + \frac{hc}{\lambda kT} \right] \frac{e^{hc/\lambda kT}}{\left(e^{hc/\lambda kT} - 1 \right)^2} = 0, \tag{1.13}$$

and hence

$$\frac{\alpha}{\lambda} = 5 \left(1 - e^{-\alpha/\lambda} \right), \tag{1.14}$$

where $\alpha = hc/(kT)$. We can solve this transcendental equation either graphically or numerically by writing $\alpha/\lambda = 5 - \varepsilon$. Inserting this value into (1.14), we obtain $5 - \varepsilon = 5 - 5e^{-5+\varepsilon}$, which leads to a suggestive approximate solution $\varepsilon \approx 5e^{-5} = 0.0337$ and hence $\alpha/\lambda = 5 - 0.0337 = 4.9663$. Since $\alpha = hc/(kT)$ and using the values $h = 6.626 \times 10^{-34}$ J s and $k = 1.3807 \times 10^{-23}$ J K^{-1}, we can write the wavelength that corresponds to the maximum of the Planck energy density (1.9) as follows:

$$\lambda_{max} = \frac{hc}{4.9663k} \frac{1}{T} = \frac{2898.9 \times 10^{-6} \text{ m K}}{T}. \tag{1.15}$$

This relation, which shows that λ_{max} decreases with increasing temperature of the body, is called *Wien's displacement law*. It can be used to determine the wavelength corresponding to the maximum intensity if the temperature of the body is known or, conversely, to determine the temperature of the radiating body if the wavelength of greatest intensity is known. This law can be used, in particular, to estimate the temperature of stars (or of glowing objects) from their radiation, as shown in part (b). From (1.15) we obtain

$$v_{max} = \frac{c}{\lambda_{max}} = \frac{4.9663}{h} kT. \tag{1.16}$$

This relation shows that the peak of the radiation spectrum occurs at a frequency that is proportional to the temperature.

(b) If the radiation emitted by the star has a maximum intensity at a wavelength of $\lambda_{max} = 446$ nm, its surface temperature is given by

$$T = \frac{2898.9 \times 10^{-6} \text{ m K}}{446 \times 10^{-9} \text{ m}} \simeq 6500 \text{ K}. \tag{1.17}$$

Using Stefan–Boltzmann's law (1.1), and assuming the star to radiate like a blackbody, we can estimate the total power per unit surface area emitted at the surface of the star:

$$\mathcal{P} = \sigma T^4 = 5.67 \times 10^{-8} \text{ W m}^{-2} \text{K}^{-4} \times (6500 \text{ K})^4 \simeq 101.2 \times 10^6 \text{ W m}^{-2}. \tag{1.18}$$

This is an enormous intensity which will decrease as it spreads over space.

(c) The wavelength of greatest intensity of the radiation emitted by a glowing tungsten filament of temperature 3300 K is

$$\lambda_{max} = \frac{2898.9 \times 10^{-6} \text{ m K}}{3300 \text{ K}} \simeq 878.45 \text{ nm}. \tag{1.19}$$

The intensity (or total power per unit surface area) radiated by the filament is given by

$$\mathcal{P} = \sigma T^4 = 5.67 \times 10^{-8} \text{ W m}^{-2} \text{K}^{-4} \times (3300 \text{ K})^4 \simeq 6.7 \times 10^6 \text{ W m}^{-2}. \tag{1.20}$$

1.2.2   Photoelectric Effect

The photoelectric effect provides a direct confirmation for the energy quantization of light. In 1887 Hertz discovered the photoelectric effect: electrons[6] were observed to be ejected from metals when irradiated with light (Figure 1.3a). Moreover, the following experimental laws were discovered prior to 1905:

- If the frequency of the incident radiation is smaller than the metal's threshold frequency— a frequency that depends on the properties of the metal—no electron can be emitted regardless of the radiation's intensity (Philip Lenard, 1902).

- No matter how low the intensity of the incident radiation, electrons will be ejected *instantly* the moment the frequency of the radiation exceeds the threshold frequency ν_0.

- At any frequency above ν_0, the number of electrons ejected increases with the intensity of the light but does not depend on the light's frequency.

- The kinetic energy of the ejected electrons depends on the frequency but not on the intensity of the beam; the kinetic energy of the ejected electron increases *linearly* with the incident frequency.

---

[6]In 1899 J. J. Thomson confirmed that the particles giving rise to the photoelectric effect (i.e., the particles ejected from the metals) are electrons.

Figure 1.3 (a) Photoelectric effect: when a metal is irradiated with light, electrons may get emitted. (b) Kinetic energy K of the electron leaving the metal when irradiated with a light of frequency ν; when $\nu < \nu_0$ no electron is ejected from the metal regardless of the intensity of the radiation.

These experimental findings cannot be explained within the context of a purely classical picture of radiation, notably the dependence of the effect on the threshold frequency. According to classical physics, any (continuous) amount of energy can be exchanged with matter. That is, since the intensity of an electromagnetic wave is proportional to the square of its amplitude, *any frequency* with sufficient intensity can supply the necessary energy to free the electron from the metal.

But what would happen when using a *weak* light source? According to classical physics, an electron would keep on absorbing energy—at a *continuous rate*—until it gained a sufficient amount; then it would leave the metal. If this argument is to hold, then when using very weak radiation, the photoelectric effect would not take place for a long time, possibly hours, until an electron gradually accumulated the necessary amount of energy. This conclusion, however, disagrees utterly with experimental observation. Experiments were conducted with a light source that was so weak it would have taken several hours for an electron to accumulate the energy needed for its ejection, and yet some electrons were observed to leave the metal *instantly*. Further experiments showed that an increase in intensity (brightness) alone can in no way dislodge electrons from the metal. But by increasing the frequency of the incident radiation beyond a certain threshold, even at very weak intensity, the emission of electrons starts immediately. These experimental facts indicate that the concept of gradual accumulation, or continuous absorption, of energy by the electron, as predicated by classical physics, is indeed erroneous.

Inspired by Planck's quantization of electromagnetic radiation, Einstein succeeded in 1905 in giving a theoretical explanation for the dependence of photoelectric emission on the frequency of the incident radiation. He assumed that light is made of corpuscles each carrying an energy $h\nu$, called *photons*. When a beam of light of frequency ν is incident on a metal, each photon transmits all its energy $h\nu$ to an electron near the surface; in the process, the photon is entirely absorbed by the electron. The electron will thus absorb energy *only* in quanta of energy $h\nu$, irrespective of the intensity of the incident radiation. If $h\nu$ is larger than the metal's *work function W*—the energy required to dislodge the electron from the metal (every metal has free electrons that move from one atom to another; the minimum energy required to free the electron from the metal is called the work function of that metal)—the electron will then be knocked out of the metal. Hence no electron can be emitted from the metal's surface unless $h\nu > W$:

$$\boxed{h\nu = W + K,}$$
(1.21)

where K represents the kinetic energy of the electron leaving the material.

Equation (1.21), which was derived by Einstein, gives the proper explanation to the experimental observation that the kinetic energy of the ejected electron *increases linearly* with the incident frequency ν, as shown in Figure 1.3b:

$$K = h\nu - W = h(\nu - \nu_0),$$
(1.22)

where $\nu_0 = W/h$ is called the threshold or cutoff frequency of the metal. Moreover, this relation shows clearly why no electron can be ejected from the metal unless $\nu > \nu_0$: since the kinetic energy cannot be negative, the photoelectric effect cannot occur when $\nu < \nu_0$ regardless of the intensity of the radiation. The ejected electrons acquire their kinetic energy from the excess energy $h(\nu - \nu_0)$ supplied by the incident radiation.

The kinetic energy of the emitted electrons can be experimentally determined as follows. The setup, which was devised by Lenard, consists of the photoelectric metal (cathode) that is placed next to an anode inside an evacuated glass tube. When light strikes the cathode's surface, the electrons ejected will be attracted to the anode, thereby generating a photoelectric current. It was found that *the magnitude of the photoelectric current thus generated is proportional to the intensity of the incident radiation, yet the speed of the electrons does not depend on the radiation's intensity, but on its frequency.* To measure the kinetic energy of the electrons, we simply need to use a varying voltage source and reverse the terminals. When the potential V across the tube is reversed, the liberated electrons will be prevented from reaching the anode; only those electrons with kinetic energy larger than $e|V|$ will make it to the negative plate and contribute to the current. We vary V until it reaches a value V_s, called the *stopping potential*, at which all of the electrons, even the most energetic ones, will be turned back before reaching the collector; hence the flow of photoelectric current ceases completely. The stopping potential V_s is connected to the electrons' kinetic energy by $e|V_s| = \frac{1}{2}m_e v^2 = K$ (in what follows, V_s will implicitly denote $|V_s|$). Thus, the relation (1.22) becomes $eV_s = h\nu - W$ or

$$\boxed{V_s = \frac{h}{e}\nu - \frac{W}{e} = \frac{hc}{e\lambda} - \frac{W}{e}.}$$
(1.23)

The shape of the plot of V_s against frequency is a straight line, much like Figure 1.3b with the slope now given by h/e. This shows that the stopping potential depends linearly on the frequency of the incident radiation.

It was Millikan who, in 1916, gave a systematic experimental confirmation to Einstein's photoelectric theory. He produced an extensive collection of photoelectric data using various metals. He verified that Einstein's relation (1.23) reproduced his data exactly. In addition, Millikan found that his empirical value for h, which he obtained by measuring the slope h/e of (1.23) (Figure 1.3b), is equal to Planck's constant to within a 0.5% experimental error.

In summary, the photoelectric effect does provide compelling evidence for the corpuscular nature of the electromagnetic radiation.

Example 1.2 (Estimation of the Planck constant)

When two ultraviolet beams of wavelengths $\lambda_1 = 80$ nm and $\lambda_2 = 110$ nm fall on a lead surface, they produce photoelectrons with maximum energies 11.390 eV and 7.154 eV, respectively.

(a) Estimate the numerical value of the Planck constant.

(b) Calculate the work function, the cutoff frequency, and the cutoff wavelength of lead.

Solution

(a) From (1.22) we can write the kinetic energies of the emitted electrons as $K_1 = hc/\lambda_1 - W$ and $K_2 = hc/\lambda_2 - W$; the difference between these two expressions is given by $K_1 - K_2 = hc(\lambda_2 - \lambda_1)/(\lambda_1\lambda_2)$ and hence

$$h = \frac{K_1 - K_2}{c} \frac{\lambda_1\lambda_2}{\lambda_2 - \lambda_1}. \tag{1.24}$$

Since 1 eV $= 1.6 \times 10^{-19}$ J, the numerical value of h follows at once:

$$h = \frac{(11.390 - 7.154) \times 1.6 \times 10^{-19} \text{ J}}{3 \times 10^8 \text{ m s}^{-1}} \times \frac{(80 \times 10^{-9} \text{ m})(110 \times 10^{-9} \text{ m})}{110 \times 10^{-9} \text{ m} - 80 \times 10^{-9} \text{ m}} \simeq 6.627 \times 10^{-34} \text{ J s}. \tag{1.25}$$

This is a very accurate result indeed.

(b) The work function of the metal can be obtained from either one of the two data

$$
\begin{aligned}
W = \frac{hc}{\lambda_1} - K_1 &= \frac{6.627 \times 10^{-34} \text{ J s} \times 3 \times 10^8 \text{ m s}^{-1}}{80 \times 10^{-9} \text{ m}} - 11.390 \times 1.6 \times 10^{-19} \text{ J} \\
&= 6.627 \times 10^{-19} \text{ J} = 4.14 \text{ eV}.
\end{aligned} \tag{1.26}
$$

The cutoff frequency and wavelength of lead are

$$\nu_0 = \frac{W}{h} = \frac{6.627 \times 10^{-19} \text{ J}}{6.627 \times 10^{-34} \text{ J s}} = 10^{15} \text{ Hz}, \qquad \lambda_0 = \frac{c}{\nu_0} = \frac{3 \times 10^8 \text{ m/s}}{10^{15} \text{ Hz}} = 300 \text{ nm}. \tag{1.27}$$

1.2.3   Compton Effect

In his 1923 experiment, Compton provided the most conclusive confirmation of the particle aspect of radiation. By scattering X-rays off free electrons, he found that the wavelength of the scattered radiation is larger than the wavelength of the incident radiation. This can be explained only by assuming that the X-ray photons behave like particles.

At issue here is to study how X-rays scatter off free electrons. According to classical physics, the incident and scattered radiation should have the same wavelength. This can be viewed as follows. Classically, since the energy of the X-ray radiation is too high to be absorbed by a free electron, the incident X-ray would then provide an oscillatory electric field which sets the electron into oscillatory motion, hence making it radiate light with the same wavelength but with an intensity I that depends on the intensity of the incident radiation I_0 (i.e., $I \propto I_0$). Neither of these two predictions of classical physics is compatible with experiment. The experimental findings of Compton reveal that the wavelength of the scattered X-radiation *increases* by an amount $\Delta\lambda$, called the wavelength shift, and that $\Delta\lambda$ depends not on the intensity of the incident radiation, but only on the scattering angle.

Figure 1.4 Compton scattering of a photon (of energy $h\nu$ and momentum $\vec{p}$) off a free, stationary electron. After collision, the photon is scattered at angle θ with energy $h\nu'$.

Compton succeeded in explaining his experimental results only after treating the incident radiation as a stream of particles—photons—colliding *elastically* with individual electrons. In this scattering process, which can be illustrated by the elastic scattering of a photon from a free[7] electron (Figure 1.4), the laws of elastic collisions can be invoked, notably the *conservation* of energy and momentum.

Consider that the incident photon, of energy $E = h\nu$ and momentum $p = h\nu/c$, collides with an electron that is initially at rest. If the photon scatters with a momentum $\vec{p}'$ at an angle[8] θ while the electron recoils with a momentum $\vec{P}_e$, the conservation of linear momentum yields

$$\vec{p} = \vec{P}_e + \vec{p}', \qquad (1.28)$$

which leads to

$$\vec{P}_e^2 = (\vec{p} - \vec{p}')^2 = p^2 + p'^2 - 2pp'\cos\theta = \frac{h^2}{c^2}\left(\nu^2 + \nu'^2 - 2\nu\nu'\cos\theta\right). \qquad (1.29)$$

Let us now turn to the energy conservation. The energies of the electron before and after the collision are given, respectively, by

$$E_0 = m_e c^2, \qquad (1.30)$$

$$E_e = \sqrt{\vec{P}_e^2 c^2 + m_e^2 c^4} = h\sqrt{\nu^2 + \nu'^2 - 2\nu\nu'\cos\theta + \frac{m_e^2 c^4}{h^2}}; \qquad (1.31)$$

in deriving this relation, we have used (1.29). Since the energies of the incident and scattered photons are given by $E = h\nu$ and $E' = h\nu'$, respectively, conservation of energy dictates that

$$E + E_0 = E' + E_e \qquad (1.32)$$

---

[7]When a metal is irradiated with high energy radiation, and at sufficiently high frequencies—as in the case of X-rays—so that $h\nu$ is much larger than the binding energies of the electrons in the metal, these electrons can be considered as free.

[8]Here θ is the angle between $\vec{p}$ and $\vec{p}'$, the photons' momenta before and after collision.

or

$$hv + m_e c^2 = hv' + h\sqrt{v^2 + v'^2 - 2vv' \cos \theta + \frac{m_e^2 c^4}{h^2}}, \qquad (1.33)$$

which in turn leads to

$$v - v' + \frac{m_e c^2}{h} = \sqrt{v^2 + v'^2 - 2vv' \cos \theta + \frac{m_e^2 c^4}{h^2}}. \qquad (1.34)$$

Squaring both sides of (1.34) and simplifying, we end up with

$$\frac{1}{v'} - \frac{1}{v} = \frac{h}{m_e c^2}(1 - \cos \theta) = \frac{2h}{m_e c^2} \sin^2 \left(\frac{\theta}{2}\right). \qquad (1.35)$$

Hence the wavelength shift is given by

$$\boxed{\Delta\lambda = \lambda' - \lambda = \frac{h}{m_e c}(1 - \cos \theta) = 2\lambda_C \sin^2 \left(\frac{\theta}{2}\right),} \qquad (1.36)$$

where $\lambda_C = h/(m_e c) = 2.426 \times 10^{-12}$ m is called the Compton wavelength of the electron. This relation, which connects the initial and final wavelengths to the scattering angle, confirms Compton's experimental observation: the wavelength shift of the X-rays depends only on the angle at which they are scattered and not on the frequency (or wavelength) of the incident photons.

In summary, the Compton effect confirms that photons behave like particles: they collide with electrons like material particles.

Example 1.3 (Compton effect)
High energy photons (γ-rays) are scattered from electrons initially at rest. Assume the photons are backscatterred and their energies are much larger than the electron's rest-mass energy, $E \gg m_e c^2$.
(a) Calculate the wavelength shift.
(b) Show that the energy of the scattered photons is half the rest mass energy of the electron, regardless of the energy of the incident photons.
(c) Calculate the electron's recoil kinetic energy if the energy of the incident photons is 150 MeV.

Solution
(a) In the case where the photons backscatter (i.e., $\theta = \pi$), the wavelength shift (1.36) becomes

$$\Delta\lambda = \lambda' - \lambda = 2\lambda_C \sin^2 \left(\frac{\pi}{2}\right) = 2\lambda_C = 4.86 \times 10^{-12} \text{ m}, \qquad (1.37)$$

since $\lambda_C = h/(m_e c) = 2.426 \times 10^{-12}$ m.
(b) Since the energy of the scattered photons E' is related to the wavelength λ' by $E' = hc/\lambda'$, equation (1.37) yields

$$E' = \frac{hc}{\lambda'} = \frac{hc}{\lambda + 2h/(m_e c)} = \frac{m_e c^2}{m_e c^2 \lambda/(hc) + 2} = \frac{m_e c^2}{m_e c^2/E + 2}, \qquad (1.38)$$

Figure 1.5 Pair production: a highly energetic photon, interacting with a nucleus, disappears and produces an electron and a positron.

where $E = hc/\lambda$ is the energy of the incident photons. If $E \gg m_e c^2$ we can approximate (1.38) by

$$E' = \frac{m_e c^2}{2} \left[1 + \frac{m_e c^2}{2E} \right]^{-1} \simeq \frac{m_e c^2}{2} - \frac{(m_e c^2)^2}{4E} \simeq \frac{m_e c^2}{2} = 0.25 \text{ MeV}. \tag{1.39}$$

(c) If $E = 150$ MeV, the kinetic energy of the recoiling electrons can be obtained from conservation of energy

$$K_e = E - E' \simeq 150 \text{ MeV} - 0.25 \text{ MeV} = 149.75 \text{ MeV}. \tag{1.40}$$

1.2.4   Pair Production

We deal here with another physical process which confirms that radiation (the photon) has corpuscular properties.

The theory of quantum mechanics that Schrödinger and Heisenberg proposed works only for nonrelativistic phenomena. This theory, which is called nonrelativistic quantum mechanics, was immensely successful in explaining a wide range of such phenomena. Combining the theory of special relativity with quantum mechanics, Dirac succeeded (1928) in extending quantum mechanics to the realm of relativistic phenomena. The new theory, called relativistic quantum mechanics, predicted the existence of a new particle, the *positron*. This particle, defined as the *antiparticle* of the electron, was predicted to have the same mass as the electron and an equal but opposite (positive) charge.

Four years after its prediction by Dirac's relativistic quantum mechanics, the positron was discovered by Anderson in 1932 while studying the trails left by cosmic rays in a cloud chamber. When high-frequency electromagnetic radiation passes through a foil, individual photons of this radiation disappear by producing a pair of particles consisting of an electron, e^-, and a positron, e^+: photon$\rightarrow e^- + e^+$. This process is called *pair production*; Anderson obtained such a process by exposing a lead foil to cosmic rays from outer space which contained highly energetic X-rays. It is useless to attempt to explain the pair production phenomenon by means of classical physics, because even nonrelativistic quantum mechanics fails utterly to account for it.

Due to charge, momentum, and energy conservation, pair production cannot occur in empty space. For the process photon$\rightarrow e^- + e^+$ to occur, the photon must interact with an external field such as the Coulomb field of an atomic nucleus to absorb some of its momentum. In the reaction depicted in Figure 1.5, an electron–positron pair is produced when the photon comes near (interacts with) a nucleus at rest; energy conservation dictates that

$$\hbar\omega = E_{e^-} + E_{e^+} + E_N = \left(m_e c^2 + k_{e^-}\right) + \left(m_e c^2 + k_{e^+}\right) + K_N$$

$$\simeq 2m_e c^2 + k_{e^-} + k_{e^+}, \tag{1.41}$$

where $\hbar\omega$ is the energy of the incident photon, $2m_e c^2$ is the sum of the rest masses of the electron and positron, and k_{e^-} and k_{e^+} are the kinetic energies of the electron and positron, respectively. As for $E_N = K_N$, it represents the recoil energy of the nucleus which is purely kinetic. Since the nucleus is very massive compared to the electron and the positron, K_N can be neglected to a good approximation. Note that the photon cannot produce an electron or a positron alone, for electric charge would not be conserved. Also, a massive object, such as the nucleus, must participate in the process to take away some of the photon's momentum.

The inverse of pair production, called pair annihilation, also occurs. For instance, when an electron and a positron collide, they *annihilate* each other and give rise to electromagnetic radiation[9]: $e^- + e^+ \rightarrow$ photon. This process explains why positrons do not last long in nature. When a positron is generated in a pair production process, its passage through matter will make it lose some of its energy and it eventually gets annihilated after colliding with an electron. The collision of a positron with an electron produces a hydrogen-like atom, called *positronium*, with a mean lifetime of about 10^{-10} s; positronium is like the hydrogen atom where the proton is replaced by the positron. Note that, unlike pair production, energy and momentum can simultaneously be conserved in pair annihilation processes without any additional (external) field or mass such as the nucleus.

The pair production process is a direct consequence of the mass–energy equation of Einstein $E = mc^2$, which states that pure energy can be converted into mass and vice versa. Conversely, pair annihilation occurs as a result of mass being converted into pure energy. All subatomic particles also have antiparticles (e.g., antiproton). Even neutral particles have antiparticles; for instance, the antineutron is the neutron's antiparticle. Although this text deals only with nonrelativistic quantum mechanics, we have included pair production and pair annihilation, which are relativistic processes, merely to illustrate how radiation interacts with matter, and also to underscore the fact that the quantum theory of Schrödinger and Heisenberg is limited to nonrelativistic phenomena only.

Example 1.4 (Minimum energy for pair production)
Calculate the minimum energy of a photon so that it converts into an electron–positron pair. Find the photon's frequency and wavelength.

Solution
The minimum energy E_{min} of a photon required to produce an electron–positron pair must be equal to the sum of rest mass energies of the electron and positron; this corresponds to the case where the kinetic energies of the electron and positron are zero. Equation (1.41) yields

$$E_{min} = 2m_e c^2 = 2 \times 0.511 \text{ MeV} = 1.02 \text{ MeV}. \tag{1.42}$$

---

[9]When an electron–positron pair annihilate, they produce at least two photons each having an energy $m_e c^2 = 0.511$ MeV.

If the photon's energy is smaller than 1.02 MeV, no pair will be produced. The photon's frequency and wavelength can be obtained at once from $E_{min} = h\nu = 2m_e c^2$ and $\lambda = c/\nu$:

$$\nu = \frac{2m_e c^2}{h} = \frac{2 \times 9.1 \times 10^{-31} \text{ kg} \times (3 \times 10^8 \text{m s}^{-1})^2}{6.63 \times 10^{-34} \text{ J s}} = 2.47 \times 10^{20} \text{ Hz}, \qquad (1.43)$$

$$\lambda = \frac{c}{\nu} = \frac{3 \times 10^8 \text{m s}^{-1}}{2.47 \times 10^{20} \text{ Hz}} = 1.2 \times 10^{-12} \text{ m}. \qquad (1.44)$$

1.3 Wave Aspect of Particles

1.3.1 de Broglie's Hypothesis: Matter Waves

As discussed above—in the photoelectric effect, the Compton effect, and the pair production effect—radiation exhibits particle-like characteristics in addition to its wave nature. In 1923 de Broglie took things even further by suggesting that this wave–particle duality is not restricted to radiation, but must be universal: *all material particles should also display a dual wave–particle behavior*. That is, the wave–particle duality present in light must also occur in matter.

So, starting from the momentum of a photon $p = h\nu/c = h/\lambda$, we can generalize this relation to any material particle[10] with nonzero rest mass: *each material particle of momentum $\vec{p}$ behaves as a group of waves* (matter waves) whose wavelength λ and wave vector $\vec{k}$ are governed by the speed and mass of the particle

$$\boxed{\lambda = \frac{h}{p}, \qquad \vec{k} = \frac{\vec{p}}{\hbar},} \qquad (1.45)$$

where $\hbar = h/2\pi$. The expression (1.45), known as the *de Broglie relation*, connects the momentum of a particle with the wavelength and wave vector of the wave corresponding to this particle.

1.3.2 Experimental Confirmation of de Broglie's Hypothesis

de Broglie's idea was confirmed experimentally in 1927 by Davisson and Germer, and later by Thomson, who obtained *interference patterns* with electrons.

1.3.2.1 Davisson–Germer Experiment

In their experiment, Davisson and Germer scattered a 54 eV monoenergetic beam of electrons from a nickel (Ni) crystal. The electron source and detector were symmetrically located with respect to the crystal's normal, as indicated in Figure 1.6; this is similar to the Bragg setup for X-ray diffraction by a grating. What Davisson and Germer found was that, although the electrons are scattered in all directions from the crystal, the intensity was a minimum at $\theta = 35°$

---

[10]In classical physics a particle is characterized by its energy E and its momentum $\vec{p}$, whereas a wave is characterized by its wavelength λ and its wave vector $\vec{k} = (2\pi/\lambda)\hat{n}$, where $\hat{n}$ is a unit vector that specifies the direction of propagation of the wave.

Figure 1.6 Davisson–Germer experiment: electrons strike the crystal's surface at an angle ϕ; the detector, symmetrically located from the electron source, measures the number of electrons scattered at an angle θ, where θ is the angle between the incident and scattered electron beams.

and a maximum at $\theta = 50°$; that is, the bulk of the electrons scatter only in well-specified directions. They showed that the pattern persisted even when the intensity of the beam was so low that the incident electrons were sent one at a time. This can only result from a constructive interference of the scattered electrons. So, instead of the diffuse distribution pattern that results from material particles, the reflected electrons formed diffraction patterns that were identical with Bragg's X-ray diffraction by a *grating*. In fact, the intensity maximum of the scattered electrons in the Davisson–Germer experiment corresponds to the first maximum ($n = 1$) of the Bragg formula,

$$n\lambda = 2d \sin \phi, \tag{1.46}$$

where d is the spacing between the Bragg planes, ϕ is the angle between the incident ray and the crystal's reflecting planes, θ is the angle between the incident and scattered beams (d is given in terms of the separation D between successive atomic layers in the crystal by $d = D \sin\theta$).

For an Ni crystal, we have $d = 0.091$ nm, since $D = 0.215$ nm. Since only one maximum is seen at $\theta = 50°$ for a mono-energetic beam of electrons of kinetic energy 54 eV, and since $2\phi + \theta = \pi$ and hence $\sin \phi = \cos(\theta/2)$ (Figure 1.6), we can obtain from (1.46) the wavelength associated with the scattered electrons:

$$\lambda = \frac{2d}{n} \sin \phi = \frac{2d}{n} \cos \frac{1}{2}\theta = \frac{2 \times 0.091 \text{ nm}}{1} \cos 25° = 0.165 \text{ nm}. \tag{1.47}$$

Now, let us look for the numerical value of λ that results from de Broglie's relation. Since the kinetic energy of the electrons is $K = 54$ eV, and since the momentum is $p = \sqrt{2m_e K}$ with $m_e c^2 = 0.511$ MeV (the rest mass energy of the electron) and $\hbar c \simeq 197.33$ eV nm, we can show that the de Broglie wavelength is

$$\lambda = \frac{h}{p} = \frac{h}{\sqrt{2m_e K}} = \frac{2\pi \hbar c}{\sqrt{2m_e c^2 K}} = 0.167 \text{ nm}, \tag{1.48}$$

which is in excellent agreement with the experimental value (1.47).

We have seen that the scattered electrons in the Davisson–Germer experiment produced interference fringes that were identical to those of Bragg's X-ray diffraction. Since the Bragg formula provided an accurate prediction of the electrons' interference fringes, the motion of an electron of momentum $\vec{p}$ must be described by means of a plane wave

$$\psi(\vec{r}, t) = Ae^{i(\vec{k}\cdot\vec{r}-\omega t)} = Ae^{i(\vec{p}\cdot\vec{r}-Et)/\hbar}, \tag{1.49}$$

Figure 1.7 Thomson experiment: diffraction of electrons through a thin film of polycrystalline material yields fringes that usually result from light diffraction.

where A is a constant, $\vec{k}$ is the wave vector of the plane wave, and ω is its angular frequency; the wave's parameters, $\vec{k}$ and ω, are related to the electron's momentum $\vec{p}$ and energy E by means of de Broglie's relations: $\vec{k} = \vec{p}/\hbar$, $\omega = E/\hbar$.

We should note that, inspired by de Broglie's hypothesis, Schrödinger constructed the theory of wave mechanics which deals with the dynamics of microscopic particles. He described the motion of particles by means of a *wave function* $\psi(\vec{r}, t)$ which corresponds to the de Broglie wave of the particle. We will deal with the physical interpretation of $\psi(\vec{r}, t)$ in the following section.

1.3.2.2   Thomson Experiment

In the Thomson experiment (Figure 1.7), electrons were diffracted through a polycrystalline thin film. Diffraction fringes were also observed. This result confirmed again the wave behavior of electrons.

The Davisson–Germer experiment has inspired others to obtain diffraction patterns with a large variety of particles. Interference patterns were obtained with bigger and bigger particles such as neutrons, protons, helium atoms, and hydrogen molecules. de Broglie wave interference of carbon 60 (C60) molecules were recently[11] observed by diffraction at a material absorption grating; these observations supported the view that each C60 molecule interferes only with itself (a C60 molecule is nearly a classical object).

1.3.3   Matter Waves for Macroscopic Objects

We have seen that microscopic particles, such as electrons, display wave behavior. What about macroscopic objects? Do they also display wave features? They surely do. Although macro-

---

[11]Markus Arndt, *et al.*, "Wave–Particle Duality of C60 Molecules", *Nature*, **V401, n6754**, 680 (Oct. 14, 1999).

scopic material particles display wave properties, the corresponding wavelengths are too small to detect; being very massive[12], macroscopic objects have extremely small wavelengths. At the microscopic level, however, the waves associated with material particles are of the same size or exceed the size of the system. Microscopic particles therefore exhibit clearly discernible wave-like aspects.

The general rule is: whenever the de Broglie wavelength of an object is in the range of, or exceeds, its size, the wave nature of the object is detectable and hence cannot be neglected. But if its de Broglie wavelength is much too small compared to its size, the wave behavior of this object is undetectable. For a quantitative illustration of this general rule, let us calculate in the following example the wavelengths corresponding to two particles, one microscopic and the other macroscopic.

Example 1.5 (Matter waves for microscopic and macroscopic systems)
Calculate the de Broglie wavelength for
    (a) a proton of kinetic energy 70 MeV kinetic energy and
    (b) a 100 g bullet moving at 900 m s^{-1}.

Solution
(a) Since the kinetic energy of the proton is $T = p^2/(2m_p)$, its momentum is $p = \sqrt{2Tm_p}$. The de Broglie wavelength is $\lambda_p = h/p = h/\sqrt{2Tm_p}$. To calculate this quantity numerically, it is more efficient to introduce the well-known quantity $\hbar c \simeq 197$ MeV fm and the rest mass of the proton $m_p c^2 = 938.3$ MeV, where c is the speed of light:

$$\lambda_p = 2\pi \frac{\hbar c}{pc} = 2\pi \frac{\hbar c}{\sqrt{2Tm_p c^2}} = 2\pi \frac{197 \text{ MeV fm}}{\sqrt{2 \times 938.3 \times 70 \text{ MeV}^2}} = 3.4 \times 10^{-15} \text{ m}. \qquad (1.50)$$

(b) As for the bullet, its de Broglie wavelength is $\lambda_b = h/p = h/(mv)$ and since $h = 6.626 \times 10^{-34}$ J s, we have

$$\lambda_b = \frac{h}{mv} = \frac{6.626 \times 10^{-34} \text{ J s}}{0.1 \text{ kg} \times 900 \text{ m s}^{-1}} = 7.4 \times 10^{-36} \text{ m}. \qquad (1.51)$$

The ratio of the two wavelengths is $\lambda_b/\lambda_p \simeq 2.2 \times 10^{-21}$. Clearly, the wave aspect of this bullet lies beyond human observational abilities. As for the wave aspect of the proton, it cannot be neglected; its de Broglie wavelength of 3.4×10^{-15} m has the same order of magnitude as the size of a typical atomic nucleus.

We may conclude that, whereas the wavelengths associated with *microscopic* systems are *finite* and display easily detectable wave-like patterns, the wavelengths associated with *macroscopic* systems are *infinitesimally small* and display no discernible wave-like behavior. So, when the wavelength approaches zero, the wave-like properties of the system disappear. In such cases of infinitesimally small wavelengths, *geometrical optics* should be used to describe the motion of the object, for the wave associated with it behaves as a *ray*.

---

[12]Very massive compared to microscopic particles. For instance, the ratio between the mass of an electron and a 100 g bullet is infinitesimal: $m_e/m_b \simeq 10^{-29}$.

Figure 1.8 The double-slit experiment with *particles*: S is a source of *bullets*; I_1 and I_2 are the intensities recorded on the screen, respectively, when only S_1 is open and then when only S_2 is open. When both slits are open, the total intensity is $I = I_1 + I_2$.

1.4   Particles versus Waves

In this section we are going to study the properties of particles and waves within the contexts of classical and quantum physics. The experimental setup to study these aspects is the *double-slit experiment*, which consists of a source S (S can be a source of material particles or of waves), a wall with two slits S_1 and S_2, and a back screen equipped with counters that record whatever arrives at it from the slits.

1.4.1   Classical View of Particles and Waves

In classical physics, particles and waves are mutually exclusive; they exhibit completely differ-ent behaviors. While the full description of a particle requires only one parameter, the position vector $\vec{r}(t)$, the complete description of a wave requires two, the amplitude and the phase. For instance, three-dimensional plane waves can be described by wave functions $\psi(\vec{r}, t)$:

$$\psi(\vec{r}, t) = Ae^{i(\vec{k}\cdot\vec{r}-\omega t)} = Ae^{i\phi}, \tag{1.52}$$

where A is the amplitude of the wave and ϕ is its phase ($\vec{k}$ is the wave vector and ω is the angular frequency). We may recall the physical meaning of ψ: the intensity of the wave is given by $I = |\psi|^2$.

(a) S is a source of streams of bullets
Consider three different experiments as displayed in Figure 1.8, in which a source S fires a stream of bullets; the bullets are assumed to be indestructible and hence arrive on the screen in identical lumps. In the first experiment, only slit S_1 is open; let $I_1(y)$ be the corresponding intensity collected on the screen (the number of bullets arriving per second at a given point y). In the second experiment, let $I_2(y)$ be the intensity collected on the screen when only S_2 is open. In the third experiments, if S_1 and S_2 are both open, the total intensity collected on the

Only slit 1 is open | Only slit 2 is open | Both slits are open

Figure 1.9 The double-slit experiment: S is a source of *waves*, I_1 and I_2 are the intensities recorded on the screen when only S_1 is open, and then when only S_2 is open, respectively. When both slits are open, the total intensity is no longer equal to the sum of I_1 and I_2; an *oscillating* term has to be added.

screen behind the two slits must be equal to the sum of I_1 and I_2:

$$I(y) = I_1(y) + I_2(y). \tag{1.53}$$

(b) S is a source of waves
Now, as depicted in Figure 1.9, S is a source of waves (e.g., light or water waves). Let I_1 be the intensity collected on the screen when only S_1 is open and I_2 be the intensity when only S_2 is open. Recall that a wave is represented by a complex function ψ, and its intensity is proportional to its amplitude (e.g., height of water or electric field) squared: $I_1 = |\psi_1|^2$, $I_2 = |\psi_2|^2$. When both slits are open, the total intensity collected on the screen displays an *interference* pattern; hence it cannot be equal to the sum of I_1 and I_2. The amplitudes, not the intensities, must add: the total amplitude ψ is the sum of ψ_1 and ψ_2; hence the total intensity is given by

$$I = |\psi_1 + \psi_2|^2 = |\psi_1|^2 + |\psi_2|^2 + (\psi_1^*\psi_2 + \psi_2^*\psi_1) = I_1 + I_2 + 2\text{Re}(\psi_1^*\psi_2)$$
$$= I_1 + I_2 + 2\sqrt{I_1 I_2}\cos\delta, \tag{1.54}$$

where δ is the phase difference between ψ_1 and ψ_2, and $2\sqrt{I_1 I_2}\cos\delta$ is an oscillating term, which is responsible for the interference pattern (Figure 1.9). So the resulting intensity distribution cannot be predicted from I_1 or from I_2 alone, for it depends on the phase δ, which cannot be measured when only one slit is open (δ can be calculated from the slits separation or from the observed intensities I_1, I_2 and I).

Conclusion: Classically, waves exhibit interference patterns, particles do not. When two non-interacting streams of particles combine in the same region of space, their intensities add; when waves combine, their amplitudes add but their intensities do not.

1.4.2 Quantum View of Particles and Waves

Let us now discuss the double-slit experiment with quantum material particles such as electrons. Figure 1.10 shows three different experiments where the source S shoots a stream of electrons,

Figure 1.10 The double-slit experiment: S is a source of *electrons*, I_1 and I_2 are the intensities recorded on the screen when only S_1 is open, and then when only S_2 is open, respectively. When both slits are open, the total intensity is equal to the sum of I_1, I_2 and an *oscillating* term.

first with only S_1 open, then with only S_2 open, and finally with both slits open. In the first two cases, the distributions of the electrons on the screen are smooth; the sum of these distributions is also smooth, a bell-shaped curve like the one obtained for classical particles (Figure 1.8).

But when both slits are open, we see a rapid variation in the distribution, an *interference pattern*. So in spite of their discreteness, the electrons seem to interfere with themselves; this means that each electron seems to have gone through both slits at once! One might ask, if an electron cannot be split, how can it appear to go through both slits at once? Note that this interference pattern has nothing to do with the intensity of the electron beam. In fact, experiments were carried out with beams so weak that the electrons were sent one at a time (i.e., each electron was sent only after the previous electron has reached the screen). In this case, if both slits were open and if we wait long enough so that sufficient impacts are collected on the screen, the interference pattern appears again.

The crucial question now is to find out the slit through which the electron went. To answer this query, an experiment can be performed to watch the electrons as they leave the slits. It consists of placing a strong light source behind the wall containing the slits, as shown in Figure 1.11. We place Geiger counters all over the screen so that whenever an electron reaches the screen we hear a click on the counter.

Since electric charges scatter light, whenever an electron passes through either of the slits, on its way to the counter, it will scatter light to our eyes. So, whenever we hear a click on the counter, we see a flash near *either S_1 or S_2 but never near both at once*. After recording the various counts with both slits open, we find out that the distribution is similar to that of classical bullets in Figure 1.8: the interference pattern has disappeared! But if we turn off the light source, the interference pattern appears again.

From this experiment we conclude that the mere act of looking at the electrons immensely affects their distribution on the screen. Clearly, electrons are very delicate: their motion gets modified when one watches them. This is the very quantum mechanical principle which states that *measurements interfere with the states of microscopic objects*. One might think of turning down the brightness (intensity) of the light source so that it is weak enough not to disturb the

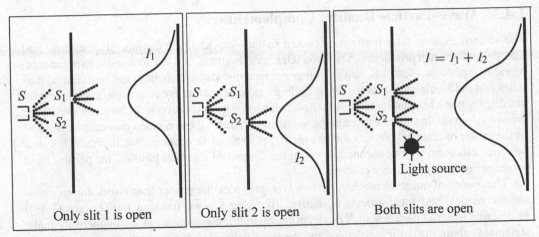

Figure 1.11 The double-slit experiment: S is a source of *electrons*. *A light source* is placed behind the wall containing S_1 and S_2. When both slits are open, the interference pattern is destroyed and the total intensity is $I = I_1 + I_2$.

electrons. We find that the light scattered from the electrons, as they pass by, does not get weaker; the same sized flash is seen, but only every once in a while. This means that, at low brightness levels, we miss some electrons: we hear the click from the counter but see no flash at all. At still lower brightness levels, we miss most of the electrons. We conclude, in this case, that some electrons went through the slits without being seen, because there were no photons around at the right moment to catch them. This process is important because it confirms that light has particle properties: light also arrives in lumps (photons) at the screen.

Two distribution profiles are compiled from this dim light source experiment, one corresponding to the electrons that were seen and the other to the electrons that were not seen (but heard on the counter). The first distribution contains no interference (i.e., it is similar to classical bullets); but the second distribution displays an interference pattern. This results from the fact that when the electrons are not seen, they display interference. When we do not see the electron, no photon has disturbed it but when we see it, a photon has disturbed it.

For the electrons that display interference, it is impossible to identify the slit that each electron had gone through. This experimental finding introduces a new fundamental concept: the microphysical world is *indeterministic*. Unlike classical physics, where we can follow accurately the particles along their trajectories, we cannot follow a microscopic particle along its motion nor can we determine its path. It is technically impossible to perform such detailed tracing of the particle's motion. Such results inspired Heisenberg to postulate the uncertainty principle, which states that *it is impossible to design an apparatus which allows us to determine the slit that the electron went through without disturbing the electron enough to destroy the interference pattern* (we shall return to this principle later).

The interference pattern obtained from the double-slit experiment indicates that electrons display both particle and wave properties. When electrons are observed or detected one by one, they behave like particles, but when they are detected after many measurements (distribution of the detected electrons), they behave like waves of wavelength $\lambda = h/p$ and display an interference pattern.

1.4.3   Wave–Particle Duality: Complementarity

The various experimental findings discussed so far—blackbody radiation, photoelectric and Compton effect, pair production, Davisson–Germer, Thomson, and the double-slit experiments—reveal that photons, electrons, and any other microscopic particles behave unlike classical particles and unlike classical waves. These findings indicate that, at the microscopic scale, nature can display particle behavior as well as wave behavior. The question now is, how can something behave as a particle and as a wave at the same time? Aren't these notions mutually exclusive? In the realm of classical physics the answer is yes, but not in quantum mechanics. This *dual* behavior can in no way be reconciled within the context of classical physics, for particles and waves are mutually exclusive entities.

The theory of quantum mechanics, however, provides the proper framework for reconciling the particle and wave aspects of matter. By using a wave function $\psi(\vec{r}, t)$ (see (1.49)) to describe material particles such as electrons, quantum mechanics can simultaneously make statements about the particle behavior and the wave behavior of microscopic systems. It combines the quantization of energy or intensity with a wave description of matter. That is, it uses both particle and wave pictures to describe the same material particle.

Our ordinary concepts of particles or waves are thus inadequate when applied to microscopic systems. These two concepts, which preclude each other in the macroscopic realm, do not strictly apply to the microphysical world. No longer valid at the microscopic scale is the notion that a wave cannot behave as a particle and vice versa. The true reality of a quantum system is that it is neither a pure particle nor a pure wave. The particle and wave aspects of a quantum system manifest themselves only when subjected to, or intruded on by, penetrating means of observation (any procedure of penetrating observation would destroy the initial state of the quantum system; for instance, the mere act of looking at an electron will knock it out of its orbit). Depending on the type of equipment used to observe an electron, the electron has the capacity to display either "grain" or wave features. As illustrated by the double-slit experiment, if we wanted to look at the particle aspect of the electron, we would need only to block one slit (or leave both slits open but introduce an observational apparatus), but if we were interested only in its wave features, we would have to leave both slits open and not intrude on it by observational tools. This means that both the "grain" and "wave" features are embedded into the electron, and by modifying the probing tool, we can suppress one aspect of the electron and keep the other. An experiment designed to isolate the particle features of a quantum system gives no information about its wave features, and vice versa. When we subject an electron to Compton scattering, we observe only its particle aspects, but when we involve it in a diffraction experiment (as in Davisson–Germer, Thomson, or the double-slit experiment), we observe its wave behavior only. So if we measure the particle properties of a quantum system, this will destroy its wave properties, and vice versa. Any measurement gives either one property or the other, but never both at once. We can get either the wave property or the particle but not both of them together.

Microscopic systems, therefore, are neither pure particles nor pure waves, they are both. The particle and wave manifestations do not contradict or preclude one another, but, as suggested by Bohr, they are just *complementary*. Both concepts are complementary in describing the true nature of microscopic systems. Being complementary features of microscopic matter, particles and waves are equally important for a complete description of quantum systems. From here comes the essence of the *complementarity principle*.

We have seen that when the *rigid* concept of *either/or* (i.e., either a particle or a wave) is indiscriminately applied or imposed on quantum systems, we get into trouble with reality.

Without the complementarity principle, quantum mechanics would not have been in a position to produce the accurate results it does.

1.4.4 Principle of Linear Superposition

How do we account mathematically for the existence of the interference pattern in the double-slit experiment with material particles such as electrons? An answer is offered by the *superposition principle*. The interference results from the superposition of the waves emitted by slits 1 and 2. If the functions $\psi_1(\vec{r}, t)$ and $\psi_2(\vec{r}, t)$, which denote the waves reaching the screen emitted respectively by slits 1 and 2, represent two physically possible states of the system, then any linear superposition

$$\psi(\vec{r}, t) = \alpha_1 \psi_1(\vec{r}, t) + \alpha_2 \psi_2(\vec{r}, t) \tag{1.55}$$

also represents a physically possible outcome of the system; α_1 and α_2 are complex constants. This is the superposition principle. The intensity produced on the screen by opening only slit 1 is $|\psi_1(\vec{r}, t)|^2$ and it is $|\psi_2(\vec{r}, t)|^2$ when only slit 2 is open. When both slits are open, the intensity is

$$\begin{aligned} |\psi(\vec{r}, t)|^2 &= |\psi_1(\vec{r}, t) + \psi_2(\vec{r}, t)|^2 \\ &= |\psi_1(\vec{r}, t)|^2 + |\psi_2(\vec{r}, t)|^2 + \psi_1^*(\vec{r}, t)\psi_2(\vec{r}, t) + \psi_1(\vec{r}, t)\psi_2^*(\vec{r}, t), \end{aligned} \tag{1.56}$$

where the asterisk denotes the complex conjugate. Note that (1.56) is not equal to the sum of $|\psi_1(\vec{r}, t)|^2$ and $|\psi_2(\vec{r}, t)|^2$; it contains an additional term $\psi_1^*(\vec{r}, t)\psi_2(\vec{r}, t) + \psi_1(\vec{r}, t)\psi_2^*(\vec{r}, t)$. This is the very term which gives rise in the case of electrons to an interference pattern similar to light waves. The interference pattern therefore results from the existence of a phase shift between $\psi_1(\vec{r}, t)$ and $\psi_2(\vec{r}, t)$. We can measure this phase shift from the interference pattern, but we can in no way measure the phases of ψ_1 and ψ_2 separately.

We can summarize the double-slit results in three principles:

- Intensities add for classical particles: $I = I_1 + I_2$.

- Amplitudes, not intensities, add for quantum particles: $\psi(\vec{r}, t) = \psi_1(\vec{r}, t) + \psi_2(\vec{r}, t)$; this gives rise to interference.

- Whenever one attempts to determine experimentally the outcome of individual events for microscopic material particles (such as trying to specify the slit through which an electron has gone), the interference pattern gets destroyed. In this case the intensities add in much the same way as for classical particles: $I = I_1 + I_2$.

1.5 Indeterministic Nature of the Microphysical World

Let us first mention two important experimental findings that were outlined above. On the one hand, the Davisson–Germer and the double-slit experiments have shown that microscopic material particles do give rise to interference patterns. To account for the interference pattern, we have seen that it is imperative to describe microscopic particles by means of waves. Waves are not localized in space. As a result, we have to give up on accuracy to describe microscopic particles, for waves give at best a probabilistic account. On the other hand, we have seen in the double-slit experiment that it is impossible to trace the motion of individual electrons; there is no experimental device that would determine the slit through which a given electron has gone. Not being able to predict single events is a stark violation of a founding principle of classical physics: predictability or determinacy. These experimental findings inspired Heisenberg to postulate the indeterministic nature of the microphysical world and Born to introduce the probabilistic interpretation of quantum mechanics.

1.5.1   Heisenberg's Uncertainty Principle

According to classical physics, given the initial conditions and the forces acting on a system, the future behavior (unique path) of this physical system can be determined exactly. That is, if the initial coordinates $\vec{r}_0$, velocity $\vec{v}_0$, and all the forces acting on the particle are known, the position $\vec{r}(t)$ and velocity $\vec{v}(t)$ are uniquely determined by means of Newton's second law. *Classical physics* is thus completely *deterministic*.

Does this deterministic view hold also for the microphysical world? Since a particle is represented within the context of quantum mechanics by means of a wave function corresponding to the particle's wave, and since wave functions cannot be localized, then a microscopic particle is somewhat spread over space and, unlike classical particles, cannot be localized in space. In addition, we have seen in the double-slit experiment that it is impossible to determine the slit that the electron went through without disturbing it. The classical concepts of exact position, exact momentum, and unique path of a particle therefore make no sense at the microscopic scale. This is the essence of Heisenberg's uncertainty principle.

In its original form, Heisenberg's uncertainty principle states that: *If the x-component of the momentum of a particle is measured with an uncertainty Δp_x, then its x-position cannot, at the same time, be measured more accurately than $\Delta x = \hbar/(2\Delta p_x)$.* The three-dimensional form of the uncertainty relations for position and momentum can be written as follows:

$$\Delta x \, \Delta p_x \geq \frac{\hbar}{2}, \qquad \Delta y \, \Delta p_y \geq \frac{\hbar}{2}, \qquad \Delta z \, \Delta p_z \geq \frac{\hbar}{2}. \qquad (1.57)$$

This principle indicates that, although it is possible to measure the momentum or position of a particle accurately, it is not possible to measure these two observables *simultaneously* to an *arbitrary* accuracy. That is, we cannot localize a microscopic particle without giving to it a rather large momentum. We cannot measure the position without disturbing it; there is no way to carry out such a measurement passively as it is bound to change the momentum. To understand this, consider measuring the position of a macroscopic object (e.g., a car) and the position of a microscopic system (e.g., an electron in an atom). On the one hand, to locate the position of a macroscopic object, you need simply to observe it; the light that strikes it and gets reflected to the detector (your eyes or a measuring device) can in no measurable way affect the motion of the object. On the other hand, to measure the position of an electron in an atom, you must use radiation of very short wavelength (the size of the atom). The energy of this radiation is high enough to change tremendously the momentum of the electron; the mere observation of the electron affects its motion so much that it can knock it entirely out of its orbit. It is therefore impossible to determine the position and the momentum simultaneously to arbitrary accuracy. If a particle were localized, its wave function would become zero everywhere else and its wave would then have a very short wavelength. According to de Broglie's relation $p = h/\lambda$, the momentum of this particle will be rather *high*. Formally, this means that if a particle is accurately localized (i.e., $\Delta x \to 0$), there will be total uncertainty about its momentum (i.e., $\Delta p_x \to \infty$). To summarize, since all quantum phenomena are described by waves, we have no choice but to accept limits on our ability to measure simultaneously any two complementary variables.

Heisenberg's uncertainty principle can be generalized to any pair of complementary, or canonically conjugate, dynamical variables: *it is impossible to devise an experiment that can measure simultaneously two complementary variables to arbitrary accuracy* (if this were ever achieved, the theory of quantum mechanics would collapse).

Energy and time, for instance, form a pair of complementary variables. Their simultaneous measurement must obey the time–energy uncertainty relation:

$$\Delta E \Delta t \geq \frac{\hbar}{2}. \tag{1.58}$$

This relation states that if we make two measurements of the energy of a system and if these measurements are separated by a time interval Δt, the measured energies will differ by an amount ΔE which can in no way be smaller than $\hbar / \Delta t$. If the time interval between the two measurements is large, the energy difference will be small. This can be attributed to the fact that, when the first measurement is carried out, the system becomes perturbed and it takes it a long time to return to its initial, unperturbed state. This expression is particularly useful in the study of decay processes, for it specifies the relationship between the mean lifetime and the energy width of the excited states.

We see that, in sharp contrast to classical physics, *quantum mechanics* is a completely *indeterministic* theory. Asking about the position or momentum of an electron, one cannot get a definite answer; only a *probabilistic* answer is possible. According to the uncertainty principle, if the position of a quantum system is well defined, its momentum will be totally undefined. In this context, the uncertainty principle has clearly brought down one of the most sacrosanct concepts of classical physics: the deterministic nature of Newtonian mechanics.

Example 1.6 (Uncertainties for microscopic and macroscopic systems)
Estimate the uncertainty in the position of (a) a neutron moving at $5 \times 10^6 \,\text{m s}^{-1}$ and (b) a 50 kg person moving at $2 \,\text{m s}^{-1}$.

Solution
(a) Using (1.57), we can write the position uncertainty as

$$\Delta x \geq \frac{\hbar}{2\Delta p} \simeq \frac{\hbar}{2m_n v} = \frac{1.05 \times 10^{-34} \,\text{J s}}{2 \times 1.65 \times 10^{-27} \,\text{kg} \times 5 \times 10^6 \,\text{m s}^{-1}} = 6.4 \times 10^{-15} \,\text{m}. \tag{1.59}$$

This distance is comparable to the size of a nucleus.
(b) The position uncertainty for the person is

$$\Delta x \geq \frac{\hbar}{2\Delta p} \simeq \frac{\hbar}{2m v} = \frac{1.05 \times 10^{-34} \,\text{J s}}{2 \times 50 \,\text{kg} \times 2 \,\text{m s}^{-1}} = 0.5 \times 10^{-36} \,\text{m}. \tag{1.60}$$

An uncertainty of this magnitude is beyond human detection; therefore, it can be neglected. The accuracy of the person's position is limited only by the uncertainties induced by the device used in the measurement. So the position and momentum uncertainties are important for microscopic systems, but negligible for macroscopic systems.

1.5.2 Probabilistic Interpretation

In quantum mechanics the state (or one of the states) of a particle is described by a *wave function* $\psi(\vec{r}, t)$ corresponding to the de Broglie wave of this particle; so $\psi(\vec{r}, t)$ describes the wave properties of a particle. As a result, when discussing quantum effects, it is suitable to use the amplitude function, ψ, whose square modulus, $|\psi|^2$, is equal to the intensity of the wave associated with this quantum effect. The intensity of a wave at a given point in space is proportional to the probability of finding, at that point, the material particle that corresponds to the wave.

In 1927 Born interpreted $|\psi|^2$ as the *probability density* and $|\psi(\vec{r}, t)|^2 d^3r$ as the probability, $dP(\vec{r}, t)$, of finding a particle at time t in the volume element d^3r located between $\vec{r}$ and $\vec{r} + d\vec{r}$:

$$|\psi(\vec{r}, t)|^2 d^3r = dP(\vec{r}, t), \qquad (1.61)$$

where $|\psi|^2$ has the dimensions of $[\text{Length}]^{-3}$. If we integrate over the entire space, we are certain that the particle is somewhere in it. Thus, the total probability of finding the particle somewhere in space must be equal to one:

$$\int_{all\ space} |\psi(\vec{r}, t)|^2 d^3r = 1. \qquad (1.62)$$

The main question now is, how does one determine the wave function ψ of a particle? The answer to this question is given by the theory of quantum mechanics, where ψ is determined by the Schrödinger equation (Chapters 3 and 4).

1.6 Atomic Transitions and Spectroscopy

Besides failing to explain blackbody radiation, the Compton, photoelectric, and pair production effects and the wave–particle duality, classical physics also fails to account for many other phenomena at the microscopic scale. In this section we consider another area where classical physics breaks down—the atom. Experimental observations reveal that atoms exist as stable, bound systems that have *discrete* numbers of energy levels. Classical physics, however, states that any such bound system must have a continuum of energy levels.

1.6.1 Rutherford Planetary Model of the Atom

After his experimental discovery of the atomic nucleus in 1911, Rutherford proposed a model in an attempt to explain the properties of the atom. Inspired by the orbiting motion of the planets around the sun, Rutherford considered the atom to consist of electrons orbiting around a positively charged massive center, the nucleus. It was soon recognized that, within the context of *classical physics*, this model suffers from two serious deficiencies: (a) atoms are *unstable* and (b) atoms radiate energy over a *continuous* range of frequencies.

The first deficiency results from the application of Maxwell's electromagnetic theory to Rutherford's model: as the electron orbits around the nucleus, it accelerates and hence radiates energy. It must therefore lose energy. The radius of the orbit should then decrease continuously (spiral motion) until the electron collapses onto the nucleus; the typical time for such a collapse is about 10^{-8} s. Second, since the frequency of the radiated energy is the same as the orbiting frequency, and as the electron orbit collapses, its orbiting frequency increases *continuously*. Thus, the spectrum of the radiation emitted by the atom should be continuous. These two conclusions completely disagree with experiment, since atoms are *stable* and radiate energy over *discrete* frequency ranges.

1.6.2 Bohr Model of the Hydrogen Atom

Combining Rutherford's planetary model, Planck's quantum hypothesis, and Einstein's photon concept, Bohr proposed in 1913 a model that gives an accurate account of the observed spectrum of the hydrogen atom as well as a convincing explanation for its stability.

Bohr assumed, as in Rutherford's model, that each atom's electron moves in an orbit around the nucleus under the influence of the electrostatic attraction of the nucleus; circular or elliptic orbits are allowed by classical mechanics. For simplicity, Bohr considered only circular orbits, and introduced several, rather arbitrary assumptions which violate classical physics but which are immensely successful in explaining many properties of the hydrogen atom:

- Instead of a *continuum* of orbits, which are possible in classical mechanics, only a *discrete* set of circular *stable* orbits, called *stationary states*, are allowed. Atoms can exist only in certain stable states with definite energies: E_1, E_2, E_3, etc.

- The allowed (stationary) orbits correspond to those for which the orbital angular momentum of the electron is an *integer multiple* of $\hbar$ ($\hbar = h/2\pi$):

$$L = n\hbar. \tag{1.63}$$

This relation is known as the *Bohr quantization rule* of the angular momentum.

- As long as an electron remains in a stationary orbit, it does not radiate electromagnetic energy. Emission or absorption of radiation can take place only when an electron jumps from one allowed orbit to another. The radiation corresponding to the electron's transition from an orbit of energy E_n to another E_m is carried out by a photon of energy

$$h\nu = E_n - E_m. \tag{1.64}$$

So an atom may emit (or absorb) radiation by having the electron jump to a lower (or higher) orbit.

In what follows we are going to apply Bohr's assumptions to the hydrogen atom. We want to provide a quantitative description of its energy levels and its spectroscopy.

1.6.2.1 Energy Levels of the Hydrogen Atom

Let us see how Bohr's quantization condition (1.63) leads to a discrete set of energies E_n and radii r_n. When the electron of the hydrogen atom moves in a *circular* orbit, the application of Newton's second law to the electron yields $F = m_e a_r = m_e v^2/r$. Since the only force[13]

---

[13] At the atomic scale, gravity has no measurable effect. The gravitational force between the hydrogen's proton and electron, $F_G = (G m_e m_p)/r^2$, is negligible compared to the electrostatic force $F_e = e^2/(4\pi \varepsilon_0 r^2)$, since $F_G/F_e = (4\pi \varepsilon_0) G m_e m_p/e^2 \simeq 10^{-40}$.

acting on the electron is the electrostatic force applied on it by the proton, we can equate the electrostatic force to the centripetal force and obtain

$$\frac{e^2}{4\pi\varepsilon_0 r^2} = m_e \frac{v^2}{r}. \tag{1.65}$$

Now, assumption (1.63) yields

$$L = m_e v r = n\hbar, \tag{1.66}$$

hence $m_e v^2/r = n^2\hbar^2/(m_e r^3)$, which when combined with (1.65) yields $e^2/(4\pi\varepsilon_0 r^2) = n^2\hbar^2/(m_e r^3)$; this relation in turn leads to a quantized expression for the radius:

$$\boxed{r_n = \left(\frac{4\pi\varepsilon_0\hbar^2}{m_e e^2}\right) n^2 = n^2 a_0,} \tag{1.67}$$

where

$$a_0 = \frac{4\pi\varepsilon_0\hbar^2}{m_e e^2} \tag{1.68}$$

is the *Bohr radius*, $a_0 = 0.053$ nm. The speed of the orbiting electron can be obtained from (1.66) and (1.67):

$$\boxed{v_n = \frac{n\hbar}{m_e r_n} = \left(\frac{e^2}{4\pi\varepsilon_0}\right)\frac{1}{n\hbar}.} \tag{1.69}$$

Note that the ratio between the speed of the electron in the first Bohr orbit, v_1, and the speed of light is equal to a dimensionless constant α, known as the *fine structure constant*:

$$\alpha = \frac{v_1}{c} = \frac{1}{4\pi\varepsilon_0}\frac{e^2}{\hbar c} = \frac{1}{137} \Rightarrow v_1 = \alpha c = \frac{3\times 10^8 \text{ m s}^{-1}}{137} \simeq 2.19\times 10^6 \text{ m s}^{-1}. \tag{1.70}$$

As for the total energy of the electron, it is given by

$$E = \frac{1}{2}m_e v^2 - \frac{1}{4\pi\varepsilon_0}\frac{e^2}{r}; \tag{1.71}$$

in deriving this relation, we have assumed that the nucleus, i.e., the proton, is infinitely heavy compared with the electron and hence it can be considered at rest; that is, the energy of the electron–proton system consists of the kinetic energy of the electron plus the electrostatic potential energy. From (1.65) we see that the kinetic energy, $\frac{1}{2}m_e v^2$, is equal to $\frac{1}{2}e^2/(4\pi\varepsilon_0 r)$, which when inserted into (1.71) leads to

$$E = -\frac{1}{2}\left(\frac{e^2}{4\pi\varepsilon_0 r}\right). \tag{1.72}$$

This equation shows that the electron circulates in an orbit of radius r with a kinetic energy equal to minus one half the potential energy (this result is the well known Virial theorem of classical mechanics). Substituting r_n of (1.67) into (1.72), we obtain

$$\boxed{E_n = -\frac{e^2}{8\pi\varepsilon_0}\frac{1}{r_n} = -\frac{m_e}{2\hbar^2}\left(\frac{e^2}{4\pi\varepsilon_0}\right)^2\frac{1}{n^2} = -\frac{\mathcal{R}}{n^2},} \tag{1.73}$$

Figure 1.12 Energy levels and transitions between them for the hydrogen atom.

known as the *Bohr energy*, where $\mathcal{R}$ is the *Rydberg constant*:

$$\mathcal{R} = \frac{m_e}{2\hbar^2} \left(\frac{e^2}{4\pi\,\varepsilon_0} \right)^2 = 13.6 \text{ eV.} \tag{1.74}$$

The energy E_n of each state of the atom is determined by the value of the quantum number n. The *negative* sign of the energy (1.73) is due to the *bound* state nature of the atom. That is, states with negative energy $E_n < 0$ correspond to bound states.

The structure of the atom's energy spectrum as given by (1.73) is displayed in Figure 1.12 (where, by convention, the energy levels are shown as horizontal lines). As n increases, the energy level separation decreases rapidly. Since n can take all integral values from $n = 1$ to $n = +\infty$, the energy spectrum of the atom contains an infinite number of discrete energy levels. In the ground state ($n = 1$), the atom has an energy $E_1 = -\mathcal{R}$ and a radius a_0. The states $n = 2, 3, 4, \ldots$ correspond to the *excited states* of the atom, since their energies are greater than the ground state energy.

When the quantum number n is very large, $n \to +\infty$, the atom's radius r_n will also be very large but the energy values go to zero, $E_n \to 0$. This means that the proton and the electron are infinitely far away from one another and hence they are no longer bound; the atom is ionized. In this case there is no restriction on the amount of kinetic energy the electron can take, for it is free. This situation is represented in Figure 1.12 by the *continuum* of positive energy states, $E_n > 0$.

Recall that in deriving (1.67) and (1.73) we have neglected the mass of the proton. If we include it, the expressions (1.67) and (1.73) become

$$r_n = \frac{4\pi\varepsilon_0\hbar^2}{\mu e^2}n^2 = \left(1 + \frac{m_e}{m_p}\right)a_0 n^2, \quad E_n = -\frac{\mu}{2\hbar^2}\left(\frac{e^2}{4\pi\varepsilon_0}\right)^2\frac{1}{n^2} = -\frac{1}{1 + m_e/m_p}\frac{\mathcal{R}}{n^2},$$

(1.75)

where $\mu = m_p m_e/(m_p + m_e) = m_e/(1 + m_e/m_p)$ is the reduced mass of the proton–electron system.

We should note that r_n and E_n of (1.75), which were derived for the hydrogen atom, can be generalized to hydrogen-like ions where all electrons save one are removed. To obtain the radius and energy of a single electron orbiting a fixed nucleus of Z protons, we need simply to replace e^2 in (1.75) by Ze^2,

$$r_n = \left(1 + \frac{m_e}{M}\right)\frac{a_0}{Z}n^2, \quad E_n = -\frac{Z^2}{1 + m_e/M}\frac{\mathcal{R}}{n^2},$$

(1.76)

where M is the mass of the nucleus; when $m_e/M \ll 1$ we can just drop the term m_e/M.

de Broglie's hypothesis and Bohr's quantization condition

The Bohr quantization condition (1.63) can be viewed as a manifestation of de Broglie's hypothesis. For the wave associated with the atom's electron to be a standing wave, the circumference of the electron's orbit must be equal to an *integral* multiple of the electron's wavelength:

$$2\pi r = n\lambda \qquad (n = 1, 2, 3, \ldots).$$

(1.77)

This relation can be reduced to (1.63) or to (1.66), provided that we make use of de Broglie's relation, $\lambda = h/p = h/(m_e v)$. That is, inserting $\lambda = h/(m_e v)$ into (1.77) and using the fact that the electron's orbital angular momentum is $L = m_e v r$, we have

$$2\pi r = n\lambda = n\frac{h}{m_e v} \quad \Longrightarrow \quad m_e v r = n\frac{h}{2\pi} \quad \Longrightarrow \quad L = n\hbar,$$

(1.78)

which is identical with Bohr's quantization condition (1.63). In essence, this condition states that the only allowed orbits for the electron are those whose circumferences are equal to integral multiples of the de Broglie wavelength. For example, in the hydrogen atom, the circumference of the electron's orbit is equal to λ when the atom is in its ground state ($n = 1$); it is equal to 2λ when the atom is in its first excited state ($n = 2$); equal to 3λ when the atom is in its second excited state ($n = 3$); and so on.

1.6.2.2   Spectroscopy of the Hydrogen Atom

Having specified the energy spectrum of the hydrogen atom, let us now study its spectroscopy. In sharp contrast to the continuous nature of the spectral distribution of the radiation emitted by glowing solid objects, the radiation emitted or absorbed by a gas displays a *discrete* spectrum distribution. When subjecting a gas to an electric discharge (or to a flame), the radiation emitted from the excited atoms of the gas discharge consists of a few sharp lines (bright lines of pure color, with darkness in between). A major success of Bohr's model lies in its ability to predict accurately the *sharpness* of the spectral lines emitted or absorbed by the atom. The model shows clearly that these discrete lines correspond to the sharply defined energy levels of the atom. The radiation emitted from the atom results from the transition of the electron from an allowed state n to another m; this radiation has a well defined (sharp) frequency v:

$$hv = E_n - E_m = \mathcal{R}\left(\frac{1}{m^2} - \frac{1}{n^2}\right). \tag{1.79}$$

For instance, the *Lyman series*, which corresponds to the emission of *ultraviolet* radiation, is due to transitions from excited states $n = 2, 3, 4, 5, \ldots$ to the ground state $n = 1$ (Figure 1.12):

$$hv_L = E_n - E_1 = \mathcal{R}\left(\frac{1}{1^2} - \frac{1}{n^2}\right) \qquad (n > 1). \tag{1.80}$$

Another transition series, the *Balmer series*, is due to transitions to the first excited state ($n = 2$):

$$hv_B = E_n - E_2 = \mathcal{R}\left(\frac{1}{2^2} - \frac{1}{n^2}\right) \qquad (n > 2). \tag{1.81}$$

The atom emits *visible* radiation as a result of the Balmer transitions. Other series are Paschen, $n \rightarrow 3$ with $n > 3$; Brackett, $n \rightarrow 4$ with $n > 4$; Pfund, $n \rightarrow 5$ with $n > 5$; and so on. They correspond to the emission of *infrared* radiation. Note that the results obtained from (1.79) are in spectacular agreement with those of experimental spectroscopy.

So far in this chapter, we have seen that when a photon passes through matter, it interacts as follows:

- If it comes in contact with an electron that is at rest, it will scatter from it like a corpuscular particle: it will impart a momentum to the electron, it will scatter and continue its travel with the speed of light but with a lower frequency (or higher wavelength). This is the Compton effect.

- If it comes into contact with an atom's electron, it will interact according to one of the following scenarios:

  - If it has enough energy, it will knock the electron completely out of the atom and then vanish, for it transmits all its energy to the electron. This is the photoelectric effect.

  - If its energy hv is not sufficient to knock out the electron altogether, it will kick the electron to a higher orbit, provided hv is equal to the energy difference between the initial and final orbits: $hv = E_n - E_m$. In the process it will transmit all its energy to the electron and then vanish. The atom will be left in an excited state. However, if $hv \neq E_n - E_m$, nothing will happen (the photon simply scatters away).

- If it comes in contact with an atomic nucleus and if its energy is sufficiently high ($hv \geq 2m_e c^2$), it will vanish by creating matter: an electron–positron pair will be produced. This is pair production.

Example 1.7 (Positronium's radius and energy spectrum)
Positronium is the bound state of an electron and a positron; it is a short-lived, hydrogen-like atom where the proton is replaced by a positron.

(a) Calculate the energy and radius expressions, E_n and r_n.

(b) Estimate the values of the energies and radii of the three lowest states.

(c) Calculate the frequency and wavelength of the electromagnetic radiation that will just ionize the positronium atom when it is in its first excited state.

Solution

(a) The radius and energy expressions of the positronium can be obtained at once from (1.75) by simply replacing the reduced mass μ with that of the electron–positron system $\mu = m_e m_e/(m_e + m_e) = \frac{1}{2} m_e$:

$$r_n = \left(\frac{8\pi \varepsilon_0 \hbar^2}{m_e e^2} \right) n^2, \qquad E_n = -\frac{m_e}{4\hbar^2} \left(\frac{e^2}{4\pi \varepsilon_0} \right)^2 \frac{1}{n^2}. \qquad (1.82)$$

We can rewrite r_n and E_n in terms of the Bohr radius, $a_0 = 4\pi \varepsilon_0 \hbar^2/(m_e e^2) = 0.053$ nm, and the Rydberg constant, $\mathcal{R} = \frac{m_e}{2\hbar^2} \left(\frac{e^2}{4\pi \varepsilon_0} \right)^2 = 13.6$ eV, as follows:

$$r_n = 2a_0 n^2, \qquad E_n = -\frac{\mathcal{R}}{2n^2}. \qquad (1.83)$$

These are related to the expressions for the hydrogen by $r_{n_{pos}} = 2r_{n_H}$ and $E_{n_{pos}} = \frac{1}{2} E_{n_H}$.

(b) The radii of the three lowest states of the positronium are given by $r_1 = 2a_0 = 0.106$ nm, $r_2 = 8a_0 = 0.424$ nm, and $r_3 = 18a_0 = 0.954$ nm. The corresponding energies are $E_1 = -\frac{1}{2}\mathcal{R} = -6.8$ eV, $E_2 = -\frac{1}{8}\mathcal{R} = -1.7$ eV, and $E_3 = -\frac{1}{18}\mathcal{R} = -0.756$ eV.

(c) Since the energy of the first excited state of the positronium is $E_2 = -1.7$ eV $= -1.7 \times 1.6 \times 10^{-19}$ J $= -2.72 \times 10^{-19}$ J, the energy of the electromagnetic radiation that will just ionize the positronium is equal to $h\nu = E_\infty - E_2 = 0 - (-2.72 \times 10^{-19}$ J$) = 2.72 \times 10^{-19}$ J $= E_{ion}$; hence the frequency and wavelength of the ionizing radiation are given by

$$\nu = \frac{E_{ion}}{h} = \frac{2.72 \times 10^{-19} \text{ J}}{6.6 \times 10^{-34} \text{ J s}} = 4.12 \times 10^{14} \text{ Hz}, \qquad (1.84)$$

$$\lambda = \frac{c}{\nu} = \frac{3 \times 10^8 \text{ m s}^{-1}}{4.12 \times 10^{14} \text{ Hz}} = 7.28 \times 10^{-7} \text{ m}. \qquad (1.85)$$

1.7 Quantization Rules

The ideas that led to successful explanations of blackbody radiation, the photoelectric effect, and the hydrogen's energy levels rest on two quantization rules: (a) the relation (1.7) that Planck postulated to explain the quantization of energy, $E = nh\nu$, and (b) the condition (1.63) that Bohr postulated to account for the quantization of the electron's orbital angular momentum, $L = n\hbar$. A number of attempts were undertaken to understand or interpret these rules. In 1916 Wilson and Sommerfeld offered a scheme that included both quantization rules as special cases. In essence, their scheme, which applies only to systems with coordinates that are periodic in time, consists in *quantizing the action variable*, $J = \oint p \, dq$, of classical mechanics:

$$\oint p \, dq = nh \qquad (n = 0, 1, 2, 3, \ldots), \qquad (1.86)$$

where n is a quantum number, p is the momentum conjugate associated with the coordinate q; the closed integral $\oint$ is taken over one period of q. This relation is known as the Wilson–Sommerfeld quantization rule.

Wilson–Sommerfeld quantization rule and Planck's quantization relation

In what follows we are going to show how the Wilson–Sommerfeld rule (1.86) leads to Planck's quantization relation $E = nh\nu$. For an illustration, consider a one-dimensional harmonic oscillator where a particle of mass m oscillates harmonically between $-a \leq x \leq a$; its classical energy is given by

$$E(x, p) = \frac{p^2}{2m} + \frac{1}{2}m\omega^2 x^2; \qquad (1.87)$$

hence $p(E, x) = \pm\sqrt{2mE - m^2\omega^2 x^2}$. At the turning points, $x_{min} = -a$ and $x_{max} = a$, the energy is purely potential: $E = V(\pm a) = \frac{1}{2}m\omega^2 a^2$; hence $a = \sqrt{2E/(m\omega^2)}$. Using $p(E, x) = \pm\sqrt{2mE - m^2\omega^2 x^2}$ and from symmetry considerations, we can write the action as

$$\oint p\, dx = 2\int_{-a}^{a} \sqrt{2mE - m^2\omega^2 x^2}\, dx = 4m\omega \int_{0}^{a} \sqrt{a^2 - x^2}\, dx. \qquad (1.88)$$

The change of variables $x = a\sin\theta$ leads to

$$\int_{0}^{a} \sqrt{a^2 - x^2}\, dx = a^2 \int_{0}^{\pi/2} \cos^2\theta\, d\theta = \frac{a^2}{2} \int_{0}^{\pi/2} (1 + \cos 2\theta)\, d\theta = \frac{\pi a^2}{4} = \frac{\pi E}{2m\omega^2}. \qquad (1.89)$$

Since $\omega = 2\pi\nu$, where ν is the frequency of oscillations, we have

$$\oint p\, dx = \frac{2\pi E}{\omega} = \frac{E}{\nu}. \qquad (1.90)$$

Inserting (1.90) into (1.86), we end up with the Planck quantization rule $E = nh\nu$, i.e.,

$$\oint p\, dx = nh \quad \Longrightarrow \quad \frac{E}{\nu} = nh \quad \Longrightarrow \quad E_n = nh\nu. \qquad (1.91)$$

We can interpret this relation as follows. From classical mechanics, we know that the motion of a mass subject to harmonic oscillations is represented in the xp phase space by a continuum of ellipses whose areas are given by $\oint p\, dx = E/\nu$, because the integral $\oint p(x)\, dx$ gives the area enclosed by the closed trajectory of the particle in the xp phase space. The condition (1.86) or (1.91) provides a mechanism for selecting, from the continuum of the oscillator's energy values, only those energies E_n for which the areas of the contours $p(x, E_n) = \sqrt{2m(E_n - V(x))}$ are equal to nh with $n = 0, 1, 2, 3, \ldots$. That is, the only allowed states of oscillation are those represented in the phase space by a series of ellipses with "quantized" areas $\oint p\, dx = nh$. Note that the area between two successive states is equal to h: $\oint p(x, E_{n+1})\, dx - \oint p(x, E_n)\, dx = h$.

This simple calculation shows that the *Planck rule for energy quantization is equivalent to the quantization of action.*

Wilson–Sommerfeld quantization rule and Bohr's quantization condition

Let us now show how the Wilson–Sommerfeld rule (1.86) leads to Bohr's quantization condition (1.63). For an electron moving in a circular orbit of radius r, it is suitable to use polar coordinates (r, φ). The action $J = \oint p\, dq$, which is expressed in Cartesian coordinates by the linear momentum p and its conjugate variable x, is characterized in polar coordinates by the orbital angular momentum L and its conjugate variable φ, the polar angle, where φ is periodic in time. That is, $J = \oint p \, dq$ is given in polar coordinates by $\int_0^{2\pi} L \, d\varphi$. In this case (1.86) becomes

$$\int_0^{2\pi} L \, d\varphi = nh. \tag{1.92}$$

For spherically symmetric potentials—as it is the case here where the electron experiences the proton's Coulomb potential—the angular momentum L is a constant of the motion. Hence (1.92) shows that angular momentum can change only in integral units of $\hbar$:

$$L \int_0^{2\pi} d\varphi = nh \quad \Longrightarrow \quad L = n\frac{h}{2\pi} = n\hbar, \tag{1.93}$$

which is identical with the Bohr quantization condition (1.63). This calculation also shows that the *Bohr quantization is equivalent to the quantization of action.* As stated above (1.78), the Bohr quantization condition (1.63) has the following physical meaning: while orbiting the nucleus, the electron moves only in well specified orbits, orbits with circumferences equal to integral multiples of the de Broglie wavelength.

Note that the Wilson–Sommerfeld quantization rule (1.86) does not tell us how to calculate the energy levels of non-periodic systems; it applies only to systems which are periodic. On a historical note, the quantization rules of Planck and Bohr have dominated quantum physics from 1900 to 1925; the quantum physics of this period is known as the "old quantum theory." The success of these quantization rules, as measured by the striking agreement of their results with experiment, gave irrefutable evidence for the quantization hypothesis of all material systems and constituted a triumph of the "old quantum theory." In spite of their quantitative success, these quantization conditions suffer from a serious inconsistency: they do not originate from a theory, they were postulated rather arbitrarily.

1.8   Wave Packets

At issue here is how to describe a particle within the context of quantum mechanics. As quantum particles jointly display particle and wave features, we need to look for a mathematical scheme that can embody them simultaneously.

In classical physics, a particle is *well localized* in space, for its position and velocity can be calculated simultaneously to arbitrary precision. As for quantum mechanics, it describes a material particle by a *wave function* corresponding to the matter wave associated with the particle (de Broglie's conjecture). Wave functions, however, depend on the *whole* space; hence they *cannot be localized.* If the wave function is made to vanish everywhere except in the neighborhood of the particle or the neighborhood of the "classical trajectory," it can then be used to describe the dynamics of the particle. That is, a particle which is localized within a certain region of space can be described by a matter wave whose amplitude is large in that region and zero outside it. This matter wave must then be *localized* around the region of space within which the particle is confined.

A *localized* wave function is called a *wave packet*. A wave packet therefore consists of a group of waves of slightly different wavelengths, with phases and amplitudes so chosen that they interfere constructively over a small region of space and destructively elsewhere. Not only are wave packets useful in the description of "isolated" particles that are confined to a certain spatial region, they also play a key role in understanding the connection between quantum mechanics and classical mechanics. The wave packet concept therefore represents a unifying mathematical tool that can cope with and embody nature's particle-like behavior and also its wave-like behavior.

1.8.1 Localized Wave Packets

Localized wave packets can be constructed by superposing, in the same region of space, waves of slightly different wavelengths, but with phases and amplitudes chosen to make the super-position constructive in the desired region and destructive outside it. Mathematically, we can carry out this superposition by means of *Fourier transforms*. For simplicity, we are going to consider a one-dimensional wave packet; this packet is intended to describe a "classical" parti-cle confined to a one-dimensional region, for instance, a particle moving along the x-axis. We can construct the packet $\psi(x, t)$ by superposing plane waves (propagating along the x-axis) of different frequencies (or wavelengths):

$$\psi(x, t) = \frac{1}{\sqrt{2\pi}} \int_{-\infty}^{+\infty} \phi(k) e^{i(kx - \omega t)} dk; \tag{1.94}$$

$\phi(k)$ is the amplitude of the wave packet.

In what follows we want to look at the form of the packet at a given time; we will deal with the time evolution of wave packets later. Choosing this time to be $t = 0$ and abbreviating $\psi(x, 0)$ by $\psi_0(x)$, we can reduce (1.94) to

$$\psi_0(x) = \frac{1}{\sqrt{2\pi}} \int_{-\infty}^{+\infty} \phi(k) e^{ikx} dk, \tag{1.95}$$

where $\phi(k)$ is the Fourier transform of $\psi_0(x)$,

$$\phi(k) = \frac{1}{\sqrt{2\pi}} \int_{-\infty}^{+\infty} \psi_0(x) e^{-ikx} dx. \tag{1.96}$$

The relations (1.95) and (1.96) show that $\phi(k)$ determines $\psi_0(x)$ and vice versa. The packet (1.95), whose form is determined by the x-dependence of $\psi_0(x)$, does indeed have the required property of localization: $|\psi_0(x)|$ peaks at $x = 0$ and vanishes far away from $x = 0$. On the one hand, as $x \to 0$ we have $e^{ikx} \to 1$; hence the waves of different frequencies interfere constructively (i.e., the various k-integrations in (1.95) add constructively). On the other hand, far away from $x = 0$ (i.e., $|x| \gg 0$) the phase e^{ikx} goes through many periods leading to violent oscillations, thereby yielding destructive interference (i.e., the various k-integrations in (1.95) add up to zero). This implies, in the language of Born's probabilistic interpretation, that the particle has a greater probability of being found near $x = 0$ and a scant chance of being found far away from $x = 0$. The same comments apply to the amplitude $\phi(k)$ as well: $\phi(k)$ peaks at $k = 0$ and vanishes far away. Figure 1.13 displays a typical wave packet that has the required localization properties we have just discussed.

In summary, the particle is represented not by a single de Broglie wave of well-defined frequency and wavelength, but by a wave packet that is obtained by adding a large number of waves of different frequencies.

The physical interpretation of the wave packet is obvious: $\psi_0(x)$ is the wave function or probability amplitude for finding the particle at position x; hence $|\psi_0(x)|^2$ gives the probability density for finding the particle at x, and $P(x) dx = |\psi_0(x)|^2 dx$ gives the probability of finding

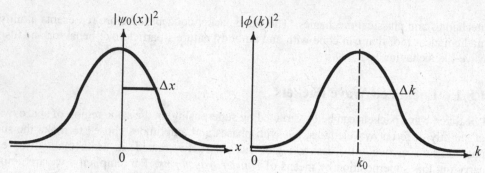

Figure 1.13 Two localized wave packets: $\psi_0(x) = (2/\pi a^2)^{1/4} e^{-x^2/a^2} e^{ik_0 x}$ and $\phi(k) = (a^2/2\pi)^{1/4} e^{-a^2(k-k_0)^2/4}$; they peak at $x = 0$ and $k = k_0$, respectively, and vanish far away.

the particle between x and $x + dx$. What about the physical interpretation of $\phi(k)$? From (1.95) and (1.96) it follows that

$$\int_{-\infty}^{+\infty} |\psi_0(x)|^2 dx = \int_{-\infty}^{+\infty} |\phi(k)|^2 dk; \qquad (1.97)$$

then if $\psi(x)$ is normalized so is $\phi(k)$, and vice versa. Thus, the function $\phi(k)$ can be interpreted most naturally, like $\psi_0(x)$, as a probability amplitude for measuring a wave vector k for a particle in the state $\phi(k)$. Moreover, while $|\phi(k)|^2$ represents the probability density for measuring k as the particle's wave vector, the quantity $P(k) dk = |\phi(k)|^2 dk$ gives the probability of finding the particle's wave vector between k and $k + dk$.

We can extract information about the particle's motion by simply expressing its corresponding matter wave in terms of the particle's energy, E, and momentum, p. Using $k = p/\hbar$, $dk = dp/\hbar$, $E = \hbar\omega$ and redefining $\tilde{\phi}(p) = \phi(k)/\sqrt{\hbar}$, we can rewrite (1.94) to (1.96) as follows:

$$\psi(x, t) = \frac{1}{\sqrt{2\pi\hbar}} \int_{-\infty}^{+\infty} \tilde{\phi}(p) e^{i(px-Et)/\hbar} dp, \qquad (1.98)$$

$$\psi_0(x) = \frac{1}{\sqrt{2\pi\hbar}} \int_{-\infty}^{+\infty} \tilde{\phi}(p) e^{ipx/\hbar} dp, \qquad (1.99)$$

$$\tilde{\phi}(p) = \frac{1}{\sqrt{2\pi\hbar}} \int_{-\infty}^{+\infty} \psi_0(x) e^{-ipx/\hbar} dx, \qquad (1.100)$$

where $E(p)$ is the total energy of the particle described by the wave packet $\psi(x, t)$ and $\tilde{\phi}(p)$ is the momentum amplitude of the packet.

In what follows we are going to illustrate the basic ideas of wave packets on a simple, instructive example: the Gaussian and square wave packets.

Example 1.8 (Gaussian and square wave packets)

(a) Find $\psi(x, 0)$ for a Gaussian wave packet $\phi(k) = A \exp\left[-a^2(k - k_0)^2/4\right]$, where A is a normalization factor to be found. Calculate the probability of finding the particle in the region $-a/2 \leq x \leq a/2$.

(b) Find $\phi(k)$ for a square wave packet $\psi_0(x) = \begin{cases} Ae^{ik_0x}, & |x| \leq a, \\ 0, & |x| > a. \end{cases}$

Find the factor A so that $\psi(x)$ is normalized.

Solution

(a) The normalization factor A is easy to obtain:

$$1 = \int_{-\infty}^{+\infty} |\phi(k)|^2 dk = |A|^2 \int_{-\infty}^{+\infty} \exp\left[-\frac{a^2}{2}(k - k_0)^2\right] dk, \qquad (1.101)$$

which, by using a change of variable $z = k - k_0$ and using the integral $\int_{-\infty}^{+\infty} e^{-a^2z^2/2} dz = \sqrt{2\pi}/a$, leads at once to $A = \sqrt{a/\sqrt{2\pi}} = [a^2/(2\pi)]^{1/4}$. Now, the wave packet corresponding to

$$\phi(k) = \left(\frac{a^2}{2\pi}\right)^{1/4} \exp\left[-\frac{a^2}{4}(k - k_0)^2\right] \qquad (1.102)$$

is

$$\psi_0(x) = \frac{1}{\sqrt{2\pi}} \int_{-\infty}^{+\infty} \phi(k) e^{ikx} dk = \frac{1}{\sqrt{2\pi}} \left(\frac{a^2}{2\pi}\right)^{1/4} \int_{-\infty}^{+\infty} e^{-a^2(k-k_0)^2/4 + ikx} dk. \qquad (1.103)$$

To carry out the integration, we need simply to rearrange the exponent's argument as follows:

$$-\frac{a^2}{4}(k - k_0)^2 + ikx = -\left[\frac{a}{2}(k - k_0) - \frac{ix}{a}\right]^2 - \frac{x^2}{a^2} + ik_0x. \qquad (1.104)$$

The introduction of a new variable $y = a(k - k_0)/2 - ix/a$ yields $dk = 2dy/a$, and when combined with (1.103) and (1.104), this leads to

$$\begin{aligned} \psi_0(x) &= \frac{1}{\sqrt{2\pi}} \left(\frac{a^2}{2\pi}\right)^{1/4} \int_{-\infty}^{+\infty} e^{-x^2/a^2} e^{ik_0x} e^{-y^2} \left(\frac{2}{a} dy\right) \\ &= \frac{1}{\sqrt{\pi}} \left(\frac{2}{\pi a^2}\right)^{1/4} e^{-x^2/a^2} e^{ik_0x} \int_{-\infty}^{+\infty} e^{-y^2} dy. \end{aligned} \qquad (1.105)$$

Since $\int_{-\infty}^{+\infty} e^{-y^2} dy = \sqrt{\pi}$, this expression becomes

$$\psi_0(x) = \left(\frac{2}{\pi a^2}\right)^{1/4} e^{-x^2/a^2} e^{ik_0x}, \qquad (1.106)$$

where e^{ik_0x} is the phase of $\psi_0(x)$; $\psi_0(x)$ is an oscillating wave with wave number k_0 modulated by a Gaussian envelope centered at the origin. We will see later that the phase factor e^{ik_0x} has real physical significance. The wave function $\psi_0(x)$ is complex, as necessitated by quantum mechanics. Note that $\psi_0(x)$, like $\phi(k)$, is normalized. Moreover, equations (1.102) and (1.106) show that the Fourier transform of a Gaussian wave packet is also a Gaussian wave packet.

The probability of finding the particle in the region $-a/2 \leq x \leq a/2$ can be obtained at once from (1.106):

$$P = \int_{-a/2}^{+a/2} |\psi_0(x)|^2 dx = \sqrt{\frac{2}{\pi a^2}} \int_{-a/2}^{+a/2} e^{-2x^2/a^2} dx = \frac{1}{\sqrt{2\pi}} \int_{-1}^{+1} e^{-z^2/2} dz \simeq \frac{2}{3}, \qquad (1.107)$$

where we have used the change of variable $z = 2x/a$.

(b) The normalization of $\psi_0(x)$ is straightforward:

$$1 = \int_{-\infty}^{+\infty} |\psi_0(x)|^2 dx = |A|^2 \int_{-a}^{a} e^{-ik_0x} e^{ik_0x} dx = |A|^2 \int_{-a}^{a} dx = 2a|A|^2; \qquad (1.108)$$

hence $A = 1/\sqrt{2a}$. The Fourier transform of $\psi_0(x)$ is

$$\phi(k) = \frac{1}{\sqrt{2\pi}} \int_{-\infty}^{+\infty} \psi_0(x) e^{-ikx} dx = \frac{1}{2\sqrt{\pi a}} \int_{-a}^{a} e^{ik_0x} e^{-ikx} dx = \frac{1}{\sqrt{\pi a}} \frac{\sin\left[(k-k_0)a\right]}{k - k_0}. \qquad (1.109)$$

1.8.2 Wave Packets and the Uncertainty Relations

We want to show here that the width of a wave packet $\psi_0(x)$ and the width of its amplitude $\phi(k)$ are not independent; they are correlated by a reciprocal relationship. As it turns out, the reciprocal relationship between the widths in the x and k spaces has a direct connection to Heisenberg's uncertainty relation.

For simplicity, let us illustrate the main ideas on the Gaussian wave packet treated in the previous example (see (1.102) and (1.106)):

$$\psi_0(x) = \left(\frac{2}{\pi a^2}\right)^{1/4} e^{-x^2/a^2} e^{ik_0x}, \qquad \phi(k) = \left(\frac{a^2}{2\pi}\right)^{1/4} e^{-a^2(k-k_0)^2/4}. \qquad (1.110)$$

As displayed in Figure 1.13, $|\psi_0(x)|^2$ and $|\phi(k)|^2$ are centered at $x = 0$ and $k = k_0$, respectively. It is convenient to define the half-widths Δx and Δk as corresponding to the half-maxima of $|\psi_0(x)|^2$ and $|\phi(k)|^2$. In this way, when x varies from 0 to $\pm\Delta x$ and k from k_0 to $k_0 \pm \Delta k$, the functions $|\psi_0(x)|^2$ and $|\phi(k)|^2$ drop to $e^{-1/2}$:

$$\frac{|\psi(\pm\Delta x, 0)|^2}{|\psi(0, 0)|^2} = e^{-1/2}, \qquad \frac{|\phi(k_0 \pm \Delta k)|^2}{|\phi(k_0)|^2} = e^{-1/2}. \qquad (1.111)$$

These equations, combined with (1.110), lead to $e^{-2\Delta x^2/a^2} = e^{-1/2}$ and $e^{-a^2\Delta k^2/2} = e^{-1/2}$, respectively, or to

$$\Delta x = \frac{a}{2}, \qquad \Delta k = \frac{1}{a}; \qquad (1.112)$$

hence

$$\Delta x \, \Delta k = \frac{1}{2}. \qquad (1.113)$$

Since $\Delta k = \Delta p/\hbar$ we have

$$\boxed{\Delta x \, \Delta p = \frac{\hbar}{2}.} \qquad (1.114)$$

This relation shows that if the packet's width is narrow in x-space, its width in momentum space must be very broad, and vice versa.

A comparison of (1.114) with Heisenberg's uncertainty relations (1.57) reveals that the Gaussian wave packet yields an *equality*, not an *inequality* relation. In fact, equation (1.114) is the *lowest limit* of Heisenberg's inequality. As a result, the Gaussian wave packet is called the *minimum uncertainty* wave packet. All other wave packets yield higher values for the product of the x and p uncertainties: $\Delta x \Delta p > \hbar/2$; for an illustration see Problem 1.11. In conclusion, the value of the uncertainties product $\Delta x \Delta p$ varies with the choice of ψ, but the lowest bound, $\hbar/2$, is provided by a Gaussian wave function. We have now seen how the wave packet concept offers a heuristic way of deriving Heisenberg's uncertainty relations; a more rigorous derivation is given in Chapter 2.

1.8.3 Motion of Wave Packets

How do wave packets evolve in time? The answer is important, for it gives an idea not only about the motion of a quantum particle in space but also about the connection between classical and quantum mechanics. Besides studying how wave packets propagate in space, we will also examine the conditions under which packets may or may not spread.

At issue here is, knowing the initial wave packet $\psi_0(x)$ or the amplitude $\phi(k)$, how do we find $\psi(x, t)$ at any later time t? This issue reduces to calculating the integral $\int \phi(k)e^{i(kx-\omega t)}dk$ in (1.94). To calculate this integral, we need to specify the angular frequency ω and the amplitude $\phi(k)$. We will see that the spreading or nonspreading of the packet is dictated by the form of the function $\omega(k)$.

1.8.3.1 Propagation of a Wave Packet without Distortion

The simplest form of the angular frequency ω is when it is *proportional* to the wave number k; this case corresponds to a *nondispersive* propagation. Since the constant of proportionality has the dimension of a velocity[14], which we denote by v_0 (i.e., $\omega = v_0 k$), the wave packet (1.94) becomes

$$\psi(x, t) = \frac{1}{\sqrt{2\pi}} \int_{-\infty}^{+\infty} \phi(k)e^{ik(x-v_0 t)}dk. \tag{1.115}$$

This relation has the same structure as (1.95), which suggests that $\psi(x,t)$ is identical with $\psi_0(x - v_0 t)$:

$$\psi(x, t) = \psi_0(x - v_0 t); \tag{1.116}$$

the form of the wave packet at time t is identical with the initial form. Therefore, when ω is proportional to k, so that $\omega = v_0 k$, the wave packet travels to the right with constant velocity v_0 *without distortion*.

However, since we are interested in wave packets that describe particles, we need to consider the more general case of *dispersive* media which transmit harmonic waves of different frequencies at different velocities. This means that ω is a *function* of k: $\omega = \omega(k)$. The form of $\omega(k)$ is determined by the requirement that the wave packet $\psi(x, t)$ describes the particle. Assuming that the amplitude $\phi(k)$ peaks at $k = k_0$, then $\phi(k) = g(k - k_0)$ is appreciably different from zero only in a narrow range $\Delta k = k - k_0$, and we can Taylor expand $\omega(k)$ about k_0:

$$\omega(k) = \omega(k_0) + (k - k_0) \frac{d\omega(k)}{dk}\bigg|_{k=k_0} + \frac{1}{2}(k - k_0)^2 \frac{d^2\omega(k)}{dk^2}\bigg|_{k=k_0} + \cdots$$

$$= \omega(k_0) + (k - k_0)v_g + (k - k_0)^2 \alpha + \cdots \tag{1.117}$$

---

[14]For propagation of light in a vacuum this constant is equal to c, the speed of light.

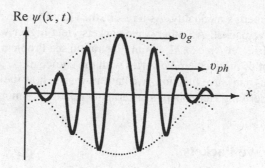

Figure 1.14 The function Re $\psi(x, t)$ of the wave packet (1.118), represented here by the solid curve contained in the dashed-curve envelope, propagates with the *group* velocity v_g along the x axis; the individual waves (not drawn here), which add up to make the solid curve, move with different *phase* velocities v_{ph}.

where $v_g = \left. \frac{d\omega(k)}{dk} \right|_{k=k_0}$ and $\alpha = \frac{1}{2} \left. \frac{d^2\omega(k)}{dk^2} \right|_{k=k_0}$.

Now, to determine $\psi(x, t)$ we need simply to substitute (1.117) into (1.94) with $\phi(k) = g(k - k_0)$. This leads to

$$\psi(x, t) = \frac{1}{\sqrt{2\pi}} e^{ik_0(x - v_{ph}t)} \int_{-\infty}^{+\infty} g(k - k_0) e^{i(k-k_0)(x-v_g t)} e^{-i(k-k_0)^2 \alpha t + \cdots} dk \qquad (1.118)$$

where[15]

$$\boxed{v_g = \frac{d\omega(k)}{dk}, \qquad v_{ph} = \frac{\omega(k)}{k};} \qquad (1.119)$$

v_{ph} and v_g are respectively the *phase* velocity and the *group* velocity. The phase velocity denotes the velocity of propagation for the phase of a single harmonic wave, $e^{ik_0(x - v_{ph}t)}$, and the group velocity represents the velocity of motion for the group of waves that make up the packet. One should not confuse the phase velocity and the group velocity; in general they are different. Only when ω is proportional to k will they be equal, as can be inferred from (1.119).

Group and phase velocities

Let us take a short detour to explain the meanings of v_{ph} and v_g. As mentioned above, when we superimpose many waves of different amplitudes and frequencies, we can obtain a wave packet or pulse which travels at the group velocity v_g; the *individual* waves that constitute the packet, however, move with different speeds; each wave moves with its own phase velocity v_{ph}. Figure 1.14 gives a qualitative illustration: the group velocity represents the velocity with which the wave packet propagates as a *whole*, where the individual waves (located inside the packet's envelope) that add up to make the packet move with different phase velocities. As shown in Figure 1.14, the wave packet has an appreciable magnitude only over a small region and falls rapidly outside this region.

The difference between the group velocity and the phase velocity can be understood quantitatively by deriving a relationship between them. A differentiation of $\omega = k v_{ph}$ (see (1.119)) with respect to k yields $d\omega/dk = v_{ph} + k(dv_{ph}/dk)$, and since $k = 2\pi/\lambda$, we have $dv_{ph}/dk =$

---

[15]In these equations we have omitted k_0 since they are valid for any choice of k_0.

$(dv_{ph}/d\lambda)(d\lambda/dk) = -(2\pi/k^2)(dv_{ph}/d\lambda)$ or $k(dv_{ph}/dk) = -\lambda(dv_{ph}/d\lambda)$; combining these relations, we obtain

$$v_g = \frac{d\omega}{dk} = v_{ph} + k\frac{dv_{ph}}{dk} = v_{ph} - \lambda\frac{dv_{ph}}{d\lambda}, \tag{1.120}$$

which we can also write as

$$v_g = v_{ph} + p\frac{dv_{ph}}{dp}, \tag{1.121}$$

since $k(dv_{ph}/dk) = (p/\hbar)(dv_{ph}/dp)(dp/dk) = p(dv_{ph}/dp)$ because $k = p/\hbar$. Equations (1.120) and (1.121) show that the group velocity may be larger or smaller than the phase velocity; it may also be equal to the phase velocity depending on the medium. If the phase velocity does not depend on the wavelength—this occurs in nondispersive media—the group and phase velocities are equal, since $dv_{ph}/d\lambda = 0$. But if v_{ph} depends on the wavelength—this occurs in dispersive media—then $dv_{ph}/d\lambda \neq 0$; hence the group velocity may be smaller or larger than the phase velocity. An example of a nondispersive medium is an inextensible string; we would expect $v_g = v_{ph}$. Water waves offer a typical dispersive medium; in Problem 1.13 we show that for deepwater waves we have $v_g = \frac{1}{2}v_{ph}$ and for surface waves we have $v_g = \frac{3}{2}v_{ph}$; see (1.212) and (1.214).

Consider the case of a particle traveling in a *constant potential* V; its total energy is $E(p) = p^2/(2m) + V$. Since the corpuscular features (energy and momentum) of a particle are connected to its wave characteristics (wave frequency and number) by the relations $E = \hbar\omega$ and $p = \hbar k$, we can rewrite (1.119) as follows:

$$v_g = \frac{dE(p)}{dp}, \qquad v_{ph} = \frac{E(p)}{p}, \tag{1.122}$$

which, when combined with $E(p) = \frac{p^2}{2m} + V$, yield

$$v_g = \frac{d}{dp}\left(\frac{p^2}{2m} + V\right) = \frac{p}{m} = v_{particle}, \qquad v_{ph} = \frac{1}{p}\left(\frac{p^2}{2m} + V\right) = \frac{p}{2m} + \frac{V}{p}. \tag{1.123}$$

The group velocity of the wave packet is thus equal to the classical velocity of the particle, $v_g = v_{particle}$. This suggests we should view the "center" of the wave packet as traveling like a classical particle that obeys the laws of classical mechanics: the center would then follow the "classical trajectory" of the particle. We now see how the wave packet concept offers a clear connection between the classical description of a particle and its quantum mechanical description. In the case of a *free* particle, an insertion of $V = 0$ into (1.123) yields

$$v_g = \frac{p}{m}, \qquad v_{ph} = \frac{p}{2m} = \frac{1}{2}v_g. \tag{1.124}$$

This shows that, while the group velocity of the wave packet corresponding to a free particle is equal to the particle's velocity, p/m, the phase velocity is half the group velocity. The expression $v_{ph} = \frac{1}{2}v_g$ is meaningless, for it states that the wave function travels at half the speed of the particle it is intended to represent. This is unphysical indeed. The phase velocity has in general no meaningful physical significance.

Time-evolution of the packet

Having taken a short detour to discuss the phase and group velocities, let us now return to our main task of calculating the packet $\psi(x, t)$ as listed in (1.118). For this, we need to decide on where to terminate the expansion (1.117) or the exponent in the integrand of (1.118). We are going to consider two separate cases corresponding to whether we terminate the exponent in (1.118) at the linear term, $(k - k_0)v_g t$, or at the quadratic term, $(k - k_0)^2 \alpha t$. These two cases are respectively known as the *linear approximation* and the *quadratic approximation*.

In the linear approximation, which is justified when $g(k - k_0)$ is narrow enough to neglect the quadratic k^2 term, $(k - k_0)^2 \alpha t \ll 1$, the wave packet (1.118) becomes

$$\psi(x, t) = \frac{1}{\sqrt{2\pi}} e^{ik_0(x - v_{ph}t)} \int_{-\infty}^{+\infty} g(k - k_0) e^{i(k - k_0)(x - v_g t)} dk. \tag{1.125}$$

This relation can be rewritten as

$$\psi(x, t) = e^{ik_0(x - v_{ph}t)} \psi_0(x - v_g t) e^{-ik_0(x - v_g t)}, \tag{1.126}$$

where ψ_0 is the initial wave packet (see (1.95))

$$\psi_0(x - v_g t) = \frac{1}{\sqrt{2\pi}} \int_{-\infty}^{+\infty} g(q) e^{i(x - v_g t)q + ik_0(x - v_g t)} dq; \tag{1.127}$$

the new variable q stands for $q = k - k_0$. Equation (1.126) leads to

$$|\psi(x, t)|^2 = |\psi_0(x - v_g t)|^2. \tag{1.128}$$

Equation (1.126) represents a wave packet whose amplitude is modulated. As depicted in Figure 1.14, the modulating wave, $\psi_0(x - v_g t)$, propagates to the right with the group velocity v_g; the modulated wave, $e^{ik_0(x - v_{ph}t)}$, represents a pure harmonic wave of constant wave number k_0 that also travels to the right with the phase velocity v_{ph}. That is, (1.126) and (1.128) represent a wave packet whose peak travels as a whole with the velocity v_g, while the individual wave propagates inside the envelope with the velocity v_{ph}. The group velocity, which gives the velocity of the packet's peak, clearly represents the velocity of the particle, since the chance of finding the particle around the packet's peak is much higher than finding it in any other region of space; the wave packet is highly localized in the neighborhood of the particle's position and vanishes elsewhere. It is therefore the group velocity, not the phase velocity, that is equal to the velocity of the particle represented by the packet. This suggests that the motion of a material particle can be described well by wave packets. By establishing a correspondence between the particle's velocity and the velocity of the wave packet's peak, we see that the wave packet concept jointly embodies the particle aspect and the wave aspect of material particles.

Now, what about the size of the wave packet in the linear approximation? Is it affected by the particle's propagation? Clearly not. This can be inferred immediately from (1.126): $\psi_0(x - v_g t)$ represents, mathematically speaking, a curve that travels to the right with a velocity v_g without deformation. This means that if the packet is initially Gaussian, it will remain Gaussian as it propagates in space without any change in its size.

To summarize, we have shown that, in the linear approximation, the wave packet propagates undistorted and undergoes a uniform translational motion. Next we are going to study the conditions under which the packet experiences deformation.

1.8.3.2 Propagation of a Wave Packet with Distortion

Let us now include the quadratic k^2 term, $(k - k_0)^2 \alpha t$, in the integrand's exponent of (1.118) and drop the higher terms. This leads to

$$\psi(x, t) = e^{ik_0(x - v_{ph}t)} f(x, t), \tag{1.129}$$

where $f(x, t)$, which represents the envelope of the packet, is given by

$$f(x, t) = \frac{1}{\sqrt{2\pi}} \int_{-\infty}^{+\infty} g(q) e^{iq(x - v_g t)} e^{-iq^2 \alpha t} \, dq, \tag{1.130}$$

with $q = k - k_0$. Were it not for the quadratic q^2 correction, $iq^2 \alpha t$, the wave packet would move uniformly without any change of shape, since similarly to (1.116), $f(x, t)$ would be given by $f(x, t) = \psi_0(x - v_g t)$.

To show how α affects the width of the packet, let us consider the Gaussian packet (1.102) whose amplitude is given by $\phi(k) = (a^2/2\pi)^{1/4} \exp[-a^2(k - k_0)^2/4]$ and whose initial width is $\Delta x_0 = a/2$ and $\Delta k = \hbar/a$. Substituting $\phi(k)$ into (1.129), we obtain

$$\psi(x, t) = \frac{1}{\sqrt{2\pi}} \left(\frac{a^2}{2\pi}\right)^{1/4} e^{ik_0(x - v_{ph}t)} \int_{-\infty}^{+\infty} \exp\left[iq(x - v_g t) - \left(\frac{a^2}{4} + i\alpha t\right) q^2\right] dq. \tag{1.131}$$

Evaluating the integral (the calculations are detailed in the following example, see Eq. (1.145)), we can show that the packet's density distribution is given by

$$|\psi(x, t)|^2 = \frac{1}{\sqrt{2\pi} \, \Delta x(t)} \exp\left\{-\frac{(x - v_g t)^2}{2 [\Delta x(t)]^2}\right\}, \tag{1.132}$$

where $\Delta x(t)$ is the width of the packet at time t:

$$\Delta x(t) = \frac{a}{2} \sqrt{1 + \frac{16\alpha^2}{a^4} t^2} = \Delta x_0 \sqrt{1 + \frac{\alpha^2 t^2}{(\Delta x_0)^4}}. \tag{1.133}$$

We see that the packet's width, which was initially given by $\Delta x_0 = a/2$, has grown by a factor of $\sqrt{1 + \alpha^2 t^2/(\Delta x_0)^4}$ after time t. Hence the wave packet is spreading; *the spreading is due to the inclusion of the quadratic q^2 term, $iq^2 \alpha t$.* Should we drop this term, the packet's width $\Delta x(t)$ would then remain constant, equal to Δx_0.

The density distribution (1.132) displays two results: (1) the center of the packet moves with the group velocity; (2) the packet's width increases linearly with time. From (1.133) we see that the packet begins to spread appreciably only when $\alpha^2 t^2/(\Delta x_0)^4 \approx 1$ or $t \approx (\Delta x_0)^2/\alpha$. In fact, if $t \ll (\Delta x_0)^2/\alpha$ the packet's spread will be negligible, whereas if $t \gg \frac{(\Delta x_0)^2}{\alpha}$ the packet's spread will be significant.

To be able to make concrete statements about the growth of the packet, as displayed in (1.133), we need to specify α; this reduces to determining the function $\omega(k)$, since $\alpha = \frac{1}{2} \frac{d^2\omega}{dk^2}\big|_{k=k_0}$. For this, let us invoke an example that yields itself to explicit calculation. In fact, the example we are going to consider—a *free* particle with a Gaussian amplitude—allows the calculations to be performed *exactly*; hence there is no need to expand $\omega(k)$.

Example 1.9 (Free particle with a Gaussian wave packet)

Determine how the wave packet corresponding to a free particle, with an initial Gaussian packet, spreads in time.

Solution

The issue here is to find out how the wave packet corresponding to a free particle with $\phi(k) = (a^2/2\pi)^{1/4} e^{-a^2(k-k_0)^2/4}$ (see (1.110)) spreads in time.

First, we need to find the form of the wave packet, $\psi(x, t)$. Substituting the amplitude $\phi(k) = (a^2/2\pi)^{1/4} e^{-a^2(k-k_0)^2/4}$ into the Fourier integral (1.94), we obtain

$$\psi(x, t) = \frac{1}{\sqrt{2\pi}} \left(\frac{a^2}{2\pi}\right)^{1/4} \int_{-\infty}^{+\infty} \exp\left[-\frac{a^2}{4}(k - k_0)^2 + i(kx - \omega t)\right] dk. \tag{1.134}$$

Since $\omega(k) = \hbar k^2/(2m)$ (the dispersion relation for a free particle), and using a change of variables $q = k - k_0$, we can write the exponent in the integrand of (1.134) as a perfect square for q:

$$-\frac{a^2}{4}(k - k_0)^2 + i\left(kx - \frac{\hbar k^2}{2m}t\right) = -\left(\frac{a^2}{4} + i\frac{\hbar t}{2m}\right)q^2 + i\left(x - \frac{\hbar k_0 t}{m}\right)q$$

$$+ ik_0\left(x - \frac{\hbar k_0 t}{2m}\right)$$

$$= -\alpha q^2 + i\left(x - \frac{\hbar k_0 t}{m}\right)q + ik_0\left(x - \frac{\hbar k_0 t}{2m}\right)$$

$$= -\alpha\left[q - \frac{i}{2\alpha}\left(x - \frac{\hbar k_0 t}{m}\right)\right]^2 - \frac{1}{4\alpha}\left(x - \frac{\hbar k_0 t}{m}\right)^2$$

$$+ ik_0\left(x - \frac{\hbar k_0 t}{2m}\right), \tag{1.135}$$

where we have used the relation $-\alpha q^2 + iyq = -\alpha[q - iy/(2\alpha)]^2 - y^2/(4\alpha)$, with $y = x - \hbar k_0 t/m$ and

$$\alpha = \frac{a^2}{4} + i\frac{\hbar t}{2m}. \tag{1.136}$$

Substituting (1.135) into (1.134) we obtain

$$\psi(x, t) = \frac{1}{\sqrt{2\pi}} \left(\frac{a^2}{2\pi}\right)^{1/4} \exp\left[ik_0\left(x - \frac{\hbar k_0 t}{2m}\right)\right] \exp\left[-\frac{1}{4\alpha}\left(x - \frac{\hbar k_0 t}{m}\right)^2\right]$$

$$\times \int_{-\infty}^{+\infty} \exp\left\{-\alpha\left[q - \frac{i}{2\alpha}\left(x - \frac{\hbar k_0 t}{m}\right)\right]^2\right\} dq. \tag{1.137}$$

Combined with the integral[16] $\int_{-\infty}^{+\infty} \exp\left[-\alpha(q - iy/(2\alpha))^2\right] dq = \sqrt{\pi/\alpha}$, (1.137) leads to

$$\psi(x, t) = \frac{1}{\sqrt{\alpha}} \left(\frac{a^2}{8\pi}\right)^{1/4} \exp\left[ik_0\left(x - \frac{\hbar k_0 t}{2m}\right)\right] \exp\left[-\frac{1}{4\alpha}\left(x - \frac{\hbar k_0 t}{m}\right)^2\right]. \tag{1.138}$$

---

[16]If β and δ are two complex numbers and if Re $\beta > 0$, we have $\int_{-\infty}^{+\infty} e^{-\beta(q+\delta)^2} dq = \sqrt{\pi/\beta}$.

Since α is a complex number (see (1.136)), we can write it in terms of its modulus and phase

$$\alpha = \frac{a^2}{4}\left(1 + i\frac{2\hbar t}{ma^2}\right) = \frac{a^2}{4}\left(1 + \frac{4\hbar^2 t^2}{m^2 a^4}\right)^{1/2} e^{i\theta}, \tag{1.139}$$

where $\theta = \tan^{-1}\left[2\hbar t/(ma^2)\right]$; hence

$$\frac{1}{\sqrt{\alpha}} = \frac{2}{a}\left(1 + \frac{4\hbar^2 t^2}{m^2 a^4}\right)^{-1/4} e^{-i\theta/2}. \tag{1.140}$$

Substituting (1.136) and (1.140) into (1.138), we have

$$\psi(x,t) = \left(\frac{2}{\pi a^2}\right)^{1/4}\left(1 + \frac{4\hbar^2 t^2}{m^2 a^4}\right)^{-1/4} e^{-i\theta/2} e^{ik_0(x - \hbar k_0 t/2m)} \exp\left[-\frac{(x - \hbar k_0 t/m)^2}{a^2 + 2i\hbar t/m}\right]. \tag{1.141}$$

Since $\left|e^{-y^2/(a^2 + 2i\hbar t/m)}\right|^2 = e^{-y^2/(a^2 - 2i\hbar t/m)} e^{-y^2/(a^2 + 2i\hbar t/m)}$, where $y = x - \hbar k_0 t/m$, and since $y^2/(a^2 - 2i\hbar t/m) + y^2/(a^2 + 2i\hbar t/m) = 2a^2 y^2/(a^4 + 4\hbar^2 t^2/m^2)$, we have

$$\left|\exp\left(-\frac{y^2}{a^2 + 2i\hbar t/m}\right)\right|^2 = \exp\left(-\frac{2a^2 y^2}{a^4 + 4\hbar^2 t^2/m^2}\right); \tag{1.142}$$

hence

$$\begin{aligned}
|\psi(x,t)|^2 &= \sqrt{\frac{2}{\pi a^2}}\left(1 + \frac{4\hbar^2 t^2}{m^2 a^4}\right)^{-1/2}\left|\exp\left[-\frac{(x - \hbar k_0 t/m)^2}{a^2 + 2i\hbar t/m}\right]\right|^2 \\
&= \sqrt{\frac{2}{\pi a^2}}\frac{1}{\gamma(t)}\exp\left\{-\frac{2}{[a\gamma(t)]^2}\left(x - \frac{\hbar k_0 t}{m}\right)^2\right\},
\end{aligned} \tag{1.143}$$

where $\gamma(t) = \sqrt{1 + 4\hbar^2 t^2/(m^2 a^4)}$.

We see that both the wave packet (1.141) and the probability density (1.143) remain Gaussian as time evolves. This can be traced to the fact that the x-dependence of the phase, $e^{ik_0 x}$, of $\psi_0(x)$ as displayed in (1.110) is *linear*. If the x-dependence of the phase were other than linear, say quadratic, the form of the wave packet would not remain Gaussian. So the phase factor $e^{ik_0 x}$, which was present in $\psi_0(x)$, allows us to account for the motion of the particle.

Since the group velocity of a free particle is $v_g = d\omega/dk = \frac{d}{dk}\left(\frac{\hbar k^2}{2m}\right)\Big|_{k_0} = \hbar k_0/m$, we can rewrite (1.141) as follows[17]:

$$\psi(x,t) = \frac{1}{\sqrt{\sqrt{2\pi}\,\Delta x(t)}} e^{-i\theta/2} e^{ik_0(x - v_g t/2)} \exp\left[-\frac{(x - v_g t)^2}{a^2 + 2i\hbar t/m}\right], \tag{1.144}$$

$$\left|\psi(x,t)\right|^2 = \frac{1}{\sqrt{2\pi}\,\Delta x(t)} \exp\left\{-\frac{(x - v_g t)^2}{2[\Delta x(t)]^2}\right\}, \tag{1.145}$$

---

[17]It is interesting to note that the harmonic wave $e^{ik_0(x - v_g t/2)}$ propagates with a phase velocity which is *half* the group velocity; as shown in (1.124), this is a property of *free* particles.

Figure 1.15 Time evolution of $|\psi(x, t)|^2$: the peak of the packet, which is centered at $x = v_g t$, moves with the speed v_g from left to right. The height of the packet, represented here by the dotted envelope, is modulated by the function $1/(\sqrt{2\pi}\,\Delta x(t))$, which goes to zero at $t \to \pm\infty$ and is equal to $\sqrt{2/\pi a^2}$ at $t = 0$. The width of the packet $\Delta x(t) = \Delta x_0 \sqrt{1 + (t/\tau)^2}$ increases linearly with time.

where[18]

$$\Delta x(t) = \frac{a}{2}\gamma(t) = \frac{a}{2}\sqrt{1 + \frac{4\hbar^2 t^2}{m^2 a^4}} \tag{1.146}$$

represents the width of the wave packet at time t. Equations (1.144) and (1.145) describe a Gaussian wave packet that is centered at $x = v_g t$ whose peak travels with the group speed $v_g = \hbar k_0/m$ and whose width $\Delta x(t)$ increases linearly with time. So, during time t, the packet's center has moved from $x = 0$ to $x = v_g t$ and its width has expanded from $\Delta x_0 = a/2$ to $\Delta x(t) = \Delta x_0 \sqrt{1 + 4\hbar^2 t^2/(m^2 a^4)}$. The wave packet therefore undergoes a distortion; although it remains Gaussian, its width broadens linearly with time whereas its height, $1/(\sqrt{2\pi}\,\Delta x(t))$, decreases with time. As depicted in Figure 1.15, the wave packet, which had a very broad width and a very small amplitude at $t \to -\infty$, becomes narrower and narrower and its amplitude larger and larger as time increases towards $t = 0$; at $t = 0$ the packet is very localized, its width and amplitude being given by $\Delta x_0 = a/2$ and $\sqrt{2/\pi a^2}$, respectively. Then, as time increases $(t > 0)$, the width of the packet becomes broader and broader, and its amplitude becomes smaller and smaller.

In the rest of this section we are going to comment on several features that are relevant not only to the Gaussian packet considered above but also to more general wave packets. First, let us begin by estimating the time at which the wave packet starts to spread out appreciably. The packet, which is initially narrow, begins to grow out noticeably only when the second term, $2\hbar t/(ma^2)$, under the square root sign of (1.146) is of order unity. For convenience, let us write

---

[18]We can derive (1.146) also from (1.111): a combination of the half-width $|\psi(\pm\Delta x,\ t)|^2 / |\psi(0,\ 0)|^2 = e^{-1/2}$ with (1.143) yields $e^{-2[\Delta x/a\gamma(t)]^2} = e^{-1/2}$, which in turn leads to (1.146).

(1.146) in the form

$$\Lambda x(t) = \Delta x_0 \sqrt{1 + \left(\frac{t}{\tau}\right)^2},$$
(1.147)

where

$$\tau = \frac{2m(\Delta x_0)^2}{\hbar}$$
(1.148)

represents a time constant that characterizes the rate of the packet's spreading. Now we can estimate the order of magnitude of τ; it is instructive to evaluate it for *microscopic* particles as well as for *macroscopic* particles. For instance, τ for an electron whose position is defined to within 10^{-10} m is given by[19] $\tau \simeq 1.7 \times 10^{-16}$ s; on the other hand, the time constant for a macroscopic particle of mass say 1 g whose position is defined to within 1 mm is of the order[20] of $\tau \simeq 2 \times 10^{25}$ s (for an illustration see Problems 1.15 and 1.16). This crude calculation suggests that the wave packets of microscopic systems very quickly undergo significant growth; as for the packets of macroscopic systems, they begin to grow out noticeably only after the system has been in motion for an absurdly long time, a time of the order of, if not much higher than, the age of the Universe itself, which is about 4.7×10^{17} s. Having estimated the times at which the packet's spread becomes appreciable, let us now shed some light on the size of the spread. From (1.147) we see that when $t \gg \tau$ the packet's spreading is significant and, conversely, when $t \ll \tau$ the spread is negligible. As the cases $t \gg \tau$ and $t \ll \tau$ correspond to microscopic and macroscopic systems, respectively, we infer that the packet's dispersion is significant for microphysical systems and negligible for macroscopic systems. In the case of macroscopic systems, the spread is there but it is too small to detect. For an illustration see Problem 1.15 where we show that the width of a 100 g object increases by an absurdly small factor of about 10^{-29} after traveling a distance of 100 m, but the width of a 25 eV electron increases by a factor of 10^9 after traveling the same distance (in a time of 3.3×10^{-5} s). Such an immense dispersion in such a short time is indeed hard to visualize classically; this motion cannot be explained by classical physics.

So the wave packets of propagating, microscopic particles are prone to spreading out very significantly in a short time. This spatial spreading seems to generate a conceptual problem: the spreading is incompatible with our expectation that the packet should remain highly localized at all times. After all, the wave packet is supposed to represent the particle and, as such, it is expected to travel without dispersion. For instance, the charge of an electron does not spread out while moving in space; the charge should remain localized inside the corresponding wave packet. In fact, whenever microscopic particles (electrons, neutrons, protons, etc.) are observed, they are always confined to small, finite regions of space; they never spread out as suggested by equation (1.146). How do we explain this apparent contradiction? The problem here has to do with the proper interpretation of the situation: we must modify the classical concepts pertaining to the meaning of the position of a particle. The wave function (1.141) cannot be identified with a material particle. The quantity $|\psi(x, t)|^2 dx$ represents the probability (Born's interpretation) of finding the particle described by the packet $\psi(x, t)$ at time t in the spatial region located between x and $x + dx$. *The material particle does not disperse (or fuzz out); yet its position cannot be known exactly.* The spreading of the matter wave, which is accompanied by a shrinkage of its height, as indicated in Figure 1.15, corresponds to a decrease

---

[19]If $\Delta x_0 = 10^{-10}$ m and since the rest mass energy of an electron is $mc^2 = 0.5$ MeV and using $\hbar c \simeq 197 \times 10^{-15}$ MeV m, we have $\tau = 2mc^2(\Delta x_0)^2/((\hbar c)c) \simeq 1.7 \times 10^{-16}$ s.

[20]Since $\hbar = 1.05 \times 10^{-34}$ J s we have $\tau = 2 \times 0.001$ kg $\times (0.001$ m$)^2/(1.05 \times 10^{-34}$ J s$) \simeq 2 \times 10^{25}$ s.

of the probability density $|\psi(x, t)|^2$ and implies in no way a growth in the size of the particle. So the wave packet gives only the probability that the particle it represents will be found at a given position. No matter how broad the packet becomes, we can show that its norm is always conserved, for it does not depend on time. In fact, as can be inferred from (1.143), the norm of the packet is equal to one:

$$\int_{-\infty}^{+\infty} |\psi(x, t)|^2 \, dx = \sqrt{\frac{2}{\pi a^2}} \frac{1}{\gamma} \int_{-\infty}^{+\infty} \exp\left\{ -\frac{2 (x - \hbar k_0 t/m)^2}{(a\gamma)^2} \right\} dx = \sqrt{\frac{2}{\pi a^2}} \frac{1}{\gamma} \sqrt{\frac{\pi a^2 \gamma^2}{2}} = 1,$$

(1.149)

since $\int_{-\infty}^{+\infty} e^{-ax^2} dx = \sqrt{\pi/a}$. This is expected, since the probability of finding the particle somewhere along the x-axis must be equal to one. The important issue here is that the norm of the packet is time independent and that its spread does not imply that the material particle becomes bloated during its motion, but simply implies a redistribution of the probability density. So, in spite of the significant spread of the packets of microscopic particles, the norms of these packets are always conserved—normalized to unity.

Besides, we should note that the example considered here is an *idealized* case, for we are dealing with a *free* particle. If the particle is subject to a potential, as in the general case, its wave packet will not spread as dramatically as that of a free particle. In fact, a varying potential can cause the wave packet to become narrow. This is indeed what happens when a measurement is performed on a microscopic system; the interaction of the system with the measuring device makes the packet very narrow, as will be seen in Chapter 3.

Let us now study how the spreading of the wave packet affects the uncertainties product $\Delta x(t)\Delta p(t)$. First, we should point out that the average momentum of the packet $\hbar k_0$ and its uncertainty $\hbar \Delta k$ do not change in time. This can be easily inferred as follows. Rewriting (1.94) in the form

$$\psi(x, t) = \frac{1}{\sqrt{2\pi}} \int_{-\infty}^{+\infty} \phi(k, 0) e^{i(kx-\omega t)} dk = \frac{1}{\sqrt{2\pi}} \int_{-\infty}^{+\infty} \phi(k, t) e^{ikx} dk,$$

(1.150)

we have

$$\phi(k, t) = e^{-i\omega(k)t} \phi(k, 0),$$

(1.151)

where $\phi(k, 0) = (a^2/2\pi)^{1/4} e^{-a^2(k-k_0)^2/4}$; hence

$$|\phi(k, t)|^2 = |\phi(k, 0)|^2.$$

(1.152)

This suggests that the widths of $\phi(k, t)$ and $\phi(k, 0)$ are equal; hence Δk remains constant and so must the momentum dispersion Δp (this is expected because the momentum of a free particle is a constant of the motion). Since the width of $\phi(k, 0)$ is given by $\Delta k = 1/a$ (see (1.112)), we have

$$\Delta p = \hbar \Delta k = \frac{\hbar}{a}.$$

(1.153)

Multiplying this relation by (1.146), we have

$$\Delta x(t)\Delta p = \frac{\hbar}{2}\sqrt{1 + \frac{4\hbar^2}{m^2 a^4} t^2},$$

(1.154)

which shows that $\Delta x(t)\Delta p \geq \hbar/2$ is satisfied at all times. Notably, when $t = 0$ we obtain the lower bound limit $\Delta x_0 \Delta p = \hbar/2$; this is the uncertainty relation for a stationary Gaussian packet (see (1.114)). As $|t|$ increases, however, we obtain an inequality, $\Delta x(t)\Delta p > \hbar/2$.

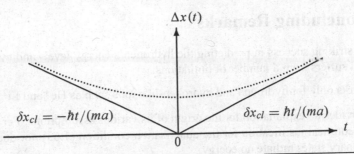

Figure 1.16 Time evolutions of the packet's width $\Delta x(t) = \Delta x_0\sqrt{1 + (\delta x_{cl}(t)/\Delta x_0)^2}$ (dotted curve) and of the classical dispersion $\delta x_{cl}(t) = \pm\hbar t/(ma)$ (solid lines). For large values of $|t|$, $\Delta x(t)$ approaches $\delta x_{cl}(t)$ and at $t = 0$, $\Delta x(0) = \Delta x_0 = a/2$.

Having shown that the width of the packet does not disperse in momentum space, let us now study the dispersion of the packet's width in x-space. Since $\Delta x_0 = a/2$ we can write (1.146) as

$$\Delta x(t) = \frac{a}{2}\sqrt{1 + \frac{4\hbar^2 t^2}{m^2 a^4}} = \Delta x_0 \sqrt{1 + \left(\frac{\delta x_{cl}(t)}{\Delta x_0}\right)^2}, \tag{1.155}$$

where the dispersion factor $\delta x_{cl}(t)/\Delta x_0$ is given by

$$\frac{\delta x_{cl}(t)}{\Delta x_0} = \pm\frac{2\hbar}{ma^2}t = \pm\frac{\hbar}{2m\Delta x_0^2}t; \tag{1.156}$$

As shown in Figure 1.16, when $|t|$ is large (i.e., $t \to \pm\infty$), we have $\Delta x(t) \to \delta x_{cl}(t)$ with

$$\delta x_{cl}(t) = \pm\frac{\hbar t}{ma} = \pm\frac{\Delta p}{m}t = \pm\Delta vt, \tag{1.157}$$

where $\Delta v = \hbar/(ma)$ represents the dispersion in velocity. This means that if a particle starts initially ($t = 0$) at $x = 0$ with a velocity dispersion equal to Δv, then Δv will remain constant but the dispersion of the particle's position will increase linearly with time: $\delta x_{cl}(t) = \hbar|t|/(ma)$ (Figure 1.16). We see from (1.155) that if $\delta x_{cl}(t)/\Delta x_0 \ll 1$, the spreading of the wave packet is negligible, but if $\delta x_{cl}(t)/\Delta x_0 \gg 1$, the wave packet will spread out without bound.

We should highlight at this level the importance of the classical limit of (1.154): in the limit $\hbar \to 0$, the product $\Delta x(t)\Delta p$ goes to zero. This means that the x and p uncertainties become negligible; that is, in the classical limit, the wave packet will propagate without spreading. In this case the center of the wave packet moves like a free particle that obeys the laws of classical mechanics. *The spread of wave packets is thus a purely quantum effect.* So when $\hbar \to 0$ all quantum effects, the spread of the packet, disappear.

We may conclude this study of wave packets by highlighting their importance:

- They provide a linkage with the Heisenberg uncertainty principle.

- They embody and unify the particle and wave features of matter waves.

- They provide a linkage between wave intensities and probabilities.

- They provide a connection between classical and quantum mechanics.

1.9   Concluding Remarks

Despite its striking success in predicting the hydrogen's energy levels and transition rates, the Bohr model suffers from a number of limitations:

- It works only for hydrogen and hydrogen-like ions such as He^+ and Li^{2+}.

- It provides no explanation for the origin of its various assumptions. For instance, it gives no theoretical justification for the quantization condition (1.63) nor does it explain why stationary states radiate no energy.

- It fails to explain why, instead of moving *continuously* from one energy level to another, the electrons *jump* from one level to the other.

The model therefore requires considerable extension to account for the electronic properties and spectra of a wide range of atoms. Even in its present limited form, Bohr's model represents a bold and major departure from classical physics: classical physics offers no justification for the existence of discrete energy states in a system such as a hydrogen atom and no justification for the quantization of the angular momentum.

In its present form, the model not only suffers from incompleteness but also lacks the ingredients of a consistent theory. It was built upon a series of ad hoc, piecemeal assumptions. These assumptions were not derived from the first principles of a more general theory, but postulated rather arbitrarily.

The formulation of the theory of quantum mechanics was largely precipitated by the need to find a theoretical foundation for Bohr's ideas as well as to explain, from first principles, a wide variety of other microphysical phenomena such as the puzzling processes discussed in this chapter. It is indeed surprising that a single theory, quantum mechanics, is powerful and rich enough to explain accurately a wide variety of phenomena taking place at the molecular, atomic, and subatomic levels.

In this chapter we have dealt with the most important experimental facts which confirmed the failure of classical physics and subsequently led to the birth of quantum mechanics. In the rest of this text we will focus on the formalism of quantum mechanics and on its application to various microphysical processes. To prepare for this task, we need first to study the mathematical tools necessary for understanding the formalism of quantum mechanics; this is taken up in Chapter 2.

1.10   Solved Problems

Numerical calculations in quantum physics can be made simpler by using the following units. First, it is convenient to express energies in units of electronvolt (eV): one eV is defined as the energy acquired by an electron passing through a potential difference of one Volt. The electronvolt unit can be expressed in terms of joules and vice versa: $1 \text{ eV} = (1.6 \times 10^{-19} \text{ C}) \times (1 \text{ V}) = 1.6 \times 10^{-19}$ J and $1 \text{ J} = 0.625 \times 10^{19}$ eV.

It is also convenient to express the masses of subatomic particles, such as the electron, proton, and neutron, in terms of their rest mass energies: $m_e c^2 = 0.511$ MeV, $m_p c^2 = 938.27$ MeV, and $m_n c^2 = 939.56$ MeV.

In addition, the quantities $\hbar c = 197.33$ MeV fm $= 197.33 \times 10^{-15}$ MeV m or $hc = 1242.37 \times 10^{-10}$ eV m are sometimes more convenient to use than $\hbar = 1.05 \times 10^{-34}$ J s.

Additionally, instead of $1/(4\pi\varepsilon_0) = 8.9 \times 10^9$ N m^2 C^{-2}, one should sometimes use the fine structure constant $\alpha = e^2/[(4\pi\varepsilon_0)\hbar c] = 1/137$.

Problem 1.1

A 45 kW broadcasting antenna emits radio waves at a frequency of 4 MHz.

(a) How many photons are emitted per second?

(b) Is the quantum nature of the electromagnetic radiation important in analyzing the radiation emitted from this antenna?

Solution

(a) The electromagnetic energy emitted by the antenna in one second is $E = 45\,000$ J. Thus, the number of photons emitted in one second is

$$n = \frac{E}{h\nu} = \frac{45\,000 \text{ J}}{6.63 \times 10^{-34} \text{ J s} \times 4 \times 10^6 \text{ Hz}} = 1.7 \times 10^{31}. \qquad (1.158)$$

(b) Since the antenna emits a huge number of photons every second, 1.7×10^{31}, the quantum nature of this radiation is unimportant. As a result, this radiation can be treated fairly accurately by the classical theory of electromagnetism.

Problem 1.2

Consider a mass–spring system where a 4 kg mass is attached to a massless spring of constant $k = 196$ N m^{-1}; the system is set to oscillate on a frictionless, horizontal table. The mass is pulled 25 cm away from the equilibrium position and then released.

(a) Use classical mechanics to find the total energy and frequency of oscillations of the system.

(b) Treating the oscillator with quantum theory, find the energy spacing between two consecutive energy levels and the total number of quanta involved. Are the quantum effects important in this system?

Solution

(a) According to classical mechanics, the frequency and the total energy of oscillations are given by

$$\nu = \frac{1}{2\pi}\sqrt{\frac{k}{m}} = \frac{1}{2\pi}\sqrt{\frac{196}{4}} = 1.11 \text{ Hz}, \qquad E = \frac{1}{2}kA^2 = \frac{196}{2}(0.25)^2 = 6.125 \text{ J}. \quad (1.159)$$

(b) The energy spacing between two consecutive energy levels is given by

$$\Delta E = h\nu = (6.63 \times 10^{-34} \text{ J s}) \times (1.11 \text{ Hz}) = 7.4 \times 10^{-34} \text{ J} \qquad (1.160)$$

and the total number of quanta is given by

$$n = \frac{E}{\Delta E} = \frac{6.125 \text{ J}}{7.4 \times 10^{-34} \text{ J}} = 8.3 \times 10^{33}. \qquad (1.161)$$

We see that the energy of one quantum, 7.4×10^{-34} J, is completely negligible compared to the total energy 6.125 J, and that the number of quanta is very large. As a result, the energy levels of the oscillator can be viewed as continuous, for it is not feasible classically to measure the spacings between them. Although the quantum effects are present in the system, they are beyond human detection. So quantum effects are negligible for macroscopic systems.

Problem 1.3

When light of a given wavelength is incident on a metallic surface, the stopping potential for the photoelectrons is 3.2 V. If a second light source whose wavelength is double that of the first is used, the stopping potential drops to 0.8 V. From these data, calculate
(a) the wavelength of the first radiation and
(b) the work function and the cutoff frequency of the metal.

Solution

(a) Using (1.23) and since the wavelength of the second radiation is double that of the first one, $\lambda_2 = 2\lambda_1$, we can write

$$V_{s_1} = \frac{hc}{e\lambda_1} - \frac{W}{e}, \tag{1.162}$$

$$V_{s_2} = \frac{hc}{e\lambda_2} - \frac{W}{e} = \frac{hc}{2e\lambda_1} - \frac{W}{e}. \tag{1.163}$$

To obtain λ_1 we have only to subtract (1.163) from (1.162):

$$V_{s_1} - V_{s_2} = \frac{hc}{e\lambda_1}\left(1 - \frac{1}{2}\right) = \frac{hc}{2e\lambda_1}. \tag{1.164}$$

The wavelength is thus given by

$$\lambda_1 = \frac{hc}{2e(V_{s_1} - V_{s_2})} = \frac{6.6 \times 10^{-34} \text{ J s} \times 3 \times 10^8 \text{m s}^{-1}}{2 \times 1.6 \times 10^{-19} \text{ C} \times (3.2 \text{ V} - 0.8 \text{ V})} = 2.6 \times 10^{-7} \text{ m}. \tag{1.165}$$

(b) To obtain the work function, we simply need to multiply (1.163) by 2 and subtract the result from (1.162), $V_{s_1} - 2V_{s_2} = W/e$, which leads to

$$W = e(V_{s_1} - 2V_{s_2}) = 1.6 \text{ eV} = 1.6 \times 1.6 \times 10^{-19} = 2.56 \times 10^{-19} \text{ J}. \tag{1.166}$$

The cutoff frequency is

$$\nu = \frac{W}{h} = \frac{2.56 \times 10^{-19} \text{ J}}{6.6 \times 10^{-34} \text{ J s}} = 3.9 \times 10^{14} \text{ Hz}. \tag{1.167}$$

Problem 1.4

(a) Estimate the energy of the electrons that we need to use in an electron microscope to resolve a separation of 0.27 nm.
(b) In a scattering of 2 eV protons from a crystal, the fifth maximum of the intensity is observed at an angle of 30°. Estimate the crystal's planar separation.

Solution

(a) Since the electron's momentum is $p = 2\pi\hbar/\lambda$, its kinetic energy is given by

$$E = \frac{p^2}{2m_e} = \frac{2\pi^2\hbar^2}{m_e\lambda^2}. \tag{1.168}$$

Since $m_e c^2 = 0.511$ MeV, $\hbar c = 197.33 \times 10^{-15}$ MeV m, and $\lambda = 0.27 \times 10^{-9}$ m, we have

$$E = \frac{2\pi^2(\hbar c)^2}{(m_e c^2)\lambda^2} = \frac{2\pi^2(197.33 \times 10^{-15} \text{ MeV m})^2}{(0.511 \text{ MeV})(0.27 \times 10^{-9} \text{ m})^2} = 20.6 \text{ eV}. \tag{1.169}$$

(b) Using Bragg's relation (1.46), $\lambda = (2d/n) \sin \phi$, where d is the crystal's planar separation, we can infer the proton's kinetic energy from (1.168):

$$E = \frac{p^2}{2m_p} = \frac{2\pi^2 \hbar^2}{m_p \lambda^2} = \frac{n^2 \pi^2 \hbar^2}{2m_p d^2 \sin^2 \phi}, \tag{1.170}$$

which leads to

$$d = \frac{n\pi \hbar}{(\sin \phi)\sqrt{2m_p E}} = \frac{n\pi \hbar c}{(\sin \phi)\sqrt{2m_p c^2 E}}. \tag{1.171}$$

Since $n = 5$ (the fifth maximum), $\phi = 30°$, $E = 2$ eV, and $m_p c^2 = 938.27$ MeV, we have

$$d = \frac{5\pi \times 197.33 \times 10^{-15} \text{ MeV m}}{(\sin 30°)\sqrt{2 \times 938.27 \text{ MeV} \times 2 \times 10^{-6} \text{ MeV}}} = 0.101 \text{ nm.} \tag{1.172}$$

Problem 1.5
A photon of energy 3 keV collides elastically with an electron initially at rest. If the photon emerges at an angle of 60°, calculate
(a) the kinetic energy of the recoiling electron and
(b) the angle at which the electron recoils.

Solution
(a) From energy conservation, we have

$$h\nu + m_e c^2 = h\nu' + (K_e + m_e c^2), \tag{1.173}$$

where $h\nu$ and $h\nu'$ are the energies of the initial and scattered photons, respectively, $m_e c^2$ is the rest mass energy of the initial electron, $(K_e + m_e c^2)$ is the total energy of the recoiling electron, and K_e is its recoil kinetic energy. The expression for K_e can immediately be inferred from (1.173):

$$K_e = h(\nu - \nu') = hc \left(\frac{1}{\lambda} - \frac{1}{\lambda'} \right) = \frac{hc}{\lambda} \frac{\lambda' - \lambda}{\lambda'} = (h\nu)\frac{\Delta\lambda}{\lambda'}, \tag{1.174}$$

where the wave shift $\Delta\lambda$ is given by (1.36):

$$\begin{aligned} \Delta\lambda &= \lambda' - \lambda = \frac{h}{m_e c}(1 - \cos\theta) = \frac{2\pi \hbar c}{m_e c^2}(1 - \cos\theta) \\ &= \frac{2\pi \times 197.33 \times 10^{-15} \text{ MeV m}}{0.511 \text{ MeV}}(1 - \cos 60°) \\ &= 0.0012 \text{ nm.} \end{aligned} \tag{1.175}$$

Since the wavelength of the incident photon is $\lambda = 2\pi \hbar c/(h\nu)$, we have $\lambda = 2\pi \times 197.33 \times 10^{-15}$ MeV m$/(0.003$ MeV$) = 0.414$ nm; the wavelength of the scattered photon is given by

$$\lambda' = \lambda + \Delta\lambda = 0.4152 \text{ nm.} \tag{1.176}$$

Now, substituting the numerical values of λ' and $\Delta\lambda$ into (1.174), we obtain the kinetic energy of the recoiling electron

$$K_e = (h\nu)\frac{\Delta\lambda}{\lambda'} = (3 \text{ keV}) \times \frac{0.0012 \text{ nm}}{0.4152 \text{ nm}} = 8.671 \text{ eV.} \tag{1.177}$$

(b) To obtain the angle at which the electron recoils, we need simply to use the conservation of the total momentum along the $x-$ and $y-$ axes:

$$p = p_e \cos \phi + p' \cos \theta, \qquad 0 = p_e \sin \phi - p' \sin \theta. \qquad (1.178)$$

These can be rewritten as

$$p_e \cos \phi = p - p' \cos \theta, \qquad p_e \sin \phi = p' \sin \theta, \qquad (1.179)$$

where p and p' are the momenta of the initial and final photons, p_e is the momentum of the recoiling electron, and θ and ϕ are the angles at which the photon and electron scatter, respectively (Figure 1.4). Taking (1.179) and dividing the second equation by the first, we obtain

$$\tan \phi = \frac{\sin \theta}{p/p' - \cos \theta} = \frac{\sin \theta}{\lambda'/\lambda - \cos \theta}, \qquad (1.180)$$

where we have used the momentum expressions of the incident photon $p = h/\lambda$ and of the scattered photon $p' = h/\lambda'$. Since $\lambda = 0.414$ nm and $\lambda' = 0.4152$ nm, the angle at which the electron recoils is given by

$$\phi = \tan^{-1} \left(\frac{\sin \theta}{\lambda'/\lambda - \cos \theta} \right) = \tan^{-1} \left(\frac{\sin 60°}{0.4152/0.414 - \cos 60°} \right) = 59.86°. \qquad (1.181)$$

Problem 1.6

Show that the maximum kinetic energy transferred to a proton when hit by a photon of energy $h\nu$ is $K_p = h\nu/[1 + m_p c^2/(2h\nu)]$, where m_p is the mass of the proton.

Solution

Using (1.35), we have

$$\frac{1}{\nu'} = \frac{1}{\nu} + \frac{h}{m_p c^2}(1 - \cos \theta), \qquad (1.182)$$

which leads to

$$h\nu' = \frac{h\nu}{1 + (h\nu/m_p c^2)(1 - \cos \theta)}. \qquad (1.183)$$

Since the kinetic energy transferred to the proton is given by $K_p = h\nu - h\nu'$, we obtain

$$K_p = h\nu - \frac{h\nu}{1 + (h\nu/m_p c^2)(1 - \cos \theta)} = \frac{h\nu}{1 + m_p c^2/[h\nu(1 - \cos \theta)]}. \qquad (1.184)$$

Clearly, the maximum kinetic energy of the proton corresponds to the case where the photon scatters backwards ($\theta = \pi$),

$$K_p = \frac{h\nu}{1 + m_p c^2/(2h\nu)}. \qquad (1.185)$$

Problem 1.7

Consider a photon that scatters from an electron at rest. If the Compton wavelength shift is observed to be triple the wavelength of the incident photon and if the photon scatters at 60°, calculate

(a) the wavelength of the incident photon,
(b) the energy of the recoiling electron, and
(c) the angle at which the electron scatters.

Solution

(a) In the case where the photons scatter at $\theta = 60°$ and since $\Delta\lambda = 3\lambda$, the wave shift relation (1.36) yields

$$3\lambda = \frac{h}{m_e c}(1 - \cos 60°),\tag{1.186}$$

which in turn leads to

$$\lambda = \frac{h}{6m_e c} = \frac{\pi \hbar c}{3m_e c^2} = \frac{3.14 \times 197.33 \times 10^{-15} \text{ MeV m}}{3 \times 0.511 \text{ MeV}} = 4.04 \times 10^{-13} \text{ m}.\tag{1.187}$$

(b) The energy of the recoiling electron can be obtained from the conservation of energy:

$$K_e = hc\left(\frac{1}{\lambda} - \frac{1}{\lambda'}\right) = \frac{3hc}{4\lambda} = \frac{3\pi \hbar c}{2\lambda} = \frac{3 \times 3.14 \times 197.33 \times 10^{-15} \text{ MeV m}}{2 \times 4.04 \times 10^{-13} \text{ m}} = 2.3 \text{ MeV}.\tag{1.188}$$

In deriving this relation, we have used the fact that $\lambda' = \lambda + \Delta\lambda = 4\lambda$.

(c) Since $\lambda' = 4\lambda$ the angle ϕ at which the electron recoils can be inferred from (1.181)

$$\phi = \tan^{-1}\left(\frac{\sin \theta}{\lambda'/\lambda - \cos \theta}\right) = \tan^{-1}\left(\frac{\sin 60°}{4 - \cos 60°}\right) = 13.9°.\tag{1.189}$$

Problem 1.8

In a double-slit experiment with a source of monoenergetic electrons, detectors are placed along a vertical screen parallel to the y-axis to monitor the diffraction pattern of the electrons emitted from the two slits. When only one slit is open, the amplitude of the electrons detected on the screen is $\psi_1(y, t) = A_1 e^{-i(ky-\omega t)}/\sqrt{1 + y^2}$, and when only the other is open the amplitude is $\psi_2(y, t) = A_2 e^{-i(ky+\pi y-\omega t)}/\sqrt{1 + y^2}$, where A_1 and A_2 are normalization constants that need to be found. Calculate the intensity detected on the screen when

(a) both slits are open and a light source is used to determine which of the slits the electron went through and

(b) both slits are open and no light source is used.

Plot the intensity registered on the screen as a function of y for cases (a) and (b).

Solution

Using the integral $\int_{-\infty}^{+\infty} dy/(1 + y^2) = \pi$, we can obtain the normalization constants at once: $A_1 = A_2 = 1/\sqrt{\pi}$; hence ψ_1 and ψ_2 become $\psi_1(y, t) = e^{-i(ky-\omega t)}/\sqrt{\pi(1 + y^2)}$, $\psi_2(y, t) = e^{-i(ky+\pi y-\omega t)}/\sqrt{\pi(1 + y^2)}$.

(a) When we use a light source to observe the electrons as they exit from the two slits on their way to the vertical screen, the total intensity recorded on the screen will be determined by a simple addition of the probability densities (or of the separate intensities):

$$I(y) = |\psi_1(y, t)|^2 + |\psi_2(y, t)|^2 = \frac{2}{\pi(1 + y^2)}.\tag{1.190}$$

As depicted in Figure 1.17a, the shape of the total intensity displays no interference pattern. Intruding on the electrons with the light source, we distort their motion.

Figure 1.17 Shape of the total intensity generated in a double slit experiment when both slits are open and (a) a light source is used to observe the electrons' motion, $I(y) = 2/\pi(1 + y^2)$, and no interference is registered; (b) no light source is used, $I(y) = 4/[\pi(1+y^2)]\cos^2(\pi y/2)$, and an interference pattern occurs.

(b) When no light source is used to observe the electrons, the motion will not be distorted and the total intensity will be determined by an addition of the amplitudes, not the intensities:

$$
\begin{aligned}
I(y) & = |\psi_1(y,t) + \psi_2(y,t)|^2 = \frac{1}{\pi(1+y^2)} \left| e^{-i(ky-\omega t)} + e^{-i(ky+\pi y-\omega t)} \right|^2 \\
& = \frac{1}{\pi(1+y^2)} \left(1 + e^{i\pi y}\right)\left(1 + e^{-i\pi y}\right) \\
& = \frac{4}{\pi(1+y^2)} \cos^2\left(\frac{\pi}{2}y\right).
\end{aligned}
\tag{1.191}
$$

The shape of this intensity does display an interference pattern which, as shown in Figure 1.17b, results from an oscillating function, $\cos^2(\pi y/2)$, modulated by $4/[\pi(1 + y^2)]$.

Problem 1.9
Consider a head-on collision between an α-particle and a lead nucleus. Neglecting the recoil of the lead nucleus, calculate the distance of closest approach of a 9.0 MeV α-particle to the nucleus.

Solution
In this head-on collision the distance of closest approach r_0 can be obtained from the conservation of energy $E_i = E_f$, where E_i is the initial energy of the system, α-particle plus the lead nucleus, when the particle and the nucleus are far from each other and thus feel no electrostatic potential between them. Assuming the lead nucleus to be at rest, E_i is simply the energy of the α-particle: $E_i = 9.0\,\text{MeV} = 9 \times 10^6 \times 1.6 \times 10^{-19}$ J.

As for E_f, it represents the energy of the system when the α-particle is at its closest distance from the nucleus. At this position, the α-particle is at rest and hence has no kinetic energy. The only energy the system has is the electrostatic potential energy between the α-particle and the lead nucleus, which has a positive charge of $82e$. Neglecting the recoil of the lead nucleus and since the charge of the α-particle is positive and equal to $2e$, we have $E_f = (2e)(82e)/(4\pi\varepsilon_0 r_0)$. The energy conservation $E_i = E_f$ or $(2e)(82e)/(4\pi\varepsilon_0 r_0) = E_i$ leads at once to

$$r_0 = \frac{(2e)(82e)}{4\pi\varepsilon_0 E_i} = 2.62 \times 10^{-14} \text{ m}, \qquad (1.192)$$

where we used the values $e = 1.6 \times 10^{-19}$ C and $1/(4\pi\varepsilon_0) = 8.9 \times 10^9$ N m^2 C^{-2}.

Problem 1.10

Considering that a quintuply ionized carbon ion, C^{5+}, behaves like a hydrogen atom, calculate

(a) the radius r_n and energy E_n for a given state n and compare them with the corresponding expressions for hydrogen,

(b) the ionization energy of C^{5+} when it is in its first excited state and compare it with the corresponding value for hydrogen, and

(c) the wavelength corresponding to the transition from state $n = 3$ to state $n = 1$; compare it with the corresponding value for hydrogen.

Solution

(a) The C^{5+} ion is generated by removing five electrons from the carbon atom. To find the expressions for r_{nC} and E_{nC} for the C^{5+} ion (which has 6 protons), we need simply to insert $Z = 6$ into (1.76):

$$r_{nC} = \frac{a_0}{6} n^2, \qquad E_{nC} = -\frac{36\mathcal{R}}{n^2}, \qquad (1.193)$$

where we have dropped the term m_e/M, since it is too small compared to one. Clearly, these expressions are related to their hydrogen counterparts by

$$r_{nC} = \frac{a_0}{6} n^2 = \frac{r_{nH}}{6}, \qquad E_{nC} = -\frac{36\mathcal{R}}{n^2} = 36 E_{nH}. \qquad (1.194)$$

(b) The ionization energy is the one needed to remove the only remaining electron of the C^{5+} ion. When the C^{5+} ion is in its first excited state, the ionization energy is

$$E_{2C} = -\frac{36\mathcal{R}}{4} = -9 \times 13.6 \text{ eV} = -122.4 \text{ eV}, \qquad (1.195)$$

which is equal to 36 times the energy needed to ionize the hydrogen atom in its first excited state: $E_{2H} = -3.4$ eV (note that we have taken $n = 2$ to correspond to the first excited state; as a result, the cases $n = 1$ and $n = 3$ will correspond to the ground and second excited states, respectively).

(c) The wavelength corresponding to the transition from state $n = 3$ to state $n = 1$ can be inferred from the relation $hc/\lambda = E_{3C} - E_{1C}$ which, when combined with $E_{1C} = -489.6$ eV and $E_{3C} = -54.4$ eV, leads to

$$\lambda = \frac{hc}{E_{3C} - E_{1C}} = \frac{2\pi\hbar c}{E_{3C} - E_{1C}} = \frac{2\pi \, 197.33 \times 10^{-9} \text{ eV m}}{-54.4 \text{ eV} + 489.6 \text{ eV}} = 2.85 \text{ nm.} \qquad (1.196)$$

Problem 1.11

(a) Find the Fourier transform for $\phi(k) = \begin{cases} A(a - |k|), & |k| \leq a, \\ 0, & |k| > a. \end{cases}$

where a is a positive parameter and A is a normalization factor to be found.

(b) Calculate the uncertainties Δx and Δp and check whether they satisfy the uncertainty principle.

Figure 1.18 The shape of the function $\phi(k)$ and its Fourier transform $\psi_0(x)$.

Solution

(a) The normalization factor A can be found at once:

$$
\begin{aligned}
1 &= \int_{-\infty}^{+\infty} |\phi(k)|^2 dk = |A|^2 \int_{-a}^{0} (a+k)^2 dk + |A|^2 \int_{0}^{a} (a-k)^2 dk \\
&= 2|A|^2 \int_{0}^{a} (a-k)^2 dk = 2|A|^2 \int_{0}^{a} \left(a^2 - 2ak + k^2\right) dk \\
&= \frac{2a^3}{3}|A|^2,
\end{aligned}
\tag{1.197}
$$

which yields $A = \sqrt{3/(2a^3)}$. The shape of $\phi(k) = \sqrt{3/(2a^3)}\,(a - |k|)$ is displayed in Figure 1.18.

Now, the Fourier transform of $\phi(k)$ is

$$
\begin{aligned}
\psi_0(x) &= \frac{1}{\sqrt{2\pi}} \int_{-\infty}^{+\infty} \phi(k) e^{ikx} dk \\
&= \frac{1}{\sqrt{2\pi}} \sqrt{\frac{3}{2a^3}} \left[\int_{-a}^{0} (a+k)\, e^{ikx} dk + \int_{0}^{a} (a-k)\, e^{ikx} dk \right] \\
&= \frac{1}{\sqrt{2\pi}} \sqrt{\frac{3}{2a^3}} \left[\int_{-a}^{0} k e^{ikx} dk - \int_{0}^{a} k e^{ikx} dk + a \int_{-a}^{a} e^{ikx} dk \right].
\end{aligned}
\tag{1.198}
$$

Using the integrations

$$
\int_{-a}^{0} k e^{ikx} dk = \frac{a}{ix} e^{-iax} + \frac{1}{x^2}\left(1 - e^{-iax}\right),
\tag{1.199}
$$

$$
\int_{0}^{a} k e^{ikx} dk = \frac{a}{ix} e^{iax} + \frac{1}{x^2}\left(e^{iax} - 1\right),
\tag{1.200}
$$

$$
\int_{-a}^{a} e^{ikx} dk = \frac{1}{ix}\left(e^{iax} - e^{-iax}\right) = \frac{2\sin(ax)}{x},
\tag{1.201}
$$

and after some straightforward calculations, we end up with

$$\psi_0(x) = \frac{4}{x^2} \sin^2\left(\frac{ax}{2}\right). \tag{1.202}$$

As shown in Figure 1.18, this wave packet is localized: it peaks at $x = 0$ and decreases gradually as x increases. We can verify that the maximum of $\psi_0(x)$ occurs at $x = 0$; writing $\psi_0(x)$ as $a^2(ax/2)^{-2}\sin^2(ax/2)$ and since $\lim_{x\to 0} \sin(bx)/(bx) \to 1$, we obtain $\psi_0(0) = a^2$.

(b) Figure 1.18a is quite suggestive in defining the half-width of $\phi(k)$: $\Delta k = a$ (hence the momentum uncertainty is $\Delta p = \hbar a$). By defining the width as $\Delta k = a$, we know with full certainty that the particle is located between $-a \le k \le a$; according to Figure 1.18a, the probability of finding the particle outside this interval is zero, for $\phi(k)$ vanishes when $|k| > a$.

Now, let us find the width Δx of $\psi_0(x)$. Since $\sin(a\pi/2a) = 1$, $\psi_0(\pi/a) = 4a^2/\pi^2$, and that $\psi_0(0) = a^2$, we can obtain from (1.202) that $\psi_0(\pi/a) = 4a^2/\pi^2 = 4/\pi^2 \psi_0(0)$, or

$$\frac{\psi_0(\pi/a)}{\psi_0(0)} = \frac{4}{\pi^2}. \tag{1.203}$$

This suggests that $\Delta x = \pi/a$: when $x = \pm\Delta x = \pm\pi/a$ the wave packet $\psi_0(x)$ drops to $4/\pi^2$ from its maximum value $\psi_0(0) = a^2$. In sum, we have $\Delta x = \pi/a$ and $\Delta k = a$; hence

$$\Delta x \, \Delta k = \pi \tag{1.204}$$

or

$$\Delta x \, \Delta p = \pi \hbar, \tag{1.205}$$

since $\Delta k = \Delta p/\hbar$. In addition to satisfying Heisenberg's uncertainty principle (1.57), this relation shows that the product $\Delta x \, \Delta p$ is higher than $\hbar/2$: $\Delta x \, \Delta p > \hbar/2$. The wave packet (1.202) therefore offers a clear illustration of the general statement outlined above; namely, only Gaussian wave packets yield the *lowest* limit to Heisenberg's uncertainty principle $\Delta x \, \Delta p = \hbar/2$ (see (1.114)). All other wave packets, such as (1.202), yield higher values for the product $\Delta x \, \Delta p$.

Problem 1.12
Calculate the group and phase velocities for the wave packet corresponding to a relativistic particle.

Solution
Recall that the energy and momentum of a relativistic particle are given by

$$E = mc^2 = \frac{m_0 c^2}{\sqrt{1 - v^2/c^2}}, \qquad p = mv = \frac{m_0 v}{\sqrt{1 - v^2/c^2}}, \tag{1.206}$$

where m_0 is the rest mass of the particle and c is the speed of light in a vacuum. Squaring and adding the expressions of E and p, we obtain $E^2 = p^2 c^2 + m_0^2 c^4$; hence

$$E = c\sqrt{p^2 + m_0^2 c^2}. \tag{1.207}$$

Using this relation along with $p^2 + m_0^2 c^2 = m_0^2 c^2/(1 - v^2/c^2)$ and (1.122), we can show that the group velocity is given as follows:

$$v_g = \frac{dE}{dp} = \frac{d}{dp}\left(c\sqrt{p^2 + m_0^2 c^2}\right) = \frac{pc}{\sqrt{p^2 + m_0^2 c^2}} = v. \tag{1.208}$$

The group velocity is thus equal to the speed of the particle, $v_g = v$.

The phase velocity can be found from (1.122) and (1.207): $v_{ph} = E/p = c\sqrt{1 + m_0^2 c^2/p^2}$ which, when combined with $p = m_0 v/\sqrt{1 - v^2/c^2}$, leads to $\sqrt{1 + m_0^2 c^2/p^2} = c/v$; hence

$$v_{ph} = \frac{E}{p} = c\sqrt{1 + \frac{m_0^2 c^2}{p^2}} = \frac{c^2}{v}. \tag{1.209}$$

This shows that the phase velocity of the wave corresponding to a relativistic particle with $m_0 \neq 0$ is larger than the speed of light, $v_{ph} = c^2/v > c$. This is indeed unphysical. The result $v_{ph} > c$ seems to violate the special theory of relativity, which states that the speed of material particles cannot exceed c. In fact, this principle is not violated because v_{ph} does not represent the velocity of the particle; the velocity of the particle is represented by the group velocity (1.208). As a result, the phase speed of a relativistic particle has no meaningful physical significance.

Finally, the product of the group and phase velocities is equal to c^2, i.e., $v_g v_{ph} = c^2$.

Problem 1.13

The angular frequency of the surface waves in a liquid is given in terms of the wave number k by $\omega = \sqrt{gk + Tk^3/\rho}$, where g is the acceleration due to gravity, ρ is the density of the liquid, and T is the surface tension (which gives an upward force on an element of the surface liquid). Find the phase and group velocities for the limiting cases when the surface waves have: (a) very large wavelengths and   (b) very small wavelengths.

Solution

The phase velocity can be found at once from (1.119):

$$v_{ph} = \frac{\omega}{k} = \sqrt{\frac{g}{k} + \frac{T}{\rho}k} = \sqrt{\frac{g\lambda}{2\pi} + \frac{2\pi T}{\rho\lambda}}, \tag{1.210}$$

where we have used the fact that $k = 2\pi/\lambda$, λ being the wavelength of the surface waves.

(a) If λ is very large, we can neglect the second term in (1.210); hence

$$v_{ph} = \sqrt{\frac{g\lambda}{2\pi}} = \sqrt{\frac{g}{k}}. \tag{1.211}$$

In this approximation the phase velocity does not depend on the nature of the liquid, since it depends on no parameter pertaining to the liquid such as its density or surface tension. This case corresponds, for instance, to deepwater waves, called gravity waves.

To obtain the group velocity, let us differentiate (1.211) with respect to k: $dv_{ph}/dk = -(1/2k)\sqrt{g/k} = -v_{ph}/2k$. A substitution of this relation into (1.120) shows that the group velocity is half the phase velocity:

$$v_g = \frac{d\omega}{dk} = v_{ph} + k\frac{dv_{ph}}{dk} = v_{ph} - \frac{1}{2}v_{ph} = \frac{1}{2}v_{ph} = \frac{1}{2}\sqrt{\frac{g\lambda}{2\pi}}. \qquad (1.212)$$

The longer the wavelength, the faster the group velocity. This explains why a strong, steady wind will produce waves of longer wavelength than those produced by a swift wind.

(b) If λ is very small, the second term in (1.210) becomes the dominant one. So, retaining only the second term, we have

$$v_{ph} = \sqrt{\frac{2\pi T}{\rho\lambda}} = \sqrt{\frac{T}{\rho}k}, \qquad (1.213)$$

which leads to $dv_{ph}/dk = \sqrt{Tk/\rho}/2k = v_{ph}/2k$. Inserting this expression into (1.120), we obtain the group velocity

$$v_g = v_{ph} + k\frac{dv_{ph}}{dk} = v_{ph} + \frac{1}{2}v_{ph} = \frac{3}{2}v_{ph}; \qquad (1.214)$$

hence the smaller the wavelength, the faster the group velocity. These are called ripple waves; they occur, for instance, when a container is subject to vibrations of high frequency and small amplitude or when a gentle wind blows on the surface of a fluid.

Problem 1.14
This problem is designed to illustrate the superposition principle and the concepts of modulated and modulating functions in a wave packet. Consider two wave functions $\psi_1(y, t) = 5y\cos 7t$ and $\psi_2(y, t) = -5y\cos 9t$, where y and t are in meters and seconds, respectively. Show that their superposition generates a wave packet. Plot it and identify the modulated and modulating functions.

Solution
Using the relation $\cos(\alpha \pm \beta) = \cos\alpha\cos\beta \mp \sin\alpha\sin\beta$, we can write the superposition of $\psi_1(y, t)$ and $\psi_2(y, t)$ as follows:

$$\begin{aligned}
\psi(y, t) &= \psi_1(y, t) + \psi_2(y, t) = 5y\cos 7t - 5y\cos 9t \\
&= 5y(\cos 8t\cos t + \sin 8t\sin t) - 5y(\cos 8t\cos t - \sin 8t\sin t) \\
&= 10y\sin t\sin 8t. \qquad (1.215)
\end{aligned}$$

The periods of $10y\sin t$ and $\sin(8t)$ are given by 2π and $2\pi/8$, respectively. Since the period of $10y\sin t$ is larger than that of $\sin 8t$, $10y\sin t$ must be the modulating function and $\sin 8t$ the modulated function. As depicted in Figure 1.19, we see that $\sin 8t$ is modulated by $10y\sin t$.

Problem 1.15
(a) Calculate the final size of the wave packet representing a free particle after traveling a distance of 100 m for the following four cases where the particle is
(i) a 25 eV electron whose wave packet has an initial width of 10^{-6} m,

Figure 1.19 Shape of the wave packet $\psi(y, t) = 10y \sin t \sin 8t$. The function $\sin 8t$, the solid curve, is modulated by $10y \sin t$, the dashed curve.

(ii) a 25 eV electron whose wave packet has an initial width of 10^{-8} m,
(iii) a 100 MeV electron whose wave packet has an initial width of 1 mm, and
(iv) a 100 g object of size 1 cm moving at a speed of 50 m s^{-1}.
(b) Estimate the times required for the wave packets of the electron in (i) and the object in (iv) to spread to 10 mm and 10 cm, respectively. Discuss the results obtained.

Solution

(a) If the initial width of the wave packet of the particle is Δx_0, the width at time t is given by

$$\Delta x(t) = \Delta x_0 \sqrt{1 + \left(\frac{\delta x}{\Delta x_0}\right)^2}, \qquad (1.216)$$

where the dispersion factor is given by

$$\frac{\delta x}{\Delta x_0} = \frac{2\hbar t}{ma^2} = \frac{\hbar t}{2m(a/2)^2} = \frac{\hbar t}{2m(\Delta x_0)^2}. \qquad (1.217)$$

(i) For the 25 eV electron, which is clearly not relativistic, the time to travel the $L = 100$ m distance is given by $t = L/v = L\sqrt{mc^2/2E}/c$, since $E = \frac{1}{2}mv^2 = \frac{1}{2}mc^2(v^2/c^2)$ or $v = c\sqrt{2E/(mc^2)}$. We can therefore write the dispersion factor as

$$\frac{\delta x}{\Delta x_0} = \frac{\hbar}{2m\Delta x_0^2}t = \frac{\hbar}{2m\Delta x_0^2}\frac{L}{c}\sqrt{\frac{mc^2}{2E}} = \frac{\hbar c L}{2mc^2\Delta x_0^2}\sqrt{\frac{mc^2}{2E}}. \qquad (1.218)$$

The numerics of this expression can be made easy by using the following quantities: $\hbar c \simeq 197 \times 10^{-15}$ MeV m, the rest mass energy of an electron is $mc^2 = 0.5$ MeV, $\Delta x_0 = 10^{-6}$ m, $E = 25$ eV $= 25 \times 10^{-6}$ MeV, and $L = 100$ m. Inserting these quantities into (1.218), we obtain

$$\frac{\delta x}{\Delta x_0} \simeq \frac{197 \times 10^{-15} \text{ MeV m} \times 100 \text{ m}}{2 \times 0.5 \text{ MeV} \times 10^{-12} \text{ m}^2}\sqrt{\frac{0.5 \text{ MeV}}{2 \times 25 \times 10^{-6} \text{ MeV}}} \simeq 2 \times 10^3; \qquad (1.219)$$

the time it takes the electron to travel the 100 m distance is given, as shown above, by

$$t = \frac{L}{c}\sqrt{\frac{mc^2}{2E}} = \frac{100 \text{ m}}{3 \times 10^8 \text{ m s}^{-1}}\sqrt{\frac{0.5 \text{ MeV}}{2 \times 25 \times 10^{-6} \text{ MeV}}} = 3.3 \times 10^{-5} \text{ s}. \qquad (1.220)$$

Using $t = 3.3 \times 10^{-5}$ s and substituting (1.219) into (1.216), we obtain

$$\Delta x(t = 3.3 \times 10^{-5} \text{ s}) = 10^{-6} \text{ m} \times \sqrt{1 + 4 \times 10^6} \simeq 2 \times 10^{-3} \text{ m} = 2 \text{ mm}. \qquad (1.221)$$

The width of the wave packet representing the electron has increased from an initial value of 10^{-6} m to 2×10^{-3} m, i.e., by a factor of about 10^3. The spread of the electron's wave packet is thus quite large.

(ii) The calculation needed here is identical to that of part (i), except the value of Δx_0 is now 10^{-8} m instead of 10^{-6} m. This leads to $\delta x / \Delta x_0 \simeq 2 \times 10^7$ and hence the width is $\Delta x(t) = 20$ cm; the width has therefore increased by a factor of about 10^7. This calculation is intended to show that *the narrower the initial wave packet, the larger the final spread*. In fact, starting in part (i) with an initial width of 10^{-6} m, the final width has increased to 2×10^{-3} m by a factor of about 10^3; but in part (ii) we started with an initial width of 10^{-8} m, and the final width has increased to 20 cm by a factor of about 10^7.

(iii) The motion of a 100 MeV electron is relativistic; hence to good approximation, its speed is equal to the speed of light, $v \simeq c$. Therefore the time it takes the electron to travel a distance of $L = 100$ m is $t \simeq L/c = 3.3 \times 10^{-7}$ s. The dispersion factor for this electron can be obtained from (1.217) where $\Delta x_0 = 10^{-3}$ m:

$$\frac{\delta x}{\Delta x_0} = \frac{\hbar L}{2mc \, \Delta x_0^2} = \frac{\hbar c L}{2mc^2 \, \Delta x_0^2} \simeq \frac{197 \times 10^{-15} \text{ MeV m} \times 100 \text{ m}}{2 \times 0.5 \text{ MeV} \times 10^{-6} \text{ m}^2} \simeq 2 \times 10^{-5}. \qquad (1.222)$$

The increase in the width of the wave packet is relatively small:

$$\Delta x(t = 3.3 \times 10^{-7} \text{ s}) = 10^{-3} \text{ m} \times \sqrt{1 + 4 \times 10^{-10}} \simeq 10^{-3} \text{ m} = \Delta x_0. \qquad (1.223)$$

So the width did not increase appreciably. We can conclude from this calculation that, when the motion of a microscopic particle is relativistic, the width of the corresponding wave packet increases by a relatively small amount.

(iv) In the case of a macroscopic object of mass $m = 0.1$ kg, the time to travel the distance $L = 100$ m is $t = L/v = 100$ m/50 m s^{-1} = 2 s. Since the size of the system is about $\Delta x_0 = 1$ cm $= 0.01$ m and $\hbar = 1.05 \times 10^{-34}$ J s, the dispersion factor for the object can be obtained from (1.217):

$$\frac{\delta x}{\Delta x_0} = \frac{\hbar t}{2m \, \Delta x_0^2} \simeq \frac{1.05 \times 10^{-34} \text{ J s} \times 2 \text{ s}}{2 \times 0.1 \text{ kg} \times 10^{-4} \text{ m}^2} \simeq 10^{-29}. \qquad (1.224)$$

Since $\delta x / \Delta x_0 = 10^{-29} \ll 1$, the increase in the width of the wave packet is utterly undetectable:

$$\Delta x(2s) = 10^{-2} \text{ m} \times \sqrt{1 + 10^{-58}} \simeq 10^{-2} \text{ m} = \Delta x_0. \qquad (1.225)$$

(b) Using (1.216) and (1.217) we obtain the expression for the time t in which the wave packet spreads to $\Delta x(t)$:

$$t = \tau \sqrt{\left(\frac{\Delta x(t)}{\Delta x_0} \right)^2 - 1}, \qquad (1.226)$$

where τ represents a time constant $\tau = 2m(\Delta x_0)^2 / \hbar$ (see (1.148)). The time constant for the electron of part (i) is given by

$$\tau = \frac{2mc^2 (\Delta x_0)^2}{\hbar c^2} \simeq \frac{2 \times 0.5 \text{ MeV} \times 10^{-12} \text{ m}^2}{197 \times 10^{-15} \text{ MeV m} \times 3 \times 10^8 \text{m s}^{-1}} = 1.7 \times 10^{-8} \text{ s}, \qquad (1.227)$$

and the time constant for the object of part (iv) is given by

$$\tau = \frac{2m(\Delta x_0)^2}{\hbar} \simeq \frac{2 \times 0.1 \text{ kg} \times 10^{-4} \text{ m}^2}{1.05 \times 10^{-34} \text{ J s}} = 1.9 \times 10^{29} \text{ s}. \tag{1.228}$$

Note that the time constant, while very small for a microscopic particle, is exceedingly large for macroscopic objects.

On the one hand, a substitution of the time constant (1.227) into (1.226) yields the time required for the electron's packet to spread to 10 mm:

$$t = 1.7 \times 10^{-8} \text{ s} \sqrt{\left(\frac{10^{-2}}{10^{-6}}\right)^2 - 1} \simeq 1.7 \times 10^{-4} \text{ s}. \tag{1.229}$$

On the other hand, a substitution of (1.228) into (1.226) gives the time required for the object to spread to 10 cm:

$$t = 1.9 \times 10^{29} \text{ s} \sqrt{\left(\frac{10^{-1}}{10^{-2}}\right)^2 - 1} \simeq 1.9 \times 10^{30} \text{ s}. \tag{1.230}$$

The result (1.229) shows that the size of the electron's wave packet grows in a matter of 1.7×10^{-4} s from 10^{-6} m to 10^{-2} m, a very large spread in a very short time. As for (1.230), it shows that the object has to be constantly in motion for about 1.9×10^{30} s for its wave packet to grow from 1 cm to 10 cm, a small spread for such an absurdly large time; this time is absurd because it is much larger than the age of the Universe, which is about 4.7×10^{17} s. We see that the spread of macroscopic objects becomes appreciable only if the motion lasts for a long, long time. However, the spread of microscopic objects is fast and large.

We can summarize these ideas in three points:

- The width of the wave packet of a nonrelativistic, microscopic particle increases substantially and quickly. The narrower the wave packet at the start, the further and the quicker it will spread.

- When the particle is microscopic and relativistic, the width corresponding to its wave packet does not increase appreciably.

- For a nonrelativistic, macroscopic particle, the width of its corresponding wave packet remains practically constant. The spread becomes appreciable only after absurdly long times, times that are larger than the lifetime of the Universe itself!

Problem 1.16

A neutron is confined in space to 10^{-14} m. Calculate the time its packet will take to spread to
   (a) four times its original size,
   (b) a size equal to the Earth's diameter, and
   (c) a size equal to the distance between the Earth and the Moon.

Solution

Since the rest mass energy of a neutron is equal to $m_n c^2 = 939.6$ MeV, we can infer the time constant for the neutron from (1.227):

$$\tau = \frac{2m_n c^2 (\Delta x_0)^2}{\hbar c^2} \simeq \frac{2 \times 939.6 \text{ MeV} \times (10^{-14} \text{ m})^2}{197 \times 10^{-15} \text{ MeV m} \times 3 \times 10^8 \text{ m s}^{-1}} = 3.2 \times 10^{-21} \text{ s}. \tag{1.231}$$

Inserting this value in (1.226) we obtain the time it takes for the neutron's packet to grow from an initial width Δx_0 to a final size $\Delta x(t)$:

$$t = \tau \sqrt{\left(\frac{\Delta x(t)}{\Delta x_0}\right)^2 - 1} = 3.2 \times 10^{-21} \text{ s} \sqrt{\left(\frac{\Delta x(t)}{\Delta x_0}\right)^2 - 1}. \qquad (1.232)$$

The calculation of t reduces to simple substitutions.

(a) Substituting $\Delta x(t) = 4\Delta x_0$ into (1.232), we obtain the time needed for the neutron's packet to expand to four times its original size:

$$t = 3.2 \times 10^{-21} \text{ s} \sqrt{16 - 1} = 1.2 \times 10^{-20} \text{ s}. \qquad (1.233)$$

(b) The neutron's packet will expand from an initial size of 10^{-14} m to 12.7×10^6 m (the diameter of the Earth) in a time of

$$t = 3.2 \times 10^{-21} \text{ s} \sqrt{\left(\frac{12.7 \times 10^6 \text{ m}}{10^{-14} \text{ m}}\right)^2 - 1} = 4.1 \text{ s}. \qquad (1.234)$$

(c) The time needed for the neutron's packet to spread from 10^{-14} m to 3.84×10^8 m (the distance between the Earth and the Moon) is

$$t = 3.2 \times 10^{-21} \text{ s} \sqrt{\left(\frac{3.84 \times 10^8 \text{ m}}{10^{-14} \text{ m}}\right)^2 - 1} = 12.3 \text{ s}. \qquad (1.235)$$

The calculations carried out in this problem show that the spread of the packets of microscopic particles is significant and occurs very fast: the size of the packet for an earthly neutron can expand to reach the Moon in a mere 12.3 s! Such an immense expansion in such a short time is indeed hard to visualize classically. One should not confuse the packet's expansion with a growth in the size of the system. As mentioned above, the spread of the wave packet does not mean that the material particle becomes bloated. It simply implies a redistribution of the probability density. In spite of the significant spread of the wave packet, the packet's norm is always conserved; as shown in (1.149) it is equal to 1.

Problem 1.17
Use the uncertainty principle to estimate: (a) the ground state radius of the hydrogen atom and (b) the ground state energy of the hydrogen atom.

Solution
(a) According to the uncertainty principle, the electron's momentum and the radius of its orbit are related by $rp \sim \hbar$; hence $p \sim \hbar/r$. To find the ground state radius, we simply need to minimize the electron–proton energy

$$E(r) = \frac{p^2}{2m_e} - \frac{e^2}{4\pi \varepsilon_0 r} = \frac{\hbar^2}{2m_e r^2} - \frac{e^2}{4\pi \varepsilon_0 r} \qquad (1.236)$$

with respect to r:

$$0 = \frac{dE}{dr} = -\frac{\hbar^2}{m_e r_0^3} + \frac{e^2}{4\pi \varepsilon_0 r_0^2}. \qquad (1.237)$$

This leads to the Bohr radius

$$r_0 = \frac{4\pi\varepsilon_0\hbar^2}{m_e e^2} = 0.053 \text{ nm}. \tag{1.238}$$

(b) Inserting (1.238) into (1.236), we obtain the Bohr energy:

$$E(r_0) = \frac{\hbar^2}{2mr_0^2} - \frac{e^2}{4\pi\varepsilon_0 r_0} = -\frac{m_e}{2\hbar^2}\left(\frac{e^2}{4\pi\varepsilon_0}\right)^2 = -13.6 \text{ eV}. \tag{1.239}$$

The results obtained for r_0 and $E(r_0)$, as shown in (1.238) and (1.239), are indeed impressively accurate given the crudeness of the approximation.

Problem 1.18

Consider the bound state of two quarks having the same mass m and interacting via a potential energy $V(r) = kr$ where k is a constant.

(a) Using the Bohr model, find the speed, the radius, and the energy of the system in the case of circular orbits. Determine also the angular frequency of the radiation generated by a transition of the system from energy state n to energy state m.

(b) Obtain numerical values for the speed, the radius, and the energy for the case of the ground state, $n = 1$, by taking a quark mass of $mc^2 = 2$ GeV and $k = 0.5$ GeV fm^{-1}.

Solution

(a) Consider the two quarks to move circularly, much like the electron and proton in a hydrogen atom; then we can write the force between them as

$$\mu\frac{v^2}{r} = \frac{dV(r)}{dr} = k, \tag{1.240}$$

where $\mu = m/2$ is the reduced mass and $V(r)$ is the potential. From the Bohr quantization condition of the orbital angular momentum, we have

$$L = \mu v r = n\hbar. \tag{1.241}$$

Multiplying (1.240) by (1.241), we end up with $\mu^2 v^3 = n\hbar k$, which yields the (quantized) speed of the relative motion for the two-quark system:

$$v_n = \left(\frac{\hbar k}{\mu^2}\right)^{1/3} n^{1/3}. \tag{1.242}$$

The radius can be obtained from (1.241), $r_n = n\hbar/(\mu v_n)$; using (1.242), this leads to

$$r_n = \left(\frac{\hbar^2}{\mu k}\right)^{1/3} n^{2/3}. \tag{1.243}$$

We can obtain the total energy of the relative motion by adding the kinetic and potential energies:

$$E_n = \frac{1}{2}\mu v_n^2 + kr_n = \frac{3}{2}\left(\frac{\hbar^2 k^2}{\mu}\right)^{1/3} n^{2/3}. \tag{1.244}$$

In deriving this relation, we have used the relations for v_n and r_n as given by (1.242) by (1.243), respectively.

The angular frequency of the radiation generated by a transition from n to m is given by

$$\omega_{nm} = \frac{E_n - E_m}{\hbar} = \frac{3}{2} \left(\frac{k^2}{\mu\hbar} \right)^{1/3} \left(n^{2/3} - m^{2/3} \right). \qquad (1.245)$$

(b) Inserting $n = 1$, $\hbar c \simeq 0.197$ GeV fm, $\mu c^2 = mc^2/2 = 1$ GeV, and $k = 0.5$ GeV fm^{-1} into (1.242) to (1.244), we have

$$v_1 = \left(\frac{\hbar c k}{(\mu c^2)^2} \right)^{1/3} c \simeq \left(\frac{0.197 \text{ GeV fm} \times 0.5 \text{ GeV fm}^{-1}}{(1 \text{ GeV})^2} \right)^{1/3} c = 0.46c, \qquad (1.246)$$

where c is the speed of light and

$$r_1 = \left(\frac{(\hbar c)^2}{\mu c^2 k} \right)^{1/3} \simeq \left(\frac{(0.197 \text{ GeV fm})^2}{1 \text{ GeV} \times 0.5 \text{ GeV fm}^{-1}} \right)^{1/3} = 0.427 \text{ fm}, \qquad (1.247)$$

$$E_1 = \frac{3}{2} \left(\frac{(\hbar c)^2 k^2}{\mu c^2} \right)^{1/3} \simeq \frac{3}{2} \left(\frac{(0.197 \text{ GeV fm})^2 (0.5 \text{ GeV fm}^{-1})^2}{1 \text{ GeV}} \right)^{1/3} = 0.32 \text{ GeV}. \quad (1.248)$$

1.11 Exercises

Exercise 1.1
Consider a metal that is being welded.
(a) How hot is the metal when it radiates most strongly at 490 nm?
(b) Assuming that it radiates like a blackbody, calculate the intensity of its radiation.

Exercise 1.2
Consider a star, a light bulb, and a slab of ice; their respective temperatures are 8500 K, 850 K, and 273.15 K.
(a) Estimate the wavelength at which their radiated energies peak.
(b) Estimate the intensities of their radiation.

Exercise 1.3
Consider a 75 W light bulb and an 850 W microwave oven. If the wavelengths of the radiation they emit are 500 nm and 150 mm, respectively, estimate the number of photons they emit per second. Are the quantum effects important in them?

Exercise 1.4
Assuming that a given star radiates like a blackbody, estimate
(a) the temperature at its surface and
(b) the wavelength of its strongest radiation,
when it emits a total intensity of 575 MW m^{-2}.

Exercise 1.5

The intensity reaching the surface of the Earth from the Sun is about $1.36\,\text{kW}\,\text{m}^{-2}$. Assuming the Sun to be a sphere (of radius 6.96×10^8 m) that radiates like a blackbody, estimate
   (a) the temperature at its surface and the wavelength of its strongest radiation, and
   (b) the total power radiated by the Sun (the Earth–Sun distance is 1.5×10^{11} m).

Exercise 1.6

   (a) Calculate: (i) the energy spacing ΔE between the ground state and the first excited state of the hydrogen atom; (ii) and the ratio $\Delta E/E_1$ between the spacing and the ground state energy.
   (b) Consider now a macroscopic system: a simple pendulum which consists of a 5 g mass attached to a 2 m long, massless and inextensible string. Calculate (i) the total energy E_1 of the pendulum when the string makes an angle of $60°$ with the vertical; (ii) the frequency of the pendulum's small oscillations and the energy ΔE of one quantum; and (iii) the ratio $\Delta E/E_1$.
   (c) Examine the sizes of the ratio $\Delta E/E_1$ calculated in parts (a) and (b) and comment on the importance of the quantum effects for the hydrogen atom and the pendulum.

Exercise 1.7

A beam of X-rays from a sulfur source ($\lambda = 53.7\,\text{nm}$) and a γ-ray beam from a Cs^{137} sample ($\lambda = 0.19\,\text{nm}$) impinge on a graphite target. Two detectors are set up at angles $30°$ and $120°$ from the direction of the incident beams.
   (a) Estimate the wavelength shifts of the X-rays and the γ-rays recorded at both detectors.
   (b) Find the kinetic energy of the recoiling electron in each of the four cases.
   (c) What percentage of the incident photon energy is lost in the collision in each of the four cases?

Exercise 1.8

It has been suggested that high energy photons might be found in cosmic radiation, as a result of the *inverse Compton effect*, i.e., a photon of visible light *gains* energy by scattering from a high energy *proton*. If the proton has a momentum of 10^{10} eV/c, find the maximum final energy of an initially yellow photon emitted by a sodium atom ($\lambda_0 = 2.1\,\text{nm}$).

Exercise 1.9

Estimate the number of photons emitted per second from a $75\,rm W$ light bulb; use 575 nm as the average wavelength of the (visible) light emitted. Is the quantum nature of this radiation important?

Exercise 1.10

A 0.7 MeV photon scatters from an electron initially at rest. If the photon scatters at an angle of $35°$, calculate
   (a) the energy and wavelength of the scattered photon,
   (b) the kinetic energy of the recoiling electron, and
   (c) the angle at which the electron recoils.

Exercise 1.11

Light of wavelength 350 nm is incident on a metallic surface of work function 1.9 eV.
   (a) Calculate the kinetic energy of the ejected electrons.
   (b) Calculate the cutoff frequency of the metal.

Exercise 1.12
Find the wavelength of the radiation that can eject electrons from the surface of a zinc sheet with a kinetic energy of 75 eV; the work function of zinc is 3.74 eV. Find also the cutoff wavelength of the metal.

Exercise 1.13
If the stopping potential of a metal when illuminated with a radiation of wavelength 480 nm is 1.2 V, find
    (a) the work function of the metal,
    (b) the cutoff wavelength of the metal, and
    (c) the maximum energy of the ejected electrons.

Exercise 1.14
Find the maximum Compton wave shift corresponding to a collision between a photon and a proton at rest.

Exercise 1.15
If the stopping potential of a metal when illuminated with a radiation of wavelength 150 nm is 7.5 V, calculate the stopping potential of the metal when illuminated by a radiation of wavelength 275 nm.

Exercise 1.16
A light source of frequency 9.5×10^{14} Hz illuminates the surface of a metal of work function 2.8 eV and ejects electrons. Calculate
    (a) the stopping potential,
    (b) the cutoff frequency, and
    (c) the kinetic energy of the ejected electrons.

Exercise 1.17
Consider a metal with a cutoff frequency of 1.2×10^{14} Hz.
    (a) Find the work function of the metal.
    (b) Find the kinetic energy of the ejected electrons when the metal is illuminated with a radiation of frequency 7×10^{14} Hz.

Exercise 1.18
A light of frequency 7.2×10^{14} Hz is incident on four different metallic surfaces of cesium, aluminum, cobalt, and platinum whose work functions are 2.14 eV, 4.08 eV, 3.9 eV, and 6.35 eV, respectively.
    (a) Which among these metals will exhibit the photoelectric effect?
    (b) For each one of the metals producing photoelectrons, calculate the maximum kinetic energy for the electrons ejected.

Exercise 1.19
Consider a metal with stopping potentials of 9 V and 4 V when illuminated by two sources of frequencies 17×10^{14} Hz and 8×10^{14} Hz, respectively.
    (a) Use these data to find a numerical value for the Planck constant.
    (b) Find the work function and the cutoff frequency of the metal.
    (c) Find the maximum kinetic energy of the ejected electrons when the metal is illuminated with a radiation of frequency 12×10^{14} Hz.

Exercise 1.20

Using energy and momentum conservation requirements, show that a free electron cannot absorb all the energy of a photon.

Exercise 1.21

Photons of wavelength 5 nm are scattered from electrons that are at rest. If the photons scatter at 60° relative to the incident photons, calculate
(a) the Compton wave shift,
(b) the kinetic energy imparted to the recoiling electrons, and
(c) the angle at which the electrons recoil.

Exercise 1.22

X-rays of wavelength 0.0008 nm collide with electrons initially at rest. If the wavelength of the scattered photons is 0.0017 nm, determine
(a) the kinetic energy of the recoiling electrons,
(b) the angle at which the photons scatter, and
(c) the angle at which the electrons recoil.

Exercise 1.23

Photons of energy 0.7 MeV are scattered from electrons initially at rest. If the energy of the scattered photons is 0.5 MeV, find
(a) the wave shift,
(b) the angle at which the photons scatter,
(c) the angle at which the electrons recoil, and
(d) the kinetic energy of the recoiling electrons.

Exercise 1.24

In a Compton scattering of photons from electrons at rest, if the photons scatter at an angle of 45° and if the wavelength of the scattered photons is 9×10^{-13} m, find
(a) the wavelength and the energy of the incident photons,
(b) the energy of the recoiling electrons and the angle at which they recoil.

Exercise 1.25

When scattering photons from electrons at rest, if the scattered photons are detected at 90° and if their wavelength is double that of the incident photons, find
(a) the wavelength of the incident photons,
(b) the energy of the recoiling electrons and the angle at which they recoil, and
(c) the energies of the incident and scattered photons.

Exercise 1.26

In scattering electrons from a crystal, the first maximum is observed at an angle of 60°. What must be the energy of the electrons that will enable us to probe as deep as 19 nm inside the crystal?

Exercise 1.27

Estimate the resolution of a microscope which uses electrons of energy 175 eV.

Exercise 1.28

What are the longest and shortest wavelengths in the Balmer and Paschen series for hydrogen?

Exercise 1.29

(a) Calculate the ground state energy of the doubly ionized lithium ion, Li^{2+}, obtained when one removes two electrons from the lithium atom.

(b) If the lithium ion Li^{2+} is bombarded with a photon and subsequently absorbs it, calculate the energy and wavelength of the photon needed to excite the Li^{2+} ion into its third excited state.

Exercise 1.30

Consider a tenfold ionized sodium ion, Na^{10+}, which is obtained by removing ten electrons from an Na atom.

(a) Calculate the orbiting speed and orbital angular momentum of the electron (with respect to the ion's origin) when the ion is in its fourth excited state.

(b) Calculate the frequency of the radiation emitted when the ion deexcites from its fourth excited state to the first excited state.

Exercise 1.31

Calculate the wavelength of the radiation needed to excite the triply ionized beryllium atom, Be^{3+}, from the ground state to its third excited state.

Exercise 1.32

According to the classical model of the hydrogen atom, an electron moving in a circular orbit of radius 0.053 nm around a proton fixed at the center is unstable, and the electron should eventually collapse into the proton. Estimate how long it would take for the electron to collapse into the proton.

Hint: Start with the classical expression for radiation from an accelerated charge

$$\frac{dE}{dt} = -\frac{2}{3}\frac{e^2 a^2}{4\pi \varepsilon_0 c^3}, \qquad E = \frac{p^2}{2m} - \frac{e^2}{4\pi \varepsilon_0 r} = -\frac{e^2}{8\pi \varepsilon_0 r},$$

where a is the *acceleration* of the electron and E is its total energy.

Exercise 1.33

Calculate the de Broglie wavelength of

(a) an electron of kinetic energy 54 eV,

(b) a proton of kinetic energy 70 MeV,

(c) a 100 g bullet moving at 1200 m s^{-1}, and

Useful data: $m_e c^2 = 0.511 \text{ MeV}$, $m_p c^2 = 938.3 \text{ MeV}$, $\hbar c \simeq 197.3 \text{ eV nm}$.

Exercise 1.34

A simple one-dimensional harmonic oscillator is a particle acted upon by a linear restoring force $F(x) = -m\omega^2 x$. Classically, the minimum energy of the oscillator is zero, because we can place it precisely at $x = 0$, its equilibrium position, while giving it zero initial velocity. Quantum mechanically, the uncertainty principle does not allow us to localize the particle precisely and simultaneously have it at rest. Using the uncertainty principle, estimate the minimum energy of the quantum mechanical oscillator.

Exercise 1.35

Consider a double-slit experiment where the waves emitted from the slits superpose on a vertical screen parallel to the y-axis. When only one slit is open, the amplitude of the wave which gets

through is $\psi_1(y, t) = e^{-y^2/32}e^{i(\omega t - ay)}$ and when only the other slit is open, the amplitude is
$\psi_2(y, t) = e^{-y^2/32}e^{i(\omega t - ay - \pi y)}$.

(a) What is the interference pattern along the y-axis with both slits open? Plot the intensity of the wave as a function of y.

(b) What would be the intensity if we put a light source behind the screen to measure which of the slits the light went through? Plot the intensity of the wave as a function of y.

Exercise 1.36

Consider the following three wave functions:

$$\psi_1(y) = A_1 e^{-y^2}, \qquad \psi_2(y) = A_2 e^{-y^2/2}, \qquad \psi_3(y) = A_3(e^{-y^2} + ye^{-y^2/2}),$$

where A_1, A_2, and A_3 are normalization constants.

(a) Find the constants A_1, A_2, and A_3 so that ψ_1, ψ_2, and ψ_3 are normalized.

(b) Find the probability that each one of the states will be in the interval $-1 < y < 1$.

Exercise 1.37

Find the Fourier transform $\phi(p)$ of the following function and plot it:

$$\psi(x) = \begin{cases} 1 - |x|, & |x| < 1, \\ 0, & |x| \geq 1. \end{cases}$$

Exercise 1.38

(a) Find the Fourier transform of $\phi(k) = Ae^{-a|k|-ibk}$, where a and b are real numbers, but a is positive.

(b) Find A so that $\psi(x)$ is normalized.

(c) Find the x and k uncertainties and calculate the uncertainty product $\Delta x \Delta p$. Does it satisfy Heisenberg's uncertainty principle?

Exercise 1.39

(a) Find the Fourier transform $\psi(x)$ of

$$\phi(p) = \begin{cases} 0, & p < -p_0, \\ A, & -p_0 < p < p_0, \\ 0, & p_0 < p, \end{cases}$$

where A is a real constant.

(b) Find A so that $\psi(x)$ is normalized and plot $\phi(p)$ and $\psi(x)$. *Hint:* The following integral might be needed: $\int_{-\infty}^{+\infty} dx \, (\sin^2(ax))/x^2 = \pi a$.

(c) Estimate the uncertainties Δp and Δx and then verify that $\Delta x \Delta p$ satisfies Heisenberg's uncertainty relation.

Exercise 1.40

Estimate the lifetime of the excited state of an atom whose natural width is 3×10^{-4} eV; you may need the value $\hbar = 6.626 \times 10^{-34}$ J s $= 4.14 \times 10^{-15}$ eV s.

Exercise 1.41

Calculate the final width of the wave packet corresponding to an 80 g bullet after traveling for 20 s; the size of the bullet is 2 cm.

Exercise 1.42
A 100 g arrow travels with a speed of 30 m s^{-1} over a distance of 50 m. If the initial size of the wave packet is 5 cm, what will be its final size?

Exercise 1.43
A 50 MeV beam of protons is fired over a distance of 10 km. If the initial size of the wave packet is 1.5×10^{-6} m, what will be the final size upon arrival?

Exercise 1.44
A 250 GeV beam of protons is fired over a distance of 1 km. If the initial size of the wave packet is 1 mm, find its final size.

Exercise 1.45
Consider an inextensible string of linear density μ (mass per unit length). If the string is subject to a tension T, the angular frequency of the string waves is given in terms of the wave number k by $\omega = k\sqrt{T/\mu}$. Find the phase and group velocities.

Exercise 1.46
The angular frequency for a wave propagating inside a waveguide is given in terms of the wave number k and the width b of the guide by $\omega = kc\left[1 - \pi^2/(b^2k^2)\right]^{-1/2}$. Find the phase and group velocities of the wave.

Exercise 1.47
Show that for those waves whose angular frequency ω and wave number k obey the dispersion relation $k^2c^2 = \omega^2 + constant$, the product of the phase and group velocities is equal to c^2, $v_g v_{ph} = c^2$, where c is the speed of light.

Exercise 1.48
How long will the wave packet of a 10 g object, initially confined to 1 mm, take to quadruple its size?

Exercise 1.49
How long will it take for the wave packet of a proton confined to 10^{-15} m to grow to a size equal to the distance between the Earth and the Sun? This distance is equal to 1.5×10^8 km.

Exercise 1.50
Assuming the wave packet representing the Moon to be confined to 1 m, how long will the packet take to reach a size triple that of the Sun? The Sun's radius is 6.96×10^5 km.

Chapter 2

Mathematical Tools of Quantum Mechanics

2.1 Introduction

We deal here with the mathematical machinery needed to study quantum mechanics. Although this chapter is mathematical in scope, no attempt is made to be mathematically complete or rigorous. We limit ourselves to those practical issues that are relevant to the formalism of quantum mechanics.

The Schrödinger equation is one of the cornerstones of the theory of quantum mechanics; it has the structure of a *linear* equation. The formalism of quantum mechanics deals with operators that are linear and wave functions that belong to an abstract Hilbert space. The mathematical properties and structure of Hilbert spaces are essential for a proper understanding of the formalism of quantum mechanics. For this, we are going to review briefly the properties of Hilbert spaces and those of linear operators. We will then consider Dirac's *bra-ket* notation.

Quantum mechanics was formulated in two different ways by Schrödinger and Heisenberg. Schrödinger's wave mechanics and Heisenberg's matrix mechanics are the representations of the general formalism of quantum mechanics in *continuous* and *discrete* basis systems, respectively. For this, we will also examine the mathematics involved in representing kets, bras, bra-kets, and operators in discrete and continuous bases.

2.2 The Hilbert Space and Wave Functions

2.2.1 The Linear Vector Space

A linear vector space consists of two sets of elements and two algebraic rules:

- a set of *vectors* ψ, ϕ, χ, ... and a set of *scalars* $a, b, c, \ldots$;

- a rule for vector *addition* and a rule for scalar *multiplication*.

(a) Addition rule
The addition rule has the properties and structure of an *abelian* group:

- If ψ and ϕ are vectors (elements) of a space, their sum, $\psi + \phi$, is also a vector of the same space.

- Commutativity: $\psi + \phi = \phi + \psi$.

- Associativity: $(\psi + \phi) + \chi = \psi + (\phi + \chi)$.

- Existence of a zero or neutral vector: for each vector ψ, there must exist a zero vector O such that: $O + \psi = \psi + O = \psi$.

- Existence of a symmetric or inverse vector: each vector ψ must have a symmetric vector $(-\psi)$ such that $\psi + (-\psi) = (-\psi) + \psi = O$.

(b) Multiplication rule

The multiplication of vectors by scalars (scalars can be real or complex numbers) has these properties:

- The product of a scalar with a vector gives another vector. In general, if ψ and ϕ are two vectors of the space, any linear combination $a\psi + b\phi$ is also a vector of the space, a and b being scalars.

- Distributivity with respect to addition:

$$a(\psi + \phi) = a\psi + a\phi, \qquad\qquad (a + b)\psi = a\psi + b\psi, \qquad\qquad (2.1)$$

- Associativity with respect to multiplication of scalars:

$$a(b\psi) = (ab)\psi \qquad\qquad (2.2)$$

- For each element ψ there must exist a unitary scalar I and a zero scalar "o" such that

$$I\psi = \psi I = \psi \quad \text{and} \quad o\psi = \psi o = o. \qquad\qquad (2.3)$$

2.2.2 The Hilbert Space

A Hilbert space $\mathcal{H}$ consists of a set of vectors $\psi, \phi, \chi, \ldots$ and a set of *scalars* $a, b, c, \ldots$ which satisfy the following *four* properties:

(a) $\mathcal{H}$ is a linear space

The properties of a linear space were considered in the previous section.

(b) $\mathcal{H}$ has a defined scalar product that is strictly positive

The scalar product of an element ψ with another element ϕ is in general a complex number, denoted by (ψ, ϕ), where $(\psi, \phi) = $ complex number. **Note:** Watch out for the order! Since the scalar product is a complex number, the quantity (ψ, ϕ) is generally not equal to (ϕ, ψ): $(\psi, \phi) = \psi^* \phi$ while $(\phi, \psi) = \phi^* \psi$. The scalar product satisfies the following properties:

- The scalar product of ψ with ϕ is equal to the complex conjugate of the scalar product of ϕ with ψ:

$$(\psi, \phi) = (\phi, \psi)^*. \qquad\qquad (2.4)$$

- The scalar product of ϕ with ψ is linear with respect to the second factor if $\psi = a\psi_1 + b\psi_2$:

$$(\phi, a\psi_1 + b\psi_2) = a(\phi, \psi_1) + b(\phi, \psi_2), \tag{2.5}$$

and antilinear with respect to the first factor if $\phi = a\phi_1 + b\phi_2$:

$$(a\phi_1 + b\phi_2, \psi) = a^*(\phi_1, \psi) + b^*(\phi_2, \psi). \tag{2.6}$$

- The scalar product of a vector ψ with itself is a positive real number:

$$(\psi, \psi) = \| \psi \|^2 \geq 0, \tag{2.7}$$

where the equality holds only for $\psi = O$.

(c) $\mathcal{H}$ is separable

There exists a Cauchy sequence $\psi_n \in \mathcal{H}$ $(n = 1, 2, \ldots)$ such that for every ψ of $\mathcal{H}$ and $\varepsilon > 0$, there exists at least one ψ_n of the sequence for which

$$\| \psi - \psi_n \| < \varepsilon. \tag{2.8}$$

(d) $\mathcal{H}$ is complete

Every Cauchy sequence $\psi_n \in \mathcal{H}$ converges to an element of $\mathcal{H}$. That is, for any ψ_n, the relation

$$\lim_{n,m \to \infty} \| \psi_n - \psi_m \| = 0, \tag{2.9}$$

defines a unique limit ψ of $\mathcal{H}$ such that

$$\lim_{n \to \infty} \| \psi - \psi_n \| = 0. \tag{2.10}$$

Remark

We should note that in a scalar product (ϕ, ψ), the second factor, ψ, belongs to the Hilbert space $\mathcal{H}$, while the first factor, ϕ, belongs to its dual Hilbert space $\mathcal{H}_d$. The distinction between $\mathcal{H}$ and $\mathcal{H}_d$ is due to the fact that, as mentioned above, the scalar product is not commutative: $(\phi, \psi) \neq (\psi, \phi)$; the order matters! From linear algebra, we know that every vector space can be associated with a dual vector space.

2.2.3 Dimension and Basis of a Vector Space

A set of N nonzero vectors $\phi_1, \phi_2, \ldots, \phi_N$ is said to be *linearly independent* if and only if the solution of the equation

$$\sum_{i=1}^{N} a_i \phi_i = 0 \tag{2.11}$$

is $a_1 = a_2 = \cdots = a_N = 0$. But if there exists a set of scalars, which are not all zero, so that one of the vectors (say ϕ_n) can be expressed as a linear combination of the others,

$$\phi_n = \sum_{i=1}^{n-1} a_i \phi_i + \sum_{i=n+1}^{N} a_i \phi_i, \tag{2.12}$$

the set $\{\phi_i\}$ is said to be *linearly dependent*.

Dimension: The *dimension* of a vector space is given by the *maximum number* of linearly independent vectors the space can have. For instance, if the maximum number of linearly independent vectors a space has is N (i.e., $\phi_1, \phi_2, \ldots, \phi_N$), this space is said to be N-dimensional. In this N-dimensional vector space, any vector ψ can be expanded as a linear combination:

$$\psi = \sum_{i=1}^{N} a_i \phi_i. \tag{2.13}$$

Basis: The *basis* of a vector space consists of a set of the maximum possible number of linearly independent vectors belonging to that space. This set of vectors, $\phi_1, \phi_2, \ldots, \phi_N$, to be denoted in short by $\{\phi_i\}$, is called the basis of the vector space, while the vectors $\phi_1, \phi_2, \ldots, \phi_N$ are called the base vectors. Although the set of these linearly independent vectors is arbitrary, it is convenient to choose them *orthonormal*; that is, their scalar products satisfy the relation $(\phi_i, \phi_j) = \delta_{ij}$ (we may recall that $\delta_{ij} = 1$ whenever $i = j$ and zero otherwise). The basis is said to be *orthonormal* if it consists of a set of orthonormal vectors. Moreover, the basis is said to be *complete* if it spans the entire space; that is, there is no need to introduce any additional base vector. The expansion coefficients a_i in (2.13) are called the *components* of the vector ψ in the basis. Each component is given by the scalar product of ψ with the corresponding base vector, $a_j = (\phi_j, \psi)$.

Examples of linear vector spaces

Let us give two examples of linear spaces that are Hilbert spaces: one having a *finite (discrete)* set of base vectors, the other an *infinite (continuous)* basis.

- The first one is the three-dimensional Euclidean vector space; the basis of this space consists of three linearly independent vectors, usually denoted by $\vec{i}, \vec{j}, \vec{k}$. Any vector of the Euclidean space can be written in terms of the base vectors as $\vec{A} = a_1\vec{i} + a_2\vec{j} + a_3\vec{k}$, where a_1, a_2, and a_3 are the components of $\vec{A}$ in the basis; each component can be determined by taking the scalar product of $\vec{A}$ with the corresponding base vector: $a_1 = \vec{i} \cdot \vec{A}, a_2 = \vec{j} \cdot \vec{A}$, and $a_3 = \vec{k} \cdot \vec{A}$. Note that the scalar product in the Euclidean space is real and hence symmetric. The norm in this space is the usual length of vectors $\| \vec{A} \| = A$. Note also that whenever $a_1\vec{i} + a_2\vec{j} + a_3\vec{k} = 0$ we have $a_1 = a_2 = a_3 = 0$ and that none of the unit vectors $\vec{i}, \vec{j}, \vec{k}$ can be expressed as a linear combination of the other two.

- The second example is the space of the entire complex functions $\psi(x)$; the dimension of this space is infinite for it has an infinite number of linearly independent basis vectors.

Example 2.1

Check whether the following sets of functions are linearly independent or dependent on the real x-axis.

(a) $f(x) = 4, g(x) = x^2, h(x) = e^{2x}$
(b) $f(x) = x, g(x) = x^2, h(x) = x^3$
(c) $f(x) = x, g(x) = 5x, h(x) = x^2$
(d) $f(x) = 2 + x^2, g(x) = 3 - x + 4x^3, h(x) = 2x + 3x^2 - 8x^3$

Solution

(a) The first set is clearly linearly independent since $a_1 f(x) + a_2 g(x) + a_3 h(x) = 4a_1 + a_2 x^2 + a_3 e^{2x} = 0$ implies that $a_1 = a_2 = a_3 = 0$ for any value of x.

(b) The functions $f(x) = x$, $g(x) = x^2$, $h(x) = x^3$ are also linearly independent since $a_1 x + a_2 x^2 + a_3 x^3 = 0$ implies that $a_1 = a_2 = a_3 = 0$ no matter what the value of x. For instance, taking $x = -1, 1, 3$, the following system of three equations

$$-a_1 + a_2 - a_3 = 0, \qquad a_1 + a_2 + a_3 = 0, \qquad 3a_1 + 9a_2 + 27a_3 = 0 \qquad (2.14)$$

yields $a_1 = a_2 = a_3 = 0$.

(c) The functions $f(x) = x$, $g(x) = 5x$, $h(x) = x^2$ are not linearly independent, since $g(x) = 5f(x) + 0 \times h(x)$.

(d) The functions $f(x) = 2 + x^2$, $g(x) = 3 - x + 4x^3$, $h(x) = 2x + 3x^2 - 8x^3$ are not linearly independent since $h(x) = 3f(x) - 2g(x)$.

Example 2.2
Are the following sets of vectors (in the three-dimensional Euclidean space) linearly independent or dependent?

(a) $\vec{A} = (3, 0, 0)$, $\vec{B} = (0, -2, 0)$, $\vec{C} = (0, 0, -1)$
(b) $\vec{A} = (6, -9, 0)$, $\vec{B} = (-2, 3, 0)$
(c) $\vec{A} = (2, 3, -1)$, $\vec{B} = (0, 1, 2)$, $\vec{C} = (0, 0, -5)$
(d) $\vec{A} = (1, -2, 3)$, $\vec{B} = (-4, 1, 7)$, $\vec{C} = (0, 10, 11)$, and $\vec{D} = (14, 3, -4)$

Solution
(a) The three vectors $\vec{A} = (3, 0, 0)$, $\vec{B} = (0, -2, 0)$, $\vec{C} = (0, 0, -1)$ are linearly independent, since

$$a_1 \vec{A} + a_2 \vec{B} + a_3 \vec{C} = 0 \Longrightarrow 3a_1 \vec{i} - 2a_2 \vec{j} - a_3 \vec{k} = 0 \qquad (2.15)$$

leads to

$$3a_1 = 0, \qquad -2a_2 = 0, \qquad -a_3 = 0, \qquad (2.16)$$

which yields $a_1 = a_2 = a_3 = 0$.

(b) The vectors $\vec{A} = (6, -9, 0)$, $\vec{B} = (-2, 3, 0)$ are linearly dependent, since the solution to

$$a_1 \vec{A} + a_2 \vec{B} = 0 \quad \Longrightarrow \quad (6a_1 - 2a_2)\vec{i} + (-9a_1 + 3a_2)\vec{j} = 0 \qquad (2.17)$$

is $a_1 = a_2/3$. The first vector is equal to -3 times the second one: $\vec{A} = -3\vec{B}$.

(c) The vectors $\vec{A} = (2, 3, -1)$, $\vec{B} = (0, 1, 2)$, $\vec{C} = (0, 0, -5)$ are linearly independent, since

$$a_1 \vec{A} + a_2 \vec{B} + a_3 \vec{C} = 0 \Longrightarrow 2a_1 \vec{i} + (3a_1 + a_2)\vec{j} + (-a_1 + 2a_2 - 5a_3)\vec{k} = 0 \qquad (2.18)$$

leads to

$$2a_1 = 0, \qquad 3a_1 + a_2 = 0, \qquad -a_1 + 2a_2 - 5a_3 = 0. \qquad (2.19)$$

The only solution of this system is $a_1 = a_2 = a_3 = 0$.

(d) The vectors $\vec{A} = (1, -2, 3)$, $\vec{B} = (-4, 1, 7)$, $\vec{C} = (0, 10, 11)$, and $\vec{D} = (14, 3, -4)$ are not linearly independent, because $\vec{D}$ can be expressed in terms of the other vectors:

$$\vec{D} = 2\vec{A} - 3\vec{B} + \vec{C}. \qquad (2.20)$$

2.2.4 Square-Integrable Functions: Wave Functions

In the case of function spaces, a "vector" element is given by a *complex function* and the *scalar product* by *integrals*. That is, the scalar product of two functions $\psi(x)$ and $\phi(x)$ is given by

$$(\psi, \phi) = \int \psi^*(x)\phi(x)\, dx. \tag{2.21}$$

If this integral *diverges*, the scalar product *does not exist*. As a result, if we want the function space to possess a scalar product, we must select only those functions for which (ψ, ϕ) is *finite*. In particular, a function $\psi(x)$ is said to be *square integrable* if the scalar product of ψ with itself,

$$(\psi, \psi) = \int |\psi(x)|^2\, dx, \tag{2.22}$$

is *finite*.

It is easy to verify that the space of square-integrable functions possesses the properties of a Hilbert space. For instance, any linear combination of square-integrable functions is also a square-integrable function and (2.21) satisfies all the properties of the scalar product of a Hilbert space.

Note that the dimension of the Hilbert space of square-integrable functions is infinite, since each wave function can be expanded in terms of an infinite number of linearly independent functions. The dimension of a space is given by the maximum number of linearly independent basis vectors required to span that space.

A good example of square-integrable functions is the *wave function* of quantum mechanics, $\psi(\vec{r}, t)$. We have seen in Chapter 1 that, according to Born's probabilistic interpretation of $\psi(\vec{r}, t)$, the quantity $|\psi(\vec{r}, t)|^2\, d^3r$ represents the probability of finding, at time t, the particle in a volume d^3r, centered around the point $\vec{r}$. The probability of finding the particle somewhere in space must then be equal to 1:

$$\int |\psi(\vec{r}, t)|^2\, d^3r = \int_{-\infty}^{+\infty} dx \int_{-\infty}^{+\infty} dy \int_{-\infty}^{+\infty} |\psi(\vec{r}, t)|^2\, dz = 1; \tag{2.23}$$

hence the wave functions of quantum mechanics are square-integrable. Wave functions satisfying (2.23) are said to be normalized or square-integrable. As wave mechanics deals with square-integrable functions, any wave function which is not square-integrable has no physical meaning in quantum mechanics.

2.3   Dirac Notation

The physical state of a system is represented in quantum mechanics by elements of a Hilbert space; these elements are called state vectors. We can represent the state vectors in different bases by means of function expansions. This is analogous to specifying an ordinary (Euclidean) vector by its components in various coordinate systems. For instance, we can represent equivalently a vector by its components in a Cartesian coordinate system, in a spherical coordinate system, or in a cylindrical coordinate system. *The meaning of a vector is, of course, independent of the coordinate system chosen to represent its components.* Similarly, the state of a microscopic system has a meaning independent of the basis in which it is expanded.

To free state vectors from coordinate meaning, Dirac introduced what was to become an invaluable notation in quantum mechanics; it allows one to manipulate the formalism of quantum mechanics with ease and clarity. He introduced the concepts of kets, bras, and bra-kets, which will be explained below.

Kets: elements of a vector space

Dirac denoted the state vector ψ by the symbol $\mid \psi \rangle$, which he called a *ket* vector, or simply a ket. Kets belong to the Hilbert (vector) space $\mathcal{H}$, or, in short, to the ket-space.

Bras: elements of a dual space

As mentioned above, we know from linear algebra that a dual space can be associated with every vector space. Dirac denoted the elements of a dual space by the symbol $\langle \mid$, which he called a bra vector, or simply a bra; for instance, the element $\langle \psi \mid$ represents a bra. *Note: For every ket $\mid \psi \rangle$ there exists a unique bra $\langle \psi \mid$ and vice versa.* Again, while kets belong to the Hilbert space $\mathcal{H}$, the corresponding bras belong to its dual (Hilbert) space $\mathcal{H}_d$.

Bra-ket: Dirac notation for the scalar product

Dirac denoted the scalar (inner) product by the symbol $\langle \mid \rangle$, which he called a a *bra-ket*. For instance, the scalar product (ϕ, ψ) is denoted by the bra-ket $\langle \phi \mid \psi \rangle$:

$$(\phi, \psi) \quad \longrightarrow \quad \langle \phi \mid \psi \rangle. \tag{2.24}$$

Note: When a ket (or bra) is multiplied by a complex number, we also get a ket (or bra).

Remark: In wave mechanics we deal with wave functions $\psi(\vec{r}, t)$, but in the more general formalism of quantum mechanics we deal with abstract kets $\mid \psi \rangle$. Wave functions, like kets, are elements of a Hilbert space. We should note that, like a wave function, a ket represents the system completely, and hence knowing $\mid \psi \rangle$ means knowing all its amplitudes in all possible representations. As mentioned above, kets are independent of any particular representation. There is no reason to single out a particular representation basis such as the representation in the position space. Of course, if we want to know the probability of finding the particle at some position in space, we need to work out the formalism within the coordinate representation. The state vector of this particle at time t will be given by the spatial wave function $\langle \vec{r}, t \mid \psi \rangle = \psi(\vec{r}, t)$. In the coordinate representation, the scalar product $\langle \phi \mid \psi \rangle$ is given by

$$\langle \phi \mid \psi \rangle = \int \phi^*(\vec{r}, t) \psi(\vec{r}, t) \, d^3r. \tag{2.25}$$

Similarly, if we are considering the three-dimensional momentum of a particle, the ket $\mid \psi \rangle$ will have to be expressed in momentum space. In this case the state of the particle will be described by a wave function $\psi(\vec{p}, t)$, where $\vec{p}$ is the momentum of the particle.

Properties of kets, bras, and bra-kets

- **Every ket has a corresponding bra**

    To every *ket* $\mid \psi \rangle$, there corresponds a unique *bra* $\langle \psi \mid$ and vice versa:

$$\mid \psi \rangle \quad \longleftrightarrow \quad \langle \psi \mid. \tag{2.26}$$

    There is a one-to-one correspondence between bras and kets:

$$a \mid \psi \rangle + b \mid \phi \rangle \quad \longleftrightarrow \quad a^* \langle \psi \mid + b^* \langle \phi \mid, \tag{2.27}$$

    where a and b are complex numbers. The following is a common notation:

$$\mid a\psi \rangle = a \mid \psi \rangle, \qquad \langle a\psi \mid = a^* \langle \psi \mid. \tag{2.28}$$

- **Properties of the scalar product**

In quantum mechanics, since the scalar product is a complex number, the ordering matters a lot. We must be careful to distinguish a scalar product from its complex conjugate; $\langle \psi \mid \phi \rangle$ is not the same thing as $\langle \phi \mid \psi \rangle$:

$$\langle \phi \mid \psi \rangle^* = \langle \psi \mid \phi \rangle. \tag{2.29}$$

This property becomes clearer if we apply it to (2.21):

$$\langle \phi \mid \psi \rangle^* = \left(\int \phi^*(\vec{r}, t) \psi(\vec{r}, t) \, d^3r \right)^* = \int \psi^*(\vec{r}, t) \phi(\vec{r}, t) \, d^3r = \langle \psi \mid \phi \rangle. \tag{2.30}$$

When $\mid \psi \rangle$ and $\mid \phi \rangle$ are real, we would have $\langle \psi \mid \phi \rangle = \langle \phi \mid \psi \rangle$. Let us list some additional properties of the scalar product:

$$\langle \psi \mid a_1 \psi_1 + a_2 \psi_2 \rangle = a_1 \langle \psi \mid \psi_1 \rangle + a_2 \langle \psi \mid \psi_2 \rangle, \tag{2.31}$$

$$\langle a_1 \phi_1 + a_2 \phi_2 \mid \psi \rangle = a_1^* \langle \phi_1 \mid \psi \rangle + a_2^* \langle \phi_2 \mid \psi \rangle, \tag{2.32}$$

$$\langle a_1 \phi_1 + a_2 \phi_2 \mid b_1 \psi_1 + b_2 \psi_2 \rangle = a_1^* b_1 \langle \phi_1 \mid \psi_1 \rangle + a_1^* b_2 \langle \phi_1 \mid \psi_2 \rangle$$
$$+ a_2^* b_1 \langle \phi_2 \mid \psi_1 \rangle + a_2^* b_2 \langle \phi_2 \mid \psi_2 \rangle. \tag{2.33}$$

- **The norm is real and positive**

For any state vector $\mid \psi \rangle$ of the Hilbert space $\mathcal{H}$, the norm $\langle \psi \mid \psi \rangle$ is real and positive; $\langle \psi \mid \psi \rangle$ is equal to zero only for the case where $\mid \psi \rangle = O$, where O is the zero vector. If the state $\mid \psi \rangle$ is normalized then $\langle \psi \mid \psi \rangle = 1$.

- **Schwarz inequality**

For any two states $\mid \psi \rangle$ and $\mid \phi \rangle$ of the Hilbert space, we can show that

$$|\langle \psi \mid \phi \rangle|^2 \leq \langle \psi \mid \psi \rangle \langle \phi \mid \phi \rangle. \tag{2.34}$$

If $\mid \psi \rangle$ and $\mid \phi \rangle$ are linearly dependent (i.e., proportional: $\mid \psi \rangle = \alpha \mid \phi \rangle$, where α is a scalar), this relation becomes an equality. The Schwarz inequality (2.34) is analogous to the following relation of the real Euclidean space

$$|\vec{A} \cdot \vec{B}|^2 \leq |\vec{A}|^2 |\vec{B}|^2. \tag{2.35}$$

- **Triangle inequality**

$$\sqrt{\langle \psi + \phi \mid \psi + \phi \rangle} \leq \sqrt{\langle \psi \mid \psi \rangle} + \sqrt{\langle \phi \mid \phi \rangle}. \tag{2.36}$$

If $\mid \psi \rangle$ and $\mid \phi \rangle$ are linearly dependent, $\mid \psi \rangle = \alpha \mid \phi \rangle$, and if the proportionality scalar α is real and positive, the triangle inequality becomes an equality. The counterpart of this inequality in Euclidean space is given by $|\vec{A} + \vec{B}| \leq |\vec{A}| + |\vec{B}|$.

- **Orthogonal states**

Two *kets*, $\mid \psi \rangle$ and $\mid \phi \rangle$, are said to be orthogonal if they have a vanishing scalar product:

$$\langle \psi \mid \phi \rangle = 0. \tag{2.37}$$

● **Orthonormal states**

Two *kets*, $|\psi\rangle$ and $|\phi\rangle$, are said to be orthonormal if they are orthogonal and if each one of them has a unit norm:

$$\langle\psi\,|\,\phi\rangle = 0, \qquad \langle\psi\,|\,\psi\rangle = 1, \qquad \langle\phi\,|\,\phi\rangle = 1. \tag{2.38}$$

● **Forbidden quantities**

If $|\psi\rangle$ and $|\phi\rangle$ belong to the same vector (Hilbert) space, products of the type $|\psi\rangle\,|\phi\rangle$ and $\langle\psi\,|\,\langle\phi\,|$ are forbidden. They are nonsensical, since $|\psi\rangle\,|\phi\rangle$ and $\langle\psi\,|\,\langle\phi\,|$ are neither kets nor bras (an explicit illustration of this will be carried out in the example below and later on when we discuss the representation in a discrete basis). If $|\psi\rangle$ and $|\phi\rangle$ belong, however, to different vector spaces (e.g., $|\psi\rangle$ belongs to a spin space and $|\phi\rangle$ to an orbital angular momentum space), then the product $|\psi\rangle\,|\phi\rangle$, written as $|\psi\rangle\otimes|\phi\rangle$, represents a tensor product of $|\psi\rangle$ and $|\phi\rangle$. Only in these typical cases are such products meaningful.

Example 2.3
(**Note:** We will see later in this chapter that kets are represented by column matrices and bras by row matrices; this example is offered earlier than it should because we need to show some concrete illustrations of the formalism.) Consider the following two kets:

$$|\psi\rangle = \begin{pmatrix} -3i \\ 2+i \\ 4 \end{pmatrix}, \qquad |\phi\rangle = \begin{pmatrix} 2 \\ -i \\ 2-3i \end{pmatrix}.$$

(a) Find the bra $\langle\phi\,|$.
(b) Evaluate the scalar product $\langle\phi\,|\,\psi\rangle$.
(c) Examine why the products $|\psi\rangle\,|\phi\rangle$ and $\langle\phi\,|\,\langle\psi\,|$ do not make sense.

Solution
(a) As will be explained later when we introduce the Hermitian adjoint of kets and bras, we want to mention that the bra $\langle\phi\,|$ can be obtained by simply taking the complex conjugate of the transpose of the ket $|\phi\rangle$:

$$\langle\phi\,| = (2 \quad i \quad 2+3i). \tag{2.39}$$

(b) The scalar product $\langle\phi\,|\,\psi\rangle$ can be calculated as follows:

$$\begin{aligned}
\langle\phi\,|\,\psi\rangle &= (2 \quad i \quad 2+3i) \begin{pmatrix} -3i \\ 2+i \\ 4 \end{pmatrix} \\
&= 2(-3i) + i(2+i) + 4(2+3i) \\
&= 7+8i. \tag{2.40}
\end{aligned}$$

(c) First, the product $|\psi\rangle\,|\phi\rangle$ cannot be performed because, from linear algebra, the product of two column matrices cannot be performed. Similarly, since two row matrices cannot be multiplied, the product $\langle\phi\,|\,\langle\psi\,|$ is meaningless.

Physical meaning of the scalar product

The scalar product can be interpreted in two ways. First, by analogy with the scalar product of ordinary vectors in the Euclidean space, where $\vec{A} \cdot \vec{B}$ represents the projection of $\vec{B}$ on $\vec{A}$, the product $\langle \phi \mid \psi \rangle$ also represents the projection of $\mid \psi \rangle$ onto $\mid \phi \rangle$. Second, in the case of normalized states and according to Born's probabilistic interpretation, the quantity $\langle \phi \mid \psi \rangle$ represents the probability amplitude that the system's state $\mid \psi \rangle$ will, after a measurement is performed on the system, be found to be in another state $\mid \phi \rangle$.

Example 2.4 (Bra-ket algebra)

Consider the states $\mid \psi \rangle = 3i \mid \phi_1 \rangle - 7i \mid \phi_2 \rangle$ and $\mid \chi \rangle = - \mid \phi_1 \rangle + 2i \mid \phi_2 \rangle$, where $\mid \phi_1 \rangle$ and $\mid \phi_2 \rangle$ are orthonormal.

(a) Calculate $\mid \psi + \chi \rangle$ and $\langle \psi + \chi \mid$.
(b) Calculate the scalar products $\langle \psi \mid \chi \rangle$ and $\langle \chi \mid \psi \rangle$. Are they equal?
(c) Show that the states $\mid \psi \rangle$ and $\mid \chi \rangle$ satisfy the Schwarz inequality.
(d) Show that the states $\mid \psi \rangle$ and $\mid \chi \rangle$ satisfy the triangle inequality.

Solution

(a) The calculation of $\mid \psi + \chi \rangle$ is straightforward:

$$\begin{aligned} \mid \psi + \chi \rangle &= \mid \psi \rangle + \mid \chi \rangle = (3i \mid \phi_1 \rangle - 7i \mid \phi_2 \rangle) + (- \mid \phi_1 \rangle + 2i \mid \phi_2 \rangle) \\ &= (-1 + 3i) \mid \phi_1 \rangle - 5i \mid \phi_2 \rangle. \end{aligned} \tag{2.41}$$

This leads at once to the expression of $\langle \psi + \chi \mid$:

$$\langle \psi + \chi \mid = (-1 + 3i)^* \langle \phi_1 \mid + (-5i)^* \langle \phi_2 \mid = (-1 - 3i) \langle \phi_1 \mid + 5i \langle \phi_2 \mid . \tag{2.42}$$

(b) Since $\langle \phi_1 \mid \phi_1 \rangle = \langle \phi_2 \mid \phi_2 \rangle = 1$, $\langle \phi_1 \mid \phi_2 \rangle = \langle \phi_2 \mid \phi_1 \rangle = 0$, and since the bras corresponding to the kets $\mid \psi \rangle = 3i \mid \phi_1 \rangle - 7i \mid \phi_2 \rangle$ and $\mid \chi \rangle = - \mid \phi_1 \rangle + 2i \mid \phi_2 \rangle$ are given by $\langle \psi \mid = -3i \langle \phi_1 \mid + 7i \langle \phi_2 \mid$ and $\langle \chi \mid = -\langle \phi_1 \mid - 2i \langle \phi_2 \mid$, the scalar products are

$$\begin{aligned} \langle \psi \mid \chi \rangle &= (-3i \langle \phi_1 \mid + 7i \langle \phi_2 \mid) (- \mid \phi_1 \rangle + 2i \mid \phi_2 \rangle) \\ &= (-3i)(-1) \langle \phi_1 \mid \phi_1 \rangle + (7i)(2i) \langle \phi_2 \mid \phi_2 \rangle \\ &= -14 + 3i, \end{aligned} \tag{2.43}$$

$$\begin{aligned} \langle \chi \mid \psi \rangle &= (-\langle \phi_1 \mid - 2i \langle \phi_2 \mid) (3i \mid \phi_1 \rangle - 7i \mid \phi_2 \rangle) \\ &= (-1)(3i) \langle \phi_1 \mid \phi_1 \rangle + (-2i)(-7i) \langle \phi_2 \mid \phi_2 \rangle \\ &= -14 - 3i. \end{aligned} \tag{2.44}$$

We see that $\langle \psi \mid \chi \rangle$ is equal to the complex conjugate of $\langle \chi \mid \psi \rangle$.

(c) Let us first calculate $\langle \psi \mid \psi \rangle$ and $\langle \chi \mid \chi \rangle$:

$$\langle \psi \mid \psi \rangle = (-3i \langle \phi_1 \mid + 7i \langle \phi_2 \mid) (3i \mid \phi_1 \rangle - 7i \mid \phi_2 \rangle) = (-3i)(3i) + (7i)(-7i) = 58, \tag{2.45}$$

$$\langle \chi \mid \chi \rangle = (-\langle \phi_1 \mid - 2i \langle \phi_2 \mid) (- \mid \phi_1 \rangle + 2i \mid \phi_2 \rangle) = (-1)(-1) + (-2i)(2i) = 5. \tag{2.46}$$

Since $\langle \psi \mid \chi \rangle = -14 + 3i$ we have $\mid \langle \psi \mid \chi \rangle \mid^2 = 14^2 + 3^2 = 205$. Combining the values of $\mid \langle \psi \mid \chi \rangle \mid^2$, $\langle \psi \mid \psi \rangle$, and $\langle \chi \mid \chi \rangle$, we see that the Schwarz inequality (2.34) is satisfied:

$$205 < (58)(5) \implies \mid \langle \psi \mid \chi \rangle \mid^2 < \langle \psi \mid \psi \rangle \langle \chi \mid \chi \rangle. \tag{2.47}$$

(d) First, let us use (2.41) and (2.42) to calculate $\langle \psi + \chi \mid \psi + \chi \rangle$:

$$
\begin{aligned}
\langle \psi + \chi \mid \psi + \chi \rangle &= [(-1 - 3i)\langle \phi_1 \mid + 5i \langle \phi_2 \mid][(-1 + 3i) \mid \phi_1 \rangle - 5i \mid \phi_2 \rangle] \\
&= (-1 - 3i)(-1 + 3i) + (5i)(-5i) \\
&= 35.
\end{aligned}
\tag{2.48}
$$

Since $\langle \psi \mid \psi \rangle = 58$ and $\langle \chi \mid \chi \rangle = 5$, we infer that the triangle inequality (2.36) is satisfied:

$$
\sqrt{35} < \sqrt{58} + \sqrt{5} \Longrightarrow \sqrt{\langle \psi + \chi \mid \psi + \chi \rangle} < \sqrt{\langle \psi \mid \psi \rangle} + \sqrt{\langle \chi \mid \chi \rangle}.
\tag{2.49}
$$

Example 2.5
Consider two states $|\psi_1\rangle = 2i|\phi_1\rangle + |\phi_2\rangle - a|\phi_3\rangle + 4|\phi_4\rangle$ and $|\psi_2\rangle = 3|\phi_1\rangle - i|\phi_2\rangle + 5|\phi_3\rangle - |\phi_4\rangle$, where $|\phi_1\rangle$, $|\phi_2\rangle$, $|\phi_3\rangle$, and $|\phi_4\rangle$ are orthonormal kets, and where a is a constant. Find the value of a so that $|\psi_1\rangle$ and $|\psi_2\rangle$ are orthogonal.

Solution
For the states $|\psi_1\rangle$ and $|\psi_2\rangle$ to be orthogonal, the scalar product $\langle \psi_2 \mid \psi_1 \rangle$ must be zero. Using the relation $\langle \psi_2 \mid = 3\langle \phi_1 \mid + i\langle \phi_2 \mid + 5\langle \phi_3 \mid - \langle \phi_4 \mid$, we can easily find the scalar product

$$
\begin{aligned}
\langle \psi_2 \mid \psi_1 \rangle &= (3\langle \phi_1 \mid + i\langle \phi_2 \mid + 5\langle \phi_3 \mid - \langle \phi_4 \mid)(2i|\phi_1\rangle + |\phi_2\rangle - a|\phi_3\rangle + 4|\phi_4\rangle) \\
&= 7i - 5a - 4.
\end{aligned}
\tag{2.50}
$$

Since $\langle \psi_2 \mid \psi_1 \rangle = 7i - 5a - 4 = 0$, the value of a is $a = (7i - 4)/5$.

2.4 Operators

2.4.1 General Definitions

Definition of an operator: An operator[1] $\hat{A}$ is a *mathematical rule* that when applied to a ket $| \psi \rangle$ transforms it into another ket $| \psi' \rangle$ of the same space and when it acts on a bra $\langle \phi |$ transforms it into another bra $\langle \phi' |$:

$$
\hat{A} \mid \psi \rangle = \mid \psi' \rangle, \qquad \langle \phi \mid \hat{A} = \langle \phi' \mid.
\tag{2.51}
$$

A similar definition applies to wave functions:

$$
\hat{A}\psi(\vec{r}) = \psi'(\vec{r}), \qquad \phi(\vec{r})\hat{A} = \phi'(\vec{r}).
\tag{2.52}
$$

Examples of operators
Here are some of the operators that we will use in this text:

- Unity operator: it leaves any ket unchanged, $\hat{I} \mid \psi \rangle = \mid \psi \rangle$.

- The gradient operator: $\vec{\nabla}\psi(\vec{r}) = (\partial\psi(\vec{r})/\partial x)\vec{i} + (\partial\psi(\vec{r})/\partial y)\vec{j} + (\partial\psi(\vec{r})/\partial z)\vec{k}$.

---

[1] The hat on $\hat{A}$ will be used throughout this text to distinguish an operator $\hat{A}$ from a complex number or a matrix A.

- The linear momentum operator: $\vec{P}\psi(\vec{r}) = -i\hbar\vec{\nabla}\psi(\vec{r})$.

- The Laplacian operator: $\nabla^2\psi(\vec{r}) = \partial^2\psi(\vec{r})/\partial x^2 + \partial^2\psi(\vec{r})/\partial y^2 + \partial^2\psi(\vec{r})/\partial z^2$.

- The parity operator: $\hat{\mathcal{P}}\psi(\vec{r}) = \psi(-\vec{r})$.

Products of operators

The product of two operators is generally not commutative:

$$\hat{A}\hat{B} \neq \hat{B}\hat{A}. \tag{2.53}$$

The product of operators is, however, associative:

$$\hat{A}\hat{B}\hat{C} = \hat{A}(\hat{B}\hat{C}) = (\hat{A}\hat{B})\hat{C}. \tag{2.54}$$

We may also write $\hat{A}^n\hat{A}^m = \hat{A}^{n+m}$. When the product $\hat{A}\hat{B}$ operates on a ket $|\psi\rangle$ (the order of application is important), the operator $\hat{B}$ acts first on $|\psi\rangle$ and then $\hat{A}$ acts on the new ket $(\hat{B}|\psi\rangle)$:

$$\hat{A}\hat{B}|\psi\rangle = \hat{A}(\hat{B}|\psi\rangle). \tag{2.55}$$

Similarly, when $\hat{A}\hat{B}\hat{C}\hat{D}$ operates on a ket $|\psi\rangle$, $\hat{D}$ acts first, then $\hat{C}$, then $\hat{B}$, and then $\hat{A}$.

When an operator $\hat{A}$ is sandwiched between a bra $\langle\phi|$ and a ket $|\psi\rangle$, it yields in general a complex number: $\langle\phi|\hat{A}|\psi\rangle$ = complex number. The quantity $\langle\phi|\hat{A}|\psi\rangle$ can also be a purely real or a purely imaginary number. **Note:** In evaluating $\langle\phi|\hat{A}|\psi\rangle$ it does not matter if one first applies $\hat{A}$ to the ket and then takes the bra-ket or one first applies $\hat{A}$ to the bra and then takes the bra-ket; that is $((\langle\phi|\hat{A})|\psi\rangle = \langle\phi|(\hat{A}|\psi\rangle))$.

Linear operators

An operator $\hat{A}$ is said to be *linear* if it obeys the distributive law and, like all operators, it commutes with constants. That is, an operator $\hat{A}$ is linear if, for any vectors $|\psi_1\rangle$ and $|\psi_2\rangle$ and any complex numbers a_1 and a_2, we have

$$\hat{A}(a_1|\psi_1\rangle + a_2|\psi_2\rangle) = a_1\hat{A}|\psi_1\rangle + a_2\hat{A}|\psi_2\rangle, \tag{2.56}$$

and

$$((\langle\psi_1|a_1 + \langle\psi_2|a_2)\hat{A} = a_1\langle\psi_1|\hat{A} + a_2\langle\psi_2|\hat{A}. \tag{2.57}$$

Remarks

- The *expectation* or *mean* value $\langle\hat{A}\rangle$ of an operator $\hat{A}$ with respect to a state $|\psi\rangle$ is defined by

$$\langle\hat{A}\rangle = \frac{\langle\psi|\hat{A}|\psi\rangle}{\langle\psi|\psi\rangle}. \tag{2.58}$$

- The quantity $|\phi\rangle\langle\psi|$ (i.e., the product of a ket with a bra) is a linear operator in Dirac's notation. To see this, when $|\phi\rangle\langle\psi|$ is applied to a ket $|\psi'\rangle$, we obtain another ket:

$$|\phi\rangle\langle\psi|\psi'\rangle = \langle\psi|\psi'\rangle|\phi\rangle, \tag{2.59}$$

since $\langle\psi|\psi'\rangle$ is a complex number.

- Products of the type $|\psi\rangle\hat{A}$ and $\hat{A}\langle\psi|$ (i.e., when an operator stands on the right of a ket or on the left of a bra) are forbidden. They are not operators, or kets, or bras; they have no mathematical or physical meanings (see equation (2.219) for an illustration).

2.4.2 Hermitian Adjoint

The Hermitian adjoint or conjugate[2], $\hat{a}^\dagger$, of a complex number α is the complex conjugate of this number: $\alpha^\dagger = \alpha^*$. The Hermitian adjoint, or simply the adjoint, $\hat{A}^\dagger$, of an operator $\hat{A}$ is defined by this relation:

$$\langle \psi \mid \hat{A}^\dagger \mid \phi \rangle = \langle \phi \mid \hat{A} \mid \psi \rangle^*. \tag{2.60}$$

Properties of the Hermitian conjugate rule
To obtain the Hermitian adjoint of any expression, we must cyclically reverse the order of the factors and make three replacements:

- Replace constants by their complex conjugates: $\alpha^\dagger = \alpha^*$.

- Replace kets (bras) by the corresponding bras (kets): $(\mid \psi \rangle)^\dagger = \langle \psi \mid$ and $((\langle \psi \mid)^\dagger = \mid \psi \rangle$.

- Replace operators by their adjoints.

Following these rules, we can write

$$(\hat{A}^\dagger)^\dagger = \hat{A}, \tag{2.61}$$
$$(a\hat{A})^\dagger = a^* \hat{A}^\dagger, \tag{2.62}$$
$$(\hat{A}^n)^\dagger = (\hat{A}^\dagger)^n, \tag{2.63}$$
$$(\hat{A} + \hat{B} + \hat{C} + \hat{D})^\dagger = \hat{A}^\dagger + \hat{B}^\dagger + \hat{C}^\dagger + \hat{D}^\dagger, \tag{2.64}$$
$$(\hat{A}\hat{B}\hat{C}\hat{D})^\dagger = \hat{D}^\dagger \hat{C}^\dagger \hat{B}^\dagger \hat{A}^\dagger, \tag{2.65}$$
$$(\hat{A}\hat{B}\hat{C}\hat{D} \mid \psi \rangle)^\dagger = \langle \psi \mid D^\dagger C^\dagger B^\dagger A^\dagger. \tag{2.66}$$

The Hermitian adjoint of the operator $\mid \psi \rangle \langle \phi \mid$ is given by

$$(\mid \psi \rangle \langle \phi \mid)^\dagger = \mid \phi \rangle \langle \psi \mid. \tag{2.67}$$

Operators act inside kets and bras, respectively, as follows:

$$\mid a\hat{A}\psi \rangle = a\hat{A} \mid \psi \rangle, \qquad \langle a\hat{A}\psi \mid = a^* \langle \psi \mid \hat{A}^\dagger. \tag{2.68}$$

Note also that $\langle a\hat{A}^\dagger \psi \mid = a^* \langle \psi \mid (\hat{A}^\dagger)^\dagger = a^* \langle \psi \mid \hat{A}$. Hence, we can also write:

$$\langle \psi \mid \hat{A} \mid \phi \rangle = \langle \hat{A}^\dagger \psi \mid \phi \rangle = \langle \psi \mid \hat{A}\phi \rangle. \tag{2.69}$$

Hermitian and skew-Hermitian operators
An operator $\hat{A}$ is said to be *Hermitian* if it is equal to its adjoint $\hat{A}^\dagger$:

$$\boxed{\hat{A} = \hat{A}^\dagger \qquad \text{or} \qquad \langle \psi \mid \hat{A} \mid \phi \rangle = \langle \phi \mid \hat{A} \mid \psi \rangle^*.} \tag{2.70}$$

---

[2] The terms "adjoint" and "conjugate" are used indiscriminately.

On the other hand, an operator $\hat{B}$ is said to be *skew-Hermitian* or *anti-Hermitian* if

$$\boxed{\hat{B}^\dagger = -\hat{B} \quad \text{or} \quad \langle \psi | \hat{B} | \phi \rangle = -\langle \phi | \hat{B} | \psi \rangle^*.} \tag{2.71}$$

Remark

The Hermitian adjoint of an operator is not, in general, equal to its complex conjugate: $\hat{A}^\dagger \neq \hat{A}^*$.

Example 2.6

(a) Discuss the hermiticity of the operators $(\hat{A} + \hat{A}^\dagger)$, $i(\hat{A} + \hat{A}^\dagger)$, and $i(\hat{A} - \hat{A}^\dagger)$.

(b) Find the Hermitian adjoint of $f(\hat{A}) = (1 + i\hat{A} + 3\hat{A}^2)(1 - 2i\hat{A} - 9\hat{A}^2)/(5 + 7\hat{A})$.

(c) Show that the expectation value of a Hermitian operator is real and that of an anti-Hermitian operator is imaginary.

Solution

(a) The operator $\hat{B} = \hat{A} + \hat{A}^\dagger$ is Hermitian regardless of whether or not $\hat{A}$ is Hermitian, since

$$\hat{B}^\dagger = (\hat{A} + \hat{A}^\dagger)^\dagger = \hat{A}^\dagger + \hat{A} = \hat{B}. \tag{2.72}$$

Similarly, the operator $i(\hat{A} - \hat{A}^\dagger)$ is also Hermitian; but $i(\hat{A} + \hat{A}^\dagger)$ is anti-Hermitian, since $[i(\hat{A} + \hat{A}^\dagger)]^\dagger = -i(\hat{A} + \hat{A}^\dagger)$.

(b) Since the Hermitian adjoint of an operator function $f(\hat{A})$ is given by $f^\dagger(\hat{A}) = f^*(\hat{A}^\dagger)$, we can write

$$\left(\frac{(1 + i\hat{A} + 3\hat{A}^2)(1 - 2i\hat{A} - 9\hat{A}^2)}{5 + 7\hat{A}} \right)^\dagger = \frac{(1 + 2i\hat{A}^\dagger - 9\hat{A}^{\dagger 2})(1 - i\hat{A}^\dagger + 3\hat{A}^{\dagger 2})}{5 + 7\hat{A}^\dagger}. \tag{2.73}$$

(c) From (2.70) we immediately infer that the expectation value of a Hermitian operator is real, for it satisfies the following property:

$$\langle \psi | \hat{A} | \psi \rangle = \langle \psi | \hat{A} | \psi \rangle^*; \tag{2.74}$$

that is, if $\hat{A}^\dagger = \hat{A}$ then $\langle \psi | \hat{A} | \psi \rangle$ is real. Similarly, for an anti-Hermitian operator, $\hat{B}^\dagger = -\hat{B}$, we have

$$\langle \psi | \hat{B} | \psi \rangle = -\langle \psi | \hat{B} | \psi \rangle^*, \tag{2.75}$$

which means that $\langle \psi | \hat{B} | \psi \rangle$ is a purely imaginary number.

2.4.3 Projection Operators

An operator $\hat{P}$ is said to be a *projection operator* if it is Hermitian and equal to its own square:

$$\hat{P}^\dagger = \hat{P}, \qquad \hat{P}^2 = \hat{P}. \tag{2.76}$$

The unit operator $\hat{I}$ is a simple example of a projection operator, since $\hat{I}^\dagger = \hat{I}$, $\hat{I}^2 = \hat{I}$.

Properties of projection operators

- The product of two commuting projection operators, $\hat{P}_1$ and $\hat{P}_2$, is also a projection operator, since

$$(\hat{P}_1\hat{P}_2)^\dagger = \hat{P}_2^\dagger\hat{P}_1^\dagger = \hat{P}_2\hat{P}_1 = \hat{P}_1\hat{P}_2 \text{ and } (\hat{P}_1\hat{P}_2)^2 = \hat{P}_1\hat{P}_2\hat{P}_1\hat{P}_2 = \hat{P}_1^2\hat{P}_2^2 = \hat{P}_1\hat{P}_2.$$
(2.77)

- The sum of two projection operators is generally not a projection operator.

- Two projection operators are said to be orthogonal if their product is zero.

- For a sum of projection operators $\hat{P}_1 + \hat{P}_2 + \hat{P}_3 + \cdots$ to be a projection operator, it is necessary and sufficient that these projection operators be mutually orthogonal (i.e., the cross-product terms must vanish).

Example 2.7
Show that the operator $| \psi \rangle \langle \psi |$ is a projection operator only when $| \psi \rangle$ is normalized.

Solution
It is easy to ascertain that the operator $| \psi \rangle \langle \psi |$ is Hermitian, since $(| \psi \rangle \langle \psi |)^\dagger = | \psi \rangle \langle \psi |$. As for the square of this operator, it is given by

$$(| \psi \rangle \langle \psi |)^2 = (| \psi \rangle \langle \psi |)(| \psi \rangle \langle \psi |) = | \psi \rangle \langle \psi | \psi \rangle \langle \psi |.$$
(2.78)

Thus, if $| \psi \rangle$ is normalized, we have $(| \psi \rangle \langle \psi |)^2 = | \psi \rangle \langle \psi |$. In sum, if the state $| \psi \rangle$ is normalized, the product of the ket $| \psi \rangle$ with the bra $\langle \psi |$ is a projection operator.

2.4.4 Commutator Algebra

The *commutator* of two operators $\hat{A}$ and $\hat{B}$, denoted by $[\hat{A}, \hat{B}]$, is defined by

$$[\hat{A}, \hat{B}] = \hat{A}\hat{B} - \hat{B}\hat{A},$$
(2.79)

and the *anticommutator* $\{\hat{A}, \hat{B}\}$ is defined by

$$\{\hat{A}, \hat{B}\} = \hat{A}\hat{B} + \hat{B}\hat{A}.$$
(2.80)

Two operators are said to commute if their commutator is equal to zero and hence $\hat{A}\hat{B} = \hat{B}\hat{A}$. Any operator commutes with itself:

$$[\hat{A}, \hat{A}] = 0.$$
(2.81)

Note that if two operators are Hermitian and their product is also Hermitian, these operators commute:

$$(\hat{A}\hat{B})^\dagger = \hat{B}^\dagger\hat{A}^\dagger = \hat{B}\hat{A},$$
(2.82)

and since $(\hat{A}\hat{B})^\dagger = \hat{A}\hat{B}$ we have $\hat{A}\hat{B} = \hat{B}\hat{A}$.

As an example, we may mention the commutators involving the x-position operator, $\hat{X}$, and the x-component of the momentum operator, $\hat{P}_x = -i\hbar\partial/\partial x$, as well as the y and the z components

$$[\hat{X}, \hat{P}_x] = i\hbar\hat{I}, \qquad [\hat{Y}, \hat{P}_y] = i\hbar\hat{I}, \qquad [\hat{Z}, \hat{P}_z] = i\hbar\hat{I}, \qquad (2.83)$$

where $\hat{I}$ is the unit operator.

Properties of commutators

Using the commutator relation (2.79), we can establish the following properties:

- Antisymmetry:

$$[\hat{A}, \hat{B}] = -[\hat{B}, \hat{A}] \qquad (2.84)$$

- Linearity:

$$[\hat{A}, \hat{B} + \hat{C} + \hat{D} + \cdots] = [\hat{A}, \hat{B}] + [\hat{A}, \hat{C}] + [\hat{A}, \hat{D}] + \cdots \qquad (2.85)$$

- Hermitian conjugate of a commutator:

$$[\hat{A}, \hat{B}]^\dagger = [\hat{B}^\dagger, \hat{A}^\dagger] \qquad (2.86)$$

- Distributivity:

$$[\hat{A}, \hat{B}\hat{C}] = [\hat{A}, \hat{B}]\hat{C} + \hat{B}[\hat{A}, \hat{C}] \qquad (2.87)$$

$$[\hat{A}\hat{B}, \hat{C}] = \hat{A}[\hat{B}, \hat{C}] + [\hat{A}, \hat{C}]\hat{B} \qquad (2.88)$$

- Jacobi identity:

$$[\hat{A},[\hat{B}, \hat{C}]] + [\hat{B}, [\hat{C}, \hat{A}]] + [\hat{C}, [\hat{A}, \hat{B}]] = 0 \qquad (2.89)$$

- By repeated applications of (2.87), we can show that

$$[\hat{A}, \hat{B}^n] = \sum_{j=0}^{n-1} \hat{B}^j[\hat{A}, \hat{B}]\hat{B}^{n-j-1} \qquad (2.90)$$

$$[\hat{A}^n, \hat{B}] = \sum_{j=0}^{n-1} \hat{A}^{n-j-1}[\hat{A}, \hat{B}]\hat{A}^j \qquad (2.91)$$

- Operators commute with scalars: an operator $\hat{A}$ commutes with any scalar b:

$$[\hat{A}, b] = 0 \qquad (2.92)$$

Example 2.8

(a) Show that the commutator of two Hermitian operators is anti-Hermitian.

(b) Evaluate the commutator $[\hat{A}, [\hat{B}, \hat{C}]\hat{D}]$.

Solution

(a) If $\hat{A}$ and $\hat{B}$ are Hermitian, we can write

$$[\hat{A}, \hat{B}]^\dagger = (\hat{A}\hat{B} - \hat{B}\hat{A})^\dagger = \hat{B}^\dagger \hat{A}^\dagger - \hat{A}^\dagger \hat{B}^\dagger = \hat{B}\hat{A} - \hat{A}\hat{B} = -[\hat{A}, \hat{B}]; \qquad (2.93)$$

that is, the commutator of $\hat{A}$ and $\hat{B}$ is anti-Hermitian: $[\hat{A}, \hat{B}]^\dagger = -[\hat{A}, \hat{B}]$.

(b) Using the distributivity relation (2.87), we have

$$
\begin{aligned}
[\hat{A}, [\hat{B}, \hat{C}]\hat{D}] &= [\hat{B}, \hat{C}][\hat{A}, \hat{D}] + [\hat{A}, [\hat{B}, \hat{C}]]\hat{D} \\
&= (\hat{B}\hat{C} - \hat{C}\hat{B})(\hat{A}\hat{D} - \hat{D}\hat{A}) + \hat{A}(\hat{B}\hat{C} - \hat{C}\hat{B})\hat{D} - (\hat{B}\hat{C} - \hat{C}\hat{B})\hat{A}\hat{D} \\
&= \hat{C}\hat{B}\hat{D}\hat{A} - \hat{B}\hat{C}\hat{D}\hat{A} + \hat{A}\hat{B}\hat{C}\hat{D} - \hat{A}\hat{C}\hat{B}\hat{D}.
\end{aligned}
\qquad (2.94)
$$

2.4.5 Uncertainty Relation between Two Operators

An interesting application of the commutator algebra is to derive a general relation giving the uncertainties product of two operators, $\hat{A}$ and $\hat{B}$. In particular, we want to give a formal derivation of Heisenberg's uncertainty relations.

Let $\langle \hat{A} \rangle$ and $\langle \hat{B} \rangle$ denote the expectation values of two Hermitian operators $\hat{A}$ and $\hat{B}$ with respect to a normalized state vector $| \psi \rangle$: $\langle \hat{A} \rangle = \langle \psi | \hat{A} | \psi \rangle$ and $\langle \hat{B} \rangle = \langle \psi | \hat{B} | \psi \rangle$. Introducing the operators $\Delta \hat{A}$ and $\Delta \hat{B}$,

$$\Delta \hat{A} = \hat{A} - \langle \hat{A} \rangle, \qquad \Delta \hat{B} = \hat{B} - \langle \hat{B} \rangle, \qquad (2.95)$$

we have $(\Delta \hat{A})^2 = \hat{A}^2 - 2\hat{A}\langle \hat{A} \rangle + \langle \hat{A} \rangle^2$ and $(\Delta \hat{B})^2 = \hat{B}^2 - 2\hat{B}\langle \hat{B} \rangle + \langle \hat{B} \rangle^2$, and hence

$$\langle \psi | (\Delta \hat{A})^2 | \psi \rangle = \langle (\Delta \hat{A})^2 \rangle = \langle \hat{A}^2 \rangle - \langle \hat{A} \rangle^2, \qquad \langle (\Delta \hat{B})^2 \rangle = \langle \hat{B}^2 \rangle - \langle \hat{B} \rangle^2, \qquad (2.96)$$

where $\langle \hat{A}^2 \rangle = \langle \psi | \hat{A}^2 | \psi \rangle$ and $\langle \hat{B}^2 \rangle = \langle \psi | \hat{B}^2 | \psi \rangle$. The *uncertainties* ΔA and ΔB are defined by

$$\boxed{\Delta A = \sqrt{\langle (\Delta \hat{A})^2 \rangle} = \sqrt{\langle \hat{A}^2 \rangle - \langle \hat{A} \rangle^2}, \qquad \Delta B = \sqrt{\langle (\Delta \hat{B})^2 \rangle} = \sqrt{\langle \hat{B}^2 \rangle - \langle \hat{B} \rangle^2}.} \qquad (2.97)$$

Let us write the action of the operators (2.95) on any state $| \psi \rangle$ as follows:

$$| \chi \rangle = \Delta \hat{A} | \psi \rangle = \left(\hat{A} - \langle \hat{A} \rangle \right) | \psi \rangle, \qquad | \phi \rangle = \Delta \hat{B} | \psi \rangle = \left(\hat{B} - \langle \hat{B} \rangle \right) | \psi \rangle. \quad (2.98)$$

The Schwarz inequality for the states $| \chi \rangle$ and $| \phi \rangle$ is given by

$$\langle \chi | \chi \rangle \langle \phi | \phi \rangle \geq |\langle \chi | \phi \rangle|^2. \qquad (2.99)$$

Since $\hat{A}$ and $\hat{B}$ are Hermitian, $\Delta \hat{A}$ and $\Delta \hat{B}$ must also be Hermitian: $\Delta \hat{A}^\dagger = \hat{A}^\dagger - \langle \hat{A} \rangle = \hat{A} - \langle \hat{A} \rangle = \Delta \hat{A}$ and $\Delta \hat{B}^\dagger = \hat{B} - \langle \hat{B} \rangle = \Delta \hat{B}$. Thus, we can show the following three relations:

$$\langle \chi | \chi \rangle = \langle \psi | (\Delta \hat{A})^2 | \psi \rangle, \qquad \langle \phi | \phi \rangle = \langle \psi | (\Delta \hat{B})^2 | \psi \rangle, \qquad \langle \chi | \phi \rangle = \langle \psi | \Delta \hat{A} \Delta \hat{B} | \psi \rangle. \qquad (2.100)$$

For instance, since $\Delta \hat{A}^{\dagger} = \Delta \hat{A}$ we have $\langle \chi \mid \chi \rangle = \langle \psi \mid \Delta \hat{A}^{\dagger} \Delta \hat{A} \mid \psi \rangle = \langle \psi \mid (\Delta \hat{A})^2 \mid \psi \rangle = \langle (\Delta \hat{A})^2 \rangle$. Hence, the Schwarz inequality (2.99) becomes

$$\langle (\Delta \hat{A})^2 \rangle \langle (\Delta \hat{B})^2 \rangle \geq \left| \langle \Delta \hat{A} \Delta \hat{B} \rangle \right|^2 . \tag{2.101}$$

Notice that the last term $\Delta \hat{A} \Delta \hat{B}$ of this equation can be written as

$$\Delta \hat{A} \Delta \hat{B} = \frac{1}{2} [\Delta \hat{A}, \ \Delta \hat{B}] + \frac{1}{2} \{\Delta \hat{A}, \ \Delta \hat{B}\} = \frac{1}{2} [\hat{A}, \ \hat{B}] + \frac{1}{2} \{\Delta \hat{A}, \ \Delta \hat{B}\}, \tag{2.102}$$

where we have used the fact that $[\Delta \hat{A}, \ \Delta \hat{B}] = [\hat{A}, \ \hat{B}]$. Since $[\hat{A}, \ \hat{B}]$ is anti-Hermitian and $\{\Delta \hat{A}, \ \Delta \hat{B}\}$ is Hermitian and since the expectation value of a Hermitian operator is real and that the expectation value of an anti-Hermitian operator is imaginary (see Example 2.6), the expectation value $\langle \Delta \hat{A} \Delta \hat{B} \rangle$ of (2.102) becomes equal to the sum of a real part $\langle \{\Delta \hat{A}, \ \Delta \hat{B}\} \rangle / 2$ and an imaginary part $\langle [\hat{A}, \ \hat{B}] \rangle / 2$; hence

$$\left| \langle \Delta \hat{A} \Delta \hat{B} \rangle \right|^2 = \frac{1}{4} \left| \langle [\hat{A}, \ \hat{B}] \rangle \right|^2 + \frac{1}{4} \left| \langle \{\Delta \hat{A}, \ \Delta \hat{B}\} \rangle \right|^2 . \tag{2.103}$$

Since the last term is a positive real number, we can infer the following relation:

$$\left| \langle \Delta \hat{A} \Delta \hat{B} \rangle \right|^2 \geq \frac{1}{4} \left| \langle [\hat{A}, \ \hat{B}] \rangle \right|^2 . \tag{2.104}$$

Comparing equations (2.101) and (2.104), we conclude that

$$\langle (\Delta \hat{A})^2 \rangle \langle (\Delta \hat{B})^2 \rangle \geq \frac{1}{4} \left| \langle [\hat{A}, \ \hat{B}] \rangle \right|^2 , \tag{2.105}$$

which (by taking its square root) can be reduced to

$$\boxed{\Delta A \Delta B \geq \frac{1}{2} \left| \langle [\hat{A}, \ \hat{B}] \rangle \right| .} \tag{2.106}$$

This uncertainty relation plays an important role in the formalism of quantum mechanics. Its application to position and momentum operators leads to the Heisenberg uncertainty relations, which represent one of the cornerstones of quantum mechanics; see the next example.

Example 2.9 (Heisenberg uncertainty relations)
Find the uncertainty relations between the components of the position and the momentum operators.

Solution
By applying (2.106) to the x-components of the position operator $\hat{X}$, and the momentum operator $\hat{P}_x$, we obtain $\Delta x \Delta p_x \geq \frac{1}{2} \mid \langle [\hat{X}, \ \hat{P}_x] \rangle \mid$. But since $[\hat{X}, \ \hat{P}_x] = i\hbar \hat{I}$, we have $\Delta x \Delta p_x \geq \hbar/2$; the uncertainty relations for the $y-$ and $z-$ components follow immediately:

$$\boxed{\Delta x \Delta p_x \geq \frac{\hbar}{2}, \qquad \Delta y \Delta p_y \geq \frac{\hbar}{2}, \qquad \Delta z \Delta p_z \geq \frac{\hbar}{2}.} \tag{2.107}$$

These are the Heisenberg uncertainty relations.

2.4.6 Functions of Operators

Let $F(\hat{A})$ be a function of an operator $\hat{A}$. If $\hat{A}$ is a linear operator, we can Taylor expand $F(\hat{A})$ in a power series of $\hat{A}$:

$$F(\hat{A}) = \sum_{n=0}^{\infty} a_n \hat{A}^n, \tag{2.108}$$

where a_n is just an expansion coefficient. As an illustration of an operator function, consider $e^{a\hat{A}}$, where a is a scalar which can be complex or real. We can expand it as follows:

$$e^{a\hat{A}} = \sum_{n=0}^{\infty} \frac{a^n}{n!} \hat{A}^n = \hat{I} + a\hat{A} + \frac{a^2}{2!} \hat{A}^2 + \frac{a^3}{3!} \hat{A}^3 + \cdots. \tag{2.109}$$

Commutators involving function operators
If $\hat{A}$ commutes with another operator $\hat{B}$, then $\hat{B}$ commutes with any operator function that depends on $\hat{A}$:

$$[\hat{A}, \hat{B}] = 0 \implies [\hat{B}, F(\hat{A})] = 0; \tag{2.110}$$

in particular, $F(\hat{A})$ commutes with $\hat{A}$ and with any other function, $G(\hat{A})$, of $\hat{A}$:

$$[\hat{A}, F(\hat{A})] = 0, \qquad [\hat{A}^n, F(\hat{A})] = 0, \qquad [F(\hat{A}), G(\hat{A})] = 0. \tag{2.111}$$

Hermitian adjoint of function operators
The adjoint of $F(\hat{A})$ is given by

$$[F(\hat{A})]^{\dagger} = F^*(\hat{A}^{\dagger}). \tag{2.112}$$

Note that if $\hat{A}$ is Hermitian, $F(\hat{A})$ is not necessarily Hermitian; $F(\hat{A})$ will be Hermitian only if F is a real function and $\hat{A}$ is Hermitian. An example is

$$(e^{\hat{A}})^{\dagger} = e^{\hat{A}^{\dagger}}, \qquad (e^{i\hat{A}})^{\dagger} = e^{-i\hat{A}^{\dagger}}, \qquad (e^{i\alpha\hat{A}})^{\dagger} = e^{-i\alpha^*\hat{A}^{\dagger}}, \tag{2.113}$$

where α is a complex number. So if $\hat{A}$ is Hermitian, an operator function which can be expanded as $F(\hat{A}) = \sum_{n=0}^{\infty} a_n \hat{A}^n$ will be Hermitian only if the expansion coefficients a_n are real numbers. But in general, $F(\hat{A})$ is not Hermitian even if $\hat{A}$ is Hermitian, since

$$F^*(\hat{A}^{\dagger}) = \sum_{n=0}^{\infty} a_n^* (\hat{A}^{\dagger})^n. \tag{2.114}$$

Relations involving function operators
Note that

$$[\hat{A}, \hat{B}] \neq 0 \implies [\hat{B}, F(\hat{A})] \neq 0; \tag{2.115}$$

in particular, $e^{\hat{A}} e^{\hat{B}} \neq e^{\hat{A}+\hat{B}}$. Using (2.109) we can ascertain that

$$e^{\hat{A}} e^{\hat{B}} = e^{\hat{A}+\hat{B}} e^{[\hat{A}, \hat{B}]/2}, \tag{2.116}$$

$$e^{\hat{A}} \hat{B} e^{-\hat{A}} = \hat{B} + [\hat{A}, \hat{B}] + \frac{1}{2!} [\hat{A}, [\hat{A}, \hat{B}]] + \frac{1}{3!} [\hat{A}, [\hat{A}, [\hat{A}, \hat{B}]]] + \cdots \tag{2.117}$$

2.4.7 Inverse and Unitary Operators

Inverse of an operator: Assuming it exists[3] the *inverse* $\hat{A}^{-1}$ of a linear operator $\hat{A}$ is defined by the relation

$$\hat{A}^{-1}\hat{A} = \hat{A}\hat{A}^{-1} = \hat{I},$$ (2.118)

where $\hat{I}$ is the unit operator, the operator that leaves any state $|\psi\rangle$ unchanged.

Quotient of two operators: Dividing an operator $\hat{A}$ by another operator $\hat{B}$ (provided that the inverse $\hat{B}^{-1}$ exists) is equivalent to multiplying $\hat{A}$ by $\hat{B}^{-1}$:

$$\frac{\hat{A}}{\hat{B}} = \hat{A}\hat{B}^{-1}.$$ (2.119)

The side on which the quotient is taken matters:

$$\frac{\hat{A}}{\hat{B}} = \hat{A}\frac{\hat{I}}{\hat{B}} = \hat{A}\hat{B}^{-1} \quad \text{and} \quad \frac{\hat{I}}{\hat{B}}\hat{A} = \hat{B}^{-1}\hat{A}.$$ (2.120)

In general, we have $\hat{A}\hat{B}^{-1} \neq \hat{B}^{-1}\hat{A}$. For an illustration of these ideas, see Problem 2.12. We may mention here the following properties about the inverse of operators:

$$\left(\hat{A}\hat{B}\hat{C}\hat{D}\right)^{-1} = \hat{D}^{-1}\hat{C}^{-1}\hat{B}^{-1}\hat{A}^{-1}, \qquad \left(\hat{A}^n\right)^{-1} = \left(\hat{A}^{-1}\right)^n.$$ (2.121)

Unitary operators: A linear operator $\hat{U}$ is said to be *unitary* if its inverse $\hat{U}^{-1}$ is equal to its adjoint $\hat{U}^\dagger$:

$$\hat{U}^\dagger = \hat{U}^{-1} \qquad \text{or} \qquad \hat{U}\hat{U}^\dagger = \hat{U}^\dagger\hat{U} = \hat{I}.$$ (2.122)

The product of two unitary operators is also unitary, since

$$(\hat{U}\hat{V})(\hat{U}\hat{V})^\dagger = (\hat{U}\hat{V})(\hat{V}^\dagger\hat{U}^\dagger) = \hat{U}(\hat{V}\hat{V}^\dagger)\hat{U}^\dagger = \hat{U}\hat{U}^\dagger = \hat{I},$$ (2.123)

or $(\hat{U}\hat{V})^\dagger = (\hat{U}\hat{V})^{-1}$. This result can be generalized to any number of operators; the product of a number of unitary operators is also unitary, since

$$
\begin{aligned}
(\hat{A}\hat{B}\hat{C}\hat{D}\cdots)(\hat{A}\hat{B}\hat{C}\hat{D}\cdots)^\dagger &= \hat{A}\hat{B}\hat{C}\hat{D}(\cdots)\hat{D}^\dagger\hat{C}^\dagger\hat{B}^\dagger\hat{A}^\dagger = \hat{A}\hat{B}\hat{C}(\hat{D}\hat{D}^\dagger)\hat{C}^\dagger\hat{B}^\dagger\hat{A}^\dagger \\
&= \hat{A}\hat{B}(\hat{C}\hat{C}^\dagger)\hat{B}^\dagger\hat{A}^\dagger = \hat{A}(\hat{B}\hat{B}^\dagger)\hat{A}^\dagger \\
&= \hat{A}\hat{A}^\dagger = \hat{I},
\end{aligned}
$$ (2.124)

or $(\hat{A}\hat{B}\hat{C}\hat{D}\cdots)^\dagger = (\hat{A}\hat{B}\hat{C}\hat{D}\cdots)^{-1}$.

Example 2.10 (Unitary operator)

What conditions must the parameter ε and the operator $\hat{G}$ satisfy so that the operator $\hat{U} = e^{i\varepsilon\hat{G}}$ is unitary?

---

[3]Not every operator has an inverse, just as in the case of matrices. The inverse of a matrix exists only when its determinant is nonzero.

Solution

Clearly, if ε is real and $\hat{G}$ is Hermitian, the operator $e^{i\varepsilon\hat{G}}$ would be unitary. Using the property $[F(\hat{A})]^{\dagger} = F^*(\hat{A}^{\dagger})$, we see that

$$(e^{i\varepsilon\hat{G}})^{\dagger} = e^{-i\varepsilon\hat{G}} = (e^{i\varepsilon\hat{G}})^{-1}, \tag{2.125}$$

that is, $\hat{U}^{\dagger} = \hat{U}^{-1}$.

2.4.8 Eigenvalues and Eigenvectors of an Operator

Having studied the properties of operators and states, we are now ready to discuss how to find the eigenvalues and eigenvectors of an operator.

A state vector $| \psi \rangle$ is said to be an *eigenvector* (also called an eigenket or eigenstate) of an operator $\hat{A}$ if the application of $\hat{A}$ to $| \psi \rangle$ gives

$$\hat{A} | \psi \rangle = a | \psi \rangle, \tag{2.126}$$

where a is a complex number, called an *eigenvalue* of $\hat{A}$. This equation is known as the *eigenvalue equation*, or *eigenvalue problem*, of the operator $\hat{A}$. Its solutions yield the eigenvalues and eigenvectors of $\hat{A}$. In Section 2.5.3 we will see how to solve the eigenvalue problem in a discrete basis.

A simple example is the eigenvalue problem for the unity operator $\hat{I}$:

$$\hat{I} | \psi \rangle = | \psi \rangle. \tag{2.127}$$

This means that all vectors are eigenvectors of $\hat{I}$ with one eigenvalue, 1. Note that

$$\hat{A} | \psi \rangle = a | \psi \rangle \implies \hat{A}^n | \psi \rangle = a^n | \psi \rangle \quad \text{and} \quad F(\hat{A}) | \psi \rangle = F(a) | \psi \rangle. \tag{2.128}$$

For instance, we have

$$\hat{A} | \psi \rangle = a | \psi \rangle \implies e^{i\hat{A}} | \psi \rangle = e^{ia} | \psi \rangle. \tag{2.129}$$

Example 2.11 (Eigenvalues of the inverse of an operator)

Show that if $\hat{A}^{-1}$ exists, the eigenvalues of $\hat{A}^{-1}$ are just the inverses of those of $\hat{A}$.

Solution

Since $\hat{A}^{-1}\hat{A} = \hat{I}$ we have on the one hand

$$\hat{A}^{-1}\hat{A} | \psi \rangle = | \psi \rangle, \tag{2.130}$$

and on the other hand

$$\hat{A}^{-1}\hat{A} | \psi \rangle = \hat{A}^{-1}(\hat{A} | \psi \rangle) = a\hat{A}^{-1} | \psi \rangle. \tag{2.131}$$

Combining the previous two equations, we obtain

$$a\hat{A}^{-1} | \psi \rangle = | \psi \rangle, \tag{2.132}$$

hence

$$\hat{A}^{-1} \mid \psi \rangle = \frac{1}{a} \mid \psi \rangle. \tag{2.133}$$

This means that $\mid \psi \rangle$ is also an eigenvector of $\hat{A}^{-1}$ with eigenvalue $1/a$. That is, if $\hat{A}^{-1}$ exists, then

$$\boxed{\hat{A} \mid \psi \rangle = a \mid \psi \rangle \quad \Longrightarrow \quad \hat{A}^{-1} \mid \psi \rangle = \frac{1}{a} \mid \psi \rangle.} \tag{2.134}$$

Some useful theorems pertaining to the eigenvalue problem

Theorem 2.1 *For a Hermitian operator, all of its eigenvalues are real and the eigenvectors corresponding to different eigenvalues are orthogonal.*

$$\text{If } \hat{A}^{\dagger} = \hat{A}, \quad \hat{A} \mid \phi_n \rangle = a_n \mid \phi_n \rangle \quad \Longrightarrow \quad a_n = \text{real number, and } \langle \phi_m \mid \phi_n \rangle = \delta_{mn}. \tag{2.135}$$

Proof of Theorem 2.1
Note that

$$\hat{A} \mid \phi_n \rangle = a_n \mid \phi_n \rangle \quad \Longrightarrow \quad \langle \phi_m \mid \hat{A} \mid \phi_n \rangle = a_n \langle \phi_m \mid \phi_n \rangle, \tag{2.136}$$

and

$$\langle \phi_m \mid \hat{A}^{\dagger} = a_m^* \langle \phi_m \mid \quad \Longrightarrow \quad \langle \phi_m \mid \hat{A}^{\dagger} \mid \phi_n \rangle = a_m^* \langle \phi_m \mid \phi_n \rangle. \tag{2.137}$$

Subtracting (2.137) from (2.136) and using the fact that $\hat{A}$ is Hermitian, $\hat{A} = \hat{A}^{\dagger}$, we have

$$(a_n - a_m^*) \langle \phi_m \mid \phi_n \rangle = 0. \tag{2.138}$$

Two cases must be considered separately:

- Case $m = n$: since $\langle \phi_n \mid \phi_n \rangle > 0$, we must have $a_n = a_n^*$; hence the eigenvalues a_n must be real.

- Case $m \neq n$: since in general $a_n \neq a_m^*$, we must have $\langle \phi_m \mid \phi_n \rangle = 0$; that is, $\mid \phi_m \rangle$ and $\mid \phi_n \rangle$ must be orthogonal.

Theorem 2.2 *The eigenstates of a Hermitian operator define a complete set of mutually orthonormal basis states. The operator is diagonal in this eigenbasis with its diagonal elements equal to the eigenvalues. This basis set is unique if the operator has no degenerate eigenvalues and not unique (in fact it is infinite) if there is any degeneracy.*

Theorem 2.3 *If two Hermitian operators, $\hat{A}$ and $\hat{B}$, commute and if $\hat{A}$ has no degenerate eigenvalue, then each eigenvector of $\hat{A}$ is also an eigenvector of $\hat{B}$. In addition, we can construct a common orthonormal basis that is made of the joint eigenvectors of $\hat{A}$ and $\hat{B}$.*

Proof of Theorem 2.3
Since $\hat{A}$ is Hermitian with no degenerate eigenvalue, to each eigenvalue of $\hat{A}$ there corresponds only one eigenvector. Consider the equation

$$\hat{A} \mid \phi_n \rangle = a_n \mid \phi_n \rangle. \tag{2.139}$$

Since $\hat{A}$ commutes with $\hat{B}$ we can write

$$\hat{B}\hat{A} \mid \phi_n\rangle = \hat{A}\hat{B} \mid \phi_n\rangle \quad \text{or} \quad \hat{A}(\hat{B} \mid \phi_n\rangle) = a_n(\hat{B} \mid \phi_n\rangle); \tag{2.140}$$

that is, $(\hat{B} \mid \phi_n\rangle)$ is an eigenvector of $\hat{A}$ with eigenvalue a_n. But since this eigenvector is unique (apart from an arbitrary phase constant), the ket $\mid \phi_n\rangle$ must also be an eigenvector of $\hat{B}$:

$$\hat{B} \mid \phi_n\rangle = b_n \mid \phi_n\rangle. \tag{2.141}$$

Since each eigenvector of $\hat{A}$ is also an eigenvector of $\hat{B}$ (and vice versa), both of these operators must have a common basis. This basis is unique; it is made of the joint eigenvectors of $\hat{A}$ and $\hat{B}$. This theorem also holds for any number of mutually commuting Hermitian operators.

Now, if a_n is a degenerate eigenvalue, we can only say that $\hat{B} \mid \phi_n\rangle$ is an eigenvector of $\hat{A}$ with eigenvalue a_n; $\mid \phi_n\rangle$ is not necessarily an eigenvector of $\hat{B}$. If one of the operators is degenerate, there exist an infinite number of orthonormal basis sets that are common to these two operators; that is, the joint basis does exist and it is not unique.

Theorem 2.4 *The eigenvalues of an anti-Hermitian operator are either purely imaginary or equal to zero.*

Theorem 2.5 *The eigenvalues of a unitary operator are complex numbers of moduli equal to one; the eigenvectors of a unitary operator that has no degenerate eigenvalues are mutually orthogonal.*

Proof of Theorem 2.5
Let $\mid \phi_n\rangle$ and $\mid \phi_m\rangle$ be eigenvectors to the unitary operator $\hat{U}$ with eigenvalues a_n and a_m, respectively. We can write

$$(\langle\phi_m \mid \hat{U}^\dagger)(\hat{U} \mid \phi_n\rangle) = a_m^* a_n \langle\phi_m \mid \phi_n\rangle. \tag{2.142}$$

Since $\hat{U}^\dagger \hat{U} = \hat{I}$ this equation can be rewritten as

$$(a_m^* a_n - 1)\langle\phi_m \mid \phi_n\rangle = 0, \tag{2.143}$$

which in turn leads to the following two cases:

- Case $n = m$: since $\langle\phi_n \mid \phi_n\rangle > 0$ then $a_n^* a_n = \mid a_n \mid^2 = 1$, and hence $\mid a_n \mid = 1$.

- Case $n \neq m$: the only possibility for this case is that $\mid \phi_m\rangle$ and $\mid \phi_n\rangle$ are orthogonal, $\langle\phi_m \mid \phi_n\rangle = 0$.

2.4.9 Infinitesimal and Finite Unitary Transformations

We want to study here how quantities such as kets, bras, operators, and scalars transform under unitary transformations. A unitary transformation is the application of a unitary operator $\hat{U}$ to one of these quantities.

2.4.9.1 Unitary Transformations

Kets $| \psi \rangle$ and bras $\langle \psi |$ transform as follows:

$$| \psi' \rangle = \hat{U} | \psi \rangle, \qquad \langle \psi' | = \langle \psi | \hat{U}^\dagger. \qquad (2.144)$$

Let us now find out how operators transform under unitary transformations. Since the transform of $\hat{A} | \psi \rangle = | \phi \rangle$ is $\hat{A}' | \psi' \rangle = | \phi' \rangle$, we can rewrite $\hat{A}' | \psi' \rangle = | \phi' \rangle$ as $\hat{A}' \hat{U} | \psi \rangle = \hat{U} | \phi \rangle = \hat{U} \hat{A} | \psi \rangle$ which, in turn, leads to $\hat{A}' \hat{U} = \hat{U} \hat{A}$. Multiplying both sides of $\hat{A}' \hat{U} = \hat{U} \hat{A}$ by $\hat{U}^\dagger$ and since $\hat{U} \hat{U}^\dagger = \hat{U}^\dagger \hat{U} = \hat{I}$, we have

$$\hat{A}' = \hat{U} \hat{A} \hat{U}^\dagger, \qquad \hat{A} = \hat{U}^\dagger \hat{A}' \hat{U}. \qquad (2.145)$$

The results reached in (2.144) and (2.145) may be summarized as follows:

$$\boxed{| \psi' \rangle = \hat{U} | \psi \rangle, \qquad \langle \psi' | = \langle \psi | \hat{U}^\dagger, \qquad \hat{A}' = \hat{U} \hat{A} \hat{U}^\dagger,} \qquad (2.146)$$

$$\boxed{| \psi \rangle = \hat{U}^\dagger | \psi' \rangle, \qquad \langle \psi | = \langle \psi' | \hat{U}, \qquad \hat{A} = \hat{U}^\dagger \hat{A}' \hat{U}.} \qquad (2.147)$$

Properties of unitary transformations

- If an operator $\hat{A}$ is Hermitian, its transformed $\hat{A}'$ is also Hermitian, since

$$\hat{A}'^{\,\dagger} = (\hat{U} \hat{A} \hat{U}^\dagger)^\dagger = \hat{U} \hat{A}^\dagger \hat{U}^\dagger = \hat{U} \hat{A} \hat{U}^\dagger = \hat{A}'. \qquad (2.148)$$

- The eigenvalues of $\hat{A}$ and those of its transformed $\hat{A}'$ are the same:

$$\hat{A} | \psi_n \rangle = a_n | \psi_n \rangle \implies \hat{A}' | \psi'_n \rangle = a_n | \psi'_n \rangle, \qquad (2.149)$$

since

$$\begin{aligned} \hat{A}' | \psi'_n \rangle &= (\hat{U} \hat{A} \hat{U}^\dagger)(\hat{U} | \psi_n \rangle) = \hat{U} \hat{A} (\hat{U}^\dagger \hat{U}) | \psi_n \rangle \\ &= \hat{U} \hat{A} | \psi_n \rangle = a_n (\hat{U} | \psi_n \rangle) = a_n | \psi'_n \rangle. \end{aligned} \qquad (2.150)$$

- Commutators that are equal to (complex) numbers remain unchanged under unitary transformations, since the transformation of $[\hat{A}, \hat{B}] = a$, where a is a complex number, is given by

$$\begin{aligned} [\hat{A}', \hat{B}'] &= [\hat{U} \hat{A} \hat{U}^\dagger, \hat{U} \hat{B} \hat{U}^\dagger] = (\hat{U} \hat{A} \hat{U}^\dagger)(\hat{U} \hat{B} \hat{U}^\dagger) - (\hat{U} \hat{B} \hat{U}^\dagger)(\hat{U} \hat{A} \hat{U}^\dagger) \\ &= \hat{U} [\hat{A}, \hat{B}] \hat{U}^\dagger = \hat{U} a \hat{U}^\dagger = a \hat{U} \hat{U}^\dagger = a \\ &= [\hat{A}, \hat{B}]. \end{aligned} \qquad (2.151)$$

- We can also verify the following general relations:

$$\hat{A} = \beta \hat{B} + \gamma \hat{C} \implies \hat{A}' = \beta \hat{B}' + \gamma \hat{C}', \qquad (2.152)$$

$$\hat{A} = \alpha \hat{B} \hat{C} \hat{D} \implies \hat{A}' = \alpha \hat{B}' \hat{C}' \hat{D}', \qquad (2.153)$$

where $\hat{A}'$, $\hat{B}'$, $\hat{C}'$, and $\hat{D}'$ are the transforms of $\hat{A}$, $\hat{B}$, $\hat{C}$, and $\hat{D}$, respectively.

- Since the result (2.151) is valid for any complex number, we can state that complex numbers, such as $\langle \psi \mid \hat{A} \mid \chi \rangle$, remain unchanged under unitary transformations, since

$$\langle \psi' \mid \hat{A}' \mid \chi' \rangle = ((\psi \mid \hat{U}^\dagger)(\hat{U}\hat{A}\hat{U}^\dagger)(\hat{U} \mid \chi)) = \langle \psi \mid (\hat{U}^\dagger\hat{U})\hat{A}(\hat{U}^\dagger U) \mid \chi \rangle = \langle \psi \mid \hat{A} \mid \chi \rangle. \tag{2.154}$$

Taking $\hat{A} = \hat{I}$ we see that scalar products of the type

$$\langle \psi' \mid \chi' \rangle = \langle \psi \mid \chi \rangle \tag{2.155}$$

are invariant under unitary transformations; notably, the norm of a state vector is conserved:

$$\langle \psi' \mid \psi' \rangle = \langle \psi \mid \psi \rangle. \tag{2.156}$$

- We can also verify that $\left(\hat{U}\hat{A}\hat{U}^\dagger\right)^n = \hat{U}\hat{A}^n\hat{U}^\dagger$ since

$$\begin{aligned}
\left(\hat{U}\hat{A}\hat{U}^\dagger\right)^n &= \left(\hat{U}\hat{A}\hat{U}^\dagger\right)\left(\hat{U}\hat{A}\hat{U}^\dagger\right)\cdots\left(\hat{U}\hat{A}\hat{U}^\dagger\right) = \hat{U}\hat{A}(\hat{U}^\dagger\hat{U})\hat{A}(\hat{U}^\dagger\hat{U})\cdots(\hat{U}^\dagger\hat{U})\hat{A}\hat{U}^\dagger \\
&= \hat{U}\hat{A}^n\hat{U}^\dagger. \tag{2.157}
\end{aligned}$$

- We can generalize the previous result to obtain the transformation of any operator function $f(\hat{A})$:

$$\hat{U}f(\hat{A})\hat{U}^\dagger = f(\hat{U}\hat{A}\hat{U}^\dagger) = f(\hat{A}'), \tag{2.158}$$

or more generally

$$\hat{U}f(\hat{A}, \hat{B}, \hat{C}, \cdots)\hat{U}^\dagger = f(\hat{U}\hat{A}\hat{U}^\dagger, \hat{U}\hat{B}\hat{U}^\dagger, \hat{U}\hat{C}\hat{U}^\dagger, \cdots) = f(\hat{A}', \hat{B}', \hat{C}', \cdots). \tag{2.159}$$

A unitary transformation does not change the physics of a system; it merely transforms one description of the system to another physically equivalent description.

In what follows we want to consider two types of unitary transformations: infinitesimal transformations and finite transformations.

2.4.9.2 Infinitesimal Unitary Transformations

Consider an operator $\hat{U}$ which depends on an infinitesimally small real parameter ε and which varies only slightly from the unity operator $\hat{I}$:

$$\hat{U}_\varepsilon(\hat{G}) = \hat{I} + i\varepsilon\hat{G}, \tag{2.160}$$

where $\hat{G}$ is called the *generator* of the infinitesimal transformation. Clearly, $\hat{U}_\varepsilon$ is a unitary transformation only when the parameter ε is real and $\hat{G}$ is Hermitian, since

$$\hat{U}_\varepsilon\hat{U}_\varepsilon^\dagger = (\hat{I} + i\varepsilon\hat{G})(\hat{I} - i\varepsilon\hat{G}^\dagger) \simeq \hat{I} + i\varepsilon(\hat{G} - \hat{G}^\dagger) = \hat{I}, \tag{2.161}$$

where we have neglected the quadratic terms in ε.

The transformation of a state vector $\mid \psi \rangle$ is

$$\mid \psi' \rangle = (\hat{I} + i\varepsilon\hat{G}) \mid \psi \rangle = \mid \psi \rangle + \delta \mid \psi \rangle, \tag{2.162}$$

where

$$\delta \mid \psi \rangle = i\varepsilon\hat{G} \mid \psi \rangle. \tag{2.163}$$

The transformation of an operator $\hat{A}$ is given by

$$\boxed{\hat{A}' = (\hat{I} + i\varepsilon\hat{G})\hat{A}(\hat{I} - i\varepsilon\hat{G}) \simeq \hat{A} + i\varepsilon[\hat{G}, \hat{A}].} \tag{2.164}$$

If $\hat{G}$ commutes with $\hat{A}$, the unitary transformation will leave $\hat{A}$ unchanged, $\hat{A}' = \hat{A}$:

$$[\hat{G}, \hat{A}] = 0 \quad \Longrightarrow \quad \hat{A}' = (\hat{I} + i\varepsilon\hat{G})\hat{A}(\hat{I} - i\varepsilon\hat{G}) = \hat{A}. \tag{2.165}$$

2.4.9.3 Finite Unitary Transformations

We can construct a *finite* unitary transformation from (2.160) by performing a succession of infinitesimal transformations in steps of ε; the application of a series of successive unitary transformations is equivalent to the application of a single unitary transformation. Denoting $\varepsilon = \alpha/N$, where N is an integer and α is a finite parameter, we can apply the same unitary transformation N times; in the limit $N \to +\infty$ we obtain

$$\hat{U}_\alpha(\hat{G}) = \lim_{N\to\infty} \prod_{k=1}^{N} \left(1 + i\frac{\alpha}{N}\hat{G}\right) = \lim_{N\to+\infty} \left(1 + i\frac{\alpha}{N}\hat{G}\right)^N = e^{i\alpha\hat{G}}, \tag{2.166}$$

where $\hat{G}$ is now the generator of the finite transformation and α is its parameter.

As shown in (2.125), $\hat{U}$ is unitary only when the parameter α is real and $\hat{G}$ is Hermitian, since

$$(e^{i\alpha\hat{G}})^\dagger = e^{-i\alpha\hat{G}} = (e^{i\alpha\hat{G}})^{-1}. \tag{2.167}$$

Using the commutation relation (2.117), we can write the transformation $\hat{A}'$ of an operator $\hat{A}$ as follows:

$$\boxed{e^{i\alpha\hat{G}}\hat{A}e^{-i\alpha\hat{G}} = \hat{A} + i\alpha[\hat{G}, \hat{A}] + \frac{(i\alpha)^2}{2!}\left[\hat{G}, [\hat{G}, \hat{A}]\right] + \frac{(i\alpha)^3}{3!}\left[\hat{G}, [\hat{G}, [\hat{G}, \hat{A}]]\right] + \cdots.}$$
$$\tag{2.168}$$

If $\hat{G}$ commutes with $\hat{A}$, the unitary transformation will leave $\hat{A}$ unchanged, $\hat{A}' = \hat{A}$:

$$[\hat{G}, \hat{A}] = 0 \Longrightarrow \hat{A}' = e^{i\alpha\hat{G}}\hat{A}e^{-i\alpha\hat{G}} = \hat{A}. \tag{2.169}$$

In Chapter 3, we will consider some important applications of infinitesimal unitary transformations to study time translations, space translations, space rotations, and conservation laws.

2.5   Representation in Discrete Bases

By analogy with the expansion of Euclidean space vectors in terms of the basis vectors, we need to express any ket $\mid \psi \rangle$ of the Hilbert space in terms of a complete set of mutually orthonormal base kets. State vectors are then represented by their components in this basis.

2.5.1 Matrix Representation of Kets, Bras, and Operators

Consider a discrete, complete, and orthonormal basis which is made of an infinite[4] set of kets $| \phi_1 \rangle, | \phi_2 \rangle, | \phi_3 \rangle, \ldots, | \phi_n \rangle$ and denote it by $\{| \phi_n \rangle\}$. Note that the basis $\{| \phi_n \rangle\}$ is discrete, yet it has an infinite number of unit vectors. In the limit $n \to \infty$, the ordering index n of the unit vectors $| \phi_n \rangle$ is *discrete* or *countable*; that is, the sequence $| \phi_1 \rangle, | \phi_2 \rangle, | \phi_3 \rangle, \ldots$ is countably infinite. As an illustration, consider the special functions, such as the Hermite, Legendre, or Laguerre polynomials, $H_n(x)$, $P_n(x)$, and $L_n(x)$. These polynomials are identified by a discrete index n and by a continuous variable x; although n varies discretely, it can be infinite.

In Section 2.6, we will consider bases that have a continuous and infinite number of base vectors; in these bases the index n increases continuously. Thus, each basis has a continuum of base vectors.

In this section the notation $\{| \phi_n \rangle\}$ will be used to abbreviate an infinitely countable set of vectors (i.e., $| \phi_1 \rangle, | \phi_2 \rangle, | \phi_3 \rangle, \ldots$) of the Hilbert space $\mathcal{H}$. The orthonormality condition of the base kets is expressed by

$$\langle \phi_n | \phi_m \rangle = \delta_{nm}, \tag{2.170}$$

where δ_{nm} is the *Kronecker delta* symbol defined by

$$\delta_{nm} = \begin{cases} 1, & n = m, \\ 0, & n \neq m. \end{cases} \tag{2.171}$$

The completeness, or closure, relation for this basis is given by

$$\sum_{n=1}^{\infty} | \phi_n \rangle \langle \phi_n | = \hat{I}, \tag{2.172}$$

where $\hat{I}$ is the unit operator; when the unit operator acts on any ket, it leaves the ket unchanged.

2.5.1.1 Matrix Representation of Kets and Bras

Let us now examine how to represent the vector $| \psi \rangle$ within the context of the basis $\{| \phi_n \rangle\}$. The completeness property of this basis enables us to expand any state vector $| \psi \rangle$ in terms of the base kets $| \phi_n \rangle$:

$$| \psi \rangle = \hat{I} | \psi \rangle = \left(\sum_{n=1}^{\infty} | \phi_n \rangle \langle \phi_n | \right) | \psi \rangle = \sum_{n=1}^{\infty} a_n | \phi_n \rangle, \tag{2.173}$$

where the coefficient a_n, which is equal to $\langle \phi_n | \psi \rangle$, represents the projection of $| \psi \rangle$ onto $| \phi_n \rangle$; a_n is the component of $| \psi \rangle$ along the vector $| \phi_n \rangle$. Recall that the coefficients a_n are complex numbers. So, within the basis $\{| \phi_n \rangle\}$, the ket $| \psi \rangle$ is represented by the set of its components, $a_1, a_2, a_3, \ldots$ along $| \phi_1 \rangle, | \phi_2 \rangle, | \phi_3 \rangle, \ldots$, respectively. Hence $| \psi \rangle$ can be represented by a *column* vector which has a countably infinite number of components:

$$| \psi \rangle \longrightarrow \begin{pmatrix} \langle \phi_1 | \psi \rangle \\ \langle \phi_2 | \psi \rangle \\ \vdots \\ \langle \phi_n | \psi \rangle \\ \vdots \end{pmatrix} = \begin{pmatrix} a_1 \\ a_2 \\ \vdots \\ a_n \\ \vdots \end{pmatrix}. \tag{2.174}$$

---

[4]Kets are elements of the Hilbert space, and the dimension of a Hilbert space is infinite.

The bra $\langle \psi \mid$ can be represented by a *row* vector:

$$
\begin{aligned}
\langle \psi \mid &\longrightarrow (\langle \psi \mid \phi_1 \rangle \ \langle \psi \mid \phi_2 \rangle \ \cdots \ \langle \psi \mid \phi_n \rangle \ \cdots) \\
&= (\langle \phi_1 \mid \psi \rangle^* \ \langle \phi_2 \mid \psi \rangle^* \ \cdots \ \langle \phi_n \mid \psi \rangle^* \ \cdots) \\
&= (a_1^* \ a_2^* \ \cdots \ a_n^* \ \cdots).
\end{aligned}
\tag{2.175}
$$

Using this representation, we see that a bra-ket $\langle \psi \mid \phi \rangle$ is a complex number equal to the matrix product of the row matrix corresponding to the bra $\langle \psi \mid$ with the column matrix corresponding to the ket $\mid \phi \rangle$:

$$
\langle \psi \mid \phi \rangle = (a_1^* \ a_2^* \ \cdots \ a_n^* \ \cdots)
\begin{pmatrix} b_1 \\ b_2 \\ \vdots \\ b_n \\ \vdots \end{pmatrix}
= \sum_n a_n^* b_n,
\tag{2.176}
$$

where $b_n = \langle \phi_n \mid \phi \rangle$. We see that, within this representation, the matrices representing $\mid \psi \rangle$ and $\langle \psi \mid$ are Hermitian adjoints of each other.

Remark

A ket $\mid \psi \rangle$ is normalized if $\langle \psi \mid \psi \rangle = \sum_n |a_n|^2 = 1$. If $\mid \psi \rangle$ is not normalized and we want to normalized it, we need simply to multiply it by a constant α so that $\langle \alpha \psi \mid \alpha \psi \rangle = |\alpha|^2 \langle \psi \mid \psi \rangle = 1$, and hence $\alpha = 1/\sqrt{\langle \psi \mid \psi \rangle}$.

Example 2.12

Consider the following two kets:

$$
\mid \psi \rangle = \begin{pmatrix} 5i \\ 2 \\ -i \end{pmatrix}, \qquad \mid \phi \rangle = \begin{pmatrix} 3 \\ 8i \\ -9i \end{pmatrix}.
$$

(a) Find $\mid \psi \rangle^*$ and $\langle \psi \mid$.
(b) Is $\mid \psi \rangle$ normalized? If not, normalize it.
(c) Are $\mid \psi \rangle$ and $\mid \phi \rangle$ orthogonal?

Solution

(a) The expressions of $\mid \psi \rangle^*$ and $\langle \psi \mid$ are given by

$$
\mid \psi \rangle^* = \begin{pmatrix} -5i \\ 2 \\ i \end{pmatrix}, \qquad \langle \psi \mid = (-5i \ \ 2 \ \ i),
\tag{2.177}
$$

where we have used the fact that $\langle \psi \mid$ is equal to the complex conjugate of the transpose of the ket $\mid \psi \rangle$. Hence, we should reiterate the important fact that $\mid \psi \rangle^* \neq \langle \psi \mid$.

(b) The norm of $\mid \psi \rangle$ is given by

$$
\langle \psi \mid \psi \rangle = (-5i \ \ 2 \ \ i) \begin{pmatrix} 5i \\ 2 \\ -i \end{pmatrix} = (-5i)(5i) + (2)(2) + (i)(-i) = 30.
\tag{2.178}
$$

Thus, $|\psi\rangle$ is not normalized. By multiplying it with $1/\sqrt{30}$, it becomes normalized:

$$|\chi\rangle = \frac{1}{\sqrt{30}}|\psi\rangle = \frac{1}{\sqrt{30}}\begin{pmatrix} 5i \\ 2 \\ -i \end{pmatrix} \implies \langle\chi\mid\chi\rangle = 1. \qquad (2.179)$$

(c) The kets $|\psi\rangle$ and $|\phi\rangle$ are not orthogonal since their scalar product is not zero:

$$\langle\psi\mid\phi\rangle = (-5i \quad 2 \quad i)\begin{pmatrix} 3 \\ 8i \\ -9i \end{pmatrix} = (-5i)(3) + (2)(8i) + (i)(-9i) = 9 + i. \qquad (2.180)$$

2.5.1.2 Matrix Representation of Operators

For each linear operator $\hat{A}$, we can write

$$\hat{A} = \hat{I}\hat{A}\hat{I} = \left(\sum_{n=1}^{\infty}|\phi_n\rangle\langle\phi_n|\right)\hat{A}\left(\sum_{m=1}^{\infty}|\phi_m\rangle\langle\phi_m|\right) = \sum_{nm}A_{nm}|\phi_n\rangle\langle\phi_m|, \qquad (2.181)$$

where A_{nm} is the nm matrix element of the operator $\hat{A}$:

$$A_{nm} = \langle\phi_n\mid\hat{A}\mid\phi_m\rangle. \qquad (2.182)$$

We see that the operator $\hat{A}$ is represented, within the basis $\{|\phi_n\rangle\}$, by a *square* matrix A (A without a hat designates a matrix), which has a countably infinite number of columns and a countably infinite number of rows:

$$A = \begin{pmatrix} A_{11} & A_{12} & A_{13} & \cdots \\ A_{21} & A_{22} & A_{23} & \cdots \\ A_{31} & A_{32} & A_{33} & \cdots \\ \vdots & \vdots & \vdots & \ddots \end{pmatrix}, \qquad (2.183)$$

For instance, the *unit* operator $\hat{I}$ is represented by the unit matrix; when the unit matrix is multiplied with another matrix, it leaves that unchanged:

$$I = \begin{pmatrix} 1 & 0 & 0 & \cdots \\ 0 & 1 & 0 & \cdots \\ 0 & 0 & 1 & \cdots \\ \vdots & \vdots & \vdots & \ddots \end{pmatrix}. \qquad (2.184)$$

In summary, *kets are represented by column vectors, bras by row vectors, and operators by square matrices.*

2.5.1.3　Matrix Representation of Some Other Operators

(a) Hermitian adjoint operation

Let us now look at the matrix representation of the Hermitian adjoint operation of an operator. First, recall that the *transpose* of a matrix A, denoted by A^T, is obtained by interchanging the rows with the columns:

$$(A^T)_{nm} = A_{mn} \quad \text{or} \quad \begin{pmatrix} A_{11} & A_{12} & A_{13} & \cdots \\ A_{21} & A_{22} & A_{23} & \cdots \\ A_{31} & A_{32} & A_{33} & \cdots \\ \vdots & \vdots & \vdots & \ddots \end{pmatrix}^T = \begin{pmatrix} A_{11} & A_{21} & A_{31} & \cdots \\ A_{12} & A_{22} & A_{32} & \cdots \\ A_{13} & A_{23} & A_{33} & \cdots \\ \vdots & \vdots & \vdots & \ddots \end{pmatrix}.$$

(2.185)

Similarly, the transpose of a column matrix is a row matrix, and the transpose of a row matrix is a column matrix:

$$\begin{pmatrix} a_1 \\ a_2 \\ \vdots \\ a_n \\ \vdots \end{pmatrix}^T = \begin{pmatrix} a_1 & a_2 & \cdots & a_n & \cdots \end{pmatrix} \quad \text{and} \quad \begin{pmatrix} a_1 & a_2 & \cdots & a_n & \cdots \end{pmatrix}^T = \begin{pmatrix} a_1 \\ a_2 \\ \vdots \\ a_n \\ \vdots \end{pmatrix}.$$

(2.186)

So a square matrix A is symmetric if it is equal to its transpose, $A^T = A$. A skew-symmetric matrix is a square matrix whose transpose equals the negative of the matrix, $A^T = -A$.

The *complex conjugate* of a matrix is obtained by simply taking the complex conjugate of all its elements: $(A^*)_{nm} = (A_{nm})^*$.

The matrix which represents the operator $\hat{A}^\dagger$ is obtained by taking the complex conjugate of the matrix transpose of A:

$$A^\dagger = (A^T)^* \quad \text{or} \quad (\hat{A}^\dagger)_{nm} = \langle \phi_n | \hat{A}^\dagger | \phi_m \rangle = \langle \phi_m | \hat{A} | \phi_n \rangle^* = A_{mn}^*;$$

(2.187)

that is,

$$\begin{pmatrix} A_{11} & A_{12} & A_{13} & \cdots \\ A_{21} & A_{22} & A_{23} & \cdots \\ A_{31} & A_{32} & A_{33} & \cdots \\ \vdots & \vdots & \vdots & \ddots \end{pmatrix}^\dagger = \begin{pmatrix} A_{11}^* & A_{21}^* & A_{31}^* & \cdots \\ A_{12}^* & A_{22}^* & A_{32}^* & \cdots \\ A_{13}^* & A_{23}^* & A_{33}^* & \cdots \\ \vdots & \vdots & \vdots & \ddots \end{pmatrix}.$$

(2.188)

If an operator $\hat{A}$ is Hermitian, its matrix satisfies this condition:

$$(A^T)^* = A \quad \text{or} \quad A_{mn}^* = A_{nm}.$$

(2.189)

The diagonal elements of a Hermitian matrix therefore must be real numbers. Note that *a Hermitian matrix must be square*.

(b) Inverse and unitary operators

A matrix has an inverse only if it is square and its determinant is nonzero; a matrix that has an inverse is called a nonsingular matrix and a matrix that has no inverse is called a singular matrix. The elements A_{nm}^{-1} of the inverse matrix A^{-1}, representing an operator $\hat{A}^{-1}$, are given by the relation

$$A_{nm}^{-1} = \frac{\text{cofactor of } A_{mn}}{\text{determinant of } A} \quad \text{or} \quad A^{-1} = \frac{B^I}{\text{determinant of } A}, \quad (2.190)$$

where B is the matrix of *cofactors* (also called the minor); the cofactor of element A_{mn} is equal to $(-1)^{m+n}$ times the determinant of the submatrix obtained from A by removing the mth row and the nth column. Note that when the matrix, representing an operator, has a determinant equal to zero, this operator does not possess an inverse. Note that $A^{-1}A = AA^{-1} = I$ where I is the unit matrix.

The inverse of a product of matrices is obtained as follows:

$$(ABC \cdots PQ)^{-1} = Q^{-1}P^{-1} \cdots C^{-1}B^{-1}A^{-1}. \quad (2.191)$$

The inverse of the inverse of a matrix is equal to the matrix itself, $(A^{-1})^{-1} = A$.

A *unitary* operator $\hat{U}$ is represented by a unitary matrix. A matrix U is said to be unitary if its inverse is equal to its adjoint:

$$U^{-1} = U^\dagger \quad \text{or} \quad U^\dagger U = I, \quad (2.192)$$

where I is the unit matrix.

Example 2.13 (Inverse of a matrix)

Calculate the inverse of the matrix $A = \begin{pmatrix} 2 & i & 0 \\ 3 & 1 & 5 \\ 0 & -i & -2 \end{pmatrix}$. Is this matrix unitary?

Solution
Since the determinant of A is $\det(A) = -4 + 16i$, we have $A^{-1} = B^T/(-4 + 16i)$, where the elements of the cofactor matrix B are given by $B_{nm} = (-1)^{n+m}$ times the determinant of the submatrix obtained from A by removing the nth row and the mth column. In this way, we have

$$B_{11} = (-1)^{1+1} \begin{vmatrix} A_{22} & A_{23} \\ A_{32} & A_{33} \end{vmatrix} = (-1)^2 \begin{vmatrix} 1 & 5 \\ -i & -2 \end{vmatrix} = -2 + 5i, \quad (2.193)$$

$$B_{12} = (-1)^{1+2} \begin{vmatrix} A_{21} & A_{23} \\ A_{31} & A_{33} \end{vmatrix} = (-1)^3 \begin{vmatrix} 3 & 5 \\ 0 & -2 \end{vmatrix} = 6, \quad (2.194)$$

$$B_{13} = (-1)^{1+3} \begin{vmatrix} A_{21} & A_{22} \\ A_{31} & A_{32} \end{vmatrix} = (-1)^4 \begin{vmatrix} 3 & 1 \\ 0 & -i \end{vmatrix} = -3i, \quad (2.195)$$

$$B_{21} = (-1)^3 \begin{vmatrix} i & 0 \\ -i & -2 \end{vmatrix} = 2i, \qquad B_{22} = (-1)^4 \begin{vmatrix} 2 & 0 \\ 0 & -2 \end{vmatrix} = -4, \quad (2.196)$$

$$B_{23} = (-1)^5 \begin{vmatrix} 2 & i \\ 0 & -i \end{vmatrix} = 2i, \qquad B_{31} = (-1)^4 \begin{vmatrix} i & 0 \\ 1 & 5 \end{vmatrix} = 5i, \quad (2.197)$$

$$B_{32} = (-1)^5 \begin{vmatrix} 2 & 0 \\ 3 & 5 \end{vmatrix} = -10, \qquad B_{33} = (-1)^6 \begin{vmatrix} 2 & i \\ 3 & 1 \end{vmatrix} = 2 - 3i, \quad (2.198)$$

and hence

$$B = \begin{pmatrix} -2+5i & 6 & -3i \\ 2i & -4 & 2i \\ 5i & -10 & 2-3i \end{pmatrix}. \tag{2.199}$$

Taking the transpose of B, we obtain

$$A^{-1} = \frac{1}{-4+16i} B^T = \frac{-1-4i}{68} \begin{pmatrix} -2+5i & 2i & 5i \\ 6 & -4 & -10 \\ -3i & 2i & 2-3i \end{pmatrix}$$

$$= \frac{1}{68} \begin{pmatrix} 22+3i & 8-2i & 20-5i \\ -6-24i & 4+16i & 10+40i \\ -12+3i & 8-2i & -14-5i \end{pmatrix}. \tag{2.200}$$

Clearly, this matrix is not unitary since its inverse is not equal to its Hermitian adjoint: $A^{-1} \neq A^{\dagger}$.

(c) Matrix representation of $| \psi \rangle \langle \psi |$
It is now easy to see that the product $| \psi \rangle \langle \psi |$ is indeed an operator, since its representation within $\{| \phi_n \rangle\}$ is a square matrix:

$$| \psi \rangle \langle \psi | = \begin{pmatrix} a_1 \\ a_2 \\ a_3 \\ \vdots \end{pmatrix} \begin{pmatrix} a_1^* & a_2^* & a_3^* & \cdots \end{pmatrix} = \begin{pmatrix} a_1 a_1^* & a_1 a_2^* & a_1 a_3^* & \cdots \\ a_2 a_1^* & a_2 a_2^* & a_2 a_3^* & \cdots \\ a_3 a_1^* & a_3 a_2^* & a_3 a_3^* & \cdots \\ \vdots & \vdots & \vdots & \ddots \end{pmatrix}. \tag{2.201}$$

(d) Trace of an operator
The trace $\text{Tr}(\hat{A})$ of an operator $\hat{A}$ is given, within an orthonormal basis $\{| \phi_n \rangle\}$, by the expression

$$\text{Tr}(\hat{A}) = \sum_n \langle \phi_n | \hat{A} | \phi_n \rangle = \sum_n A_{nn}; \tag{2.202}$$

we will see later that the trace of an operator does not depend on the basis. The trace of a matrix is equal to the sum of its diagonal elements:

$$\text{Tr} \begin{pmatrix} A_{11} & A_{12} & A_{13} & \cdots \\ A_{21} & A_{22} & A_{23} & \cdots \\ A_{31} & A_{32} & A_{33} & \cdots \\ \vdots & \vdots & \vdots & \ddots \end{pmatrix} = A_{11} + A_{22} + A_{33} + \cdots. \tag{2.203}$$

Properties of the trace
We can ascertain that

$$\text{Tr}(\hat{A}^{\dagger}) = (\text{Tr}(\hat{A}))^*, \tag{2.204}$$

$$\text{Tr}(\alpha \hat{A} + \beta \hat{B} + \gamma \hat{C} + \cdots) = \alpha \text{Tr}(\hat{A}) + \beta \text{Tr}(\hat{B}) + \gamma \text{Tr}(\hat{C}) + \cdots, \tag{2.205}$$

and the trace of a product of operators is invariant under the cyclic permutations of these operators:

$$\text{Tr}(\hat{A}\hat{B}\hat{C}\hat{D}\hat{E}) = \text{Tr}(\hat{E}\hat{A}\hat{B}\hat{C}\hat{D}) = \text{Tr}(\hat{D}\hat{E}\hat{A}\hat{B}\hat{C}) = \text{Tr}(\hat{C}\hat{D}\hat{E}\hat{A}\hat{B}) = \cdots. \tag{2.206}$$

Example 2.14

(a) Show that $\text{Tr}(\hat{A}\hat{B}) = \text{Tr}(\hat{B}\hat{A})$.

(b) Show that the trace of a commutator is always zero.

(c) Illustrate the results shown in (a) and (b) on the following matrices:

$$A = \begin{pmatrix} 8-2i & 4i & 0 \\ 1 & 0 & 1-i \\ -8 & i & 6i \end{pmatrix}, \qquad B = \begin{pmatrix} -i & 2 & 1-i \\ 6 & 1+i & 3i \\ 1 & 5+7i & 0 \end{pmatrix}.$$

Solution

(a) Using the definition of the trace,

$$\text{Tr}(\hat{A}\hat{B}) = \sum_n \langle \phi_n \mid \hat{A}\hat{B} \mid \phi_n \rangle, \tag{2.207}$$

and inserting the unit operator between $\hat{A}$ and $\hat{B}$ we have

$$\text{Tr}(\hat{A}\hat{B}) = \sum_n \langle \phi_n \mid \hat{A} \left(\sum_m \mid \phi_m \rangle \langle \phi_m \mid \right) \hat{B} \mid \phi_n \rangle = \sum_{nm} \langle \phi_n \mid \hat{A} \mid \phi_m \rangle \langle \phi_m \mid \hat{B} \mid \phi_n \rangle$$

$$= \sum_{nm} A_{nm} B_{mn}. \tag{2.208}$$

On the other hand, since $\text{Tr}(\hat{A}\hat{B}) = \sum_n \langle \phi_n \mid \hat{A}\hat{B} \mid \phi_n \rangle$, we have

$$\text{Tr}(\hat{B}\hat{A}) = \sum_m \langle \phi_m \mid \hat{B} \left(\sum_n \mid \phi_n \rangle \langle \phi_n \mid \right) \hat{A} \mid \phi_m \rangle = \sum_m \langle \phi_m \mid \hat{B} \mid \phi_n \rangle \langle \phi_n \mid \hat{A} \mid \phi_m \rangle$$

$$= \sum_{nm} B_{mn} A_{nm}. \tag{2.209}$$

Comparing (2.208) and (2.209), we see that $\text{Tr}(\hat{A}\hat{B}) = \text{Tr}(\hat{B}\hat{A})$.

(b) Since $\text{Tr}(\hat{A}\hat{B}) = \text{Tr}(\hat{B}\hat{A})$ we can infer at once that the trace of any commutator is always zero:

$$\text{Tr}([\hat{A}, \hat{B}]) = \text{Tr}(\hat{A}\hat{B}) - \text{Tr}(\hat{B}\hat{A}) = 0. \tag{2.210}$$

(c) Let us verify that the traces of the products AB and BA are equal. Since

$$AB = \begin{pmatrix} -2+16i & 12 & -6-10i \\ 1-2i & 14+2i & 1-i \\ 20i & -59+31i & -11+8i \end{pmatrix}, \qquad BA = \begin{pmatrix} -8 & 5+i & 8+4i \\ 49-35i & -3+24i & -16 \\ 13+5i & 4i & 12+2i \end{pmatrix}, \tag{2.211}$$

we have

$$\text{Tr}(AB) = \text{Tr} \begin{pmatrix} -2+16i & 12 & -6-10i \\ 1-2i & 14+2i & 1-i \\ 20i & -59+31i & -11+8i \end{pmatrix} = 1+26i, \tag{2.212}$$

$$\text{Tr}(BA) = \text{Tr} \begin{pmatrix} -8 & 5+i & 8+4i \\ 49-35i & -3+24i & -16 \\ 13+5i & 4i & 12+2i \end{pmatrix} = 1+26i = \text{Tr}(AB). \tag{2.213}$$

This leads to $\text{Tr}(AB) - \text{Tr}(BA) = (1+26i) - (1+26i) = 0$ or $\text{Tr}([A, B]) = 0$.

2.5.1.4   Matrix Representation of Several Other Quantities

(a) Matrix representation of $| \phi \rangle = \hat{A} | \psi \rangle$
The relation $| \phi \rangle = \hat{A} | \psi \rangle$ can be cast into the algebraic form $\hat{I} | \phi \rangle = \hat{I} \hat{A} \hat{I} | \psi \rangle$ or

$$\left(\sum_n | \phi_n \rangle \langle \phi_n | \right) | \phi \rangle = \left(\sum_n | \phi_n \rangle \langle \phi_n | \right) \hat{A} \left(\sum_m | \phi_m \rangle \langle \phi_m | \right) | \psi \rangle, \qquad (2.214)$$

which in turn can be written as

$$\sum_n b_n | \phi_n \rangle = \sum_{nm} a_m | \phi_n \rangle \langle \phi_n | \hat{A} | \phi_m \rangle = \sum_{nm} a_m A_{nm} | \phi_n \rangle, \qquad (2.215)$$

where $b_n = \langle \phi_n | \phi \rangle$, $A_{nm} = \langle \phi_n | \hat{A} | \phi_m \rangle$, and $a_m = \langle \phi_m | \psi \rangle$. It is easy to see that (2.215) yields $b_n = \sum_m A_{nm} a_m$; hence the matrix representation of $| \phi \rangle = \hat{A} | \psi \rangle$ is given by

$$\begin{pmatrix} b_1 \\ b_2 \\ b_3 \\ \vdots \end{pmatrix} = \begin{pmatrix} A_{11} & A_{12} & A_{13} & \cdots \\ A_{21} & A_{22} & A_{23} & \cdots \\ A_{31} & A_{32} & A_{33} & \cdots \\ \vdots & \vdots & \vdots & \ddots \end{pmatrix} \begin{pmatrix} a_1 \\ a_2 \\ a_3 \\ \vdots \end{pmatrix}. \qquad (2.216)$$

(b) Matrix representation of $\langle \phi | \hat{A} | \psi \rangle$
As for $\langle \phi | \hat{A} | \psi \rangle$ we have

$$\begin{aligned} \langle \phi | \hat{A} | \psi \rangle &= \langle \phi | \hat{I} \hat{A} \hat{I} | \psi \rangle = \langle \phi | \left(\sum_{n=1}^{\infty} | \phi_n \rangle \langle \phi_n | \right) \hat{A} \left(\sum_{m=1}^{\infty} | \phi_m \rangle \langle \phi_m | \right) | \psi \rangle \\ &= \sum_{nm} \langle \phi | \phi_n \rangle \langle \phi_n | \hat{A} | \phi_m \rangle \langle \phi_m | \psi \rangle \\ &= \sum_{nm} b_n^* A_{nm} a_m. \end{aligned} \qquad (2.217)$$

This is a complex number; its matrix representation goes as follows:

$$\langle \phi | \hat{A} | \psi \rangle \longrightarrow \begin{pmatrix} b_1^* & b_2^* & b_3^* & \cdots \end{pmatrix} \begin{pmatrix} A_{11} & A_{12} & A_{13} & \cdots \\ A_{21} & A_{22} & A_{23} & \cdots \\ A_{31} & A_{32} & A_{33} & \cdots \\ \vdots & \vdots & \vdots & \ddots \end{pmatrix} \begin{pmatrix} a_1 \\ a_2 \\ a_3 \\ \vdots \end{pmatrix}. \qquad (2.218)$$

Remark
It is now easy to see explicitly why products of the type $| \psi \rangle | \phi \rangle$, $\langle \psi | \langle \phi |$, $\hat{A} \langle \psi |$, or $| \psi \rangle \hat{A}$ are forbidden. They cannot have matrix representations; they are nonsensical. For instance, $| \psi \rangle | \phi \rangle$ is represented by the product of two column matrices:

$$| \psi \rangle | \phi \rangle \longrightarrow \begin{pmatrix} \langle \phi_1 | \psi \rangle \\ \langle \phi_2 | \psi \rangle \\ \vdots \end{pmatrix} \begin{pmatrix} \langle \phi_1 | \phi \rangle \\ \langle \phi_2 | \phi \rangle \\ \vdots \end{pmatrix}. \qquad (2.219)$$

This product is clearly not possible to perform, for the product of two matrices is possible only when the number of columns of the first is equal to the number of rows of the second; in (2.219) the first matrix has one single column and the second an infinite number of rows.

2.5.1.5 Properties of a Matrix A

- Real if $A = A^*$ or $A_{mn} = A_{mn}^*$

- Imaginary if $A = -A^*$ or $A_{mn} = -A_{mn}^*$

- Symmetric if $A = A^T$ or $A_{mn} = A_{nm}$

- Antisymmetric if $A = -A^T$ or $A_{mn} = -A_{nm}$ with $A_{mm} = 0$

- Hermitian if $A = A^\dagger$ or $A_{mn} = A_{nm}^*$

- Anti-Hermitian if $A = -A^\dagger$ or $A_{mn} = -A_{nm}^*$

- Orthogonal if $A^T = A^{-1}$ or $AA^T = I$ or $(AA^T)_{mn} = \delta_{mn}$

- Unitary if $A^\dagger = A^{-1}$ or $AA^\dagger = I$ or $(AA^\dagger)_{mn} = \delta_{mn}$

Example 2.15

Consider a matrix A (which represents an operator $\hat{A}$), a ket $| \psi \rangle$, and a bra $\langle \phi |$:

$$A = \begin{pmatrix} 5 & 3+2i & 3i \\ -i & 3i & 8 \\ 1-i & 1 & 4 \end{pmatrix}, \quad | \psi \rangle = \begin{pmatrix} -1+i \\ 3 \\ 2+3i \end{pmatrix}, \quad \langle \phi | = (6 \quad -i \quad 5).$$

(a) Calculate the quantities $A | \psi \rangle$, $\langle \phi | A$, $\langle \phi | A | \psi \rangle$, and $| \psi \rangle \langle \phi |$.

(b) Find the complex conjugate, the transpose, and the Hermitian conjugate of A, $| \psi \rangle$, and $\langle \phi |$.

(c) Calculate $\langle \phi | \psi \rangle$ and $\langle \psi | \phi \rangle$; are they equal? Comment on the differences between the complex conjugate, Hermitian conjugate, and transpose of kets and bras.

Solution

(a) The calculations are straightforward:

$$A | \psi \rangle = \begin{pmatrix} 5 & 3+2i & 3i \\ -i & 3i & 8 \\ 1-i & 1 & 4 \end{pmatrix} \begin{pmatrix} -1+i \\ 3 \\ 2+3i \end{pmatrix} = \begin{pmatrix} -5+17i \\ 17+34i \\ 11+14i \end{pmatrix}, \tag{2.220}$$

$$\langle \phi | A = (6 \quad -i \quad 5) \begin{pmatrix} 5 & 3+2i & 3i \\ -i & 3i & 8 \\ 1-i & 1 & 4 \end{pmatrix} = (34-5i \quad 26+12i \quad 20+10i), \tag{2.221}$$

$$\langle \phi | A | \psi \rangle = (6 \quad -i \quad 5) \begin{pmatrix} 5 & 3+2i & 3i \\ -i & 3i & 8 \\ 1-i & 1 & 4 \end{pmatrix} \begin{pmatrix} -1+i \\ 3 \\ 2+3i \end{pmatrix} = 59+155i, \tag{2.222}$$

$$| \psi \rangle \langle \phi | = \begin{pmatrix} -1+i \\ 3 \\ 2+3i \end{pmatrix} (6 \quad -i \quad 5) = \begin{pmatrix} -6+6i & 1+i & -5+5i \\ 18 & -3i & 15 \\ 12+18i & 3-2i & 10+15i \end{pmatrix}. \tag{2.223}$$

(b) To obtain the complex conjugate of A, $\mid \psi \rangle$, and $\langle \phi \mid$, we need simply to take the complex conjugate of their elements:

$$A^* = \begin{pmatrix} 5 & 3-2i & -3i \\ i & -3i & 8 \\ 1+i & 1 & 4 \end{pmatrix}, \quad \mid \psi \rangle^* = \begin{pmatrix} -1-i \\ 3 \\ 2-3i \end{pmatrix}, \quad \langle \phi \mid^* = \begin{pmatrix} 6 & i & 5 \end{pmatrix}.$$

(2.224)

For the transpose of A, $\mid \psi \rangle$, and $\langle \phi \mid$, we simply interchange columns with rows:

$$A^T = \begin{pmatrix} 5 & -i & 1-i \\ 3+2i & 3i & 1 \\ 3i & 8 & 4 \end{pmatrix}, \quad \mid \psi \rangle^T = \begin{pmatrix} -1+i & 3 & 2+3i \end{pmatrix}, \quad \langle \phi \mid^T = \begin{pmatrix} 6 \\ -i \\ 5 \end{pmatrix}.$$

(2.225)

The Hermitian conjugate can be obtained by taking the complex conjugates of the transpose expressions calculated above: $A^\dagger = (A^T)^*, \mid \psi \rangle^\dagger = (\mid \psi \rangle^T)^* = \langle \psi \mid, \langle \phi \mid^\dagger = (\langle \phi \mid^T)^* = \mid \phi \rangle$:

$$A^\dagger = \begin{pmatrix} 5 & i & 1+i \\ 3-2i & -3i & 1 \\ -3i & 8 & 4 \end{pmatrix}, \quad \langle \psi \mid = \begin{pmatrix} -1-i & 3 & 2-3i \end{pmatrix}, \quad \mid \phi \rangle = \begin{pmatrix} 6 \\ i \\ 5 \end{pmatrix}.$$

(2.226)

(c) Using the kets and bras above, we can easily calculate the needed scalar products:

$$\langle \phi \mid \psi \rangle = \begin{pmatrix} 6 & -i & 5 \end{pmatrix} \begin{pmatrix} -1+i \\ 3 \\ 2+3i \end{pmatrix} = 6(-1+i)+(-i)(3)+5(2+3i) = 4+18i, \quad (2.227)$$

$$\langle \psi \mid \phi \rangle = \begin{pmatrix} -1-i & 3 & 2-3i \end{pmatrix} \begin{pmatrix} 6 \\ i \\ 5 \end{pmatrix} = 6(-1-i)+(i)(3)+5(2-3i) = 4-18i. \quad (2.228)$$

We see that $\langle \phi \mid \psi \rangle$ and $\langle \psi \mid \phi \rangle$ are not equal; they are complex conjugates of each other:

$$\langle \psi \mid \phi \rangle = \langle \phi \mid \psi \rangle^* = 4 - 18i. \quad (2.229)$$

Remark

We should underscore the importance of the differences between $\mid \psi \rangle^*, \mid \psi \rangle^T$, and $\mid \psi \rangle^\dagger$. Most notably, we should note (from equations (2.224)–(2.226)) that $\mid \psi \rangle^*$ is a ket, while $\mid \psi \rangle^T$ and $\mid \psi \rangle^\dagger$ are bras. Additionally, we should note that $\langle \phi \mid^*$ is a bra, while $\langle \phi \mid^T$ and $\langle \phi \mid^\dagger$ are kets.

2.5.2 Change of Bases and Unitary Transformations

In a Euclidean space, a vector $\vec{A}$ may be represented by its components in different coordinate systems or in different bases. The transformation from one basis to the other is called a change of basis. The components of $\vec{A}$ in a given basis can be expressed in terms of the components of $\vec{A}$ in another basis by means of a *transformation* matrix.

Similarly, state vectors and operators of quantum mechanics may also be represented in different bases. In this section we are going to study how to transform from one basis to another. That is, knowing the components of kets, bras, and operators in a basis $\{\mid \phi_n \rangle\}$, how does one determine the corresponding components in a different basis $\{| \phi'_n \rangle\}$? Assuming that $\{| \phi_n \rangle\}$ and $\{| \phi'_n \rangle\}$ are two different bases, we can expand each ket $| \phi_n \rangle$ of the old basis in terms of the new basis $\{| \phi'_n \rangle\}$ as follows:

$$| \phi_n \rangle = \left(\sum_m | \phi'_m \rangle \langle \phi'_m | \right) | \phi_n \rangle = \sum_m U_{mn} | \phi'_m \rangle, \tag{2.230}$$

where

$$U_{mn} = \langle \phi'_m | \phi_n \rangle. \tag{2.231}$$

The matrix U, providing the transformation from the old basis $\{| \phi_n \rangle\}$ to the new basis $\{| \phi'_n \rangle\}$, is given by

$$U = \begin{pmatrix} \langle \phi'_1 | \phi_1 \rangle & \langle \phi'_1 | \phi_2 \rangle & \langle \phi'_1 | \phi_3 \rangle \\ \langle \phi'_2 | \phi_1 \rangle & \langle \phi'_2 | \phi_2 \rangle & \langle \phi'_2 | \phi_3 \rangle \\ \langle \phi'_3 | \phi_1 \rangle & \langle \phi'_3 | \phi_2 \rangle & \langle \phi'_3 | \phi_3 \rangle \end{pmatrix}. \tag{2.232}$$

Example 2.16 (Unitarity of the transformation matrix)
Let U be a transformation matrix which connects two *complete* and *orthonormal* bases $\{| \phi_n \rangle\}$ and $\{| \phi'_n \rangle\}$. Show that U is *unitary*.

Solution
For this we need to prove that $\hat{U}\hat{U}^\dagger = \hat{I}$, which reduces to showing that $\langle \phi_m | \hat{U}\hat{U}^\dagger | \phi_n \rangle = \delta_{mn}$. This goes as follows:

$$\langle \phi_m | \hat{U}\hat{U}^\dagger | \phi_n \rangle = \langle \phi_m | \hat{U} \left(\sum_l | \phi_l \rangle \langle \phi_l | \right) \hat{U}^\dagger | \phi_n \rangle = \sum_l U_{ml} U^*_{nl}, \tag{2.233}$$

where $U_{ml} = \langle \phi_m | \hat{U} | \phi_l \rangle$ and $U^*_{nl} = \langle \phi_l | \hat{U}^\dagger | \phi_n \rangle = \langle \phi_n | \hat{U} | \phi_l \rangle^*$. According to (2.231), $U_{ml} = \langle \phi'_m | \phi_l \rangle$ and $U^*_{nl} = \langle \phi_l | \phi'_n \rangle$; we can thus rewrite (2.233) as

$$\sum_l U_{ml} U^*_{nl} = \sum_l \langle \phi'_m | \phi_l \rangle \langle \phi_l | \phi'_n \rangle = \langle \phi'_m | \phi'_n \rangle = \delta_{mn}. \tag{2.234}$$

Combining (2.233) and (2.234), we infer $\langle \phi_m | \hat{U}\hat{U}^\dagger | \phi_n \rangle = \delta_{mn}$, or $\hat{U}\hat{U}^\dagger = \hat{I}$.

2.5.2.1 Transformations of Kets, Bras, and Operators

The components $\langle \phi'_n | \psi \rangle$ of a state vector $| \psi \rangle$ in a new basis $\{| \phi'_n \rangle\}$ can be expressed in terms of the components $\langle \phi_n | \psi \rangle$ of $| \psi \rangle$ in an old basis $\{| \phi_n \rangle\}$ as follows:

$$\langle \phi'_m | \psi \rangle = \langle \phi'_m | \hat{I} | \psi \rangle = \langle \phi'_m | \left(\sum_n | \phi_n \rangle \langle \phi_n | \right) | \psi \rangle = \sum_n U_{mn} \langle \phi_n | \psi \rangle. \tag{2.235}$$

This relation, along with its complex conjugate, can be generalized into

$$| \psi_{new} \rangle = \hat{U} | \psi_{old} \rangle, \qquad \langle \psi_{new} | = \langle \psi_{old} | \hat{U}^\dagger. \tag{2.236}$$

Let us now examine how operators transform when we change from one basis to another. The matrix elements $A'_{mn} = \langle \phi'_m \mid \hat{A} \mid \phi'_n \rangle$ of an operator $\hat{A}$ in the new basis can be expressed in terms of the old matrix elements, $A_{jl} = \langle \phi_j \mid \hat{A} \mid \phi_l \rangle$, as follows:

$$A'_{mn} = \langle \phi'_m \mid \left(\sum_j \mid \phi_j \rangle \langle \phi_j \mid \right) \hat{A} \left(\sum_l \mid \phi_l \rangle \langle \phi_l \mid \right) \mid \phi'_n \rangle = \sum_{jl} U_{mj} A_{jl} U^*_{nl}; \qquad (2.237)$$

that is,

$$\hat{A}_{new} = \hat{U} \hat{A}_{old} \hat{U}^\dagger \quad \text{or} \quad \hat{A}_{old} = \hat{U}^\dagger \hat{A}_{new} \hat{U}. \qquad (2.238)$$

We may summarize the results of the change of basis in the following relations:

$$\boxed{\mid \psi_{new} \rangle = \hat{U} \mid \psi_{old} \rangle, \qquad \langle \psi_{new} \mid = \langle \psi_{old} \mid \hat{U}^\dagger, \qquad \hat{A}_{new} = \hat{U} \hat{A}_{old} \hat{U}^\dagger,} \qquad (2.239)$$

or

$$\boxed{\mid \psi_{old} \rangle = \hat{U}^\dagger \mid \psi_{new} \rangle, \qquad \langle \psi_{old} \mid = \langle \psi_{new} \mid \hat{U}, \qquad \hat{A}_{old} = \hat{U}^\dagger \hat{A}_{new} \hat{U}.} \qquad (2.240)$$

These relations are similar to the ones we derived when we studied unitary transformations; see (2.146) and (2.147).

Example 2.17
Show that the operator $\hat{U} = \sum_n \mid \phi'_n \rangle \langle \phi_n \mid$ satisfies all the properties discussed above.

Solution
First, note that $\hat{U}$ is unitary:

$$\hat{U}\hat{U}^\dagger = \sum_{nl} \mid \phi'_n \rangle \langle \phi_n \mid \phi_l \rangle \langle \phi'_l \mid = \sum_{nl} \mid \phi'_n \rangle \langle \phi'_l \mid \delta_{nl} = \sum_n \mid \phi'_n \rangle \langle \phi'_n \mid = \hat{I}. \qquad (2.241)$$

Second, the action of $\hat{U}$ on a ket of the old basis gives the corresponding ket from the new basis:

$$\hat{U} \mid \phi_m \rangle = \sum_n \mid \phi'_n \rangle \langle \phi_n \mid \phi_m \rangle = \sum_n \mid \phi'_n \rangle \delta_{nm} = \mid \phi'_m \rangle. \qquad (2.242)$$

We can also verify that the action $\hat{U}^\dagger$ on a ket of the new basis gives the corresponding ket from the old basis:

$$\hat{U}^\dagger \mid \phi'_m \rangle = \sum_l \mid \phi_l \rangle \langle \phi'_l \mid \phi'_m \rangle = \sum_l \mid \phi_l \rangle \delta_{lm} = \mid \phi_m \rangle. \qquad (2.243)$$

How does a trace transform under unitary transformations? Using the cyclic property of the trace, $\text{Tr}(\hat{A}\hat{B}\hat{C}) = \text{Tr}(\hat{C}\hat{A}\hat{B}) = \text{Tr}(\hat{B}\hat{C}\hat{A})$, we can ascertain that

$$\text{Tr}(\hat{A}') = \text{Tr}(\hat{U}\hat{A}\hat{U}^\dagger) = \text{Tr}(\hat{U}^\dagger \hat{U} \hat{A}) = \text{Tr}(\hat{A}), \qquad (2.244)$$

$$\text{Tr}\left(\mid \phi_n\rangle\langle\phi_m\mid\right) = \sum_l \langle\phi_l\mid\phi_n\rangle\langle\phi_m\mid\phi_l\rangle = \sum_l \langle\phi_m\mid\phi_l\rangle\langle\phi_l\mid\phi_n\rangle$$

$$= \langle\phi_m\mid\left(\sum_l \mid\phi_l\rangle\langle\phi_l\mid\right)\mid\phi_n\rangle = \langle\phi_m\mid\phi_n\rangle = \delta_{mn}, \qquad (2.245)$$

$$\text{Tr}\left(\mid\phi'_m\rangle\langle\phi_n\mid\right) = \langle\phi_n\mid\phi'_m\rangle. \qquad (2.246)$$

Example 2.18 (The trace is base independent)
Show that the trace of an operator does not depend on the basis in which it is expressed.

Solution
Let us show that the trace of an operator $\hat{A}$ in a basis $\{\mid\phi_n\rangle\}$ is equal to its trace in another basis $\{\mid\phi'_n\rangle\}$. First, the trace of $\hat{A}$ in the basis $\{\mid\phi_n\rangle\}$ is given by

$$\text{Tr}(\hat{A}) = \sum_n \langle\phi_n\mid\hat{A}\mid\phi_n\rangle \qquad (2.247)$$

and in $\{\mid\phi'_n\rangle\}$ by

$$\text{Tr}(\hat{A}) = \sum_n \langle\phi'_n\mid\hat{A}\mid\phi'_n\rangle. \qquad (2.248)$$

Starting from (2.247) and using the completeness of the other basis, $\{\mid\phi'_n\rangle\}$, we have

$$\text{Tr}(\hat{A}) = \sum_n \langle\phi_n\mid\hat{A}\mid\phi_n\rangle = \sum_n \langle\phi_n\mid\left(\sum_m \mid\phi'_m\rangle\langle\phi'_m\mid\right)\hat{A}\mid\phi_n\rangle$$

$$= \sum_{nm} \langle\phi_n\mid\phi'_m\rangle\langle\phi'_m\mid\hat{A}\mid\phi_n\rangle. \qquad (2.249)$$

All we need to do now is simply to interchange the positions of the numbers (scalars) $\langle\phi_n\mid\phi'_m\rangle$ and $\langle\phi'_m\mid\hat{A}\mid\phi_n\rangle$:

$$\text{Tr}(\hat{A}) = \sum_m \langle\phi'_m\mid\hat{A}\left(\sum_n \mid\phi_n\rangle\langle\phi_n\mid\right)\mid\phi'_m\rangle = \sum_m \langle\phi'_m\mid\hat{A}\mid\phi'_m\rangle. \qquad (2.250)$$

From (2.249) and (2.250) we see that

$$\text{Tr}(\hat{A}) = \sum_n \langle\phi_n\mid\hat{A}\mid\phi_n\rangle = \sum_n \langle\phi'_n\mid\hat{A}\mid\phi'_n\rangle. \qquad (2.251)$$

2.5.3 Matrix Representation of the Eigenvalue Problem

At issue here is to work out the matrix representation of the eigenvalue problem (2.126) and then solve it. That is, we want to find the eigenvalues a and the eigenvectors $\mid\psi\rangle$ of an operator $\hat{A}$ such that

$$\hat{A}\mid\psi\rangle = a\mid\psi\rangle, \qquad (2.252)$$

where a is a complex number. Inserting the unit operator between $\hat{A}$ and $| \psi \rangle$ and multiplying by $\langle \phi_m |$, we can cast the eigenvalue equation in the form

$$\langle \phi_m | \hat{A} \left(\sum_n | \phi_n \rangle \langle \phi_n | \right) | \psi \rangle = a \langle \phi_m | \left(\sum_n | \phi_n \rangle \langle \phi_n | \right) | \psi \rangle, \qquad (2.253)$$

or

$$\sum_n A_{mn} \langle \phi_n | \psi \rangle = a \sum_n \langle \phi_n | \psi \rangle \delta_{nm}, \qquad (2.254)$$

which can be rewritten as

$$\sum_n [A_{mn} - a \delta_{nm}] \langle \phi_n | \psi \rangle = 0, \qquad (2.255)$$

with $A_{mn} = \langle \phi_m | \hat{A} | \phi_n \rangle$.

This equation represents an infinite, homogeneous system of equations for the coefficients $\langle \phi_n | \psi \rangle$, since the basis $\{| \phi_n \rangle\}$ is made of an infinite number of base kets. This system of equations can have nonzero solutions only if its determinant vanishes:

$$\det (A_{mn} - a \delta_{nm}) = 0. \qquad (2.256)$$

The problem that arises here is that this determinant corresponds to a matrix with an infinite number of columns and rows. To solve (2.256) we need to truncate the basis $\{| \phi_n \rangle\}$ and assume that it contains only N terms, where N must be large enough to guarantee convergence. In this case we can reduce (2.256) to the following Nth degree determinant:

$$\begin{vmatrix} A_{11} - a & A_{12} & A_{13} & \cdots & A_{1N} \\ A_{21} & A_{22} - a & A_{23} & \cdots & A_{2N} \\ A_{31} & A_{32} & A_{33} - a & \cdots & A_{3N} \\ \vdots & \vdots & \vdots & \ddots & \vdots \\ A_{N1} & A_{N2} & A_{N3} & \cdots & A_{NN} - a \end{vmatrix} = 0. \qquad (2.257)$$

This is known as the *secular* or *characteristic equation*. The solutions of this equation yield the N *eigenvalues* $a_1, a_2, a_3, \ldots, a_N$, since it is an Nth order equation in a. The set of these N eigenvalues is called the spectrum of $\hat{A}$. Knowing the set of eigenvalues $a_1, a_2, a_3, \ldots, a_N$, we can easily determine the corresponding set of *eigenvectors* $| \phi_1 \rangle, | \phi_2 \rangle, \ldots, | \phi_N \rangle$. For each eigenvalue a_m of $\hat{A}$, we can obtain from the "secular" equation (2.257) the N components $\langle \phi_1 | \psi \rangle, \langle \phi_2 | \psi \rangle, \langle \phi_3 | \psi \rangle, \ldots, \langle \phi_N | \psi \rangle$ of the corresponding eigenvector $| \phi_m \rangle$.

If a number of different eigenvectors (two or more) have the same eigenvalue, this eigenvalue is said to be *degenerate*. The order of degeneracy is determined by the number of linearly independent eigenvectors that have the same eigenvalue. For instance, if an eigenvalue has five different eigenvectors, it is said to be fivefold degenerate.

In the case where the set of eigenvectors $| \phi_n \rangle$ of $\hat{A}$ is complete and orthonormal, this set can be used as a basis. In this basis the matrix representing the operator $\hat{A}$ is diagonal,

$$A = \begin{pmatrix} a_1 & 0 & 0 & \cdots \\ 0 & a_2 & 0 & \cdots \\ 0 & 0 & a_3 & \cdots \\ \vdots & \vdots & \vdots & \ddots \end{pmatrix}, \qquad (2.258)$$

the diagonal elements being the eigenvalues a_n of $\hat{A}$, since

$$\langle \phi_m \mid \hat{A} \mid \phi_n \rangle = a_n \langle \phi_m \mid \phi_n \rangle = a_n \delta_{mn}. \tag{2.259}$$

Note that the trace and determinant of a matrix are given, respectively, by the sum and product of the *eigenvalues*:

$$\mathrm{Tr}(A) = \sum_n a_n = a_1 + a_2 + a_3 + \cdots, \tag{2.260}$$

$$\det(A) = \prod_n a_n = a_1 a_2 a_3 \cdots. \tag{2.261}$$

Properties of determinants

Let us mention several useful properties that pertain to determinants. The determinant of a product of matrices is equal to the product of their determinants:

$$\det(ABCD\cdots) = \det(A) \cdot \det(B) \cdot \det(C) \cdot \det(D)\cdots, \tag{2.262}$$

$$\det(A^*) = (\det(A))^*, \qquad \det(A^\dagger) = (\det(A))^*, \tag{2.263}$$

$$\det(A^T) = \det(A), \qquad \det(A) = e^{\mathrm{Tr}(\ln A)}. \tag{2.264}$$

Some theorems pertaining to the eigenvalue problem

Here is a list of useful theorems (the proofs are left as exercises):

- The eigenvalues of a symmetric matrix are real; the eigenvectors form an orthonormal basis.

- The eigenvalues of an antisymmetric matrix are purely imaginary or zero.

- The eigenvalues of a Hermitian matrix are real; the eigenvectors form an orthonormal basis.

- The eigenvalues of a skew-Hermitian matrix are purely imaginary or zero.

- The eigenvalues of a unitary matrix have absolute value equal to one.

- If the eigenvalues of a square matrix are not degenerate (distinct), the corresponding eigenvectors form a basis (i.e., they form a linearly independent set).

Example 2.19 (Eigenvalues and eigenvectors of a matrix)

Find the eigenvalues and the normalized eigenvectors of the matrix

$$A = \begin{pmatrix} 7 & 0 & 0 \\ 0 & 1 & -i \\ 0 & i & -1 \end{pmatrix}.$$

Solution

To find the eigenvalues of A, we simply need to solve the secular equation $\det(A - aI) = 0$:

$$0 = \begin{vmatrix} 7-a & 0 & 0 \\ 0 & 1-a & -i \\ 0 & i & -1-a \end{vmatrix} = (7-a)\left[-(1-a)(1+a) + i^2\right] = (7-a)(a^2 - 2).$$

(2.265)

The eigenvalues of A are thus given by

$$a_1 = 7, \quad a_2 = \sqrt{2}, \quad a_3 = -\sqrt{2}.$$

(2.266)

Let us now calculate the eigenvectors of A. To find the eigenvector corresponding to the first eigenvalue, $a_1 = 7$, we need to solve the matrix equation

$$\begin{pmatrix} 7 & 0 & 0 \\ 0 & 1 & -i \\ 0 & i & -1 \end{pmatrix} \begin{pmatrix} x \\ y \\ z \end{pmatrix} = 7 \begin{pmatrix} x \\ y \\ z \end{pmatrix} \implies \begin{matrix} 7x & = & 7x \\ y - iz & = & 7y \; ; \\ iy - z & = & 7z \end{matrix}$$

(2.267)

this yields $x = 1$ (because the eigenvector is normalized) and $y = z = 0$. So the eigenvector corresponding to $a_1 = 7$ is given by the column matrix

$$|a_1\rangle = \begin{pmatrix} 1 \\ 0 \\ 0 \end{pmatrix}.$$

(2.268)

This eigenvector is normalized since $\langle a_1 \mid a_1 \rangle = 1$.

The eigenvector corresponding to the second eigenvalue, $a_2 = \sqrt{2}$, can be obtained from the matrix equation

$$\begin{pmatrix} 7 & 0 & 0 \\ 0 & 1 & -i \\ 0 & i & -1 \end{pmatrix} \begin{pmatrix} x \\ y \\ z \end{pmatrix} = \sqrt{2} \begin{pmatrix} x \\ y \\ z \end{pmatrix} \implies \begin{matrix} (7 - \sqrt{2})x & = & 0 \\ (1 - \sqrt{2})y - iz & = & 0 \; ; \\ iy - (1 + \sqrt{2})z & = & 0 \end{matrix}$$

(2.269)

this yields $x = 0$ and $z = i(\sqrt{2} - 1)y$. So the eigenvector corresponding to $a_2 = \sqrt{2}$ is given by the column matrix

$$|a_2\rangle = \begin{pmatrix} 0 \\ y \\ i(\sqrt{2} - 1)y \end{pmatrix}.$$

(2.270)

The value of the variable y can be obtained from the normalization condition of $|a_2\rangle$:

$$1 = \langle a_2 \mid a_2 \rangle = \begin{pmatrix} 0 & y^* & -i(\sqrt{2} - 1)y^* \end{pmatrix} \begin{pmatrix} 0 \\ y \\ i(\sqrt{2} - 1)y \end{pmatrix} = 2(2 - \sqrt{2}) \mid y \mid^2.$$

(2.271)

Taking only the positive value of y (a similar calculation can be performed easily if one is interested in the negative value of y), we have $y = 1/\sqrt{2(2 - \sqrt{2})}$; hence the eigenvector (2.270) becomes

$$|a_2\rangle = \begin{pmatrix} 0 \\ \frac{1}{\sqrt{2(2-\sqrt{2})}} \\ \frac{i(\sqrt{2}-1)}{\sqrt{2(2-\sqrt{2})}} \end{pmatrix}.$$

(2.272)

Following the same procedure that led to (2.272), we can show that the third eigenvector is given by

$$| a_3 \rangle = \begin{pmatrix} 0 \\ y \\ -i(1 + \sqrt{2})y \end{pmatrix} ; \tag{2.273}$$

its normalization leads to $y = 1/\sqrt{2(2 + \sqrt{2})}$ (we have considered only the positive value of y); hence

$$| a_3 \rangle = \begin{pmatrix} 0 \\ \frac{1}{\sqrt{2(2+\sqrt{2})}} \\ -\frac{i(1+\sqrt{2})}{\sqrt{2(2+\sqrt{2})}} \end{pmatrix} . \tag{2.274}$$

2.6   Representation in Continuous Bases

In this section we are going to consider the representation of state vectors, bras, and operators in *continuous* bases. After presenting the general formalism, we will consider two important applications: representations in the *position* and *momentum* spaces.

In the previous section we saw that the representations of kets, bras, and operators in a discrete basis are given by discrete matrices. We will show here that these quantities are represented in a *continuous* basis by *continuous matrices*, that is, by noncountable infinite matrices.

2.6.1   General Treatment

The orthonormality condition of the base *kets* of the continuous basis $| \chi_k \rangle$ is expressed not by the usual discrete Kronecker delta as in (2.170) but by Dirac's *continuous delta* function:

$$\langle \chi_k | \chi_{k'} \rangle = \delta(k' - k), \tag{2.275}$$

where k and k' are continuous parameters and where $\delta(k' - k)$ is the Dirac delta function (see Appendix A), which is defined by

$$\delta(x) = \frac{1}{2\pi} \int_{-\infty}^{+\infty} e^{ikx} dk. \tag{2.276}$$

As for the completeness condition of this continuous basis, it is not given by a discrete sum as in (2.172), but by an integral over the continuous variable

$$\int_{-\infty}^{+\infty} dk \ | \chi_k \rangle \langle \chi_k | = \hat{I}, \tag{2.277}$$

where $\hat{I}$ is the unit operator.

Every state vector $| \psi \rangle$ can be expanded in terms of the complete set of basis *kets* $| \chi_k \rangle$:

$$| \psi \rangle = \hat{I} | \psi \rangle = \left(\int_{-\infty}^{+\infty} dk \ | \chi_k \rangle \langle \chi_k | \right) | \psi \rangle = \int_{-\infty}^{+\infty} dk \, b(k) \ | \chi_k \rangle, \tag{2.278}$$

where $b(k)$, which is equal to $\langle \chi_k \mid \psi \rangle$, represents the projection of $\mid \psi \rangle$ on $\mid \chi_k \rangle$.

The norm of the discrete base *kets* is finite ($\langle \phi_n \mid \phi_n \rangle = 1$), but the norm of the continuous base *kets* is infinite; a combination of (2.275) and (2.276) leads to

$$\langle \chi_k \mid \chi_k \rangle = \delta(0) = \frac{1}{2\pi} \int_{-\infty}^{+\infty} dk \longrightarrow \infty. \tag{2.279}$$

This implies that the kets $\mid \chi_k \rangle$ are not square integrable and hence are not elements of the Hilbert space; recall that the space spanned by square-integrable functions is a Hilbert space. Despite the divergence of the norm of $\mid \chi_k \rangle$, the set $\mid \chi_k \rangle$ does constitute a valid basis of vectors that span the Hilbert space, since for any state vector $\mid \psi \rangle$, the scalar product $\langle \chi_k \mid \psi \rangle$ is finite.

The Dirac delta function

Before dealing with the representation of kets, bras, and operators, let us make a short detour to list some of the most important properties of the Dirac delta function (for a more detailed presentation, see Appendix A):

$$\delta(x) = 0, \qquad \text{for} \qquad x \neq 0, \tag{2.280}$$

$$\int_a^b f(x)\delta(x - x_0) \, dx = \begin{cases} f(x_0) & \text{if } a < x_0 < b, \\ 0 & \text{elsewhere,} \end{cases} \tag{2.281}$$

$$\int_{-\infty}^{\infty} f(x)\frac{d^n \delta(x - a)}{dx^n} \, dx = (-1)^n \left. \frac{d^n f(x)}{dx^n} \right|_{x=a}, \tag{2.282}$$

$$\delta(\vec{r} - \vec{r}\,') = \delta(x - x')\delta(y - y')\delta(z - z') = \frac{1}{r^2 \sin\theta}\delta(r - r')\delta(\theta - \theta')\delta(\varphi - \varphi'). \tag{2.283}$$

Representation of kets, bras, and operators

The representation of kets, bras, and operators can be easily inferred from the study that was carried out in the previous section, for the case of a discrete basis. For instance, the ket $\mid \psi \rangle$ is represented by a single column matrix which has a continuous (noncountable) and infinite number of components (rows) $b(k)$:

$$\mid \psi \rangle \longrightarrow \begin{pmatrix} \vdots \\ \langle \chi_k \mid \psi \rangle \\ \vdots \end{pmatrix}. \tag{2.284}$$

The bra $\langle \psi \mid$ is represented by a single row matrix which has a continuous (noncountable) and infinite number of components (columns):

$$\langle \psi \mid \longrightarrow (\cdots\cdots \quad \langle \psi \mid \chi_k \rangle \quad \cdots\cdots). \tag{2.285}$$

Operators are represented by square continuous matrices whose rows and columns have continuous and infinite numbers of components:

$$\hat{A} \longrightarrow \begin{pmatrix} \ddots & \vdots & \ddots \\ \cdots & A(k, k') & \cdots \\ \ddots & \vdots & \ddots \end{pmatrix}. \tag{2.286}$$

As an application, we are going to consider the representations in the position and momentum bases.

2.6.2 Position Representation

In the position representation, the basis consists of an infinite set of vectors $\{|\vec{r}\rangle\}$ which are eigenkets to the position operator $\hat{\vec{R}}$:

$$\hat{\vec{R}} \, |\vec{r}\rangle = \vec{r} \, |\vec{r}\rangle, \tag{2.287}$$

where $\vec{r}$ (without a hat), the position vector, is the eigenvalue of the operator $\hat{\vec{R}}$. The orthonormality and completeness conditions are respectively given by

$$\langle \vec{r} \, | \vec{r}\,' \rangle = \delta(\vec{r} - \vec{r}\,') \quad = \quad \delta(x - x')\delta(y - y')\delta(z - z'), \tag{2.288}$$

$$\int d^3 r \, |\vec{r}\rangle\langle \vec{r}\, | \quad = \quad \hat{I}, \tag{2.289}$$

since, as discussed in Appendix A, the three-dimensional delta function is given by

$$\delta(\vec{r} - \vec{r}\,') = \frac{1}{(2\pi)^3} \int d^3 k \, e^{i\vec{k}\cdot(\vec{r} - \vec{r}\,')}. \tag{2.290}$$

So every state vector $|\psi\rangle$ can be expanded as follows:

$$|\psi\rangle = \int d^3 r \, |\vec{r}\rangle\langle \vec{r} \, | \psi\rangle \equiv \int d^3 r \, \psi(\vec{r}) \, |\vec{r}\rangle, \tag{2.291}$$

where $\psi(\vec{r})$ denotes the components of $|\psi\rangle$ in the $\{|\vec{r}\rangle\}$ basis:

$$\langle \vec{r} \, | \psi\rangle = \psi(\vec{r}). \tag{2.292}$$

This is known as the *wave function* for the state vector $|\psi\rangle$. Recall that, according to the probabilistic interpretation of Born, the quantity $|\langle \vec{r} \, | \psi\rangle|^2 \, d^3 r$ represents the probability of finding the system in the volume element $d^3 r$.

The scalar product between two state vectors, $|\psi\rangle$ and $|\phi\rangle$, can be expressed in this form:

$$\langle \phi \, | \psi\rangle = \langle \phi \, | \left(\int d^3 r \, |\vec{r}\rangle\langle \vec{r}\, | \right) | \psi\rangle = \int d^3 r \, \phi^*(\vec{r})\psi(\vec{r}). \tag{2.293}$$

Since $\hat{\vec{R}} \, |\vec{r}\rangle = \vec{r} \, |\vec{r}\rangle$ we have

$$\langle \vec{r}\,' \, | \hat{\vec{R}}^n | \vec{r}\rangle = \vec{r}^{\,n}\delta(\vec{r}\,' - \vec{r}). \tag{2.294}$$

Note that the operator $\hat{\vec{R}}$ is Hermitian, since

$$\langle \phi \, | \hat{\vec{R}} \, | \psi\rangle = \int d^3 r \, \vec{r}\langle \phi \, | \vec{r}\rangle\langle \vec{r} \, | \psi\rangle = \left[\int d^3 r \, \vec{r}\langle \psi \, | \vec{r}\rangle\langle \vec{r} \, | \phi\rangle \right]^*$$

$$= \langle \psi \, | \hat{\vec{R}} \, | \phi\rangle^*. \tag{2.295}$$

2.6.3 Momentum Representation

The basis $\{|\vec{p}\rangle\}$ of the momentum representation is obtained from the eigenkets of the momentum operator $\hat{\vec{P}}$:

$$\hat{\vec{P}} \mid \vec{p}\rangle = \vec{p} \mid \vec{p}\rangle, \tag{2.296}$$

where $\vec{p}$ is the momentum vector. The algebra relevant to this representation can be easily inferred from the position representation. The orthonormality and completeness conditions of the momentum space basis $\mid \vec{p}\rangle$ are given by

$$\langle \vec{p} \mid \vec{p}\,'\rangle = \delta(\vec{p} - \vec{p}\,') \quad \text{and} \quad \int d^3p \mid \vec{p}\rangle\langle \vec{p} \mid = \hat{I}. \tag{2.297}$$

Expanding $\mid \psi\rangle$ in this basis, we obtain

$$\mid \psi\rangle = \int d^3p \mid \vec{p}\rangle\langle \vec{p} \mid \psi\rangle = \int d^3p\, \Psi(\vec{p}) \mid \vec{p}\rangle, \tag{2.298}$$

where the expansion coefficient $\Psi(\vec{p})$ represents the *momentum space wave function*. The quantity $\mid \Psi(\vec{p}) \mid^2 d^3p$ is the probability of finding the system's momentum in the volume element d^3p located between $\vec{p}$ and $\vec{p} + d\vec{p}$.

By analogy with (2.293) the scalar product between two states is given in the momentum space by

$$\langle \phi \mid \psi\rangle = \langle \phi \mid \left(\int d^3p \mid \vec{p}\rangle\langle \vec{p} \mid \right) \mid \psi\rangle = \int d^3p\, \Phi^*(\vec{p})\Psi(\vec{p}). \tag{2.299}$$

Since $\hat{\vec{P}} \mid \vec{p}\rangle = \vec{p} \mid \vec{p}\rangle$ we have

$$\langle \vec{p}\,' \mid \hat{\vec{P}}^n \mid \vec{p}\rangle = \vec{p}^{\,n}\delta(\vec{p}\,' - \vec{p}). \tag{2.300}$$

2.6.4 Connecting the Position and Momentum Representations

Let us now study how to establish a connection between the position and the momentum representations. By analogy with the foregoing study, when changing from the $\{|\vec{r}\rangle\}$ basis to the $\{|\vec{p}\rangle\}$ basis, we encounter the *transformation* function $\langle \vec{r} \mid \vec{p}\rangle$.

To find the expression for the transformation function $\langle \vec{r} \mid \vec{p}\rangle$, let us establish a connection between the position and momentum representations of the state vector $\mid \psi\rangle$:

$$\langle \vec{r} \mid \psi\rangle = \langle \vec{r} \mid \left(\int d^3p \mid \vec{p}\rangle\langle \vec{p} \mid \right) \mid \psi\rangle = \int d^3p\, \langle \vec{r} \mid \vec{p}\rangle\Psi(\vec{p}); \tag{2.301}$$

that is,

$$\psi(\vec{r}) = \int d^3p\, \langle \vec{r} \mid \vec{p}\rangle\Psi(\vec{p}). \tag{2.302}$$

Similarly, we can write

$$\Psi(\vec{p}) = \langle \vec{p} \mid \psi\rangle = \langle \vec{p} \mid \int d^3r \mid \vec{r}\rangle\langle \vec{r} \mid \psi\rangle = \int d^3r\, \langle \vec{p} \mid \vec{r}\rangle\psi(\vec{r}). \tag{2.303}$$

The last two relations imply that $\psi(\vec{r})$ and $\Psi(\vec{p})$ are to be viewed as Fourier transforms of each other. In quantum mechanics the Fourier transform of a function $f(\vec{r})$ is given by

$$f(\vec{r}) = \frac{1}{(2\pi\hbar)^{3/2}} \int d^3p \, e^{i\vec{p}\cdot\vec{r}/\hbar} g(\vec{p}); \tag{2.304}$$

notice the presence of Planck's constant. Hence the function $\langle \vec{r} \mid \vec{p} \rangle$ is given by

$$\boxed{\langle \vec{r} \mid \vec{p} \rangle = \frac{1}{(2\pi\hbar)^{3/2}} e^{i\vec{p}\cdot\vec{r}/\hbar}.} \tag{2.305}$$

This function transforms from the momentum to the position representation. The function corresponding to the inverse transformation, $\langle \vec{p} \mid \vec{r} \rangle$, is given by

$$\boxed{\langle \vec{p} \mid \vec{r} \rangle = \langle \vec{r} \mid \vec{p} \rangle^* = \frac{1}{(2\pi\hbar)^{3/2}} e^{-i\vec{p}\cdot\vec{r}/\hbar}.} \tag{2.306}$$

The quantity $|\langle \vec{r} \mid \vec{p} \rangle|^2$ represents the probability density of finding the particle in a region around $\vec{r}$ where its momentum is equal to $\vec{p}$.

Remark

If the position wave function

$$\psi(\vec{r}) = \frac{1}{(2\pi\hbar)^{3/2}} \int d^3p \, e^{i\vec{p}\cdot\vec{r}/\hbar} \Psi(\vec{p}) \tag{2.307}$$

is normalized (i.e., $\int d^3r \, \psi(\vec{r})\psi^*(\vec{r}) = 1$), its Fourier transform

$$\Psi(\vec{p}) = \frac{1}{(2\pi\hbar)^{3/2}} \int d^3r \, e^{-i\vec{p}\cdot\vec{r}/\hbar} \psi(\vec{r}) \tag{2.308}$$

must also be normalized, since

$$\begin{aligned}
\int d^3p \, \Psi^*(\vec{p})\Psi(\vec{p}) &= \int d^3p \Psi^*(\vec{p}) \left[\frac{1}{(2\pi\hbar)^{3/2}} \int d^3r \, e^{-i\vec{p}\cdot\vec{r}/\hbar} \psi(\vec{r}) \right] \\
&= \int d^3r \, \psi(\vec{r}) \left[\frac{1}{(2\pi\hbar)^{3/2}} \int d^3p \, \Psi^*(\vec{p}) e^{-i\vec{p}\cdot\vec{r}/\hbar} \right] \\
&= \int d^3r \, \psi(\vec{r})\psi^*(\vec{r}) \\
&= 1.
\end{aligned} \tag{2.309}$$

This result is known as *Parseval's theorem*.

2.6.4.1 Momentum Operator in the Position Representation

To determine the form of the momentum operator $\hat{\vec{P}}$ in the position representation, let us calculate $\langle \vec{r} \mid \hat{\vec{P}} \mid \psi \rangle$:

$$\begin{aligned}
\langle \vec{r} \mid \hat{\vec{P}} \mid \psi \rangle &= \int \langle \vec{r} \mid \hat{\vec{P}} \mid \vec{p} \rangle \langle \vec{p} \mid \psi \rangle d^3p = \int \vec{p} \langle \vec{r} \mid \vec{p} \rangle \langle \vec{p} \mid \psi \rangle d^3p \\
&= \frac{1}{(2\pi\hbar)^{3/2}} \int \vec{p} \, e^{i\vec{p}\cdot\vec{r}/\hbar} \Psi(\vec{p}) d^3p, \tag{2.310}
\end{aligned}$$

where we have used the relation $\int \mid \vec{p} \rangle \langle \vec{p} \mid d^3p = \hat{I}$ along with Eq. (2.305). Now, since $\vec{p}\, e^{i\vec{p}\cdot\vec{r}/\hbar} = -i\hbar\vec{\nabla} e^{i\vec{p}\cdot\vec{r}/\hbar}$, and using Eq. (2.305) again, we can rewrite (2.310) as

$$
\begin{aligned}
\langle \vec{r} \mid \hat{\vec{P}} \mid \psi \rangle &= -i\hbar\vec{\nabla}\left(\frac{1}{(2\pi\hbar)^{3/2}}\int e^{i\vec{p}\cdot\vec{r}/\hbar}\Psi(\vec{p})d^3p\right) \\
&= -i\hbar\vec{\nabla}\left(\int \langle \vec{r} \mid \vec{p} \rangle\langle \vec{p} \mid \psi \rangle d^3p\right) \\
&= -i\hbar\vec{\nabla}\langle \vec{r} \mid \psi \rangle.
\end{aligned}
\tag{2.311}
$$

Thus, $\hat{\vec{P}}$ is given in the position representation by

$$
\boxed{\hat{\vec{P}} = -i\hbar\vec{\nabla}.}
\tag{2.312}
$$

Its Cartesian components are

$$
\boxed{\hat{P}_x = -i\hbar\frac{\partial}{\partial x}, \qquad \hat{P}_y = -i\hbar\frac{\partial}{\partial y}, \qquad \hat{P}_z = -i\hbar\frac{\partial}{\partial z}.}
\tag{2.313}
$$

Note that the form of the momentum operator (2.312) can be derived by simply applying the gradient operator $\vec{\nabla}$ on a *plane* wave function $\psi(\vec{r}, t) = Ae^{i(\vec{p}\cdot\vec{r}-Et)/\hbar}$:

$$
-i\hbar\vec{\nabla}\psi(\vec{r}, t) = \vec{p}\psi(\vec{r}, t) = \hat{\vec{P}}\psi(\vec{r}, t).
\tag{2.314}
$$

It is easy to verify that $\hat{\vec{P}}$ is Hermitian (see equation (2.378)).

Now, since $\hat{\vec{P}} = -i\hbar\vec{\nabla}$, we can write the Hamiltonian operator $\hat{H} = \hat{\vec{P}}^2/(2m) + \hat{V}$ in the position representation as follows:

$$
\boxed{\hat{H} = -\frac{\hbar^2}{2m}\nabla^2 + \hat{V}(\vec{r}) = -\frac{\hbar^2}{2m}\left(\frac{\partial^2}{\partial x^2} + \frac{\partial^2}{\partial y^2} + \frac{\partial^2}{\partial z^2}\right) + \hat{V}(\vec{r}),}
\tag{2.315}
$$

where ∇^2 is the Laplacian operator; it is given in Cartesian coordinates by $\nabla^2 = \partial^2/\partial x^2 + \partial^2/\partial y^2 + \partial^2/\partial z^2$.

2.6.4.2 Position Operator in the Momentum Representation

The form of the position operator $\hat{\vec{R}}$ in the momentum representation can be easily inferred from the representation of $\hat{\vec{P}}$ in the position space. In momentum space the position operator can be written as follows:

$$
\hat{R}_j = i\hbar\frac{\partial}{\partial p_j} \qquad (j = x, y, z)
\tag{2.316}
$$

or

$$
\boxed{\hat{X} = i\hbar\frac{\partial}{\partial p_x}, \qquad \hat{Y} = i\hbar\frac{\partial}{\partial p_y}, \qquad \hat{Z} = i\hbar\frac{\partial}{\partial p_z}.}
\tag{2.317}
$$

2.6.4.3 Important Commutation Relations

Let us now calculate the commutator $[\hat{R}_j, \hat{P}_k]$ in the position representation. As the separate actions of $\hat{X}\hat{P}_x$ and $\hat{P}_x\hat{X}$ on the wave function $\psi(\vec{r})$ are given by

$$\hat{X}\hat{P}_x\psi(\vec{r}) = -i\hbar x\frac{\partial\psi(\vec{r})}{\partial x}, \qquad (2.318)$$

$$\hat{P}_x\hat{X}\psi(\vec{r}) = -i\hbar\frac{\partial}{\partial x}(x\psi(\vec{r})) = -i\hbar\psi(\vec{r}) - i\hbar x\frac{\partial\psi(\vec{r})}{\partial x}, \qquad (2.319)$$

we have

$$\begin{aligned}
[\hat{X}, \hat{P}_x]\psi(\vec{r}) &= \hat{X}\hat{P}_x\psi(\vec{r}) - \hat{P}_x\hat{X}\psi(\vec{r}) = -i\hbar x\frac{\partial\psi(\vec{r})}{\partial x} + i\hbar\psi(\vec{r}) + i\hbar x\frac{\partial\psi(\vec{r})}{\partial x} \\
&= i\hbar\psi(\vec{r})
\end{aligned} \qquad (2.320)$$

or

$$[\hat{X}, \hat{P}_x] = i\hbar. \qquad (2.321)$$

Similar relations can be derived at once for the y and the z components:

$$\boxed{[\hat{X}, \hat{P}_x] = i\hbar, \qquad [\hat{Y}, \hat{P}_Y] = i\hbar, \qquad [\hat{Z}, \hat{P}_Z] = i\hbar.} \qquad (2.322)$$

We can verify that

$$[\hat{X}, \hat{P}_y] = [\hat{X}, \hat{P}_z] = [\hat{Y}, \hat{P}_x] = [\hat{Y}, \hat{P}_z] = [\hat{Z}, \hat{P}_x] = [\hat{Z}, \hat{P}_y] = 0, \qquad (2.323)$$

since the x, y, z degrees of freedom are independent; the previous two relations can be grouped into

$$\boxed{[\hat{R}_j, \hat{P}_k] = i\hbar\delta_{jk}, \qquad [\hat{R}_j, \hat{R}_k] = 0, \qquad [\hat{P}_j, \hat{P}_k] = 0 \qquad (j, k = x, y, z).} \qquad (2.324)$$

These relations are often called the *canonical commutation relations*.

Now, from (2.321) we can show that (for the proof see Problem 2.8 on page 139)

$$\boxed{[\hat{X}^n, \hat{P}_x] = i\hbar n\hat{X}^{n-1}, \qquad [\hat{X}, \hat{P}_x^n] = i\hbar n\hat{P}_x^{n-1}.} \qquad (2.325)$$

Following the same procedure that led to (2.320), we can obtain a more general commutation relation of $\hat{P}_x$ with an arbitrary function $f(\hat{X})$:

$$\boxed{[f(\hat{X}), \hat{P}_x] = i\hbar\frac{df(\hat{X})}{d\hat{X}} \implies [\hat{\vec{P}}, F(\hat{\vec{R}})] = -i\hbar\vec{\nabla}F(\hat{\vec{R}})),} \qquad (2.326)$$

where F is a function of the operator $\hat{\vec{R}}$.

The explicit form of operators thus depends on the representation adopted. We have seen, however, that the *commutation relations for operators are representation independent*. In particular, the commutator $[\hat{R}_j, \hat{P}_k]$ is given by $i\hbar\delta_{jk}$ in the position and the momentum representations; see the next example.

Example 2.20 (Commutators are representation independent)
Calculate the commutator $[\hat{X}, \hat{P}]$ in the momentum representation and verify that it is equal to $i\hbar$.

Solution
As the operator $\hat{X}$ is given in the momentum representation by $\hat{X} = i\hbar\partial/\partial p$, we have

$$
\begin{aligned}
[\hat{X}, \hat{P}]\psi(p) &= \hat{X}\hat{P}\psi(p) - \hat{P}\hat{X}\psi(p) = i\hbar\frac{\partial}{\partial p}(p\psi(p)) - i\hbar p\frac{\partial\psi(p)}{\partial p} \\
&= i\hbar\psi(p) + i\hbar p\frac{\partial\psi(p)}{\partial p} - i\hbar p\frac{\partial\psi(p)}{\partial p} = i\hbar\psi(p).
\end{aligned}
\tag{2.327}
$$

Thus, the commutator $[\hat{X}, \hat{P}]$ is given in the *momentum representation* by

$$
[\hat{X}, \hat{P}] = \left[i\hbar\frac{\partial}{\partial p}, \hat{P}\right] = i\hbar.
\tag{2.328}
$$

The commutator $[\hat{X}, \hat{P}]$ was also shown to be equal to $i\hbar$ in the *position representation* (see equation (2.321)):

$$
[\hat{X}, \hat{P}] = -\left[\hat{X}, i\hbar\frac{\partial}{\partial p_x}\right] = i\hbar.
\tag{2.329}
$$

2.6.5 Parity Operator

The *space reflection* about the origin of the coordinate system is called an *inversion* or a *parity* operation. This transformation is *discrete*. The parity operator $\hat{P}$ is defined by its action on the kets $|\vec{r}\rangle$ of the position space:

$$
\hat{P}|\vec{r}\rangle = |-\vec{r}\rangle, \qquad \langle\vec{r}|\hat{P}^{\dagger} = \langle-\vec{r}|,
\tag{2.330}
$$

such that

$$
\hat{P}\psi(\vec{r}) = \psi(-\vec{r}).
\tag{2.331}
$$

The parity operator is Hermitian, $\hat{P}^{\dagger} = \hat{P}$, since

$$
\begin{aligned}
\int d^3r\, \phi^*(\vec{r})\left[\hat{P}\psi(\vec{r})\right] &= \int d^3r\, \phi^*(\vec{r})\psi(-\vec{r}) = \int d^3r\, \phi^*(-\vec{r})\psi(\vec{r}) \\
&= \int d^3r\, \left[\hat{P}\phi(\vec{r})\right]^* \psi(\vec{r}).
\end{aligned}
\tag{2.332}
$$

From the definition (2.331), we have

$$
\hat{P}^2\psi(\vec{r}) = \hat{P}\psi(-\vec{r}) = \psi(\vec{r});
\tag{2.333}
$$

hence $\hat{P}^2$ is equal to the unity operator:

$$
\hat{P}^2 = \hat{I} \quad \text{or} \quad \hat{P} = \hat{P}^{-1}.
\tag{2.334}
$$

The parity operator is therefore *unitary*, since its Hermitian adjoint is equal to its inverse:

$$\hat{P}^\dagger = \hat{P}^{-1}. \tag{2.335}$$

Now, since $\hat{P}^2 = \hat{I}$, the eigenvalues of $\hat{P}$ are $+1$ or -1 with the corresponding eigenstates

$$\hat{P}\psi_+(\vec{r}) = \psi_+(-\vec{r}) = \psi_+(\vec{r}), \qquad \hat{P}\psi_-(\vec{r}) = \psi_-(-\vec{r}) = -\psi_-(\vec{r}). \tag{2.336}$$

The eigenstate $|\psi_+\rangle$ is said to be *even* and $|\psi_-\rangle$ is *odd*. Therefore, the eigenfunctions of the parity operator have *definite parity*: they are either even or odd.

Since $|\psi_+\rangle$ and $|\psi_-\rangle$ are joint eigenstates of the same Hermitian operator $\hat{P}$ but with different eigenvalues, these eigenstates must be orthogonal:

$$\langle\psi_+ \mid \psi_-\rangle = \int d^3r\, \psi_+^*(-\vec{r})\psi_-(-\vec{r}) \equiv -\int d^3r\, \psi_+^*(\vec{r})\psi_-(\vec{r}) = -\langle\psi_+ \mid \psi_-\rangle; \tag{2.337}$$

hence $\langle\psi_+ \mid \psi_-\rangle$ is zero. The states $|\psi_+\rangle$ and $|\psi_-\rangle$ form a complete set since any function can be written as $\psi(\vec{r}) = \psi_+(\vec{r}) + \psi_-(\vec{r})$, which leads to

$$\psi_+(\vec{r}) = \frac{1}{2}\left[\psi(\vec{r}) + \psi(-\vec{r})\right], \qquad \psi_-(\vec{r}) = \frac{1}{2}\left[\psi(\vec{r}) - \psi(-\vec{r})\right]. \tag{2.338}$$

Since $\hat{P}^2 = I$ we have

$$\hat{P}^n = \left\{ \begin{array}{ll} \hat{P} & \text{when } n \text{ is odd,} \\ \hat{I} & \text{when } n \text{ is even.} \end{array} \right. \tag{2.339}$$

Even and odd operators

An operator $\hat{A}$ is said to be *even* if it obeys the condition

$$\hat{P}\hat{A}\hat{P} = \hat{A} \tag{2.340}$$

and an operator $\hat{B}$ is *odd* if

$$\hat{P}\hat{B}\hat{P} = -\hat{B}. \tag{2.341}$$

We can easily verify that even operators commute with the parity operator $\hat{P}$ and that odd operators anticommute with $\hat{P}$:

$$\hat{A}\hat{P} = (\hat{P}\hat{A}\hat{P})\hat{P} = \hat{P}\hat{A}\hat{P}^2 = \hat{P}\hat{A}, \tag{2.342}$$

$$\hat{B}\hat{P} = -(\hat{P}\hat{B}\hat{P})\hat{P} = -\hat{P}\hat{B}\hat{P}^2 = -\hat{P}\hat{B}. \tag{2.343}$$

The fact that even operators commute with the parity operator has very useful consequences. Let us examine the following two important cases depending on whether an even operator has nondegenerate or degenerate eigenvalues:

- If an even operator is Hermitian and none of its eigenvalues is degenerate, then this operator has the same eigenvectors as those of the parity operator. And since the eigenvectors of the parity operator are either even or odd, the eigenvectors of an even, Hermitian, and nondegenerate operator must also be either even or odd; they are said to have a *definite parity*. This property will have useful applications when we solve the Schrödinger equation for even Hamiltonians.

- If the even operator has a degenerate spectrum, its eigenvectors do not necessarily have a definite parity.

What about the parity of the position and momentum operators, $\hat{\vec{R}}$ and $\hat{\vec{P}}$? We can easily show that both of them are odd, since they anticommute with the parity operator:

$$\hat{P}\hat{\vec{R}} = -\hat{\vec{R}}\hat{P}, \qquad \hat{P}\hat{\vec{P}} = -\hat{\vec{P}}\hat{P}; \tag{2.344}$$

hence

$$\hat{P}\hat{\vec{R}}\hat{P}^{\dagger} = -\hat{\vec{R}}, \qquad \hat{P}\hat{\vec{P}}\hat{P}^{\dagger} = -\hat{\vec{P}}, \tag{2.345}$$

since $\hat{P}\hat{P}^{\dagger} = 1$. For instance, to show that $\hat{\vec{R}}$ anticommutes with $\hat{P}$, we need simply to look at the following relations:

$$\hat{P}\hat{\vec{R}} \mid \vec{r} \rangle = \vec{r}\hat{P} \mid \vec{r} \rangle = \vec{r} \mid -\vec{r} \rangle, \tag{2.346}$$

$$\hat{\vec{R}}\hat{P} \mid \vec{r} \rangle = \hat{\vec{R}} \mid -\vec{r} \rangle = -\vec{r} \mid -\vec{r} \rangle. \tag{2.347}$$

If the operators $\hat{A}$ and $\hat{B}$ are even and odd, respectively, we can verify that

$$\hat{P}\hat{A}^{n}\hat{P} = \hat{A}^{n}, \qquad \hat{P}\hat{B}^{n}\hat{P} = (-1)^{n}\hat{B}^{n}. \tag{2.348}$$

These relations can be shown as follows:

$$\begin{aligned} \hat{P}\hat{A}^{n}\hat{P} &= (\hat{P}\hat{A}\hat{P})(\hat{P}\hat{A}\hat{P})\cdots(\hat{P}\hat{A}\hat{P}) = \hat{A}^{n}, \tag{2.349}\\ \hat{P}\hat{B}^{n}\hat{P} &= (\hat{P}\hat{B}\hat{P})(\hat{P}\hat{B}\hat{P})\cdots(\hat{P}\hat{B}\hat{P}) = (-1)^{n}\hat{B}^{n}. \tag{2.350} \end{aligned}$$

2.7 Matrix and Wave Mechanics

In this chapter we have so far worked out the mathematics pertaining to quantum mechanics in two different representations: *discrete* basis systems and *continuous* basis systems. The theory of quantum mechanics deals in essence with solving the following eigenvalue problem:

$$\hat{H} \mid \psi \rangle = E \mid \psi \rangle, \tag{2.351}$$

where $\hat{H}$ is the Hamiltonian of the system. This equation is general and does not depend on any coordinate system or representation. But to solve it, we need to represent it in a given basis system. The complexity associated with solving this eigenvalue equation will then vary from one basis to another.

In what follows we are going to examine the representation of this eigenvalue equation in a *discrete* basis and then in a *continuous* basis.

2.7.1 Matrix Mechanics

The representation of quantum mechanics in a *discrete* basis yields a *matrix* eigenvalue problem. That is, the representation of (2.351) in a discrete basis $\{\mid \phi_n \rangle\}$ yields the following matrix eigenvalue equation (see (2.257)):

$$\begin{vmatrix} H_{11} - E & H_{12} & H_{13} & \cdots & H_{1N} \\ H_{21} & H_{22} - E & H_{23} & \cdots & H_{2N} \\ H_{31} & H_{32} & H_{33} - E & \cdots & H_{3N} \\ \vdots & \vdots & \vdots & \ddots & \vdots \\ H_{N1} & H_{N2} & H_{N3} & \cdots & H_{NN} - E \end{vmatrix} = 0. \qquad (2.352)$$

This is an Nth order equation in E; its solutions yield the energy spectrum of the system: E_1, E_2, E_3, ..., E_N. Knowing the set of eigenvalues E_1, E_2, E_3, ..., E_N, we can easily determine the corresponding set of eigenvectors $| \phi_1 \rangle, | \phi_2 \rangle, ..., | \phi_N \rangle$.

The diagonalization of the Hamiltonian matrix (2.352) of a system yields the energy spectrum as well as the state vectors of the system. This procedure, which was worked out by Heisenberg, involves only matrix quantities and matrix eigenvalue equations. This formulation of quantum mechanics is known as *matrix mechanics*.

The starting point of Heisenberg, in his attempt to find a theoretical foundation to Bohr's ideas, was the atomic transition relation, $v_{mn} = (E_m - E_n)/h$, which gives the frequencies of the radiation associated with the electron's transition from orbit m to orbit n. The frequencies v_{mn} can be arranged in a square matrix, where the mn element corresponds to the transition from the mth to the nth quantum state.

We can also construct matrices for other dynamical quantities related to the transition $m \rightarrow n$. In this way, every physical quantity is represented by a matrix. For instance, we represent the energy levels by an energy matrix, the position by a position matrix, the momentum by a momentum matrix, the angular momentum by an angular momentum matrix, and so on. In calculating the various physical magnitudes, one has thus to deal with the algebra of matrix quantities. So, within the context of matrix mechanics, one deals with noncommuting quantities, for the product of matrices does not commute. This is an essential feature that distinguishes matrix mechanics from classical mechanics, where all the quantities commute. Take, for instance, the position and momentum quantities. While commuting in classical mechanics, $px = xp$, they do not commute within the context of matrix mechanics; they are related by the commutation relation $[\hat{X}, \hat{P}_x] = i\hbar$. The same thing applies for the components of angular momentum. We should note that the role played by the commutation relations within the context of matrix mechanics is similar to the role played by Bohr's quantization condition in atomic theory. Heisenberg's matrix mechanics therefore requires the introduction of some mathematical machinery—linear vector spaces, Hilbert space, commutator algebra, and matrix algebra—that is entirely different from the mathematical machinery of classical mechanics. Here lies the justification for having devoted a somewhat lengthy section, Section 2.5, to study the matrix representation of quantum mechanics.

2.7.2 Wave Mechanics

Representing the formalism of quantum mechanics in a *continuous* basis yields an eigenvalue problem not in the form of a matrix equation, as in Heisenberg's formulation, but in the form of a *differential equation*. The representation of the eigenvalue equation (2.351) in the *position* space yields

$$\langle \vec{r} \mid \hat{H} \mid \psi \rangle = E \langle \vec{r} \mid \psi \rangle. \qquad (2.353)$$

As shown in (2.315), the Hamiltonian is given in the position representation by $-\hbar^2 \nabla^2/(2m) + \hat{V}(\vec{r})$, so we can rewrite (2.353) in a more familiar form:

$$-\frac{\hbar^2}{2m} \nabla^2 \psi(\vec{r}) + \hat{V}(\vec{r})\psi(\vec{r}) = E\psi(\vec{r}), \qquad (2.354)$$

where $\langle \vec{r} \mid \psi \rangle = \psi(\vec{r})$ is the *wave function* of the system. This differential equation is known as the *Schrödinger equation* (its origin will be discussed in Chapter 3). Its solutions yield the energy spectrum of the system as well as its wave function. This formulation of quantum mechanics in *the position representation* is called *wave mechanics*.

Unlike Heisenberg, Schödinger took an entirely different starting point in his quest to find a theoretical justification for Bohr's ideas. He started from the de Broglie particle–wave hypothesis and extended it to the electrons orbiting around the nucleus. Schrödinger aimed at finding an equation that describes the motion of the electron within an atom. Here the focus is on the *wave* aspect of the electron. We can show, as we did in Chapter 1, that the Bohr quantization condition, $L = n\hbar$, is equivalent to the de Broglie relation, $\lambda = 2\pi\hbar/p$. To establish this connection, we need simply to make three assumptions: (a) the wavelength of the wave associated with the orbiting electron is connected to the electron's linear momentum p by $\lambda = 2\pi\hbar/p$, (b) the electron's orbit is circular, and (c) the circumference of the electron's orbit is an integer multiple of the electron's wavelength, i.e., $2\pi r = n\lambda$. This leads at once to $2\pi r = n \times (2\pi\hbar/p)$ or $n\hbar = rp \equiv L$. This means that, for every orbit, there is only one wavelength (or one wave) associated with the electron while revolving in that orbit. This wave can be described by means of a *wave function*. So Bohr's quantization condition implies, in essence, a uniqueness of the wave function for each orbit of the electron. In Chapter 3 we will show how Schrödinger obtained his differential equation (2.354) to describe the motion of an electron in an atom.

2.8 Concluding Remarks

Historically, the matrix formulation of quantum mechanics was worked out by Heisenberg shortly before Schrödinger introduced his wave theory. The equivalence between the matrix and wave formulations was proved a few years later by using the theory of unitary transformations. Different in form, yet identical in contents, wave mechanics and matrix mechanics achieve the same goal: finding the energy spectrum and the states of quantum systems.

The matrix formulation has the advantage of greater (formal) generality, yet it suffers from a number of disadvantages. On the conceptual side, it offers no visual idea about the structure of the atom; it is less intuitive than wave mechanics. On the technical side, it is difficult to use in some problems of relative ease such as finding the stationary states of atoms. Matrix mechanics, however, becomes powerful and practical in solving problems such as the harmonic oscillator or in treating the formalism of angular momentum.

But most of the efforts of quantum mechanics focus on solving the Schrödinger equation, not the Heisenberg matrix eigenvalue problem. So in the rest of this text we deal mostly with wave mechanics. Matrix mechanics is used only in a few problems, such as the harmonic oscillator, where it is more suitable than Schrödinger's wave mechanics.

In wave mechanics we need only to specify the potential in which the particle moves; the Schrödinger equation takes care of the rest. That is, knowing $\hat{V}(\vec{r})$, we can in principle solve equation (2.354) to obtain the various energy levels of the particle and their corresponding wave functions. The complexity we encounter in solving the differential equation depends entirely on the form of the potential; the simpler the potential the easier the solution. Exact solutions of the Schrödinger equation are possible only for a few idealized systems; we deal with such systems in Chapters 4 and 6. However, exact solutions are generally not possible, for real systems do not yield themselves to exact solutions. In such cases one has to resort to approximate solutions. We deal with such approximate treatments in Chapters 9 and 10; Chapter 9 deals with time-independent potentials and Chapter 10 with time-dependent potentials.

Before embarking on the applications of the Schrödinger equation, we need first to lay down the theoretical foundations of quantum mechanics. We take up this task in Chapter 3, where we deal with the postulates of the theory as well as their implications; the postulates are the bedrock on which the theory is built.

2.9   Solved Problems

Problem 2.1
Consider the states $| \psi \rangle = 9i | \phi_1 \rangle + 2 | \phi_2 \rangle$ and $| \chi \rangle = -\frac{i}{\sqrt{2}} | \phi_1 \rangle + \frac{1}{\sqrt{2}} | \phi_2 \rangle$, where the two vectors $| \phi_1 \rangle$ and $| \phi_2 \rangle$ form a complete and orthonormal basis.

   (a) Calculate the operators $| \psi \rangle \langle \chi |$ and $| \chi \rangle \langle \psi |$. Are they equal?
   (b) Find the Hermitian conjugates of $| \psi \rangle, | \chi \rangle, | \psi \rangle \langle \chi |$, and $| \chi \rangle \langle \psi |$.
   (c) Calculate $\text{Tr}(| \psi \rangle \langle \chi |)$ and $\text{Tr}(| \chi \rangle \langle \psi |)$. Are they equal?
   (d) Calculate $| \psi \rangle \langle \psi |$ and $| \chi \rangle \langle \chi |$ and the traces $\text{Tr}(| \psi \rangle \langle \psi |)$ and $\text{Tr}(| \chi \rangle \langle \chi |)$. Are they projection operators?

Solution
   (a) The bras corresponding to $| \psi \rangle = 9i | \phi_1 \rangle + 2 | \phi_2 \rangle$ and $| \chi \rangle = -i | \phi_1 \rangle / \sqrt{2} + | \phi_2 \rangle / \sqrt{2}$ are given by $\langle \psi | = -9i \langle \phi_1 | + 2 \langle \phi_2 |$ and $\langle \chi | = \frac{i}{\sqrt{2}} \langle \phi_1 | + \frac{1}{\sqrt{2}} \langle \phi_2 |$, respectively. Hence we have

$$
\begin{aligned}
| \psi \rangle \langle \chi | &= \frac{1}{\sqrt{2}} (9i | \phi_1 \rangle + 2 | \phi_2 \rangle) (i \langle \phi_1 | + \langle \phi_2 |) \\
&= \frac{1}{\sqrt{2}} (-9 | \phi_1 \rangle \langle \phi_1 | + 9i | \phi_1 \rangle \langle \phi_2 | + 2i | \phi_2 \rangle \langle \phi_1 | + 2 | \phi_2 \rangle \langle \phi_2 |),
\end{aligned}
$$

(2.355)

$$
| \chi \rangle \langle \psi | = \frac{1}{\sqrt{2}} (-9 | \phi_1 \rangle \langle \phi_1 | - 2i | \phi_1 \rangle \langle \phi_2 | - 9i | \phi_2 \rangle \langle \phi_1 | + 2 | \phi_2 \rangle \langle \phi_2 |).
$$
(2.356)

As expected, $| \psi \rangle \langle \chi |$ and $| \chi \rangle \langle \psi |$ are not equal; they would be equal only if the states $| \psi \rangle$ and $| \chi \rangle$ were proportional and the proportionality constant real.

   (b) To find the Hermitian conjugates of $| \psi \rangle, | \chi \rangle, | \psi \rangle \langle \chi |$, and $| \chi \rangle \langle \psi |$, we need simply to replace the factors with their respective complex conjugates, the bras with kets, and the kets with bras:

$$
| \psi \rangle^\dagger = \langle \psi | = -9i \langle \phi_1 | + 2 \langle \phi_2 |, \qquad | \chi \rangle^\dagger = \langle \chi | = \frac{1}{\sqrt{2}} (i \langle \phi_1 | + \langle \phi_2 |),
$$
(2.357)

$$(|\psi\rangle\langle\chi|)^\dagger = |\chi\rangle\langle\psi| = \frac{1}{\sqrt{2}}(-9|\phi_1\rangle\langle\phi_1| - 2i|\phi_1\rangle\langle\phi_2|$$
$$- 9i|\phi_2\rangle\langle\phi_1| + 2|\phi_2\rangle\langle\phi_2|), \tag{2.358}$$

$$(|\chi\rangle\langle\psi|)^\dagger = |\psi\rangle\langle\chi| = \frac{1}{\sqrt{2}}(-9|\phi_1\rangle\langle\phi_1| + 9i|\phi_1\rangle\langle\phi_2|$$
$$+ 2i|\phi_2\rangle\langle\phi_1| + 2|\phi_2\rangle\langle\phi_2|). \tag{2.359}$$

(c) Using the property $\text{Tr}(AB) = \text{Tr}(BA)$ and since $\langle\phi_1|\phi_1\rangle = \langle\phi_2|\phi_2\rangle = 1$ and $\langle\phi_1|\phi_2\rangle = \langle\phi_2|\phi_1\rangle = 0$, we obtain

$$\text{Tr}(|\psi\rangle\langle\chi|) = \text{Tr}(\langle\chi|\psi\rangle) = \langle\chi|\psi\rangle$$
$$= \left(\frac{i}{\sqrt{2}}\langle\phi_1| + \frac{1}{\sqrt{2}}\langle\phi_2|\right)(9i|\phi_1\rangle + 2|\phi_2\rangle) = -\frac{7}{\sqrt{2}}, \tag{2.360}$$

$$\text{Tr}(|\chi\rangle\langle\psi|) = \text{Tr}(\langle\psi|\chi\rangle) = \langle\psi|\chi\rangle$$
$$= (-9i\langle\phi_1| + 2\langle\phi_2|)\left(-\frac{i}{\sqrt{2}}|\phi_1\rangle + \frac{1}{\sqrt{2}}|\phi_2\rangle\right) = -\frac{7}{\sqrt{2}}$$
$$= \text{Tr}(|\psi\rangle\langle\chi|). \tag{2.361}$$

The traces $\text{Tr}(|\psi\rangle\langle\chi|)$ and $\text{Tr}(|\chi\rangle\langle\psi|)$ are equal only because the scalar product of $|\psi\rangle$ and $|\chi\rangle$ is a real number. Were this product a complex number, the traces would be different; in fact, they would be the complex conjugate of one another.

(d) The expressions $|\psi\rangle\langle\psi|$ and $|\chi\rangle\langle\chi|$ are

$$|\psi\rangle\langle\psi| = (9i|\phi_1\rangle + 2|\phi_2\rangle)(-9i\langle\phi_1| + 2\langle\phi_2|)$$
$$= 81|\phi_1\rangle\langle\phi_1| + 18i|\phi_1\rangle\langle\phi_2| - 18i|\phi_2\rangle\langle\phi_1| + 4|\phi_2\rangle\langle\phi_2|,$$
$$\tag{2.362}$$

$$|\chi\rangle\langle\chi|) = \frac{1}{2}(|\phi_1\rangle\langle\phi_1| - i|\phi_1\rangle\langle\phi_2| + i|\phi_2\rangle\langle\phi_1| + |\phi_2\rangle\langle\phi_2|)$$
$$= \frac{1}{2}(1 - i|\phi_1\rangle\langle\phi_2| + i|\phi_2\rangle\langle\phi_1|). \tag{2.363}$$

In deriving (2.363) we have used the fact that the basis is complete, $|\phi_1\rangle\langle\phi_1| + |\phi_2\rangle\langle\phi_2| = 1$. The traces $\text{Tr}(|\psi\rangle\langle\psi|)$ and $\text{Tr}(|\chi\rangle\langle\chi|)$ can then be calculated immediately:

$$\text{Tr}(|\psi\rangle\langle\psi|) = \langle\psi|\psi\rangle = (-9i\langle\phi_1| + 2\langle\phi_2|)(9i|\phi_1\rangle + 2|\phi_2\rangle) = 85, \tag{2.364}$$

$$\text{Tr}(|\chi\rangle\langle\chi|) = \langle\chi|\chi\rangle = \frac{1}{2}(i\langle\phi_1| + \langle\phi_2|)(-i|\phi_1\rangle + |\phi_2\rangle) = 1. \tag{2.365}$$

So $|\chi\rangle$ is normalized but $|\psi\rangle$ is not. Since $|\chi\rangle$ is normalized, we can easily ascertain that $|\chi\rangle\langle\chi|$ is a projection operator, because it is Hermitian, $(|\chi\rangle\langle\chi|)^\dagger = |\chi\rangle\langle\chi|$, and equal to its own square:

$$(|\chi\rangle\langle\chi|)^2 = |\chi\rangle\langle\chi|\chi\rangle\langle\chi| = (\langle\chi|\chi\rangle)|\chi\rangle\langle\chi| = |\chi\rangle\langle\chi|. \tag{2.366}$$

As for $|\psi\rangle\langle\psi|$, although it is Hermitian, it cannot be a projection operator since $|\psi\rangle$ is not normalized. That is, $|\psi\rangle\langle\psi|$ is not equal to its own square:

$$(|\psi\rangle\langle\psi|)^2 = |\psi\rangle\langle\psi|\psi\rangle\langle\psi| = (\langle\psi|\psi\rangle)|\psi\rangle\langle\psi| = 85|\psi\rangle\langle\psi|. \tag{2.367}$$

Problem 2.2

(a) Find a complete and orthonormal basis for a space of the trigonometric functions of the form $\psi(\theta) = \sum_{n=0}^{N} a_n \cos(n\theta)$.

(b) Illustrate the results derived in (a) for the case $N = 5$; find the basis vectors.

Solution

(a) Since $\cos(n\theta) = \frac{1}{2}\left(e^{in\theta} + e^{-in\theta}\right)$, we can write $\sum_{n=0}^{N} a_n \cos(n\theta)$ as

$$\frac{1}{2}\sum_{n=0}^{N} a_n \left(e^{in\theta} + e^{-in\theta}\right) = \frac{1}{2}\left[\sum_{n=0}^{N} a_n e^{in\theta} + \sum_{n=-N}^{0} a_{-n} e^{in\theta}\right] = \sum_{n=-N}^{N} C_n e^{in\theta}, \qquad (2.368)$$

where $C_n = a_n/2$ for $n > 0$, $C_n = a_{-n}/2$ for $n < 0$, and $C_0 = a_0$. Since any trigonometric function of the form $\psi(x) = \sum_{n=0}^{N} a_n \cos(n\theta)$ can be expressed in terms of the functions $\phi_n(\theta) = e^{in\theta}/\sqrt{2\pi}$, we can try to take the set $\phi_n(\theta)$ as a basis. As this set is complete, let us see if it is orthonormal. The various functions $\phi_n(\theta)$ are indeed orthonormal, since their scalar products are given by

$$\langle \phi_m \mid \phi_n \rangle = \int_{-\pi}^{\pi} \phi_m^*(\theta)\phi_n(\theta)d\theta = \frac{1}{2\pi}\int_{-\pi}^{\pi} e^{i(n-m)\theta}\, d\theta = \delta_{nm}. \qquad (2.369)$$

In deriving this result, we have considered two cases: $n = m$ and $n \neq m$. First, the case $n = m$ is obvious, since $\langle \phi_n \mid \phi_n \rangle = \frac{1}{2\pi}\int_{-\pi}^{\pi} d\theta = 1$. On the other hand, when $n \neq m$ we have

$$\langle \phi_m \mid \phi_n \rangle = \frac{1}{2\pi}\int_{-\pi}^{\pi} e^{i(n-m)\theta}d\theta = \frac{1}{2\pi}\frac{e^{i(n-m)\pi} - e^{-i(n-m)\pi}}{i(n-m)} = \frac{2i\sin((n-m)\pi)}{2i\pi(n-m)} = 0,$$
$$(2.370)$$

since $\sin((n-m)\pi) = 0$. So the functions $\phi_n(\theta) = e^{in\theta}/\sqrt{2\pi}$ form a complete and orthonormal basis. From (2.368) we see that the basis has $2N + 1$ functions $\phi_n(\theta)$; hence the dimension of this space of functions is equal to $2N + 1$.

(b) In the case where $N = 5$, the dimension of the space is equal to 11, for the basis has 11 vectors: $\phi_{-5}(\theta) = e^{-5i\theta}/\sqrt{2\pi}$, $\phi_{-4}(\theta) = e^{-4i\theta}/\sqrt{2\pi}$, ..., $\phi_0(\theta) = 1/\sqrt{2\pi}$, ..., $\phi_4(\theta) = e^{4i\theta}/\sqrt{2\pi}$, $\phi_5(\theta) = e^{5i\theta}/\sqrt{2\pi}$.

Problem 2.3

(a) Show that the sum of two projection operators cannot be a projection operator unless their product is zero.

(b) Show that the product of two projection operators cannot be a projection operator unless they commute.

Solution

Recall that an operator $\hat{P}$ is a projection operator if it satisfies $\hat{P}^\dagger = \hat{P}$ and $\hat{P}^2 = \hat{P}$.

(a) If two operators $\hat{A}$ and $\hat{B}$ are projection operators and if $\hat{A}\hat{B} = \hat{B}\hat{A}$, we want to show that $(\hat{A} + \hat{B})^\dagger = \hat{A} + \hat{B}$ and that $(\hat{A} + \hat{B})^2 = \hat{A} + \hat{B}$. First, the hermiticity is easy to ascertain since $\hat{A}$ and $\hat{B}$ are both Hermitian: $(\hat{A} + \hat{B})^\dagger = \hat{A} + \hat{B}$. Let us now look at the square of $(\hat{A} + \hat{B})$; since $\hat{A}^2 = \hat{A}$ and $\hat{B}^2 = \hat{B}$, we can write

$$(\hat{A} + \hat{B})^2 = \hat{A}^2 + \hat{B}^2 + (\hat{A}\hat{B} + \hat{B}\hat{A}) = \hat{A} + \hat{B} + (\hat{A}\hat{B} + \hat{B}\hat{A}). \qquad (2.371)$$

Clearly, only when the product of $\hat{A}$ and $\hat{B}$ is zero will their sum be a projection operator.

(b) At issue here is to show that if two operators $\hat{A}$ and $\hat{B}$ are projection operators and if they commute, $[\hat{A}, \hat{B}] = 0$, their product is a projection operator. That is, we need to show that $(\hat{A}\hat{B})^{\dagger} = \hat{A}\hat{B}$ and $(\hat{A}\hat{B})^2 = \hat{A}\hat{B}$. Again, since $\hat{A}$ and $\hat{B}$ are Hermitian and since they commute, we see that $(\hat{A}\hat{B})^{\dagger} = \hat{B}\hat{A} = \hat{A}\hat{B}$. As for the square of $\hat{A}\hat{B}$, we have

$$(\hat{A}\hat{B})^2 = (\hat{A}\hat{B})(\hat{A}\hat{B}) = \hat{A}(\hat{B}\hat{A})\hat{B} = \hat{A}(\hat{A}\hat{B})\hat{B} = \hat{A}^2\hat{B}^2 = \hat{A}\hat{B}, \qquad (2.372)$$

hence the product $\hat{A}\hat{B}$ is a projection operator.

Problem 2.4

Consider a state $|\psi\rangle = \frac{1}{\sqrt{2}}|\phi_1\rangle + \frac{1}{\sqrt{5}}|\phi_2\rangle + \frac{1}{\sqrt{10}}|\phi_3\rangle$ which is given in terms of three orthonormal eigenstates $|\phi_1\rangle$, $|\phi_2\rangle$ and $|\phi_3\rangle$ of an operator $\hat{B}$ such that $\hat{B}|\phi_n\rangle = n^2|\phi_n\rangle$. Find the expectation value of $\hat{B}$ for the state $|\psi\rangle$.

Solution

Using Eq (2.58), we can write the expectation value of $\hat{B}$ for the state $|\psi\rangle$ as $\langle\hat{B}\rangle = \langle\psi | \hat{B} | \psi\rangle / \langle\psi | \psi\rangle$ where

$$\langle\psi | \psi\rangle = \left(\frac{1}{\sqrt{2}}\langle\phi_1| + \frac{1}{\sqrt{5}}\langle\phi_2| + \frac{1}{\sqrt{10}}\langle\phi_3|\right)\left(\frac{1}{\sqrt{2}}|\phi_1\rangle + \frac{1}{\sqrt{5}}|\phi_2\rangle + \frac{1}{\sqrt{10}}|\phi_3\rangle\right)$$

$$= \frac{8}{10} \qquad (2.373)$$

and

$$\langle\psi | \hat{B} | \psi\rangle = \left(\frac{1}{\sqrt{2}}\langle\phi_1| + \frac{1}{\sqrt{5}}\langle\phi_2| + \frac{1}{\sqrt{10}}\langle\phi_3|\right)\hat{B}\left(\frac{1}{\sqrt{2}}|\phi_1\rangle + \frac{1}{\sqrt{5}}|\phi_2\rangle + \frac{1}{\sqrt{10}}|\phi_3\rangle\right)$$

$$= \frac{1}{2} + \frac{2^2}{5} + \frac{3^2}{10}$$

$$= \frac{22}{10}. \qquad (2.374)$$

Hence, the expectation value of $\hat{B}$ is given by

$$\langle\hat{B}\rangle = \frac{\langle\psi | \hat{B} | \psi\rangle}{\langle\psi | \psi\rangle} = \frac{22/10}{8/10} = \frac{11}{4}. \qquad (2.375)$$

Problem 2.5

(a) Study the hermiticity of these operators: $\hat{X}$, d/dx, and id/dx. What about the complex conjugate of these operators? Are the Hermitian conjugates of the position and momentum operators equal to their complex conjugates?

(b) Use the results of (a) to discuss the hermiticity of the operators $e^{\hat{X}}$, $e^{d/dx}$, and $e^{id/dx}$.

(c) Find the Hermitian conjugate of the operator $\hat{X}d/dx$.

(d) Use the results of (a) to discuss the hermiticity of the components of the angular momentum operator (Chapter 5): $\hat{L}_x = -i\hbar\left(\hat{Y}\partial/\partial z - \hat{Z}\partial/\partial y\right)$, $\hat{L}_y = -i\hbar\left(\hat{Z}\partial/\partial x - \hat{X}\partial/\partial z\right)$, $\hat{L}_z = -i\hbar\left(\hat{X}\partial/\partial y - \hat{Y}\partial/\partial x\right)$.

Solution

(a) Using (2.69) and (2.70), and using the fact that the eigenvalues of $\hat{X}$ are real (i.e., $\hat{X}^* = \hat{X}$ or $x^* - x$), we can verify that $\hat{X}$ is Hermitian (i.e., $\hat{X}^\dagger = \hat{X}$) since

$$\langle \psi \mid \hat{X}\psi \rangle = \int_{-\infty}^{+\infty} \psi^*(x)\,(x\psi(x))\,dx = \int_{-\infty}^{+\infty} (x\psi(x)^*)\,\psi(x)\,dx$$

$$= \int_{-\infty}^{+\infty} (x\psi(x))^*\,\psi(x)\,dx = \langle \hat{X}\psi \mid \psi \rangle. \tag{2.376}$$

Now, since $\psi(x)$ vanishes as $x \to \pm\infty$, an integration by parts leads to

$$\langle \psi \mid \frac{d}{dx}\psi \rangle = \int_{-\infty}^{+\infty} \psi^*(x) \left(\frac{d\psi(x)}{dx} \right) dx = \psi^*(x)\psi(x)\Big|_{x=-\infty}^{x=+\infty} - \int_{-\infty}^{+\infty} \left(\frac{d\psi^*(x)}{dx} \right) \psi(x)\,dx$$

$$= -\int_{-\infty}^{+\infty} \left(\frac{d\psi(x)}{dx} \right)^* \psi(x)\,dx = -\langle \frac{d}{dx}\psi \mid \psi \rangle. \tag{2.377}$$

So, d/dx is anti-Hermitian: $(d/dx)^\dagger = -d/dx$. Since d/dx is anti-Hermitian, id/dx must be Hermitian, since $(id/dx)^\dagger = -i\,(-d/dx) = id/dx$. The results derived above are

$$\boxed{\hat{X}^\dagger = \hat{X}, \qquad \left(\frac{d}{dx} \right)^\dagger = -\frac{d}{dx}, \qquad \left(i\frac{d}{dx} \right)^\dagger = i\frac{d}{dx}.} \tag{2.378}$$

From this relation we see that the momentum operator $\hat{P} = -i\hbar d/dx$ is Hermitian: $\hat{P}^\dagger = \hat{P}$. We can also infer that, although the momentum operator is Hermitian, its complex conjugate is not equal to $\hat{P}$, since $\hat{P}^* = (-i\hbar d/dx)^* = i\hbar d/dx = -\hat{P}$. We may group these results into the following relation:

$$\boxed{\hat{X}^\dagger = \hat{X}, \quad \hat{X}^* = \hat{X}, \qquad \hat{P}^\dagger = \hat{P}, \quad \hat{P}^* = -\hat{P}.} \tag{2.379}$$

(b) Using the relations $(e^{\hat{A}})^\dagger = e^{\hat{A}^\dagger}$ and $(e^{i\hat{A}})^\dagger = e^{-i\hat{A}^\dagger}$ derived in (2.113), we infer

$$\boxed{(e^{\hat{X}})^\dagger = e^{\hat{X}}, \qquad (e^{d/dx})^\dagger = e^{-d/dx}, \qquad (e^{id/dx})^\dagger = e^{id/dx}.} \tag{2.380}$$

(c) Since $\hat{X}$ is Hermitian and d/dx is anti-Hermitian, we have

$$\left(\hat{X}\frac{d}{dx} \right)^\dagger = \left(\frac{d}{dx} \right)^\dagger (\hat{X})^\dagger = -\frac{d}{dx}\hat{X}, \tag{2.381}$$

where $d\hat{X}/dx$ is given by

$$\frac{d}{dx}\left(\hat{X}\psi(x) \right) = \left(1 + x\frac{d}{dx} \right)\psi(x); \tag{2.382}$$

hence

$$\left(\hat{X}\frac{d}{dx} \right)^\dagger = -\hat{X}\frac{d}{dx} - 1. \tag{2.383}$$

(d) From the results derived in (a), we infer that the operators $\hat{Y}$, $\hat{Z}$, $i\partial/\partial x$, and $i\partial/\partial y$ are Hermitian. We can verify that $\hat{L}_x$ is also Hermitian:

$$\hat{L}_x^\dagger = -i\hbar \left(\frac{\partial}{\partial z}\hat{Y} - \frac{\partial}{\partial y}\hat{Z} \right) = -i\hbar \left(\hat{Y}\frac{\partial}{\partial z} - \hat{Z}\frac{\partial}{\partial y} \right) = \hat{L}_x; \qquad (2.384)$$

in deriving this relation, we used the fact that the y and z degrees of freedom commute (i.e., $\partial\hat{Y}/\partial z = \hat{Y}\partial/\partial z$ and $\partial\hat{Z}/\partial y = \hat{Z}\partial/\partial y$), for they are independent. Similarly, the hermiticity of $\hat{L}_y = -i\hbar\left(\hat{Z}\partial/\partial x - \hat{X}\partial/\partial z\right)$ and $\hat{L}_z = -i\hbar\left(\hat{X}\partial/\partial y - \hat{Y}\partial/\partial x\right)$ is obvious.

Problem 2.6

(a) Show that the operator $\hat{A} = i(\hat{X}^2 + 1)d/dx + i\hat{X}$ is Hermitian.

(b) Find the state $\psi(x)$ for which $\hat{A}\psi(x) = 0$ and normalize it.

(c) Calculate the probability of finding the particle (represented by $\psi(x)$) in the region: $-1 \le x \le 1$.

Solution

(a) From the previous problem we know that $\hat{X}^\dagger = \hat{X}$ and $(d/dx)^\dagger = -d/dx$. We can thus infer the Hermitian conjugate of $\hat{A}$:

$$\begin{aligned} \hat{A}^\dagger &= -i\left(\frac{d}{dx}\right)^\dagger (\hat{X}^2)^\dagger - i\left(\frac{d}{dx}\right)^\dagger - i\hat{X}^\dagger = i\left(\frac{d}{dx}\right)(\hat{X}^2) + i\left(\frac{d}{dx}\right) - i\hat{X} \\ &= i\hat{X}^2\frac{d}{dx} + i\left[\frac{d}{dx}, \hat{X}^2\right] + i\frac{d}{dx} - i\hat{X}. \end{aligned} \qquad (2.385)$$

Using the relation $[\hat{B}, \hat{C}^2] = \hat{C}[\hat{B}, \hat{C}] + [\hat{B}, \hat{C}]\hat{C}$ along with $[d/dx, \hat{X}] = 1$, we can easily evaluate the commutator $[d/dx, \hat{X}^2]$:

$$\left[\frac{d}{dx}, \hat{X}^2\right] = \hat{X}\left[\frac{d}{dx}, \hat{X}\right] + \left[\frac{d}{dx}, \hat{X}\right]\hat{X} = 2\hat{X}. \qquad (2.386)$$

A combination of (2.385) and (2.386) shows that $\hat{A}$ is Hermitian:

$$\hat{A}^\dagger = i(\hat{X}^2 + 1)\frac{d}{dx} + i\hat{X} = \hat{A}. \qquad (2.387)$$

(b) The state $\psi(x)$ for which $\hat{A}\psi(x) = 0$, i.e.,

$$i(\hat{X}^2 + 1)\frac{d\psi(x)}{dx} + i\hat{X}\psi(x) = 0, \qquad (2.388)$$

corresponds to

$$\frac{d\psi(x)}{dx} = -\frac{x}{x^2 + 1}\psi(x). \qquad (2.389)$$

The solution to this equation is given by

$$\psi(x) = \frac{B}{\sqrt{x^2 + 1}}. \qquad (2.390)$$

Since $\int_{-\infty}^{+\infty} dx/(x^2 + 1) = \pi$ we have

$$1 = \int_{-\infty}^{+\infty} |\psi(x)|^2 \, dx = B^2 \int_{-\infty}^{+\infty} \frac{dx}{x^2 + 1} = B^2 \pi, \qquad (2.391)$$

which leads to $B = 1/\sqrt{\pi}$ and hence $\psi(x) = \frac{1}{\sqrt{\pi(x^2+1)}}$.

(c) Using the integral $\int_{-1}^{+1} dx/(x^2 + 1) = \pi/2$, we can obtain the probability immediately:

$$P = \int_{-1}^{+1} |\psi(x)|^2 \, dx = \frac{1}{\pi} \int_{-1}^{+1} \frac{dx}{x^2 + 1} = \frac{1}{2}. \qquad (2.392)$$

Problem 2.7

Discuss the conditions for these operators to be unitary: (a) $(1 + i\hat{A})/(1 - i\hat{A})$,
(b) $(\hat{A} + i\hat{B})/\sqrt{\hat{A}^2 + \hat{B}^2}$.

Solution

An operator $\hat{U}$ is unitary if $\hat{U}\hat{U}^\dagger = \hat{U}^\dagger\hat{U} = \hat{I}$ (see (2.122)).

(a) Since

$$\left(\frac{1 + i\hat{A}}{1 - i\hat{A}}\right)^\dagger = \frac{1 - i\hat{A}^\dagger}{1 + i\hat{A}^\dagger}, \qquad (2.393)$$

we see that if $\hat{A}$ is Hermitian, the expression $(1 + i\hat{A})/(1 - i\hat{A})$ is unitary:

$$\left(\frac{1 + i\hat{A}}{1 - i\hat{A}}\right)^\dagger \frac{1 + i\hat{A}}{1 - i\hat{A}} = \frac{1 - i\hat{A}}{1 + i\hat{A}}\frac{1 + i\hat{A}}{1 - i\hat{A}} = \hat{I}. \qquad (2.394)$$

(b) Similarly, if $\hat{A}$ and $\hat{B}$ are Hermitian and commute, the expression $(\hat{A} + i\hat{B})/\sqrt{\hat{A}^2 + \hat{B}^2}$ is unitary:

$$\left(\frac{\hat{A} + i\hat{B}}{\sqrt{\hat{A}^2 + \hat{B}^2}}\right)^\dagger \frac{\hat{A} + i\hat{B}}{\sqrt{\hat{A}^2 + \hat{B}^2}} = \frac{\hat{A} - i\hat{B}}{\sqrt{\hat{A}^2 + \hat{B}^2}} \frac{\hat{A} + i\hat{B}}{\sqrt{\hat{A}^2 + \hat{B}^2}} = \frac{\hat{A}^2 + \hat{B}^2 + i(\hat{A}\hat{B} - \hat{B}\hat{A})}{\hat{A}^2 + \hat{B}^2}$$

$$= \frac{\hat{A}^2 + \hat{B}^2}{\hat{A}^2 + \hat{B}^2} = \hat{I}. \qquad (2.395)$$

Problem 2.8

(a) Using the commutator $[\hat{X}, \hat{p}] = i\hbar$, show that $[\hat{X}^m, \hat{P}] = im\hbar\hat{X}^{m-1}$, with $m > 1$. Can you think of a direct way to get to the same result?

(b) Use the result of (a) to show the general relation $[F(\hat{X}), \hat{P}] = i\hbar dF(\hat{X})/d\hat{X}$, where $F(\hat{X})$ is a differentiable operator function of $\hat{X}$.

Solution

(a) Let us attempt a proof by induction. Assuming that $[\hat{X}^m, \hat{P}] = im\hbar\hat{X}^{m-1}$ is valid for $m = k$ (note that it holds for $n = 1$; i.e., $[\hat{X}, \hat{P}] = i\hbar$),

$$[\hat{X}^k, \hat{P}] = ik\hbar\hat{X}^{k-1}, \tag{2.396}$$

let us show that it holds for $m = k + 1$:

$$[\hat{X}^{k+1}, \hat{P}] = [\hat{X}^k\hat{X}, \hat{P}] = \hat{X}^k[\hat{X}, \hat{P}] + [\hat{X}^k, \hat{P}]\hat{X}, \tag{2.397}$$

where we have used the relation $[\hat{A}\hat{B}, \hat{C}] = \hat{A}[\hat{B}, \hat{C}] + [\hat{A}, \hat{C}]\hat{B}$. Now, since $[\hat{X}, \hat{P}] = i\hbar$ and $[\hat{X}^k, \hat{P}] = ik\hbar\hat{X}^{k-1}$, we rewrite (2.397) as

$$[\hat{X}^{k+1}, \hat{P}] = i\hbar\hat{X}^k + (ik\hbar\hat{X}^{k-1})\hat{X} = i\hbar(k + 1)\hat{X}^k. \tag{2.398}$$

So this relation is valid for any value of k, notably for $k = m - 1$:

$$\boxed{[\hat{X}^m, \hat{P}] = im\hbar\hat{X}^{m-1}.} \tag{2.399}$$

In fact, it is easy to arrive at this result *directly* through brute force as follows. Using the relation $[\hat{A}^n, \hat{B}] = \hat{A}^{n-1}[\hat{A}, \hat{B}] + [\hat{A}^{n-1}, \hat{B}]\hat{A}$ along with $[\hat{X}, \hat{P}_x] = i\hbar$, we can obtain

$$[\hat{X}^2, \hat{P}_x] = \hat{X}[\hat{X}, \hat{P}_x] + [\hat{X}, \hat{P}_x]\hat{X} = 2i\hbar\hat{X}, \tag{2.400}$$

which leads to

$$[\hat{X}^3, \hat{P}_x] = \hat{X}^2[\hat{X}, \hat{P}_x] + [\hat{X}^2, \hat{P}_x]\hat{X} = 3i\hat{X}^2\hbar; \tag{2.401}$$

this in turn leads to

$$[\hat{X}^4, \hat{P}_x] = \hat{X}^3[\hat{X}, \hat{P}_x] + [\hat{X}^3, \hat{P}_x]\hat{X} = 4i\hat{X}^3\hbar. \tag{2.402}$$

Continuing in this way, we can get to any power of $\hat{X}$: $[\hat{X}^m, \hat{P}] = im\hbar\hat{X}^{m-1}$.

A more direct and simpler method is to apply the commutator $[\hat{X}^m, \hat{P}]$ on some wave function $\psi(x)$:

$$\begin{aligned}
[\hat{X}^m, \hat{P}_x]\psi(x) &= \left(\hat{X}^m\hat{P}_x - \hat{P}_x\hat{X}^m\right)\psi(x) \\
&= x^m\left(-i\hbar\frac{d\psi(x)}{dx}\right) + i\hbar\frac{d}{dx}\left(x^m\psi(x)\right) \\
&= x^m\left(-i\hbar\frac{d\psi(x)}{dx}\right) + im\hbar x^{m-1}\psi(x) - x^m\left(-i\hbar\frac{d\psi(x)}{dx}\right) \\
&= im\hbar x^{m-1}\psi(x).
\end{aligned} \tag{2.403}$$

Since $[\hat{X}^m, \hat{P}_x]\psi(x) = im\hbar x^{m-1}\psi(x)$ we see that $[\hat{X}^m, \hat{P}] = im\hbar\hat{X}^{m-1}$.

(b) Let us Taylor expand $F(\hat{X})$ in powers of $\hat{X}$, $F(\hat{X}) = \sum_k a_k\hat{X}^k$, and insert this expression into $[F(\hat{X}), \hat{P}]$:

$$\left[F(\hat{X}), \hat{P}\right] = \left[\sum_k a_k\hat{X}^k, \hat{P}\right] = \sum_k a_k[\hat{X}^k, \hat{P}], \tag{2.404}$$

where the commutator $[\hat{X}^k, \hat{P}]$ is given by (2.396). Thus, we have

$$\left[F(\hat{X}), \hat{P}\right] = i\hbar \sum_k k a_k \hat{X}^{k-1} = i\hbar \frac{d(\sum_k a_k \hat{X}^k)}{d\hat{X}} = i\hbar \frac{dF(\hat{X})}{d\hat{X}}. \qquad (2.405)$$

A much simpler method again consists in applying the commutator $\left[F(\hat{X}), \hat{P}\right]$ on some wave function $\psi(x)$. Since $F(\hat{X})\psi(x) = F(x)\psi(x)$, we have

$$\begin{aligned}
\left[F(\hat{X}), \hat{P}\right]\psi(x) &= F(\hat{X})\hat{P}\psi(x) + i\hbar \frac{d}{dx}(F(x)\psi(x)) \\
&= F(\hat{X})\hat{P}\psi(x) - \left(-i\hbar \frac{d\psi(x)}{dx}\right)F(x) + i\hbar \frac{dF(x)}{dx}\psi(x) \\
&= F(\hat{X})\hat{P}\psi(x) - F(\hat{X})\hat{P}\psi(x) + i\hbar \frac{dF(x)}{dx}\psi(x) \\
&= i\hbar \frac{dF(x)}{dx}\psi(x). \qquad (2.406)
\end{aligned}$$

Since $\left[F(\hat{X}), \hat{P}\right]\psi(x) = i\hbar \frac{dF(x)}{dx}\psi(x)$ we see that $\left[F(\hat{X}), \hat{P}\right] = i\hbar \frac{dF(\hat{X})}{d\hat{X}}$.

Problem 2.9

Consider the matrices $A = \begin{pmatrix} 7 & 0 & 0 \\ 0 & 1 & -i \\ 0 & i & -1 \end{pmatrix}$ and $B = \begin{pmatrix} 1 & 0 & 3 \\ 0 & 2i & 0 \\ i & 0 & -5i \end{pmatrix}$.

(a) Are A and B Hermitian? Calculate AB and BA and verify that $\text{Tr}(AB) = \text{Tr}(BA)$; then calculate $[A, B]$ and verify that $\text{Tr}([A, B]) = 0$.

(b) Find the eigenvalues and the normalized eigenvectors of A. Verify that the sum of the eigenvalues of A is equal to the value of $\text{Tr}(A)$ calculated in (a) and that the three eigenvectors form a basis.

(c) Verify that $U^\dagger A U$ is diagonal and that $U^{-1} = U^\dagger$, where U is the matrix formed by the normalized eigenvectors of A.

(d) Calculate the inverse of $A' = U^\dagger A U$ and verify that A'^{-1} is a diagonal matrix whose eigenvalues are the inverse of those of A'.

Solution

(a) Taking the Hermitian adjoints of the matrices A and B (see (2.188))

$$A^\dagger = \begin{pmatrix} 7 & 0 & 0 \\ 0 & 1 & -i \\ 0 & i & -1 \end{pmatrix}, \qquad B^\dagger = \begin{pmatrix} 1 & 0 & -i \\ 0 & -2i & 0 \\ 3 & 0 & 5i \end{pmatrix}, \qquad (2.407)$$

we see that A is Hermitian and B is not. Using the products

$$AB = \begin{pmatrix} 7 & 0 & 21 \\ 1 & 2i & -5 \\ -i & -2 & 5i \end{pmatrix}, \qquad BA = \begin{pmatrix} 7 & 3i & -3 \\ 0 & 2i & 2 \\ 7i & 5 & 5i \end{pmatrix}, \qquad (2.408)$$

we can obtain the commutator

$$[A, B] = \begin{pmatrix} 0 & -3i & 24 \\ 1 & 0 & -7 \\ -8i & -7 & 0 \end{pmatrix}. \tag{2.409}$$

From (2.408) we see that

$$\text{Tr}(AB) = 7 + 2i + 5i = 7 + 7i = \text{Tr}(BA). \tag{2.410}$$

That is, the cyclic permutation of matrices leaves the trace unchanged; see (2.206). On the other hand, (2.409) shows that the trace of the commutator $[A, B]$ is zero: $\text{Tr}([A, B]) = 0+0+0 = 0$.

(b) The eigenvalues and eigenvectors of A were calculated in Example 2.19 (see (2.266), (2.268), (2.272), (2.274)). We have $a_1 = 7$, $a_2 = \sqrt{2}$, and $a_3 = -\sqrt{2}$:

$$| a_1 \rangle = \begin{pmatrix} 1 \\ 0 \\ 0 \end{pmatrix}, \quad | a_2 \rangle = \begin{pmatrix} 0 \\ \frac{1}{\sqrt{2(2-\sqrt{2})}} \\ \frac{i(\sqrt{2}-1)}{\sqrt{2(2-\sqrt{2})}} \end{pmatrix}, \quad | a_3 \rangle = \begin{pmatrix} 0 \\ \frac{1}{\sqrt{2(2+\sqrt{2})}} \\ -\frac{i(1+\sqrt{2})}{\sqrt{2(2+\sqrt{2})}} \end{pmatrix}. \tag{2.411}$$

One can easily verify that the eigenvectors $| a_1 \rangle$, $| a_2 \rangle$, and $| a_3 \rangle$ are mutually orthogonal: $\langle a_i | a_j \rangle = \delta_{ij}$ where $i, j = 1, 2, 3$. Since the set of $| a_1 \rangle$, $| a_2 \rangle$, and $| a_3 \rangle$ satisfy the completeness condition

$$\sum_{j=1}^{3} | a_j \rangle \langle a_j | = \begin{pmatrix} 1 & 0 & 0 \\ 0 & 1 & 0 \\ 0 & 0 & 1 \end{pmatrix}, \tag{2.412}$$

and since they are orthonormal, they form a complete and orthonormal basis.

(c) The columns of the matrix U are given by the eigenvectors (2.411):

$$U = \begin{pmatrix} 1 & 0 & 0 \\ 0 & \frac{1}{\sqrt{2(2-\sqrt{2})}} & \frac{1}{\sqrt{2(2+\sqrt{2})}} \\ 0 & \frac{i(\sqrt{2}-1)}{\sqrt{2(2-\sqrt{2})}} & -\frac{i(1+\sqrt{2})}{\sqrt{2(2+\sqrt{2})}} \end{pmatrix}. \tag{2.413}$$

We can show that the product $U^\dagger A U$ is diagonal where the diagonal elements are the eigenvalues of the matrix A; $U^\dagger A U$ is given by

$$\begin{pmatrix} 1 & 0 & 0 \\ 0 & \frac{1}{\sqrt{2(2-\sqrt{2})}} & -\frac{i(\sqrt{2}-1)}{\sqrt{2(2-\sqrt{2})}} \\ 0 & \frac{1}{\sqrt{2(2+\sqrt{2})}} & \frac{i(1+\sqrt{2})}{\sqrt{2(2+\sqrt{2})}} \end{pmatrix} \begin{pmatrix} 7 & 0 & 0 \\ 0 & 1 & -i \\ 0 & i & -1 \end{pmatrix} \begin{pmatrix} 1 & 0 & 0 \\ 0 & \frac{1}{\sqrt{2(2-\sqrt{2})}} & \frac{1}{\sqrt{2(2+\sqrt{2})}} \\ 0 & \frac{i(\sqrt{2}-1)}{\sqrt{2(2-\sqrt{2})}} & -\frac{i(1+\sqrt{2})}{\sqrt{2(2+\sqrt{2})}} \end{pmatrix}$$

$$= \begin{pmatrix} 7 & 0 & 0 \\ 0 & \sqrt{2} & 0 \\ 0 & 0 & -\sqrt{2} \end{pmatrix}. \tag{2.414}$$

We can also show that $U^\dagger U = 1$:

$$\begin{pmatrix} 1 & 0 & 0 \\ 0 & \dfrac{1}{\sqrt{2(2-\sqrt{2})}} & -\dfrac{i(\sqrt{2}-1)}{\sqrt{2(2-\sqrt{2})}} \\ 0 & \dfrac{1}{\sqrt{2(2+\sqrt{2})}} & \dfrac{i(1+\sqrt{2})}{\sqrt{2(2+\sqrt{2})}} \end{pmatrix} \begin{pmatrix} 1 & 0 & 0 \\ 0 & \dfrac{1}{\sqrt{2(2-\sqrt{2})}} & \dfrac{1}{\sqrt{2(2+\sqrt{2})}} \\ 0 & \dfrac{i(\sqrt{2}-1)}{\sqrt{2(2-\sqrt{2})}} & -\dfrac{i(1+\sqrt{2})}{\sqrt{2(2+\sqrt{2})}} \end{pmatrix} = \begin{pmatrix} 1 & 0 & 0 \\ 0 & 1 & 0 \\ 0 & 0 & 1 \end{pmatrix}.$$

(2.415)

This implies that the matrix U is unitary: $U^\dagger = U^{-1}$. Note that, from (2.413), we have $\mid \det(U) \mid = \mid -i \mid = 1$.

(d) Using (2.414) we can verify that the inverse of $A' = U^\dagger AU$ is a diagonal matrix whose elements are given by the inverse of the diagonal elements of A':

$$A' = \begin{pmatrix} 7 & 0 & 0 \\ 0 & \sqrt{2} & 0 \\ 0 & 0 & -\sqrt{2} \end{pmatrix} \implies A'^{-1} = \begin{pmatrix} \frac{1}{7} & 0 & 0 \\ 0 & \frac{1}{\sqrt{2}} & 0 \\ 0 & 0 & -\frac{1}{\sqrt{2}} \end{pmatrix}.$$

(2.416)

Problem 2.10

Consider a particle whose Hamiltonian matrix is $H = \begin{pmatrix} 2 & i & 0 \\ -i & 1 & 1 \\ 0 & 1 & 0 \end{pmatrix}$.

(a) Is $\mid \lambda \rangle = \begin{pmatrix} i \\ 7i \\ -2 \end{pmatrix}$ an eigenstate of H? Is H Hermitian?

(b) Find the energy eigenvalues, a_1, a_2, and a_3, and the normalized energy eigenvectors, $\mid a_1 \rangle$, $\mid a_2 \rangle$, and $\mid a_3 \rangle$, of H.

(c) Find the matrix corresponding to the operator obtained from the ket-bra product of the first eigenvector $P = \mid a_1 \rangle \langle a_1 \mid$. Is P a projection operator? Calculate the commutator $[P, H]$ firstly by using commutator algebra and then by using matrix products.

Solution

(a) The ket $\mid \lambda \rangle$ is an eigenstate of H only if the action of the Hamiltonian on $\mid \lambda \rangle$ is of the form $H \mid \lambda \rangle = b \mid \lambda \rangle$, where b is constant. This is not the case here:

$$H \mid \lambda \rangle = \begin{pmatrix} 2 & i & 0 \\ -i & 1 & 1 \\ 0 & 1 & 0 \end{pmatrix} \begin{pmatrix} i \\ 7i \\ -2 \end{pmatrix} = \begin{pmatrix} -7+2i \\ -1+7i \\ 7i \end{pmatrix}.$$

(2.417)

Using the definition of the Hermitian adjoint of matrices (2.188), it is easy to ascertain that H is Hermitian:

$$H^\dagger = \begin{pmatrix} 2 & i & 0 \\ -i & 1 & 1 \\ 0 & 1 & 0 \end{pmatrix} = H.$$

(2.418)

(b) The energy eigenvalues can be obtained by solving the secular equation

$$0 = \begin{vmatrix} 2-a & i & 0 \\ -i & 1-a & 1 \\ 0 & 1 & -a \end{vmatrix} = (2-a)[(1-a)(-a)-1] - i(-i)(-a)$$

$$= -(a-1)(a-1-\sqrt{3})(a-1+\sqrt{3}),$$

(2.419)

which leads to

$$a_1 = 1, \quad a_2 = 1 - \sqrt{3}, \quad a_3 = 1 + \sqrt{3}. \tag{2.420}$$

To find the eigenvector corresponding to the first eigenvalue, $a_1 = 1$, we need to solve the matrix equation

$$\begin{pmatrix} 2 & i & 0 \\ -i & 1 & 1 \\ 0 & 1 & 0 \end{pmatrix} \begin{pmatrix} x \\ y \\ z \end{pmatrix} = \begin{pmatrix} x \\ y \\ z \end{pmatrix} \implies \begin{array}{rcl} x + iy & = & 0 \\ -ix + z & = & 0 \\ y - z & = & 0 \end{array} \tag{2.421}$$

which yields $x = 1, y = z = i$. So the eigenvector corresponding to $a_1 = 1$ is

$$| a_1 \rangle = \begin{pmatrix} 1 \\ i \\ i \end{pmatrix}. \tag{2.422}$$

This eigenvector is not normalized since $\langle a_1 \mid a_1 \rangle = 1 + (i^*)(i) + (i^*)(i) = 3$. The normalized $| a_1 \rangle$ is therefore

$$| a_1 \rangle = \frac{1}{\sqrt{3}} \begin{pmatrix} 1 \\ i \\ i \end{pmatrix}. \tag{2.423}$$

Solving (2.421) for the other two energy eigenvalues, $a_2 = 1 - \sqrt{3}$, $a_3 = 1 + \sqrt{3}$, and normalizing, we end up with

$$| a_2 \rangle = \frac{1}{\sqrt{6(2 - \sqrt{3})}} \begin{pmatrix} i(2 - \sqrt{3}) \\ 1 - \sqrt{3} \\ 1 \end{pmatrix}, \quad | a_3 \rangle = \frac{1}{\sqrt{6(2 + \sqrt{3})}} \begin{pmatrix} i(2 + \sqrt{3}) \\ 1 + \sqrt{3} \\ 1 \end{pmatrix}. \tag{2.424}$$

(c) The operator P is given by

$$P = | a_1 \rangle \langle a_1 | = \frac{1}{3} \begin{pmatrix} 1 \\ i \\ i \end{pmatrix} \begin{pmatrix} 1 & -i & -i \end{pmatrix} = \frac{1}{3} \begin{pmatrix} 1 & -i & -i \\ i & 1 & 1 \\ i & 1 & 1 \end{pmatrix}. \tag{2.425}$$

Since this matrix is Hermitian and since the square of P is equal to P,

$$P^2 = \frac{1}{9} \begin{pmatrix} 1 & -i & -i \\ i & 1 & 1 \\ i & 1 & 1 \end{pmatrix} \begin{pmatrix} 1 & -i & -i \\ i & 1 & 1 \\ i & 1 & 1 \end{pmatrix} = \frac{1}{3} \begin{pmatrix} 1 & -i & -i \\ i & 1 & 1 \\ i & 1 & 1 \end{pmatrix} = P, \tag{2.426}$$

so P is a projection operator. Using the relations $H | a_1 \rangle = | a_1 \rangle$ and $\langle a_1 | H = \langle a_1 |$ (because H is Hermitian), and since $P = | a_1 \rangle \langle a_1 |$, we can evaluate algebraically the commutator $[P, H]$ as follows:

$$[P, H] = PH - HP = | a_1 \rangle \langle a_1 | H - H | a_1 \rangle \langle a_1 | = | a_1 \rangle \langle a_1 | - | a_1 \rangle \langle a_1 | = 0. \tag{2.427}$$

We can reach the same result by using the matrices of H and P:

$$\begin{aligned} [P, H] &= \frac{1}{3} \begin{pmatrix} 1 & -i & -i \\ i & 1 & 1 \\ i & 1 & 1 \end{pmatrix} \begin{pmatrix} 2 & i & 0 \\ -i & 1 & 1 \\ 0 & 1 & 0 \end{pmatrix} - \frac{1}{3} \begin{pmatrix} 2 & i & 0 \\ -i & 1 & 1 \\ 0 & 1 & 0 \end{pmatrix} \begin{pmatrix} 1 & -i & -i \\ i & 1 & 1 \\ i & 1 & 1 \end{pmatrix} \\ &= \begin{pmatrix} 0 & 0 & 0 \\ 0 & 0 & 0 \\ 0 & 0 & 0 \end{pmatrix}. \end{aligned} \tag{2.428}$$

Problem 2.11

Consider the matrices $A = \begin{pmatrix} 0 & 0 & i \\ 0 & 1 & 0 \\ -i & 0 & 0 \end{pmatrix}$ and $B = \begin{pmatrix} 2 & i & 0 \\ 3 & 1 & 5 \\ 0 & -i & -2 \end{pmatrix}$.

(a) Check if A and B are Hermitian and find the eigenvalues and eigenvectors of A. Any degeneracies?

(b) Verify that $\mathrm{Tr}(AB) = \mathrm{Tr}(BA)$, $\det(AB) = \det(A)\det(B)$, and $\det(B^\dagger) = (\det(B))^*$.

(c) Calculate the commutator $[A, B]$ and the anticommutator $\{A, B\}$.

(d) Calculate the inverses A^{-1}, B^{-1}, and $(AB)^{-1}$. Verify that $(AB)^{-1} = B^{-1}A^{-1}$.

(e) Calculate A^2 and infer the expressions of A^{2n} and A^{2n+1}. Use these results to calculate the matrix of e^{xA}.

Solution

(a) The matrix A is Hermitian but B is not. The eigenvalues of A are $a_1 = -1$ and $a_2 = a_3 = 1$ and its normalized eigenvectors are

$$| a_1 \rangle = \frac{1}{\sqrt{2}} \begin{pmatrix} 1 \\ 0 \\ i \end{pmatrix}, \qquad | a_2 \rangle = \frac{1}{\sqrt{2}} \begin{pmatrix} 1 \\ 0 \\ -i \end{pmatrix}, \qquad | a_3 \rangle = \begin{pmatrix} 0 \\ 1 \\ 0 \end{pmatrix}. \qquad (2.429)$$

Note that the eigenvalue 1 is doubly degenerate, since the two eigenvectors $| a_2 \rangle$ and $| a_3 \rangle$ correspond to the same eigenvalue $a_2 = a_3 = 1$.

(b) A calculation of the products (AB) and (BA) reveals that the traces $\mathrm{Tr}(AB)$ and $\mathrm{Tr}(BA)$ are equal:

$$\mathrm{Tr}(AB) = \mathrm{Tr} \begin{pmatrix} 0 & 1 & -2i \\ 3 & 1 & 5 \\ -2i & 1 & 0 \end{pmatrix} = 1,$$

$$\mathrm{Tr}(BA) = \mathrm{Tr} \begin{pmatrix} 0 & i & 2i \\ -5i & 1 & 3i \\ 2i & -i & 0 \end{pmatrix} = 1 = \mathrm{Tr}(AB). \qquad (2.430)$$

From the matrices A and B, we have $\det(A) = i(i) = -1$, $\det(B) = -4 + 16i$. We can thus write

$$\det(AB) = \det \begin{pmatrix} 0 & 1 & -2i \\ 3 & 1 & 5 \\ -2i & 1 & 0 \end{pmatrix} = 4 - 16i = (-1)(-4 + 16i) = \det(A)\det(B). \quad (2.431)$$

On the other hand, since $\det(B) = -4 + 16i$ and $\det(B^\dagger) = -4 - 16i$, we see that $\det(B^\dagger) = -4 - 16i = (-4 + 16i)^* = (\det(B))^*$.

(c) The commutator $[A, B]$ is given by

$$AB - BA = \begin{pmatrix} 0 & 1 & -2i \\ 3 & 1 & 5 \\ -2i & 1 & 0 \end{pmatrix} - \begin{pmatrix} 0 & i & 2i \\ -5i & 1 & 3i \\ 2i & -i & 0 \end{pmatrix} = \begin{pmatrix} 0 & 1-i & -4i \\ 3+5i & 0 & 5-3i \\ -4i & 1+i & 0 \end{pmatrix}$$

$$(2.432)$$

and the anticommutator $\{A, B\}$ by

$$AB + BA = \begin{pmatrix} 0 & 1 & -2i \\ 3 & 1 & 5 \\ -2i & 1 & 0 \end{pmatrix} + \begin{pmatrix} 0 & i & 2i \\ -5i & 1 & 3i \\ 2i & -i & 0 \end{pmatrix} = \begin{pmatrix} 0 & 1+i & 0 \\ 3-5i & 2 & 5+3i \\ 0 & 1-i & 0 \end{pmatrix}.$$

$$(2.433)$$

(d) A calculation similar to (2.200) leads to the inverses of A, B, and AB:

$$A^{-1} = \begin{pmatrix} 0 & 0 & i \\ 0 & 1 & 0 \\ -i & 0 & 0 \end{pmatrix}, \qquad B^{-1} = \frac{1}{68} \begin{pmatrix} 22+3i & 8-2i & 20-5i \\ -6-24i & 4+16i & 10+40i \\ -12+3i & 8-2i & -14-5i \end{pmatrix}, \qquad (2.434)$$

$$(AB)^{-1} = \frac{1}{68} \begin{pmatrix} -5-20i & 8-2i & -3+22i \\ 40-10i & 4+16i & 24-6i \\ -5+14i & 8-2i & -3-12i \end{pmatrix}. \qquad (2.435)$$

From (2.434) it is now easy to verify that the product $B^{-1}A^{-1}$ is equal to $(AB)^{-1}$:

$$B^{-1}A^{-1} = \frac{1}{68} \begin{pmatrix} -5-20i & 8-2i & -3+22i \\ 40-10i & 4+16i & 24-6i \\ -5+14i & 8-2i & -3-12i \end{pmatrix} = (AB)^{-1}. \qquad (2.436)$$

(e) Since

$$A^2 = \begin{pmatrix} 0 & 0 & i \\ 0 & 1 & 0 \\ -i & 0 & 0 \end{pmatrix}\begin{pmatrix} 0 & 0 & i \\ 0 & 1 & 0 \\ -i & 0 & 0 \end{pmatrix} = \begin{pmatrix} 1 & 0 & 0 \\ 0 & 1 & 0 \\ 0 & 0 & 1 \end{pmatrix} = I, \qquad (2.437)$$

we can write $A^3 = A$, $A^4 = I$, $A^5 = A$, and so on. We can generalize these relations to any value of n: $A^{2n} = I$ and $A^{2n+1} = A$:

$$A^{2n} = \begin{pmatrix} 1 & 0 & 0 \\ 0 & 1 & 0 \\ 0 & 0 & 1 \end{pmatrix} = I, \qquad A^{2n+1} = \begin{pmatrix} 0 & 0 & i \\ 0 & 1 & 0 \\ -i & 0 & 0 \end{pmatrix} = A. \qquad (2.438)$$

Since $A^{2n} = I$ and $A^{2n+1} = A$, we can write

$$e^{xA} = \sum_{n=0}^{\infty} \frac{x^n A^n}{n!} = \sum_{n=0}^{\infty} \frac{x^{2n} A^{2n}}{(2n)!} + \sum_{n=0}^{\infty} \frac{x^{2n+1} A^{2n+1}}{(2n+1)!} = I \sum_{n=0}^{\infty} \frac{x^{2n}}{(2n)!} + A \sum_{n=0}^{\infty} \frac{x^{2n+1}}{(2n+1)!}.$$

$$(2.439)$$

The relations

$$\sum_{n=0}^{\infty} \frac{x^{2n}}{(2n)!} = \cosh x, \qquad \sum_{n=0}^{\infty} \frac{x^{2n+1}}{(2n+1)!} = \sinh x, \qquad (2.440)$$

lead to

$$e^{xA} = I \cosh x + A \sinh x = \begin{pmatrix} 1 & 0 & 0 \\ 0 & 1 & 0 \\ 0 & 0 & 1 \end{pmatrix} \cosh x + \begin{pmatrix} 0 & 0 & i \\ 0 & 1 & 0 \\ -i & 0 & 0 \end{pmatrix} \sinh x$$

$$= \begin{pmatrix} \cosh x & 0 & i \sinh x \\ 0 & \cosh x + \sinh x & 0 \\ -i \sinh x & 0 & \cosh x \end{pmatrix}. \qquad (2.441)$$

Problem 2.12

Consider two matrices: $A = \begin{pmatrix} 0 & i & 2 \\ 0 & 1 & 0 \\ -i & 0 & 0 \end{pmatrix}$ and $B = \begin{pmatrix} 2 & i & 0 \\ 3 & 1 & 5 \\ 0 & -i & -2 \end{pmatrix}$. Calculate $A^{-1} B$

and $B A^{-1}$. Are they equal?

Solution

As mentioned above, a calculation similar to (2.200) leads to the inverse of A:

$$A^{-1} = \begin{pmatrix} 0 & 0 & i \\ 0 & 1 & 0 \\ 1/2 & -i/2 & 0 \end{pmatrix}. \tag{2.442}$$

The products $A^{-1} B$ and $B A^{-1}$ are given by

$$A^{-1}B = \begin{pmatrix} 0 & 0 & i \\ 0 & 1 & 0 \\ 1/2 & -i/2 & 0 \end{pmatrix} \begin{pmatrix} 2 & i & 0 \\ 3 & 1 & 5 \\ 0 & -i & -2 \end{pmatrix} = \begin{pmatrix} 0 & 1 & -2i \\ 3 & 1 & 5 \\ 1-3i/2 & 0 & -5i/2 \end{pmatrix}, \tag{2.443}$$

$$BA^{-1} = \begin{pmatrix} 2 & i & 0 \\ 3 & 1 & 5 \\ 0 & -i & -2 \end{pmatrix} \begin{pmatrix} 0 & 0 & i \\ 0 & 1 & 0 \\ 1/2 & -i/2 & 0 \end{pmatrix} = \begin{pmatrix} 0 & i & 2i \\ 5/2 & 1-5i/2 & 3i \\ -1 & 0 & 0 \end{pmatrix}. \tag{2.444}$$

We see that $A^{-1} B$ and $B A^{-1}$ are not equal.

Remark

We should note that the quotient B/A of two matrices A and B is equal to the product $B A^{-1}$ and not $A^{-1}B$; that is:

$$\frac{B}{A} = BA^{-1} = \frac{\begin{pmatrix} 2 & i & 0 \\ 3 & 1 & 5 \\ 0 & -i & -2 \end{pmatrix}}{\begin{pmatrix} 0 & i & 2 \\ 0 & 1 & 0 \\ -i & 0 & 0 \end{pmatrix}} = \begin{pmatrix} 0 & i & 2i \\ 5/2 & 1-5i/2 & 3i \\ -1 & 0 & 0 \end{pmatrix}. \tag{2.445}$$

Problem 2.13

Consider the matrices $A = \begin{pmatrix} 0 & 1 & 0 \\ 1 & 0 & 1 \\ 0 & 1 & 0 \end{pmatrix}$ and $B = \begin{pmatrix} 1 & 0 & 0 \\ 0 & 0 & 0 \\ 0 & 0 & -1 \end{pmatrix}$.

(a) Find the eigenvalues and normalized eigenvectors of A and B. Denote the eigenvectors of A by $| a_1 \rangle$, $| a_2 \rangle$, $| a_3 \rangle$ and those of B by $| b_1 \rangle$, $| b_2 \rangle$, $| b_3 \rangle$. Are there any degenerate eigenvalues?

(b) Show that each of the sets $| a_1 \rangle$, $| a_2 \rangle$, $| a_3 \rangle$ and $| b_1 \rangle$, $| b_2 \rangle$, $| b_3 \rangle$ forms an orthonormal and complete basis, i.e., show that $\langle a_j | a_k \rangle = \delta_{jk}$ and $\sum_{j=1}^{3} | a_j \rangle \langle a_j | = I$, where I is the 3×3 unit matrix; then show that the same holds for $| b_1 \rangle$, $| b_2 \rangle$, $| b_3 \rangle$.

(c) Find the matrix U of the transformation from the basis $\{| a \rangle\}$ to $\{| b \rangle\}$. Show that $U^{-1} = U^\dagger$. Verify that $U^\dagger U = I$. Calculate how the matrix A transforms under U, i.e., calculate $A' = U A U^\dagger$.

Solution

(a) It is easy to verify that the eigenvalues of A are $a_1 = 0$, $a_2 = \sqrt{2}$, $a_3 = -\sqrt{2}$ and their corresponding normalized eigenvectors are

$$| a_1 \rangle = \frac{1}{\sqrt{2}} \begin{pmatrix} -1 \\ 0 \\ 1 \end{pmatrix}, \quad | a_2 \rangle = \frac{1}{2} \begin{pmatrix} 1 \\ \sqrt{2} \\ 1 \end{pmatrix}, \quad | a_3 \rangle = \frac{1}{2} \begin{pmatrix} 1 \\ -\sqrt{2} \\ 1 \end{pmatrix}. \quad (2.446)$$

The eigenvalues of B are $b_1 = 1$, $b_2 = 0$, $b_3 = -1$ and their corresponding normalized eigenvectors are

$$| b_1 \rangle = \begin{pmatrix} 1 \\ 0 \\ 0 \end{pmatrix}, \quad | b_2 \rangle = \begin{pmatrix} 0 \\ 1 \\ 0 \end{pmatrix}, \quad | b_3 \rangle = \begin{pmatrix} 0 \\ 0 \\ 1 \end{pmatrix}. \quad (2.447)$$

None of the eigenvalues of A and B are degenerate.

(b) The set $| a_1 \rangle$, $| a_2 \rangle$, $| a_3 \rangle$ is indeed complete because the sum of $| a_1 \rangle \langle a_1 |$, $| a_2 \rangle \langle a_2 |$, and $| a_3 \rangle \langle a_3 |$ as given by

$$| a_1 \rangle \langle a_1 | = \frac{1}{2} \begin{pmatrix} -1 \\ 0 \\ 1 \end{pmatrix} (-1 \quad 0 \quad 1) = \frac{1}{2} \begin{pmatrix} 1 & 0 & -1 \\ 0 & 0 & 0 \\ -1 & 0 & 1 \end{pmatrix}, \quad (2.448)$$

$$| a_2 \rangle \langle a_2 | = \frac{1}{4} \begin{pmatrix} 1 \\ \sqrt{2} \\ 1 \end{pmatrix} (1 \quad \sqrt{2} \quad 1) = \frac{1}{4} \begin{pmatrix} 1 & \sqrt{2} & 1 \\ \sqrt{2} & 2 & \sqrt{2} \\ 1 & \sqrt{2} & 1 \end{pmatrix}, \quad (2.449)$$

$$| a_3 \rangle \langle a_3 | = \frac{1}{4} \begin{pmatrix} 1 \\ -\sqrt{2} \\ 1 \end{pmatrix} (1 \quad -\sqrt{2} \quad 1) = \frac{1}{4} \begin{pmatrix} 1 & -\sqrt{2} & 1 \\ -\sqrt{2} & 2 & -\sqrt{2} \\ 1 & -\sqrt{2} & 1 \end{pmatrix}, \quad (2.450)$$

is equal to unity:

$$\sum_{j=1}^{3} | a_j \rangle \langle a_j | = \frac{1}{2} \begin{pmatrix} 1 & 0 & -1 \\ 0 & 0 & 0 \\ -1 & 0 & 1 \end{pmatrix} + \frac{1}{4} \begin{pmatrix} 1 & \sqrt{2} & 1 \\ \sqrt{2} & 2 & \sqrt{2} \\ 1 & \sqrt{2} & 1 \end{pmatrix}$$

$$+ \frac{1}{4} \begin{pmatrix} 1 & -\sqrt{2} & 1 \\ -\sqrt{2} & 2 & -\sqrt{2} \\ 1 & -\sqrt{2} & 1 \end{pmatrix}$$

$$= \begin{pmatrix} 1 & 0 & 0 \\ 0 & 1 & 0 \\ 0 & 0 & 1 \end{pmatrix}. \quad (2.451)$$

The states $| a_1 \rangle$, $| a_2 \rangle$, $| a_3 \rangle$ are orthonormal, since $\langle a_1 | a_2 \rangle = \langle a_1 | a_3 \rangle = \langle a_3 | a_2 \rangle = 0$ and $\langle a_1 | a_1 \rangle = \langle a_2 | a_2 \rangle = \langle a_3 | a_3 \rangle = 1$. Following the same procedure, we can ascertain that

$$| b_1 \rangle \langle b_1 | + | b_2 \rangle \langle b_2 | + | b_3 \rangle \langle b_3 | = \begin{pmatrix} 1 & 0 & 0 \\ 0 & 1 & 0 \\ 0 & 0 & 1 \end{pmatrix}. \quad (2.452)$$

We can verify that the states $| b_1 \rangle, | b_2 \rangle, | b_3 \rangle$ are orthonormal, since $\langle b_1 | b_2 \rangle = \langle b_1 | b_3 \rangle = \langle b_3 | b_2 \rangle = 0$ and $\langle b_1 | b_1 \rangle = \langle b_2 | b_2 \rangle = \langle b_3 | b_3 \rangle = 1$.

(c) The elements of the matrix U, corresponding to the transformation from the basis $\{| a \rangle\}$ to $\{| b \rangle\}$, are given by $U_{jk} = \langle b_j | a_k \rangle$ where $j, k = 1, 2, 3$:

$$U = \begin{pmatrix} \langle b_1 | a_1 \rangle & \langle b_1 | a_2 \rangle & \langle b_1 | a_3 \rangle \\ \langle b_2 | a_1 \rangle & \langle b_2 | a_2 \rangle & \langle b_2 | a_3 \rangle \\ \langle b_3 | a_1 \rangle & \langle b_3 | a_2 \rangle & \langle b_3 | a_3 \rangle \end{pmatrix}, \tag{2.453}$$

where the elements $\langle b_j | a_k \rangle$ can be calculated from (2.446) and (2.447):

$$U_{11} = \langle b_1 | a_1 \rangle = \frac{1}{\sqrt{2}} \begin{pmatrix} 1 & 0 & 0 \end{pmatrix} \begin{pmatrix} -1 \\ 0 \\ 1 \end{pmatrix} = -\frac{\sqrt{2}}{2}, \tag{2.454}$$

$$U_{12} = \langle b_1 | a_2 \rangle = \frac{1}{2} \begin{pmatrix} 1 & 0 & 0 \end{pmatrix} \begin{pmatrix} 1 \\ \sqrt{2} \\ 1 \end{pmatrix} = \frac{1}{2}, \tag{2.455}$$

$$U_{13} = \langle b_1 | a_3 \rangle = \frac{1}{2} \begin{pmatrix} 1 & 0 & 0 \end{pmatrix} \begin{pmatrix} 1 \\ -\sqrt{2} \\ 1 \end{pmatrix} = \frac{1}{2}, \tag{2.456}$$

$$U_{21} = \langle b_2 | a_1 \rangle = \frac{1}{\sqrt{2}} \begin{pmatrix} 0 & 1 & 0 \end{pmatrix} \begin{pmatrix} -1 \\ 0 \\ 1 \end{pmatrix} = 0, \tag{2.457}$$

$$U_{22} = \langle b_2 | a_2 \rangle = \frac{1}{2} \begin{pmatrix} 0 & 1 & 0 \end{pmatrix} \begin{pmatrix} 1 \\ \sqrt{2} \\ 1 \end{pmatrix} = \frac{\sqrt{2}}{2}, \tag{2.458}$$

$$U_{23} = \langle b_2 | a_3 \rangle = \frac{1}{2} \begin{pmatrix} 0 & 1 & 0 \end{pmatrix} \begin{pmatrix} 1 \\ -\sqrt{2} \\ 1 \end{pmatrix} = -\frac{\sqrt{2}}{2}, \tag{2.459}$$

$$U_{31} = \langle b_3 | a_1 \rangle = \frac{1}{\sqrt{2}} \begin{pmatrix} 0 & 0 & 1 \end{pmatrix} \begin{pmatrix} -1 \\ 0 \\ 1 \end{pmatrix} = \frac{\sqrt{2}}{2}, \tag{2.460}$$

$$U_{32} = \langle b_3 | a_2 \rangle = \frac{1}{2} \begin{pmatrix} 0 & 0 & 1 \end{pmatrix} \begin{pmatrix} 1 \\ \sqrt{2} \\ 1 \end{pmatrix} = \frac{1}{2}, \tag{2.461}$$

$$U_{33} = \langle b_3 | a_3 \rangle = \frac{1}{2} \begin{pmatrix} 0 & 0 & 1 \end{pmatrix} \begin{pmatrix} 1 \\ -\sqrt{2} \\ 1 \end{pmatrix} = \frac{1}{2}. \tag{2.462}$$

Collecting these elements, we obtain

$$U = \frac{1}{2} \begin{pmatrix} -\sqrt{2} & 1 & 1 \\ 0 & \sqrt{2} & -\sqrt{2} \\ \sqrt{2} & 1 & 1 \end{pmatrix}. \tag{2.463}$$

Calculating the inverse of U as we did in (2.200), we see that it is equal to its Hermitian adjoint:

$$U^{-1} = \frac{1}{2}\begin{pmatrix} -\sqrt{2} & 0 & \sqrt{2} \\ 1 & \sqrt{2} & 1 \\ 1 & -\sqrt{2} & 1 \end{pmatrix} = U^{\dagger}. \tag{2.464}$$

This implies that the matrix U is unitary. The matrix A transforms as follows:

$$A' = UAU^{\dagger} = \frac{1}{4}\begin{pmatrix} -\sqrt{2} & 1 & 1 \\ 0 & \sqrt{2} & -\sqrt{2} \\ \sqrt{2} & 1 & 1 \end{pmatrix}\begin{pmatrix} 0 & 1 & 0 \\ 1 & 0 & 1 \\ 0 & 1 & 0 \end{pmatrix}\begin{pmatrix} -\sqrt{2} & 0 & \sqrt{2} \\ 1 & \sqrt{2} & 1 \\ 1 & -\sqrt{2} & 1 \end{pmatrix}$$

$$= \frac{1}{2}\begin{pmatrix} 1-\sqrt{2} & -1 & 1 \\ -1 & -2 & 1 \\ 1 & 1 & 1+\sqrt{2} \end{pmatrix}. \tag{2.465}$$

Problem 2.14

Calculate the following expressions involving Dirac's delta function:

(a) $\int_{-5}^{5} \cos(3x)\delta(x - \pi/3)\,dx$

(b) $\int_{0}^{10} \left[e^{2x-7} + 4\right]\delta(x + 3)\,dx$

(c) $\left[2\cos^2(3x) - \sin(x/2)\right]\delta(x + \pi)$

(d) $\int_{0}^{\pi} \cos(3\theta)\delta'''(\theta - \pi/2)\,d\theta$

(e) $\int_{2}^{9} \left(x^2 - 5x + 2\right)\delta[2(x - 4)]\,dx$.

Solution

(a) Since $x = \pi/3$ lies within the interval $(-5,\ 5)$, equation (2.281) yields

$$\int_{-5}^{5} \cos(3x)\delta(x - \pi/3)\,dx = \cos\left(3\frac{\pi}{3}\right) = -1. \tag{2.466}$$

(b) Since $x = -3$ lies outside the interval $(0,\ 10)$, Eq (2.281) yields at once

$$\int_{0}^{10} \left[e^{2x-7} + 4\right]\delta(x + 3)\,dx = 0. \tag{2.467}$$

(c) Using the relation $f(x)\delta(x - a) = f(a)\delta(x - a)$ which is listed in Appendix A, we have

$$\left[2\cos^2(3x) - \sin(x/2)\right]\delta(x + \pi) = \left[2\cos^2(3(-\pi)) - \sin((-\pi)/2)\right]\delta(x + \pi)$$

$$= 3\delta(x + \pi). \tag{2.468}$$

(d) Inserting $n = 3$ into Eq (2.282) and since $\cos'''(3\theta) = 27\sin(3\theta)$, we obtain

$$\int_{0}^{\pi} \cos(3\theta)\delta'''(\theta - \pi/2)\,d\theta = (-1)^3 \cos'''(3\pi/2) = (-1)^3\,27\sin(3\pi/2)$$

$$= 27. \tag{2.469}$$

(e) Since $\delta[2(x-4)] = (1/2)\delta(x-4)$, we have

$$\int_2^9 \left(x^2 - 5x + 2\right) \delta[2(x-4)], dx \quad - \quad \frac{1}{2}\int_2^9 \left(x^2 \quad 5x \quad 2\right) \delta(x-4)\, dx$$

$$= \frac{1}{2}\left(4^2 - 5 \times 4 + 2\right) = -1. \qquad (2.470)$$

Problem 2.15
Consider a system whose Hamiltonian is given by $\hat{H} = \alpha\, (|\,\phi_1\rangle\langle\phi_2\,| + |\,\phi_2\rangle\langle\phi_1\,|)$, where α is a real number having the dimensions of energy and $|\,\phi_1\rangle,\,|\,\phi_2\rangle$ are normalized eigenstates of a Hermitian operator $\hat{A}$ that has no degenerate eigenvalues.

(a) Is $\hat{H}$ a projection operator? What about $\alpha^{-2}\hat{H}^2$?

(b) Show that $|\,\phi_1\rangle$ and $|\,\phi_2\rangle$ are not eigenstates of $\hat{H}$.

(c) Calculate the commutators $[\hat{H},\,|\,\phi_1\rangle\langle\phi_1\,|]$ and $[\hat{H},\,|\,\phi_2\rangle\langle\phi_2\,|]$ then find the relation that may exist between them.

(d) Find the normalized eigenstates of $\hat{H}$ and their corresponding energy eigenvalues.

(e) Assuming that $|\,\phi_1\rangle$ and $|\,\phi_2\rangle$ form a complete and orthonormal basis, find the matrix representing $\hat{H}$ in the basis. Find the eigenvalues and eigenvectors of the matrix and compare the results with those derived in (d).

Solution
(a) Since $|\,\phi_1\rangle$ and $|\,\phi_2\rangle$ are eigenstates of $\hat{A}$ and since $\hat{A}$ is Hermitian, they must be orthogonal, $\langle\phi_1\,|\,\phi_2\rangle = 0$ (instance of Theorem 2.1). Now, since $|\,\phi_1\rangle$ and $|\,\phi_2\rangle$ are both normalized and since $\langle\phi_1\,|\,\phi_2\rangle = 0$, we can reduce $\hat{H}^2$ to

$$\begin{aligned}
\hat{H}^2 &= \alpha^2\,(|\,\phi_1\rangle\langle\phi_2\,| + |\,\phi_2\rangle\langle\phi_1\,|)\,(|\,\phi_1\rangle\langle\phi_1\,| + |\,\phi_2\rangle\langle\phi_2\,|) \\
&= \alpha^2\,(|\,\phi_1\rangle\langle\phi_2\,| + |\,\phi_2\rangle\langle\phi_1\,|), \qquad (2.471)
\end{aligned}$$

which is different from $\hat{H}$; hence $\hat{H}$ is not a projection operator. The operator $\alpha^{-2}\hat{H}^2$ is a projection operator since it is both Hermitian and equal to its own square. Using (2.471) we can write

$$\begin{aligned}
(\alpha^{-2}\hat{H}^2)^2 &= (|\,\phi_1\rangle\langle\phi_2\,| + |\,\phi_2\rangle\langle\phi_1\,|)\,(|\,\phi_1\rangle\langle\phi_2\,| + |\,\phi_2\rangle\langle\phi_1\,|) \\
&= |\,\phi_1\rangle\langle\phi_1\,| + |\,\phi_2\rangle\langle\phi_2\,| = \alpha^{-2}\hat{H}^2. \qquad (2.472)
\end{aligned}$$

(b) Since $|\,\phi_1\rangle$ and $|\,\phi_2\rangle$ are both normalized, and since $\langle\phi_1\,|\,\phi_2\rangle = 0$, we have

$$\hat{H}\,|\,\phi_1\rangle = \alpha\,|\,\phi_1\rangle\langle\phi_2\,|\,\phi_1\rangle + \alpha\,|\,\phi_2\rangle\langle\phi_1\,|\,\phi_1\rangle = \alpha\,|\,\phi_2\rangle, \qquad (2.473)$$

$$\hat{H}\,|\,\phi_2\rangle = \alpha\,|\,\phi_1\rangle; \qquad (2.474)$$

hence $|\,\phi_1\rangle$ and $|\,\phi_2\rangle$ are not eigenstates of $\hat{H}$. In addition, we have

$$\langle\phi_1\,|\,\hat{H}\,|\,\phi_1\rangle = \langle\phi_2\,|\,\hat{H}\,|\,\phi_2\rangle = 0. \qquad (2.475)$$

(c) Using the relations derived above, $\hat{H}\,|\,\phi_1\rangle = \alpha\,|\,\phi_2\rangle$ and $\hat{H}\,|\,\phi_2\rangle = \alpha\,|\,\phi_1\rangle$, we can write

$$[\hat{H},\,|\,\phi_1\rangle\langle\phi_1\,|] = \alpha\,(|\,\phi_2\rangle\langle\phi_1\,| - |\,\phi_1\rangle\langle\phi_2\,|), \qquad (2.476)$$

$$[\hat{H}, |\phi_2\rangle\langle\phi_2|] = \alpha \left(|\phi_1\rangle\langle\phi_2| - |\phi_2\rangle\langle\phi_1|\right); \tag{2.477}$$

hence

$$[\hat{H}, |\phi_1\rangle\langle\phi_1|] = -[\hat{H}, |\phi_2\rangle\langle\phi_2|]. \tag{2.478}$$

(d) Consider a general state $|\psi\rangle = \lambda_1 |\phi_1\rangle + \lambda_2 |\phi_2\rangle$. Applying $\hat{H}$ to this state, we get

$$\begin{aligned}
\hat{H} |\psi\rangle &= \alpha \left(|\phi_1\rangle\langle\phi_2| + |\phi_2\rangle\langle\phi_1|\right) \left(\lambda_1 |\phi_1\rangle + \lambda_2 |\phi_2\rangle\right) \\
&= \alpha \left(\lambda_2 |\phi_1\rangle + \lambda_1 |\phi_2\rangle\right).
\end{aligned} \tag{2.479}$$

Now, since $|\psi\rangle$ is normalized, we have

$$\langle\psi \mid \psi\rangle = |\lambda_1|^2 + |\lambda_2|^2 = 1. \tag{2.480}$$

The previous two equations show that $|\lambda_1| = |\lambda_2| = 1/\sqrt{2}$ and that $\lambda_1 = \pm\lambda_2$. Hence the eigenstates of the system are:

$$|\psi_\pm\rangle = \frac{1}{\sqrt{2}} \left(|\phi_1\rangle \pm |\phi_2\rangle\right). \tag{2.481}$$

The corresponding eigenvalues are $\pm\alpha$:

$$\hat{H} |\psi_\pm\rangle = \pm\alpha |\psi_\pm\rangle. \tag{2.482}$$

(e) Since $\langle\phi_1 \mid \phi_2\rangle = \langle\phi_2 \mid \phi_1\rangle = 0$ and $\langle\phi_1 \mid \phi_1\rangle = \langle\phi_2 \mid \phi_2\rangle = 1$, we can verify that $H_{11} = \langle\phi_1 | \hat{H} | \phi_1\rangle = 0$, $H_{22} = \langle\phi_2 | \hat{H} | \phi_2\rangle = 0$, $H_{12} = \langle\phi_1 | \hat{H} | \phi_2\rangle = \alpha$, $H_{21} = \langle\phi_2 | \hat{H} | \phi_1\rangle = \alpha$. The matrix of $\hat{H}$ is thus given by

$$H = \alpha \begin{pmatrix} 0 & 1 \\ 1 & 0 \end{pmatrix}. \tag{2.483}$$

The eigenvalues of this matrix are equal to $\pm\alpha$ and the corresponding eigenvectors are $\frac{1}{\sqrt{2}} \begin{pmatrix} 1 \\ \pm 1 \end{pmatrix}$. These results are indeed similar to those derived in (d).

Problem 2.16

Consider the matrices $A = \begin{pmatrix} 1 & 0 & 0 \\ 0 & 7 & -3i \\ 0 & 3i & 5 \end{pmatrix}$ and $B = \begin{pmatrix} 0 & -i & 3i \\ -i & 0 & i \\ 3i & i & 0 \end{pmatrix}$.

(a) Check the hermiticity of A and B.

(b) Find the eigenvalues of A and B; denote the eigenvalues of A by a_1, a_2, and a_3. Explain why the eigenvalues of A are real and those of B are imaginary.

(c) Calculate $\text{Tr}(A)$ and $\det(A)$. Verify $\text{Tr}(A) = a_1 + a_2 + a_3$, $\det(A) = a_1 a_2 a_3$.

Solution

(a) Matrix A is Hermitian but B is anti-Hermitian:

$$A^\dagger = \begin{pmatrix} 1 & 0 & 0 \\ 0 & 7 & -3i \\ 0 & 3i & 5 \end{pmatrix} = A, \qquad B^\dagger = \begin{pmatrix} 0 & i & -3i \\ i & 0 & -i \\ -3i & -i & 0 \end{pmatrix} = -B. \tag{2.484}$$

(b) The eigenvalues of A are $a_1 = 6 - \sqrt{10}$, $a_2 = 1$, and $a_3 = 6 + \sqrt{10}$ and those of B are $b_1 = -i\left(3 + \sqrt{17}\right)/2$, $b_2 = 3i$, and $b_3 = i\left(-3 + \sqrt{17}\right)/2$. The eigenvalues of A are real and those of B are imaginary. This is expected since, as shown in (2.74) and (2.75), the expectation values of Hermitian operators are real and those of anti-Hermitian operators are imaginary.

(c) A direct calculation of the trace and the determinant of A yields $\text{Tr}(A) = 1 + 7 + 5 = 13$ and $\det(A) = (7)(5) - (3i)(-3i) = 26$. Adding and multiplying the eigenvalues $a_1 = 6 - \sqrt{10}$, $a_2 = 1$, $a_3 = 6 + \sqrt{10}$, we have $a_1 + a_2 + a_3 = 6 - \sqrt{10} + 1 + 6 + \sqrt{10} = 13$ and $a_1 a_2 a_3 = (6 - \sqrt{10})(1)(6 + \sqrt{10}) = 26$. This confirms the results (2.260) and (2.261):

$$\text{Tr}(A) = a_1 + a_2 + a_3 = 13, \qquad \det(A) = a_1 a_2 a_3 = 26. \tag{2.485}$$

Problem 2.17
Consider a one-dimensional particle which moves along the x-axis and whose Hamiltonian is $\hat{H} = -\mathcal{E}d^2/dx^2 + 16\mathcal{E}\hat{X}^2$, where $\mathcal{E}$ is a real constant having the dimensions of energy.

(a) Is $\psi(x) = Ae^{-2x^2}$, where A is a normalization constant that needs to be found, an eigenfunction of $\hat{H}$? If yes, find the energy eigenvalue.

(b) Calculate the probability of finding the particle anywhere along the negative x-axis.

(c) Find the energy eigenvalue corresponding to the wave function $\phi(x) = 2x\psi(x)$.

(d) Specify the parities of $\phi(x)$ and $\psi(x)$. Are $\phi(x)$ and $\psi(x)$ orthogonal?

Solution
(a) The integral $\int_{-\infty}^{+\infty} e^{-4x^2} dx = \sqrt{\pi}/2$ allows us to find the normalization constant:

$$1 = \int_{-\infty}^{+\infty} |\psi(x)|^2 \, dx = A^2 \int_{-\infty}^{+\infty} e^{-4x^2} dx = A^2 \frac{\sqrt{\pi}}{2}; \tag{2.486}$$

this leads to $A = \sqrt{2/\sqrt{\pi}}$ and hence $\psi(x) = \sqrt{2/\sqrt{\pi}}\, e^{-2x^2}$. Since the first and second derivatives of $\psi(x)$ are given by

$$\psi'(x) = \frac{d\psi(x)}{dx} = -4x\psi(x), \qquad \psi''(x) = \frac{d^2\psi(x)}{dx^2} = (16x^2 - 4)\psi(x), \tag{2.487}$$

we see that $\psi(x)$ is an eigenfunction of $\hat{H}$ with an energy eigenvalue equal to $4\mathcal{E}$:

$$\hat{H}\psi(x) = -\mathcal{E}\frac{d^2\psi(x)}{dx^2} + 16\mathcal{E}x^2\psi(x) = -\mathcal{E}(16x^2 - 4)\psi(x) + 16\mathcal{E}x^2\psi(x) = 4\mathcal{E}\psi(x). \tag{2.488}$$

(b) Since $\int_{-\infty}^{0} e^{-4x^2} dx = \sqrt{\pi}/4$, the probability of finding the particle anywhere along the negative x-axis is equal to $\frac{1}{2}$:

$$\int_{-\infty}^{0} |\psi(x)|^2 \, dx = \frac{2}{\sqrt{\pi}} \int_{-\infty}^{0} e^{-4x^2} dx = \frac{1}{2}. \tag{2.489}$$

This is expected, since this probability is half the total probability, which in turn is equal to one.

(c) Since the second derivative of $\phi(x) = 2x\psi(x)$ is $\phi''(x) = 4\psi'(x) + 2x\psi''(x) = 8x(-3 + 4x^2)\psi(x) = 4(-3 + 4x^2)\phi(x)$, we see that $\phi(x)$ is an eigenfunction of $\hat{H}$ with an energy eigenvalue equal to $12\mathcal{E}$:

$$\hat{H}\phi(x) = -\mathcal{E}\frac{d^2\phi(x)}{dx^2} + 16\mathcal{E}x^2\phi(x) = -4\mathcal{E}(-3 + 4x^2)\phi(x) + 16\mathcal{E}x^2\phi(x) = 12\mathcal{E}\phi(x).$$

(2.490)

(d) The wave functions $\psi(x)$ and $\phi(x)$ are even and odd, respectively, since $\psi(-x) = \psi(x)$ and $\phi(-x) = -\phi(x)$; hence their product is an odd function. Therefore, they are orthogonal, since the symmetric integration of an odd function is zero:

$$\langle \phi \mid \psi \rangle = \int_{-\infty}^{+\infty} \phi^*(x)\psi(x)\,dx = \int_{-\infty}^{+\infty} \phi(x)\psi(x)\,dx = \int_{+\infty}^{-\infty} \phi(-x)\psi(-x)(-dx)$$

$$= -\int_{-\infty}^{+\infty} \phi(x)\psi(x)\,dx = 0.$$

(2.491)

Problem 2.18

(a) Find the eigenvalues and the eigenfunctions of the operator $\hat{A} = -d^2/dx^2$; restrict the search for the eigenfunctions to those complex functions that vanish everywhere except in the region $0 < x < a$.

(b) Normalize the eigenfunction and find the probability in the region $0 < x < a/2$.

Solution

(a) The eigenvalue problem for $-d^2/dx^2$ consists of solving the differential equation

$$-\frac{d^2\psi(x)}{dx^2} = \alpha\psi(x)$$

(2.492)

and finding the eigenvalues α and the eigenfunction $\psi(x)$. The most general solution to this equation is

$$\psi(x) = Ae^{ibx} + Be^{-ibx},$$

(2.493)

with $\alpha = b^2$. Using the boundary conditions of $\psi(x)$ at $x = 0$ and $x = a$, we have

$$\psi(0) = A + B = 0 \implies B = -A, \qquad \psi(a) = Ae^{iba} + Be^{-iba} = 0.$$

(2.494)

A substitution of $B = -A$ into the second equation leads to $A\left(e^{iba} - e^{-iba}\right) = 0$ or $e^{iba} = e^{-iba}$ which leads to $e^{2iba} = 1$. Thus, we have $\sin 2ba = 0$ and $\cos 2ba = 1$, so $ba = n\pi$. The eigenvalues are then given by $\alpha_n = n^2\pi^2/a^2$ and the corresponding eigenvectors by $\psi_n(x) = A\left(e^{in\pi x/a} - e^{-in\pi x/a}\right)$; that is,

$$\alpha_n = \frac{n^2\pi^2}{a^2}, \qquad \psi_n(x) = C_n \sin\left(\frac{n\pi x}{a}\right).$$

(2.495)

So the eigenvalue spectrum of the operator $\hat{A} = -d^2/dx^2$ is discrete, because the eigenvalues and eigenfunctions depend on a discrete number n.

(b) The normalization of $\psi_n(x)$,

$$1 = C_n^2 \int_0^a \sin^2\left(\frac{n\pi x}{a}\right) dx = \frac{C_n^2}{2} \int_0^a \left[1 - \cos\left(\frac{2n\pi x}{a}\right)\right] dx = \frac{C_n^2}{2}a,$$

(2.496)

yields $C_n = \sqrt{2/a}$ and hence $\psi_n(x) = \sqrt{2/a}\sin(n\pi x/a)$. The probability in the region $0 < x < a/2$ is given by

$$\frac{2}{a}\int_0^{a/2}\sin^2\left(\frac{n\pi x}{a}\right)dx = \frac{1}{a}\int_0^{a/2}\left[1 - \cos\left(\frac{2n\pi x}{a}\right)\right]dx = \frac{1}{2}. \qquad (2.497)$$

This is expected since the total probability is 1: $\int_0^a |\psi_n(x)|^2\,dx = 1$.

2.10 Exercises

Exercise 2.1
Consider the two states $|\psi\rangle = i|\phi_1\rangle + 3i|\phi_2\rangle - |\phi_3\rangle$ and $|\chi\rangle = |\phi_1\rangle - i|\phi_2\rangle + 5i|\phi_3\rangle$, where $|\phi_1\rangle$, $|\phi_2\rangle$ and $|\phi_3\rangle$ are orthonormal.
   (a) Calculate $\langle\psi|\psi\rangle$, $\langle\chi|\chi\rangle$, $\langle\psi|\chi\rangle$, $\langle\chi|\psi\rangle$, and infer $\langle\psi+\chi|\psi+\chi\rangle$. Are the scalar products $\langle\psi|\chi\rangle$ and $\langle\chi|\psi\rangle$ equal?
   (b) Calculate $|\psi\rangle\langle\chi|$ and $|\chi\rangle\langle\psi|$. Are they equal? Calculate their traces and compare them.
   (c) Find the Hermitian conjugates of $|\psi\rangle$, $|\chi\rangle$, $|\psi\rangle\langle\chi|$, and $|\chi\rangle\langle\psi|$.

Exercise 2.2
Consider two states $|\psi_1\rangle = |\phi_1\rangle + 4i|\phi_2\rangle + 5|\phi_3\rangle$ and $|\psi_2\rangle = b|\phi_1\rangle + 4|\phi_2\rangle - 3i|\phi_3\rangle$, where $|\phi_1\rangle$, $|\phi_2\rangle$, and$|\phi_3\rangle$ are orthonormal kets, and where b is a constant. Find the value of b so that $|\psi_1\rangle$ and $|\psi_2\rangle$ are orthogonal.

Exercise 2.3
If $|\phi_1\rangle$, $|\phi_2\rangle$, and $|\phi_3\rangle$ are orthonormal, show that the states $|\psi\rangle = i|\phi_1\rangle + 3i|\phi_2\rangle - |\phi_3\rangle$ and $|\chi\rangle = |\phi_1\rangle - i|\phi_2\rangle + 5i|\phi_3\rangle$ satisfy
   (a) the triangle inequality and
   (b) the Schwarz inequality.

Exercise 2.4
Find the constant α so that the states $|\psi\rangle = \alpha|\phi_1\rangle + 5|\phi_2\rangle$ and $|\chi\rangle = 3\alpha|\phi_1\rangle - 4|\phi_2\rangle$ are orthogonal; consider $|\phi_1\rangle$ and $|\phi_2\rangle$ to be orthonormal.

Exercise 2.5
If $|\psi\rangle = |\phi_1\rangle + |\phi_2\rangle$ and $|\chi\rangle = |\phi_1\rangle - |\phi_2\rangle$, prove the following relations (note that $|\phi_1\rangle$ and $|\phi_2\rangle$ are not orthonormal):
   (a) $\langle\psi|\psi\rangle + \langle\chi|\chi\rangle = 2\langle\phi_1|\phi_1\rangle + 2\langle\phi_2|\phi_2\rangle$,
   (b) $\langle\psi|\psi\rangle - \langle\chi|\chi\rangle = 2\langle\phi_1|\phi_2\rangle + 2\langle\phi_2|\phi_1\rangle$.

Exercise 2.6
Consider a state which is given in terms of three orthonormal vectors $|\phi_1\rangle$, $|\phi_2\rangle$, and $|\phi_3\rangle$ as follows:

$$|\psi\rangle = \frac{1}{\sqrt{15}}|\phi_1\rangle + \frac{1}{\sqrt{3}}|\phi_2\rangle + \frac{1}{\sqrt{5}}|\phi_3\rangle,$$

where $|\phi_n\rangle$ are eigenstates to an operator $\hat{B}$ such that: $\hat{B}|\phi_n\rangle = (3n^2 - 1)|\phi_n\rangle$ with $n = 1, 2, 3$.
   (a) Find the norm of the state $|\psi\rangle$.
   (b) Find the expectation value of $\hat{B}$ for the state $|\psi\rangle$.
   (c) Find the expectation value of $\hat{B}^2$ for the state $|\psi\rangle$.

Exercise 2.7

Are the following sets of functions linearly independent or dependent?
(a) $4e^x, e^x, 5e^x$
(b) $\cos x, e^{ix}, 3 \sin x$
(c) $7, x^2, 9x^4, e^{-x}$

Exercise 2.8

Are the following sets of functions linearly independent or dependent on the positive x-axis?
(a) $x, x + 2, x + 5$
(b) $\cos x, \cos 2x, \cos 3x$
(c) $\sin^2 x, \cos^2 x, \sin 2x$
(d) $x, (x - 1)^2, (x + 1)^2$
(e) $\sinh^2 x, \cosh^2 x, 1$

Exercise 2.9

Are the following sets of vectors linearly independent or dependent over the complex field?
(a) $(2, -3, 0), (0, 0, 1), (2i, i, -i)$
(b) $(0, 4, 0), (i, -3i, i), (2, 0, 1)$
(c) $(i, 1, 2), (3, i, -1), (-i, 3i, 5i)$

Exercise 2.10

Are the following sets of vectors (in the three-dimensional Euclidean space) linearly independent or dependent?
(a) $(4, 5, 6), (1, 2, 3), (7, 8, 9)$
(b) $(1, 0, 0), (0, -5, 0), (0, 0, \sqrt{7})$
(c) $(5, 4, 1), (2, 0, -2), (0, 6, -1)$

Exercise 2.11

Show that if $\hat{A}$ is a projection operator, the operator $1 - \hat{A}$ is also a projection operator.

Exercise 2.12

Show that $| \psi \rangle \langle \psi | / \langle \psi | \psi \rangle$ is a projection operator, regardless of whether $| \psi \rangle$ is normalized or not.

Exercise 2.13

In the following expressions, where $\hat{A}$ is an operator, specify the nature of each expression (i.e., specify whether it is an operator, a bra, or a ket); then find its Hermitian conjugate.
(a) $\langle \phi | \hat{A} | \psi \rangle \langle \psi |$
(b) $\hat{A} | \psi \rangle \langle \phi |$
(c) $\langle \phi | \hat{A} | \psi \rangle | \psi \rangle \langle \phi | \hat{A}$
(d) $\langle \psi | \hat{A} | \phi \rangle | \phi \rangle + i \hat{A} | \psi \rangle$
(e) $\left(| \phi \rangle \langle \phi | \hat{A} \right) - i \left(\hat{A} | \psi \rangle \langle \psi | \right)$

Exercise 2.14

Consider a two-dimensional space where a Hermitian operator $\hat{A}$ is defined by $\hat{A} | \phi_1 \rangle = | \phi_1 \rangle$ and $\hat{A} | \phi_2 \rangle = - | \phi_2 \rangle$; $| \phi_1 \rangle$ and $| \phi_2 \rangle$ are orthonormal.
(a) Do the states $| \phi_1 \rangle$ and $| \phi_2 \rangle$ form a basis?
(b) Consider the operator $\hat{B} = | \phi_1 \rangle \langle \phi_2 |$. Is $\hat{B}$ Hermitian? Show that $\hat{B}^2 = 0$.

(c) Show that the products $\hat{B}\hat{B}^{\dagger}$ and $\hat{B}^{\dagger}\hat{B}$ are projection operators.

(d) Show that the operator $\hat{B}\hat{B}^{\dagger} - \hat{B}^{\dagger}\hat{B}$ is unitary.

(e) Consider $\hat{C} = \hat{B}\hat{B}^{\dagger} + \hat{B}^{\dagger}\hat{B}$. Show that $\hat{C} \mid \phi_1\rangle = -\mid \phi_1\rangle$ and $\hat{C} \mid \phi_2\rangle = \mid \phi_2\rangle$.

Exercise 2.15

Prove the following two relations:

(a) $e^{\hat{A}}e^{\hat{B}} = e^{\hat{A}+\hat{B}}e^{[\hat{A},\hat{B}]/2}$,

(b) $e^{\hat{A}}\hat{B}e^{-\hat{A}} = \hat{B} + [\hat{A},\ \hat{B}] + \frac{1}{2!}[\hat{A}, [\hat{A},\ \hat{B}]] + \frac{1}{3!}[\hat{A}, [\hat{A}, [\hat{A},\ \hat{B}]]] + \cdots$.

Hint: To prove the first relation, you may consider defining an operator function $\hat{F}(t) = e^{\hat{A}t}e^{\hat{B}t}$, where t is a parameter, $\hat{A}$ and $\hat{B}$ are t-independent operators, and then make use of $[\hat{A}, G(\hat{B})] = [\hat{A}, \hat{B}]dG(\hat{B})/d\hat{B}$, where $G(\hat{B})$ is a function depending on the operator $\hat{B}$.

Exercise 2.16

(a) Verify that the matrix

$$\begin{pmatrix} \cos\theta & \sin\theta \\ -\sin\theta & \cos\theta \end{pmatrix}$$

is unitary.

(b) Find its eigenvalues and the corresponding normalized eigenvectors.

Exercise 2.17

Consider the following three matrices:

$$A = \begin{pmatrix} 0 & 1 & 0 \\ 1 & 0 & 1 \\ 0 & 1 & 0 \end{pmatrix}, \quad B = \begin{pmatrix} 0 & -i & 0 \\ i & 0 & -i \\ 0 & i & 0 \end{pmatrix}, \quad C = \begin{pmatrix} 1 & 0 & 0 \\ 0 & 0 & 0 \\ 0 & 0 & -1 \end{pmatrix}.$$

(a) Calculate the commutators $[A,\ B]$, $[B,\ C]$, and $[C,\ A]$.

(b) Show that $A^2 + B^2 + 2C^2 = 4I$, where I is the unity matrix.

(c) Verify that $\text{Tr}(ABC) = \text{Tr}(BCA) = \text{Tr}(CAB)$.

Exercise 2.18

Consider the following two matrices:

$$A = \begin{pmatrix} 3 & i & 1 \\ -1 & -i & 2 \\ 4 & 3i & 1 \end{pmatrix}, \quad B = \begin{pmatrix} 2i & 5 & -3 \\ -i & 3 & 0 \\ 7i & 1 & i \end{pmatrix}.$$

Verify the following relations:

(a) $\det(AB) = \det(A)\det(B)$,

(b) $\det(A^T) = \det(A)$,

(c) $\det(A^{\dagger}) = (\det(A))^*$, and

(d) $\det(A^*) = (\det(A))^*$.

Exercise 2.19

Consider the matrix

$$A = \begin{pmatrix} 0 & i \\ -i & 0 \end{pmatrix}.$$

(a) Find the eigenvalues and the normalized eigenvectors for the matrix A.

(b) Do these eigenvectors form a basis (i.e., is this basis complete and orthonormal)?

(c) Consider the matrix U which is formed from the normalized eigenvectors of A. Verify that U is unitary and that it satisfies

$$U^\dagger AU = \begin{pmatrix} \lambda_1 & 0 \\ 0 & \lambda_2 \end{pmatrix},$$

where λ_1 and λ_2 are the eigenvalues of A.

(d) Show that $e^{xA} = \cosh x + A \sinh x$.

Exercise 2.20

Using the bra-ket algebra, show that $\text{Tr}(\hat{A}\hat{B}\hat{C}) = \text{Tr}(\hat{C}\hat{A}\hat{B}) = \text{Tr}(\hat{B}\hat{C}\hat{A})$ where $\hat{A}, \hat{B}, \hat{C}$ are operators.

Exercise 2.21

For any two kets $| \psi \rangle$ and $| \phi \rangle$ that have finite norm, show that $\text{Tr}(| \psi \rangle \langle \phi |) = \langle \phi | \psi \rangle$.

Exercise 2.22

Consider the matrix $A = \begin{pmatrix} 0 & 0 & -1+i \\ 0 & 3 & 0 \\ -1-i & 0 & 0 \end{pmatrix}$.

(a) Find the eigenvalues and normalized eigenvectors of A. Denote the eigenvectors of A by $| a_1 \rangle, | a_2 \rangle, | a_3 \rangle$. Any degenerate eigenvalues?

(b) Show that the eigenvectors $| a_1 \rangle, | a_2 \rangle, | a_3 \rangle$ form an orthonormal and complete basis, i.e., show that $\sum_{j=1}^{3} | a_j \rangle \langle a_j | = I$, where I is the 3×3 unit matrix, and that $\langle a_j | a_k \rangle = \delta_{jk}$.

(c) Find the matrix corresponding to the operator obtained from the ket-bra product of the first eigenvector $P = | a_1 \rangle \langle a_1 |$. Is P a projection operator?

Exercise 2.23

In a three-dimensional vector space, consider the operator whose matrix, in an orthonormal basis $\{| 1 \rangle, | 2 \rangle, | 3 \rangle\}$, is

$$A = \begin{pmatrix} 0 & 0 & 1 \\ 0 & -1 & 0 \\ 1 & 0 & 0 \end{pmatrix}.$$

(a) Is A Hermitian? Calculate its eigenvalues and the corresponding normalized eigenvectors. Verify that the eigenvectors corresponding to the two nondegenerate eigenvalues are orthonormal.

(b) Calculate the matrices representing the projection operators for the two nondegenerate eigenvectors found in part (a).

Exercise 2.24

Consider two operators $\hat{A}$ and $\hat{B}$ whose matrices are

$$A = \begin{pmatrix} 1 & 3 & 0 \\ 1 & 0 & 1 \\ 0 & -1 & 1 \end{pmatrix}, \qquad B = \begin{pmatrix} 1 & 0 & -2 \\ 0 & 0 & 0 \\ -2 & 0 & 4 \end{pmatrix}.$$

(a) Are $\hat{A}$ and $\hat{B}$ Hermitian?

(b) Do $\hat{A}$ and $\hat{B}$ commute?

(c) Find the eigenvalues and eigenvectors of $\hat{A}$ and $\hat{B}$.

(d) Are the eigenvectors of each operator orthonormal?

(e) Verify that $\hat{U}^{\dagger}\hat{B}\hat{U}$ is diagonal, $\hat{U}$ being the matrix of the normalized eigenvectors of $\hat{B}$.

(f) Verify that $\hat{U}^{-1} = \hat{U}^{\dagger}$.

Exercise 2.25

Consider an operator $\hat{A}$ so that $[\hat{A}, \hat{A}^{\dagger}] = 1$.

(a) Evaluate the commutators $[\hat{A}^{\dagger}\hat{A}, \hat{A}]$ and $[\hat{A}^{\dagger}\hat{A}, \hat{A}^{\dagger}]$.

(b) If the actions of $\hat{A}$ and $\hat{A}^{\dagger}$ on the states $\{|a\rangle\}$ are given by $\hat{A}|a\rangle = \sqrt{a}|a-1\rangle$ and $\hat{A}^{\dagger}|a\rangle = \sqrt{a+1}|a+1\rangle$ and if $\langle a'|a\rangle = \delta_{a'a}$, calculate $\langle a|\hat{A}|a+1\rangle$, $\langle a+1|\hat{A}^{\dagger}|a\rangle$ and $\langle a|\hat{A}^{\dagger}\hat{A}|a\rangle$ and $\langle a|\hat{A}\hat{A}^{\dagger}|a\rangle$.

(c) Calculate $\langle a|(\hat{A} + \hat{A}^{\dagger})^2|a\rangle$ and $\langle a|(\hat{A} - \hat{A}^{\dagger})^2|a\rangle$.

Exercise 2.26

Consider a 4×4 matrix

$$A = \begin{pmatrix} 0 & \sqrt{1} & 0 & 0 \\ 0 & 0 & \sqrt{2} & 0 \\ 0 & 0 & 0 & \sqrt{3} \\ 0 & 0 & 0 & 0 \end{pmatrix}.$$

(a) Find the matrices of $A^{\dagger}$, $N = A^{\dagger}A$, $H = N + \frac{1}{2}I$ (where I is the unit matrix), $B = A + A^{\dagger}$, and $C = i(A - A^{\dagger})$.

(b) Find the matrices corresponding to the commutators $[A^{\dagger}, A]$, $[B, C]$, $[N, B]$, and $[N, C]$.

(c) Find the matrices corresponding to B^2, C^2, $[N, B^2 + C^2]$, $[H, A^{\dagger}]$, $[H, A]$, and $[H, N]$.

(d) Verify that $\det(ABC) = \det(A)\det(B)\det(C)$ and $\det(C^{\dagger}) = (\det(C))^*$.

Exercise 2.27

If $\hat{A}$ and $\hat{B}$ commute, and if $|\psi_1\rangle$ and $|\psi_2\rangle$ are two eigenvectors of $\hat{A}$ with different eigenvalues ($\hat{A}$ is Hermitian), show that

(a) $\langle\psi_1|\hat{B}|\psi_2\rangle$ is zero and

(b) $\hat{B}|\psi_1\rangle$ is also an eigenvector to $\hat{A}$ with the same eigenvalue as $|\psi_1\rangle$; i.e., if $\hat{A}|\psi_1\rangle = a_1|\psi_1\rangle$, show that $\hat{A}(\hat{B}|\psi_1\rangle) = a_1\hat{B}|\psi_1\rangle$.

Exercise 2.28

Let A and B be two $n \times n$ matrices. Assuming that B^{-1} exists, show that $[A, B^{-1}] = -B^{-1}[A, B]B^{-1}$.

Exercise 2.29

Consider a physical system whose Hamiltonian H and an operator A are given, in a three-dimensional space, by the matrices

$$H = \hbar\omega \begin{pmatrix} 1 & 0 & 0 \\ 0 & -1 & 0 \\ 0 & 0 & -1 \end{pmatrix}, \qquad A = a \begin{pmatrix} 1 & 0 & 0 \\ 0 & 0 & 1 \\ 0 & 1 & 0 \end{pmatrix}.$$

(a) Are H and A Hermitian?

(b) Show that H and A commute. Give a basis of eigenvectors common to H and A.

Exercise 2.30

(a) Using $[\hat{X}, \hat{P}] = i\hbar$, show that $[\hat{X}^2, \hat{P}] = 2i\hbar\hat{X}$ and $[\hat{X}, \hat{P}^2] = 2i\hbar\hat{P}$.

(b) Show that $[\hat{X}^2, \hat{P}^2] = 2i\hbar(i\hbar + 2\hat{P}\hat{X})$.

(c) Calculate the commutator $[\hat{X}^2, \hat{P}^3]$.

Exercise 2.31

Discuss the hermiticity of the commutators $[\hat{X}, \hat{P}]$, $[\hat{X}^2, \hat{P}]$ and $[\hat{X}, \hat{P}^2]$.

Exercise 2.32

(a) Evaluate the commutator $[\hat{X}^2, d/dx]$ by operating it on a wave function.

(b) Using $[\hat{X}, \hat{P}] = i\hbar$, evaluate the commutator $[\hat{X}\hat{P}^2, \hat{P}\hat{X}^2]$ in terms of a linear combination of $\hat{X}^2\hat{P}^2$ and $\hat{X}\hat{P}$.

Exercise 2.33

Show that $[\hat{X}, \hat{P}^n] = i\hbar\hat{X}\hat{P}^{n-1}$.

Exercise 2.34

Evaluate the commutators $[e^{i\hat{X}}, \hat{P}]$, $[e^{i\hat{X}^2}, \hat{P}]$, and $[e^{i\hat{X}}, \hat{P}^2]$.

Exercise 2.35

Consider the matrix

$$A = \begin{pmatrix} 0 & 0 & -1 \\ 0 & 1 & 0 \\ -1 & 0 & 0 \end{pmatrix}.$$

(a) Find the eigenvalues and the normalized eigenvectors of A.

(b) Do these eigenvectors form a basis (i.e., is this basis complete and orthonormal)?

(c) Consider the matrix U which is formed from the normalized eigenvectors of A. Verify that U is unitary and that it satisfies the relation

$$U^\dagger A U = \begin{pmatrix} \lambda_1 & 0 & 0 \\ 0 & \lambda_2 & 0 \\ 0 & 0 & \lambda_3 \end{pmatrix},$$

where λ_1, λ_2, and λ_3 are the eigenvalues of A.

(d) Show that $e^{xA} = \cosh x + A \sinh x$.

Hint: $\cosh x = \sum_{n=0}^{\infty} x^{2n}/(2n)!$ and $\sinh x = \sum_{n=0}^{\infty} x^{2n+1}/(2n+1)!$.

Exercise 2.36

(a) If $[\hat{A}, \hat{B}] = c$, where c is a number, prove the following two relations: $e^{\hat{A}}\hat{B}e^{-\hat{A}} = \hat{B}+c$ and $e^{\hat{A}+\hat{B}} = e^{\hat{A}}e^{\hat{B}}e^{-c/2}$.

(b) Now if $[\hat{A}, \hat{B}] = c\hat{B}$, where c is again a number, show that $e^{\hat{A}}\hat{B}e^{-\hat{A}} = e^c\hat{B}$.

Exercise 2.37

Consider the matrix

$$A = \frac{1}{2}\begin{pmatrix} 2 & 0 & 0 \\ 0 & 3 & -1 \\ 0 & -1 & 3 \end{pmatrix}.$$

(a) Find the eigenvalues of A and their corresponding eigenvectors.

(b) Consider the basis which is constructed from the three eigenvectors of A. Using matrix algebra, verify that this basis is both orthonormal and complete.

Exercise 2.38

(a) Specify the condition that must be satisfied by a matrix A so that it is both unitary and Hermitian.

(b) Consider the three matrices

$$M_1 = \begin{pmatrix} 0 & 1 \\ 1 & 0 \end{pmatrix}, \qquad M_2 = \begin{pmatrix} 0 & -i \\ i & 0 \end{pmatrix}, \qquad M_3 = \begin{pmatrix} 1 & 0 \\ 0 & -1 \end{pmatrix}.$$

Calculate the inverse of each matrix. Do they satisfy the condition derived in (a)?

Exercise 2.39

Consider the two matrices

$$A = \frac{1}{\sqrt{2}} \begin{pmatrix} 1 & i \\ i & 1 \end{pmatrix}, \qquad B = \frac{1}{2} \begin{pmatrix} 1+i & 1-i \\ 1-i & 1+i \end{pmatrix}.$$

(a) Are these matrices Hermitian?

(b) Calculate the inverses of these matrices.

(c) Are these matrices unitary?

(d) Verify that the determinants of A and B are of the form $e^{i\theta}$. Find the corresponding values of θ.

Exercise 2.40

Show that the transformation matrix representing a $90°$ counterclockwise rotation about the z-axis of the basis vectors $(\vec{i}, \vec{j}, \vec{k})$ is given by

$$U = \begin{pmatrix} 0 & -1 & 0 \\ 1 & 0 & 0 \\ 0 & 0 & 1 \end{pmatrix}.$$

Exercise 2.41

Show that the transformation matrix representing a $90°$ clockwise rotation about the y-axis of the basis vectors $(\vec{i}, \vec{j}, \vec{k})$ is given by

$$U = \begin{pmatrix} 0 & 0 & -1 \\ 0 & 1 & 0 \\ 1 & 0 & 0 \end{pmatrix}.$$

Exercise 2.42

Show that the operator $(\hat{X}\hat{P} + \hat{P}\hat{X})^2$ is equal to $(\hat{X}^2\hat{P}^2 + \hat{P}^2\hat{X}^2)$ plus a term of the order of $\hbar^2$.

Exercise 2.43

Consider the two matrices $A = \begin{pmatrix} 4 & i & 7 \\ 1 & 0 & 1 \\ 0 & 1 & -i \end{pmatrix}$ and $B = \begin{pmatrix} 1 & 1 & 1 \\ 0 & i & 0 \\ -i & 0 & i \end{pmatrix}$. Calculate the products $B^{-1} A$ and $A B^{-1}$. Are they equal? What is the significance of this result?

Exercise 2.44

Use the relations listed in Appendix A to evaluate the following integrals involving Dirac's delta function:

(a) $\int_0^\pi \sin(3x) \cos^2(4x) \delta(x - \pi/2) \, dx$.

(b) $\int_{-2}^2 e^{7x+2} \delta(5x) \, dx$.

(c) $\int_{-2\pi}^{2\pi} \sin(\theta/2) \delta''(\theta + \pi) \, d\theta$.

(d) $\int_0^{2\pi} \cos^2 \theta \delta[(\theta - \pi)/4] \, d\theta$.

Exercise 2.45

Use the relations listed in Appendix A to evaluate the following expressions:

(a) $\int_0^5 (3x^2 + 2) \delta(x - 1) \, dx$.

(b) $(2x^5 - 4x^3 + 1) \delta(x + 2)$.

(c) $\int_0^\infty (5x^3 - 7x^2 - 3) \delta(x^2 - 4) \, dx$.

Exercise 2.46

Use the relations listed in Appendix A to evaluate the following expressions:

(a) $\int_3^7 e^{6x-2} \delta(-4x) \, dx$.

(b) $\cos(2\theta) \sin(\theta) \delta(\theta^2 - \pi^2/4)$.

(c) $\int_{-1}^1 e^{5x-1} \delta'''(x) \, dx$.

Exercise 2.47

If the position and momentum operators are denoted by $\hat{\vec{R}}$ and $\hat{\vec{P}}$, respectively, show that $\hat{\mathcal{P}}^\dagger \hat{\vec{R}}{}^n \hat{\mathcal{P}} = (-1)^n \hat{\vec{R}}{}^n$ and $\hat{\mathcal{P}}^\dagger \hat{\vec{P}}{}^n \hat{\mathcal{P}} = (-1)^n \hat{\vec{P}}{}^n$, where $\hat{\mathcal{P}}$ is the parity operator and n is an integer.

Exercise 2.48

Consider an operator

$$\hat{A} = | \phi_1 \rangle \langle \phi_1 | + | \phi_2 \rangle \langle \phi_2 | + | \phi_3 \rangle \langle \phi_3 | -i | \phi_1 \rangle \langle \phi_2 |$$
$$- | \phi_1 \rangle \langle \phi_3 | +i | \phi_2 \rangle \langle \phi_1 | - | \phi_3 \rangle \langle \phi_1 |,$$

where $| \phi_1 \rangle, | \phi_2 \rangle$, and $| \phi_3 \rangle$ form a complete and orthonormal basis.

(a) Is $\hat{A}$ Hermitian? Calculate $\hat{A}^2$; is it a projection operator?

(b) Find the 3×3 matrix representing $\hat{A}$ in the $| \phi_1 \rangle, | \phi_2 \rangle, | \phi_3 \rangle$ basis.

(c) Find the eigenvalues and the eigenvectors of the matrix.

Exercise 2.49

The Hamiltonian of a two-state system is given by

$$\hat{H} = E (| \phi_1 \rangle \langle \phi_1 | - | \phi_2 \rangle \langle \phi_2 | - i | \phi_1 \rangle \langle \phi_2 | + i | \phi_2 \rangle \langle \phi_1 |),$$

where $| \phi_1 \rangle, | \phi_2 \rangle$ form a complete and orthonormal basis; E is a real constant having the dimensions of energy.

(a) Is $\hat{H}$ Hermitian? Calculate the trace of $\hat{H}$.

(b) Find the matrix representing $\hat{H}$ in the $| \phi_1 \rangle, | \phi_2 \rangle$ basis and calculate the eigenvalues and the eigenvectors of the matrix. Calculate the trace of the matrix and compare it with the result you obtained in (a).

(c) Calculate $[\hat{H}, | \phi_1 \rangle \langle \phi_1 |]$, $[\hat{H}, | \phi_2 \rangle \langle \phi_2 |]$, and $[\hat{H}, | \phi_1 \rangle \langle \phi_2 |]$.

Exercise 2.50
Consider a particle which is confined to move along the positive x-axis and whose Hamiltonian is $\hat{H} = \mathcal{E}d^2/dx^2$, where $\mathcal{E}$ is a positive real constant having the dimensions of energy.

(a) Find the wave function that corresponds to an energy eigenvalue of $9\mathcal{E}$ (make sure that the function you find is finite everywhere along the positive x-axis and is square integrable). Normalize this wave function.

(b) Calculate the probability of finding the particle in the region $0 \le x \le 15$.

(c) Is the wave function derived in (a) an eigenfunction of the operator $\hat{A} = d/dx - 7$?

(d) Calculate the commutator $[\hat{H},\ \hat{A}]$.

Exercise 2.51
Consider the wave functions:

$$\psi(x, y) = \sin 2x \cos 5x, \qquad \phi(x, y) = e^{-2(x^2+y^2)}, \qquad \chi(x, y) = e^{-i(x+y)}.$$

(a) Verify if any of the wave functions is an eigenfunction of $\hat{A} = \partial/\partial x + \partial/\partial y$.

(b) Find out if any of the wave functions is an eigenfunction of $\hat{B} = \partial^2/\partial x^2 + \partial^2/\partial y^2 + 1$.

(c) Calculate the actions of $\hat{A}\hat{B}$ and $\hat{B}\hat{A}$ on each one of the wave functions and infer $[\hat{A},\ \hat{B}]$.

Exercise 2.52
(a) Is the state $\psi(\theta, \phi) = e^{-3i\phi} \cos \theta$ an eigenfunction of $\hat{A}_\phi = \partial/\partial\phi$ or of $\hat{B}_\theta = \partial/\partial\theta$?

(b) Are $\hat{A}_\phi$ and $\hat{B}_\theta$ Hermitian?

(c) Evaluate the expressions $\langle \psi \mid \hat{A}_\phi \mid \psi \rangle$ and $\langle \psi \mid \hat{B}_\theta \mid \psi \rangle$.

(d) Find the commutator $[\hat{A}_\phi,\ \hat{B}_\theta]$.

Exercise 2.53
Consider an operator $\hat{A} = (\hat{X}d/dx + 2)$.

(a) Find the eigenfunction of $\hat{A}$ corresponding to a zero eigenvalue. Is this function normalizable?

(b) Is the operator $\hat{A}$ Hermitian?

(c) Calculate $[\hat{A},\ \hat{X}]$, $[\hat{A},\ d/dx]$, $[\hat{A},\ d^2/dx^2]$, $[\hat{X},\ [\hat{A},\ \hat{X}]]$, and $[d/dx,\ [\hat{A},\ d/dx]]$.

Exercise 2.54
If $\hat{A}$ and $\hat{B}$ are two Hermitian operators, find their respective eigenvalues such that $\hat{A}^2 = 2\hat{I}$ and $\hat{B}^4 = \hat{I}$, where $\hat{I}$ is the unit operator.

Exercise 2.55
Consider the Hilbert space of two-variable complex functions $\psi(x, y)$. A permutation operator is defined by its action on $\psi(x, y)$ as follows: $\hat{\pi} \psi(x, y) = \psi(y, x)$.

(a) Verify that the operator $\hat{\pi}$ is linear and Hermitian.

(b) Show that $\hat{\pi}^2 = \hat{I}$. Find the eigenvalues and show that the eigenfunctions of $\hat{\pi}$ are given by

$$\psi_+(x, y) = \frac{1}{2}[\psi(x, y) + \psi(y, x)] \quad \text{and} \quad \psi_-(x, y) = \frac{1}{2}[\psi(x, y) - \psi(y, x)].$$

Chapter 3

Postulates of Quantum Mechanics

3.1 Introduction

The formalism of quantum mechanics is based on a number of postulates. These postulates are in turn based on a wide range of experimental observations; the underlying physical ideas of these experimental observations have been briefly mentioned in Chapter 1. In this chapter we present a formal discussion of these postulates, and how they can be used to extract quantitative information about microphysical systems.

These postulates cannot be derived; they result from experiment. They represent the minimal set of assumptions needed to develop the theory of quantum mechanics. But how does one find out about the validity of these postulates? Their validity cannot be determined directly; only an indirect inferential statement is possible. For this, one has to turn to the theory built upon these postulates: if the theory works, the postulates will be valid; otherwise they will make no sense. Quantum theory not only works, but works extremely well, and this represents its experimental justification. It has a very penetrating qualitative as well as quantitative prediction power; this prediction power has been verified by a rich collection of experiments. So the accurate prediction power of quantum theory gives irrefutable evidence to the validity of the postulates upon which the theory is built.

3.2 The Basic Postulates of Quantum Mechanics

According to classical mechanics, the state of a particle is specified, at any time t, by two fundamental dynamical variables: the position $\vec{r}(t)$ and the momentum $\vec{p}(t)$. Any other physical quantity, relevant to the system, can be calculated in terms of these two dynamical variables. In addition, knowing these variables at a time t, we can predict, using for instance Hamilton's equations $dx/dt = \partial H/\partial p$ and $dp/dt = -\partial H/\partial x$, the values of these variables at any later time t'.

The quantum mechanical counterparts to these ideas are specified by postulates, which enable us to understand:

- how a quantum state is described mathematically at a given time t,

- how to calculate the various physical quantities from this quantum state, and

- knowing the system's state at a time t, how to find the state at any later time t'; that is, how to describe the time evolution of a system.

The answers to these questions are provided by the following set of five postulates.

Postulate 1: State of a system

The state of any physical system is specified, at each time t, by a state vector $|\psi(t)\rangle$ in a Hilbert space $\mathcal{H}$; $|\psi(t)\rangle$ contains (and serves as the basis to extract) all the needed information about the system. Any superposition of state vectors is also a state vector.

Postulate 2: Observables and operators

To every physically measurable quantity A, called an observable or dynamical variable, there corresponds a linear Hermitian operator $\hat{A}$ whose eigenvectors form a complete basis.

Postulate 3: Measurements and eigenvalues of operators

The measurement of an observable A may be represented formally by the action of $\hat{A}$ on a state vector $|\psi(t)\rangle$. The only possible result of such a measurement is one of the eigenvalues a_n (which are real) of the operator $\hat{A}$. If the result of a measurement of A on a state $|\psi(t)\rangle$ is a_n, the state of the system *immediately after* the measurement changes to $|\psi_n\rangle$:

$$\hat{A}|\psi(t)\rangle = a_n|\psi_n\rangle, \tag{3.1}$$

where $a_n = \langle\psi_n|\psi(t)\rangle$. **Note:** a_n is the component of $|\psi(t)\rangle$ when projected[1] onto the eigenvector $|\psi_n\rangle$.

Postulate 4: Probabilistic outcome of measurements

- **Discrete spectra**: When measuring an observable A of a system in a state $|\psi\rangle$, the probability of obtaining one of the nondegenerate eigenvalues a_n of the corresponding operator $\hat{A}$ is given by

$$P_n(a_n) = \frac{|\langle\psi_n|\psi\rangle|^2}{\langle\psi|\psi\rangle} = \frac{|a_n|^2}{\langle\psi|\psi\rangle}, \tag{3.2}$$

where $|\psi_n\rangle$ is the eigenstate of $\hat{A}$ with eigenvalue a_n. If the eigenvalue a_n is m-degenerate, P_n becomes

$$P_n(a_n) = \frac{\sum_{j=1}^m |\langle\psi_n^j|\psi\rangle|^2}{\langle\psi|\psi\rangle} = \frac{\sum_{j=1}^m |a_n^{(j)}|^2}{\langle\psi|\psi\rangle}. \tag{3.3}$$

The act of measurement changes the state of the system from $|\psi\rangle$ to $|\psi_n\rangle$. If the system is already in an eigenstate $|\psi_n\rangle$ of $\hat{A}$, a measurement of A yields with certainty the corresponding eigenvalue a_n: $\hat{A}|\psi_n\rangle = a_n|\psi_n\rangle$.

- **Continuous spectra**: The relation (3.2), which is valid for discrete spectra, can be extended to determine the probability density that a measurement of $\hat{A}$ yields a value between a and $a + da$ on a system which is initially in a state $|\psi\rangle$:

$$\frac{dP(a)}{da} = \frac{|\psi(a)|^2}{\langle\psi|\psi\rangle} = \frac{|\psi(a)|^2}{\int_{-\infty}^{+\infty} |\psi(a')|^2 da'}; \tag{3.4}$$

for instance, the probability density for finding a particle between x and $x + dx$ is given by $dP(x)/dx = |\psi(x)|^2/\langle\psi|\psi\rangle$.

---

[1] To see this, we need only to expand $|\psi(t)\rangle$ in terms of the eigenvectors of $\hat{A}$ which form a complete basis: $|\psi(t)\rangle = \sum_n |\psi_n\rangle\langle\psi_n|\psi(t)\rangle = \sum_n a_n|\psi_n\rangle$.

Postulate 5: Time evolution of a system

The time evolution of the state vector $|\psi(t)\rangle$ of a system is governed by the time-dependent *Schrödinger equation*

$$i\hbar \frac{\partial |\psi(t)\rangle}{\partial t} = \hat{H} |\psi(t)\rangle, \tag{3.5}$$

where $\hat{H}$ is the Hamiltonian operator corresponding to the total energy of the system.

Remark

These postulates fall into two categories:

- The first four describe the system at a given time.

- The fifth shows how this description evolves in time.

In the rest of this chapter we are going to consider the physical implications of each one of the four postulates. Namely, we shall look at the state of a quantum system and its interpretation, the physical observables, measurements in quantum mechanics, and finally the time evolution of quantum systems.

3.3 The State of a System

To describe a system in quantum mechanics, we use a mathematical entity (a complex function) belonging to a Hilbert space, the state vector $|\psi(t)\rangle$, which contains all the information we need to know about the system and from which all needed physical quantities can be computed. As discussed in Chapter 2, the state vector $|\psi(t)\rangle$ may be represented in two ways:

- A wave function $\psi(\vec{r}, t)$ in the position space: $\psi(\vec{r}, t) = \langle \vec{r} | \psi(t) \rangle$.

- A momentum wave function $\Psi(\vec{p}, t)$ in the momentum space: $\Psi(\vec{p}, t) = \langle \vec{p} | \psi(t) \rangle$.

So, for instance, to describe the state of a one-dimensional particle in quantum mechanics we use a complex function $\psi(x, t)$ instead of two real real numbers (x, p) in classical physics.

The wave functions to be used are only those that correspond to physical systems. What are the mathematical requirements that a wave function must satisfy to represent a physical system? Wave functions $\psi(x)$ that are physically acceptable must, along with their first derivatives $d\psi(x)/dx$, be *finite*, *continuous*, and *single-valued everywhere*. As will be discussed in Chapter 4, we will examine the underlying physics behind the continuity conditions of $\psi(x)$ and $d\psi(x)/dx$ (we will see that $\psi(x)$ and $d\psi(x)/dx$ must be be continuous because the probability density and the linear momentum are continuous functions of x).

3.3.1 Probability Density

What about the physical meaning of a wave function? Only the square of its norm, $|\psi(\vec{r}, t)|^2$, has meaning. According to Born's probabilistic interpretation, the square of the norm of $\psi(\vec{r}, t)$,

$$P(\vec{r}, t) = |\psi(\vec{r}, t)|^2, \tag{3.6}$$

represents a position probability density; that is, the quantity $|\psi(\vec{r}, t)|^2 d^3r$ represents the probability of finding the particle at time t in a volume element d^3r located between $\vec{r}$ and $\vec{r} + d\vec{r}$. Therefore, the total probability of finding the system somewhere in space is equal to 1:

$$\int |\psi(\vec{r}, t)|^2 d^3r = \int_{-\infty}^{+\infty} dx \int_{-\infty}^{+\infty} dy \int_{-\infty}^{+\infty} |\psi(\vec{r}, t)|^2 dz = 1. \qquad (3.7)$$

A wave function $\psi(\vec{r}, t)$ satisfying this relation is said to be *normalized*. We may mention that $\psi(\vec{r})$ has the physical dimensions of $1/\sqrt{L^3}$, where L is a length. Hence, the physical dimensions of $|\psi(\vec{r})|^2$ is $1/L^3$: $[|\psi(\vec{r})|^2] = 1/L^3$.

Note that the wave functions $\psi(\vec{r}, t)$ and $e^{i\alpha} \psi(\vec{r}, t)$, where α is a real number, represent the same state.

Example 3.1 (Physical and unphysical wave functions)
Which among the following functions represent physically acceptable wave functions: $f(x) = 3 \sin \pi x$, $g(x) = 4 - |x|$, $h^2(x) = 5x$, and $e(x) = x^2$.

Solution
Among these functions only $f(x) = 3 \sin \pi x$ represents a physically acceptable wave function, since $f(x)$ and its derivative are finite, continuous, single-valued everywhere, and integrable.

The other functions cannot be wave functions, since $g(x) = 4 - |x|$ is not continuous, not finite, and not square integrable; $h^2(x) = 5x$ is neither finite nor square integrable; and $e(x) = x^2$ is neither finite nor square integrable.

3.3.2   The Superposition Principle

The state of a system does not have to be represented by a *single* wave function; it can be represented by a *superposition* of two or more wave functions. An example from the macroscopic world is a vibrating string; its state can be represented by a single wave or by the superposition (linear combination) of many waves.

If $\psi_1(\vec{r}, t)$ and $\psi_2(\vec{r}, t)$ separately satisfy the Schrödinger equation, then the wave function $\psi(\vec{r}, t) = \alpha_1 \psi_1(\vec{r}, t) + \alpha_2 \psi_2(\vec{r}, t)$ also satisfies the Schrödinger equation, where α_1 and α_2 are complex numbers. The Schrödinger equation is a linear equation. So in general, according to the superposition principle, the linear superposition of many wave functions (which describe the various permissible physical states of a system) gives a new wave function which represents a possible physical state of the system:

$$|\psi\rangle = \sum_i \alpha_i |\psi_i\rangle, \qquad (3.8)$$

where the α_i are complex numbers. The quantity

$$P = \left| \sum_i \alpha_i |\psi_i\rangle \right|^2, \qquad (3.9)$$

represents the probability for this superposition. If the states $|\psi_i\rangle$ are mutually *orthonormal*, the probability will be equal to the sum of the individual probabilities:

$$P = \left| \sum_i a_i |\psi_i\rangle \right|^2 = \sum_i |a_i|^2 = P_1 + P_2 + P_3 + \cdots, \tag{3.10}$$

where $P_i = |a_i|^2$; P_i is the probability of finding the system in the state $|\psi_i\rangle$.

Example 3.2

Consider a system whose state is given in terms of an orthonormal set of three vectors: $|\phi_1\rangle$, $|\phi_2\rangle$, $|\phi_3\rangle$ as

$$|\psi\rangle = \frac{\sqrt{3}}{3}|\phi_1\rangle + \frac{2}{3}|\phi_2\rangle + \frac{\sqrt{2}}{3}|\phi_3\rangle.$$

(a) Verify that $|\psi\rangle$ is normalized. Then, calculate the probability of finding the system in any one of the states $|\phi_1\rangle$, $|\phi_2\rangle$, and $|\phi_3\rangle$. Verify that the total probability is equal to one.

(b) Consider now an ensemble of 810 identical systems, each one of them in the state $|\psi\rangle$. If measurements are done on all of them, how many systems will be found in each of the states $|\phi_1\rangle$, $|\phi_2\rangle$, and $|\phi_3\rangle$?

Solution

(a) Using the orthonormality condition $\langle \phi_j | \phi_k \rangle = \delta_{jk}$ where $j, k = 1, 2, 3$, we can verify that $|\psi\rangle$ is normalized:

$$\langle \psi | \psi \rangle = \frac{1}{3}\langle \phi_1 | \phi_1 \rangle + \frac{4}{9}\langle \phi_2 | \phi_2 \rangle + \frac{2}{9}\langle \phi_3 | \phi_3 \rangle = \frac{1}{3} + \frac{4}{9} + \frac{2}{9} = 1. \tag{3.11}$$

Since $|\psi\rangle$ is normalized, the probability of finding the system in $|\phi_1\rangle$ is given by

$$P_1 = |\langle \phi_1 | \psi \rangle|^2 = \left| \frac{\sqrt{3}}{3}\langle \phi_1 | \phi_1 \rangle + \frac{2}{3}\langle \phi_1 | \phi_2 \rangle + \frac{\sqrt{2}}{3}\langle \phi_1 | \phi_3 \rangle \right|^2 = \frac{1}{3}, \tag{3.12}$$

since $\langle \phi_1 | \phi_1 \rangle = 1$ and $\langle \phi_1 | \phi_2 \rangle = \langle \phi_1 | \phi_3 \rangle = 0$.

Similarly, from the relations $\langle \phi_2 | \phi_2 \rangle = 1$ and $\langle \phi_2 | \phi_1 \rangle = \langle \phi_2 | \phi_3 \rangle = 0$, we obtain the probability of finding the system in $|\phi_2\rangle$:

$$P_2 = |\langle \phi_2 | \psi \rangle|^2 = \left| \frac{2}{3}\langle \phi_2 | \phi_2 \rangle \right|^2 = \frac{4}{9}. \tag{3.13}$$

As for $\langle \phi_3 | \phi_3 \rangle = 1$ and $\langle \phi_3 | \phi_1 \rangle = \langle \phi_3 | \phi_2 \rangle = 0$, they lead to the probability of finding the system in $|\phi_3\rangle$:

$$P_3 = |\langle \phi_3 | \psi \rangle|^2 = \left| \frac{\sqrt{2}}{3}\langle \phi_3 | \phi_3 \rangle \right|^2 = \frac{2}{9}. \tag{3.14}$$

As expected, the total probability is equal to one:

$$P = P_1 + P_2 + P_3 = \frac{1}{3} + \frac{4}{9} + \frac{2}{9} = 1. \tag{3.15}$$

(b) The number of systems that will be found in the state $|\phi_1\rangle$ is

$$N_1 = 810 \times P_1 = \frac{810}{3} = 270.$$ (3.16)

Likewise, the number of systems that will be found in states $|\phi_2\rangle$ and $|\phi_3\rangle$ are given, respectively, by

$$N_2 = 810 \times P_2 = \frac{810 \times 4}{9} = 360, \qquad N_3 = 810 \times P_3 = \frac{810 \times 2}{9} = 180.$$ (3.17)

3.4 Observables and Operators

An observable is a dynamical variable that can be measured; the dynamical variables encountered most in classical mechanics are the position, linear momentum, angular momentum, and energy. How do we mathematically represent these and other variables in quantum mechanics?

According to the second postulate, a *Hermitian operator* is associated with every *physical observable*. In the preceding chapter, we have seen that the position representation of the linear momentum operator is given in one-dimensional space by $\hat{P} = -i\hbar\partial/\partial x$ and in three-dimensional space by $\hat{\vec{P}} = -i\hbar\vec{\nabla}$.

In general, any function, $f(\vec{r}, \vec{p})$, which depends on the position and momentum variables, $\vec{r}$ and $\vec{p}$, can be "*quantized*" or made into a function of operators by replacing $\vec{r}$ and $\vec{p}$ with their corresponding operators:

$$f(\vec{r}, \vec{p}) \quad \longrightarrow \quad F(\hat{\vec{R}}, \hat{\vec{P}}) = f(\hat{\vec{R}}, -i\hbar\vec{\nabla}),$$ (3.18)

or $f(x, p) \to F(\hat{X}, -i\hbar\partial/\partial x)$. For instance, the operator corresponding to the Hamiltonian

$$H = \frac{1}{2m}\vec{p}^{\,2} + V(\vec{r}, t)$$ (3.19)

is given in the position representation by

$$\hat{H} = -\frac{\hbar^2}{2m}\nabla^2 + V(\hat{\vec{R}}, t),$$ (3.20)

where ∇^2 is the Laplacian operator; it is given in Cartesian coordinates by: $\nabla^2 = \partial^2/\partial x^2 + \partial^2/\partial y^2 + \partial^2/\partial z^2$.

Since the momentum operator $\hat{\vec{P}}$ is Hermitian, and if the potential $V(\hat{\vec{R}}, t)$ is a real function, the Hamiltonian (3.19) is Hermitian. We saw in Chapter 2 that the eigenvalues of Hermitian operators are real. Hence, the spectrum of the Hamiltonian, which consists of the entire set of its eigenvalues, is real. This spectrum can be discrete, continuous, or a mixture of both. In the case of *bound* states, the Hamiltonian has a *discrete* spectrum of values and a *continuous* spectrum for *unbound* states. In general, an operator will have bound or unbound spectra in the same manner that the corresponding classical variable has bound or unbound orbits. As for $\hat{\vec{R}}$ and $\hat{\vec{P}}$, they have continuous spectra, since r and p may take a continuum of values.

Table 3.1 Some observables and their corresponding operators.

Observable	Corresponding operator
$\vec{r}$	$\hat{\vec{R}}$
$\vec{p}$	$\hat{\vec{P}} = -i\hbar\vec{\nabla}$
$T = \frac{p^2}{2m}$	$\hat{T} = -\frac{\hbar^2}{2m}\nabla^2$
$E = \frac{p^2}{2m} + V(\vec{r}, t)$	$\hat{H} = -\frac{\hbar^2}{2m}\nabla^2 + \hat{V}(\hat{\vec{R}}, t)$
$\vec{L} = \vec{r} \times \vec{p}$	$\hat{\vec{L}} = -i\hbar\hat{\vec{R}} \times \vec{\nabla}$

According to Postulate 5, the total energy E for time-dependent systems is associated to the operator

$$\hat{H} = i\hbar\frac{\partial}{\partial t}. \tag{3.21}$$

This can be seen as follows. The wave function of a free particle of momentum $\vec{p}$ and total energy E is given by $\psi(\vec{r}, t) = Ae^{i(\vec{p}\cdot\vec{r}-Et)/\hbar}$, where A is a constant. The time derivative of $\psi(\vec{r}, t)$ yields

$$i\hbar\frac{\partial\psi(\vec{r}, t)}{\partial t} = E\psi(\vec{r}, t). \tag{3.22}$$

Let us look at the eigenfunctions and eigenvalues of the momentum operator $\hat{\vec{P}}$. The eigenvalue equation

$$-i\hbar\vec{\nabla}\psi(\vec{r}) = \vec{p}\psi(\vec{r}) \tag{3.23}$$

yields the eigenfunction $\psi(\vec{r})$ corresponding to the eigenvalue $\vec{p}$ such that $|\psi(\vec{r})|^2d^3r$ is the probability of finding the particle with a momentum $\vec{p}$ in the volume element d^3r centered about $\vec{r}$. The solution to the eigenvalue equation (3.23) is

$$\psi(\vec{r}) = Ae^{i\vec{p}\cdot\vec{r}/\hbar}, \tag{3.24}$$

where A is a normalization constant. Since $\vec{p} = \hbar\vec{k}$ is the eigenvalue of the operator $\hat{\vec{P}}$, the eigenfunction (3.24) reduces to $\psi(\vec{r}) = Ae^{i\vec{k}\cdot\vec{r}}$; hence the eigenvalue equation (3.23) becomes

$$\hat{\vec{P}}\psi(\vec{r}) = \hbar\vec{k}\psi(\vec{r}). \tag{3.25}$$

To summarize, there is a one-to-one correspondence between observables and operators (Table 3.1).

Example 3.3 (Orbital angular momentum)
Find the operator representing the classical orbital angular momentum.

Solution
The classical expression for the orbital angular momentum of a particle whose position and linear momentum are $\vec{r}$ and $\vec{p}$ is given by $\vec{L} = \vec{r} \times \vec{p} = l_x\vec{i} + l_y\vec{j} + l_z\vec{k}$, where $l_x = yp_z - zp_y$, $l_y = zp_x - xp_z$, $l_z = xp_y - yp_x$.

To find the operator representing the classical angular momentum, we need simply to replace $\vec{r}$ and $\vec{p}$ with their corresponding operators $\hat{\vec{R}}$ and $\hat{\vec{P}} = -i\hbar\vec{\nabla}$: $\hat{\vec{L}} = -i\hbar\hat{\vec{R}} \times \vec{\nabla}$. This leads to

$$\hat{L}_x = \hat{Y}\hat{P}_z - \hat{Z}\hat{P}_y = -i\hbar\left(\hat{Y}\frac{\partial}{\partial z} - \hat{Z}\frac{\partial}{\partial y}\right), \tag{3.26}$$

$$\hat{L}_y = \hat{Z}\hat{P}_x - \hat{X}\hat{P}_z = -i\hbar\left(\hat{Z}\frac{\partial}{\partial x} - \hat{X}\frac{\partial}{\partial Z}\right), \tag{3.27}$$

$$\hat{L}_z = \hat{X}\hat{P}_y - \hat{Y}\hat{P}_x = -i\hbar\left(\hat{X}\frac{\partial}{\partial y} - \hat{Y}\frac{\partial}{\partial x}\right). \tag{3.28}$$

Recall that in classical mechanics the position and momentum components commute, $xp_x = p_x x$, and so do the components of the angular momentum, $l_x l_y = l_y l_x$. In quantum mechanics, however, this is not the case, since $\hat{X}\hat{P}_x = \hat{P}_x\hat{X} + i\hbar$ and, as will be shown in Chapter 5, $\hat{L}_x\hat{L}_y = \hat{L}_y\hat{L}_x + i\hbar\hat{L}_z$, and so on.

3.5   Measurement in Quantum Mechanics

Quantum theory is about the results of measurement; it says nothing about what might happen in the physical world outside the context of measurement. So the emphasis is on measurement.

3.5.1   How Measurements Disturb Systems

In classical physics it is possible to perform measurements on a system without disturbing it significantly. In quantum mechanics, however, the measurement process perturbs the system significantly. While carrying out measurements on classical systems, this perturbation does exist, but it is small enough that it can be neglected. In atomic and subatomic systems, however, the act of measurement induces nonnegligible or significant disturbances.

As an illustration, consider an experiment that measures the position of a hydrogenic electron. For this, we need to bombard the electron with electromagnetic radiation (photons). If we want to determine the position accurately, the wavelength of the radiation must be sufficiently short. Since the electronic orbit is of the order of 10^{-10} m, we must use a radiation whose wavelength is smaller than 10^{-10} m. That is, we need to bombard the electron with photons of energies higher than

$$h\nu = h\frac{c}{\lambda} = h\frac{3 \times 10^8}{10^{-10}} \sim 10^4 \text{ eV}. \tag{3.29}$$

When such photons strike the electron, not only will they perturb it, they will knock it completely off its orbit; recall that the ionization energy of the hydrogen atom is about 13.5 eV. Thus, the mere act of measuring the position of the electron disturbs it appreciably.

Let us now discuss the general concept of measurement in quantum mechanics. *The act of measurement generally changes the state of the system.* In theory we can represent the measuring device by an operator so that, after carrying out the measurement, the system will be in one of the eigenstates of the operator. Consider a system which is in a state $|\psi\rangle$. Before measuring an observable A, the state $|\psi\rangle$ can be represented by a linear superposition of eigenstates $|\psi_n\rangle$

of the corresponding operator $\hat{A}$:

$$|\psi\rangle = \sum_n |\psi_n\rangle\langle\psi_n|\psi\rangle = \sum_n a_n |\psi_n\rangle. \tag{3.30}$$

According to Postulate 4, the act of measuring A changes the state of the system from $|\psi\rangle$ to one of the eigenstates $|\psi_n\rangle$ of the operator $\hat{A}$, and the result obtained is the eigenvalue a_n. The only *exception* to this rule is when the *system is already in one of the eigenstates of the observable being measured*. For instance, if the system is in the eigenstate $|\psi_n\rangle$, a measurement of the observable A yields with certainty (i.e., with probability = 1) the value a_n without changing the state $|\psi_n\rangle$.

Before a measurement, we do not know in advance with certainty in which eigenstate, among the various states $|\psi_n\rangle$, a system will be after the measurement; only a probabilistic outcome is possible. Postulate 4 states that the probability of finding the system in one particular nondegenerate eigenstate $|\psi_n\rangle$ is given by

$$P_n = \frac{|\langle\psi_n|\psi\rangle|^2}{\langle\psi|\psi\rangle}. \tag{3.31}$$

Note that the wave function does not predict the results of individual measurements; it instead determines the probability distribution, $P = |\psi|^2$, over measurements on many identical systems in the same state.

Finally, we may state that quantum mechanics is the mechanics applicable to objects for which measurements necessarily interfere with the state of the system. Quantum mechanically, we cannot ignore the effects of the measuring equipment on the system, for they are important. In general, certain measurements cannot be performed without major disturbances to other properties of the quantum system. In conclusion, *it is the effects of the interference by the equipment on the system which is the essence of quantum mechanics.*

3.5.2 Expectation Values

The expectation value $\langle\hat{A}\rangle$ of $\hat{A}$ with respect to a state $|\psi\rangle$ is defined by

$$\langle\hat{A}\rangle = \frac{\langle\psi|\hat{A}|\psi\rangle}{\langle\psi|\psi\rangle}. \tag{3.32}$$

For instance, the energy of a system is given by the expectation value of the Hamiltonian: $E = \langle\hat{H}\rangle = \langle\psi|\hat{H}|\psi\rangle/\langle\psi|\psi\rangle$.

In essence, the expectation value $\langle\hat{A}\rangle$ represents the average result of measuring $\hat{A}$ on the state $|\psi\rangle$. To see this, using the complete set of eigenvectors $|\psi_n\rangle$ of $\hat{A}$ as a basis (i.e., $\hat{A}$ is diagonal in ψ_n), we can rewrite $\langle\hat{A}\rangle$ as follows:

$$\langle\hat{A}\rangle = \frac{1}{\langle\psi|\psi\rangle}\sum_{nm}\langle\psi|\psi_m\rangle\langle\psi_m|\hat{A}|\psi_n\rangle\langle\psi_n|\psi\rangle = \sum_n a_n\frac{|\langle\psi_n|\psi\rangle|^2}{\langle\psi|\psi\rangle}, \tag{3.33}$$

where we have used $\langle\psi_m|\hat{A}|\psi_n\rangle = a_n\delta_{nm}$. Since the quantity $|\langle\psi_n|\psi\rangle|^2/\langle\psi|\psi\rangle$ gives the probability P_n of finding the value a_n after measuring the observable A, we can indeed interpret $\langle\hat{A}\rangle$ as an *average* of a series of measurements of A:

$$\langle\hat{A}\rangle = \sum_n a_n\frac{|\langle\psi_n|\psi\rangle|^2}{\langle\psi|\psi\rangle} = \sum_n a_n P_n. \tag{3.34}$$

That is, the expectation value of an observable is obtained by adding all permissible eigenvalues a_n, with each a_n multiplied by the corresponding probability P_n.

The relation (3.34), which is valid for *discrete* spectra, can be extended to a *continuous* distribution of probabilities $P(a)$ as follows:

$$\langle \hat{A} \rangle = \frac{\int_{-\infty}^{+\infty} a \, |\psi(a)|^2 da}{\int_{-\infty}^{+\infty} |\psi(a)|^2 da} = \int_{-\infty}^{+\infty} a \, dP(a). \tag{3.35}$$

The expectation value of an observable can be obtained physically as follows: prepare a very large number of *identical* systems each in the *same* state $|\psi\rangle$. The observable A is then measured on all these identical systems; the results of these measurements are $a_1, a_2, \ldots, a_n, \ldots$; the corresponding probabilities of occurrence are $P_1, P_2, \ldots, P_n, \ldots$. The average value of all these repeated measurements is called the expectation value of $\hat{A}$ with respect to the state $|\psi\rangle$.

Note that the process of obtaining different results when measuring the same observable on many identically prepared systems is contrary to classical physics, where these measurements must give the same outcome. In quantum mechanics, however, we can predict only the probability of obtaining a certain value for an observable.

Example 3.4
Consider a system whose state is given in terms of a complete and orthonormal set of five vectors $|\phi_1\rangle, |\phi_2\rangle, |\phi_3\rangle, |\phi_4\rangle, |\phi_5\rangle$ as follows:

$$|\psi\rangle = \frac{1}{\sqrt{19}}|\phi_1\rangle + \frac{2}{\sqrt{19}}|\phi_2\rangle + \sqrt{\frac{2}{19}}|\phi_3\rangle + \sqrt{\frac{3}{19}}|\phi_4\rangle + \sqrt{\frac{5}{19}}|\phi_5\rangle,$$

where $|\phi_n\rangle$ are eigenstates to the system's Hamiltonian, $\hat{H}|\phi_n\rangle = n\varepsilon_0|\phi_n\rangle$ with $n = 1, 2, 3, 4, 5$, and where ε_0 has the dimensions of energy.

(a) If the energy is measured on a large number of identical systems that are all initially in the same state $|\psi\rangle$, what values would one obtain and with what probabilities?

(b) Find the average energy of one such system.

Solution
First, note that $|\psi\rangle$ is not normalized:

$$\langle \psi | \psi \rangle = \sum_{n=1}^{5} a_n^2 \langle \phi_n | \phi_n \rangle = \sum_{n=1}^{5} a_n^2 = \frac{1}{19} + \frac{4}{19} + \frac{2}{19} + \frac{3}{19} + \frac{5}{19} = \frac{15}{19}, \tag{3.36}$$

since $\langle \phi_j | \phi_k \rangle = \delta_{jk}$ with $j, k = 1, 2, 3, 4, 5$.

(a) Since $E_n = \langle \phi_n | \hat{H} | \phi_n \rangle = n\varepsilon_0$ ($n = 1, 2, 3, 4, 5$), the various measurements of the energy of the system yield the values $E_1 = \varepsilon_0$, $E_2 = 2\varepsilon_0$, $E_3 = 3\varepsilon_0$, $E_4 = 4\varepsilon_0$, $E_5 = 5\varepsilon_0$ with the following probabilities:

$$P_1(E_1) = \frac{|\langle \phi_1 | \psi \rangle|^2}{\langle \psi | \psi \rangle} = \left| \frac{1}{\sqrt{19}} \langle \phi_1 | \phi_1 \rangle \right|^2 \times \frac{19}{15} = \frac{1}{15}, \tag{3.37}$$

$$P_2(E_2) = \frac{|\langle \phi_2 | \psi \rangle|^2}{\langle \psi | \psi \rangle} = \left| \frac{2}{\sqrt{19}} \langle \phi_2 | \phi_2 \rangle \right|^2 \times \frac{19}{15} = \frac{4}{15}, \tag{3.38}$$

$$P_3(E_3) = \frac{|\langle\phi_3|\psi\rangle|^2}{\langle\psi|\psi\rangle} = \left|\sqrt{\frac{2}{19}}\langle\phi_3|\phi_3\rangle\right|^2 \times \frac{19}{15} = \frac{2}{15}, \tag{3.39}$$

$$P_4(E_4) = \frac{|\langle\phi_4|\psi\rangle|^2}{\langle\psi|\psi\rangle} = \left|\sqrt{\frac{3}{19}}\langle\phi_4|\phi_4\rangle\right|^2 \times \frac{19}{15} = \frac{3}{15}, \tag{3.40}$$

and

$$P_5(E_5) = \frac{|\langle\phi_5|\psi\rangle|^2}{\langle\psi|\psi\rangle} = \left|\sqrt{\frac{5}{19}}\langle\phi_5|\phi_5\rangle\right|^2 \times \frac{19}{15} = \frac{5}{15}, \tag{3.41}$$

(b) The average energy of a system is given by

$$E = \sum_{j=1}^{5} P_j E_j = \frac{1}{15}\varepsilon_0 + \frac{8}{15}\varepsilon_0 + \frac{6}{15}\varepsilon_0 + \frac{12}{15}\varepsilon_0 + \frac{25}{15}\varepsilon_0 = \frac{52}{15}\varepsilon_0. \tag{3.42}$$

This energy can also be obtained from the expectation value of the Hamiltonian:

$$\begin{aligned}
E &= \frac{\langle\psi|\hat{H}|\psi\rangle}{\langle\psi|\psi\rangle} = \frac{19}{15}\sum_{n=1}^{5} a_n^2\langle\phi_n|\hat{H}|\phi_n\rangle = \frac{19}{15}\left(\frac{1}{19} + \frac{8}{19} + \frac{6}{19} + \frac{12}{19} + \frac{25}{19}\right)\varepsilon_0 \\
&= \frac{52}{15}\varepsilon_0,
\end{aligned} \tag{3.43}$$

where the values of the coefficients a_n^2 are listed in (3.36).

3.5.3 Complete Sets of Commuting Operators (CSCO)

Two observables A and B are said to be *compatible* when their corresponding operators commute, $[\hat{A}, \hat{B}] = 0$; observables corresponding to noncommuting operators are said to be *noncompatible*.

In what follows we are going to consider the task of measuring two observables A and B on a given system. Since the act of measurement generally *perturbs* the system, the result of measuring A and B therefore depends on the *order* in which they are carried out. Measuring A first and then B leads[2] in general to results that are different from those obtained by measuring B first and then A. How does this take place?

If $\hat{A}$ and $\hat{B}$ do not commute and if the system is in an eigenstate $|\psi_n^{(a)}\rangle$ of $\hat{A}$, a measurement of A yields with certainty a value a_n, since $\hat{A}|\psi_n^{(a)}\rangle = a_n|\psi_n^{(a)}\rangle$. Then, when we measure B, the state of the system will be left in one of the eigenstates of B. If we measure A again, we will find a value which will be different from a_n. What is this new value? We cannot answer this question with certainty: only a probabilistic outcome is possible. For this, we need to expand the eigenstates of B in terms of those of A, and thus provide a probabilistic answer as to the value of measuring A. So if $\hat{A}$ and $\hat{B}$ do not commute, *they cannot be measured simultaneously; the order in which they are measured matters.*

---

[2]The act of measuring A first and then B is represented by the action of product $\hat{B}\hat{A}$ of their corresponding operators on the state vector.

What happens when A and B commute? We can show that the results of their measurements will not depend on the order in which they are carried out. Before showing this, let us mention a useful theorem.

Theorem 3.1 *If two observables are compatible, their corresponding operators possess a set of common (or simultaneous) eigenstates (this theorem holds for both degenerate and nondegenerate eigenstates).*

Proof
We provide here a proof for the nondegenerate case only. If $|\psi_n\rangle$ is a nondegenerate eigenstate of $\hat{A}$, $\hat{A}|\psi_n\rangle = a_n|\psi_n\rangle$, we have

$$\langle\psi_m|[\hat{A}, \hat{B}]|\psi_n\rangle = (a_m - a_n)\langle\psi_m|\hat{B}|\psi_n\rangle = 0, \tag{3.44}$$

since $\hat{A}$ and $\hat{B}$ commute. So $\langle\psi_m|\hat{B}|\psi_n\rangle$ must vanish unless $a_n = a_m$. That is,

$$\langle\psi_m|\hat{B}|\psi_n\rangle = \langle\psi_n|\hat{B}|\psi_n\rangle \propto \delta_{nm}. \tag{3.45}$$

Hence the $|\psi_n\rangle$ are *joint* or *simultaneous* eigenstates of $\hat{A}$ and $\hat{B}$ (this completes the proof).
Denoting the simultaneous eigenstate of $\hat{A}$ and $\hat{B}$ by $|\psi_{n_1}^{(a)}, \psi_{n_2}^{(b)}\rangle$, we have

$$\hat{A}|\psi_{n_1}^{(a)}, \psi_{n_2}^{(b)}\rangle = a_{n_1}|\psi_{n_1}^{(a)}, \psi_{n_2}^{(b)}\rangle, \tag{3.46}$$

$$\hat{B}|\psi_{n_1}^{(a)}, \psi_{n_2}^{(b)}\rangle = b_{n_2}|\psi_{n_1}^{(a)}, \psi_{n_2}^{(b)}\rangle. \tag{3.47}$$

Theorem 3.1 can be generalized to the case of many mutually compatible observables A, B, C, ... These compatible observables possess a *complete* set of *joint eigenstates*

$$|\psi_n\rangle = |\psi_{n_1}^{(a)}, \psi_{n_2}^{(b)}, \psi_{n_3}^{(c)}, \ldots\rangle. \tag{3.48}$$

The completeness and orthonormality conditions of this set are

$$\sum_{n_1}\sum_{n_2}\sum_{n_3}\cdots|\psi_{n_1}^{(a)}, \psi_{n_2}^{(b)}, \psi_{n_3}^{(c)}, \ldots\rangle\langle\psi_{n_1}^{(a)}, \psi_{n_2}^{(b)}, \psi_{n_3}^{(c)}, \ldots| = 1; \tag{3.49}$$

$$\langle\psi_{n'}|\psi_n\rangle = \delta_{n'n} = \delta_{n_1'n_1}\delta_{n_2'n_2}\delta_{n_3'n_3}\cdots. \tag{3.50}$$

Let us now show why, when two observables A and B are compatible, the order in which we carry out their measurements is irrelevant. Measuring A first, we would find a value a_n and would leave the system in an eigenstate of A. According to Theorem 3.1, this eigenstate is also an eigenstate of B. Thus a measurement of B yields with certainty b_n without affecting the state of the system. In this way, if we measure A again, we obtain with certainty the same initial value a_n. Similarly, another measurement of B will yield b_n and will leave the system in the same joint eigenstate of A and B. Thus, if two observables A and B are compatible, and if the system is initially in an eigenstate of one of their operators, their measurements not only yield precise values (eigenvalues) but they will not depend on the order in which the measurements were performed. In this case, A and B are said to be *simultaneously measurable*. So compatible observables can be measured simultaneously with arbitrary accuracy; noncompatible observables cannot.

What happens if an operator, say $\hat{A}$, has *degenerate* eigenvalues? The specification of one eigenvalue does not uniquely determine the state of the system. Among the degenerate eigenstates of $\hat{A}$, only a subset of them are also eigenstates of $\hat{B}$. Thus, the set of states that are joint eigenstates of both $\hat{A}$ and $\hat{B}$ is not complete. To resolve the degeneracy, we can introduce a third operator $\hat{C}$ which commutes with both $\hat{A}$ and $\hat{B}$; then we can construct a set of joint eigenstates of $\hat{A}$, $\hat{B}$, and $\hat{C}$ that is complete. If the degeneracy persists, we may introduce a fourth operator $\hat{D}$ that commutes with the previous three and then look for their joint eigenstates which form a complete set. Continuing in this way, we will ultimately exhaust all the operators (that is, there are no more independent operators) which commute with each other. When that happens, we have then obtained a *complete set of commuting operators* (CSCO). Only then will the state of the system be specified unambiguously, for the joint eigenstates of the CSCO are determined uniquely and will form a complete set (recall that a complete set of eigenvectors of an operator is called a basis). We should, at this level, state the following definition.

Definition: A set of Hermitian operators, $\hat{A}$, $\hat{B}$, $\hat{C}$, ..., is called a CSCO if the operators mutually commute and if the set of their common eigenstates is complete and not degenerate (i.e., unique).

The complete commuting set may sometimes consist of only one operator. *Any operator with nondegenerate eigenvalues constitutes, all by itself, a CSCO.* For instance, the position operator $\hat{X}$ of a one-dimensional, spinless particle provides a complete set. Its momentum operator $\hat{P}$ is also a complete set; together, however, $\hat{X}$ and $\hat{P}$ cannot form a CSCO, for they do not commute. In three-dimensional problems, the three-coordinate position operators $\hat{X}$, $\hat{Y}$, and $\hat{Z}$ form a CSCO; similarly, the components of the momentum operator $\hat{P}_x$, $\hat{P}_y$, and $\hat{P}_z$ also form a CSCO. In the case of spherically symmetric three-dimensional potentials, the set $\hat{H}$, $\hat{\vec{L}}^2$, $\hat{L}_z$ forms a CSCO. Note that in this case of spherical symmetry, we need three operators to form a CSCO because $\hat{H}$, $\hat{\vec{L}}^2$, and $\hat{L}_z$ are all degenerate; hence the complete and unique determination of the wave function cannot be achieved with one operator or with two.

In summary, when a given operator, say $\hat{A}$, is degenerate, the wave function cannot be determined uniquely unless we introduce one or more additional operators so as to form a complete commuting set.

3.5.4 Measurement and the Uncertainty Relations

We have seen in Chapter 2 that the uncertainty condition pertaining to the measurement of any two observables A and B is given by

$$\Delta A \Delta B \geq \frac{1}{2}|\langle[\hat{A}, \hat{B}]\rangle|, \tag{3.51}$$

where $\Delta A = \sqrt{\langle \hat{A}^2 \rangle - \langle \hat{A} \rangle^2}$.

Let us illustrate this on the joint measurement of the position and momentum observables. Since these observables are not compatible, their simultaneous measurement with infinite accuracy is not possible; that is, since $[\hat{X}, \hat{P}] = i\hbar$ there exists no state which is a simultaneous eigenstate of $\hat{X}$ and $\hat{P}$. For the case of the position and momentum operators, the relation (3.51) yields

$$\Delta x \Delta p \geq \frac{\hbar}{2}. \tag{3.52}$$

This condition shows that the position and momentum of a microscopic system cannot be measured with infinite accuracy both at once. If the position is measured with an uncertainty Δx, the uncertainty associated with its momentum measurement cannot be smaller than $\hbar/2\Delta x$. This is due to *the interference between the two measurements*. If we measure the position first, we perturb the system by changing its state to an eigenstate of the position operator; then the measurement of the momentum throws the system into an eigenstate of the momentum operator.

Another interesting application of the uncertainty relation (3.51) is to the orbital angular momentum of a particle. Since its components satisfy the commutator $[\hat{L}_x, \hat{L}_y] = i\hbar \hat{L}_z$, we obtain

$$\Delta L_x \Delta L_y \geq \frac{1}{2}\hbar |\langle \hat{L}_z \rangle|. \tag{3.53}$$

We can obtain the other two inequalities by means of a cyclic permutation of x, y, and z. If $\langle \hat{L}_z \rangle = 0$, $\hat{L}_x$ and $\hat{L}_y$ will have sharp values simultaneously. This occurs when the particle is in an s state. In fact, when a particle is in an s state, we have $\langle \hat{L}_x \rangle = \langle \hat{L}_y \rangle = \langle \hat{L}_z \rangle = 0$; hence all the components of orbital angular momentum will have sharp values simultaneously.

3.6 Time Evolution of the System's State

3.6.1 Time Evolution Operator

We want to examine here how quantum states evolve in time. That is, given the initial state $|\psi(t_0)\rangle$, how does one find the state $|\psi(t)\rangle$ at any later time t? The two states can be related by means of a linear operator $\hat{U}(t, t_0)$ such that

$$|\psi(t)\rangle = \hat{U}(t, t_0)|\psi(t_0)\rangle \qquad (t > t_0); \tag{3.54}$$

$\hat{U}(t, t_0)$ is known as the *time evolution operator* or *propagator*. From (3.54), we infer that

$$\hat{U}(t_0, t_0) = \hat{I}, \tag{3.55}$$

where $\hat{I}$ is the unit (identity) operator.

The issue now is to find $\hat{U}(t, t_0)$. For this, we need simply to substitute (3.54) into the time-dependent Schrödinger equation (3.5):

$$i\hbar \frac{\partial}{\partial t} \left(\hat{U}(t, t_0)|\psi(t_0)\rangle \right) = \hat{H} \left(\hat{U}(t, t_0)|\psi(t_0)\rangle \right) \tag{3.56}$$

or

$$\frac{\partial \hat{U}(t, t_0)}{\partial t} = -\frac{i}{\hbar} \hat{H}\hat{U}(t, t_0). \tag{3.57}$$

The integration of this differential equation depends on whether or not the Hamiltonian depends on time. If it does not depend on time, and taking into account the initial condition (3.55), we can easily ascertain that the integration of (3.57) leads to

$$\boxed{\hat{U}(t, t_0) = e^{-i(t-t_0)\hat{H}/\hbar} \qquad \text{and} \qquad |\psi(t)\rangle = e^{-i(t-t_0)\hat{H}/\hbar}|\psi(t_0)\rangle.} \tag{3.58}$$

We will show in Section 3.7 that the operator $\hat{U}(t, t_0) = e^{-i(t-t_0)\hat{H}/\hbar}$ represents a finite time translation.

If, on the other hand, $\hat{H}$ depends on time the integration of (3.57) becomes less trivial. We will deal with this issue in Chapter 10 when we look at time-dependent potentials or at the time-dependent perturbation theory. In this chapter, and in all chapters up to Chapter 10, we will consider only Hamiltonians that do not depend on time.

Note that $\hat{U}(t, t_0)$ is a unitary operator, since

$$\hat{U}(t, t_0)\hat{U}^\dagger(t, t_0) = \hat{U}(t, t_0)\hat{U}^{-1}(t, t_0) = e^{-i(t-t_0)\hat{H}/\hbar} e^{i(t-t_0)\hat{H}/\hbar} = \hat{I} \qquad (3.59)$$

or $\hat{U}^\dagger = \hat{U}^{-1}$.

3.6.2 Stationary States: Time-Independent Potentials

In the position representation, the time-dependent Schrödinger equation (3.5) for a particle of mass m moving in a time-dependent potential $\hat{V}(\vec{r}, t)$ can be written as follows:

$$i\hbar \frac{\partial \Psi(\vec{r}, t)}{\partial t} = -\frac{\hbar^2}{2m} \nabla^2 \Psi(\vec{r}, t) + \hat{V}(\vec{r}, t)\Psi(\vec{r}, t). \qquad (3.60)$$

Now, let us consider the particular case of *time-independent* potentials: $\hat{V}(\vec{r}, t) = \hat{V}(\vec{r})$. In this case the Hamiltonian operator will also be time independent, and hence the Schrödinger equation will have solutions that are *separable*, i.e., solutions that consist of a product of two functions, one depending only on $\vec{r}$ and the other only on time:

$$\Psi(\vec{r}, t) = \psi(\vec{r}) f(t). \qquad (3.61)$$

Substituting (3.61) into (3.60) and dividing both sides by $\psi(\vec{r}) f(t)$, we obtain

$$i\hbar \frac{1}{f(t)} \frac{df(t)}{dt} = \frac{1}{\psi(\vec{r})} \left[-\frac{\hbar^2}{2m} \nabla^2 \psi(\vec{r}) + \hat{V}(\vec{r})\psi(\vec{r}) \right]. \qquad (3.62)$$

Since the left-hand side depends only on time and the right-hand side depends only on $\vec{r}$, both sides must be equal to a constant; this constant, which we denote by E, has the dimensions of energy. We can therefore break (3.62) into two separate differential equations, one depending on time only,

$$i\hbar \frac{df(t)}{dt} = Ef(t), \qquad (3.63)$$

and the other on the space variable $\vec{r}$,

$$\left[-\frac{\hbar^2}{2m} \nabla^2 + \hat{V}(\vec{r}) \right] \psi(\vec{r}) = E\psi(\vec{r}). \qquad (3.64)$$

This equation is known as the *time-independent* Schrödinger equation for a particle of mass m moving in a time-independent potential $\hat{V}(\vec{r})$.

The solutions to (3.63) can be written as $f(t) = e^{-iEt/\hbar}$; hence the state (3.61) becomes

$$\Psi(\vec{r}, t) = \psi(\vec{r}) e^{-iEt/\hbar}. \qquad (3.65)$$

This particular solution of the Schrödinger equation (3.60) for a *time-independent potential* is called a *stationary state*. Why is this state called *stationary*? The reason is obvious: the probability density is stationary, i.e., it does not depend on time:

$$|\Psi(\vec{r}, t)|^2 = |\psi(\vec{r}) e^{-iEt/\hbar}|^2 = |\psi(\vec{r})|^2. \qquad (3.66)$$

Note that such a state has a precise value for the energy, $E = \hbar\omega$.

In summary, stationary states, which are given by the solutions of (3.64), exist only for time-independent potentials. The set of energy levels that are solutions to this equation are called the *energy spectrum* of the system. The states corresponding to discrete and continuous spectra are called *bound* and *unbound* states, respectively. We will consider these questions in detail in Chapter 4.

The most general solution to the time-dependent Schrödinger equation (3.60) can be written as an expansion in terms of the stationary states $\psi_n(\vec{r}) \exp(-i E_n t/\hbar)$:

$$\Psi(\vec{r}, t) = \sum_n c_n \psi_n(\vec{r}) \exp\left(-\frac{i E_n t}{\hbar}\right), \tag{3.67}$$

where $c_n = \langle \psi_n | \Psi(t = 0) \rangle = \int \psi_n^*(\vec{r}) \psi(\vec{r}) \, d^3 r$. The general solution (3.67) is not a stationary state, because a linear superposition of stationary states is not necessarily a stationary state.

Remark

The time-dependent and time-independent Schrödinger equations are given in one dimension by (see (3.60) and (3.64))

$$i\hbar \frac{\partial \Psi(x, t)}{\partial t} = -\frac{\hbar^2}{2m} \frac{\partial^2 \Psi(x, t)}{\partial x^2} + \hat{V}(x, t)\Psi(x, t), \tag{3.68}$$

$$-\frac{\hbar^2}{2m} \frac{d^2 \psi(x)}{dx^2} + \hat{V}(x)\psi(x) = E\psi(x). \tag{3.69}$$

3.6.3 Schrödinger Equation and Wave Packets

Can we derive the Schrödinger equation (3.5) formally from first principles? No, we cannot; we can only postulate it. What we can do, however, is to provide an educated guess on the formal steps leading to it. *Wave packets* offer the formal tool to achieve that. We are going to show how to start from a wave packet and end up with the Schrödinger equation.

As seen in Chapter 1, the wave packet representing a particle of energy E and momentum p moving in a potential V is given by

$$\begin{aligned}
\Psi(x, t) &= \frac{1}{\sqrt{2\pi\hbar}} \int_{-\infty}^{+\infty} \tilde{\phi}(p) \exp\left[\frac{i}{\hbar}(px - Et)\right] dp \\
&= \frac{1}{\sqrt{2\pi\hbar}} \int_{-\infty}^{+\infty} \tilde{\phi}(p) \exp\left[\frac{i}{\hbar}\left(px - \left(\frac{p^2}{2m} + V\right)t\right)\right] dp;
\end{aligned} \tag{3.70}$$

recall that wave packets unify the corpuscular (E and p) and the wave (k and ω) features of particles: $k = p/\hbar$, $\hbar\omega = E = p^2/(2m) + V$. A partial time derivative of (3.70) yields

$$i\hbar \frac{\partial}{\partial t} \Psi(x, t) = \frac{1}{\sqrt{2\pi\hbar}} \int_{-\infty}^{+\infty} \tilde{\phi}(p) \left(\frac{p^2}{2m} + V\right) \exp\left[\frac{i}{\hbar}\left(px - \left(\frac{p^2}{2m} + V\right)t\right)\right] dp. \tag{3.71}$$

Since $p^2/(2m) = -(\hbar^2/2m)\partial^2/\partial x^2$ and assuming that V is constant, we can take the term $-(\hbar^2/2m)\partial^2/\partial x^2 + V$ outside the integral sign, for it does not depend on p:

$$i\hbar\frac{\partial}{\partial t}\Psi(x,t) = \left(-\frac{\hbar^2}{2m}\frac{\partial^2}{\partial x^2} + V\right)\frac{1}{\sqrt{2\pi\hbar}}\int_{-\infty}^{+\infty}\tilde{\phi}(p)\exp\left[\frac{i}{\hbar}\left(px - \left(\frac{p^2}{2m} + V\right)t\right)\right]dp.$$

(3.72)

This can be written as

$$i\hbar\frac{\partial}{\partial t}\Psi(x,t) = \left[-\frac{\hbar^2}{2m}\frac{\partial^2}{\partial x^2} + V\right]\Psi(x,t).$$

(3.73)

Now, since this equation is valid for spatially varying potentials $V = V(x)$, we see that we have ended up with the Schrödinger equation (3.68).

3.6.4 The Conservation of Probability

Since the Hamiltonian operator is Hermitian, we can show that the norm $\langle\Psi(t)|\Psi(t)\rangle$, which is given by

$$\langle\Psi(t)|\Psi(t)\rangle = \int |\Psi(\vec{r},t)|^2\, d^3r,$$

(3.74)

is time independent. This means, if $|\Psi(t)\rangle$ is normalized, it stays normalized for all subsequent times. This is a direct consequence of the hermiticity of $\hat{H}$.

To prove that $\langle\Psi(t)|\Psi(t)\rangle$ is constant, we need simply to show that its time derivative is zero. First, the time derivative of $\langle\Psi(t)|\Psi(t)\rangle$ is

$$\frac{d}{dt}\langle\Psi(t)|\Psi(t)\rangle = \left(\frac{d}{dt}\langle\Psi(t)|\right)|\Psi(t)\rangle + \langle\Psi(t)|\left(\frac{d|\Psi(t)\rangle}{dt}\right),$$

(3.75)

where $d|\Psi(t)\rangle/dt$ and $d\langle\Psi(t)|/dt$ can be obtained from (3.5):

$$\frac{d}{dt}|\Psi(t)\rangle = -\frac{i}{\hbar}\hat{H}|\Psi(t)\rangle,$$

(3.76)

$$\frac{d}{dt}\langle\Psi(t)| = \frac{i}{\hbar}\langle\Psi(t)|\hat{H}^\dagger = \frac{i}{\hbar}\langle\Psi(t)|\hat{H}.$$

(3.77)

Inserting these two equations into (3.75), we end up with

$$\frac{d}{dt}\langle\Psi(t)|\Psi(t)\rangle = \left(\frac{i}{\hbar} - \frac{i}{\hbar}\right)\langle\Psi(t)|\hat{H}|\Psi(t)\rangle = 0.$$

(3.78)

Thus, the probability density $\langle\Psi|\Psi\rangle$ does not evolve in time.

In what follows we are going to calculate the probability density in the position representation. For this, we need to invoke the time-dependent Schrödinger equation

$$i\hbar\frac{\partial\Psi(\vec{r},t)}{\partial t} = -\frac{\hbar^2}{2m}\nabla^2\Psi(\vec{r},t) + \hat{V}(\vec{r},t)\Psi(\vec{r},t)$$

(3.79)

and its complex conjugate

$$-i\hbar\frac{\partial\Psi^*(\vec{r},t)}{\partial t} = -\frac{\hbar^2}{2m}\nabla^2\Psi^*(\vec{r},t) + \hat{V}(\vec{r},t)\Psi^*(\vec{r},t).$$

(3.80)

Multiplying both sides of (3.79) by $\Psi^*(\vec{r}, t)$ and both sides of (3.80) by $\Psi(\vec{r}, t)$, and subtracting the two resulting equations, we obtain

$$i\hbar \frac{\partial}{\partial t} \left[\Psi^*(\vec{r}, t)\Psi(\vec{r}, t) \right] = -\frac{\hbar^2}{2m} \left[\Psi^*(\vec{r}, t)\nabla^2\Psi(\vec{r}, t) - \Psi\nabla^2\Psi^* \right]. \qquad (3.81)$$

We can rewrite this equation as

$$\boxed{\frac{\partial \rho(\vec{r}, t)}{\partial t} + \vec{\nabla} \cdot \vec{J} = 0,} \qquad (3.82)$$

where $\rho(\vec{r}, t)$ and $\vec{J}$ are given by

$$\boxed{\rho(\vec{r}, t) = \Psi^*(\vec{r}, t)\Psi(\vec{r}, t), \qquad \vec{J}(\vec{r}, t) = \frac{i\hbar}{2m} \left(\Psi\vec{\nabla}\Psi^* - \Psi^*\vec{\nabla}\Psi \right);} \qquad (3.83)$$

$\rho(\vec{r}, t)$ is called the *probability density*, while $\vec{J}(\vec{r}, t)$ is the *probability current density*, or simply the *current density*, or even the particle density flux. By analogy with charge conservation in electrodynamics, equation (3.82) is interpreted as the *conservation of probability*.

Let us find the relationship between the density operators $\hat{\rho}(t)$ and $\hat{\rho}(t_0)$. Since $|\Psi(t)\rangle = \hat{U}(t, t_0)|\Psi(t_0)\rangle$ and $\langle\Psi(t)| = \langle\Psi(t_0)|\hat{U}^\dagger(t, t_0)$, we have

$$\hat{\rho}(t) = |\Psi(t)\rangle\langle\Psi(t)| = \hat{U}(t, t_0)|\Psi(0)\rangle\langle\Psi(0)|\hat{U}^\dagger(t, t_0). \qquad (3.84)$$

This is known as the *density operator* for the state $|\Psi(t)\rangle$. Hence knowing $\hat{\rho}(t_0)$ we can calculate $\hat{\rho}(t)$ as follows:

$$\hat{\rho}(t) = \hat{U}(t, t_0)\hat{\rho}(t_0)\hat{U}^\dagger(t, t_0). \qquad (3.85)$$

3.6.5   Time Evolution of Expectation Values

We want to look here at the time dependence of the expectation value of a linear operator; if the state $|\Psi(t)\rangle$ is normalized, the expectation value is given by

$$\langle\hat{A}\rangle = \langle\Psi(t)|\hat{A}|\Psi(t)\rangle. \qquad (3.86)$$

Using (3.76) and (3.77), we can write $d\langle\hat{A}\rangle/dt$ as follows:

$$\frac{d}{dt}\langle\hat{A}\rangle = \frac{1}{i\hbar}\langle\Psi(t)|\hat{A}\hat{H} - \hat{H}\hat{A}|\Psi(t)\rangle + \langle\Psi(t)|\frac{\partial A}{\partial t}|\Psi(t)\rangle \qquad (3.87)$$

or

$$\boxed{\frac{d}{dt}\langle\hat{A}\rangle = \frac{1}{i\hbar}\langle[\hat{A}, \hat{H}]\rangle + \langle\frac{\partial\hat{A}}{\partial t}\rangle.} \qquad (3.88)$$

Two important results stem from this relation. First, if the observable A does not depend explicitly on time, the term $\partial\hat{A}/\partial t$ will vanish, so the rate of change of the expectation value of $\hat{A}$ is given by $\langle[\hat{A}, \hat{H}]\rangle/i\hbar$. Second, besides not depending explicitly on time, if the observable A commutes with the Hamiltonian, the quantity $d\langle\hat{A}\rangle/dt$ will then be zero; hence the expectation value $\langle \hat{A} \rangle$ will be constant in time. So if $\hat{A}$ commutes with the Hamiltonian and is not dependent on time, the observable A is said to be a *constant of the motion*; that is, *the expectation value of an operator that does not depend on time and that commutes with the Hamiltonian is constant in time*:

$$\text{If} \quad [\hat{H}, \hat{A}] = 0 \quad \text{and} \quad \frac{\partial \hat{A}}{\partial t} = 0 \implies \frac{d\langle \hat{A} \rangle}{dt} = 0 \implies \langle \hat{A} \rangle = \text{constant.} \tag{3.89}$$

For instance, we can verify that the energy, the linear momentum, and the angular momentum of an isolated system are conserved: $d\langle \hat{H} \rangle / dt = 0$, $d\langle \hat{\vec{P}} \rangle / dt = 0$, and $d\langle \hat{\vec{L}} \rangle / dt = 0$. This implies that the expectation values of $\hat{H}$, $\hat{\vec{P}}$, and $\hat{\vec{L}}$ are constant. Recall from classical physics that the conservation of energy, linear momentum, and angular momentum are consequences of the following symmetries, respectively: homogeneity of time, homogeneity of space, and isotropy of space. We will show in the following section that these symmetries are associated, respectively, with invariances in time translation, space translation, and space rotation.

As an example, let us consider the time evolution of the expectation value of the density operator $\hat{\rho}(t) = |\Psi(t)\rangle\langle\Psi(t)|$; see (3.84). From (3.5), which leads to $\partial|\Psi(t)\rangle/\partial t = (1/i\hbar)\hat{H}|\Psi(t)\rangle$ and $\partial\langle\Psi(t)|/\partial t = -(1/i\hbar)\langle\Psi(t)|\hat{H}$, we have

$$\frac{\partial \hat{\rho}(t)}{\partial t} = \frac{1}{i\hbar}\hat{H}|\Psi(t)\rangle\langle\Psi(t)| - \frac{1}{i\hbar}|\Psi(t)\rangle\langle\Psi(t)|\hat{H} = -\frac{1}{i\hbar}[\hat{\rho}(t), \hat{H}]. \tag{3.90}$$

A substitution of this relation into (3.88) leads to

$$\frac{d}{dt}\langle\hat{\rho}(t)\rangle = \frac{1}{i\hbar}\langle[\hat{\rho}(t), \hat{H}]\rangle + \langle\frac{\partial\hat{\rho}(t)}{\partial t}\rangle = \frac{1}{i\hbar}\langle[\hat{\rho}(t), \hat{H}]\rangle - \frac{1}{i\hbar}\langle[\hat{\rho}(t), \hat{H}]\rangle = 0. \tag{3.91}$$

So the density operator is a constant of the motion. In fact, we can easily show that

$$\begin{aligned}
\langle[\hat{\rho}(t), \hat{H}]\rangle &= \langle\Psi(t)|[|\Psi(t)\rangle\langle\Psi(t)|, \hat{H}]|\Psi(t)\rangle \\
&= \langle\Psi(t)|\Psi(t)\rangle\langle\Psi(t)|\hat{H}|\Psi(t)\rangle - \langle\Psi(t)|\hat{H}|\Psi(t)\rangle\langle\Psi(t)|\Psi(t)\rangle \\
&= 0,
\end{aligned} \tag{3.92}$$

which, when combined with (3.90), yields $\langle\partial\hat{\rho}(t)/\partial t\rangle = 0$.

Finally, we should note that the constants of motion are nothing but observables that can be measured simultaneously with the energy to arbitrary accuracy. If a system has a complete set of commuting operators (CSCO), the number of these operators is given by the total number of constants of the motion.

3.7 Symmetries and Conservation Laws

We are interested here in symmetries that leave the Hamiltonian of an *isolated* system invariant. We will show that for each such symmetry there corresponds an observable which is a constant of the motion. The invariance principles relevant to our study are the time translation invariance and the space translation invariance. We may recall from classical physics that whenever a system is invariant under space translations, its total momentum is conserved; and whenever it is invariant under rotations, its total angular momentum is also conserved.

To prepare the stage for symmetries and conservation laws in quantum mechanics, we are going to examine the properties of infinitesimal and finite unitary transformations that are most essential to these invariance principles.

3.7.1 Infinitesimal Unitary Transformations

In Chapter 2 we saw that the transformations of a state vector $|\psi\rangle$ and an operator $\hat{A}$ under an infinitesimal unitary transformation $U_\varepsilon(\hat{G}) = \hat{I} + i\varepsilon\hat{G}$ are given by

$$
\begin{aligned}
|\psi'\rangle &= (\hat{I} + i\varepsilon\hat{G})|\psi\rangle = |\psi\rangle + \delta|\psi\rangle, \\
\hat{A}' &= (\hat{I} + i\varepsilon\hat{G})\hat{A}(\hat{I} - i\varepsilon\hat{G}) \simeq \hat{A} + i\varepsilon[\hat{G}, \hat{A}],
\end{aligned}
$$

(3.93)

(3.94)

where ε and $\hat{G}$ are called the parameter and the generator of the transformation, respectively.

Let us consider two important applications of infinitesimal unitary transformations: time and space translations.

3.7.1.1 Time Translations: $\hat{G} = \hat{H}/\hbar$

The application of $\hat{U}_{\delta t}(\hat{H}) = \hat{I} + (i/\hbar)\delta t\,\hat{H}$ on a state $|\psi(t)\rangle$ gives

$$
\left(\hat{I} + \frac{i}{\hbar}\,\delta t\,\hat{H}\right)|\psi(t)\rangle = |\psi(t)\rangle + \left(\frac{i}{\hbar}\delta t\right)\hat{H}|\psi(t)\rangle.
$$

(3.95)

Since $\hat{H}|\psi(t)\rangle = i\hbar\partial|\psi(t)\rangle/\partial t$ we have

$$
\boxed{\left(\hat{I} + \frac{i}{\hbar}\,\delta t\,\hat{H}\right)|\psi(t)\rangle = |\psi(t)\rangle - \delta t\frac{\partial|\psi(t)\rangle}{\partial t} \simeq |\psi(t - \delta t)\rangle,}
$$

(3.96)

because $|\psi(t)\rangle - \delta t\,\partial|\psi(t)\rangle/\partial t$ is nothing but the first-order Taylor expansion of $|\psi(t-\delta t)\rangle$. We conclude from (3.96) that the application of $\hat{U}_{\delta t}(\hat{H})$ to $|\psi(t)\rangle$ generates a state $|\psi(t-\delta t)\rangle$ which consists simply of a *time translation* of $|\psi(t)\rangle$ by an amount equal to δt. *The Hamiltonian in* $(\hat{I}+(i/\hbar)\delta t\,\hat{H})$ *is thus the generator of infinitesimal time translations.* Note that this translation preserves the shape of the state $|\psi(t)\rangle$, for its overall shape is merely translated in time by δt.

3.7.1.2 Spatial Translations: $\hat{G} = \hat{P}_x/\hbar$

The application of $\hat{U}_\varepsilon(\hat{P}_x) = \hat{I} + (i/\hbar)\varepsilon\hat{P}_x$ to $\psi(x)$ gives

$$
\left(\hat{I} + \frac{i}{\hbar}\varepsilon\hat{P}_x\right)\psi(x) = \psi(x) + \left(\frac{i}{\hbar}\varepsilon\right)\hat{P}_x\psi(x).
$$

(3.97)

Since $\hat{P}_x = -i\hbar\partial/\partial x$ and since the first-order Taylor expansion of $\psi(x + \varepsilon)$ is given by $\psi(x + \varepsilon) = \psi(x) + \varepsilon\partial\psi(x)/\partial x$, we have

$$
\boxed{\left(\hat{I} + \frac{i}{\hbar}\varepsilon\hat{P}_x\right)\psi(x) = \psi(x) + \varepsilon\frac{\partial\psi(x)}{\partial x} \simeq \psi(x + \varepsilon).}
$$

(3.98)

So, when $\hat{U}_\varepsilon(\hat{P}_x)$ acts on a wave function, it translates it spatially by an amount equal to ε.

Using $[\hat{X}, \hat{P}_x] = i\hbar$ we infer from (3.94) that the position operator $\hat{X}$ transforms as follows:

$$
\hat{X}' = \left(\hat{I} + \frac{i}{\hbar}\varepsilon\hat{P}_x\right)\hat{X}\left(\hat{I} - \frac{i}{\hbar}\varepsilon\hat{P}_x\right) \simeq \hat{X} + \frac{i}{\hbar}\varepsilon[\hat{P}_x, \hat{X}] = \hat{X} + \varepsilon.
$$

(3.99)

The relations (3.98) and (3.99) show that the *linear momentum operator in* $(\hat{I} + (i/\hbar)\varepsilon\hat{P}_x)$ *is a generator of infinitesimal spatial translations.*

3.7.2 Finite Unitary Transformations

In Chapter 2 we saw that a *finite* unitary transformation can be constructed by performing a succession of infinitesimal transformations. For instance, by applying a single infinitesimal time translation N times in steps of τ/N, we can generate a finite time translation

$$\hat{U}_\tau(\hat{H}) = \lim_{N\to+\infty} \prod_{k=1}^{N} \left(\hat{I} + \frac{i}{\hbar}\frac{\tau}{N}\hat{H}\right) = \lim_{N\to+\infty}\left(\hat{I} + \frac{i}{\hbar}\tau\hat{H}\right)^N = \exp\left(\frac{i}{\hbar}\tau\hat{H}\right), \quad (3.100)$$

where the Hamiltonian is the generator of finite time translations. We should note that the time evolution operator $\hat{U}(t, t_0) = e^{-i(t-t_0)\hat{H}/\hbar}$, displayed in (3.58), represents a finite unitary transformation where $\hat{H}$ is the generator of the time translation.

By analogy with (3.96) we can show that the application of $\hat{U}_\tau(\hat{H})$ to $|\psi(t)\rangle$ yields

$$\hat{U}_\tau(\hat{H})|\psi(t)\rangle = \exp\left(\frac{i}{\hbar}\tau\hat{H}\right)|\psi(t)\rangle = |\psi(t-\tau)\rangle, \quad (3.101)$$

where $|\psi(t-\tau)\rangle$ is merely a time translation of $|\psi(t)\rangle$.

Similarly, we can infer from (3.98) that the application of $\hat{U}_{\vec{a}}(\hat{\vec{P}}) = \exp(i\vec{a}\cdot\hat{\vec{P}}/\hbar)$ to a wave function causes it to be translated in space by a vector $\vec{a}$:

$$\hat{U}_{\vec{a}}(\hat{\vec{P}})\psi(\vec{r}) = \exp\left(\frac{i}{\hbar}\vec{a}\cdot\hat{\vec{P}}\right)\psi(\vec{r}) = \psi(\vec{r}+\vec{a}). \quad (3.102)$$

To calculate the transformed position vector operator $\hat{\vec{R}}'$, let us invoke a relation we derived in Chapter 2:

$$\hat{A}' = e^{i\alpha\hat{G}}\hat{A}e^{-i\alpha\hat{G}} = \hat{A} + i\alpha[\hat{G}, \hat{A}] + \frac{(i\alpha)^2}{2!}[\hat{G}, [\hat{G}, \hat{A}]] + \frac{(i\alpha)^3}{3!}[\hat{G}, [\hat{G}, [\hat{G}, \hat{A}]]] + \cdots. \quad (3.103)$$

An application of this relation to the spatial translation operator $\hat{U}_{\vec{a}}(\hat{\vec{P}})$ yields

$$\hat{\vec{R}}' = \exp\left(\frac{i}{\hbar}\vec{a}\cdot\hat{\vec{P}}\right)\hat{\vec{R}}\exp\left(-\frac{i}{\hbar}\vec{a}\cdot\hat{\vec{P}}\right) = \hat{\vec{R}} + \frac{i}{\hbar}[\vec{a}\cdot\hat{\vec{P}}, \hat{\vec{R}}] = \hat{\vec{R}} + \vec{a}. \quad (3.104)$$

In deriving this, we have used the fact that $[\vec{a}\cdot\hat{\vec{P}}, \hat{\vec{R}}] = -i\hbar\vec{a}$ and that the other commutators are zero, notably $[\vec{a}\cdot\hat{\vec{P}}, [\vec{a}\cdot\hat{\vec{P}}, \hat{\vec{R}}]] = 0$. From (3.102) and (3.104), we see that *the linear momentum in* $\exp(i\vec{a}\cdot\hat{\vec{P}}/\hbar)$ *is a generator of finite spatial translations*.

3.7.3 Symmetries and Conservation Laws

We want to show here that every invariance principle of $\hat{H}$ is connected with a conservation law.

The Hamiltonian of a system transforms under a unitary transformation $e^{i\alpha\hat{G}}$ as follows; see (3.103):

$$\hat{H}' = e^{i\alpha\hat{G}}\hat{H}e^{-i\alpha\hat{G}} = \hat{H} + i\alpha[\hat{G}, \hat{H}] + \frac{(i\alpha)^2}{2!}[\hat{G}, [\hat{G}, \hat{H}]] + \frac{(i\alpha)^3}{3!}[\hat{G}, [\hat{G}, [\hat{G}, \hat{H}]]] + \cdots. \quad (3.105)$$

If $\hat{H}$ commutes with $\hat{G}$, it also commutes with the unitary transformation $\hat{U}_\alpha(\hat{G}) = e^{i\alpha\hat{G}}$. In this case we may infer two important conclusions. On the one hand, there is an *invariance principle*: the Hamiltonian is invariant under the transformation $\hat{U}_\alpha(\hat{G})$, since

$$\hat{H}' = e^{i\alpha\hat{G}}\hat{H}e^{-i\alpha\hat{G}} = e^{i\alpha\hat{G}}e^{-i\alpha\hat{G}}\hat{H} = \hat{H}. \tag{3.106}$$

On the other hand, if in addition to $[\hat{G}, \hat{H}] = 0$, the operator $\hat{G}$ does not depend on time explicitly, there is a *conservation law*: equation (3.88) shows that $\hat{G}$ is a *constant of the motion*, since

$$\frac{d}{dt}\langle\hat{G}\rangle = \frac{1}{i\hbar}\langle[\hat{G}, \hat{H}]\rangle + \langle\frac{\partial\hat{G}}{\partial t}\rangle = 0. \tag{3.107}$$

We say that $\hat{G}$ is conserved.

So whenever the Hamiltonian is invariant under a unitary transformation, the generator of the transformation is conserved. We may say, in general, that *for every invariance symmetry of the Hamiltonian, there corresponds a conservation law*.

3.7.3.1 Conservation of Energy and Linear Momentum

Let us consider two interesting applications pertaining to the invariance of the Hamiltonian of an *isolated* system with respect to time translations and to space translations. First, let us consider time translations. As shown in (3.58), time translations are generated in the case of time-independent Hamiltonians by the evolution operator $\hat{U}(t, t_0) = e^{-i(t-t_0)\hat{H}/\hbar}$. Since $\hat{H}$ commutes with the generator of the time translation (which is given by $\hat{H}$ itself), it is invariant under time translations. As $\hat{H}$ is invariant under time translations, the energy of an isolated system is conserved. We should note that if the system is invariant under time translations, this means there is a symmetry of time homogeneity. Time homogeneity implies that the time-displaced state $\psi(t - \tau)$, like $\psi(t)$, satisfies the Schrödinger equation.

The second application pertains to the spatial translations, or to transformations under $\hat{U}_{\vec{a}}(\hat{\vec{P}}) = \exp(i\vec{a} \cdot \hat{\vec{P}}/\hbar)$, of an isolated system. The linear momentum is invariant under $\hat{U}_{\vec{a}}(\hat{\vec{P}})$ and the position operator transforms according to (3.104):

$$\hat{\vec{P}}' = \hat{\vec{P}}, \qquad \hat{\vec{R}}' = \hat{\vec{R}} + \vec{a}. \tag{3.108}$$

For instance, since the Hamiltonian of a free particle does not depend on the coordinates, it commutes with the linear momentum $[\hat{H}, \hat{\vec{P}}] = 0$. The Hamiltonian is then invariant under spatial translations, since

$$\hat{H}' = \exp\left(\frac{i}{\hbar}\vec{a} \cdot \hat{\vec{P}}\right)\hat{H}\exp\left(-\frac{i}{\hbar}\vec{a} \cdot \hat{\vec{P}}\right) = \exp\left(\frac{i}{\hbar}\vec{a} \cdot \hat{\vec{P}}\right)\exp\left(-\frac{i}{\hbar}\vec{a} \cdot \hat{\vec{P}}\right)\hat{H} = \hat{H}. \tag{3.109}$$

Since $[\hat{H}, \hat{\vec{P}}] = 0$ and since the linear momentum operator does not depend explicitly on time, we infer from (3.88) that $\hat{\vec{P}}$ is a constant of the motion, since

$$\frac{d}{dt}\langle\hat{\vec{P}}\rangle = \frac{1}{i\hbar}\langle[\hat{\vec{P}}, \hat{H}]\rangle + \langle\frac{\partial\hat{\vec{P}}}{\partial t}\rangle = 0. \tag{3.110}$$

So if $[\hat{H}, \hat{\vec{P}}] = 0$ the *Hamiltonian will be invariant under spatial translations and the linear momentum will be conserved*. A more general case where the linear momentum is a constant of the motion is provided by an isolated system, for its total linear momentum is conserved. Note that the invariance of the system under spatial translations means there is a symmetry of spatial homogeneity. The requirement for the homogeneity of space implies that the spatially displaced wave function $\psi(\vec{r} + \vec{a})$, much like $\psi(\vec{r})$, satisfies the Schrödinger equation.

In summary, the symmetry of time homogeneity gives rise to the conservation of energy, whereas the symmetry of space homogeneity gives rise to the conservation of linear momentum.

In Chapter 7 we will see that the symmetry of space isotropy, or the invariance of the Hamiltonian with respect to space rotations, leads to conservation of the angular momentum.

Parity operator

The unitary transformations we have considered so far, time translations and space translations, are *continuous*. We may consider now a *discrete* unitary transformation, the *parity*. As seen in Chapter 2, the parity transformation consists of an inversion or reflection through the origin of the coordinate system:

$$\hat{\mathcal{P}}\psi(\vec{r}) = \psi(-\vec{r}). \tag{3.111}$$

If the parity operator commutes with the system's Hamiltonian,

$$[\hat{H}, \hat{\mathcal{P}}] = 0, \tag{3.112}$$

the *parity will be conserved*, and hence a constant of the motion. In this case the Hamiltonian and the parity operator have simultaneous eigenstates. For instance, we will see in Chapter 4 that the wave functions of a particle moving in a symmetric potential, $\hat{V}(\vec{r}) = \hat{V}(-\vec{r})$, have definite parities: they can be only even or odd. Similarly, we can ascertain that the parity of an isolated system is a constant of the motion.

3.8 Connecting Quantum to Classical Mechanics

3.8.1 Poisson Brackets and Commutators

To establish a connection between quantum mechanics and classical mechanics, we may look at the time evolution of observables.

Before describing the time evolution of a dynamical variable within the context of classical mechanics, let us review the main ideas of the mathematical tool relevant to this description, the *Poisson bracket*. The Poisson bracket between two dynamical variables A and B is defined in terms of the generalized coordinates q_i and the momenta p_i of the system:

$$\{A, B\} = \sum_j \left(\frac{\partial A}{\partial q_j} \frac{\partial B}{\partial p_j} - \frac{\partial A}{\partial p_j} \frac{\partial B}{\partial q_j} \right). \tag{3.113}$$

Since the variables q_i are independent of p_i, we have $\partial q_j / \partial p_k = 0$, $\partial p_j / \partial q_k = 0$; thus we can show that

$$\{q_j, q_k\} = \{p_j, p_k\} = 0, \qquad \{q_j, p_k\} = \delta_{jk}. \tag{3.114}$$

Using (3.113) we can easily infer the following properties of the Poisson brackets:

- Antisymmetry

$$\{A, B\} = -\{B, A\} \tag{3.115}$$

- Linearity

$$\{A, \alpha B + \beta C + \gamma D + \cdots\} = \alpha\{A, B\} + \beta\{A, C\} + \gamma\{A, D\} + \cdots \quad (3.116)$$

- Complex conjugate

$$\{A, B\}^* = \{A^*, B^*\} \quad (3.117)$$

- Distributivity

$$\{A, BC\} = \{A, B\}C + B\{A, C\}, \qquad \{AB, C\} = A\{B, C\} + \{A, C\}B \quad (3.118)$$

- Jacobi identity

$$\{A, \{B, C\}\} + \{B, \{C, A\}\} + \{C, \{A, B\}\} = 0 \quad (3.119)$$

- Using $df^n(x)/dx = nf^{n-1}(x)df(x)/dx$, we can show that

$$\{A, B^n\} = nB^{n-1}\{A, B\}, \qquad \{A^n, B\} = nA^{n-1}\{A, B\} \quad (3.120)$$

These properties are similar to the properties of the quantum mechanical commutators seen in Chapter 2.

The total time derivative of a dynamical variable A is given by

$$\frac{dA}{dt} = \sum_j \left(\frac{\partial A}{\partial q_j}\frac{\partial q_j}{\partial t} + \frac{\partial A}{\partial p_j}\frac{\partial p_j}{\partial t} \right) + \frac{\partial A}{\partial t} = \sum_j \left(\frac{\partial A}{\partial q_j}\frac{\partial H}{\partial p_j} - \frac{\partial A}{\partial p_j}\frac{\partial H}{\partial p_j} \right) + \frac{\partial A}{\partial t}; \quad (3.121)$$

in deriving this relation we have used the Hamilton equations of classical mechanics:

$$\frac{dq_j}{dt} = \frac{\partial H}{\partial p_j}, \qquad \frac{dp_j}{dt} = -\frac{\partial H}{\partial q_j}, \quad (3.122)$$

where H is the Hamiltonian of the system. The total time evolution of a dynamical variable A is thus given by the following equation of motion:

$$\frac{dA}{dt} = \{A, H\} + \frac{\partial A}{\partial t}. \quad (3.123)$$

Note that if A does not depend explicitly on time, its time evolution is given simply by $dA/dt = \{A, H\}$. If $dA/dt = 0$ or $\{A, H\} = 0$, A is said to be a *constant of the motion*.

Comparing the classical relation (3.123) with its quantum mechanical counterpart (3.88),

$$\frac{d}{dt}\langle \hat{A} \rangle = \frac{1}{i\hbar}\langle [\hat{A}, \hat{H}] \rangle + \langle \frac{\partial \hat{A}}{\partial t} \rangle, \quad (3.124)$$

we see that they are identical only if we identify the Poisson bracket $\{A, H\}$ with the commutator $[\hat{A}, \hat{H}]/(i\hbar)$. We may thus infer the following general rule. The Poisson bracket of any pair of classical variables can be obtained from the commutator between the corresponding pair of quantum operators by dividing it by $i\hbar$:

$$\boxed{\frac{1}{i\hbar}[\hat{A}, \hat{B}] \longrightarrow \{A, B\}_{classical}.} \quad (3.125)$$

Note that the expressions of classical mechanics can be derived from their quantum counterparts, but the opposite is not possible. That is, dividing quantum mechanical expressions by $i\hbar$ leads to their classical analog, but multiplying classical mechanical expressions by $i\hbar$ doesn't necessarily lead to their quantum counterparts.

Example 3.5

(a) Evaluate the Poisson bracket $\{x, p\}$ between the position, x, and momentum, p, variables.

(b) Compare the commutator $\left[\hat{X}, \hat{P}\right]$ with Poisson bracket $\{x, p\}$ calculated in Part (a).

Solution

(a) Applying the general relation

$$\{A, B\} = \sum_j \left(\frac{\partial A}{\partial x_j} \frac{\partial B}{\partial p_j} - \frac{\partial A}{\partial p_j} \frac{\partial B}{\partial x_j} \right) \tag{3.126}$$

to x and p, we can readily evaluate the given Poisson bracket:

$$
\begin{aligned}
\{x, p\} &= \frac{\partial(x)}{\partial x} \frac{\partial(p)}{\partial p} - \frac{\partial(x)}{\partial p} \frac{\partial(p)}{\partial x} \\
&= \frac{\partial(x)}{\partial x} \frac{\partial(p)}{\partial p} \\
&= 1.
\end{aligned}
\tag{3.127}
$$

(b) Using the fact that $[\hat{X}, \hat{P}] = i\hbar$, we see that

$$\frac{1}{i\hbar}[\hat{X}, \hat{P}] = 1, \tag{3.128}$$

which is equal to the Poisson bracket (3.127); that is,

$$\frac{1}{i\hbar}[\hat{X}, \hat{P}] = \{x, p\}_{classical} = 1. \tag{3.129}$$

This result is in agreement with Eq. (3.125).

3.8.2 The Ehrenfest Theorem

If quantum mechanics is to be more general than classical mechanics, it must contain classical mechanics as a limiting case. To illustrate this idea, let us look at the time evolution of the expectation values of the position and momentum operators, $\hat{\vec{R}}$ and $\hat{\vec{P}}$, of a particle moving in a potential $\hat{V}(\vec{r})$, and then compare these relations with their classical counterparts.

Since the position and the momentum observables do not depend explicitly on time, within the context of wave mechanics, the terms $\langle \partial \hat{\vec{R}}/\partial t \rangle$ and $\langle \partial \hat{\vec{P}}/\partial t \rangle$ are zero. Hence, inserting

$\hat{H} = \hat{\vec{P}}^2/(2m) + \hat{V}(\hat{\vec{R}}, t)$ into (3.88) and using the fact that $\hat{\vec{R}}$ commutes with $\hat{V}(\hat{\vec{R}}, t)$, we can write

$$\frac{d}{dt}\langle\hat{\vec{R}}\rangle = \frac{1}{i\hbar}\langle[\hat{\vec{R}}, \hat{H}]\rangle = \frac{1}{i\hbar}\langle[\hat{\vec{R}}, \frac{\hat{\vec{P}}^2}{2m} + \hat{V}(\hat{\vec{R}}, t)]\rangle = \frac{1}{2im\hbar}\langle[\hat{\vec{R}}, \hat{\vec{P}}^2]\rangle. \tag{3.130}$$

Since

$$[\hat{\vec{R}}, \hat{\vec{P}}^2] = 2i\hbar\hat{\vec{P}}, \tag{3.131}$$

we have

$$\boxed{\frac{d}{dt}\langle\hat{\vec{R}}\rangle = \frac{1}{m}\langle\hat{\vec{P}}\rangle.} \tag{3.132}$$

As for $d\langle\hat{\vec{P}}\rangle/dt$, we can infer its expression from a treatment analogous to $d\langle\hat{\vec{R}}\rangle/dt$. Using

$$[\hat{\vec{P}}, \hat{V}(\hat{\vec{R}}, t)] = -i\hbar\vec{\nabla}\hat{V}(\vec{R}, t), \tag{3.133}$$

we can write

$$\boxed{\frac{d}{dt}\langle\hat{\vec{P}}\rangle = \frac{1}{i\hbar}\langle[\hat{\vec{P}}, \hat{V}(\hat{\vec{R}}, t)]\rangle = -\langle\vec{\nabla}\hat{V}(\hat{\vec{R}}, t)\rangle.} \tag{3.134}$$

The two relations (3.132) and (3.134), expressing the time evolution of the expectation values of the position and momentum operators, are known as the *Ehrenfest theorem*, or Ehrenfest equations. Their respective forms are reminiscent of the Hamilton–Jacobi equations of classical mechanics,

$$\frac{d\vec{r}}{dt} = \frac{\vec{p}}{m}, \qquad\qquad \frac{d\vec{p}}{dt} = -\vec{\nabla}V(\vec{r}), \tag{3.135}$$

which reduce to Newton's equation of motion for a *classical* particle of mass m, position $\vec{r}$, and momentum $\vec{p}$:

$$\frac{d\vec{p}}{dt} = m\frac{d^2\vec{r}}{dt^2} = -\vec{\nabla}V(\vec{r}). \tag{3.136}$$

Notice $\hbar$ has completely disappeared in the Ehrenfest equations (3.132) and (3.134). These two equations certainly establish a connection between quantum mechanics and classical mechanics. We can, within this context, view the center of the wave packet as moving like a classical particle when subject to a potential $V(\vec{r})$.

3.8.3 Quantum Mechanics and Classical Mechanics

In Chapter 1 we focused mainly on those experimental observations which confirm the failure of classical physics at the microscopic level. We should bear in mind, however, that classical physics works perfectly well within the realm of the macroscopic world. Thus, if the theory of quantum mechanics is to be considered more general than classical physics, it must yield accurate results not only on the microscopic scale but at the classical limit as well.

How does one decide on when to use classical or quantum mechanics to describe the motion of a given system? That is, how do we know when a classical description is good enough or when a quantum description becomes a must? The answer is provided by comparing the size of those quantities of the system that have the dimensions of an action with the Planck constant, h. Since, as shown in (3.125), the quantum relations are characterized by h, we can state that if the value of the action of a system is too large compared to h, this system can be accurately described by means of classical physics. Otherwise, the use of a quantal description becomes unavoidable. One should recall that, for microscopic systems, the size of action variables is of the order of h; for instance, the angular momentum of the hydrogen atom is $L = n\hbar$, where n is finite.

Another equivalent way of defining the classical limit is by means of "*length*." Since $\lambda = h/p$ the classical domain can be specified by the limit $\lambda \to 0$. This means that, when the de Broglie wavelength of a system is too small compared to its size, the system can be described accurately by means of classical physics.

In summary, the classical limit can be described as the limit $h \to 0$ or, equivalently, as the limit $\lambda \to 0$. In these limits the results of quantum mechanics should be similar to those of classical physics:

$$\lim_{h \to 0} \text{Quantum Mechanics} \longrightarrow \text{Classical Mechanics}, \tag{3.137}$$

$$\lim_{\lambda \to 0} \text{Quantum Mechanics} \longrightarrow \text{Classical Mechanics}. \tag{3.138}$$

Classical mechanics can thus be regarded as the short wavelength limit of quantum mechanics. In this way, quantum mechanics contains classical mechanics as a limiting case. So, in the limit of $h \to 0$ or $\lambda \to 0$, quantum dynamical quantities should have, as proposed by Bohr, a one-to-one correspondence with their classical counterparts. This is the essence of the *correspondence principle*.

But how does one reconcile, in the classical limit, the probabilistic nature of quantum mechanics with the determinism of classical physics? The answer is quite straightforward: quantum *fluctuations* must become negligible or even vanish when $h \to 0$, for Heisenberg's uncertainty principle would acquire the status of *certainty*; when $h \to 0$, the fluctuations in the position and momentum will vanish, $\Delta x \to 0$ and $\Delta p \to 0$. Thus, the position and momentum can be measured simultaneously with arbitrary accuracy. This implies that the probabilistic assessments of dynamical quantities by quantum mechanics must give way to exact calculations (these ideas will be discussed further when we study the WKB method in Chapter 9).

So, for those cases where the action variables of a system are too large compared to h (or, equivalently, when the lengths of this system are too large compared to its de Broglie wavelength), quantum mechanics gives the same results as classical mechanics.

In the rest of this text, we will deal with the various applications of the Schrödinger equation. We start, in Chapter 4, with the simple case of one-dimensional systems and later on consider more realistic systems.

3.9 Solved Problems

Problem 3.1

A particle of mass m, which moves freely inside an infinite potential well of length a, has the following initial wave function at $t = 0$:

$$\psi(x, 0) = \frac{A}{\sqrt{a}} \sin\left(\frac{\pi x}{a}\right) + \sqrt{\frac{3}{5a}} \sin\left(\frac{3\pi x}{a}\right) + \frac{1}{\sqrt{5a}} \sin\left(\frac{5\pi x}{a}\right),$$

where A is a real constant.

(a) Find A so that $\psi(x, 0)$ is normalized.

(b) If measurements of the energy are carried out, what are the values that will be found and what are the corresponding probabilities? Calculate the average energy.

(c) Find the wave function $\psi(x, t)$ at any later time t.

(d) Determine the probability of finding the system at a time t in the state $\varphi(x, t) = \sqrt{2/a} \sin(5\pi x/a) \exp(-iE_5 t/\hbar)$; then determine the probability of finding it in the state $\chi(x, t) = \sqrt{2/a} \sin(2\pi x/a) \exp(-iE_2 t/\hbar)$.

Solution

Since the functions

$$\phi_n(x) = \sqrt{\frac{2}{a}} \sin\left(\frac{n\pi x}{a}\right) \tag{3.139}$$

are orthonormal,

$$\langle \phi_n | \phi_m \rangle = \int_0^a \phi_n^*(x)\phi_m(x)\, dx = \frac{2}{a} \int_0^a \sin\left(\frac{n\pi x}{a}\right) \sin\left(\frac{m\pi x}{a}\right) dx = \delta_{nm}, \tag{3.140}$$

it is more convenient to write $\psi(x, 0)$ in terms of $\phi_n(x)$:

$$\begin{aligned}
\psi(x, 0) &= \frac{A}{\sqrt{a}} \sin\left(\frac{\pi x}{a}\right) + \sqrt{\frac{3}{5a}} \sin\left(\frac{3\pi x}{a}\right) + \frac{1}{\sqrt{5a}} \sin\left(\frac{5\pi x}{a}\right) \\
&= \frac{A}{\sqrt{2}} \phi_1(x) + \sqrt{\frac{3}{10}} \phi_3(x) + \frac{1}{\sqrt{10}} \phi_5(x).
\end{aligned} \tag{3.141}$$

(a) Since $\langle \phi_n | \phi_m \rangle = \delta_{nm}$ the normalization of $\psi(x, 0)$ yields

$$1 = \langle \psi | \psi \rangle = \frac{A^2}{2} + \frac{3}{10} + \frac{1}{10}, \tag{3.142}$$

or $A = \sqrt{6/5}$; hence

$$\psi(x, 0) = \sqrt{\frac{3}{5}} \phi_1(x) + \sqrt{\frac{3}{10}} \phi_3(x) + \frac{1}{\sqrt{10}} \phi_5(x). \tag{3.143}$$

(b) Since the second derivative of (3.139) is given by $d^2\phi_n(x)/dx^2 = -(n^2\pi^2/a^2)\phi_n(x)$, and since the Hamiltonian of a free particle is $\hat{H} = -(\hbar^2/2m)d^2/dx^2$, the expectation value of $\hat{H}$ with respect to $\phi_n(x)$ is

$$E_n = \langle \phi_n | \hat{H} | \phi_n \rangle = -\frac{\hbar^2}{2m} \int_0^a \phi_n^*(x) \frac{d^2\phi_n(x)}{dx^2} dx = \frac{n^2\pi^2\hbar^2}{2ma^2}. \tag{3.144}$$

If a measurement is carried out on the system, we would obtain $E_n = n^2\pi^2\hbar^2/(2ma^2)$ with a corresponding probability of $P_n(E_n) = |\langle \phi_n | \psi \rangle|^2$. Since the initial wave function (3.143) contains only three eigenstates of $\hat{H}$, $\phi_1(x)$, $\phi_3(x)$, and $\phi_5(x)$, the results of the energy measurements along with the corresponding probabilities are

$$E_1 = \langle \phi_1 | \hat{H} | \phi_1 \rangle = \frac{\pi^2\hbar^2}{2ma^2}, \qquad P_1(E_1) = |\langle \phi_1 | \psi \rangle|^2 = \frac{3}{5}, \tag{3.145}$$

$$E_3 = \langle \phi_3 | \hat{H} | \phi_3 \rangle = \frac{9\pi^2\hbar^2}{2ma^2}, \qquad P_3(E_3) = |\langle \phi_3 | \psi \rangle|^2 = \frac{3}{10}, \tag{3.146}$$

$$E_5 = \langle \phi_5 | \hat{H} | \phi_5 \rangle = \frac{25\pi^2\hbar^2}{2ma^2}, \qquad P_5(E_5) = |\langle \phi_5 | \psi \rangle|^2 = \frac{1}{10}. \tag{3.147}$$

The average energy is

$$E = \sum_n P_n E_n = \frac{3}{5}E_1 + \frac{3}{10}E_3 + \frac{1}{10}E_5 = \frac{29\pi^2\hbar^2}{10ma^2}. \tag{3.148}$$

(c) As the initial state $\psi(x, 0)$ is given by (3.143), the wave function $\psi(x, t)$ at any later time t is

$$\psi(x, t) = \sqrt{\frac{3}{5}}\phi_1(x)e^{-iE_1t/\hbar} + \sqrt{\frac{3}{10}}\phi_3(x)e^{-iE_3t/\hbar} + \frac{1}{\sqrt{10}}\phi_5(x)e^{-iE_5t/\hbar}, \tag{3.149}$$

where the expressions of E_n are listed in (3.144) and $\phi_n(x)$ in (3.139).

(d) First, let us express $\varphi(x, t)$ in terms of $\phi_n(x)$:

$$\varphi(x, t) = \sqrt{\frac{2}{a}} \sin\left(\frac{5\pi x}{a}\right) e^{-iE_5t/\hbar} = \phi_5(x)e^{-iE_5t/\hbar}. \tag{3.150}$$

The probability of finding the system at a time t in the state $\varphi(x, t)$ is

$$P = |\langle\varphi|\psi\rangle|^2 = \left|\int_0^a \varphi^*(x, t)\psi(x, t)\, dx\right|^2 = \frac{1}{10}\left|\int_0^a \phi_5^*(x)\phi_5(x)\, dx\right|^2 = \frac{1}{10}, \tag{3.151}$$

since $\langle\varphi|\phi_1\rangle = \langle\varphi|\phi_3\rangle = 0$ and $\langle\varphi|\phi_5\rangle = \exp(iE_5t/\hbar)$.

Similarly, since $\chi(x, t) = \sqrt{2/a}\sin(2\pi x/a)\exp(-iE_2t/\hbar) = \phi_2(x)\exp(-iE_2t/\hbar)$, we can easily show that the probability for finding the system in the state $\chi(x, t)$ is zero:

$$P = |\langle\chi|\psi\rangle|^2 = \left|\int_0^a \chi^*(x, t)\psi(x, t)\, dx\right|^2 = 0, \tag{3.152}$$

since $\langle\chi|\phi_1\rangle = \langle\chi|\phi_3\rangle = \langle\chi|\phi_5\rangle = 0$.

Problem 3.2

A particle of mass m, which moves freely inside an infinite potential well of length a, is initially in the state $\psi(x, 0) = \sqrt{3/5a}\sin(3\pi x/a) + (1/\sqrt{5a})\sin(5\pi x/a)$.

(a) Find $\psi(x, t)$ at any later time t.

(b) Calculate the probability density $\rho(x, t)$ and the current density, $\vec{J}(x, t)$.

(c) Verify that the probability is conserved, i.e., $\partial\rho/\partial t + \vec{\nabla}\cdot\vec{J}(x, t) = 0$.

Solution

(a) Since $\psi(x, 0)$ can be expressed in terms of $\phi_n(x) = \sqrt{2/a}\sin(n\pi x/a)$ as

$$\psi(x, 0) = \sqrt{\frac{3}{5a}}\sin\left(\frac{3\pi x}{a}\right) + \frac{1}{\sqrt{5a}}\sin\left(\frac{5\pi x}{a}\right) = \sqrt{\frac{3}{10}}\phi_3(x) + \frac{1}{\sqrt{10}}\phi_5(x), \tag{3.153}$$

we can write

$$\begin{aligned}
\psi(x, t) &= \sqrt{\frac{3}{5a}}\sin\left(\frac{3\pi x}{a}\right)e^{-iE_3t/\hbar} + \frac{1}{\sqrt{5a}}\sin\left(\frac{5\pi x}{a}\right)e^{-iE_5t/\hbar} \\
&= \sqrt{\frac{3}{10}}\phi_3(x)e^{-iE_3t/\hbar} + \frac{1}{\sqrt{10}}\phi_5(x)e^{-iE_5t/\hbar},
\end{aligned} \tag{3.154}$$

where the expressions for E_n are listed in (3.144): $E_n = n^2 \pi^2 \hbar^2 / (2ma^2)$.

(b) Since $\rho(x, t) = \psi^*(x, t)\psi(x, t)$, where $\psi(x, t)$ is given by (3.154), we can write

$$\rho(x, t) = \frac{3}{10}\phi_3^2(x) + \frac{\sqrt{3}}{10}\phi_3(x)\phi_5(x)\left[e^{i(E_3 - E_5)t/\hbar} + e^{-i(E_3 - E_5)t/\hbar}\right] + \frac{1}{10}\phi_5^2(x). \quad (3.155)$$

From (3.144) we have $E_3 - E_5 = 9E_1 - 25E_1 = -16E_1 = -8\pi^2\hbar^2/(ma^2)$. Thus, $\rho(x, t)$ becomes

$$\begin{aligned}
\rho(x, t) &= \frac{3}{10}\phi_3^2(x) + \frac{\sqrt{3}}{5}\phi_3(x)\phi_5(x)\cos\left(\frac{16E_1 t}{\hbar}\right) + \frac{1}{10}\phi_5^2(x) \\
&= \frac{3}{5a}\sin^2\left(\frac{3\pi x}{a}\right) + \frac{2\sqrt{3}}{5a}\sin\left(\frac{3\pi x}{a}\right)\sin\left(\frac{5\pi x}{a}\right)\cos\left(\frac{16E_1 t}{\hbar}\right) \\
&\quad + \frac{1}{5a}\sin^2\left(\frac{5\pi x}{a}\right).
\end{aligned} \quad (3.156)$$

Since the system is one-dimensional, the action of the gradient operator on $\psi(x, t)$ and $\psi^*(x, t)$ is given by $\vec{\nabla}\psi(x, t) = (d\psi(x, t)/dx)\vec{i}$ and $\vec{\nabla}\psi^*(x, t) = (d\psi^*(x, t)/dx)\vec{i}$. We can thus write the current density $\vec{J}(x, t) = (i\hbar/2m)\left(\psi(x, t)\vec{\nabla}\psi^*(x, t) - \psi^*(x, t)\vec{\nabla}\psi(x, t)\right)$ as

$$\vec{J}(x, t) = \frac{i\hbar}{2m}\left(\psi(x, t)\frac{d\psi^*(x, t)}{dx} - \psi^*(x, t)\frac{d\psi(x, t)}{dx}\right)\vec{i}. \quad (3.157)$$

Using (3.154) we have

$$\frac{d\psi(x, t)}{dx} = \frac{3\pi}{a}\sqrt{\frac{3}{5a}}\cos\left(\frac{3\pi x}{a}\right)e^{-iE_3 t/\hbar} + \frac{5\pi}{a}\frac{1}{\sqrt{5a}}\cos\left(\frac{5\pi x}{a}\right)e^{-iE_5 t/\hbar}, \quad (3.158)$$

$$\frac{d\psi^*(x, t)}{dx} = \frac{3\pi}{a}\sqrt{\frac{3}{5a}}\cos\left(\frac{3\pi x}{a}\right)e^{iE_3 t/\hbar} + \frac{5\pi}{a}\frac{1}{\sqrt{5a}}\cos\left(\frac{5\pi x}{a}\right)e^{iE_5 t/\hbar}. \quad (3.159)$$

A straightforward calculation yields

$$\begin{aligned}
\psi\frac{d\psi^*}{dx} - \psi^*\frac{d\psi}{dx} &= -2i\pi\frac{\sqrt{3}}{5a^2}\left[5\sin\left(\frac{3\pi x}{a}\right)\cos\left(\frac{5\pi x}{a}\right) - 3\sin\left(\frac{5\pi x}{a}\right)\cos\left(\frac{3\pi x}{a}\right)\right] \\
&\quad \times \sin\left(\frac{E_3 - E_5}{\hbar}t\right).
\end{aligned} \quad (3.160)$$

Inserting this into (3.157) and using $E_3 - E_5 = -16E_1$, we have

$$\vec{J}(x, t) = -\frac{\pi\hbar}{m}\frac{\sqrt{3}}{5a^2}\left[5\sin\left(\frac{3\pi x}{a}\right)\cos\left(\frac{5\pi x}{a}\right) - 3\sin\left(\frac{5\pi x}{a}\right)\cos\left(\frac{3\pi x}{a}\right)\right]\sin\left(\frac{16E_1 t}{\hbar}\right)\vec{i}. \quad (3.161)$$

(c) Performing the time derivative of (3.156) and using the expression $32\sqrt{3}E_1/(5a\hbar) = 16\pi^2\hbar\sqrt{3}/(5ma^3)$, since $E_1 = \pi^2\hbar^2/(2ma^2)$, we obtain

$$\begin{aligned}
\frac{\partial\rho}{\partial t} &= -\frac{32\sqrt{3}E_1}{5a\hbar}\sin\left(\frac{3\pi x}{a}\right)\sin\left(\frac{5\pi x}{a}\right)\sin\left(\frac{16E_1 t}{\hbar}\right) \\
&= -\frac{16\pi^2\hbar\sqrt{3}}{5ma^3}\sin\left(\frac{3\pi x}{a}\right)\sin\left(\frac{5\pi x}{a}\right)\sin\left(\frac{16E_1 t}{\hbar}\right),
\end{aligned} \quad (3.162)$$

Now, taking the divergence of (3.161), we end up with

$$\vec{\nabla} \cdot \vec{J}(x,t) = \frac{dJ(x,t)}{dx} = \frac{16\pi^2 \hbar \sqrt{3}}{5ma^3} \sin\left(\frac{3\pi x}{a}\right) \sin\left(\frac{5\pi x}{a}\right) \sin\left(\frac{16E_1 t}{\hbar}\right). \qquad (3.163)$$

The addition of (3.162) and (3.163) confirms the conservation of probability:

$$\frac{\partial \rho}{\partial t} + \vec{\nabla} \cdot \vec{J}(x,t) = 0. \qquad (3.164)$$

Problem 3.3

Consider a one-dimensional particle which is confined within the region $0 \le x \le a$ and whose wave function is $\Psi(x,t) = \sin(\pi x/a) \exp(-i\omega t)$.

(a) Find the potential $V(x)$.

(b) Calculate the probability of finding the particle in the interval $a/4 \le x \le 3a/4$.

Solution

(a) Since the first time derivative and the second x derivative of $\Psi(x,t)$ are given by $\partial \Psi(x,t)/\partial t = -i\omega \Psi(x,t)$ and $\partial^2 \Psi(x,t)/\partial x^2 = -(\pi^2/a^2)\Psi(x,t)$, the Schrödinger equation (3.68) yields

$$i\hbar(-i\omega)\Psi(x,t) = \frac{\hbar^2}{2m}\frac{\pi^2}{a^2}\Psi(x,t) + \hat{V}(x,t)\Psi(x,t). \qquad (3.165)$$

Hence $V(x,t)$ is time independent and given by $V(x) = \hbar\omega - \hbar^2\pi^2/(2ma^2)$.

(b) The probability of finding the particle in the interval $a/4 \le x \le 3a/4$ can be obtained from (3.4):

$$P = \frac{\int_{a/4}^{3a/4} |\psi(x)|^2 dx}{\int_0^a |\psi(x)|^2 dx} = \frac{\int_{a/4}^{3a/4} \sin^2(\pi x/a)\, dx}{\int_0^a \sin^2(\pi x/a)\, dx} = \frac{2+\pi}{2\pi} = 0.82 \qquad (3.166)$$

Problem 3.4

A system is initially in the state $|\psi_0\rangle = [\sqrt{2}|\phi_1\rangle + \sqrt{3}|\phi_2\rangle + |\phi_3\rangle + |\phi_4\rangle]/\sqrt{7}$, where $|\phi_n\rangle$ are eigenstates of the system's Hamiltonian such that $\hat{H}|\phi_n\rangle = n^2 \mathcal{E}_0 |\phi_n\rangle$.

(a) If energy is measured, what values will be obtained and with what probabilities?

(b) Consider an operator $\hat{A}$ whose action on $|\phi_n\rangle$ is defined by $\hat{A}|\phi_n\rangle = (n+1)a_0|\phi_n\rangle$. If A is measured, what values will be obtained and with what probabilities?

(c) Suppose that a measurement of the energy yields $4\mathcal{E}_0$. If we measure A immediately afterwards, what value will be obtained?

Solution

(a) A measurement of the energy yields $E_n = \langle \phi_n | \hat{H} | \phi_n \rangle = n^2 \mathcal{E}_0$, that is

$$E_1 = \mathcal{E}_0, \quad E_2 = 4\mathcal{E}_0, \quad E_3 = 9\mathcal{E}_0, \quad E_4 = 16\mathcal{E}_0. \qquad (3.167)$$

Since $|\psi_0\rangle$ is normalized, $\langle \psi_0 | \psi_0 \rangle = (2+3+1+1)/7 = 1$, and using (3.2), we can write the probabilities corresponding to (3.167) as $P(E_n) = |\langle \phi_n | \psi_0 \rangle|^2 / \langle \psi_0 | \psi_0 \rangle = |\langle \phi_n | \psi_0 \rangle|^2$; hence, using the fact that $\langle \phi_n | \phi_m \rangle = \delta_{nm}$, we have

$$P(E_1) = \left| \sqrt{\frac{2}{7}} \langle \phi_1 | \phi_1 \rangle \right|^2 = \frac{2}{7}, \qquad P(E_2) = \left| \sqrt{\frac{3}{7}} \langle \phi_2 | \phi_2 \rangle \right|^2 = \frac{3}{7}, \qquad (3.168)$$

$$P(E_3) = \left| \frac{1}{\sqrt{7}} \langle \phi_3 | \phi_3 \rangle \right|^2 = \frac{1}{7}, \qquad P(E_4) = \left| \frac{1}{\sqrt{7}} \langle \phi_4 | \phi_4 \rangle \right|^2 = \frac{1}{7}. \qquad (3.169)$$

(b) Similarly, a measurement of the observable $\hat{A}$ yields $a_n = \langle \phi_n | \hat{A} | \phi_n \rangle = (n+1)a_0$; that is,

$$a_1 = 2a_0, \quad a_2 = 3a_0, \quad a_3 = 4a_0, \quad a_4 = 5a_0. \qquad (3.170)$$

Again, using (3.2) and since $| \psi_0 \rangle$ is normalized, we can ascertain that the probabilities corresponding to the values (3.170) are given by $P(a_n) = |\langle \phi_n | \psi_0 \rangle|^2 / \langle \psi_0 | \psi_0 \rangle = |\langle \phi_n | \psi_0 \rangle|^2$, or

$$P(a_1) = \left| \sqrt{\frac{2}{7}} \langle \phi_1 | \phi_1 \rangle \right|^2 = \frac{2}{7}, \qquad P(a_2) = \left| \sqrt{\frac{3}{7}} \langle \phi_2 | \phi_2 \rangle \right|^2 = \frac{3}{7}, \qquad (3.171)$$

$$P(a_3) = \left| \frac{1}{\sqrt{7}} \langle \phi_3 | \phi_3 \rangle \right|^2 = \frac{1}{7}, \qquad P(a_4) = \left| \frac{1}{\sqrt{7}} \langle \phi_4 | \phi_4 \rangle \right|^2 = \frac{1}{7}. \qquad (3.172)$$

(c) An energy measurement that yields $4\mathcal{E}_0$ implies that the system is left in the state $|\phi_2\rangle$. A measurement of the observable A immediately afterwards leads to

$$\langle \phi_2 | \hat{A} | \phi_2 \rangle = 3a_0 \langle \phi_2 | \phi_2 \rangle = 3a_0. \qquad (3.173)$$

Problem 3.5

(a) Assuming that the system of Problem 3.4 is initially in the state $|\phi_3\rangle$, what values for the energy and the observable A will be obtained if we measure: (i)H first then A, (ii) A first then H?

(b) Compare the results obtained in (i) and (ii) and infer whether $\hat{H}$ and $\hat{A}$ are compatible. Calculate $[\hat{H}, \hat{A}]|\phi_3\rangle$.

Solution

(a) (i) The measurement of H first then A is represented by $\hat{A}\hat{H}|\phi_3\rangle$. Using the relations $\hat{H}|\phi_n\rangle = n^2 \mathcal{E}_0 |\phi_n\rangle$ and $\hat{A}|\phi_n\rangle = na_0|\phi_{n+1}\rangle$, we have

$$\hat{A}\hat{H}|\phi_3\rangle = 9\mathcal{E}_0 \hat{A}|\phi_3\rangle = 27\mathcal{E}_0 a_0|\phi_4\rangle. \qquad (3.174)$$

(ii) Measuring A first and then H, we will obtain

$$\hat{H}\hat{A}|\phi_3\rangle = 3a_0 \hat{H}|\phi_4\rangle = 48\mathcal{E}_0 a_0|\phi_4\rangle. \qquad (3.175)$$

(b) Equations (3.174) and (3.175) show that the actions of $\hat{A}\hat{H}$ and $\hat{H}\hat{A}$ yield different results. This means that $\hat{H}$ and $\hat{A}$ do not commute; hence they are not compatible. We can thus write

$$[\hat{H}, \hat{A}]|\phi_3\rangle = (48 - 27)\mathcal{E}_0 a_0|\phi_4\rangle = 17\mathcal{E}_0 a_0|\phi_4\rangle. \qquad (3.176)$$

Problem 3.6

Consider a physical system whose Hamiltonian H and initial state $|\psi_0\rangle$ are given by

$$H = \mathcal{E} \begin{pmatrix} 0 & i & 0 \\ -i & 0 & 0 \\ 0 & 0 & -1 \end{pmatrix}, \qquad |\psi_0\rangle = \frac{1}{\sqrt{5}} \begin{pmatrix} 1-i \\ 1-i \\ 1 \end{pmatrix},$$

where $\mathcal{E}$ has the dimensions of energy.

(a) What values will we obtain when measuring the energy and with what probabilities?

(b) Calculate $\langle \hat{H} \rangle$, the expectation value of the Hamiltonian.

Solution

(a) The results of the energy measurement are given by the eigenvalues of H. A diagonalization of H yields a nondegenerate eigenenergy $E_1 = \mathcal{E}$ and a doubly degenerate value $E_2 = E_3 = -\mathcal{E}$ whose respective eigenvectors are given by

$$|\phi_1\rangle = \frac{1}{\sqrt{2}} \begin{pmatrix} 1 \\ -i \\ 0 \end{pmatrix}, \qquad |\phi_2\rangle = \frac{1}{\sqrt{2}} \begin{pmatrix} -i \\ 1 \\ 0 \end{pmatrix}, \qquad |\phi_3\rangle = \begin{pmatrix} 0 \\ 0 \\ 1 \end{pmatrix}; \qquad (3.177)$$

these eigenvectors are orthogonal since H is Hermitian. Note that the initial state $|\psi_0\rangle$ can be written in terms of $|\phi_1\rangle$, $|\phi_2\rangle$, and $|\phi_3\rangle$ as follows:

$$|\psi_0\rangle = \frac{1}{\sqrt{5}} \begin{pmatrix} 1-i \\ 1-i \\ 1 \end{pmatrix} = \sqrt{\frac{2}{5}}|\phi_1\rangle + \sqrt{\frac{2}{5}}|\phi_2\rangle + \frac{1}{\sqrt{5}}|\phi_3\rangle. \qquad (3.178)$$

Since $|\phi_1\rangle$, $|\phi_2\rangle$, and $|\phi_3\rangle$ are orthonormal, the probability of measuring $E_1 = \mathcal{E}$ is given by

$$P_1(E_1) = |\langle \phi_1 | \psi_0 \rangle|^2 = \left| \sqrt{\frac{2}{5}} \langle \phi_1 | \phi_1 \rangle \right|^2 = \frac{2}{5}. \qquad (3.179)$$

Now, since the other eigenvalue is doubly degenerate, $E_2 = E_3 = -\mathcal{E}$, the probability of measuring $-\mathcal{E}$ can be obtained from (3.3):

$$P_2(E_2) = |\langle \phi_2 | \psi_0 \rangle|^2 + |\langle \phi_3 | \psi_0 \rangle|^2 = \frac{2}{5} + \frac{1}{5} = \frac{3}{5}. \qquad (3.180)$$

(b) From (3.179) and (3.180), we have

$$\langle \hat{H} \rangle = P_1 E_1 + P_2 E_2 = \frac{2}{5}\mathcal{E} - \frac{3}{5}\mathcal{E} = -\frac{1}{5}\mathcal{E}. \qquad (3.181)$$

We can obtain the same result by calculating the expectation value of $\hat{H}$ with respect to $|\psi_0\rangle$. Since $\langle \psi_0 | \psi_0 \rangle = 1$, we have $\langle \hat{H} \rangle = \langle \psi_0 | \hat{H} | \psi_0 \rangle / \langle \psi_0 | \psi_0 \rangle = \langle \psi_0 | \hat{H} | \psi_0 \rangle$:

$$\langle \hat{H} \rangle = \langle \psi_0 | \hat{H} | \psi_0 \rangle = \frac{\mathcal{E}}{5} \begin{pmatrix} 1+i & 1+i & 1 \end{pmatrix} \begin{pmatrix} 0 & i & 0 \\ -i & 0 & 0 \\ 0 & 0 & -1 \end{pmatrix} \begin{pmatrix} 1-i \\ 1-i \\ 1 \end{pmatrix} = -\frac{1}{5}\mathcal{E}. \qquad (3.182)$$

Problem 3.7

Consider a system whose Hamiltonian H and an operator A are given by the matrices

$$H = \mathcal{E}_0 \begin{pmatrix} 1 & -1 & 0 \\ -1 & 1 & 0 \\ 0 & 0 & -1 \end{pmatrix}, \quad A = a \begin{pmatrix} 0 & 4 & 0 \\ 4 & 0 & 1 \\ 0 & 1 & 0 \end{pmatrix},$$

where $\mathcal{E}_0$ has the dimensions of energy.

(a) If we measure the energy, what values will we obtain?

(b) Suppose that when we measure the energy, we obtain a value of $-\mathcal{E}_0$. Immediately afterwards, we measure A. What values will we obtain for A and what are the probabilities corresponding to each value?

(c) Calculate the uncertainty ΔA.

Solution

(a) The possible energies are given by the eigenvalues of H. A diagonalization of H yields three nondegenerate eigenenergies $E_1 = 0$, $E_2 = -\mathcal{E}_0$, and $E_3 = 2\mathcal{E}_0$. The respective eigenvectors are

$$|\phi_1\rangle = \frac{1}{\sqrt{2}} \begin{pmatrix} 1 \\ 1 \\ 0 \end{pmatrix}, \quad |\phi_2\rangle = \begin{pmatrix} 0 \\ 0 \\ 1 \end{pmatrix}, \quad |\phi_3\rangle = \frac{1}{\sqrt{2}} \begin{pmatrix} -1 \\ 1 \\ 0 \end{pmatrix}; \quad (3.183)$$

these eigenvectors are orthonormal.

(b) If a measurement of the energy yields $-\mathcal{E}_0$, this means that the system is left in the state $|\phi_2\rangle$. When we measure the next observable, A, the system is in the state $|\phi_2\rangle$. The result we obtain for A is given by any of the eigenvalues of A. A diagonalization of A yields three nondegenerate values: $a_1 = -\sqrt{17}a$, $a_2 = 0$, and $a_3 = \sqrt{17}a$; their respective eigenvectors are given by

$$|a_1\rangle = \frac{1}{\sqrt{34}} \begin{pmatrix} 4 \\ -\sqrt{17} \\ 1 \end{pmatrix}, \quad |a_2\rangle = \frac{1}{\sqrt{17}} \begin{pmatrix} 1 \\ 0 \\ -4 \end{pmatrix}, \quad |a_3\rangle = \frac{1}{\sqrt{2}} \begin{pmatrix} 4 \\ \sqrt{17} \\ 1 \end{pmatrix}.$$

$$(3.184)$$

Thus, when measuring A on a system which is in the state $|\phi_2\rangle$, the probability of finding $-\sqrt{17}a$ is given by

$$P_1(a_1) = |\langle a_1|\phi_2\rangle|^2 = \left| \frac{1}{\sqrt{34}} \begin{pmatrix} 4 & -\sqrt{17} & 1 \end{pmatrix} \begin{pmatrix} 0 \\ 0 \\ 1 \end{pmatrix} \right|^2 = \frac{1}{34}. \quad (3.185)$$

Similarly, the probabilities of measuring 0 and $\sqrt{17}a$ are

$$P_2(a_2) = |\langle a_2|\phi_2\rangle|^2 = \left| \frac{1}{\sqrt{17}} \begin{pmatrix} 1 & 0 & -4 \end{pmatrix} \begin{pmatrix} 0 \\ 0 \\ 1 \end{pmatrix} \right|^2 = \frac{16}{17}, \quad (3.186)$$

$$P_3(a_3) = |\langle a_3|\phi_2\rangle|^2 = \left| \frac{1}{\sqrt{34}} \begin{pmatrix} 4 & \sqrt{17} & 1 \end{pmatrix} \begin{pmatrix} 0 \\ 0 \\ 1 \end{pmatrix} \right|^2 = \frac{1}{34}. \quad (3.187)$$

(c) Since the system, when measuring A is in the state $|\phi_2\rangle$, the uncertainty ΔA is given by $\Delta A = \sqrt{\langle\phi_2|A^2|\phi_2\rangle - \langle\phi_2|A|\phi_2\rangle^2}$, where

$$\langle\phi_2|A|\phi_2\rangle = a \begin{pmatrix} 0 & 0 & 1 \end{pmatrix} \begin{pmatrix} 0 & 4 & 0 \\ 4 & 0 & 1 \\ 0 & 1 & 0 \end{pmatrix} \begin{pmatrix} 0 \\ 0 \\ 1 \end{pmatrix} = 0, \tag{3.188}$$

$$\langle\phi_2|A^2|\phi_2\rangle = a^2 \begin{pmatrix} 0 & 0 & 1 \end{pmatrix} \begin{pmatrix} 0 & 4 & 0 \\ 4 & 0 & 1 \\ 0 & 1 & 0 \end{pmatrix} \begin{pmatrix} 0 & 4 & 0 \\ 4 & 0 & 1 \\ 0 & 1 & 0 \end{pmatrix} \begin{pmatrix} 0 \\ 0 \\ 1 \end{pmatrix} = a^2. \tag{3.189}$$

Thus we have $\Delta A = a$.

Problem 3.8

Consider a system whose state and two observables are given by

$$|\psi(t)\rangle = \begin{pmatrix} -1 \\ 2 \\ 1 \end{pmatrix}, \qquad A = \frac{1}{\sqrt{2}} \begin{pmatrix} 0 & 1 & 0 \\ 1 & 0 & 1 \\ 0 & 1 & 0 \end{pmatrix}, \qquad B = \begin{pmatrix} 1 & 0 & 0 \\ 0 & 0 & 0 \\ 0 & 0 & -1 \end{pmatrix}.$$

(a) What is the probability that a measurement of A at time t yields -1?

(b) Let us carry out a set of two measurements where B is measured first and then, immediately afterwards, A is measured. Find the probability of obtaining a value of 0 for B and a value of 1 for A.

(c) Now we measure A first then, immediately afterwards, B. Find the probability of obtaining a value of 1 for A and a value of 0 for B.

(d) Compare the results of (b) and (c). Explain.

(e) Which among the sets of operators $\{\hat{A}\}$, $\{\hat{B}\}$, and $\{\hat{A}, \hat{B}\}$ form a complete set of commuting operators (CSCO)?

Solution

(a) A measurement of A yields any of the eigenvalues of A which are given by $a_1 = -1$, $a_2 = 0$, $a_3 = 1$; the respective (normalized) eigenstates are

$$|a_1\rangle = \frac{1}{2} \begin{pmatrix} -1 \\ \sqrt{2} \\ -1 \end{pmatrix}, \qquad |a_2\rangle = \frac{1}{\sqrt{2}} \begin{pmatrix} -1 \\ 0 \\ 1 \end{pmatrix}, \qquad |a_3\rangle = \frac{1}{2} \begin{pmatrix} 1 \\ \sqrt{2} \\ 1 \end{pmatrix}. \tag{3.190}$$

The probability of obtaining $a_1 = -1$ is

$$P(-1) = \frac{|\langle a_1|\psi(t)\rangle|^2}{\langle\psi(t)|\psi(t)\rangle} = \frac{1}{6} \left| \frac{1}{2} \begin{pmatrix} -1 & \sqrt{2} & -1 \end{pmatrix} \begin{pmatrix} -1 \\ 2 \\ 1 \end{pmatrix} \right|^2 = \frac{1}{3}, \tag{3.191}$$

where we have used the fact that $\langle\psi(t)|\psi(t)\rangle = \begin{pmatrix} -1 & 2 & 1 \end{pmatrix} \begin{pmatrix} -1 \\ 2 \\ 1 \end{pmatrix} = 6$.

(b) A measurement of B yields a value which is equal to any of the eigenvalues of B: $b_1 = -1$, $b_2 = 0$, and $b_3 = 1$; their corresponding eigenvectors are

$$|b_1\rangle = \begin{pmatrix} 0 \\ 0 \\ 1 \end{pmatrix}, \qquad |b_2\rangle = \begin{pmatrix} 0 \\ 1 \\ 0 \end{pmatrix}, \qquad |b_3\rangle = \begin{pmatrix} 1 \\ 0 \\ 0 \end{pmatrix}. \tag{3.192}$$

Since the system was in the state $|\psi(t)\rangle$, the probability of obtaining the value $b_2 = 0$ for B is

$$P(b_2) = \frac{|\langle b_2|\psi(t)\rangle|^2}{\langle\psi(t)|\psi(t)\rangle} = \frac{1}{6}\left|(\,0\quad 1\quad 0\,)\begin{pmatrix}-1\\2\\1\end{pmatrix}\right|^2 = \frac{2}{3}. \tag{3.193}$$

We deal now with the measurement of the other observable, A. The observables A and B do not have common eigenstates, since they do not commute. After measuring B (the result is $b_2 = 0$), the system is left, according to Postulate 3, in a state $|\phi\rangle$ which can be found by projecting $|\psi(t)\rangle$ onto $|b_2\rangle$:

$$|\phi\rangle = |b_2\rangle\langle b_2|\psi(t)\rangle = \begin{pmatrix}0\\1\\0\end{pmatrix}(\,0\quad 1\quad 0\,)\begin{pmatrix}-1\\2\\1\end{pmatrix} = \begin{pmatrix}0\\2\\0\end{pmatrix}. \tag{3.194}$$

The probability of finding 1 when we measure A is given by

$$P(a_3) = \frac{|\langle a_3|\phi\rangle|^2}{\langle\phi|\phi\rangle} = \frac{1}{4}\left|\frac{1}{2}(\,1\quad \sqrt{2}\quad 1\,)\begin{pmatrix}0\\2\\0\end{pmatrix}\right|^2 = \frac{1}{2}, \tag{3.195}$$

since $\langle\phi|\phi\rangle = 4$. In summary, when measuring B then A, the probability of finding a value of 0 for B and 1 for A is given by the product of the probabilities (3.193) and (3.195):

$$P(b_2, a_3) = P(b_2)P(a_3) = \frac{2}{3}\frac{1}{2} = \frac{1}{3}. \tag{3.196}$$

(c) Next we measure A first then B. Since the system is in the state $|\psi(t)\rangle$, the probability of measuring $a_3 = 1$ for A is given by

$$P'(a_3) = \frac{|\langle a_3|\psi(t)\rangle|^2}{\langle\psi(t)|\psi(t)\rangle} = \frac{1}{6}\left|\frac{1}{2}(\,1\quad \sqrt{2}\quad 1\,)\begin{pmatrix}-1\\2\\1\end{pmatrix}\right|^2 = \frac{1}{3}, \tag{3.197}$$

where we have used the expression (3.190) for $|a_3\rangle$.

We then proceed to the measurement of B. The state of the system just after measuring A (with a value $a_3 = 1$) is given by a projection of $|\psi(t)\rangle$ onto $|a_3\rangle$:

$$|\chi\rangle = |a_3\rangle\langle a_3|\psi(t)\rangle = \frac{1}{4}\begin{pmatrix}1\\\sqrt{2}\\1\end{pmatrix}(\,1\quad \sqrt{2}\quad 1\,)\begin{pmatrix}-1\\2\\1\end{pmatrix} = \frac{\sqrt{2}}{2}\begin{pmatrix}1\\\sqrt{2}\\1\end{pmatrix}. \tag{3.198}$$

So the probability of finding a value of $b_2 = 0$ when measuring B is given by

$$P'(b_2) = \frac{|\langle b_2|\chi\rangle|^2}{\langle\chi|\chi\rangle} = \frac{1}{2}\left|\frac{\sqrt{2}}{2}(\,0\quad 1\quad 0\,)\begin{pmatrix}1\\\sqrt{2}\\1\end{pmatrix}\right|^2 = \frac{1}{2}, \tag{3.199}$$

since $\langle\chi|\chi\rangle = 2$.

So when measuring A then B, the probability of finding a value of 1 for A and 0 for B is given by the product of the probabilities (3.199) and (3.197):

$$P(a_3, b_2) = P'(a_3)P'(b_2) = \frac{1}{3}\frac{1}{2} = \frac{1}{6}. \tag{3.200}$$

(d) The probabilities $P(b_2, a_3)$ and $P(a_3, b_2)$, as shown in (3.196) and (3.200), are different. This is expected, since A and B do not commute. The result of the successive measurements of A and B therefore depends on the order in which they are carried out. The probability of obtaining 0 for B then 1 for A is equal to $\frac{1}{3}$. On the other hand, the probability of obtaining 1 for A then 0 for B is equal to $\frac{1}{6}$. However, if the observables A and B commute, the result of the measurements will not depend on the order in which they are carried out (this idea is illustrated in the following solved problem).

(e) As stated in the text, any operator with non-degenerate eigenvalues constitutes, all by itself, a CSCO. Hence each of $\{\hat{A}\}$ and $\{\hat{B}\}$ forms a CSCO, since their eigenvalues are not degenerate. However, the set $\{\hat{A}, \hat{B}\}$ does not form a CSCO since the opertators $\{\hat{A}\}$ and $\{\hat{B}\}$ do not commute.

Problem 3.9
Consider a system whose state and two observables A and B are given by

$$|\psi(t)\rangle = \frac{1}{6}\begin{pmatrix} 1 \\ 0 \\ 4 \end{pmatrix}, \qquad A = \frac{1}{\sqrt{2}}\begin{pmatrix} 2 & 0 & 0 \\ 0 & 1 & i \\ 0 & -i & 1 \end{pmatrix}, \qquad B = \begin{pmatrix} 1 & 0 & 0 \\ 0 & 0 & -i \\ 0 & i & 0 \end{pmatrix}.$$

(a) We perform a measurement where A is measured first and then, immediately afterwards, B is measured. Find the probability of obtaining a value of 0 for A and a value of 1 for B.

(b) Now we measure B first then, immediately afterwards, A. Find the probability of obtaining a value of 1 for B and a value of 0 for A.

(c) Compare the results of (b) and (c). Explain.

(d) Which among the sets of operators $\{\hat{A}\}$, $\{\hat{B}\}$, and $\{\hat{A}, \hat{B}\}$ form a complete set of commuting operators (CSCO)?

Solution
(a) A measurement of A yields any of the eigenvalues of A which are given by $a_1 = 0$ (not degenerate) and $a_2 = a_3 = 2$ (doubly degenerate); the respective (normalized) eigenstates are

$$|a_1\rangle = \frac{1}{\sqrt{2}}\begin{pmatrix} 0 \\ -i \\ 1 \end{pmatrix}, \qquad |a_2\rangle = \frac{1}{\sqrt{2}}\begin{pmatrix} 0 \\ i \\ 1 \end{pmatrix}, \qquad |a_3\rangle = \begin{pmatrix} 1 \\ 0 \\ 0 \end{pmatrix}. \tag{3.201}$$

The probability that a measurement of A yields $a_1 = 0$ is given by

$$P(a_1) = \frac{|\langle a_1|\psi(t)\rangle|^2}{\langle\psi(t)|\psi(t)\rangle} = \frac{36}{17}\left|\frac{1}{\sqrt{2}}\frac{1}{6}\begin{pmatrix} 0 & i & 1 \end{pmatrix}\begin{pmatrix} 1 \\ 0 \\ 4 \end{pmatrix}\right|^2 = \frac{8}{17}, \tag{3.202}$$

where we have used the fact that $\langle\psi(t)|\psi(t)\rangle = \frac{1}{36}\begin{pmatrix} 1 & 0 & 4 \end{pmatrix}\begin{pmatrix} 1 \\ 0 \\ 4 \end{pmatrix} = \frac{17}{36}.$

Since the system was initially in the state $|\psi(t)\rangle$, after a measurement of A yields $a_1 = 0$, the system is left, as mentioned in Postulate 3, in the following state:

$$|\phi\rangle = |a_1\rangle\langle a_1|\psi(t)\rangle = \frac{1}{2}\frac{1}{6}\begin{pmatrix} 0 \\ -i \\ 1 \end{pmatrix}\begin{pmatrix} 0 & i & 1 \end{pmatrix}\begin{pmatrix} 1 \\ 0 \\ 4 \end{pmatrix} = \frac{1}{3}\begin{pmatrix} 0 \\ -i \\ 1 \end{pmatrix}. \tag{3.203}$$

As for the measurement of B, we obtain any of the eigenvalues $b_1 = -1$, $b_2 = b_3 = 1$; their corresponding eigenvectors are

$$|b_1\rangle = \frac{1}{\sqrt{2}}\begin{pmatrix} 0 \\ i \\ 1 \end{pmatrix}, \qquad |b_2\rangle = \frac{1}{\sqrt{2}}\begin{pmatrix} 0 \\ -i \\ 1 \end{pmatrix}, \qquad |b_3\rangle = \begin{pmatrix} 1 \\ 0 \\ 0 \end{pmatrix}. \tag{3.204}$$

Since the system is now in the state $|\phi\rangle$, the probability of obtaining the (doubly degenerate) value $b_2 = b_3 = 1$ for B is

$$\begin{aligned} P(b_2) &= \frac{|\langle b_2|\phi\rangle|^2}{\langle\phi|\phi\rangle} + \frac{|\langle b_3|\phi\rangle|^2}{\langle\phi|\phi\rangle} \\ &= \frac{1}{2}\left|\frac{1}{\sqrt{2}}\begin{pmatrix} 0 & i & 1 \end{pmatrix}\begin{pmatrix} 0 \\ -i \\ 1 \end{pmatrix}\right|^2 + \frac{1}{2}\left|\begin{pmatrix} 1 & 0 & 0 \end{pmatrix}\begin{pmatrix} 0 \\ -i \\ 1 \end{pmatrix}\right|^2 \\ &= 1. \end{aligned} \tag{3.205}$$

The reason $P(b_2) = 1$ is because the new state $|\phi\rangle$ is an eigenstate of B; in fact $|\phi\rangle = \sqrt{2}/3|b_2\rangle$.

In sum, when measuring A then B, the probability of finding a value of 0 for A and 1 for B is given by the product of the probabilities (3.202) and (3.205):

$$P(a_1, b_2) = P(a_1)P(b_2) = \frac{8}{17}. \tag{3.206}$$

(b) Next we measure B first then A. Since the system is in the state $|\psi(t)\rangle$ and since the value $b_2 = b_3 = 1$ is doubly degenerate, the probability of measuring 1 for B is given by

$$\begin{aligned} P'(b_2) &= \frac{|\langle b_2|\psi(t)\rangle|^2}{\langle\psi(t)|\psi(t)\rangle} + \frac{|\langle b_3|\psi(t)\rangle|^2}{\langle\psi(t)|\psi(t)\rangle} \\ &= \frac{36}{17}\frac{1}{36}\left[\left|\frac{1}{\sqrt{2}}\begin{pmatrix} 0 & i & 1 \end{pmatrix}\begin{pmatrix} 1 \\ 0 \\ 4 \end{pmatrix}\right|^2 + \left|\begin{pmatrix} 1 & 0 & 0 \end{pmatrix}\begin{pmatrix} 1 \\ 0 \\ 4 \end{pmatrix}\right|^2\right] \\ &= \frac{9}{17}. \end{aligned} \tag{3.207}$$

We now proceed to the measurement of A. The state of the system immediately after measuring B (with a value $b_2 = b_3 = 1$) is given by a projection of $|\psi(t)\rangle$ onto $|b_2\rangle$, and $|b_3\rangle$

$$\begin{aligned} |\chi\rangle &= |b_2\rangle\langle b_2|\psi(t)\rangle + |b_3\rangle\langle b_3|\psi(t)\rangle \\ &= \frac{1}{12}\begin{pmatrix} 0 \\ -i \\ 1 \end{pmatrix}\begin{pmatrix} 0 & i & 1 \end{pmatrix}\begin{pmatrix} 1 \\ 0 \\ 4 \end{pmatrix} + \frac{1}{6}\begin{pmatrix} 1 \\ 0 \\ 0 \end{pmatrix}\begin{pmatrix} 1 & 0 & 0 \end{pmatrix}\begin{pmatrix} 1 \\ 0 \\ 4 \end{pmatrix} \\ &= \frac{1}{6}\begin{pmatrix} 1 \\ -2i \\ 2i \end{pmatrix}. \end{aligned} \tag{3.208}$$

So the probability of finding a value of $a_1 = 0$ when measuring A is given by

$$P'(a_1) = \frac{|\langle a_1 | \chi \rangle|^2}{\langle \chi | \chi \rangle} = \frac{36}{9} \left| \frac{1}{6\sqrt{2}} (0 \quad i \quad 1) \begin{pmatrix} 1 \\ -2i \\ 2i \end{pmatrix} \right|^2 = \frac{8}{9}, \qquad (3.209)$$

since $\langle \chi | \chi \rangle = \frac{9}{36}$.

Therefore, when measuring B then A, the probability of finding a value of 1 for B and 0 for A is given by the product of the probabilities (3.207) and (3.209):

$$P(b_2, a_3) = P'(b_2) P'(a_1) = \frac{9}{17} \frac{8}{9} = \frac{8}{17}. \qquad (3.210)$$

(c) The probabilities $P(a_1, b_2)$ and $P(b_2, a_1)$, as shown in (3.206) and (3.210), are equal. This is expected since A and B do commute. The result of the successive measurements of A and B does not depend on the order in which they are carried out.

(d) Neither $\{\hat{A}\}$ nor $\{\hat{B}\}$ forms a CSCO since their eigenvalues are degenerate. The set $\{\hat{A}, \hat{B}\}$, however, does form a CSCO since the opertators $\{\hat{A}\}$ and $\{\hat{B}\}$ commute. The set of eigenstates that are common to $\{\hat{A}, \hat{B}\}$ are given by

$$|a_2, b_1\rangle = \frac{1}{\sqrt{2}} \begin{pmatrix} 0 \\ i \\ 1 \end{pmatrix}, \qquad |a_1, b_2\rangle = \frac{1}{\sqrt{2}} \begin{pmatrix} 0 \\ -i \\ 1 \end{pmatrix}, \qquad |a_3, b_3\rangle = \begin{pmatrix} 1 \\ 0 \\ 0 \end{pmatrix}. \qquad (3.211)$$

Problem 3.10

Consider a physical system which has a number of observables that are represented by the following matrices:

$$A = \begin{pmatrix} 5 & 0 & 0 \\ 0 & 1 & 2 \\ 0 & 2 & 1 \end{pmatrix}, \quad B = \begin{pmatrix} 1 & 0 & 0 \\ 0 & 0 & 3 \\ 0 & 3 & 0 \end{pmatrix}, \quad C = \begin{pmatrix} 0 & 3 & 0 \\ 3 & 0 & 2 \\ 0 & 2 & 0 \end{pmatrix}, \quad D = \begin{pmatrix} 1 & 0 & 0 \\ 0 & 0 & -i \\ 0 & i & 0 \end{pmatrix}.$$

(a) Find the results of the measurements of these observables.

(b) Which among these observables are compatible? Give a basis of eigenvectors common to these observables.

(c) Which among the sets of operators $\{\hat{A}\}$, $\{\hat{B}\}$, $\{\hat{C}\}$, $\{\hat{D}\}$ and their various combinations, such as $\{\hat{A}, \hat{B}\}$, $\{\hat{A}, \hat{C}\}$, $\{\hat{B}, \hat{C}\}$, $\{\hat{A}, \hat{D}\}$, $\{\hat{A}, \hat{B}, \hat{C}\}$, form a complete set of commuting operators (CSCO)?

Solution

(a) The measurements of A, B, C and D yield $a_1 = -1$, $a_2 = 3$, $a_3 = 5$, $b_1 = -3$, $b_2 = 1$, $b_3 = 3$, $c_1 = -1/\sqrt{2}$, $c_2 = 0$, $c_3 = 1/\sqrt{2}$, $d_1 = -1$, $d_2 = d_3 = 1$; the respective eigenvectors of A, B, C and D are

$$|a_1\rangle = \frac{1}{\sqrt{2}} \begin{pmatrix} 0 \\ -1 \\ 1 \end{pmatrix}, \qquad |a_2\rangle = \frac{1}{\sqrt{2}} \begin{pmatrix} 0 \\ 1 \\ 1 \end{pmatrix}, \qquad |a_3\rangle = \begin{pmatrix} 1 \\ 0 \\ 0 \end{pmatrix}, \qquad (3.212)$$

$$|b_1\rangle = \frac{1}{\sqrt{2}} \begin{pmatrix} 0 \\ -1 \\ 1 \end{pmatrix}, \qquad |b_2\rangle = \begin{pmatrix} 1 \\ 0 \\ 0 \end{pmatrix}, \qquad |b_3\rangle = \frac{1}{\sqrt{2}} \begin{pmatrix} 0 \\ 1 \\ 1 \end{pmatrix}, \qquad (3.213)$$

$$|c_1\rangle = \frac{1}{\sqrt{26}} \begin{pmatrix} 3 \\ -\sqrt{13} \\ 2 \end{pmatrix}, \quad |c_2\rangle = \frac{1}{\sqrt{13}} \begin{pmatrix} 2 \\ 0 \\ -3 \end{pmatrix}, \quad |c_3\rangle = \frac{1}{\sqrt{26}} \begin{pmatrix} 3 \\ \sqrt{13} \\ 2 \end{pmatrix}, \quad (3.214)$$

$$|d_1\rangle = \frac{1}{\sqrt{2}} \begin{pmatrix} 0 \\ i \\ 1 \end{pmatrix}, \quad |d_2\rangle = \begin{pmatrix} 1 \\ 0 \\ 0 \end{pmatrix}, \quad |d_3\rangle = \frac{1}{\sqrt{2}} \begin{pmatrix} 0 \\ 1 \\ i \end{pmatrix}. \quad (3.215)$$

(b) We can verify that, among the observables $A, B, C,$ and D, only A and B are compatible, since the matrices A and B commute; the rest do not commute with one another (neither A nor B commutes with C or D; C and D do not commute).

From (3.212) and (3.213) we see that the three states $|a_1, b_1\rangle, |a_2, b_3\rangle, |a_3, b_2\rangle$,

$$|a_1, b_1\rangle = \frac{1}{\sqrt{2}} \begin{pmatrix} 0 \\ -1 \\ 1 \end{pmatrix}, \quad |a_2, b_3\rangle = \frac{1}{\sqrt{2}} \begin{pmatrix} 0 \\ 1 \\ 1 \end{pmatrix}, \quad |a_3, b_2\rangle = \begin{pmatrix} 1 \\ 0 \\ 0 \end{pmatrix}, \quad (3.216)$$

form a common, complete basis for A and B, since $\hat{A}|a_n, b_m\rangle = a_n|a_n, b_m\rangle$ and $\hat{B}|a_n, b_m\rangle = b_m|a_n, b_m\rangle$.

(c) First, since the eigenvalues of the operators $\{\hat{A}\}, \{\hat{B}\},$ and $\{\hat{C}\}$ are all nondegenerate, each one of $\{\hat{A}\}, \{\hat{B}\},$ and $\{\hat{C}\}$ forms separately a CSCO. Additionally, since two eigenvalues of $\{\hat{D}\}$ are degenerate ($d_2 = d_3 = 1$), the operator $\{\hat{D}\}$ does not form a CSCO.

Now, among the various combinations $\{\hat{A}, \hat{B}\}, \{\hat{A}, \hat{C}\}, \{\hat{B}, \hat{C}\}, \{\hat{A}, \hat{D}\},$ and $\{\hat{A}, \hat{B}, \hat{C}\}$, only $\{\hat{A}, \hat{B}\}$ forms a CSCO, because $\{\hat{A}\}$ and $\{\hat{B}\}$ are the only operators that commute; the set of their joint eigenvectors are given by $|a_1, b_1\rangle, |a_2, b_3\rangle, |a_3, b_2\rangle$.

Problem 3.11

Consider a system whose initial state $|\psi(0)\rangle$ and Hamiltonian are given by

$$|\psi(0)\rangle = \frac{1}{5} \begin{pmatrix} 3 \\ 0 \\ 4 \end{pmatrix}, \quad H = \begin{pmatrix} 3 & 0 & 0 \\ 0 & 0 & 5 \\ 0 & 5 & 0 \end{pmatrix}.$$

(a) If a measurement of the energy is carried out, what values would we obtain and with what probabilities?

(b) Find the state of the system at a later time t; you may need to expand $|\psi(0)\rangle$ in terms of the eigenvectors of H.

(c) Find the total energy of the system at time $t = 0$ and any later time t; are these values different?

(d) Does $\{\hat{H}\}$ form a complete set of commuting operators?

Solution

(a) A measurement of the energy yields the values $E_1 = -5, E_2 = 3, E_3 = 5$; the respective (orthonormal) eigenvectors of these values are

$$|\phi_1\rangle = \frac{1}{\sqrt{2}} \begin{pmatrix} 0 \\ -1 \\ 1 \end{pmatrix}, \quad |\phi_2\rangle = \begin{pmatrix} 1 \\ 0 \\ 0 \end{pmatrix}, \quad |\phi_3\rangle = \frac{1}{\sqrt{2}} \begin{pmatrix} 0 \\ 1 \\ 1 \end{pmatrix}. \quad (3.217)$$

The probabilities of finding the values $E_1 = -5$, $E_2 = 3$, $E_3 = 5$ are given by

$$P(E_1) = |\langle \phi_1 | \psi(0) \rangle|^2 = \left| \frac{1}{5\sqrt{2}} \begin{pmatrix} 0 & -1 & 1 \end{pmatrix} \begin{pmatrix} 3 \\ 0 \\ 4 \end{pmatrix} \right|^2 = \frac{8}{25},$$ (3.218)

$$P(E_2) = |\langle \phi_2 | \psi(0) \rangle|^2 = \left| \frac{1}{5} \begin{pmatrix} 1 & 0 & 0 \end{pmatrix} \begin{pmatrix} 3 \\ 0 \\ 4 \end{pmatrix} \right|^2 = \frac{9}{25},$$ (3.219)

$$P(E_3) = |\langle \phi_3 | \psi(0) \rangle|^2 = \left| \frac{1}{5\sqrt{2}} \begin{pmatrix} 0 & 1 & 1 \end{pmatrix} \begin{pmatrix} 3 \\ 0 \\ 4 \end{pmatrix} \right|^2 = \frac{8}{25}.$$ (3.220)

(b) To find $|\psi(t)\rangle$ we need to expand $|\psi(0)\rangle$ in terms of the eigenvectors (3.217):

$$|\psi(0)\rangle = \frac{1}{5} \begin{pmatrix} 3 \\ 0 \\ 4 \end{pmatrix} = \frac{2\sqrt{2}}{5} |\phi_1\rangle + \frac{3}{5} |\phi_2\rangle + \frac{2\sqrt{2}}{5} |\phi_3\rangle);$$ (3.221)

hence

$$|\psi(t)\rangle = \frac{2\sqrt{2}}{5} e^{-iE_1 t} |\phi_1\rangle + \frac{3}{5} e^{-iE_2 t} |\phi_2\rangle + \frac{2\sqrt{2}}{5} e^{-iE_3 t} |\phi_3\rangle = \frac{1}{5} \begin{pmatrix} 3e^{-3it} \\ -4i \sin 5t \\ 4 \cos 5t \end{pmatrix}.$$ (3.222)

(c) We can calculate the energy at time $t = 0$ in three quite different ways. The first method uses the bra-ket notation. Since $\langle \psi(0) | \psi(0) \rangle = 1$, $\langle \phi_n | \phi_m \rangle = \delta_{nm}$ and since $\hat{H} | \phi_n \rangle = E_n | \phi_n \rangle$, we have

$$E(0) = \langle \psi(0) | \hat{H} | \psi(0) \rangle = \frac{8}{25} \langle \phi_1 | \hat{H} | \phi_1 \rangle + \frac{9}{25} \langle \phi_2 | \hat{H} | \phi_2 \rangle + \frac{8}{25} \langle \phi_3 | \hat{H} | \phi_3 \rangle$$

$$= \frac{8}{25}(-5) + \frac{9}{25}(3) + \frac{8}{25}(5) = \frac{27}{25}.$$ (3.223)

The second method uses matrix algebra:

$$E(0) = \langle \psi(0) | \hat{H} | \psi(0) \rangle = \frac{1}{25} \begin{pmatrix} 3 & 0 & 4 \end{pmatrix} \begin{pmatrix} 3 & 0 & 0 \\ 0 & 0 & 5 \\ 0 & 5 & 0 \end{pmatrix} \begin{pmatrix} 3 \\ 0 \\ 4 \end{pmatrix} = \frac{27}{25}.$$ (3.224)

The third method uses the probabilities:

$$E(0) = \sum_{n=1}^{2} P(E_n) E_n = \frac{8}{25}(-5) + \frac{9}{25}(3) + \frac{8}{25}(5) = \frac{27}{25}.$$ (3.225)

The energy at a time t is

$$E(t) = \langle \psi(t) | \hat{H} | \psi(t) \rangle = \frac{8}{25} e^{iE_1 t} e^{-iE_1 t} \langle \phi_1 | \hat{H} | \phi_1 \rangle + \frac{9}{25} e^{iE_2 t} e^{-iE_2 t} \langle \phi_2 | \hat{H} | \phi_2 \rangle$$

$$+ \frac{8}{25} e^{iE_3 t} e^{-iE_3 t} \langle \phi_3 | \hat{H} | \phi_3 \rangle = \frac{8}{25}(-5) + \frac{9}{25}(3) + \frac{8}{25}(5) = \frac{27}{25} = E(0).$$ (3.226)

As expected, $E(t) = E(0)$ since $d\langle \hat{H} \rangle / dt = 0$.

(d) Since none of the eigenvalues of $\hat{H}$ is degenerate, the eigenvectors $|\phi_1\rangle, |\phi_2\rangle, |\phi_3\rangle$ form a compete (orthonormal) basis. Thus $\{\hat{H}\}$ forms a complete set of commuting operators.

Problem 3.12

(a) Calculate the Poisson bracket between the x and y components of the classical orbital angular momentum.

(b) Calculate the commutator between the x and y components of the orbital angular momentum operator.

(c) Compare the results obtained in (a) and (b).

Solution

(a) Using the definition (3.113) we can write the Poisson bracket $\{l_x, l_y\}$ as

$$\{l_x, l_y\} = \sum_{j=1}^{3} \left(\frac{\partial l_x}{\partial q_j} \frac{\partial l_y}{\partial p_j} - \frac{\partial l_x}{\partial p_j} \frac{\partial l_y}{\partial q_j} \right), \tag{3.227}$$

where $q_1 = x$, $q_2 = y$, $q_3 = z$, $p_1 = p_x$, $p_2 = p_y$, and $p_3 = p_z$. Since $l_x = yp_z - zp_y$, $l_y = zp_x - xp_z$, $l_z = xp_y - yp_x$, the only partial derivatives that survive are $\partial l_x / \partial z = -p_y$, $\partial l_y / \partial p_z = -x$, $\partial l_x / \partial p_z = y$, and $\partial l_y / \partial z = p_x$. Thus, we have

$$\{l_x, l_y\} = \frac{\partial l_x}{\partial z} \frac{\partial l_y}{\partial p_z} - \frac{\partial l_x}{\partial p_z} \frac{\partial l_y}{\partial z} = xp_y - yp_x = l_z. \tag{3.228}$$

(b) The components of $\hat{\vec{L}}$ are listed in (3.26) to (3.28): $\hat{L}_x = \hat{Y}\hat{P}_z - \hat{Z}\hat{P}_y$, $\hat{L}_y = \hat{Z}\hat{P}_x - \hat{X}\hat{P}_z$, and $\hat{L}_z = \hat{X}\hat{P}_y - \hat{Y}\hat{P}_x$. Since $\hat{X}$, $\hat{Y}$, and $\hat{Z}$ mutually commute and so do $\hat{P}_x$, $\hat{P}_y$, and $\hat{P}_z$, we have

$$\begin{aligned}
[\hat{L}_x, \hat{L}_y] &= [\hat{Y}\hat{P}_z - \hat{Z}\hat{P}_y, \ \hat{Z}\hat{P}_x - \hat{X}\hat{P}_z] \\
&= [\hat{Y}\hat{P}_z, \hat{Z}\hat{P}_x] - [\hat{Y}\hat{P}_z, \hat{X}\hat{P}_z] - [\hat{Z}\hat{P}_y, \hat{Z}\hat{P}_x] + [\hat{Z}\hat{P}_y, \hat{X}\hat{P}_z] \\
&= \hat{Y}[\hat{P}_z, \hat{Z}]\hat{P}_x + \hat{X}[\hat{Z}, \hat{P}_z]\hat{P}_y = i\hbar(\hat{X}\hat{P}_y - \hat{Y}\hat{P}_x) \\
&= i\hbar\hat{L}_z.
\end{aligned} \tag{3.229}$$

(c) A comparison of (3.228) and (3.229) shows that

$$\{l_x, l_y\} = l_z \longrightarrow [\hat{L}_x, \hat{L}_y] = i\hbar\hat{L}_z. \tag{3.230}$$

Problem 3.13

Consider a charged oscillator, of positive charge q and mass m, which is subject to an oscillating electric field $E_0 \cos \omega t$; the particle's Hamiltonian is $\hat{H} = P^2/(2m) + k\hat{X}^2/2 + qE_0\hat{X} \cos \omega t$.

(a) Calculate $d\langle \hat{X} \rangle / dt$, $d\langle \hat{P} \rangle / dt$, $d\langle \hat{H} \rangle / dt$.

(b) Solve the equation for $d\langle \hat{X} \rangle / dt$ and obtain $\langle \hat{X} \rangle (t)$ such that $\langle \hat{X} \rangle (0) = x_0$.

Solution

(a) Since the position operator $\hat{X}$ does not depend explicitly on time (i.e., $\partial \hat{X}/\partial t = 0$), equation (3.88) yields

$$\frac{d}{dt}\langle \hat{X} \rangle = \frac{1}{i\hbar}\langle [\hat{X}, \hat{H}] \rangle = \frac{1}{i\hbar}\langle \left[\hat{X}, \frac{\hat{P}^2}{2m} \right] \rangle = \frac{\langle \hat{P} \rangle}{m}. \tag{3.231}$$

Now, since $[\hat{P}, \hat{X}] = -i\hbar$, $[\hat{P}, \hat{X}^2] = -2i\hbar\hat{X}$ and $\partial \hat{P}/\partial t = 0$, we have

$$\frac{d}{dt}\langle \hat{P} \rangle = \frac{1}{i\hbar}\langle [\hat{P}, \hat{H}] \rangle = \frac{1}{i\hbar}\langle \left[\hat{P}, \frac{1}{2}k\hat{X}^2 + q E_0 \hat{X} \cos \omega t \right] \rangle = -k\langle \hat{X} \rangle - q E_0 \cos \omega t,$$
$$\tag{3.232}$$

$$\frac{d}{dt}\langle \hat{H} \rangle = \frac{1}{i\hbar}\langle [\hat{H}, \hat{H}] \rangle + \langle \frac{\partial \hat{H}}{\partial t} \rangle = \langle \frac{\partial \hat{H}}{\partial t} \rangle = -q E_0 \omega \langle \hat{X} \rangle \sin \omega t. \tag{3.233}$$

(b) To find $\langle \hat{X} \rangle$ we need to take a time derivative of (3.231) and then make use of (3.232):

$$\frac{d^2}{dt^2}\langle \hat{X} \rangle = \frac{1}{m}\frac{d}{dt}\langle \hat{P} \rangle = -\frac{k}{m}\langle \hat{X} \rangle - \frac{q E_0}{m} \cos \omega t. \tag{3.234}$$

The solution of this equation is

$$\langle \hat{X} \rangle(t) = \langle \hat{X} \rangle(0) \cos \left(\sqrt{\frac{k}{m}}t \right) - \frac{q E_0}{m\omega} \sin \omega t + A, \tag{3.235}$$

where A is a constant which can be determined from the initial conditions; since $\langle \hat{X} \rangle(0) = x_0$ we have $A = 0$, and hence

$$\langle \hat{X} \rangle(t) = x_0 \cos \left(\sqrt{\frac{k}{m}}t \right) - \frac{q E_0}{m\omega} \sin \omega t. \tag{3.236}$$

Problem 3.14
Consider a one-dimensional free particle of mass m whose position and momentum at time $t = 0$ are given by x_0 and p_0, respectively.
(a) Calculate $\langle \hat{P} \rangle(t)$ and show that $\langle \hat{X} \rangle(t) = p_0 t^2/m + x_0$.
(b) Show that $d\langle \hat{X}^2 \rangle/dt = 2\langle \hat{P}\hat{X} \rangle/m + i\hbar/m$ and $d\langle \hat{P}^2 \rangle/dt = 0$.
(c) Show that the position and momentum fluctuations are related by $d^2(\Delta x)^2/dt^2 = 2(\Delta p)^2/m^2$ and that the solution to this equation is given by $(\Delta x)^2 = (\Delta p)_0^2 t^2/m^2 + (\Delta x)_0^2$ where $(\Delta x)_0$ and $(\Delta p)_0$ are the initial fluctuations.

Solution
(a) From the Ehrenfest equations $d\langle \hat{P} \rangle/dt = \langle [\hat{P}, \hat{V}(x, t)] \rangle/i\hbar$ as shown in (3.134), and since for a free particle $\hat{V}(x, t) = 0$, we see that $d\langle \hat{P} \rangle/dt = 0$. As expected this leads to $\langle \hat{P} \rangle(t) = p_0$, since the linear momentum of a free particle is conserved. Inserting $\langle \hat{P} \rangle = p_0$ into Ehrenfest's other equation $d\langle \hat{X} \rangle/dt = \langle \hat{P} \rangle/m$ (see (3.132)), we obtain

$$\frac{d\langle \hat{X} \rangle}{dt} = \frac{1}{m}p_0. \tag{3.237}$$

The solution of this equation with the initial condition $\langle \hat{X} \rangle(0) = x_0$ is

$$\langle \hat{X} \rangle(t) = \frac{p_0}{m}t + x_0. \tag{3.238}$$

(b) First, the proof of $d\langle \hat{P}^2 \rangle/dt = 0$ is straightforward. Since $[\hat{P}^2, \hat{H}] = [\hat{P}^2, \hat{P}^2/2m] = 0$ and $\partial\hat{P}^2/\partial t = 0$ (the momentum operator does not depend on time), (3.124) yields

$$\frac{d}{dt}\langle \hat{P}^2 \rangle = \frac{1}{i\hbar}\langle [\hat{P}^2, \hat{H}] \rangle + \langle \frac{\partial \hat{P}^2}{\partial t} \rangle = 0. \tag{3.239}$$

For $d\langle \hat{X}^2 \rangle/dt$ we have

$$\frac{d}{dt}\langle \hat{X}^2 \rangle = \frac{1}{i\hbar}\langle [\hat{X}^2, \hat{H}] \rangle = \frac{1}{2im\hbar}\langle [\hat{X}^2, \hat{P}^2] \rangle, \tag{3.240}$$

since $\partial\hat{X}^2/\partial t = 0$. Using $[\hat{X}, \hat{P}] = i\hbar$, we obtain

$$\begin{aligned}
[\hat{X}^2, \hat{P}^2] &= \hat{P}[\hat{X}^2, \hat{P}] + [\hat{X}^2, \hat{P}]\hat{P} \\
&= \hat{P}\hat{X}[\hat{X}, \hat{P}] + \hat{P}[\hat{X}, \hat{P}]\hat{X} + \hat{X}[\hat{X}, \hat{P}]\hat{P} + [\hat{X}, \hat{P}]\hat{X}\hat{P} \\
&= 2i\hbar(\hat{P}\hat{X} + \hat{X}\hat{P}) = 2i\hbar(2\hat{P}\hat{X} + i\hbar);
\end{aligned} \tag{3.241}$$

hence

$$\frac{d}{dt}\langle \hat{X}^2 \rangle = \frac{2}{m}\langle \hat{P}\hat{X} \rangle + \frac{i\hbar}{m}. \tag{3.242}$$

(c) As the position fluctuation is given by $(\Delta x)^2 = \langle \hat{X}^2 \rangle - \langle \hat{X} \rangle^2$, we have

$$\frac{d(\Delta x)^2}{dt} = \frac{d\langle \hat{X}^2 \rangle}{dt} - 2\langle \hat{X} \rangle\frac{d\langle \hat{X} \rangle}{dt} = \frac{2}{m}\langle \hat{P}\hat{X} \rangle + \frac{i\hbar}{m} - \frac{2}{m}\langle \hat{X} \rangle\langle \hat{P} \rangle. \tag{3.243}$$

In deriving this expression we have used (3.242) and $d\langle \hat{X} \rangle/dt = \langle \hat{P} \rangle/m$. Now, since $d(\langle \hat{X} \rangle\langle \hat{P} \rangle)/dt = \langle \hat{P} \rangle d\langle \hat{X} \rangle/dt = \langle \hat{P} \rangle^2/m$ and

$$\frac{d\langle \hat{P}\hat{X} \rangle}{dt} = \frac{1}{i\hbar}\langle [\hat{P}\hat{X}, \hat{H}] \rangle = \frac{1}{2im\hbar}\langle [\hat{P}\hat{X}, \hat{P}^2] \rangle = \frac{1}{m}\langle \hat{P}^2 \rangle, \tag{3.244}$$

we can write the second time derivative of (3.243) as follows:

$$\frac{d^2(\Delta x)^2}{dt^2} = \frac{2}{m}\left(\frac{d\langle \hat{P}\hat{X} \rangle}{dt} - \frac{d\langle \hat{X} \rangle\langle \hat{P} \rangle}{dt} \right) = \frac{2}{m^2}\left(\langle \hat{P}^2 \rangle - \langle \hat{P} \rangle^2 \right) = \frac{2}{m^2}(\Delta p)_0^2, \tag{3.245}$$

where $(\Delta p)_0^2 = \langle \hat{P}^2 \rangle - \langle \hat{P} \rangle^2 = \langle \hat{P}^2 \rangle_0 - \langle \hat{P} \rangle_0^2$; the momentum of the free particle is a constant of the motion. We can verify that the solution of the differential equation (3.245) is given by

$$(\Delta x)^2 = \frac{1}{m^2}(\Delta p)_0^2 t^2 + (\Delta x)_0^2. \tag{3.246}$$

This fluctuation is similar to the spreading of a Gaussian wave packet we derived in Chapter 1.

3.10 Exercises

Exercise 3.1

A particle in an infinite potential box with walls at $x = 0$ and $x = a$ (i.e., the potential is infinite for $x < 0$ and $x > a$ and zero in between) has the following wave function at some initial time:

$$\psi(x) = \frac{1}{\sqrt{5a}} \sin\left(\frac{\pi x}{a}\right) + \frac{2}{\sqrt{5a}} \sin\left(\frac{3\pi x}{a}\right).$$

(a) Find the possible results of the measurement of the system's energy and the corresponding probabilities.

(b) Find the form of the wave function after such a measurement.

(c) If the energy is measured again immediately afterwards, what are the relative probabilities of the possible outcomes?

Exercise 3.2

Let $\psi_n(x)$ denote the orthonormal stationary states of a system corresponding to the energy E_n. Suppose that the normalized wave function of the system at time $t = 0$ is $\psi(x, 0)$ and suppose that a measurement of the energy yields the value E_1 with probability 1/2, E_2 with probability 3/8, and E_3 with probability 1/8.

(a) Write the most general expansion for $\psi(x, 0)$ consistent with this information.

(b) What is the expansion for the wave function of the system at time t, $\psi(x, t)$?

(c) Show that the expectation value of the Hamiltonian does not change with time.

Exercise 3.3

Consider a neutron which is confined to an infinite potential well of width $a = 8$ fm. At time $t = 0$ the neutron is assumed to be in the state

$$\Psi(x, 0) = \sqrt{\frac{4}{7a}} \sin\left(\frac{\pi x}{a}\right) + \sqrt{\frac{2}{7a}} \sin\left(\frac{2\pi x}{a}\right) + \sqrt{\frac{8}{7a}} \sin\left(\frac{3\pi x}{a}\right).$$

(a) If an energy measurement is carried out on the system, what are the values that will be found for the energy and with what probabilities? Express your answer in MeV (the mass of the neutron is $mc^2 \simeq 939$ MeV, $\hbar c \simeq 197$ MeV fm).

(b) If this measurement is repeated on many identical systems, what is the average value of the energy that will be found? Again, express your answer in MeV.

(c) Using the uncertainty principle, estimate the order of magnitude of the neutron's speed in this well as a function of the speed of light c.

Exercise 3.4

Consider the dimensionless harmonic oscillator Hamiltonian

$$\hat{H} = \frac{1}{2}\hat{P}^2 + \frac{1}{2}\hat{X}^2, \quad \text{with} \quad \hat{P} = -i\frac{d}{dx}.$$

(a) Show that the two wave functions $\psi_0(x) = e^{-x^2/2}$ and $\psi_1(x) = xe^{-x^2/2}$ are eigenfunctions of $\hat{H}$ with eigenvalues 1/2 and 3/2, respectively.

(b) Find the value of the coefficient α such that $\psi_2(x) = (1 + \alpha x^2) e^{-x^2/2}$ is orthogonal to $\psi_0(x)$. Then show that $\psi_2(x)$ is an eigenfunction of $\hat{H}$ with eigenvalue 5/2.

Exercise 3.5

Consider that the wave function of a dimensionless harmonic oscillator, whose Hamiltonian is $\hat{H} = \frac{1}{2}\hat{P}^2 + \frac{1}{2}\hat{X}^2$, is given at time $t = 0$ by

$$\psi(x, 0) = \frac{1}{\sqrt{8\pi}}\phi_0(x) + \frac{1}{\sqrt{18\pi}}\phi_2(x) = \frac{1}{\sqrt{8\pi}}e^{-x^2/2} + \frac{1}{\sqrt{18\pi}}\left(1 - 2x^2\right)e^{-x^2/2}.$$

(a) Find the expression of the oscillator's wave function at any later time t.

(b) Calculate the probability P_0 to find the system in an eigenstate of energy $1/2$ and the probability P_2 of finding the system in an eigenstate of energy $5/2$.

(c) Calculate the probability density, $\rho(x, t)$, and the current density, $\vec{J}(x, t)$.

(d) Verify that the probability is conserved; that is, show that $\partial\rho/\partial t + \vec{\nabla} \cdot \vec{J}(x, t) = 0$.

Exercise 3.6

A particle of mass m, in an infinite potential well of length a, has the following initial wave function at $t = 0$:

$$\psi(x, 0) = \sqrt{\frac{3}{5a}}\sin\left(\frac{3\pi x}{a}\right) + \frac{1}{\sqrt{5a}}\sin\left(\frac{5\pi x}{a}\right), \qquad (3.247)$$

and an energy spectrum $E_n = -\hbar^2\pi^2 n^2/(2ma^2)$.

Find $\psi(x, t)$ at any later time t, then calculate $\frac{\partial\rho}{\partial t}$ and the probability current density vector $\vec{J}(x, t)$ and verify that $\frac{\partial\rho}{\partial t} + \vec{\nabla} \cdot \vec{J}(x, t) = 0$. Recall that $\rho = \psi^*(x, t)\psi(x, t)$ and $\vec{J}(x, t) = \frac{i\hbar}{2m}\left(\psi(x, t)\vec{\nabla}\psi^*(x, t) - \psi^*(x, t)\vec{\nabla}\psi(x, t)\right)$.

Exercise 3.7

Consider a system whose initial state at $t = 0$ is given in terms of a complete and orthonormal set of three vectors: $|\phi_1\rangle$, $|\phi_2\rangle$, $|\phi_3\rangle$ as follows: $|\psi(0)\rangle = 1/\sqrt{3}|\phi_1\rangle + A|\phi_2\rangle + 1/\sqrt{6}|\phi_3\rangle$, where A is a real constant.

(a) Find A so that $|\psi(0)\rangle$ is normalized.

(b) If the energies corresponding to $|\phi_1\rangle$, $|\phi_2\rangle$, $|\phi_3\rangle$ are given by E_1, E_2, and E_3, respectively, write down the state of the system $|\psi(t)\rangle$ at any later time t.

(c) Determine the probability of finding the system at a time t in the state $|\phi_3\rangle$.

Exercise 3.8

The components of the initial state $|\psi_i\rangle$ of a quantum system are given in a complete and orthonormal basis of three states $|\phi_1\rangle$, $|\phi_2\rangle$, $|\phi_3\rangle$ by

$$\langle\phi_1|\psi_i\rangle = \frac{i}{\sqrt{3}}, \quad \langle\phi_2|\psi_i\rangle = \sqrt{\frac{2}{3}}, \quad \langle\phi_3|\psi_i\rangle = 0.$$

Calculate the probability of finding the system in a state $|\psi_f\rangle$ whose components are given in the same basis by

$$\langle\phi_1|\psi_f\rangle = \frac{1+i}{\sqrt{3}}, \quad \langle\phi_2|\psi_f\rangle = \frac{1}{\sqrt{6}}, \quad \langle\phi_3|\psi_f\rangle = \frac{1}{\sqrt{6}}.$$

Exercise 3.9

(a) Evaluate the Poisson bracket $\{x^2, p^2\}$.

(b) Express the commutator $\left[\hat{X}^2, \hat{P}^2\right]$ in terms of $\hat{X}\hat{P}$ plus a constant in $\hbar^2$.

(c) Find the classical limit of $\left[\hat{x}^2, \hat{p}^2\right]$ for this expression and then compare it with the result of part (a).

Exercise 3.10

A particle bound in a one-dimensional potential has a wave function

$$\psi(x) = \begin{cases} Ae^{5ikx}\cos(3\pi x/a), & -a/2 \leq x \leq a/2, \\ 0, & |x| > a/2. \end{cases}$$

(a) Calculate the constant A so that $\psi(x)$ is normalized.

(b) Calculate the probability of finding the particle between $x = 0$ and $x = a/4$.

Exercise 3.11

(a) Show that any component of the momentum operator of a particle is compatible with its kinetic energy operator.

(b) Show that the momentum operator is compatible with the Hamiltonian operator only if the potential operator is constant in space coordinates.

Exercise 3.12

Consider a physical system whose Hamiltonian H and an operator A are given by

$$H = \mathcal{E}_0 \begin{pmatrix} -2 & 0 & 0 \\ 0 & 1 & 0 \\ 0 & 0 & 1 \end{pmatrix}, \quad A = a_0 \begin{pmatrix} 5 & 0 & 0 \\ 0 & 0 & 2 \\ 0 & 2 & 0 \end{pmatrix},$$

where $\mathcal{E}_0$ has the dimensions of energy.

(a) Do H and A commute? If yes, give a basis of eigenvectors common to H and A.

(b) Which among the sets of operators $\{\hat{H}\}$, $\{\hat{A}\}$, $\{\hat{H}, \hat{A}\}$, $\{\hat{H}^2, \hat{A}\}$ form a complete set of commuting operators (CSCO)?

Exercise 3.13

Show that the momentum and the total energy can be measured simultaneously only when the potential is constant everywhere.

Exercise 3.14

The initial state of a system is given in terms of four orthonormal energy eigenfunctions $|\phi_1\rangle$, $|\phi_2\rangle$, $|\phi_3\rangle$, and $|\phi_4\rangle$ as follows:

$$|\psi_0\rangle = |\psi(t = 0)\rangle = \frac{1}{\sqrt{3}}|\phi_1\rangle + \frac{1}{2}|\phi_2\rangle + \frac{1}{\sqrt{6}}|\phi_3\rangle + \frac{1}{2}|\phi_4\rangle.$$

(a) If the four kets $|\phi_1\rangle$, $|\phi_2\rangle$, $|\phi_3\rangle$, and $|\phi_4\rangle$ are eigenvectors to the Hamiltonian $\hat{H}$ with energies E_1, E_2, E_3, and E_4, respectively, find the state $|\psi(t)\rangle$ at any later time t.

(b) What are the possible results of measuring the energy of this system and with what probability will they occur?

(c) Find the expectation value of the system's Hamiltonian at $t = 0$ and $t = 10$ s.

Exercise 3.15

The complete set expansion of an initial wave function $\psi(x, 0)$ of a system in terms of orthonormal energy eigenfunctions $\phi_n(x)$ of the system has three terms, $n = 1, 2, 3$. The measurement of energy on the system represented by $\psi(x, 0)$ gives three values, E_1 and E_2 with probability $1/4$ and E_3 with probability $1/2$.

(a) Write down $\psi(x, 0)$ in terms of $\phi_1(x)$, $\phi_2(x)$, and $\phi_3(x)$.

(b) Find $\psi(x, 0)$ at any later time t, i.e., find $\psi(x, t)$.

Exercise 3.16

Consider a system whose Hamiltonian H and an operator A are given by the matrices

$$H = \mathcal{E}_0 \begin{pmatrix} 0 & -i & 0 \\ i & 0 & 2i \\ 0 & -2i & 0 \end{pmatrix}, \quad A = a_0 \begin{pmatrix} 0 & -i & 0 \\ i & 1 & 1 \\ 0 & 1 & 0 \end{pmatrix}.$$

(a) If we measure energy, what values will we obtain?

(b) Suppose that when we measure energy, we obtain a value of $\sqrt{5}\mathcal{E}_0$. Immediately afterwards, we measure A. What values will we obtain for A and what are the probabilities corresponding to each value?

(c) Calculate the expectation value $\langle A \rangle$.

Exercise 3.17

Consider a physical system whose Hamiltonian and initial state are given by

$$H = \mathcal{E}_0 \begin{pmatrix} 1 & -1 & 0 \\ -1 & 1 & 0 \\ 0 & 0 & -1 \end{pmatrix}, \quad |\psi_0\rangle = \frac{1}{\sqrt{6}} \begin{pmatrix} 1 \\ 1 \\ 2 \end{pmatrix},$$

where $\mathcal{E}_0$ has the dimensions of energy.

(a) What values will we obtain when measuring the energy and with what probabilities?

(b) Calculate the expectation value of the Hamiltonian $\langle \hat{H} \rangle$.

Exercise 3.18

Consider a system whose state $|\psi(t)\rangle$ and two observables A and B are given by

$$|\psi(t)\rangle = \begin{pmatrix} 5 \\ 1 \\ 3 \end{pmatrix}, \quad A = \frac{1}{\sqrt{2}} \begin{pmatrix} 2 & 0 & 0 \\ 0 & 1 & 1 \\ 0 & 1 & 1 \end{pmatrix}, \quad B = \begin{pmatrix} 1 & 0 & 0 \\ 0 & 0 & 1 \\ 0 & 1 & 0 \end{pmatrix}.$$

(a) We perform a measurement where A is measured first and then B immediately afterwards. Find the probability of obtaining a value of $\sqrt{2}$ for A and a value of -1 for B.

(b) Now we measure B first and then A immediately afterwards. Find the probability of obtaining a value of -1 for B and a value of $\sqrt{2}$ for A.

(c) Compare the results of (a) and (b). Explain.

Exercise 3.19

Consider a system whose state $|\psi(t)\rangle$ and two observables A and B are given by

$$|\psi(t)\rangle = \frac{1}{\sqrt{3}} \begin{pmatrix} -i \\ 2 \\ 0 \end{pmatrix}, \quad A = \frac{1}{\sqrt{2}} \begin{pmatrix} 1 & i & 1 \\ -i & 0 & 0 \\ 1 & 0 & 0 \end{pmatrix}, \quad B = \begin{pmatrix} 3 & 0 & 0 \\ 0 & 1 & i \\ 0 & -i & 0 \end{pmatrix}.$$

(a) Are A and B compatible? Which among the sets of operators $\{\hat{A}\}$, $\{\hat{B}\}$, and $\{\hat{A}, \hat{B}\}$ form a complete set of commuting operators?

(b) Measuring A first and then B immediately afterwards, find the probability of obtaining a value of -1 for A and a value of 3 for B.

(c) Now, measuring B first then A immediately afterwards, find the probability of obtaining 3 for B and -1 for A. Compare this result with the probability obtained in (b).

Exercise 3.20

Consider a physical system which has a number of observables that are represented by the following matrices:

$$A = \begin{pmatrix} 1 & 0 & 0 \\ 0 & 0 & 1 \\ 0 & 1 & 0 \end{pmatrix}, \quad B = \begin{pmatrix} 0 & 0 & -1 \\ 0 & 0 & i \\ -1 & -i & 4 \end{pmatrix}, \quad C = \begin{pmatrix} 2 & 0 & 0 \\ 0 & 1 & 3 \\ 0 & 3 & 1 \end{pmatrix}.$$

(a) Find the results of the measurements of the compatible observables.

(b) Which among these observables are compatible? Give a basis of eigenvectors common to these observables.

(c) Which among the sets of operators $\{\hat{A}\}$, $\{\hat{B}\}$, $\{\hat{C}\}$, $\{\hat{A}, \hat{B}\}$, $\{\hat{A}, \hat{C}\}$, $\{\hat{B}, \hat{C}\}$ form a complete set of commuting operators?

Exercise 3.21

Consider a system which is initially in a state $|\psi(0)\rangle$ and having a Hamiltonian $\hat{H}$, where

$$|\psi(0)\rangle = \begin{pmatrix} 4 - i \\ -2 + 5i \\ 3 + 2i \end{pmatrix}, \quad H = \frac{1}{\sqrt{2}} \begin{pmatrix} 0 & -i & 0 \\ i & 3 & 3 \\ 0 & 3 & 0 \end{pmatrix}.$$

(a) If a measurement of H is carried out, what values will we obtain and with what probabilities?

(b) Find the state of the system at a later time t; you may need to expand $|\psi(0)\rangle$ in terms of the eigenvectors of H.

(c) Find the total energy of the system at time $t = 0$ and any later time t; are these values different?

(d) Does $\hat{H}$ form a complete set of commuting operators?

Exercise 3.22

Consider a particle which moves in a scalar potential $V(\vec{r}) = V_x(x) + V_y(y) + V_z(z)$.

(a) Show that the Hamiltonian of this particle can be written as $\hat{H} = \hat{H}_x + \hat{H}_y + \hat{H}_z$, where $\hat{H}_x = p_x^2/(2m) + V_x(x)$, and so on.

(b) Do $\hat{H}_x$, $\hat{H}_y$, and $\hat{H}_z$ form a complete set of commuting operators?

Exercise 3.23

Consider a system whose Hamiltonian is $H = \mathcal{E} \begin{pmatrix} 0 & -i \\ i & 0 \end{pmatrix}$, where $\mathcal{E}$ is a real constant with the dimensions of energy.

(a) Find the eigenenergies, E_1 and E_2, of H.

(b) If the system is initially (i.e., $t = 0$) in the state $|\psi_0\rangle = \begin{pmatrix} 1 \\ 0 \end{pmatrix}$, find the probability so that a measurement of energy at $t = 0$ yields: (i) E_1, and (ii) E_2.

(c) Find the average value of the energy $\langle \hat{H} \rangle$ and the energy uncertainty $\sqrt{\langle \hat{H}^2 \rangle - \langle \hat{H} \rangle^2}$.

(d) Find the state $|\psi(t)\rangle$.

Exercise 3.24

Prove the relation

$$\frac{d}{dt}\langle \hat{A}\hat{B} \rangle = \langle \frac{\partial \hat{A}}{\partial t}\hat{B} \rangle + \langle \hat{A}\frac{\partial \hat{B}}{\partial t} \rangle + \frac{1}{i\hbar}\langle [\hat{A}, \hat{H}]\hat{B} \rangle + \frac{1}{i\hbar}\langle \hat{A}[\hat{B}, \hat{H}] \rangle.$$

Exercise 3.25

Consider a particle of mass m which moves under the influence of gravity; the particle's Hamiltonian is $\hat{H} = \hat{P}_z^2/(2m) - mg\hat{Z}$, where g is the acceleration due to gravity, $g = 9.8\,\mathrm{m\,s^{-2}}$.

(a) Calculate $d\langle \hat{Z} \rangle/dt, d\langle \hat{P}_z \rangle/dt, d\langle \hat{H} \rangle/dt$.

(b) Solve the equation $d\langle \hat{Z} \rangle/dt$ and obtain $\langle \hat{Z} \rangle(t)$, such that $\langle \hat{Z} \rangle(0) = h$ and $\langle \hat{P}_z \rangle(0) = 0$. Compare the result with the classical relation $z(t) = -\frac{1}{2}gt^2 + h$.

Exercise 3.26

Calculate $d\langle \hat{X} \rangle/dt, d\langle \hat{P}_x \rangle/dt, d\langle \hat{H} \rangle/dt$ for a particle with $\hat{H} = \hat{P}_x^2/(2m) + \frac{1}{2}m\omega^2\hat{X}^2 + V_0\hat{X}^3$.

Exercise 3.27

Consider a system whose initial state at $t = 0$ is given in terms of a complete and orthonormal set of four vectors $|\phi_1\rangle, |\phi_2\rangle, |\phi_3\rangle, |\phi_4\rangle$ as follows:

$$|\psi(0)\rangle = \frac{A}{\sqrt{12}}|\phi_1\rangle + \frac{1}{\sqrt{6}}|\phi_2\rangle + \frac{2}{\sqrt{12}}|\phi_3\rangle + \frac{1}{2}|\phi_4\rangle,$$

where A is a real constant.

(a) Find A so that $|\psi(0)\rangle$ is normalized.

(b) If the energies corresponding to $|\phi_1\rangle, |\phi_2\rangle, |\phi_3\rangle, |\phi_4\rangle$ are given by E_1, E_2, E_3, and E_4, respectively, write down the state of the system $|\psi(t)\rangle$ at any later time t.

(c) Determine the probability of finding the system at a time t in the state $|\phi_2\rangle$.

Chapter 4

One-Dimensional Problems

4.1 Introduction

After presenting the formalism of quantum mechanics in the previous two chapters, we are now well equipped to apply it to the study of physical problems. Here we apply the Schrödinger equation to *one-dimensional* problems. These problems are interesting since there exist many physical phenomena whose motion is one-dimensional. The application of the Schrödinger equation to *one-dimensional* problems enables us to compare the predictions of classical and quantum mechanics in a simple setting. In addition to being simple to solve, one-dimensional problems will be used to illustrate some nonclassical effects.

The Schrödinger equation describing the dynamics of a microscopic particle of mass m in a one-dimensional time-independent potential $V(x)$ is given by

$$-\frac{\hbar^2}{2m}\frac{d^2\psi(x)}{dx^2} + V(x)\psi(x) = E\psi(x), \tag{4.1}$$

where E is the total energy of the particle. The solutions of this equation yield the allowed energy eigenvalues E_n and the corresponding wave functions $\psi_n(x)$. To solve this partial differential equation, we need to specify the potential $V(x)$ as well as the boundary conditions; the boundary conditions can be obtained from the physical requirements of the system.

We have seen in the previous chapter that the solutions of the Schrödinger equation for time-independent potentials are stationary,

$$\Psi(x,t) = \psi(x)e^{-iEt/\hbar}, \tag{4.2}$$

for the probability density does not depend on time. Recall that the state $\psi(x)$ has the physical dimensions of $1/\sqrt{L}$, where L is a length. Hence, the physical dimension of $|\psi(x)|^2$ is $1/L$: $[|\psi(x)|^2] = 1/L$.

We begin by examining some general properties of one-dimensional motion and discussing the symmetry character of the solutions. Then, in the rest of the chapter, we apply the Schrödinger equation to various one-dimensional potentials: the free particle, the potential step, the finite and infinite potential wells, and the harmonic oscillator. We conclude by showing how to solve the Schrödinger equation numerically.

Figure 4.1 Shape of a general potential.

4.2   Properties of One-Dimensional Motion

To study the dynamic properties of a single particle moving in a one-dimensional potential, let us consider a potential $V(x)$ that is general enough to allow for the illustration of all the desired features. One such potential is displayed in Figure 4.1; it is finite at $x \rightarrow \pm\infty$, $V(-\infty) = V_1$ and $V(+\infty) = V_2$ with V_1 smaller than V_2, and it has a minimum, V_{min}. In particular, we want to study the conditions under which discrete and continuous spectra occur. As the character of the states is completely determined by the size of the system's energy, we will be considering separately the cases where the energy is smaller and larger than the potential.

4.2.1   Discrete Spectrum (Bound States)

Bound states occur whenever the particle cannot move to infinity. That is, the particle is *confined* or *bound* at all energies to move within a *finite* and limited region of space which is delimited by two classical turning points. The Schrödinger equation in this region admits only solutions that are *discrete*. The infinite square well potential and the harmonic oscillator are typical examples that display bound states.

In the potential of Figure 4.1, the motion of the particle is bounded between the classical turning points x_1 and x_2 when the particle's energy lies between V_{min} and V_1:

$$V_{min} < E < V_1. \tag{4.3}$$

The states corresponding to this energy range are called *bound* states. They are defined as states whose wave functions are finite (or zero) at $x \rightarrow \pm\infty$; usually the bound states have energies smaller than the potential $E < V$. For the bound states to exist, the potential $V(x)$ must have at least one minimum which is lower than V_1 (i.e., $V_{min} < V_1$). The energy spectra of bound states are discrete. We need to use the boundary conditions[1] to find the wave function and the energy.

Let us now list two theorems that are important to the study of bound states.

---

[1] Since the Schrödinger equation is a *second-order* differential equation, only *two* boundary conditions are required to solve it.

Theorem 4.1 *In a one-dimensional problem the energy levels of a bound state system are discrete and not degenerate.*

Theorem 4.2 *The wave function $\psi_n(x)$ of a one-dimensional bound state system has n nodes (i.e., $\psi_n(x)$ vanishes n times) if $n = 0$ corresponds to the ground state and $(n - 1)$ nodes if $n = 1$ corresponds to the ground state.*

4.2.2 Continuous Spectrum (Unbound States)

Unbound states occur in those cases where the motion of the system is not confined; a typical example is the *free* particle. For the potential displayed in Figure 4.1 there are two energy ranges where the particle's motion is infinite: $V_1 < E < V_2$ and $E > V_2$.

- Case $V_1 < E < V_2$

    In this case the particle's motion is infinite only towards $x = -\infty$; that is, the particle can move between $x = x_3$ and $x \to -\infty$, x_3 being a classical turning point. The energy spectrum is continuous and none of the energy eigenvalues is degenerate. The nondegeneracy can be shown to result as follows. Since the Schrödinger equation (4.1) is a second-order differential equation, it has, for this case, two linearly independent solutions, but only one is physically acceptable. The solution is oscillatory for $x \leq x_3$ and rapidly decaying for $x > x_3$ so that it is finite (zero) at $x \to +\infty$, since divergent solutions are unphysical.

- Case $E > V_2$

    The energy spectrum is continuous and the particle's motion is infinite in both directions of x (i.e., towards $x \to \pm\infty$). All the energy levels of this spectrum are doubly degenerate. To see this, note that the general solution to (4.1) is a linear combination of two independent oscillatory solutions, one moving to the left and the other to the right. In the previous nondegenerate case only one solution is retained, since the other one diverges as $x \to +\infty$ and it has to be rejected.

In contrast to bound states, unbound states cannot be normalized and we cannot use boundary conditions.

4.2.3 Mixed Spectrum

Potentials that confine the particle for only some energies give rise to mixed spectra; the motion of the particle for such potentials is confined for some energy values only. For instance, for the potential displayed in Figure 4.1, if the energy of the particle is between $V_{min} < E < V_1$, the motion of the particle is confined (bound) and its spectrum is discrete, but if $E > V_2$, the particle's motion is unbound and its spectrum is continuous (if $V_1 < E < V_2$, the motion is unbound only along the $x = -\infty$ direction). Other typical examples where mixed spectra are encountered are the finite square well potential and the Coulomb or molecular potential.

4.2.4   Symmetric Potentials and Parity

Most of the potentials that are encountered at the microscopic level are symmetric (or even) with respect to space inversion, $\hat{V}(-x) = \hat{V}(x)$. This symmetry introduces considerable simplifications in the calculations.

When $\hat{V}(x)$ is even, the corresponding Hamiltonian, $\hat{H}(x) = -(\hbar^2/2m)d^2/dx^2 + \hat{V}(x)$, is also even. We saw in Chapter 2 that even operators commute with the parity operator; hence they can have a common eigenbasis.

Let us consider the following two cases pertaining to degenerate and nondegenerate spectra of this Hamiltonian:

- **Nondegenerate spectrum**

  First we consider the particular case where the eigenvalues of the Hamiltonian corresponding to this symmetric potential are not degenerate. According to Theorem 4.1, this Hamiltonian describes bound states. We saw in Chapter 2 that a nondegenerate, even operator has the same eigenstates as the parity operator. Since the eigenstates of the parity operator have definite parity, *the bound eigenstates of a particle moving in a one-dimensional symmetric potential have definite parity; they are either even or odd*:

$$\boxed{\hat{V}(-x) = \hat{V}(x) \quad \Longrightarrow \quad \psi(-x) = \pm\psi(x).}$$

$$(4.4)$$

- **Degenerate spectrum**

  If the spectrum of the Hamiltonian corresponding to a symmetric potential is degenerate, the eigenstates are expressed only in terms of even and odd states. That is, the eigenstates do not have definite parity.

Summary: The various properties of the one-dimensional motion discussed in this section can be summarized as follows:

- The energy spectrum of a bound state system is discrete and nondegenerate.

- The bound state wave function $\psi_n(x)$ has: (a) n nodes if $n = 0$ corresponds to the ground state and (b) $(n-1)$ nodes if $n = 1$ corresponds to the ground state.

- The bound state eigenfunctions in an even potential have definite parity.

- The eigenfunctions of a degenerate spectrum in an even potential do not have definite parity.

4.3   The Free Particle: Continuous States

This is the simplest one-dimensional problem; it corresponds to $V(x) = 0$ for any value of x. In this case the Schrödinger equation is given by

$$-\frac{\hbar^2}{2m}\frac{d^2\psi(x)}{dx^2} = E\psi(x) \quad \Longrightarrow \quad \left(\frac{d^2}{dx^2} + k^2\right)\psi(x) = 0,$$

$$(4.5)$$

where $k^2 = 2mE/\hbar^2$, k being the wave number. The most general solution to (4.5) is a combination of two linearly independent *plane waves* $\psi_+(x) = e^{ikx}$ and $\psi_-(x) = e^{-ikx}$:

$$\psi_k(x) = A_+e^{ikx} + A_-e^{-ikx},$$

$$(4.6)$$

where A_+ and A_- are two arbitrary constants. The complete wave function is thus given by the stationary state

$$\Psi_k(x,t) = A_+ e^{i(kx-\omega t)} + A_- e^{-i(kx+\omega t)} = A_+ e^{i(kx-\hbar k^2 t/2m)} + A_- e^{-i(kx+\hbar k^2 t/2m)}, \quad (4.7)$$

since $\omega = E/\hbar = \hbar k^2/2m$. The first term, $\Psi_+(x,t) = A_+ e^{i(kx-\omega t)}$, represents a wave traveling to the right, while the second term, $\Psi_-(x,t) = A_- e^{-i(kx+\omega t)}$, represents a wave traveling to the left. The intensities of these waves are given by $|A_+|^2$ and $|A_-|^2$, respectively. We should note that the waves $\Psi_+(x,t)$ and $\Psi_-(x,t)$ are associated, respectively, with a free particle traveling to the right and to the left with *well-defined* momenta and energy: $p_\pm = \pm\hbar k$, $E_\pm = \hbar^2 k^2/2m$. We will comment on the physical implications of this in a moment. Since there are no boundary conditions, there are no restrictions on k or on E; all values yield solutions to the equation.

The free particle problem is simple to solve mathematically, yet it presents a number of physical subtleties. Let us discuss briefly three of these subtleties. First, the probability densities corresponding to either solutions

$$P_\pm(x,t) = |\Psi_\pm(x,t)|^2 = |A_\pm|^2 \qquad (4.8)$$

are constant, for they depend neither on x nor on t. This is due to the complete loss of information about the position and time for a state with definite values of momentum, $p_\pm = \pm\hbar k$, and energy, $E_\pm = \hbar^2 k^2/2m$. This is a consequence of Heisenberg's uncertainty principle: when the momentum and energy of a particle are known exactly, $\Delta p = 0$ and $\Delta E = 0$, there must be total uncertainty about its position and time: $\Delta x \longrightarrow \infty$ and $\Delta t \longrightarrow \infty$. The second subtlety pertains to an apparent discrepancy between the speed of the wave and the speed of the particle it is supposed to represent. The speed of the plane waves $\Psi_\pm(x,t)$ is given by

$$v_{wave} = \frac{\omega}{k} = \frac{E}{\hbar k} = \frac{\hbar^2 k^2/2m}{\hbar k} = \frac{\hbar k}{2m}. \qquad (4.9)$$

On the other hand, the classical speed of the particle[2] is given by

$$v_{classical} = \frac{p}{m} = \frac{\hbar k}{m} = 2v_{wave}. \qquad (4.10)$$

This means that the particle travels twice as fast as the wave that represents it! Third, the wave function is not normalizable:

$$\int_{-\infty}^{+\infty} \Psi_\pm^*(x,t)\Psi_\pm(x,t)\,dx = |A_\pm|^2 \int_{-\infty}^{+\infty} dx \to \infty. \qquad (4.11)$$

The solutions $\Psi_\pm(x,t)$ are thus unphysical; physical wave functions must be square integrable. The problem can be traced to this: a free particle cannot have sharply defined momenta and energy.

In view of the three subtleties outlined above, the solutions of the Schrödinger equation (4.5) that are physically acceptable cannot be plane waves. Instead, we can construct physical

---

[2]The classical speed can be associated with the flux (or current density) which, as shown in Chapter 3, is $J_+ = i\hbar \frac{1}{2m}(\Psi_+ \frac{\partial \Psi_+^*}{\partial x} - \Psi_+^* \frac{\partial \Psi_+}{\partial x}) = \frac{\hbar k}{m} = \frac{p}{m}$, where use was made of $A_+ = 1$.

solutions by means of a linear superposition of plane waves. The answer is provided by *wave packets*, which we have seen in Chapter 1:

$$\psi(x, t) = \frac{1}{\sqrt{2\pi}} \int_{-\infty}^{+\infty} \phi(k) e^{i(kx - \omega t)} dk, \tag{4.12}$$

where $\phi(k)$, the amplitude of the wave packet, is given by the Fourier transform of $\psi(x, 0)$ as

$$\phi(k) = \frac{1}{\sqrt{2\pi}} \int_{-\infty}^{+\infty} \psi(x, 0) e^{-ikx} dx. \tag{4.13}$$

The wave packet solution cures and avoids all the subtleties raised above. First, the momentum, the position and the energy of the particle are no longer known exactly; only probabilistic outcomes are possible. Second, as shown in Chapter 1, the wave packet (4.12) and the particle travel with the same speed $v_g = p/m$, called the *group* speed or the speed of the whole packet. Third, the wave packet (4.12) is normalizable.

To summarize, a free particle cannot be represented by a single (monochromatic) plane wave; it has to be represented by a wave packet. The physical solutions of the Schrödinger equation are thus given by wave packets, not by stationary solutions.

4.4   The Potential Step

Another simple problem consists of a particle that is free everywhere, but beyond a particular point, say $x = 0$, the potential increases sharply (i.e., it becomes repulsive or attractive). A potential of this type is called a potential step (see Figure 4.2):

$$V(x) = \begin{cases} 0, & x < 0, \\ V_0, & x \geq 0. \end{cases} \tag{4.14}$$

In this problem we try to analyze the dynamics of a flux of particles (all having the same mass m and moving with the same velocity) moving from left to right. We are going to consider two cases, depending on whether the energy of the particles is larger or smaller than V_0.

(a) Case $E > V_0$

The particles are free for $x < 0$ and feel a repulsive potential V_0 that starts at $x = 0$ and stays flat (constant) for $x > 0$. Let us analyze the dynamics of this flux of particles classically and then quantum mechanically.

Classically, the particles approach the potential step or barrier from the left with a constant momentum $\sqrt{2mE}$. As the particles enter the region $x \geq 0$, where the potential now is $V = V_0$, they slow down to a momentum $\sqrt{2m(E - V_0)}$; they will then conserve this momentum as they travel to the right. Since the particles have sufficient energy to penetrate into the region $x \geq 0$, there will be *total transmission*: all the particles will emerge to the right with a smaller kinetic energy $E - V_0$. This is then a simple *scattering* problem in one dimension.

Quantum mechanically, the dynamics of the particle is regulated by the Schrödinger equation, which is given in these two regions by

$$\left(\frac{d^2}{dx^2} + k_1^2 \right) \psi_1(x) = 0 \qquad (x < 0), \tag{4.15}$$

Figure 4.2 Potential step and propagation directions of the incident, reflected, and transmitted waves, plus their probability densities $|\psi(x)|^2$ when $E > V_0$ and $E < V_0$.

$$\left(\frac{d^2}{dx^2} + k_2^2\right)\psi_2(x) = 0 \qquad (x \geq 0), \tag{4.16}$$

where $k_1^2 = 2mE/\hbar^2$ and $k_2^2 = 2m(E - V_0)/\hbar^2$. The most general solutions to these two equations are plane waves:

$$\psi_1(x) = Ae^{ik_1x} + Be^{-ik_1x} \qquad (x < 0), \tag{4.17}$$

$$\psi_2(x) = Ce^{ik_2x} + De^{-ik_2x} \qquad (x \geq 0), \tag{4.18}$$

where Ae^{ik_1x} and Ce^{ik_2x} represent waves moving in the positive x-direction, but Be^{-ik_1x} and De^{-ik_2x} correspond to waves moving in the negative x-direction. We are interested in the case where the particles are initially incident on the potential step from the left: they can be reflected or transmitted at $x = 0$. Since no wave is reflected from the region $x > 0$ to the left, the constant D must vanish. Since we are dealing with stationary states, the complete wave function is thus given by

$$\Psi(x, t) = \begin{cases} \psi_1(x)e^{-i\omega t} = Ae^{i(k_1x-\omega t)} + Be^{-i(k_1x+\omega t)} & x < 0 \\ \psi_2(x)e^{-i\omega t} = Ce^{i(k_2x-\omega t)} & x \geq 0, \end{cases} \tag{4.19}$$

where $A\exp[i(k_1x - \omega t)]$, $B\exp[-i(k_1x + \omega t)]$, and $C\exp[i(k_2x - \omega t)]$ represent the *incident*, the *reflected*, and the *transmitted* waves, respectively; they travel to the right, the left, and the right (Figure 4.2). Note that the probability density $|\psi(x)|^2$ shown in the lower left plot of Figure 4.2 is a straight line for $x > 0$, since $|\psi_2(x)|^2 = |C\exp i(k_2x - \omega t)|^2 = |C|^2$.

Let us now evaluate the *reflection* and *transmission coefficients*, R and T, as defined by

$$R = \left|\frac{\text{reflected current density}}{\text{incident current density}}\right| = \left|\frac{J_{reflected}}{J_{incident}}\right|, \qquad T = \left|\frac{J_{transmitted}}{J_{incident}}\right|; \tag{4.20}$$

R represents the ratio of the reflected to the incident beams and T the ratio of the transmitted to the incident beams. To calculate R and T, we need to find $J_{incident}$, $J_{reflected}$, and $J_{transmitted}$. Since the incident wave is $\psi_i(x) = Ae^{ik_1 x}$, the incident current density (or incident flux) is given by

$$J_{incident} = \frac{i\hbar}{2m}\left(\psi_i(x)\frac{d\psi_i^*(x)}{dx} - \psi_i^*(x)\frac{d\psi_i(x)}{dx}\right) = \frac{\hbar k_1}{m}|A|^2. \qquad (4.21)$$

Similarly, since the reflected and transmitted waves are $\psi_r(x) = Be^{-ik_1 x}$ and $\psi_t(x) = Ce^{ik_2 x}$, we can verify that the reflected and transmitted fluxes are

$$J_{reflected} = -\frac{\hbar k_1}{m}|B|^2, \qquad\qquad J_{transmitted} = \frac{\hbar k_2}{m}|C|^2. \qquad (4.22)$$

A combination of (4.20) to (4.22) yields

$$R = \frac{|B|^2}{|A|^2}, \qquad\qquad T = \frac{k_2}{k_1}\frac{|C|^2}{|A|^2}. \qquad (4.23)$$

Thus, the calculation of R and T is reduced to determining the constants B and C. For this, we need to use the boundary conditions of the wave function at $x = 0$. Since both the wave function and its first derivative are continuous at $x = 0$,

$$\psi_1(0) = \psi_2(0), \qquad \frac{d\psi_1(0)}{dx} = \frac{d\psi_2(0)}{dx}, \qquad (4.24)$$

equations (4.17) and (4.18) yield

$$A + B = C, \qquad k_1(A - B) = k_2 C; \qquad (4.25)$$

hence

$$B = \frac{k_1 - k_2}{k_1 + k_2}A, \qquad C = \frac{2k_1}{k_1 + k_2}A. \qquad (4.26)$$

As for the constant A, it can be determined from the normalization condition of the wave function, but we don't need it here, since R and T are expressed in terms of ratios. A combination of (4.23) with (4.26) leads to

$$R = \frac{(k_1 - k_2)^2}{(k_1 + k_2)^2} = \frac{(1 - \mathcal{K})^2}{(1 + \mathcal{K})^2}, \qquad T = \frac{4k_1 k_2}{(k_1 + k_2)^2} = \frac{4\mathcal{K}}{(1 + \mathcal{K})^2}, \qquad (4.27)$$

where $\mathcal{K} = k_2/k_1 = \sqrt{1 - V_0/E}$. The sum of R and T is equal to 1, as it should be.

In contrast to classical mechanics, which states that none of the particles get reflected, equation (4.27) shows that the quantum mechanical reflection coefficient R is not zero: there are particles that get reflected in spite of their energies being higher than the step V_0. This effect must be attributed to the *wavelike behavior* of the particles.

From (4.27) we see that as E gets smaller and smaller, T also gets smaller and smaller so that when $E = V_0$ the transmission coefficient T becomes zero and $R = 1$. On the other hand, when $E \gg V_0$, we have $\mathcal{K} = \sqrt{1 - V_0/E} \simeq 1$; hence $R = 0$ and $T = 1$. This is expected since, when the incident particles have very high energies, the potential step is so weak that it produces no noticeable effect on their motion.

Remark: physical meaning of the boundary conditions

Throughout this chapter, we will encounter at numerous times the use of the boundary conditions of the wave function and its first derivative as in Eq (4.24). What is the underlying physics behind these continuity conditions? We can make two observations:

- Since the probability density $|\psi(x)|^2$ of finding the particle in any small region varies continuously from one point to another, the wave function $\psi(x)$ must, therefore, be a continuous function of x; thus, as shown in (4.24), we must have $\psi_1(0) = \psi_2(0)$.

- Since the linear momentum of the particle, $\hat{P}\psi(x) = -i\hbar d\psi(x)/dx$, must be a continuous function of x as the particle moves from left to right, the first derivative of the wave function, $d\psi(x)/dx$, must also be a continuous function of x, notably at $x = 0$. Hence, as shown in (4.24), we must have $d\psi_1(0)/dx = d\psi_2(0)/dx$.

(b) Case $E < V_0$

Classically, the particles arriving at the potential step from the left (with momenta $p = \sqrt{2mE}$) will come to a stop at $x = 0$ and then all will bounce back to the left with the magnitudes of their momenta unchanged. None of the particles will make it into the right side of the barrier $x = 0$; there is total reflection of the particles. So the motion of the particles is reversed by the potential barrier.

Quantum mechanically, the picture will be somewhat different. In this case, the Schrödinger equation and the wave function in the region $x < 0$ are given by (4.15) and (4.17), respectively. But for $x > 0$ the Schrödinger equation is given by

$$\left(\frac{d^2}{dx^2} - k_2'^2\right)\psi_2(x) = 0 \quad (x \geq 0), \tag{4.28}$$

where $k_2'^2 = 2m(V_0 - E)/\hbar^2$. This equation's solution is

$$\psi_2(x) = Ce^{-k_2'x} + De^{k_2'x} \quad (x \geq 0). \tag{4.29}$$

Since the wave function must be finite everywhere, and since the term $e^{k_2'x}$ diverges when $x \to \infty$, the constant D has to be zero. Thus, the complete wave function is

$$\Psi(x, t) = \begin{cases} Ae^{i(k_1x - \omega t)} + Be^{-i(k_1x + \omega t)}, & x < 0, \\ Ce^{-k_2'x}e^{-i\omega t}, & x \geq 0. \end{cases} \tag{4.30}$$

Let us now evaluate, as we did in the previous case, the reflected and the transmitted coefficients. First we should note that the transmitted coefficient, which corresponds to the transmitted wave function $\psi_t(x) = Ce^{-k_2'x}$, is zero since $\psi_t(x)$ is a *purely real* function ($\psi_t^*(x) = \psi_t(x)$) and therefore

$$J_{transmitted} = \frac{\hbar}{2im}\left(\psi_t(x)\frac{d\psi_t(x)}{dx} - \psi_t(x)\frac{d\psi_t(x)}{dx}\right) = 0. \tag{4.31}$$

Hence, the reflected coefficient R must be equal to 1. We can obtain this result by applying the continuity conditions at $x = 0$ for (4.17) and (4.29):

$$B = \frac{k_1 - ik_2'}{k_1 + ik_2'}A, \qquad C = \frac{2k_1}{k_1 + ik_2'}A. \tag{4.32}$$

Thus, the reflected coefficient is given by

$$R = \frac{|B|^2}{|A|^2} = \frac{k_1^2 + k_2'^2}{k_1^2 + k_2'^2} = 1. \tag{4.33}$$

We therefore have total reflection, as in the classical case.

There is, however, a difference with the classical case: while none of the particles can be found classically in the region $x > 0$, quantum mechanically there is a *nonzero probability* that the wave function penetrates this *classically forbidden* region. To see this, note that the relative probability density

$$P(x) = |\psi_t(x)|^2 = |C|^2 e^{-2k_2'x} = \frac{4k_1^2|A|^2}{k_1^2 + k_2'^2} e^{-2k_2'x} \tag{4.34}$$

is appreciable near $x = 0$ and falls exponentially to small values as x becomes large; the behavior of the probability density is shown in Figure 4.2.

4.5 The Potential Barrier and Well

Consider a beam of particles of mass m that are sent from the left on a potential barrier

$$V(x) = \begin{cases} 0, & x < 0, \\ V_0, & 0 \le x \le a, \\ 0, & x > a. \end{cases} \tag{4.35}$$

This potential, which is repulsive, supports no bound states (Figure 4.3). We are dealing here, as in the case of the potential step, with a one-dimensional *scattering* problem.

Again, let us consider the following two cases which correspond to the particle energies being respectively larger and smaller than the potential barrier.

4.5.1 The Case $E > V_0$

Classically, the particles that approach the barrier from the left at constant momentum, $p_1 = \sqrt{2mE}$, as they enter the region $0 \le x \le a$ will slow down to a momentum $p_2 = \sqrt{2m(E - V_0)}$. They will maintain the momentum p_2 until they reach the point $x = a$. Then, as soon as they pass beyond the point $x = a$, they will accelerate to a momentum $p_3 = \sqrt{2mE}$ and maintain this value in the entire region $x > a$. Since the particles have enough energy to cross the barrier, none of the particles will be reflected back; all the particles will emerge on the right side of $x = a$: *total transmission*.

It is easy to infer the quantum mechanical study from the treatment of the potential step presented in the previous section. We need only to mention that the wave function will display an oscillatory pattern in all three regions; its amplitude reduces every time the particle enters a new region (see Figure 4.3):

$$\psi(x) = \begin{cases} \psi_1(x) = Ae^{ik_1x} + Be^{-ik_1x}, & x \le 0, \\ \psi_2(x) = Ce^{ik_2x} + De^{-ik_2x}, & 0 < x < a, \\ \psi_3(x) = Ee^{ik_1x}, & x \ge a, \end{cases} \tag{4.36}$$

where $k_1 = \sqrt{2mE/\hbar^2}$ and $k_2 = \sqrt{2m(E - V_0)/\hbar^2}$. The constants B, C, D, and E can be obtained in terms of A from the boundary conditions: $\psi(x)$ and $d\psi/dx$ must be continuous at $x = 0$ and $x = a$, respectively:

Figure 4.3 Potential barrier and propagation directions of the incident, reflected, and transmitted waves, plus their probability densities $|\psi(x)|^2$ when $E > V_0$ and $E < V_0$.

$$\psi_1(0) = \psi_2(0), \qquad \frac{d\psi_1(0)}{dx} = \frac{d\psi_2(0)}{dx}, \tag{4.37}$$

$$\psi_2(a) = \psi_3(a), \qquad \frac{d\psi_2(a)}{dx} = \frac{d\psi_3(a)}{dx}. \tag{4.38}$$

These equations yield

$$A + B = C + D, \qquad ik_1(A - B) = ik_2(C - D), \tag{4.39}$$

$$Ce^{ik_2a} + De^{-ik_2a} = Ee^{ik_1a}, \qquad ik_2\left(Ce^{ik_2a} - De^{-ik_2a}\right) = ik_1 Ee^{ik_1a}. \tag{4.40}$$

Solving for E, we obtain

$$\begin{aligned} E &= 4k_1k_2 Ae^{-ik_1a}[(k_1 + k_2)^2 e^{-ik_2a} - (k_1 - k_2)^2 e^{ik_2a}]^{-1} \\ &= 4k_1k_2 Ae^{-ik_1a}\left[4k_1k_2 \cos(k_2a) - 2i\left(k_1^2 + k_2^2\right)\sin(k_2a)\right]^{-1}. \end{aligned} \tag{4.41}$$

The transmission coefficient is thus given by

$$\begin{aligned} T &= \frac{k_1|E|^2}{k_1|A|^2} = \left[1 + \frac{1}{4}\left(\frac{k_1^2 - k_2^2}{k_1 k_2}\right)^2 \sin^2(k_2a)\right]^{-1} \\ &= \left[1 + \frac{V_0^2}{4E(E - V_0)} \sin^2\left(a\sqrt{2mV_0/\hbar^2}\sqrt{E/V_0 - 1}\right)\right]^{-1}, \end{aligned} \tag{4.42}$$

Figure 4.4 Transmission coefficients for a potential barrier, $T_B(\varepsilon) = \frac{4\varepsilon(\varepsilon-1)}{4\varepsilon(\varepsilon-1)+\sin^2(\lambda\sqrt{\varepsilon-1})}$, and for a potential well, $T_W(\varepsilon) = \frac{4\varepsilon(\varepsilon+1)}{4\varepsilon(\varepsilon+1)+\sin^2(\lambda\sqrt{\varepsilon+1})}$.

because

$$\left(\frac{k_1^2 - k_2^2}{k_1 k_2}\right)^2 = \frac{V_0^2}{E(E - V_0)}. \tag{4.43}$$

Using the notation $\lambda = a\sqrt{2mV_0/\hbar^2}$ and $\varepsilon = E/V_0$, we can rewrite T as

$$T = \left[1 + \frac{1}{4\varepsilon(\varepsilon - 1)} \sin^2(\lambda\sqrt{\varepsilon - 1})\right]^{-1}. \tag{4.44}$$

Similarly, we can show that

$$R = \frac{\sin^2(\lambda\sqrt{\varepsilon - 1})}{4\varepsilon(\varepsilon - 1) + \sin^2(\lambda\sqrt{\varepsilon - 1})} = \left[1 + \frac{4\varepsilon(\varepsilon - 1)}{\sin^2(\lambda\sqrt{\varepsilon - 1})}\right]^{-1}. \tag{4.45}$$

Special cases

- If $E \gg V_0$, and hence $\varepsilon \gg 1$, the transmission coefficient T becomes asymptotically equal to unity, $T \simeq 1$, and $R \simeq 0$. So, at very high energies and weak potential barrier, the particles would not feel the effect of the barrier; we have total transmission.

- We also have total transmission when $\sin(\lambda\sqrt{\varepsilon - 1}) = 0$ or $\lambda\sqrt{\varepsilon - 1} = n\pi$. As shown in Figure 4.4, the total transmission, $T(\varepsilon_n) = 1$, occurs whenever $\varepsilon_n = E_n/V_0 = n^2\pi^2\hbar^2/(2ma^2V_0) + 1$ or whenever the incident energy of the particle is $E_n = V_0 + n^2\pi^2\hbar^2/(2ma^2)$ with $n = 1, 2, 3, \dots$. The maxima of the transmission coefficient coincide with the energy eigenvalues of the infinite square well potential; these are known as resonances. This resonance phenomenon, which does not occur in classical physics, results from a constructive interference between the incident and the reflected waves. This phenomenon is observed experimentally in a number of cases such as when scattering low-energy ($E \sim 0.1$ eV) electrons off noble atoms (known as the *Ramsauer–Townsend effect*, a consequence of symmetry of noble atoms) and neutrons off nuclei.

- In the limit $\varepsilon \to 1$ we have $\sin(\lambda\sqrt{\varepsilon - 1}) \sim \lambda\sqrt{\varepsilon - 1}$, hence (4.44) and (4.45) become

$$T = \left(1 + \frac{ma^2 V_0}{2\hbar^2}\right)^{-1}, \qquad R = \left(1 + \frac{2\hbar^2}{ma^2 V_0}\right)^{-1}. \tag{4.46}$$

The potential well ($V_0 < 0$)

The transmission coefficient (4.44) was derived for the case where $V_0 > 0$, i.e., for a *barrier* potential. Following the same procedure that led to (4.44), we can show that the transmission coefficient for a finite *potential well*, $V_0 < 0$, is given by

$$T_W = \left[1 + \frac{1}{4\varepsilon(\varepsilon + 1)} \sin^2(\lambda\sqrt{\varepsilon + 1})\right]^{-1}, \tag{4.47}$$

where $\varepsilon = E/|V_0|$ and $\lambda = a\sqrt{2m|V_0|/\hbar^2}$. Notice that there is total transmission whenever $\sin(\lambda\sqrt{\varepsilon + 1}) = 0$ or $\lambda\sqrt{\varepsilon + 1} = n\pi$. As shown in Figure 4.4, the total transmission, $T_W(\varepsilon_n) = 1$, occurs whenever $\varepsilon_n = E_n/|V_0| = n^2\pi^2\hbar^2/(2ma^2 V_0) - 1$ or whenever the incident energy of the particle is $E_n = n^2\pi^2\hbar^2/(2ma^2) - |V_0|$ with $n = 1, 2, 3, \ldots$. We will study in more detail the *symmetric* potential well in Section 4.7.

4.5.2 The Case $E < V_0$: **Tunneling**

Classically, we would expect total reflection: every particle that arrives at the barrier ($x = 0$) will be reflected back; no particle can penetrate the barrier, where it would have a negative kinetic energy.

We are now going to show that the quantum mechanical predictions differ sharply from their classical counterparts, for the wave function is not zero beyond the barrier. The solutions of the Schrödinger equation in the three regions yield expressions that are similar to (4.36) except that $\psi_2(x) = Ce^{ik_2 x} + De^{-ik_2 x}$ should be replaced with $\psi_2(x) = Ce^{k_2 x} + De^{-k_2 x}$:

$$\psi(x) = \begin{cases} \psi_1(x) = Ae^{ik_1 x} + Be^{-ik_1 x}, & x \le 0, \\ \psi_2(x) = Ce^{k_2 x} + De^{-k_2 x}, & 0 < x < a, \\ \psi_3(x) = Ee^{ik_1 x}, & x \ge a, \end{cases} \tag{4.48}$$

where $k_1^2 = 2mE/\hbar^2$ and $k_2^2 = 2m(V_0 - E)/\hbar^2$. The behavior of the probability density corresponding to this wave function is expected, as displayed in Figure 4.3, to be oscillatory in the regions $x < 0$ and $x > a$, and exponentially decaying for $0 \le x \le a$.

To find the reflection and transmission coefficients,

$$R = \frac{|B|^2}{|A|^2}, \qquad T = \frac{|E|^2}{|A|^2}, \tag{4.49}$$

we need only to calculate B and E in terms of A. The continuity conditions of the wave function and its derivative at $x = 0$ and $x = a$ yield

$$A + B = C + D, \tag{4.50}$$
$$ik_1(A - B) = k_2(C - D), \tag{4.51}$$
$$Ce^{k_2 a} + De^{-k_2 a} = Ee^{ik_1 a}, \tag{4.52}$$
$$k_2\left(Ce^{k_2 a} - De^{-k_2 a}\right) = ik_1 Ee^{ik_1 a}. \tag{4.53}$$

The last two equations lead to the following expressions for C and D:

$$C = \frac{E}{2}\left(1 + i\frac{k_1}{k_2}\right)e^{(ik_1 - k_2)a}, \qquad D = \frac{E}{2}\left(1 - i\frac{k_1}{k_2}\right)e^{(ik_1 + k_2)a}. \qquad (4.54)$$

Inserting these two expressions into the two equations (4.50) and (4.51) and dividing by A, we can show that these two equations reduce, respectively, to

$$1 + \frac{B}{A} = \frac{E}{A}e^{ik_1 a}\left[\cosh(k_2 a) - i\frac{k_1}{k_2}\sinh(k_2 a)\right], \qquad (4.55)$$

$$1 - \frac{B}{A} = \frac{E}{A}e^{ik_1 a}\left[\cosh(k_2 a) + i\frac{k_2}{k_1}\sinh(k_2 a)\right]. \qquad (4.56)$$

Solving these two equations for B/A and E/A, we obtain

$$\frac{B}{A} = -i\frac{k_1^2 + k_2^2}{k_1 k_2}\sinh(k_2 a)\left[2\cosh(k_2 a) + i\frac{k_2^2 - k_1^2}{k_1 k_2}\sinh(k_2 a)\right]^{-1}, \qquad (4.57)$$

$$\frac{E}{A} = 2e^{-ik_1 a}\left[2\cosh(k_2 a) + i\frac{k_2^2 - k_1^2}{k_1 k_2}\sinh(k_2 a)\right]^{-1}. \qquad (4.58)$$

Thus, the coefficients R and T become

$$R = \left(\frac{k_1^2 + k_2^2}{k_1 k_2}\right)^2 \sinh^2(k_2 a)\left[4\cosh^2(k_2 a) + \left(\frac{k_2^2 - k_1^2}{k_1 k_2}\right)^2 \sinh^2(k_2 a)\right]^{-1}, \qquad (4.59)$$

$$T = \frac{|E|^2}{|A|^2} = 4\left[4\cosh^2(k_2 a) + \left(\frac{k_2^2 - k_1^2}{k_1 k_2}\right)^2 \sinh^2(k_2 a)\right]^{-1}. \qquad (4.60)$$

We can rewrite R in terms of T as

$$R = \frac{1}{4}T\left(\frac{k_1^2 + k_2^2}{k_1 k_2}\right)^2 \sinh^2(k_2 a). \qquad (4.61)$$

Since $\cosh^2(k_2 a) = 1 + \sinh^2(k_2 a)$ we can reduce (4.60) to

$$T = \left[1 + \frac{1}{4}\left(\frac{k_1^2 + k_2^2}{k_1 k_2}\right)^2 \sinh^2(k_2 a)\right]^{-1}. \qquad (4.62)$$

Note that T is *finite*. This means that the probability for the transmission of the particles into the region $x \geq a$ is *not zero* (in classical physics, however, the particle can in no way make it into the $x \geq 0$ region). This is a purely quantum mechanical effect which is due to the *wave aspect* of microscopic objects; it is known as the *tunneling effect*: *quantum mechanical objects can tunnel through classically impenetrable barriers*. This *barrier penetration* effect has important applications in various branches of modern physics ranging from particle and nuclear physics to semiconductor devices. For instance, radioactive decays and charge transport in electronic devices are typical examples of the tunneling effect.

Now since

$$\left(\frac{k_1^2 + k_2^2}{k_1 k_2}\right)^2 = \left(\frac{V_0}{\sqrt{E(V_0 - E)}}\right)^2 = \frac{V_0^2}{E(V_0 - E)}, \tag{4.63}$$

we can rewrite (4.61) and (4.62) as follows:

$$R = \frac{1}{4} \frac{V_0^2 T}{E(V_0 - E)} \sinh^2\left(\frac{a}{\hbar}\sqrt{2m(V_0 - E)}\right), \tag{4.64}$$

$$T = \left[1 + \frac{1}{4} \frac{V_0^2}{E(V_0 - E)} \sinh^2\left(\frac{a}{\hbar}\sqrt{2m(V_0 - E)}\right)\right]^{-1}, \tag{4.65}$$

or

$$R = \frac{T}{4\varepsilon(1-\varepsilon)} \sinh^2\left(\lambda\sqrt{1-\varepsilon}\right), \tag{4.66}$$

$$T = \left[1 + \frac{1}{4\varepsilon(1-\varepsilon)} \sinh^2\left(\lambda\sqrt{1-\varepsilon}\right)\right]^{-1}, \tag{4.67}$$

where $\lambda = a\sqrt{2mV_0/\hbar^2}$ and $\varepsilon = E/V_0$.

Special cases

- If $E \ll V_0$, hence $\varepsilon \ll 1$ or $\lambda\sqrt{1-\varepsilon} \gg 1$, we may approximate $\sinh\left(\lambda\sqrt{1-\varepsilon}\right) \simeq \frac{1}{2}\exp\left(\lambda\sqrt{1-\varepsilon}\right)$. We can thus show that the transmission coefficient (4.67) becomes asymptotically equal to

$$T \simeq \left\{\frac{1}{4\varepsilon(1-\varepsilon)}\left[\frac{1}{2}e^{\lambda\sqrt{1-\varepsilon}}\right]^2\right\}^{-1} = 16\varepsilon(1-\varepsilon)e^{-2\lambda\sqrt{1-\varepsilon}}$$

$$= \frac{16E}{V_0}\left(1 - \frac{E}{V_0}\right)e^{-(2a/\hbar)\sqrt{2m(V_0 - E)}}. \tag{4.68}$$

  This shows that the transmission coefficient is not zero, as it would be classically, but has a finite value. So, quantum mechanically, there is a finite tunneling beyond the barrier, $x > a$.

- When $E \simeq V_0$, hence $\varepsilon \simeq 1$, we can verify that (4.66) and (4.67) lead to the relations (4.46).

- Taking the classical limit $\hbar \to 0$, the coefficients (4.66) and (4.67) reduce to the classical result: $R \to 1$ and $T \to 0$.

4.5.3 The Tunneling Effect

In general, the tunneling effect consists of the propagation of a particle through a region where the particle's energy is smaller than the potential energy $E < V(x)$. Classically this region, defined by $x_1 < x < x_2$ (Figure 4.5a), is forbidden to the particle where its kinetic energy

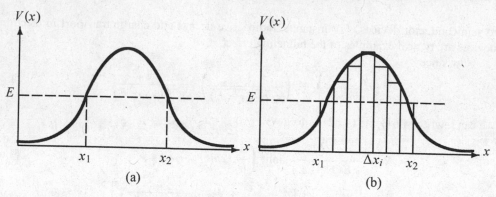

Figure 4.5 (a) Tunneling though a potential barrier. (b) Approximation of a smoothly varying potential $V(x)$ by square barriers.

would be negative; the points $x = x_1$ and $x = x_2$ are known as the *classical turning points*. Quantum mechanically, however, since particles display wave features, the quantum waves can tunnel through the barrier.

As shown in the square barrier example, the particle has a finite probability of tunneling through the barrier. In this case we managed to find an analytical expression (4.67) for the tunneling probability only because we dealt with a simple square potential. Analytic expressions cannot be obtained for potentials with arbitrary spatial dependence. In such cases one needs approximations. The Wentzel–Kramers–Brillouin (WKB) method (Chapter 9) provides one of the most useful approximation methods. We will show that the transmission coefficient for a barrier potential $V(x)$ is given by

$$T \sim \exp\left\{-\frac{2}{\hbar}\int_{x_1}^{x_2} dx\sqrt{2m\left[V(x) - E\right]}\right\}. \tag{4.69}$$

We can obtain this relation by means of a crude approximation. For this, we need simply to take the classically forbidden region $x_1 < x < x_2$ (Figure 4.5b) and divide it into a series of small intervals Δx_i. If Δx_i is small enough, we may approximate the potential $V(x_i)$ at each point x_i by a square potential barrier. Thus, we can use (4.68) to calculate the transmission probability corresponding to $V(x_i)$:

$$T_i \sim \exp\left[-\frac{2\Delta x_i}{\hbar}\sqrt{2m(V(x_i) - E)}\right]. \tag{4.70}$$

The transmission probability for the general potential of Figure 4.5, where we divided the region $x_1 < x < x_2$ into a very large number of small intervals Δx_i, is given by

$$
\begin{aligned}
T &\sim \lim_{N\to\infty}\prod_{i=1}^{N}\exp\left[-\frac{2\Delta x_i}{\hbar}\sqrt{2m(V(x_i) - E)}\right] \\
&= \exp\left[-\frac{2}{\hbar}\lim_{\Delta x_i\to 0}\sum_i \Delta x_i\sqrt{2m(V(x_i) - E)}\right] \\
&\longrightarrow \exp\left[-\frac{2}{\hbar}\int_{x_1}^{x_2} dx\sqrt{2m\left[V(x) - E\right]}\right].
\end{aligned} \tag{4.71}
$$

The approximation leading to this relation is valid, as will be shown in Chapter 9, only if the potential $V(x)$ is a smooth, slowly varying function of x.

4.6 The Infinite Square Well Potential

4.6.1 The Asymmetric Square Well

Consider a particle of mass m confined to move inside an infinitely deep asymmetric potential well

$$V(x) = \begin{cases} +\infty, & x < 0, \\ 0, & 0 \leq x \leq a, \\ +\infty, & x > a. \end{cases} \qquad (4.72)$$

Classically, the particle remains confined inside the well, moving at constant momentum $p = \pm\sqrt{2mE}$ back and forth as a result of repeated reflections from the walls of the well.

Quantum mechanically, we expect this particle to have only bound state solutions and a discrete nondegenerate energy spectrum. Since $V(x)$ is infinite outside the region $0 \leq x \leq a$, the wave function of the particle must be zero outside the boundary. Hence we can look for solutions only inside the well

$$\frac{d^2 \psi(x)}{dx^2} + k^2 \psi(x) = 0, \qquad (4.73)$$

with $k^2 = 2mE/\hbar^2$; the solutions are

$$\psi(x) = A'e^{ikx} + B'e^{-ikx} \quad \Longrightarrow \quad \psi(x) = A\sin(kx) + B\cos(kx). \qquad (4.74)$$

The wave function vanishes at the walls, $\psi(0) = \psi(a) = 0$: the condition $\psi(0) = 0$ gives $B = 0$, while $\psi(a) = A\sin(ka) = 0$ gives

$$k_n a = n\pi \qquad (n = 1, 2, 3, \cdots). \qquad (4.75)$$

This condition determines the energy

$$\boxed{E_n = \frac{\hbar^2}{2m}k_n^2 = \frac{\hbar^2\pi^2}{2ma^2}n^2 \qquad (n = 1, 2, 3, \cdots).} \qquad (4.76)$$

The energy is *quantized*; only certain values are permitted. This is expected since the states of a particle which is confined to a limited region of space are *bound states* and the energy spectrum is *discrete*. This is in sharp contrast to classical physics where the energy of the particle, given by $E = p^2/(2m)$, takes any value; the classical energy evolves *continuously*.

As it can be inferred from (4.76), we should note that the energy between adjacent levels is not constant:

$$E_{n+1} - E_n = 2n + 1, \qquad (4.77)$$

which leads to

$$\frac{E_{n+1} - E_n}{E_n} = \frac{(n+1)^2 - n^2}{n^2} = \frac{2n+1}{n^2}. \qquad (4.78)$$

In the classical limit $n \to \infty$,

$$\lim_{n\to\infty} \frac{E_{n+1} - E_n}{E_n} = \lim_{n\to\infty} \frac{2n+1}{n^2} = 0, \qquad (4.79)$$

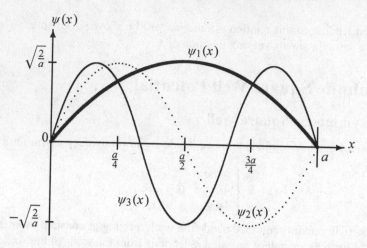

Figure 4.6 Three lowest states of an infinite potential well, $\psi_n(x) = \sqrt{2/a}\sin(n\pi x/a)$; the states $\psi_{2n+1}(x)$ and $\psi_{2n}(x)$ are even and odd, respectively, with respect to $x = a/2$.

the levels become so close together as to be practically indistinguishable.

Since $B = 0$ and $k_n = n\pi/a$, (4.74) yields $\psi_n(x) = A\sin(n\pi x/a)$. We can choose the constant A so that $\psi_n(x)$ is normalized:

$$1 = \int_0^a |\psi_n(x)|^2 dx = |A|^2 \int_0^a \sin^2\left(\frac{n\pi}{a}x\right) dx \quad\Longrightarrow\quad A = \sqrt{\frac{2}{a}}, \qquad (4.80)$$

hence

$$\boxed{\psi_n(x) = \sqrt{\frac{2}{a}}\sin\left(\frac{n\pi}{a}x\right) \qquad (n = 1, 2, 3, \cdots).} \qquad (4.81)$$

The first few functions are plotted in Figure 4.6.

The solution of the time-independent Schrödinger equation has thus given us the energy (4.76) and the wave function (4.81). There is then an infinite sequence of discrete energy levels corresponding to the positive integer values of the *quantum number* n. It is clear that $n = 0$ yields an uninteresting result: $\psi_0(x) = 0$ and $E_0 = 0$; later, we will examine in more detail the physical implications of this. So, the lowest energy, or *ground state* energy, corresponds to $n = 1$; it is $E_1 = \hbar^2\pi^2/(2ma^2)$. As will be explained later, this is called the *zero-point energy*, for there exists no state with zero energy. The states corresponding to $n = 2, 3, 4, \ldots$ are called *excited states*; their energies are given by $E_n = n^2 E_1$. As mentioned in Theorem 4.2, each function $\psi_n(x)$ has $(n-1)$ nodes. Figure 4.6 shows that the functions $\psi_{2n+1}(x)$ are even and the functions $\psi_{2n}(x)$ are odd with respect to the center of the well; we will study this in Section 4.6.2 when we consider the symmetric potential well. Note that none of the energy levels is degenerate (there is only one eigenfunction for each energy level) and that the wave functions corresponding to different energy levels are orthogonal:

$$\int_0^a \psi_m^*(x)\psi_n(x)\, dx = \delta_{mn}. \qquad (4.82)$$

Since we are dealing with stationary states and since $E_n = n^2 E_1$, the most general solutions of the time-dependent Schrödinger equation are given by

$$\Psi(x, t) = \sum_{n=1}^{\infty} \psi_n(x) e^{-iE_n t/\hbar} - \sqrt{\frac{2}{a}} \sum_{n=1}^{\infty} \sin\left(\frac{n\pi x}{a}\right) e^{-in^2 E_1 t/\hbar}. \tag{4.83}$$

Zero-point energy

Let us examine why there is no state with zero energy for a square well potential. If the particle has zero energy, it will be at rest inside the well, and this violates Heisenberg's uncertainty principle. By *localizing* or confining the particle to a limited region in space, it will acquire a *finite* momentum leading to a minimum kinetic energy. That is, the localization of the particle's motion to $0 \le x \le a$ implies a position uncertainty of order $\Delta x \sim a$ which, according to the uncertainty principle, leads to a minimum momentum uncertainty $\Delta p \sim \hbar/a$ and this in turn leads to a minimum kinetic energy of order $\hbar^2/(2ma^2)$. This is in qualitative agreement with the exact value $E_1 = \pi^2 \hbar^2/(2ma^2)$. In fact, as will be shown in (4.216), an accurate evaluation of Δp_1 leads to a zero-point energy which is equal to E_1.

Note that, as the momentum uncertainty is inversely proportional to the width of the well, $\Delta p \sim \hbar/a$, if the width decreases (i.e., the particle's position is confined further and further), the uncertainty on $\hat{P}$ will increase. This makes the particle move faster and faster, so the zero-point energy will also increase. Conversely, if the width of the well increases, the zero-point energy decreases, but it will never vanish.

The zero-point energy therefore reflects the necessity of a *minimum motion* of a particle due to localization. The zero-point energy occurs in all bound state potentials. In the case of binding potentials, the lowest energy state has an energy which is higher than the minimum of the potential energy. This is in sharp contrast to classical mechanics, where the lowest possible energy is equal to the minimum value of the potential energy, with zero kinetic energy. In quantum mechanics, however, the lowest state does not minimize the potential alone, but applies to the sum of the kinetic and potential energies, and this leads to a finite ground state or zero-point energy. This concept has far-reaching physical consequences in the realm of the microscopic world. For instance, without the zero-point motion, atoms would not be stable, for the electrons would fall into the nuclei. Also, it is the zero-point energy which prevents helium from freezing at very low temperatures.

The following example shows that the zero-point energy is also present in macroscopic systems, but it is infinitesimally small. In the case of microscopic systems, however, it has a nonnegligible size.

Example 4.1 (Zero-point energy)

To illustrate the idea that the zero-point energy gets larger by going from macroscopic to microscopic systems, calculate the zero-point energy for a particle in an infinite potential well for the following three cases:

(a) a 100 g ball confined on a 5 m long line,
(b) an oxygen atom confined to a 2×10^{-10} m lattice, and
(c) an electron confined to a 10^{-10} m atom.

Solution

(a) The zero-point energy of a 100 g ball that is confined to a 5 m long line is

$$E = \frac{\hbar^2 \pi^2}{2ma^2} \simeq \frac{10 \times 10^{-68} \text{ J}}{2 \times 0.1 \times 25} \simeq 2 \times 10^{-68} \text{ J} = 1.25 \times 10^{-49} \text{ eV}. \tag{4.84}$$

This energy is too small to be detected, much less measured, by any known experimental technique.

(b) For the zero-point energy of an oxygen atom confined to a 2×10^{-10} m lattice, since the oxygen atom has 16 nucleons, its mass is of the order of $m \simeq 16 \times 1.6 \times 10^{-27}$ kg $\simeq 26 \times 10^{-27}$ kg, so we have

$$E = \frac{10^{-67} \text{ J}}{2 \times 26 \times 10^{-27} \times 4 \times 10^{-20}} \simeq 0.5 \times 10^{-22} \text{ J} \simeq 3 \times 10^{-4} \text{ eV}. \tag{4.85}$$

(c) The zero-point energy of an electron ($m \sim 10^{-30}$ kg) that is confined to an atom ($a \sim 1 \times 10^{-10}$ m) is

$$E = \frac{10^{-67} \text{ J}}{2 \times 10^{-30} \times 10^{-20}} \simeq 5 \times 10^{-18} \text{ J} \simeq 30 \text{ eV}. \tag{4.86}$$

This energy is important at the atomic scale, for the binding energy of a hydrogen electron is about 14 eV. So the zero-point energy is negligible for macroscopic objects, but important for microscopic systems.

4.6.2 The Symmetric Potential Well

What happens if the potential (4.72) is translated to the left by a distance of $a/2$ to become symmetric?

$$V(x) = \begin{cases} +\infty, & x < -a/2, \\ 0, & -a/2 \le x \le a/2, \\ +\infty, & x > a/2. \end{cases} \tag{4.87}$$

First, we would expect the energy spectrum (4.76) to remain unaffected by this translation, since the Hamiltonian is invariant under spatial translations; as it contains only a kinetic part, it commutes with the particle's momentum, $[\hat{H}, \hat{P}] = 0$. The energy spectrum is discrete and nondegenerate.

Second, earlier in this chapter we saw that for symmetric potentials, $V(-x) = V(x)$, the wave function of bound states must be either even or odd. The wave function corresponding to the potential (4.87) can be written as follows:

$$\psi_n(x) = \sqrt{\frac{2}{a}} \sin\left[\frac{n\pi}{a}\left(x + \frac{a}{2}\right)\right] = \begin{cases} \sqrt{\frac{2}{a}} \cos(\frac{n\pi}{a} x) & (n = 1, 3, 5, 7, \cdots), \\ \sqrt{\frac{2}{a}} \sin(\frac{n\pi}{a} x) & (n = 2, 4, 6, 8, \cdots). \end{cases} \tag{4.88}$$

That is, the wave functions corresponding to odd quantum numbers $n = 1, 3, 5, \ldots$ are symmetric, $\psi(-x) = \psi(x)$, and those corresponding to even numbers $n = 2, 4, 6, \ldots$ are antisymmetric, $\psi(-x) = -\psi(x)$.

4.7 The Finite Square Well Potential

Consider a particle of mass m moving in the following symmetric potential:

$$V(x) = \begin{cases} V_0, & x < -a/2, \\ 0, & -a/2 \le x \le a/2, \\ V_0, & x > a/2. \end{cases} \tag{4.89}$$

Figure 4.7 Finite square well potential and propagation directions of the incident, reflected and transmitted waves when $E > V_0$ and $0 < E < V_0$.

The two physically interesting cases are $E > V_0$ and $E < V_0$ (see Figure 4.7). We expect the solutions to yield a *continuous* doubly-degenerate energy spectrum for $E > V_0$ and a *discrete* nondegenerate spectrum for $0 < E < V_0$.

4.7.1 The Scattering Solutions $(E > V_0)$

Classically, if the particle is initially incident from left with constant momentum $\sqrt{2m(E - V_0)}$, it will speed up to $\sqrt{2mE}$ between $-a/2 \le x \le a/2$ and then slow down to its initial momentum in the region $x > a$. All the particles that come from the left will be transmitted, none will be reflected back; therefore $T = 1$ and $R = 0$.

Quantum mechanically, and as we did for the step and barrier potentials, we can verify that we get a *finite* reflection coefficient. The solution is straightforward to obtain; just follow the procedure outlined in the previous two sections. The wave function has an oscillating pattern in all three regions (see Figure 4.7).

4.7.2 The Bound State Solutions $(0 < E < V_0)$

Classically, when $E < V_0$ the particle is completely confined to the region $-a/2 \le x \le a/2$; it will bounce back and forth between $x = -a/2$ and $x = a/2$ with constant momentum $p = \sqrt{2mE}$.

Quantum mechanically, the solutions are particularly interesting for they are expected to yield a *discrete* energy spectrum and wave functions that decay in the two regions $x < -a/2$ and $x > a/2$, but oscillate in $-a/2 \le x \le a/2$. In these three regions, the Schrödinger equation can be written as

$$\left(\frac{d^2}{dx^2} - k_1^2\right)\psi_1(x) = 0 \qquad \left(x < -\frac{1}{2}a\right), \tag{4.90}$$

$$\left(\frac{d^2}{dx^2} + k_2^2\right)\psi_2(x) = 0 \qquad \left(-\frac{1}{2}a \le x \le \frac{1}{2}a\right), \tag{4.91}$$

$$\left(\frac{d^2}{dx^2} - k_1^2\right)\psi_3(x) = 0 \qquad \left(x > \frac{1}{2}a\right), \tag{4.92}$$

where $k_1^2 = 2m(V_0 - E)/\hbar^2$ and $k_2^2 = 2mE/\hbar^2$. Eliminating the physically unacceptable solutions which grow exponentially for large values of $|x|$, we can write the solution to this Schrödinger equation in the regions $x < -a/2$ and $x > a/2$ as follows:

$$\psi_1(x) = Ae^{k_1x} \qquad \left(x < -\frac{1}{2}a\right), \tag{4.93}$$

$$\psi_3(x) = De^{-k_1x} \qquad \left(x > \frac{1}{2}a\right). \tag{4.94}$$

As mentioned in (4.4), since the bound state eigenfunctions of symmetric one-dimensional Hamiltonians are either even or odd under space inversion, the solutions of (4.90) to (4.92) are then either antisymmetric (odd)

$$\psi_a(x) = \begin{cases} Ae^{k_1x}, & x < -a/2, \\ C\sin(k_2x), & -a/2 \le x \le a/2, \\ De^{-k_1x}, & x > a/2, \end{cases} \tag{4.95}$$

or symmetric (even)

$$\psi_s(x) = \begin{cases} Ae^{k_1x}, & x < -a/2, \\ B\cos(k_2x), & -a/2 \le x \le a/2, \\ De^{-k_1x}, & x > a/2. \end{cases} \tag{4.96}$$

To determine the eigenvalues, we need to use the continuity conditions at $x = \pm a/2$. The continuity of the logarithmic derivative, $(1/\psi_a(x))d\psi_a(x)/dx$, of $\psi_a(x)$ at $x = \pm a/2$ yields

$$k_2 \cot\left(\frac{k_2a}{2}\right) = -k_1. \tag{4.97}$$

Similarly, the continuity of $(1/\psi_s(x))d\psi_s(x)/dx$ at $x = \pm a/2$ gives

$$k_2 \tan\left(\frac{k_2a}{2}\right) = k_1. \tag{4.98}$$

The transcendental equations (4.97) and (4.98) cannot be solved directly; we can solve them either graphically or numerically. To solve these equations graphically, we need only to rewrite them in the following suggestive forms:

$$-\alpha_n \cot\alpha_n = \sqrt{R^2 - \alpha_n^2} \qquad \text{(for odd states)}, \tag{4.99}$$

$$\alpha_n \tan\alpha_n = \sqrt{R^2 - \alpha_n^2} \qquad \text{(for even states)}, \tag{4.100}$$

where $\alpha_n^2 = (k_2a/2)^2 = ma^2E_n/(2\hbar^2)$ and $R^2 = ma^2V_0/(2\hbar^2)$; these equations are obtained by inserting $k_1 = \sqrt{2m(V_0 - E)/\hbar^2}$ and $k_2 = \sqrt{2mE/\hbar^2}$ into (4.97) and (4.98). The left-hand sides of (4.99) and (4.100) consist of trigonometric functions; the right-hand sides consist of a circle of radius R. The solutions are given by the points where the circle $\sqrt{R^2 - \alpha_n^2}$ intersects the functions $-\alpha_n \cot\alpha_n$ and $\alpha_n \tan\alpha_n$ (Figure 4.8). The solutions form a *discrete* set. As illustrated in Figure 4.8, the intersection of the small circle with the curve $\alpha_n \tan\alpha_n$ yields only one bound state, $n = 0$, whereas the intersection of the larger circle with $\alpha_n \tan\alpha_n$ yields two bound states, $n = 0, 2$, and its intersection with $-\alpha_n \cot \alpha_n$ yields two other bound states, $n = 1, 3$.

The number of solutions depends on the size of R, which in turn depends on the depth V_0 and the width a of the well, since $R = \sqrt{ma^2 V_0/(2\hbar^2)}$. The deeper and broader the well, the larger the value of R, and hence the greater the number of bound states. Note that there is always at least one bound state (i.e., one intersection) no matter how small V_0 is. When

$$0 < R < \frac{\pi}{2} \quad \text{or} \quad 0 < V_0 < \left(\frac{\pi}{2}\right)^2 \frac{2\hbar^2}{ma^2}, \tag{4.101}$$

there is only one bound state corresponding to $n = 0$ (see Figure 4.8); this state—the ground state—is even. Then, and when

$$\frac{\pi}{2} < R < \pi \quad \text{or} \quad \left(\frac{\pi}{2}\right)^2 \frac{2\hbar^2}{ma^2} < V_0 < \pi^2 \frac{2\hbar^2}{ma^2}, \tag{4.102}$$

there are two bound states: an even state (the ground state) corresponding to $n = 0$ and the first odd state corresponding to $n = 1$. Now, if

$$\pi < R < \frac{3\pi}{2} \quad \text{or} \quad \pi^2 \frac{2\hbar^2}{ma^2} < V_0 < \left(\frac{3\pi}{2}\right)^2 \frac{2\hbar^2}{ma^2}, \tag{4.103}$$

there exist three bound states: the ground state (even state), $n = 0$, the first excited state (odd state), corresponding to $n = 1$, and the second excited state (even state), which corresponds to $n = 2$. In general, the well width at which n states are allowed is given by

$$R = \frac{n\pi}{2} \quad \text{or} \quad V_0 = \left(\frac{\pi}{2}\right)^2 \frac{2\hbar^2}{ma^2} n^2. \tag{4.104}$$

The spectrum, therefore, consists of a set of alternating even and odd states: the lowest state, the ground state, is even, the next state (first excited sate) is odd, and so on.

In the limiting case $V_0 \to \infty$, the circle's radius R also becomes infinite, and hence the function $\sqrt{R^2 - \alpha_n^2}$ will cross $-\alpha_n \cot \alpha_n$ and $\alpha_n \tan \alpha_n$ at the asymptotes $\alpha_n = n\pi/2$, because when $V_0 \to \infty$ both $\tan \alpha_n$ and $\cot \alpha_n$ become infinite:

$$\tan \alpha_n \to \infty \quad \implies \quad \alpha_n = \frac{2n+1}{2}\pi \quad (n = 0, 1, 2, 3, \cdots), \tag{4.105}$$

$$\cot \alpha_n \to \infty \quad \implies \quad \alpha_n = n\pi \quad (n = 1, 2, 3, \cdots). \tag{4.106}$$

Combining these two cases, we obtain

$$\alpha_n = \frac{n\pi}{2} \quad (1, 2, 3, \ldots). \tag{4.107}$$

Since $\alpha_n^2 = ma^2 E_n/(2\hbar^2)$ we see that we recover the energy expression for the infinite well:

$$\alpha_n = \frac{n\pi}{2} \longrightarrow E_n = \frac{\pi^2 \hbar^2}{2ma^2} n^2. \tag{4.108}$$

Figure 4.8 Graphical solutions for the finite square well potential: they are given by the intersections of $\sqrt{R^2 - \alpha_n^2}$ with $\alpha_n \tan \alpha_n$ and $-\alpha_n \cot \alpha_n$, where $\alpha_n^2 = ma^2 E_n/(2\hbar^2)$ and $R^2 = ma^2 V_0/(2\hbar^2)$.

Example 4.2

Find the number of bound states and the corresponding energies for the finite square well potential when: (a) $R = 1$ (i.e., $\sqrt{ma^2 V_0/(2\hbar^2)} = 1$), and (b) $R = 2$.

Solution

(a) From Figure 4.8, when $R = \sqrt{ma^2 V_0/(2\hbar^2)} = 1$, there is only one bound state since $\alpha_n \leq R$. This bound state corresponds to $n = 0$. The corresponding energy is given by the intersection of $\alpha_0 \tan \alpha_0$ with $\sqrt{1 - \alpha_0^2}$:

$$\alpha_0 \tan \alpha_0 = \sqrt{1 - \alpha_0^2} \implies \alpha_0^2(1 + \tan^2 \alpha_0) = 1 \implies \cos^2 \alpha_0 = \alpha_0^2. \tag{4.109}$$

The solution of $\cos^2 \alpha_0 = \alpha_0^2$ is given numerically by $\alpha_0 = 0.739\,09$. Thus, the corresponding energy is determined by the relation $\sqrt{ma^2 E_0/(2\hbar^2)} = 0.739\,09$, which yields $E_0 \simeq 1.1\hbar^2/(ma^2)$.

(b) When $R = 2$ there are two bound states resulting from the intersections of $\sqrt{4 - \alpha_0^2}$ with $\alpha_0 \tan \alpha_0$ and $-\alpha_1 \cot \alpha_1$; they correspond to $n = 0$ and $n = 1$, respectively. The numerical solutions of the corresponding equations

$$\alpha_0 \tan \alpha_0 = \sqrt{4 - \alpha_0^2} \implies 4 \cos^2 \alpha_0 = \alpha_0^2, \tag{4.110}$$

$$-\alpha_1 \cot \alpha_1 = \sqrt{4 - \alpha_1^2} \implies 4 \sin^2 \alpha_1 = \alpha_1^2, \tag{4.111}$$

yield $\alpha_0 \simeq 1.03$ and $\alpha_1 \simeq 1.9$, respectively. The corresponding energies are

$$\alpha_0 = \sqrt{\frac{ma^2 E_0}{2\hbar^2}} \simeq 1.03 \implies E_0 \simeq \frac{2.12\hbar^2}{ma^2}, \tag{4.112}$$

$$\alpha_1 = \sqrt{\frac{ma^2 E_1}{2\hbar^2}} \simeq 1.9 \implies E_1 \simeq \frac{7.22\hbar^2}{ma^2}. \tag{4.113}$$

4.8 The Harmonic Oscillator

The harmonic oscillator is one of those few problems that are important to all branches of physics. It provides a useful model for a variety of vibrational phenomena that are encountered, for instance, in classical mechanics, electrodynamics, statistical mechanics, solid state, atomic, nuclear, and particle physics. In quantum mechanics, it serves as an invaluable tool to illustrate the basic concepts and the formalism.

The Hamiltonian of a particle of mass m which oscillates with an angular frequency ω under the influence of a one-dimensional harmonic potential is

$$\hat{H} = \frac{\hat{P}^2}{2m} + \frac{1}{2}m\omega^2\hat{X}^2. \tag{4.114}$$

The problem is how to find the energy eigenvalues and eigenstates of this Hamiltonian. This problem can be studied by means of two separate methods. The first method, called the *analytic method*, consists in solving the time-independent Schrödinger equation (TISE) for the Hamiltonian (4.114). The second method, called the *ladder* or *algebraic method*, does not deal with solving the Schrödinger equation, but deals instead with operator algebra involving operators known as the *creation* and *annihilation* or *ladder* operators; this method is in essence a matrix formulation, because it expresses the various quantities in terms of matrices. In our presentation, we are going to adopt the second method, for it is more straightforward, more elegant and much simpler than solving the Schrödinger equation. Unlike the examples seen up to now, solving the Schrödinger equation for the potential $V(x) = \frac{1}{2}m\omega x^2$ is no easy job. Before embarking on the second method, let us highlight the main steps involved in the first method.

Brief outline of the analytic method
This approach consists in using the power series method to solve the following differential (Schrödinger) equation:

$$-\frac{\hbar^2}{2m}\frac{d^2\psi(x)}{dx^2} + \frac{1}{2}m\omega^2 x^2\psi(x) = E\psi(x), \tag{4.115}$$

which can be reduced to

$$\frac{d^2\psi(x)}{dx^2} + \left(\frac{2mE}{\hbar^2} - \frac{x^2}{x_0^4}\right)\psi(x) = 0, \tag{4.116}$$

where $x_0 = \sqrt{\hbar/(m\omega)}$ is a constant that has the dimensions of length; it sets the length scale of the oscillator, as will be seen later. The solutions of differential equations like (4.116) have been worked out by our mathematician colleagues well before the arrival of quantum mechanics (the solutions are expressed in terms of some special functions, the Hermite polynomials). The occurrence of the term $x^2\psi(x)$ in (4.116) suggests trying a Gaussian type solution[3]: $\psi(x) =$

---

[3] Solutions of the type $\psi(x) = f(x)\exp(x^2/2x_0^2)$ are physically unacceptable, for they diverge when $x \longrightarrow \pm\infty$.

$f(x) \exp(-x^2/2x_0^2)$, where $f(x)$ is a function of x. Inserting this trial function into (4.116), we obtain a differential equation for $f(x)$. This new differential equation can be solved by expanding $f(x)$ out in a power series (i.e., $f(x) = \sum_{n=0}^{\infty} a_n x^n$, where a_n is just a coefficient), which when inserted into the differential equation leads to a recursion relation. By demanding the power series of $f(x)$ to terminate at some finite value of n (because the wave function $\psi(x)$ has to be finite everywhere, notably when $x \longrightarrow \pm\infty$), the recursion relation yields an expression for the energy eigenvalues which are discrete or quantized:

$$E_n = \left(n + \frac{1}{2}\right)\hbar\omega \qquad (n = 0, 1, 2, 3, \ldots). \tag{4.117}$$

After some calculations, we can show that the wave functions that are physically acceptable and that satisfy (4.116) are given by

$$\psi_n(x) = \frac{1}{\sqrt{\sqrt{\pi}\, 2^n n!\, x_0}} e^{-x^2/2x_0^2} H_n\left(\frac{x}{x_0}\right), \tag{4.118}$$

where $H_n(y)$ are nth order polynomials called *Hermite polynomials*:

$$H_n(y) = (-1)^n e^{y^2} \frac{d^n}{dy^n} e^{-y^2}. \tag{4.119}$$

From this relation it is easy to calculate the first few polynomials:

$$\begin{aligned}
&H_0(y) = 1, && H_1(y) = 2y, \\
&H_2(y) = 4y^2 - 2, && H_3(y) = 8y^3 - 12y, \\
&H_4(y) = 16y^4 - 48y^2 + 12, && H_5(y) = 32y^5 - 160y^3 + 120y.
\end{aligned} \tag{4.120}$$

We will deal with the physical interpretations of the harmonic oscillator results when we study the second method.

Algebraic method

Let us now show how to solve the harmonic oscillator eigenvalue problem using the algebraic method. For this, we need to rewrite the Hamiltonian (4.114) in terms of the two Hermitian, dimensionless operators $\hat{p} = \hat{P}/\sqrt{m\hbar\omega}$ and $\hat{q} = \hat{X}\sqrt{m\omega/\hbar}$:

$$\hat{H} = \frac{\hbar\omega}{2}(\hat{p}^2 + \hat{q}^2), \tag{4.121}$$

and then introduce two non-Hermitian, dimensionless operators:

$$\hat{a} = \frac{1}{\sqrt{2}}(\hat{q} + i\hat{p}), \qquad \hat{a}^\dagger = \frac{1}{\sqrt{2}}(\hat{q} - i\hat{p}). \tag{4.122}$$

The physical meaning of the operators $\hat{a}$ and $\hat{a}^\dagger$ will be examined later. Note that

$$\hat{a}^\dagger \hat{a} = \frac{1}{2}(\hat{q} - i\hat{p})(\hat{q} + i\hat{p}) = \frac{1}{2}(\hat{q}^2 + \hat{p}^2 + i\hat{q}\hat{p} - i\hat{p}\hat{q}) = \frac{1}{2}(\hat{q}^2 + \hat{p}^2) + \frac{i}{2}[\hat{q}, \hat{p}], \tag{4.123}$$

where, using $[\hat{X}, \hat{P}] = i\hbar$, we can verify that the commutator between $\hat{q}$ and $\hat{p}$ is

$$[\hat{q}, \hat{p}] = \left[\sqrt{\frac{m\omega}{\hbar}}\hat{X}, \frac{1}{\sqrt{\hbar m\omega}}\hat{P}\right] = \frac{1}{\hbar}[\hat{X}, \hat{P}] = i; \tag{4.124}$$

hence

$$\hat{a}^\dagger \hat{a} = \frac{1}{2}(\hat{q}^2 + \hat{p}^2) - \frac{1}{2} \qquad (4.125)$$

or

$$\frac{1}{2}(\hat{q}^2 + \hat{p}^2) = \hat{a}^\dagger \hat{a} + \frac{1}{2}. \qquad (4.126)$$

Inserting (4.126) into (4.121) we obtain

$$\hat{H} = \hbar\omega \left(\hat{a}^\dagger \hat{a} + \frac{1}{2} \right) = \hbar\omega \left(\hat{N} + \frac{1}{2} \right) \qquad \text{with} \quad \hat{N} = \hat{a}^\dagger \hat{a}, \qquad (4.127)$$

where $\hat{N}$ is known as the *number* operator or *occupation number* operator, which is clearly Hermitian.

Let us now derive the commutator $[\hat{a}, \hat{a}^\dagger]$. Since $[\hat{X}, \hat{P}] = i\hbar$ we have $[\hat{q}, \hat{p}] = \frac{1}{\hbar}[\hat{X}, \hat{P}] = i$; hence

$$[\hat{a}, \hat{a}^\dagger] = \frac{1}{2}[\hat{q} + i\hat{p}, \hat{q} - i\hat{p}] = -i[\hat{q}, \hat{p}] = 1 \qquad (4.128)$$

or

$$[\hat{a}, \hat{a}^\dagger] = 1. \qquad (4.129)$$

4.8.1 Energy Eigenvalues

Note that $\hat{H}$ as given by (4.127) commutes with $\hat{N}$, since $\hat{H}$ is linear in $\hat{N}$. Thus, $\hat{H}$ and $\hat{N}$ can have a set of joint eigenstates, to be denoted by $| n \rangle$:

$$\hat{N} | n \rangle = n | n \rangle \qquad (4.130)$$

and

$$\hat{H} | n \rangle = E_n | n \rangle; \qquad (4.131)$$

the states $| n \rangle$ are called energy eigenstates. Combining (4.127) and (4.131), we obtain the energy eigenvalues at once:

$$E_n = \left(n + \frac{1}{2} \right) \hbar\omega. \qquad (4.132)$$

We will show later that n is a *positive integer*; it cannot have negative values.

The physical meaning of the operators $\hat{a}$, $\hat{a}^\dagger$, and $\hat{N}$ can now be clarified. First, we need the following two commutators that can be extracted from (4.129) and (4.127):

$$[\hat{a}, \hat{H}] = \hbar\omega\hat{a}, \qquad [\hat{a}^\dagger, \hat{H}] = -\hbar\omega\hat{a}^\dagger. \qquad (4.133)$$

These commutation relations along with (4.131) lead to

$$\hat{H}(\hat{a} | n \rangle) = (\hat{a}\hat{H} - \hbar\omega\hat{a}) | n \rangle = (E_n - \hbar\omega)(\hat{a} | n \rangle), \qquad (4.134)$$

$$\hat{H}(\hat{a}^\dagger | n \rangle) = (\hat{a}^\dagger \hat{H} + \hbar\omega\hat{a}^\dagger) | n \rangle = (E_n + \hbar\omega)(\hat{a}^\dagger | n \rangle). \qquad (4.135)$$

Thus, $\hat{a} | n \rangle$ and $\hat{a}^\dagger | n \rangle$ are eigenstates of $\hat{H}$ with eigenvalues $(E_n - \hbar\omega)$ and $(E_n + \hbar\omega)$, respectively. So the actions of $\hat{a}$ and $\hat{a}^\dagger$ on $| n \rangle$ generate new energy states that are lower and higher by one unit of $\hbar\omega$, respectively. As a result, $\hat{a}$ and $\hat{a}^\dagger$ are respectively known as the *lowering* and *raising* operators, or the *annihilation* and *creation* operators; they are also known as the *ladder* operators.

Let us now find out how the operators $\hat{a}$ and $\hat{a}^\dagger$ act on the energy eigenstates $\mid n \rangle$. Since $\hat{a}$ and $\hat{a}^\dagger$ do not commute with $\hat{N}$, the states $\mid n \rangle$ are eigenstates neither to $\hat{a}$ nor to $\hat{a}^\dagger$. Using (4.129) along with $[\hat{A}\hat{B}, \hat{C}] = \hat{A}[\hat{B}, \hat{C}] + [\hat{A}, \hat{C}]\hat{B}$, we can show that

$$\boxed{[\hat{N}, \hat{a}] = -\hat{a}, \qquad [\hat{N}, \hat{a}^\dagger] = \hat{a}^\dagger;} \tag{4.136}$$

hence $\hat{N}\hat{a} = \hat{a}(\hat{N} - 1)$ and $\hat{N}\hat{a}^\dagger = \hat{a}^\dagger(\hat{N} + 1)$. Combining these relations with (4.130), we obtain

$$\hat{N}\left(\hat{a} \mid n \rangle\right) = \hat{a}(\hat{N} - 1) \mid n \rangle = (n - 1)\left(\hat{a} \mid n \rangle\right), \tag{4.137}$$

$$\hat{N}\left(\hat{a}^\dagger \mid n \rangle\right) = \hat{a}^\dagger(\hat{N} + 1) \mid n \rangle = (n + 1)(\hat{a}^\dagger \mid n \rangle). \tag{4.138}$$

These relations reveal that $\hat{a} \mid n \rangle$ and $\hat{a}^\dagger \mid n \rangle$ are eigenstates of $\hat{N}$ with eigenvalues $(n - 1)$ and $(n + 1)$, respectively. This implies that when $\hat{a}$ and $\hat{a}^\dagger$ operate on $\mid n \rangle$, respectively, they decrease and increase n by one unit. That is, while the action of $\hat{a}$ on $\mid n \rangle$ generates a new state $\mid n - 1 \rangle$ (i.e., $\hat{a} \mid n \rangle \sim \mid n - 1 \rangle$), the action of $\hat{a}^\dagger$ on $\mid n \rangle$ generates $\mid n + 1 \rangle$. Hence from (4.137) we can write

$$\hat{a} \mid n \rangle = c_n \mid n - 1 \rangle, \tag{4.139}$$

where c_n is a constant to be determined from the requirement that the states $\mid n \rangle$ be normalized for all values of n. On the one hand, (4.139) yields

$$\left(\langle n \mid \hat{a}^\dagger\right) \cdot \left(\hat{a} \mid n \rangle\right) = \langle n \mid \hat{a}^\dagger \hat{a} \mid n \rangle = |c_n|^2 \langle n - 1 \mid n - 1 \rangle = |c_n|^2 \tag{4.140}$$

and, on the other hand, (4.130) gives

$$\left(\langle n \mid \hat{a}^\dagger\right) \cdot \left(\hat{a} \mid n \rangle\right) = \langle n \mid \hat{a}^\dagger \hat{a} \mid n \rangle = n\langle n \mid n \rangle = n. \tag{4.141}$$

When combined, the last two equations yield

$$|c_n|^2 = n. \tag{4.142}$$

This implies that n, which is equal to the norm of $\hat{a} \mid n \rangle$ (see (4.141)), *cannot be negative*, $n \geq 0$, since the norm is a positive quantity. Substituting (4.142) into (4.139) we end up with

$$\boxed{\hat{a} \mid n \rangle = \sqrt{n} \mid n - 1 \rangle.} \tag{4.143}$$

This equation shows that repeated applications of the operator $\hat{a}$ on $\mid n \rangle$ generate a sequence of eigenvectors $\mid n - 1 \rangle$, $\mid n - 2 \rangle$, $\mid n - 3 \rangle$, …. Since $n \geq 0$ and since $\hat{a} \mid 0 \rangle = 0$, this sequence has to terminate at $n = 0$; this is true if we start with an integer value of n. But if we start with a noninteger n, the sequence will not terminate; hence it leads to eigenvectors with negative values of n. But as shown above, since n cannot be negative, we conclude that n has to be a *nonnegative integer*.

Now, we can easily show, as we did for (4.143), that

$$\hat{a}^\dagger \mid n \rangle = \sqrt{n+1} \mid n+1 \rangle. \tag{4.144}$$

This implies that repeated applications of $\hat{a}^\dagger$ on $\mid n \rangle$ generate an infinite sequence of eigenvectors $\mid n+1 \rangle$, $\mid n+2 \rangle$, $\mid n+3 \rangle$, ... Since n is a positive integer, the energy spectrum of a harmonic oscillator as specified by (4.132) is therefore *discrete*:

$$E_n = \left(n + \frac{1}{2} \right) \hbar\omega \qquad (n = 0, 1, 2, 3, \ldots). \tag{4.145}$$

This expression is similar to the one obtained from the first method (see Eq. (4.117)). The energy spectrum of the harmonic oscillator consists of energy levels that are equally spaced: $E_{n+1} - E_n = \hbar\omega$. This is Planck's famous equidistant energy idea—the energy of the radiation emitted by the oscillating charges (from the inside walls of the cavity) must come only in bundles (quanta) that are integral multiples of $\hbar\omega$—which, as mentioned in Chapter 1, led to the birth of quantum mechanics.

As expected for bound states of one-dimensional potentials, the energy spectrum is both discrete and nondegenerate. Once again, as in the case of the infinite square well potential, we encounter the zero-point energy phenomenon: the lowest energy eigenvalue of the oscillator is not zero but is instead equal to $E_0 = \hbar\omega/2$. It is called the *zero-point energy* of the oscillator, for it corresponds to $n = 0$. The zero-point energy of bound state systems cannot be zero, otherwise it would violate the uncertainty principle. For the harmonic oscillator, for instance, the classical minimum energy corresponds to $x = 0$ and $p = 0$; there would be no oscillations in this case. This would imply that we know simultaneously and with absolute precision both the position and the momentum of the system. This would contradict the uncertainty principle.

4.8.2 Energy Eigenstates

The algebraic or operator method can also be used to determine the energy eigenvectors. First, using (4.144), we see that the various eigenvectors can be written in terms of the ground state $\mid 0 \rangle$ as follows:

$$\mid 1 \rangle = \hat{a}^\dagger \mid 0 \rangle, \tag{4.146}$$

$$\mid 2 \rangle = \frac{1}{\sqrt{2}} \hat{a}^\dagger \mid 1 \rangle = \frac{1}{\sqrt{2!}} \left(\hat{a}^\dagger \right)^2 \mid 0 \rangle, \tag{4.147}$$

$$\mid 3 \rangle = \frac{1}{\sqrt{3}} \hat{a}^\dagger \mid 2 \rangle = \frac{1}{\sqrt{3!}} \left(\hat{a}^\dagger \right)^3 \mid 0 \rangle, \tag{4.148}$$

$$\vdots$$

$$\mid n \rangle = \frac{1}{\sqrt{n}} \hat{a}^\dagger \mid n-1 \rangle = \frac{1}{\sqrt{n!}} \left(\hat{a}^\dagger \right)^n \mid 0 \rangle. \tag{4.149}$$

So, to find any excited eigenstate $\mid n \rangle$, we need simply to operate $\hat{a}^\dagger$ on $\mid 0 \rangle$ n successive times.

Note that any set of kets $\mid n \rangle$ and $\mid n' \rangle$, corresponding to different eigenvalues, must be orthogonal, $\langle n' \mid n \rangle \sim \delta_{n',n}$, since $\hat{H}$ is Hermitian and none of its eigenstates is degenerate.

Moreover, the states $| 0 \rangle, | 1 \rangle, | 2 \rangle, | 3 \rangle, \ldots, | n \rangle, \ldots$ are *simultaneous eigenstates* of $\hat{H}$ and $\hat{N}$; the set $\{| n \rangle\}$ constitutes an orthonormal and complete basis:

$$\langle n' \mid n \rangle = \delta_{n',n}, \qquad \sum_{n=0}^{+\infty} | n \rangle \langle n \mid = 1. \tag{4.150}$$

4.8.3 Energy Eigenstates in Position Space

Let us now determine the harmonic oscillator wave function in the position representation.

Equations (4.146) to (4.149) show that, knowing the ground state wave function, we can determine any other eigenstate by successive applications of the operator $a^\dagger$ on the ground state. So let us first determine the ground state wave function in the position representation.

The operator $\hat{p}$, defined by $\hat{p} = \hat{P}/\sqrt{m\hbar\omega}$, is given in the position space by

$$\hat{p} = -\frac{i\hbar}{\sqrt{m\hbar\omega}} \frac{d}{dx} = -ix_0 \frac{d}{dx}, \tag{4.151}$$

where, as mentioned above, $x_0 = \sqrt{\hbar/(m\omega)}$ is a constant that has the dimensions of length; it sets the length scale of the oscillator. We can easily show that the annihilation and creation operators $\hat{a}$ and $\hat{a}^\dagger$, defined in (4.122), can be written in the position representation as

$$\hat{a} = \frac{1}{\sqrt{2}} \left(\frac{\hat{X}}{x_0} + x_0 \frac{d}{dx} \right) = \frac{1}{\sqrt{2}x_0} \left(\hat{X} + x_0^2 \frac{d}{dx} \right), \tag{4.152}$$

$$\hat{a}^\dagger = \frac{1}{\sqrt{2}} \left(\frac{\hat{X}}{x_0} - x_0 \frac{d}{dx} \right) = \frac{1}{\sqrt{2}x_0} \left(\hat{X} - x_0^2 \frac{d}{dx} \right). \tag{4.153}$$

Using (4.152) we can write the equation $\hat{a} | 0 \rangle = 0$ in the position space as

$$\langle x | \hat{a} | 0 \rangle = \frac{1}{\sqrt{2}x_0} \langle x | \hat{X} + x_0^2 \frac{d}{dx} | 0 \rangle = \frac{1}{\sqrt{2}x_0} \left(x \, \psi_0(x) + x_0^2 \frac{d\psi_0(x)}{dx} \right) = 0; \tag{4.154}$$

hence

$$\frac{d\psi_0(x)}{dx} = -\frac{x}{x_0^2} \psi_0(x), \tag{4.155}$$

where $\psi_0(x) = \langle x \mid 0 \rangle$ represents the ground state wave function. The solution of this differential equation is

$$\psi_0(x) = A \exp\left(-\frac{x^2}{2x_0^2} \right), \tag{4.156}$$

where A is a constant that can be determined from the normalization condition

$$1 = \int_{-\infty}^{+\infty} dx \, |\psi_0(x)|^2 = A^2 \int_{-\infty}^{+\infty} dx \, \exp\left(-\frac{x^2}{x_0^2} \right) = A^2 \sqrt{\pi} x_0; \tag{4.157}$$

hence $A = (m\omega/(\pi\hbar))^{1/4} = 1/\sqrt{\sqrt{\pi}x_0}$. The normalized ground state wave function is then given by

$$\boxed{\psi_0(x) = \frac{1}{\sqrt{\sqrt{\pi}x_0}}\exp\left(-\frac{x^2}{2x_0^2}\right).}$$

(4.158)

This is a Gaussian function.

We can then obtain the wave function of any excited state by a series of applications of $\hat{a}^\dagger$ on the ground state. For instance, the first excited state is obtained by one single application of the operator $\hat{a}^\dagger$ of (4.153) on the ground state:

$$\langle x \mid 1 \rangle = \langle x|\hat{a}^\dagger \mid 0 \rangle = \frac{1}{\sqrt{2}x_0}\left(x - x_0^2\frac{d}{dx}\right)\langle x \mid 0 \rangle$$

$$= \frac{1}{\sqrt{2}x_0}\left(x - x_0^2\left(-\frac{x}{x_0^2}\right)\right)\psi_0(x) = \frac{\sqrt{2}}{x_0}x\psi_0(x)$$

(4.159)

or

$$\psi_1(x) = \frac{\sqrt{2}}{x_0}x\psi_0(x) = \sqrt{\frac{2}{\sqrt{\pi}x_0^3}}x\exp\left(-\frac{x^2}{2x_0^2}\right).$$

(4.160)

As for the eigenstates of the second and third excited states, we can obtain them by applying $\hat{a}^\dagger$ on the ground state twice and three times, respectively:

$$\langle x \mid 2 \rangle = \frac{1}{\sqrt{2!}}\langle x|\left(a^\dagger\right)^2 \mid 0 \rangle = \frac{1}{\sqrt{2!}}\left(\frac{1}{\sqrt{2}x_0}\right)^2\left(x - x_0^2\frac{d}{dx}\right)^2\psi_0(x),$$

(4.161)

$$\langle x \mid 3 \rangle = \frac{1}{\sqrt{3!}}\langle x|\left(a^\dagger\right)^3 \mid 0 \rangle = \frac{1}{\sqrt{3!}}\left(\frac{1}{\sqrt{2}x_0}\right)^3\left(x - x_0^2\frac{d}{dx}\right)^3\psi_0(x)$$

(4.162)

or

$$\psi_2(x) = \frac{1}{\sqrt{2\sqrt{\pi}x_0}}\left(\frac{2x^2}{x_0^2} - 1\right)\exp\left(-\frac{x^2}{2x_0^2}\right), \quad \psi_3(x) = \frac{1}{\sqrt{3\sqrt{\pi}x_0}}\left(\frac{2x^3}{x_0^3} - \frac{3x}{x_0}\right)\exp\left(-\frac{x^2}{2x_0^2}\right).$$

(4.163)

Similarly, using (4.149), (4.153), and (4.158), we can easily infer the energy eigenstate for the nth excited state:

$$\langle x \mid n \rangle = \frac{1}{\sqrt{n!}}\langle x|\left(a^\dagger\right)^n \mid 0 \rangle = \frac{1}{\sqrt{n!}}\left(\frac{1}{\sqrt{2}x_0}\right)^n\left(x - x_0^2\frac{d}{dx}\right)^n\psi_0(x),$$

(4.164)

which in turn can be rewritten as

$$\boxed{\psi_n(x) = \frac{1}{\sqrt{\sqrt{\pi}2^n n!}}\frac{1}{x_0^{n+1/2}}\left(x - x_0^2\frac{d}{dx}\right)^n\exp\left(-\frac{x^2}{2x_0^2}\right).}$$

(4.165)

In summary, by successive applications of $\hat{a}^\dagger = (\hat{X} - x_0^2 d/dx)/(\sqrt{2}x_0)$ on $\psi_0(x)$, we can find the wave function of any excited state $\psi_n(x)$.

Oscillator wave functions and the Hermite polynomials

At this level, we can show that the wave function (4.165) derived from the algebraic method is similar to the one obtained from the first method (4.118). To see this, we simply need to use the following operator identity:

$$e^{-x^2/2}\left(x - \frac{d}{dx}\right)e^{x^2/2} = -\frac{d}{dx} \quad \text{or} \quad e^{-x^2/2x_0^2}\left(x - x_0^2\frac{d}{dx}\right)e^{x^2/2x_0^2} = -x_0^2\frac{d}{dx}. \quad (4.166)$$

An application of this operator n times leads at once to

$$e^{-x^2/2x_0^2}\left(x - x_0^2\frac{d}{dx}\right)^n e^{x^2/2x_0^2} = (-1)^n(x_0^2)^n\frac{d^n}{dx^n}, \quad (4.167)$$

which can be shown to yield

$$\left(x - x_0^2\frac{d}{dx}\right)^n e^{-x^2/2x_0^2} = (-1)^n(x_0^2)^n e^{x^2/2x_0^2}\frac{d^n}{dx^n}e^{-x^2/x_0^2}. \quad (4.168)$$

We can now rewrite the right-hand side of this equation as follows:

$$
\begin{aligned}
(-1)^n(x_0^2)^n e^{x^2/2x_0^2}\frac{d^n}{dx^n}e^{-x^2/x_0^2} &= x_0^n e^{-x^2/2x_0^2}\left[(-1)^n e^{x^2/x_0^2}\frac{d^n}{d(x/x_0)^n}e^{-x^2/x_0^2}\right] \\
&= x_0^n e^{-x^2/2x_0^2}\left[(-1)^n e^{y^2}\frac{d^n}{dy^n}e^{-y^2}\right] \\
&= x_0^n e^{-x^2/2x_0^2}H_n(y), \quad (4.169)
\end{aligned}
$$

where $y = x/x_0$ and where $H_n(y)$ are the Hermite polynomials listed in (4.119):

$$H_n(y) = (-1)^n e^{y^2}\frac{d^n}{dy^n}e^{-y^2}. \quad (4.170)$$

Note that the polynomials $H_{2n}(y)$ are even and $H_{2n+1}(y)$ are odd, since $H_n(-y) = (-1)^n H_n(y)$. Inserting (4.169) into (4.168), we obtain

$$\left(x - x_0^2\frac{d}{dx}\right)^n e^{-x^2/2x_0^2} = x_0^n e^{-x^2/2x_0^2}H_n\left(\frac{x}{x_0}\right); \quad (4.171)$$

substituting this equation into (4.165), we can write the oscillator wave function in terms of the Hermite polynomials as follows:

$$\boxed{\psi_n(x) = \frac{1}{\sqrt{\sqrt{\pi}2^n n!x_0}}e^{-x^2/2x_0^2}H_n\left(\frac{x}{x_0}\right).} \quad (4.172)$$

This wave function is identical with the one obtained from the first method (see Eq. (4.118)).

Remark

This wave function is either even or odd depending on n; in fact, the functions $\psi_{2n}(x)$ are even (i.e., $\psi_{2n}(-x) = \psi_{2n}(x)$) and $\psi_{2n+1}(x)$ are odd (i.e., $\psi_{2n+1}(-x) = -\psi_{2n+1}(x)$) since, as can be inferred from Eq (4.120), the Hermite polynomials $H_{2n}(x)$ are even and $H_{2n+1}(x)$ are odd. This is expected because, as mentioned in Section 4.2.4, the wave functions of even one-dimensional potentials have definite parity. Figure 4.9 displays the shapes of the first few wave functions.

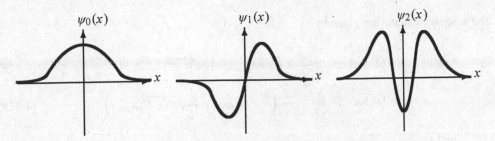

Figure 4.9 Shapes of the first three wave functions of the harmonic oscillator.

4.8.4 The Matrix Representation of Various Operators

Here we look at the matrix representation of several operators in the N-space. In particular, we focus on the representation of the operators $\hat{a}$, $\hat{a}^\dagger$, $\hat{X}$, and $\hat{P}$. First, since the states $|n\rangle$ are joint eigenstates of $\hat{H}$ and $\hat{N}$, it is easy to see from (4.130) and (4.132) that $\hat{H}$ and $\hat{N}$ are represented within the $\{|n\rangle\}$ basis by infinite diagonal matrices:

$$\langle n'|\hat{N}|n\rangle = n\delta_{n',n}, \qquad \langle n'|\hat{H}|n\rangle = \hbar\omega\left(n + \frac{1}{2}\right)\delta_{n',n}; \qquad (4.173)$$

that is,

$$\hat{N} = \begin{pmatrix} 0 & 0 & 0 & \cdots \\ 0 & 1 & 0 & \cdots \\ 0 & 0 & 2 & \cdots \\ \vdots & \vdots & \vdots & \ddots \end{pmatrix}, \qquad \hat{H} = \frac{\hbar\omega}{2}\begin{pmatrix} 1 & 0 & 0 & \cdots \\ 0 & 3 & 0 & \cdots \\ 0 & 0 & 5 & \cdots \\ \vdots & \vdots & \vdots & \ddots \end{pmatrix}. \qquad (4.174)$$

As for the operators $\hat{a}$, $\hat{a}^\dagger$, $\hat{X}$, $\hat{P}$, none of them are diagonal in the N-representation, since they do not commute with $\hat{N}$. The matrix elements of $\hat{a}$ and $\hat{a}^\dagger$ can be obtained from (4.143) and (4.144):

$$\langle n'|\hat{a}|n\rangle = \sqrt{n}\delta_{n',n-1}, \qquad \langle n'|\hat{a}^\dagger|n\rangle = \sqrt{n+1}\delta_{n',n+1}; \qquad (4.175)$$

that is,

$$\hat{a} = \begin{pmatrix} 0 & \sqrt{1} & 0 & 0 & \cdots \\ 0 & 0 & \sqrt{2} & 0 & \cdots \\ 0 & 0 & 0 & \sqrt{3} & \cdots \\ 0 & 0 & 0 & 0 & \cdots \\ \vdots & \vdots & \vdots & \vdots & \ddots \end{pmatrix}, \qquad \hat{a}^\dagger = \begin{pmatrix} 0 & 0 & 0 & 0 & \cdots \\ \sqrt{1} & 0 & 0 & 0 & \cdots \\ 0 & \sqrt{2} & 0 & 0 & \cdots \\ 0 & 0 & \sqrt{3} & 0 & \cdots \\ \vdots & \vdots & \vdots & \vdots & \ddots \end{pmatrix}. \qquad (4.176)$$

Now, let us find the N-representation of the position and momentum operators, $\hat{X}$ and $\hat{P}$. From (4.122) we can show that $\hat{X}$ and $\hat{P}$ are given in terms of $\hat{a}$ and $\hat{a}^\dagger$ as follows:

$$\hat{X} = \sqrt{\frac{\hbar}{2m\omega}}\left(\hat{a} + \hat{a}^\dagger\right), \qquad \hat{P} = i\sqrt{\frac{m\hbar\omega}{2}}\left(\hat{a}^\dagger - \hat{a}\right). \qquad (4.177)$$

Their matrix elements are given by

$$\langle n' | \hat{X} | n \rangle = \sqrt{\frac{\hbar}{2m\omega}} \left(\sqrt{n}\delta_{n',n-1} + \sqrt{n+1}\delta_{n',n+1} \right), \tag{4.178}$$

$$\langle n' | \hat{P} | n \rangle = i\sqrt{\frac{m\hbar\omega}{2}} \left(-\sqrt{n}\delta_{n',n-1} + \sqrt{n+1}\delta_{n',n+1} \right), \tag{4.179}$$

in particular

$$\langle n | \hat{X} | n \rangle = \langle n | \hat{P} | n \rangle = 0. \tag{4.180}$$

The matrices corresponding to $\hat{X}$ and $\hat{P}$ are thus given by

$$\hat{X} = \sqrt{\frac{\hbar}{2m\omega}} \begin{pmatrix} 0 & \sqrt{1} & 0 & 0 & \cdots \\ \sqrt{1} & 0 & \sqrt{2} & 0 & \cdots \\ 0 & \sqrt{2} & 0 & \sqrt{3} & \cdots \\ 0 & 0 & \sqrt{3} & 0 & \cdots \\ \vdots & \vdots & \vdots & \vdots & \ddots \end{pmatrix}, \tag{4.181}$$

$$\hat{P} = i\sqrt{\frac{m\hbar\omega}{2}} \begin{pmatrix} 0 & -\sqrt{1} & 0 & 0 & \cdots \\ \sqrt{1} & 0 & -\sqrt{2} & 0 & \cdots \\ 0 & \sqrt{2} & 0 & -\sqrt{3} & \cdots \\ 0 & 0 & \sqrt{3} & 0 & \cdots \\ \vdots & \vdots & \vdots & \vdots & \ddots \end{pmatrix}. \tag{4.182}$$

As mentioned in Chapter 2, the momentum operator is Hermitian, but not equal to its own complex conjugate: (4.182) shows that $\hat{P}^{\dagger} = \hat{P}$ and $\hat{P}^{*} = -\hat{P}$. As for $\hat{X}$, however, it is both Hermitian and equal to its complex conjugate: from (4.181) we have that $\hat{X}^{\dagger} = \hat{X}^{*} = \hat{X}$.

Finally, we should mention that the eigenstates $|n\rangle$ are represented by infinite column matrices; the first few states can be written as

$$|0\rangle = \begin{pmatrix} 1 \\ 0 \\ 0 \\ 0 \\ \vdots \end{pmatrix}, \quad |1\rangle = \begin{pmatrix} 0 \\ 1 \\ 0 \\ 0 \\ \vdots \end{pmatrix}, \quad |2\rangle = \begin{pmatrix} 0 \\ 0 \\ 1 \\ 0 \\ \vdots \end{pmatrix}, \quad |3\rangle = \begin{pmatrix} 0 \\ 0 \\ 0 \\ 1 \\ \vdots \end{pmatrix}, \dots . \tag{4.183}$$

The set of states $\{|n\rangle\}$ forms indeed a complete and orthonormal basis.

4.8.5 Expectation Values of Various Operators

Let us evaluate the expectation values for $\hat{X}^2$ and $\hat{P}^2$ in the N-representation:

$$\hat{X}^2 = \frac{\hbar}{2m\omega} \left(\hat{a}^2 + \hat{a}^{\dagger 2} + \hat{a}\hat{a}^{\dagger} + \hat{a}^{\dagger}\hat{a} \right) = \frac{\hbar}{2m\omega} \left(\hat{a}^2 + \hat{a}^{\dagger 2} + 2\hat{a}^{\dagger}\hat{a} + 1 \right), \tag{4.184}$$

$$\hat{P}^2 = -\frac{m\hbar\omega}{2} \left(\hat{a}^2 + \hat{a}^{\dagger 2} - \hat{a}\hat{a}^{\dagger} - \hat{a}^{\dagger}\hat{a} \right) = -\frac{m\hbar\omega}{2} \left(\hat{a}^2 + \hat{a}^{\dagger 2} - 2\hat{a}^{\dagger}\hat{a} - 1 \right), \tag{4.185}$$

where we have used the fact that $\hat{a}\hat{a}^\dagger + \hat{a}^\dagger\hat{a} = 2\hat{a}^\dagger\hat{a} + 1$. Since the expectation values of $\hat{a}^2$ and $\hat{a}^{\dagger 2}$ are zero, $\langle n \mid \hat{a}^2 \mid n \rangle = \langle n \mid \hat{a}^{\dagger 2} \mid n \rangle = 0$, and $\langle n \mid \hat{a}^\dagger\hat{a} \mid n \rangle = n$, we have

$$\langle n \mid \hat{a}\hat{a}^\dagger + \hat{a}^\dagger\hat{a} \mid n \rangle = \langle n \mid 2\hat{a}^\dagger\hat{a} + 1 \mid n \rangle = 2n + 1; \tag{4.186}$$

hence

$$\langle n \mid \hat{X}^2 \mid n \rangle = \frac{\hbar}{2m\omega} \langle n \mid \hat{a}\hat{a}^\dagger + \hat{a}^\dagger\hat{a} \mid n \rangle = \frac{\hbar}{2m\omega}(2n+1), \tag{4.187}$$

$$\langle n \mid \hat{P}^2 \mid n \rangle = \frac{m\hbar\omega}{2} \langle n \mid \hat{a}\hat{a}^\dagger + \hat{a}^\dagger\hat{a} \mid n \rangle = \frac{m\hbar\omega}{2}(2n+1). \tag{4.188}$$

Comparing (4.187) and (4.188) we see that the expectation values of the potential and kinetic energies are equal and are also equal to half the total energy:

$$\boxed{\frac{m\omega^2}{2} \langle n \mid \hat{X}^2 \mid n \rangle = \frac{1}{2m} \langle n \mid \hat{P}^2 \mid n \rangle = \frac{1}{2} \langle n \mid \hat{H} \mid n \rangle.} \tag{4.189}$$

This result is known as the *Virial theorem*.

We can now easily calculate the product $\Delta x \Delta p$ from (4.187) and (4.188). Since $\langle \hat{X} \rangle = \langle \hat{P} \rangle = 0$ we have

$$\Delta x = \sqrt{\langle \hat{X}^2 \rangle - \langle \hat{X} \rangle^2} = \sqrt{\langle \hat{X}^2 \rangle} = \sqrt{\frac{\hbar}{2m\omega}(2n+1)}, \tag{4.190}$$

$$\Delta p = \sqrt{\langle \hat{P}^2 \rangle - \langle \hat{P} \rangle^2} = \sqrt{\langle \hat{P}^2 \rangle} = \sqrt{\frac{m\hbar\omega}{2}(2n+1)}; \tag{4.191}$$

hence

$$\Delta x \Delta p = \left(n + \frac{1}{2}\right)\hbar \implies \Delta x \Delta p \geq \frac{\hbar}{2}, \tag{4.192}$$

since $n \geq 0$; this is the Heisenberg uncertainty principle.

4.9 Numerical Solution of the Schrödinger Equation

In this section we are going to show how to solve a one-dimensional Schrödinger equation numerically. The numerical solutions provide an idea about the properties of stationary states.

4.9.1 Numerical Procedure

We want to solve the following equation numerically:

$$-\frac{\hbar^2}{2m}\frac{d^2\psi}{dx^2} + V(x)\psi(x) = E\psi(x) \implies \frac{d^2\psi}{dx^2} + k^2\psi(x) = 0, \tag{4.193}$$

where $k^2 = 2m[E - V(x)]/\hbar^2$.

First, divide the x-axis into a set of equidistant points with a spacing of $h_0 = \Delta x$, as shown in Figure 4.10a. The wave function $\psi(x)$ can be approximately described by its values at the

Figure 4.10 (a) Discretization of the wave function. (b) If the energy E used in the computation is too high (too low), the wave function will diverge as $x \to \pm\infty$; but at the appropriate value of E, the wave function converges to the correct values.

points of the grid (i.e., $\psi_0 = \psi(x = 0)$, $\psi_1 = \psi(h_0)$, $\psi_2 = \psi(2h_0)$, $\psi_3 = \psi(3h_0)$, and so on). The first derivative of ψ can then be approximated by

$$\frac{d\psi}{dx} \approx \frac{\psi_{n+1} - \psi_n}{h_0}. \tag{4.194}$$

An analogous approximation for the second derivative is actually a bit tricky. There are several methods to calculate it, but a very efficient procedure is called the *Numerov algorithm* (which is described in standard numerical analysis textbooks). In short, the second derivative is approximated by the so-called three-point difference formula:

$$\frac{\psi_{n+1} - 2\psi_n + \psi_{n-1}}{h_0^2} = \psi_n'' + \frac{h_0^2}{12}\psi_n'''' + 0(h_0^4). \tag{4.195}$$

From (4.193) we have

$$\psi_n'''' = \frac{d^2}{dx^2}(-k^2\psi)\bigg|_{x=x_n} = -\frac{(k^2\psi)_{n+1} - 2(k^2\psi)_n + (k^2\psi)_{n-1}}{h_0^2}. \tag{4.196}$$

Using $\psi_n'' = -k_n^2\psi_n$ and substituting (4.196) into (4.195) we can show that

$$\psi_{n+1} = \frac{2\left(1 - \frac{5}{12}h_0^2k_n^2\right)\psi_n - \left(1 + \frac{1}{12}h_0^2k_{n-1}^2\right)\psi_{n-1}}{1 + \frac{1}{12}h_0^2k_{n+1}^2}. \tag{4.197}$$

We can thus assign arbitrary values for ψ_0 and ψ_1; this is equivalent to providing the starting (or initial) values for $\psi(x)$ and $\psi'(x)$. Knowing ψ_0 and ψ_1, we can use (4.197) to calculate ψ_2, then ψ_3, then ψ_4, and so on. The solution of a linear equation, equation (4.197), for either ψ_{n+1} or ψ_{n-1} yields a recursion relation for integrating either *forward* or *backward* in x with a local error $O(h_0^6)$. In this way, the solution depends on two arbitrary constants, ψ_0 and ψ_1, as it should for any second-order differential equation (i.e., there are two linearly independent solutions).

The *boundary conditions* play a crucial role in solving any Schrödinger equation. Every boundary condition gives a linear homogeneous equation satisfied by the wave function or its derivative. For example, in the case of the infinite square well potential and the harmonic oscillator, the conditions $\psi(x_{min}) = 0$, $\psi(x_{max}) = 0$ are satisfied as follows:

- Infinite square well: $\psi(-a/2) = \psi(a/2) = 0$

- Harmonic oscillator: $\psi(-\infty) = \psi(+\infty) = 0$

4.9.2 Algorithm

To solve the Schrödinger equation with the boundary conditions $\psi(x_{min}) = \psi(x_{max}) = 0$, you may proceed as follows. Suppose you want to find the wave function, $\psi^{(n)}(x)$, and the energy E_n for the nth excited[4] state of a system:

- Take $\psi_0 = 0$ and choose ψ_1 (any small number you like), because the value of ψ_1 must be very close to that of ψ_0.

- Choose a *trial* energy E_n.

- With this value of the energy, E_n, together with ψ_0 and ψ_1, you can calculate iteratively the wave function at different values of x; that is, you can calculate $\psi_2, \psi_3, \psi_4, \ldots$ How? You need simply to inject $\psi_0 = 0$, ψ_1, and E_n into (4.197) and proceed incrementally to calculate ψ_2; then use ψ_1 and ψ_2 to calculate ψ_3; then use ψ_2 and ψ_3 to calculate ψ_4; and so on till you end up with the value of the wave function at $x_n = nh_0$, $\psi_n = \psi(nh_0)$.

- Next, you need to check whether the ψ_n you obtained is zero or not. If ψ_n is zero, this means that you have made the right choice for the trial energy. This value E_n can then be taken as a possible eigenenergy for the system; at this value of E_n, the wave function converges to the correct value (dotted curve in Figure 4.10b). Of course, it is highly unlikely to have chosen the correct energy from a first trial. In this case you need to proceed as follows. If the value of ψ_n obtained is a nonzero positive number or if it diverges, this means that the trial E_n you started with is larger than the correct eigenvalue (Figure 4.10b); on the other hand, if ψ_n is a negative nonzero number, this means that the E_n you started with is less than the true energy. If the ψ_n you end up with is a positive nonzero number, you need to start all over again with a smaller value of the energy. But if the ψ_n you end up with is negative, you need to start again with a larger value of E. You can continue in this way, improving every time, till you end up with a zero value for ψ_n. Note that in practice there is no way to get ψ_n exactly equal to zero. You may stop the procedure the moment ψ_n is sufficiently small; that is, you stop the iteration at the desired accuracy, say at 10^{-8} of its maximum value.

Example 4.3 (Numerical solution of the Schrödinger equation)
A proton is subject to a harmonic oscillator potential $V(x) = m\omega^2 x^2/2$, $\omega = 5.34 \times 10^{21} s^{-1}$.
   (a) Find the exact energies of the five lowest states (express them in MeV).
   (b) Solve the Schrödinger equation *numerically* and find the energies of the five lowest states and compare them with the exact results obtained in (a). Note: You may use these quantities: rest mass energy of the proton $mc^2 \simeq 10^3$ MeV, $\hbar c \simeq 200$ MeV fm, and $\hbar\omega \simeq 3.5$ MeV.

---

[4]We have denoted here the wave function of the nth excited state by $\psi^{(n)}(x)$ to distinguish it from the value of the wave function at $x_n = nh_0$, $\psi_n = \psi(nh_0)$.

Table 4.1 Exact and numerical energies for the five lowest states of the harmonic oscillator.

n	E_n^{Exact} (MeV)	$E_n^{Numeric}$ (MeV)
0.0	1.750 000	1.749 999 999 795
1.0	5.250 000	5.249 999 998 112
2.0	8.750 000	8.749 999 992 829
3.0	12.250 000	12.249 999 982 320
4.0	15.750 000	15.749 999 967 590

Solution

(a) The exact energies can be calculated at once from $E_n = \hbar\omega(n + \frac{1}{2}) \simeq 3.5(n + \frac{1}{2})$ MeV. The results for the five lowest states are listed in Table 4.1.

(b) To obtain the numerical values, we need simply to make use of the Numerov relation (4.197), where $k_n^2(x) = 2m(E_n - \frac{1}{2}m\omega^2 x^2)/\hbar^2$. The numerical values involved here can be calculated as follows:

$$\frac{m^2\omega^2}{\hbar^2} = \frac{(mc^2)^2(\hbar\omega)^2}{(\hbar c)^4} \simeq \frac{(10^3\,\text{MeV})^2(3.5\,\text{MeV})^2}{(200\,\text{MeV fm})^4} = 7.66 \times 10^{-4}\,\text{fm}^{-3}, \quad (4.198)$$

$$\frac{2m}{\hbar^2} = \frac{2mc^2}{(\hbar c)^2} \simeq \frac{2 \times 10^3\,\text{MeV}}{(200\,\text{MeV fm})^2} = 0.05\,\text{MeV}^{-1}\,\text{fm}^{-2}. \quad (4.199)$$

The boundary conditions for the harmonic oscillator imply that the wave function vanishes at $x = \pm\infty$, i.e., at $x_{min} = -\infty$ and $x_{max} = \infty$. How does one deal with infinities within a computer program? For this, we need to choose the numerical values of x_{min} and x_{max} in a way that the wave function would not feel the "edge" effects. That is, we simply need to assign numerical values to x_{min} and x_{max} so that they are far away from the turning points $x_{Left} = -\sqrt{2E_n/(m\omega^2)}$ and $x_{Right} = \sqrt{2E_n/(m\omega^2)}$, respectively. For instance, in the case of the ground state, where $E_0 = 1.75$ MeV, we have $x_{Left} = -3.38$ fm and $x_{Right} = 3.38$ fm; we may then take $x_{min} = -20$ fm and $x_{max} = 20$ fm. The wave function should be practically zero at $x = \pm 20$ fm.

To calculate the energies numerically for the five lowest states, a C++ computer code has been prepared (see Appendix C). The numerical results generated by this code are listed in Table 4.1; they are in excellent agreement with the exact results. Figure 4.11 displays the wave functions obtained from this code for the five lowest states for the proton moving in a harmonic oscillator potential (these plotted wave functions are normalized).

4.10 Solved Problems

Problem 4.1

A particle moving in one dimension is in a stationary state whose wave function

$$\psi(x) = \begin{cases} 0, & x < -a, \\ A(1 + \cos\frac{\pi x}{a}), & -a \leq x \leq a, \\ 0, & x > a, \end{cases}$$

Figure 4.11 Wave functions $\psi_n(x)$ of the five lowest states of a harmonic oscillator potential in terms of x, where the x-axis values are in fm (obtained from the C++ code of Appendix C).

where A and a are real constants.

    (a) Is this a physically acceptable wave function? Explain.
    (b) Find the magnitude of A so that $\psi(x)$ is normalized.
    (c) Evaluate Δx and Δp. Verify that $\Delta x \Delta p \geq \hbar/2$.
    (d) Find the classically allowed region.

Solution

    (a) Since $\psi(x)$ is square integrable, single-valued, continuous, and has a continuous first derivative, it is indeed physically acceptable.

    (b) Normalization of $\psi(x)$: using the relation $\cos^2 y = (1 + \cos 2y)/2$, we have

$$
\begin{aligned}
1 &= \int_{-\infty}^{+\infty} |\psi(x)|^2 dx = A^2 \int_{-a}^{a} dx \left[1 + 2\cos\frac{\pi x}{a} + \cos^2\left(\frac{\pi x}{a}\right) \right] \\
&= A^2 \int_{-a}^{a} dx \left[\frac{3}{2} + 2\cos\frac{\pi x}{a} + \frac{1}{2}\cos\frac{2\pi x}{a} \right] \\
&= \frac{3}{2} A^2 \int_{-a}^{a} dx = 3a A^2;
\end{aligned}
\tag{4.200}
$$

hence $A = 1/\sqrt{3a}$.

    (c) As $\psi(x)$ is even, we have $\langle \hat{X} \rangle = \int_{-a}^{+a} \psi^*(x) x \psi(x)\, dx = 0$, since the symmetric integral of an odd function (i.e., $\psi^*(x) x \psi(x)$ is odd) is zero. On the other hand, we also have $\langle \hat{P} \rangle = 0$ because $\psi(x)$ is real and even. We can thus write

$$
\Delta x = \sqrt{\langle \hat{X}^2 \rangle}, \qquad \Delta p = \sqrt{\langle \hat{P}^2 \rangle},
\tag{4.201}
$$

since $\Delta A = \sqrt{\langle \hat{A}^2 \rangle - \langle \hat{A} \rangle^2}$. The calculations of $\langle \hat{X}^2 \rangle$ and $\langle \hat{P}^2 \rangle$ are straightforward:

$$
\langle \hat{X}^2 \rangle = -\hbar^2 \int_{-a}^{a} \psi^*(x) x^2 \psi(x)\, dx = \frac{1}{3a} \int_{-a}^{a} \left[x^2 + 2x^2 \cos\left(\frac{\pi x}{a}\right) + x^2 \cos^2\left(\frac{\pi x}{a}\right) \right] dx
$$

$$= \frac{a^2}{6\pi^2} \left(2\pi^2 - 15\right), \tag{4.202}$$

$$
\begin{aligned}
\langle \hat{P}^2 \rangle &= -\hbar^2 \int_{-a}^{+a} \psi(x) \frac{d^2 \psi(x)}{dx^2} \, dx = \frac{\pi^2 \hbar^2}{a^2} A^2 \int_{-a}^{a} \left[\cos \frac{\pi x}{a} + \cos^2 \left(\frac{\pi x}{a}\right)\right] dx \\
&= \frac{\pi^2 \hbar^2}{3a^3} \int_{-a}^{a} \left[\frac{1}{2} + \cos \frac{\pi x}{a} + \frac{1}{2} \cos \frac{2\pi x}{a}\right] dx = \frac{\pi^2 \hbar^2}{3a^2};
\end{aligned} \tag{4.203}
$$

hence $\Delta x = a\sqrt{1/3 - 5/(2\pi^2)}$ and $\Delta p = \pi \hbar/(\sqrt{3}a)$. We see that the uncertainties product

$$\Delta x \, \Delta p = \frac{\pi \hbar}{3} \sqrt{1 - \frac{15}{2\pi^2}} \tag{4.204}$$

satisfies Heisenberg's uncertainty principle, $\Delta x \, \Delta p > \hbar/2$.

(d) Since $d\psi^2/dx^2$ is zero at the inflection points, we have

$$\frac{d^2 \psi}{dx^2} = -\frac{\pi^2}{a^2} A \cos \frac{\pi x}{a} = 0. \tag{4.205}$$

This relation holds when $x = \pm a/2$; hence the classically allowed region is defined by the interval between the inflection points $-a/2 \leq x \leq a/2$. That is, since $\psi(x)$ decays exponentially for $x > a/2$ and for $x < -a/2$, the energy of the system must be smaller than the potential. Classically, the system cannot be found in this region.

Problem 4.2

Consider a particle of mass m moving freely between $x = 0$ and $x = a$ inside an infinite square well potential.

(a) Calculate the expectation values $\langle \hat{X} \rangle_n$, $\langle \hat{P} \rangle_n$, $\langle \hat{X}^2 \rangle_n$, and $\langle \hat{P}^2 \rangle_n$, and compare them with their classical counterparts.

(b) Calculate the uncertainties product $\Delta x_n \Delta p_n$.

(c) Use the result of (b) to estimate the zero-point energy.

Solution

(a) Since $\psi_n(x) = \sqrt{2/a} \sin(n\pi x/a)$ and since it is a real function, we have $\langle \psi_n | \hat{P} | \psi_n \rangle = 0$ because for any real function $\phi(x)$ the integral $\langle \hat{P} \rangle = -i\hbar \int \phi^*(x)(d\phi(x)/dx) \, dx$ is imaginary and this contradicts the fact that $\langle \hat{P} \rangle$ has to be real. On the other hand, the expectation values of $\hat{X}$, $\hat{X}^2$, and $\hat{P}^2$ are

$$
\begin{aligned}
\langle \psi_n | \hat{X} | \psi_n \rangle &= \int_0^a \psi_n^*(x) x \psi_n(x) \, dx = \frac{2}{a} \int_0^a x \sin^2 \left(\frac{n\pi x}{a}\right) dx \\
&= \frac{1}{a} \int_0^a x \left[1 - \cos \left(\frac{2n\pi x}{a}\right)\right] dx = \frac{a}{2},
\end{aligned} \tag{4.206}
$$

$$
\begin{aligned}
\langle \psi_n | \hat{X}^2 | \psi_n \rangle &= \frac{2}{a} \int_0^a x^2 \sin^2 \left(\frac{n\pi x}{a}\right) dx = \frac{1}{a} \int_0^a x^2 \left[1 - \cos \left(\frac{2n\pi x}{a}\right)\right] dx \\
&= \frac{a^2}{3} - \frac{1}{a} \int_0^a x^2 \cos \left(\frac{2n\pi x}{a}\right) dx
\end{aligned}
$$

$$= \frac{a^2}{3} - \frac{1}{2n\pi}x^2 \sin\left(\frac{2n\pi x}{a}\right)\Big|_{x=0}^{x=a} + \frac{1}{n\pi}\int_0^a x \sin\left(\frac{2n\pi x}{a}\right) dx$$

$$= \frac{a^2}{3} - \frac{a^2}{2n^2\pi^2}, \tag{4.207}$$

$$\langle\psi_n|\hat{P}^2|\psi_n\rangle = -\hbar^2 \int_0^a \psi_n^*(x)\frac{d^2\psi_n(x)}{dx^2} dx = \frac{n^2\pi^2\hbar^2}{a^2}\int_0^a |\psi_n(x)|^2 dx = \frac{n^2\pi^2\hbar^2}{a^2}. \tag{4.208}$$

In deriving the previous three expressions, we have used integrations by parts. Since $E_n = n^2\pi^2\hbar^2/(2ma^2)$, we may write

$$\langle\psi_n|\hat{P}^2|\psi_n\rangle = \frac{n^2\pi^2\hbar^2}{a^2} = 2mE_n. \tag{4.209}$$

To calculate the classical average values $x_{av}, p_{av}, x_{av}^2, p_{av}^2$, it is easy first to infer that $p_{av} = 0$ and $p_{av}^2 = 2mE$, since the particle moves to the right with *constant* momentum $p = mv$ and to the left with $p = -mv$. As the particle moves at constant speed, we have $x = vt$, hence

$$x_{av} = \frac{1}{T}\int_0^T x(t)\, dt = \frac{v}{T}\int_0^T t\, dt = v\frac{T}{2} = \frac{a}{2}, \tag{4.210}$$

$$x_{av}^2 = \frac{1}{T}\int_0^T x^2(t)\, dt = \frac{v^2}{T}\int_0^T t^2 dt = \frac{1}{3}v^2T^2 = \frac{a^2}{3}, \tag{4.211}$$

where T is half[5] of the period of the motion, with $a = vT$.

We conclude that, while the average classical and quantum expressions for x, p and p^2 are identical, a comparison of (4.207) and (4.211) yields

$$\langle\psi_n|\hat{X}^2|\psi_n\rangle = \frac{a^2}{3} - \frac{a^2}{2n^2\pi^2} = x_{av}^2 - \frac{a^2}{2n^2\pi^2}, \tag{4.212}$$

so that in the limit of large quantum numbers, the quantum expression $\langle\psi_n|\hat{X}^2|\psi_n\rangle$ matches with its classical counterpart x_{av}^2: $\lim_{n\to\infty}\langle\psi_n|\hat{X}^2|\psi_n\rangle = a^2/3 = x_{av}^2$.

(b) The position and the momentum uncertainties can be calculated from (4.206) to (4.208):

$$\Delta x_n = \sqrt{\langle\psi_n|\hat{X}^2|\psi_n\rangle - \langle\psi_n|\hat{X}|\psi_n\rangle^2} = \sqrt{\frac{a^2}{3} - \frac{a^2}{2n^2\pi^2} - \frac{a^2}{4}} = a\sqrt{\frac{1}{12} - \frac{1}{2n^2\pi^2}}, \tag{4.213}$$

$$\Delta p_n = \sqrt{\langle\psi_n|\hat{P}^2|\psi_n\rangle - \langle\psi_n|\hat{P}|\psi_n\rangle^2} = \sqrt{\langle\psi_n|\hat{P}^2|\psi_n\rangle} = \frac{n\pi\hbar}{a}, \tag{4.214}$$

hence

$$\Delta x_n \Delta p_n = n\pi\hbar\sqrt{\frac{1}{12} - \frac{1}{2n^2\pi^2}}. \tag{4.215}$$

(c) Equation (4.214) shows that the momentum uncertainty for the ground state is not zero, but

$$\Delta p_1 = \frac{\pi\hbar}{a}. \tag{4.216}$$

---

[5] We may parameterize the other half of the motion by $x = -vt$, which when inserted in (4.210) and (4.211), where the variable t varies between $-T$ and 0, the integrals would yield the same results, namely $x_{av} = a/2$ and $x_{av}^2 = a^2/3$, respectively.

This leads to a nonzero kinetic energy. Therefore, the lowest value of the particle's kinetic energy is of the order of $E_{min} \sim (\Delta p_1)^2/(2m) \sim \pi^2\hbar^2/(2ma^2)$. This value, which is in full agreement with the ground state energy, $E_1 = \pi^2\hbar^2/(2ma^2)$, is the zero-point energy of the particle.

Problem 4.3

An electron is moving freely inside a one-dimensional infinite potential box with walls at $x = 0$ and $x = a$. If the electron is initially in the ground state ($n = 1$) of the box and if we *suddenly* quadruple the size of the box (i.e., the right-hand side wall is moved instantaneously from $x = a$ to $x = 4a$), calculate the probability of finding the electron in:

   (a) the ground state of the new box and
   (b) the first excited state of the new box.

Solution

Initially, the electron is in the ground state of the box $x = 0$ and $x = a$; its energy and wave function are

$$E_1 = \frac{\pi^2\hbar^2}{2ma^2}, \qquad \phi_1(x) = \sqrt{\frac{2}{a}}\sin\left(\frac{\pi x}{a}\right). \tag{4.217}$$

(a) Once in the new box, $x = 0$ and $x = 4a$, the ground state energy and wave function of the electron are

$$E_1' = \frac{\pi^2\hbar^2}{2m(4a)^2} = \frac{\pi^2\hbar^2}{32ma^2}, \qquad \psi_1(x) = \frac{1}{\sqrt{2a}}\sin\left(\frac{\pi x}{4a}\right). \tag{4.218}$$

The probability of finding the electron in $\psi_1(x)$ is

$$P(E_1') = |\langle\psi_1|\phi_1\rangle|^2 = \left|\int_0^a \psi_1^*(x)\phi_1(x)dx\right|^2 = \frac{1}{a^2}\left|\int_0^a \sin\left(\frac{\pi x}{4a}\right)\sin\left(\frac{\pi x}{a}\right)dx\right|^2; \tag{4.219}$$

the upper limit of the integral sign is a (and not $4a$) because $\phi_1(x)$ is limited to the region between 0 and a. Using the relation $\sin a \sin b = \frac{1}{2}\cos(a - b) - \frac{1}{2}\cos(a + b)$, we have $\sin(\pi x/4a)\sin(\pi x/a) = \frac{1}{2}\cos(3\pi x/4a) - \frac{1}{2}\cos(5\pi x/4a)$; hence

$$\begin{aligned}
P(E_1') &= \frac{1}{a^2}\left|\frac{1}{2}\int_0^a \cos\left(\frac{3\pi x}{4a}\right)dx - \frac{1}{2}\int_0^a \cos\left(\frac{5\pi x}{4a}\right)dx\right|^2 \\
&= \frac{128}{15^2\pi^2} = 0.058 = 5.8\%.
\end{aligned} \tag{4.220}$$

(b) If the electron is in the first excited state of the new box, its energy and wave function are

$$E_2' = \frac{\pi^2\hbar^2}{8ma^2}, \qquad \psi_2(x) = \frac{1}{\sqrt{2a}}\sin\left(\frac{\pi x}{2a}\right). \tag{4.221}$$

The corresponding probability is

$$\begin{aligned}
P(E_2') &= |\langle\psi_2|\phi_1\rangle|^2 = \left|\int_0^a \psi_2^*(x)\phi_1(x)dx\right|^2 = \frac{1}{a^2}\left|\int_0^a \sin\left(\frac{\pi x}{2a}\right)\sin\left(\frac{\pi x}{a}\right)dx\right|^2 \\
&= \frac{16}{9\pi^2} = 0.18 = 18\%.
\end{aligned} \tag{4.222}$$

Problem 4.4

Consider a particle of mass m subject to an attractive delta potential $V(x) = -V_0\,\delta(x)$, where $V_0 > 0$ (V_0 has the dimensions of Energy×Distance).

(a) In the case of negative energies, show that this particle has only one bound state; find the binding energy and the wave function.

(b) Calculate the probability of finding the particle in the interval $-a \leq x \leq a$.

(c) What is the probability that the particle remains bound when V_0 is (i) halved suddenly, (ii) quadrupled suddenly?

(d) Study the scattering case (i.e., $E > 0$) and calculate the reflection and transmission coefficients as a function of the wave number k.

Solution

(a) Let us consider first the bound state case $E < 0$. We can write the Schrödinger equation as follows:

$$\frac{d^2\psi(x)}{dx^2} + \frac{2mV_0}{\hbar^2}\delta(x)\psi(x) + \frac{2mE}{\hbar^2}\psi(x) = 0. \tag{4.223}$$

Since $\delta(x)$ vanishes for $x \neq 0$, this equation becomes

$$\frac{d^2\psi(x)}{dx^2} + \frac{2mE}{\hbar^2}\psi(x) = 0. \tag{4.224}$$

The bound solutions require that $\psi(x)$ vanishes at $x = \pm\infty$; these bound solutions are given by

$$\psi(x) = \begin{cases} \psi_-(x) = Ae^{kx}, & x < 0, \\ \psi_+(x) = Be^{-kx}, & x > 0, \end{cases} \tag{4.225}$$

where $k = \sqrt{2m|E|}/\hbar$. Since $\psi(x)$ is continuous at $x = 0$, $\psi_-(0) = \psi_+(0)$, we have $A = B$. Thus, the wave function is given by $\psi(x) = Ae^{-k|x|}$; note that $\psi(x)$ is even.

The energy can be obtained from the discontinuity condition of the first derivative of the wave function, which in turn can be obtained by integrating (4.223) from $-\varepsilon$ to $+\varepsilon$,

$$\int_{-\varepsilon}^{+\varepsilon} dx \frac{d^2\psi(x)}{dx^2} + \frac{2mV_0}{\hbar^2}\int_{-\varepsilon}^{+\varepsilon}\delta(x)\psi(x)dx + \frac{2mE}{\hbar^2}\int_{-\varepsilon}^{+\varepsilon}\psi(x)dx = 0, \tag{4.226}$$

and then letting $\varepsilon \to 0$. Using the facts that

$$\int_{-\varepsilon}^{+\varepsilon} dx \frac{d^2\psi(x)}{dx^2} = \left.\frac{d\psi(x)}{dx}\right|_{x=+\varepsilon} - \left.\frac{d\psi(x)}{dx}\right|_{x=-\varepsilon} = \left.\frac{d\psi_+(x)}{dx}\right|_{x=+\varepsilon} - \left.\frac{d\psi_-(x)}{dx}\right|_{x=-\varepsilon} \tag{4.227}$$

and that $\int_{-\varepsilon}^{+\varepsilon}\psi(x)dx = 0$ (because $\psi(x)$ is even), we can rewrite (4.226) as follows:

$$\lim_{\varepsilon\to 0}\left(\left.\frac{d\psi_+(x)}{dx}\right|_{x=+\varepsilon} - \left.\frac{d\psi_-(x)}{dx}\right|_{x=-\varepsilon}\right)_{\varepsilon=0} + \frac{2mV_0}{\hbar^2}\psi(0) = 0, \tag{4.228}$$

since the wave function is continuous at $x = 0$, but its first derivative is not. Substituting (4.225) into (4.228) and using $A = B$, we obtain

$$(-2kA) + \frac{2mV_0}{\hbar^2}A = 0 \tag{4.229}$$

or $k = mV_0/\hbar^2$. But since $k = \sqrt{2m|E|/\hbar^2}$, we have $mV_0/\hbar^2 = \sqrt{2m|E|/\hbar^2}$, and since the energy is negative, we conclude that $E = -mV_0^2/(2\hbar^2)$. There is, therefore, only one bound state solution. As for the excited states, all of them are unbound. We may normalize $\psi(x)$,

$$
\begin{aligned}
1 &= \int_{-\infty}^{\infty} \psi^*(x)\psi(x)\, dx = A^2 \int_{-\infty}^{0} \exp(2kx)\, dx + A^2 \int_{0}^{\infty} \exp(-2kx)\, dx \\
&= 2A^2 \int_{0}^{\infty} \exp(-2kx)\, dx = \frac{A^2}{k},
\end{aligned}
\tag{4.230}
$$

hence $A = \sqrt{k}$. The normalized wave function is thus given by $\psi(x) = \sqrt{k}\,e^{-k|x|}$. So the energy and normalized wave function of the bound state are given by

$$
E = -\frac{mV_0^2}{2\hbar^2}, \qquad \psi(x) = \sqrt{\frac{mV_0}{\hbar^2}} \exp\left(-\frac{mV_0}{\hbar^2}|x|\right).
\tag{4.231}
$$

(b) Since the wave function $\psi(x) = \sqrt{k}\,e^{-k|x|}$ is normalized, the probability of finding the particle in the interval $-a \le x \le a$ is given by

$$
\begin{aligned}
P &= \frac{\int_{-a}^{a} |\psi(x)|^2\, dx}{\int_{-\infty}^{\infty} |\psi(x)|^2\, dx} = \int_{-a}^{a} |\psi(x)|^2\, dx = k \int_{-a}^{a} e^{-2k|x|}\, dx \\
&= k \int_{-a}^{0} e^{2kx}\, dx + k \int_{0}^{a} e^{-2kx}\, dx = 2k \int_{0}^{a} e^{-2kx}\, dx \\
&= 1 - e^{-2ka} = 1 - e^{-2mV_0 a/\hbar^2}.
\end{aligned}
\tag{4.232}
$$

(c) If the strength of the potential changed suddenly from V_0 to V_1, the wave function will be given by $\psi_1(x) = \sqrt{mV_1/\hbar^2}\exp(-mV_1|x|/\hbar^2)$. The probability that the particle remains in the bound state $\psi_1(x)$ is

$$
\begin{aligned}
P &= |\langle \psi_1 | \psi \rangle|^2 = \left| \int_{-\infty}^{\infty} \psi_1^*(x)\psi(x)\, dx \right|^2 \\
&= \left| \frac{m}{\hbar^2}\sqrt{V_0 V_1} \int_{-\infty}^{\infty} \exp\left(-\frac{m(V_0 + V_1)}{\hbar^2}|x|\right) dx \right|^2 \\
&= \left| 2\frac{m}{\hbar^2}\sqrt{V_0 V_1} \int_{0}^{\infty} \exp\left(-\frac{m(V_0 + V_1)}{\hbar^2}x\right) dx \right|^2 = \frac{4V_0 V_1}{(V_0 + V_1)^2}.
\end{aligned}
\tag{4.233}
$$

(i) In the case where the strength of the potential is halved, $V_1 = \frac{1}{2}V_0$, the probability that the particle remains bound is

$$
P = \frac{2V_0^2}{(V_0 + \frac{1}{2}V_0)^2} = \frac{8}{9} = 89\%.
\tag{4.234}
$$

(ii) When the strength is quadrupled, $V_1 = 4V_0$, the probability is given by

$$
P = \frac{16V_0^2}{(5V_0)^2} = \frac{16}{25} = 64\%.
\tag{4.235}
$$

(d) The case $E > 0$ corresponds to a free motion and the energy levels represent a continuum. The solution of the Schrödinger equation for $E > 0$ is given by

$$\psi(x) = \begin{cases} \psi_-(x) = Ae^{ikx} + Be^{-ikx}, & x < 0, \\ \psi_+(x) = Ce^{ikx}, & x > 0, \end{cases} \quad (4.236)$$

where $k = \sqrt{2mE}/\hbar$; this corresponds to a plane wave incident from the left together with a reflected wave in the region $x < 0$, and only a transmitted wave for $x > 0$.

The values of the constants A and B are to be found from the continuity relations. On the one hand, the continuity of $\psi(x)$ at $x = 0$ yields

$$A + B = C \quad (4.237)$$

and, on the other hand, substituting (4.236) into (4.228), we end up with

$$ik(C - A + B) + \frac{2mV_0}{\hbar^2}C = 0. \quad (4.238)$$

Solving (4.237) and (4.238) for B/A and C/A, we find

$$\frac{B}{A} = \frac{-1}{1 + \frac{ik\hbar^2}{mV_0}}, \qquad \frac{C}{A} = \frac{1}{1 - \frac{imV_0}{\hbar^2 k}}. \quad (4.239)$$

Thus, the reflection and transmission coefficients are

$$R = \left|\frac{B}{A}\right|^2 = \frac{1}{1 + \frac{\hbar^4 k^2}{m^2 V_0^2}} = \frac{1}{1 + \frac{2\hbar^2 E}{mV_0^2}}, \qquad T = \left|\frac{C}{A}\right|^2 = \frac{1}{1 + \frac{m^2 V_0^2}{\hbar^4 k^2}} = \frac{1}{1 + \frac{mV_0^2}{2\hbar^2 E}}, \quad (4.240)$$

with $R + T = 1$.

Problem 4.5

A particle of mass m is subject to an attractive double-delta potential $V(x) = -V_0\delta(x - a) - V_0\delta(x + a)$, where $V_0 > 0$. Consider only the case of negative energies.
  (a) Obtain the wave functions of the bound states.
  (b) Derive the eigenvalue equations.
  (c) Specify the number of bound states and the limit on their energies. Is the ground state an even state or an odd state?
  (d) Estimate the ground state energy for the limits $a \to 0$ and $a \to \infty$.

Solution
  (a) The Schrödinger equation for this problem is

$$\frac{d^2\psi(x)}{dx^2} + \frac{2mV_0}{\hbar^2}[\delta(x - a) + \delta(x + a)]\psi(x) + \frac{2mE}{\hbar^2}\psi(x) = 0. \quad (4.241)$$

For $x \neq \pm a$ this equation becomes

$$\frac{d^2\psi(x)}{dx^2} + \frac{2mE}{\hbar^2}\psi(x) = 0 \quad \text{or} \quad \frac{d^2\psi(x)}{dx^2} - k^2\psi(x) = 0, \quad (4.242)$$

Even wave function   Odd wave function

Figure 4.12 Shapes of the even and odd wave functions for $V(x) = -V_0\delta(x-a) - V_0\delta(x+a)$.

where $k^2 = -2mE/\hbar^2 = 2m|E|/\hbar^2$, since this problem deals only with the bound states $E < 0$.

Since the potential is symmetric, $V(-x) = V(x)$, the wave function is either even or odd; we will denote the even states by $\psi_+(x)$ and the odd states by $\psi_-(x)$. The bound state solutions for $E < 0$ require that $\psi_\pm(x)$ vanish at $x = \pm\infty$:

$$\psi_\pm(x) = \begin{cases} Ae^{-kx}, & x > a, \\ \frac{B}{2}\left(e^{kx} \pm e^{-kx}\right), & -a < x < a, \\ \pm Ae^{kx}, & x < -a; \end{cases} \tag{4.243}$$

hence

$$\psi_+(x) = \begin{cases} Ae^{-kx}, \\ B\cosh kx, \\ Ae^{kx}, \end{cases} \qquad \psi_-(x) = \begin{cases} Ae^{-kx}, & x > a, \\ B\sinh kx, & -a < x < a, \\ -Ae^{kx}, & x < -a. \end{cases} \tag{4.244}$$

The shapes of $\psi_\pm(x)$ are displayed in Figure 4.12.

(b) As for the energy eigenvalues, they can be obtained from the boundary conditions. The continuity condition at $x = a$ of $\psi_+(x)$ leads to

$$Ae^{-ka} = B\cosh ka \tag{4.245}$$

and that of $\psi_-(x)$ leads to

$$Ae^{-ka} = B\sinh ka. \tag{4.246}$$

To obtain the discontinuity condition for the first derivative of $\psi_+(x)$ at $x = a$, we need to integrate (4.241):

$$\lim_{\varepsilon\to 0}\left[\psi'_+(a+\varepsilon) - \psi'_+(a-\varepsilon)\right] + \frac{2mV_0}{\hbar^2}\psi_+(a) = 0; \tag{4.247}$$

hence

$$-kAe^{-ka} - kB\sinh ka + \frac{2mV_0}{\hbar^2}Ae^{-ka} = 0 \implies A\left(\frac{2mV_0}{k\hbar^2} - 1\right)e^{-ka} = B\sinh ka. \tag{4.248}$$

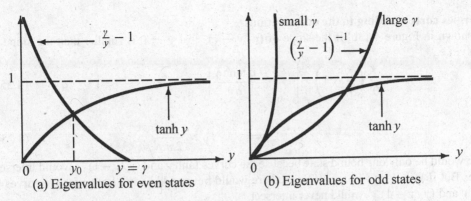

Figure 4.13 Graphical solutions of the eigenvalue equations for the even states and the odd states for the double-delta potential $V(x) = -V_0\delta(x-a) - V_0\delta(x+a)$.

Similarly, the continuity of the first derivative of $\psi_-(x)$ at $x = a$ yields

$$-kAe^{-ka} - kB\cosh ka + \frac{2mV_0}{\hbar^2}Ae^{-ka} = 0 \implies A\left(\frac{2mV_0}{k\hbar^2} - 1\right)e^{-ka} = B\cosh ka.$$
$$(4.249)$$

Dividing (4.248) by (4.245) we obtain the eigenvalue equation for the even solutions:

$$\frac{2mV_0}{k\hbar^2} - 1 = \tanh ka \implies \tanh y = \frac{\gamma}{y} - 1, \qquad (4.250)$$

where $y = ka$ and $\gamma = 2maV_0/\hbar^2$. The eigenvalue equation for the odd solutions can be obtained by dividing (4.249) by (4.246):

$$\frac{2mV_0}{k\hbar^2} - 1 = \coth ka \implies \coth y = \frac{\gamma}{y} - 1 \implies \tanh y = \left(\frac{\gamma}{y} - 1\right)^{-1}, \quad (4.251)$$

because $\coth y = 1/\tanh y$.

To obtain the energy eigenvalues for the even and odd solutions, we need to solve the transcendental equations (4.250) and (4.251). These equations can be solved graphically. In what follows, let us determine the upper and lower limits of the energy for both the even and odd solutions.

(c) To find the number of bound states and the limits on the energy, let us consider the even and odd states separately.

Energies corresponding to the even solutions
There is only one bound state, since the curves $\tanh y$ and $\gamma/y - 1$ intersect only once (Figure 4.13a); we call this point $y = y_0$. When $y = \gamma$ we have $\gamma/y - 1 = 0$, while $\tanh y > 0$. Therefore $y_0 < \gamma$. On the other hand, since $\tanh y_0 < 1$ we have $\gamma/y_0 - 1 < 1$ or $y_0 > \gamma/2$. We conclude then that $\gamma/2 < y_0 < \gamma$ or

$$\frac{\gamma}{2} < y_0 < \gamma \implies -\frac{2mV_0^2}{\hbar^2} < E_{even} < -\frac{mV_0^2}{2\hbar^2}. \qquad (4.252)$$

In deriving this relation, we have used the fact that $\gamma^2/4 < y_0^2 < \gamma^2$ where $\gamma = 2maV_0/\hbar^2$ and $y_0^2 = k_0^2a^2 = -2ma^2E_{even}/\hbar^2$. So there is always one even bound state, the ground state, whose energy lies within the range specified by (4.252).

Energies corresponding to the odd solutions

As shown in Figure 4.13b, if the slope of $(\gamma/y - 1)^{-1}$ at $y = 0$ is smaller than the slope of $\tanh y$, i.e.,

$$\frac{d}{dy}\left(\frac{\gamma}{y} - 1\right)^{-1}\bigg|_{y=0} < \frac{d\tanh y}{dy}\bigg|_{y=0} \quad\Longrightarrow\quad \frac{1}{\gamma} < 1 \tag{4.253}$$

or

$$\gamma > 1 \quad\Longrightarrow\quad V_0 > \frac{\hbar^2}{2ma}, \tag{4.254}$$

there would be only one bound state because the curves $\tanh y$ and $(\gamma/y-1)^{-1}$ would intersect once. But if $\gamma < 1$ or $V_0 < \hbar^2/(2ma)$, there would be no odd bound states, for the curves of $\tanh y$ and $(\gamma/y - 1)^{-1}$ would never intersect.

Note that if $y = \gamma/2$ we have $(\gamma/y - 1)^{-1} = 1$. Thus the intersection of $\tanh y$ and $(\gamma/y-1)^{-1}$, if it takes place at all, has to take place for $y < \gamma/2$. That is, the odd bound states occur only when

$$y < \frac{\gamma}{2} \quad\Longrightarrow\quad E_{odd} > -\frac{mV_0^2}{2\hbar^2}. \tag{4.255}$$

A comparison of (4.252) and (4.255) shows that the energies corresponding to even states are smaller than those of odd states:

$$E_{even} < E_{odd}. \tag{4.256}$$

Thus, the even bound state is the ground state. Using this result, we may infer (a) if $\gamma < 1$ there are no odd bound states, but there is always one even bound state, the ground state; (b) if $\gamma > 1$ there are two bound states: the ground state (even) and the first excited state (odd).

We may summarize these results as follows:

$$\boxed{\text{If } \gamma < 1 \quad\text{or}\quad V_0 < \frac{\hbar^2}{2ma} \longrightarrow \text{ there is only one bound state.}} \tag{4.257}$$

$$\boxed{\text{If } \gamma > 1 \quad\text{or}\quad V_0 > \frac{\hbar^2}{2ma} \longrightarrow \text{ there are two bound states .}} \tag{4.258}$$

(d) In the limit $a \longrightarrow 0$ we have $y \longrightarrow 0$ and $\gamma \longrightarrow 0$; hence the even transcendental equation $\tanh y = \gamma/y - 1$ reduces to $y \simeq \gamma/y - 1$ or $y = \gamma$, which in turn leads to $y^2 = (ka)^2 = \gamma^2$ or $-2ma^2 E_{even}/\hbar^2 = (2maV_0/\hbar^2)^2$:

$$E_{even} = -\frac{2mV_0^2}{\hbar^2}. \tag{4.259}$$

Note that in the limit $a \longrightarrow 0$, the potential $V(x) = -V_0\delta(x - a) - V_0\delta(x + a)$ reduces to $V(x) = -2V_0\delta(x)$. We can see that the ground state energy (4.231) of the single-delta potential is identical with (4.259) provided we replace V_0 in (4.231) by $2V_0$.

In the limit $a \longrightarrow \infty$, we have $y \longrightarrow \infty$ and $\gamma \longrightarrow \infty$; hence $\tanh y = \gamma/y - 1$ reduces to $1 \simeq \gamma/y-1$ or $y = \gamma/2$. This leads to $y^2 = (ka)^2 = \gamma^2/4$ or $-2ma^2 E_{even}/\hbar^2 = (maV_0/\hbar^2)^2$:

$$E_{even} = -\frac{mV_0^2}{2\hbar^2}. \tag{4.260}$$

This relation is identical with that of the single-delta potential (4.231).

Problem 4.6
Consider a particle of mass m subject to the potential

$$V(x) = \begin{cases} \infty, & x \le 0, \\ -V_0\delta(x-a), & x > 0, \end{cases}$$

where $V_0 > 0$. Discuss the existence of bound states in terms of the size of a.

Solution
The Schrödinger equation for $x > 0$ is

$$\frac{d^2\psi(x)}{dx^2} + \left[\frac{2mV_0}{\hbar^2}\delta(x-a) - k^2\right]\psi(x) = 0, \tag{4.261}$$

where $k^2 = -2mE/\hbar^2$, since we are looking here at the bound states only, $E < 0$. The solutions of this equation are

$$\psi(x) = \begin{cases} \psi_1(x) = Ae^{kx} + Be^{-kx}, & 0 < x < a, \\ \psi_2(x) = Ce^{-kx}, & x > a. \end{cases} \tag{4.262}$$

The energy eigenvalues can be obtained from the boundary conditions. As the wave function vanishes at $x = 0$, we have

$$\psi_1(0) = 0 \implies A + B = 0 \implies B = -A. \tag{4.263}$$

The continuity condition at $x = a$ of $\psi(x)$, $\psi_1(a) = \psi_2(a)$, leads to

$$Ae^{ka} - Ae^{-ka} = Ce^{-ka}. \tag{4.264}$$

To obtain the discontinuity condition for the first derivative of $\psi(x)$ at $x = a$, we need to integrate (4.261):

$$\lim_{\varepsilon \to a}\left[\psi_2'(a+\varepsilon) - \psi_1'(a-\varepsilon)\right] + \frac{2mV_0}{\hbar^2}\psi_2(a) = 0 \tag{4.265}$$

or

$$-kCe^{-ka} - kAe^{ka} - kAe^{-ka} + \frac{2mV_0}{\hbar^2}Ce^{-ka} = 0 \tag{4.266}$$

Substituting $Ce^{-ka} = Ae^{ka} - Ae^{-ka}$ or (4.264) into (4.266) we have

$$-kAe^{ka} + kAe^{-ka} - kAe^{ka} - kAe^{-ka} + \frac{2mV_0}{\hbar^2}\left(Ae^{ka} - Ae^{-ka}\right) = 0. \tag{4.267}$$

From this point on, we can proceed in two different, yet equivalent, ways. These two methods differ merely in the way we exploit (4.267). For completeness of the presentation, let us discuss both methods.

First method
The second and fourth terms of (4.267) cancel each other, so we can reduce it to

$$-kAe^{ka} - kAe^{ka} + \frac{2mV_0}{\hbar^2}\left(Ae^{ka} - Ae^{-ka}\right) = 0, \tag{4.268}$$

(a) Case where $a < \frac{\hbar^2}{2mV_0}$ (b) Case where $a > \frac{\hbar^2}{2mV_0}$

Figure 4.14 Graphical solutions of $f(k) = g(k)$ or $k = (mV_0/\hbar^2)(1 - e^{-2ka})$. If the slope of $g(k)$ is smaller than 1, i.e., $a < \hbar^2/(2mV_0)$, no bound state will exist, but if the slope of $g(k)$ is greater than 1, i.e., $a > \frac{\hbar^2}{2mV_0}$, there will be only one bound state.

which in turn leads to the following transcendental equation:

$$k = \frac{mV_0}{\hbar^2}\left(1 - e^{-2ka}\right). \tag{4.269}$$

The energy eigenvalues are given by the intersection of the curves $f(k) = k$ and $g(k) = mV_0(1 - e^{-2ka})/\hbar^2$. As the slope of $f(k)$ is equal to 1, if the slope of $g(k)$ at $k = 0$ is smaller than 1 (i.e., $a < \hbar^2/(2mV_0)$), there will be no bound states (Figure 4.14a). But if the slope of $g(k)$ is greater than 1 (i.e., $a > \hbar^2/(2mV_0)$),

$$\left.\frac{dg(k)}{dk}\right|_{k=0} > 1 \quad \text{or} \quad a > \frac{\hbar^2}{2mV_0}, \tag{4.270}$$

and there will be one bound state (Figure 4.14b).

Second method

We simply combine the first and second terms of (4.267) to generate $-2kA\sinh(ka)$; the third and fourth terms yield $-2kA\cosh(ka)$; and the fifth and sixth terms lead to $2A(2mV_0/\hbar^2)\sinh ka$. Hence

$$-2kA\sinh ka - 2kA\cosh ka + (2A)\frac{2mV_0}{\hbar^2}\sinh ka = 0, \tag{4.271}$$

which leads to

$$\gamma \coth \gamma = \frac{2mV_0}{\hbar^2}a - \gamma, \tag{4.272}$$

where $\gamma = ka$. The energy eigenvalues are given by the intersection of the curves $h(\gamma) = \gamma \coth \gamma$ and $u(\gamma) = 2mV_0a/\hbar^2 - \gamma$. As displayed in Figure 4.15a, if $a < 2mV_0/\hbar^2$, no bound state solution will exist, since the curves of $h(\gamma)$ and $u(\gamma)$ do not intersect. But if $a > 2mV_0/\hbar^2$, the curves intersect only once; hence there will be one bound state (Figure 4.15b).

We may summarize the results as follows:

$$a < \frac{\hbar^2}{2mV_0} \quad \Longrightarrow \quad \text{no bound states,} \tag{4.273}$$

$$a > \frac{\hbar^2}{2mV_0} \quad \Longrightarrow \quad \text{one bound state.} \tag{4.274}$$

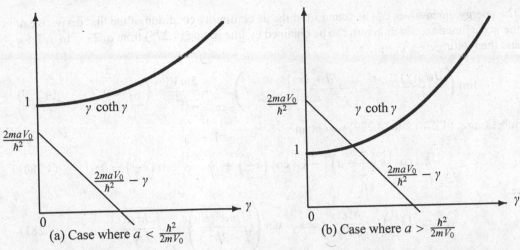

Figure 4.15 Graphical solutions of $h(\gamma) = u(\gamma)$, with $\gamma = ka$, $h(\gamma) = \gamma \coth \gamma$, and $u(\gamma) = 2mV_0 a/\hbar^2 - \gamma$. If $a < 2mV_0/\hbar^2$ there is no bound state. If $a > 2mV_0/\hbar^2$ there is one bound state.

Problem 4.7

A particle of mass m, besides being confined to move inside an infinite square well potential of size a with walls at $x = 0$ and $x = a$, is subject to a delta potential of strength V_0

$$V(x) = \begin{cases} V_0 \delta(x - a/2), & 0 < x < a, \\ \infty, & \text{elsewhere,} \end{cases}$$

where $V_0 > 0$. Show how to calculate the energy levels of the system in terms of V_0 and a.

Solution

The Schrödinger equation

$$\frac{d^2\psi(x)}{dx^2} + \frac{2mV_0}{\hbar^2}\delta\left(x - \frac{a}{2}\right)\psi(x) + \frac{2mE}{\hbar^2}\psi(x) = 0 \qquad (4.275)$$

can be written for $x \neq a/2$ as

$$\frac{d^2\psi(x)}{dx^2} + \frac{2mE}{\hbar^2}\psi(x) = 0. \qquad (4.276)$$

The solutions of this equation must vanish at $x = 0$ and $x = a$:

$$\psi(x) = \begin{cases} \psi_L(x) = A \sin kx, & 0 \leq x < a/2, \\ \psi_R(x) = B \sin k(x - a), & a/2 < x \leq a, \end{cases} \qquad (4.277)$$

where $k = \sqrt{2m|E|}/\hbar$. The continuity of $\psi(x)$ at $x = a/2$, $\psi_L(a/2) = \psi_R(a/2)$, leads to $A \sin(a/2) = -B \sin(a/2)$; hence $B = -A$. The wave function is thus given by

$$\psi(x) = \begin{cases} \psi_L(x) = A \sin kx, & 0 \leq x < a/2, \\ \psi_R(x) = -A \sin k(x - a), & a/2 < x \leq a, \end{cases} \qquad (4.278)$$

The energy eigenvalues can be found from the discontinuity condition of the first derivative of the wave function, which in turn can be obtained by integrating (4.275) from $a/2 - \varepsilon$ to $a/2 + \varepsilon$ and then letting $\varepsilon \to 0$:

$$\lim_{\varepsilon \to 0} \left(\left. \frac{d\psi_R(x)}{dx} \right|_{x=a/2+\varepsilon} - \left. \frac{d\psi_L(x)}{dx} \right|_{x=a/2-\varepsilon} \right)_{\varepsilon=0} + \frac{2mV_0}{\hbar^2} \psi \left(\frac{a}{2} \right) = 0. \qquad (4.279)$$

Substituting (4.278) into (4.279) we obtain

$$-kA \cos \left[k \left(\frac{a}{2} - a \right) \right] - kA \cos \left(k\frac{a}{2} \right) + A\frac{2mV_0}{\hbar^2} \sin \left(k\frac{a}{2} \right) = 0 \qquad (4.280)$$

or

$$\tan \left(k\frac{a}{2} \right) = \frac{\hbar^2 k}{mV_0} \implies \tan \left(\sqrt{\frac{ma^2|E|}{2\hbar^2}} \right) = \sqrt{\frac{2\hbar^2|E|}{mV_0^2}}. \qquad (4.281)$$

This is a transcendental equation for the energy; its solutions, which can be obtained numerically or graphically, yield the values of E.

Problem 4.8
Using the uncertainty principle, show that the lowest energy of an oscillator is $\hbar\omega/2$.

Solution
The motion of the particle is confined to the region $-a/2 \leq x \leq a/2$; that is, $\Delta x \simeq a$. Then as a result of the uncertainty principle, the lowest value of this particle's momentum is $\hbar/(2\Delta x) \simeq \hbar/(2a)$. The total energy as a function of a is

$$E(a) \simeq \frac{1}{2m} \left(\frac{\hbar}{2a} \right)^2 + \frac{1}{2}m\omega^2 a^2. \qquad (4.282)$$

The minimization of E with respect to a,

$$0 = \left. \frac{dE}{da} \right|_{a=a_0} = -\frac{\hbar^2}{4ma_0^3} + m\omega^2 a_0, \qquad (4.283)$$

gives $a_0 = \sqrt{\hbar/2m\omega}$ and hence $E(a_0) \simeq \hbar\omega/2$; this is equal to the exact value of the oscillator's zero-point energy.

Problem 4.9
Find the energy levels of a particle of mass m moving in a one-dimensional potential:

$$V(x) = \begin{cases} +\infty, & x \leq 0, \\ \frac{1}{2}m\omega^2 x^2, & x > 0. \end{cases}$$

Solution
This is an asymmetric harmonic oscillator potential in which the particle moves only in the region $x > 0$. The only acceptable solutions are those for which the wave function vanishes at $x = 0$. These solutions must be those of an ordinary (symmetric) harmonic oscillator that have odd parity, since the wave functions corresponding to the symmetric harmonic oscillator are

either even (n even) or odd (n odd), and only the odd solutions vanish at the origin, $\psi_{2n+1}(0) = 0$ ($n = 0, 1, 2, 3, \ldots$). Therefore, the energy levels of this asymmetric potential must be given by those corresponding to the odd n energy levels of the symmetric potential, i.e.,

$$E_n = \left[(2n+1) + \frac{1}{2}\right]\hbar\omega = \left(2n + \frac{3}{2}\right)\hbar\omega \qquad (n = 0, 1, 2, 3, \ldots). \tag{4.284}$$

Problem 4.10

Consider the box potential

$$V(x) = \begin{cases} 0, & 0 < x < a, \\ \infty, & \text{elsewhere.} \end{cases}$$

(a) Estimate the energies of the ground state as well as those of the first and the second excited states for (i) an electron enclosed in a box of size $a = 10^{-10}$ m (express your answer in electron volts; you may use these values: $\hbar c = 200$ MeV fm, $m_e c^2 = 0.5$ MeV); (ii) a 1 g metallic sphere which is moving in a box of size $a = 10$ cm (express your answer in joules).

(b) Discuss the importance of the quantum effects for both of these two systems.

(c) Use the uncertainty principle to estimate the velocities of the electron and the metallic sphere.

Solution

The energy of a particle of mass m in a box having perfectly rigid walls is given by

$$E_n = \frac{n^2 h^2}{8ma^2}, \qquad n = 1, 2, 3, \ldots, \tag{4.285}$$

where a is the size of the box.

(a) (i) For the electron in the box of size 10^{-10} m, we have

$$\begin{aligned} E_n &= \frac{\hbar^2 c^2}{m_e c^2 a^2} \frac{4\pi^2 n^2}{8} \equiv \frac{4 \times 10^4 \,(\text{MeV fm})^2}{0.5 \,\text{MeV} \times 10^{10} \,\text{fm}^2} \frac{\pi^2}{2} n^2 \\ &= 4\pi^2 n^2 \,\text{eV} \simeq 39 n^2 \,\text{eV}. \end{aligned} \tag{4.286}$$

Hence $E_1 = 39$ eV, $E_2 = 156$ eV, and $E_3 = 351$ eV.

(ii) For the sphere in the box of side 10 cm we have

$$E_n = \frac{(6.6 \times 10^{-34} \,\text{J s})^2}{10^{-3} \,\text{kg} \times 10^{-2} \,\text{m}^2} n^2 = 43.6 \times 10^{-63} n^2 \,\text{J} \tag{4.287}$$

Hence $E_1 = 43.6 \times 10^{-63}$ J, $E_2 = 174.4 \times 10^{-63}$ J, and $E_3 = 392.4 \times 10^{-63}$ J.

(b) The differences between the energy levels are

$$(E_2 - E_1)_{electron} = 117 \,\text{eV}, \qquad (E_3 - E_2)_{electron} = 195 \,\text{eV}, \tag{4.288}$$

$$(E_2 - E_1)_{sphere} = 130.8 \times 10^{-63} \,\text{J}, \qquad (E_3 - E_2)_{sphere} = 218 \times 10^{-63} \,\text{J}. \tag{4.289}$$

These results show that:

- The spacings between the energy levels of the electron are quite large; the levels are far apart from each other. Thus, the quantum effects are important.

- The energy levels of the sphere are practically indistinguishable; the spacings between the levels are negligible. The energy spectrum therefore forms a continuum; hence the quantum effects are not noticeable for the sphere.

(c) According to the uncertainty principle, the speed is proportional to $v \sim \hbar/(ma)$. For the electron, the typical distances are atomic, $a \simeq 10^{-10}$ m; hence

$$v \sim \frac{\hbar c}{mc^2 a} c \sim \frac{200 \, \text{MeV fm}}{0.5 \, \text{MeV} \times 10^5 \, \text{fm}} c \simeq 4 \times 10^{-3} c = 1.2 \times 10^6 \text{m s}^{-1}, \qquad (4.290)$$

where c is the speed of light. The electron therefore moves quite fast; this is expected since we have confined the electron to move within a small region.

For the sphere, the typical distances are in the range of 1 cm:

$$v \sim \frac{\hbar}{ma} \sim \frac{6.6 \times 10^{-34} \, \text{J s}}{10^{-3} \, \text{kg} \times 10^{-2} \, \text{m}} \simeq 6.6 \times 10^{-29} \text{m s}^{-1} \qquad (4.291)$$

At this speed the sphere is practically at rest.

Problem 4.11

(a) Verify that the matrices representing the operators $\hat{X}$ and $\hat{P}$ in the N-space for a harmonic oscillator obey the correct commutation relation $[\hat{X}, \hat{P}] = i\hbar$.

(b) Show that the energy levels of the harmonic oscillator can be obtained by inserting the matrices of $\hat{X}$ and $\hat{P}$ into the Hamiltonian $\hat{H} = \hat{P}^2/(2m) + \frac{1}{2}m\omega^2 \hat{X}^2$.

Solution

(a) Using the matrices of $\hat{X}$ and $\hat{P}$ in (4.181) and (4.182), we obtain

$$\hat{X}\hat{P} = i\frac{\hbar}{2} \begin{pmatrix} 1 & 0 & -\sqrt{2} & \cdots \\ 0 & 1 & 0 & \cdots \\ \sqrt{2} & 0 & 1 & \cdots \\ \vdots & \vdots & \vdots & \ddots \end{pmatrix}, \qquad \hat{P}\hat{X} = i\frac{\hbar}{2} \begin{pmatrix} -1 & 0 & -\sqrt{2} & \cdots \\ 0 & -1 & 0 & \cdots \\ \sqrt{2} & 0 & -1 & \cdots \\ \vdots & \vdots & \vdots & \ddots \end{pmatrix}; \quad (4.292)$$

hence

$$\hat{X}\hat{P} - \hat{P}\hat{X} = i\hbar \begin{pmatrix} 1 & 0 & 0 & \cdots \\ 0 & 1 & 0 & \cdots \\ 0 & 0 & 1 & \cdots \\ \vdots & \vdots & \vdots & \ddots \end{pmatrix}, \qquad (4.293)$$

or $[\hat{X}, \hat{P}] = i\hbar I$, where I is the unit matrix.

(b) Again, using the matrices of $\hat{X}$ and $\hat{P}$ in (4.181) and (4.182), we can verify that

$$\hat{X}^2 = \frac{\hbar}{2m\omega} \begin{pmatrix} 1 & 0 & \sqrt{2} & \cdots \\ 0 & 3 & 0 & \cdots \\ \sqrt{2} & 0 & 5 & \cdots \\ \vdots & \vdots & \vdots & \ddots \end{pmatrix}, \qquad \hat{P}^2 = -\frac{m\hbar\omega}{2} \begin{pmatrix} -1 & 0 & \sqrt{2} & \cdots \\ 0 & -3 & 0 & \cdots \\ \sqrt{2} & 0 & -5 & \cdots \\ \vdots & \vdots & \vdots & \ddots \end{pmatrix};$$

$$(4.294)$$

hence

$$\frac{\hat{p}^2}{2m} + \frac{1}{2}m\omega^2\hat{X}^2 = \frac{\hbar\omega}{2}\begin{pmatrix} 1 & 0 & 0 & \cdots \\ 0 & 3 & 0 & \cdots \\ 0 & 0 & 5 & \cdots \\ \vdots & \vdots & \vdots & \ddots \end{pmatrix}.$$

(4.295)

The form of this matrix is similar to the result we obtain from an analytical treatment, $E_n = \hbar\omega(2n+1)/2$, since

$$H_{n'n} = \langle n'|\hat{H}|n\rangle = \frac{\hbar\omega}{2}(2n+1)\delta_{n'n}.$$

(4.296)

Problem 4.12

Calculate the probability of finding a particle in the classically forbidden region of a harmonic oscillator for the states $n = 0, 1, 2, 3, 4$. Are these results compatible with their classical counterparts?

Solution

The classical turning points are defined by $E_n = V(x_n)$ or by $\hbar\omega(n + \frac{1}{2}) = \frac{1}{2}m\omega^2 x_n^2$; that is, $x_n = \pm\sqrt{\hbar/(m\omega)}\sqrt{2n+1}$. Thus, the probability of finding a particle in the classically forbidden region for a state $\psi_n(x)$ is

$$P_n = \int_{-\infty}^{-|x_n|} |\psi_n(x)|^2\, dx + \int_{|x_n|}^{+\infty} |\psi_n(x)|^2\, dx = 2\int_{|x_n|}^{+\infty} |\psi_n(x)|^2\, dx,$$

(4.297)

where $\psi_n(x)$ is given in (4.172), $\psi_n(x) = 1/\sqrt{\sqrt{\pi}2^n n! x_0}\, e^{-x^2/2x_0^2} H_n(x/x_0)$, where x_0 is given by $x_0 = \sqrt{\hbar/(m\omega)}$. Using the change of variable $y = x/x_0$, we can rewrite P_n as

$$P_n = \frac{2}{\sqrt{\pi}2^n n!} \int_{\sqrt{2n+1}}^{+\infty} e^{-y^2} H_n^2(y)\, dy,$$

(4.298)

where the Hermite polynomials $H_n(y)$ are listed in (4.120). The integral in (4.298) can be evaluated only numerically. Using the numerical values

$$\int_1^\infty e^{-y^2}\, dy = 0.1394, \qquad \int_{\sqrt{3}}^\infty y^2 e^{-y^2}\, dy = 0.0495,$$

(4.299)

$$\int_{\sqrt{5}}^\infty \left(4y^2 - 2\right)^2 e^{-y^2}\, dy = 0.6740, \qquad \int_{\sqrt{7}}^\infty \left(8y^3 - 12y\right)^2 e^{-y^2}\, dy = 3.6363,$$

(4.300)

$$\int_{\sqrt{9}}^\infty \left(16y^4 - 48y^2 + 12\right)^2 e^{-y^2}\, dx = 26.86,$$

(4.301)

we obtain

$$P_0 = 0.1573, \qquad P_1 = 0.1116, \qquad P_2 = 0.095\,069,$$

(4.302)

$$P_3 = 0.085\,48, \qquad P_4 = 0.078\,93.$$

(4.303)

This shows that the probability decreases as n increases, so it would be very small for very large values of n. It is therefore unlikely to find the particle in the classically forbidden region when the particle is in a very highly excited state. This is what we expect, since the classical approximation is recovered in the limit of high values of n.

Problem 4.13

Consider a particle of mass m moving in the following potential

$$V(x) = \begin{cases} \infty, & x \leq 0, \\ -V_0, & 0 < x < a, \\ 0, & x \geq a, \end{cases}$$

where $V_0 > 0$.

(a) Find the wave function.

(b) Show how to obtain the energy eigenvalues from a graph.

(c) Calculate the minimum value of V_0 (in terms of m, a, and $\hbar$) so that the particle will have one bound state; then calculate it for two bound states. From these two results, try to obtain the lowest value of V_0 so that the system has n bound states.

Solution

(a) As shown in Figure 4.16, the wave function in the region $x < 0$ is zero, $\psi(x) = 0$. In the region $x > 0$ the Schrödinger equation for the bound state solutions, $-V_0 < E < 0$, is given by

$$\left(\frac{d^2}{dx^2} + k_1^2\right)\psi_1(x) = 0 \qquad (0 < x < a), \tag{4.304}$$

$$\left(\frac{d^2}{dx^2} - k_2^2\right)\psi_2(x) = 0 \qquad (x > a), \tag{4.305}$$

where $k_1^2 = 2m(V_0 + E)/\hbar^2$ and $k_2^2 = -2mE/\hbar^2$. On one hand, the solution of (4.304) is oscillatory, $\psi_1(x) = A\sin k_1 x + B\cos k_1 x$, but since $\psi_1(0) = 0$ we must have $B = 0$. On the other hand, eliminating the physically unacceptable solutions which grow exponentially for large values of x, the solution of (4.305) is $\psi_2(x) = Ce^{-k_2 x}$. Thus, the wave function is given by

$$\psi(x) = \begin{cases} 0, & x < 0, \\ \psi_1(x) = A\sin k_1 x, & 0 < x < 0, \\ \psi_2(x) = Ce^{-k_2 x}, & x > a. \end{cases} \tag{4.306}$$

(b) To determine the eigenvalues, we need to use the boundary conditions at $x = a$. The condition $\psi_1(a) = \psi_2(a)$ yields

$$A\sin k_1 a = Ce^{-k_2 a}, \tag{4.307}$$

while the continuity of the first derivative, $\psi_1'(a) = \psi_2{'}(a)$, leads to

$$Ak_1\cos k_1 a = -Ck_2 e^{-k_2 a}. \tag{4.308}$$

Dividing (4.308) by (4.307) we obtain

$$k_1 a\cot k_1 a = -k_2 a. \tag{4.309}$$

Since $k_1^2 = 2m(V_0 + E)/\hbar^2$ and $k_2^2 = -2mE/\hbar^2$, we have

$$(k_1 a)^2 + (k_2 a)^2 = \gamma^2, \tag{4.310}$$

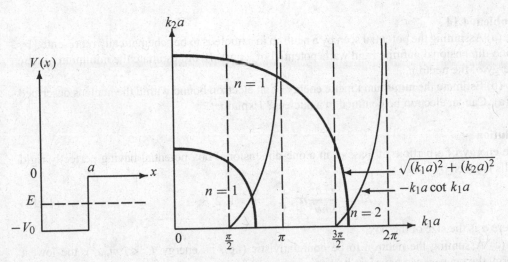

Figure 4.16 Potential $V(x)$ (left curve); the energy levels of $V(x)$ are given graphically by the intersection of the circular curve $\sqrt{(k_1a)^2 + (k_2a)^2}$ with $-k_1a \cot k_1a$ (right curve).

where $\gamma = \sqrt{2mV_0}a/\hbar$.

The transcendental equations (4.309) and (4.310) can be solved graphically. As shown in Figure 4.16, the energy levels are given by the intersection of the circular curve $(k_1a)^2 + (k_2a)^2 = \gamma^2$ with $k_1a \cot k_1a = -k_2a$.

(c) If $\pi/2 < \gamma < 3\pi/2$ there will be only one bound state, the ground state $n = 1$, for there is only one crossing between the curves $(k_1a)^2 + (k_2a)^2 = \gamma^2$ and $k_1a \cot k_1a = -k_2a$. The lowest value of V_0 that yields a single bound state is given by the relation $\gamma = \pi/2$, which leads to $2ma^2 V_0/\hbar^2 = \pi^2/4$ or to

$$V_0 = \frac{\pi^2\hbar^2}{8ma^2}. \tag{4.311}$$

Similarly, if $3\pi/2 < \gamma < 5\pi/2$ there will be two crossings between $(k_1a)^2 + (k_2a)^2 = \gamma^2$ and $k_1a \cot k_1a = -k_2a$. Thus, there will be two bound states: the ground state, $n = 1$, and the first excited state, $n = 2$. The lowest value of V_0 that yields two bound states corresponds to $2ma^2 V_0/\hbar^2 = 9\pi^2/4$ or to

$$V_0 = \frac{9\pi^2\hbar^2}{8ma^2}. \tag{4.312}$$

We may thus infer the following general result. If $n\pi - \pi/2 < \gamma < n\pi + \pi/2$, there will be n crossings and hence n bound states:

$$n\pi - \frac{\pi}{2} < \frac{\sqrt{2mV_0}}{\hbar}a < n\pi + \frac{\pi}{2} \implies \text{there are } n \text{ bound states.} \tag{4.313}$$

The lowest value of V_0 giving n bound states is

$$V_0 = \frac{\pi^2\hbar^2}{8ma^2}(2n - 1)^2. \tag{4.314}$$

Problem 4.14

(a) Assuming the potential seen by a neutron in a nucleus to be schematically represented by a one-dimensional, infinite rigid walls potential of length 10 fm, estimate the minimum kinetic energy of the neutron.

(b) Estimate the minimum kinetic energy of an electron bound within the nucleus described in (a). Can an electron be confined in a nucleus? Explain.

Solution

The energy of a particle of mass m in a one-dimensional box potential having perfectly rigid walls is given by

$$E_n = \frac{\pi^2 \hbar^2}{2ma^2} n^2, \qquad n = 1, 2, 3, \ldots, \qquad (4.315)$$

where a is the size of the box.

(a) Assuming the neutron to be nonrelativistic (i.e., its energy $E \ll m_n c^2$), the lowest energy the neutron can have in a box of size $a = 10$ fm is

$$E_{min} = \frac{\pi^2 \hbar^2}{2m_n a^2} = \frac{\pi^2 (\hbar^2 c^2)}{2(m_n c^2) a^2} \simeq 2.04 \, \text{MeV}, \qquad (4.316)$$

where we have used the fact that the rest mass energy of a neutron is $m_n c^2 \simeq 939.57$ MeV and $\hbar c \simeq 197.3$ MeV fm. Indeed, we see that $E_{min} \ll m_n c^2$.

(b) The minimum energy of a (nonrelativistic) electron moving in a box of size $a = 10$ fm is given by

$$E_{min} = \frac{\pi^2 \hbar^2}{2m_e a^2} = \frac{\pi^2 (\hbar^2 c^2)}{2(m_e c^2) a^2} \simeq 3755.45 \, \text{MeV}. \qquad (4.317)$$

The rest mass energy of an electron is $m_e c^2 \simeq 0.511$ MeV, so this electron is ultra-relativistic since $E_{min} \gg m_e c^2$. It implies that an electron with this energy cannot be confined within such a nucleus.

Problem 4.15

(a) Calculate the expectation value of the operator $\hat{X}^4$ in the N-representation with respect to the state $\mid n \rangle$ (i.e., $\langle n \mid \hat{X}^4 \mid n \rangle$).

(b) Use the result of (a) to calculate the energy E_n for a particle whose Hamiltonian is $\hat{H} = \hat{P}^2/(2m) + \frac{1}{2} m \omega^2 \hat{X}^2 - \lambda \hat{X}^4$.

Solution

(a) Since $\sum_{m=0}^{\infty} \mid m \rangle \langle m \mid = 1$ we can write the expectation value of $\hat{X}^4$ as

$$\langle n \mid \hat{X}^4 \mid n \rangle = \sum_{m=0}^{\infty} \langle n \mid \hat{X}^2 \mid m \rangle \langle m \mid \hat{X}^2 \mid n \rangle = \sum_{m=0}^{\infty} \left| \langle m \mid \hat{X}^2 \mid n \rangle \right|^2. \qquad (4.318)$$

Now since

$$\hat{X}^2 = \frac{\hbar}{2m\omega} \left(\hat{a}^2 + \hat{a}^{\dagger 2} + \hat{a}\hat{a}^{\dagger} + \hat{a}^{\dagger}\hat{a} \right) = \frac{\hbar}{2m\omega} \left(\hat{a}^2 + \hat{a}^{\dagger 2} + 2\hat{a}^{\dagger}\hat{a} + 1 \right), \qquad (4.319)$$

the only terms $\langle m | \hat{X}^2 | n \rangle$ that survive are

$$\langle n | \hat{X}^2 | n \rangle = \frac{\hbar}{2m\omega} \langle n | 2\hat{a}^\dagger \hat{a} + 1 | n \rangle = \frac{\hbar}{2m\omega}(2n+1), \tag{4.320}$$

$$\langle n-2 | \hat{X}^2 | n \rangle = \frac{\hbar}{2m\omega} \langle n-2 | \hat{a}^2 | n \rangle = \frac{\hbar}{2m\omega}\sqrt{n(n-1)}, \tag{4.321}$$

$$\langle n+2 | \hat{X}^2 | n \rangle = \frac{\hbar}{2m\omega} \langle n+2 | \hat{a}^{\dagger 2} | n \rangle = \frac{\hbar}{2m\omega}\sqrt{(n+1)(n+2)}. \tag{4.322}$$

Thus

$$
\begin{aligned}
\langle n | \hat{X}^4 | n \rangle &= \left| \langle n | \hat{X}^2 | n \rangle \right|^2 + \left| \langle n-2 | \hat{X}^2 | n \rangle \right|^2 + \left| \langle n+2 | \hat{X}^2 | n \rangle \right|^2 \\
&= \frac{\hbar^2}{4m^2\omega^2} \left[(2n+1)^2 + n(n-1) + (n+1)(n+2) \right] \\
&= \frac{\hbar^2}{4m^2\omega^2} \left(6n^2 + 6n + 3 \right).
\end{aligned}
\tag{4.323}
$$

(b) Using (4.323), and since the Hamiltonian can be expressed in terms of the harmonic oscillator, $\hat{H} = \hat{H}_{HO} - \lambda \hat{X}^4$, we immediately obtain the particle energy:

$$E_n = \langle n | \hat{H}_{HO} | n \rangle - \lambda \langle n | \hat{X}^4 | n \rangle = \hbar\omega \left(n + \frac{1}{2} \right) - \frac{\lambda \hbar^2}{4m^2\omega^2} \left(6n^2 + 6n + 3 \right). \tag{4.324}$$

Problem 4.16
Find the energy levels and the wave functions of two harmonic oscillators of masses m_1 and m_2, having identical frequencies ω, and coupled by the interaction $\frac{1}{2}k(\hat{X}_1 - \hat{X}_2)^2$.

Solution
This problem reduces to finding the eigenvalues for the Hamiltonian

$$
\begin{aligned}
\hat{H} &= \hat{H}_1 + \hat{H}_2 + \frac{1}{2}K(\hat{X}_1 - \hat{X}_2)^2 \\
&= \frac{1}{2m_1}\hat{P}_1^2 + \frac{1}{2}m_1\omega^2\hat{X}_1^2 + \frac{1}{2m_2}\hat{P}_2^2 + \frac{1}{2}m_2\omega^2\hat{X}_2^2 + \frac{1}{2}K(\hat{X}_1 - \hat{X}_2)^2.
\end{aligned}
\tag{4.325}
$$

This is a two-particle problem. As in classical mechanics, it is more convenient to describe the dynamics of a two-particle system in terms of the center of mass (CM) and relative motions. For this, let us introduce the following operators:

$$\hat{P} = \hat{p}_1 + \hat{p}_2, \qquad \hat{X} = \frac{m_1\hat{x}_1 + m_2\hat{x}_2}{M}, \tag{4.326}$$

$$\hat{p} = \frac{m_2\hat{p}_1 - m_1\hat{p}_2}{M}, \qquad \hat{x} = \hat{x}_1 - \hat{x}_2, \tag{4.327}$$

where $M = m_1 + m_2$ and $\mu = m_1 m_2/(m_1 + m_2)$ is the reduced mass; $\hat{P}$ and $\hat{X}$ pertain to the CM; $\hat{p}$ and $\hat{x}$ pertain to the relative motion. These relations lead to

$$\hat{p}_1 = \frac{m_1}{M}\hat{P} + \hat{p}, \qquad \hat{p}_2 = \frac{m_2}{M}\hat{P} - \hat{p}, \tag{4.328}$$

$$\hat{x}_1 = \frac{m_2}{M}\hat{x} + \hat{X}, \qquad \hat{x}_2 = -\frac{m_1}{M}\hat{x} + \hat{X}. \tag{4.329}$$

Note that the sets (X, P) and (x, p) are conjugate variables separately: $[\hat{X}, \hat{P}] = i\hbar$, $[\hat{x}, \hat{p}] = i\hbar$, $[\hat{X}, \hat{p}] = [\hat{x}, \hat{P}] = 0$. Taking $\hat{p}_1$, $\hat{p}_2$, $\hat{x}_1$, and $\hat{x}_2$ of (4.328) and (4.329) and inserting them into (4.325), we obtain

$$
\begin{aligned}
\hat{H} &= \frac{1}{2m_1}\left(\frac{m_1}{M}\hat{P} + \hat{p}\right)^2 + \frac{1}{2}m_1\omega^2\left(\frac{m_2}{M}\hat{x} + \hat{X}\right)^2 \\
&\quad + \frac{1}{2m_2}\left(\frac{m_2}{M}\hat{P} - \hat{p}\right)^2 + \frac{1}{2}m_2\omega^2\left(-\frac{m_1}{M}\hat{x} + \hat{X}\right)^2 + \frac{1}{2}K\hat{x}^2 \\
&= \hat{H}_{CM} + \hat{H}_{rel},
\end{aligned}
\tag{4.330}
$$

where

$$
\hat{H}_{CM} = \frac{1}{2M}\hat{P}^2 + \frac{1}{2}M\omega^2\hat{X}^2, \qquad \hat{H}_{rel} = \frac{1}{2\mu}\hat{p}^2 + \frac{1}{2}\mu\Omega^2\hat{x}^2, \tag{4.331}
$$

with $\Omega^2 = \omega^2 + k/\mu$. We have thus reduced the Hamiltonian of these two coupled harmonic oscillators to the sum of two independent harmonic oscillators, one with frequency ω and mass M and the other of mass μ and frequency $\Omega = \sqrt{\omega^2 + k/\mu}$. That is, by introducing the CM and relative motion variables, we have managed to eliminate the coupled term from the Hamiltonian.

The energy levels of this two-oscillator system can be inferred at once from the suggestive Hamiltonians of (4.331):

$$
E_{n_1 n_2} = \hbar\omega\left(n_1 + \frac{1}{2}\right) + \hbar\Omega\left(n_2 + \frac{1}{2}\right). \tag{4.332}
$$

The states of this two-particle system are given by the product of the two states $|N\rangle = |n_1\rangle|n_2\rangle$; hence the total wave function, $\psi_n(X, x)$, is equal to the product of the center of mass wave function, $\psi_{n_1}(X)$, and the wave function of the relative motion, $\psi_{n_2}(x)$: $\psi_n(X, x) = \psi_{n_1}(X)\psi_{n_2}(x)$. Note that both of these wave functions are harmonic oscillator functions whose forms can be found in (4.172):

$$
\psi_n(X, x) = \frac{1}{\sqrt{\pi}\sqrt{2^{n_1}2^{n_2}n_1!n_2!x_{0_1}x_{0_2}}}e^{-X^2/2x_{0_1}^2}e^{-x^2/2x_{0_2}^2}H_{n_1}\left(\frac{X}{x_{0_1}}\right)H_{n_2}\left(\frac{x}{x_{0_2}}\right), \tag{4.333}
$$

where $n = (n_1, n_2)$, $x_{0_1} = \sqrt{\hbar/(M\omega)}$, and $x_{0_2} = \sqrt{\hbar/(\mu\Omega)}$.

Problem 4.17

Consider a particle of mass m and charge q moving under the influence of a one-dimensional harmonic oscillator potential. Assume it is placed in a constant electric field $\mathcal{E}$. The Hamiltonian of this particle is therefore given by $\hat{H} = \hat{P}^2/(2m) + \frac{1}{2}m\omega^2\hat{X}^2 - q\mathcal{E}\hat{X}$. Derive the energy expression and the wave function of the nth excited state.

Solution

To find the eigenenergies of the Hamiltonian

$$
\hat{H} = \frac{1}{2m}\hat{P}^2 + \frac{1}{2}m\omega^2\hat{X}^2 - q\mathcal{E}\hat{X}, \tag{4.334}
$$

it is convenient to use the change of variable $y = \hat{X} - q\mathcal{E}/(m\omega^2)$. Thus the Hamiltonian becomes

$$
\hat{H} = \frac{1}{2m}\hat{P}^2 + \frac{1}{2}m\omega^2\hat{y}^2 - \frac{q^2\mathcal{E}^2}{2m\omega^2}. \tag{4.335}
$$

Since the term $q^2\mathcal{E}^2/(2m\omega^2)$ is a mere constant and $\hat{P}^2/(2m) + \frac{1}{2}m\omega^2\hat{y}^2 = \hat{H}_{HO}$ has the structure of a harmonic oscillator Hamiltonian, we can easily infer the energy levels:

$$E_n = \langle n \mid \hat{H} \mid n \rangle = \hbar\omega\left(n + \frac{1}{2}\right) - \frac{q^2\mathcal{E}^2}{2m\omega^2}. \tag{4.336}$$

The wave function is given by $\psi_n(y) = \psi_n(x - q\mathcal{E}/(m\omega^2))$, where $\psi_n(y)$ is given in (4.172):

$$\psi_n(y) = \frac{1}{\sqrt{\sqrt{\pi}\,2^n n! x_0}}e^{-y^2/2x_0^2} H_n\left(\frac{y}{x_0}\right). \tag{4.337}$$

Problem 4.18

Consider a particle of mass m that is bouncing vertically and elastically on a smooth reflecting floor in the Earth's gravitational field

$$V(z) = \begin{cases} mgz, & z > 0, \\ +\infty, & z \le 0, \end{cases}$$

where g is a constant (the acceleration due to gravity). Find the energy levels and wave function of this particle.

Solution

We need to solve the Schrödinger equation with the boundary conditions $\psi(0) = 0$ and $\psi(+\infty) = 0$:

$$-\frac{\hbar^2}{2m}\frac{d^2\psi(z)}{dz^2} + mgz\,\psi(z) = E\psi(z) \implies \frac{d^2\psi(z)}{dz^2} - \frac{2m}{\hbar^2}(mgz - E)\,\psi(z) = 0. \tag{4.338}$$

With the change of variable $x = (\hbar^2/(2m^2g))^{2/3}(2m/\hbar^2)(mgz - E)$, we can reduce this equation to

$$\frac{d^2\phi(x)}{dx^2} - x\phi(x) = 0. \tag{4.339}$$

This is a standard differential equation; its solution (which vanishes at $x \to = +\infty$, i.e., $\phi(+\infty) = 0$) is given by

$$\phi(x) = B\mathrm{Ai}(x) \quad \text{where} \quad \mathrm{Ai}(x) = \frac{1}{\pi}\int_0^\infty \cos\left(\frac{1}{3}t^3 + xt\right)dt, \tag{4.340}$$

where $\mathrm{Ai}(x)$ is called the *Airy function*.

When $z = 0$ we have $x = -(2/(mg^2\hbar^2))^{1/3}E$. The boundary condition $\psi(0) = 0$ yields $\phi[-(2/(mg^2\hbar^2))^{1/3}E] = 0$ or $\mathrm{Ai}[-(2/(mg^2\hbar^2))^{1/3}E] = 0$. The Airy function has zeros only at certain values of R_n: $\mathrm{Ai}(R_n) = 0$ with $n = 0, 1, 2, 3, \ldots$. The roots R_n of the Airy function can be found in standard tables. For instance, the first few roots are $R_0 = -2.338$, $R_1 = -4.088$, $R_2 = -5.521$, $R_3 = -6.787$.

The boundary condition $\psi(0) = 0$ therefore gives a *discrete* set of energy levels which can be expressed in terms of the roots of the Airy function:

$$\mathrm{Ai}\left[-\left(\frac{2}{mg^2\hbar^2}\right)^{1/3}E\right] = 0 \implies -\left(\frac{2}{mg^2\hbar^2}\right)^{1/3}E_n = R_n; \tag{4.341}$$

hence

$$E_n = -\left(\frac{1}{2}mg^2\hbar^2\right)^{1/3} R_n, \qquad \psi_n(z) = B_n \text{Ai}\left[-\left(\frac{2m^2g^2}{\hbar^2}\right)^{1/3} z - R_n\right]. \qquad (4.342)$$

The first few energy levels are

$$E_0 = 2.338\left(\frac{1}{2}mg^2\hbar^2\right)^{1/3}, \qquad E_1 = 4.088\left(\frac{1}{2}mg^2\hbar^2\right)^{1/3}, \qquad (4.343)$$

$$E_2 = 5.521\left(\frac{1}{2}mg^2\hbar^2\right)^{1/3}. \qquad E_3 = 6.787\left(\frac{1}{2}mg^2\hbar^2\right)^{1/3}. \qquad (4.344)$$

4.11 Exercises

Exercise 4.1
A particle of mass m is subjected to a potential

$$V(x) = \begin{cases} 0, & |x| < a/2, \\ \infty, & |x| > a/2. \end{cases}$$

(a) Find the ground, first, and second excited state wave functions.
(b) Find expressions for E_1, E_2, and E_3.
(c) Plot the probability densities $P_2(x, t)$ and $P_3(x, t)$.
(d) Find $\langle X \rangle_2$, $\langle X \rangle_3$, $\langle P \rangle_2$, and $\langle P \rangle_3$.
(e) Evaluate $\Delta x \Delta p$ for the states $\psi_2(x, t)$ and $\psi_3(x, t)$.

Exercise 4.2
Consider a system whose wave function at $t = 0$ is

$$\psi(x, 0) = \frac{3}{\sqrt{30}}\phi_0(x) + \frac{4}{\sqrt{30}}\phi_1(x) + \frac{1}{\sqrt{6}}\phi_4(x),$$

where $\phi_n(x)$ is the wave function of the nth excited state of an infinite square well potential of width a and whose energy is $E_n = \pi^2\hbar^2 n^2/(2ma^2)$.
(a) Find the average energy of this system.
(b) Find the state $\psi(x, t)$ at a later time t and the average value of the energy. Compare the result with the value obtained in (a).

Exercise 4.3
An electron with a kinetic energy of 10 eV at large negative values of x is moving from left to right along the x-axis. The potential energy is

$$V(x) = \begin{cases} 0 & (x \leq 0), \\ 20\,\text{eV} & (x > 0). \end{cases}$$

(a) Write the time-independent Schrödinger equation in the regions $x \leq 0$ and $x > 0$.
(b) Describe the shapes for $\psi(x)$ for $x \leq 0$ and $x > 0$.

(c) Calculate the electron wavelength (in meters) in $-20\,\text{m} < x < -10\,\text{m}$ and $x > 10\,\text{m}$.

(d) Write down the boundary conditions at $x = 0$.

(e) Calculate the ratio of the probabilities for finding the electron near $x = 10^{-10}\,\text{m}$ and $x = 0$.

Exercise 4.4

A particle is moving in the potential well

$$V(x) = \begin{cases} 0, & -a \le x \le -b, \\ V_0, & -b \le x \le b, \\ 0, & b \le x \le a, \\ +\infty & \text{elsewhere} \end{cases},$$

where V_0 is positive. In this problem consider $E < V_0$. Let $\psi_1(x)$ and $\psi_2(x)$ represent the two lowest energy solutions of the Schrödinger equation; call their energies E_1 and E_2, respectively.

(a) Calculate E_1 and E_2 in units of eV for the case where $mc^2 = 1\,\text{GeV}$, $a = 10^{-14}\,\text{m}$, and $b = 0.4 \times 10^{-14}\,\text{m}$; take $\hbar c \simeq 200\,\text{MeV fm}$.

(b) A particular solution of the Schrödinger equation can be constructed by superposing $\psi_1(x)e^{iE_1t/\hbar}$ and $\psi_2(x)e^{iE_2t/\hbar}$. Construct a wave packet ψ which at $t = 0$ is (almost) entirely to the left-hand side of the well and describe its motion in time; find the period of oscillations between the two terms of ψ.

Exercise 4.5

A particle moves in the potential

$$V(x) = \frac{\hbar^2}{2m} \left[\frac{4}{225} \sinh^2 x - \frac{2}{5} \cosh x \right].$$

(a) Sketch $V(x)$ and locate the position of the two minima.

(b) Show that $\psi(x) = (1 + 4\cosh x)\exp\left(-\frac{2}{15}\cosh x\right)$ is a solution of the time-independent Schrödinger equation for the particle. Find the corresponding energy level and indicate its position on the sketch of $V(x)$.

(c) Sketch $\psi(x)$ and show that it has the proper behavior at the classical turning points and in the classically forbidden regions.

Exercise 4.6

Show that for a particle of mass m which moves in a one-dimensional infinite potential well of length a, the uncertainties product $\Delta x_n \Delta p_n$ is given by $\Delta x_n \Delta p_n \simeq n\pi\hbar/\sqrt{12}$.

Exercise 4.7

A particle of mass m is moving in an infinite potential well

$$V(x) = \begin{cases} V_0, & 0 < x < a, \\ \infty, & \text{elsewhere.} \end{cases}$$

(a) Solve the Schrödinger equation and find the energy levels and the corresponding normalized wave functions.

(b) Calculate $\langle \hat{X} \rangle_5$, $\langle \hat{P} \rangle_5$, $\langle \hat{X}^2 \rangle_5$, and $\langle \hat{P}^2 \rangle_5$ for the fourth excited state and infer the value of $\Delta x \Delta p$.

Exercise 4.8

Consider the potential step

$$V(x) = \begin{cases} 6\,\text{eV}, & x < 0, \\ 0, & x > 0. \end{cases}$$

(a) An electron of energy 8 eV is moving from left to right in this potential. Calculate the probability that the electron will (i) continue moving along its initial direction after reaching the step and (ii) get reflected at the potential step.

(b) Now suppose the electron is moving from right to left with an energy 3 eV. (i) Estimate the order of magnitude of the distance the electron can penetrate the barrier. (ii) Repeat part (i) for a 70 kg person initially moving at 4 m s^{-1} and running into a wall which can be represented by a potential step of height equal to four times this person's energy before reaching the step.

Exercise 4.9

Consider a system whose wave function at time $t = 0$ is given by

$$\psi(x, 0) = \frac{5}{\sqrt{50}}\phi_0(x) + \frac{4}{\sqrt{50}}\phi_1(x) + \frac{3}{\sqrt{50}}\phi_2(x),$$

where $\phi_n(x)$ is the wave function of the nth excited state for a harmonic oscillator of energy $E_n = \hbar\omega(n + 1/2)$.

(a) Find the average energy of this system.

(b) Find the state $\psi(x, t)$ at a later time t and the average value of the energy; compare the result with the value obtained in (a).

(c) Find the expectation value of the operator $\hat{X}$ with respect to the state $\psi(x, t)$ (i.e., find $\langle \psi(x, t) | \hat{X} | \psi(x, t) \rangle$).

Exercise 4.10

Calculate $\langle n | \hat{X}^2 | m \rangle$ and $\langle m | \hat{X}^4 | n \rangle$ in the N-representation; $| n \rangle$ and $| m \rangle$ are harmonic oscillator states.

Exercise 4.11

Consider the dimensionless Hamiltonian $\hat{H} = \frac{1}{2}\hat{P}^2 + \frac{1}{2}\hat{X}^2$, with $\hat{P} = -id/dx$.

(a) Show that the wave functions $\psi_0(x) = e^{-x^2/2}/\sqrt{\sqrt{\pi}}$ and $\psi_1(x) = \sqrt{2/\sqrt{\pi}}\,xe^{-x^2/2}$ are eigenfunctions of $\hat{H}$ with eigenvalues $1/2$ and $3/2$, respectively.

(b) Find the values of the coefficients α and β such that

$$\psi_2(x) = \frac{1}{\sqrt{2\sqrt{\pi}}}\left(\alpha x^2 - 1\right)e^{-x^2/2} \quad \text{and} \quad \psi_3(x) = \frac{1}{\sqrt{6\sqrt{\pi}}}x\left(1 + \beta x^2\right)e^{-x^2/2}$$

are orthogonal to $\psi_0(x)$ and $\psi_1(x)$, respectively. Then show that $\psi_2(x)$ and $\psi_3(x)$ are eigenfunctions of $\hat{H}$ with eigenvalues $5/2$ and $7/2$, respectively.

Exercise 4.12

Consider the dimensionless Hamiltonian $\hat{H} = \frac{1}{2}\hat{P}^2 + \frac{1}{2}\hat{X}^2$ (with $\hat{P} = -id/dx$) whose wave function at time $t = 0$ is given by

$$\Psi(x, 0) = \frac{1}{\sqrt{2}}\psi_0(x) + \frac{1}{\sqrt{8}}\psi_1(x) + \frac{1}{\sqrt{10}}\psi_2(x),$$

where $\psi_0(x) = \frac{1}{\sqrt{\sqrt{\pi}}}e^{-x^2/2}$, $\psi_1(x) = \sqrt{\frac{2}{\sqrt{\pi}}}xe^{-x^2/2}$, and $\psi_2(x) = \frac{1}{\sqrt{2\sqrt{\pi}}}(2x^2 - 1)e^{-x^2/2}$.

(a) Calculate $\Delta x_n \Delta p_n$ for $n = 0, 1$ where $\Delta x_n = \sqrt{\langle\psi_n|\hat{X}^2|\psi_n\rangle - \langle\psi_n|\hat{X}|\psi_n\rangle^2}$.

(b) Calculate $\hat{a}^\dagger \psi_0(x)$, $\hat{a}\psi_0(x)$, $\hat{a}^\dagger \psi_1(x)$, $\hat{a}\psi_1(x)$, and $\hat{a}\psi_2(x)$, where the operators $\hat{a}^\dagger$ and $\hat{a}$ are defined by $\hat{a} = (\hat{X} + d/dx)/\sqrt{2}$ and $\hat{a}^\dagger = (\hat{X} - d/dx)/\sqrt{2}$.

Exercise 4.13
Consider a particle of mass m that is moving in a one-dimensional infinite potential well with walls at $x = 0$ and $x = a$ which is initially (i.e., at $t = 0$) in the state

$$\psi(x, 0) = \frac{1}{\sqrt{2}}[\phi_1(x) + \phi_3(x)],$$

where $\phi_1(x)$ and $\phi_3(x)$ are the ground and second excited states, respectively.
(a) What is the state vector $\psi(x, t)$ for $t > 0$ in the Schrödinger picture.
(b) Find the expectation values $\langle\hat{X}\rangle$, $\langle\hat{P}\rangle$, $\langle\hat{X}^2\rangle$, and $\langle\hat{P}^2\rangle$ with respect to $|\psi\rangle$.
(c) Evaluate $\Delta x \Delta p$ and verify that it satisfies the uncertainty principle.

Exercise 4.14
If the state of a particle moving in a one-dimensional harmonic oscillator is given by

$$|\psi\rangle = \frac{1}{\sqrt{17}}|0\rangle + \frac{3}{\sqrt{17}}|1\rangle - \frac{2}{\sqrt{17}}|2\rangle - \sqrt{\frac{3}{17}}|3\rangle,$$

where $|n\rangle$ represents the normalized nth energy eigenstate, find the expectation values of the number operator, $\hat{N}$, and of the Hamiltonian operator.

Exercise 4.15
Find the number of bound states and the corresponding energies for the finite square well potential when (a) $R = 7$ (i.e., $\sqrt{ma^2 V_0/(2\hbar^2)} = 7$) and (b) $R = 3\pi$.

Exercise 4.16
A ball of mass $m = 0.2\,\text{kg}$ bouncing on a table located at $z = 0$ is subject to the potential

$$V(z) = \begin{cases} V_0 & (z < 0), \\ mgz & (z > 0), \end{cases}$$

where $V_0 = 3\,\text{J}$ and g is the acceleration due to gravity.
(a) Describe the spectrum of possible energies (i.e., continuous, discrete, or nonexistent) as E increases from large negative values to large positive values.
(b) Estimate the order of magnitude for the lowest energy state.
(c) Describe the general shapes of the wave functions $\psi_0(z)$ and $\psi_1(z)$ corresponding to the lowest two energy states and sketch the corresponding probability densities.

Exercise 4.17
Consider a particle of mass m moving in a one-dimensional harmonic oscillator potential, with $\hat{X} = \sqrt{\hbar/(2m\omega)}(\hat{a} + \hat{a}^\dagger)$ and $\hat{P} = i\sqrt{m\hbar\omega/2}(\hat{a}^\dagger - \hat{a})$.
(a) Calculate the product of the uncertainties in position and momentum for the particle in the fifth excited state, i.e., $(\Delta X \Delta P)_5$.
(b) Compare the result of (a) with the uncertainty product when the particle is in its lowest energy state. Explain why the two uncertainty products are different.

Exercise 4.18

A particle of mass m in an infinite potential well of length a has the following initial wave function at $t = 0$:

$$\psi(x, 0) = \frac{2}{\sqrt{7a}} \sin\left(\frac{\pi x}{a}\right) + \sqrt{\frac{6}{7a}} \sin\left(\frac{2\pi x}{a}\right) + \frac{2}{\sqrt{7a}} \sin\left(\frac{3\pi x}{a}\right).$$

(a) If we measure energy, what values will we find and with what probabilities? Calculate the average energy.

(b) Find the wave function $\psi(x, t)$ at any later time t. Determine the probability of finding the particle at a time t in the state $\varphi(x, t) = 1/\sqrt{a} \sin(3\pi x/a) \exp(-i E_3 t/\hbar)$.

(c) Calculate the probability density $\rho(x, t)$ and the current density $\vec{J}(x, t)$. Verify that $\partial\rho/\partial t + \vec{\nabla} \cdot \vec{J}(x, t) = 0$.

Exercise 4.19

Consider a particle in an infinite square well whose wave function is given by

$$\psi(x) = \begin{cases} Ax(a^2 - x^2), & 0 < x < a, \\ 0, & \text{elsewhere,} \end{cases}$$

where A is a real constant.

(a) Find A so that $\psi(x)$ is normalized.

(b) Calculate the position and momentum uncertainties, Δx and Δp, and the product $\Delta x \Delta p$.

(c) Calculate the probability of finding $5^2\pi^2\hbar^2/(2ma^2)$ for a measurement of the energy.

Exercise 4.20

The relativistic expression for the energy of a free particle is $E^2 = m_0^2 c^4 + p^2 c^2$.

(a) Write down the corresponding relativistic Schrödinger equation, by quantizing this energy expression (i.e., replacing E and p with their corresponding operators). This equation is called the Klein–Gordon equation.

(b) Find the solutions corresponding to a free particle moving along the x-axis.

Exercise 4.21

(a) Write down the classical (gravitational) energy E_c of a particle of mass m at rest a height h_0 above the ground (take the zero potential energy to be located at the ground level).

(b) Use the uncertainty principle to estimate the ground state energy E_0 of the particle introduced in (a); note that the particle is subject to gravity. Compare E_0 to E_c.

(c) If $h_0 = 3$ m obtain the numerical values of E_c and the quantum mechanical correction $(E_0 - E_c)$ for a neutron and then for a particle of mass $m = 0.01$ kg. Comment on the importance of the quantum correction in both cases.

Exercise 4.22

Find the energy levels and the wave functions of two noninteracting particles of masses m_1 and m_2 that are moving in a common infinite square well potential

$$V(x_i) = \begin{cases} 0, & 0 \leq x_i \leq a, \\ +\infty, & \text{elsewhere,} \end{cases}$$

where x_i is the position of the ith particle (i.e., x_i denotes $x = x_1$ or x_2).

Exercise 4.23

A particle of mass m is subject to a repulsive delta potential $V(x) = V_0\delta(x)$, where $V_0 > 0$ (V_0 has the dimensions of Energy×Distance). Find the reflection and transmission coefficients, R and T.

Exercise 4.24

A particle of mass m is scattered by a double-delta potential $V(x) = V_0\delta(x-a) + V_0\delta(x+a)$, where $V_0 > 0$.

(a) Find the transmission coefficient for the particle at an energy $E > 0$.

(b) When V_0 is very large (i.e., $V_0 \to \infty$), find the energies corresponding to the resonance case (i.e., $T = 1$) and compare them with the energies of an infinite square well potential having a width of $2a$.

Exercise 4.25

A particle of mass m is subject to an antisymmetric delta potential $V(x) = V_0\delta(x + a) - V_0\delta(x - a)$, where $V_0 > 0$.

(a) Show that there is always one and only one bound state, and find the expression that gives its energy.

(b) Find the transmission coefficient T.

Exercise 4.26

A particle of mass m is subject to a delta potential

$$V(x) = \begin{cases} \infty, & x \leq 0, \\ V_0\delta(x - a), & x > 0, \end{cases}$$

where $V_0 > 0$.

(a) Find the wave functions corresponding to the cases $0 < x < a$ and $x > a$.

(b) Find the transmission coefficient.

Exercise 4.27

A particle of mass m, besides being confined to move in an infinite square well potential of size $2a$ with walls at $x = -a$ and $x = a$, is subject to an attractive delta potential

$$V(x) = \begin{cases} V_0\delta(x), & -a < x < a, \\ \infty, & \text{elsewhere,} \end{cases}$$

where $V_0 > 0$.

(a) Find the particle's wave function corresponding to *even* solutions when $E > 0$.

(b) Find the energy levels corresponding to *even* solutions.

Exercise 4.28

A particle of mass m, besides being confined to move in an infinite square well potential of size $2a$ with walls at $x = -a$ and $x = a$, is subject to an attractive delta potential

$$V(x) = \begin{cases} V_0\delta(x), & -a < x < a, \\ \infty, & \text{elsewhere,} \end{cases}$$

where $V_0 > 0$.

(a) Find the particle's wave function corresponding to *odd* solutions when $E > 0$.

(b) Find the energy levels corresponding to *odd* solutions.

Exercise 4.29

Consider a particle of mass m that is moving under the influence of an attractive delta potential

$$V(x) = \begin{cases} -V_0\delta(x), & x > -a, \\ \infty, & x < -a, \end{cases}$$

where $V_0 > 0$. Discuss the existence of bound states in terms of V_0 and a.

Exercise 4.30

Consider a system of two identical harmonic oscillators (with an angular frequency ω).

(a) Find the energy levels when the oscillators are independent (non-interacting).

(b) Find the energy levels when the oscillators are coupled by an interaction $-\lambda\hat{X}_1\hat{X}_2$, where λ is a constant.

(c) Assuming that $\lambda \ll m\omega^2$ (weak coupling limit), find an approximate value to first order in $\lambda/m\omega^2$ for the energy expression derived in part (b).

Exercise 4.31

A particle is initially in its ground state in an infinite one-dimensional potential box with sides at $x = 0$ and $x = a$. If the wall of the box at $x = a$ is *suddenly* moved to $x = 3a$, calculate the probability of finding the particle in

(a) the ground state of the new box and

(b) the first excited state of the new box.

(c) Now, calculate the probability of finding the particle in the first excited state of the new box, assuming the particle was initially in the first excited state of the old box.

Exercise 4.32

A particle is initially in its ground state in a one-dimensional harmonic oscillator potential, $\hat{V}(x) = \frac{1}{2}k\hat{x}^2$. If the spring constant is *suddenly* doubled, calculate the probability of finding the particle in the ground state of the new potential.

Exercise 4.33

Consider an electron in an infinite potential well

$$V(x) = \begin{cases} 0, & 0 < x < a, \\ +\infty, & \text{elsewhere}, \end{cases}$$

where $a = 10^{-10}$ m.

(a) Calculate the energy levels of the three lowest states (the results should be expressed in eV) and the corresponding wavelengths of the electron.

(b) Calculate the frequency of the radiation that would cause the electron to jump from the ground to the third excited energy level.

(c) When the electron de-excites, what are the frequencies of the emitted photons?

(d) Specify the probability densities for all these three states and plot them.

Exercise 4.34

Consider an electron which is confined to move in an infinite square well of width $a = 10^{-10}m$.

(a) Find the exact energies of the 11 lowest states (express them in eV).

(b) Solve the Schrödinger equation *numerically* and find the energies of the 11 lowest states and compare them with the exact results obtained in (a). Plot the wave functions of the five lowest states.

Chapter 5

Angular Momentum

5.1 Introduction

After treating one-dimensional problems in Chapter 4, we now should deal with three-dimensional problems. However, the study of three-dimensional systems such as atoms cannot be undertaken unless we first cover the formalism of angular momentum. The current chapter, therefore, serves as an essential prelude to Chapter 6.

Angular momentum is as important in classical mechanics as in quantum mechanics. It is particularly useful for studying the dynamics of systems that move under the influence of *spherically symmetric*, or *central*, potentials, $V(\vec{r}) = V(r)$, for the orbital angular momenta of these systems are *conserved*. For instance, as mentioned in Chapter 1, one of the cornerstones of Bohr's model of the hydrogen atom (where the electron moves in the proton's Coulomb potential, a central potential) is based on the quantization of angular momentum. Additionally, angular momentum plays a critical role in the description of molecular rotations, the motion of electrons in atoms, and the motion of nucleons in nuclei. The quantum theory of angular momentum is thus a prerequisite for studying molecular, atomic, and nuclear systems.

In this chapter we are going to consider the general formalism of angular momentum. We will examine the various properties of the angular momentum operator, and then focus on determining its eigenvalues and eigenstates. Finally, we will apply this formalism to the determination of the eigenvalues and eigenvectors of the spin and orbital angular momenta.

5.2 Orbital Angular Momentum

In classical physics the angular momentum of a particle with momentum $\vec{p}$ and position $\vec{r}$ is defined by

$$\vec{L} = \vec{r} \times \vec{p} = (yp_z - zp_y)\vec{i} + (zp_x - xp_z)\vec{j} + (xp_y - yp_x)\vec{k}. \tag{5.1}$$

The orbital angular momentum operator $\hat{\vec{L}}$ can be obtained at once by replacing $\vec{r}$ and $\vec{p}$ by the corresponding operators in the position representation, $\hat{\vec{R}}$ and $\hat{\vec{P}} = -i\hbar\vec{\nabla}$:

$$\boxed{\hat{\vec{L}} = \hat{\vec{R}} \times \hat{\vec{P}} = -i\hbar\hat{\vec{R}} \times \vec{\nabla}.} \tag{5.2}$$

The Cartesian components of $\hat{\vec{L}}$ are

$$\hat{L}_x = \hat{Y}\hat{P}_z - \hat{Z}\hat{P}_y = -i\hbar\left(\hat{Y}\frac{\partial}{\partial z} - \hat{Z}\frac{\partial}{\partial y}\right), \tag{5.3}$$

$$\hat{L}_y = \hat{Z}\hat{P}_x - \hat{X}\hat{P}_z = -i\hbar\left(\hat{Z}\frac{\partial}{\partial x} - \hat{X}\frac{\partial}{\partial z}\right), \tag{5.4}$$

$$\hat{L}_z = \hat{X}\hat{P}_y - \hat{Y}\hat{P}_x = -i\hbar\left(\hat{X}\frac{\partial}{\partial y} - \hat{Y}\frac{\partial}{\partial x}\right). \tag{5.5}$$

Clearly, angular momentum does not exist in a one-dimensional space. We should mention that the components $\hat{L}_x$, $\hat{L}_y$, $\hat{L}_z$, and the square of $\hat{\vec{L}}$,

$$\hat{\vec{L}}^2 = \hat{L}_x^2 + \hat{L}_y^2 + \hat{L}_z^2, \tag{5.6}$$

are all Hermitian.

Commutation relations

Since $\hat{X}$, $\hat{Y}$, and $\hat{Z}$ mutually commute and so do $\hat{P}_x$, $\hat{P}_y$, and $\hat{P}_z$, and since $[\hat{X}, \hat{P}_x] = i\hbar$, $[\hat{Y}, \hat{P}_y] = i\hbar$, $[\hat{Z}, \hat{P}_z] = i\hbar$, we have

$$\begin{aligned}
[\hat{L}_x, \hat{L}_y] &= [\hat{Y}\hat{P}_z - \hat{Z}\hat{P}_y, \ \hat{Z}\hat{P}_x - \hat{X}\hat{P}_z] \\
&= [\hat{Y}\hat{P}_z, \hat{Z}\hat{P}_x] - [\hat{Y}\hat{P}_z, \hat{X}\hat{P}_z] - [\hat{Z}\hat{P}_y, \hat{Z}\hat{P}_x] + [\hat{Z}\hat{P}_y, \hat{X}\hat{P}_z] \\
&= \hat{Y}[\hat{P}_z, \hat{Z}]\hat{P}_x + \hat{X}[\hat{Z}, \hat{P}_z]\hat{P}_y = i\hbar(\hat{X}\hat{P}_y - \hat{Y}\hat{P}_x) \\
&= i\hbar\hat{L}_z.
\end{aligned} \tag{5.7}$$

A similar calculation yields the other two commutation relations; but it is much simpler to infer them from (5.7) by means of a *cyclic permutation* of the xyz components, $x \to y \to z \to x$:

$$\boxed{[\hat{L}_x, \hat{L}_y] = i\hbar\hat{L}_z, \qquad [\hat{L}_y, \hat{L}_z] = i\hbar\hat{L}_x, \qquad [\hat{L}_z, \hat{L}_x] = i\hbar\hat{L}_y.} \tag{5.8}$$

As mentioned in Chapter 3, since $\hat{L}_x$, $\hat{L}_y$, and $\hat{L}_z$ do not commute, we cannot measure them simultaneously to arbitrary accuracy.

Note that the commutation relations (5.8) were derived by expressing the orbital angular momentum in the *position representation*, but since these are operator relations, they must be valid in any representation. In the following section we are going to consider the general formalism of angular momentum, a formalism that is restricted to no particular representation.

Example 5.1

(a) Calculate the commutators $[\hat{X}, \ \hat{L}_x]$, $[\hat{X}, \ \hat{L}_y]$, and $[\hat{X}, \ \hat{L}_z]$.

(b) Calculate the commutators: $[\hat{P}_x, \ \hat{L}_x]$, $[\hat{P}_x, \ \hat{L}_y]$, and $[\hat{P}_x, \ \hat{L}_z]$.

(c) Use the results of (a) and (b) to calculate $[\hat{X}, \ \hat{\vec{L}}^2]$ and $[\hat{P}_x, \ \hat{\vec{L}}^2]$.

Solution

(a) The only nonzero commutator which involves $\hat{X}$ and the various components of $\hat{L}_x$, $\hat{L}_y$, $\hat{L}_z$ is $[\hat{X}, \hat{P}_x] = i\hbar$. Having stated this result, we can easily evaluate the needed commutators. First, since $\hat{L}_x = \hat{Y}\hat{P}_z - \hat{Z}\hat{P}_y$ involves no $\hat{P}_x$, the operator $\hat{X}$ commutes separately with $\hat{Y}$, $\hat{P}_z$, $\hat{Z}$, and $\hat{P}_y$; hence

$$[\hat{X}, \hat{L}_x] = [\hat{X}, \hat{Y}\hat{P}_z - \hat{Z}\hat{P}_y] = 0. \tag{5.9}$$

The evaluation of the other two commutators is straightforward:

$$[\hat{X}, \hat{L}_y] = [\hat{X}, \hat{Z}\hat{P}_x - \hat{X}\hat{P}_z] = [\hat{X}, \hat{Z}\hat{P}_x] = \hat{Z}[\hat{X}, \hat{P}_x] = i\hbar\hat{Z}, \tag{5.10}$$

$$[\hat{X}, \hat{L}_z] = [\hat{X}, \hat{X}\hat{P}_y - \hat{Y}\hat{P}_x] = -[\hat{X}, \hat{Y}\hat{P}_x] = -\hat{Y}[\hat{X}, \hat{P}_x] = -i\hbar\hat{Y}. \tag{5.11}$$

(b) The only commutator between $\hat{P}_x$ and the components of $\hat{L}_x$, $\hat{L}_y$, $\hat{L}_z$ that survives is again $[\hat{P}_x, \hat{X}] = -i\hbar$. We may thus infer

$$[\hat{P}_x, \hat{L}_x] = [\hat{P}_x, \hat{Y}\hat{P}_z - \hat{Z}\hat{P}_y] = 0, \tag{5.12}$$

$$[\hat{P}_x, \hat{L}_y] = [\hat{P}_x, \hat{Z}\hat{P}_x - \hat{X}\hat{P}_z] = -[\hat{P}_x, \hat{X}\hat{P}_z] = -[\hat{P}_x, \hat{X}]\hat{P}_z = i\hbar\hat{P}_z, \tag{5.13}$$

$$[\hat{P}_x, \hat{L}_z] = [\hat{P}_x, \hat{X}\hat{P}_y - \hat{Y}\hat{P}_x] = [\hat{P}_x, \hat{X}\hat{P}_y] = [\hat{P}_x, \hat{X}]\hat{P}_y = -i\hbar\hat{P}_y. \tag{5.14}$$

(c) Using the commutators derived in (a) and (b), we infer

$$[\hat{X}, \hat{\vec{L}}^2] = [\hat{X}, \hat{L}_x^2] + [\hat{X}, \hat{L}_y^2] + [\hat{X}, \hat{L}_z^2]$$
$$= 0 + \hat{L}_y[\hat{X}, \hat{L}_y] + [\hat{X}, \hat{L}_y]\hat{L}_y + \hat{L}_z[\hat{X}, \hat{L}_z] + [\hat{X}, \hat{L}_z]\hat{L}_z$$
$$= i\hbar(\hat{L}_y\hat{Z} + \hat{Z}\hat{L}_y - \hat{L}_z\hat{Y} - \hat{Y}\hat{L}_y), \tag{5.15}$$

$$[\hat{P}_x, \hat{\vec{L}}^2] = [\hat{P}_x, \hat{L}_x^2] + [\hat{P}_x, \hat{L}_y^2] + [\hat{P}_x, \hat{L}_z^2]$$
$$= 0 + \hat{L}_y[\hat{P}_x, \hat{L}_y] + [\hat{P}_x, \hat{L}_y]\hat{L}_y + \hat{L}_z[\hat{P}_x, \hat{L}_z] + [\hat{P}_x, \hat{L}_z]\hat{L}_z$$
$$= i\hbar(\hat{L}_y\hat{P}_z + \hat{P}_z\hat{L}_y - \hat{L}_z\hat{P}_y - \hat{P}_y\hat{L}_y). \tag{5.16}$$

5.3 General Formalism of Angular Momentum

Let us now introduce a more general angular momentum operator $\hat{\vec{J}}$ that is defined by its three components $\hat{J}_x$, $\hat{J}_y$, and $\hat{J}_z$, which satisfy the following commutation relations:

$$\boxed{[\hat{J}_x, \hat{J}_y] = i\hbar\hat{J}_z, \qquad [\hat{J}_y, \hat{J}_z] = i\hbar\hat{J}_x, \qquad [\hat{J}_z, \hat{J}_x] = i\hbar\hat{J}_y,} \tag{5.17}$$

or equivalently by

$$\hat{\vec{J}} \times \hat{\vec{J}} = i\hbar\hat{\vec{J}}. \tag{5.18}$$

Since $\hat{J}_x$, $\hat{J}_y$, and $\hat{J}_z$ do not mutually commute, they cannot be simultaneously diagonalized; that is, they do not possess common eigenstates. The square of the angular momentum,

$$\hat{\vec{J}}^2 = \hat{J}_x^2 + \hat{J}_y^2 + \hat{J}_z^2, \tag{5.19}$$

is a scalar operator; hence it commutes with $\hat{J}_x$, $\hat{J}_y$, and $\hat{J}_z$:

$$[\hat{\vec{J}}^2, \hat{J}_k] = 0, \tag{5.20}$$

where k stands for x, y, and z. For instance, in the the case $k = x$ we have

$$\begin{aligned}
[\hat{\vec{J}}^2, \hat{J}_x] &= [\hat{J}_x^2, \hat{J}_x] + \hat{J}_y[\hat{J}_y, \hat{J}_x] + [\hat{J}_y, \hat{J}_x]\hat{J}_y + \hat{J}_z[\hat{J}_z, \hat{J}_x] + [\hat{J}_z, \hat{J}_x]\hat{J}_z \\
&= \hat{J}_y(-i\hbar\hat{J}_z) + (-i\hbar\hat{J}_z)\hat{J}_y + \hat{J}_z(i\hbar\hat{J}_y) + (i\hbar\hat{J}_y)\hat{J}_z \\
&= 0,
\end{aligned} \tag{5.21}$$

because $[\hat{J}_x^2, \hat{J}_x] = 0$, $[\hat{J}_y, \hat{J}_x] = -i\hbar\hat{J}_z$, and $[\hat{J}_z, \hat{J}_x] = i\hbar\hat{J}_y$. We should note that the operators $\hat{J}_x$, $\hat{J}_y$, $\hat{J}_z$, and $\hat{\vec{J}}^2$ are all Hermitian; their eigenvalues are real.

Eigenstates and eigenvalues of the angular momentum operator

Since $\hat{\vec{J}}^2$ commutes with $\hat{J}_x$, $\hat{J}_y$ and $\hat{J}_z$, each component of $\vec{J}$ can be separately diagonalized (hence it has simultaneous eigenfunctions) with $\hat{\vec{J}}^2$. But since the components $\hat{J}_x$, $\hat{J}_y$, and $\hat{J}_z$ do not mutually commute, we can choose only one of them to be simultaneously diagonalized with $\hat{\vec{J}}^2$. By convention we choose $\hat{J}_z$. There is nothing special about the z-direction; we can just as well take $\hat{\vec{J}}^2$ and $\hat{J}_x$ or $\hat{\vec{J}}^2$ and $\hat{J}_y$.

Let us now look for the joint eigenstates of $\hat{\vec{J}}^2$ and $\hat{J}_z$ and their corresponding eigenvalues. Denoting the joint eigenstates by $| \alpha, \beta \rangle$ and the eigenvalues of $\hat{\vec{J}}^2$ and $\hat{J}_z$ by $\hbar^2\alpha$ and $\hbar\beta$, respectively, we have

$$\hat{\vec{J}}^2 | \alpha, \beta \rangle = \hbar^2\alpha | \alpha, \beta \rangle, \tag{5.22}$$
$$\hat{J}_z | \alpha, \beta \rangle = \hbar\beta | \alpha, \beta \rangle. \tag{5.23}$$

The factor $\hbar$ is introduced so that α and β are dimensionless; recall that the angular momentum has the dimensions of $\hbar$ and that the physical dimensions of $\hbar$ are: $[\hbar] =$ energy $\times$ time. For simplicity, we will assume that these eigenstates are orthonormal:

$$\langle \alpha', \beta' | \alpha, \beta \rangle = \delta_{\alpha', \alpha}\delta_{\beta', \beta}. \tag{5.24}$$

Now we need to introduce *raising* and *lowering* operators $\hat{J}_+$ and $\hat{J}_-$, just as we did when we studied the harmonic oscillator in Chapter 4:

$$\boxed{\hat{J}_\pm = \hat{J}_x \pm i\hat{J}_y.} \tag{5.25}$$

This leads to

$$\hat{J}_x = \frac{1}{2}(\hat{J}_+ + \hat{J}_-), \qquad \hat{J}_y = \frac{1}{2i}(\hat{J}_+ - \hat{J}_-); \tag{5.26}$$

hence

$$\hat{J}_x^2 = \frac{1}{4}(\hat{J}_+^2 + \hat{J}_+\hat{J}_- + \hat{J}_-\hat{J}_+ + \hat{J}_-^2), \qquad \hat{J}_y^2 = -\frac{1}{4}(\hat{J}_+^2 - \hat{J}_+\hat{J}_- - \hat{J}_-\hat{J}_+ + \hat{J}_-^2). \tag{5.27}$$

Using (5.17) we can easily obtain the following commutation relations:

$$[\hat{\vec{J}}^2, \hat{J}_\pm] = 0, \qquad [\hat{J}_+, \hat{J}_-] = 2\hbar\hat{J}_z, \qquad [\hat{J}_z, \hat{J}_\pm] = \pm\hbar\hat{J}_\pm. \tag{5.28}$$

In addition, $\hat{J}_+$ and $\hat{J}_-$ satisfy

$$\hat{J}_+\hat{J}_- = \hat{J}_x^2 + \hat{J}_y^2 + \hbar.\hat{J}_z = \hat{J}^2 - \hat{J}_z^2 + \hbar\hat{J}_z, \tag{5.29}$$

$$\hat{J}_-\hat{J}_+ = \hat{J}_x^2 + \hat{J}_y^2 - \hbar\hat{J}_z = \hat{J}^2 - \hat{J}_z^2 - \hbar\hat{J}_z. \tag{5.30}$$

These relations lead to

$$\boxed{\hat{J}^2 = \hat{J}_\pm\hat{J}_\mp + \hat{J}_z^2 \mp \hbar\hat{J}_z,} \tag{5.31}$$

which in turn yield

$$\boxed{\hat{J}^2 = \frac{1}{2}(\hat{J}_+\hat{J}_- + \hat{J}_-\hat{J}_+) + \hat{J}_z^2.} \tag{5.32}$$

Let us see how $\hat{J}_\pm$ operate on $| \alpha, \beta \rangle$. First, since $\hat{J}_\pm$ do not commute with $\hat{J}_z$, the kets $| \alpha, \beta \rangle$ are not eigenstates of $\hat{J}_\pm$. Using the relations (5.28) we have

$$\hat{J}_z(\hat{J}_\pm | \alpha, \beta\rangle) = (\hat{J}_\pm\hat{J}_z \pm \hbar\hat{J}_\pm) | \alpha, \beta\rangle = \hbar(\beta \pm 1)(\hat{J}_\pm | \alpha, \beta\rangle); \tag{5.33}$$

hence the ket $(\hat{J}_\pm | \alpha, \beta\rangle)$ is an eigenstate of $\hat{J}_z$ with eigenvalues $\hbar(\beta \pm 1)$. Now since $\hat{J}_z$ and $\hat{J}^2$ commute, $(\hat{J}_\pm | \alpha, \beta\rangle)$ must also be an eigenstate of $\hat{J}^2$. The eigenvalue of $\hat{J}^2$ when acting on $\hat{J}_\pm | \alpha, \beta\rangle$ can be determined by making use of the commutator $[\hat{J}^2, \hat{J}_\pm] = 0$. The state $(\hat{J}_\pm | \alpha, \beta\rangle)$ is also an eigenstate of $\hat{J}^2$ with eigenvalue $\hbar^2\alpha$:

$$\hat{J}^2(\hat{J}_\pm | \alpha, \beta\rangle) = \hat{J}_\pm\hat{J}^2 | \alpha, \beta\rangle = \hbar^2\alpha(\hat{J}_\pm | \alpha, \beta\rangle). \tag{5.34}$$

From (5.33) and (5.34) we infer that when $\hat{J}_\pm$ acts on $| \alpha, \beta \rangle$, it does not affect the first quantum number α, but it raises or lowers the second quantum number β by one unit. That is, $\hat{J}_\pm | \alpha, \beta \rangle$ is proportional to $| \alpha, \beta \pm 1 \rangle$:

$$\hat{J}_\pm | \alpha, \beta \rangle = C_{\alpha\beta}^\pm | \alpha, \beta \pm 1 \rangle. \tag{5.35}$$

We will determine the constant $C_{\alpha\beta}^\pm$ later on.

Note that, for a given eigenvalue α of $\hat{J}^2$, there exists an *upper limit* for the quantum number β. This is due to the fact that the operator $\hat{J}^2 - \hat{J}_z^2$ is positive, since the matrix elements of $\hat{J}^2 - \hat{J}_z^2 = \hat{J}_x^2 + \hat{J}_y^2$ are ≥ 0; we can therefore write

$$\langle \alpha, \beta | \hat{J}^2 - \hat{J}_z^2 | \alpha, \beta \rangle = \hbar^2(\alpha - \beta^2) \geq 0, \quad \Longrightarrow \quad \alpha \geq \beta^2. \tag{5.36}$$

Since β has an upper limit β_{max}, there must exist a state $| \alpha, \beta_{max} \rangle$ which cannot be raised further:

$$\hat{J}_+ | \alpha, \beta_{max} \rangle = 0. \tag{5.37}$$

Using this relation along with $\hat{J}_-\hat{J}_+ = \hat{J}^2 - \hat{J}_z^2 - \hbar\hat{J}_z$, we see that $\hat{J}_-\hat{J}_+ | \alpha, \beta_{max} \rangle = 0$ or

$$(\hat{J}^2 - \hat{J}_z^2 - \hbar\hat{J}_z) | \alpha, \beta_{max} \rangle = \hbar^2(\alpha - \beta_{max}^2 - \beta_{max}) | \alpha, \beta_{max} \rangle; \tag{5.38}$$

hence

$$\alpha = \beta_{max}(\beta_{max} + 1).\tag{5.39}$$

After n successive applications of $\hat{J}_-$ on $|\,\alpha,\ \beta_{max}\rangle$, we must be able to reach a state $|\,\alpha,\ \beta_{min}\rangle$ which cannot be lowered further:

$$\hat{J}_-\,|\,\alpha,\ \beta_{min}\rangle = 0.\tag{5.40}$$

Using $\hat{J}_+\hat{J}_- = \hat{\vec{J}}^2 - \hat{J}_z^2 + \hbar\hat{J}_z$, and by analogy with (5.38) and (5.39), we infer that

$$\alpha = \beta_{min}(\beta_{min} - 1).\tag{5.41}$$

Comparing (5.39) and (5.41) we obtain

$$\beta_{max} = -\beta_{min}.\tag{5.42}$$

Since β_{min} was reached by n applications of $\hat{J}_-$ on $|\,\alpha,\ \beta_{max}\rangle$, it follows that

$$\beta_{max} = \beta_{min} + n,\tag{5.43}$$

and since $\beta_{min} = -\beta_{max}$ we conclude that

$$\beta_{max} = \frac{n}{2}.\tag{5.44}$$

Hence β_{max} can be integer or half-odd-integer, depending on n being even or odd.

It is now appropriate to introduce the notation j and m to denote β_{max} and β, respectively:

$$j = \beta_{max} = \frac{n}{2}, \qquad m = \beta;\tag{5.45}$$

hence the eigenvalue of $\hat{\vec{J}}^2$ is given by

$$\alpha = j(j + 1).\tag{5.46}$$

Now since $\beta_{min} = -\beta_{max}$, and with n positive, we infer that the allowed values of m lie between $-j$ and $+j$:

$$\boxed{-j \le m \le j.}\tag{5.47}$$

The results obtained thus far can be summarized as follows: the eigenvalues of $\hat{\vec{J}}^2$ and J_z corresponding to the joint eigenvectors $|\,j,\ m\rangle$ are given, respectively, by $\hbar^2 j(j + 1)$ and $\hbar m$:

$$\boxed{\hat{\vec{J}}^2\,|\,j,\ m\rangle = \hbar^2 j(j + 1)\,|\,j,\ m\rangle \quad \text{and} \quad \hat{J}_z\,|\,j,\ m\rangle = \hbar m\,|\,j,\ m\rangle,}\tag{5.48}$$

where $j = 0, 1/2, 1, 3/2, \ldots$ and $m = -j, -(j-1), \ldots, j-1, j$. So for each j there are $2j+1$ values of m. For example, if $j = 1$ then m takes the three values $-1, 0, 1$; and if $j = 5/2$ then m takes the six values $-5/2, -3/2, -1/2, 1/2, 3/2, 5/2$. The values of j are either integer or half-integer. We see that the spectra of the angular momentum operators $\hat{\vec{J}}^2$ and $\hat{J}_z$ are discrete. Since the eigenstates corresponding to different angular momenta are orthogonal, and since the angular momentum spectra are discrete, the orthonormality condition is

$$\boxed{\langle j',\ m' \,|\, j,\ m\rangle = \delta_{j',j}\delta_{m',m}.}\tag{5.49}$$

Let us now determine the eigenvalues of $\hat{J}_\pm$ within the $\{|\, j,\ m\rangle\}$ basis; $|\, j,\ m\rangle$ is not an eigenstate of $\hat{J}_\pm$. We can rewrite equation (5.35) as

$$\hat{J}_\pm \,|\, j,\ m\rangle = C^{\pm}_{jm} \,|\, j,\ m\pm 1\rangle. \tag{5.50}$$

We are going to derive C^+_{jm} and then infer C^-_{jm}. Since $|\, j,\ m\rangle$ is normalized, we can use (5.50) to obtain the following two expressions:

$$(\hat{J}_+ \,|\, j,\ m\rangle)^\dagger \,(\hat{J}_+ \,|\, j,\ m\rangle) = |C^+_{jm}|^2 \langle j,\ m+1 \,|\, j,\ m+1\rangle = |C^+_{jm}|^2, \tag{5.51}$$

$$\left|C^+_{jm}\right|^2 = \langle j,\ m |\hat{J}_- \hat{J}_+ \,|\, j,\ m\rangle. \tag{5.52}$$

But since $\hat{J}_- \hat{J}_+$ is equal to $(\hat{J}^2 - \hat{J}^2_z - \hbar \hat{J}_z)$, and assuming the arbitrary phase of C^+_{jm} to be zero, we conclude that

$$C^+_{jm} = \sqrt{\langle j,\ m |\hat{J}^2 - \hat{J}^2_z - \hbar \hat{J}_z \,|\, j,\ m\rangle} = \hbar\sqrt{j(j+1) - m(m+1)}. \tag{5.53}$$

By analogy with C^+_{jm} we can easily infer the expression for C^-_{jm}:

$$C^-_{jm} = \hbar\sqrt{j(j+1) - m(m-1)}. \tag{5.54}$$

Thus, the eigenvalue equations for $\hat{J}_+$ and $\hat{J}_-$ are given by

$$\boxed{\hat{J}_\pm \,|\, j,\ m\rangle = \hbar\sqrt{j(j+1) - m(m\pm 1)} \,\,|\, j,\ m\pm 1\rangle} \tag{5.55}$$

or

$$\boxed{\hat{J}_\pm \,|\, j,\ m\rangle = \hbar\sqrt{(j\mp m)(j\pm m+1)} \,\,|\, j,\ m\pm 1\rangle,} \tag{5.56}$$

which in turn leads to the two relations:

$$\hat{J}_x \,|\, j,\ m\rangle = \frac{1}{2}(\hat{J}_+ + \hat{J}_-) \,|\, j,\ m\rangle$$
$$= \frac{\hbar}{2}\left[\sqrt{(j-m)(j+m+1)} \,\,|\, j,\ m+1\rangle + \sqrt{(j+m)(j-m+1)} \,\,|\, j,\ m-1\rangle\right], \tag{5.57}$$

$$\hat{J}_y \,|\, j,\ m\rangle = \frac{1}{2i}(\hat{J}_+ - \hat{J}_-) \,|\, j,\ m\rangle$$
$$= \frac{\hbar}{2i}\left[\sqrt{(j-m)(j+m+1)} \,\,|\, j,\ m+1\rangle - \sqrt{(j+m)(j-m+1)} \,\,|\, j,\ m-1\rangle\right]. \tag{5.58}$$

The expectation values of $\hat{J}_x$ and $\hat{J}_y$ are therefore zero:

$$\langle j,\ m \,|\, \hat{J}_x \,|\, j,\ m\rangle = \langle j,\ m \,|\, \hat{J}_y \,|\, j,\ m\rangle = 0 \tag{5.59}$$

We will show later in (5.208) that the expectation values $\langle j,\ m \,|\, \hat{J}^2_x \,|\, j,\ m\rangle$ and $\langle j,\ m \,|\, \hat{J}^2_y \,|\, j,\ m\rangle$ are equal and given by

$$\boxed{\langle \hat{J}^2_x\rangle = \langle \hat{J}^2_y\rangle = \frac{1}{2}\left[\langle j,\ m \,|\, \hat{J}^2 \,|\, j,\ m\rangle - \langle j,\ m \,|\, \hat{J}^2_z \,|\, j,\ m\rangle\right] = \frac{\hbar^2}{2}\left[j(j+1) - m^2\right].} \tag{5.60}$$

Example 5.2

Calculate $[\hat{J}_x^2, \hat{J}_y]$, $[\hat{J}_z^2, \hat{J}_y]$, and $[\hat{\vec{J}}^2, \hat{J}_y]$; then show $\langle j, m \mid \hat{J}_x^2 \mid j, m \rangle = \langle j, m \mid \hat{J}_y^2 \mid j, m \rangle$.

Solution

Since $[\hat{J}_x, \hat{J}_y] = i\hbar \hat{J}_z$ and $[\hat{J}_z, \hat{J}_x] = i\hbar \hat{J}_y$, we have

$$[\hat{J}_x^2, \hat{J}_y] = \hat{J}_x[\hat{J}_x, \hat{J}_y] + [\hat{J}_x, \hat{J}_y]\hat{J}_x = i\hbar(\hat{J}_x\hat{J}_z + \hat{J}_z\hat{J}_x) = i\hbar(2\hat{J}_x\hat{J}_z + i\hbar\hat{J}_y). \tag{5.61}$$

Similarly, since $[\hat{J}_z, \hat{J}_y] = -i\hbar \hat{J}_x$ and $[\hat{J}_z, \hat{J}_x] = i\hbar \hat{J}_y$, we have

$$[\hat{J}_z^2, \hat{J}_y] = \hat{J}_z[\hat{J}_z, \hat{J}_y] + [\hat{J}_z, \hat{J}_y]\hat{J}_z = -i\hbar(\hat{J}_z\hat{J}_x + \hat{J}_x\hat{J}_z) = -i\hbar(2\hat{J}_x\hat{J}_z + i\hbar\hat{J}_y). \tag{5.62}$$

The previous two expressions yield

$$[\hat{\vec{J}}^2, \hat{J}_y] = [\hat{J}_x^2 + \hat{J}_y^2 + \hat{J}_z^2, \hat{J}_y] = [\hat{J}_x^2, \hat{J}_y] + [\hat{J}_z^2, \hat{J}_y]$$
$$= i\hbar(2\hat{J}_x\hat{J}_z + i\hbar\hat{J}_y) - i\hbar(2\hat{J}_x\hat{J}_z + i\hbar\hat{J}_y) = 0. \tag{5.63}$$

Since we have

$$\hat{J}_x^2 = \frac{1}{4}(\hat{J}_+^2 + \hat{J}_+\hat{J}_- + \hat{J}_-\hat{J}_+ + J_-^2), \qquad \hat{J}_y^2 = -\frac{1}{4}(\hat{J}_+^2 - \hat{J}_+\hat{J}_- - \hat{J}_-\hat{J}_+ + J_-^2), \tag{5.64}$$

and since $\langle j, m \mid \hat{J}_+^2 \mid j, m \rangle = \langle j, m \mid \hat{J}_-^2 \mid j, m \rangle = 0$, we can write

$$\langle j, m \mid \hat{J}_x^2 \mid j, m \rangle = \frac{1}{4}\langle j, m \mid \hat{J}_+\hat{J}_- + \hat{J}_-\hat{J}_+ \mid j, m \rangle = \langle j, m \mid \hat{J}_y^2 \mid j, m \rangle. \tag{5.65}$$

5.4 Matrix Representation of Angular Momentum

The formalism of the previous section is general and independent of any particular representation. There are many ways to represent the angular momentum operators and their eigenstates. In this section we are going to discuss the matrix representation of angular momentum where eigenkets and operators will be represented by column vectors and square matrices, respectively. This is achieved by expanding states and operators in a discrete basis. We will see later how to represent the orbital angular momentum in the position representation.

Since $\hat{\vec{J}}^2$ and $\hat{J}_z$ commute, the set of their common eigenstates $\{\mid j, m \rangle\}$ can be chosen as a basis; this basis is discrete, orthonormal, and complete. For a given value of j, the orthonormalization condition for this base is given by (5.49), and the completeness condition is expressed by

$$\sum_{m=-j}^{+j} \mid j, m \rangle \langle j, m \mid = \hat{I}, \tag{5.66}$$

where $\hat{I}$ is the unit matrix. The operators $\hat{\vec{J}}^2$ and $\hat{J}_z$ are diagonal in the basis given by their joint eigenstates

$$\langle j', m' \mid \hat{\vec{J}}^2 \mid j, m \rangle = \hbar^2 j(j+1)\delta_{j', j}\delta_{m', m}, \tag{5.67}$$

$$\langle j', m' \mid \hat{J}_z \mid j, m \rangle = \hbar m \delta_{j', j}\delta_{m', m}. \tag{5.68}$$

Thus, the matrices representing $\hat{J}^2$ and $\hat{J}_z$ in the $\{| j, m\rangle\}$ eigenbasis are diagonal, their diagonal elements being equal to $\hbar^2 j(j+1)$ and $\hbar m$, respectively.

Now since the operators $\hat{J}_\pm$ do not commute with $\hat{J}_z$, they are represented in the $\{| j, m\rangle\}$ basis by matrices that are not diagonal:

$$\boxed{\langle j', m' | \hat{J}_\pm | j, m\rangle = \hbar\sqrt{j(j+1) - m(m \pm 1)}\, \delta_{j', j}\delta_{m', m\pm 1}.} \qquad (5.69)$$

We can infer the matrices of $\hat{J}_x$ and $\hat{J}_y$ from (5.57) and (5.58):

$$\langle j', m' | \hat{J}_x | j, m\rangle = \frac{\hbar}{2}\left[\sqrt{j(j+1) - m(m+1)}\delta_{m', m+1}\right.$$
$$\left. + \sqrt{j(j+1) - m(m-1)}\delta_{m', m-1}\right]\delta_{j', j}, \qquad (5.70)$$

$$\langle j', m' | \hat{J}_y | j, m\rangle = \frac{\hbar}{2i}\left[\sqrt{j(j+1) - m(m+1)}\delta_{m', m+1}\right.$$
$$\left. - \sqrt{j(j+1) - m(m-1)}\delta_{m', m-1}\right]\delta_{j', j}. \qquad (5.71)$$

Example 5.3 (Angular momentum $j = 1$)

Consider the case where $j = 1$.

(a) Find the matrices representing the operators $\hat{J}^2$, $\hat{J}_z$, $\hat{J}_\pm$, $\hat{J}_x$, and $\hat{J}_y$.

(b) Find the joint eigenstates of $\hat{J}^2$ and $\hat{J}_z$ and verify that they form an orthonormal and complete basis.

(c) Use the matrices of $\hat{J}_x$, $\hat{J}_y$ and $\hat{J}_z$ to calculate $[\hat{J}_x, \hat{J}_y]$, $[\hat{J}_y, \hat{J}_z]$, and $[\hat{J}_z, \hat{J}_x]$.

(d) Verify that $\hat{J}_z^3 = \hbar^2 \hat{J}_z$ and $\hat{J}_\pm^3 = 0$.

Solution

(a) For $j = 1$ the allowed values of m are $-1, 0, 1$. The joint eigenstates of $\hat{J}^2$ and $\hat{J}_z$ are $| 1, -1\rangle$, $| 1, 0\rangle$, and $| 1, 1\rangle$. The matrix representations of the operators $\hat{J}^2$ and $\hat{J}_z$ can be inferred from (5.67) and (5.68):

$$\hat{J}^2 = \begin{pmatrix} \langle 1, 1 | \hat{J}^2 | 1, 1\rangle & \langle 1, 1 | \hat{J}^2 | 1, 0\rangle & \langle 1, 1 | \hat{J}^2 | 1, -1\rangle \\ \langle 1, 0 | \hat{J}^2 | 1, 1\rangle & \langle 1, 0 | \hat{J}^2 | 1, 0\rangle & \langle 1, 0 | \hat{J}^2 | 1, -1\rangle \\ \langle 1, -1 | \hat{J}^2 | 1, 1\rangle & \langle 1, -1 | \hat{J}^2 | 1, 0\rangle & \langle 1, -1 | \hat{J}^2 | 1, -1\rangle \end{pmatrix}$$

$$= 2\hbar^2 \begin{pmatrix} 1 & 0 & 0 \\ 0 & 1 & 0 \\ 0 & 0 & 1 \end{pmatrix}, \qquad (5.72)$$

$$\hat{J}_z = \hbar \begin{pmatrix} 1 & 0 & 0 \\ 0 & 0 & 0 \\ 0 & 0 & -1 \end{pmatrix}. \qquad (5.73)$$

Similarly, using (5.69), we can ascertain that the matrices of $\hat{J}_+$ and $\hat{J}_-$ are given by

$$\hat{J}_- = \hbar\sqrt{2} \begin{pmatrix} 0 & 0 & 0 \\ 1 & 0 & 0 \\ 0 & 1 & 0 \end{pmatrix}, \qquad \hat{J}_+ = \hbar\sqrt{2} \begin{pmatrix} 0 & 1 & 0 \\ 0 & 0 & 1 \\ 0 & 0 & 0 \end{pmatrix}. \qquad (5.74)$$

The matrices for $\hat{J}_x$ and $\hat{J}_y$ in the $\{| j, m\rangle\}$ basis result immediately from the relations $\hat{J}_x = (\hat{J}_+ + \hat{J}_-)/2$ and $\hat{J}_y = i(\hat{J}_- - \hat{J}_+)/2$:

$$\hat{J}_x = \frac{\hbar}{\sqrt{2}} \begin{pmatrix} 0 & 1 & 0 \\ 1 & 0 & 1 \\ 0 & 1 & 0 \end{pmatrix}, \qquad \hat{J}_y = \frac{\hbar}{\sqrt{2}} \begin{pmatrix} 0 & -i & 0 \\ i & 0 & -i \\ 0 & i & 0 \end{pmatrix}. \tag{5.75}$$

(b) The joint eigenvectors of $\hat{J}^2$ and $\hat{J}_z$ can be obtained as follows. The matrix equation of $\hat{J}_z | j, m\rangle = m\hbar | j, m\rangle$ is

$$\hbar \begin{pmatrix} 1 & 0 & 0 \\ 0 & 0 & 0 \\ 0 & 0 & -1 \end{pmatrix} \begin{pmatrix} a \\ b \\ c \end{pmatrix} = m\hbar \begin{pmatrix} a \\ b \\ c \end{pmatrix} \implies \begin{array}{l} \hbar a = m\hbar a \\ 0 = m\hbar b \\ -\hbar c = m\hbar c \end{array}. \tag{5.76}$$

The normalized solutions to these equations for $m = 1, 0, -1$ are respectively given by $a = 1$, $b = c = 0$; $a = 0, b = 1, c = 0$; and $a = b = 0, c = 1$; that is,

$$| 1, 1\rangle = \begin{pmatrix} 1 \\ 0 \\ 0 \end{pmatrix}, \qquad | 1, 0\rangle = \begin{pmatrix} 0 \\ 1 \\ 0 \end{pmatrix}, \qquad | 1, -1\rangle = \begin{pmatrix} 0 \\ 0 \\ 1 \end{pmatrix}. \tag{5.77}$$

We can verify that these vectors are orthonormal:

$$\langle 1, m' | 1, m\rangle = \delta_{m', m} \qquad (m', m = -1, 0, 1). \tag{5.78}$$

We can also verify that they are complete:

$$\sum_{m=-1}^{1} | 1, m\rangle\langle 1, m| = \begin{pmatrix} 0 \\ 0 \\ 1 \end{pmatrix} (0\ 0\ 1) + \begin{pmatrix} 0 \\ 1 \\ 0 \end{pmatrix} (0\ 1\ 0) + \begin{pmatrix} 1 \\ 0 \\ 0 \end{pmatrix} (1\ 0\ 0)$$

$$= \begin{pmatrix} 1 & 0 & 0 \\ 0 & 1 & 0 \\ 0 & 0 & 1 \end{pmatrix}. \tag{5.79}$$

(c) Using the matrices (5.75) we have

$$\hat{J}_x \hat{J}_y = \frac{\hbar^2}{2} \begin{pmatrix} 0 & 1 & 0 \\ 1 & 0 & 1 \\ 0 & 1 & 0 \end{pmatrix} \begin{pmatrix} 0 & -i & 0 \\ i & 0 & -i \\ 0 & i & 0 \end{pmatrix} = \frac{\hbar^2}{2} \begin{pmatrix} i & 0 & -i \\ 0 & 0 & 0 \\ i & 0 & -i \end{pmatrix}, \tag{5.80}$$

$$\hat{J}_y \hat{J}_x = \frac{\hbar^2}{2} \begin{pmatrix} 0 & -i & 0 \\ i & 0 & -i \\ 0 & i & 0 \end{pmatrix} \begin{pmatrix} 0 & 1 & 0 \\ 1 & 0 & 1 \\ 0 & 1 & 0 \end{pmatrix} = \frac{\hbar^2}{2} \begin{pmatrix} -i & 0 & -i \\ 0 & 0 & 0 \\ i & 0 & i \end{pmatrix}; \tag{5.81}$$

hence

$$\hat{J}_x \hat{J}_y - \hat{J}_y \hat{J}_x = \frac{\hbar^2}{2} \begin{pmatrix} 2i & 0 & 0 \\ 0 & 0 & 0 \\ 0 & 0 & -2i \end{pmatrix} = i\hbar^2 \begin{pmatrix} 1 & 0 & 0 \\ 0 & 0 & 0 \\ 0 & 0 & -1 \end{pmatrix} = i\hbar \hat{J}_z, \tag{5.82}$$

where the matrix of $\hat{J}_z$ is given by (5.73). A similar calculation leads to $[\hat{J}_y, \hat{J}_z] = i\hbar \hat{J}_x$ and $[\hat{J}_z, \hat{J}_x] = i\hbar \hat{J}_y$.

Figure 5.1 Geometrical representation of the angular momentum $\hat{\vec{J}}$: the vector $\hat{\vec{J}}$ rotates along the surface of a cone about its axis; the cone's height is equal to $m\hbar$, the projection of $\hat{\vec{J}}$ on the cone's axis. The tip of $\hat{\vec{J}}$ lies, within the $J_z J_{xy}$ plane, on a circle of radius $\hbar\sqrt{j(j+1)}$.

(d) The calculation of $\hat{J}_z^3$ and $\hat{J}_\pm^3$ is straightforward:

$$\hat{J}_z^3 = \hbar^3 \left[\begin{pmatrix} 1 & 0 & 0 \\ 0 & 0 & 0 \\ 0 & 0 & -1 \end{pmatrix} \right]^3 = \hbar^3 \begin{pmatrix} 1 & 0 & 0 \\ 0 & 0 & 0 \\ 0 & 0 & -1 \end{pmatrix} = \hbar^2 \hat{J}_z, \tag{5.83}$$

$$\hat{J}_+^3 = 2\hbar^3\sqrt{2} \left[\begin{pmatrix} 0 & 1 & 0 \\ 0 & 0 & 1 \\ 0 & 0 & 0 \end{pmatrix} \right]^3 = 2\hbar^3\sqrt{2} \begin{pmatrix} 0 & 0 & 0 \\ 0 & 0 & 0 \\ 0 & 0 & 0 \end{pmatrix} = 0, \tag{5.84}$$

and

$$\hat{J}_-^3 = 2\hbar^3\sqrt{2} \left[\begin{pmatrix} 0 & 0 & 0 \\ 1 & 0 & 0 \\ 0 & 1 & 0 \end{pmatrix} \right]^3 = 2\hbar^3\sqrt{2} \begin{pmatrix} 0 & 0 & 0 \\ 0 & 0 & 0 \\ 0 & 0 & 0 \end{pmatrix} = 0. \tag{5.85}$$

5.5 Geometrical Representation of Angular Momentum

At issue here is the relationship between the angular momentum and its z-component; this relation can be represented geometrically as follows. For a fixed value of j, the total angular momentum $\hat{\vec{J}}$ may be represented by a vector whose length, as displayed in Figure 5.1, is given by $\sqrt{\langle \hat{\vec{J}}^2 \rangle} = \hbar\sqrt{j(j+1)}$ and whose z-component is $\langle \hat{J}_z \rangle = \hbar m$. Since $\hat{J}_x$ and $\hat{J}_y$ are separately undefined, only their sum $\hat{J}_x^2 + \hat{J}_y^2 = \hat{J}^2 - \hat{J}_z^2$, which lies within the xy plane, is well defined.

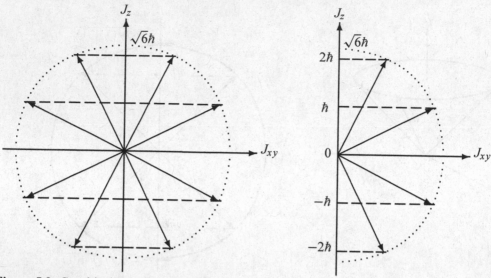

Figure 5.2 Graphical representation of the angular momentum $j = 2$ for the state $| 2, m \rangle$ with $m = -2, -1, 0, 1, 2$. The radius of the circle is $\hbar\sqrt{2(2+1)} = \sqrt{6}\hbar$.

In classical terms, we can think of $\vec{J}$ as representable graphically by a vector, whose endpoint lies on a circle of radius $\hbar\sqrt{j(j+1)}$, rotating along the surface of a cone of half-angle

$$\theta = \cos^{-1}\left(\frac{m}{\sqrt{j(j+1)}}\right),$$
(5.86)

such that its projection along the z-axis is always $m\hbar$. Notice that, as the values of the quantum number m are limited to $m = -j, -j + 1, \ldots, j - 1, j$, the angle θ is quantized; the only possible values of θ consist of a discrete set of $2j + 1$ values:

$$\theta = \cos^{-1}\left(\frac{-j}{\sqrt{j(j+1)}}\right), \cos^{-1}\left(\frac{-j+1}{\sqrt{j(j+1)}}\right), \ldots, \cos^{-1}\left(\frac{j-1}{\sqrt{j(j+1)}}\right),$$

$$\cos^{-1}\left(\frac{j}{\sqrt{j(j+1)}}\right).$$
(5.87)

Since all orientations of $\vec{\hat{J}}$ on the surface of the cone are equally likely, the projection of $\vec{\hat{J}}$ on both the x and y axes average out to zero:

$$\langle \hat{J}_x \rangle = \langle \hat{J}_y \rangle = 0,$$
(5.88)

where $\langle \hat{J}_x \rangle$ stands for $\langle j, m | \hat{J}_x | j, m \rangle$.

As an example, Figure 5.2 shows the graphical representation for the $j = 2$ case. As specified in (5.87), θ takes only a discrete set of values. In this case where $j = 2$, the angle θ takes only five values corresponding respectively to $m = -2, -1, 0, 1, 2$; they are given by

$$\theta = -35.26°, \ -65.91°, \ 90°, \ 65.91°, \ 35.26°.$$
(5.89)

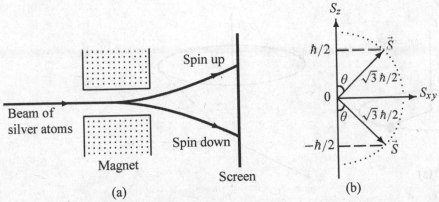

Figure 5.3 (a) Stern–Gerlach experiment: when a beam of silver atoms passes through an inhomogeneous magnetic field, it splits into two distinct components corresponding to spin-up and spin-down. (b) Graphical representation of spin $\frac{1}{2}$: the tip of $\vec{S}$ lies on a circle of radius $|\vec{S}| = \sqrt{3}\hbar/2$ so that its projection on the z-axis takes only two values, $\pm\hbar/2$, with $\theta = 54.73°$.

5.6 Spin Angular Momentum

5.6.1 Experimental Evidence of the Spin

The existence of spin was confirmed experimentally by Stern and Gerlach in 1922 using silver (Ag) atoms. Silver has 47 electrons; 46 of them form a spherically symmetric charge distribution and the 47th electron occupies a 5s orbital. If the silver atom were in its ground state, its total orbital angular momentum would be zero: $l = 0$ (since the fifth shell electron would be in a 5s state). In the Stern–Gerlach experiment, a beam of silver atoms passes through an inhomogeneous (nonuniform) magnetic field. If, for argument's sake, the field were along the z-direction, we would expect classically to see on the screen a *continuous* band that is symmetric about the undeflected direction, $z = 0$. According to Schrödinger's wave theory, however, if the atoms had an orbital angular momentum l, we would expect the beam to split into an *odd* (*discrete*) number of $2l + 1$ components. Suppose the beam's atoms were in their ground state $l = 0$, there would be only one spot on the screen, and if the fifth shell electron were in a 5p state ($l = 1$), we would expect to see three spots. Experimentally, however, the beam behaves according to the predictions of neither classical physics nor Schrödinger's wave theory. Instead, *it splits into two distinct components* as shown in Figure 5.3a. This result was also observed for hydrogen atoms in their ground state ($l = 0$), where no splitting is expected.

To solve this puzzle, Goudsmit and Uhlenbeck postulated in 1925 that, in addition to its orbital angular momentum, the electron possesses an *intrinsic* angular momentum which, unlike the orbital angular momentum, has nothing to do with the spatial degrees of freedom. By analogy with the motion of the Earth, which consists of an orbital motion around the Sun and an internal rotational or *spinning* motion about its axis, the electron or, for that matter, any other microscopic particle may also be considered to have some sort of internal or intrinsic spinning motion. This intrinsic degree of freedom was given the suggestive name of *spin* angular momentum. One has to keep in mind, however, that the electron remains thus far a structureless or pointlike particle; hence caution has to be exercised when trying to link the electron's spin to an internal spinning motion. The spin angular momentum of a particle does not depend on

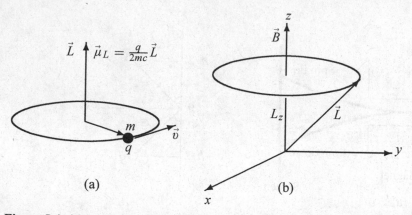

Figure 5.4 (a) Orbital magnetic dipole moment of a positive charge q. (b) When an external magnetic field is applied, the orbital magnetic moment precesses about it.

its spatial degrees of freedom. The spin, an intrinsic degree of freedom, is a purely quantum mechanical concept with no classical analog. Unlike the orbital angular momentum, *the spin cannot be described by a differential operator.*

From the classical theory of electromagnetism, an *orbital magnetic dipole moment* is generated with the orbital motion of a particle of charge q:

$$\vec{\mu}_L = \frac{q}{2mc}\vec{L}, \tag{5.90}$$

where $\vec{L}$ is the orbital angular momentum of the particle, m is its mass, and c is the speed of light. As shown in Figure 5.4a, if the charge q is positive, $\vec{\mu}_L$ and $\vec{L}$ will be in the same direction; for a negative charge such as an electron ($q = -e$), the magnetic dipole moment $\vec{\mu}_L = -e\vec{L}/(2m_e c)$ and the orbital angular momentum will be in opposite directions. Similarly, if we follow a classical analysis and picture the electron as a spinning spherical charge, then we obtain an intrinsic or *spin magnetic dipole moment* $\vec{\mu}_S = -e\vec{S}/(2m_e c)$. This classical derivation of $\vec{\mu}_S$ is quite erroneous, since the electron cannot be viewed as a spinning sphere; in fact, it turns out that the electron's spin magnetic moment is twice its classical expression. Although the spin magnetic moment cannot be derived classically, as we did for the orbital magnetic moment, it can still be postulated by analogy with (5.90):

$$\vec{\mu}_S = -g_s \frac{e}{2m_e c}\vec{S}, \tag{5.91}$$

where g_s is called the Landé factor or the gyromagnetic ratio of the electron; its experimental value is $g_s \simeq 2$ (this factor can be calculated using Dirac's relativistic theory of the electron).

When the electron is placed in a magnetic field $\vec{B}$ and if the field is inhomogeneous, a force will be exerted on the electron's intrinsic dipole moment; the direction and magnitude of the force depend on the relative orientation of the field and the dipole. This force tends to align $\vec{\mu}_S$ along $\vec{B}$, producing a precessional motion of $\vec{\mu}_S$ around $\vec{B}$ (Figure 5.4b). For instance, if $\vec{\mu}_S$ is parallel to $\vec{B}$, the electron will move in the direction in which the field increases; conversely, if $\vec{\mu}_S$ is antiparallel to $\vec{B}$, the electron will move in the direction in which the field decreases. For hydrogen-like atoms (such as silver) that are in the ground state, the orbital angular momentum will be zero; hence the dipole moment of the atom will be entirely due to the spin of the electron.

The atomic beam will therefore deflect according to the orientation of the electron's spin. Since, experimentally, the beam splits into two components, the electron's spin must have only two possible orientations relative to the magnetic field, either parallel or antiparallel.

By analogy with the orbital angular momentum of a particle, which is characterized by two quantum numbers—the orbital number l and the azimuthal number m_l (with $m_l = -l, -l + 1,$ $\ldots, l - 1, l$)—the spin angular momentum is also characterized by two quantum numbers, the spin s and its projection m_s on the z-axis (the direction of the magnetic field), where $m_s = -s,$ $-s+1, \ldots, s-1, s$. Since only two components were observed in the Stern–Gerlach experiment, we must have $2s + 1 = 2$. The quantum numbers for the electron must then be given by $s = \frac{1}{2}$ and $m_s = \pm \frac{1}{2}$.

In nature it turns out that every fundamental particle has a specific spin. Some particles have integer spins $s = 0, 1, 2, \ldots$ (the pi mesons have spin $s = 0$, the photons have spin $s = 1$, and so on) and others have half-odd-integer spins $s = \frac{1}{2}, \frac{3}{2}, \frac{5}{2}, \ldots$ (the electrons, protons, and neutrons have spin $s = \frac{1}{2}$, the deltas have spin $s = \frac{3}{2}$, and so on). We will see in Chapter 8 that particles with *half-odd-integer* spins are called *fermions* (quarks, electrons, protons, neutrons, etc.) and those with *integer* spins are called *bosons* (pions, photons, gravitons, etc.).

Besides confirming the existence of spin and measuring it, the Stern–Gerlach experiment offers a number of other important uses to quantum mechanics. First, by showing that a beam splits into a *discrete* set of components rather than a continuous band, it provides additional confirmation for the quantum hypothesis on the discrete character of the microphysical world. The Stern–Gerlach experiment also turns out to be an invaluable technique for preparing a quantum state. Suppose we want to prepare a beam of spin-up atoms; we simply pass an unpolarized beam through an inhomogeneous magnet, then collect the desired component and discard (or block) the other. The Stern–Gerlach experiment can also be used to determine the total angular momentum of an atom which, in the case where $l \neq 0$, is given by the sum of the orbital and spin angular momenta: $\vec{J} = \vec{L} + \vec{S}$. The addition of angular momenta is covered in Chapter 7.

5.6.2 General Theory of Spin

The theory of spin is identical to the general theory of angular momentum (Section 5.3). By analogy with the vector angular momentum $\vec{\hat{J}}$, the spin is also represented by a vector operator $\vec{\hat{S}}$ whose components $\hat{S}_x$, $\hat{S}_y$, $\hat{S}_z$ obey the same commutation relations as $\hat{J}_x$, $\hat{J}_y$, $\hat{J}_z$:

$$[\hat{S}_x, \hat{S}_y] = i\hbar \hat{S}_z, \qquad [\hat{S}_y, \hat{S}_z] = i\hbar \hat{S}_x, \qquad [\hat{S}_z, \hat{S}_x] = i\hbar \hat{S}_y. \tag{5.92}$$

In addition, $\vec{\hat{S}}^2$ and $\hat{S}_z$ commute; hence they have common eigenvectors:

$$\vec{\hat{S}}^2 \mid s, m_s \rangle = \hbar^2 s(s + 1) \mid s, m_s \rangle, \qquad \hat{S}_z \mid s, m_s \rangle = \hbar m_s \mid s, m_s \rangle, \tag{5.93}$$

where $m_s = -s, -s + 1, \ldots, -s + 1, s$. Similarly, we have

$$\hat{S}_\pm \mid s, m_s \rangle = \hbar \sqrt{s(s + 1) - m_s(m_s \pm 1)} \mid s, m_s \pm 1 \rangle, \tag{5.94}$$

where $\hat{S}_\pm = \hat{S}_x \pm i\hat{S}_y$, and

$$\langle \hat{S}_x^2 \rangle = \langle \hat{S}_y^2 \rangle = \frac{1}{2}(\langle \vec{\hat{S}}^2 \rangle - \langle \hat{S}_z^2 \rangle) = \frac{\hbar^2}{2} \left[s(s + 1) - m_s^2 \right], \tag{5.95}$$

where $\langle \hat{A} \rangle$ denotes $\langle \hat{A} \rangle = \langle s, m_s | \hat{A} | s, m_s \rangle$.

The spin states form an orthonormal and complete basis

$$\langle s', m_s' \mid s, m_s \rangle = \delta_{s', s} \delta_{m_s', m_s}, \qquad \sum_{m_s=-s}^{s} | s, m_s \rangle \langle s, m_s \mid= I, \qquad (5.96)$$

where I is the unit matrix.

5.6.3 Spin $1/2$ and the Pauli Matrices

For a particle with spin $\frac{1}{2}$ the quantum number m_s takes only two values: $m_s = -\frac{1}{2}$ and $\frac{1}{2}$. The particle can thus be found in either of the following two states: $| s, m_s \rangle = \left| \frac{1}{2}, \frac{1}{2} \right\rangle$ and $\left| \frac{1}{2}, -\frac{1}{2} \right\rangle$. The eigenvalues of $\hat{S}^2$ and $\hat{S}_z$ are given by

$$\hat{S}^2 \left| \frac{1}{2}, \pm\frac{1}{2} \right\rangle = \frac{3}{4}\hbar^2 \left| \frac{1}{2}, \pm\frac{1}{2} \right\rangle, \qquad \hat{S}_z \left| \frac{1}{2}, \pm\frac{1}{2} \right\rangle = \pm\frac{\hbar}{2} \left| \frac{1}{2}, \pm\frac{1}{2} \right\rangle. \qquad (5.97)$$

Hence the spin may be represented graphically, as shown in Figure 5.3b, by a vector of length $|\vec{S}| = \sqrt{3}\hbar/2$, whose endpoint lies on a circle of radius $\sqrt{3}\hbar/2$, rotating along the surface of a cone with half-angle

$$\theta = \cos^{-1}\left(\frac{|m_s|}{\sqrt{s(s+1)}} \right) = \cos^{-1}\left(\frac{\hbar/2}{\sqrt{3}\hbar/2} \right) = \cos^{-1}\left(\frac{1}{\sqrt{3}} \right) = 54.73°. \qquad (5.98)$$

The projection of $\hat{\vec{S}}$ on the z-axis is restricted to two values only: $\pm\hbar/2$ corresponding to spin-up and spin-down.

Let us now study the matrix representation of the spin $s = \frac{1}{2}$. Using (5.67) and (5.68) we can represent the operators $\hat{S}^2$ and $\hat{S}_z$ within the $\{| s, m_s \rangle\}$ basis by the following matrices:

$$\hat{S}^2 = \begin{pmatrix} \langle \frac{1}{2}, \frac{1}{2} | \hat{S}^2 | \frac{1}{2}, \frac{1}{2} \rangle & \langle \frac{1}{2}, \frac{1}{2} | \hat{S}^2 | \frac{1}{2}, -\frac{1}{2} \rangle \\ \langle \frac{1}{2}, -\frac{1}{2} | \hat{S}^2 | \frac{1}{2}, \frac{1}{2} \rangle & \langle \frac{1}{2}, -\frac{1}{2} | \hat{S}^2 | \frac{1}{2}, -\frac{1}{2} \rangle \end{pmatrix} = \frac{3\hbar^2}{4} \begin{pmatrix} 1 & 0 \\ 0 & 1 \end{pmatrix}, \qquad (5.99)$$

$$\hat{S}_z = \frac{\hbar}{2} \begin{pmatrix} 1 & 0 \\ 0 & -1 \end{pmatrix}. \qquad (5.100)$$

The matrices of $\hat{S}_+$ and $\hat{S}_-$ can be inferred from (5.69):

$$\hat{S}_+ = \hbar \begin{pmatrix} 0 & 1 \\ 0 & 0 \end{pmatrix}, \qquad \hat{S}_- = \hbar \begin{pmatrix} 0 & 0 \\ 1 & 0 \end{pmatrix}, \qquad (5.101)$$

and since $\hat{S}_x = \frac{1}{2}(\hat{S}_+ + \hat{S}_-)$ and $\hat{S}_y = \frac{i}{2}(\hat{S}_- - \hat{S}_+)$, we have

$$\hat{S}_x = \frac{\hbar}{2} \begin{pmatrix} 0 & 1 \\ 1 & 0 \end{pmatrix}, \qquad \hat{S}_y = \frac{\hbar}{2} \begin{pmatrix} 0 & -i \\ i & 0 \end{pmatrix}. \qquad (5.102)$$

The joint eigenvectors of $\hat{S}^2$ and $\hat{S}_z$ are expressed in terms of two-element column matrices, known as *spinors*:

$$\left| \frac{1}{2}, \frac{1}{2} \right\rangle = \begin{pmatrix} 1 \\ 0 \end{pmatrix}, \qquad \left| \frac{1}{2}, -\frac{1}{2} \right\rangle = \begin{pmatrix} 0 \\ 1 \end{pmatrix}. \qquad (5.103)$$

It is easy to verify that these eigenvectors form a basis that is complete,

$$\sum_{m_s=-\frac{1}{2}}^{\frac{1}{2}} \left| \frac{1}{2}, m_s \right\rangle \left\langle \frac{1}{2}, m_s \right| = \begin{pmatrix} 0 \\ 1 \end{pmatrix} (0 \; 1) + \begin{pmatrix} 1 \\ 0 \end{pmatrix} (1 \; 0) = \begin{pmatrix} 1 & 0 \\ 0 & 1 \end{pmatrix}, \qquad (5.104)$$

and orthonormal,

$$\left\langle \frac{1}{2}, \frac{1}{2} \middle| \frac{1}{2}, \frac{1}{2} \right\rangle = (1 \; 0) \begin{pmatrix} 1 \\ 0 \end{pmatrix} = 1, \qquad (5.105)$$

$$\left\langle \frac{1}{2}, -\frac{1}{2} \middle| \frac{1}{2}, -\frac{1}{2} \right\rangle = (0 \; 1) \begin{pmatrix} 0 \\ 1 \end{pmatrix} = 1, \qquad (5.106)$$

$$\left\langle \frac{1}{2}, \frac{1}{2} \middle| \frac{1}{2}, -\frac{1}{2} \right\rangle = \left\langle \frac{1}{2}, -\frac{1}{2} \middle| \frac{1}{2}, \frac{1}{2} \right\rangle = 0. \qquad (5.107)$$

Let us now find the eigenvectors of $\hat{S}_x$ and $\hat{S}_y$. First, note that the basis vectors $|s, m_s\rangle$ are eigenvectors of neither $\hat{S}_x$ nor $\hat{S}_y$; their eigenvectors can, however, be expressed in terms of $|s, m_s\rangle$ as follows:

$$| \psi_x \rangle_\pm = \frac{1}{\sqrt{2}} \left[\left| \frac{1}{2}, \frac{1}{2} \right\rangle \pm \left| \frac{1}{2}, -\frac{1}{2} \right\rangle \right], \qquad (5.108)$$

$$| \psi_y \rangle_\pm = \frac{1}{\sqrt{2}} \left[\left| \frac{1}{2}, \frac{1}{2} \right\rangle \pm i \left| \frac{1}{2}, -\frac{1}{2} \right\rangle \right]. \qquad (5.109)$$

The eigenvalue equations for $\hat{S}_x$ and $\hat{S}_y$ are thus given by

$$\hat{S}_x | \psi_x \rangle_\pm = \pm \frac{\hbar}{2} | \psi_x \rangle_\pm, \qquad \hat{S}_y | \psi_y \rangle_\pm = \pm \frac{\hbar}{2} | \psi_y \rangle_\pm. \qquad (5.110)$$

Pauli matrices

When $s = \frac{1}{2}$ it is convenient to introduce the *Pauli matrices* $\sigma_x, \sigma_y, \sigma_z$, which are related to the spin vector as follows:

$$\hat{\vec{S}} = \frac{\hbar}{2} \vec{\sigma}. \qquad (5.111)$$

Using this relation along with (5.100) and (5.102), we have

$$\sigma_x = \begin{pmatrix} 0 & 1 \\ 1 & 0 \end{pmatrix}, \qquad \sigma_y = \begin{pmatrix} 0 & -i \\ i & 0 \end{pmatrix}, \qquad \sigma_z = \begin{pmatrix} 1 & 0 \\ 0 & -1 \end{pmatrix}. \qquad (5.112)$$

These matrices satisfy the following two properties:

$$\sigma_j^2 = \hat{I} \qquad (j = x, y, z), \qquad (5.113)$$

$$\sigma_j \sigma_k + \sigma_k \sigma_j = 0 \qquad (j \neq k), \qquad (5.114)$$

where the subscripts j and k refer to x, y, z, and $\hat{I}$ is the 2×2 unit matrix. These two equations are equivalent to the anticommutation relation

$$\{\sigma_j, \sigma_k\} = 2\hat{I}\delta_{j,k}. \qquad (5.115)$$

We can verify that the Pauli matrices satisfy the commutation relations

$$[\sigma_j, \sigma_k] = 2i \, \varepsilon_{jkl}\sigma_l, \tag{5.116}$$

where ε_{jkl} is the antisymmetric tensor (also known as the Levi–Civita tensor)

$$\varepsilon_{jkl} = \begin{cases} 1 & \text{if } jkl \text{ is an even permutation of } x, y, z, \\ -1 & \text{if } jkl \text{ is an odd permutation of } x, y, z, \\ 0 & \text{if any two indices among } j, k, l \text{ are equal.} \end{cases} \tag{5.117}$$

We can condense the relations (5.113), (5.114), and (5.116) into

$$\sigma_j\sigma_k = \delta_{j,k} + i \sum_l \varepsilon_{jkl}\sigma_l. \tag{5.118}$$

Using this relation we can verify that, for any two vectors $\vec{A}$ and $\vec{B}$ which commute with $\vec{\sigma}$, we have

$$(\vec{\sigma} \cdot \vec{A})(\vec{\sigma} \cdot \vec{B}) = (\vec{A} \cdot \vec{B})\hat{I} + i\vec{\sigma} \cdot (\vec{A} \times \vec{B}), \tag{5.119}$$

where $\hat{I}$ is the unit matrix. The Pauli matrices are Hermitian, traceless, and have determinants equal to -1:

$$\sigma_j^{\dagger} = \sigma_j, \qquad \text{Tr}(\sigma_j) = 0, \qquad \det(\sigma_j) = -1 \qquad (j = x, \ y, \ z). \tag{5.120}$$

Using the relation $\sigma_x\sigma_y = i\sigma_z$ along with $\sigma_z^2 = \hat{I}$, we obtain

$$\sigma_x\sigma_y\sigma_z = i\hat{I}. \tag{5.121}$$

From the commutation relations (5.116) we can show that

$$e^{i\alpha\sigma_j} = I \cos\alpha + i\sigma_j \sin\alpha \qquad (j = x, y, z), \tag{5.122}$$

where I is the unit matrix and α is an arbitrary real constant.

Remarks

- Since the spin does not depend on the spatial degrees of freedom, the components $\hat{S}_x$, $\hat{S}_y$, $\hat{S}_z$ of the spin operator commute with all the spatial operators, notably the orbital angular momentum $\hat{\vec{L}}$, the position and the momentum operators $\hat{\vec{R}}$ and $\hat{\vec{P}}$:

$$[\hat{S}_j, \hat{L}_k] = 0, \qquad [\hat{S}_j, \hat{R}_k] = 0, \qquad [\hat{S}_j, \hat{P}_k] = 0 \qquad (j, k = x, y, z). \tag{5.123}$$

- The total wave function $| \Psi \rangle$ of a system with spin consists of a product of two parts: a spatial part $\psi(\vec{r})$ and a spin part $| s, m_s \rangle$:

$$| \Psi \rangle = | \psi \rangle \, | s, m_s \rangle. \tag{5.124}$$

This product of the space and spin degrees of freedom is not a product in the usual sense, but a direct or tensor product as discussed in Chapter 7. We will show in Chapter 6 that the four quantum numbers n, l, m_l, and m_s are required to completely describe the state of an electron moving in a central field; its wave function is

$$\Psi_{nlm_lm_s}(\vec{r}) = \psi_{nlm_l}(\vec{r}) \, | s, m_s \rangle. \tag{5.125}$$

Since the spin operator does not depend on the spatial degrees of freedom, it acts only on the spin part $| s, m_s \rangle$ and leaves the spatial wave function, $\psi_{nlm_l}(\vec{r})$, unchanged; conversely, the spatial operators $\hat{\vec{L}}$, $\hat{\vec{R}}$, and $\hat{\vec{P}}$ act on the spatial part and not on the spin part. For spin $\frac{1}{2}$ particles, the total wave function corresponding to spin-up and spin-down cases are respectively expressed in terms of the spinors:

$$\Psi_{nlm_l \frac{1}{2}}(\vec{r}) = \psi_{nlm_l}(\vec{r}) \begin{pmatrix} 1 \\ 0 \end{pmatrix} = \begin{pmatrix} \psi_{nlm_l}(\vec{r}) \\ 0 \end{pmatrix}, \tag{5.126}$$

$$\Psi_{nlm_l - \frac{1}{2}}(\vec{r}) = \psi_{nlm_l}(\vec{r}) \begin{pmatrix} 0 \\ 1 \end{pmatrix} = \begin{pmatrix} 0 \\ \psi_{nlm_l}(\vec{r}) \end{pmatrix}. \tag{5.127}$$

Example 5.4

Find the energy levels of a spin $s = \frac{3}{2}$ particle whose Hamiltonian is given by

$$\hat{H} = \frac{\alpha}{\hbar^2}(\hat{S}_x^2 + \hat{S}_y^2 - 2\hat{S}_z^2) - \frac{\beta}{\hbar}\hat{S}_z;$$

α and β are constants. Are these levels degenerate?

Solution

Rewriting $\hat{H}$ in the form,

$$\hat{H} = \frac{\alpha}{\hbar^2}\left(\hat{S}^2 - 3\hat{S}_z^2\right) - \frac{\beta}{\hbar}\hat{S}_z, \tag{5.128}$$

we see that $\hat{H}$ is diagonal in the $\{|s, m\rangle\}$ basis:

$$E_m = \langle s, m \mid \hat{H} \mid s, m \rangle = \frac{\alpha}{\hbar^2}\left[\hbar^2 s(s+1) - 3\hbar^2 m^2\right] - \frac{\beta}{\hbar}\hbar m = \frac{15}{4}\alpha - m(3\alpha m + \beta), \tag{5.129}$$

where the quantum number m takes any of the four values $m = -\frac{3}{2}, -\frac{1}{2}, \frac{1}{2}, \frac{3}{2}$. Since E_m depends on m, the energy levels of this particle are nondegenerate.

5.7 Eigenfunctions of Orbital Angular Momentum

We now turn to the coordinate representation of the angular momentum. In this section, we are going to work within the spherical coordinate system. Let us denote the joint eigenstates of $\hat{\vec{L}}^2$ and $\hat{L}_z$ by $| l, m \rangle$:

$$\boxed{\hat{\vec{L}}^2 \mid l, m \rangle = \hbar^2 l(l+1) \mid l, m \rangle,} \tag{5.130}$$

$$\boxed{\hat{L}_z \mid l, m \rangle = \hbar m \mid l, m \rangle.} \tag{5.131}$$

The operators $\hat{L}_z$, $\hat{\vec{L}}^2$, $\hat{L}_\pm$, whose Cartesian components are listed in Eqs (5.3) to (5.5), can be expressed in terms of spherical coordinates (Appendix B) as follows:

$$\hat{L}_z = -i\hbar \frac{\partial}{\partial \varphi}, \tag{5.132}$$

$$\hat{\vec{L}}^2 = -\hbar^2 \left[\frac{1}{\sin \theta} \frac{\partial}{\partial \theta} \left(\sin \theta \frac{\partial}{\partial \theta} \right) + \frac{1}{\sin^2 \theta} \frac{\partial^2}{\partial \varphi^2} \right], \tag{5.133}$$

$$\hat{L}_\pm = \hat{L}_x \pm i\hat{L}_y = \pm \hbar e^{\pm i\varphi} \left[\frac{\partial}{\partial \theta} \pm i \frac{\cos \theta}{\sin \theta} \frac{\partial}{\partial \varphi} \right]. \tag{5.134}$$

Since the operators $\hat{L}_z$ and $\hat{\vec{L}}$ depend only on the angles θ and φ, their eigenstates depend only on θ and φ. Denoting their joint eigenstates by

$$\langle \theta \varphi \mid l, m \rangle = Y_{lm}(\theta, \varphi), \tag{5.135}$$

where[1] $Y_{lm}(\theta, \varphi)$ are continuous functions of θ and φ, we can rewrite the eigenvalue equations (5.130) and (5.131) as follows:

$$\hat{\vec{L}}^2 Y_{lm}(\theta, \varphi) = \hbar^2 l(l+1) Y_{lm}(\theta, \varphi), \tag{5.136}$$

$$\hat{L}_z Y_{lm}(\theta, \varphi) = m\hbar Y_{lm}(\theta, \varphi). \tag{5.137}$$

Since $\hat{L}_z$ depends only on φ, as shown in (5.132), the previous two equations suggest that the eigenfunctions $Y_{lm}(\theta, \varphi)$ are separable:

$$Y_{lm}(\theta, \varphi) = \Theta_{lm}(\theta) \Phi_m(\varphi). \tag{5.138}$$

We ascertain that

$$\hat{L}_\pm Y_{lm}(\theta, \varphi) = \hbar \sqrt{l(l+1) - m(m \pm 1)} \, Y_{l\,m\pm1}(\theta, \varphi). \tag{5.139}$$

5.7.1 Eigenfunctions and Eigenvalues of $\hat{L}_z$

Inserting (5.138) into (5.137) we obtain $\hat{L}_z \Theta_{lm}(\theta) \Phi_m(\varphi) = m\hbar \Theta_{lm}(\theta) \Phi_m(\varphi)$. Now since $\hat{L}_z = -i\hbar \partial/\partial \varphi$, we have

$$-i\hbar \Theta_{lm}(\theta) \frac{\partial \Phi_m(\varphi)}{\partial \varphi} = m\hbar \Theta_{lm}(\theta) \Phi_m(\varphi), \tag{5.140}$$

which reduces to

$$-i \frac{\partial \Phi_m(\varphi)}{\partial \varphi} = m\Phi_m(\varphi). \tag{5.141}$$

The normalized solutions of this equation are given by

$$\Phi_m(\varphi) = \frac{1}{\sqrt{2\pi}} e^{im\varphi}, \tag{5.142}$$

---

[1]For notational consistency throughout this text, we will insert a comma between l and m in $Y_{lm}(\theta, \varphi)$ whenever m is negative.

where $1/\sqrt{2\pi}$ is the normalization constant,

$$\int_0^{2\pi} d\varphi \, \Phi_{m'}^*(\varphi)\Phi_m(\varphi) = \delta_{m', m} \tag{5.143}$$

For $\Phi_m(\varphi)$ to be single-valued, it must be periodic in φ with period 2π, $\Phi_m(\varphi+2\pi) = \Phi_m(\varphi)$; hence

$$e^{im(\varphi+2\pi)} = e^{im\varphi}. \tag{5.144}$$

This relation shows that the expectation value of $\hat{L}_z$, $l_z = \langle l, \, m \mid \hat{L}_z \mid l \, m\rangle$, is restricted to a *discrete* set of values

$$l_z = m\hbar, \qquad m = 0, \pm 1, \pm 2, \pm 3, \ldots. \tag{5.145}$$

Thus, the values of m vary from $-l$ to l:

$$m = -l, -(l-1), -(l-2), \ldots, 0, 1, 2, \ldots, l-2, l-1, l. \tag{5.146}$$

Hence the quantum number l must also be an integer. This is expected since the orbital angular momentum must have integer values.

5.7.2 Eigenfunctions of $\hat{L}^2$

Let us now focus on determining the eigenfunctions $\Theta_{lm}(\theta)$ of $\hat{L}^2$. We are going to follow two methods. The first method involves differential equations and gives $\Theta_{lm}(\theta)$ in terms of the well-known associated Legendre functions. The second method is algebraic; it deals with the operators $\hat{L}_\pm$ and enables an explicit construction of $Y_{lm}(\theta, \varphi)$, the spherical harmonics.

5.7.2.1 First Method for Determining the Eigenfunctions of $\hat{L}^2$

We begin by applying $\hat{L}^2$ of (5.133) to the eigenfunctions

$$Y_{lm}(\theta, \varphi) = \frac{1}{\sqrt{2\pi}}\Theta_{lm}(\theta)e^{im\varphi}. \tag{5.147}$$

This gives

$$\hat{L}^2 Y_{lm}(\theta, \varphi) = \frac{-\hbar^2}{\sqrt{2\pi}}\left[\frac{1}{\sin\theta}\frac{\partial}{\partial\theta}\left(\sin\theta\frac{\partial}{\partial\theta}\right) + \frac{1}{\sin^2\theta}\frac{\partial^2}{\partial\varphi^2}\right]\Theta_{lm}(\theta)e^{im\varphi}$$
$$= \frac{\hbar^2 l(l+1)}{\sqrt{2\pi}}\Theta_{lm}(\theta)e^{im\varphi}, \tag{5.148}$$

which, after eliminating the φ-dependence, reduces to

$$\frac{1}{\sin\theta}\frac{d}{d\theta}\left(\sin\theta\frac{d\Theta_{lm}(\theta)}{d\theta}\right) + \left[l(l+1) - \frac{m^2}{\sin^2\theta}\right]\Theta_{lm}(\theta) = 0. \tag{5.149}$$

This equation is known as the *Legendre differential equation*. Its solutions can be expressed in terms of the *associated Legendre functions* $P_l^m(\cos\theta)$:

$$\Theta_{lm}(\theta) = C_{lm} P_l^m(\cos\theta), \tag{5.150}$$

which are defined by

$$P_l^m(x) = (1 - x^2)^{|m|/2} \frac{d^{|m|}}{dx^{|m|}} P_l(x). \tag{5.151}$$

This shows that

$$P_l^{-m}(x) = P_l^m(x), \tag{5.152}$$

where $P_l(x)$ is the lth *Legendre polynomial* which is defined by the *Rodrigues formula*

$$P_l(x) = \frac{1}{2^l l!} \frac{d^l}{dx^l} (x^2 - 1)^l. \tag{5.153}$$

We can obtain at once the first few Legendre polynomials:

$$P_0(x) = 1, \qquad P_1(x) = \frac{1}{2} \frac{d(x^2 - 1)}{dx} = x, \tag{5.154}$$

$$P_2(x) = \frac{1}{8} \frac{d^2(x^2 - 1)^2}{dx^2} = \frac{1}{2}(3x^2 - 1), \qquad P_3(x) = \frac{1}{48} \frac{d^3(x^2 - 1)^3}{dx^3} = \frac{1}{2}(5x^3 - 3x),$$

$$\tag{5.155}$$

$$P_4(x) = \frac{1}{8}(35x^4 - 30x^2 + 3), \qquad P_5(x) = \frac{1}{8}(63x^5 - 70x^3 + 15x). \tag{5.156}$$

The Legendre polynomials satisfy the following closure or completeness relation:

$$\frac{1}{2} \sum_{l=0}^{\infty} (2l + 1) P_l(x') P_l(x) = \delta(x - x'). \tag{5.157}$$

From (5.153) we can infer at once

$$P_l(-x) = (-1)^l P_l(x). \tag{5.158}$$

A similar calculation leads to the first few associated Legendre functions:

$$P_1^1(x) = \sqrt{1 - x^2}, \tag{5.159}$$

$$P_2^1(x) = 3x\sqrt{1 - x^2}, \qquad P_2^2(x) = 3(1 - x^2), \tag{5.160}$$

$$P_3^1(x) = \frac{3}{2}(5x^2 - 1)\sqrt{1 - x^2}, \quad P_3^2(x) = 15x(1 - x^2), \quad P_3^3(x) = 15(1 - x^2)^{3/2}, \tag{5.161}$$

where $P_l^0(x) = P_l(x)$, with $l = 0, 1, 2, 3, \ldots$. The first few expressions for the associated Legendre functions and the Legendre polynomials are listed in Table 5.1. Note that

$$P_l^m(-x) = (-1)^{l+m} P_l^m(x). \tag{5.162}$$

The constant C_{lm} of (5.150) can be determined from the orthonormalization condition

$$\langle l', m' \mid l, m \rangle = \int_0^{2\pi} d\varphi \int_0^{\pi} d\theta \sin\theta \langle l', m' | \theta\varphi \rangle \langle \theta\varphi \mid l, m \rangle = \delta_{l', l} \delta_{m', m}, \tag{5.163}$$

which can be written as

$$\boxed{\int_0^{2\pi} d\varphi \int_0^{\pi} d\theta \sin\theta \, Y_{l'm'}^*(\theta, \varphi) Y_{lm}(\theta, \varphi) = \delta_{l', l} \delta_{m', m}.} \tag{5.164}$$

Table 5.1 First few Legendre polynomials and associated Legendre functions.

Legendre polynomials	Associated Legendre functions
$P_0(\cos\theta) = 1$	$P_1^1(\cos\theta) = \sin\theta$
$P_1(\cos\theta) = \cos\theta$	$P_2^1(\cos\theta) = 3\cos\theta\sin\theta$
$P_2(\cos\theta) = \frac{1}{2}(3\cos^2\theta - 1)$	$P_2^2(\cos\theta) = 3\sin^2\theta$
$P_3(\cos\theta) = \frac{1}{2}(5\cos^3\theta - 3\cos\theta)$	$P_3^1(\cos\theta) = \frac{3}{2}\sin\theta(5\cos^2\theta - 1)$
$P_4(\cos\theta) = \frac{1}{8}(35\cos^4\theta - 30\cos^2\theta + 3)$	$P_3^2(\cos\theta) = 15\sin^2\theta\cos\theta$
$P_5(\cos\theta) = \frac{1}{8}(63\cos^5\theta - 70\cos^3\theta + 15\cos\theta)$	$P_3^3(\cos\theta) = 15\sin^3\theta$

This relation is known as the normalization condition of spherical harmonics. Using the form (5.147) for $Y_{lm}(\theta, \varphi)$, we obtain

$$\int_0^{2\pi} d\varphi \int_0^{\pi} d\theta \, \sin\theta \, |Y_{lm}(\theta, \varphi)|^2 = \frac{|C_{lm}|^2}{2\pi} \int_0^{2\pi} d\varphi \int_0^{\pi} d\theta \, \sin\theta |P_l^m(\cos\theta)|^2 = 1. \quad (5.165)$$

From the theory of associated Legendre functions, we have

$$\int_0^{\pi} d\theta \, \sin\theta \, P_l^m(\cos\theta) P_{l'}^m(\cos\theta) = \frac{2}{2l+1} \frac{(l+m)!}{(l-m)!} \delta_{l,l'}, \quad (5.166)$$

which is known as the normalization condition of associated Legendre functions. A combination of the previous two relations leads to an expression for the coefficient C_{lm}:

$$C_{lm} = (-1)^m \sqrt{\left(\frac{2l+1}{2}\right) \frac{(l-m)!}{(l+m)!}} \qquad (m \geq 0). \quad (5.167)$$

Inserting this equation into (5.150), we obtain the eigenfunctions of $\hat{L}^2$:

$$\Theta_{lm}(\theta) = (-1)^m \sqrt{\left(\frac{2l+1}{2}\right) \frac{(l-m)!}{(l+m)!}} \, P_l^m(\cos\theta). \quad (5.168)$$

Finally, the joint eigenfunctions, $Y_{lm}(\theta, \varphi)$, of $\hat{L}^2$ and $\hat{J}_z$ can be obtained by substituting (5.142) and (5.168) into (5.138):

$$\boxed{Y_{lm}(\theta, \varphi) = (-1)^m \sqrt{\left(\frac{2l+1}{4\pi}\right) \frac{(l-m)!}{(l+m)!}} \, P_l^m(\cos\theta) e^{im\varphi} \qquad (m \geq 0).} \quad (5.169)$$

These are called the *normalized spherical harmonics*.

5.7.2.2 Second Method for Determining the Eigenfunctions of $\hat{L}^2$

The second method deals with a direct construction of $Y_{lm}(\theta, \varphi)$; it starts with the case $m = l$ (this is the maximum value of m). By analogy with the general angular momentum algebra developed in the previous section, the action of $\hat{L}_+$ on Y_{ll} gives zero,

$$\langle \theta \, \varphi | \hat{L}_+ | l, l \rangle = \hat{L}_+ Y_{ll}(\theta, \varphi) = 0, \quad (5.170)$$

since Y_{ll} cannot be raised further as $Y_{ll} = Y_{lm_{max}}$.

Using the expression (5.134) for $\hat{L}_+$ in the spherical coordinates, we can rewrite (5.170) as follows:

$$\frac{\hbar e^{i\varphi}}{\sqrt{2\pi}}\left[\frac{\partial}{\partial\theta} + i\cot\theta\frac{\partial}{\partial\varphi}\right]\Theta_{ll}(\theta)e^{il\varphi} = 0, \tag{5.171}$$

which leads to

$$\frac{1}{\Theta_{ll}}\frac{\partial\Theta_{ll}(\theta)}{\partial\theta} = l\cot\theta. \tag{5.172}$$

The solution to this differential equation is of the form

$$\Theta_{ll}(\theta) = C_l\sin^l\theta, \tag{5.173}$$

where C_l is a constant to be determined from the normalization condition (5.164) of $Y_{ll}(\theta, \varphi)$:

$$Y_{ll}(\theta, \varphi) = \frac{C_l}{\sqrt{2\pi}}e^{il\varphi}\sin^l\theta. \tag{5.174}$$

We can ascertain that C_l is given by

$$C_l = \frac{(-1)^l}{2^l l!}\sqrt{\frac{(2l+1)!}{2}}. \tag{5.175}$$

The action of $\hat{L}_-$ on $Y_{ll}(\theta, \varphi)$ is given, on the one hand, by

$$\hat{L}_- Y_{ll}(\theta, \varphi) = \hbar\sqrt{2l}\,Y_{l,l-1}(\theta, \varphi) \tag{5.176}$$

and, on the other hand, by

$$\hat{L}_- Y_{ll}(\theta, \varphi) = \hbar\frac{(-1)^l}{2^l\,l!}\sqrt{\frac{(2l+1)!}{4\pi}}e^{i(l-1)\varphi}(\sin\theta)^{1-l}\frac{d}{d(\cos\theta)}[(\sin\theta)^{2l}], \tag{5.177}$$

where we have used the spherical coordinate form (5.134).

Similarly, we can show that the action of $\hat{L}_-^{l-m}$ on $Y_{ll}(\theta, \varphi)$ is given, on the one hand, by

$$\hat{L}_-^{l-m} Y_{ll}(\theta, \varphi) = \hbar^{l-m}\sqrt{\frac{(2l)!(l+m)!}{(l-m)!}}Y_{lm}(\theta, \varphi) \tag{5.178}$$

and, on the other hand, by

$$\hat{L}_-^{l-m} Y_{ll}(\theta, \varphi) = \hbar^{l-m}\frac{(-1)^l}{2^l l!}\sqrt{\frac{(2l)!(2l+1)!}{4\pi}}e^{im\varphi}\frac{1}{\sin^m\theta}\frac{d^{l-m}}{d(\cos\theta)^{l-m}}(\sin\theta)^{2l}, \tag{5.179}$$

where $m \geq 0$. Equating the previous two relations, we obtain the expression of the spherical harmonic $Y_{lm}(\theta, \varphi)$ for $m \geq 0$:

$$\boxed{Y_{lm}(\theta, \varphi) = \frac{(-1)^l}{2^l l!}\sqrt{\left(\frac{2l+1}{4\pi}\right)\frac{(l+m)!}{(l-m)!}}e^{im\varphi}\frac{1}{\sin^m\theta}\frac{d^{l-m}}{d(\cos\theta)^{l-m}}(\sin\theta)^{2l}.} \tag{5.180}$$

5.7.3   Properties of the Spherical Harmonics

Since the spherical harmonics $Y_{lm}(\theta, \varphi)$ are joint eigenfunctions of $\hat{\vec{L}}^2$ and $\hat{L}_z$ and are ortho-normal (5.164), they constitute an orthonormal basis in the Hilbert space of square-integrable functions of θ and φ. The completeness relation is given by

$$\sum_{m=-l}^{l} |l, m\rangle\langle l, m| = 1 \tag{5.181}$$

or

$$\sum_m \langle \theta\varphi \mid l, m\rangle\langle l, m \mid \theta'\varphi'\rangle = \sum_m Y_{lm}^*(\theta', \varphi')Y_{lm}(\theta, \varphi) = \delta(\cos\theta - \cos\theta')\delta(\varphi - \varphi')$$

$$= \frac{\delta(\theta - \theta')}{\sin\theta}\delta(\varphi - \varphi'). \tag{5.182}$$

Let us mention some essential properties of the spherical harmonics. First, the spherical harmonics are complex functions; their complex conjugate is given by

$$\boxed{[Y_{lm}(\theta, \varphi)]^* = (-1)^m Y_{l,-m}(\theta, \varphi).} \tag{5.183}$$

We can verify that $Y_{lm}(\theta, \varphi)$ is an eigenstate of the parity operator $\hat{P}$ with an eigenvalue $(-1)^l$:

$$\hat{P}Y_{lm}(\theta, \varphi) = Y_{lm}(\pi - \theta, \varphi + \pi) = (-1)^l Y_{lm}(\theta, \varphi), \tag{5.184}$$

since a spatial reflection about the origin, $\vec{r}\,' = -\vec{r}$, corresponds to $r' = r$, $\theta' = \pi - \theta$, and $\varphi' = \pi + \varphi$, which leads to $P_l^m(\cos\theta') = P_l^m(-\cos\theta) = (-1)^{l+m}P_l^m(\cos\theta)$ and $e^{im\varphi'} = e^{im\pi}e^{im\varphi} = (-1)^m e^{im\varphi}$.

We can establish a connection between the spherical harmonics and the Legendre polynomials by simply taking $m = 0$. Then equation (5.180) yields

$$Y_{l0}(\theta, \varphi) = \frac{(-1)^l}{2^l l!}\sqrt{\frac{2l+1}{4\pi}}\frac{d^l}{d(\cos\theta)^l}(\sin\theta)^{2l} = \sqrt{\frac{2l+1}{4\pi}}P_l(\cos\theta), \tag{5.185}$$

with

$$P_l(\cos\theta) = \frac{1}{2^l l!}\frac{d^l}{d(\cos\theta)^l}(\cos^2\theta - 1)^l. \tag{5.186}$$

From the expression of Y_{lm}, we can verify that

$$Y_{lm}(0, \varphi) = \sqrt{\frac{2l+1}{4\pi}}\delta_{m,0}. \tag{5.187}$$

The expressions for the spherical harmonics corresponding to $l = 0$, $l = 1$, and $l = 2$ are listed in Table 5.2.

Spherical harmonics in Cartesian coordinates
Note that $Y_{lm}(\theta, \varphi)$ can also be expressed in terms of the Cartesian coordinates. For this, we need only to substitute

$$\sin\theta\cos\varphi = \frac{x}{r}, \qquad \sin\theta\sin\varphi = \frac{y}{r}, \qquad \cos\theta = \frac{z}{r} \tag{5.188}$$

Table 5.2 Spherical harmonics and their expressions in Cartesian coordinates.

$Y_{lm}(\theta, \varphi)$	$Y_{lm}(x, y, z)$
$Y_{00}(\theta, \varphi) = \frac{1}{\sqrt{4\pi}}$	$Y_{00}(x, y, z) = \frac{1}{\sqrt{4\pi}}$
$Y_{10}(\theta, \varphi) = \sqrt{\frac{3}{4\pi}} \cos\theta$	$Y_{10}(x, y, z) = \sqrt{\frac{3}{4\pi}} \frac{z}{r}$
$Y_{1,\pm1}(\theta, \varphi) = \mp\sqrt{\frac{3}{8\pi}} e^{\pm i\varphi} \sin\theta$	$Y_{1,\pm1}(x, y, z) = \mp\sqrt{\frac{3}{8\pi}} \frac{x\pm iy}{r}$
$Y_{20}(\theta, \varphi) = \sqrt{\frac{5}{16\pi}}(3\cos^2\theta - 1)$	$Y_{20}(x, y, z) = \sqrt{\frac{5}{16\pi}} \frac{3z^2 - r^2}{r^2}$
$Y_{2,\pm1}(\theta, \varphi) = \mp\sqrt{\frac{15}{8\pi}} e^{\pm i\varphi} \sin\theta\cos\theta$	$Y_{2,\pm1}(x, y, z) = \mp\sqrt{\frac{15}{8\pi}} \frac{(x\pm iy)z}{r^2}$
$Y_{2,\pm2}(\theta, \varphi) = \sqrt{\frac{15}{32\pi}} e^{\pm 2i\varphi} \sin^2\theta$	$Y_{2,\pm2}(x, y, z) = \sqrt{\frac{15}{32\pi}} \frac{x^2 - y^2 \pm 2ixy}{r^2}$

in the expression for $Y_{lm}(\theta, \varphi)$.

As an illustration, let us show how to derive the Cartesian expressions for Y_{10} and $Y_{1,\pm1}$. Substituting $\cos\theta = z/r$ into $Y_{10}(\theta, \varphi) = \sqrt{3/4\pi} \cos\theta \, Y_{10}$, we have

$$Y_{10}(x, y, z) = \sqrt{\frac{3}{4\pi}} \frac{z}{r} = \sqrt{\frac{3}{4\pi}} \frac{z}{\sqrt{x^2 + y^2 + z^2}}. \tag{5.189}$$

Using $\sin\theta \cos\varphi = x/r$ and $\sin\theta \sin\varphi = y/r$, we obtain

$$\frac{x \pm iy}{r} = \sin\theta\cos\varphi \pm i\sin\theta\sin\varphi = \sin\theta \, e^{\pm i\varphi}, \tag{5.190}$$

which, when substituted into $Y_{1\pm1}(\theta, \varphi) = \mp\sqrt{3/8\pi} \sin\theta \, e^{\pm i\varphi}$, leads to

$$Y_{1,\pm1}(x, y, z) = \mp\sqrt{\frac{3}{8\pi}} \frac{x \pm iy}{r}. \tag{5.191}$$

Following the same procedure, we can derive the Cartesian expressions of the remaining harmonics; for a listing, see Table 5.2.

Example 5.5 (Application of ladder operators to spherical harmonics)
   (a) Use the relation $Y_{l0}(\theta, \varphi) = \sqrt{(2l + 1)/4\pi} \, P_l(\cos\theta)$ to find the expression of $Y_{30}(\theta, \varphi)$.
   (b) Find the expression of Y_{30} in Cartesian coordinates.
   (c) Use the expression of $Y_{30}(\theta, \varphi)$ to infer those of $Y_{3,\pm1}(\theta, \varphi)$.

Solution
   (a) From Table 5.1 we have $P_3(\cos\theta) = \frac{1}{2}(5\cos^3\theta - 3\cos\theta)$; hence

$$Y_{30}(\theta, \varphi) = \sqrt{\frac{7}{4\pi}} P_3(\cos\theta) = \sqrt{\frac{7}{16\pi}}(5\cos^3\theta - 3\cos\theta). \tag{5.192}$$

(b) Since $\cos\theta = z/r$, we have $5\cos^3\theta - 3\cos\theta = 5\cos\theta(5\cos^2\theta-3) = z(5z^2-3r^2)/r^3$; hence

$$Y_{30}(x,y,z) = \sqrt{\frac{7}{16\pi}}\frac{z}{r^3}(5z^2 - 3r^2).\tag{5.193}$$

(c) To find Y_{31} from Y_{30}, we need to apply the ladder operator $\hat{L}_+$ on Y_{30} in two ways: first, algebraically

$$\hat{L}_+Y_{30} = \hbar\sqrt{3(3+1)-0}\,Y_{31} = 2\hbar\sqrt{3}\,Y_{31}\tag{5.194}$$

and hence

$$Y_{31} = \frac{1}{2\hbar\sqrt{3}}\hat{L}_+Y_{30};\tag{5.195}$$

then we use the differential form (5.134) of $\hat{L}_+$:

$$\begin{aligned}\hat{L}_+Y_{30}(\theta,\varphi) &= \hbar e^{i\varphi}\left[\frac{\partial}{\partial\theta} + i\frac{\cos\theta}{\sin\theta}\frac{\partial}{\partial\varphi}\right]Y_{30}(\theta,\varphi)\\ &= \hbar\sqrt{\frac{7}{16\pi}}e^{i\varphi}\left[\frac{\partial}{\partial\theta} + i\frac{\cos\theta}{\sin\theta}\frac{\partial}{\partial\varphi}\right](5\cos^3\theta - 3\cos\theta)\\ &= -3\hbar\sqrt{\frac{7}{16\pi}}\sin\theta(5\cos^2\theta - 1)e^{i\varphi}.\end{aligned}\tag{5.196}$$

Inserting (5.196) into (5.195) we end up with

$$Y_{31} = \frac{1}{2\hbar\sqrt{3}}\hat{L}_+Y_{30} = -\sqrt{\frac{21}{64\pi}}\sin\theta(5\cos^2\theta - 1)e^{i\varphi}.\tag{5.197}$$

Now, to find $Y_{3,-1}$ from Y_{30}, we also need to apply $\hat{L}_-$ on Y_{30} in two ways:

$$\hat{L}_-Y_{30} = \hbar\sqrt{3(3+1)-0}\,Y_{3,-1} = 2\hbar\sqrt{3}\,Y_{3,-1}\tag{5.198}$$

and hence

$$Y_{3,-1} = \frac{1}{2\hbar\sqrt{3}}\hat{L}_-Y_{30};\tag{5.199}$$

then we use the differential form (5.134) of $\hat{L}_-$:

$$\begin{aligned}\hat{L}_-Y_{30}(\theta,\varphi) &= -\hbar e^{-i\varphi}\left[\frac{\partial}{\partial\theta} - i\frac{\cos\theta}{\sin\theta}\frac{\partial}{\partial\varphi}\right]Y_{30}(\theta,\varphi)\\ &= -\hbar\sqrt{\frac{7}{16\pi}}e^{-i\varphi}\left[\frac{\partial}{\partial\theta} - i\frac{\cos\theta}{\sin\theta}\frac{\partial}{\partial\varphi}\right](5\cos^3\theta - 3\cos\theta)\\ &= 3\hbar\sqrt{\frac{7}{16\pi}}\sin\theta(5\cos^2\theta - 1)e^{-i\varphi}.\end{aligned}\tag{5.200}$$

Inserting (5.200) into (5.199), we obtain

$$Y_{3,-1} = \frac{1}{2\hbar\sqrt{3}}\hat{L}_-Y_{30} = \sqrt{\frac{21}{64\pi}}\sin\theta(5\cos^2\theta - 1)e^{-i\varphi}.\tag{5.201}$$

5.8 Solved Problems

Problem 5.1

(a) Show that $\Delta J_x \Delta J_y = \hbar^2[j(j+1)-m^2]/2$, where $\Delta J_x = \sqrt{\langle \hat{J}_x^2 \rangle - \langle \hat{J}_x \rangle^2}$ and the same for ΔJ_y.

(b) Show that this relation is consistent with $\Delta J_x \Delta J_y \geq (\hbar/2) |\langle \hat{J}_z \rangle| = \hbar^2 m/2$.

Solution

(a) First, note that $\langle \hat{J}_x \rangle$ and $\langle \hat{J}_y \rangle$ are zero, since

$$\langle \hat{J}_x \rangle = \frac{1}{2}\langle j,\, m \mid \hat{J}_+ \mid j,\, m \rangle + \frac{1}{2}\langle j,\, m \mid \hat{J}_- \mid j,\, m \rangle = 0. \tag{5.202}$$

As for $\langle \hat{J}_x^2 \rangle$ and $\langle \hat{J}_y^2 \rangle$, they are given by

$$\langle \hat{J}_x^2 \rangle = \frac{1}{4}\langle (\hat{J}_+ + \hat{J}_-)^2 \rangle = \frac{1}{4}\langle \hat{J}_+^2 + \hat{J}_+\hat{J}_- + \hat{J}_-\hat{J}_+ + J_-^2 \rangle, \tag{5.203}$$

$$\langle \hat{J}_y^2 \rangle = \frac{1}{4}\langle (\hat{J}_+ - \hat{J}_-)^2 \rangle = -\frac{1}{4}\langle \hat{J}_+^2 - \hat{J}_+\hat{J}_- - \hat{J}_-\hat{J}_+ + J_-^2 \rangle. \tag{5.204}$$

Since $\langle \hat{J}_+^2 \rangle = \langle \hat{J}_-^2 \rangle = 0$, we see that

$$\langle \hat{J}_x^2 \rangle = \frac{1}{4}\langle \hat{J}_+\hat{J}_- + \hat{J}_-\hat{J}_+ \rangle = \langle \hat{J}_y^2 \rangle. \tag{5.205}$$

Using the fact that

$$\langle \hat{J}_x^2 \rangle + \langle \hat{J}_y^2 \rangle = \langle \hat{J}^2 \rangle - \langle \hat{J}_z^2 \rangle \tag{5.206}$$

along with $\langle \hat{J}_x^2 \rangle = \langle \hat{J}_y^2 \rangle$, we see that

$$\langle \hat{J}_x^2 \rangle = \langle \hat{J}_y^2 \rangle = \frac{1}{2}[\langle \hat{J}^2 \rangle - \langle \hat{J}_z^2 \rangle]. \tag{5.207}$$

Now, since $\mid j,\, m \rangle$ is a joint eigenstate of $\hat{J}^2$ and $\hat{J}_z$ with eigenvalues $j(j+1)\hbar^2$ and $m\hbar$, we can easily see that the expressions of $\langle \hat{J}_x^2 \rangle$ and $\langle \hat{J}_y^2 \rangle$ are given by

$$\boxed{\langle \hat{J}_x^2 \rangle = \langle \hat{J}_y^2 \rangle = \frac{1}{2}[\langle \hat{J}^2 \rangle - \langle \hat{J}_z^2 \rangle] = \frac{\hbar^2}{2}\left[j(j+1) - m^2 \right].} \tag{5.208}$$

Hence $\Delta J_x \Delta J_y$ is given by

$$\Delta J_x \Delta J_y = \sqrt{\langle \hat{J}_x^2 \rangle \langle \hat{J}_y^2 \rangle} = \frac{\hbar^2}{2}[j(j+1) - m^2]. \tag{5.209}$$

(b) Since $j \geq m$ (because $m = -j, -j+1, \ldots, j-1, j$), we have

$$j(j+1) - m^2 \geq m(m+1) - m^2 = m, \tag{5.210}$$

from which we infer that $\Delta J_x \Delta J_y \geq \hbar^2 m/2$, or

$$\Delta J_x \Delta J_y \geq \frac{\hbar}{2}|\langle \hat{J}_z \rangle|. \tag{5.211}$$

Problem 5.2

Find the energy levels of a particle which is free except that it is constrained to move on the surface of a sphere of radius r.

Solution

This system consists of a particle that is constrained to move on the surface of a sphere but free from the influence of any other potential; it is called a *rigid rotator*. Since $V = 0$ the energy of this system is purely kinetic; the Hamiltonian of the rotator is

$$\hat{H} = \frac{\hat{L}^2}{2I}, \tag{5.212}$$

where $I = mr^2$ is the moment of inertia of the particle with respect to the origin. In deriving this relation, we have used the fact that $H = p^2/2m = (rp)^2/2mr^2 = L^2/2I$, since $L = |\vec{r} \times \vec{p}| = rp$.

The wave function of the system is clearly independent of the radial degree of freedom, for it is constant. The Schrödinger equation is thus given by

$$\hat{H}\psi(\theta, \varphi) = \frac{\hat{L}^2}{2I}\psi(\theta, \varphi) = E\psi(\theta, \varphi). \tag{5.213}$$

Since the eigenstates of $\hat{L}^2$ are the spherical harmonics $Y_{lm}(\theta, \varphi)$, the corresponding energy eigenvalues are given by

$$E_l = \frac{\hbar^2}{2I}l(l + 1), \qquad l = 0, 1, 2, 3, \dots, \tag{5.214}$$

and the Schrödinger equation by

$$\frac{\hat{L}^2}{2I}Y_{lm}(\theta, \varphi) = \frac{\hbar^2}{2I}l(l + 1)Y_{lm}(\theta, \varphi). \tag{5.215}$$

Note that the energy levels do not depend on the azimuthal quantum number m. This means that there are $(2l + 1)$ eigenfunctions $Y_{l-l}, Y_{l-l+1}, \dots, Y_{ll-1}, Y_{ll}$ corresponding to the same energy. Thus, every energy level E_l is $(2l + 1)$-fold degenerate. This is due to the fact that the rotator's Hamiltonian, $\hat{L}^2/2I$, commutes with $\hat{L}$. That is, the Hamiltonian is independent of the orientation of $\hat{L}$ in space; hence the energy spectrum does not depend on the component of $\hat{L}$ in any particular direction.

Problem 5.3

Find the rotational energy levels of a diatomic molecule.

Solution

Consider two molecules of masses m_1 and m_2 separated by a constant distance $\vec{r}$. Let r_1 and r_2 be their distances from the center of mass, i.e., $m_1r_1 = m_2r_2$. The moment of inertia of the diatomic molecule is

$$I = m_1r_1^2 + m_2r_2^2 \equiv \mu r^2, \tag{5.216}$$

where $r = |\vec{r}_1 - \vec{r}_2|$ and where μ is their reduced mass, $\mu = m_1 m_2 / (m_1 + m_2)$. The total angular momentum is given by

$$|\vec{L}| = m_1 r_1 r_1 \omega + m_2 r_2 r_2 \omega = I \omega = \mu r^2 \omega \tag{5.217}$$

and the Hamiltonian by

$$\hat{H} = \frac{\hat{L}^2}{2I} = \frac{\hat{L}^2}{2\mu r^2}. \tag{5.218}$$

The corresponding eigenvalue equation

$$\hat{H} \, | \, l, \, m \rangle = \frac{\hat{L}^2}{2\mu r^2} \, | \, l, \, m \rangle = \frac{l(l+1)\hbar^2}{2\mu r^2} \, | \, l, \, m \rangle, \tag{5.219}$$

shows that the eigenenergies are $(2l + 1)$-fold degenerate and given by

$$E_l = \frac{l(l+1)\hbar^2}{2\mu r^2}. \tag{5.220}$$

Problem 5.4

(a) Find the eigenvalues and eigenstates of the spin operator $\vec{S}$ of an electron in the direction of a unit vector $\vec{n}$; assume that $\vec{n}$ lies in the xz plane.

(b) Find the probability of measuring $\hat{S}_z = +\hbar/2$.

Solution

(a) In this question we want to solve

$$\vec{n} \cdot \vec{S} | \lambda \rangle = \frac{\hbar}{2} \lambda | \lambda \rangle, \tag{5.221}$$

where $\vec{n}$ is given by $\vec{n} = (\sin \theta \, \vec{i} + \cos \theta \, \vec{k})$, because it lies in the xz plane, with $0 \le \theta \le \pi$. We can thus write

$$\vec{n} \cdot \vec{S} = (\sin \theta \, \vec{i} + \cos \theta \, \vec{k}) \cdot (S_x \vec{i} + S_y \vec{j} + S_z \vec{k}) = S_x \sin \theta + S_z \cos \theta. \tag{5.222}$$

Using the spin matrices

$$\hat{S}_x = \frac{\hbar}{2} \begin{pmatrix} 0 & 1 \\ 1 & 0 \end{pmatrix}, \qquad \hat{S}_y = \frac{\hbar}{2} \begin{pmatrix} 0 & -i \\ i & 0 \end{pmatrix}, \qquad \hat{S}_z = \frac{\hbar}{2} \begin{pmatrix} 1 & 0 \\ 0 & -1 \end{pmatrix}, \tag{5.223}$$

we can write (5.222) in the following matrix form:

$$\vec{n} \cdot \vec{S} = \frac{\hbar}{2} \begin{pmatrix} 0 & 1 \\ 1 & 0 \end{pmatrix} \sin \theta + \frac{\hbar}{2} \begin{pmatrix} 1 & 0 \\ 0 & -1 \end{pmatrix} \cos \theta = \frac{\hbar}{2} \begin{pmatrix} \cos \theta & \sin \theta \\ \sin \theta & -\cos \theta \end{pmatrix}. \tag{5.224}$$

The diagonalization of this matrix leads to the following secular equation:

$$-\frac{\hbar^2}{4} (\cos \theta - \lambda)(\cos \theta + \lambda) - \frac{\hbar^2}{4} \sin^2 \theta = 0, \tag{5.225}$$

which in turn leads as expected to the eigenvalues $\lambda = \pm 1$.

The eigenvector corresponding to $\lambda = 1$ can be obtained from

$$\frac{\hbar}{2} \begin{pmatrix} \cos\theta & \sin\theta \\ \sin\theta & -\cos\theta \end{pmatrix} \begin{pmatrix} a \\ b \end{pmatrix} = \frac{\hbar}{2} \begin{pmatrix} a \\ b \end{pmatrix}. \tag{5.226}$$

This matrix equation can be reduced to a single equation

$$a \sin\frac{1}{2}\theta = b \cos\frac{1}{2}\theta. \tag{5.227}$$

Combining this equation with the normalization condition $|a|^2 + |b|^2 = 1$, we infer that $a = \cos\frac{1}{2}\theta$ and $b = \sin\frac{1}{2}\theta$; hence the eigenvector corresponding to $\lambda = 1$ is

$$|\lambda_+\rangle = \begin{pmatrix} \cos(\theta/2) \\ \sin(\theta/2) \end{pmatrix}. \tag{5.228}$$

Proceeding in the same way, we can easily obtain the eigenvector for $\lambda = -1$:

$$|\lambda_-\rangle = \begin{pmatrix} -\sin(\theta/2) \\ \cos(\theta/2) \end{pmatrix}. \tag{5.229}$$

(b) Let us write $|\lambda_\pm\rangle$ of (5.228) and (5.229) in terms of the spin-up and spin-down eigenvectors, $\left|\frac{1}{2}, \frac{1}{2}\right\rangle = \begin{pmatrix} 1 \\ 0 \end{pmatrix}$ and $\left|\frac{1}{2}, -\frac{1}{2}\right\rangle = \begin{pmatrix} 0 \\ 1 \end{pmatrix}$:

$$|\lambda_+\rangle = \cos\frac{1}{2}\theta \left|\frac{1}{2}, \frac{1}{2}\right\rangle + \sin\frac{1}{2}\theta \left|\frac{1}{2}, -\frac{1}{2}\right\rangle, \tag{5.230}$$

$$|\lambda_-\rangle = -\sin\frac{1}{2}\theta \left|\frac{1}{2}, \frac{1}{2}\right\rangle + \cos\frac{1}{2}\theta \left|\frac{1}{2}, -\frac{1}{2}\right\rangle. \tag{5.231}$$

We see that the probability of measuring $\hat{S}_z = +\hbar/2$ is given by

$$\left|\left\langle \frac{1}{2}, \frac{1}{2} \Big| \lambda_+ \right\rangle\right|^2 = \cos^2\frac{1}{2}\theta. \tag{5.232}$$

Problem 5.5

(a) Find the eigenvalues and eigenstates of the spin operator $\vec{S}$ of an electron in the direction of a unit vector $\vec{n}$, where $\vec{n}$ is *arbitrary*.

(b) Find the probability of measuring $\hat{S}_z = -\hbar/2$.

(c) Assuming that the eigenvectors of the spin calculated in (a) correspond to $t = 0$, find these eigenvectors at time t.

Solution

(a) We need to solve

$$\vec{n} \cdot \vec{S} |\lambda\rangle = \frac{\hbar}{2} \lambda |\lambda\rangle, \tag{5.233}$$

where $\vec{n}$, a unit vector pointing along an *arbitrary* direction, is given in spherical coordinates by

$$\vec{n} = (\sin\theta\cos\varphi)\vec{i} + (\sin\theta\sin\varphi)\vec{j} + (\cos\theta)\vec{k}, \tag{5.234}$$

with $0 \le \theta \le \pi$ and $0 \le \varphi \le 2\pi$. We can thus write

$$
\begin{aligned}
\vec{n}\cdot\vec{S} &= (\sin\theta\cos\varphi\,\vec{i} + \sin\theta\sin\varphi\,\vec{j} + \cos\theta\,\vec{k})\cdot(S_x\vec{i} + S_y\vec{j} + S_z\vec{k}) \\
&= S_x\sin\theta\cos\varphi + S_y\sin\theta\sin\varphi + S_z\cos\theta.
\end{aligned}
\tag{5.235}
$$

Using the spin matrices, we can write this equation in the following matrix form:

$$
\begin{aligned}
\vec{n}\cdot\vec{S} &= \frac{\hbar}{2}\begin{pmatrix} 0 & 1 \\ 1 & 0 \end{pmatrix}\sin\theta\cos\varphi + \frac{\hbar}{2}\begin{pmatrix} 0 & -i \\ i & 0 \end{pmatrix}\sin\theta\sin\varphi + \frac{\hbar}{2}\begin{pmatrix} 1 & 0 \\ 0 & -1 \end{pmatrix}\cos\theta \\
&= \frac{\hbar}{2}\begin{pmatrix} \cos\theta & \sin\theta\,(\cos\varphi - i\sin\varphi) \\ \sin\theta\,(\cos\varphi + i\sin\varphi) & -\cos\theta \end{pmatrix} \\
&= \frac{\hbar}{2}\begin{pmatrix} \cos\theta & e^{-i\varphi}\sin\theta \\ e^{i\varphi}\sin\theta & -\cos\theta \end{pmatrix}.
\end{aligned}
\tag{5.236}
$$

Diagonalization of this matrix leads to the secular equation

$$-\frac{\hbar^2}{4}(\cos\theta - \lambda)(\cos\theta + \lambda) - \frac{\hbar^2}{4}\sin^2\theta = 0, \tag{5.237}$$

which in turn leads to the eigenvalues $\lambda = \pm 1$.

The eigenvector corresponding to $\lambda = 1$ can be obtained from

$$\frac{\hbar}{2}\begin{pmatrix} \cos\theta & e^{-i\varphi}\sin\theta \\ e^{i\varphi}\sin\theta & -\cos\theta \end{pmatrix}\begin{pmatrix} a \\ b \end{pmatrix} = \frac{\hbar}{2}\begin{pmatrix} a \\ b \end{pmatrix}, \tag{5.238}$$

which leads to

$$a\cos\theta + be^{-i\varphi}\sin\theta = a \tag{5.239}$$

or

$$a(1 - \cos\theta) = be^{-i\varphi}\sin\theta. \tag{5.240}$$

Using the relations $1 - \cos\theta = 2\sin^2\frac{1}{2}\theta$ and $\sin\theta = 2\cos\frac{1}{2}\theta\sin\frac{1}{2}\theta$, we have

$$b = a\tan\frac{1}{2}\theta e^{i\varphi}. \tag{5.241}$$

Combining this equation with the normalization condition $|a|^2 + |b|^2 = 1$, we obtain $a = \cos\frac{1}{2}\theta$ and $b = e^{i\varphi}\sin\frac{1}{2}\theta$. Thus, the eigenvector corresponding to $\lambda = 1$ is

$$|\lambda_+\rangle = \begin{pmatrix} \cos(\theta/2) \\ e^{i\varphi}\sin(\theta/2) \end{pmatrix}. \tag{5.242}$$

A similar treatment leads to the eigenvector for $\lambda = -1$:

$$|\lambda_-\rangle = \begin{pmatrix} -\sin(\theta/2) \\ e^{i\varphi}\cos(\theta/2) \end{pmatrix}. \tag{5.243}$$

(b) Write $|\lambda_-\rangle$ of (5.243) in terms of $|\frac{1}{2}, \frac{1}{2}\rangle = \begin{pmatrix} 1 \\ 0 \end{pmatrix}$ and $|\frac{1}{2}, -\frac{1}{2}\rangle = \begin{pmatrix} 0 \\ 1 \end{pmatrix}$:

$$|\lambda_+\rangle = \cos \frac{1}{2}\theta \left|\frac{1}{2}, \frac{1}{2}\right\rangle + e^{i\varphi} \sin \frac{1}{2}\theta \left|\frac{1}{2}, -\frac{1}{2}\right\rangle, \qquad (5.244)$$

$$|\lambda_-\rangle = -\sin \frac{1}{2}\theta \left|\frac{1}{2}, \frac{1}{2}\right\rangle + e^{i\varphi} \cos \frac{1}{2}\theta \left|\frac{1}{2}, -\frac{1}{2}\right\rangle. \qquad (5.245)$$

We can then obtain the probability of measuring $\hat{S}_z = -\hbar/2$:

$$\left|\left\langle\frac{1}{2}, -\frac{1}{2}\Big|\lambda_-\right\rangle\right|^2 = \cos^2 \frac{1}{2}\theta. \qquad (5.246)$$

(c) The spin's eigenstates at time t are given by

$$|\lambda_+(t)\rangle = e^{-iE_+t/\hbar} \cos \frac{1}{2}\theta \left|\frac{1}{2}, \frac{1}{2}\right\rangle + e^{i(\varphi-E_-t/\hbar)} \sin \frac{1}{2}\theta \left|\frac{1}{2}, -\frac{1}{2}\right\rangle, \qquad (5.247)$$

$$|\lambda_-(t)\rangle = -e^{-iE_+t/\hbar} \sin \frac{1}{2}\theta \left|\frac{1}{2}, \frac{1}{2}\right\rangle + e^{i(\varphi-E_-t/\hbar)} \cos \frac{1}{2}\theta \left|\frac{1}{2}, -\frac{1}{2}\right\rangle, \qquad (5.248)$$

where $E_\pm$ are the energy eigenvalues corresponding to the spin-up and spin-down states, respectively.

Problem 5.6

The Hamiltonian of a system is $\hat{H} = \varepsilon \vec{\sigma} \cdot \vec{n}$, where ε is a constant having the dimensions of energy, $\vec{n}$ is an arbitrary unit vector, and σ_x, σ_y, and σ_z are the Pauli matrices.

(a) Find the energy eigenvalues and normalized eigenvectors of $\hat{H}$.

(b) Find a transformation matrix that diagonalizes $\hat{H}$.

Solution

(a) Using the Pauli matrices $\sigma_x = \begin{pmatrix} 0 & 1 \\ 1 & 0 \end{pmatrix}$, $\sigma_y = \begin{pmatrix} 0 & -i \\ i & 0 \end{pmatrix}$, $\sigma_z = \begin{pmatrix} 1 & 0 \\ 0 & -1 \end{pmatrix}$

and the expression of an arbitrary unit vector in spherical coordinates $\vec{n} = (\sin\theta \cos\varphi)\vec{i} + (\sin\theta \sin\varphi)\vec{j} + (\cos\theta)\vec{k}$, we can rewrite the Hamiltonian

$$\hat{H} = \varepsilon \vec{\sigma} \cdot \vec{n} = \varepsilon \left(\sigma_x \sin\theta \cos\varphi + \sigma_y \sin\theta \sin\varphi + \sigma_z \cos\theta \right) \qquad (5.249)$$

in the following matrix form:

$$\hat{H} = \varepsilon \begin{pmatrix} \cos\theta & \exp(-i\varphi)\sin\theta \\ \exp(i\varphi)\sin\theta & -\cos\theta \end{pmatrix}. \qquad (5.250)$$

The eigenvalues of $\hat{H}$ are obtained by solving the secular equation $\det(H - E) = 0$, or

$$(\varepsilon\cos\theta - E)(-\varepsilon\cos\theta - E) - \varepsilon^2 \sin^2\theta = 0, \qquad (5.251)$$

which yields two eigenenergies $E_1 = \varepsilon$ and $E_2 = -\varepsilon$.

The energy eigenfunctions are obtained from

$$\varepsilon \begin{pmatrix} \cos\theta & \exp(-i\varphi)\sin\theta \\ \exp(i\varphi)\sin\theta & -\cos\theta \end{pmatrix} \begin{pmatrix} x \\ y \end{pmatrix} = E \begin{pmatrix} x \\ y \end{pmatrix}. \tag{5.252}$$

For the case $E = E_1 = \varepsilon$, this equation yields

$$(\cos\theta - 1)x + y\sin\theta\exp(-i\varphi) = 0, \tag{5.253}$$

which in turn leads to

$$\frac{x}{y} = \frac{\sin\theta\exp(-i\varphi)}{1 - \cos\theta} = \frac{\cos\theta/2\exp(-i\varphi/2)}{\sin\theta/2\exp(i\varphi/2)}; \tag{5.254}$$

hence

$$\begin{pmatrix} x_1 \\ y_1 \end{pmatrix} = \begin{pmatrix} \exp(-i\varphi/2)\cos(\theta/2) \\ \exp(i\varphi/2)\sin(\theta/2) \end{pmatrix}; \tag{5.255}$$

this vector is normalized. Similarly, in the case where $E = E_2 = -\varepsilon$, we can show that the second normalized eigenvector is

$$\begin{pmatrix} x_2 \\ y_2 \end{pmatrix} = \begin{pmatrix} -\exp(-i\varphi/2)\sin(\theta/2) \\ \exp(i\varphi/2)\cos(\theta/2) \end{pmatrix}. \tag{5.256}$$

(b) A transformation $\hat{U}$ that diagonalizes $\hat{H}$ can be obtained from the two eigenvectors obtained in part (a): $U_{11} = x_1$, $U_{21} = y_1$, $U_{12} = x_2$, $U_{22} = y_2$. That is,

$$U = \begin{pmatrix} \exp(-i\varphi/2)\cos(\theta/2) & -\exp(-i\varphi/2)\sin(\theta/2) \\ \exp(i\varphi/2)\sin(\theta/2) & \exp(i\varphi/2)\cos(\theta/2) \end{pmatrix}. \tag{5.257}$$

Note that this matrix is unitary, since $U^\dagger = U^{-1}$ and $\det(U) = 1$. We can ascertain that

$$\hat{U}\hat{H}\hat{U}^\dagger = \begin{pmatrix} \varepsilon & 0 \\ 0 & -\varepsilon \end{pmatrix}. \tag{5.258}$$

Problem 5.7

Consider a system of total angular momentum $j = 1$. As shown in (5.73) and (5.75), the operators $\hat{J}_x$, $\hat{J}_y$, and $\hat{J}_z$ are given by

$$\hat{J}_x = \frac{\hbar}{\sqrt{2}} \begin{pmatrix} 0 & 1 & 0 \\ 1 & 0 & 1 \\ 0 & 1 & 0 \end{pmatrix}, \quad \hat{J}_y = \frac{\hbar}{\sqrt{2}} \begin{pmatrix} 0 & -i & 0 \\ i & 0 & -i \\ 0 & i & 0 \end{pmatrix}, \quad \hat{J}_z = \hbar \begin{pmatrix} 1 & 0 & 0 \\ 0 & 0 & 0 \\ 0 & 0 & -1 \end{pmatrix}. \tag{5.259}$$

(a) What are the possible values when measuring $\hat{J}_x$?

(b) Calculate $\langle \hat{J}_z \rangle$, $\langle \hat{J}_z^2 \rangle$, and ΔJ_z if the system is in the state $j_x = -\hbar$.

(c) Repeat (b) for $\langle \hat{J}_y \rangle$, $\langle \hat{J}_y^2 \rangle$, and ΔJ_y.

(d) If the system were initially in state $| \psi \rangle = \frac{1}{\sqrt{14}} \begin{pmatrix} -\sqrt{3} \\ 2\sqrt{2} \\ \sqrt{3} \end{pmatrix}$, what values will one obtain

when measuring $\hat{J}_x$ and with what probabilities?

Solution

(a) According to Postulate 2 of Chapter 3, the results of the measurements are given by the eigenvalues of the measured quantity. Here the eigenvalues of $\hat{J}_x$, which are obtained by diagonalizing the matrix J_x, are $j_x = -\hbar, 0$, and $\hbar$; the respective (normalized) eigenstates are

$$
|-1\rangle = \frac{1}{2}\begin{pmatrix} -1 \\ \sqrt{2} \\ -1 \end{pmatrix}, \qquad |0\rangle = \frac{1}{\sqrt{2}}\begin{pmatrix} -1 \\ 0 \\ 1 \end{pmatrix}, \qquad |1\rangle = \frac{1}{2}\begin{pmatrix} 1 \\ \sqrt{2} \\ 1 \end{pmatrix}. \qquad (5.260)
$$

(b) If the system is in the state $j_x = -\hbar$, its eigenstate is given by $|-1\rangle$. In this case $\langle \hat{J}_z \rangle$ and $\langle \hat{J}_z^2 \rangle$ are given by

$$
\langle -1|\hat{J}_z|-1\rangle = \frac{\hbar}{4}\begin{pmatrix} -1 & \sqrt{2} & -1 \end{pmatrix}\begin{pmatrix} 1 & 0 & 0 \\ 0 & 0 & 0 \\ 0 & 0 & -1 \end{pmatrix}\begin{pmatrix} -1 \\ \sqrt{2} \\ -1 \end{pmatrix} = 0, \qquad (5.261)
$$

$$
\langle -1|\hat{J}_z^2|-1\rangle = \frac{\hbar^2}{4}\begin{pmatrix} -1 & \sqrt{2} & -1 \end{pmatrix}\begin{pmatrix} 1 & 0 & 0 \\ 0 & 0 & 0 \\ 0 & 0 & 1 \end{pmatrix}\begin{pmatrix} -1 \\ \sqrt{2} \\ -1 \end{pmatrix} = \frac{\hbar^2}{2}. \qquad (5.262)
$$

Thus, the uncertainty ΔJ_z is given by

$$
\Delta J_z = \sqrt{\langle -1|\hat{J}_z^2|-1\rangle - \langle -1|\hat{J}_z|-1\rangle^2} = \sqrt{\frac{\hbar^2}{2}} = \frac{\hbar}{\sqrt{2}}. \qquad (5.263)
$$

(c) Following the same procedure in (b), we have

$$
\langle -1|\hat{J}_y|-1\rangle = \frac{\hbar}{4\sqrt{2}}\begin{pmatrix} -1 & \sqrt{2} & -1 \end{pmatrix}\begin{pmatrix} 0 & -i & 0 \\ i & 0 & -i \\ 0 & i & 0 \end{pmatrix}\begin{pmatrix} -1 \\ \sqrt{2} \\ -1 \end{pmatrix} = 0, \qquad (5.264)
$$

$$
\langle -1|\hat{J}_y^2|-1\rangle = \frac{\hbar^2}{8}\begin{pmatrix} -1 & \sqrt{2} & -1 \end{pmatrix}\begin{pmatrix} 1 & 0 & -1 \\ 0 & 2 & 0 \\ -1 & 0 & 1 \end{pmatrix}\begin{pmatrix} -1 \\ \sqrt{2} \\ -1 \end{pmatrix} = \frac{\hbar^2}{2}; \qquad (5.265)
$$

hence

$$
\Delta J_y = \sqrt{\langle -1|\hat{J}_y^2|-1\rangle - \langle -1|\hat{J}_y|-1\rangle^2} = \frac{\hbar}{\sqrt{2}}. \qquad (5.266)
$$

(d) We can express $|\psi\rangle = \frac{1}{\sqrt{14}}\begin{pmatrix} -\sqrt{3} \\ 2\sqrt{2} \\ \sqrt{3} \end{pmatrix}$ in terms of the eigenstates (5.260) as

$$
\frac{1}{\sqrt{14}}\begin{pmatrix} -\sqrt{3} \\ 2\sqrt{2} \\ \sqrt{3} \end{pmatrix} = \sqrt{\frac{2}{7}}\frac{1}{2}\begin{pmatrix} -1 \\ \sqrt{2} \\ -1 \end{pmatrix} + \sqrt{\frac{3}{7}}\frac{1}{\sqrt{2}}\begin{pmatrix} -1 \\ 0 \\ 1 \end{pmatrix} + \sqrt{\frac{2}{7}}\frac{1}{2}\begin{pmatrix} 1 \\ \sqrt{2} \\ 1 \end{pmatrix} \qquad (5.267)
$$

or

$$
|\psi\rangle = \sqrt{\frac{2}{7}}|-1\rangle + \sqrt{\frac{3}{7}}|0\rangle + \sqrt{\frac{2}{7}}|1\rangle. \qquad (5.268)
$$

A measurement of $\hat{J}_x$ on a system initially in the state (5.268) yields a value $j_x = -\hbar$ with probability

$$P_{-1} = |\langle -1 \mid \psi \rangle|^2 = \left| \sqrt{\frac{2}{7}} \langle -1| - 1\rangle + \sqrt{\frac{3}{7}} \langle -1|0\rangle + \sqrt{\frac{2}{7}} \langle -1|1\rangle \right|^2 = \frac{2}{7}, \qquad (5.269)$$

since $\langle -1|0\rangle = \langle -1|1\rangle = 0$ and $\langle -1| - 1\rangle = 1$, and the values $j_x = 0$ and $j_x = \hbar$ with the respective probabilities

$$P_0 = |\langle 0 \mid \psi \rangle|^2 = \left| \sqrt{\frac{3}{7}} \langle 0|0\rangle \right|^2 = \frac{3}{7}, \qquad P_1 = |\langle 1 \mid \psi \rangle|^2 = \left| \sqrt{\frac{2}{7}} \langle 1|1\rangle \right|^2 = \frac{2}{7}. \qquad (5.270)$$

Problem 5.8

Consider a particle of total angular momentum $j = 1$. Find the matrix for the component of $\vec{J}$ along a unit vector with arbitrary direction $\vec{n}$. Find its eigenvalues and eigenvectors.

Solution

Since $\vec{J} = J_x \vec{i} + J_y \vec{j} + J_z \vec{k}$ and $\vec{n} = (\sin\theta \cos\varphi)\vec{i} + (\sin\theta \sin\varphi)\vec{j} + (\cos\theta)\vec{k}$, the component of $\vec{J}$ along $\vec{n}$ is

$$\vec{n} \cdot \vec{J} = J_x \sin\theta \cos\varphi + J_y \sin\theta \sin\varphi + J_z \cos\theta, \qquad (5.271)$$

with $0 \leq \theta \leq \pi$ and $0 \leq \varphi \leq 2\pi$; the matrices of $\hat{J}_x$, $\hat{J}_y$, and $\hat{J}_z$ are given by (5.259). We can therefore write this equation in the following matrix form:

$$\vec{n} \cdot \vec{J} = \frac{\hbar}{\sqrt{2}} \begin{pmatrix} 0 & 1 & 0 \\ 1 & 0 & 1 \\ 0 & 1 & 0 \end{pmatrix} \sin\theta \cos\varphi + \frac{\hbar}{\sqrt{2}} \begin{pmatrix} 0 & -i & 0 \\ i & 0 & -i \\ 0 & i & 0 \end{pmatrix} \sin\theta \sin\varphi$$

$$+ \hbar \begin{pmatrix} 1 & 0 & 0 \\ 0 & 0 & 0 \\ 0 & 0 & -1 \end{pmatrix} \cos\theta = \frac{\hbar}{\sqrt{2}} \begin{pmatrix} \sqrt{2}\cos\theta & e^{-i\varphi}\sin\theta & 0 \\ e^{i\varphi}\sin\theta & 0 & e^{-i\varphi}\sin\theta \\ 0 & e^{i\varphi}\sin\theta & -\sqrt{2}\cos\theta \end{pmatrix}. \qquad (5.272)$$

The diagonalization of this matrix leads to the eigenvalues $\lambda_1 = -\hbar$, $\lambda_2 = 0$, and $\lambda_3 = \hbar$; the corresponding eigenvectors are given by

$$|\lambda_1\rangle = \frac{1}{2} \begin{pmatrix} (1 - \cos\theta)e^{-i\varphi} \\ -\frac{2}{\sqrt{2}}\sin\theta \\ (1 + \cos\theta)e^{i\varphi} \end{pmatrix}, \qquad |\lambda_2\rangle = \frac{1}{\sqrt{2}} \begin{pmatrix} -e^{-i\varphi}\sin\theta \\ \sqrt{2}\cos\theta \\ e^{i\varphi}\sin\theta \end{pmatrix}, \qquad (5.273)$$

$$|\lambda_3\rangle = \frac{1}{2} \begin{pmatrix} (1 + \cos\theta)e^{-i\varphi} \\ \frac{2}{\sqrt{2}}\sin\theta \\ (1 - \cos\theta)e^{i\varphi} \end{pmatrix}. \qquad (5.274)$$

Problem 5.9

Consider a system which is initially in the state

$$\psi(\theta, \varphi) = \frac{1}{\sqrt{5}} Y_{1,-1}(\theta, \varphi) + \sqrt{\frac{3}{5}} Y_{10}(\theta, \varphi) + \frac{1}{\sqrt{5}} Y_{11}(\theta, \varphi).$$

(a) Find $\langle \psi \mid \hat{L}_+ \mid \psi \rangle$.

(b) If $\hat{L}_z$ were measured what values will one obtain and with what probabilities?

(c) If after measuring $\hat{L}_z$ we find $l_z = -\hbar$, calculate the uncertainties ΔL_x and ΔL_y and their product $\Delta L_x \Delta L_y$.

Solution

(a) Let us use a lighter notation for $\mid \psi \rangle$: $\mid \psi \rangle = \frac{1}{\sqrt{5}} \mid 1, \ -1 \rangle + \sqrt{\frac{3}{5}} \mid 1, \ 0 \rangle + \frac{1}{\sqrt{5}} \mid 1, \ 1 \rangle$.

From (5.56) we can write $\hat{L}_+ \mid l, \ m \rangle = \hbar \sqrt{l(l+1) - m(m+1)} \mid l, \ m+1 \rangle$; hence the only terms that survive in $\langle \psi \mid \hat{L}_+ \mid \psi \rangle$ are

$$\langle \psi \mid \hat{L}_+ \mid \psi \rangle = \frac{\sqrt{3}}{5} \langle 1, \ 0 | \hat{L}_+ \mid 1, \ -1 \rangle + \frac{\sqrt{3}}{5} \langle 1, \ 1 | \hat{L}_+ \mid 1, \ 0 \rangle = \frac{2\sqrt{6}}{5} \hbar, \qquad (5.275)$$

since $\langle 1, \ 0 | \hat{L}_+ \mid 1, \ -1 \rangle = \langle 1, \ 1 | \hat{L}_+ \mid 1, \ 0 \rangle = \sqrt{2} \hbar$.

(b) If $\hat{L}_z$ were measured, we will find three values $l_z = -\hbar$, 0, and $\hbar$. The probability of finding the value $l_z = -\hbar$ is

$$P_{-1} = |\langle 1, \ -1 \mid \psi \rangle|^2 = \left| \frac{1}{\sqrt{5}} \langle 1, \ -1 \mid 1, \ -1 \rangle + \sqrt{\frac{3}{5}} \langle 1, \ -1 \mid 1, \ 0 \rangle + \frac{1}{\sqrt{5}} \langle 1, \ -1 \mid 1, \ 1 \rangle \right|^2$$

$$= \frac{1}{5}, \qquad (5.276)$$

since $\langle 1, \ -1 \mid 1, \ 0 \rangle = \langle 1, \ -1 \mid 1, \ 1 \rangle = 0$ and $\langle 1, \ -1 \mid 1, \ -1 \rangle = 1$. Similarly, we can verify that the probabilities of measuring $l_z = 0$ and $\hbar$ are respectively given by

$$P_0 = |\langle 1, \ 0 \mid \psi \rangle|^2 = \left| \sqrt{\frac{3}{5}} \langle 1, \ 0 \mid 1, \ 0 \rangle \right|^2 = \frac{3}{5}, \qquad (5.277)$$

$$P_1 = |\langle 1, \ 1 \mid \psi \rangle|^2 = \left| \sqrt{\frac{1}{5}} \langle 1, \ 1 \mid 1, \ 1 \rangle \right|^2 = \frac{1}{5}. \qquad (5.278)$$

(c) After measuring $l_z = -\hbar$, the system will be in the eigenstate $\mid lm \rangle = \mid 1, \ -1 \rangle$, that is, $\psi(\theta, \ \varphi) = Y_{1,-1}(\theta, \ \varphi)$. We need first to calculate the expectation values of $\hat{L}_x$, $\hat{L}_y$, $\hat{L}_x^2$, and $\hat{L}_y^2$ using $\mid 1, \ -1 \rangle$. Symmetry requires that $\langle 1, \ -1 \mid \hat{L}_x \mid 1, \ -1 \rangle = \langle 1, \ -1 \mid \hat{L}_y \mid 1, \ -1 \rangle = 0$. The expectation values of $\hat{L}_x^2$ and $\hat{L}_y^2$ are equal, as shown in (5.60); they are given by

$$\langle \hat{L}_x^2 \rangle = \langle \hat{L}_y^2 \rangle = \frac{1}{2} [\langle \hat{L}^2 \rangle - \langle \hat{L}_z^2 \rangle] = \frac{\hbar^2}{2} \left[l(l+1) - m^2 \right] = \frac{\hbar^2}{2}; \qquad (5.279)$$

in this relation, we have used the fact that $l = 1$ and $m = -1$. Hence

$$\Delta L_x = \sqrt{\langle \hat{L}_x^2 \rangle} = \frac{\hbar}{\sqrt{2}} = \Delta L_y, \qquad (5.280)$$

and the uncertainties product $\Delta L_x \Delta L_y$ is given by

$$\Delta L_x \Delta L_y = \sqrt{\langle \hat{L}_x^2 \rangle \langle \hat{L}_y^2 \rangle} = \frac{\hbar^2}{2}. \qquad (5.281)$$

Problem 5.10
Find the angle between the angular momentum $l = 4$ and the z-axis for all possible orientations.

Solution
Since $m_l = 0, \pm 1, \pm 2, \ldots, \pm l$ and the angle between the orbital angular momentum l and the z-axis is $\cos\theta_{m_l} = m_l / \sqrt{l(l+1)}$ we have

$$\theta_{m_l} = \cos^{-1}\left[\frac{m_l}{\sqrt{l(l+1)}}\right] = \cos^{-1}\left[\frac{m_l}{2\sqrt{5}}\right]; \tag{5.282}$$

hence

$$\theta_0 = \cos^{-1}(0) = 90°, \tag{5.283}$$

$$\theta_1 = \cos^{-1}\left[\frac{1}{2\sqrt{5}}\right] = 77.08°, \qquad \theta_2 = \cos^{-1}\left[\frac{2}{2\sqrt{5}}\right] = 63.43°, \tag{5.284}$$

$$\theta_3 = \cos^{-1}\left[\frac{3}{2\sqrt{5}}\right] = 47.87°, \qquad \theta_4 = \cos^{-1}\left[\frac{4}{2\sqrt{5}}\right] = 26.57°. \tag{5.285}$$

The angles for the remaining quantum numbers $m_4 = -1, -2, -3, -4$ can be inferred at once from the relation

$$\theta_{-m_l} = 180° - \theta_{m_l}, \tag{5.286}$$

hence

$$\theta_{-1} = 180° - 77.08° = 102.92°, \qquad \theta_{-2} = 180° - 63.43° = 116.57°, \tag{5.287}$$

$$\theta_{-3} = 180° - 47.87° = 132.13°, \qquad \theta_{-4} = 180° - 26.57° = 153.43°. \tag{5.288}$$

Problem 5.11
Using $[\hat{X}, \hat{P}] = i\hbar$, calculate the various commutation relations between the following operators[2]

$$\hat{T}_1 = \frac{1}{4}(\hat{P}^2 - \hat{X}^2), \qquad \hat{T}_2 = \frac{1}{4}(\hat{X}\hat{P} + \hat{P}\hat{X}), \qquad \hat{T}_3 = \frac{1}{4}(\hat{P}^2 + \hat{X}^2).$$

Solution
The operators $\hat{T}_1$, $\hat{T}_2$, and $\hat{T}_3$ can be viewed as describing some sort of *collective* vibrations; $\hat{T}_3$ has the structure of a harmonic oscillator Hamiltonian. The first commutator can be calculated as follows:

$$[\hat{T}_1, \hat{T}_2] = \frac{1}{4}[\hat{P}^2 - \hat{X}^2, \hat{T}_2] = \frac{1}{4}[\hat{P}^2, \hat{T}_2] - \frac{1}{4}[\hat{X}^2, \hat{T}_2], \tag{5.289}$$

where, using the commutation relation $[\hat{X}, \hat{P}] = i\hbar$, we have

$$[\hat{P}^2, \hat{T}_2] = \frac{1}{4}[\hat{P}^2, \hat{X}\hat{P}] + \frac{1}{4}[\hat{P}^2, \hat{P}\hat{X}]$$

$$= \frac{1}{4}\hat{P}[\hat{P}, \hat{X}\hat{P}] + \frac{1}{4}[\hat{P}, \hat{X}\hat{P}]\hat{P} + \frac{1}{4}\hat{P}[\hat{P}, \hat{P}\hat{X}] + \frac{1}{4}[\hat{P}, \hat{P}\hat{X}]\hat{P}$$

$$= \frac{1}{4}\hat{P}[\hat{P}, \hat{X}]\hat{P} + \frac{1}{4}[\hat{P}, \hat{X}]\hat{P}^2 + \frac{1}{4}\hat{P}^2[\hat{P}, \hat{X}] + \frac{1}{4}\hat{P}[\hat{P}, \hat{X}]\hat{P}$$

---

[2]N. Zettili and F. Villars, *Nucl. Phys.*, **A469**, 77 (1987).

$$= -\frac{i\hbar}{4}\hat{P}^2 - \frac{i\hbar}{4}\hat{P}^2 - \frac{i\hbar}{4}\hat{P}^2 - \frac{i\hbar}{4}\hat{P}^2 = -i\hbar\hat{P}^2, \tag{5.290}$$

$$[\hat{X}^2, \hat{T}_2] = \frac{1}{4}[\hat{X}^2, \hat{X}\hat{P}] + \frac{1}{4}[\hat{X}^2, \hat{P}\hat{X}]$$

$$= \frac{1}{4}\hat{X}[\hat{X}, \hat{X}\hat{P}] + \frac{1}{4}[\hat{X}, \hat{X}\hat{P}]\hat{X} + \frac{1}{4}\hat{X}[\hat{X}, \hat{P}\hat{X}] + \frac{1}{4}[\hat{X}, \hat{P}\hat{X}]\hat{X}$$

$$= \frac{1}{4}\hat{X}^2[\hat{X}, \hat{P}] + \frac{1}{4}\hat{X}[\hat{X}, \hat{P}]\hat{X} + \frac{1}{4}\hat{X}[\hat{X}, \hat{P}]\hat{X} + \frac{1}{4}[\hat{X}, \hat{P}]\hat{X}^2$$

$$= \frac{i\hbar}{4}\hat{X}^2 + \frac{i\hbar}{4}\hat{X}^2 + \frac{i\hbar}{4}\hat{X}^2 + \frac{i\hbar}{4}\hat{X}^2 = i\hbar\hat{X}^2; \tag{5.291}$$

hence

$$[\hat{T}_1, \hat{T}_2] = \frac{1}{4}[\hat{P}^2 - \hat{X}^2, \hat{T}_2] = -\frac{1}{4}(i\hbar\hat{P}^2 + i\hbar\hat{X}^2) = -i\hbar\hat{T}_3. \tag{5.292}$$

The second commutator is calculated as follows:

$$[\hat{T}_2, \hat{T}_3] = \frac{1}{4}[\hat{T}_2, \hat{P}^2 + \hat{X}^2] = \frac{1}{4}[\hat{T}_2, \hat{P}^2] + \frac{1}{4}[\hat{T}_2, \hat{X}^2], \tag{5.293}$$

where $[\hat{T}_2, \hat{P}^2]$ and $[\hat{T}_2, \hat{X}^2]$ were calculated in (5.290) and (5.291):

$$[\hat{T}_2, \hat{P}^2] = i\hbar\hat{P}^2, \qquad [\hat{T}_2, \hat{X}^2] = -i\hbar\hat{X}^2. \tag{5.294}$$

Thus, we have

$$[\hat{T}_2, \hat{T}_3] = \frac{1}{4}(i\hbar\hat{P}^2 - i\hbar\hat{X}^2) = i\hbar\hat{T}_1. \tag{5.295}$$

The third commutator is

$$[\hat{T}_3, \hat{T}_1] = \frac{1}{4}[\hat{T}_3, \hat{P}^2 - \hat{X}^2] = \frac{1}{4}[\hat{T}_3, \hat{P}^2] - \frac{1}{4}[\hat{T}_3, \hat{X}^2], \tag{5.296}$$

where

$$[\hat{T}_3, \hat{P}^2] = \frac{1}{4}[\hat{P}^2, \hat{P}^2] + \frac{1}{4}[\hat{X}^2, \hat{P}^2] = \frac{1}{4}[\hat{X}^2, \hat{P}^2] = \frac{1}{4}\hat{X}[\hat{X}, \hat{P}^2] + \frac{1}{4}[\hat{X}, \hat{P}^2]\hat{X}$$

$$= \frac{1}{4}\hat{X}\hat{P}[\hat{X}, \hat{P}] + \frac{1}{4}\hat{X}[\hat{X}, \hat{P}]\hat{P} + \frac{1}{4}\hat{P}[\hat{X}, \hat{P}]\hat{X} + \frac{1}{4}[\hat{X}, \hat{P}]\hat{P}\hat{X}$$

$$= \frac{i\hbar}{4}(2\hat{X}\hat{P} + 2\hat{P}\hat{X}) = \frac{i\hbar}{2}(\hat{X}\hat{P} + \hat{P}\hat{X}), \tag{5.297}$$

$$[\hat{T}_3, \hat{X}^2] = \frac{1}{4}[\hat{P}^2, \hat{X}^2] + \frac{1}{4}[\hat{X}^2, \hat{X}^2] = \frac{1}{4}[\hat{P}^2, \hat{X}^2] = -\frac{i\hbar}{2}(\hat{X}\hat{P} + \hat{P}\hat{X}); \tag{5.298}$$

hence

$$[\hat{T}_3, \hat{T}_1] = \frac{1}{4}[\hat{T}_3, \hat{P}^2] - \frac{1}{4}[\hat{T}_3, \hat{X}^2] = \frac{i\hbar}{8}(\hat{X}\hat{P} + \hat{P}\hat{X}) + \frac{i\hbar}{8}(\hat{X}\hat{P} + \hat{P}\hat{X})$$

$$= \frac{i\hbar}{4}(\hat{X}\hat{P} + \hat{P}\hat{X}) = i\hbar\hat{T}_2. \tag{5.299}$$

In sum, the commutation relations between $\hat{T}_1$, $\hat{T}_2$, and $\hat{T}_3$ are

$$[\hat{T}_1, \hat{T}_2] = -i\hbar\hat{T}_3, \qquad [\hat{T}_2, \hat{T}_3] = i\hbar\hat{T}_1, \qquad [\hat{T}_3, \hat{T}_1] = i\hbar\hat{T}_2. \tag{5.300}$$

These relations are similar to those of ordinary angular momentum, save for the minus sign in $[\hat{T}_1, \hat{T}_2] = -i\hbar\hat{T}_3$.

Problem 5.12

Consider a particle whose wave function is

$$\psi(x, y, z) = \frac{1}{4\sqrt{\pi}} \frac{2z^2 - x^2 - y^2}{r^2} + \sqrt{\frac{3}{\pi}} \frac{xz}{r^2}.$$

(a) Calculate $\hat{L}^2 \psi(x, y, z)$ and $\hat{L}_z \psi(x, y, z)$. Find the total angular momentum of this particle.

(b) Calculate $\hat{L}_+ \psi(x, y, z)$ and $\langle \psi \mid \hat{L}_+ \mid \psi \rangle$.

(c) If a measurement of the z-component of the orbital angular momentum is carried out, find the probabilities corresponding to finding the results 0, $\hbar$, and $-\hbar$.

(d) What is the probability of finding the particle at the position $\theta = \pi/3$ and $\varphi = \pi/2$ within $d\theta = 0.03$ rad and $d\varphi = 0.03$ rad?

Solution

(a) Since $Y_{20}(x, y, z) = \sqrt{5/16\pi}\,(3z^2 - r^2)/r^2$ and $Y_{2,\pm1}(x, y, z) = \mp\sqrt{15/8\pi}\,(x \pm iy)z/r^2$, we can write

$$\frac{2z^2 - x^2 - y^2}{r^2} = \frac{3z^2 - r^2}{r^2} = \sqrt{\frac{16\pi}{5}}\,Y_{20} \qquad \text{and} \qquad \frac{xz}{r^2} = \sqrt{\frac{2\pi}{15}}\,(Y_{2,-1} - Y_{21}); \quad (5.301)$$

hence

$$\psi(x, y, z) = \frac{1}{4\sqrt{\pi}}\sqrt{\frac{16\pi}{5}}\,Y_{20} + \sqrt{\frac{3}{\pi}}\sqrt{\frac{2\pi}{15}}\,(Y_{2,-1} - Y_{21}) = \frac{1}{\sqrt{5}}\,Y_{20} + \sqrt{\frac{2}{5}}\,(Y_{2,-1} - Y_{21}).$$

$$(5.302)$$

Having expressed ψ in terms of the spherical harmonics, we can now easily write

$$\hat{L}^2 \psi(x, y, z) = \frac{1}{\sqrt{5}}\hat{L}^2 Y_{20} + \sqrt{\frac{2}{5}}\hat{L}^2\,(Y_{2,-1} - Y_{21}) = 6\hbar^2 \psi(x, y, z) \qquad (5.303)$$

and

$$\hat{L}_z \psi(x, y, z) = \frac{1}{\sqrt{5}}\hat{L}_z Y_{20} + \sqrt{\frac{2}{5}}\hat{L}_z\,(Y_{2,-1} - Y_{21}) = -\hbar\sqrt{\frac{2}{5}}\hat{L}_z\,(Y_{2,-1} + Y_{21}). \qquad (5.304)$$

This shows that $\psi(x, y, z)$ is an eigenstate of $\hat{L}^2$ with eigenvalue $6\hbar^2$; $\psi(x, y, z)$ is, however, not an eigenstate of $\hat{L}_z$. Thus the total angular momentum of the particle is

$$\sqrt{\langle \psi \mid \hat{L}^2 \mid \psi \rangle} = \sqrt{6}\hbar. \qquad (5.305)$$

(b) Using the relation $\hat{L}_+ Y_{lm} = \hbar\sqrt{l(l+1) - m(m+1)}\,Y_{l\,m+1}$, we have

$$\hat{L}_+ \psi(x, y, z) = \frac{1}{\sqrt{5}}\hat{L}_+ Y_{20} + \sqrt{\frac{2}{5}}\hat{L}_+\,(Y_{2,-1} - Y_{21}) = \hbar\sqrt{\frac{6}{5}}Y_{21} + \hbar\sqrt{\frac{2}{5}}\,\left(\sqrt{6}Y_{20} - 2Y_{22}\right);$$

$$(5.306)$$

hence

$$\langle \psi \mid \hat{L}_+ \mid \psi \rangle = \left[\frac{1}{\sqrt{5}} \langle 2, 0 \mid + \sqrt{\frac{2}{5}} (\langle 2, -1 \mid - \langle 2, 1 \mid) \right]$$

$$\times \left[\hbar \sqrt{\frac{6}{5}} Y_{21} + \hbar \sqrt{\frac{2}{5}} \left(\sqrt{6} Y_{20} - 2 Y_{22} \right) \right]$$

$$= 0. \tag{5.307}$$

(c) Since $\mid \psi \rangle = (1/\sqrt{5}) Y_{20} + \sqrt{2/5}(Y_{2,-1} - Y_{21})$, a calculation of $\langle \psi \mid \hat{L}_z \mid \psi \rangle$ yields

$$\langle \psi \mid \hat{L}_z \mid \psi \rangle = 0, \qquad \text{with probability} \qquad P_0 = \frac{1}{5}, \tag{5.308}$$

$$\langle \psi \mid \hat{L}_z \mid \psi \rangle = -\hbar, \qquad \text{with probability} \qquad P_{-1} = \frac{2}{5}, \tag{5.309}$$

$$\langle \psi \mid \hat{L}_z \mid \psi \rangle = \hbar, \qquad \text{with probability} \qquad P_1 = \frac{2}{5}. \tag{5.310}$$

(d) Since $\psi(x, y, z) = (1/4\sqrt{\pi})(2z^2 - x^2 - y^2)/r^2 + \sqrt{3/\pi} xz/r^2$ can be written in terms of the spherical coordinates as

$$\psi(\theta, \varphi) = \frac{1}{4\sqrt{\pi}}(3\cos\theta^2 - 1) + \sqrt{\frac{3}{\pi}} \sin\theta \cos\theta \cos\varphi, \tag{5.311}$$

the probability of finding the particle at the position θ and φ is

$$P(\theta, \varphi) = |\psi(\theta, \varphi)|^2 \sin\theta d\theta d\varphi = \left[\frac{1}{4\sqrt{\pi}}(3\cos^2\theta - 1) + \sqrt{\frac{3}{\pi}} \sin\theta \cos\theta \cos\varphi \right]^2 \sin\theta d\theta d\varphi; \tag{5.312}$$

hence

$$P\left(\frac{\pi}{3}, \frac{\pi}{2}\right) = \left[\frac{1}{4\sqrt{\pi}} \left(3\cos^2\frac{\pi}{3} - 1\right) + 0 \right]^2 (0.03)^2 \sin\frac{\pi}{3} = 9.7 \times 10^{-7}. \tag{5.313}$$

Problem 5.13

Consider a particle of spin $s = 3/2$.

(a) Find the matrices representing the operators $\hat{S}_z$, $\hat{S}_x$, $\hat{S}_y$, $\hat{S}_x^2$, and $\hat{S}_y^2$ within the basis of $\hat{S}^2$ and $\hat{S}_z$.

(b) Find the energy levels of this particle when its Hamiltonian is given by

$$\hat{H} = \frac{\varepsilon_0}{\hbar^2}(\hat{S}_x^2 - \hat{S}_y^2) - \frac{\varepsilon_0}{\hbar} \hat{S}_z,$$

where ε_0 is a constant having the dimensions of energy. Are these levels degenerate?

(c) If the system was initially in an eigenstate $\mid \psi_0 \rangle = \begin{pmatrix} 1 \\ 0 \\ 0 \\ 0 \end{pmatrix}$, find the state of the system at time t.

Solution

(a) Following the same procedure that led to (5.73) and (5.75), we can verify that for $s = \frac{3}{2}$ we have

$$S_z = \frac{\hbar}{2} \begin{pmatrix} 3 & 0 & 0 & 0 \\ 0 & 1 & 0 & 0 \\ 0 & 0 & -1 & 0 \\ 0 & 0 & 0 & -3 \end{pmatrix}, \tag{5.314}$$

$$\hat{S}_- = \hbar \begin{pmatrix} 0 & 0 & 0 & 0 \\ \sqrt{3} & 0 & 0 & 0 \\ 0 & 2 & 0 & 0 \\ 0 & 0 & \sqrt{3} & 0 \end{pmatrix}, \qquad \hat{S}_+ = \hbar \begin{pmatrix} 0 & \sqrt{3} & 0 & 0 \\ 0 & 0 & 2 & 0 \\ 0 & 0 & 0 & \sqrt{3} \\ 0 & 0 & 0 & 0 \end{pmatrix}, \tag{5.315}$$

which, when combined with $\hat{S}_x = (\hat{S}_+ + \hat{S}_-)/2$ and $\hat{S}_y = i(\hat{S}_- - \hat{S}_+)/2$, lead to

$$\hat{S}_x = \frac{\hbar}{2} \begin{pmatrix} 0 & \sqrt{3} & 0 & 0 \\ \sqrt{3} & 0 & 2 & 0 \\ 0 & 2 & 0 & \sqrt{3} \\ 0 & 0 & \sqrt{3} & 0 \end{pmatrix}, \qquad \hat{S}_y = \frac{i\hbar}{2} \begin{pmatrix} 0 & -\sqrt{3} & 0 & 0 \\ \sqrt{3} & 0 & -2 & 0 \\ 0 & 2 & 0 & -\sqrt{3} \\ 0 & 0 & \sqrt{3} & 0 \end{pmatrix}. \tag{5.316}$$

Thus, we have

$$\hat{S}_x^2 = \frac{\hbar^2}{4} \begin{pmatrix} 3 & 0 & 2\sqrt{3} & 0 \\ 0 & 7 & 0 & 2\sqrt{3} \\ 2\sqrt{3} & 0 & 7 & 0 \\ 0 & 2\sqrt{3} & 0 & 3 \end{pmatrix}, \qquad \hat{S}_y^2 = \frac{\hbar^2}{4} \begin{pmatrix} 3 & 0 & -2\sqrt{3} & 0 \\ 0 & 7 & 0 & -2\sqrt{3} \\ -2\sqrt{3} & 0 & 7 & 0 \\ 0 & -2\sqrt{3} & 0 & 3 \end{pmatrix}. \tag{5.317}$$

(b) The Hamiltonian is then given by

$$H = \frac{\varepsilon_0}{\hbar^2}(\hat{S}_x^2 - \hat{S}_y^2) - \frac{\varepsilon_0}{\hbar}\hat{S}_z = \frac{1}{2}\varepsilon_0 \begin{pmatrix} -3 & 0 & 2\sqrt{3} & 0 \\ 0 & -1 & 0 & 2\sqrt{3} \\ 2\sqrt{3} & 0 & 1 & 0 \\ 0 & 2\sqrt{3} & 0 & 3 \end{pmatrix}. \tag{5.318}$$

The diagonalization of this Hamiltonian yields the following energy values:

$$E_1 = -\frac{5}{2}\varepsilon_0, \qquad E_2 = -\frac{3}{2}\varepsilon_0, \qquad E_3 = \frac{3}{2}\varepsilon_0, \qquad E_4 = \frac{5}{2}\varepsilon_0. \tag{5.319}$$

The corresponding normalized eigenvectors are given by

$$|1\rangle = \frac{1}{2}\begin{pmatrix} -\sqrt{3} \\ 0 \\ 1 \\ 0 \end{pmatrix}, \quad |2\rangle = \frac{1}{2}\begin{pmatrix} 0 \\ -\sqrt{3} \\ 0 \\ 1 \end{pmatrix}, \quad |3\rangle = \frac{1}{\sqrt{12}}\begin{pmatrix} \sqrt{3} \\ 0 \\ 3 \\ 0 \end{pmatrix}, \quad |4\rangle = \frac{1}{2}\begin{pmatrix} 0 \\ 1 \\ 0 \\ \sqrt{3} \end{pmatrix}. \tag{5.320}$$

None of the energy levels is degenerate.

(c) Since the initial state $|\psi_0\rangle$ can be written in terms of the eigenvectors (5.320) as follows:

$$|\psi_0\rangle = \begin{pmatrix} 1 \\ 0 \\ 0 \\ 0 \end{pmatrix} = -\frac{\sqrt{3}}{2}|1\rangle + \frac{1}{2}|3\rangle; \tag{5.321}$$

the eigenfunction at a later time t is given by

$$| \psi(t) \rangle = -\frac{\sqrt{3}}{2} | 1 \rangle e^{-i E_1 t/\hbar} + \frac{1}{2} | 3 \rangle e^{-i E_3 t/\hbar}$$

$$= -\frac{\sqrt{3}}{4} \begin{pmatrix} -\sqrt{3} \\ 0 \\ 1 \\ 0 \end{pmatrix} \exp\left[\frac{5 i \varepsilon_0 t}{2\hbar} \right] + \frac{1}{2\sqrt{12}} \begin{pmatrix} \sqrt{3} \\ 0 \\ 3 \\ 0 \end{pmatrix} \exp\left[-\frac{3 i \varepsilon_0 t}{2\hbar} \right].$$

(5.322)

5.9 Exercises

Exercise 5.1
(a) Show the following commutation relations:

$$[\hat{Y}, \hat{L}_y] = 0, \qquad [\hat{Y}, \hat{L}_z] = i\hbar \hat{X}, \qquad [\hat{Y}, \hat{L}_x] = -i\hbar \hat{Z},$$

$$[\hat{Z}, \hat{L}_z] = 0, \qquad [\hat{Z}, \hat{L}_x] = i\hbar \hat{Y}, \qquad [\hat{Z}, \hat{L}_y] = -i\hbar \hat{X}.$$

(b) Using a cyclic permutation of xyz, apply the results of (a) to infer expressions for $[\hat{X}, \hat{L}_x], [\hat{X}, \hat{L}_y]$, and $[\hat{X}, \hat{L}_z]$.

(c) Use the results of (a) and (b) to calculate $[\hat{R}^2, \hat{L}_x], [\hat{R}^2, \hat{L}_y]$, and $[\hat{R}^2, \hat{L}_x]$, where $\hat{R}^2 = \hat{X}^2 + \hat{Y}^2 + \hat{Z}^2$.

Exercise 5.2
(a) Show the following commutation relations:

$$[\hat{P}_y, \hat{L}_y] = 0, \qquad [\hat{P}_y, \hat{L}_z] = i\hbar \hat{P}_x, \qquad [\hat{P}_y, \hat{L}_x] = -i\hbar \hat{P}_z,$$

$$[\hat{P}_z, \hat{L}_z] = 0, \qquad [\hat{P}_z, \hat{L}_x] = i\hbar \hat{P}_y, \qquad [\hat{P}_z, \hat{L}_y] = -i\hbar \hat{P}_x.$$

(b) Use the results of (a) to infer by means of a cyclic permutation the expressions for $[\hat{P}_x, \hat{L}_x], [\hat{P}_x, \hat{L}_y]$, and $[\hat{P}_x, \hat{L}_z]$.

(c) Use the results of (a) and (b) to calculate $[\hat{P}^2, \hat{L}_x], [\hat{P}^2, \hat{L}_y]$, and $[\hat{P}^2, \hat{L}_z]$, where $\hat{P}^2 = \hat{P}_x^2 + \hat{P}_y^2 + \hat{P}_z^2$.

Exercise 5.3
If $\hat{L}_\pm$ and $\hat{R}_\pm$ are defined by $\hat{L}_\pm = \hat{L}_x \pm i\hat{L}_y$ and $\hat{R}_\pm = \hat{X} \pm i\hat{Y}$, prove the following commutators: (a) $[\hat{L}_\pm, \hat{R}_\pm] = \pm 2\hbar \hat{Z}$ and (b) $[\hat{L}_\pm, \hat{R}_\mp] = 0$.

Exercise 5.4
If $\hat{L}_\pm$ and $\hat{R}_\pm$ are defined by $\hat{L}_\pm = \hat{L}_x \pm i\hat{L}_y$ and $\hat{R}_\pm = \hat{X} \pm i\hat{Y}$, prove the following commutators: (a) $[\hat{L}_\pm, \hat{Z}] = \mp \hbar \hat{R}_\pm$, (b) $[\hat{L}_z, \hat{R}_\pm] = \pm \hbar \hat{R}_\pm$, and (c) $[\hat{L}_z, \hat{Z}] = 0$.

Exercise 5.5
Prove the following two relations: $\hat{R} \cdot \hat{L} = 0$ and $\hat{P} \cdot \hat{L} = 0$.

Exercise 5.6

The Hamiltonian due to the interaction of a particle of spin $\vec{S}$ with a magnetic field $\vec{B}$ is given by $\hat{H} = -\vec{S} \cdot \vec{B}$ where $\vec{S}$ is the spin. Calculate the commutator $[\vec{S}, \hat{H}]$.

Exercise 5.7

Prove the following relation:

$$[\hat{L}_z, \cos\varphi] = i\hbar \sin\varphi,$$

where φ is the azimuthal angle.

Exercise 5.8

Prove the following relation:

$$[\hat{L}_z, \sin(2\varphi)] = 2i\hbar \left(\sin^2\varphi - \cos^2\varphi \right),$$

where φ is the azimuthal angle. *Hint:* $[\hat{A}, \hat{B}\hat{C}] = \hat{B}[\hat{A}, \hat{C}] + [\hat{A}, \hat{B}]\hat{C}$.

Exercise 5.9

Using the properties of $\hat{J}_+$ and $\hat{J}_-$, calculate $|j, \pm j\rangle$ and $|j, \pm m\rangle$ as functions of the action of $\hat{J}_\pm$ on the states $|j, \pm m\rangle$ and $|j, \pm j\rangle$, respectively.

Exercise 5.10

Consider the operator $\hat{A} = \frac{1}{2}(\hat{J}_x\hat{J}_y + \hat{J}_y\hat{J}_x)$.

(a) Calculate the expectation value of $\hat{A}$ and $\hat{A}^2$ with respect to the state $|j, m\rangle$.

(b) Use the result of (a) to find an expression for $\hat{A}^2$ in terms of: $\hat{J}^4, \hat{J}^2, \hat{J}_z^2, \hat{J}_+^4, \hat{J}_-^4$.

Exercise 5.11

Consider the wave function

$$\psi(\theta, \varphi) = 3\sin\theta \cos\theta e^{i\varphi} - 2(1 - \cos^2\theta)e^{2i\varphi}.$$

(a) Write $\psi(\theta, \varphi)$ in terms of the spherical harmonics.

(b) Write the expression found in (a) in terms of the Cartesian coordinates.

(c) Is $\psi(\theta, \varphi)$ an eigenstate of $\hat{L}^2$ or $\hat{L}_z$?

(d) Find the probability of measuring $2\hbar$ for the z-component of the orbital angular momentum.

Exercise 5.12

Show that $\hat{L}_z(\cos^2\varphi - \sin^2\varphi + 2i\sin\varphi\cos\varphi) = 2\hbar^{2i\varphi}$, where φ is the azimuthal angle.

Exercise 5.13

Find the expressions for the spherical harmonics $Y_{30}(\theta, \varphi)$ and $Y_{3,\pm1}(\theta, \varphi)$,

$$Y_{30}(\theta, \varphi) = \sqrt{7/16\pi}\,(5\cos^3\theta - 3\cos\theta), \quad Y_{3,\pm1}(\theta, \varphi) = \mp\sqrt{21/64\pi}\,\sin\theta(5\cos^2\theta - 1)e^{\pm i\varphi},$$

in terms of the Cartesian coordinates x, y, z.

Exercise 5.14

(a) Show that the following expectation values between $|lm\rangle$ states satisfy the relations $\langle\hat{L}_x\rangle = \langle\hat{L}_y\rangle = 0$ and $\langle\hat{L}_x^2\rangle = \langle\hat{L}_y^2\rangle = \frac{1}{2}[l(l+1)\hbar^2 - m^2\hbar^2]$.

(b) Verify the inequality $\Delta L_x \Delta L_y \geq \hbar^2 m/2$, where $\Delta L_x = \sqrt{\langle L_x^2\rangle - \langle L_x\rangle^2}$.

Exercise 5.15

A particle of mass m is fixed at one end of a rigid rod of negligible mass and length R. The other end of the rod rotates in the xy plane about a bearing located at the origin, whose axis is in the z-direction.

(a) Write the system's total energy in terms of its angular momentum L.

(b) Write down the time-independent Schrödinger equation of the system. *Hint:* In spherical coordinates, only φ varies.

(c) Solve for the possible energy levels of the system, in terms of m and the moment of inertia $I = mR^2$.

(d) Explain why there is no zero-point energy.

Exercise 5.16

Consider a system which is described by the state

$$\psi(\theta, \varphi) = \sqrt{\frac{3}{8}} Y_{11}(\theta, \varphi) + \sqrt{\frac{1}{8}} Y_{10}(\theta, \varphi) + A Y_{1,-1}(\theta, \varphi),$$

where A is a real constant

(a) Calculate A so that $| \psi \rangle$ is normalized.

(b) Find $\hat{L}_+ \psi(\theta, \varphi)$.

(c) Calculate the expectation values of $\hat{L}_x$ and $\hat{L}^2$ in the state $| \psi \rangle$.

(d) Find the probability associated with a measurement that gives zero for the z-component of the angular momentum.

(e) Calculate $\langle \Phi | \hat{L}_z | \psi \rangle$ and $\langle \Phi | \hat{L}_- | \psi \rangle$ where

$$\Phi(\theta, \varphi) = \sqrt{\frac{8}{15}} Y_{11}(\theta, \varphi) + \sqrt{\frac{4}{15}} Y_{10}(\theta, \varphi) + \sqrt{\frac{3}{15}} Y_{2,-1}(\theta, \varphi).$$

Exercise 5.17

(a) Using the commutation relations of angular momentum, verify the validity of the (Jacobi) identity: $[\hat{J}_x, [\hat{J}_y, \hat{J}_z]] + [\hat{J}_y, [\hat{J}_z, \hat{J}_x]] + [\hat{J}_z, [\hat{J}_x, \hat{J}_y]] = 0$.

(b) Prove the following identity: $[\hat{J}_x^2, \hat{J}_y^2] = [\hat{J}_y^2, \hat{J}_z^2] = [\hat{J}_z^2, \hat{J}_x^2]$.

(c) Calculate the expressions of $\hat{L}_- \hat{L}_+ Y_{lm}(\theta, \varphi)$ and $\hat{L}_+ \hat{L}_- Y_{lm}(\theta, \varphi)$, and then infer the commutator $[\hat{L}_+ \hat{L}_-, \hat{L}_- \hat{L}_+] Y_{lm}(\theta, \varphi)$.

Exercise 5.18

Consider a particle whose wave function is given by $\psi(x, y, z) = A[(x + z)y + z^2]/r^2 - A/3$, where A is a constant.

(a) Is ψ an eigenstate of $\hat{L}^2$? If yes, what is the corresponding eigenvalue? Is it also an eigenstate of $\hat{L}_z$?

(b) Find the constant A so that ψ is normalized.

(c) Find the relative probabilities for measuring the various values of $\hat{L}_z$ and $\hat{L}^2$, and then calculate the expectation values of $\hat{L}_z$ and $\hat{L}^2$.

(d) Calculate $\hat{L}_\pm | \psi \rangle$ and then infer $\langle \psi | \hat{L}_\pm | \psi \rangle$.

Exercise 5.19

Consider a system which is in the state

$$\psi(\theta, \varphi) = \sqrt{\frac{2}{13}} Y_{3,-3} + \sqrt{\frac{3}{13}} Y_{3,-2} + \sqrt{\frac{3}{13}} Y_{30} + \sqrt{\frac{3}{13}} Y_{3,2} + \sqrt{\frac{2}{13}} Y_{33}.$$

(a) If $\hat{L}_z$ were measured, what values will one obtain and with what probabilities?

(b) If after a measurement of $\hat{L}_z$ we find $l_z = 2\hbar$, calculate the uncertainties ΔL_x and ΔL_y and their product $\Delta L_x \Delta L_y$.

(c) Find $\langle \psi \mid \hat{L}_x \mid \psi \rangle$ and $\langle \psi \mid \hat{L}_y \mid \psi \rangle$.

Exercise 5.20

(a) Calculate the energy eigenvalues of an axially symmetric rotator and find the degeneracy of each energy level (i.e., for each value of the azimuthal quantum number m, find how many states $|l\ m\rangle$ correspond to the same energy). We may recall that the Hamiltonian of an axially symmetric rotator is given by

$$\hat{H} = \frac{\hat{L}_x^2 + \hat{L}_y^2}{2I_1} + \frac{\hat{L}_z^2}{2I_2},$$

where I_1 and I_2 are the moments of inertia.

(b) From part (a) infer the energy eigenvalues for the various levels of $l = 3$.

(c) In the case of a rigid rotator (i.e., $I_1 = I_2 = I$), find the energy expression and the corresponding degeneracy relation.

(d) Calculate the orbital quantum number l and the corresponding energy degeneracy for a rigid rotator where the magnitude of the total angular momentum is $\sqrt{56}\hbar$.

Exercise 5.21

Consider a system of total angular momentum $j = 1$. We are interested here in the measurement of $\hat{J}_y$; its matrix is given by

$$\hat{J}_y = \frac{\hbar}{\sqrt{2}} \begin{pmatrix} 0 & -i & 0 \\ i & 0 & -i \\ 0 & i & 0 \end{pmatrix}.$$

(a) What are the possible values will we obtain when measuring $\hat{J}_y$?

(b) Calculate $\langle \hat{J}_z \rangle$, $\langle \hat{J}_z^2 \rangle$, and ΔJ_z if the system is in the state $j_y = \hbar$.

(c) Repeat (b) for $\langle \hat{J}_x \rangle$, $\langle \hat{J}_x^2 \rangle$, and ΔJ_x.

Exercise 5.22

Calculate $Y_{3,\pm 2}(\theta, \varphi)$ by applying the ladder operators $\hat{L}_\pm$ on $Y_{3,\pm 1}(\theta, \varphi)$.

Exercise 5.23

Consider a system of total angular momentum $j = 1$. We want to carry out measurements on

$$\hat{J}_z = \hbar \begin{pmatrix} 1 & 0 & 0 \\ 0 & 0 & 0 \\ 0 & 0 & -1 \end{pmatrix}.$$

(a) What are the possible values will we obtain when measuring $\hat{J}_z$?

(b) Calculate $\langle \hat{J}_x \rangle$, $\langle \hat{J}_x^2 \rangle$, and ΔJ_x if the system is in the state $j_z = -\hbar$.

(c) Repeat (b) for $\langle \hat{J}_y \rangle$, $\langle \hat{J}_y^2 \rangle$, and ΔJ_y.

Exercise 5.24

Consider a system which is in the state

$$\psi(x, y, z) = \frac{1}{4\sqrt{\pi}} \frac{z}{r} + \frac{1}{\sqrt{3\pi}} \frac{x}{r}.$$

(a) Express $\psi(x, y, z)$ in terms of the spherical harmonics then calculate $\hat{L}^2 \psi(x, y, z)$ and $\hat{L}_z \psi(x, y, z)$. Is $\psi(x, y, z)$ an eigenstate of $\hat{L}^2$ or $\hat{L}_z$?

(b) Calculate $\hat{L}_\pm \psi(x, y, z)$ and $\langle \psi \mid \hat{L}_\pm \mid \psi \rangle$.

(c) If a measurement of the z-component of the orbital angular momentum is carried out, find the probabilities corresponding to finding the results 0, $\hbar$, and $-\hbar$.

Exercise 5.25

Consider a system whose wave function is given by

$$\psi(\theta, \varphi) = \frac{1}{2} Y_{00}(\theta, \varphi) + \frac{1}{\sqrt{3}} Y_{11}(\theta, \varphi) + \frac{1}{2} Y_{1,-1}(\theta, \varphi) + \frac{1}{\sqrt{6}} Y_{22}(\theta, \varphi).$$

(a) Is $\psi(\theta, \varphi)$ normalized?

(b) Is $\psi(\theta, \varphi)$ an eigenstate of $\hat{L}^2$ or $\hat{L}_z$?

(c) Calculate $\hat{L}_\pm \psi(\theta, \varphi)$ and $\langle \psi \mid \hat{L}_\pm \mid \psi \rangle$.

(d) If a measurement of the z-component of the orbital angular momentum is carried out, find the probabilities corresponding to finding the results 0, $\hbar$, $-\hbar$, and $2\hbar$.

Exercise 5.26

Using the expression of $\hat{L}_-$ in spherical coordinates, prove the following two commutators: $[\hat{L}_-, e^{-i\varphi} \sin\theta] = 0$ and $[\hat{L}_-, \cos\theta] = \hbar e^{-i\varphi} \sin\theta$.

Exercise 5.27

Consider a particle whose angular momentum is $l = 1$.

(a) Find the eigenvalues and eigenvectors, $|1, m_x\rangle$, of $\hat{L}_x$.

(b) Express the state $|1, m_x = 1\rangle$ as a linear superposition of the eigenstates of $\hat{L}_z$. Hint: you need first to find the eigenstates of L_x and find which of them corresponds to the eigenvalue $m_x = 1$; this eigenvector will be expanded in the z basis.

(c) What is the probability of measuring $m_z = 1$ when the particle is in the eigenstate $|1, m_x = 1\rangle$? What about the probability corresponding to measuring $m_z = 0$?

(d) Suppose that a measurement of the z-component of angular momentum is performed and that the result $m_z = 1$ is obtained. Now we measure the x-component of angular momentum. What are the possible results and with what probabilities?

Exercise 5.28

Consider a system which is given in the following angular momentum eigenstates $| l, m \rangle$:

$$| \psi \rangle = \frac{1}{\sqrt{7}} | 1, -1 \rangle + A | 1, 0 \rangle + \sqrt{\frac{2}{7}} | 1, 1 \rangle,$$

where A is a real constant

(a) Calculate A so that $| \psi \rangle$ is normalized.

(b) Calculate the expectation values of $\hat{L}_x$, $\hat{L}_y$, $\hat{L}_z$, and $\hat{L}^2$ in the state $| \psi \rangle$.

(c) Find the probability associated with a measurement that gives $1\hbar$ for the z-component of the angular momentum.

(d) Calculate $\langle 1, m|\hat{L}_+^2 \mid \psi \rangle$ and $\langle 1, m|\hat{L}_-^2 \mid \psi \rangle$.

Exercise 5.29
Consider a particle of angular momentum $j = 3/2$.
(a) Find the matrices representing the operators $\hat{J}^2$, $\hat{J}_x$, $\hat{J}_y$, and $\hat{J}_z$ in the $\{|\frac{3}{2}, m\rangle\}$ basis.
(b) Using these matrices, show that $\hat{J}_x$, $\hat{J}_y$, $\hat{J}_z$ satisfy the commutator $[\hat{J}_x, \hat{J}_y] = i\hbar\hat{J}_z$.

(c) Calculate the mean values of $\hat{J}_x$ and $\hat{J}_x^2$ with respect to the state $\begin{pmatrix} 0 \\ 0 \\ 1 \\ 0 \end{pmatrix}$.

(d) Calculate $\Delta J_x \Delta J_y$ with respect to the state

$$\begin{pmatrix} 0 \\ 0 \\ 1 \\ 0 \end{pmatrix}$$

and verify that this product satisfies Heisenberg's uncertainty principle.

Exercise 5.30
Consider the Pauli matrices

$$\sigma_x = \begin{pmatrix} 0 & 1 \\ 1 & 0 \end{pmatrix}, \qquad \sigma_y = \begin{pmatrix} 0 & -i \\ i & 0 \end{pmatrix}, \qquad \sigma_z = \begin{pmatrix} 1 & 0 \\ 0 & -1 \end{pmatrix}.$$

(a) Verify that $\sigma_x^2 = \sigma_y^2 = \sigma_z^2 = I$, where I is the unit matrix

$$I = \begin{pmatrix} 1 & 0 \\ 0 & 1 \end{pmatrix}.$$

(b) Calculate the commutators $[\sigma_x, \sigma_y]$, $[\sigma_x, \sigma_z]$, and $[\sigma_y, \sigma_z]$.
(c) Calculate the anticommutator $\sigma_x\sigma_y + \sigma_y\sigma_x$.
(d) Show that $e^{i\theta\sigma_y} = I\cos\theta + i\sigma_y\sin\theta$, where I is the unit matrix.
(e) Derive an expression for $e^{i\theta\sigma_z}$ by analogy with the one for σ_y.

Exercise 5.31
Consider a spin $\frac{3}{2}$ particle whose Hamiltonian is given by

$$\hat{H} = \frac{\varepsilon_0}{\hbar^2}(\hat{S}_x^2 - \hat{S}_y^2) - \frac{\varepsilon_0}{\hbar^2}\hat{S}_z^2,$$

where ε_0 is a constant having the dimensions of energy.
(a) Find the matrix of the Hamiltonian and diagonalize it to find the energy levels.
(b) Find the eigenvectors and verify that the energy levels are doubly degenerate.

Exercise 5.32
Find the energy levels of a spin $\frac{5}{2}$ particle whose Hamiltonian is given by

$$\hat{H} = \frac{\varepsilon_0}{\hbar^2}(\hat{S}_x^2 + \hat{S}_y^2) + \frac{\varepsilon_0}{\hbar}\hat{S}_z,$$

where ε_0 is a constant having the dimensions of energy. Are the energy levels degenerate?

Exercise 5.33

Consider an electron whose spin direction is located in the xy plane.

(a) Find the eigenvalues (call them λ_1, λ_2) and eigenstates ($| \lambda_1\rangle$, $| \lambda_2\rangle$) of the electron's spin operator $\vec{S}$.

(b) Assuming that the initial state of the electron is given by

$$| \psi_0\rangle = \frac{1}{3} | \lambda_1\rangle + \frac{2\sqrt{2}}{3} | \lambda_2\rangle,$$

find the probability of obtaining a value of $\hat{S} = -\hbar/2$ after measuring the spin of the electron.

Exercise 5.34

(a) Find the eigenvalues (call them λ_1, λ_2) and eigenstates ($| \lambda_1\rangle$, $| \lambda_2\rangle$) of the spin operator $\vec{S}$ of an electron when $\vec{S}$ is pointing along an arbitrary unit vector $\vec{n}$ that lies within the yz plane.

(b) Assuming that the initial state of the electron is given by

$$| \psi_0\rangle = \frac{1}{2} | \lambda_1\rangle + \frac{\sqrt{3}}{2} | \lambda_2\rangle,$$

find the probability of obtaining a value of $\hat{S} = \hbar/2$ after measuring the spin of the electron.

Exercise 5.35

Consider a particle of spin $\frac{3}{2}$. Find the matrix for the component of the spin along a unit vector with arbitrary direction $\vec{n}$. Find its eigenvalues and eigenvectors. *Hint:*

$$\vec{n} = (\sin\theta\cos\varphi)\vec{i} + (\sin\theta\sin\varphi)\vec{j} + (\cos\theta)\vec{k}.$$

Exercise 5.36

Show that $[\hat{J}_x\hat{J}_y, \ \hat{J}_z] + [\hat{J}_x, \ \hat{J}_y\hat{J}_z] = i\hbar\left(\hat{J}_x^2 - 2\hat{J}_y^2 + \hat{J}_z^2\right).$

Exercise 5.37

Find the eigenvalues of the operators $\hat{L}^2$ and $\hat{L}_z$ for each of the following states:

(a) $Y_{21}(\theta, \varphi)$,

(b) $Y_{3,-2}(\theta, \varphi)$,

(c) $\frac{1}{\sqrt{2}}[Y_{33}(\theta, \varphi) + Y_{3,-3}(\theta, \varphi)]$, and

(d) $Y_{40}(\theta, \varphi)$.

Exercise 5.38

Use the following general relations:

$$| \psi_x\rangle_\pm = \frac{1}{\sqrt{2}}\left[\left|\frac{1}{2}, \frac{1}{2}\right\rangle \pm \left|\frac{1}{2}, -\frac{1}{2}\right\rangle\right], \qquad | \psi_y\rangle_\pm = \frac{1}{\sqrt{2}}\left[\left|\frac{1}{2}, \frac{1}{2}\right\rangle \pm i\left|\frac{1}{2}, -\frac{1}{2}\right\rangle\right]$$

to verify the following eigenvalue equations:

$$\hat{S}_x | \psi_x\rangle_\pm = \pm\frac{\hbar}{2}| \psi_x\rangle_\pm \qquad \text{and} \qquad \hat{S}_y | \psi_y\rangle_\pm = \pm\frac{\hbar}{2}| \psi_y\rangle_\pm.$$

Chapter 6

Three-Dimensional Problems

6.1 Introduction

In this chapter we examine how to solve the Schrödinger equation for spinless particles moving in three-dimensional potentials. We carry out this study in two different coordinate systems: the Cartesian system and the spherical system.

First, working within the context of Cartesian coordinates, we study the motion of a particle in different potentials: the free particle, a particle in a (three-dimensional) rectangular potential, and a particle in a harmonic oscillator potential. This study is going to be a simple generalization of the one-dimensional problems presented in Chapter 4. Unlike the one-dimensional case, three-dimensional problems often exhibit degeneracy, which occurs whenever the potential displays symmetry.

Second, using spherical coordinates, we describe the motion of a particle in spherically symmetric potentials. After presenting a general treatment, we consider several applications ranging from the free particle and the isotropic harmonic oscillator to the hydrogen atom. We conclude the chapter by calculating the energy levels of a hydrogen atom when placed in a constant magnetic field; this gives rise to the Zeeman effect.

6.2 3D Problems in Cartesian Coordinates

We examine here how to extend Schrödinger's theory of one-dimensional problems (Chapter 4) to three dimensions.

6.2.1 General Treatment: Separation of Variables

The time-dependent Schrödinger equation for a spinless particle of mass m moving under the influence of a three-dimensional potential is

$$-\frac{\hbar^2}{2m}\vec{\nabla}^2\Psi(x,y,z,t) + \hat{V}(x,y,z,t)\Psi(x,y,z) = i\hbar\frac{\partial\Psi(x,y,z,t)}{\partial t}, \qquad (6.1)$$

where $\vec{\nabla}^2$ is the Laplacian, $\vec{\nabla}^2 = \partial^2/\partial x^2 + \partial^2/\partial y^2 + \partial^2/\partial z^2$. As seen in Chapter 4, the wave function of a particle moving in a time-independent potential can be written as a product of spatial and time components:

$$\Psi(x, y, z, t) = \psi(x, y, z)e^{-iEt/\hbar}, \tag{6.2}$$

where $\psi(x, y, z)$ is the solution to the time-independent Schrödinger equation:

$$-\frac{\hbar^2}{2m}\vec{\nabla}^2\psi(x, y, z) + \hat{V}(x, y, z)\psi(x, y, z) = E\psi(x, y, z), \tag{6.3}$$

which is of the form $\hat{H}\psi = E\psi$.

This partial differential equation is generally difficult to solve. But, for those cases where the potential $\hat{V}(x, y, z)$ separates into the sum of three independent, one-dimensional terms (which should not be confused with a vector)

$$V(x, y, z) = V_x(x) + V_y(y) + V_z(z), \tag{6.4}$$

we can solve (6.3) by means of the technique of *separation of variables*. This technique consists of separating the three-dimensional Schrödinger equation (6.3) into three independent one-dimensional Schrödinger equations. Let us examine how to achieve this. Note that (6.3), in conjunction with (6.4), can be written as

$$\left[\hat{H}_x + \hat{H}_y + \hat{H}_z\right]\psi(x, y, z) = E\psi(x, y, z), \tag{6.5}$$

where $\hat{H}_x$ is given by

$$\hat{H}_x = -\frac{\hbar^2}{2m}\frac{\partial^2}{\partial x^2} + V_x(x); \tag{6.6}$$

the expressions for $\hat{H}_y$ and $\hat{H}_z$ are analogous.

As $\hat{V}(x, y, z)$ separates into three independent terms, we can also write $\psi(x, y, z)$ as a product of three functions of a single variable each:

$$\psi(x, y, z) = X(x)Y(y)Z(z). \tag{6.7}$$

Substituting (6.7) into (6.5) and dividing by $X(x)Y(y)Z(z)$, we obtain

$$\left[-\frac{\hbar^2}{2m}\frac{1}{X}\frac{d^2X}{dx^2} + V_x(x)\right] + \left[-\frac{\hbar^2}{2m}\frac{1}{Y}\frac{d^2Y}{dy^2} + V_y(y)\right]$$

$$+ \left[-\frac{\hbar^2}{2m}\frac{1}{Z}\frac{d^2Z}{dz^2} + V_z(z)\right] = E. \tag{6.8}$$

Since each expression in the square brackets depends on only one of the variables x, y, z, and since the sum of these three expressions is equal to a constant, E, each separate expression must then be equal to a constant such that the sum of these three constants is equal to E. For instance, the x-dependent expression is given by

$$\left[-\frac{\hbar^2}{2m}\frac{d^2}{dx^2} + V_x(x)\right]X(x) = E_x X(x). \tag{6.9}$$

Similar equations hold for the y and z coordinates, with

$$E_x + E_y + E_z = E. \tag{6.10}$$

The separation of variables technique consists in essence of reducing the three-dimensional Schrödinger equation (6.3) into three separate one-dimensional equations (6.9).

6.2.2 The Free Particle

In the simple case of a free particle, the Schrödinger equation (6.3) reduces to three equations similar to (6.9) with $V_x = 0$, $V_y = 0$, and $V_z = 0$. The x-equation can be obtained from (6.9):

$$\frac{d^2 X(x)}{dx^2} = -k_x^2 X(x), \tag{6.11}$$

where $k_x^2 = 2m E_x/\hbar^2$, and hence $E_x = \hbar^2 k_x^2/(2m)$. As shown in Chapter 4, the normalized solutions to (6.11) are plane waves

$$X(x) = \frac{1}{\sqrt{2\pi}} e^{ik_x x}. \tag{6.12}$$

Thus, the solution to the three-dimensional Schrödinger equation (6.3) is given by

$$\psi_{\vec{k}}(x, y, z) = (2\pi)^{-3/2} e^{ik_x x} e^{ik_y y} e^{ik_z z} = (2\pi)^{-3/2} e^{i\vec{k}\cdot\vec{r}}, \tag{6.13}$$

where $\vec{k}$ and $\vec{r}$ are the wave and position vectors of the particle, respectively. As for the total energy E, it is equal to the sum of the eigenvalues of the three one-dimensional equations (6.11):

$$E = E_x + E_y + E_z = \frac{\hbar^2}{2m} \left(k_x^2 + k_y^2 + k_z^2 \right) = \frac{\hbar^2}{2m} \vec{k}^2. \tag{6.14}$$

Note that, since the energy (6.14) depends only on the magnitude of $\vec{k}$, all different orientations of $\vec{k}$ (obtained by varying k_x, k_y, k_z) subject to the condition

$$|\vec{k}| = \sqrt{k_x^2 + k_y^2 + k_z^2} = constant \tag{6.15}$$

generate different eigenfunctions (6.13) without a change in the energy. As the total number of orientations of $\vec{k}$ which preserve its magnitude is infinite, the energy of a free particle is *infinitely degenerate*.

Note that the solutions to the time-dependent Schrödinger equation (6.1) are obtained by substituting (6.13) into (6.2):

$$\Psi_{\vec{k}}(\vec{r}, t) = \psi(\vec{r}) e^{-i\omega t} = (2\pi)^{-3/2} e^{i(\vec{k}\cdot\vec{r} - \omega t)}, \tag{6.16}$$

where $\omega = E/\hbar$; this represents a propagating wave with wave vector $\vec{k}$. The orthonormality condition of this wave function is expressed by

$$\int \Psi_{\vec{k}'}^*(\vec{r}, t) \Psi_{\vec{k}}(\vec{r}, t) \, d^3 r = \int \psi_{\vec{k}'}^*(\vec{r}) \psi_{\vec{k}}(\vec{r}) \, d^3 r = (2\pi)^{-3} \int e^{i(\vec{k}-\vec{k}')\cdot\vec{r}} \, d^3 r = \delta(\vec{k} - \vec{k}'), \tag{6.17}$$

which can be written in Dirac's notation as

$$\langle \Psi_{\vec{k}'}(t) | \Psi_{\vec{k}}(t) \rangle = \langle \psi_{\vec{k}'} | \psi_{\vec{k}} \rangle = \delta(\vec{k} - \vec{k}'). \tag{6.18}$$

The free particle can be represented, as seen in Chapter 3, by a wave packet (a superposition of wave functions corresponding to the various wave vectors):

$$\Psi(\vec{r}, t) = (2\pi)^{-3/2} \int A(\vec{k}, t) \Psi_{\vec{k}}(\vec{r}, t) \, d^3 k = (2\pi)^{-3/2} \int A(\vec{k}, t) e^{i(\vec{k}\cdot\vec{r} - \omega t)} \, d^3 k, \tag{6.19}$$

where $A(\vec{k}, t)$ is the Fourier transform of $\Psi(\vec{r}, t)$:

$$A(\vec{k}, t) = (2\pi)^{-3/2} \int \Psi(\vec{r}, t) e^{-i(\vec{k}\cdot\vec{r} - \omega t)} d^3 r. \tag{6.20}$$

As seen in Chapters 1 and 4, the position of the particle can be represented classically by the center of the wave packet.

6.2.3 The Box Potential

We are going to begin with the rectangular box potential, which has no symmetry, and then consider the cubic potential, which displays a great deal of symmetry, since the xyz axes are equivalent.

6.2.3.1 The Rectangular Box Potential

Consider first the case of a spinless particle of mass m confined in a *rectangular* box of sides a, b, c:

$$V(x, y, z) = \begin{cases} 0, & 0 < x < a, \ 0 < y < b, \ 0 < z < c, \\ \infty, & \text{elsewhere,} \end{cases} \tag{6.21}$$

which can be written as $V(x, y, z) = V_x(x) + V_y(y) + V_z(z)$, with

$$V_x(x) = \begin{cases} 0, & 0 < x < a, \\ \infty, & \text{elsewhere;} \end{cases} \tag{6.22}$$

the potentials $V_y(y)$ and $V_z(z)$ have similar forms.

The wave function $\psi(x, y, z)$ must vanish at the walls of the box. We have seen in Chapter 4 that the solutions for this potential are of the form

$$X(x) = \sqrt{\frac{2}{a}} \sin\left(\frac{n_x \pi}{a} x\right), \qquad n_x = 1, 2, 3, \ldots, \tag{6.23}$$

and the corresponding energy eigenvalues are

$$E_{n_x} = \frac{\hbar^2 \pi^2}{2ma^2} n_x^2. \tag{6.24}$$

From these expressions we can write the normalized three-dimensional eigenfunctions and their corresponding energies:

$$\boxed{\psi_{n_x n_y n_z}(x, y, z) = \sqrt{\frac{8}{abc}} \sin\left(\frac{n_x \pi}{a} x\right) \sin\left(\frac{n_y \pi}{b} y\right) \sin\left(\frac{n_z \pi}{c} z\right),} \tag{6.25}$$

$$\boxed{E_{n_x n_y n_z} = \frac{\hbar^2 \pi^2}{2m} \left(\frac{n_x^2}{a^2} + \frac{n_y^2}{b^2} + \frac{n_z^2}{c^2}\right).} \tag{6.26}$$

Table 6.1 Energy levels and their degeneracies for the cubic potential, with $E_1 = \frac{\pi^2\hbar^2}{2mL^2}$.

$E_{n_x n_y n_z}/E_1$	(n_x, n_y, n_z)	g_n
3	(111)	1
6	(211), (121), (112)	3
9	(221), (212), (122)	3
11	(311), (131), (113)	3
12	(222)	1
14	(321), (312), (231), (213), (132), (123)	6

6.2.3.2 The Cubic Potential

For the simpler case of a *cubic* box of side L, the energy expression can be inferred from (6.26) by substituting $a = b = c = L$:

$$E_{n_x n_y n_z} = \frac{\hbar^2 \pi^2}{2mL^2}(n_x^2 + n_y^2 + n_z^2), \qquad n_x, n_y, n_z = 1, 2, 3, \ldots. \qquad (6.27)$$

The ground state corresponds to $n_x = n_y = n_z = 1$; its energy is given by

$$E_{111} = \frac{3\pi^2\hbar^2}{2mL^2} = 3E_1, \qquad (6.28)$$

where, as shown in Chapter 4, $E_1 = \pi^2\hbar^2/(2mL^2)$ is the zero-point energy of a particle in a one-dimensional box. Thus, the zero-point energy for a particle in a three-dimensional box is three times that in a one-dimensional box. The factor 3 can be viewed as originating from the fact that we are confining the particle symmetrically in all three dimensions.

The first excited state has three possible sets of quantum numbers $(n_x, n_y, n_z) = (2, 1, 1)$, $(1, 2, 1), (1, 1, 2)$ corresponding to three different states $\psi_{211}(x, y, z)$, $\psi_{121}(x, y, z)$, and $\psi_{112}(x, y, z)$, where

$$\psi_{211}(x, y, z) = \sqrt{\frac{8}{L^3}} \sin\left(\frac{2\pi}{L}x\right) \sin\left(\frac{\pi}{L}y\right) \sin\left(\frac{\pi}{L}z\right); \qquad (6.29)$$

the expressions for $\psi_{121}(x, y, z)$ and $\psi_{112}(x, y, z)$ can be inferred from $\psi_{211}(x, y, z)$. Notice that all three states have the same energy:

$$E_{211} = E_{121} = E_{112} = 6\frac{\pi^2\hbar^2}{2mL^2} = 6E_1. \qquad (6.30)$$

The first excited state is thus threefold *degenerate*.

Degeneracy occurs only when there is a symmetry in the problem. For the present case of a particle in a cubic box, there is a great deal of symmetry, since all three dimensions are equivalent. Note that for the rectangular box, there is no degeneracy since the three dimensions are not equivalent. Moreover, degeneracy did not exist when we treated one-dimensional problems in Chapter 4, for they give rise to only one quantum number.

The second excited state also has three different states, and hence it is threefold degenerate; its energy is equal to $9E_1$: $E_{221} = E_{212} = E_{122} = 9E_1$.

The energy spectrum is shown in Table 6.1, where every nth level is characterized by its energy, its quantum numbers, and its degeneracy g_n.

6.2.4 The Harmonic Oscillator

We are going to begin with the anisotropic oscillator, which displays no symmetry, and then consider the isotropic oscillator where the xyz axes are all equivalent.

6.2.4.1 The Anisotropic Oscillator

Consider a particle of mass m moving in a three-dimensional anisotropic oscillator potential

$$\hat{V}(\hat{x}, \hat{y}, \hat{z}) = \frac{1}{2}m\omega_x^2 \hat{X}^2 + \frac{1}{2}m\omega_y^2 \hat{Y}^2 + \frac{1}{2}m\omega_z^2 \hat{Z}^2. \tag{6.31}$$

Its Schrödinger equation separates into three equations similar to (6.9):

$$-\frac{\hbar^2}{2m}\frac{d^2 X(x)}{dx^2} + \frac{1}{2}m\omega_x x^2 X(x) = E_x X(x), \tag{6.32}$$

with similar equations for $Y(y)$ and $Z(z)$. The eigenenergies corresponding to the potential (6.31) can be expressed as

$$E_{n_x n_y n_z} = E_{n_x} + E_{n_y} + E_{n_z} = \left(n_x + \frac{1}{2}\right)\hbar\omega_x + \left(n_y + \frac{1}{2}\right)\hbar\omega_y + \left(n_z + \frac{1}{2}\right)\hbar\omega_z, \tag{6.33}$$

with $n_x, n_y, n_z = 0, 1, 2, 3, \ldots$. The corresponding stationary states are

$$\psi_{n_x n_y n_z}(x, y, z) = X_{n_x}(x)Y_{n_y}(y)Z_{n_z}(z), \tag{6.34}$$

where $X_{n_x}(x)$, $Y_{n_y}(y)$, and $Z_{n_z}(z)$ are one-dimensional harmonic oscillator wave functions. These states are not degenerate, because the potential (6.31) has no symmetry (it is anisotropic).

6.2.4.2 The Isotropic Harmonic Oscillator

Consider now an *isotropic* harmonic oscillator potential. Its energy eigenvalues can be inferred from (6.33) by substituting $\omega_x = \omega_y = \omega_z = \omega$,

$$E_{n_x n_y n_z} = \left(n_x + n_y + n_z + \frac{3}{2}\right)\hbar\omega. \tag{6.35}$$

Since the energy depends on the sum of n_x, n_y, n_z, any set of quantum numbers having the same sum will represent states of equal energy.

The ground state, whose energy is $E_{000} = 3\hbar\omega/2$, is not degenerate. The first excited state is threefold degenerate, since there are three different states, $\psi_{100}, \psi_{010}, \psi_{001}$, that correspond to the same energy $5\hbar\omega/2$. The second excited state is sixfold degenerate; its energy is $7\hbar\omega/2$.

In general, we can show that the degeneracy g_n of the nth excited state, which is equal to the number of ways the nonnegative integers n_x, n_y, n_z may be chosen to total to n, is given by

$$g_n = \frac{1}{2}(n+1)(n+2), \tag{6.36}$$

where $n = n_x + n_y + n_z$. Table 6.2 displays the first few energy levels along with their degeneracies.

Table 6.2 Energy levels and their degeneracies for an isotropic harmonic oscillator.

n	$2E_n/(\hbar\omega)$	$(n_x n_y n_z)$	g_n
0	3	(000)	1
1	5	(100), (010), (001)	3
2	7	(200), (020), (002)	6
		(110), (101), (011)	
3	9	(300), (030), (003)	10
		(210), (201), (021)	
		(120), (102), (012)	
		(111)	

Example 6.1 (Degeneracy of a harmonic oscillator)

Show how to derive the degeneracy relation (6.36).

Solution

For a fixed value of n, the degeneracy g_n is given by the number of ways of choosing n_x, n_y, and n_z so that $n = n_x + n_y + n_z$.

For a fixed value of n_x, the number of ways of choosing n_y and n_z so that $n_y + n_z = n - n_x$ is given by $(n - n_x + 1)$; this can be shown as follows. For a given value of n_x, the various permissible values of (n_y, n_z) are given by $(n_y, n_z) = (0, n-n_x), (1, n-n_x-1), (2, n-n_x-2), (3, n - n_x - 3), \ldots, (n - n_x - 3, 3), (n - n_x - 2, 2), (n - n_x - 1, 1),$ and $(n - n_x, 0)$. In all, there are $(n - n_x + 1)$ sets of (n_y, n_z) so that $n_y + n_z = n - n_x$. Now, since the values of n_x can vary from 0 to n, the degeneracy is then given by

$$g_n = \sum_{n_x=0}^{n}(n - n_x + 1) = (n+1)\sum_{n_x=0}^{n}1 - \sum_{n_x=0}^{n}n_x = (n+1)^2 - \frac{1}{2}n(n+1) = \frac{1}{2}(n+1)(n+2).$$

(6.37)

A more primitive way of calculating this series is to use Gauss's method: simply write the series $\sum_{n_x=0}^{n}(n - n_x + 1)$ in the following two equivalent forms:

$$g_n = (n+1) + n + (n-1) + (n-2) + \cdots + 4 + 3 + 2 + 1,$$

(6.38)

$$g_n = 1 + 2 + 3 + 4 + \cdots + (n-2) + (n-1) + n + (n+1).$$

(6.39)

Since both of these two series contain $(n + 1)$ terms each, a term by term addition of these relations yields

$$\begin{aligned} 2g_n &= (n+2) + (n+2) + (n+2) + \cdots + (n+2) + (n+2) + (n+2) \\ &= (n+1)(n+2); \end{aligned}$$

(6.40)

hence $g_n = \frac{1}{2}(n + 1)(n + 2)$.

6.3 3D Problems in Spherical Coordinates

6.3.1 Central Potential: General Treatment

In this section we study the structure of the Schrödinger equation for a particle of mass[1] M moving in a *spherically symmetric* potential

$$V(\vec{r}) = V(r),\tag{6.41}$$

which is also known as the *central* potential.

The time-independent Schrödinger equation for this particle, of momentum $-i\hbar\vec{\nabla}$ and position vector $\vec{r}$, is

$$\left[-\frac{\hbar^2}{2M}\nabla^2 + V(r)\right]\psi(\vec{r}) = E\psi(\vec{r}).\tag{6.42}$$

Since the Hamiltonian is spherically symmetric, we are going to use the spherical coordinates (r, θ, φ) which are related to their Cartesian counterparts by

$$x = r\sin\theta\cos\varphi, \qquad y = r\sin\theta\sin\varphi, \qquad z = r\cos\theta.\tag{6.43}$$

The Laplacian ∇^2 separates into a radial part ∇_r^2 and an angular part ∇_Ω^2 as follows (see Chapter 5):

$$\nabla^2 = \nabla_r^2 - \frac{1}{\hbar^2 r^2}\nabla_\Omega^2 = \frac{1}{r^2}\frac{\partial}{\partial r}\left(r^2\frac{\partial}{\partial r}\right) - \frac{1}{\hbar^2 r^2}\hat{\vec{L}}^2 = \frac{1}{r}\frac{\partial^2}{\partial r^2}r - \frac{1}{\hbar^2 r^2}\hat{\vec{L}}^2,\tag{6.44}$$

where $\hat{\vec{L}}$ is the orbital angular momentum with

$$\hat{\vec{L}}^2 = -\hbar^2\left[\frac{1}{\sin\theta}\frac{\partial}{\partial\theta}\left(\sin\theta\frac{\partial}{\partial\theta}\right) + \frac{1}{\sin^2\theta}\frac{\partial^2}{\partial\varphi^2}\right].\tag{6.45}$$

In spherical coordinates the Schrödinger equation therefore takes the form

$$\left[-\frac{\hbar^2}{2M}\frac{1}{r}\frac{\partial^2}{\partial r^2}r + \frac{1}{2Mr^2}\hat{\vec{L}}^2 + V(r)\right]\psi(\vec{r}) = E\psi(\vec{r}).\tag{6.46}$$

The first term of this equation can be viewed as the *radial* kinetic energy

$$-\frac{\hbar^2}{2M}\frac{1}{r}\frac{\partial^2}{\partial r^2}r = \frac{\hat{P}_r^2}{2M},\tag{6.47}$$

since the radial momentum operator is given by the Hermitian form[2]

$$\hat{P}_r = \frac{1}{2}\left[\left(\frac{\vec{r}}{r}\right)\cdot\hat{\vec{P}} + \hat{\vec{P}}\cdot\left(\frac{\vec{r}}{r}\right)\right] = -i\hbar\left(\frac{\partial}{\partial r} + \frac{1}{r}\right) \equiv -i\hbar\frac{1}{r}\frac{\partial}{\partial r}r.\tag{6.48}$$

---

[1]Throughout this section we will designate the mass of the particle by a capital M to avoid any confusion with the azimuthal quantum number m.

[2]Note that we can show that the commutator between the position operator, $\hat{r}$, and the radial momentum operator, $\hat{p}_r$, is given by: $[\hat{r}, \hat{p}_r] = i\hbar$ (the proof is left as an exercise).

The second term $\hat{L}^2/(2Mr^2)$ of (6.46) can be identified with the *rotational* kinetic energy, for this term is generated from a "pure" rotation of the particle about the origin (i.e., no change in the radial variable r, where Mr^2 is its moment of inertia with respect to the origin).

Now, since $\hat{L}^2$ as shown in (6.45) does not depend on r, it commutes with both $\hat{V}(r)$ and the radial kinetic energy; hence it also commutes with the Hamiltonian $\hat{H}$. In addition, since $\hat{L}_z$ commutes with $\hat{L}^2$, the three operators $\hat{H}$, $\hat{L}^2$, and L_z mutually commute:

$$[\hat{H}, \hat{L}^2] = [\hat{H}, \hat{L}_z] = 0. \tag{6.49}$$

Thus $\hat{H}$, $\hat{L}^2$, and $\hat{L}_z$ have common eigenfunctions. We have seen in Chapter 5 that the simultaneous eigenfunctions of $\hat{L}^2$ and $\hat{L}_z$ are given by the spherical harmonics $Y_{lm}(\theta, \varphi)$:

$$\hat{L}^2 Y_{lm}(\theta, \varphi) = l(l+1)\hbar^2 Y_{lm}(\theta, \varphi), \tag{6.50}$$
$$\hat{L}_z Y_{lm}(\theta, \varphi) = m\hbar Y_{lm}(\theta, \varphi). \tag{6.51}$$

Since the Hamiltonian in (6.46) is a sum of a radial part and an angular part, we can look for solutions that are products of a radial part and an angular part, where the angular part is simply the spherical harmonic $Y_{lm}(\theta, \varphi)$:

$$\psi(\vec{r}) = \langle \vec{r} \mid nlm \rangle = \psi_{nlm}(r, \theta, \varphi) = R_{nl}(r)Y_{lm}(\theta, \varphi). \tag{6.52}$$

Note that the orbital angular momentum of a system moving in a central potential is conserved, since, as shown in (6.49), it commutes with the Hamiltonian.

The radial wave function $R_{nl}(r)$ has yet to be found. The quantum number n is introduced to identify the eigenvalues of $\hat{H}$:

$$\hat{H} \mid nlm \rangle = E_n \mid nlm \rangle. \tag{6.53}$$

Substituting (6.52) into (6.46) and using the fact that $\psi_{nlm}(r, \theta, \varphi)$ is an eigenfunction of $\hat{L}^2$ with eigenvalue $l(l+1)\hbar^2$, then dividing through by $R_{nl}(r)Y_{lm}(\theta, \varphi)$ and multiplying by $2Mr^2$, we end up with an equation where the radial and angular degrees of freedom are separated:

$$\left[-\hbar^2 \frac{r}{R_{nl}} \frac{\partial^2}{\partial r^2}(r R_{nl}) + 2Mr^2(V(r) - E)\right] + \left[\frac{\hat{L}^2 Y_{lm}(\theta, \varphi)}{Y_{lm}(\theta, \varphi)}\right] = 0. \tag{6.54}$$

The terms inside the first square bracket are independent of θ and φ and those of the second are independent of r. They must then be separately equal to constants and their sum equal to zero. The second square bracket is nothing but (6.50), the eigenvalue equation of $\hat{L}^2$; hence it is equal to $l(l+1)\hbar^2$. As for the first bracket, it must be equal to $-l(l+1)\hbar^2$; this leads to an equation known as the *radial equation* for a central potential:

$$-\frac{\hbar^2}{2M}\frac{d^2}{dr^2}(r R_{nl}(r)) + \left[V(r) + \frac{l(l+1)\hbar^2}{2Mr^2}\right](r R_{nl}(r)) = E_n (r R_{nl}(r)). \tag{6.55}$$

Note that (6.55), which gives the energy levels of the system, does not depend on the azimuthal quantum number m. Thus, the energy E_n is $(2l+1)$-fold *degenerate*. This is due to the fact that, for a given l, there are $(2l + 1)$ different eigenfunctions ψ_{nlm} (i.e., $\psi_{nl\,-l}$, $\psi_{nl\,-l+1}$, ..., $\psi_{nl\,l-1}$, $\psi_{nl\,l}$) which correspond to the same eigenenergy E_n. This degeneracy property is peculiar to central potentials.

Note that (6.55) has the structure of a one-dimensional equation in r,

$$-\frac{\hbar^2}{2M}\frac{d^2 U_{nl}(r)}{dr^2} + \left[V(r) + \frac{l(l+1)\hbar^2}{2Mr^2}\right]U_{nl}(r) = E_n U_{nl}(r), \tag{6.56}$$

or

$$\boxed{-\frac{\hbar^2}{2M}\frac{d^2 U_{nl}(r)}{dr^2} + V_{eff}(r)U_{nl}(r) = E_n U_{nl}(r),} \tag{6.57}$$

whose solutions give the energy levels of the system. The wave function $U_{nl}(r)$ is given by

$$U_{nl}(r) = r R_{nl}(r) \tag{6.58}$$

and the potential by

$$V_{eff}(r) = V(r) + \frac{l(l+1)\hbar^2}{2Mr^2}, \tag{6.59}$$

which is known as the *effective* or *centrifugal* potential, where $V(r)$ is the central potential and $l(l+1)\hbar^2/2Mr^2$ is a repulsive or centrifugal potential, associated with the orbital angular momentum, which tends to repel the particle away from the center. As will be seen later, in the case of atoms, $V(r)$ is the Coulomb potential resulting from the attractive forces between the electrons and the nucleus. Notice that although (6.57) has the structure of a one-dimensional eigenvalue equation, it differs from the one-dimensional Schrödinger equation in one major aspect: the variable r cannot have negative values, for it varies from $r = 0$ to $r \to +\infty$. We must therefore require the wave function $\psi_{nlm}(r, \theta, \varphi)$ to be finite for all values of r between 0 and ∞, notably for $r = 0$. But if $R_{nl}(0)$ is finite, $r R_{nl}(r)$ must vanish at $r = 0$, i.e.,

$$\lim_{r \to 0}[r R_{nl}(r)] = U_{nl}(0) = 0. \tag{6.60}$$

Thus, to make the radial equation (6.57) equivalent to a one-dimensional eigenvalue problem, we need to assume that the particle's potential is given by the effective potential $V_{eff}(r)$ for $r > 0$ and by an infinite potential for $r \leq 0$.

For the eigenvalue equation (6.57) to describe bound states, the potential $V(r)$ must be attractive (i.e., negative) because $l(l+1)\hbar^2/(2Mr^2)$ is repulsive. Figure 6.1 shows that, as l increases, the depth of $V_{eff}(r)$ decreases and its minimum moves farther away from the origin. The farther the particle from the origin, the less bound it will be. This is due to the fact that as the particle's angular momentum increases, the particle becomes less and less bound.

In summary, we want to emphasize the fact that, for spherically symmetric potentials, the Schrödinger equation (6.46) reduces to a trivial angular equation (6.50) for $\hat{L}^2$ and to a one-dimensional radial equation (6.57).

Remark

When a particle has orbital and spin degrees of freedom, its total wave function $\mid \Psi \rangle$ consists of a product of two parts: a spatial part, $\psi(\vec{r})$, and a spin part, $\mid s, m_s \rangle$; that is, $\mid \Psi \rangle = \mid \psi \rangle \mid s, m_s \rangle$. In the case of an electron moving in a central field, besides the quantum numbers

Figure 6.1 The effective potential $V_{eff}(r) = V(r)+\hbar^2 l(l+1)/(2Mr^2)$ corresponding to several values of l: $l = 0, 1, 2, 3$; $V(r)$ is an attractive central potential, while $\hbar^2 l(l+1)/(2Mr^2)$ is a repulsive (centrifugal) potential.

n, l, m_l, a complete description of its state would require a fourth quantum number, the spin quantum number m_s: $\mid nlm_l m_s \rangle =\mid nlm_l \rangle \mid s, m_s \rangle$; hence

$$\Psi_{n,l,m_l,m_s}(\vec{r}) = \psi_{nlm_l}(\vec{r}) \mid s, m_s \rangle = R_{nl}(r)Y_{lm_l}(\theta, \varphi) \mid s, m_s \rangle. \tag{6.61}$$

Since the spin does not depend on the spatial degrees of freedom, the spin operator does not act on the spatial wave function $\psi_{nlm_l}(\vec{r})$ but acts only on the spin part $\mid s, m_s \rangle$; conversely, $\hat{L}$ acts only the spatial part.

6.3.2 The Free Particle in Spherical Coordinates

In what follows we want to apply the general formalism developed above to study the motion of a free particle of mass M and energy $E_k = \hbar^2 k^2/(2M)$, where k is the wave number ($k = |\vec{k}|$). The Hamiltonian $\hat{H} = -\hbar^2 \nabla^2/(2M)$ of a free particle commutes with $\hat{L}^2$ and $\hat{L}_z$. Since $V(r) = 0$ the Hamiltonian of a free particle is *rotationally invariant*. The free particle can then be viewed as a special case of central potentials. We have shown above that the radial and angular parts of the wave function can be separated, $\psi_{klm}(r, \theta, \varphi) = \langle r\theta\varphi \mid klm \rangle = R_{kl}(r)Y_{lm}(\theta, \varphi)$.

The radial equation for a free particle is obtained by setting $V(r) = 0$ in (6.55):

$$-\frac{\hbar^2}{2M}\frac{1}{r}\frac{d^2}{dr^2}(r R_{kl}(r)) + \frac{l(l+1)\hbar^2}{2Mr^2} R_{kl}(r) = E_k R_{kl}(r), \tag{6.62}$$

which can be rewritten as

$$-\frac{1}{r}\frac{d^2}{dr^2}(r R_{kl}(r)) + \frac{l(l+1)}{r^2} R_{kl}(r) = k^2 R_{kl}(r), \tag{6.63}$$

where $k^2 = 2ME_k/\hbar^2$.

Table 6.3 First few spherical Bessel and Neumann functions.

Bessel functions $j_l(r)$	Neumann functions $n_l(r)$
$j_0(r) = \frac{\sin r}{r}$	$n_0(r) = -\frac{\cos r}{r}$
$j_1(r) = \frac{\sin r}{r^2} - \frac{\cos r}{r}$	$n_1(r) = -\frac{\cos r}{r^2} - \frac{\sin r}{r}$
$j_2(r) = \left(\frac{3}{r^3} - \frac{1}{r}\right)\sin r - \frac{3\cos r}{r^2}$	$n_2(r) = -\left(\frac{3}{r^3} - \frac{1}{r}\right)\cos r - \frac{3}{r^2}\sin r$

Using the change of variable $\rho = kr$, we can reduce this equation to

$$\frac{d^2\mathcal{R}_l(\rho)}{d\rho^2} + \frac{2}{\rho}\frac{d\mathcal{R}_l(\rho)}{d\rho} + \left[1 - \frac{l(l+1)}{\rho^2}\right]\mathcal{R}_l(\rho) = 0, \tag{6.64}$$

where $\mathcal{R}_l(\rho) = \mathcal{R}_l(kr) = R_{kl}(r)$. This differential equation is known as the *spherical Bessel equation*. The general solutions to this equation are given by an independent linear combination of the *spherical Bessel functions* $j_l(\rho)$ and the *spherical Neumann functions* $n_l(\rho)$:

$$\mathcal{R}_l(\rho) = A_l j_l(\rho) + B_l n_l(\rho), \tag{6.65}$$

where $j_l(\rho)$ and $n_l(\rho)$ are given by

$$j_l(\rho) = (-\rho)^l \left(\frac{1}{\rho}\frac{d}{d\rho}\right)^l \frac{\sin\rho}{\rho}, \qquad n_l(\rho) = -(-\rho)^l \left(\frac{1}{\rho}\frac{d}{d\rho}\right)^l \frac{\cos\rho}{\rho}. \tag{6.66}$$

The first few spherical Bessel and Neumann functions are listed in Table 6.3 and their shapes are displayed in Figure 6.2.

Expanding $\sin\rho/\rho$ and $\cos\rho/\rho$ in a power series of ρ, we see that the functions $j_l(\rho)$ and $n_l(\rho)$ reduce for small values of ρ (i.e., near the origin) to

$$j_l(\rho) \simeq \frac{2^l l!}{(2l+1)!}\rho^l, \qquad n_l(\rho) \simeq -\frac{(2l)!}{2^l l!}\rho^{-l-1}, \qquad \rho \ll 1, \tag{6.67}$$

and for large values of ρ to

$$j_l(\rho) \simeq \frac{1}{\rho}\sin\left(\rho - \frac{l\pi}{2}\right), \qquad n_l(\rho) \simeq -\frac{1}{\rho}\cos\left(\rho - \frac{l\pi}{2}\right), \qquad \rho \gg 1. \tag{6.68}$$

Since the Neumann functions $n_l(\rho)$ diverge at the origin, and since the wave functions ψ_{klm} are required to be finite everywhere in space, the functions $n_l(\rho)$ are unacceptable solutions to the problem. Hence only the spherical Bessel functions $j_l(kr)$ contribute to the eigenfunctions of the free particle:

$$\psi_{klm}(r, \theta, \psi) = j_l(kr)Y_{lm}(\theta, \varphi), \tag{6.69}$$

where $k = \sqrt{2ME_k}/\hbar$. As shown in Figure 6.2, the amplitude of the wave functions becomes smaller and smaller as r increases. At large distances, the wave functions are represented by spherical waves.

Figure 6.2 Spherical Bessel functions $j_l(r)$ and spherical Neumann functions $n_l(r)$; only the Bessel functions are finite at the origin.

Note that, since the index k in $E_k = \hbar^2 k^2/(2M)$ varies *continuously*, the energy spectrum of a free particle is *infinitely degenerate*. This is because all orientations of $\vec{k}$ in space correspond to the same energy.

Remark

We have studied the free particle within the context of Cartesian and spherical coordinate systems. Whereas the energy is given in both coordinate systems by the same expression, $E_k = \hbar^2 k^2/(2M)$, the wave functions are given in Cartesian coordinates by *plane waves* $e^{i\vec{k}\cdot\vec{r}}$ (see (6.13)) and in spherical coordinates by *spherical waves* $j_l(kr)Y_{lm}(\theta, \varphi)$ (see (6.69)). We can, however, show that both sets of wave functions are equivalent, since we can express a plane wave $e^{i\vec{k}\cdot\vec{r}}$ in terms of spherical wave states $j_l(kr)Y_{lm}(\theta, \varphi)$. In particular, we can generate plane waves from a linear combination of spherical states that have the same k but different values of l and m:

$$e^{i\vec{k}\cdot\vec{r}} = \sum_{l=0}^{\infty} \sum_{m=-l}^{l} a_{lm} \, j_l(kr)Y_{lm}(\theta, \varphi). \tag{6.70}$$

The problem therefore reduces to finding the expansion coefficients a_{lm}. For instance, in the case where $\vec{k}$ is along the z-axis, $m = 0$, we can show that

$$e^{i\vec{k}\cdot\vec{r}} = e^{ikr\cos\theta} = \sum_{l=0}^{\infty} i^l \, (2l + 1)j_l \, (kr) P_l(\cos\theta), \tag{6.71}$$

where $P_l(\cos\theta)$ are the Legendre polynomials, with $Y_{l0}(\theta, \varphi) \sim P_l(\cos\theta)$. The wave functions $\psi_{klm}(r, \theta, \varphi) = j_l(kr)Y_{lm}(\theta, \varphi)$ describe a free particle of energy E_k, with angular momentum l, but they give no information on the linear momentum $\vec{p}$ (ψ_{klm} is an eigenstate of $\hat{H}$, $\hat{\vec{L}}^2$, and $\hat{L}_z$ but not of $\hat{\vec{P}}$). On the other hand, the plane wave $e^{i\vec{k}\cdot\vec{r}}$ which is an eigenfunction of $\hat{H}$ and $\hat{\vec{P}}$, but not of $\hat{\vec{L}}^2$ nor of $\hat{L}_z$, gives no information about the particle's angular momentum. That is, plane waves describe states with well-defined linear momenta but poorly defined angular momenta. Conversely, spherical waves describe states with well-defined angular momenta but poorly defined linear momenta.

6.3.3   The Spherical Square Well Potential

Consider now the problem of a particle of mass M in an attractive square well potential

$$V(r) = \begin{cases} -V_0, & r < a, \\ 0, & r > a. \end{cases} \tag{6.72}$$

Let us consider the cases $0 < r < a$ and $r > a$ separately.

6.3.3.1   Case where $0 < r < a$

Inside the well, $0 < r < a$, the time-independent Schrödinger equation for this particle can be obtained from (6.55):

$$-\frac{\hbar^2}{2M}\frac{1}{r}\frac{d^2}{dr^2}(r R_l(r)) + \frac{l(l+1)\hbar^2}{2Mr^2}R_l(r) = (E + V_0)R_l(r). \tag{6.73}$$

Using the change of variable $\rho = k_1 r$, where k_1 is now given by $k_1 = \sqrt{2M(E + V_0)}/\hbar$, we see that (6.73) reduces to the spherical Bessel differential equation (6.64). As in the case of a free particle, the radial wave function must be finite everywhere, and is given as follows in terms of the spherical Bessel functions $j_l(k_1 r)$:

$$R_l(r) = A j_l(k_1 r) = A j_l\left(\frac{\sqrt{2M(E + V_0)}}{\hbar}r\right), \qquad \text{for} \qquad r < a, \tag{6.74}$$

where A is a normalization constant.

6.3.3.2   Case where $r > a$

Outside the well, $r > a$, the particle moves freely; its Schrödinger equation is (6.62):

$$-\frac{\hbar^2}{2M}\frac{1}{r}\frac{d^2}{dr^2}(r R_{kl}(r)) + \frac{l(l+1)\hbar^2}{2Mr^2}R_{kl}(r) = E_k R_{kl}(r) \qquad (r > a). \tag{6.75}$$

Two possibilities arise here, depending on whether the energy is negative or positive.

- The negative energy case corresponds to bound states (i.e., to a discrete energy spectrum). The general solutions of (6.75) are similar to those of (6.63), but k is now an imaginary number; that is, we must replace k by ik_2 and, hence, the solutions are given by linear combinations of $j_l(ik_2 r)$ and $n_l(ik_2 r)$:

$$R_l(ik_2 r) = B\left[j_l(ik_2 r) \pm n_l(ik_2 r)\right], \tag{6.76}$$

where B is a normalization constant, with $k_2 = \sqrt{-2ME}/\hbar$. **Note:** Linear combinations of $j_l(\rho)$ and $n_l(\rho)$ can be expressed in terms of the *spherical Hankel functions* of the first kind, $h_l^{(1)}(\rho)$, and the second kind, $h_l^{(2)}(\rho)$, as follows:

$$h_l^{(1)}(\rho) = j_l(\rho) + i n_l(\rho), \tag{6.77}$$

$$h_l^{(2)}(\rho) = j_l(\rho) - i n_l(\rho) = \left(h_l^{(1)}(\rho)\right)^*. \tag{6.78}$$

The first few spherical Hankel functions of the first kind are

$$h_0^{(1)}(\rho) = -i\frac{e^{i\rho}}{\rho}, \quad h_1^{(1)}(\rho) = -\left(\frac{1}{\rho} + \frac{i}{\rho^2}\right)e^{i\rho}, \quad h_2^{(1)}(\rho) = \left(\frac{i}{\rho} - \frac{3}{\rho^2} - \frac{3i}{\rho^3}\right)e^{i\rho}.$$

(6.79)

The asymptotic behavior of the Hankel functions when $\rho \to \infty$ can be inferred from (6.68):

$$h_l^{(1)}(\rho) \to -\frac{i}{\rho}e^{i(\rho - l\pi/2)}, \qquad h_l^{(2)}(\rho) \to \frac{i}{\rho}e^{-i(\rho - l\pi/2)}. \qquad (6.80)$$

The solutions that need to be retained in (6.76) must be finite everywhere. As can be inferred from Eq (6.80), only the Hankel functions of the first kind $h_l^{(1)}(ik_2r)$ are finite at large values of r (the functions $h_l^{(2)}(ik_2r)$ diverge for large values of r). Thus, the wave functions outside the well that are physically meaningful are those expressed in terms of the Hankel functions of the first kind (see (6.76)):

$$R_l(ik_2r) = Bh_l^{(1)}\left(i\frac{\sqrt{-2ME}}{\hbar}r\right) = Bj_l\left(i\frac{\sqrt{-2ME}}{\hbar}r\right) + iBn_l\left(i\frac{\sqrt{-2ME}}{\hbar}r\right).$$

(6.81)

The continuity of the radial function and its derivative at $r = a$ yields

$$\frac{1}{h_l^{(1)}(ik_2r)}\frac{dh_l^{(1)}(ik_2r)}{dr}\bigg|_{r=a} = \frac{1}{j_l(k_1r)}\frac{dj_l(k_1r)}{dr}\bigg|_{r=a}. \qquad (6.82)$$

For the $l = 0$ states, this equation reduces to

$$-k_2 = k_1 \cot(k_1a). \qquad (6.83)$$

This continuity condition is analogous to the transcendental equation we obtained in Chapter 4 when we studied the one-dimensional finite square well potential.

- The positive energy case corresponds to the continuous spectrum (unbound or scattering states), where the solution is asymptotically oscillatory. The solution consists of a linear combination of $j_l(k'r)$ and $n_l(k'r)$, where $k' = \sqrt{2ME}/\hbar$. Since the solution must be finite everywhere, the continuity condition at $r = a$ determines the coefficients of the linear combination. The particle can move freely to infinity with a finite kinetic energy $E = \hbar^2k'^2/(2M)$.

6.3.4 The Isotropic Harmonic Oscillator

The radial Schrödinger equation for a particle of mass M in an isotropic harmonic oscillator potential

$$V(r) = \frac{1}{2}M\omega^2r^2 \qquad (6.84)$$

is obtained from (6.57):

$$-\frac{\hbar^2}{2M}\frac{d^2U_{nl}(r)}{dr^2} + \left[\frac{1}{2}M\omega^2r^2 + \frac{l(l+1)\hbar^2}{2Mr^2}\right]U_{nl}(r) = EU_{nl}(r). \qquad (6.85)$$

We are going to solve this equation by examining the behavior of the solutions at the asymptotic limits (at very small and very large values of r). On the one hand, when $r \to 0$, the E and $M\omega^2 r^2/2$ terms become too small compared to the $l(l+1)\hbar^2/2Mr^2$ term. Hence, when $r \to 0$, Eq. (6.85) reduces to

$$-\frac{\hbar^2}{2M}\frac{d^2U(r)}{dr^2} + \frac{l(l+1)\hbar^2}{2Mr^2}U(r) = 0; \tag{6.86}$$

the solutions of this equation are of the form $U(r) \sim r^{l+1}$. On the other hand, when $r \to \infty$, the E and $l(l+1)\hbar^2/2Mr^2$ terms become too small compared to the $M\omega^2 r^2/2$ term; hence, the asymptotic form of (6.85) when $r \to \infty$ is

$$-\frac{\hbar^2}{2M}\frac{d^2U(r)}{dr^2} + \frac{1}{2}M\omega^2 r^2 U(r) = 0, \tag{6.87}$$

which admits solutions of type $U(r) \sim e^{-M\omega r^2/2\hbar}$. Combining (6.86) and (6.87), we can write the solutions of (6.85) as

$$U(r) = f(r)r^{l+1}e^{-M\omega r^2/2\hbar}, \tag{6.88}$$

where $f(r)$ is a function of r. Substituting this expression into (6.85), we obtain an equation for $f(r)$:

$$\frac{d^2 f(r)}{dr^2} + 2\left(\frac{l+1}{r} - \frac{M\omega}{\hbar}r\right)\frac{df(r)}{dr} + \left[\frac{2ME}{\hbar^2} - (2l+3)\frac{M\omega}{\hbar}\right]f(r) = 0. \tag{6.89}$$

Let us try a power series solution

$$f(r) = \sum_{n=0}^{\infty} a_n r^n = a_0 + a_1 r + a_2 r^2 + \cdots + a_n r^n + \cdots. \tag{6.90}$$

Substituting this function into (6.89), we obtain

$$\sum_{n=0}^{\infty}\left\{n(n-1)a_n r^{n-2} + 2\left(\frac{l+1}{r} - \frac{M\omega}{\hbar}r\right)na_n r^{n-1}\right.$$
$$\left. + \left[\frac{2ME}{\hbar^2} - (2l+3)\frac{M\omega}{\hbar}\right]a_n r^n\right\} = 0, \tag{6.91}$$

which in turn reduces to

$$\sum_{n=0}^{\infty}\left\{n(n+2l+1)a_n r^{n-2} + \left[-\frac{2M\omega}{\hbar}n + \frac{2ME}{\hbar^2} - (2l+3)\frac{M\omega}{\hbar}\right]a_n r^n\right\} = 0. \tag{6.92}$$

For this equation to hold, the coefficients of the various powers of r must vanish separately. For instance, when $n = 0$ the coefficient of r^{-2} is indeed zero:

$$0 \cdot (2l+1)a_0 = 0. \tag{6.93}$$

Note that a_0 need not be zero for this equation to hold. The coefficient of r^{-1} corresponds to $n = 1$ in (6.92); for this coefficient to vanish, we must have

$$1 \cdot (2l+2)a_1 = 0. \tag{6.94}$$

Since $(2l + 2)$ cannot be zero, because the quantum number l is a positive integer, a_1 must vanish.

The coefficient of r^n results from the relation

$$\sum_{n=0}^{\infty} \left\{ (n+2)(n+2l+3)a_{n+2} + \left[\frac{2ME}{\hbar^2} - \frac{M\omega}{\hbar}(2n+2l+3) \right] a_n \right\} r^n = 0, \qquad (6.95)$$

which leads to the recurrence formula

$$(n+2)(n+2l+3)a_{n+2} = \left[\frac{-2ME}{\hbar^2} + \frac{M\omega}{\hbar}(2n+2l+3) \right] a_n. \qquad (6.96)$$

This recurrence formula shows that all coefficients a_n corresponding to odd values of n are zero, since $a_1 = 0$ (see (6.94)). The function $f(r)$ must therefore contain only even powers of r:

$$f(r) = \sum_{n=0}^{\infty} a_{2n} r^{2n} = \sum_{n'=0,2,4,\cdots}^{\infty} a_{n'} r^{n'}, \qquad (6.97)$$

where all coefficients a_{2n}, with $n \geq 1$, are proportional to a_0.

Now note that when $n \to +\infty$ the function $f(r)$ diverges, for it behaves asymptotically like e^{r^2}. To obtain a finite solution, we must require the series (6.97) to stop at a maximum power $r^{n'}$; hence it must be *polynomial*. For this, we require $a_{n'+2}$ to be zero. Thus, setting $a_{n'+2} = 0$ into the recurrence formula (6.96) and since $a_{n'} \neq 0$, we obtain at once the *quantization condition*

$$2\frac{M}{\hbar^2} E_{n'l} - \frac{M\omega}{\hbar}(2n'+2l+3) = 0, \qquad (6.98)$$

or

$$E_{n'l} = \left(n' + l + \frac{3}{2} \right) \hbar\omega, \qquad (6.99)$$

where n' is even (see (6.97)). Denoting n' by $2N$, where $N = 0, 1, 2, 3, \ldots$, we rewrite this energy expression as

$$\boxed{E_n = \left(n + \frac{3}{2} \right) \hbar\omega \qquad (n = 0, 1, 2, 3, \ldots),} \qquad (6.100)$$

where $n = n' + l = 2N + l$.

The ground state, whose energy is $E_0 = \frac{3}{2}\hbar\omega$, is not degenerate; the first excited state, $E_1 = \frac{5}{2}\hbar\omega$, is threefold degenerate; and the second excited state, $E_2 = \frac{7}{2}\hbar\omega$, is sixfold degenerate (Table 6.4). As shown in the following example, the degeneracy relation for the nth level is given by

$$\boxed{g_n = \frac{1}{2}(n+1)(n+2).} \qquad (6.101)$$

This expression is in agreement with (6.36) obtained for an isotropic harmonic oscillator in Cartesian coordinates.

Finally, since the radial wave function is given by $R_{nl}(r) = U_{nl}(r)/r$, where $U_{nl}(r)$ is listed in (6.88) with $f(r)$ being a polynomial in r^{2l} of degree $(n-l)/2$, the total wave function for the isotropic harmonic oscillator is

$$\psi_{nlm}(r, \theta, \varphi) = R_{nl}(r)Y_{lm}(\theta, \varphi) = \frac{U_{nl}(r)}{r}Y_{lm}(\theta, \varphi) = r^l f(r)Y_{lm}(\theta, \varphi)e^{-M\omega r^2/2\hbar}, \qquad (6.102)$$

Table 6.4 Energy levels E_n and degeneracies g_n for an isotropic harmonic oscillator.

n	E_n	Nl	m	g_n
0	$\frac{3}{2}\hbar\omega$	0 0	0	1
1	$\frac{5}{2}\hbar\omega$	0 1	$\pm 1, 0$	3
2	$\frac{7}{2}\hbar\omega$	1 0	0	6
		0 2	$\pm 2, \pm 1, 0$	
3	$\frac{9}{2}\hbar\omega$	1 1	$\pm 1, 0$	10
		0 3	$\pm 3, \pm 2, \pm 1, 0$	

where l takes only odd or only even values. For instance, the ground state corresponds to $(n, l, m) = (0, 0, 0)$; its wave function is

$$\psi_{000}(r, \theta, \varphi) = R_{00}(r)Y_{00}(\theta, \varphi) = \frac{2}{\sqrt{\sqrt{\pi}}}\left(\frac{M\omega}{\hbar}\right)^{3/4} e^{-M\omega r^2/2\hbar} Y_{00}(\theta, \varphi). \tag{6.103}$$

The (n, l, m) configurations of the first, second, and third excited states can be determined as follows. The first excited state has three degenerate states: $(1, 1, m)$ with $m = -1, 0, 1$. The second excited states has 6 degenerate states: $(2, 0, 0)$ and $(2, 2, m)$ with $m = -2, -1, 0, 1, 2$. The third excited state has 10 degenerate states: $(3, 1, m)$ with $m = -1, 0, 1$ and $(3, 3, m)$ where $m = -3, -2, -1, 0, 1, 2, 3$. Some of these wave functions are given by

$$\psi_{11m}(r, \theta, \varphi) = R_{11}(r)Y_{1m}(\theta, \varphi) = \sqrt{\frac{8}{3\sqrt{\pi}}}\left(\frac{M\omega}{\hbar}\right)^{5/4} re^{-M\omega r^2/2\hbar} Y_{1m}(\theta, \varphi), \tag{6.104}$$

$$\psi_{200}(r, \theta, \varphi) = R_{20}(r)Y_{00}(\theta, \varphi) = \sqrt{\frac{8}{3\sqrt{\pi}}}\left(\frac{M\omega}{\hbar}\right)^{3/4}\left(\frac{3}{2} - \frac{M\omega}{\hbar}r^2\right)e^{-M\omega r^2/2\hbar} Y_{00}(\theta, \varphi), \tag{6.105}$$

$$\psi_{31m}(r, \theta, \varphi) = R_{31}(r)Y_{1m}(\theta, \varphi) = \frac{4}{\sqrt{15\sqrt{\pi}}}\left(\frac{M\omega}{\hbar}\right)^{7/4} r^2 e^{-M\omega r^2/2\hbar} Y_{1m}(\theta, \varphi). \tag{6.106}$$

Example 6.2 (Degeneracy relation for an isotropic oscillator)
Prove the degeneracy relation (6.101) for an isotropic harmonic oscillator.

Solution
Since $n = 2N + l$ the quantum numbers n and l must have the same parity. Also, since the isotropic harmonic oscillator is spherically symmetric, its states have definite parity[3]. In addition, since the parity of the states corresponding to a central potential is given by $(-1)^l$, the

---

[3] Recall from Chapter 4 that if the potential of a system is symmetric, $V(x) = V(-x)$, the states of the system must be either odd or even.

quantum number l (hence n) can take only even or only odd values. Let us consider separately the cases when n is even or odd.

First, when n is even the degeneracy g_n of the nth excited state is given by

$$g_n = \sum_{l=0,2,4,\ldots}^{n}(2l+1) = \sum_{l=0,2,4,\ldots}^{n}1 + 2\sum_{l=0,2,4,\ldots}^{n}l = \frac{1}{2}(n+2) + \frac{n(n+2)}{2} = \frac{1}{2}(n+1)(n+2). \quad (6.107)$$

A more explicit way of obtaining this series consists of writing it in the following two equivalent forms:

$$g_n = 1+5+9+13+\cdots+(2n-7)+(2n-3)+(2n+1), \quad (6.108)$$
$$g_n = (2n+1)+(2n-3)+(2n-7)+(2n-11)+\cdots+13+9+5+1. \quad (6.109)$$

We then add them, term by term, to get

$$2g_n = (2n+2)+(2n+2)+(2n+2)+(2n+2)+\cdots+(2n+2) = (2n+2)\left(\frac{n}{2}+1\right). \quad (6.110)$$

This relation yields $g_n = \frac{1}{2}(n+1)(n+2)$, which proves (6.101) when n is even.

Second, when n is odd, a similar treatment leads to

$$g_n = \sum_{l=1,3,5,7,\ldots}^{n}(2l+1) = \sum_{l=1,3,5,7,\ldots}^{n}1 + 2\sum_{l=1,3,5,7,\ldots}^{n}l = \frac{1}{2}(n+1)+\frac{1}{2}(n+1)^2 = \frac{1}{2}(n+1)(n+2), \quad (6.111)$$

which proves (6.101) when n is odd. Note that this degeneracy relation is, as expected, identical with the degeneracy expression (6.36) obtained for a harmonic oscillator in Cartesian coordinates.

6.3.5 The Hydrogen Atom

The hydrogen atom consists of an electron and a proton. For simplicity, we will ignore their spins. The wave function then depends on six coordinates $\vec{r}_e(x_e, y_e, z_e)$ and $\vec{r}_p(x_p, y_p, z_p)$, where $\vec{r}_e$ and $\vec{r}_p$ are the electron and proton position vectors, respectively. According to the probabilistic interpretation of the wave function, the quantity $|\Psi(\vec{r}_e, \vec{r}_p, t)|^2 d^3 r_e\, d^3 r_p$ represents the probability that a simultaneous measurement of the electron and proton positions at time t will result in the electron being in the volume element $d^3 r_e$ and the proton in $d^3 r_p$.

The time-dependent Schrödinger equation for the hydrogen atom is given by

$$\left[-\frac{\hbar^2}{2m_p}\nabla_p^2 - \frac{\hbar^2}{2m_e}\nabla_e^2 + V(r)\right]\Psi(\vec{r}_e, \vec{r}_p, t) = i\hbar\frac{\partial}{\partial t}\Psi(\vec{r}_e, \vec{r}_p, t), \quad (6.112)$$

where ∇_p^2 and ∇_e^2 are the Laplacians with respect to the proton and the electron degrees of freedom, with $\nabla_p^2 = \partial^2/\partial x_p^2 + \partial^2/\partial y_p^2 + \partial^2/\partial z_p^2$ and $\nabla_e^2 = \partial^2/\partial x_e^2 + \partial^2/\partial y_e^2 + \partial^2/\partial z_e^2$, and where $V(r)$ is the potential (interaction) between the electron and the proton. This interaction, which depends only on the distance that separates the electron and the proton $\vec{r} = \vec{r}_e - \vec{r}_p$, is given by the Coulomb potential:

$$V(r) = -\frac{e^2}{r}. \quad (6.113)$$

Note: Throughout this text, we will be using the CGS units for the Coulomb potential where it is given by $V(r) = -e^2/r$ (in the MKS units, however, it is given by $V(r) = -e^2/(4\pi\varepsilon_0 r)$).

Since V does not depend on time, the solutions of (6.112) are stationary; hence, they can be written as follows:

$$\Psi(\vec{r}_e, \vec{r}_p, t) = \chi(\vec{r}_e, \vec{r}_p)e^{-iEt/\hbar}, \tag{6.114}$$

where E is the total energy of the electron–proton system. Substituting this into (6.112), we obtain the time-independent Schrödinger equation for the hydrogen atom:

$$\left[-\frac{\hbar^2}{2m_p}\nabla_p^2 - \frac{\hbar^2}{2m_e}\nabla_e^2 - \frac{e^2}{|\vec{r}_e - \vec{r}_p|}\right]\chi(\vec{r}_e, \vec{r}_p) = E\chi(\vec{r}_e, \vec{r}_p). \tag{6.115}$$

6.3.5.1 Separation of the Center of Mass Motion

Since V depends only on the relative distance r between the electron and proton, instead of the coordinates $\vec{r}_e$ and $\vec{r}_p$ (position vectors of the electron and proton), it is more appropriate to use the coordinates of the center of mass, $\vec{R} = X\vec{i} + Y\vec{j} + Z\vec{k}$, and the relative coordinates of the electron with respect to the proton, $\vec{r} = x\vec{i} + y\vec{j} + z\vec{k}$. The transformation from $\vec{r}_e, \vec{r}_p$ to $\vec{R}, \vec{r}$ is given by

$$\vec{R} = \frac{m_e\vec{r}_e + m_p\vec{r}_p}{m_e + m_p}, \qquad \vec{r} = \vec{r}_e - \vec{r}_p. \tag{6.116}$$

We can verify that the Laplacians ∇_e^2 and ∇_p^2 are related to

$$\nabla_R^2 = \frac{\partial^2}{\partial X^2} + \frac{\partial^2}{\partial Y^2} + \frac{\partial^2}{\partial Z^2}, \qquad \nabla_r^2 = \frac{\partial^2}{\partial x^2} + \frac{\partial^2}{\partial y^2} + \frac{\partial^2}{\partial z^2} \tag{6.117}$$

as follows:

$$\frac{1}{m_e}\nabla_e^2 + \frac{1}{m_p}\nabla_p^2 = \frac{1}{M}\nabla_R^2 + \frac{1}{\mu}\nabla_r^2, \tag{6.118}$$

where

$$M = m_e + m_p, \qquad \mu = \frac{m_e m_p}{m_e + m_p} \tag{6.119}$$

are the total and reduced masses, respectively The time-independent Schrödinger equation (6.115) then becomes

$$\left[-\frac{\hbar^2}{2M}\nabla_R^2 - \frac{\hbar^2}{2\mu}\nabla_r^2 + V(r)\right]\Psi_E(\vec{R}, \vec{r}) = E\,\Psi_E(\vec{R}, \vec{r}), \tag{6.120}$$

where $\Psi_E(\vec{R}, \vec{r}) = \chi(\vec{r}_e, \vec{r}_p)$. Let us now solve this equation by the separation of variables; that is, we look for solutions of the form

$$\Psi_E(\vec{R}, \vec{r}) = \Phi(\vec{R})\psi(\vec{r}), \tag{6.121}$$

where $\Phi(\vec{R})$ and $\psi(\vec{r})$ are the wave functions of the CM and of the relative motions, respectively. Substituting this wave function into (6.120) and dividing by $\Phi(\vec{R})\psi(\vec{r})$, we obtain

$$\left[-\frac{\hbar^2}{2M}\frac{1}{\Phi(\vec{R})}\nabla_R^2\Phi(\vec{R})\right] + \left[-\frac{\hbar^2}{2\mu}\frac{1}{\psi(\vec{r})}\nabla_r^2\psi(\vec{r}) + V(r)\right] = E. \tag{6.122}$$

The first bracket depends only on $\vec{R}$ whereas the second bracket depends only on $\vec{r}$. Since $\vec{R}$ and $\vec{r}$ are independent vectors, the two expressions of the left hand side of (6.122) must be separately constant. Thus, we can reduce (6.122) to the following two separate equations:

$$-\frac{\hbar^2}{2M}\nabla_R^2\Phi(\vec{R}) = E_R\Phi(\vec{R}), \qquad (6.123)$$

$$-\frac{\hbar^2}{2\mu}\nabla_r^2\psi(\vec{r}) + V(r)\psi(\vec{r}) = E_r\psi(\vec{r}), \qquad (6.124)$$

with the condition

$$E_R + E_r = E. \qquad (6.125)$$

We have thus reduced the Schrödinger equation (6.120), which involves two variables $\vec{R}$ and $\vec{r}$, into two separate equations (6.123) and (6.124) each involving a single variable. Note that equation (6.123) shows that the center of mass moves like a free particle of mass M. The solution to this kind of equation was examined earlier in this chapter; it has the form

$$\Phi(\vec{R}) = (2\pi)^{-3/2}e^{i\vec{k}\cdot\vec{R}}, \qquad (6.126)$$

where $\vec{k}$ is the wave vector associated with the center of mass. The constant $E_R = \hbar^2k^2/(2M)$ gives the kinetic energy of the center of mass in the lab system (the total mass M is located at the origin of the center of mass coordinate system).

The second equation (6.124) represents the Schrödinger equation of a fictitious particle of mass μ moving in the central potential $-e^2/r$.

We should note that the total wave function $\Psi_E(\vec{R},\vec{r}) = \Phi(\vec{R})\psi(\vec{r})$ is seldom used. When the hydrogen problem is mentioned, this implicitly refers to $\psi(\vec{r})$ and E_r. That is, the hydrogen wave function and energy are taken to be given by $\psi(\vec{r})$ and E_r, not by Ψ_E and E.

6.3.5.2 Solution of the Radial Equation for the Hydrogen Atom

The Schrödinger equation (6.124) for the relative motion has the form of an equation for a central potential. The wave function $\psi(\vec{r})$ that is a solution to this equation is a product of an angular part and a radial part. The angular part is given by the spherical harmonic $Y_{lm}(\theta,\varphi)$. The radial part $R(r)$ can be obtained by solving the following radial equation:

$$-\frac{\hbar^2}{2\mu}\frac{d^2U(r)}{dr^2} + \left[\frac{l(l+1)\hbar^2}{2\mu r^2} - \frac{e^2}{r}\right]U(r) = EU(r), \qquad (6.127)$$

where $U(r) = rR(r)$. To solve this radial equation, we are going to consider first its asymptotic solutions and then attempt a power series solution.

(a) Asymptotic behavior of the radial wave function
For very small values of r, (6.127) reduces to

$$-\frac{d^2U(r)}{dr^2} + \frac{l(l+1)}{r^2}U(r) = 0, \qquad (6.128)$$

whose solutions are of the form

$$U(r) = Ar^{l+1} + Br^{-l}, \qquad (6.129)$$

where A and B are constants. Since $U(r)$ vanishes at $r = 0$, the second term r^{-l}, which diverges at $r = 0$, must be discarded. Thus, for small r, the solution is

$$U(r) \sim r^{l+1}. \tag{6.130}$$

Now, in the limit of very large values of r, we can approximate (6.127) by

$$\frac{d^2 U(r)}{dr^2} + \frac{2\mu E}{\hbar^2} U(r) = 0. \tag{6.131}$$

Note that, for bound state solutions, which correspond to the states where the electron and the proton are bound together, the energy E must be negative. Hence the solutions to this equation are of the form $U(r) \sim e^{\pm \lambda r}$ where $\lambda = \sqrt{2\mu(-E)}/\hbar$. Only the minus sign solution is physically acceptable, since $e^{\lambda r}$ diverges for large values of r. So, for large values of r, $U(r)$ behaves like

$$U(r) \longrightarrow e^{-\lambda r}. \tag{6.132}$$

The solutions to (6.127) can be obtained by combining (6.130) and (6.132):

$$U(r) = r^{l+1} f(r) e^{-\lambda r}, \tag{6.133}$$

where $f(r)$ is an r-dependent function. Substituting (6.133) into (6.127) we end up with a differential equation that determines the form of $f(r)$:

$$\frac{d^2 f}{dr^2} + 2\left(\frac{l+1}{r} - \lambda\right)\frac{df}{dr} + 2\left[\frac{-\lambda(l+1) + \mu e^2/\hbar^2}{r}\right] f(r) = 0. \tag{6.134}$$

(b) Power series solutions for the radial equation
As in the case of the three-dimensional harmonic oscillator, let us try a power series solution for (6.134):

$$f(r) = \sum_{k=0}^{\infty} b_k r^k, \tag{6.135}$$

which, when inserted into (6.134), yields

$$\sum_{k=0}^{\infty} \left\{ k(k + 2l + 1) b_k r^{k-2} + 2\left[-\lambda(k + l + 1) + \frac{\mu e^2}{\hbar^2}\right] b_k r^{k-1} \right\} = 0. \tag{6.136}$$

This equation leads to the following recurrence relation (by changing k to $k - 1$ in the last term):

$$k(k + 2l + 1) b_k = 2\left[\lambda(k + l) - \frac{\mu e^2}{\hbar^2}\right] b_{k-1}. \tag{6.137}$$

In the limit of large values of k, the ratio of successive coefficients,

$$\frac{b_k}{b_{k-1}} = \frac{2\left[\lambda(k + l) - \mu e^2/\hbar^2\right]}{k(k + 2l + 1)}, \tag{6.138}$$

is of the order of

$$\frac{b_k}{b_{k-1}} \longrightarrow \frac{2\lambda}{k}. \tag{6.139}$$

This is the behavior of an exponential series, since the ratio of successive coefficients of the relation $e^{2x} = \sum_{k=0}^{\infty} (2x)^k / k!$ is given by

$$\frac{2^k}{k!} \frac{(k-1)!}{2^{k-1}} = \frac{2}{k}. \tag{6.140}$$

That is, the asymptotic behavior of (6.135) is

$$f(r) = \sum_{k=0}^{\infty} b_k r^k \longrightarrow e^{2\lambda r}; \tag{6.141}$$

hence the radial solution (6.133) becomes

$$U(r) = r^{l+1} e^{2\lambda r} e^{-\lambda r} = r^{l+1} e^{\lambda r}. \tag{6.142}$$

But this contradicts (6.133): for large values of r, the asymptotic behavior of the physically acceptable radial function (6.133) is given by $e^{-\lambda r}$ while that of (6.142) by $e^{\lambda r}$; the form (6.142) is thus physically unacceptable.

(c) Energy quantization
To obtain physically acceptable solutions, the series (6.135) *must terminate at a certain power N*; hence the function $f(r)$ becomes a *polynomial* of order N:

$$f(r) = \sum_{k=0}^{N} b_k r^k. \tag{6.143}$$

This requires that all coefficients $b_{N+1}, b_{N+2}, b_{N+3}, \ldots$ have to vanish. When $b_{N+1} = 0$ the recurrence formula (6.137) yields

$$\lambda(N + l + 1) - \frac{\mu e^2}{\hbar^2} = 0. \tag{6.144}$$

Since $\lambda = \sqrt{-2\mu E/\hbar^2}$ and using the notation

$$n = N + l + 1, \tag{6.145}$$

where n is known as the *principal* quantum number and N as the *radial* quantum number, we can infer the energy

$$E_n = -\frac{\mu e^4}{2\hbar^2} \frac{1}{n^2}, \tag{6.146}$$

which in turn can be written as

$$\boxed{E_n = -\frac{\mu e^4}{2\hbar^2} \frac{1}{n^2} = -\frac{e^2}{2a_0} \frac{1}{n^2},} \tag{6.147}$$

because (from Bohr theory of the hydrogen atom) the Bohr radius is given by $a_0 = \hbar^2/(\mu e^2)$ and hence $\mu/\hbar^2 = 1/(e^2 a_0)$. Note that we can write λ in terms of a_0 as follows:

$$\lambda = \sqrt{-2\frac{\mu}{\hbar^2} E_n} = \sqrt{2\frac{1}{e^2 a_0} \frac{e^2}{2a_0 n^2}} = \frac{1}{na_0}. \tag{6.148}$$

Since $N = 0, 1, 2, 3, \ldots$, the allowed values of n are nonzero integers, $n = l+1, l+2, l+3,$ $\ldots$. For a given value of n, the orbital quantum number l can have values only between 0 and $n-1$ (i.e., $l = 0, 1, 2, \ldots, n-1$).

Remarks

- Note that (6.147) is similar to the energy expression obtained from the Bohr quantization condition, discussed in Chapter 1. It can be rewritten in terms of the Rydberg constant $\mathcal{R} = m_e e^4 / (2\hbar^2)$ as follows:

$$E_n = -\frac{m_p}{m_p + m_e} \frac{\mathcal{R}}{n^2}, \tag{6.149}$$

where $\mathcal{R} = 13.6\,\text{eV}$. Since the ratio m_e/m_p is very small ($m_e/m_p \ll 1$), we can approximate this expression by

$$E_n = -\left(1 + \frac{m_e}{m_p}\right)^{-1} \frac{\mathcal{R}}{n^2} \simeq -\left(1 - \frac{m_e}{m_p}\right) \frac{\mathcal{R}}{n^2}. \tag{6.150}$$

So, if we consider the proton to be infinitely more massive than the electron, we recover the energy expression as derived by Bohr: $E_n = -\mathcal{R}/n^2$.

- **Energy of hydrogen-like atoms:** How does one obtain the energy of an atom or ion with a nuclear charge Ze but which has only one electron[4]? Since the Coulomb potential felt by the single electron due to the charge Ze is given by $V(r) = -Ze^2/r$, the energy of the electron can be inferred from (6.147) by simply replacing e^2 with Ze^2:

$$E_n = -\frac{m_e(Ze^2)^2}{2\hbar^2} \frac{1}{n^2} = -\frac{Z^2 E_0}{n^2}, \tag{6.151}$$

where $E_0 = e^2/(2a_0) = 13.6\,\text{eV}$; in deriving this relation, we have assumed that the mass of the nucleus is infinitely large compared to the electronic mass.

(d) Radial wave functions of the hydrogen atom

The radial wave function $R_{nl}(r)$ can be obtained by inserting (6.143) into (6.133),

$$R_{nl}(r) = \frac{1}{r}U_{nl}(r) = A_{nl}r^l e^{-\lambda r} \sum_{k=0}^{N} b_k r^k = A_{nl}r^l e^{-r/na_0} \sum_{k=0}^{N} b_k r^k, \tag{6.152}$$

since, as shown in (6.148), $\lambda = 1/(na_0)$; A_{nl} is a normalization constant.

How does one determine the expression of $R_{nl}(r)$? This issue reduces to obtaining the form of the polynomial $r^l \sum_{k=0}^{N} b_k r^k$ and the normalization constant A_{nl}. For this, we are going to explore two methods: the first approach follows a straightforward calculation and the second makes use of special functions.

---

[4]For instance, $Z = 1$ refers to H, $Z = 2$ to He$^+$, $Z = 3$ to Li^{2+}, $Z = 4$ to Be^{3+}, $Z = 5$ to B^{4+}, $Z = 6$ to C^{5+}, and so on.

(i) First approach: straightforward calculation of $R_{nl}(r)$.

This approach consists of a straightforward construction of $R_{nl}(r)$; we are going to show how to construct only the first few expressions. For instance, if $n = 1$ and $l = 0$ then $N = 0$. Since $N = n - l - 1$ and $\lambda = 1/(na_0)$ we can write (6.152) as

$$R_{10}(r) = A_{10}e^{-r/a_0} \sum_{k=0}^{0} b_k r^k = A_{10}b_0 e^{-r/a_0}, \tag{6.153}$$

where $A_{10}b_0$ can be obtained from the normalization of $R_{10}(r)$: using $\int_0^\infty x^n e^{-ax}\, dx = n!/a^{n+1}$, we have

$$1 = \int_0^\infty r^2 |R_{10}(r)|^2\, dr = A_{10}^2 b_0^2 \int_0^\infty r^2 e^{-2r/a_0}\, dr = A_{10}^2 b_0^2 \frac{a_0^3}{4}; \tag{6.154}$$

hence $A_{10} = 1$ and $b_0 = 2\,(a_0)^{-3/2}$. Thus, $R_{10}(r)$ is given by

$$R_{10}(r) = 2\,(a_0)^{-3/2}\, e^{-r/a_0}. \tag{6.155}$$

Next, let us find $R_{20}(r)$. Since $n = 2, l = 0$ we have $N = 2 - 0 - 1 = 1$ and

$$R_{20}(r) = A_{20}e^{-r/2a_0} \sum_{k=0}^{1} b_k r^k = A_{20}(b_0 + b_1 r)e^{-r/2a_0}. \tag{6.156}$$

From (6.138) we can express b_1 in terms of b_0 as

$$b_1 = \frac{2\lambda(k+l) - 2/a_0}{k(k+2l+1)} b_0 = -\frac{1}{2a_0} b_0 = -\frac{1}{a_0\sqrt{a_0^3}}, \tag{6.157}$$

because $\lambda = 1/(2a_0)$, $k = 1$, and $l = 0$. So, substituting (6.157) into (6.156) and normalizing, we get $A_{20} = 1/(2\sqrt{2})$; hence

$$R_{20}(r) = \frac{1}{\sqrt{2a_0^3}} \left(1 - \frac{r}{2a_0}\right) e^{-r/2a_0}. \tag{6.158}$$

Continuing in this way, we can obtain the expression of any radial wave function $R_{nl}(r)$; note that, knowing $b_0 = 2\,(a_0)^{-3/2}$, we can use the recursion relation (6.138) to obtain all other coefficients $b_2, b_3, \ldots$.

(ii) Second approach: determination of $R_{nl}(r)$ by means of special functions

The polynomial $r^l \sum_{k=0}^{N} b_k r^k$ present in (6.152) is a polynomial of degree $N + l$ or $n - 1$ since $n = N + l + 1$. This polynomial, which is denoted by $L_k^N(r)$, is known as the *associated Laguerre polynomial*; it is a solution to the Schrödinger equation (6.134). The solutions to differential equations of the form (6.134) were studied by Laguerre long before the birth of quantum mechanics. The associated Laguerre polynomial is defined, in terms of the *Laguerre polynomials* of order k, $L_k(r)$, by

$$L_k^N(r) = \frac{d^N}{dr^N} L_k(r), \tag{6.159}$$

where

$$L_k(r) = e^r \frac{d^k}{dr^k} (r^k e^{-r}). \tag{6.160}$$

Table 6.5 First few Laguerre polynomials and associated Laguerre polynomials.

Laguerre polynomials $L_k(r)$	Associated Laguerre polynomials $L_k^N(r)$
$L_0 = 1$	
$L_1 = 1 - r$	$L_1^1 = -1$
$L_2 = 2 - 4r + r^2$	$L_2^1 = -4 + 2r, \quad L_2^2 = 2$
$L_3 = 6 - 18r + 9r^2 - r^3$	$L_3^1 = -18 + 18r - 3r^2, \ L_3^2 = 18 - 6r, \ L_3^3 = -6$
$L_4 = 24 - 96r + 72r^2 - 16r^3 + r^4$	$L_4^1 = -96 + 144r - 48r^2 + 4r^3$
	$L_4^2 = 144 - 96r + 12r^2, \ L_4^3 = 24r - 96, \ L_4^4 = 24$
$L_5 = 120 - 600r + 600r^2 - 200r^3$	$L_5^1 = -600 + 1200r - 600r^2 + 100r^3 - 5r^4$
$\quad + 25r^4 - r^5$	$L_5^2 = 1200 - 1200r + 300r^2 - 20r^3$
	$L_5^3 = -1200 + 600r - 60r^2, \ L_5^4 = 600 - 120r$
	$L_5^5 = -120$

The first few Laguerre polynomials are listed in Table 6.5.

We can verify that $L_k(r)$ and $L_k^N(r)$ satisfy the following differential equations:

$$r\frac{d^2 L_k(r)}{dr^2} + (1-r)\frac{dL_k(r)}{dr} + kL_k(r) = 0, \tag{6.161}$$

$$r\frac{d^2 L_k^N(r)}{dr^2} + (N+1-r)\frac{dL_k^N(r)}{dr} + (k-N)L_k^N(r) = 0. \tag{6.162}$$

This last equation is identical to the hydrogen atom radial equation (6.134). The proof goes as follows. Using a change of variable

$$\rho = 2\lambda r = 2\frac{\sqrt{-2\mu E}}{\hbar}r, \tag{6.163}$$

along with the fact that $a_0 = \hbar^2/(\mu e^2)$ (Bohr radius), we can show that (6.134) reduces to

$$\rho\frac{d^2 g(\rho)}{d\rho^2} + [(2l+1)+1-\rho]\frac{dg(\rho)}{d\rho} + [(n+l)-(2l+1)]g(\rho) = 0, \tag{6.164}$$

where $f(r) = g(\rho)$. In deriving (6.164), we have used the fact that $1/\lambda a_0 = n$ (see (6.148)). Note that equations (6.162) and (6.164) are identical; the solutions to (6.134) are thus given by the associated Laguerre polynomials $L_{n+l}^{2l+1}(2\lambda r)$.

The radial wave function of the hydrogen atom is then given by

$$R_{nl}(r) = N_{nl}\left(\frac{2r}{na_0}\right)^l e^{-r/na_0} L_{n+l}^{2l+1}\left(\frac{2r}{na_0}\right), \tag{6.165}$$

where N_{nl} is a constant obtained by normalizing the radial function $R_{nl}(r)$:

$$\int_0^\infty r^2 R_{nl}^2(r)\, dr = 1. \tag{6.166}$$

Table 6.6 The first few radial wave functions $R_{nl}(r)$ of the hydrogen atom.

$$R_{10}(r) = 2a_0^{-3/2} e^{-r/a_0} \qquad\qquad R_{21}(r) = \frac{1}{\sqrt{6a_0^3}} \frac{r}{2a_0} e^{-r/2a_0}$$

$$R_{20}(r) = \frac{1}{\sqrt{2a_0^3}} \left(1 - \frac{r}{2a_0}\right) e^{-r/2a_0} \qquad R_{31}(r) = \frac{8}{9\sqrt{6a_0^3}} \left(1 - \frac{r}{6a_0}\right) \left(\frac{r}{3a_0}\right) e^{-r/3a_0}$$

$$R_{30}(r) = \frac{2}{3\sqrt{3a_0^3}} \left(1 - \frac{2r}{3a_0} + \frac{2r^2}{27a_0^2}\right) e^{-r/3a_0} \qquad R_{32}(r) = \frac{4}{9\sqrt{30a_0^3}} \left(\frac{r}{3a_0}\right)^2 e^{-r/3a_0}$$

Using the normalization condition of the associated Laguerre functions

$$\int_0^\infty e^{-\rho} \rho^{2l} \left[L_{n+l}^{2l+1}(\rho)\right]^2 \rho^2 d\rho = \frac{2n[(n+l)!]^3}{(n-l-1)!}, \tag{6.167}$$

where $\rho = 2\lambda r = 2r/(na_0)$, we can show that N_{nl} is given by

$$N_{nl} = -\left(\frac{2}{na_0}\right)^{3/2} \sqrt{\frac{(n-l-1)!}{2n[(n+l)!]^3}}. \tag{6.168}$$

The wave functions of the hydrogen atom are given by

$$\boxed{\psi_{nlm}(r, \theta, \varphi) = R_{nl}(r) Y_{lm}(\theta, \varphi),} \tag{6.169}$$

where the radial functions $R_{nl}(r)$ are

$$\boxed{R_{nl}(r) = -\left(\frac{2}{na_0}\right)^{3/2} \sqrt{\frac{(n-l-1)!}{2n[(n+l)!]^3}} \left(\frac{2r}{na_0}\right)^l e^{-r/na_0} L_{n+l}^{2l+1}\left(\frac{2r}{na_0}\right).} \tag{6.170}$$

The first few radial wave functions are listed in Table 6.6; as shown in (6.155) and (6.158), they are identical with those obtained from a straightforward construction of $R_{nl}(r)$. The shapes of some of these radial functions are plotted in Figure 6.3.

(e) Properties of the radial wave functions of hydrogen
The radial wave functions of the hydrogen atom behave as follows (see Figure 6.3):

- They behave like r^l for small r.

- They decrease exponentially at large r, since L_{n+l}^{2l+1} is dominated by the highest power, r^{n-l-1}.

- Each function $R_{nl}(r)$ has $n-l-1$ radial nodes, since $L_{n+l}^{2l+1}(\rho)$ is a polynomial of degree $n-l-1$.

Figure 6.3 The first few radial wave functions $R_{nl}(r)$ for hydrogen; the radial length is in units of the Bohr radius $a_0 = \hbar^2/(\mu e^2)$. Notice that $R_{nl}(r)$ has $(n - l - 1)$ nodes.

Table 6.7 Hydrogen energy levels and their degeneracies when the electron's spin is ignored.

n	l	Orbitals	m	g_n	E_n
1	0	s	0	1	$-e^2/(2a_0)$
2	0	s	0	4	$-e^2/(8a_0)$
	1	p	$-1, 0, 1$		
3	0	s	0	9	$-e^2/(18a_0)$
	1	p	$-1, 0, 1$		
	2	d	$-2, -1, 0, 1, 2$		
4	0	s	0	16	$-e^2/(32a_0)$
	1	p	$-1, 0, 1$		
	2	d	$-2, -1, 0, 1, 2$		
	3	f	$-3, -2, -1, 0, 1, 2, 3$		
5	0	s	0	25	$-e^2/(50a_0)$
	1	p	$-1, 0, 1$		
	2	d	$-2, -1, 0, 1, 2$		
	3	f	$-3, -2, -1, 0, 1, 2, 3$		
	4	g	$-4, -3, -2, -1, 0, 1, 2, 3, 4$		

6.3.5.3 Degeneracy of the Bound States of Hydrogen

Besides being independent of m, which is a property of central potentials (see (6.55)), the energy levels (6.147) are also independent of l. This *additional degeneracy* in l is not a property of central potentials, but a *particular feature of the Coulomb potential*. In the case of central potentials, the energy E usually depends on two quantum numbers: one radial, n, and the other orbital, l, giving E_{nl}.

The total quantum number n takes only nonzero values $1, 2, 3, \ldots$. As displayed in Table 6.7, for a given n, the quantum l number may vary from 0 to $n - 1$; and for each l, m can take $(2l + 1)$ values: $m = -l, -l+1, \ldots, l-1, l$. The degeneracy of the state n, which is specified by the total number of different states associated with n, is then given by (see Example 6.3 on page 364)

$$g_n = \sum_{l=0}^{n-1}(2l + 1) = n^2. \tag{6.171}$$

Remarks

- The state of every hydrogenic electron is specified by three quantum numbers (n, l, m), called the single-particle state or *orbital*, $|nlm\rangle$. According to the spectroscopic notation, the states corresponding to the respective numerical values $l = 0, 1, 2, 3, 4, 5, \ldots$ are called the s, p, d, f, g, h, $\ldots$ states; the letters s, p, d, f refer to sharp, principal, diffuse, and fundamental labels, respectively (as the letters g, h, $\ldots$ have yet to be assigned labels, the reader is free to guess how to refer to them!). Hence, as shown in Table 6.7, for a given n an s-state has 1 orbital $|n00\rangle$, a p-state has 3 orbitals $|n1m\rangle$ corresponding to $m = -1, 0, 1$, a d-state has 5 orbitals $|n2m\rangle$ corresponding to $m = -2, -1, 0, 1, 2$, and so on.

- If we take into account the spin of the electron, the state of every electron will be specified by four quantum numbers (n, l, m_l, m_s), where $m_s = \pm\frac{1}{2}$ is the z-component of the spin of the electron. Hence the complete wave function of the hydrogen atom must be equal to the product of a space part or orbital $\psi_{nlm_l}(r, \theta, \varphi) = R_{nl}(r)Y_{lm_l}(\theta, \varphi)$, and a spin part $\left|\frac{1}{2}, m_s\right\rangle$:

$$\Psi_{nlm_lm_s}(\vec{r}) = \psi_{nlm_l}(r, \theta, \varphi)\left|\frac{1}{2}, \pm\frac{1}{2}\right\rangle = R_{nl}(r)Y_{lm_l}(\theta, \varphi)\left|\frac{1}{2}, \pm\frac{1}{2}\right\rangle. \tag{6.172}$$

Using the spinors from Chapter 5 we can write the spin-up wave function as

$$\Psi_{nlm_l\frac{1}{2}}(\vec{r}) = \psi_{nlm_l}(r, \theta, \varphi)\left|\frac{1}{2}, \frac{1}{2}\right\rangle = \psi_{nlm_l}\begin{pmatrix}1\\0\end{pmatrix} = \begin{pmatrix}\psi_{nlm_l}\\0\end{pmatrix}, \tag{6.173}$$

and the spin-down wave function as

$$\Psi_{nlm_l-\frac{1}{2}}(\vec{r}) = \psi_{nlm_l}(r, \theta, \varphi)\left|\frac{1}{2}, -\frac{1}{2}\right\rangle = \psi_{nlm_l}\begin{pmatrix}0\\1\end{pmatrix} = \begin{pmatrix}0\\\psi_{nlm_l}\end{pmatrix}. \tag{6.174}$$

For instance, the spin-up and spin-down *ground state* wave functions of hydrogen are given by

$$\Psi_{100\frac{1}{2}}(\vec{r}) = \begin{pmatrix}\psi_{100}\\0\end{pmatrix} = \begin{pmatrix}(1/\sqrt{\pi})a_0^{-3/2}e^{-r/a_0}\\0\end{pmatrix}, \tag{6.175}$$

$$\Psi_{100-\frac{1}{2}}(\vec{r}) = \begin{pmatrix}0\\\psi_{100}\end{pmatrix} = \begin{pmatrix}0\\(1/\sqrt{\pi})a_0^{-3/2}e^{-r/a_0}\end{pmatrix}. \tag{6.176}$$

- When spin is included the degeneracy of the hydrogen's energy levels is given by

$$2\sum_{l=0}^{n-1}(2l+1) = 2n^2, \tag{6.177}$$

since, in addition to the degeneracy (6.171), each level is doubly degenerate with respect to the spin degree of freedom. For instance, the ground state of hydrogen is doubly degenerate since $\Psi_{100\frac{1}{2}}(\vec{r})$ and $\Psi_{100-\frac{1}{2}}(\vec{r})$ correspond to the same energy -13.6 eV. Similarly, the first excited state is eightfold degenerate $(2(2)^2 = 8)$ because the eight states $\Psi_{200\pm\frac{1}{2}}(\vec{r})$, $\Psi_{211\pm\frac{1}{2}}(\vec{r})$, $\Psi_{210\pm\frac{1}{2}}(\vec{r})$, and $\Psi_{21-1\pm\frac{1}{2}}(\vec{r})$ correspond to the same energy $-13.6/4$ eV $= -3.4$ eV.

6.3.5.4 Probabilities and Averages

When a hydrogen atom is in the stationary state $\psi_{nlm}(r, \theta, \varphi)$, the quantity $|\psi_{nlm}(r, \theta, \varphi)|^2 d^3r$ represents the probability of finding the electron in the volume element d^3r, where

$d^3r = r^2 \sin\theta \, dr \, d\theta \, d\varphi$. The probability of finding the electron in a spherical shell located between r and $r + dr$ (i.e., a shell of thickness dr) is given by

$$P_{nl}(r)\, dr = \left(\int_0^\pi \sin\theta \, d\theta \int_0^{2\pi} d\varphi \, |\psi_{nlm}(r,\theta,\varphi)|^2 \right) r^2 dr$$

$$= |R_{nl}(r)|^2 r^2 dr \int_0^\pi \sin\theta \, d\theta \int_0^{2\pi} Y_{lm}^*(\theta,\varphi) Y_{lm}(\theta,\varphi) \, d\varphi$$

$$= |R_{nl}(r)|^2 r^2 dr. \tag{6.178}$$

If we integrate this quantity between $r = 0$ and $r = a$, we obtain the probability of finding the electron in a sphere of radius a centered about the origin. Hence integrating between $r = 0$ and $r = \infty$, we would obtain 1, which is the probability of finding the electron somewhere in space.

Let us now specify the average values of the various powers of r. Since $\psi_{nlm}(r,\theta,\varphi) = R_{nl}(r)Y_{lm}(\theta,\varphi)$, we can see that the average of r^k is independent of the azimuthal quantum number m:

$$\langle nlm|r^k|nlm \rangle = \int r^k |\psi_{nlm}(r,\theta,\varphi)|^2 r^2 \sin\theta \, dr \, d\theta \, d\varphi$$

$$= \int_0^\infty r^{k+2} |R_{nl}(r)|^2 dr \int_0^\pi \sin\theta \, d\theta \int_0^{2\pi} Y_{lm}^*(\theta,\varphi) Y_{lm}(\theta,\varphi) \, d\varphi$$

$$= \int_0^\infty r^{k+2} |R_{nl}(r)|^2 dr$$

$$= \langle nl \mid r^k|nl \rangle. \tag{6.179}$$

Using the properties of Laguerre polynomials, we can show that (Problem 6.2, page 370)

$$\langle nl| \, r \, |nl \rangle = \frac{1}{2}\left[3n^2 - l(l+1)\right]a_0, \tag{6.180}$$

$$\langle nl| \, r^2 \, | nl \rangle = \frac{1}{2}n^2\left[5n^2 + 1 - 3l\,(l+1)\right]a_0^2, \tag{6.181}$$

$$\langle nl|r^{-1}|nl \rangle = \frac{1}{n^2 a_0}, \tag{6.182}$$

$$\langle nl|r^{-2}|nl \rangle = \frac{2}{n^3(2l+1)a_0^2}, \tag{6.183}$$

where a_0 is the Bohr radius, $a_0 = \hbar^2/(\mu e^2)$. The averages (6.180) to (6.183) can be easily derived from Kramers' recursion relation (Problem 6.3, page 371):

$$\frac{k+1}{n^2}\langle nl|r^k|nl \rangle - (2k+1)a_0\langle nl|r^{k-1}|nl \rangle + \frac{ka_0^2}{4}\left[(2l+1)^2 - k^2\right]\langle nl|r^{k-2}|nl \rangle = 0. \tag{6.184}$$

Equations (6.180) and (6.182) reveal that $1/\langle r \rangle$ and $\langle 1/r \rangle$ are not equal, but are of the same order of magnitude:

$$\langle r \rangle \sim n^2 a_0. \tag{6.185}$$

This relation is in agreement with the expression obtained from the Bohr theory of hydrogen: the quantized radii of circular orbits for the hydrogen atom are given by $r_n = n^2 a_0$. We will show in Problem 6.6 page 375 that the Bohr radii for circular orbits give the locations where the probability density of finding the electron reaches its maximum.

Next, using the expression (6.182) for $\langle r^{-1} \rangle$, we can obtain the average value of the Coulomb potential

$$\langle V(r) \rangle = -e^2 \left\langle \frac{1}{r} \right\rangle = -\frac{e^2}{a_0} \frac{1}{n^2}, \tag{6.186}$$

which, as specified by (6.147), is equal to twice the total energy:

$$E_n = \frac{1}{2} \langle V(r) \rangle = -\frac{e^2}{2a_0} \frac{1}{n^2}. \tag{6.187}$$

This is known as the Virial theorem, which states that if $V(\alpha r) = \alpha^n V(r)$, the average expressions of the kinetic and potential energies are related by

$$\langle T \rangle = \frac{n}{2} \langle V(r) \rangle. \tag{6.188}$$

For instance, in the case of a Coulomb potential $V(\alpha r) = \alpha^{-1} V(r)$, we have $\langle T \rangle = -\frac{1}{2} \langle V \rangle$; hence $E = -\frac{1}{2} \langle V \rangle + \langle V \rangle = \frac{1}{2} \langle V \rangle$.

Example 6.3 (Degeneracy relation for the hydrogen atom)
Prove the degeneracy relation (6.171) for the hydrogen atom.

Solution
The energy $E_n = -e^2/(2a_0 n^2)$ of the hydrogen atom (6.147) does not depend on the orbital quantum number l or on the azimuthal number m; it depends only on the principal quantum number n. For a given n, the orbital number l can take $n - 1$ values: $l = 0, 1, 2, 3, \ldots, n - 1$; while for each l, the azimuthal number m takes $2l + 1$ values: $m = -l, -l + 1, \ldots, l - 1, l$. Thus, for each n, there exist g_n different wave functions $\psi_{nlm}(\vec{r})$, which correspond to the same energy E_n, with

$$g_n = \sum_{l=0}^{n-1} (2l + 1) = 2 \sum_{l=0}^{n-1} l + \sum_{l=0}^{n-1} 1 = n(n - 1) + n = n^2. \tag{6.189}$$

Another way of finding this result consists of writing $\sum_{l=0}^{n-1} (2l + 1)$ in the following two equivalent forms:

$$g_n = 1 + 3 + 5 + 7 + \cdots + (2n - 7) + (2n - 5) + (2n - 3) + (2n - 1), \tag{6.190}$$
$$g_n = (2n - 1) + (2n - 3) + (2n - 5) + (2n - 7) + \cdots + 7 + 5 + 3 + 1, \tag{6.191}$$

and then add them, term by term:

$$2g_n = (2n) + (2n) + (2n) + (2n) + \cdots + (2n) + (2n) + (2n) + (2n). \tag{6.192}$$

Since there are n terms (because l can take n values: $l = 0, 1, 2, 3, \ldots, n - 1$), we have $2g_n = n(2n)$; hence $g_n = n^2$.

6.3.6 Effect of Magnetic Fields on Central Potentials

As discussed earlier (6.55), the energy levels of a particle in a central potential do not depend on the azimuthal quantum number m. This degeneracy can be lifted if we place the particle in a uniform magnetic field $\vec{B}$ (if $\vec{B}$ is uniform, its spatial derivatives vanish).

6.3.6.1 Effect of a Magnetic Field on a Charged Particle

Consider a particle of mass μ and charge q which, besides moving in a central potential $V(r)$, is subject to a uniform magnetic field $\vec{B}$.

From the theory of classical electromagnetism, the vector potential corresponding to a uniform magnetic field may be written as $\vec{A} = \frac{1}{2}(\vec{B} \times \vec{r})$ since, using the relation $\vec{\nabla} \times (\vec{C} \times \vec{D}) = \vec{C}(\vec{\nabla} \cdot \vec{D}) - \vec{D}(\vec{\nabla} \cdot \vec{C}) + (\vec{D} \cdot \vec{\nabla})\vec{C} - (\vec{C} \cdot \vec{\nabla})\vec{D}$, we have

$$\vec{\nabla} \times \vec{A} = \frac{1}{2}\vec{\nabla} \times (\vec{B} \times \vec{r}) = \frac{1}{2}\left[\vec{B}(\vec{\nabla} \cdot \vec{r}) - (\vec{B} \cdot \vec{\nabla})\vec{r}\right] = \frac{1}{2}\left[3\vec{B} - \vec{B}\right] = \vec{B}, \qquad (6.193)$$

where we have used $\vec{\nabla} \cdot \vec{B} = 0$, $(\vec{r} \cdot \vec{\nabla})\vec{B} = 0$, $\vec{\nabla} \cdot \vec{r} = 3$, and $(\vec{B} \cdot \vec{\nabla})\vec{r} = \vec{B}$. When the charge is placed in a magnetic field $\vec{B}$, its linear momentum becomes $\vec{p} \to \vec{p} - (q/c)\vec{A}$, where c is the speed of light. The Hamiltonian of the particle is thus given by (see (6.124))

$$\hat{H} = \frac{1}{2\mu}\left(\vec{p} - \frac{q}{c}\vec{A}\right)^2 + V(r) = \hat{H}_0 - \frac{q}{2\mu c}\left(\vec{p} \cdot \vec{A} + \vec{A} \cdot \vec{p}\right) + \frac{q^2}{2\mu c^2}\vec{A}^2, \qquad (6.194)$$

where $\hat{H}_0 = \vec{p}^{\,2}/(2\mu) + V(r)$ is the Hamiltonian of the particle when the magnetic field $\vec{B}$ is not present. The term $\vec{p} \cdot \vec{A}$ can be calculated by analogy with the commutator $[\hat{p}, \hat{F}(x)] = -i\hbar d\hat{F}(x)/dx$:

$$(\vec{p} \cdot \vec{A}) \mid \psi \rangle = -i\hbar(\vec{\nabla} \cdot \vec{A}) \mid \psi \rangle - i\hbar\vec{A} \cdot \vec{\nabla} \mid \psi \rangle = -i\hbar(\vec{\nabla} \cdot \vec{A}) \mid \psi \rangle + \vec{A} \cdot \vec{p} \mid \psi \rangle. \qquad (6.195)$$

We see that, whenever $\vec{\nabla} \cdot \vec{A} = 0$ is valid (the Coulomb gauge), $\vec{A} \cdot \vec{p}$ is equal to $\vec{p} \cdot \vec{A}$:

$$\vec{p} \cdot \vec{A} - \vec{A} \cdot \vec{p} = -i\hbar(\vec{\nabla} \cdot \vec{A}) = 0 \qquad \Longrightarrow \qquad \vec{A} \cdot \vec{p} = \vec{p} \cdot \vec{A}. \qquad (6.196)$$

On the other hand, since $\vec{A} = \frac{1}{2}(\vec{B} \times \vec{r})$, we have

$$\vec{A} \cdot \vec{p} = \frac{1}{2}(\vec{B} \times \vec{r}) \cdot \vec{p} = \frac{1}{2}\vec{B} \cdot (\vec{r} \times \vec{p}) = \frac{1}{2}\vec{B} \cdot \vec{L}, \qquad (6.197)$$

where $\hat{\vec{L}}$ is the orbital angular momentum operator of the particle. Now, a combination of (6.196) and (6.197) leads to $\vec{p} \cdot \vec{A} = \vec{A} \cdot \vec{p} = \frac{1}{2}\vec{B} \cdot \hat{\vec{L}}$ which, when inserted in the Hamiltonian (6.194), yields

$$\hat{H} = \hat{H}_0 - \frac{q}{\mu c}\vec{A} \cdot \vec{p} + \frac{q^2}{2\mu c^2}\vec{A}^2 = \hat{H}_0 - \frac{q}{2\mu c}\vec{B} \cdot \hat{\vec{L}} + \frac{q^2}{2\mu c^2}\vec{A}^2 = \hat{H}_0 - \vec{\mu}_L \cdot \vec{B} + \frac{q^2}{2\mu c^2}\vec{A}^2, \qquad (6.198)$$

where

$$\vec{\mu}_L = \frac{q}{2\mu c}\vec{L} = \frac{\mu_B}{\hbar}\vec{L} \qquad (6.199)$$

is called the *orbital magnetic dipole moment* of the charge q and $\mu_B = q\hbar/(2\mu c)$ is known as the *Bohr magneton*; as mentioned in Chapter 5, $\vec{\mu}_L$ is due to the orbiting motion of the charge about the center of the potential. The term $-\vec{\mu}_L \cdot \vec{B}$ in (6.198) represents the energy resulting from the interaction between the particle's orbital magnetic dipole moment $\vec{\mu}_L = q\vec{L}/(2\mu c)$ and the magnetic field $\vec{B}$. We should note that if the charge q had an intrinsic spin $\vec{S}$, its spinning motion would give rise to a magnetic dipole moment $\vec{\mu}_S = q\vec{S}/(2\mu c)$ which, when interacting with an external magnetic field $\vec{B}$, would in turn generate an energy term $-\vec{\mu}_S \cdot \vec{B}$ that must be added to the Hamiltonian. This issue will be discussed further in Chapter 7.

Finally, using the relation $(\vec{C} \times \vec{D}) \cdot (\vec{E} \times \vec{F}) = (\vec{C} \cdot \vec{E})(\vec{D} \cdot \vec{F}) - (\vec{C} \cdot \vec{F})(\vec{D} \cdot \vec{E})$, and since $\vec{A} = \frac{1}{2}(\vec{B} \times \vec{r})$, we have

$$\vec{A}^2 = \frac{1}{4}(\vec{B} \times \vec{r}) \cdot (\vec{B} \times \vec{r}) = \frac{1}{4}\left[B^2 r^2 - (\vec{B} \cdot \vec{r})^2\right]. \tag{6.200}$$

We can thus write (6.198) as

$$\hat{H} = \frac{1}{2\mu}\vec{p}^2 + V(r) - \frac{q}{2\mu c}\vec{B} \cdot \hat{\vec{L}} + \frac{q^2}{8\mu c^2}\left[B^2 r^2 - (\vec{B} \cdot \vec{r})^2\right]. \tag{6.201}$$

This is the Hamiltonian of a particle of mass μ and charge q moving in a central potential $V(r)$ under the influence of a uniform magnetic field $\vec{B}$.

6.3.6.2 The Normal Zeeman Effect ($\vec{S} = 0$)

When a hydrogen atom is placed in an external uniform magnetic field, its energy levels get shifted. This energy shift is known as the *Zeeman effect*.

In this study *we ignore the spin* of the hydrogen's electron. The Zeeman effect *without the spin* of the electron is called the *normal* Zeeman effect. When the spin of the electron is considered, we get what is called the *anomalous* Zeeman effect, to be examined in Chapter 9 since its study requires familiarity with the formalisms of addition of angular momenta and perturbation theory, which will be studied in Chapters 7 and 9, respectively.

For simplicity, we take $\vec{B}$ along the z-direction: $\vec{B} = B\hat{z}$. The Hamiltonian of the hydrogen atom when subject to such a magnetic field can be obtained from (6.201) by replacing q with the electron's charge $q \to -e$,

$$\hat{H} = \frac{1}{2\mu}\vec{p}^2 - \frac{e^2}{r} + \frac{e}{2\mu c}B\hat{L}_z + \frac{e^2 B^2}{8\mu c^2}\left(x^2 + y^2\right) = \hat{H}_0 + \frac{e}{2\mu c}B\hat{L}_z + \frac{e^2 B^2}{8\mu c^2}\left(x^2 + y^2\right), \tag{6.202}$$

where $\hat{H}_0 = \vec{p}^2/(2\mu) - e^2/r$ is the atom's Hamiltonian in the absence of a magnetic field. We can ignore the quadratic term $e^2 B^2(x^2 + y^2)/(8\mu c^2)$; it is too small for a one-electron atom even when the field $\vec{B}$ is strong; then (6.202) reduces to

$$\hat{H} = \hat{H}_0 + \frac{B\mu_B}{\hbar}\hat{L}_z, \tag{6.203}$$

where $\mu_B = e\hbar/(2\mu c) = 9.2740 \times 10^{-24}\,\text{J T}^{-1} = 5.7884 \times 10^{-5}\,\text{eV T}^{-1}$ is the Bohr magneton; the electron's orbital magnetic dipole moment, which results from the orbiting motion of the electron about the proton, would be given by $\vec{\mu}_L = -e\vec{B}/(2\mu c)$. Since $\hat{H}_0$ commutes with $\hat{L}_z$, the operators $\hat{H}$, $\hat{L}_z$, and $\hat{H}_0$ mutually commute; hence they possess a set of common eigenfunctions: $\psi_{nlm}(r, \theta, \varphi) = R_{nl}(r)Y_{lm}(\theta, \varphi)$. The eigenvalues of (6.203) are

$$E_{nlm} = \langle nlm \mid \hat{H} \mid nlm \rangle = \langle nlm \mid \hat{H}_0 \mid nlm \rangle + \frac{B\mu_B}{\hbar}\langle nlm \mid \hat{L}_z \mid nlm \rangle \tag{6.204}$$

Figure 6.4 Normal Zeeman effect in hydrogen. (Left) When $\vec{B} = 0$ the energy levels are degenerate with respect to l and m. (Right) When $\vec{B} \neq 0$ the degeneracy with respect to m is removed, but the degeneracy with respect to l persists; $\mu_B = e\hbar/(2\mu c)$.

or

$$E_{nlm} = E_n^0 + m\mu_B B = E_n^0 + m\hbar\omega_L, \tag{6.205}$$

where E_n^0 are the hydrogen's energy levels $E_n^0 = -\mu e^4/(2\hbar^2 n^2)$ (6.147) and ω_L is called the *Larmor frequency*:

$$\omega_L = \frac{eB}{2\mu c}. \tag{6.206}$$

So when a hydrogen atom is placed in a uniform magnetic field, and if we ignore the spin of the electron, the atom's spherical symmetry will be broken: each level with angular momentum l will split into $(2l + 1)$ equally spaced levels (Figure 6.4), where the spacing is given by $\Delta E = \hbar\omega_L = B\mu_B$; the spacing is independent of l. This *equidistant* splitting of the levels is known as the *normal* Zeeman effect. The splitting leads to transitions which are restricted by the selection rule: $\Delta m = -1, 0, 1$. Transitions $m' = 0 \longrightarrow m = 0$ are not allowed.

The normal Zeeman effect has removed the degeneracy of the levels only partially; the degeneracy with respect to l remains. For instance, as shown in Figure 6.4, the following levels are still degenerate: $E_{nlm} = E_{200} = E_{210}$, $E_{32,-1} = E_{31,-1}$, $E_{300} = E_{310} = E_{320}$, and $E_{321} = E_{311}$. That is, the degeneracies of the levels corresponding to the same n and m but different values of l are not removed by the normal Zeeman effect: $E_{nl'm} = E_{nlm}$ with $l' \neq l$.

The results of the normal Zeeman effect, which show that each energy level splits into an *odd* number of $(2l + 1)$ equally spaced levels, disagree with the experimental observations. For instance, every level in the hydrogen atom actually splits into an *even* number of levels. This suggests that the angular momentum is not integer but half-integer. This disagreement is due to the simplifying assumption where the spin of the electron was ignored. A proper treatment, which includes the electron spin, confirms that the angular momentum is not purely orbital but includes a spin component as well. This leads to the splitting of each level into an *even*[5] number of $(2j+1)$ unequally spaced energy levels. This effect, known as the *anomalous* Zeeman effect, is in full agreement with experimental findings.

6.4   Concluding Remarks

An important result that needs to be highlighted in this chapter is the solution of the Schrödinger equation for the hydrogen atom. Unlike Bohr's semiclassical model, which is founded on piecemeal assumptions, we have seen how the Schrödinger equation yields the energy levels systematically and without ad hoc arguments, the quantization of the energy levels comes out naturally as a by-product of the formalism, not as an unjustified assumption: it is a consequence of the boundary conditions which require the wave function to be finite as $r \to \infty$; see (6.144) and (6.147). So we have seen that by solving a single differential equation—the Schrödinger equation—we obtain all that we need to know about the hydrogen atom. As such, the Schrödinger equation has delivered on the promise made in Chapter 1: namely, a theory that avoids the undesired aspects of Bohr's model—its hand-waving, ad hoc assumptions—while preserving its good points (i.e., the expressions for the energy levels, the radii, and the transition relations).

6.5   Solved Problems

Problem 6.1

Consider a spinless particle of mass m which is moving in a three-dimensional potential

$$V(x, y, z) = \begin{cases} \frac{1}{2}m\omega^2 z^2, & 0 < x < a, \ 0 < y < a, \\ \infty, & \text{elsewhere.} \end{cases}$$

(a) Write down the total energy and the total wave function of this particle.

(b) Assuming that $\hbar\omega > 3\pi^2\hbar^2/(2ma^2)$, find the energies and the corresponding degeneracies for the ground state and first excited state.

(c) Assume now that, in addition to the potential $V(x, y, z)$, this particle also has a negative electric charge $-q$ and that it is subjected to a constant electric field ϵ directed along the z-axis. The Hamiltonian along the z-axis is thus given by

$$\hat{H}_z = -\frac{\hbar^2}{2m}\frac{\partial^2}{\partial z^2} + \frac{1}{2}m\omega^2 z^2 - q\epsilon z.$$

Derive the energy expression E_{n_z} for this particle and also its total energy $E_{n_x n_y n_z}$. Then find the energies and the corresponding degeneracies for the ground state and first excited state.

Solution

(a) This three-dimensional potential consists of three independent one-dimensional potentials: (i) a potential well along the x-axis, (ii) a potential well along the y-axis, and (iii) a

---

[5]When spin is included, the electron's total angular momentum j would be half-integer; $(2j + 1)$ is then an even number.

harmonic oscillator along the z-axis. The energy must then be given by

$$E_{n_x n_y n_z} = \frac{\pi^2 \hbar^2}{2ma^2} \left(n_x^2 + n_y^2 \right) + \hbar\omega \left(n_z + \frac{1}{2} \right) \tag{6.207}$$

and the wave function by

$$\psi_{n_x n_y n_z}(x, y, z) = X_{n_x}(x) Y_{n_y}(y) Z_{n_z}(z) = \frac{2}{a} \sin \left(\frac{\pi n_x}{a} x \right) \sin \left(\frac{\pi n_y}{a} y \right) Z_{n_z}, \tag{6.208}$$

where $Z_{n_z}(z)$ is the wave function of a harmonic oscillator which, as shown in Chapter 4, is given in terms of the Hermite polynomial $H_{n_z} \left(\frac{z}{z_0} \right)$ by

$$Z_{n_z}(z) = \frac{1}{\sqrt{\sqrt{\pi} 2^{n_z} n_z! z_0}} e^{-z^2/2z_0^2} H_{n_z} \left(\frac{z}{z_0} \right), \tag{6.209}$$

with $z_0 = \sqrt{\pi \hbar/(m\omega)}$.

(b) The energy of the ground state is given by

$$E_{110} = \frac{\pi^2 \hbar^2}{ma^2} + \frac{\hbar\omega}{2} \tag{6.210}$$

and the energy of the first excited state is given by

$$E_{120} = E_{210} = \frac{5\pi^2 \hbar^2}{2ma^2} + \frac{\hbar\omega}{2}. \tag{6.211}$$

Note that, while the ground state is not degenerate, the first excited state is twofold degenerate. We should also mention that, since $\hbar\omega > 3\pi^2 \hbar^2/(2ma^2)$, we have $E_{120} < E_{111}$, or

$$E_{111} = \frac{\pi^2 \hbar^2}{ma^2} + \frac{3\hbar\omega}{2} = E_{120} + \hbar\omega - \frac{3\pi^2 \hbar^2}{2ma^2}, \tag{6.212}$$

and hence the first excited state is given by E_{120} and not by E_{111}.

(c) To obtain the energies for

$$\hat{H}_z = -\frac{\hbar^2}{2m} \frac{\partial^2}{\partial z^2} + \frac{1}{2} m\omega^2 z^2 - q\epsilon z, \tag{6.213}$$

we need simply to make the change of variable $\lambda = z - q\epsilon/(m\omega^2)$; hence $dz = d\lambda$. The Hamiltonian $\hat{H}_z$ then reduces to

$$\hat{H}_z = -\frac{\hbar^2}{2m} \frac{\partial^2}{\partial \lambda^2} + \frac{1}{2} m\omega^2 \lambda^2 - \frac{q^2 \epsilon^2}{2m\omega^2}. \tag{6.214}$$

This suggestive form implies that the energy eigenvalues of $\hat{H}_z$ are those of a harmonic oscillator that are shifted downwards by an amount equal to $q^2 \epsilon^2/(2m\omega^2)$:

$$E_{n_z} = \langle n_z | \hat{H}_z | n_z \rangle = \hbar\omega \left(n_z + \frac{1}{2} \right) - \frac{q^2 \epsilon^2}{2m\omega^2}. \tag{6.215}$$

As a result, the total energy is now given by

$$E_{n_x n_y n_z} = \frac{\pi^2 \hbar^2}{2ma^2} \left(n_x^2 + n_y^2 \right) + \hbar\omega \left(n_z + \frac{1}{2} \right) - \frac{q^2 \epsilon^2}{2m\omega^2}. \tag{6.216}$$

The energies of the ground and first excited states are

$$E_{110} = \frac{\pi^2 \hbar^2}{ma^2} + \frac{\hbar\omega}{2} - \frac{q^2 \epsilon^2}{2m\omega^2}, \qquad E_{120} = E_{210} = \frac{5\pi^2 \hbar^2}{2ma^2} + \frac{\hbar\omega}{2} - \frac{q^2 \epsilon^2}{2m\omega^2}. \tag{6.217}$$

Problem 6.2

Show how to obtain the expressions of: (a) $\langle nl|r^{-2}|nl\rangle$ and (b) $\langle nl|r^{-1}|nl\rangle$; that is, prove (6.183) and (6.182).

Solution

The starting point is the radial equation (6.127),

$$-\frac{\hbar^2}{2\mu} \frac{d^2 U_{nl}(r)}{dr^2} + \left[\frac{l(l+1)\hbar^2}{2\mu r^2} - \frac{e^2}{r} \right] U_{nl}(r) = E_n U_{nl}(r), \tag{6.218}$$

which can be rewritten as

$$\frac{U_{nl}''(r)}{U_{nl}(r)} = \frac{l(l+1)}{r^2} - \frac{2\mu e^2}{\hbar^2} \frac{1}{r} + \frac{\mu^2 e^4}{\hbar^4 n^2}, \tag{6.219}$$

where $U_{nl}(r) = r R_{nl}(r)$, $U_{nl}''(r) = d^2 U_{nl}(r)/dr^2$, and $E_n = -\mu e^4/(2\hbar^2 n^2)$.

(a) To find $\langle r^{-2} \rangle_{nl}$, let us treat the orbital quantum number l as a continuous variable and take the first l derivative of (6.219):

$$\frac{\partial}{\partial l} \left[\frac{U_{nl}''(r)}{U_{nl}(r)} \right] = \frac{2l+1}{r^2} - \frac{2\mu^2 e^4}{\hbar^4 n^3}, \tag{6.220}$$

where we have the fact that n depends on l since, as shown in (6.145), $n = N + l + 1$; thus $\partial n/\partial l = 1$. Now since $\int_0^\infty U_{nl}^2(r)\, dr = \int_0^\infty r^2 R_{nl}^2(r)\, dr = 1$, multiplying both sides of (6.220) by $U_{nl}^2(r)$ and integrating over r we get

$$\int_0^\infty U_{nl}^2(r) \frac{\partial}{\partial l} \left[\frac{U_{nl}''(r)}{U_{nl}(r)} \right] dr = (2l+1) \int_0^\infty U_{nl}^2(r) \frac{1}{r^2}\, dr - \frac{2\mu^2 e^4}{\hbar^4 n^3} \int_0^\infty U_{nl}^2(r)\, dr, \tag{6.221}$$

or

$$\int_0^\infty U_{nl}^2(r) \frac{\partial}{\partial l} \left[\frac{U_{nl}''(r)}{U_{nl}(r)} \right] dr = (2l+1) \left\langle nl \left| \frac{1}{r^2} \right| nl \right\rangle - \frac{2\mu^2 e^4}{\hbar^4 n^3}. \tag{6.222}$$

The left-hand side of this relation is equal to zero, since

$$\int_0^\infty U_{nl}^2(r) \frac{\partial}{\partial l} \left[\frac{U_{nl}''(r)}{U_{nl}(r)} \right] dr = \int_0^\infty U_{nl}(r) \frac{\partial U_{nl}''(r)}{\partial l}\, dr - \int_0^\infty U_{nl}''(r) \frac{\partial U_{nl}(r)}{\partial l}\, dr = 0. \tag{6.223}$$

We may therefore rewrite (6.222) as

$$(2l+1) \left\langle nl \left| \frac{1}{r^2} \right| nl \right\rangle = \frac{2\mu^2 e^4}{\hbar^4 n^3}; \tag{6.224}$$

hence

$$\left\langle nl \left| \frac{1}{r^2} \right| nl \right\rangle = \frac{2}{n^3(2l+1)a_0^2},$$ (6.225)

since $a_0 = \hbar^2/(\mu e^2)$.

(b) To find $\langle r^{-1} \rangle_{nl}$ we need now to treat the electron's charge e as a continuous variable in (6.219). The first e-derivative of (6.219) yields

$$\frac{\partial}{\partial e} \left[\frac{U_{nl}''(r)}{U_{nl}(r)} \right] = -\frac{4\mu e}{\hbar^2} \frac{1}{r} + \frac{4\mu^2 e^3}{\hbar^4 n^2}.$$ (6.226)

Again, since $\int_0^\infty U_{nl}^2(r)\, dr = 1$, multiplying both sides of (6.226) by $U_{nl}^2(r)$ and integrating over r we obtain

$$\int_0^\infty U_{nl}^2(r) \frac{\partial}{\partial e} \left[\frac{U_{nl}''(r)}{U_{nl}(r)} \right] dr = -\frac{4\mu e}{\hbar^2} \int_0^\infty U_{nl}^2(r) \frac{1}{r} dr + \frac{4\mu^2 e^3}{\hbar^4 n^2} \int_0^\infty U_{nl}^2(r)\, dr,$$ (6.227)

or

$$\int_0^\infty U_{nl}^2(r) \frac{\partial}{\partial e} \left[\frac{U_{nl}''(r)}{U_{nl}(r)} \right] dr = -\frac{4\mu e}{\hbar^2} \left\langle nl \left| \frac{1}{r} \right| nl \right\rangle + \frac{4\mu^2 e^3}{\hbar^4 n^2}.$$ (6.228)

As shown in (6.223), the left-hand side of this is equal to zero. Thus, we have

$$\frac{4\mu e}{\hbar^2} \left\langle nl \left| \frac{1}{r} \right| nl \right\rangle = \frac{4\mu^2 e^3}{\hbar^4 n^2} \quad \Longrightarrow \quad \left\langle nl \left| \frac{1}{r} \right| nl \right\rangle = \frac{1}{n^2 a_0},$$ (6.229)

since $a_0 = \hbar^2/(\mu e^2)$.

Problem 6.3

(a) Use Kramers' recursion rule (6.184) to obtain expressions (6.180) to (6.182) for $\langle nl|r^{-1}|nl \rangle$, $\langle nl|r|nl \rangle$, and $\langle nl|r^2|nl \rangle$.

(b) Using (6.225) for $\langle nl|r^{-2}|nl \rangle$ and combining it with Kramers' rule, obtain the expression for $\langle nl|r^{-3}|nl \rangle$.

(c) Repeat (b) to obtain the expression for $\langle nl|r^{-4}|nl \rangle$.

Solution

(a) First, to obtain $\langle nl \left| r^{-1} \right| nl \rangle$, we need simply to insert $k = 0$ into Kramers' recursion rule (6.184):

$$\frac{1}{n^2} \left\langle nl \left| r^0 \right| nl \right\rangle - a_0 \left\langle nl \left| r^{-1} \right| nl \right\rangle = 0;$$ (6.230)

hence

$$\left\langle nl \left| \frac{1}{r} \right| nl \right\rangle = \frac{1}{n^2 a_0}.$$ (6.231)

Second, an insertion of $k = 1$ into (6.184) leads to the relation for $\langle nl|r|nl \rangle$:

$$\frac{2}{n^2} \langle nl|r|nl \rangle - 3a_0 \left\langle nl \left| r^0 \right| nl \right\rangle + \frac{a_0^2}{4} \left[(2l+1)^2 - 1 \right] \left\langle nl \left| r^{-1} \right| nl \right\rangle = 0,$$ (6.232)

and since $\langle nl \left| r^{-1} \right| nl \rangle = 1/(n^2 a_0)$, we have

$$\langle nl|r|nl \rangle = \frac{1}{2} \left[3n^2 - l(l+1) \right] a_0.$$ (6.233)

Third, substituting $k = 2$ into (6.184) we get

$$\frac{3}{n^2} \left\langle nl \left| r^2 \right| nl \right\rangle - 5a_0 \left\langle nl \left| r \right| nl \right\rangle + \frac{a_0^2}{2} \left[(2l + 1)^2 - 4 \right] \left\langle nl \left| r^0 \right| nl \right\rangle = 0, \tag{6.234}$$

which when combined with $\langle nl \left| r \right| nl \rangle = \frac{1}{2} \left[3n^2 - l(l + 1) \right] a_0$ yields

$$\boxed{\left\langle nl \left| r^2 \right| nl \right\rangle = \frac{1}{2} n^2 \left[5n^2 + 1 - 3l(l + 1) \right] a_0^2.} \tag{6.235}$$

We can continue in this way to obtain any positive power of r: $\langle nl \left| r^k \right| nl \rangle$.

(b) Inserting $k = -1$ into Kramers' rule,

$$0 + a_0 \left\langle nl \left| r^{-2} \right| nl \right\rangle - \frac{1}{4} \left[(2l + 1)^2 - 1 \right] a_0^2 \left\langle nl \left| r^{-3} \right| nl \right\rangle, \tag{6.236}$$

we obtain

$$\left\langle nl \left| \frac{1}{r^3} \right| nl \right\rangle = \frac{1}{l(l + 1)a_0} \left\langle nl \left| \frac{1}{r^2} \right| nl \right\rangle, \tag{6.237}$$

where the expression for $\langle nl \left| r^{-2} \right| nl \rangle$ is given by (6.225); thus, we have

$$\boxed{\left\langle nl \left| \frac{1}{r^3} \right| nl \right\rangle = \frac{2}{n^3 l(l + 1)(2l + 1)a_0^3}.} \tag{6.238}$$

(c) To obtain the expression for $\langle nl \left| r^{-4} \right| nl \rangle$ we need to substitute $k = -2$ into Kramers' rule:

$$-\frac{1}{n^2} \left\langle nl \left| r^{-2} \right| nl \right\rangle + 3a_0 \left\langle nl \left| r^{-3} \right| nl \right\rangle - \frac{a_0^2}{2} \left[(2l + 1)^2 - 4 \right] \left\langle nl \left| r^{-4} \right| nl \right\rangle = 0. \tag{6.239}$$

Inserting (6.225) and (6.238) for $\langle nl \left| r^{-2} \right| nl \rangle$ and $\langle nl \left| r^{-3} \right| nl \rangle$, we obtain

$$\boxed{\left\langle nl \left| \frac{1}{r^4} \right| nl \right\rangle = \frac{4 \left[3n^2 - l(l + 1) \right]}{n^5 l(l + 1)(2l + 1) \left[(2l + 1)^2 - 4 \right] a_0^4}.} \tag{6.240}$$

We can continue in this way to obtain any negative power of r: $\langle nl \left| r^{-k} \right| nl \rangle$.

Problem 6.4

An electron is trapped inside an infinite spherical well $V(r) = \begin{cases} 0, & r < a, \\ +\infty, & r > a. \end{cases}$

(a) Using the radial Schrödinger equation, determine the bound eigenenergies and the corresponding normalized radial wave functions for the case where the orbital angular momentum of the electron is zero (i.e., $l = 0$).

(b) Show that the lowest energy state for $l = 7$ lies above the second lowest energy state for $l = 0$.

(c) Calculate the probability of finding the electron in a sphere of radius $a/2$, and then in a spherical shell of thickness $a/2$ situated between $r = a$ and $r = 3a/2$.

Solution

(a) Since $V(r) = 0$ in the region $r \le a$, the radial Schrödinger equation (6.57) becomes

$$-\frac{\hbar^2}{2m} \left[\frac{d^2 U_{nl}(r)}{dr^2} - \frac{l(l+1)}{r^2} U_{nl}(r) \right] = E U_{nl}(r), \tag{6.241}$$

where $U_{nl}(r) = r R_{nl}(r)$. For the case where $l = 0$, this equation reduces to

$$\frac{d^2 U_{n0}(r)}{dr^2} = -k_n^2 U_{n0}(r), \tag{6.242}$$

where $k_n^2 = 2m E_n/\hbar^2$. The general solution to this differential equation is given by

$$U_{n0}(r) = A \cos(k_n r) + B \sin(k_n r) \tag{6.243}$$

or

$$R_{n0}(r) = \frac{1}{r} \left(A \cos(k_n r) + B \sin(k_n r) \right). \tag{6.244}$$

Since $R_{n0}(r)$ is finite at the origin or $U_{n0}(0) = 0$, the coefficient A must be zero. In addition, since the potential is infinite at $r = a$ (rigid wall), the radial function $R_{n0}(a)$ must vanish:

$$R_{n0}(a) = B \frac{\sin k_n a}{a} = 0; \tag{6.245}$$

hence $ka = n\pi$, $n = 1, 2, 3, \ldots$. This relation leads to

$$E_n = \frac{\hbar^2 \pi^2}{2ma^2} n^2. \tag{6.246}$$

The normalization of the radial wave function $R(r)$, $\int_0^a |R_{n0}(r)|^2 r^2 dr = 1$, leads to

$$1 = |B|^2 \int_0^a \frac{1}{r^2} \sin^2(k_n r) r^2 dr = \frac{|B|^2}{k_n} \int_0^{k_n a} \sin^2 \rho \, d\rho = \frac{|B|^2}{k_n} \left(\frac{\rho}{2} - \frac{\sin 2\rho}{4} \right) \Big|_{\rho=0}^{\rho=k_n a}$$

$$= \frac{1}{2} |B|^2 a; \tag{6.247}$$

hence $B = \sqrt{2/a}$. The normalized radial wave function is thus given by

$$R_{n0}(r) = \sqrt{\frac{2}{a}} \frac{1}{r} \sin \left(\sqrt{\frac{2m E_n}{\hbar^2}} r \right). \tag{6.248}$$

(b) For $l = 7$ we have

$$E_1(l = 7) > V_{eff}(l = 7) = \frac{56\hbar^2}{2ma^2} = \frac{28\hbar^2}{ma^2}. \tag{6.249}$$

The second lowest state for $l = 0$ is given by the 3s state; its energy is

$$E_2(l = 0) = \frac{2\pi^2 \hbar^2}{ma^2}, \tag{6.250}$$

since $n = 2$. We see that

$$E_1(l = 7) > E_2(l = 0). \tag{6.251}$$

(c) Since the probability of finding the electron in the sphere of radius a is equal to 1, the probability of finding it in a sphere of radius $a/2$ is equal to $1/2$.

As for the probability of finding the electron in the spherical shell between $r = a$ and $r = 3a/2$, it is equal to zero, since the electron cannot tunnel through the infinite potential from $r < a$ to $r > a$.

Problem 6.5

Find the $l = 0$ energy and wave function of a particle of mass m that is subject to the following central potential $V(r) = \begin{cases} 0, & a < r < b, \\ \infty, & \text{elsewhere.} \end{cases}$

Solution

This particle moves between two concentric, hard spheres of radii $r = a$ and $r = b$. The $l = 0$ radial equation between $a < r < b$ can be obtained from (6.57):

$$\frac{d^2 U_{n0}(r)}{dr^2} + k^2 U_{n0}(r) = 0, \tag{6.252}$$

where $U_{n0}(r) = r R_{n0}(r)$ and $k^2 = 2mE/\hbar^2$. Since the solutions of this equation must satisfy the condition $U_{n0}(a) = 0$, we may write

$$U_{n0}(r) = A \sin[k(r - a)]; \tag{6.253}$$

the radial wave function is zero elsewhere, i.e., $U_{n0}(r) = 0$ for $0 < r < a$ and $r > b$.

Moreover, since the radial function must vanish at $r = b$, $U_{n0}(b) = 0$, we have

$$A \sin[k(b - a)] = 0 \implies k(b - a) = n\pi, \qquad n = 1, 2, 3, \dots. \tag{6.254}$$

Coupled with the fact that $k^2 = 2mE/(\hbar^2)$, this condition leads to the energy

$$E_n = \frac{\hbar^2 k^2}{2m} = \frac{\pi^2 \hbar^2}{2m(a - b)^2}, \qquad n = 1, 2, 3, \dots. \tag{6.255}$$

We can normalize the radial function (6.253) to obtain the constant A:

$$1 = \int_a^b r^2 R_{n0}^2(r) dr = \int_a^b U_{n0}^2(r) dr = A^2 \int_a^b \sin^2[k(r - a)] dr$$

$$= \frac{A^2}{2} \int_a^b \{1 - \cos[2k(r - a)]\} \, dr = \frac{b - a}{2} A^2; \tag{6.256}$$

hence $A = \sqrt{2/(b - a)}$. Since $k_n = n\pi/(b - a)$ the normalized radial function is given by

$$R_{n0}(r) = \frac{1}{r} U_{n0}(r) = \begin{cases} \sqrt{\frac{2}{b-a}} \frac{1}{r} \sin[\frac{n\pi(r-a)}{b-a}], & a < r < b, \\ 0, & \text{elsewhere.} \end{cases} \tag{6.257}$$

To obtain the total wave function $\psi_{nlm}(\vec{r})$, we need simply to divide the radial function by a $1/\sqrt{4\pi}$ factor, because in this case of $l = 0$ the wave function $\psi_{n00}(r)$ depends on no angular degrees of freedom, it depends only on the radius:

$$\psi_{n00}(r) = \frac{1}{\sqrt{4\pi}} R_{n0}(r) = \begin{cases} \sqrt{\frac{2}{4\pi(b-a)}} \frac{1}{r} \sin[\frac{n\pi(r-a)}{b-a}], & a < r < b, \\ 0, & \text{elsewhere.} \end{cases} \tag{6.258}$$

Problem 6.6
(a) For the following cases, calculate the value of r at which the radial probability density of the hydrogen atom reaches its maximum: (i) $n = 1, l = 0, m = 0$; (ii) $n = 2, l = 1, m = 0$; (iii) $l = n - 1, m = 0$.

(b) Compare the values obtained with the Bohr radius for circular orbits.

Solution
(a) Since the radial wave function for $n = 1$ and $l = 0$ is $R_{10}(r) = 2a_0^{-3/2}e^{-r/a_0}$, the probability density is given by

$$P_{10}(r) = r^2|R_{10}(r)|^2 = \frac{4}{a_0^3}r^2 e^{-2r/a_0}. \tag{6.259}$$

(i) The maximum of $P_{10}(r)$ occurs at r_1:

$$\frac{dP_{10}(r)}{dr}\Big|_{r=r_1} = 0 \implies 2r_1 - \frac{2r_1^2}{a_0} = 0 \implies r_1 = a_0. \tag{6.260}$$

(ii) Similarly, since $R_{21}(r) = 1/(2\sqrt{6}a_0^{5/2})re^{-r/2a_0}$, we have

$$P_{21}(r) = r^2|R_{21}(r)|^2 = \frac{1}{24a_0^5}r^4 e^{-r/a_0}. \tag{6.261}$$

The maximum of the probability density is given by

$$\frac{dP_{21}(r)}{dr}\Big|_{r=r_2} = 0 \implies 4r_2^3 - \frac{r_2^4}{a_0} = 0 \implies r_2 = 4a_0. \tag{6.262}$$

(iii) The radial function for $l = n - 1$ can be obtained from (6.170):

$$R_{n(n-1)}(r) = -\left(\frac{2}{na_0}\right)^{3/2} \frac{1}{\sqrt{2n[(2n-1)!]^3}} \left(\frac{2r}{na_0}\right)^{(n-1)} e^{-r/na_0} L_{2n-1}^{2n-1}\left(\frac{2r}{na_0}\right). \tag{6.263}$$

From (6.159) and (6.160) we can verify that the associated Laguerre polynomial L_{2n-1}^{2n-1} is a constant, $L_{2n-1}^{2n-1}(y) = -(2n-1)!$. We can thus write $R_{n(n-1)}(r)$ as $R_{n(n-1)}(r) = A_n r^{n-1}e^{-r/na_0}$, where A_n is a constant. Hence the probability density is given by

$$P_{n(n-1)}(r) = r^2|R_{n(n-1)}(r)|^2 = A_n^2 r^{2n}e^{-2r/na_0}. \tag{6.264}$$

The maximum of the probability density is given by

$$\frac{dP_{n(n-1)}(r)}{dr}\Big|_{r=r_n} = 0 \implies 2nr_n^{2n-1} - \frac{2r_n^{2n}}{na_0} = 0 \implies r_n = n^2a_0. \tag{6.265}$$

Figure 6.5 The probability density $P_{21}(r) = r^4 e^{-r/a_0}/(24a_0^5)$ is asymmetric about its maximum $r_2 = 4a_0$; the average of r is $\langle r_{21} \rangle = 5a_0$ and the width of the probability density is $\Delta r_{21} = \sqrt{5}a_0$.

(b) The values of r_n displayed in (6.260), (6.262), and (6.265) are nothing but the Bohr radii for circular orbits, $r_n = n^2 a_0$. The Bohr radius $r_n = n^2 a_0$ gives the position of maximum probability density for an electron in a hydrogen atom.

Problem 6.7

(a) Calculate the expectation value $\langle r \rangle_{21}$ for the hydrogen atom and compare it with the value r at which the radial probability density reaches its maximum for the state $n = 2, l = 1$.

(b) Calculate the width of the probability density distribution for r.

Solution

(a) Since $R_{21}(r) = re^{-r/2a_0}/\sqrt{24a_0^5}$ the average value of r in the state $R_{21}(r)$ is

$$\langle r \rangle_{21} = \frac{1}{24a_0^5} \int_0^\infty r^5 e^{-r/a_0} dr = \frac{a_0}{24} \int_0^\infty u^5 e^{-u} du = \frac{120a_0}{24} = 5a_0; \qquad (6.266)$$

in deriving this relation we have made use of $\int_0^\infty x^n e^{-x} dx = n!$.

The value r at which the radial probability density reaches its maximum for the state $n = 2$, $l = 1$ is given by $r_2 = 4a_0$, as shown in (6.262).

What makes the results $r_2 = 4a_0$ and $\langle r \rangle_{21} = 5a_0$ different? The reason that $\langle r \rangle_{21}$ is different from r_2 can be attributed to the fact that the probability density $P_{21}(r)$ is asymmetric about its maximum, as shown in Figure 6.5. Although the most likely location of the electron is at $r_0 = 4a_0$, the average value of the measurement of its location is $\langle r \rangle_{21} = 5a_0$.

(b) The width of the probability distribution is given by $\Delta r = \sqrt{\langle r^2 \rangle_{21} - \langle r \rangle_{21}^2}$, where the expectation value of r^2 is

$$\langle r^2 \rangle_{21} = \int_0^\infty r^4 R_{21}^2(r) dr = \frac{1}{24a_0^5} \int_0^\infty r^6 \exp\left(-\frac{1}{a_0}r\right) dr = \frac{6!a_0^7}{24a_0^5} = 30a_0^2. \qquad (6.267)$$

Thus, the width of the probability distribution shown in Figure 6.5 is given by

$$\Delta r_{21} = \sqrt{\langle r^2 \rangle_0 - \langle r \rangle_0^2} = \sqrt{30a_0^2 - (5a_0)^2} = \sqrt{5}a_0. \qquad (6.268)$$

Problem 6.8

The operators associated with the radial component of the momentum p_r and the radial coordinate r are denoted by $\hat{P}_r$ and $\hat{R}$, respectively. Their actions on a radial wave function $\psi(r)$ are given by $\hat{P}_r \psi(\vec{r}) = -i\hbar (1/r)(\partial/\partial r)(r \psi(\vec{r}))$ and $\hat{R} \psi(\vec{r}) = r \psi(\vec{r})$.

(a) Find the commutator $[\hat{P}_r, \hat{R}]$ and $\Delta P_r \Delta r$, where $\Delta r = \sqrt{\langle \hat{R}^2 \rangle - \langle \hat{R} \rangle^2}$ and $\Delta P_r = \sqrt{\langle \hat{P}_r^2 \rangle - \langle \hat{P}_r \rangle^2}$.

(b) Show that $\hat{P}_r^2 = -(\hbar^2/r)(\partial^2/\partial r^2)r$.

Solution

(a) Since $\hat{R} \psi(\vec{r}) = r \psi(\vec{r})$ and

$$\hat{P}_r \psi(\vec{r}) = -i\hbar \frac{1}{r} \frac{\partial}{\partial r} (r \psi(\vec{r})) = -i\hbar \frac{1}{r} \psi(\vec{r}) - i\hbar \frac{\partial \psi(\vec{r})}{\partial r}, \tag{6.269}$$

and since

$$\hat{P}_r (\hat{R} \psi(\vec{r})) = -i\hbar \frac{1}{r} \frac{\partial}{\partial r} \left(r^2 \psi(\vec{r}) \right) = -2i\hbar \psi(\vec{r}) - i\hbar r \frac{\partial \psi(\vec{r})}{\partial r}, \tag{6.270}$$

the action of the commutator $[\hat{P}_r, \hat{R}]$ on a function $\psi(\vec{r})$ is given by

$$[\hat{P}_r, \hat{R}]\psi(\vec{r}) = -i\hbar \left[\frac{1}{r} \frac{\partial}{\partial r} r, \; \hat{R} \right] \psi(\vec{r}) = -i\hbar \frac{1}{r} \frac{\partial}{\partial r} \left(r^2 \psi(\vec{r}) \right) + i\hbar \frac{\partial}{\partial r} (r \psi(\vec{r}))$$

$$= -2i\hbar \psi(\vec{r}) - i\hbar r \frac{\partial \psi(\vec{r})}{\partial r} + i\hbar \psi(\vec{r}) + i\hbar r \frac{\partial \psi(\vec{r})}{\partial r}$$

$$= -i\hbar \psi(\vec{r}). \tag{6.271}$$

Thus, we have

$$[\hat{P}_r, \; \hat{R}] = -i\hbar. \tag{6.272}$$

Using the uncertainty relation for a pair of operators $\hat{A}$ and $\hat{B}$, $\Delta A \Delta B \geq \frac{1}{2}|\langle[\hat{A}, \hat{B}]\rangle|$, we can write

$$\Delta P_r \Delta r \geq \frac{1}{2} \left| \langle [\hat{P}_r, \hat{R}] \rangle \right|, \tag{6.273}$$

or

$$\Delta P_r \Delta r \geq \frac{\hbar}{2}. \tag{6.274}$$

(b) The action of $\hat{P}_r^2$ on $\psi(\vec{r})$ gives

$$\hat{P}_r^2 \psi(\vec{r}) = -\hbar^2 \frac{1}{r} \frac{\partial}{\partial r} \left[r \frac{1}{r} \frac{\partial}{\partial r} (r \psi) \right] = -\hbar^2 \frac{1}{r} \frac{\partial^2}{\partial r^2} (r \psi(\vec{r})); \tag{6.275}$$

hence

$$\hat{P}_r^2 = -\hbar^2 \frac{1}{r} \frac{\partial^2}{\partial r^2} (r). \tag{6.276}$$

Problem 6.9

Find the number of s bound states for a particle of mass m moving in a delta potential $V(r) = -V_0 \delta(r - a)$ where $V_0 > 0$. Discuss the existence of bound states in terms of the size of a. Find the normalized wave function of the bound state(s).

Solution

The $l = 0$ radial equation can be obtained from (6.57):

$$\frac{d^2 U_{n0}(r)}{dr^2} + \left[\frac{2m V_0}{\hbar^2}\delta(r - a) - k^2\right] U_{n0}(r) = 0, \tag{6.277}$$

where $U_{nl}(r) = U_{n0}(r) = r R_{n0}(r)$ and $k^2 = -2m E/\hbar^2$, since we are looking here at the bound states only, $E < 0$. The solutions of this equation are

$$U_{n0}(r) = \begin{cases} U_{n0_1}(r) = Ae^{kr} + Be^{-kr}, & 0 < r < a, \\ U_{n0_2}(r) = Ce^{-kr}, & r > a. \end{cases} \tag{6.278}$$

The energy eigenvalues can be obtained from the boundary conditions. As the wave function vanishes at $r = 0$, $U_{n0}(0) = 0$, we have $A + B = 0$ or $B = -A$; hence $U_{n0_1}(r) = D \sinh kr$:

$$U_{n0}(r) = D \sinh kr, \qquad\qquad 0 < r < a, \tag{6.279}$$

with $D = 2A$. The continuity condition at $r = a$ of $U_{n0}(r)$, $U_{n0_1}(a) = U_{n0_2}(a)$, leads to

$$D \sinh ka = Ce^{-ka}. \tag{6.280}$$

To obtain the discontinuity condition for the first derivative of $U_{n0}(r)$ at $r = a$, we need to integrate (6.277):

$$\lim_{\varepsilon \to a} \left[U'_{n0_2}(a + \varepsilon) - U'_{n0_1}(a - \varepsilon) \right] + \frac{2m V_0}{\hbar^2} U_{n0_2}(a) = 0 \tag{6.281}$$

or

$$-kCe^{-ka} - kD \cosh ka + \frac{2m V_0}{\hbar^2} Ce^{-ka} = 0. \tag{6.282}$$

Taking $Ce^{-ka} = D \sinh ka$, as given by (6.280), and substituting it into (6.282), we get

$$-k \sinh ka - k \cosh ka + \frac{2m V_0}{\hbar^2} \sinh ka = 0; \tag{6.283}$$

hence

$$\gamma \coth \gamma = \frac{2m V_0}{\hbar^2} a - \gamma, \tag{6.284}$$

where $\gamma = ka$.

The energy eigenvalues are given by the intersection of the curves $f(\gamma) = \gamma \coth \gamma$ and $g(\gamma) = 2m V_0 a/\hbar^2 - \gamma$. As shown in Figure 6.6, if $a < \hbar^2/(2m V_0)$ then no bound state solution can exist, since the curves of $f(\gamma)$ and $g(\gamma)$ do not intersect. But if $a > \hbar^2/(2m V_0)$ the curves intersect only once; hence there is one bound state. We can summarize these results as follows:

$$a < \frac{\hbar^2}{2m V_0} \implies \text{no bound states}, \tag{6.285}$$

$$a > \frac{\hbar^2}{2m V_0} \implies \text{only one bound state.} \tag{6.286}$$

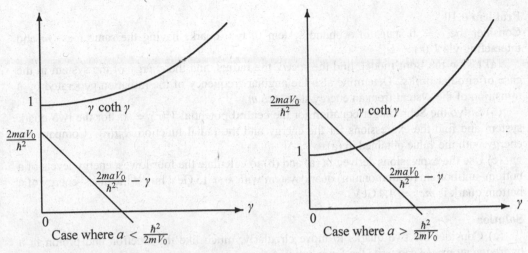

Figure 6.6 Graphical solutions of $f(\gamma) = g(\gamma)$, with $\gamma = ka$, $f(\gamma) = \gamma \coth \gamma$, and $g(\gamma) = 2m V_0 a/\hbar^2 - \gamma$. If $a < \hbar^2/(2m V_0)$ there is no bound state. If $a > \hbar^2/(2m V_0)$ there is one bound state.

The radial wave function is given by

$$R_{n0}(r) = \frac{1}{r} U_{n0}(r) = \begin{cases} (D/r) \sinh kr, & 0 < r < a, \\ (C/r)e^{-kr}, & r > a. \end{cases} \tag{6.287}$$

The normalization of this function yields

$$1 = \int_0^\infty r^2 R_{n0}^2(r)\, dr = \int_0^\infty U_{n0}^2(r)\, dr$$

$$= D^2 \int_0^a \sinh^2 kr\, dr + C^2 \int_a^\infty e^{-2kr}\, dr = \frac{D^2}{2} \int_0^a [\cosh 2kr - 1]\, dr + \frac{C^2}{2k} e^{-2ka}$$

$$= D^2 \left[\frac{1}{4k} \sinh 2ka - \frac{a}{2} \right] + \frac{C^2}{2k} e^{-2ka}. \tag{6.288}$$

From (6.280) we have $Ce^{-ka} = D \sinh ka$, so we can rewrite this relation as

$$1 = D^2 \left[\frac{1}{4k} \sinh 2ka - \frac{a}{2} \right] + \frac{D^2}{2k} \sinh^2 ka = D^2 \left[\frac{\sinh 2ka + 2 \sinh^2 ka}{4k} - \frac{a}{2} \right]; \tag{6.289}$$

hence

$$D = \frac{2\sqrt{k}}{\sqrt{\sinh 2ka + 2 \sinh^2 ka - 2ak}}. \tag{6.290}$$

The normalized wave function is thus given by $\psi_{nlm}(r) = \psi_{n00}(r) = (1/\sqrt{4\pi}) R_{n0}(r)$ or

$$\psi_{n00}(r) = \frac{\sqrt{k}}{\sqrt{\pi \sinh 2ka + 2\pi \sinh^2 ka - 2\pi ak}} \begin{cases} (1/r) \sinh(kr), & 0 < r < a, \\ (1/r) \sinh(ka)e^{-k(r-a)}, & r > a. \end{cases} \tag{6.291}$$

Problem 6.10

Consider the $l = 0$ states of a bound system of two quarks having the same mass m and interacting via $V(r) = kr$.

(a) Using the Bohr model, find the speed, the radius, and the energy of the system in the case of circular orbits. Determine also the angular frequency of the radiation generated by a transition of the system from an energy state n to m.

(b) Solve the Schrödinger equation for the central potential $V(r) = kr$ for the two-quark system and find the expressions for the energy and the radial function $R_{nl}(r)$. Compare the energy with the value obtained in (a).

(c) Use the expressions derived in (a) and (b) to calculate the four lowest energy levels of a bottom–antibottom (bottomonium) quark system with $k = 15\,\text{GeV fm}^{-1}$; the mass–energy of a bottom quark is $mc^2 = 4.4\,\text{GeV}$.

Solution

(a) Consider the two quarks to move circularly, much like the electron and proton in a hydrogen atom; we can write the force between them as

$$\mu \frac{v^2}{r} = \frac{dV(r)}{dr} = k, \tag{6.292}$$

where $\mu = m/2$ is the reduced mass. From the Bohr quantization condition of the orbital angular momentum, we have

$$L = \mu v r = n\hbar, \tag{6.293}$$

Multiplying (6.292) by (6.293), we end up with $\mu^2 v^3 = n\hbar k$ which yields the speed of the relative motion of the two-quark system:

$$v_n = \left(\frac{n\hbar k}{\mu^2}\right)^{1/3}. \tag{6.294}$$

The radius can be obtained from (6.293), $r_n = n\hbar/(\mu v_n)$; using (6.294) this leads to

$$r_n = \left(\frac{n^2\hbar^2}{\mu k}\right)^{1/3}. \tag{6.295}$$

We can obtain the total energy of the relative motion by adding the kinetic and potential energies:

$$E_n = \frac{1}{2}\mu v_n^2 + k r_n = \frac{3}{2}\left(\frac{n^2\hbar^2 k^2}{\mu}\right)^{1/3}. \tag{6.296}$$

In deriving this we have used the relations for v_n and r_n as given by (6.294) and (6.295), respectively. The angular frequency of the radiation generated by a transition from n to m is given by

$$\omega_{nm} = \frac{E_n - E_m}{\hbar} = \frac{3}{2\hbar}\left(\frac{k^2}{\mu\hbar}\right)^{1/3}\left(n^{2/3} - m^{2/3}\right). \tag{6.297}$$

(b) The radial equation is given by (6.57):

$$-\frac{\hbar^2}{2\mu}\frac{d^2 U_{nl}(r)}{dr^2} + \left[kr + \frac{l(l+1)\hbar^2}{2Mr^2}\right]U_{nl}(r) = E_n U_{nl}(r), \tag{6.298}$$

where $U_{nl}(r) = r R_{nl}(r)$. Since we are dealing with $l = 0$, we have

$$\frac{\hbar^2}{2\mu} \frac{d^2 U_{n0}(r)}{dr^2} + kr U_{n0}(r) = F_{nl} U_{n0}(r),$$ (6.299)

which can be reduced to

$$\frac{d^2 U_{n0}(r)}{dr^2} - \frac{2\mu k}{\hbar^2} \left(r - \frac{E}{k} \right) U_{n0}(r) = 0.$$ (6.300)

Making the change of variable $x = (2\mu k/\hbar^2)^{1/3}(r - E/k)$, we can rewrite (6.300) as

$$\frac{d^2 \phi_n(x)}{dx^2} - x\phi_n(x) = 0.$$ (6.301)

We have already studied the solutions of this equation in Chapter 4; they are given by the Airy functions Ai(x): $\phi(x) = B\text{Ai}(x)$. The bound state energies result from the zeros of Ai(x). The boundary conditions on U_{nl} of (6.301) are $U_{nl}(r = 0) = 0$ and $U_{nl}(r \to +\infty) = 0$. The second condition is satisfied by the Airy functions, since Ai$(x \to +\infty) = 0$. The first condition corresponds to $\phi[-(2\mu k/\hbar^2)^{1/3} E/k] = 0$ or to Ai$[-(2\mu k/\hbar^2)^{1/3} E/k] = \text{Ai}(R_n) = 0$, where R_n are the zeros of the Airy function.

The boundary condition $U_{nl}(r = 0) = 0$ then yields a discrete set of energy levels which can be expressed in terms of the Airy roots as follows:

$$\text{Ai}\left[-\left(\frac{2\mu k}{\hbar^2} \right)^{1/3} \frac{E}{k} \right] = 0 \quad \Longrightarrow \quad -\left(\frac{2\mu k}{\hbar^2} \right)^{1/3} \frac{E_n}{k} = R_n;$$ (6.302)

hence

$$E_n = -\left(\frac{\hbar^2 k^2}{2\mu} \right)^{1/3} R_n.$$ (6.303)

The radial function of the system is given by $R_{n0}(r) = (1/r)U_{n0}(r) = (B_n/r)\text{Ai}(x)$ or

$$R_{n0}(r) = \frac{B_n}{r}\text{Ai}(x) = \frac{B_n}{r}\text{Ai}\left[\left(\frac{2\mu k}{\hbar^2} \right)^{1/3} r + R_n \right].$$ (6.304)

The energy expression (6.303) has the same structure as the energy (6.296) derived from the Bohr model $E_n^B = \frac{3}{2}(n^2\hbar^2 k^2/\mu)^{1/3}$; the ratio of the two expressions is

$$\frac{E_n}{E_n^B} = -\frac{2}{3} \frac{R_n}{(2n^2)^{1/3}}.$$ (6.305)

(c) In the following calculations we will be using $k = 15\,\text{GeV fm}^{-1}$, $\mu c^2 = mc^2/2 = 2.2\,\text{GeV}$, and $\hbar c = 197.3\,\text{MeV fm}$. The values of the four lowest energy levels corresponding to the expression $E_n^B = \frac{3}{2}(n^2\hbar^2 k^2/\mu)^{1/3}$, derived from the Bohr model, are

$$E_1^B = \frac{3}{2}\left(\frac{\hbar^2 k^2}{\mu} \right)^{1/3} = 2.38\,\text{GeV}, \qquad E_2^B = 2^{2/3}E_1^B = 3.77\,\text{GeV},$$ (6.306)

$$E_3^B = 3^{2/3}E_1^B = 4.95\,\text{GeV}, \qquad E_4^B = 4^{2/3}E_1^B = 5.99\,\text{GeV}.$$ (6.307)

Let us now calculate the exact energy levels. As mentioned in Chapter 4, the first few roots of the Airy function are given by $R_1 = -2.338$, $R_2 = -4.088$, $R_3 = -5.521$, $R_4 = -6.787$, so we can immediately obtain the first few energy levels:

$$E_1 = \left(\frac{\hbar^2 k^2}{2\mu}\right)^{1/3} R_1 = 2.94\,\text{GeV}, \qquad E_2 = \left(\frac{\hbar^2 k^2}{2\mu}\right)^{1/3} R_2 = 5.14\,\text{GeV}, \qquad (6.308)$$

$$E_3 = \left(\frac{\hbar^2 k^2}{2\mu}\right)^{1/3} R_3 = 6.95\,\text{GeV}, \qquad E_4 = \left(\frac{\hbar^2 k^2}{2\mu}\right)^{1/3} R_4 = 8.54\,\text{GeV}. \qquad (6.309)$$

Problem 6.11

Consider a system of two spinless particles of reduced mass μ that is subject to a finite, central potential well

$$V(r) = \begin{cases} -V_0, & 0 \le r \le a, \\ 0, & r > a, \end{cases}$$

where V_0 is positive. The purpose of this problem is to show how to find the minimum value of V_0 so that the potential well has one $l = 0$ bound state.

(a) Find the solution of the radial Schrödinger equation in both regions, $0 \le r \le a$ and $r > a$, in the case where the particle has zero angular momentum and its energy is located in the range $-V_0 < E < 0$.

(b) Show that the continuity condition of the radial function at $r = a$ can be reduced to a transcendental equation in E.

(c) Use this continuity condition to find the minimum values of V_0 so that the system has one, two, and three bound states.

(d) Obtain the results of (c) from a graphical solution of the transcendental equation derived in (b).

(e) Use the expression obtained in (c) to estimate a numerical value of V_0 for a deuteron nucleus with $a = 2 \times 10^{-15}$ m; a deuteron nucleus consists of a neutron and a proton.

Solution

(a) When $l = 0$ and $-V_0 < E < 0$ the radial equation (6.56),

$$-\frac{\hbar^2}{2\mu}\frac{d^2 U_{nl}(r)}{dr^2} + \left[\frac{l(l+1)\hbar^2}{2\mu r^2} + V(r)\right] U_{nl}(r) = E_n U_{nl}(r), \qquad (6.310)$$

can be written inside the well, call it region (1), as

$$U_n''(r)_1 + k_1^2 U_n(r)_1 = 0, \qquad 0 \le r \le a, \qquad (6.311)$$

and outside the well, call it region (2), as

$$U_n''(r)_2 - k_2^2 U_n(r)_2 = 0, \qquad r > a, \qquad (6.312)$$

where $U_n''(r) = d^2 U_n(r)/dr^2$, $U_n(r)_1 = r R_n(r)_1$, $U_n(r)_2 = r R_n(r)_2$, $k_1 = \sqrt{2\mu(V_0 + E)/\hbar^2}$ and $k_2 = \sqrt{-2\mu E/\hbar^2}$. Since $U_n(r)_1$ must vanish at $r = 0$, while $U_n(r)_2$ has to be finite at $r \longrightarrow \infty$, the respective solutions of (6.311) and (6.312) are given by

$$U_n(r)_1 = A \sin(k_1 r), \qquad 0 \le r \le a, \qquad (6.313)$$
$$U_n(r)_2 = B e^{-k_2 r}, \qquad r > a. \qquad (6.314)$$

The corresponding radial functions are

$$R_n(r)_1 = A\frac{\sin(k_1 r)}{r}, \qquad R_n(r)_2 = B\frac{e^{-k_2 r}}{r}. \qquad (6.315)$$

(b) Since the logarithmic derivative of the radial function is continuous at $r = a$, we can write

$$\frac{R'_n(a)_1}{R_n(a)_1} = \frac{R'_n(a)_2}{R_n(a)_2}. \qquad (6.316)$$

From (6.315) we have

$$\frac{R'_n(a)_1}{R_n(a)_1} = k_1 \cot(k_1 a) - \frac{1}{a}, \qquad \frac{R'_n(a)_2}{R_n(a)_2} = -k_2 - \frac{1}{a}. \qquad (6.317)$$

Substituting (6.317) into (6.316) we obtain

$$-k_1 \cot(k_1 a) = k_2 \qquad (6.318)$$

or

$$\sqrt{\frac{2\mu}{\hbar^2}(V_0 + E)} \cot\left[\sqrt{\frac{2\mu}{\hbar^2}(V_0 + E)}a\right] = -\sqrt{-\frac{2\mu E}{\hbar^2}}, \qquad (6.319)$$

since $k_1 = \sqrt{2\mu(V_0 + E)/\hbar^2}$ and $k_2 = \sqrt{-2\mu E/\hbar^2}$.

(c) In the limit $E \to 0$, the system has very few bound states; in this limit, equation (6.319) becomes

$$\sqrt{\frac{2\mu V_0}{\hbar^2}} \cot\left(\sqrt{\frac{2\mu V_0}{\hbar^2}}a\right) = 0, \qquad (6.320)$$

which leads to $a\sqrt{2\mu V_{0n}/\hbar^2} = (2n + 1)\pi/2$; hence

$$V_{0n} = \frac{\pi^2 \hbar^2}{8\mu a^2}(2n + 1)^2, \qquad n = 0, 1, 2, 3, \ldots. \qquad (6.321)$$

Thus, the minimum values of V_0 corresponding to one, two, and three bound states are respectively

$$V_{00} = \frac{\pi^2 \hbar^2}{8\mu a^2}, \qquad V_{01} = \frac{9\pi^2 \hbar^2}{8\mu a^2}, \qquad V_{02} = \frac{25\pi^2 \hbar^2}{8\mu a^2}. \qquad (6.322)$$

(d) Using the notation $\alpha = ak_1$ and $\beta = ak_2$ we can, on the one hand, write

$$\alpha^2 + \beta^2 = \frac{2\mu a^2 V_0}{\hbar^2}, \qquad (6.323)$$

and, on the other hand, reduce the transcendental equation (6.318) to

$$-\alpha \cot \alpha = \beta, \qquad (6.324)$$

since $k_1 = \sqrt{2\mu(V_0 + E)/\hbar^2}$ and $k_2 = \sqrt{-2\mu E/\hbar^2}$.

Figure 6.7 Graphical solutions for the finite, spherical square well potential: they are given by the intersection of the circle $\alpha^2 + \beta^2 = 2\mu a^2 V_0/\hbar^2$ with the curve of $-\alpha \cot \alpha$, where $\alpha^2 = 2\mu a^2(V_0 + E)/\hbar^2$ and $\beta^2 = -2\mu a^2 E/\hbar^2$, with $-V_0 < E < 0$.

As shown in Figure 6.7, when $\pi/2 < \alpha < 3\pi/2$, which in the limit of $E \to 0$ leads to

$$\frac{\pi^2 \hbar^2}{8\mu a^2} < V_0 < \frac{9\pi^2 \hbar^2}{8\mu a^2}, \tag{6.325}$$

there exists only one bound state, since the circle intersects only once with the curve $-\alpha \cot \alpha$. Similarly, there are two bound states if $3\pi/2 < \alpha < 5\pi/2$ or

$$\frac{9\pi^2 \hbar^2}{8\mu a^2} < V_0 < \frac{25\pi^2 \hbar^2}{8\mu a^2}, \tag{6.326}$$

and three bound states if $5\pi/2 < \alpha < 7\pi/2$:

$$\frac{25\pi^2 \hbar^2}{8\mu a^2} < V_0 < \frac{49\pi^2 \hbar^2}{8\mu a^2}. \tag{6.327}$$

(e) Since $m_p c^2 \simeq 938\,\text{MeV}$ and $m_n c^2 \simeq 940\,\text{MeV}$, the reduced mass of the deuteron is given by $\mu c^2 = (m_p c^2)(m_n c^2)/(m_p c^2 + m_n c^2) \simeq 469.5\,\text{MeV}$. Since $a = 2 \times 10^{-15}\,\text{m}$ the minimum value of V_0 corresponding to one bound state is

$$V_0 = \frac{\pi^2 \hbar^2}{8\mu a^2} = \frac{\pi^2 (\hbar c)^2}{8(\mu c^2)a^2} = \frac{\pi^2 (197\,\text{MeV fm})^2}{8(469.5\,\text{MeV})(2 \times 10^{-15}\,\text{m})^2} \simeq 25.5\,\text{MeV}. \tag{6.328}$$

Problem 6.12

Calculate $\langle nl \mid \hat{P}^4 \mid nl \rangle$ in a stationary state $\mid nl \rangle$ of the hydrogen atom.

Solution

To calculate $\langle nl \mid \hat{P}^4 \mid nl \rangle$ we may consider expressing $\hat{P}^4$ in terms of the hydrogen's Hamiltonian. Since $\hat{H} = \hat{P}^2/(2m_e) - e^2/r$ we have $\hat{P}^2 = 2m_e(\hat{H} + e^2/r)$; hence

$$\langle nl \mid \hat{P}^4 \mid nl \rangle = (2m_e)^2 \left\langle nl \left| \left(\hat{H} + \frac{e^2}{r} \right)^2 \right| nl \right\rangle$$

$$= (2m_e)^2 \left\langle nl \left| \hat{H}^2 + \hat{H}\frac{e^2}{r} + \frac{e^2}{r}\hat{H} + \frac{e^4}{r^2} \right| nl \right\rangle$$

$$= (2m_e)^2 \left[E_n^2 + E_n \left\langle nl \left| \frac{e^2}{r} \right| nl \right\rangle + \left\langle nl \left| \frac{e^2}{r} \right| nl \right\rangle E_n + \left\langle nl \left| \frac{e^4}{r^2} \right| nl \right\rangle \right],$$

$$(6.329)$$

where we have used the fact that $\mid nl \rangle$ is an eigenstate of $\hat{H}$: $\hat{H} \mid nl \rangle = E_n \mid nl \rangle$ with $E_n = -e^2/(2a_0 n^2) = -13.6\,\text{eV}/n^2$. The expectation values of $1/r$ and $1/r^2$ are given by (6.182) and (6.183), $\langle nl|r^{-1}|nl \rangle = 1/(n^2 a_0)$ and $\langle nl|r^{-2}|nl \rangle = 2/[n^3(2l+1)a_0^2]$; we can thus rewrite (6.329) as

$$\langle nl \mid \hat{P}^4 \mid nl \rangle = (2m_e)^2 \left[E_n^2 + 2E_n \left\langle nl \left| \frac{e^2}{r} \right| nl \right\rangle + \left\langle nl \left| \frac{e^4}{r^2} \right| nl \right\rangle \right]$$

$$= (2m_e E_n)^2 \left[1 + \frac{2e^2}{E_n} \frac{1}{n^2 a_0} + \frac{e^4}{E_n^2} \frac{2}{n^3(2l+1)a_0^2} \right]$$

$$= (2m_e E_n)^2 \left[1 - 4 + \frac{8n}{2l+1} \right];$$

$$(6.330)$$

in deriving the last relation we have used $E_n = -e^2/(2a_0 n^2)$. Now, since $a_0 = \hbar^2/(m_e e^2)$, the energy E_n becomes $E_n = -e^2/(2a_0 n^2) = -m_e e^4/(2\hbar^2 n^2)$ which, when inserted into (6.330), leads to

$$\boxed{\langle nl \mid \hat{P}^4 \mid nl \rangle = \frac{m_e^4 e^8}{\hbar^4 n^4} \left[\frac{8n}{2l+1} - 3 \right].} \qquad (6.331)$$

6.6 Exercises

Exercise 6.1
A spinless particle of mass m is confined to move in the xy plane under the influence of a harmonic oscillator potential $\hat{V}(x, y) = \frac{1}{2}m\omega^2 (\hat{x}^2 + \hat{y}^2)$ for all values of x and y.

(a) Show that the Hamiltonian $\hat{H}$ of this particle can be written as a sum of two familiar one-dimensional Hamiltonians, $\hat{H}_x$ and $\hat{H}_y$. Then show that $\hat{H}$ commutes with $\hat{L}_z = \hat{X}\hat{P}_y - \hat{Y}\hat{P}_x$.

(b) Find the expression for the energy levels $E_{n_x n_y}$.

(c) Find the energies of the four lowest states and their corresponding degeneracies.

(d) Find the degeneracy g_n of the nth excited state as a function of the quantum number n $(n = n_x + n_y)$.

(e) If the state vector of the nth excited state is $|n\rangle = |n_x\rangle|n_y\rangle$ or

$$\langle xy|n \rangle = \langle x|n_x \rangle \langle y|n_y \rangle = \psi_{n_x}(x)\psi_{n_y}(y),$$

calculate the expectation value of the operator $\hat{A} = \hat{x}^4 + \hat{y}^2$ in the state $|n\rangle$ as a function of the quantum numbers n_x and n_y.

Exercise 6.2

A particle of mass m moves in the xy plane in the potential

$$V(x, y) = \begin{cases} \frac{1}{2}m\omega^2 y^2 & \text{for all } y \text{ and } 0 < x < a, \\ +\infty, & \text{elsewhere.} \end{cases}$$

(a) Write down the time-independent Schrödinger equation for this particle and reduce it to a set of familiar one-dimensional equations.

(b) Find the normalized eigenfunctions and the eigenenergies.

Exercise 6.3

A particle of mass m moves in the xy plane in a two-dimensional rectangular well

$$V(x, y) = \begin{cases} 0, & 0 < x < a, \quad 0 < y < b, \\ +\infty, & \text{elsewhere.} \end{cases}$$

By reducing the time-independent Schrödinger equation to a set of more familiar one-dimensional equations, find the normalized wave functions and the energy levels of this particle.

Exercise 6.4

Consider an anisotropic three-dimensional harmonic oscillator potential

$$V(x, y, z) = \frac{1}{2}m(\omega_x^2 x^2 + \omega_y^2 y^2 + \omega_z^2 z^2).$$

(a) Evaluate the energy levels in terms of ω_x, ω_y, and ω_z.

(b) Calculate $[\hat{H}, \hat{L}_z]$. Do you expect the wave functions to be eigenfunctions of $\hat{L}^2$?

(c) Find the three lowest levels for the case $\omega_x = \omega_y = 2\omega_z/3$, and determine the degeneracy of each level.

Exercise 6.5

Consider a spinless particle of mass m which is confined to move under the influence of a three-dimensional potential

$$\hat{V}(x, y, z) = \begin{cases} 0 & \text{for } 0 < x < a, \, 0 < y < a, \, 0 < z < b, \\ +\infty & \text{elsewhere.} \end{cases}$$

(a) Find the expression for the energy levels $E_{n_x n_y n_z}$ and their corresponding wave functions.

(b) If $a = 2b$ find the energies of the five lowest states and their degeneracies.

Exercise 6.6

A particle of mass m moves in the three-dimensional potential

$$V(x, y, z) = \begin{cases} \frac{1}{2}m\omega^2 z^2 & \text{for } 0 < x < a, \, 0 < y < a, \text{ and } z > 0, \\ +\infty, & \text{elsewhere.} \end{cases}$$

(a) Write down the time-independent Schrödinger equation for this particle and reduce it to a set of familiar one-dimensional equations; then find the normalized wave function $\psi_{n_x n_y n_z}(x, y, z)$.

(b) Find the allowed eigenenergies of this particle and show that they can be written as: $E_{n_x n_y n_z} = E_{n_x n_y} + E_{n_z}$.

(c) Find the four lowest energy levels in the xy plane (i.e., $E_{n_x n_y}$) and their corresponding degeneracies.

Exercise 6.7

A particle of mass m moves in the potential $V(x, y, z) = V_1(x, y) + V_2(z)$ where

$$V_1(x, y) = \frac{1}{2}m\omega^2\left(x^2 + y^2\right), \qquad V_2(z) = \begin{cases} 0, & 0 \le z \le a, \\ +\infty, & \text{elsewhere.} \end{cases}$$

(a) Calculate the energy levels and the wave function of this particle.

(b) Let us now turn off $V_2(z)$ (i.e., m is subject only to $V_1(x, y)$). Calculate the degeneracy g_n of the nth energy level (note that $n = n_x + n_y$).

Exercise 6.8

Consider a muonic atom which consists of a nucleus that has Z protons (no neutrons) and a negative muon moving around it; the muon's charge is $-e$ and its mass is 207 times the mass of the electron, $m_{\mu-} = 207 m_e$. For a muonic atom with $Z = 6$, calculate

(a) the radius of the first Bohr orbit,

(b) the energy of the ground, first, and second excited states, and

(c) the frequency associated with the transitions $n_i = 2 \to n_f = 1$, $n_i = 3 \to n_f = 1$, and $n_i = 3 \to n_f = 2$.

Exercise 6.9

A hydrogen atom has the wave function $\Psi_{nlm}(\vec{r})$, where $n = 4$, $l = 3$, $m = 3$.

(a) What is the magnitude of the orbital angular momentum of the electron around the proton?

(b) What is the angle between the orbital angular momentum vector and the z-axis? Can this angle be reduced by changing n or m if l is held constant? What is the physical significance of this result?

(c) Sketch the shapes of the radial function and of the probability of finding the electron a distance r from the proton.

Exercise 6.10

An electron in a hydrogen atom is in the energy eigenstate

$$\psi_{2,1,-1}(r, \theta, \varphi) = Nre^{-r/2a_0}Y_{1,-1}(\theta, \varphi).$$

(a) Find the normalization constant, N.

(b) What is the probability per unit volume of finding the electron at $r = a_0$, $\theta = 45°$, $\varphi = 60°$?

(c) What is the probability per unit radial interval (dr) of finding the electron at $r = 2a_0$? (One must take an integral over θ and φ at $r = 2a_0$.)

(d) If the measurements of $\hat{L}^2$ and $\hat{L}_z$ were carried out, what will be the results?

Exercise 6.11

Consider a hydrogen atom which is in its ground state; the ground state wave function is given by

$$\Psi(r, \theta, \varphi) = \frac{1}{\sqrt{\pi a_0^3}} e^{-r/a_0},$$

where a_0 is the Bohr radius.

(a) Find the most probable distance between the electron and the proton when the hydrogen atom is in its ground state.

(b) Find the average distance between the electron and the proton.

Exercise 6.12

Consider a hydrogen atom whose state at time $t = 0$ is given by

$$\Psi(\vec{r}, 0) = \frac{1}{\sqrt{2}} \phi_{300}(\vec{r}) + \frac{1}{\sqrt{3}} \phi_{311}(\vec{r}) + \frac{1}{\sqrt{6}} \phi_{322}(\vec{r}).$$

(a) What is the time-dependent wave function?

(b) If a measurement of the energy were carried out, what values could be found and with what probabilities?

(c) Repeat part (b) for $\hat{L}^2$ and $\hat{L}_z$. That is, if a measurement of $\hat{L}^2$ and $\hat{L}_z$ were carried out, what values could be found and with what probabilities?

Exercise 6.13

The wave function of an electron in a hydrogen atom is given by

$$\psi_{21 m_l m_s}(r, \theta, \varphi) = R_{21}(r) \left[\frac{1}{\sqrt{3}} Y_{10}(\theta, \varphi) \left| \frac{1}{2}, \frac{1}{2} \right\rangle + \sqrt{\frac{2}{3}} Y_{11}(\theta, \varphi) \left| \frac{1}{2}, -\frac{1}{2} \right\rangle \right],$$

where $\left| \frac{1}{2}, \pm\frac{1}{2} \right\rangle$ are the spin state vectors.

(a) Is this wave function an eigenfunction of $\hat{J}_z$, the z-component of the electron's total angular momentum? If yes, find the eigenvalue. (*Hint:* For this, you need to calculate $\hat{J}_z \psi_{21 m_l m_s}$.)

(b) If you measure the z-component of the electron's spin angular momentum, what values will you obtain? What are the corresponding probabilities?

(c) If you measure $\hat{J}^2$, what values will you obtain? What are the corresponding probabilities?

Exercise 6.14

Consider a hydrogen atom whose state at time $t = 0$ is given by

$$\Psi(\vec{r}, 0) = A\phi_{200}(\vec{r}) + \frac{1}{\sqrt{5}} \phi_{311}(\vec{r}) + \frac{1}{\sqrt{3}} \phi_{422}(\vec{r}),$$

where A is a normalization constant.

(a) Find A so that the state is normalized.

(b) Find the state of this atom at any later time t.

(c) If a measurement of the energy were carried out, what values would be found and with what probabilities?

(d) Find the mean energy of the atom.

Exercise 6.15

Calculate the width of the probability density distribution for r for the hydrogen atom in its ground state: $\Delta r = \sqrt{\langle r^2 \rangle_{10} - \langle r \rangle_{10}^2}$.

Exercise 6.16

Consider a hydrogen atom whose wave function is given at time $t = 0$ by

$$\psi(\vec{r}, 0) = \frac{A}{\sqrt{\pi}} \left(\frac{1}{a_0}\right)^{3/2} e^{-r/a_0} + \frac{1}{\sqrt{2\pi}} \left(\frac{z - \sqrt{2}x}{r}\right) R_{21}(r),$$

where A is a real constant, a_0 is the Bohr radius, and $R_{21}(r)$ is the radial wave function: $R_{21}(r) = 1/\sqrt{6}(1/a_0)^{3/2}(r/2a_0)e^{-r/2a_0}$.

   (a) Write down $\psi(\vec{r}, 0)$ in terms of $\sum_{nlm} \phi_{nlm}(\vec{r})$ where $\phi_{nlm}(\vec{r})$ is the hydrogen wave function $\phi_{nlm}(\vec{r}) = R_{nl}(r) Y_{lm}(\theta, \varphi)$.

   (b) Find A so that $\psi(\vec{r}, 0)$ is normalized. (Recall that $\int \phi_{n'l'm'}^*(\vec{r}) \phi_{nlm}(\vec{r}) d^3 r = \delta_{n', n} \delta_{l', l} \delta_{m', m}$.)

   (c) Write down the wave function $\psi(\vec{r}, t)$ at any later time t.

   (d) Is $\psi(\vec{r}, 0)$ an eigenfunction of $\hat{L}^2$ and $\hat{L}_z$? If yes, what are the eigenvalues?

   (e) If a measurement of the energy is made, what value could be found and with what probability?

   (f) What is the probability that a measurement of $\hat{L}_z$ yields $1\hbar$?

   (g) Find the mean value of r in the state $\psi(\vec{r}, 0)$.

Exercise 6.17

Consider a pendulum undergoing small harmonic oscillations (with angular frequency $\omega = \sqrt{g/l}$, where g is the acceleration due to gravity and l is the length of the pendulum). Show that the quantum energy levels and the corresponding degeneracies of the pendulum are given by $E_n = (n + 1)\hbar\omega$ and $g_n = n + 1$, respectively.

Exercise 6.18

Consider a proton that is trapped inside an infinite central potential well

$$V(r) = \begin{cases} -V_0, & 0 < r < a, \\ +\infty, & r \geq a, \end{cases}$$

where $V_0 = 5104.34$ MeV and $a = 10$ fm.

   (a) Find the energy and the (normalized) radial wave function of this particle for the s states (i.e., $l = 0$).

   (b) Find the number of bound states that have energies lower than zero; you may use the values $mc^2 = 938$ MeV and $\hbar c = 197$ MeV fm.

   (c) Calculate the energies of the levels that lie just below and just above the zero-energy level; express your answer in MeV.

Exercise 6.19

Consider the function $\psi(\vec{r}) = -A(x + iy) e^{-r/2a_0}$, where a_0 is the Bohr radius and A is a real constant.

   (a) Is $\psi(\vec{r})$ an eigenfunction to $\hat{L}^2$ and $\hat{L}_z$? If yes, write $\psi(\vec{r})$ in terms of $R_{nl}(r) Y_{lm}(\theta, \varphi)$ and find the values of the quantum numbers n, m, l; $R_{nl}(r)$ are the radial wave functions of the hydrogen atom.

   (b) Find the constant A so that $\psi(\vec{r})$ is normalized.

   (c) Find the mean value of r and the most probable value of r in this state.

Exercise 6.20

The wave function of a hydrogen-like atom at time $t = 0$ is

$$\Psi(\vec{r}, 0) = \frac{1}{\sqrt{11}}\left[\sqrt{3}\psi_{2,1,-1}(\vec{r}) - \psi_{2,1,0}(\vec{r}) + \sqrt{5}\psi_{2,1,1}(\vec{r}) + \sqrt{2}\psi_{3,1,1}(\vec{r})\right],$$

where $\psi_{nlm}(\vec{r})$ is a normalized eigenfunction (i.e., $\psi_{nlm}(\vec{r}) = R_{nl}(r)Y_{lm}(\theta, \varphi)$).

(a) What is the time-dependent wave function?

(b) If a measurement of energy is made, what values could be found and with what probabilities?

(c) What is the probability for a measurement of $\hat{L}_z$ which yields $-1\hbar$?

Exercise 6.21

Using the fact that the radial momentum operator is given by $\hat{p}_r = -i\hbar\frac{1}{r}\frac{\partial}{\partial r}r$, calculate the commutator $[\hat{r}, \hat{p}_r]$ between the position operator, $\hat{r}$, and the radial momentum operator.

Exercise 6.22

Calculate $\Delta r \Delta p_r$ with respect to the state

$$\psi_{2,1,0}(\vec{r}) = \frac{1}{\sqrt{6}}\left(\frac{1}{a_0}\right)^{3/2}\frac{r}{2a_0}e^{-r/2a_0}\, Y_{1,0}(\theta, \varphi),$$

and verify that $\Delta r \Delta p_r$ satisfies the Heisenberg uncertainty principle.

Chapter 7

Rotations and Addition of Angular Momenta

In this chapter we deal with rotations, the properties of addition of angular momenta, and the properties of tensor operators.

7.1 Rotations in Classical Physics

A rotation is defined by an angle of rotation and an axis about which the rotation is performed. Knowing the rotation matrix R, we can determine how vectors transform under rotations; in a three-dimensional space, a vector $\vec{A}$ becomes $\vec{A}'$ when rotated: $\vec{A}' = R\vec{A}$. For instance, a rotation over an angle ϕ about the z-axis transforms the components A_x, A_y, A_z of the vector $\vec{A}$ into A'_x, A'_y, A'_z:

$$\begin{pmatrix} A'_x \\ A'_y \\ A'_z \end{pmatrix} = \begin{pmatrix} \cos\phi & \sin\phi & 0 \\ -\sin\phi & \cos\phi & 0 \\ 0 & 0 & 1 \end{pmatrix} \begin{pmatrix} A_x \\ A_y \\ A_z \end{pmatrix} \tag{7.1}$$

or

$$\vec{A}' = R_z(\phi)\vec{A}, \tag{7.2}$$

where

$$R_z(\phi) = \begin{pmatrix} \cos\phi & \sin\phi & 0 \\ -\sin\phi & \cos\phi & 0 \\ 0 & 0 & 1 \end{pmatrix}. \tag{7.3}$$

Similarly, the rotation matrices about the $x-$ and $y-$ axes are given by

$$R_x(\phi) = \begin{pmatrix} 1 & 0 & 0 \\ 0 & \cos\phi & \sin\phi \\ 0 & -\sin\phi & \cos\phi \end{pmatrix}, \qquad R_y(\phi) = \begin{pmatrix} \cos\phi & 0 & \sin\phi \\ 0 & 1 & 0 \\ -\sin\phi & 0 & \cos\phi \end{pmatrix}. \tag{7.4}$$

From classical physics we know that while rotations about the same axis commute, rotations about different axes do not. From (7.4) we can verify that $R_x(\phi)R_y(\phi) \neq R_y(\phi)R_x(\phi)$. In fact, using (7.4) we can have

$$R_x(\phi)R_y(\phi) = \begin{pmatrix} \cos\phi & 0 & \sin\phi \\ -\sin^2\phi & \cos\phi & \cos\phi\sin\phi \\ -\cos\phi\sin\phi & -\sin\phi & \cos^2\phi \end{pmatrix}, \tag{7.5}$$

$$R_y(\phi)R_x(\phi) = \begin{pmatrix} \cos\phi & -\sin^2\phi & \cos\phi\sin\phi \\ 0 & \cos\phi & \sin\phi \\ -\sin\phi & -\sin\phi\cos\phi & \cos^2\phi \end{pmatrix}; \tag{7.6}$$

hence $R_x(\phi)R_y(\phi) - R_y(\phi)R_x(\phi)$ is given by

$$\begin{pmatrix} 0 & \sin^2\phi & \sin\phi - \cos\phi\sin\phi \\ -\sin^2\phi & 0 & \cos\phi\sin\phi - \sin\phi \\ \sin\phi - \cos\phi\sin\phi & \cos\phi\sin\phi - \sin\phi & 0 \end{pmatrix}. \tag{7.7}$$

In the case of infinitesimal rotations of angle δ about the $x-$ $y-$, $z-$ axes, and using $\cos\delta \simeq 1 - \delta^2/2$ and $\sin\delta \simeq \delta$, we can reduce (7.7) to

$$R_x(\delta)R_y(\delta) - R_y(\delta)R_x(\delta) = \begin{pmatrix} 0 & \delta^2 & 0 \\ -\delta^2 & 0 & 0 \\ 0 & 0 & 0 \end{pmatrix}, \tag{7.8}$$

which, when combined with $R_z(\delta^2)$ of (7.3),

$$R_z(\delta) = \begin{pmatrix} 1 - \frac{\delta^2}{2} & \delta & 0 \\ -\delta & 1 - \frac{\delta^2}{2} & 0 \\ 0 & 0 & 1 \end{pmatrix} \implies R_z(\delta^2) = \begin{pmatrix} 1 & \delta^2 & 0 \\ -\delta^2 & 1 & 0 \\ 0 & 0 & 1 \end{pmatrix}, \tag{7.9}$$

leads to

$$R_x(\delta)R_y(\delta) - R_y(\delta)R_x(\delta) = R_z(\delta^2) - 1 = \begin{pmatrix} 1 & \delta^2 & 0 \\ -\delta^2 & 1 & 0 \\ 0 & 0 & 1 \end{pmatrix} - \begin{pmatrix} 1 & 0 & 0 \\ 0 & 1 & 0 \\ 0 & 0 & 1 \end{pmatrix}. \tag{7.10}$$

We will show later that this relation can be used to derive the commutation relations between the components of the angular momentum (7.26).

The rotation matrices R are orthogonal, i.e.,

$$RR^T = R^T R = 1, \tag{7.11}$$

where R^T is the transpose of the matrix R. In addition, the orthogonal matrices conserve the magnitude of vectors:

$$|\vec{A}'| = |\vec{A}|, \tag{7.12}$$

since $\vec{A}' = \hat{R}\vec{A}$ yields $\vec{A}'^2 = \vec{A}^2$ or $A_x'^2 + A_y'^2 + A_z'^2 = A_x^2 + A_y^2 + A_z^2$.

It is easy to show that the matrices of orthogonal rotations form a (nonabelian) group and that they satisfy this relation

$$\det(R) = 1. \tag{7.13}$$

This group is called the *special* three-dimensional orthogonal group, $SO(3)$, because the rotation group is a special case of a more general group, the group of three-dimensional orthogonal transformations, $O(3)$, which consist of both rotations and reflections and for which

$$\det(R) = \pm 1. \tag{7.14}$$

The group $SO(3)$ transforms a vector $\vec{A}$ into another vector $\vec{A}'$ while conserving the size of its length.

7.2 Rotations in Quantum Mechanics

In this section we study the relationship between the angular momentum and the rotation operator and then study the properties as well as the representation of the rotation operator. The connection is analogous to that between the linear momentum operator and translations. We will see that the angular momentum operator acts as a generator for rotations.

A rotation is specified by an angle and by a unit vector $\vec{n}$ about which the rotation is performed. Knowing the rotation operator $\hat{R}$, we can determine how state vectors and operators transform under rotations; as shown in Chapter 2, a state $| \psi \rangle$ and an operator $\hat{A}$ transform according to

$$| \psi' \rangle = \hat{R} | \psi \rangle, \qquad \hat{A}' = \hat{R}\hat{A}\hat{R}^{\dagger}. \tag{7.15}$$

The problem reduces then to finding $\hat{R}$. We may now consider infinitesimal as well as finite rotations.

7.2.1 Infinitesimal Rotations

Consider a rotation of the coordinates of a *spinless* particle over an *infinitesimal* angle $\delta\phi$ about the z-axis. Denoting this rotation by the operator $\hat{R}_z(\delta\phi)$, we have

$$\hat{R}_z(\delta\phi)\psi(r, \theta, \phi) = \psi(r, \theta, \phi - \delta\phi). \tag{7.16}$$

Taylor expanding the wave function to the first order in $\delta\phi$, we obtain

$$\psi(r, \theta, \phi - \delta\phi) \simeq \psi(r, \theta, \phi) - \delta\phi\frac{\delta\psi}{\delta\phi} = \left(1 - \delta\phi\frac{\delta}{\delta\phi}\right)\psi(r, \theta, \phi). \tag{7.17}$$

Comparing (7.16) and (7.17) we see that $\hat{R}_z(\delta\phi)$ is given by

$$\hat{R}_z(\delta\phi) = 1 - \delta\phi\frac{\delta}{\delta\phi}. \tag{7.18}$$

Since the z-component of the orbital angular momentum is

$$\hat{L}_z = -i\hbar\frac{\delta}{\delta\phi}, \tag{7.19}$$

we can rewrite (7.18) as

$$\boxed{\hat{R}_z(\delta\phi) = 1 - \frac{i}{\hbar}\delta\phi\hat{L}_z.} \tag{7.20}$$

We may generalize this relation to a rotation of angle $\delta\phi$ about an arbitrary axis whose direction is given by the unit vector $\vec{n}$:

$$\hat{R}(\delta\phi) = 1 - \frac{i}{\hbar}\delta\phi\,\vec{n}\cdot\hat{\vec{L}}.$$

(7.21)

This is the operator corresponding to an *infinitesimal* rotation of angle $\delta\phi$ about $\vec{n}$ for a *spinless* system. The orbital angular momentum is thus the *generator* of infinitesimal spatial rotations.

Rotations and the commutation relations

We can show that the relation (7.10) leads to the commutation relations of angular momentum $[\hat{L}_x,\ \hat{L}_y] = i\hbar\hat{L}_z$. The operators corresponding to infinitesimal rotations of angle δ about the x and y axes can be inferred from (7.20):

$$\hat{R}_x(\delta) = 1 - \frac{i\delta}{\hbar}\hat{L}_x - \frac{\delta^2}{2\hbar^2}\hat{L}_x^2, \qquad \hat{R}_y(\delta) = 1 - \frac{i\delta}{\hbar}\hat{L}_y - \frac{\delta^2}{2\hbar^2}\hat{L}_y^2,$$

(7.22)

where we have extended the expansions to the second power in δ. On the one hand, the following useful relation can be obtained from (7.22):

$$
\begin{aligned}
\hat{R}_x(\delta)\hat{R}_y(\delta) - \hat{R}_y(\delta)\hat{R}_x(\delta) &= \left(1 - \frac{i\delta}{\hbar}\hat{L}_x - \frac{\delta^2}{2\hbar^2}\hat{L}_x^2\right)\left(1 - \frac{i\delta}{\hbar}\hat{L}_y - \frac{\delta^2}{2\hbar^2}\hat{L}_y^2\right) \\
&\quad - \left(1 - \frac{i\delta}{\hbar}\hat{L}_y - \frac{\delta^2}{2\hbar^2}\hat{L}_y^2\right)\left(1 - \frac{i\delta}{\hbar}\hat{L}_x - \frac{\delta^2}{2\hbar^2}\hat{L}_x^2\right) \\
&= -\frac{\delta^2}{\hbar^2}\left(\hat{L}_x\hat{L}_y - \hat{L}_y\hat{L}_x\right) \\
&= -\frac{\delta^2}{\hbar^2}[\hat{L}_x,\ \hat{L}_y],
\end{aligned}
$$

(7.23)

where we have kept only terms up to the second power in δ; the terms in δ cancel out automatically.

On the other hand, according to (7.10), we have

$$R_x(\delta)R_y(\delta) - R_y(\delta)R_x(\delta) = R_z(\delta^2) - 1.$$

(7.24)

Since $\hat{R}_z(\delta^2) = 1 - (i\delta^2/\hbar)\hat{L}_z$ this relations leads to

$$R_x(\delta)R_y(\delta) - R_y(\delta)R_x(\delta) = R_z(\delta^2) - 1 = -\frac{i\delta^2}{\hbar}\hat{L}_z.$$

(7.25)

Finally, equating (7.23) and (7.25), we end up with

$$[\hat{L}_x,\ \hat{L}_y] = i\hbar\hat{L}_z.$$

(7.26)

Similar calculations for $R_y(\delta)R_z(\delta) - R_z(\delta)R_y(\delta)$ and $R_z(\delta)R_x(\delta) - R_x(\delta)R_z(\delta)$ lead to the other two commutation relations $[\hat{L}_y,\ \hat{L}_z] = i\hbar\hat{L}_x$ and $[\hat{L}_z,\ \hat{L}_x] = i\hbar\hat{L}_y$.

7.2.2 Finite Rotations

The operator $\hat{R}_z(\phi)$ corresponding to a rotation (of the coordinates of a spinless particle) over a *finite* angle ϕ about the z-axis can be constructed in terms of the infinitesimal rotation operator (7.20) as follows. We divide the angle ϕ into N infinitesimal angles $\delta\phi$: $\phi = N\delta\phi$. The rotation over the finite angle ϕ can thus be viewed as a series of N consecutive infinitesimal rotations, each over the angle $\delta\phi$, about the z-axis, applied consecutively one after the other:

$$\hat{R}_z(\phi) = \hat{R}_z(N\delta\phi) = (R_z(\delta\phi))^N = \left(1 - i\frac{\delta\phi}{\hbar}\hat{L}_z\right)^N. \tag{7.27}$$

Since $\delta\phi = \phi/N$, and if $\delta\phi$ is infinitesimally small, we have

$$\hat{R}_z(\phi) = \lim_{N\to\infty}\prod_{k=1}^{N}\left(1 - \frac{i}{\hbar}\frac{\phi}{N}\hat{n}\cdot\vec{L}\right) = \lim_{N\to\infty}\left(1 - \frac{i}{\hbar}\frac{\phi}{N}\hat{L}_z\right)^N, \tag{7.28}$$

or

$$\boxed{\hat{R}_z(\phi) = e^{-i\phi\hat{L}_z/\hbar}.} \tag{7.29}$$

We can generalize this result to infer the rotation operator $\hat{R}_n(\phi)$ corresponding to a rotation over a finite angle ϕ around an axis $\vec{n}$:

$$\boxed{\hat{R}_n(\phi) = e^{-i\phi\vec{n}\cdot\hat{\vec{L}}/\hbar},} \tag{7.30}$$

where $\vec{L}$ is the orbital angular momentum. This operator represents the rotation of the coordinates of a spinless particle over an angle ϕ about an axis $\vec{n}$.

The discussion that led to (7.30) was carried out for a spinless system. A more general study for a system with spin would lead to a relation similar to (7.30):

$$\boxed{\hat{R}_n(\phi) = e^{-\frac{i}{\hbar}\phi\vec{n}\cdot\hat{\vec{J}}},} \tag{7.31}$$

where $\hat{\vec{J}}$ is the total angular momentum operator; this is known as the *rotation operator*. For instance, the rotation operator $\vec{R}_x(\phi)$ of a rotation through an angle ϕ about the x-axis is given by

$$\hat{R}_x(\phi) = e^{-i\phi\hat{J}_x/\hbar}. \tag{7.32}$$

The properties of $\hat{R}_n(\phi)$ are determined by those of the operators $\hat{J}_x$, $\hat{J}_y$, $\hat{J}_z$.

Remark

The Hamiltonian of a particle in a *central potential*, $\hat{H} = \hat{P}^2/(2m) + \hat{V}(r)$, is *invariant under spatial rotations* since, as shown in Chapter 6, it commutes with the orbital angular momentum:

$$[\hat{H},\ \hat{\vec{L}}] = 0 \quad\implies\quad \left[\hat{H},\ e^{-i\phi\vec{n}\cdot\hat{\vec{L}}/\hbar}\right] = 0. \tag{7.33}$$

Due to this symmetry of space isotropy or rotational invariance, the *orbital angular momentum is conserved*[1]. So, in the case of particles moving in central potentials, the orbital angular momentum is a constant of the motion.

---

[1] In classical physics when a system is invariant under rotations, its total angular momentum is also conserved.

7.2.3 Properties of the Rotation Operator

The rotation operators constitute a representation of the rotation group and satisfy the following properties:

- The product of any two rotation operators is another rotation operator:

$$\hat{R}_{n_1} \hat{R}_{n_2} = \hat{R}_{n_3}. \tag{7.34}$$

- The associative law holds for rotation operators:

$$\left(\hat{R}_{n_1} \hat{R}_{n_2} \right) \hat{R}_{n_3} = \hat{R}_{n_1} \left(\hat{R}_{n_2} \hat{R}_{n_3} \right). \tag{7.35}$$

- The identity operator (corresponding to no rotation) satisfies the relation

$$\hat{I} \hat{R}_n = \hat{R}_n \hat{I} = \hat{R}_n. \tag{7.36}$$

From (7.31) we see that for each rotation operator $\hat{R}_n$, there exists an inverse operator $\hat{R}_n^{-1}$ so that

$$\hat{R}_n \hat{R}_n^{-1} = \hat{R}_n^{-1} \hat{R}_n = \hat{I}. \tag{7.37}$$

The operator $\hat{R}_{-n}$, which is equal to $\hat{R}_n^{-1}$, corresponds to a rotation in the opposite sense to $\hat{R}_n$.

In sharp contrast to the translation group[2] in three dimensions, the rotation group is not commutative (nonabelian). *The product of two rotation operators depends on the order in which they are performed*:

$$\hat{R}_{n_1}(\phi) \hat{R}_{n_2}(\theta) \neq \hat{R}_{n_2}(\theta) \hat{R}_{n_1}(\phi); \tag{7.38}$$

this is due to the fact that the commutator $[\vec{n}_1 \cdot \hat{\vec{J}}, \vec{n}_2 \cdot \hat{\vec{J}}]$ is not zero. In this way, the rotation group is in general nonabelian.

But if the two rotations were performed about the *same* axis, the corresponding operators would *commute*:

$$\hat{R}_n(\phi) \hat{R}_n(\theta) = \hat{R}_n(\theta) \hat{R}_n(\phi) = \hat{R}_n(\phi + \theta). \tag{7.39}$$

Note that, since the angular momentum operator $\hat{J}$ is Hermitian, equation (7.31) yields

$$\boxed{\hat{R}_n^{\dagger}(\phi) = \hat{R}_n^{-1}(\phi) = \hat{R}_n(-\phi) = e^{i\phi\hat{n}\cdot\hat{\vec{J}}/\hbar};} \tag{7.40}$$

hence the rotation operator (7.31) is unitary:

$$\boxed{\hat{R}_n^{\dagger}(\phi) = \hat{R}_n^{-1}(\phi) \qquad \Longrightarrow \qquad \hat{R}_n^{\dagger}(\phi) \hat{R}_n(\phi) = \hat{I}.} \tag{7.41}$$

The operator $\hat{R}_n(\phi)$ therefore conserves the scalar product of kets, notably the norm of vectors. For instance, using

$$| \psi' \rangle = \hat{R}_n(\phi) | \psi \rangle, \qquad | \chi' \rangle = \hat{R}_n(\phi) | \chi \rangle, \tag{7.42}$$

along with (7.41), we can show that $\langle \chi' | \psi' \rangle = \langle \chi | \psi \rangle$, since

$$\langle \chi' | \psi' \rangle = \langle \chi | \hat{R}_n^{\dagger}(\phi) \hat{R}_n(\phi) | \psi \rangle = \langle \chi | \psi \rangle. \tag{7.43}$$

---

[2]The linear momenta $\hat{P}_i$ and $\hat{P}_j$—which are the generators of translation—commute even when $i \neq j$; hence the translation group is said to be abelian.

7.2.4 Euler Rotations

It is known from classical mechanics that an arbitrary rotation of a rigid body can be expressed in terms of three consecutive rotations, called the Euler rotations. In quantum mechanics, instead of expressing the rotation operator $\hat{R}_n(\phi) = e^{-i\phi \hat{n} \cdot \hat{J}/\hbar}$ in terms of a rotation through an angle ϕ about an arbitrary axis $\hat{n}$, it is more convenient to parameterize it, as in classical mechanics, in terms of the three *Euler angles* (α, β, γ) where $0 \leq \alpha \leq 2\pi$, $0 \leq \beta \leq \pi$, and $0 \leq \gamma \leq 2\pi$. The Euler rotations transform the space-fixed set of axes xyz into a new set $x'y'z'$, having the same origin O, by means of three consecutive counterclockwise rotations:

- First, rotate the space-fixed $Oxyz$ system through an angle α about the z-axis; this rotation transforms the $Oxyz$ system into $Ouvz$: $Oxyz \longrightarrow Ouvz$.

- Second, rotate the uvz system through an angle β about the v-axis; this rotation transforms the $Ouvz$ system into $Owvz'$: $Ouvz \longrightarrow Owvz'$.

- Third, rotate the wvz' system through an angle γ about the z'-axis; this rotation transforms the $Owvz'$ system into $Ox'y'z'$: $Owvz' \longrightarrow Ox'y'z'$.

The operators representing these three rotations are given by $\hat{R}_z(\alpha)$, $\hat{R}_v(\beta)$, and $\hat{R}_{z'}(\gamma)$, respectively. Using (7.31) we can represent these three rotations by

$$\hat{R}(\alpha, \beta, \gamma) = \hat{R}_{z'}(\gamma)\hat{R}_v(\beta)\hat{R}_z(\alpha) = \exp\left[-i\gamma J_{z'}/\hbar\right]\exp\left[-i\beta J_v/\hbar\right]\exp\left[-i\alpha J_z/\hbar\right]. \quad (7.44)$$

The form of this operator is rather inconvenient, for it includes rotations about axes belonging to different systems (i.e., z', v, and z); this form would be most convenient were we to express (7.44) as a product of three rotations about the space-fixed axes x, y, z. So let us express $\hat{R}_{z'}(\gamma)$ and $\hat{R}_v(\beta)$ in terms of rotations about the x, y, z axes. Since the first Euler rotation described above, $\hat{R}_z(\alpha)$, transforms the operator $\hat{J}_y$ into $\hat{J}_v$, i.e., $\hat{J}_v = \hat{R}_z(\alpha)\hat{J}_y\hat{R}_z(-\alpha)$ by (7.15), we have

$$\hat{R}_v(\beta) = \hat{R}_z(\alpha)\hat{R}_y(\beta)\hat{R}_z(-\alpha) = e^{-i\alpha J_z/\hbar}e^{-i\beta J_y/\hbar}e^{i\alpha J_z/\hbar}. \quad (7.45)$$

Here $\hat{J}_{z'}$ is obtained from $\hat{J}_z$ by the consecutive application of the second and third Euler rotations, $\hat{J}_{z'} = \hat{R}_v(\beta)\hat{R}_z(\alpha)\hat{J}_z\hat{R}_z(-\alpha)\hat{R}_v(-\beta)$; hence

$$\hat{R}_{z'}(\gamma) = \hat{R}_v(\beta)\hat{R}_z(\alpha)\hat{R}_z(\gamma)\hat{R}_z(-\alpha)\hat{R}_v(-\beta). \quad (7.46)$$

Since $\hat{R}_v(-\beta) = \hat{R}_z(\alpha)\hat{R}_y(-\beta)\hat{R}_z(-\alpha)$, substituting (7.45) into (7.46) we obtain

$$\begin{aligned}
\hat{R}_{z'}(\gamma) &= \left[\hat{R}_z(\alpha)\hat{R}_y(\beta)\hat{R}_z(-\alpha)\right]\hat{R}_z(\alpha)\hat{R}_z(\gamma)\hat{R}_z(-\alpha)\left[\hat{R}_z(\alpha)\hat{R}_y(-\beta)\hat{R}_z(-\alpha)\right] \\
&= \hat{R}_z(\alpha)\hat{R}_y(\beta)\hat{R}_z(\gamma)\hat{R}_y(-\beta)\hat{R}_z(-\alpha) \\
&= e^{-i\alpha J_z/\hbar}e^{-i\beta J_y/\hbar}e^{-i\gamma J_z/\hbar}e^{i\beta J_y/\hbar}e^{i\alpha J_z/\hbar}, \quad (7.47)
\end{aligned}$$

where we used the fact that $\hat{R}_z(-\alpha)\hat{R}_z(\alpha) = e^{-i\alpha J_z/\hbar}e^{i\alpha J_z/\hbar} = 1$.

Finally, inserting (7.45) and (7.47) into (7.44) and simplifying (i.e., using $\hat{R}_z(-\alpha)\hat{R}_z(\alpha) = 1$ and $\hat{R}_y(-\beta)\hat{R}_y(\beta) = 1$), we end up with a product of three rotations about the space-fixed axes y and z:

$$\boxed{\hat{R}(\alpha, \beta, \gamma) = \hat{R}_z(\alpha)\hat{R}_y(\beta)\hat{R}_z(\gamma) = e^{-i\alpha J_z/\hbar}e^{-i\beta J_y/\hbar}e^{-i\gamma J_z/\hbar}.} \quad (7.48)$$

The inverse transformation of (7.48) is obtained by taking three rotations in reverse order over the angles $(-\gamma, -\beta, -\alpha)$:

$$\hat{R}^{-1}(\alpha, \beta, \gamma) = \hat{R}_z(-\gamma)\hat{R}_y(-\beta)\hat{R}_z(-\alpha) = \hat{R}^\dagger(\alpha, \beta, \gamma) = e^{i\gamma J_z/\hbar}e^{i\beta J_y/\hbar}e^{i\alpha J_z/\hbar}. \qquad (7.49)$$

7.2.5   Representation of the Rotation Operator

The rotation operator $\hat{R}(\alpha, \beta, \gamma)$ as given by (7.48) implies that its properties are determined by the algebraic properties of the angular momentum operators $\hat{J}_x, \hat{J}_y, \hat{J}_z$. Since $\hat{R}(\alpha, \beta, \gamma)$ commutes with $\hat{J}^2$, we may look for a representation of $\hat{R}(\alpha, \beta, \gamma)$ in the basis spanned by the eigenvectors of $\hat{J}^2$ and J_z, i.e., the $| j, m \rangle$ states.

From (7.48), we see that $\hat{J}^2$ commutes with the rotation operator, $[\hat{J}^2, \hat{R}(\alpha, \beta, \gamma)] = 0$; thus, the total angular momentum is conserved under rotations

$$\hat{J}^2 \hat{R}(\alpha, \beta, \gamma) | j, m \rangle = \hat{R}(\alpha, \beta, \gamma)\hat{J}^2 | j, m \rangle = j(j+1)\hat{R}(\alpha, \beta, \gamma) | j, m \rangle. \qquad (7.50)$$

However, the z-component of the angular momentum changes under rotations, unless the axis of rotation is along the z-axis. That is, when $\hat{R}(\alpha, \beta, \gamma)$ acts on the state $| j, m \rangle$, we end up with a new state having the same j but with a different value of m:

$$\hat{R}(\alpha, \beta, \gamma) | j, m \rangle = \sum_{m'=-j}^{j} | j, m' \rangle\langle j, m' | \hat{R}(\alpha, \beta, \gamma) | j, m \rangle$$

$$= \sum_{m'=-j}^{j} D^{(j)}_{m'm}(\alpha, \beta, \gamma) | j, m' \rangle, \qquad (7.51)$$

where

$$D^{(j)}_{m'm}(\alpha, \beta, \gamma) = \langle j, m' | \hat{R}(\alpha, \beta, \gamma) | j, m \rangle. \qquad (7.52)$$

These are the matrix elements of $\hat{R}(\alpha, \beta, \gamma)$ for the $| j, m \rangle$ states; $D^{(j)}_{m'm}(\alpha, \beta, \gamma)$ is the amplitude of $| j, m' \rangle$ when $| j, m \rangle$ is rotated. The rotation operator is thus represented by a $(2j+1) \times (2j+1)$ square matrix in the $\{| j, m \rangle\}$ basis. The matrix of $D^{(j)}(\alpha, \beta, \gamma)$ is known as the *Wigner D-matrix* and its elements $D^{(j)}_{m'm}(\alpha, \beta, \gamma)$ as the *Wigner functions*. This matrix representation is often referred to as the $(2j+1)$-dimensional *irreducible representation* of the rotation operator $\hat{R}(\alpha, \beta, \gamma)$.

Since $| j, m \rangle$ is an eigenstate of J_z, it is also an eigenstate of the rotation operator $e^{i\alpha J_z/\hbar}$, because

$$e^{i\alpha J_z/\hbar} | j, m \rangle = e^{i\alpha m} | j, m \rangle. \qquad (7.53)$$

We may thus rewrite (7.52) as

$$D^{(j)}_{m'm}(\alpha, \beta, \gamma) = e^{-i(m'\alpha+m\gamma)}d^{(j)}_{m'm}(\beta), \qquad (7.54)$$

where

$$d^{(j)}_{m'm}(\beta) = \langle j, m' | e^{-i\beta \hat{J}_y/\hbar} | j, m \rangle. \qquad (7.55)$$

This shows that only the middle rotation operator, $e^{-i\beta \hat{J}_y/\hbar}$, mixes states with different values of m. Determining the matrix elements $D_{m'm}^{(j)}(\alpha, \beta, \gamma)$ therefore reduces to evaluation of the quantities $d_{m'm}^{(j)}(\beta)$.

A general expression of $d_{m'm}^{(j)}(\beta)$, called the *Wigner formula*, is given by the following explicit expression:

$$d_{m'm}^{(j)}(\beta) = \sum_k (-1)^{k+m'-m} \frac{\sqrt{(j+m)!(j-m)!(j+m')!(j-m')!}}{(j-m'-k)!(j+m-k)!(k+m'-m)!k!}$$

$$\times \left(\cos\frac{\beta}{2}\right)^{2j+m-m'-2k} \left(\sin\frac{\beta}{2}\right)^{m'-m+2k}. \tag{7.56}$$

The summation over k is taken such that none of the arguments of factorials in the denominator are negative.

We should note that, since the D-function $D_{m'm}^{(j)}(\alpha, \beta, \gamma)$ is a joint eigenfunction of $\vec{J}^2$ and J_z, we have

$$\hat{J}^2 D_{m'm}^{(j)}(\alpha, \beta, \gamma) = j(j+1)\hbar^2 D_{m'm}^{(j)}(\alpha, \beta, \gamma), \tag{7.57}$$

$$\hat{J}_z D_{m'm}^{(j)}(\alpha, \beta, \gamma) = \hbar m D_{m'm}^{(j)}(\alpha, \beta, \gamma), \tag{7.58}$$

$$\hat{J}_\pm D_{m'm}^{(j)}(\alpha, \beta, \gamma) = \hbar \sqrt{(j \pm m)(j \mp m + 1)} D_{m'm\pm1}^{(j)}(\alpha, \beta, \gamma). \tag{7.59}$$

Properties of the D-functions

We now list some of the most useful properties of the rotation matrices. The complex conjugate of the D-functions can be expressed as

$$\left[D_{m'm}^{(j)}(\alpha, \beta, \gamma) \right]^* = \langle j, m' \mid \hat{R}(\alpha, \beta, \gamma) \mid j, m \rangle^* = \langle j \, m \mid \hat{R}^\dagger(\alpha, \beta, \gamma) \mid j, m' \rangle$$

$$= \langle j \, m \mid \hat{R}^{-1}(\alpha, \beta, \gamma) \mid j, m' \rangle$$

$$= D_{mm'}^{(j)}(-\gamma, -\beta, -\alpha). \tag{7.60}$$

We can easily show that

$$\left[D_{m'm}^{(j)}(\alpha, \beta, \gamma) \right]^* = (-1)^{m'-m} D_{-m'-m}^{(j)}(\alpha, \beta, \gamma) = D_{mm'}^{(j)}(-\gamma, -\beta, -\alpha). \tag{7.61}$$

The D-functions satisfy the following unitary relations:

$$\sum_m \left[D_{km}^{(j)}(\alpha, \beta, \gamma) \right]^* D_{k'm}^{(j)}(\alpha, \beta, \gamma) = \delta_{k, k'}, \tag{7.62}$$

$$\sum_m \left[D_{mk}^{(j)}(\alpha, \beta, \gamma) \right]^* D_{mk'}^{(j)}(\alpha, \beta, \gamma) = \delta_{k, k'}, \tag{7.63}$$

since

$$\sum_m \left[D^{(j)}_{mk}(\alpha, \beta, \gamma) \right]^* D^{(j)}_{mk'}(\alpha, \beta, \gamma) = \sum_m \langle j\ k | \hat{R}^{-1}(\alpha, \beta, \gamma)\ |\ j,\ m \rangle \langle j\ m\ |\ R(\alpha, \beta, \gamma) | j\ k' \rangle$$

$$= \langle j\ k | \hat{R}^{-1}(\alpha, \beta, \gamma)\ \hat{R}(\alpha, \beta, \gamma))(|j\ k' \rangle$$

$$= \langle j\ k | j\ k' \rangle$$

$$= \delta_{k, k'}. \tag{7.64}$$

From (7.55) we can show that the d-functions satisfy the following relations:

$$d^{(j)}_{m'm}(\pi) = (-1)^{j-m} \delta_{m', -m}, \qquad d^{(j)}_{m'm}(0) = \delta_{m', m}. \tag{7.65}$$

Since $d^j_{m'm}$ are elements of a unitary real matrix, the matrix $d^{(j)}(\beta)$ must be orthogonal. We may thus write

$$\boxed{d^{(j)}_{m'm}(\beta) = \left(d^{(j)}_{m'm}(\beta) \right)^{-1} = d^{(j)}_{mm'}(-\beta)} \tag{7.66}$$

and

$$\boxed{d^{(j)}_{m'm}(\beta) = (-1)^{m'-m} d^{(j)}_{mm'}(\beta) = (-1)^{m'-m} d^{(j)}_{-m'-m}(\beta).} \tag{7.67}$$

The unitary matrices $D^{(j)}$ form a $(2j + 1)$ dimensional irreducible representation of the $SO(3)$ group.

7.2.6 Rotation Matrices and the Spherical Harmonics

In the case where the angular momentum operator $\hat{\vec{J}}$ is purely orbital (i.e., the values of j are integer, $j = l$), there exists a connection between the D-functions and the spherical harmonics $Y_{lm}(\theta, \varphi)$. The operator $\hat{R}(\alpha, \beta, \gamma)$ when applied to a vector $|\vec{r}\rangle$ pointing in the direction (θ, φ) would generate a vector $|\vec{r}'\rangle$ along a new direction (θ', φ'):

$$|\vec{r}'\rangle = \hat{R}(\alpha, \beta, \gamma)\ |\vec{r}\rangle. \tag{7.68}$$

An expansion in terms of $|l,\ m'\rangle$ and a multiplication by $\langle l,\ m\ |$ leads to

$$\langle l,\ m\ |\ \vec{r}'\rangle = \sum_{m'} \langle l,\ m | \hat{R}(\alpha, \beta, \gamma)\ |\ l,\ m'\rangle \langle l,\ m'\ |\ \vec{r}\rangle, \tag{7.69}$$

or to

$$Y^*_{lm}(\theta', \varphi') = \sum_{m'} D^{(l)}_{m\ m'}(\alpha, \beta, \gamma) Y^*_{lm'}(\theta, \varphi), \tag{7.70}$$

since $\langle l,\ m\ |\ \vec{r}'\rangle = Y^*_{lm}(\theta', \varphi')$ and $\langle l,\ m'\ |\ \vec{r}\rangle = Y^*_{lm'}(\theta, \varphi)$.

In the case where the vector $\vec{r}$ is along the z-axis, we have $\theta = 0$; hence $m' = 0$. From Chapter 5, $Y^*_{l0}(0, \varphi)$ is given by

$$Y^*_{lm'}(0, \varphi) = \sqrt{\frac{2l + 1}{4\pi}} \delta_{m', 0}. \tag{7.71}$$

We can thus reduce (7.70) to

$$Y^*_{lm}(\beta, \alpha) = D^{(l)}_{m\ 0}(\alpha, \beta, \gamma) Y^*_{l0}(0, \varphi) = \sqrt{\frac{2l + 1}{4\pi}} D^{(l)}_{m\ 0}(\alpha, \beta, \gamma), \tag{7.72}$$

or to

$$D_{m0}^{(l)}(\alpha, \beta, \gamma) = \sqrt{\frac{4\pi}{2l+1}} Y_{lm}^*(\beta, \alpha).$$ (7.73)

This means that a rotation through the Euler angles (α, β, γ) of the vector $\vec{r}$, when it is along the z-axis, produces a vector $\vec{r}\,'$ whose azimuthal and polar angles are given by β and α, respectively. Similarly, we can show that

$$D_{0m}^{(l)}(\gamma, \beta, \alpha) = \sqrt{\frac{4\pi}{2l+1}} Y_{lm}(\beta, \alpha)$$ (7.74)

and

$$D_{00}^{(l)}(0, \theta, 0) = P_l(\cos \theta),$$ (7.75)

where $P_l(\cos \theta)$ is the Legendre polynomial.

We are now well equipped to derive the theorem for the addition of spherical harmonics. Let (θ, φ) be the polar coordinates of the vector $\vec{r}$ with respect to the space-fixed x, y, z system and let (θ', φ') be its polar coordinates with respect to the rotated system x', y', z'; taking the complex conjugate of (7.70) we obtain

$$Y_{lm}(\theta', \varphi') = \sum_{m'} \left[D_{m\,m'}^{(l)}(\alpha, \beta, \gamma) \right]^* Y_{lm'}(\theta, \varphi).$$ (7.76)

For the case $m = 0$, since (from Chapter 5)

$$Y_{l0}(\theta', \varphi') = \sqrt{\frac{2l+1}{4\pi}} P_l(\cos \theta')$$ (7.77)

and since from (7.74)

$$\left[D_{0m'}^{(l)}(\alpha, \beta, \gamma) \right]^* = \sqrt{\frac{4\pi}{2l+1}} Y_{lm'}^*(\beta, \gamma),$$ (7.78)

we can reduce (7.76) to

$$\sqrt{\frac{2l+1}{4\pi}} P_l(\cos \theta') = \sum_{m'} \sqrt{\frac{4\pi}{2l+1}} Y_{lm'}^*(\beta, \gamma) Y_{lm'}(\theta, \varphi),$$ (7.79)

or to

$$P_l(\cos \theta') = \frac{4\pi}{2l+1} \sum_{m'} Y_{lm'}^*(\beta, \gamma) Y_{lm'}(\theta, \varphi).$$ (7.80)

Integrals involving the D-functions

Let ω denote the Euler angles; hence

$$\int d\omega = \int_0^\pi \sin \beta \, d\beta \int_0^{2\pi} d\alpha \int_0^{2\pi} d\gamma.$$ (7.81)

Using the relation

$$\int D_{m'm}^{(j)}(\omega) \, d\omega = \int_0^\pi d_{m'm}^{(j)}(\beta) \sin \beta d\beta \int_0^{2\pi} e^{-im'\alpha} d\alpha \int_0^{2\pi} e^{-im\gamma} d\gamma$$

$$= 8\pi^2 \delta_{j,0} \delta_{m',0} \delta_{m,0},$$ (7.82)

we may write

$$\int D_{mk}^{(j)*}(\omega) D_{m'k'}^{(j')}(\omega) \, d\omega = (-1)^{m-k} \int D_{-m-k}^{(j)}(\omega) D_{m'k'}^{(j')}(\omega) \, d\omega$$

$$= (-1)^{m-k} \int_0^\pi d_{-m-k}^{(j)}(\beta) d_{m'k'}^{(j')}(\beta) \sin \beta d\beta$$

$$\times \int_0^{2\pi} e^{-i(m'-m)\alpha} \, d\alpha \int_0^{2\pi} e^{-i(k'-k)\gamma} \, d\gamma$$

$$= \frac{8\pi^2}{2j+1} \delta_{j,\,j'} \delta_{m,\,m'} \delta_{k,\,k'}. \tag{7.83}$$

Example 7.1
Find the rotation matrices $d^{(1/2)}$ and $D^{(1/2)}$ corresponding to $j = \frac{1}{2}$.

Solution
On the one hand, since the matrix of $\hat{J}_y$ for $j = \frac{1}{2}$ (Chapter 5) is given by

$$\hat{J}_y = \frac{\hbar}{2} \begin{pmatrix} 0 & -i \\ i & 0 \end{pmatrix} = \frac{\hbar}{2} \sigma_y, \tag{7.84}$$

and since the square of the Pauli matrix σ_y is equal to the unit matrix, $\sigma_y^2 = 1$, the even and odd powers of σ_y are given by

$$\sigma_y^{2n} = \begin{pmatrix} 1 & 0 \\ 0 & 1 \end{pmatrix}, \qquad \sigma_y^{2n+1} = \begin{pmatrix} 0 & -i \\ i & 0 \end{pmatrix} = \sigma_y. \tag{7.85}$$

On the other hand, since the rotation operator

$$\hat{R}_y(\beta) = e^{-i\beta \hat{J}_y/\hbar} = e^{-i\beta\sigma_y/2} \tag{7.86}$$

can be written as

$$e^{-i\beta\sigma_y/2} = \sum_{n=0}^\infty \frac{(-i)^{2n}}{(2n)!} \left(\frac{\beta}{2}\right)^{2n} \sigma^{2n} + \sum_{n=0}^\infty \frac{(-i)^{2n+1}}{(2n+1)!} \left(\frac{\beta}{2}\right)^{2n+1} \sigma_y^{2n+1}, \tag{7.87}$$

a substitution of (7.85) into (7.87) yields

$$e^{-i\beta\sigma_y/2} = \begin{pmatrix} 1 & 0 \\ 0 & 1 \end{pmatrix} \sum_{n=0}^\infty \frac{(-1)^n}{(2n)!} \left(\frac{\beta}{2}\right)^{2n} - i\sigma_y \sum_{n=0}^\infty \frac{(-1)^n}{(2n+1)!} \left(\frac{\beta}{2}\right)^{2n+1}$$

$$= \begin{pmatrix} 1 & 0 \\ 0 & 1 \end{pmatrix} \cos\left(\frac{\beta}{2}\right) + \begin{pmatrix} 0 & -1 \\ 1 & 0 \end{pmatrix} \sin\left(\frac{\beta}{2}\right); \tag{7.88}$$

hence

$$d^{(1/2)}(\beta) = e^{-i\beta \hat{J}_y/\hbar} = \begin{pmatrix} d_{\frac{1}{2}\frac{1}{2}}^{(1/2)} & d_{\frac{1}{2}-\frac{1}{2}}^{(1/2)} \\ d_{-\frac{1}{2}\frac{1}{2}}^{(1/2)} & d_{-\frac{1}{2}-\frac{1}{2}}^{(1/2)} \end{pmatrix} = \begin{pmatrix} \cos(\beta/2) & -\sin(\beta/2) \\ \sin(\beta/2) & \cos(\beta/2) \end{pmatrix}. \tag{7.89}$$

Since as shown in (7.54) $D^{(j)}_{m'm}(\alpha, \beta, \gamma) = e^{-i(m'\alpha + m\gamma)} d^{(j)}_{m'm}(\beta)$, we have

$$D^{(1/2)}(\alpha, \beta, \gamma) = \begin{pmatrix} e^{-i(\alpha+\gamma)/2} \cos(\beta/2) & e^{-i(\alpha-\gamma)/2} \sin(\beta/2) \\ e^{i(\alpha-\gamma)/2} \sin(\beta/2) & e^{i(\alpha+\gamma)/2} \cos(\beta/2) \end{pmatrix}. \qquad (7.90)$$

7.3 Addition of Angular Momenta

The addition of angular momenta is encountered in all areas of modern physics. Mastering its techniques is essential for an understanding of the various subatomic phenomena. For instance, the total angular momentum of the electron in a hydrogen atom consists of two parts, an orbital part $\hat{\vec{L}}$, which is due to the orbiting motion of the electron around the proton, and a spin part $\hat{\vec{S}}$, which is due to the spinning motion of the electron about itself. The properties of the hydrogen atom cannot be properly discussed without knowing how to add the orbital and spin parts of the electron's total angular momentum.

In what follows we are going to present the formalism of angular momentum addition and then consider some of its most essential applications.

7.3.1 Addition of Two Angular Momenta: General Formalism

In this section we present the general formalism corresponding to the problem of adding two *commuting* angular momenta.

Consider two angular momenta $\hat{\vec{J}}_1$ and $\hat{\vec{J}}_2$ which belong to different subspaces 1 and 2; $\hat{\vec{J}}_1$ and $\hat{\vec{J}}_2$ may refer to two distinct particles or to two different properties of the same particle[3]. The latter case may refer to the orbital and spin angular momenta of the same particle. Assuming that the spin–orbit coupling is sufficiently weak, then the space and spin degrees of freedom of the electron evolve *independently* of each other.

The components of $\hat{\vec{J}}_1$ and $\hat{\vec{J}}_2$ satisfy the usual commutation relations of angular momentum:

$$\left[\hat{J}_{1x}, \hat{J}_{1y}\right] = i\hbar\hat{J}_{1z}, \qquad \left[\hat{J}_{1y}, \hat{J}_{1z}\right] = i\hbar\hat{J}_{1x}, \qquad \left[\hat{J}_{1z}, \hat{J}_{1x}\right] = i\hbar\hat{J}_{1y}, \qquad (7.91)$$

$$\left[\hat{J}_{2x}, \hat{J}_{2y}\right] = i\hbar\hat{J}_{2z}, \qquad \left[\hat{J}_{2y}, \hat{J}_{2z}\right] = i\hbar\hat{J}_{2x}, \qquad \left[\hat{J}_{2z}, \hat{J}_{2x}\right] = i\hbar\hat{J}_{2y}. \qquad (7.92)$$

Since $\hat{\vec{J}}_1$, and $\hat{\vec{J}}_2$ belong to different spaces, their components commute:

$$\left[\hat{J}_{1j}, \hat{J}_{2k}\right] = 0, \qquad (j, k = x, y, z). \qquad (7.93)$$

---

[3] Throughout this section we shall use the labels 1 and 2 to refer to quantities relevant to the two particles or the two subspaces.

Now, denoting the joint eigenstates of $\hat{\vec{J}}_1^2$ and $\hat{J}_{1_z}$ by $|\, j_1,\ m_1\rangle$ and those of $\hat{\vec{J}}_2^2$ and $\hat{J}_{2_z}$ by $|\, j_2,\ m_2\rangle$, we have

$$\hat{\vec{J}}_1^2 \,|\, j_1;\ m_1\rangle = j_1(j_1+1)\hbar^2 \,|\, j_1,\ m_1\rangle, \tag{7.94}$$

$$\hat{J}_{1_z} \,|\, j_1,\ m_1\rangle = m_1\hbar \,|\, j_1,\ m_1\rangle, \tag{7.95}$$

$$\hat{\vec{J}}_2^2 \,|\, j_2,\ m_2\rangle = j_2(j_2+1)\hbar^2 \,|\, j_2,\ m_2\rangle, \tag{7.96}$$

$$\hat{J}_{2_z} \,|\, j_2,\ m_2\rangle = m_2\hbar \,|\, j_2,\ m_2\rangle. \tag{7.97}$$

The dimensions of the spaces to which $\hat{\vec{J}}_1$ and $\hat{\vec{J}}_2$ belong are given by $(2j_1+1)$ and $(2j_2+1)$, respectively[4]. The operators $\hat{\vec{J}}_1^2$ and $\hat{J}_{1_z}$ are represented within the $\{|\, j_1,\ m_1\rangle\}$ basis by square matrices of dimension $(2j_1+1) \times (2j_1+1)$, while $\hat{\vec{J}}_2^2$ and $\hat{J}_{2_z}$ are representation by square matrices of dimension $(2j_2+1) \times (2j_2+1)$ within the $\{|\, j_2,\ m_2\rangle\}$ basis.

Consider now the two particles (or two subspaces) 1 and 2 together. The four operators $\hat{\vec{J}}_1^2$, $\hat{\vec{J}}_2^2$, $\hat{J}_{1_z}$, $\hat{J}_{2_z}$ form a complete set of commuting operators; they can thus be jointly diagonalized by the same states. Denoting their joint eigenstates by $|\, j_1, j_2;\ m_1, m_2\rangle$, we can write them as *direct products* of $|\, j_1,\ m_1\rangle$, and $|\, j_2,\ m_2\rangle$

$$|\, j_1, j_2;\ m_1, m_2\rangle = |\, j_1,\ m_1\rangle \,|\, j_2,\ m_2\rangle, \tag{7.98}$$

because the coordinates of $\hat{\vec{J}}_1$ and $\hat{\vec{J}}_2$ are *independent*. We can thus rewrite (7.94)–(7.97) as

- $\hat{\vec{J}}_1^2 \,|\, j_1, j_2;\ m_1, m_2\rangle = j_1(j_1+1)\hbar^2 \,|\, j_1, j_2;\ m_1, m_2\rangle,$ (7.99)

$$\hat{J}_{1_z} \,|\, j_1, j_2;\ m_1, m_2\rangle = m_1\hbar \,|\, j_1, j_2;\ m_1, m_2\rangle, \tag{7.100}$$

$$\hat{\vec{J}}_2^2 \,|\, j_1, j_2;\ m_1, m_2\rangle = j_2(j_2+1)\hbar^2 \,|\, j_1, j_2;\ m_1, m_2\rangle, \tag{7.101}$$

$$\hat{J}_{2_z} \,|\, j_1, j_2;\ m_1, m_2\rangle = m_2\hbar \,|\, j_1, j_2;\ m_1, m_2\rangle. \tag{7.102}$$

The kets $|\, j_1, j_2;\ m_1, m_2\rangle$ form a complete and orthonormal basis. Using

$$\sum_{m_1 m_2} |\, j_1, j_2;\ m_1, m_2\rangle\langle j_1, j_2;\ m_1, m_2\,| = \left(\sum_{m_1} |\, j_1,\ m_1\rangle\langle j_1,\ m_1\,|\right)\left(\sum_{m_2} |\, j_2,\ m_2\rangle\langle j_2,\ m_2\,|\right), \tag{7.103}$$

and since $\{|\, j_1,\ m_1\rangle\}$ and $\{|\, j_2,\ m_2\rangle\}$ are complete (i.e., $\sum_{m_1} |\, j_1,\ m_1\rangle\langle j_1,\ m_1\,| = 1$) and orthonormal (i.e., $\langle j_1',\ m_1' \,|\, j_1,\ m_1\rangle = \delta_{j_1', j_1}\delta_{m_1', m_1}$ and similarly for $|\, j_2,\ m_2\rangle$), we see that the basis $\{|\, j_1, j_2;\ m_1, m_2\rangle\}$ is complete,

$$\sum_{m_1=-j_1}^{j_1} \sum_{m_2=-j_2}^{j_2} |\, j_1, j_2;\ m_1, m_2\rangle\langle j_1, j_2;\ m_1, m_2\,| = 1, \tag{7.104}$$

and orthonormal,

$$\langle j_1', j_2';\ m_1', m_2' \,|\, j_1, j_2;\ m_1, m_2\rangle = \langle j_1',\ m_1' \,|\, j_1,\ m_1\rangle\langle j_2',\ m_2' \,|\, j_2,\ m_2\rangle$$
$$= \delta_{j_1', j_1}\delta_{j_2', j_2}\delta_{m_1', m_1}\delta_{m_2', m_2}. \tag{7.105}$$

---

[4]This is due to the fact that the number of basis vectors spanning the spaces to which $\hat{\vec{J}}_1$ and $\hat{\vec{J}}_2$ belong are equal to $(2j_1+1)$ and $(2j_2+1)$, respectively; these vectors are $|j_1, -j_1\rangle, |j_1, -j_1+1\rangle, \ldots, |j_1, j_1-1\rangle, |j_1, j_1\rangle$ and $|j_2, -j_2\rangle, |j_2, -j_2+1\rangle, \ldots, |j_2, j_2-1\rangle, |j_2, j_2\rangle$.

The basis $\{| j_1, j_2; m_1, m_2\rangle\}$ clearly spans the total space which is made of subspaces 1 and 2. From (7.98) we see that the dimension N of this space is equal to the product of the dimensions of the two subspaces spanned by $\{| j_1, m_1\rangle\}$ and $\{| j_2, m_2\rangle\}$:

$$N = (2j_1 + 1) \times (2j_2 + 1). \tag{7.106}$$

We can now introduce the *step* operators $\hat{J}_{1\pm} = \hat{J}_{1x} \pm i\hat{J}_{1y}$ and $\hat{J}_{2\pm} = \hat{J}_{2x} \pm i\hat{J}_{2y}$; their actions on $|j_1 j_2; m_1 m_2\rangle$ are given by

$$\hat{J}_{1\pm} | j_1, j_2; m_1, m_2\rangle = \hbar\sqrt{(j_1 \mp m_1)(j_1 \pm m_1 + 1)} | j_1, j_2; m_1 \pm 1, m_2\rangle, \tag{7.107}$$

$$\hat{J}_{2\pm} | j_1, j_2; m_1, m_2\rangle = \hbar\sqrt{(j_2 \mp m_2)(j_2 \pm m_2 + 1)} | j_1, j_2; m_1, m_2 \pm 1\rangle. \tag{7.108}$$

The problem of adding two angular momenta, $\hat{\vec{J}}_1$ and $\hat{\vec{J}}_2$,

$$\hat{\vec{J}} = \hat{\vec{J}}_1 + \hat{\vec{J}}_2, \tag{7.109}$$

consists of finding the eigenvalues and eigenvectors of $\hat{\vec{J}}^2$ and $\hat{J}_z$ in terms of the eigenvalues and eigenvectors of $\hat{\vec{J}}_1^2$, $\hat{\vec{J}}_2^2$, $\hat{J}_{1z}$, and $\hat{J}_{2z}$. Since the matrices of $\hat{\vec{J}}_1$ and $\hat{\vec{J}}_2$ have in general different dimensions, the addition specified by (7.109) is not an addition of matrices; it is a symbolic addition.

By adding (7.91) and (7.92), we can easily ascertain that the components of $\hat{\vec{J}}$ satisfy the commutation relations of angular momentum:

$$\left[\hat{J}_x, \hat{J}_y\right] = i\hbar\hat{J}_z, \qquad \left[\hat{J}_y, \hat{J}_z\right] = i\hbar\hat{J}_x, \qquad \left[\hat{J}_z, \hat{J}_x\right] = i\hbar\hat{J}_y. \tag{7.110}$$

Note that $\hat{\vec{J}}_1^2$, $\hat{\vec{J}}_2^2$, $\hat{\vec{J}}^2$, $\hat{J}_z$ jointly commute; this can be ascertained from the relation:

$$\hat{\vec{J}}^2 = \hat{\vec{J}}_1^2 + \hat{\vec{J}}_2^2 + 2\hat{J}_{1z}\hat{J}_{2z} + \hat{J}_{1+}\hat{J}_{2+} + \hat{J}_{1-}\hat{J}_{2-}, \tag{7.111}$$

which leads to

$$\left[\hat{\vec{J}}^2, \hat{\vec{J}}_1^2\right] = \left[\hat{\vec{J}}^2, \hat{\vec{J}}_2^2\right] = 0, \tag{7.112}$$

and to

$$\left[\hat{\vec{J}}^2, \hat{J}_z\right] = \left[\hat{\vec{J}}_1^2, \hat{J}_z\right] = \left[\hat{\vec{J}}_2^2, \hat{J}_z\right] = 0. \tag{7.113}$$

But in spite of the fact that $\left[\hat{\vec{J}}^2, \hat{J}_z\right] = 0$, the operators $\hat{J}_{1z}$ and $\hat{J}_{2z}$ do not commute separately with $\hat{\vec{J}}^2$:

$$\left[\hat{\vec{J}}^2, \hat{J}_{1z}\right] \neq 0, \qquad \left[\hat{\vec{J}}^2, \hat{J}_{2z}\right] \neq 0. \tag{7.114}$$

Now, since $\hat{\vec{J}}_1^2$, $\hat{\vec{J}}_2^2$, $\hat{\vec{J}}^2$, $\hat{J}_z$ form a complete set of commuting operators, they can be diagonalized simultaneously by the same states; designating these joint eigenstates by $|j_1, j_2; j, m\rangle$, we have

$$\hat{\vec{J}}_1^2 | j_1, j_2; j, m\rangle = j_1(j_1 + 1)\hbar^2 | j_1, j_2; j, m\rangle, \tag{7.115}$$

$$\hat{\vec{J}}_2^2 | j_1, j_2; j, m\rangle = j_2(j_2 + 1)\hbar^2 | j_1, j_2; j, m\rangle, \tag{7.116}$$

$$\hat{\vec{J}}^2 | j_1, j_2; j, m\rangle = j(j + 1)\hbar^2 | j_1, j_2; j, m\rangle, \tag{7.117}$$

$$\hat{J}_z | j_1, j_2; j, m\rangle = m\hbar | j_1, j_2; j, m\rangle. \tag{7.118}$$

For every j, the number m has $(2j + 1)$ allowed values: $m = -j, -j + 1, \ldots, j - 1, j$.

Since j_1 and j_2 are usually fixed, *we will be using, throughout the rest of this chapter, the shorthand notation* $| j, m \rangle$ *to abbreviate* $| j_1, j_2; j, m \rangle$. The set of vectors $\{| j, m \rangle\}$ form a complete and orthonormal basis:

$$\sum_j \sum_{m=-j}^{j} | j, m \rangle \langle j, m | = 1, \tag{7.119}$$

$$\langle j', m' | j, m \rangle = \delta_{j, j'} \delta_{m', m}. \tag{7.120}$$

The space where the total angular momentum $\hat{\vec{J}}$ operates is spanned by the basis $\{| j, m \rangle\}$; this space is known as a *product space*. It is important to know that this space is the same as the one spanned by $\{| j_1, j_2; m_1, m_2 \rangle\}$; that is, the space which includes both subspaces 1 and 2. So the dimension of the space which is spanned by the basis $\{| j, m \rangle\}$ is also equal to $N = (2j_1 + 1) \times (2j_2 + 1)$ as specified by (7.106).

The issue now is to find the transformation that connects the bases $\{| j_1, j_2; m_1, m_2 \rangle\}$ and $\{| j, m \rangle\}$.

7.3.1.1 Transformation between Bases: Clebsch–Gordan Coefficients

Let us now return to the addition of $\hat{\vec{J}}_1$ and $\hat{\vec{J}}_2$. This problem consists in essence of obtaining the eigenvalues of $\hat{\vec{J}}^2$ and $\hat{J}_z$ and of expressing the states $| j, m \rangle$ in terms of $| j_1, j_2; m_1, m_2 \rangle$. We should mention that $| j, m \rangle$ is the state in which $\hat{\vec{J}}^2$ and $\hat{\vec{J}}_z$ have fixed values, $j(j + 1)$ and m, but in general not a state in which the values of $\hat{\vec{J}}_{1z}$ and $\hat{\vec{J}}_{2z}$ are fixed; as for $| j_1, j_2; m_1, m_2 \rangle$, it is the state in which $\hat{\vec{J}}_1^2, \hat{\vec{J}}_2^2, \hat{\vec{J}}_{1z}$, and $\hat{\vec{J}}_{2z}$ have fixed values.

The $\{| j_1, j_2; m_1, m_2 \rangle\}$ and $\{| j, m \rangle\}$ bases can be connected by means of a transformation as follows. Inserting the identity operator as a sum over the complete basis $| j_1, j_2; m_1, m_2 \rangle$, we can write

$$| j, m \rangle = \left(\sum_{m_1=-j_1}^{j_1} \sum_{m_2=-j_2}^{j_2} | j_1, j_2; m_1, m_2 \rangle \langle j_1, j_2; m_1, m_2 | \right) | j, m \rangle$$

$$= \sum_{m_1 m_2} \langle j_1, j_2; m_1, m_2 | j, m \rangle | j_1, j_2; m_1, m_2 \rangle, \tag{7.121}$$

where we have used the normalization condition (7.104); since the bases $\{| j_1, j_2; m_1, m_2 \rangle\}$ and $\{| j, m \rangle\}$ are both *normalized*, this transformation must be *unitary*. The coefficients $\langle j_1, j_2; m_1, m_2 | j, m \rangle$, which depend only on the quantities j_1, j_2, j, m_1, m_2, and m, *are the matrix elements of the unitary transformation which connects the* $\{| j, m \rangle\}$ and $\{| j_1, j_2; m_1, m_2 \rangle\}$ bases. These coefficients are called the *Clebsch–Gordan coefficients*.

The problem of angular momentum addition reduces then to finding the Clebsch–Gordan coefficients $\langle j_1, j_2; m_1, m_2 | j, m \rangle$. These coefficients are taken to be *real by convention*; hence

$$\boxed{\langle j_1, j_2; m_1, m_2 | j, m \rangle = \langle j, m | j_1, j_2; m_1, m_2 \rangle.} \tag{7.122}$$

Using (7.104) and (7.120) we can infer the orthonormalization relation for the Clebsch–Gordan coefficients:

$$\sum_{m_1 m_2} \langle j', m' | j_1, j_2; m_1, m_2 \rangle \langle j_1, j_2; m_1, m_2 | j, m \rangle = \delta_{j', j} \delta_{m', m}, \tag{7.123}$$

and since the Clebsch–Gordan coefficients are real, this relation can be rewritten as

$$\sum_{m_1 m_2} \langle j_1, j_2; m_1, m_2 \mid j', m' \rangle \langle j_1, j_2; m_1, m_2 \mid j, m \rangle = \delta_{j', j} \delta_{m', m}, \qquad (7.124)$$

which leads to

$$\sum_{m_1 m_2} \langle j_1, j_2; m_1, m_2 \mid j, m \rangle^2 = 1. \qquad (7.125)$$

Likewise, we have

$$\sum_{j} \sum_{m=-j}^{j} \langle j_1, j_2; m_1', m_2' \mid j, m \rangle \langle j_1, j_2; m_1, m_2 \mid j, m \rangle = \delta_{m_1', m_1} \delta_{m_2', m_2} \qquad (7.126)$$

and, in particular,

$$\sum_{j} \sum_{m} \langle j_1, j_2; m_1, m_2 \mid j, m \rangle^2 = 1. \qquad (7.127)$$

7.3.1.2 Eigenvalues of $\hat{J}^2$ and $\hat{J}_z$

Let us study how to find the eigenvalues of $\hat{J}^2$ and $\hat{J}_z$ in terms of those of $\hat{J}_1^2$, $\hat{J}_2^2$, $\hat{J}_{1_z}$, and $\hat{J}_{2_z}$; that is, obtain j and m in terms of j_1, j_2, m_1 and m_2. First, since $\hat{J}_z = \hat{J}_{1_z} + \hat{J}_{2_z}$, we have $m = m_1 + m_2$. Now, to find j in terms of j_1 and j_2, we proceed as follows. Since the maximum values of m_1 and m_2 are $m_{1max} = j_1$ and $m_{2max} = j_2$, we have $m_{max} = m_{1max} + m_{2max} = j_1 + j_2$; but since $|m| \leq j$, then $j_{max} = j_1 + j_2$.

Next, to find the minimum value j_{min} of j, we need to use the fact that there are a total of $(2j_1 + 1) \times (2j_2 + 1)$ eigenkets $\mid j, m \rangle$. To each value of j there correspond $(2j + 1)$ eigenstates $\mid j, m \rangle$, so we have

$$\sum_{j=j_{min}}^{j_{max}} (2j + 1) = (2j_1 + 1)(2j_2 + 1), \qquad (7.128)$$

which leads to (see Example 7.2, page 408, for the proof)

$$j_{min}^2 = (j_1 - j_2)^2 \qquad \Longrightarrow \qquad j_{min} = |j_1 - j_2|. \qquad (7.129)$$

Hence the allowed values of j are located within the range

$$\boxed{|j_1 - j_2| \leq j \leq j_1 + j_2.} \qquad (7.130)$$

This expression can also be inferred from the well-known triangle relation[5]. So the allowed values of j proceed in integer steps according to

$$\boxed{j = |j_1 - j_2|, \ |j_1 - j_2| + 1, \ \ldots, j_1 + j_2 - 1, \ j_1 + j_2.} \qquad (7.131)$$

---

[5] The length of the sum of two classical vectors, $\vec{A} + \vec{B}$, must be located between the sum and the difference of the lengths of the two vectors, $A + B$ and $|A - B|$, i.e., $|A - B| \leq |\vec{A} + \vec{B}| \leq A + B$.

Thus, for every j the allowed values of m are located within the range $-j \leq m \leq j$.

Note that the coefficient $\langle j_1, j_2; m_1, m_2 \mid j, m \rangle$ vanishes unless $m_1 + m_2 = m$. This can be seen as follows: since $\hat{J}_z = \hat{J}_{1_z} + \hat{J}_{2_z}$, we have

$$\langle j_1, j_2; m_1, m_2 \mid \hat{J}_z - \hat{J}_{1_z} - \hat{J}_{2_z} \mid j, m \rangle = 0, \tag{7.132}$$

and since $\hat{J}_z \mid j, m \rangle = m\hbar \mid j, m \rangle$, $\langle j_1, j_2; m_1, m_2 \mid \hat{J}_{1_z} = m_1\hbar \langle j_1, j_2; m_1, m_2 \mid$, and $\langle j_1, j_2; m_1, m_2 \mid \hat{J}_{2z} = m_2\hbar \langle j_1, j_2; m_1, m_2 \mid$, we can write

$$(m - m_1 - m_2) \langle j_1, j_2; m_1, m_2 \mid j, m \rangle = 0, \tag{7.133}$$

which shows that $\langle j_1, j_2; m_1, m_2 \mid j, m \rangle$ is not zero only when $m - m_1 - m_2 = 0$.

$$\boxed{\text{If} \qquad m_1 + m_2 \neq m \quad \Longrightarrow \quad \langle j_1, j_2; m_1, m_2 \mid j, m \rangle = 0.} \tag{7.134}$$

So, for the Clebsch–Gordan coefficient $\langle j_1, j_2; m_1, m_2 \mid j, m \rangle$ not to be zero, we must simultaneously have

$$\boxed{m_1 + m_2 = m \qquad \text{and} \qquad \mid j_1 - j_2 \mid \leq j \leq j_1 + j_2.} \tag{7.135}$$

These are known as the *selection rules* for the Clebsch–Gordan coefficients.

Example 7.2

Starting from $\sum_{j=j_{min}}^{j_{max}} (2j + 1) = (2j_1 + 1)(2j_2 + 1)$, prove (7.129).

Solution

Let us first work on the left-hand side of

$$\sum_{j=j_{min}}^{j_{max}} (2j + 1) = (2j_1 + 1)(2j_2 + 1). \tag{7.136}$$

Since $j_{max} = j_1 + j_2$ we can write the left-hand side of this equation as an arithmetic sum which has $(j_{max} - j_{min} + 1) = [(j_1 + j_2 + 1) - j_{min}]$ terms:

$$\sum_{j=j_{min}}^{j_{max}} (2j + 1) = (2j_{min} + 1) + (2j_{min} + 3) + (2j_{min} + 5) + \cdots + [2(j_1 + j_2) + 1]. \tag{7.137}$$

To calculate this sum, we simply write it in the following two equivalent ways:

$$S = (2j_{min} + 1) + (2j_{min} + 3) + (2j_{min} + 5) + \cdots + [2(j_1 + j_2) + 1], \tag{7.138}$$
$$S = [2(j_1 + j_2) + 1] + [2(j_1 + j_2) - 1] + [2(j_1 + j_2) - 3] + \cdots + (2j_{min} + 1). \tag{7.139}$$

Adding these two series term by term, we obtain

$$2S = 2[(j_1 + j_2 + 1) + j_{min}] + 2[(j_1 + j_2 + 1) + j_{min}] + \cdots + 2[(j_1 + j_2 + 1) + j_{min}]. \tag{7.140}$$

Since this expression has $(j_{max} - j_{min} + 1) = [(j_1 + j_2 + 1) - j_{min}]$ terms, we have

$$2S = 2[(j_1 + j_2 + 1) + j_{min}][(j_1 + j_2 + 1) - j_{min}]; \tag{7.141}$$

hence

$$S = [(j_1 + j_2 + 1) + j_{min}][(j_1 + j_2 + 1) - j_{min}] = (j_1 + j_2 + 1)^2 - j_{min}^2. \qquad (7.142)$$

Now, equating this expression with the right-hand side of (7.136), we obtain

$$(j_1 + j_2 + 1)^2 - j_{min}^2 = (2j_1 + 1)(2j_2 + 1), \qquad (7.143)$$

which in turn leads to

$$j_{min}^2 = (j_1 - j_2)^2. \qquad (7.144)$$

7.3.2 Calculation of the Clebsch–Gordan Coefficients

First, we should point out that the Clebsch–Gordan coefficients corresponding to the two limiting cases where $m_1 = j_1$, $m_2 = j_2$, $j = j_1 + j_2$, $m = j_1 + j_2$ and $m_1 = -j_1$, $m_2 = -j_2$, $j = j_1 + j_2$, $m = -(j_1 + j_2)$ are equal to one:

$$\langle j_1, j_2; j_1, j_2 | (j_1 + j_2), (j_1 + j_2) \rangle = 1, \quad \langle j_1, j_2; -j_1, -j_2 | (j_1 + j_2), -(j_1 + j_2) \rangle = 1. \qquad (7.145)$$

These results can be inferred from (7.121), since $|(j_1+j_2), (j_1+j_2)\rangle$, and $|(j_1+j_2), -(j_1+j_2)\rangle$ have one element each:

$$|(j_1 + j_2), (j_1 + j_2)\rangle = \langle j_1, j_2; j_1, j_2 | (j_1 + j_2), (j_1 + j_2) \rangle |j_1, j_2; j_1, j_2 \rangle, \qquad (7.146)$$

$$|(j_1+j_2), -(j_1+j_2)\rangle = \langle j_1, j_2; -j_1, -j_2 | (j_1+j_2), -(j_1+j_2) \rangle |j_1, j_2; -j_1, -j_2 \rangle, \qquad (7.147)$$

where $|(j_1 + j_2), (j_1 + j_2)\rangle$, $|(j_1 + j_2), -(j_1 + j_2)\rangle$, $|j_1, j_2; j_1, j_2 \rangle$, and $|j_1, j_2; -j_1, -j_2 \rangle$ are all normalized.

The calculations of the other coefficients are generally more involved than the two limiting cases mentioned above. For this, we need to derive the recursion relations between the matrix elements of the unitary transformation between the $\{| j, m \rangle\}$ and $\{|j_1, j_2; m_1, m_2 \rangle\}$ bases, since, when j_1, j_2 and j are fixed, the various Clebsch–Gordan coefficients are related to one another by means of recursion relations. To find the recursion relations, we need to evaluate the matrix elements $\langle j_1, j_2; m_1, m_2 | \hat{J}_\pm | j, m \rangle$ in two different ways. First, allow $\hat{J}_\pm$ to act to the right, i.e., on $| j, m \rangle$:

$$\langle j_1, j_2; m_1, m_2 | \hat{J}_\pm | j, m \rangle = \hbar\sqrt{(j \mp m)(j \pm m + 1)}\langle j_1, j_2; m_1, m_2 | j, m \pm 1 \rangle. \qquad (7.148)$$

Second, make $\hat{J}_\pm = \hat{J}_{1\pm} + \hat{J}_{2\pm}$ act to the left[6], i.e., on $\langle j_1, j_2; m_1, m_2 |$:

$$\langle j_1, j_2; m_1, m_2 | \hat{J}_\pm | j, m \rangle = \hbar\sqrt{(j_1 \pm m_1)(j_1 \mp m_1 + 1)}\langle j_1, j_2; m_1 \mp 1, m_2 | j, m \rangle$$
$$+ \hbar\sqrt{(j_2 \pm m_2)(j_2 \mp m_2 + 1)}\langle j_1, j_2; m_1, m_2 \mp 1 | j, m \rangle. \qquad (7.149)$$

---

[6]Recall that $\langle j_1, j_2; m_1, m_2 | \hat{J}_{1\pm} = \hbar\sqrt{(j_1 \pm m_1)(j_1 \mp m_1 + 1)}\langle j_1, j_2; m_1 \mp 1, m_2 |$.

Equating (7.148) and (7.149) we obtain the desired recursion relations for the Clebsch–Gordan coefficients:

$$
\begin{aligned}
\sqrt{(j \mp m)(j \pm m + 1)} \; & \langle j_1, j_2; m_1, m_2 \mid j, m \pm 1 \rangle \\
&= \sqrt{(j_1 \pm m_1)(j_1 \mp m_1 + 1)} \langle j_1, j_2; m_1 \mp 1, m_2 \mid j, m \rangle \\
&\quad + \sqrt{(j_2 \pm m_2)(j_2 \mp m_2 + 1)} \langle j_1, j_2; m_1, m_2 \mp 1 \mid j, m \rangle .
\end{aligned}
$$

$$(7.150)$$

These relations, together with the orthonormalization relation (7.125), determine all Clebsch–Gordan coefficients for any given values of j_1, j_2, and j. To see this, let us substitute $m_1 = j_1$ and $m = j$ into the lower part of (7.150). Since m_2 can be equal only to $m_2 = j - j_1 - 1$, we obtain

$$
\sqrt{2j} \langle j_1, j_2; j_1, (j - j_1 - 1) \mid j, j - 1 \rangle = \sqrt{(j_2 - j + j_1 + 1)(j_2 + j - j_1)}
$$
$$
\times \langle j_1, j_2; j_1, (j - j_1) \mid j, j \rangle .
$$

$$(7.151)$$

Thus, knowing $\langle j_1, j_2; j_1, (j - j_1) \mid j, j \rangle$, we can determine $\langle j_1, j_2; j_1, (j - j_1 - 1) \mid j, j - 1 \rangle$. In addition, substituting $m_1 = j_1$, $m = j - 1$ and $m_2 = j - j_1$ into the upper part of (7.150), we end up with

$$
\sqrt{2j} \, \langle j_1, j_2; j_1, (j - j_1) \mid j, j \rangle = \sqrt{2j_1} \langle j_1, j_2; (j_1 - 1), (j - j_1) \mid j, j - 1 \rangle
$$
$$
+ \sqrt{(j_2 + j - j_1)(j_2 - j + j_1 + 1)} \langle j_1, j_2; j_1, (j - j_1 - 1) \mid j, j - 1 \rangle .
$$

$$(7.152)$$

Thus knowing $\langle j_1, j_2; j_1, (j - j_1) \mid j, j \rangle$ and $\langle j_1, j_2; j_1, (j - j_1 - 1) \mid j, j - 1 \rangle$, we can determine $\langle j_1, j_2; (j_1 - 1), (j - j_1) \mid j, j - 1 \rangle$. Repeated application of the recursion relation (7.150) will determine all the other Clebsch–Gordan coefficients, provided we know only one of them: $\langle j_1, j_2; j_1, (j - j_1) \mid j, j \rangle$. As for the absolute value of this coefficient, it can be determined from the normalization condition (7.124). Thus, the recursion relation (7.150), in conjunction with the normalization condition (7.124), determines all the Clebsch–Gordan coefficients except for a sign. But how does one determine this sign?

The convention, known as the phase convention, is to consider $\langle j_1, j_2; j_1, (j - j_1) \mid j, j \rangle$ to be *real* and *positive*. This phase convention implies that

$$
\langle j_1, j_2; m_1, m_2 \mid j, m \rangle = (-1)^{j - j_1 - j_2} \langle j_2, j_1; m_2, m_1 \mid j, m \rangle ;
$$

$$(7.153)$$

hence

$$
\langle j_1, j_2; m_1, m_2 \mid j, m \rangle = (-1)^{j - j_1 - j_2} \langle j_1, j_2; -m_1, -m_2 \mid j, -m \rangle
$$
$$
= \langle j_2, j_1; -m_2, -m_1 \mid j, -m \rangle .
$$

$$(7.154)$$

Note that, since all the Clebsch–Gordan coefficients are obtained from a single coefficient $\langle j_1, j_2; j_1, (j - j_1) \mid j, j \rangle$, and since this coefficient is real, all other Clebsch–Gordan coefficients must also be real numbers.

Following the same method that led to (7.150) from $\langle j_1, j_2; m_1, m_2 \mid J_\pm \mid j, m \rangle$, we can show that a calculation of $\langle j_1, j_2; m_1, m_2 \mid J_\pm \mid j\, m \mp 1 \rangle$ leads to the following recursion relation:

$$\sqrt{(j \mp m + 1)(j \pm m)}\, \langle j_1, j_2; m_1, m_2 \mid j, m \rangle$$
$$= \sqrt{(j_1 \pm m_1)(j_1 \mp m_1 + 1)}\langle j_1, j_2; m_1 \mp 1, m_2 \mid j, m \mp 1 \rangle$$
$$+ \sqrt{(j_2 \pm m_2)(j_2 \mp m_2 + 1)}\langle j_1, j_2; m_1, m_2 \mp 1 \mid j, m \mp 1 \rangle.$$
$$(7.155)$$

We can use the recursion relations (7.150) and (7.155) to obtain the values of the various Clebsch–Gordan coefficients. For instance, if we insert $m_1 = j_1$, $m_2 = j_2 - 1$, $j = j_1 + j_2$, and $m = j_1 + j_2$ into the lower sign of (7.150), we obtain

$$\langle j_1, j_2; j_1, (j_2 - 1) \mid (j_1 + j_2), (j_1 + j_2 - 1) \rangle = \sqrt{\frac{j_2}{j_1 + j_2}}. \qquad (7.156)$$

Similarly, a substitution of $m_1 = j_1 - 1$, $m_2 = j_2$, $j = j_1 + j_2$, and $m = j_1 + j_2$ into the lower sign of (7.150) leads to

$$\langle j_1, j_2; (j_1 - 1), j_2 \mid (j_1 + j_2), (j_1 + j_2 - 1) \rangle = \sqrt{\frac{j_1}{j_1 + j_2}}. \qquad (7.157)$$

We can also show that

$$\langle j, 1; m, 0 \mid j, m \rangle = \frac{m}{\sqrt{j(j + 1)}}, \qquad \langle j, 0; m, 0 \mid j, m \rangle = 1. \qquad (7.158)$$

Example 7.3
(a) Find the Clebsch–Gordan coefficients associated with the coupling of the spins of the electron and the proton of a hydrogen atom in its ground state.
(b) Find the transformation matrix which is formed by the Clebsch–Gordan coefficients. Verify that this matrix is unitary.

Solution
In their ground states the proton and electron have no orbital angular momenta. Thus, the total angular momentum of the atom is obtained by simply adding the spins of the proton and electron.

This is a simple example to illustrate the general formalism outlined in this section. Since $j_1 = \frac{1}{2}$ and $j_2 = \frac{1}{2}$, j has two possible values $j = 0, 1$. When $j = 0$, there is only a single state $\mid j, m \rangle = \mid 0, 0 \rangle$; this is called the *spin singlet*. On the other hand, there are three possible values of $m = -1, 0, 1$ for the case $j = 1$; this corresponds to a *spin triplet state* $\mid 1, -1 \rangle$, $\mid 1, 0 \rangle$, $\mid 1, 1 \rangle$.

From (7.121), we can express the states $\mid j, m \rangle$ in terms of $\mid \frac{1}{2}, \frac{1}{2}; m_1, m_2 \rangle$ as follows:

$$\mid j, m \rangle = \sum_{m_1=-1/2}^{1/2} \sum_{m_2=-1/2}^{1/2} \langle \tfrac{1}{2}, \tfrac{1}{2}; m_1, m_2 \mid j, m \rangle \mid \tfrac{1}{2}, \tfrac{1}{2}; m_1, m_2 \rangle, \qquad (7.159)$$

which, when applied to the two cases $j = 0$ and $j = 1$, leads to

$$| 0, 0 \rangle = \langle \tfrac{1}{2}, \tfrac{1}{2}; \tfrac{1}{2}, -\tfrac{1}{2} | 0, 0 \rangle \left| \tfrac{1}{2}, \tfrac{1}{2}; \tfrac{1}{2}, -\tfrac{1}{2} \right\rangle + \langle \tfrac{1}{2}, \tfrac{1}{2}; -\tfrac{1}{2}, \tfrac{1}{2} | 0, 0 \rangle \left| \tfrac{1}{2}, \tfrac{1}{2}; -\tfrac{1}{2}, \tfrac{1}{2} \right\rangle,$$
(7.160)

$$| 1, 1 \rangle = \langle \tfrac{1}{2}, \tfrac{1}{2}; \tfrac{1}{2}, \tfrac{1}{2} | 1, 1 \rangle \left| \tfrac{1}{2}, \tfrac{1}{2}; \tfrac{1}{2}, \tfrac{1}{2} \right\rangle,$$
(7.161)

$$| 1, 0 \rangle = \langle \tfrac{1}{2}, \tfrac{1}{2}; \tfrac{1}{2}, -\tfrac{1}{2} | 1, 0 \rangle \left| \tfrac{1}{2}, \tfrac{1}{2}; \tfrac{1}{2}, -\tfrac{1}{2} \right\rangle + \langle \tfrac{1}{2}, \tfrac{1}{2}; -\tfrac{1}{2}, \tfrac{1}{2} | 1, 0 \rangle \left| \tfrac{1}{2}, \tfrac{1}{2}; -\tfrac{1}{2}, \tfrac{1}{2} \right\rangle,$$
(7.162)

$$| 1, -1 \rangle = \langle \tfrac{1}{2}, \tfrac{1}{2}; -\tfrac{1}{2}, -\tfrac{1}{2} | 1, -1 \rangle \left| \tfrac{1}{2}, \tfrac{1}{2}; -\tfrac{1}{2}, -\tfrac{1}{2} \right\rangle.$$
(7.163)

To calculate the Clebsch–Gordan coefficients involved in (7.160)–(7.163), we are going to adopt two separate approaches: the first approach uses the recursion relations (7.150) and (7.155), while the second uses the algebra of angular momentum.

First approach: using the recursion relations
First, to calculate the two coefficients $\langle \tfrac{1}{2}, \tfrac{1}{2}; \pm\tfrac{1}{2}, \mp\tfrac{1}{2} | 0, 0 \rangle$ involved in (7.160), we need, on the one hand, to substitute $j = 0, m = 0, m_1 = m_2 = \tfrac{1}{2}$ into the upper sign relation of (7.150):

$$\langle \tfrac{1}{2}, \tfrac{1}{2}; -\tfrac{1}{2}, \tfrac{1}{2} | 0, 0 \rangle = -\langle \tfrac{1}{2}, \tfrac{1}{2}; \tfrac{1}{2}, -\tfrac{1}{2} | 0, 0 \rangle.$$
(7.164)

On the other hand, the substitution of $j = 0$ and $m = 0$ into (7.125) yields

$$\langle \tfrac{1}{2}, \tfrac{1}{2}; -\tfrac{1}{2}, \tfrac{1}{2} | 0, 0 \rangle^2 + \langle \tfrac{1}{2}, \tfrac{1}{2}; \tfrac{1}{2}, -\tfrac{1}{2} | 0, 0 \rangle^2 = 1$$
(7.165)

Combining (7.164) and (7.165) we end up with

$$\langle \tfrac{1}{2}, \tfrac{1}{2}; \tfrac{1}{2}, -\tfrac{1}{2} | 0, 0 \rangle = \pm \frac{1}{\sqrt{2}}.$$
(7.166)

The sign of $\langle \tfrac{1}{2}, \tfrac{1}{2}; \tfrac{1}{2}, -\tfrac{1}{2} | 0, 0 \rangle$ has to be positive because, according to the phase convention, the coefficient $\langle j_1, j_2; j_1, (j - j_1) | j, j \rangle$ is positive; hence

$$\langle \tfrac{1}{2}, \tfrac{1}{2}; \tfrac{1}{2}, -\tfrac{1}{2} | 0, 0 \rangle = \frac{1}{\sqrt{2}}.$$
(7.167)

As for $\langle \tfrac{1}{2}, \tfrac{1}{2}; -\tfrac{1}{2}, \tfrac{1}{2} | 0, 0 \rangle$, its value can be inferred from (7.164) and (7.167):

$$\langle \tfrac{1}{2}, \tfrac{1}{2}; -\tfrac{1}{2}, \tfrac{1}{2} | 0, 0 \rangle = -\frac{1}{\sqrt{2}}.$$
(7.168)

Second, the calculation of the coefficients involved in (7.161) to (7.163) goes as follows. The orthonormalization relation (7.125) leads to

$$\langle \tfrac{1}{2}, \tfrac{1}{2}; \tfrac{1}{2}, \tfrac{1}{2} | 1, 1 \rangle^2 = 1, \qquad \langle \tfrac{1}{2}, \tfrac{1}{2}; -\tfrac{1}{2}, -\tfrac{1}{2} | 1, -1 \rangle^2 = 1,$$
(7.169)

and since $\langle \frac{1}{2}, \frac{1}{2}; \frac{1}{2}, \frac{1}{2} \mid 1, 1 \rangle$ and $\langle \frac{1}{2}, \frac{1}{2}; -\frac{1}{2}, -\frac{1}{2} \mid 1, -1 \rangle$ are both real and positive, we have

$$\langle \frac{1}{2}, \frac{1}{2}; \frac{1}{2}, \frac{1}{2} \mid 1, 1 \rangle = 1, \qquad\qquad \langle \frac{1}{2}, \frac{1}{2}; -\frac{1}{2} - \frac{1}{2} \mid 1, -1 \rangle = 1. \qquad (7.170)$$

As for the coefficients $\langle \frac{1}{2}, \frac{1}{2}; \frac{1}{2}, -\frac{1}{2} \mid 1, 0 \rangle$ and $\langle \frac{1}{2}, \frac{1}{2}; -\frac{1}{2}, \frac{1}{2} \mid 1, 0 \rangle$, they can be extracted by setting $j = 1, m = 0, m_1 = \frac{1}{2}, m_2 = -\frac{1}{2}$ and $j = 1, m = 0, m_1 = -\frac{1}{2}, m_2 = \frac{1}{2}$, respectively, into the lower sign case of (7.155):

$$\sqrt{2} \langle \frac{1}{2}, \frac{1}{2}; -\frac{1}{2}, \frac{1}{2} \mid 1, 0 \rangle = \langle \frac{1}{2}, \frac{1}{2}; \frac{1}{2}, \frac{1}{2} \mid 1, 1 \rangle, \qquad (7.171)$$

$$\sqrt{2} \langle \frac{1}{2}, \frac{1}{2}; \frac{1}{2}, -\frac{1}{2} \mid 1, 0 \rangle = \langle \frac{1}{2}, \frac{1}{2}; \frac{1}{2}, \frac{1}{2} \mid 1, 1 \rangle. \qquad (7.172)$$

Combining (7.170) with (7.171) and (7.172), we find

$$\langle \frac{1}{2}, \frac{1}{2}; \frac{1}{2}, -\frac{1}{2} \mid 1, 0 \rangle = \langle \frac{1}{2}, \frac{1}{2}; -\frac{1}{2}, \frac{1}{2} \mid 1, 0 \rangle = \frac{1}{\sqrt{2}}. \qquad (7.173)$$

Finally, substituting the Clebsch–Gordan coefficients (7.167), (7.168) into (7.160) and (7.170), and substituting (7.173) into (7.161) to (7.163), we end up with

$$\mid 0, 0 \rangle = -\frac{1}{\sqrt{2}} \left| \frac{1}{2}, \frac{1}{2}; -\frac{1}{2}, \frac{1}{2} \right\rangle + \frac{1}{\sqrt{2}} \left| \frac{1}{2}, \frac{1}{2}; \frac{1}{2}, -\frac{1}{2} \right\rangle, \qquad (7.174)$$

$$\mid 1, 1 \rangle = \left| \frac{1}{2}, \frac{1}{2}; \frac{1}{2}, \frac{1}{2} \right\rangle, \qquad (7.175)$$

$$\mid 1, 0 \rangle = \frac{1}{\sqrt{2}} \left| \frac{1}{2}, \frac{1}{2}; -\frac{1}{2}, \frac{1}{2} \right\rangle + \frac{1}{\sqrt{2}} \left| \frac{1}{2}, \frac{1}{2}; \frac{1}{2}, -\frac{1}{2} \right\rangle, \qquad (7.176)$$

$$\mid 1, -1 \rangle = \left| \frac{1}{2}, \frac{1}{2}; -\frac{1}{2}, -\frac{1}{2} \right\rangle. \qquad (7.177)$$

Note that the singlet state $\mid 0, 0 \rangle$ is antisymmetric, whereas the triplet states $\mid 1, -1 \rangle, \mid 1, 0 \rangle$, and $\mid 1, 1 \rangle$ are symmetric.

Second approach: using angular momentum algebra
Beginning with $j = 1$, and since $\mid 1, 1 \rangle$ and $\mid \frac{1}{2}, \frac{1}{2}; \frac{1}{2}, \frac{1}{2} \rangle$ are both normalized, equation (7.161) leads to

$$\langle \frac{1}{2}, \frac{1}{2}; \frac{1}{2}, \frac{1}{2} \mid 1, 1 \rangle^2 = 1. \qquad (7.178)$$

From the phase convention, which states that $\langle j_1, j_2; j, (j - j_1) \mid j, j \rangle$ must be positive, we see that $\langle \frac{1}{2}, \frac{1}{2}; \frac{1}{2}, \frac{1}{2} \mid 1, 1 \rangle = 1$, and hence

$$\mid 1, 1 \rangle = \left| \frac{1}{2}, \frac{1}{2}; \frac{1}{2}, \frac{1}{2} \right\rangle. \qquad (7.179)$$

Now, to find the Clebsch–Gordan coefficients in $\mid 1, 0 \rangle$, we simply apply $\hat{J}_-$ on $\mid 1, 1 \rangle$:

$$\hat{J}_- \mid 1, 1 \rangle = (\hat{J}_{1-} + \hat{J}_{2-}) \left| \frac{1}{2}, \frac{1}{2}; \frac{1}{2}, \frac{1}{2} \right\rangle, \qquad (7.180)$$

which leads to

$$|1, 0\rangle = \frac{1}{\sqrt{2}}\left|\frac{1}{2}, \frac{1}{2}; -\frac{1}{2}, \frac{1}{2}\right\rangle + \frac{1}{\sqrt{2}}\left|\frac{1}{2}, \frac{1}{2}; \frac{1}{2}, -\frac{1}{2}\right\rangle, \tag{7.181}$$

hence $\langle\frac{1}{2}, \frac{1}{2}; -\frac{1}{2}, \frac{1}{2} \mid 1, 0\rangle = 1/\sqrt{2}$ and $\langle\frac{1}{2}, \frac{1}{2}; \frac{1}{2}, -\frac{1}{2} \mid 1, 0\rangle = 1/\sqrt{2}$. Next, applying $\hat{J}_-$ on (7.181), we get

$$|1, -1\rangle = \left|\frac{1}{2}, \frac{1}{2}; -\frac{1}{2}, -\frac{1}{2}\right\rangle. \tag{7.182}$$

Finally, to find $|0, 0\rangle$, we proceed in two steps: first, since

$$|0, 0\rangle = a\left|\frac{1}{2}, \frac{1}{2}; -\frac{1}{2}, \frac{1}{2}\right\rangle + b\left|\frac{1}{2}, \frac{1}{2}; \frac{1}{2}, -\frac{1}{2}\right\rangle, \tag{7.183}$$

where $a = \langle\frac{1}{2}, \frac{1}{2}; -\frac{1}{2}, \frac{1}{2} \mid 0, 0\rangle$ and $b = \langle\frac{1}{2}, \frac{1}{2}; \frac{1}{2}, -\frac{1}{2} \mid 0, 0\rangle$, a combination of (7.181) with (7.183) leads to

$$\langle 0, 0 \mid 1, 0\rangle = \frac{a}{\sqrt{2}} + \frac{b}{\sqrt{2}} = 0; \tag{7.184}$$

second, since $|0, 0\rangle$ is normalized, we have

$$\langle 0, 0 \mid 0, 0\rangle = a^2 + b^2 = 1. \tag{7.185}$$

Combining (7.184) and (7.185), and since $\langle\frac{1}{2}, \frac{1}{2}; \frac{1}{2}, -\frac{1}{2} \mid 0, 0\rangle$ must be positive, we obtain $a = \langle\frac{1}{2}, \frac{1}{2}; -\frac{1}{2}, \frac{1}{2} \mid 0, 0\rangle = -1/\sqrt{2}$ and $b = \langle\frac{1}{2}, \frac{1}{2}; \frac{1}{2}, -\frac{1}{2} \mid 0, 0\rangle = 1/\sqrt{2}$. Inserting these values into (7.183) we obtain

$$|0, 0\rangle = -\frac{1}{\sqrt{2}}\left|\frac{1}{2}, \frac{1}{2}; -\frac{1}{2}, \frac{1}{2}\right\rangle + \frac{1}{\sqrt{2}}\left|\frac{1}{2}, \frac{1}{2}; \frac{1}{2}, -\frac{1}{2}\right\rangle. \tag{7.186}$$

(b) Writing (7.174) to (7.177) in a matrix form:

$$\begin{pmatrix} |0, 0\rangle \\ |1, 1\rangle \\ |1, 0\rangle \\ |1, -1\rangle \end{pmatrix} = \begin{pmatrix} 0 & 1/\sqrt{2} & -1/\sqrt{2} & 0 \\ 1 & 0 & 0 & 0 \\ 0 & 1/\sqrt{2} & 1/\sqrt{2} & 0 \\ 0 & 0 & 0 & 1 \end{pmatrix} \begin{pmatrix} |\frac{1}{2}, \frac{1}{2}; \frac{1}{2}, \frac{1}{2}\rangle \\ |\frac{1}{2}, \frac{1}{2}; \frac{1}{2}, -\frac{1}{2}\rangle \\ |\frac{1}{2}, \frac{1}{2}; -\frac{1}{2}, \frac{1}{2}\rangle \\ |\frac{1}{2}, \frac{1}{2}; -\frac{1}{2}, -\frac{1}{2}\rangle \end{pmatrix}, \tag{7.187}$$

we see that the elements of the transformation matrix

$$U = \begin{pmatrix} 0 & 1/\sqrt{2} & -1/\sqrt{2} & 0 \\ 1 & 0 & 0 & 0 \\ 0 & 1/\sqrt{2} & 1/\sqrt{2} & 0 \\ 0 & 0 & 0 & 1 \end{pmatrix}, \tag{7.188}$$

which connects the $\{|j, m\rangle\}$ vectors to their $\{|j_1, j_2; m_1, m_2\rangle\}$ counterparts, are given by the Clebsch–Gordan coefficients derived above. Inverting (7.187) we obtain

$$\begin{pmatrix} |\frac{1}{2}, \frac{1}{2}; \frac{1}{2}, \frac{1}{2}\rangle \\ |\frac{1}{2}, \frac{1}{2}; \frac{1}{2}, -\frac{1}{2}\rangle \\ |\frac{1}{2}, \frac{1}{2}; -\frac{1}{2}, \frac{1}{2}\rangle \\ |\frac{1}{2}, \frac{1}{2}; -\frac{1}{2}, -\frac{1}{2}\rangle \end{pmatrix} = \begin{pmatrix} 0 & 1 & 0 & 0 \\ 1/\sqrt{2} & 0 & 1/\sqrt{2} & 0 \\ -1/\sqrt{2} & 0 & 1/\sqrt{2} & 0 \\ 0 & 0 & 0 & 1 \end{pmatrix} \begin{pmatrix} |0, 0\rangle \\ |1, 1\rangle \\ |1, 0\rangle \\ |1, -1\rangle \end{pmatrix}. \tag{7.189}$$

From (7.187) and (7.189) we see that the transformation matrix U is unitary; this is expected since $U^{-1} = U^{\dagger}$.

7.3.3 Coupling of Orbital and Spin Angular Momenta

We consider here an important application of the formalism of angular momenta addition to the coupling of an orbital and a spin angular momentum: $\hat{\vec{J}} = \hat{\vec{L}} + \hat{\vec{S}}$. In particular, we want to find Clebsch–Gordan coefficients associated with this coupling for a spin $s = \frac{1}{2}$ particle. In this case we have: $j_1 = l$ (integer), $m_1 = m_l$, $j_2 = s = \frac{1}{2}$, and $m_2 = m_s = \pm\frac{1}{2}$. The allowed values of j as given by (7.130) are located within the interval $|l - \frac{1}{2}| \le j \le |l + \frac{1}{2}|$. If $l = 0$ the problem would be obvious: the particle would have only spin and no orbital angular momentum. But if $l > 0$ then j can take only two possible values $j = l \pm \frac{1}{2}$. There are $2(l + 1)$ states $\{|\, l + \frac{1}{2}, m\rangle\}$ corresponding to the case $j = l + 1/2$ and $2l$ states $\{|\, l - \frac{1}{2}, m\rangle\}$ corresponding to $j = l - \frac{1}{2}$. Let us study in detail each one of these two cases.

Case $j = l + 1/2$
Applying the relation (7.121) to the case where $j = l + \frac{1}{2}$, we have

$$
\begin{aligned}
\left| l + \frac{1}{2}, m \right\rangle &= \sum_{m_l=-l}^{l} \sum_{m_2=-1/2}^{1/2} \left\langle l, \frac{1}{2}; \, m_l, m_2 \middle| l + \frac{1}{2}, m \right\rangle \left| l, \frac{1}{2}; \, m_l, m_2 \right\rangle \\
&= \sum_{m_l} \left\langle l, \frac{1}{2}; \, m_l, -\frac{1}{2} \middle| l + \frac{1}{2}, m \right\rangle \left| l, \frac{1}{2}; \, m_l, -\frac{1}{2} \right\rangle \\
&\quad + \sum_{m_l} \left\langle l, \frac{1}{2}; \, m_l, \frac{1}{2} \middle| l + \frac{1}{2}, m \right\rangle \left| l, \frac{1}{2}; \, m_l, \frac{1}{2} \right\rangle.
\end{aligned}
\tag{7.190}
$$

Using the selection rule $m_l + m_2 = m$ or $m_l = m - m_2$, we can rewrite (7.190) as follows:

$$
\begin{aligned}
\left| l + \frac{1}{2}, m \right\rangle &= \left\langle l, \frac{1}{2}; \, m + \frac{1}{2}, -\frac{1}{2} \middle| l + \frac{1}{2}, m \right\rangle \left| l, \frac{1}{2}; \, m + \frac{1}{2}, -\frac{1}{2} \right\rangle \\
&\quad + \left\langle l, \frac{1}{2}; \, m - \frac{1}{2}, \frac{1}{2} \middle| l + \frac{1}{2}, m \right\rangle \left| l, \frac{1}{2}; \, m - \frac{1}{2}, \frac{1}{2} \right\rangle.
\end{aligned}
\tag{7.191}
$$

We need now to calculate $\langle l, \frac{1}{2}; \, m + \frac{1}{2}, -\frac{1}{2} | l + \frac{1}{2}, m \rangle$ and $\langle l, \frac{1}{2}; \, m - \frac{1}{2}, \frac{1}{2} | l + \frac{1}{2}, m \rangle$. We begin with the calculation of $\langle l, \frac{1}{2}; \, m + \frac{1}{2}, -\frac{1}{2} | l + \frac{1}{2}, m \rangle$. Substituting $j = l + \frac{1}{2}$, $j_1 = l$, $j_2 = \frac{1}{2}$, $m_1 + m = \frac{1}{2}$, $m_2 = -\frac{1}{2}$ into the upper sign case of (7.155), we obtain

$$
\sqrt{\left(l - m + \frac{3}{2} \right) \left(l + m + \frac{1}{2} \right)} \left\langle l, \frac{1}{2}; \, m + \frac{1}{2}, -\frac{1}{2} \middle| l + \frac{1}{2}, m \right\rangle
$$

$$
= \sqrt{\left(l + m + \frac{1}{2} \right) \left(l - m + \frac{1}{2} \right)} \left\langle l, \frac{1}{2}; \, m - \frac{1}{2}, -\frac{1}{2} \middle| l + \frac{1}{2}, m - 1 \right\rangle
\tag{7.192}
$$

or

$$
\left\langle l, \frac{1}{2}; \, m + \frac{1}{2}, -\frac{1}{2} \middle| l + \frac{1}{2}, m \right\rangle = \sqrt{\frac{l - m + 1/2}{l - m + 3/2}} \left\langle l, \frac{1}{2}; \, m - \frac{1}{2}, -\frac{1}{2} \middle| l + \frac{1}{2}, m - 1 \right\rangle.
\tag{7.193}
$$

By analogy with $\langle l, \frac{1}{2}; \ m + \frac{1}{2}, -\frac{1}{2} | l + \frac{1}{2}, \ m \rangle$ we can express the Clebsch–Gordan coefficient $\langle l, \frac{1}{2}; \ m + \frac{1}{2}, -\frac{1}{2} | l + \frac{1}{2}, \ m - 1 \rangle$ in terms of $\langle l, \frac{1}{2}; \ m - \frac{3}{2}, -\frac{1}{2} | l + \frac{1}{2}, \ m - 2 \rangle$:

$$\left\langle l, \frac{1}{2}; \ m + \frac{1}{2}, -\frac{1}{2} \middle| l + \frac{1}{2}, \ m \right\rangle = \sqrt{\frac{l - m + 1/2}{l - m + 3/2}} \sqrt{\frac{l - m + 3/2}{l - m + 5/2}}$$

$$\times \left\langle l, \frac{1}{2}; \ m - \frac{3}{2}, -\frac{1}{2} \middle| l + \frac{1}{2}, m - 2 \right\rangle. \qquad (7.194)$$

We can continue this procedure until m reaches its lowest values, $-l - \frac{1}{2}$:

$$\left\langle l, \frac{1}{2}; \ m + \frac{1}{2}, -\frac{1}{2} \middle| l + \frac{1}{2}, m \right\rangle = \sqrt{\frac{l - m + 1/2}{l - m + 3/2}} \sqrt{\frac{l - m + 3/2}{l - m + 5/2}}$$

$$\times \cdots \times \sqrt{\frac{2l}{2l + 1}} \left\langle l, \frac{1}{2}; \ -l, -\frac{1}{2} \middle| + \frac{1}{2}, -l - \frac{1}{2} \right\rangle,$$

$$\qquad (7.195)$$

or

$$\left\langle l, \frac{1}{2}; \ m + \frac{1}{2}, -\frac{1}{2} \middle| l + \frac{1}{2}, \ m \right\rangle = \sqrt{\frac{l - m + 1/2}{2l + 1}} \left\langle l, \frac{1}{2}; \ -l, -\frac{1}{2} \middle| l + \frac{1}{2}, \ -l - \frac{1}{2} \right\rangle. \quad (7.196)$$

From (7.125) we can easily obtain $\left\langle l, -\frac{1}{2}; \ -l, -\frac{1}{2} \middle| l + \frac{1}{2}, -l - \frac{1}{2} \right\rangle^2 = 1$, and since this coefficient is real we have $\left\langle l, -\frac{1}{2}; \ -l, -\frac{1}{2} \middle| l + \frac{1}{2}, -l - \frac{1}{2} \right\rangle = 1$. Inserting this value into (7.196) we end up with

$$\boxed{\left\langle l, \frac{1}{2}; \ m + \frac{1}{2}, -\frac{1}{2} \middle| l + \frac{1}{2}, \ m \right\rangle = \sqrt{\frac{l - m + 1/2}{2l + 1}}.} \qquad (7.197)$$

Now we turn to the calculation of the second coefficient, $\langle l, \frac{1}{2}; \ m - \frac{1}{2}, \frac{1}{2} | l + \frac{1}{2} \ m \rangle$, involved in (7.191). We can perform this calculation in two different ways. The first method consists of following the same procedure adopted above to find $\langle l, \frac{1}{2}; \ m + \frac{1}{2}, -\frac{1}{2} | l + \frac{1}{2}, m \rangle$. For this, we need only to substitute $j = l + \frac{1}{2}, j_1 = l, j_2 = \frac{1}{2}, m_1 = m - \frac{1}{2}, m_2 = \frac{1}{2}$ in the lower sign case of (7.155) and work our way through. A second, simpler method consists of substituting (7.197) into (7.191) and then calculating the norm of the resulting equation:

$$1 = \frac{l - m + 1/2}{2l + 1} + \left\langle l, \frac{1}{2}; \ m - \frac{1}{2}, \frac{1}{2} \middle| l + \frac{1}{2}, \ m \right\rangle^2, \qquad (7.198)$$

where we have used the facts that the three kets $|l + \frac{1}{2}, m\rangle$ and $\left| l, \frac{1}{2}; \ m \pm \frac{1}{2}, \mp \frac{1}{2} \right\rangle$ are normalized. Again, since $\langle l, \frac{1}{2}; \ m - \frac{1}{2}, \frac{1}{2} | l + \frac{1}{2}, m \rangle$ is real, (7.198) leads to

$$\boxed{\left\langle l, \frac{1}{2}; \ m - \frac{1}{2}, \frac{1}{2} \middle| l + \frac{1}{2}, \ m \right\rangle = \sqrt{\frac{l + m + 1/2}{2l + 1}}.} \qquad (7.199)$$

A combination of (7.191), (7.197), and (7.199) yields

$$\left|l+\frac{1}{2}, m\right\rangle = \sqrt{\frac{l-m+1/2}{2l+1}}\left|l,\frac{1}{2}; m+\frac{1}{2},-\frac{1}{2}\right\rangle + \sqrt{\frac{l+m+1/2}{2l+1}}\left|l,\frac{1}{2}; m-\frac{1}{2},\frac{1}{2}\right\rangle,$$

(7.200)

where the possible values of m are given by

$$m = -l - \frac{1}{2}, -l + \frac{1}{2}, -l + \frac{3}{2}, \ldots, l - \frac{3}{2}, l - \frac{1}{2}, l + \frac{1}{2}.$$

(7.201)

Case $j = l - 1/2$

There are $2l$ states, $\{|l - \frac{1}{2}, m\rangle\}$, corresponding to $j = l - \frac{1}{2}$; these are $\left|l - \frac{1}{2}, -l + \frac{1}{2}\right\rangle$, $\left|l - \frac{1}{2}, -l + \frac{3}{2}\right\rangle, \ldots, \left|l - \frac{1}{2}, l - \frac{1}{2}\right\rangle$. Using (7.121) we write any state $\left|l - \frac{1}{2}, m\right\rangle$ as

$$\left|l - \frac{1}{2}, m\right\rangle = \left\langle l, \frac{1}{2}; m + \frac{1}{2}, -\frac{1}{2}\middle|l - \frac{1}{2}, m\right\rangle\left|l, \frac{1}{2}; m + \frac{1}{2}; -\frac{1}{2}\right\rangle$$
$$+ \left\langle l, \frac{1}{2}; m - \frac{1}{2}, \frac{1}{2}\middle|l - \frac{1}{2}, m\right\rangle\left|l, \frac{1}{2}; m - \frac{1}{2}, \frac{1}{2}\right\rangle.$$

(7.202)

The two Clebsch–Gordan coefficients involved in this equation can be calculated by following the same method that we adopted above for the case $j = l + \frac{1}{2}$. Thus, we can ascertain that $|l - \frac{1}{2}, m\rangle$ is given by

$$\left|l - \frac{1}{2}, m\right\rangle = \sqrt{\frac{l+m+1/2}{2l+1}}\left|l,\frac{1}{2}; m+\frac{1}{2},-\frac{1}{2}\right\rangle - \sqrt{\frac{l-m+1/2}{2l+1}}\left|l,\frac{1}{2}; m-\frac{1}{2},\frac{1}{2}\right\rangle,$$

(7.203)

where

$$m = -l + \frac{1}{2}, -l + \frac{3}{2}, \ldots, l - \frac{3}{2}, l - \frac{1}{2}.$$

(7.204)

We can combine (7.200) and (7.203) into

$$\boxed{\left|l \pm \frac{1}{2}, m\right\rangle = \sqrt{\frac{l \mp m + \frac{1}{2}}{2l+1}}\left|l,\frac{1}{2}; m+\frac{1}{2},-\frac{1}{2}\right\rangle \pm \sqrt{\frac{l \pm m + \frac{1}{2}}{2l+1}}\left|l,\frac{1}{2}; m-\frac{1}{2},\frac{1}{2}\right\rangle.}$$

(7.205)

Illustration on a particle with $l = 1$

As an illustration of the formalism worked out above, we consider the particular case of $l = 1$. Inserting $l = 1$ and $m = \frac{3}{2}, \frac{1}{2}, -\frac{1}{2}, -\frac{3}{2}$ into the upper sign of (7.205), we obtain

$$\left|\frac{3}{2}, \frac{3}{2}\right\rangle = \left|1, \frac{1}{2}; 1, \frac{1}{2}\right\rangle,$$

(7.206)

$$\left|\frac{3}{2}, \frac{1}{2}\right\rangle = \sqrt{\frac{2}{3}}\left|1, \frac{1}{2}; 0, \frac{1}{2}\right\rangle + \frac{1}{\sqrt{3}}\left|1, \frac{1}{2}; 1, -\frac{1}{2}\right\rangle,$$

(7.207)

$$\left|\frac{3}{2}, -\frac{1}{2}\right\rangle = \frac{1}{\sqrt{3}}\left|1, \frac{1}{2}; -1, \frac{1}{2}\right\rangle + \sqrt{\frac{2}{3}}\left|1, \frac{1}{2}; 0, -\frac{1}{2}\right\rangle,$$

(7.208)

$$\left|\frac{3}{2}, -\frac{3}{2}\right\rangle = \left|1, \frac{1}{2}; -1, -\frac{1}{2}\right\rangle.$$

(7.209)

Similarly, an insertion of $l = 1$ and $m = \frac{1}{2}, -\frac{1}{2}$ into the lower sign of (7.205) yields

$$\left|\frac{1}{2}, \frac{1}{2}\right\rangle = \sqrt{\frac{2}{3}}\left|1, \frac{1}{2}; 1, -\frac{1}{2}\right\rangle - \frac{1}{\sqrt{3}}\left|1, \frac{1}{2}; 0, \frac{1}{2}\right\rangle, \qquad (7.210)$$

$$\left|\frac{1}{2}, -\frac{1}{2}\right\rangle = \frac{1}{\sqrt{3}}\left|1, \frac{1}{2}; 0, -\frac{1}{2}\right\rangle - \sqrt{\frac{2}{3}}\left|1, \frac{1}{2}; -1, \frac{1}{2}\right\rangle. \qquad (7.211)$$

Spin–orbit functions

The eigenfunctions of the particle's total angular momentum $\hat{\vec{J}} = \hat{\vec{L}} + \hat{\vec{S}}$ may be represented by the direct product of the eigenstates of the orbital and spin angular momenta, $\left|l, m - \frac{1}{2}\right\rangle$ and $\left|\frac{1}{2}, \frac{1}{2}\right\rangle$. From (7.205) we have

$$\left|l \pm \frac{1}{2}, m\right\rangle = \sqrt{\frac{l \mp m + \frac{1}{2}}{2l+1}}\left|l, m+\frac{1}{2}\right\rangle\left|\frac{1}{2}, -\frac{1}{2}\right\rangle \pm \sqrt{\frac{l \pm m + \frac{1}{2}}{2l+1}}\left|l, m-\frac{1}{2}\right\rangle\left|\frac{1}{2}, \frac{1}{2}\right\rangle. \qquad (7.212)$$

If this particle moves in a central potential, its complete wave function consists of a space part, $\langle r\theta\varphi|n, l, m \pm \frac{1}{2}\rangle = R_{nl}(r)Y_{l,m\pm\frac{1}{2}}$, and a spin part, $\left|\frac{1}{2}, \pm\frac{1}{2}\right\rangle$:

$$\Psi_{n,l,j=l\pm\frac{1}{2},m} = R_{nl}(r)\left[\sqrt{\frac{l \mp m + \frac{1}{2}}{2l+1}}Y_{l,m+\frac{1}{2}}\left|\frac{1}{2}, -\frac{1}{2}\right\rangle \pm \sqrt{\frac{l \pm m + \frac{1}{2}}{2l+1}}Y_{l,m-\frac{1}{2}}\left|\frac{1}{2}, \frac{1}{2}\right\rangle\right],$$
$$(7.213)$$

Using the spinor representation for the spin part, $\left|\frac{1}{2}, \frac{1}{2}\right\rangle = \begin{pmatrix} 1 \\ 0 \end{pmatrix}$ and $\left|\frac{1}{2}, -\frac{1}{2}\right\rangle = \begin{pmatrix} 0 \\ 1 \end{pmatrix}$, we can write (7.213) as follows:

$$\Psi_{n,l,j=l\pm\frac{1}{2},m}(r,\theta,\varphi) = \frac{R_{nl}(r)}{\sqrt{2l+1}}\begin{pmatrix} \pm\sqrt{l \pm m + \frac{1}{2}}Y_{l,m-\frac{1}{2}}(\theta,\varphi) \\ \sqrt{l \mp m + \frac{1}{2}}Y_{l,m+\frac{1}{2}}(\theta,\varphi) \end{pmatrix}, \qquad (7.214)$$

where m is half-integer. The states (7.213) and (7.214) are simultaneous eigenfunctions of $\hat{J}^2$, $\hat{L}^2$, $\hat{S}^2$, and $\hat{J}_z$ with eigenvalues $\hbar^2 j(j+1)$, $\hbar^2 l(l+1)$, $\hbar^2 s(s+1) = 3\hbar^2/4$, and $\hbar m$, respectively. The wave functions $\Psi_{n,l,j=l\pm\frac{1}{2},m}(r,\theta,\varphi)$ are eigenstates of $\hat{\vec{L}} \cdot \hat{\vec{S}}$ as well, since

$$\hat{\vec{L}} \cdot \hat{\vec{S}}|nljm\rangle = \frac{1}{2}\left(\hat{J}^2 - \hat{L}^2 - \hat{S}^2\right)|nljm\rangle$$

$$= \frac{\hbar^2}{2}\left[j(j+1) - l(l+1) - s(s+1)\right]|nljm\rangle. \qquad (7.215)$$

Here j takes only two values, $j = l \pm \frac{1}{2}$, so we have

$$\langle nljm|\hat{\vec{L}} \cdot \hat{\vec{S}}|nljm\rangle = \frac{\hbar^2}{2}\left[j(j+1) - l(l+1) - \frac{3}{4}\right] = \begin{cases} \frac{1}{2}l\hbar^2, & j = l + \frac{1}{2}, \\ -\frac{1}{2}(l+1)\hbar^2, & j = l - \frac{1}{2}. \end{cases}$$
$$(7.216)$$

7.3.4   Addition of More Than Two Angular Momenta

The formalism for adding two angular momenta may be generalized to those cases where we add three or more angular momenta. For instance, to add three mutually commuting angular momenta $\hat{\vec{J}} = \hat{\vec{J}}_1 + \hat{\vec{J}}_2 + \hat{\vec{J}}_3$, we may follow any of these three methods. (a) Add $\hat{\vec{J}}_1$ and $\hat{\vec{J}}_2$ to obtain $\hat{\vec{J}}_{12} = \hat{\vec{J}}_1 + \hat{\vec{J}}_2$, and then add $\hat{\vec{J}}_{12}$ to $\hat{\vec{J}}_3$: $\hat{\vec{J}} = \hat{\vec{J}}_{12} + \hat{\vec{J}}_3$. (b) Add $\hat{\vec{J}}_2$ and $\hat{\vec{J}}_3$ to form $\hat{\vec{J}}_{23} = \hat{\vec{J}}_2 + \hat{\vec{J}}_3$, and then add $\hat{\vec{J}}_{23}$ to $\hat{\vec{J}}_1$: $\hat{\vec{J}} = \hat{\vec{J}}_1 + \hat{\vec{J}}_{23}$. (c) Add $\hat{\vec{J}}_1$ and $\hat{\vec{J}}_3$ to form $\hat{\vec{J}}_{13} = \hat{\vec{J}}_1 + \hat{\vec{J}}_3$, and then add $\hat{\vec{J}}_{13}$ to $\hat{\vec{J}}_2$: $\hat{\vec{J}} = \hat{\vec{J}}_2 + \hat{\vec{J}}_{13}$.

Considering the first method and denoting the eigenstates of $\hat{J}_1^2$ and $\hat{J}_{1z}$ by $| j_1, m_1 \rangle$, those of $\hat{J}_2^2$, and $\hat{J}_{2z}$ by $| j_2, m_2 \rangle$, and those of $\hat{J}_3^2$ and $\hat{J}_{3z}$ by $| j_3, m_3 \rangle$, we may express the joint eigenstates $| j_{12}, j, m \rangle$ of $\hat{J}_1^2$, $\hat{J}_2^2$, $\hat{J}_3^2$, $\hat{J}_{12}^2$, $\hat{J}^2$ and J_z in terms of the states

$$| j_1, j_2, j_3; m_1, m_2, m_3 \rangle = | j_1, m_1 \rangle \, | j_2, m_2 \rangle | j_3, m_3 \rangle \tag{7.217}$$

as follows. First, the coupling of $\hat{\vec{J}}_1$ and $\hat{\vec{J}}_2$ leads to

$$| j_{12}, m_{12} \rangle = \sum_{m_1=-j_1}^{j_1} \sum_{m_2=-j_2}^{j_2} \langle j_1, j_2; m_1, m_2 \mid j_{12}, m_{12} \rangle | j_1, j_2; m_1, m_2 \rangle, \tag{7.218}$$

where $m_{12} = m_1 + m_2$ and $|j_1 - j_2| \le j_{12} \le |j_1 + j_2|$. Then, adding $\hat{\vec{J}}_{12}$ and $\hat{\vec{J}}_3$, the state $| j_{12}, j, m \rangle$ is given by

$$\sum_{m_{12}=-j_{12}}^{j_{12}} \sum_{m_3=-j_3}^{j_3} \langle j_1, j_2; m_1, m_2 \mid j_{12}, m_{12} \rangle \langle j_{12}, j_3; m_{12}, m_3 | j_{12}, j, m \rangle | j_1, j_2, j_3; m_1, m_2, m_3 \rangle,$$
$$\tag{7.219}$$

with $m = m_{12} + m_3$ and $|j_{12} - j_3| \le j \le |j_{12} + j_3|$; the Clebsch–Gordan coefficients $\langle j_1, j_2; m_1, m_2 | j_{12}, m_{12} \rangle$ and $\langle j_{12}, j_3; m_{12}, m_3 | j_{12}, j, m \rangle$ correspond to the coupling of $\hat{\vec{J}}_1$ and $\hat{\vec{J}}_2$ and of $\hat{\vec{J}}_{12}$ and $\hat{\vec{J}}_3$, respectively. The calculation of these coefficients is similar to that of two angular momenta. For instance, in Problem 7.4, page 438, we will see how to add three spins and how to calculate the corresponding Clebsch–Gordan coefficients.

We should note that the addition of $\hat{\vec{J}}_1$, $\hat{\vec{J}}_2$, and $\hat{\vec{J}}_3$ in essence consists of constructing the eigenvectors $| j_{12}, j, m \rangle$ in terms of the $(2j_1+1)(2j_2+1)(2j_3+1)$ states $| j_1, j_2, j_3; m_1, m_2, m_3 \rangle$. We may then write

$$\hat{J}_\pm | j_{12}, j, m \rangle = \hbar \sqrt{j(j+1) - m(m \pm 1)} | j_{12}, j, m \pm 1 \rangle, \tag{7.220}$$

$$\hat{J}_{1\pm} \, | j_1, j_2, j_3; m_1, m_2, m_3 \rangle = \hbar \sqrt{j_1(j_1+1) - m_1(m_1 \pm 1)} | j_1, j_2, j_3; (m_1 \pm 1), m_2, m_3 \rangle, \tag{7.221}$$

$$\hat{J}_{2\pm} \, | j_1, j_2, j_3; m_1, m_2, m_3 \rangle = \hbar \sqrt{j_2(j_2+1) - m_2(m_2 \pm 1)} | j_1, j_2, j_3; m_1, (m_2 \pm 1), m_3 \rangle, \tag{7.222}$$

$$\hat{J}_{3\pm} \, | j_1, j_2, j_3; m_1, m_2, m_3 \rangle = \hbar \sqrt{j_3(j_3+1) - m_3(m_3 \pm 1)} | j_1, j_2, j_3; m_1, m_2, (m_3 \pm 1) \rangle. \tag{7.223}$$

The foregoing method can be generalized to the coupling of more than three angular momenta: $\hat{\vec{J}} = \hat{\vec{J}}_1 + \hat{\vec{J}}_2 + \hat{\vec{J}}_3 + \cdots + \hat{\vec{J}}_N$. Each time we couple two angular momenta, we reduce the problem to the coupling of $(N - 1)$ angular momenta. For instance, we may start by adding $\vec{J}_1$ and $\vec{J}_2$ to generate $\vec{J}_{12}$; we are then left with $(N - 1)$ angular momenta. Second, by adding $\vec{J}_{12}$ and $\vec{J}_3$ to form $\vec{J}_{123}$, we are left with $(N - 2)$ angular momenta. Third, an addition of $\vec{J}_{123}$ and $\vec{J}_4$ leaves us with $(N - 3)$ angular momenta, and so on. We may continue in this way till we add all given angular momenta.

7.3.5   Rotation Matrices for Coupling Two Angular Momenta

We want to find out how to express the rotation matrix associated with an angular momentum $\vec{J}$ in terms of the rotation matrices corresponding to $\vec{J}_1$ and $\vec{J}_2$ such that $\vec{J} = \vec{J}_1 + \vec{J}_2$. That is, knowing the rotation matrices $d^{(j_1)}(\beta)$ and $d^{(j_2)}(\beta)$, how does one calculate $d^{(j)}_{mm'}(\beta)$?

Since

$$d^{(j)}_{m'm}(\beta) = \langle j, m' \mid \hat{R}_y(\beta) \mid j, m \rangle, \tag{7.224}$$

where

$$\mid j, m \rangle = \sum_{m_1 m_2} \langle j_1, j_2; m_1, m_2 \mid j, m \rangle \mid j_1, j_2; m_1, m_2 \rangle, \tag{7.225}$$

$$\mid j, m' \rangle = \sum_{m'_1 m'_2} \langle j_1, j_2; m'_1, m'_2 \mid j, m' \rangle \mid j_1, j_2; m'_1, m'_2 \rangle, \tag{7.226}$$

and since the Clebsch–Gordan coefficients are real,

$$\langle j, m' \mid = \sum_{m'_1 m'_2} \langle j_1, j_2; m'_1, m'_2 \mid j, m' \rangle \langle j_1, j_2; m'_1, m'_2 \mid, \tag{7.227}$$

we can rewrite (7.224) as

$$d^{(j)}_{m'm}(\beta) = \sum_{m_1 m_2} \sum_{m'_1 m'_2} \langle j_1, j_2; m_1, m_2 \mid j, m \rangle \langle j_1, j_2; m'_1, m'_2 \mid j, m' \rangle$$

$$\times \langle j_1, j_2; m'_1, m'_2 \mid \hat{R}_y(\beta) \mid j_1, j_2; m_1, m_2 \rangle. \tag{7.228}$$

Since $\hat{R}_y(\beta) = \exp[-\beta \hat{J}_y/\hbar] = \exp[-\beta \hat{J}_{1y}/\hbar] \exp[-\beta \hat{J}_{2y}/\hbar]$, because $\hat{J}_y = \hat{J}_{1y} + \hat{J}_{2y}$, and since $\langle j_1, j_2; m'_1, m'_2 \mid = \langle j_1, m'_1 \mid \langle j_2, m'_2 \mid$ and $\mid j_1, j_2; m_1, m_2 \rangle = \mid j_1, m_1 \rangle \mid j_2, m_2 \rangle$, we have.

$$d^{(j)}_{m'm}(\beta) = \sum_{m_1 m_2} \sum_{m'_1 m'_2} \langle j_1, j_2; m_1, m_2 \mid j, m \rangle \langle j_1, j_2; m'_1, m'_2 \mid j, m' \rangle$$

$$\times \langle j_1, m'_1 \mid \exp\left[-\frac{i}{\hbar}\beta \hat{J}_{1y}\right] \mid j_1, m_1 \rangle \langle j_2, m'_2 \mid \exp\left[-\frac{i}{\hbar}\beta \hat{J}_{2y}\right] \mid j_2, m_2 \rangle, \tag{7.229}$$

or

$$d^{(j)}_{m'm}(\beta) = \sum_{m_1 m_2} \sum_{m'_1 m'_2} \langle j_1, j_2; m_1, m_2 \mid j, m \rangle \langle j_1, j_2; m'_1, m'_2 \mid j, m' \rangle d^{(j_1)}_{m'_1 m_1}(\beta) d^{(j_2)}_{m'_2 m_2}(\beta), \tag{7.230}$$

with

$$d_{m'_1 m_1}^{(j_1)}(\beta) = \langle j_1, m'_1 \mid \exp\left[-\frac{i}{\hbar}\beta \hat{J}_{1_y}\right] \mid j_1, m_1 \rangle, \tag{7.231}$$

$$d_{m'_2 m_2}^{(j_2)}(\beta) = \langle j_2, m'_2 \mid \exp\left[-\frac{i}{\hbar}\beta \hat{J}_{2_y}\right] \mid j_2, m_2 \rangle. \tag{7.232}$$

From (7.54) we have

$$d_{m'm}^{(j)}(\beta) = e^{i(m'\alpha + m\gamma)} D_{m'm}^{(j)}(\alpha, \beta, \gamma); \tag{7.233}$$

hence can rewrite (7.230) as

$$D_{m'm}^{(j)}(\alpha, \beta, \gamma) = \sum_{m_1 m_2} \sum_{m'_1 m'_2} \langle j_1, j_2; m_1, m_2 \mid j, m \rangle \langle j_1, j_2; m'_1, m'_2 \mid j, m' \rangle D_{m'_1 m_1}^{(j_1)}(\alpha, \beta, \gamma) D_{m'_2 m_2}^{(j_2)}(\alpha, \beta, \gamma), \tag{7.234}$$

since $m = m_1 + m'_1$ and $m' = m_2 + m'_2$.

Now, let us see how to express the product of the rotation matrices $d^{(j_1)}(\beta)$ and $d^{(j_2)}(\beta)$ in terms of $d_{mm'}^{(j)}(\beta)$. Sandwiching both sides of

$$\exp\left[-\frac{i}{\hbar}\beta \hat{J}_{1_y}\right] \exp\left[-\frac{i}{\hbar}\beta \hat{J}_{2_y}\right] = \exp\left[-\frac{i}{\hbar}\beta \hat{J}_y\right] \tag{7.235}$$

between

$$\mid j_1, j_2; m_1, m_2 \rangle = \sum_{jm} \langle j_1, j_2; m_1, m_2 \mid j, m \rangle \mid j, m \rangle \tag{7.236}$$

and

$$\langle j_1, j_2; m'_1, m'_2 \mid = \sum_{jm'} \langle j_1, j_2; m'_1, m'_2 \mid j, m' \rangle \langle j, m' \mid, \tag{7.237}$$

and since $\langle j_1, j_2; m'_1, m'_2 \mid = \langle j_1, m'_1 \mid \langle j_2, m'_2 \mid$ and $\mid j_1, j_2; m_1, m_2 \rangle = \mid j_1, m_1 \rangle \mid j_2, m_2 \rangle$, we have

$$\langle j_1, m'_1 \mid \exp\left[-\frac{i}{\hbar}\beta \hat{J}_{1_y}\right] \mid j_1, m_1 \rangle \langle j_2, m'_2 \mid \exp\left[-\frac{i}{\hbar}\beta \hat{J}_{2_y}\right] \mid j_2, m_2 \rangle$$

$$= \sum_{jmm'} \langle j_1, j_2; m_1, m_2 \mid j, m \rangle \langle j_1, j_2; m'_1, m'_2 \mid j, m' \rangle \langle j, m' \mid \hat{R}_y(\beta) \mid j, m \rangle \tag{7.238}$$

or

$$d_{m'_1 m_1}^{(j_1)}(\beta) d_{m'_2 m_2}^{(j_2)}(\beta) = \sum_{|j_1 - j_2|}^{|j_1 + j_2|} \sum_{mm'} \langle j_1, j_2; m_1, m_2 \mid j, m \rangle \langle j_1, j_2; m'_1, m'_2 \mid j, m' \rangle d_{m'm}^{(j)}(\beta). \tag{7.239}$$

Following the same procedure that led to (7.234), we can rewrite (7.239) as

$$D_{m'_1 m_1}^{(j_1)}(\alpha, \beta, \gamma) D_{m'_2 m_2}^{(j_2)}(\alpha, \beta, \gamma) = \sum_{jmm'} \langle j_1, j_2; m_1, m_2 \mid j, m \rangle \langle j_1, j_2; m'_1, m'_2 \mid j, m' \rangle D_{m'm}^{(j)}(\alpha, \beta, \gamma). \tag{7.240}$$

This relation is known as the *Clebsch–Gordan series*.

The relation (7.240) has an important application: the derivation of an integral involving three spherical harmonics. When j_1 and j_2 are both integers (i.e., $j_1 = l_1$ and $j_2 = l_2$) and m_1 and m_2 are both zero (hence $m = 0$), equation (7.240) finds a useful application:

$$D_{m_1'0}^{(l_1)}(\alpha, \beta, \gamma) D_{m_2'0}^{(l_2)}(\alpha, \beta, \gamma) = \sum_{lm'} \langle l_1, l_2; 0, 0 \mid l, 0 \rangle \langle l_1, l_2; m_1', m_2' \mid l, m' \rangle D_{m'0}^{(l)}(\alpha, \beta, \gamma).$$

(7.241)

Since the expressions of $D_{m_1'0}^{(l_1)}$, $D_{m_2'0}^{(l_2)}$, and $D_{m'0}^{(l)}$ can be inferred from (7.73), notably

$$D_{m'0}^{(l)}(\alpha, \beta, 0) = \sqrt{\frac{4\pi}{2l + 1}} Y_{lm'}^*(\beta, \alpha),$$

(7.242)

we can reduce (7.241) to

$$Y_{l_1 m_1}(\beta, \alpha) Y_{l_2 m_2}(\beta, \alpha) = \sum_{lm} \sqrt{\frac{(2l_1 + 1)(2l_2 + 1)}{4\pi(2l + 1)}} \langle l_1, l_2; 0, 0 \mid l, 0 \rangle \langle l_1, l_2; m_1, m_2 \mid l, m \rangle Y_{lm}(\beta, \alpha),$$

(7.243)

where we have removed the primes and taken the complex conjugate. Multiplying both sides by $Y_{lm}^*(\beta, \alpha)$ and integrating over α and β, we obtain the following frequently used integral:

$$\int_0^{2\pi} d\alpha \int_0^{\pi} Y_{lm}^*(\beta, \alpha) Y_{l_1 m_1}(\beta, \alpha) Y_{l_2 m_2}(\beta, \alpha) \sin\beta \, d\beta = \sqrt{\frac{(2l_1+1)(2l_2+1)}{4\pi(2l+1)}} \langle l_1, l_2; 0, 0 \mid l, 0 \rangle$$
$$\times \langle l_1, l_2; m_1, m_2 \mid l, m \rangle.$$

(7.244)

7.3.6 Isospin

The ideas presented above—spin and the addition of angular momenta—find some interesting applications to other physical quantities. For instance, in the field of nuclear physics, the quantity known as *isotopic spin* can be represented by a set of operators which not only obey the same algebra as the components of angular momentum, but also couple in the same way as ordinary angular momenta.

Since the nuclear force does not depend on the electric charge, we can consider the proton and the neutron to be separate manifestations (states) of the same particle, the *nucleon*. The nucleon may thus be found in two different states: a proton and a neutron. In this way, as the protons and neutrons are identical particles with respect to the nuclear force, we will need an additional quantum number (or label) to indicate whether the nucleon is a proton or a neutron. Due to its formal analogy with ordinary spin, this label is called the *isotopic spin* or, in short, the *isospin*. If we take the isospin quantum number to be $\frac{1}{2}$, its z-component will then be represented by a quantum number having the values $\frac{1}{2}$ and $-\frac{1}{2}$. The difference between a proton and a neutron then becomes analogous to the difference between spin-up and spin-down particles.

The fundamental difference between ordinary spin and the isospin is that, unlike the spin, the isospin has nothing to do with rotations or spinning in the coordinate space, it hence cannot be coupled with the angular momenta of the nucleons. Nucleons can thus be distinguished by $\langle \hat{t}_3 \rangle = \pm\frac{1}{2}$, where $\hat{t}_3$ is the third or z-component of the isospin vector operator $\hat{t}$.

7.3.6.1 Isospin Algebra

Due to the formal analogy between the isospin and the spin, their formalisms have similar structures from a mathematical viewpoint. The algebra obeyed by the components $\hat{t}_1$, $\hat{t}_2$, $\hat{t}_3$ of the isospin operator $\hat{\vec{t}}$ can thus be inferred from the properties and commutation relations of the spin operator. For instance, the components of the isospin operator can be constructed from the Pauli matrices $\vec{\tau}$ in the same way as we did for the angular momentum operators of spin $\frac{1}{2}$ particles:

$$\hat{\vec{t}} = \frac{1}{2}\vec{\tau}, \qquad (7.245)$$

with

$$\tau_1 = \begin{pmatrix} 0 & 1 \\ 1 & 0 \end{pmatrix}, \quad \tau_2 = \begin{pmatrix} 0 & -i \\ i & 0 \end{pmatrix}, \quad \tau_3 = \begin{pmatrix} 1 & 0 \\ 0 & -1 \end{pmatrix}. \qquad (7.246)$$

The components $\hat{t}_1$, $\hat{t}_2$, $\hat{t}_3$ obey the same commutation relations as those of angular momentum:

$$\boxed{[\hat{t}_1, \hat{t}_2] = i\hat{t}_3, \qquad [\hat{t}_2, \hat{t}_3] = i\hat{t}_1, \qquad [\hat{t}_3, \hat{t}_1] = i\hat{t}_2.} \qquad (7.247)$$

So the nucleon can be found in two different states: when $\hat{t}_3$ acts on a nucleon state, it gives the eignvalues $\pm\frac{1}{2}$. By convention the $\hat{t}_3$ of a proton is taken to be $\hat{t}_3 = +\frac{1}{2}$ and that of a neutron is $\hat{t}_3 = -\frac{1}{2}$. Denoting the proton and neutron states, respectively, by $|p\rangle$ and $|n\rangle$,

$$|p\rangle = \left|t = \frac{1}{2}, t_3 = \frac{1}{2}\right\rangle = \begin{pmatrix} 1 \\ 0 \end{pmatrix}, \qquad |n\rangle = \left|t = \frac{1}{2}, t_3 = -\frac{1}{2}\right\rangle = \begin{pmatrix} 0 \\ 1 \end{pmatrix}, \qquad (7.248)$$

we have

$$\hat{t}_3 \,|\, p\rangle = \hat{t}_3 \left|\frac{1}{2}, \frac{1}{2}\right\rangle = \frac{1}{2}\left|\frac{1}{2}, \frac{1}{2}\right\rangle, \qquad (7.249)$$

$$\hat{t}_3 \,|\, n\rangle = \hat{t}_3 \left|\frac{1}{2}, -\frac{1}{2}\right\rangle = -\frac{1}{2}\left|\frac{1}{2}, -\frac{1}{2}\right\rangle. \qquad (7.250)$$

We can write (7.249) and (7.250), respectively, as

$$\frac{1}{2}\begin{pmatrix} 1 & 0 \\ 0 & -1 \end{pmatrix}\begin{pmatrix} 1 \\ 0 \end{pmatrix} = \frac{1}{2}\begin{pmatrix} 1 \\ 0 \end{pmatrix}, \qquad (7.251)$$

$$\frac{1}{2}\begin{pmatrix} 1 & 0 \\ 0 & -1 \end{pmatrix}\begin{pmatrix} 0 \\ 1 \end{pmatrix} = -\frac{1}{2}\begin{pmatrix} 0 \\ 1 \end{pmatrix}. \qquad (7.252)$$

By analogy with angular momentum, denoting the joint eignstates of $\hat{\vec{t}}^2$ and $\hat{t}_3$ by $|t, t_3\rangle$, we have

$$\boxed{\hat{\vec{t}}^2 \,|\, t, t_3\rangle = t(t+1) \,|\, t, t_3\rangle, \qquad \hat{t}_3 \,|\, t, t_3\rangle = t_3 \,|\, t, t_3\rangle.} \qquad (7.253)$$

We can also introduce the *raising and lowering* isospin operators:

$$\hat{t}_+ = \hat{t}_1 + i\hat{t}_2 = \frac{1}{2}(\tau_1 + i\tau_2) = \begin{pmatrix} 0 & 1 \\ 0 & 0 \end{pmatrix}, \qquad (7.254)$$

$$\hat{t}_- = \hat{t}_1 - i\hat{t}_2 = \frac{1}{2}(\tau_1 - i\tau_2) = \begin{pmatrix} 0 & 0 \\ 1 & 0 \end{pmatrix} \qquad (7.255)$$

hence

$$\hat{t}_\pm \mid t, \, t_3 \rangle = \sqrt{t(t+1) - t_3(t_3 \pm 1)} \mid t, \, t_3 \pm 1 \rangle. \qquad (7.256)$$

Note that $\hat{t}_+$ and $\hat{t}_-$ are operators which, when acting on a nucleon state, convert neutron states into proton states and proton states into neutron states, respectively:

$$\hat{t}_+ \mid n \rangle = \mid p \rangle, \qquad\qquad \hat{t}_- \mid p \rangle = \mid n \rangle. \qquad (7.257)$$

We can also define a *charge operator*

$$\hat{Q} = e \left(\hat{t}_3 + \frac{1}{2} \right), \qquad (7.258)$$

where e is the charge of the proton, with

$$\hat{Q} \mid p \rangle = e \mid p \rangle, \qquad\qquad \hat{Q} \mid n \rangle = 0. \qquad (7.259)$$

We should mention that strong interactions conserve isospin. For instance, a reaction like

$$d + d \to \alpha + \pi^0 \qquad (7.260)$$

is forbidden since the isospin is not conserved, because the isospins of d and α are both zero and the isospin of the pion is equal to one (i.e., $T(d) = T(\alpha) = 0$, but $T(\pi) = 1$); this leads to isospin zero for $(d + d)$ and isospin one for $(\alpha + \pi^0)$. The reaction was confirmed experimentally to be forbidden, since its cross-section is negligibly small. However, reactions such as

$$p + p \to d + \pi^+, \quad p + n \to d + \pi^0 \qquad (7.261)$$

are allowed, since they conserve isospin.

7.3.6.2 Addition of Two Isospins

We should note that the isospins of different nucleons can be added in the same way as adding angular momenta. For a nucleus consisting of several nucleons, the total isospin is given by the vector sum of the isospins of all individual nucleons: $\vec{\hat{T}} = \sum_i^A \vec{\hat{t}}_i$. For instance, the total isospin of a system of two nucleons can be obtained by coupling their isospins $\vec{\hat{t}}_1$ and $\vec{\hat{t}}_2$:

$$\vec{\hat{T}} = \vec{\hat{t}}_1 + \vec{\hat{t}}_2. \qquad (7.262)$$

Denoting the joint eigenstates of $\hat{t}_1^2, \hat{t}_2^2, \hat{T}^2$, and $\hat{T}_3$ by $\mid T, \, N \rangle$, we have:

$$\vec{\hat{T}}^2 \mid T, \, N \rangle = T(T+1) \mid T, \, N \rangle, \qquad \hat{T}_3 \mid T, \, N \rangle = N \mid T, \, N \rangle. \qquad (7.263)$$

Similarly, if we denote the joint eigenstates of $\hat{t}_1^2, \hat{t}_2^2, \hat{t}_{1_3}$, and $\hat{t}_{2_3}$ by $\mid t_1, t_2; \, n_1, n_2 \rangle$, we have

$$\hat{t}_1^2 \mid t_1, t_2; \, n_1, n_2 \rangle = t_1(t_1 + 1) \mid t_1, t_2; \, n_1, n_2 \rangle, \qquad (7.264)$$

$$\hat{t}_2^2 \mid t_1, t_2; \, n_1, n_2 \rangle = t_2(t_2 + 1) \mid t_1, t_2; \, n_1, n_2 \rangle, \qquad (7.265)$$

$$\hat{t}_{1_3} \mid t_1, t_2; \, n_1, n_2 \rangle = n_1 \mid t_1, t_2; \, n_1, n_2 \rangle, \qquad (7.266)$$

$$\hat{t}_{2_3} \mid t_1, t_2; \, n_1, n_2 \rangle = n_2 \mid t_1, t_2; \, n_1, n_2 \rangle, \qquad (7.267)$$

The matrix elements of the unitary transformation connecting the $\{|T,\ N\rangle\}$ and $\{|t_1, t_2;\ n_1, n_2\rangle\}$ bases,

$$|T,\ N\rangle = \sum_{n_1, n_2} \langle t_1, t_2;\ n_1 n_2 | T,\ N\rangle | t_1, t_2;\ n_1, n_2\rangle, \tag{7.268}$$

are given by the coefficients $\langle t_1, t_2;\ n_1 n_2 | T,\ N\rangle$; these coefficients can be calculated in the same way as the Clebsch–Gordan coefficients; see the next example.

Example 7.4

Find the various states corresponding to a two-nucleon system.

Solution

Let $\vec{\hat{T}}$ be the total isospin vector operator of the two-nucleon system:

$$\vec{\hat{T}} = \vec{\hat{t}}_1 + \vec{\hat{t}}_2. \tag{7.269}$$

This example is similar to adding two spin $\frac{1}{2}$ angular momenta. Thus, the values of T are 0 and 1. The case $T = 0$ corresponds to a singlet state:

$$| 0,\ 0\rangle = \frac{1}{\sqrt{2}} \left[| p\rangle_1 | n\rangle_2 - | n\rangle_1 | p\rangle_2 \right], \tag{7.270}$$

where $| p\rangle_1$ means that nucleon 1 is a proton, $| n\rangle_2$ means that nucleon 2 is a neutron, and so on. This state, which is an antisymmetric isospin state, describes a bound $(p\text{-}n)$ system such as the ground state of deuterium $(T = 0)$.

The case $T = 1$ corresponds to the triplet states $|1,\ N\rangle$ with $N = 1, 0, -1$:

$$| 1,\ 1\rangle = | p\rangle_1 | p\rangle_2, \tag{7.271}$$

$$| 1,\ 0\rangle = \frac{1}{\sqrt{2}} \left[| p\rangle_1 | n\rangle_2 + | n\rangle_1 | p\rangle_2 \right], \tag{7.272}$$

$$| 1,\ -1\rangle = | n\rangle_1 | n\rangle_2. \tag{7.273}$$

The state $| 1,\ 1\rangle$ corresponds to the case where both nucleons are protons $(p\text{-}p)$ and $| 1,\ -1\rangle$ corresponds to the case where both nucleons are neutrons $(n\text{-}n)$.

7.4 Scalar, Vector, and Tensor Operators

In this section we study how operators transform under rotations. Operators corresponding to various physical quantities can be classified as scalars, vectors, and tensors as a result of their behavior under rotations.

Consider an operator $\hat{A}$, which can be a scalar, a vector, or a tensor. The transformation of $\hat{A}$ under a rotation of infinitesimal angle $\delta\theta$ about an axis $\vec{n}$ is[7]

$$\hat{A}' = \hat{R}_n^\dagger(\delta\theta)\, \hat{A}\, \hat{R}_n(\delta\theta), \tag{7.274}$$

---

[7] The expectation value of an operator $\hat{A}$ with respect to the rotated state $| \psi'\rangle = \hat{R}_n(\delta\theta) | \psi\rangle$ is given by $\langle \psi' | \hat{A} | \psi'\rangle = \langle \psi | \hat{R}_n^\dagger(\delta\theta) \hat{A} \hat{R}_n(\delta\theta) | \psi\rangle = \langle \psi | \hat{A}' | \psi\rangle$.

where $\hat{R}_n(\delta\theta)$ can be inferred from (7.20)

$$\hat{R}_n(\delta\theta) = 1 - \frac{i}{\hbar}\delta\theta\vec{n}\cdot\hat{\vec{J}}. \tag{7.275}$$

Substituting (7.275) into (7.274) and keeping terms up to first order in $\delta\theta$, we obtain

$$\hat{A}' = \hat{A}' - \frac{i}{\hbar}\delta\theta\,[\hat{A},\,\vec{n}\cdot\hat{\vec{J}}]. \tag{7.276}$$

In the rest of this section we focus on the application of this relation to scalar, vector, and tensor operators.

7.4.1   Scalar Operators

Since scalar operators are invariant under rotations (i.e., $\hat{A}' = \hat{A}$), equation (7.276) implies that they commute with the angular momentum

$$[\hat{A},\,\hat{J}_k] = 0 \qquad (k = x, y, z). \tag{7.277}$$

This is also true for pseudo-scalars. A pseudo-scalar is defined by the product of a vector $\vec{A}$ and a pseudo-vector or axial vector $\vec{B}\times\vec{C}$: $\vec{A}\cdot(\vec{B}\times\vec{C})$.

7.4.2   Vector Operators

On the one hand, a vector operator $\hat{\vec{A}}$ transforms according to (7.276):

$$\hat{\vec{A}}' = \hat{\vec{A}} - \frac{i}{\hbar}\delta\theta\,[\hat{\vec{A}},\,\vec{n}\cdot\hat{\vec{J}}]. \tag{7.278}$$

On the other hand, from the classical theory of rotations, when a vector $\hat{\vec{A}}$ is rotated through an angle $\delta\theta$ around an axis $\vec{n}$, it is given by

$$\hat{\vec{A}}' = \hat{\vec{A}} + \delta\theta\,\vec{n}\times\hat{\vec{A}}. \tag{7.279}$$

Comparing (7.278) and (7.279), we obtain

$$[\hat{\vec{A}},\,\vec{n}\cdot\hat{\vec{J}}] = i\hbar\vec{n}\times\hat{\vec{A}}. \tag{7.280}$$

The jth component of this equation is given by

$$[\hat{\vec{A}},\,\vec{n}\cdot\hat{\vec{J}}]_j = i\hbar(\vec{n}\times\hat{\vec{A}})_j \qquad (j = x, y, z), \tag{7.281}$$

which in the case of $j = x, y, z$ leads to

$$\left[\hat{A}_x,\,\hat{J}_x\right] = \left[\hat{A}_y,\,\hat{J}_y\right] = \left[\hat{A}_z,\,\hat{J}_z\right] = 0, \tag{7.282}$$

$$\left[\hat{A}_x,\,\hat{J}_y\right] = i\hbar\hat{A}_z, \qquad \left[\hat{A}_y,\,\hat{J}_z\right] = i\hbar\hat{A}_x, \qquad \left[\hat{A}_z,\,\hat{J}_x\right] = i\hbar\hat{A}_y, \tag{7.283}$$

$$\left[\hat{A}_x,\,\hat{J}_z\right] = -i\hbar\hat{A}_y, \qquad \left[\hat{A}_y,\,\hat{J}_x\right] = -i\hbar\hat{A}_z, \qquad \left[\hat{A}_z,\,\hat{J}_y\right] = -i\hbar\hat{A}_x. \tag{7.284}$$

Some interesting applications of (7.280) correspond to the cases where the vector operator $\hat{\vec{A}}$ is either the angular momentum, the position, or the linear momentum operator. Let us consider these three cases separately. First, substituting $\hat{\vec{A}} = \hat{\vec{J}}$ into (7.280), we recover the usual angular momentum commutation relations:

$$[\hat{J}_x, \hat{J}_y] = i\hbar\hat{J}_z, \qquad [\hat{J}_y, \hat{J}_z] = i\hbar\hat{J}_x, \qquad [\hat{J}_z, \hat{J}_x] = i\hbar\hat{J}_y. \qquad (7.285)$$

Second, in the case of a spinless particle (i.e., $\hat{\vec{J}} = \hat{\vec{L}}$), and if $\hat{\vec{A}}$ is equal to the position operator, $\hat{\vec{A}} = \hat{\vec{R}}$, then (7.280) will yield the following relations:

$$\left[\hat{x}, \hat{L}_x\right] = 0, \qquad \left[\hat{x}, \hat{L}_y\right] = i\hbar\hat{z}, \qquad \left[\hat{x}, \hat{L}_z\right] = -i\hbar\hat{y}, \qquad (7.286)$$

$$\left[\hat{y}, \hat{L}_y\right] = 0, \qquad \left[\hat{y}, \hat{L}_z\right] = i\hbar\hat{x}, \qquad \left[\hat{y}, \hat{L}_x\right] = -i\hbar\hat{z}, \qquad (7.287)$$

$$\left[\hat{z}, \hat{L}_z\right] = 0, \qquad \left[\hat{z}, \hat{L}_x\right] = i\hbar\hat{y}, \qquad \left[\hat{z}, \hat{L}_y\right] = -i\hbar\hat{x}. \qquad (7.288)$$

Third, if $\hat{\vec{J}} = \hat{\vec{L}}$ and if $\hat{\vec{A}}$ is equal to the momentum operator, $\hat{\vec{A}} = \hat{\vec{P}}$, then (7.280) will lead to

$$\left[\hat{P}_x, \hat{L}_x\right] = 0, \qquad \left[\hat{P}_x, \hat{L}_y\right] = i\hbar\hat{P}_z, \qquad \left[\hat{P}_x, \hat{L}_z\right] = -i\hbar\hat{P}_y, \qquad (7.289)$$

$$\left[\hat{P}_y, \hat{L}_y\right] = 0, \qquad \left[\hat{P}_y, \hat{L}_z\right] = i\hbar\hat{P}_x, \qquad \left[\hat{P}_y, \hat{L}_x\right] = -i\hbar\hat{P}_z, \qquad (7.290)$$

$$\left[\hat{P}_z, \hat{L}_z\right] = 0, \qquad \left[\hat{P}_z, \hat{L}_x\right] = i\hbar\hat{P}_y, \qquad \left[\hat{P}_z, \hat{L}_y\right] = -i\hbar\hat{P}_x. \qquad (7.291)$$

Now, introducing the operators

$$\hat{A}_\pm = \hat{A}_x \pm i\hat{A}_y, \qquad (7.292)$$

and using the relations (7.282) to (7.284), we can show that

$$\left[\hat{J}_x, \hat{A}_\pm\right] = \mp\hbar\hat{A}_z, \qquad \left[\hat{J}_y, \hat{A}_\pm\right] = -i\hbar\hat{A}_z, \qquad \left[\hat{J}_z, \hat{A}_\pm\right] = \pm\hbar\hat{A}_\pm. \qquad (7.293)$$

These relations in turn can be shown to lead to

$$\left[\hat{J}_\pm, \hat{A}_\pm\right] = 0, \qquad \left[\hat{J}_\pm, \hat{A}_\mp\right] = \pm 2\hbar\hat{A}_z. \qquad (7.294)$$

Let us introduce the *spherical components* $\hat{A}_{-1}, \hat{A}_0, \hat{A}_1$ of the vector operator $\hat{\vec{A}}$; they are defined in terms of the Cartesian coordinates $\hat{A}_x, \hat{A}_y, \hat{A}_z$ as follows:

$$\hat{A}_{\pm 1} = \mp\frac{1}{\sqrt{2}}(\hat{A}_x \pm \hat{A}_y), \qquad \hat{A}_0 = \hat{A}_z. \qquad (7.295)$$

For the particular case where $\hat{\vec{A}}$ is equal to the position vector $\hat{\vec{R}}$, we can express the components $\hat{R}_q$ (where $q = -1, 0, 1$),

$$\hat{R}_{\pm 1} = \mp\frac{1}{\sqrt{2}}(\hat{x} \pm \hat{y}), \qquad \hat{R}_0 = \hat{z}, \qquad (7.296)$$

in terms of the spherical coordinates (recall that $\hat{R}_1 = \hat{x} = r \sin\theta \cos\phi$, $\hat{R}_2 = \hat{y} = r \sin\theta \sin\phi$, and $\hat{R}_3 = \hat{z} = r \cos\theta$) as follows:

$$\hat{R}_{\pm 1} = \mp \frac{1}{\sqrt{2}} r e^{\pm i\phi} \sin\theta, \qquad\qquad \hat{R}_0 = r \cos\theta. \qquad (7.297)$$

Using the relations (7.282) to (7.284) and (7.292) to (7.294), we can ascertain that

$$\boxed{\left[\hat{J}_z, \hat{A}_q\right] = \hbar q \hat{A}_q \qquad\qquad (q = -1, 0, 1),} \qquad (7.298)$$

$$\boxed{\left[\hat{J}_\pm, \hat{A}_q\right] = \hbar \sqrt{2 - q(q \pm 1)} \hat{A}_{q \pm 1} \qquad (q = -1, 0, 1).} \qquad (7.299)$$

7.4.3   Tensor Operators: Reducible and Irreducible Tensors

In general, a tensor of rank k has 3^k components, where 3 denotes the dimension of the space. For instance, a tensor such as

$$\hat{T}_{ij} = A_i B_j \qquad (i, j = x, y, z), \qquad (7.300)$$

which is equal to the product of the components of two vectors $\vec{A}$ and $\vec{B}$, is a *second-rank* tensor; this tensor has 3^2 components.

7.4.3.1   Reducible Tensors

A Cartesian tensor $\hat{T}_{ij}$ can be decomposed into three parts:

$$\hat{T}_{ij} = \hat{T}_{ij}^{(0)} + \hat{T}_{ij}^{(1)} + \hat{T}_{ij}^{(2)}, \qquad (7.301)$$

with

$$\hat{T}_{ij}^{(0)} = \frac{1}{3} \delta_{i,j} \sum_{i=1}^{3} \hat{T}_{ii}, \qquad (7.302)$$

$$\hat{T}_{ij}^{(1)} = \frac{1}{2} (\hat{T}_{ij} - \hat{T}_{ji}) \qquad (i \neq j), \qquad (7.303)$$

$$\hat{T}_{ij}^{(2)} = \frac{1}{2} (\hat{T}_{ij} + \hat{T}_{ji}) - \hat{T}_{ij}^{(0)}. \qquad (7.304)$$

Notice that if we add equations (7.302), (7.303), and (7.304), we end up with an identity relation: $\hat{T}_{ij} = \hat{T}_{ij}$.

The term $\hat{T}_{ij}^{(0)}$ has only one component and transforms like a scalar under rotations. The second term $\hat{T}_{ij}^{(1)}$ is an antisymmetric tensor of rank 1 which has three independent components; it transforms like a vector. The third term $\hat{T}_{ij}^{(2)}$ is a symmetric second-rank tensor with zero trace, and hence has five independent components; $\hat{T}_{ij}^{(2)}$ cannot be reduced further to tensors of lower rank. These five components define an irreducible second-rank tensor.

In general, any tensor of rank k can be decomposed into tensors of lower rank that are expressed in terms of linear combinations of its 3^k components. However, there always remain

$(2k + 1)$ components that behave as a tensor of rank k which cannot be reduced further. These $(2k + 1)$ components are symmetric and traceless with respect to any two indices; they form the components of an *irreducible tensor* of rank k.

Equations (7.301) to (7.304) show how to decompose a Cartesian tensor operator, $\hat{T}_{ij}$, into a sum of irreducible *spherical* tensor operators $\hat{T}_{ij}^{(0)}, T_{ij}^{(1)}, T_{ij}^{(2)}$. Cartesian tensors are not very suitable for studying transformations under rotations, because they are reducible whenever their rank exceeds 1. In problems that display spherical symmetry, such as those encountered in subatomic physics, spherical tensors are very useful simplifying tools. It is therefore interesting to consider irreducible spherical tensor operators.

7.4.3.2 Irreducible Spherical Tensors

Let us now focus only on the representation of irreducible tensor operators in spherical coordinates. An irreducible spherical tensor operator of rank k (k is integer) is a set of $(2k + 1)$ operators $T_q^{(k)}$, with $q = -k, \ldots, k$, which transform in the same way as angular momentum under a rotation of axes. For example, the case $k = 1$ corresponds to a vector. The quantities $T_q^{(1)}$ are related to the components of the vector $\hat{\vec{A}}$ as follows (see (7.295)):

$$\hat{T}_{\pm 1}^{(1)} = \mp \frac{1}{\sqrt{2}}(\hat{A}_x \pm \hat{A}_y), \qquad \hat{T}_0^{(1)} = \hat{A}_z. \tag{7.305}$$

In what follows we are going to study some properties of spherical tensor operators and then determine how they transform under rotations.

First, let us look at the various commutation relations of spherical tensors with the angular momentum operator. Since a vector operator is a tensor of rank 1, we can rewrite equations (7.298) to (7.299), respectively, as follows:

$$\left[\hat{J}_z, \hat{T}_q^{(1)}\right] = \hbar q \hat{T}_q^{(1)} \qquad (q = -1, 0, 1), \tag{7.306}$$

$$\left[\hat{J}_\pm, \hat{T}_q^{(1)}\right] = \hbar \sqrt{1(1+1) - q(q \pm 1)} \hat{T}_{q \pm 1}^{(1)}, \tag{7.307}$$

where we have adopted the notation $\hat{A}_q = \hat{T}_q^{(1)}$. We can easily generalize these two relations to any spherical tensor of rank k, $\hat{T}_q^{(k)}$, and obtain these commutators:

$$\boxed{\left[\hat{J}_z, \hat{T}_q^{(k)}\right] = \hbar q \hat{T}_q^{(k)} \qquad (q = -k, -k+1, \ldots, k-1, k),} \tag{7.308}$$

$$\boxed{\left[\hat{J}_\pm, \hat{T}_q^{(k)}\right] = \hbar \sqrt{k(k+1) - q(q \pm 1)} \hat{T}_{q \pm 1}^{(k)}.} \tag{7.309}$$

Using the relations

$$\langle k, q' \mid \hat{J}_z \mid k, q \rangle = \hbar q \langle k, q' \mid k, q \rangle = \hbar q \delta_{q', q}, \tag{7.310}$$

$$\langle k, q' \mid \hat{J}_\pm \mid k, q \rangle = \hbar \sqrt{k(k+1) - q(q \pm 1)} \delta_{q', q \pm 1}, \tag{7.311}$$

along with (7.308) and (7.309), we can write

$$\sum_{q'=-k}^{k} \hat{T}_{q'}^{(k)} \langle k, q' \mid \hat{J}_z \mid k, q \rangle = \hbar q \hat{T}_q^{(k)} = \left[\hat{J}_z, \hat{T}_q^{(k)}\right], \tag{7.312}$$

$$\sum_{q'=-k}^{k} \hat{T}_{q'}^{(k)} \langle k, q' \mid \hat{J}_{\pm} \mid k, q \rangle = \hbar \sqrt{k(k+1) - q(q \pm 1)} \hat{T}_{q\pm 1}^{(k)} = \left[\hat{J}_{\pm}, \ \hat{T}_{q}^{(k)} \right]. \quad (7.313)$$

The previous two relations can be combined into

$$\boxed{\left[\hat{\vec{J}}, \ \hat{T}_{q}^{(k)} \right] = \sum_{q'=-k}^{k} \hat{T}_{q'}^{(k)} \langle k, q' \mid \hat{\vec{J}} \mid k, q \rangle} \quad (7.314)$$

or

$$\boxed{\left[\vec{n} \cdot \hat{\vec{J}}, \ \hat{T}_{q}^{(k)} \right] = \sum_{q'=-k}^{k} \hat{T}_{q'}^{(k)} \langle k, q' \mid \vec{n} \cdot \hat{\vec{J}} \mid k, q \rangle.} \quad (7.315)$$

Having determined the commutation relations of the tensor operators with the angular momentum (7.315), we are now well equipped to study how irreducible spherical tensor operators transform under rotations. Using (7.276) we can write the transformation relation of a spherical tensor $T_q^{(k)}$ under an infinitesimal rotation as follows:

$$\hat{R}_n^{\dagger}(\delta\theta) \hat{T}_q^{(k)} \hat{R}_n(\delta\theta) = T_q^{(k)} + \frac{i}{\hbar} \delta\theta \left[\vec{n} \cdot \hat{\vec{J}}, \ \hat{T}_q^{(k)} \right]. \quad (7.316)$$

Inserting (7.315) into (7.316), we obtain

$$\hat{R}^{\dagger}(\delta\theta) \hat{T}_q^{(k)} \hat{R}(\delta\theta) = \sum_{q'=-k}^{k} \hat{T}_{q'}^{(k)} \langle k, q' \mid 1 + \frac{i}{\hbar} \delta\theta \, \vec{n} \cdot \hat{\vec{J}} \mid k, q \rangle = \sum_{q'} \hat{T}_{q'}^{(k)} \langle k, q' \mid e^{i\delta\theta \, \vec{n} \cdot \hat{\vec{J}}/\hbar} \mid k, q \rangle.$$

$$(7.317)$$

This result also holds for finite rotations

$$\boxed{\hat{R}^{\dagger}(\alpha, \beta, \gamma) \hat{T}_q^{(k)} \hat{R}(\alpha, \beta, \gamma) = \sum_{q'=-k}^{k} \hat{T}_{q'}^{(k)} \langle k, q' \mid \hat{R}^{\dagger}(\alpha, \beta, \gamma) \mid k, q \rangle = \sum_{q'} \hat{T}_{q'}^{(k)} D_{q'q}^{(k)}(\alpha, \beta, \gamma).}$$

$$(7.318)$$

7.4.4   Wigner–Eckart Theorem for Spherical Tensor Operators

Taking the matrix elements of (7.308) between eigenstates of $\hat{\vec{J}}^2$ and $\hat{J}_z$, we find

$$\langle j', m' \mid \left[\hat{J}_z, \ \hat{T}_q^{(k)} \right] - \hbar q \hat{T}_q^{(k)} \mid j, m \rangle = 0 \quad (7.319)$$

or

$$(m' - m - q)\langle j', m' \mid \hat{T}_q^{(k)} \mid j, m \rangle = 0. \quad (7.320)$$

This implies that $\langle j', m' \mid \hat{T}_q^{(k)} \mid j, m \rangle$ vanishes unless $m' = m + q$. This property suggests that the quantity $\langle j', m' \mid \hat{T}_q^{(k)} \mid j, m \rangle$ must be proportional to the Clebsch–Gordan coefficient $\langle j', m' \mid j, k; m, q \rangle$; hence (7.320) leads to

$$(m' - m - q)\langle j', m' \mid j, k; m, q \rangle = 0. \quad (7.321)$$

Now, taking the matrix elements of (7.309) between $| j, m \rangle$ and $|j', m' \rangle$, we obtain

$$\sqrt{(j' \pm m')(j' \mp m' + 1)} \langle j', m' \mp 1| \hat{T}_q^{(k)} | j, m \rangle$$
$$= \sqrt{(j \mp m)(j \pm m + 1)} \langle j', m' | \hat{T}_q^{(k)} | j, m \pm 1 \rangle$$
$$+ \sqrt{(k \mp q)(k \pm q + 1)} \langle j', m' | \hat{T}_{q \pm 1}^{(k)} | j, m \rangle. \tag{7.322}$$

This equation has a structure which is identical to the recursion relation (7.150). For instance, substituting $j = j', m = m', j_1 = j, m_1 = m, j_2 = k, m_2 = q$ into (7.150), we end up with

$$\sqrt{(j' \pm m')(j' \mp m' + 1)} \langle j', m' \mp 1 | j, k; m, q \rangle$$
$$= \sqrt{(j \mp m)(j \pm m + 1)} \langle j', m' | j, k; m \pm 1, q \rangle$$
$$+ \sqrt{(k \mp q)(k \pm q + 1)} \langle j', m' | j, k; m, q \pm 1 \rangle. \tag{7.323}$$

A comparison of (7.320) with (7.321) and (7.322) with (7.323) suggests that the dependence of $\langle j', m' | T_q^{(k)} | j, m \rangle$ on m', m, q is through a Clebsch–Gordan coefficient. The dependence, however, of $\langle j', m' | T_q^{(k)} | j, m \rangle$ on j', j, k has yet to be determined.

We can now state the **Wigner–Eckart theorem:** *The matrix elements of spherical tensor operators $\hat{T}_q^{(k)}$ with respect to angular momentum eigenstates $| j, m \rangle$ are given by*

$$\boxed{\langle j', m' | \hat{T}_q^{(k)} | j, m \rangle = \langle j, k; m, q | j', m' \rangle \langle j' \| \hat{T}^{(k)} \| j \rangle.} \tag{7.324}$$

The factor $\langle j' \| \hat{T}^{(k)} \| j \rangle$, which depends only on j', j, k, is called the *reduced matrix element* of the tensor $\hat{T}_q^{(k)}$ (note that the double bars notation is used to distinguish the reduced matrix elements, $\langle j' \| \hat{T}^{(k)} \| j \rangle$, from the matrix elements, $\langle j', m' | \hat{T}_q^{(k)} | j, m \rangle$). The theorem implies that the matrix elements $\langle j', m' | \hat{T}_q^{(k)} | j, m \rangle$ are written as the product of two terms: a Clebsch–Gordan coefficient $\langle j, k; m, q | j', q' \rangle$—which depends on the geometry of the system (i.e., the orientation of the system with respect to the z-axis), but not on its dynamics (i.e., j', j, k)—and a dynamical factor, the reduced matrix element, which does not depend on the orientation of the system in space (m', q, m). The quantum numbers m', m, q—which specify the projections of the angular momenta $\hat{\vec{J}}'$, $\hat{\vec{J}}$, and $\vec{k}$ onto the z-axis—give the orientation of the system in space, for they specify its orientation with respect to the z-axis. As for j', j, k, they are related to the dynamics of the system, not to its orientation in space.

Wigner–Eckart theorem for a scalar operator
The simplest application of the Wigner–Eckart theorem is when dealing with a scalar operator $\hat{B}$. As seen above, a scalar is a tensor of rank $k = 0$; hence $q = 0$ as well; thus, equation (7.324) yields

$$\langle j', m' | \hat{B} | j, m \rangle = \langle j, 0; m, 0 | j', m' \rangle \langle j' \| \hat{B} \| j \rangle = \langle j' \| \hat{B} \| j \rangle \delta j' j \, \delta m' m, \tag{7.325}$$

since $\langle j, 0; m, 0 | j', m' \rangle = \delta j' j \, \delta m' m$.

Wigner–Eckart theorem for a vector operator
As shown in (7.305), a vector is a tensor of rank 1: $T^{(1)} = A^{(1)} = \hat{\vec{A}}$, with $A_0^{(1)} = A_0 = A_z$ and $A_{\pm 1}^{(1)} = A_{\pm 1} = \mp(\hat{A}_x \pm \hat{A}_y)/\sqrt{2}$. An application of (7.324) to the q-component of a vector

operator $\hat{A}$ leads to

$$\langle j', m'|\hat{A}_q \mid j, m\rangle = \langle j, 1; m, q|j', m'\rangle\langle j' \parallel \hat{\vec{A}} \parallel j\rangle. \tag{7.326}$$

For instance, in the case of the angular momentum $\hat{\vec{J}}$, we have

$$\langle j', m'|\hat{J}_q \mid j, m\rangle = \langle j, 1; m, q|j', m'\rangle\langle j' \parallel \hat{\vec{J}} \parallel j\rangle. \tag{7.327}$$

Applying this relation to the component $\hat{J}_0$,

$$\langle j', m'|\hat{J}_0 \mid j, m\rangle = \langle j, 1; m, 0|j', m'\rangle\langle j' \parallel \hat{\vec{J}} \parallel j\rangle. \tag{7.328}$$

Since $\langle j', m'|\hat{J}_0 \mid j, m\rangle = \hbar m\, \delta j' j\, \delta m' m$ and the coefficient $\langle j, 1; m, 0 \mid j, m\rangle$ is equal to $\langle j, 1; m, 0 \mid j, m\rangle = m/\sqrt{j(j+1)}$, we have

$$\hbar m\, \delta j' j\, \delta m' m = \frac{m}{\sqrt{j(j+1)}}\langle j' \parallel \hat{\vec{J}} \parallel j\rangle \implies \langle j' \parallel \hat{\vec{J}} \parallel j\rangle = \hbar\sqrt{j(j+1)}\delta_{j', j}. \tag{7.329}$$

Due to the selection rules imposed by the Clebsch–Gordan coefficients, we see from (7.326) that *a spin zero particle cannot have a dipole moment*. Since $\langle 0, 1; 0, q|0, 0\rangle = 0$, we have $\langle 0, 0|\hat{L}_q \mid 0, 0\rangle = \langle 0, 1; 0, q|0, 0\rangle\langle 0 \parallel \hat{\vec{L}} \parallel 0\rangle = 0$; the dipole moment is $\hat{\vec{\mu}} = -q\hat{\vec{L}}/(2mc)$. Similarly, *a spin $\frac{1}{2}$ particle cannot have a quadrupole moment*, because as $\langle \frac{1}{2}, 2; m, q|\frac{1}{2}, m'\rangle = 0$, we have $\langle \frac{1}{2}, m'|\hat{T}_q^{(2)}|\frac{1}{2}, m\rangle = \langle \frac{1}{2}, 2; m, q|\frac{1}{2}m'\rangle\langle \frac{1}{2} \parallel \hat{T}^{(2)} \parallel \frac{1}{2}\rangle = 0$.

Wigner–Eckart theorem for a scalar product $\hat{\vec{J}} \cdot \hat{\vec{A}}$

On the one hand, since $\hat{\vec{J}} \cdot \hat{\vec{A}} = \hat{J}_0\hat{A}_0 - \hat{J}_{+1}\hat{A}_{-1} - \hat{J}_{-1}\hat{A}_{+1}$ and since $\hat{J}_0 \mid j, m\rangle = \hbar m \mid j, m\rangle$ and $\hat{J}_{\pm 1} \mid j, m\rangle = (\hbar/2)\sqrt{j(j+1) - m(m \pm 1)}|jm \pm 1\rangle$, we have

$$\langle j, m|\hat{\vec{J}} \cdot \hat{\vec{A}} \mid j, m\rangle = \hbar m\langle j, m \mid \hat{A}_0 \mid j, m\rangle - \frac{\hbar}{2}\sqrt{j(j+1) - m(m+1)}\langle j, m+1|\hat{A}_{+1} \mid j, m\rangle$$
$$+ \frac{\hbar}{2}\sqrt{j(j+1) - m(m-1)}\langle j, m-1|\hat{A}_{+1} \mid j, m\rangle. \tag{7.330}$$

On the other hand, from the Wigner–Eckart theorem (7.324) we have $\langle j, m \mid \hat{A}_0 \mid j, m\rangle = \langle j, 1; m, 0 \mid j, m\rangle\langle j \parallel \hat{\vec{A}} \parallel j\rangle$, $\langle j, m+1|\hat{A}_{+1} \mid j, m\rangle = \langle j, 1; m, 1|j, m+1\rangle\langle j \parallel \hat{\vec{A}} \parallel j\rangle$ and $\langle j, m-1|\hat{A}_{-1} \mid j, m\rangle = \langle j, 1; m, -1|j, m-1\rangle\langle j \parallel \hat{\vec{A}} \parallel j\rangle$; substituting these terms into (7.330) we obtain

$$\langle j, m|\hat{\vec{J}} \cdot \hat{\vec{A}} \mid j, m\rangle = \Bigg[\hbar m\langle j, 1; m, 0 \mid j, m\rangle$$
$$- \frac{\hbar}{2}\langle j, 1; m, 1|j, m+1\rangle\sqrt{j(j+1) - m(m+1)}$$
$$+ \frac{\hbar}{2}\langle j, 1; m, -1|j, m-1\rangle\sqrt{j(j+1) - m(m-1)} \Bigg] \langle j \parallel \hat{\vec{A}} \parallel j\rangle. \tag{7.331}$$

When $\hat{\vec{A}} = \hat{\vec{J}}$ this relation leads to

$$
\langle j, m | \hat{\vec{J}}^2 | j, m \rangle = \Big[\hbar m \langle j, 1; m, 0 | j, m \rangle
$$
$$
- \frac{\hbar}{2} \langle j, 1; m, 1 | j, m+1 \rangle \sqrt{j(j+1) - m(m+1)}
$$
$$
+ \frac{\hbar}{2} \langle j, 1; m, -1 | j, m-1 \rangle \sqrt{j(j+1) - m(m-1)} \Big] \langle j \| \hat{\vec{J}} \| j \rangle.
$$
(7.332)

We are now equipped to obtain a relation between the matrix elements of a vector operator $\hat{\vec{A}}$ and the matrix elements of the scalar operator $\hat{\vec{J}} \cdot \hat{\vec{A}}$; this relation is useful in the calculation of the hydrogen's energy corrections due to the Zeeman effect (see Chapter 9). For this, we need to calculate two ratios: the first is between (7.326) and (7.327)

$$
\frac{\langle j, m' | \hat{A}_q | j, m \rangle}{\langle j, m' | \hat{J}_q | j, m \rangle} = \frac{\langle j \| \hat{\vec{A}} \| j \rangle}{\langle j \| \hat{\vec{J}} \| j \rangle}
$$
(7.333)

and the second is between (7.331) and (7.332)

$$
\frac{\langle j, m | \hat{\vec{J}} \cdot \hat{\vec{A}} | j, m \rangle}{\langle j, m | \hat{\vec{J}}^2 | j, m \rangle} = \frac{\langle j \| \hat{\vec{A}} \| j \rangle}{\langle j \| \hat{\vec{J}} \| j \rangle} \implies \frac{\langle j m | \hat{\vec{J}} \cdot \hat{\vec{A}} | j, m \rangle}{\hbar^2 j(j+1)^2} = \frac{\langle j \| \hat{\vec{A}} \| j \rangle}{\langle j \| \hat{\vec{J}} \| j \rangle},
$$
(7.334)

since $\langle j m | \hat{\vec{J}}^2 | j, m \rangle = \hbar^2 j(j+1)$. Equating (7.333) and (7.334) we obtain

$$
\boxed{ \langle j, m' | \hat{A}_q | j, m \rangle = \frac{\langle j, m | \hat{\vec{J}} \cdot \hat{\vec{A}} | j, m \rangle}{\hbar^2 j(j+1)} \langle j, m' | \hat{J}_q | j, m \rangle. }
$$
(7.335)

An important application of this relation pertains to the case where the vector operator $\hat{\vec{A}}$ is a spin angular momentum $\hat{\vec{S}}$. Since

$$
\hat{\vec{J}} \cdot \hat{\vec{S}} = (\hat{\vec{L}} + \hat{\vec{S}}) \cdot \hat{\vec{S}} = \hat{\vec{L}} \cdot \hat{\vec{S}} + \hat{S}^2 = \frac{(\hat{\vec{L}} + \hat{\vec{S}})^2 - \hat{L}^2 - \hat{S}^2}{2} + \hat{S}^2 = \frac{\hat{J}^2 - \hat{L}^2 - \hat{S}^2}{2} + \hat{S}^2
$$
$$
= \frac{\hat{J}^2 - \hat{L}^2 + \hat{S}^2}{2},
$$
(7.336)

and since $| j, m \rangle$ is a joint eigenstate of $\hat{J}^2$, $\hat{L}^2$, $\hat{S}^2$ and $\hat{J}_z$ with eigenvalues $\hbar^2 j(j+1)$, $\hbar^2 l(l+1)$, $\hbar^2 s(s+1)$, and $\hbar m$, respectively, the matrix element of $\hat{S}_z$ then becomes easy to calculate from (7.335):

$$
\langle j, m | \hat{S}_z | j, m \rangle = \frac{\langle j, m | \hat{\vec{J}} \cdot \hat{\vec{S}} | j, m \rangle}{\hbar^2 j(j+1)} \langle j, m | \hat{J}_z | j, m \rangle = \frac{j(j+1) - l(l+1) + s(s+1)}{2j(j+1)} \hbar m.
$$
(7.337)

7.5   Solved Problems

Problem 7.1

(a) Show how $\hat{J}_x$ and $\hat{J}_y$ transform under a rotation of (finite) angle α about the z-axis. Using these results, determine how the angular momentum operator $\vec{\hat{J}}$ transform under the rotation.

(b) Show how a vector operator $\vec{\hat{A}}$ transforms under a rotation of angle α about the y-axis.

(c) Show that $e^{i\pi \hat{J}_z/\hbar} e^{i\alpha \hat{J}_y/\hbar} e^{-i\pi \hat{J}_z/\hbar} = e^{-i\alpha \hat{J}_y/\hbar}$.

Solution

(a) The operator corresponding to a rotation of angle α about the z-axis is given by $\hat{R}_z(\alpha) = e^{-i\alpha \hat{J}_z/\hbar}$. Under this rotation, an operator $\hat{B}$ transforms like $\hat{B}' = \hat{R}_z^\dagger \hat{B} \hat{R}_z = e^{i\alpha \hat{J}_z/\hbar} \hat{B} e^{-i\alpha \hat{J}_z/\hbar}$. Using the relation

$$e^{\hat{A}} \hat{B} e^{-\hat{A}} = \hat{B} + \left[\hat{A}, \hat{B}\right] + \frac{1}{2!}\left[\hat{A}, \left[\hat{A}, \hat{B}\right]\right] + \frac{1}{3!}\left[\hat{A}, \left[\hat{A}, \left[\hat{A}, \hat{B}\right]\right]\right] + \cdots, \qquad (7.338)$$

along with the commutation relations $\left[\hat{J}_z, \hat{J}_y\right] = -i\hbar \hat{J}_x$ and $\left[\hat{J}_z, \hat{J}_x\right] = i\hbar \hat{J}_y$, we have

$$\begin{aligned}
e^{i\alpha \hat{J}_z/\hbar} \hat{J}_x e^{-i\alpha \hat{J}_z/\hbar} &= \hat{J}_x + \frac{i\alpha}{\hbar}\left[\hat{J}_z, \hat{J}_x\right] - \frac{\alpha^2}{2!\hbar^2}\left[\hat{J}_z, \left[\hat{J}_z, \hat{J}_x\right]\right] \\
&\quad - \frac{i\alpha^3}{3!\hbar^3}\left[\hat{J}_z, \left[\hat{J}_z, \left[\hat{J}_z, \hat{J}_x\right]\right]\right] + \cdots \\
&= \hat{J}_x - \alpha \hat{J}_y - \frac{\alpha^2}{2!}\hat{J}_x + \frac{\alpha^3}{3!}\hat{J}_y + \frac{\alpha^4}{4!}\hat{J}_x - \frac{\alpha^5}{5!}\hat{J}_y + \cdots \\
&= \hat{J}_x\left(1 - \frac{\alpha^2}{2!} + \frac{\alpha^4}{4!} + \cdots\right) - \hat{J}_y\left(\alpha - \frac{\alpha^3}{3!} + \frac{\alpha^5}{5!} - \cdots\right) \\
&= \hat{J}_x \cos\alpha - \hat{J}_y \sin\alpha. \qquad (7.339)
\end{aligned}$$

Similarly, we can show that

$$e^{i\alpha \hat{J}_z/\hbar} \hat{J}_y e^{-i\alpha \hat{J}_z/\hbar} = \hat{J}_y \cos\alpha + \hat{J}_x \sin\alpha. \qquad (7.340)$$

As $\hat{J}_z$ is invariant under an arbitrary rotation about the z-axis ($e^{i\alpha \hat{J}_z/\hbar} \hat{J}_z e^{-i\alpha \hat{J}_z/\hbar} = \hat{J}_z$), we can condense equations (7.339) and (7.340) into a single matrix relation:

$$e^{i\alpha \hat{J}_z/\hbar} \vec{\hat{J}} e^{-i\alpha \hat{J}_z/\hbar} = \begin{pmatrix} \cos\alpha & -\sin\alpha & 0 \\ \sin\alpha & \cos\alpha & 0 \\ 0 & 0 & 1 \end{pmatrix} \begin{pmatrix} \hat{J}_x \\ \hat{J}_y \\ \hat{J}_z \end{pmatrix}. \qquad (7.341)$$

(b) Using the commutation relations $\left[\hat{J}_y, \hat{A}_x\right] = -i\hbar \hat{A}_z$ and $\left[\hat{J}_y, \hat{A}_z\right] = i\hbar \hat{A}_x$ (see (7.282) to (7.284)) along with (7.338), we have

$$e^{i\alpha \hat{J}_y/\hbar} \hat{A}_x e^{-i\alpha \hat{J}_y/\hbar} = \hat{A}_x + \frac{i\alpha}{\hbar}\left[\hat{J}_y, \hat{A}_x\right] - \frac{\alpha^2}{2!\hbar^2}\left[\hat{J}_y, \left[\hat{J}_y, \hat{A}_x\right]\right]$$

$$- \frac{i\alpha^3}{3!\hbar^3} \left[\hat{J}_y, \left[\hat{J}_y, \left[\hat{J}_y, \hat{A}_x \right] \right] \right] + \cdots$$

$$= \hat{A}_x + \alpha \hat{A}_z - \frac{\alpha^2}{2!} \hat{A}_x - \frac{\alpha^3}{3!} \hat{A}_z + \frac{\alpha^4}{4!} \hat{A}_x + \frac{\alpha^5}{5!} \hat{J}_z + \cdots$$

$$= \hat{A}_x \left(1 - \frac{\alpha^2}{2!} + \frac{\alpha^4}{4!} + \cdots \right) + \hat{A}_z \left(\alpha - \frac{\alpha^3}{3!} + \frac{\alpha^5}{5!} + \cdots \right)$$

$$= \hat{A}_x \cos \alpha + \hat{A}_z \sin \alpha. \qquad (7.342)$$

Similarly, we can show that

$$\hat{A}'_z = e^{i\alpha \hat{J}_y/\hbar} \hat{A}_z e^{-i\alpha \hat{J}_y/\hbar} = -\hat{A}_x \sin \alpha + \hat{A}_z \cos \alpha. \qquad (7.343)$$

Also, since $\hat{A}_y$ is invariant under an arbitrary rotation about the y-axis, we may combine equations (7.342) and (7.343) to find the vector operator $\hat{\vec{A}}'$ obtained by rotating $\hat{\vec{A}}$ through an angle α about the y-axis:

$$\hat{\vec{A}}' = e^{i\alpha \hat{J}_y/\hbar} \hat{\vec{A}} e^{-i\alpha \hat{J}_y/\hbar} = \begin{pmatrix} \cos \alpha & 0 & \sin \alpha \\ 0 & 1 & 0 \\ -\sin \alpha & 0 & \cos \alpha \end{pmatrix} \begin{pmatrix} \hat{A}_x \\ \hat{A}_y \\ \hat{A}_z \end{pmatrix}. \qquad (7.344)$$

(c) Expanding $e^{i\alpha \hat{J}_y/\hbar}$ and then using (7.340), we obtain

$$e^{i\pi \hat{J}_z/\hbar} e^{i\alpha \hat{J}_y/\hbar} e^{-i\pi \hat{J}_z/\hbar} = \sum_{n=0}^{\infty} \frac{(i\alpha/\hbar)^n}{n!} e^{i\pi \hat{J}_z/\hbar} \left(\hat{J}_y \right)^n e^{-i\pi \hat{J}_z/\hbar}$$

$$= \sum_{n=0}^{\infty} \frac{(i\alpha/\hbar)^n}{n!} \left(\hat{J}_y \cos \pi + \hat{J}_x \sin \pi \right)^n = \sum_{n=0}^{\infty} \frac{(-i\alpha/\hbar)^n}{n!} \left(\hat{J}_y \right)^n$$

$$= e^{-i\alpha \hat{J}_y/\hbar}. \qquad (7.345)$$

Problem 7.2

Use the Pauli matrices $\sigma_x = \begin{pmatrix} 0 & 1 \\ 1 & 0 \end{pmatrix}$, $\sigma_y = \begin{pmatrix} 0 & -i \\ i & 0 \end{pmatrix}$, and $\sigma_z = \begin{pmatrix} 1 & 0 \\ 0 & -1 \end{pmatrix}$, to show that

(a) $e^{-i\alpha \sigma_x} = I \cos \alpha - i \sigma_x \sin \alpha$, where I is the unit matrix,

(b) $e^{i\alpha \sigma_x} \sigma_z e^{-i\alpha \sigma_x} = \sigma_z \cos(2\alpha) + \sigma_y \sin(2\alpha)$.

Solution

(a) Using the expansion

$$e^{-i\alpha \sigma_x} = \sum_{n=0}^{\infty} \frac{(-i)^{2n}}{(2n)!} (\alpha)^{2n} \sigma_x^{2n} + \sum_{n=0}^{\infty} \frac{(-i)^{2n+1}}{(2n+1)!} (\alpha)^{2n+1} \sigma_x^{2n+1}, \qquad (7.346)$$

and since $\sigma_x^2 = 1$, $\sigma_x^{2n} = I$, and $\sigma_x^{2n+1} = \sigma_x$, where I is the unit matrix, we have

$$e^{-i\alpha \sigma_x} = \begin{pmatrix} 1 & 0 \\ 0 & 1 \end{pmatrix} \sum_{n=0}^{\infty} \frac{(-1)^n}{(2n)!} (\alpha)^{2n} - i \sigma_x \sum_{n=0}^{\infty} \frac{(-1)^n}{(2n+1)!} (\alpha)^{2n+1}$$

$$= I \cos \alpha - i \sigma_x \sin \alpha. \qquad (7.347)$$

(b) From (7.347) we can write

$$e^{i\alpha\sigma_x}\sigma_z e^{-i\alpha\sigma_x} = (\cos\alpha + i\sigma_x\sin\alpha)\sigma_z(\cos\alpha - i\sigma_x\sin\alpha)$$
$$= \sigma_z\cos^2\alpha + \sigma_x\sigma_z\sigma_x\sin^2\alpha + i[\sigma_x, \sigma_z]\sin\alpha\cos\alpha,$$

$$(7.348)$$

which, when using the facts that $\sigma_x\sigma_z = -\sigma_z\sigma_x$, $\sigma_x^2 = I$, and $[\sigma_x, \sigma_z] = -2i\sigma_y$, reduces to

$$e^{i\alpha\sigma_x}\sigma_z e^{-i\alpha\sigma_x} = \sigma_z\cos^2\alpha - \sigma_z\sigma_x^2\sin^2\alpha + 2\sigma_y\sin\alpha\cos\alpha$$
$$= \sigma_z(\cos^2\alpha - \sin^2\alpha) + \sigma_y\sin(2\alpha)$$
$$= \sigma_z\cos(2\alpha) + \sigma_y\sin(2\alpha).$$

$$(7.349)$$

Problem 7.3

Find the Clebsch–Gordan coefficients associated with the addition of two angular momenta $j_1 = 1$ and $j_2 = 1$.

Solution

The addition of $j_1 = 1$ and $j_2 = 1$ is encountered, for example, in a two-particle system where the angular momenta of both particles are orbital.

The allowed values of the total angular momentum are between $|j_1 - j_2| \le j \le j_1 + j_2$; hence $j = 0, 1, 2$. To calculate the relevant Clebsch–Gordan coefficients, we need to find the basis vectors $\{|j, m\rangle\}$, which are common eigenvectors of $\hat{J}_1^2$, $\hat{J}_2^2$, $\hat{J}^2$ and $\hat{J}_z$, in terms of $\{|1, 1; m_1, m_2\rangle\}$.

Eigenvectors $|j, m\rangle$ associated with $j = 2$
The state $|2, 2\rangle$ is simply given by

$$\boxed{|2, 2\rangle = |1, 1; 1, 1\rangle;}$$

$$(7.350)$$

the corresponding Clebsch–Gordan coefficient is thus given by $\langle 1, 1; 1, 1 | 2, 2\rangle = 1$.

As for $|2, 1\rangle$, it can be found by applying J_- to $|2, 2\rangle$ and $(J_{1-} + J_{2-})$ to $|1, 1; 1, 1\rangle$, and then equating the two results

$$J_-|2, 2\rangle = (J_{1-} + J_{2-})|1, 1; 1, 1\rangle.$$

$$(7.351)$$

This leads to

$$2\hbar|2, 1\rangle = \sqrt{2}\hbar\left(|1, 1; 1, 0\rangle + |1, 1; 0, 1\rangle\right)$$

$$(7.352)$$

or to

$$\boxed{|2, 1\rangle = \frac{1}{\sqrt{2}}\left(|1, 1; 1, 0\rangle + |1, 1; 0, 1\rangle\right);}$$

$$(7.353)$$

hence $\langle 1, 1; 1, 0 | 2, 1\rangle = \langle 1, 1; 0, 1 | 2, 1\rangle = 1/\sqrt{2}$. Using (7.353), we can find $|2, 0\rangle$ by applying J_- to $|2, 1\rangle$ and $(J_{1-} + J_{2-})$ to $[|1, 1; 1, 0\rangle + |1, 1; 0, 1\rangle]$:

$$J_-|2, 1\rangle = \frac{1}{\sqrt{2}}\hbar(J_{1-} + J_{2-})[|1, 1; 1, 0\rangle + |1, 1; 0, 1\rangle],$$

$$(7.354)$$

which leads to

$$| 2, 0\rangle = \frac{1}{\sqrt{6}} \left(|1, 1;\ 1, -1\rangle + 2|1, 1;\ 0, 0\rangle + |1, 1;\ -1, 1\rangle \right); \qquad (7.355)$$

hence $\langle 1, 1;\ 1, -1\ |\ 2,\ 0\rangle = \langle 1, 1;\ -1, 1\ |\ 2,\ 0\rangle = 1/\sqrt{6}$ and $\langle 1, 1;\ 0, 0\ |\ 2,\ 0\rangle = 2/\sqrt{6}$.

Similarly, by repeated applications of J_- and $(J_{1-} + J_{2-})$, we can show that

$$| 2, -1\rangle = \frac{1}{\sqrt{2}} \left(|1, 1;\ 0, -1\rangle + |1, 1;\ -1, 0\rangle \right), \qquad (7.356)$$

$$| 2, -2\rangle = |1, 1;\ -1, -1\rangle, \qquad (7.357)$$

with $\langle 1, 1;\ 0, -1|2, -1\rangle = \langle 1, 1;\ -1, 0|2, -1\rangle = 1/\sqrt{2}$ and $\langle 1, 1;\ -1, -1|2, -2\rangle = 1$.

Eigenvectors $| j,\ m\rangle$ associated with $j = 1$

The relation

$$|1,\ m\rangle = \sum_{m_1=-1}^{1} \sum_{m_2=-1}^{1} \langle 1, 1;\ m_1, m_2|1,\ m\rangle|1, 1;\ m_1, m_2\rangle \qquad (7.358)$$

leads to

$$|1,\ 1\rangle = a|1, 1;\ 1, 0\rangle + b|1, 1;\ 0, 1\rangle, \qquad (7.359)$$

where $a = \langle 1, 1;\ 1, 0\ |\ 1,\ 1\rangle$ and $b = \langle 1, 1;\ 0, 1|1,\ 1\rangle$. Since $|\ 1,\ 1\rangle$, $|1, 1;\ 1, 0\rangle$ and $|1, 1;\ 0, 1\rangle$ are all normalized, and since $|1, 1;\ 1, 0\rangle$ is orthogonal to $|1, 1;\ 0, 1\rangle$ and a and b are real, we have

$$\langle 1, 1\ |\ 1,\ 1\rangle = a^2 + b^2 = 1. \qquad (7.360)$$

Now, since $\langle 2, 1\ |\ 1,\ 1\rangle = 0$, equations (7.353) and (7.359) yield

$$\langle 2, 1\ |\ 1,\ 1\rangle = \frac{a}{\sqrt{2}} + \frac{b}{\sqrt{2}} = 0. \qquad (7.361)$$

A combination of (7.360) and (7.361) leads to $a = -b = \pm 1/\sqrt{2}$. The signs of a and b have yet to be found. The phase convention mandates that coefficients like $\langle j_1, j_2;\ j_1, (j - j_1)|j,\ j\rangle$ must be positive. Thus, we have $a = 1/\sqrt{2}$ and $b = -1/\sqrt{2}$, which when inserted into (7.359) give

$$| 1,\ 1\rangle = \frac{1}{\sqrt{2}} \left(|1, 1;\ 1, 0\rangle - |1, 1;\ 0, 1\rangle \right). \qquad (7.362)$$

This yields $\langle 1, 1;\ 1, 0\ |\ 1,\ 1\rangle = \frac{1}{2}$ and $\langle 1, 1;\ 0, 1\ |\ 1,\ 1\rangle = -\frac{1}{2}$.

To find $|1,\ 0\rangle$ we proceed as we did above when we obtained the states $|\ 2,\ 1\rangle$, $|\ 2,\ 0\rangle$, ..., $|2, -2\rangle$ by repeatedly applying J_- on $|\ 2,\ 2\rangle$. In this way, the application of J_- on $|\ 1,\ 1\rangle$ and $(J_{1-} + J_{2-})$ on $[|1, 1;\ 1, 0\rangle - |1, 1;\ 0, 1\rangle]$,

$$J_-\ |\ 1,\ 1\rangle = \frac{1}{2}\ (J_{1-} + J_{2-})\ [|1, 1;\ 1, 0\rangle - |1, 1;\ 0, 1\rangle] \qquad (7.363)$$

gives

$$\sqrt{2}\hbar\ |\ 1,\ 0\rangle = \frac{\sqrt{2}\hbar}{2}\ [|1, 1;\ 1, -1\rangle - |1, 1;\ -1, 1\rangle], \qquad (7.364)$$

or

$$\boxed{|\, 1,\; 0\rangle = \frac{1}{\sqrt{2}}\Big(|1,1;\; 1,-1\rangle - |1,1;\; -1,1\rangle\Big),}$$

(7.365)

with $\langle 1,1;\; 1,-1\,|\,1,\,0\rangle = \frac{1}{\sqrt{2}}$ and $\langle 1,1;\; -1,1\,|\,1,\,0\rangle = -1/\sqrt{2}$.

Similarly, we can show that

$$\boxed{|\, 1,\; -1\rangle = \frac{1}{\sqrt{2}}\Big(|1,1;\; 0,-1\rangle - |1,1;\; -1,0\rangle\Big);}$$

(7.366)

hence $\langle 1,1;\; 0,-1\,|\,1,\,-1\rangle = 1/\sqrt{2}$ and $\langle 1,1;\; -1,0\,|\,1,\,-1\rangle = -1/\sqrt{2}$.

Eigenvector $|0,\; 0\rangle$ associated with $j = 0$

Since

$$|0,\; 0\rangle = a|1,1;\; 1,-1\rangle + b|1,1;\; 0,0\rangle + c|1,1;\; -1,1\rangle,$$

(7.367)

where $a = \langle 1,1;\; 1,-1\,|\,0,\,0\rangle$, $b = \langle 1,1;\; 0,0\,|\,0,\,0\rangle$, and $c = \langle 1,1;\; -1,1\,|\,0,\,0\rangle$ are real, and since the states $|0,\;0\rangle, |1,1;\; 1,-1\rangle, |1,1;\; 0,0\rangle$, and $|1,1;\; -1,1\rangle$ are normal, we have

$$\langle 0,\; 0\,|\,0,\,0\rangle = a^2 + b^2 + c^2 = 1.$$

(7.368)

Now, combining (7.355), (7.365), and (7.367), we obtain

$$\langle 2,\; 0\,|\,0,\,0\rangle = \frac{a}{\sqrt{6}} + \frac{2b}{\sqrt{6}} + \frac{c}{\sqrt{6}} = 0,$$

(7.369)

$$\langle 1,\; 0\,|\,0,\,0\rangle = \frac{a}{\sqrt{2}} - \frac{c}{\sqrt{2}} = 0.$$

(7.370)

Since a is by convention positive, we can show that the solutions of (7.368), (7.369), and (7.370) are given by $a = 1/\sqrt{3}, b = -1/\sqrt{3}, c = 1/\sqrt{3}$, and consequently

$$\boxed{|0,\; 0\rangle = \frac{1}{\sqrt{3}}\Big(|1,1;\; 1,-1\rangle - |1,1;\; 0,0\rangle + |1,1;\; -1,1\rangle\Big),}$$

(7.371)

with $\langle 1,1;\; 1,-1\,|\,0,\,0\rangle = \langle 1,1;\; -1,1\,|\,0,\,0\rangle = 1/\sqrt{3}$ and $\langle 1,1;\; 0,0\,|\,0,\,0\rangle = -1/\sqrt{3}$.

Note that while the quintuplet states $|2,\; m\rangle$ (with $m = \pm 2, \pm 1, 0$) and the singlet state $|0,\; 0\rangle$ are symmetric, the triplet states $|1,\; m\rangle$ (with $m = \pm 1, 0$) are antisymmetric under space inversion.

Problem 7.4

(a) Find the total spin of a system of three spin $\frac{1}{2}$ particles and derive the corresponding Clebsch–Gordan coefficients.

(b) Consider a system of three nonidentical spin $\frac{1}{2}$ particles whose Hamiltonian is given by $\hat{H} = -\epsilon_0(\vec{S}_1 \cdot \vec{S}_3 + \vec{S}_2 \cdot \vec{S}_3)/\hbar^2$. Find the system's energy levels and their degeneracies.

Solution

(a) To add $j_1 = \frac{1}{2}, j_2 = \frac{1}{2}$, and $j_3 = \frac{1}{2}$, we begin by coupling j_1 and j_2 to form $j_{12} = j_1 + j_2$, where $|j_1 - j_2| \le j_{12} \le |j_1 + j_2|$; hence $j_{12} = 0, 1$. Then we add j_{12} and j_3; this leads to $|j_{12} - j_3| \le j \le |j_{12} + j_3|$ or $j = \frac{1}{2}, \frac{3}{2}$.

We are going to denote the joint eigenstates of $\hat{J}_1^2$, $\hat{J}_2^2$, $\hat{J}_3^2$, $\hat{J}_{12}^2$, $\hat{J}^2$, and J_z by $|j_{12}, j, m\rangle$ and the joint eigenstates of $\hat{J}_1^2$, $\hat{J}_2^2$, $\hat{J}_3^2$, $\hat{J}_{1_z}$, $\hat{J}_{2_z}$, and $\hat{J}_{3_z}$ by $| j_1, j_2, j_3; m_1, m_2, m_3\rangle$; since $j_1 = j_2 = j_3 = \frac{1}{2}$ and $m_1 = \pm\frac{1}{2}$, $m_2 = \pm\frac{1}{2}$, $m_3 = \pm\frac{1}{2}$, we will be using throughout this problem the lighter notation $|j_1, j_2, j_3; \pm, \pm, \pm\rangle$ to abbreviate $| \frac{1}{2}, \frac{1}{2}, \frac{1}{2}; \pm\frac{1}{2}, \pm\frac{1}{2}, \pm\frac{1}{2}\rangle$.

In total there are eight states $|j_{12}, j, m\rangle$ since $(2j_1 + 1)(2j_2 + 1)(2j_3 + 1) = 8$. Four of these correspond to the subspace $j = \frac{3}{2}$: $| 1, \frac{3}{2}, \frac{3}{2}\rangle$, $| 1, \frac{3}{2}, \frac{1}{2}\rangle$, $| 1, \frac{3}{2}, -\frac{1}{2}\rangle$, and $| 1, \frac{3}{2}, -\frac{3}{2}\rangle$. The remaining four belong to the subspace $j = \frac{1}{2}$: $| 0, \frac{1}{2}, \frac{1}{2}\rangle$, $| 0, \frac{1}{2}, -\frac{1}{2}\rangle$, $| 1, \frac{1}{2}, \frac{1}{2}\rangle$, and $| 1, \frac{1}{2}, -\frac{1}{2}\rangle$. To construct the states $|j_{12}, j, m\rangle$ in terms of $|j_1, j_2, j_3; \pm, \pm, \pm\rangle$, we are going to consider the two subspaces $j = \frac{3}{2}$ and $j = \frac{1}{2}$ separately.

Subspace $j = \frac{3}{2}$

First, the states $| 1, \frac{3}{2}, \frac{3}{2}\rangle$ and $| 1, \frac{3}{2}, -\frac{3}{2}\rangle$ are clearly given by

$$\left|1, \frac{3}{2}, \frac{3}{2}\right\rangle = |j_1, j_2, j_3; +, +, +\rangle, \qquad \left|1, \frac{3}{2}, -\frac{3}{2}\right\rangle = |j_1, j_2, j_3; -, -, -\rangle. \tag{7.372}$$

To obtain $| 1, \frac{3}{2}, \frac{1}{2}\rangle$, we need to apply, on the one hand, $\hat{J}_-$ on $| 1, \frac{3}{2}, \frac{3}{2}\rangle$ (see (7.220)),

$$\hat{J}_-\left|1, \frac{3}{2}, \frac{3}{2}\right\rangle = \hbar\sqrt{\frac{3}{2}\left(\frac{3}{2}+1\right) - \frac{3}{2}\left(\frac{3}{2}-1\right)}\left|1, \frac{3}{2}, \frac{1}{2}\right\rangle = \hbar\sqrt{3}\left|1, \frac{3}{2}, \frac{1}{2}\right\rangle, \tag{7.373}$$

and, on the other hand, apply $(\hat{J}_{1-} + \hat{J}_{2-} + \hat{J}_{3-})$ on $|j_1, j_2, j_3; +, +, +\rangle$ (see (7.221) to (7.223)). This yields

$$(\hat{J}_{1-} + \hat{J}_{2-} + \hat{J}_{3-})|j_1, j_2, j_3; +, +, +\rangle = \hbar\Big(|j_1, j_2, j_3; -, +, +\rangle + |j_1, j_2, j_3; +, -, +\rangle$$
$$+ |j_1, j_2, j_3; +, +, -\rangle\Big), \tag{7.374}$$

since $\sqrt{\frac{1}{2}(\frac{1}{2} + 1) - \frac{1}{2}(\frac{1}{2} - 1)} = 1$. Equating (7.373) and (7.374) we infer

$$\left|1, \frac{3}{2}, \frac{1}{2}\right\rangle = \frac{1}{\sqrt{3}}\Big(|j_1, j_2, j_3; -, +, +\rangle + |j_1, j_2, j_3; +, -, +\rangle + |j_1, j_2, j_3; +, +, -\rangle\Big). \tag{7.375}$$

Following the same method—applying $\hat{J}_-$ on $| 1, \frac{3}{2}, \frac{1}{2}\rangle$ and $(\hat{J}_{1-} + \hat{J}_{2-} + \hat{J}_{3-})$ on the right-hand side of (7.375) and then equating the two results—we find

$$\left|1, \frac{3}{2}, -\frac{1}{2}\right\rangle = \frac{1}{\sqrt{3}}\Big(|j_1, j_2, j_3; +, -, -\rangle + |j_1, j_2, j_3; -, +, -\rangle + |j_1, j_2, j_3; -, -, +\rangle\Big). \tag{7.376}$$

Subspace $j = \frac{1}{2}$

We can write $| 0, \frac{1}{2}, \frac{1}{2}\rangle$ as a linear combination of $|j_1, j_2, j_3; +, +, -\rangle$ and $|j_1, j_2, j_3; -, +, +\rangle$:

$$\left|0, \frac{1}{2}, \frac{1}{2}\right\rangle = \alpha|j_1, j_2, j_3; +, +, -\rangle + \beta|j_1, j_2, j_3; -, +, +\rangle. \tag{7.377}$$

Since $| \, 0, \frac{1}{2}, \frac{1}{2} \rangle$ is normalized, while $|j_1, j_2, j_3; +, +, -\rangle$ and $|j_1, j_2, j_3; -, +, +\rangle$ are ortho-normal, and since the Clebsch–Gordan coefficients, such as α and β, are real numbers, equation (7.377) yields

$$\alpha^2 + \beta^2 = 1. \tag{7.378}$$

On the other hand, since $\langle 1, \frac{3}{2}, \frac{1}{2} \, | \, 0, \frac{1}{2}, \frac{1}{2} \rangle = 0$, a combination of (7.375) and (7.377) leads to

$$\frac{1}{\sqrt{3}} (\alpha + \beta) = 0 \quad \Longrightarrow \quad \alpha = -\beta. \tag{7.379}$$

A substitution of $\alpha = -\beta$ into (7.378) yields $\alpha = -\beta = 1/\sqrt{2}$, and substituting this into (7.377) we obtain

$$\left| 0, \frac{1}{2}, \frac{1}{2} \right\rangle = \frac{1}{\sqrt{2}} \left(|j_1, j_2, j_3; +, +, -\rangle - |j_1, j_2, j_3; -, +, +\rangle \right). \tag{7.380}$$

Following the same procedure that led to (7.375)—applying $\hat{J}_-$ on the left-hand side of (7.380) and $(\hat{J}_{1-} + \hat{J}_{2-} + \hat{J}_{3-})$ on the right-hand side and then equating the two results—we find

$$\left| 0, \frac{1}{2}, -\frac{1}{2} \right\rangle = \frac{1}{\sqrt{2}} \left(- |j_1, j_2, j_3; +, -, -\rangle + |j_1, j_2, j_3; -, -, +\rangle \right). \tag{7.381}$$

Now, to find $| \, 1, \frac{1}{2}, \frac{1}{2} \rangle$, we may write it as a linear combination of $|j_1, j_2, j_3; +, +, -\rangle$, $|j_1, j_2, j_3; +, -, +\rangle$, and $|j_1, j_2, j_3; -, +, +\rangle$:

$$\left| 1, \frac{1}{2}, \frac{1}{2} \right\rangle = \alpha |j_1, j_2, j_3; +, +, -\rangle + \beta |j_1, j_2, j_3; +, -, +\rangle + \gamma |j_1, j_2, j_3; -, +, +\rangle. \tag{7.382}$$

This state is orthogonal to $| \, 0, \frac{1}{2}, \frac{1}{2} \rangle$, and hence $\alpha = \gamma$; similarly, since this state is also orthogonal to $| \, 1, \frac{3}{2}, \frac{1}{2} \rangle$, we have $\alpha + \beta + \gamma = 0$, and hence $2\alpha + \beta = 0$ or $\beta = -2\alpha = -2\gamma$. Now, since all the states of (7.382) are orthonormal, we have $\alpha^2 + \beta^2 + \gamma^2 = 1$, which when combined with $\beta = -2\alpha = -2\gamma$ leads to $\alpha = \gamma = -1/\sqrt{6}$ and $\beta = 2/\sqrt{6}$. We may thus write (7.382) as

$$\left| 1, \frac{1}{2}, \frac{1}{2} \right\rangle = \frac{1}{\sqrt{6}} \left(- |j_1, j_2, j_3; +, +, -\rangle + 2|j_1, j_2, j_3; +, -, +\rangle - |j_1, j_2, j_3; -, +, +\rangle \right). \tag{7.383}$$

Finally, applying $\hat{J}_-$ on the left-hand side of (7.383) and $(\hat{J}_{1-} + \hat{J}_{2-} + \hat{J}_{3-})$ on the right-hand side and equating the two results, we find

$$\left| 1, \frac{1}{2}, -\frac{1}{2} \right\rangle = \frac{1}{\sqrt{6}} \left(|j_1, j_2, j_3; +, -, -\rangle - 2|j_1, j_2, j_3; -, +, -\rangle + |j_1, j_2, j_3; -, -, +\rangle \right). \tag{7.384}$$

(b) Since we have three different (nonidentical) particles, their spin angular momenta mutually commute. We may thus write their Hamiltonian as $\hat{H} = -(\epsilon_0/\hbar^2)(\hat{\vec{S}}_1 + \hat{\vec{S}}_2) \cdot \hat{\vec{S}}_3$. Due to this suggestive form of $\hat{H}$, it is appropriate, as shown in (a), to start by coupling $\hat{\vec{S}}_1$ with $\hat{\vec{S}}_2$ to obtain $\hat{\vec{S}}_{12} = \hat{\vec{S}}_1 + \hat{\vec{S}}_2$, and then add $\hat{\vec{S}}_{12}$ to $\hat{\vec{S}}_3$ to generate the total spin: $\hat{\vec{S}} = \hat{\vec{S}}_{12} + \hat{\vec{S}}_3$. We may thus write $\hat{H}$ as

$$\hat{H} = -\frac{\epsilon_0}{\hbar^2} \left(\hat{\vec{S}}_1 + \hat{\vec{S}}_2 \right) \cdot \hat{\vec{S}}_3 = -\frac{\epsilon_0}{\hbar^2} \hat{\vec{S}}_{12} \cdot \hat{\vec{S}}_3 = -\frac{\epsilon_0}{2\hbar^2} \left(\hat{S}^2 - \hat{S}_{12}^2 - \hat{S}_3^2 \right), \tag{7.385}$$

since $\hat{\vec{S}}_{12} \cdot \hat{\vec{S}}_3 = \frac{1}{2}[(\hat{\vec{S}}_{12} + \hat{\vec{S}}_3)^2 - \hat{S}_{12}^2 - \hat{S}_3^2]$. Since the operators $\hat{H}$, $\hat{S}^2$, $\hat{S}_{12}^2$, and $\hat{S}_3^2$ mutually commute, we may select as their joint eigenstates the kets $|s_{12}, s, m\rangle$; we have seen in (a) how to construct these states. The eigenvalues of $\hat{H}$ are thus given by

$$\hat{H}|s_{12}, s, m\rangle = -\frac{\epsilon_0}{2\hbar^2}\left(\hat{S}^2 - \hat{S}_{12}^2 - \hat{S}_3^2\right)|s_{12}, s, m\rangle$$

$$= -\frac{\epsilon_0}{2}\left[s(s+1) - s_{12}(s_{12}+1) - \frac{3}{4}\right]|s_{12}, s, m\rangle, \qquad (7.386)$$

since $s_3 = \frac{1}{2}$ and $\hat{S}_3^2|s_{12}, s, m\rangle = \hbar^2 s_3(s_3+1)|s_{12}, s, m\rangle = (3\hbar^2/4)|s_{12}, s, m\rangle$.

As shown in (7.386), the energy levels of this system are degenerate with respect to m, since they depend on the quantum numbers s and s_{12} but not on m:

$$E_{s_{12},s} = -\frac{\epsilon_0}{2}\left[s(s+1) - s_{12}(s_{12}+1) - \frac{3}{4}\right]. \qquad (7.387)$$

For instance, the energy $E_{s_{12},s} = E_{1,3/2} = -\epsilon_0/2$ is fourfold degenerate, since it corresponds to four different states: $|s_{12}, s, m\rangle = |1, \frac{3}{2}, \pm\frac{3}{2}\rangle$ and $|1, \frac{3}{2}, \pm\frac{1}{2}\rangle$. Similarly, the energy $E_{0,1/2} = 0$ is twofold degenerate; the corresponding states are $|0, \frac{1}{2}, \pm\frac{1}{2}\rangle$. Finally, the energy $E_{1,1/2} = \epsilon_0$ is also twofold degenerate since it corresponds to $|1, \frac{1}{2}, \pm\frac{1}{2}\rangle$.

Problem 7.5

Consider a system of four nonidentical spin $\frac{1}{2}$ particles. Find the possible values of the total spin S of this system and specify the number of angular momentum eigenstates, corresponding to each value of S.

Solution

First, we need to couple two spins at a time: $\hat{\vec{S}}_{12} = \hat{\vec{S}}_1 + \hat{\vec{S}}_2$ and $\hat{\vec{S}}_{34} = \hat{\vec{S}}_3 + \hat{\vec{S}}_4$. Then we couple $\hat{\vec{S}}_{12}$ and $\hat{\vec{S}}_{34}$: $\hat{\vec{S}} = \hat{\vec{S}}_{12} + \hat{\vec{S}}_{34}$. From Problem 7.4, page 438, we have $s_{12} = 0, 1$ and $s_{34} = 0, 1$. In total there are 16 states $|sm\rangle$ since $(2s_1+1)(2s_2+1)(2s_3+1)(2s_4+1) = 2^4 = 16$.

Since $s_{12} = 0, 1$ and $s_{34} = 0, 1$, the coupling of $\hat{\vec{S}}_{12}$ and $\hat{\vec{S}}_{34}$ yields the following values for the total spin s:

- When $s_{12} = 0$ and $s_{34} = 0$ we have only one possible value, $s = 0$, and hence only one eigenstate, $|sm\rangle = |0, 0\rangle$.

- When $s_{12} = 1$ and $s_{34} = 0$, we have $s = 1$; there are three eigenstates: $|s m\rangle = |1, \pm1\rangle$, and $|1, 0\rangle$.

- When $s_{12} = 0$ and $s_{34} = 1$, we have $s = 1$; there are three eigenstates: $|sm\rangle = |1, \pm1\rangle$, and $|1, 0\rangle$.

- When $s_{12} = 1$ and $s_{34} = 1$ we have $s = 0, 1, 2$; we have here nine eigenstates (see Problem 7.3, page 436): $|0, 0\rangle, |1, \pm1\rangle, |1, 0\rangle, |2, \pm2\rangle, |2, \pm1\rangle$, and $|2, 0\rangle$.

In conclusion, the possible values of the total spin when coupling four $\frac{1}{2}$ spins are $s = 0, 1, 2$; the value $s = 0$ occurs twice, $s = 1$ three times, and $s = 2$ only once.

Problem 7.6

Work out the coupling of the isospins of a pion–nucleon system and infer the various states of this system.

Solution

Since the isospin of a pion meson is 1 and that of a nucleon is $\frac{1}{2}$, the total isospin of a pion–nucleon system can be obtained by coupling the isospins $t_1 = 1$ and $t_2 = \frac{1}{2}$. The various values of the total isospin lie in the range $|t_1 - t_2| < T < t_1 + t_2$; hence they are given by $T = \frac{3}{2}, \frac{1}{2}$.

The coupling of the isospins $t_1 = 1$ and $t_2 = \frac{1}{2}$ is analogous to the addition of an orbital angular momentum $l = 1$ and a spin $\frac{1}{2}$; the expressions pertaining to this coupling are listed in (7.206) to (7.211). Note that there are three different π-mesons:

$$| 1, 1 \rangle = | \pi^+ \rangle, \qquad | 1, 0 \rangle = | \pi^0 \rangle, \qquad | 1, -1 \rangle = | \pi^- \rangle, \qquad (7.388)$$

and two nucleons, a proton and a neutron:

$$\left| \frac{1}{2}, \frac{1}{2} \right\rangle = | p \rangle, \qquad \left| \frac{1}{2}, -\frac{1}{2} \right\rangle = | n \rangle. \qquad (7.389)$$

By analogy with (7.206) to (7.211) we can write the states corresponding to $T = \frac{3}{2}$ as

$$\left| \frac{3}{2}, \frac{3}{2} \right\rangle = | 1, 1 \rangle \left| \frac{1}{2}, \frac{1}{2} \right\rangle = | \pi^+ \rangle | p \rangle, \qquad (7.390)$$

$$\left| \frac{3}{2}, \frac{1}{2} \right\rangle = \sqrt{\frac{2}{3}} | 1, 0 \rangle \left| \frac{1}{2}, \frac{1}{2} \right\rangle + \frac{1}{\sqrt{3}} | 1, 1 \rangle \left| \frac{1}{2}, -\frac{1}{2} \right\rangle = \sqrt{\frac{2}{3}} | \pi^0 \rangle | p \rangle + \frac{1}{\sqrt{3}} | \pi^+ \rangle | n \rangle, \qquad (7.391)$$

$$\left| \frac{3}{2}, -\frac{1}{2} \right\rangle = \frac{1}{\sqrt{3}} | 1, -1 \rangle \left| \frac{1}{2}, \frac{1}{2} \right\rangle + \sqrt{\frac{2}{3}} | 1, 0 \rangle \left| \frac{1}{2}, -\frac{1}{2} \right\rangle = \frac{1}{\sqrt{3}} | \pi^- \rangle | p \rangle + \sqrt{\frac{2}{3}} | \pi^0 \rangle | n \rangle, \qquad (7.392)$$

$$\left| \frac{3}{2}, -\frac{3}{2} \right\rangle = | 1, -1 \rangle \left| \frac{1}{2}, -\frac{1}{2} \right\rangle = | \pi^- \rangle | n \rangle, \qquad (7.393)$$

and those corresponding to $T = \frac{1}{2}$ as

$$\left| \frac{1}{2}, \frac{1}{2} \right\rangle = \sqrt{\frac{2}{3}} | 1, 1 \rangle \left| \frac{1}{2}, -\frac{1}{2} \right\rangle - \frac{1}{\sqrt{3}} | 1, 0 \rangle \left| \frac{1}{2}, \frac{1}{2} \right\rangle = \sqrt{\frac{2}{3}} | \pi^+ \rangle | n \rangle - \frac{1}{\sqrt{3}} | \pi^0 \rangle | p \rangle, \qquad (7.394)$$

$$\left| \frac{1}{2}, -\frac{1}{2} \right\rangle = \frac{1}{\sqrt{3}} | 1, 0 \rangle \left| \frac{1}{2}, -\frac{1}{2} \right\rangle - \sqrt{\frac{2}{3}} | 1, -1 \rangle \left| \frac{1}{2}, \frac{1}{2} \right\rangle = \frac{1}{\sqrt{3}} | \pi^0 \rangle | n \rangle - \sqrt{\frac{2}{3}} | \pi^- \rangle | p \rangle. \qquad (7.395)$$

Problem 7.7

(a) Calculate the expression of $\langle 2, 0 | Y_{10} | 1, 0 \rangle$.

(b) Use the result of (a) along with the Wigner–Eckart theorem to calculate the reduced matrix element $\langle 2 \| Y_1 \| 1 \rangle$.

Solution

(a) Since

$$\langle 2, 0 \mid Y_{10} \mid 1, 0 \rangle = \int_0^\pi \sin\theta \, d\theta \int_0^{2\pi} Y_{20}^*(\theta, \varphi) Y_{10}(\theta, \varphi) Y_{10}(\theta, \varphi) \, d\varphi, \qquad (7.396)$$

and using the relations $Y_{20}(\theta, \varphi) = \sqrt{5/(16\pi)}(3\cos^2\theta - 1)$ and $Y_{10}(\theta, \varphi) = \sqrt{3/(4\pi)}\cos\theta$, we have

$$\begin{aligned}
\langle 2, 0 \mid Y_{10} \mid 1, 0 \rangle &= \frac{3}{4\pi}\sqrt{\frac{5}{16\pi}} \int_0^\pi \cos^2\theta(3\cos^2\theta - 1)\sin\theta \, d\theta \int_0^{2\pi} d\varphi \\
&= \frac{3}{2}\sqrt{\frac{5}{16\pi}} \int_0^\pi \cos^2\theta(3\cos^2\theta - 1)\sin\theta \, d\theta. \qquad (7.397)
\end{aligned}$$

The change of variables $x = \cos\theta$ leads to

$$\begin{aligned}
\langle 2, 0 \mid Y_{10} \mid 1, 0 \rangle &= \frac{3}{2}\sqrt{\frac{5}{16\pi}} \int_0^\pi \cos^2\theta(3\cos^3\theta - 1)\sin\theta \, d\theta \\
&= \frac{3}{2}\sqrt{\frac{5}{16\pi}} \int_{-1}^1 x^2(3x^2 - 1) \, dx = \frac{1}{\sqrt{5\pi}}. \qquad (7.398)
\end{aligned}$$

(b) Applying the Wigner–Eckart theorem to $\langle 2, 0 \mid Y_{10} \mid 1, 0 \rangle$ and using the Clebsch–Gordan coefficient $\langle 1, 1; 0, 0 \mid 2, 0 \rangle = 2/\sqrt{6}$, we have

$$\langle 2, 0 \mid Y_{10} \mid 1, 0 \rangle = \langle 1, 1; 0, 0 \mid 2, 0 \rangle \langle 2 \parallel Y_1 \parallel 1 \rangle = \frac{2}{\sqrt{6}}\langle 2 \parallel Y_1 \parallel 1 \rangle. \qquad (7.399)$$

Finally, we may obtain $\langle 2 \parallel Y_1 \parallel 1 \rangle$ from (7.398) and (7.399):

$$\langle 2 \parallel Y_1 \parallel 1 \rangle = \sqrt{\frac{3}{10\pi}}. \qquad (7.400)$$

Problem 7.8

(a) Find the reduced matrix elements associated with the spherical harmonic $Y_{kq}(\theta, \varphi)$.

(b) Calculate the dipole transitions $\langle n'l'm' \mid \vec{r} \mid nlm \rangle$.

Solution

On the one hand, an application of the Wigner–Eckart theorem to Y_{kq} yields

$$\langle l', m' \mid Y_{kq} \mid l, m \rangle = \langle l, k; m, q \mid l', m' \rangle \langle l' \parallel Y^{(k)} \parallel l \rangle \qquad (7.401)$$

and, on the other hand, a straightforward evaluation of

$$\begin{aligned}
\langle l', m' \mid Y_{kq} \mid l, m \rangle &= \int_0^{2\pi} d\varphi \int_0^\pi \sin\theta \, d\theta \langle l', m \mid \theta\varphi \rangle Y_{kq}(\theta, \varphi) \langle \theta\varphi \mid l, m \rangle \\
&= \int_0^{2\pi} d\varphi \int_0^\pi \sin\theta \, d\theta \, Y_{l'm'}^*(\theta, \varphi) Y_{kq}(\theta, \varphi) Y_{lm}(\theta, \varphi) \qquad (7.402)
\end{aligned}$$

can be inferred from the triple integral relation (7.244):

$$\langle l', m' \mid Y_{kq} \mid l, m \rangle = \sqrt{\frac{(2l+1)(2k+1)}{4\pi(2l'+1)}} \langle l, k; \; 0, 0 | l', \; 0 \rangle \langle l, k; \; m, q \mid l', \; m' \rangle. \quad (7.403)$$

We can then combine (7.401) and (7.403) to obtain the reduced matrix element

$$\langle l' \parallel Y^{(k)} \parallel l \rangle = \sqrt{\frac{(2l+1)(2k+1)}{4\pi(2l'+1)}} \langle l, k; \; 0, 0 | l', \; 0 \rangle. \quad (7.404)$$

(b) To calculate $\langle n'l'm' \mid \vec{r} \mid nlm \rangle$ it is more convenient to express the vector $\vec{r}$ in terms of the spherical components $\vec{r} = (r_{-1}, r_0, r_1)$, which are given in terms of the Cartesian coordinates x, y, z as follows:

$$r_1 = -\frac{x+iy}{\sqrt{2}} = \frac{r}{\sqrt{2}} e^{i\varphi} \sin\theta, \qquad r_0 = z = r\cos\theta, \qquad r_{-1} = \frac{x-iy}{\sqrt{2}} = \frac{r}{\sqrt{2}} e^{-i\varphi} \sin\theta,$$
$$(7.405)$$

which in turn may be condensed into a single relation

$$r_q = \sqrt{\frac{4\pi}{3}} r Y_{1q}(\theta, \varphi), \qquad\qquad q = 1, 0, -1. \quad (7.406)$$

Next we may write $\langle n'l'm' \mid r_q \mid nlm \rangle$ in terms of a radial part and an angular part:

$$\langle n'l'm' \mid r_q \mid nlm \rangle = \sqrt{\frac{4\pi}{3}} \langle n'l' \mid r_q|nl \rangle \langle l', m'|Y_{1q}(\theta, \varphi) \mid l, m \rangle. \quad (7.407)$$

The calculation of the radial part, $\langle n'l' \mid r_q \mid nl \rangle = \int_0^\infty r^3 R_{n'l'}^*(r) R_{nl}^*(r)\, dr$, is straightforward and is of no concern to us here; see Chapter 6 for its calculation. As for the angular part $\langle l', m' \mid Y_{1q}(\theta, \varphi) \mid l, m \rangle$, we can infer its expression from (7.403)

$$\langle l', m' \mid Y_{1q} \mid l, m \rangle = \sqrt{\frac{3(2l+1)}{4\pi(2l'+1)}} \langle l, 1; \; 0, 0 | l', \; 0 \rangle \langle l, 1; \; m, q \mid l', \; m' \rangle. \quad (7.408)$$

The Clebsch–Gordan coefficients $\langle l, 1; \; m, q \mid l', \; m' \rangle$ vanish unless $m' = m + q$ and $l - 1 \le l' \le l + 1$ or $\Delta m = m' - m = q = 1, 0, -1$ and $\Delta l = l' - l = 1, 0, -1$. Notice that the case $\Delta l = 0$ is ruled out from the parity selection rule; so, the only permissible values of l' and l are those for which $\Delta l = l' - l = \pm 1$. Obtaining the various relevant Clebsch–Gordan coefficients from standard tables, we can ascertain that the only terms of (7.408) that survive are

$$\langle l+1, \; m+1|Y_{11} \mid l, \; m \rangle = \sqrt{\frac{3(l+m+1)(l+m+2)}{8\pi(2l+1)(2l+3)}}, \quad (7.409)$$

$$\langle l-1, \; m+1|Y_{11} \mid l, \; m \rangle = \sqrt{\frac{3(l-m-1)(l-m)}{8\pi(2l+1)(2l+3)}}, \quad (7.410)$$

$$\langle l+1, \; m|Y_{10} \mid l, \; m \rangle = \sqrt{\frac{3[(l+1)^2 - m^2]}{4\pi(2l+1)(2l+3)}}, \quad (7.411)$$

$$\langle l-1,\ m|Y_{10}\ |\ l,\ m\rangle = \sqrt{\frac{3(l^2-m^2)}{4\pi\,(2l+1)(2l-1)}},\qquad (7.412)$$

$$\langle l+1,\ m-1|Y_{1-1}\ |\ l,\ m\rangle = \sqrt{\frac{3(l-m+1)(l-m+2)}{8\pi\,(2l+1)(2l+3)}},\qquad (7.413)$$

$$\langle l-1,\ m-1|Y_{1-1}\ |\ l,\ m\rangle = \sqrt{\frac{3(l+m)(l+m-1)}{8\pi\,(2l+1)(2l-1)}}.\qquad (7.414)$$

Problem 7.9
Find the rotation matrix $d^{(1)}$ corresponding to $j=1$.

Solution
To find the matrix of $d^{(1)}(\beta) = e^{-i\beta\hat{J}_y/\hbar}$ for $j=1$, we need first to find the matrix representation of $\hat{J}_y$ within the joint eigenstates $\{|\ j,\ m\rangle\}$ of $\hat{J}^2$ and $\hat{J}_z$. Since the basis of $j=1$ consists of three states $|\ 1,\ -1\rangle, |\ 1,\ 0\rangle, |\ 1,\ 1\rangle$, the matrix representing $\hat{J}_y$ within this basis is given by

$$J_y = \frac{\hbar}{2}\begin{pmatrix} \langle 1,\ 1\ |\ \hat{J}_y\ |\ 1,\ 1\rangle & \langle 1,\ 1\ |\ \hat{J}_y\ |\ 1,\ 0\rangle & \langle 1,\ 1\ |\ \hat{J}_y\ |\ 1,\ -1\rangle \\ \langle 1,\ 0\ |\ \hat{J}_y\ |\ 1,\ 1\rangle & \langle 1,\ 0\ |\ \hat{J}_y\ |\ 1,\ 0\rangle & \langle 1,\ 0\ |\ \hat{J}_y\ |\ 1,\ -1\rangle \\ \langle 1,\ -1\ |\ \hat{J}_y\ |\ 1,\ 1\rangle & \langle 1,\ -1\ |\ \hat{J}_y\ |\ 1,\ 0\rangle & \langle 1,\ -1\ |\ \hat{J}_y\ |\ 1,\ -1\rangle \end{pmatrix}$$

$$= \frac{i\hbar}{\sqrt{2}}\begin{pmatrix} 0 & -1 & 0 \\ 1 & 0 & -1 \\ 0 & 1 & 0 \end{pmatrix}.\qquad (7.415)$$

We can easily verify that $J_y^3 = J_y$:

$$J_y^2 = \frac{\hbar^2}{2}\begin{pmatrix} 1 & 0 & -1 \\ 0 & 2 & 0 \\ -1 & 0 & 1 \end{pmatrix},\qquad J_y^3 = \frac{i\hbar^3}{\sqrt{2}}\begin{pmatrix} 0 & -1 & 0 \\ 1 & 0 & -1 \\ 0 & 1 & 0 \end{pmatrix} = \hbar^2 J_y.\qquad (7.416)$$

We can thus infer

$$J_y^{2n} = \hbar^{2n-2} J_y^2 \quad (n>0),\qquad J_y^{2n+1} = \hbar^{2n} J_y.\qquad (7.417)$$

Combining these two relations with

$$e^{-i\beta\hat{J}_y/\hbar} = \sum_{n=0}^{\infty}\frac{1}{n!}\left(-\frac{i\beta}{\hbar}\right)^n J_y^n$$

$$= \sum_{n=0}^{\infty}\frac{1}{(2n)!}\left(-\frac{i\beta}{\hbar}\right)^{2n} J_y^{2n} + \sum_{n=0}^{\infty}\frac{1}{(2n+1)!}\left(-\frac{i\beta}{\hbar}\right)^{2n+1} J_y^{2n+1},$$

$$\qquad (7.418)$$

we obtain

$$e^{-i\beta\hat{J}_y/\hbar} = \hat{I} + \left(\frac{\hat{J}_y}{\hbar}\right)^2\sum_{n=1}^{\infty}\frac{(-1)^n}{(2n)!}(\beta)^{2n} - i\frac{\hat{J}_y}{\hbar}\sum_{n=0}^{\infty}\frac{(-1)^n}{(2n+1)!}\beta^{2n+1}$$

$$= \hat{I} + \left(\frac{\hat{J}_y}{\hbar}\right)^2 \left[\sum_{n=0}^{\infty} \frac{(-1)^n}{(2n)!}(\beta)^{2n} - 1\right] - i\frac{\hat{J}_y}{\hbar}\sum_{n=0}^{\infty}\frac{(-1)^n}{(2n+1)!}\beta^{2n+1},$$

(7.419)

where $\hat{I}$ is the 3×3 unit matrix. Using the relations $\sum_{n=0}^{\infty}[(-1)^n/(2n)!](\beta)^{2n} = \cos\beta$ and $\sum_{n=0}^{\infty}[(-1)^n/(2n+1)!]\beta^{2n+1} = \sin\beta$, we may write

$$e^{-i\beta\hat{J}_y/\hbar} = \hat{I} + \left(\frac{\hat{J}_y}{\hbar}\right)^2[\cos\beta - 1] - i\frac{\hat{J}_y}{\hbar}\sin\beta.$$

(7.420)

Inserting now the matrix expressions for J_y and J_y^2 as listed in (7.415) and (7.416), we obtain

$$e^{-i\beta\hat{J}_y/\hbar} = \hat{I} + \frac{1}{2}\begin{pmatrix} 1 & 0 & -1 \\ 0 & 2 & 0 \\ -1 & 0 & 1 \end{pmatrix}(\cos\beta - 1) - i\frac{i}{\sqrt{2}}\begin{pmatrix} 0 & -1 & 0 \\ 1 & 0 & -1 \\ 0 & 1 & 0 \end{pmatrix}\sin\beta \quad (7.421)$$

or

$$d^{(1)}(\beta) = \begin{pmatrix} d_{11}^{(1)} & d_{1,0}^{(1)} & d_{1-1}^{(1)} \\ d_{01}^{(1)} & d_{00}^{(1)} & d_{0-1}^{(1)} \\ d_{-11}^{(1)} & d_{-10}^{(1)} & d_{-1-1}^{(1)} \end{pmatrix} = \begin{pmatrix} \frac{1}{2}(1+\cos\beta) & -\frac{1}{\sqrt{2}}\sin\beta & \frac{1}{2}(1-\cos\beta) \\ \frac{1}{\sqrt{2}}\sin\beta & \cos\beta & -\frac{1}{\sqrt{2}}\sin\beta \\ \frac{1}{2}(1-\cos\beta) & \frac{1}{\sqrt{2}}\sin\beta & \frac{1}{2}(1+\cos\beta) \end{pmatrix}.$$

(7.422)

Since $\frac{1}{2}(1+\cos\beta) = \cos^2(\beta/2)$ and $\frac{1}{2}(1-\cos\beta) = \sin^2(\beta/2)$, we have

$$d^{(1)}(\beta) = e^{-i\beta J_y/\hbar} = \begin{pmatrix} \cos^2(\beta/2) & -\frac{1}{\sqrt{2}}\sin(\beta) & \sin^2(\beta/2) \\ \frac{1}{\sqrt{2}}\sin(\beta) & \cos(\beta) & -\frac{1}{\sqrt{2}}\sin(\beta) \\ \sin^2(\beta/2) & \frac{1}{\sqrt{2}}\sin(\beta) & \cos^2(\beta/2) \end{pmatrix}.$$

(7.423)

This method becomes quite intractable when attempting to derive the matrix of $d^{(j)}(\beta)$ for large values of j. In Problem 7.10 we are going to present a simpler method for deriving $d^{(j)}(\beta)$ for larger values of j; this method is based on the addition of angular momenta.

Problem 7.10

(a) Use the relation

$$d_{mm'}^{(j)}(\beta) = \sum_{m_1 m_2}\sum_{m_1' m_2'}\langle j_1, j_2; m_1, m_2 \mid j, m\rangle\langle j_1, j_2; m_1', m_2' \mid j, m'\rangle d_{m_1 m_1'}^{(j_1)}(\beta)d_{m_2 m_2'}^{(j_2)}(\beta),$$

for the case where $j_1 = 1$ and $j_2 = \frac{1}{2}$ along with the Clebsch–Gordan coefficients derived in (7.206) to (7.209), and the matrix elements of $d^{(1/2)}(\beta)$ and $d^{(1)}(\beta)$, which are given by (7.89) and (7.423), respectively, to find the expressions of the matrix elements of $d_{\frac{3}{2}\frac{3}{2}}^{(3/2)}(\beta)$, $d_{\frac{3}{2}\frac{1}{2}}^{(3/2)}(\beta)$, $d_{\frac{3}{2}-\frac{1}{2}}^{(3/2)}(\beta)$, $d_{\frac{3}{2}-\frac{3}{2}}^{(3/2)}(\beta)$, $d_{\frac{1}{2}\frac{1}{2}}^{(3/2)}(\beta)$, and $d_{\frac{1}{2}-\frac{1}{2}}^{(3/2)}(\beta)$.

(b) Use the six expressions derived in (a) to infer the matrix of $d^{(3/2)}(\beta)$.

Solution

(a) Using $\langle 1, \frac{1}{2}; 1, \frac{1}{2} \mid \frac{3}{2}, \frac{3}{2} \rangle = 1$, $d_{11}^{(1)}(\beta) = \cos^2(\beta/2)$ and $d_{\frac{1}{2}\frac{1}{2}}^{(1/2)}(\beta) = \cos(\beta/2)$, we have

$$d_{\frac{3}{2}\frac{3}{2}}^{(3/2)}(\beta) = \left\langle 1, \frac{1}{2}; 1, \frac{1}{2} \left| \frac{3}{2}, \frac{3}{2} \right. \right\rangle \left\langle 1, \frac{1}{2}; 1, \frac{1}{2} \left| \frac{3}{2}, \frac{3}{2} \right. \right\rangle d_{11}^{(1)}(\beta) d_{\frac{1}{2}\frac{1}{2}}^{(1/2)}(\beta) = \cos^3\left(\frac{\beta}{2}\right). \quad (7.424)$$

Similarly, since $\langle 1, \frac{1}{2}; 0, \frac{1}{2} \mid \frac{3}{2}, \frac{1}{2} \rangle = \sqrt{2/3}$, $\langle 1, \frac{1}{2}; 1, -\frac{1}{2} \mid \frac{3}{2}, \frac{1}{2} \rangle = 1/\sqrt{3}$, and since $d_{10}^{(1)}(\beta) = -(1/\sqrt{2})\sin(\beta)$ and $d_{\frac{1}{2}-\frac{1}{2}}^{(1/2)}(\beta) = -\sin(\beta/2)$, we have

$$d_{\frac{3}{2}\frac{1}{2}}^{(3/2)}(\beta) = \left\langle 1, \frac{1}{2}; 1, \frac{1}{2} \left| \frac{3}{2}, \frac{3}{2} \right. \right\rangle \left\langle 1, \frac{1}{2}; 0, \frac{1}{2} \left| \frac{3}{2}, \frac{1}{2} \right. \right\rangle d_{10}^{(1)}(\beta) d_{\frac{1}{2}\frac{1}{2}}^{(1/2)}(\beta)$$

$$+ \left\langle 1, \frac{1}{2}; 1, \frac{1}{2} \left| \frac{3}{2}, \frac{3}{2} \right. \right\rangle \left\langle 1, \frac{1}{2}; 1, -\frac{1}{2} \left| \frac{3}{2}, \frac{1}{2} \right. \right\rangle d_{11}^{(1)}(\beta) d_{\frac{1}{2}-\frac{1}{2}}^{(1/2)}(\beta)$$

$$= -\frac{1}{\sqrt{3}} \sin\beta \cos\left(\frac{\beta}{2}\right) - \frac{1}{\sqrt{3}} \cos^2\left(\frac{\beta}{2}\right) \sin\left(\frac{\beta}{2}\right)$$

$$= -\sqrt{3} \sin\left(\frac{\beta}{2}\right) \cos^2\left(\frac{\beta}{2}\right). \quad (7.425)$$

To calculate $d_{\frac{3}{2}-\frac{1}{2}}^{(3/2)}(\beta)$, we need to use the coefficients $\langle 1, \frac{1}{2}; 0, -\frac{1}{2} \mid \frac{3}{2}, -\frac{1}{2} \rangle = \sqrt{2/3}$ and $\langle 1, \frac{1}{2}; -1, \frac{1}{2} \mid \frac{3}{2}, -\frac{1}{2} \rangle = 1/\sqrt{3}$ along with $d_{1,-1}^{(1)}(\beta) = \sin^2(\beta/2)$:

$$d_{\frac{3}{2}-\frac{1}{2}}^{(3/2)}(\beta) = \left\langle 1, \frac{1}{2}; 1, \frac{1}{2} \left| \frac{3}{2}, \frac{3}{2} \right. \right\rangle \left\langle 1, \frac{1}{2}; -1, \frac{1}{2} \left| \frac{3}{2}, -\frac{1}{2} \right. \right\rangle d_{1-1}^{(1)}(\beta) d_{\frac{1}{2}\frac{1}{2}}^{(1/2)}(\beta)$$

$$+ \left\langle 1, \frac{1}{2}; 1, \frac{1}{2} \left| \frac{3}{2}, \frac{3}{2} \right. \right\rangle \left\langle 1, \frac{1}{2}; 0, -\frac{1}{2} \left| \frac{3}{2}, -\frac{1}{2} \right. \right\rangle d_{10}^{(1)}(\beta) d_{\frac{1}{2}-\frac{1}{2}}^{(1/2)}(\beta)$$

$$= \frac{1}{\sqrt{3}} \sin^2\left(\frac{\beta}{2}\right) \cos\left(\frac{\beta}{2}\right) + \frac{1}{\sqrt{3}} \sin\beta \sin\left(\frac{\beta}{2}\right)$$

$$= \sqrt{3} \sin^2\left(\frac{\beta}{2}\right) \cos\left(\frac{\beta}{2}\right). \quad (7.426)$$

For $d_{\frac{3}{2}-\frac{3}{2}}^{(3/2)}(\beta)$ we have

$$d_{\frac{3}{2}-\frac{3}{2}}^{(3/2)}(\beta) = \left\langle 1, \frac{1}{2}; 1, \frac{1}{2} \left| \frac{3}{2}, \frac{3}{2} \right. \right\rangle \left\langle 1, \frac{1}{2}; -1, -\frac{1}{2} \left| \frac{3}{2}, -\frac{3}{2} \right. \right\rangle d_{1-1}^{(1)}(\beta) d_{\frac{1}{2}-\frac{1}{2}}^{(1/2)}(\beta) = -\sin^3\left(\frac{\beta}{2}\right),$$

$$(7.427)$$

because $\langle 1, \frac{1}{2}; -1, -\frac{1}{2} \mid \frac{3}{2}, -\frac{3}{2} \rangle = 1$, $d_{1-1}^{(1)}(\beta) = \sin^2(\beta/2)$, and $d_{\frac{1}{2}-\frac{1}{2}}^{(1/2)}(\beta) = -\sin(\beta/2)$.

To calculate $d_{\frac{1}{2}\frac{1}{2}}^{(3/2)}(\beta)$, we need to use the coefficients $\langle 1, \frac{1}{2}; 0, \frac{1}{2} \mid \frac{3}{2}, \frac{1}{2} \rangle = \sqrt{2/3}$ and $\langle 1, \frac{1}{2}; 1, -\frac{1}{2} \mid \frac{3}{2}, \frac{1}{2} \rangle = 1/\sqrt{3}$ along with $d_{1-1}^{(1)}(\beta) = \sin^2(\beta/2)$:

$$d_{\frac{1}{2}\frac{1}{2}}^{(3/2)}(\beta) = \left\langle 1, \frac{1}{2}; 1, -\frac{1}{2} \left| \frac{3}{2}, \frac{1}{2} \right. \right\rangle \left\langle 1, \frac{1}{2}; 1, -\frac{1}{2} \left| \frac{3}{2}, \frac{1}{2} \right. \right\rangle d_{11}^{(1)}(\beta) d_{-\frac{1}{2}-\frac{1}{2}}^{(1/2)}(\beta)$$

$$+ \left\langle 1, \frac{1}{2}; 0, \frac{1}{2} \left| \frac{3}{2}, \frac{1}{2} \right. \right\rangle \left\langle 1, \frac{1}{2}; 1, -\frac{1}{2} \left| \frac{3}{2}, \frac{1}{2} \right. \right\rangle d_{01}^{(1)}(\beta) d_{\frac{1}{2}-\frac{1}{2}}^{(1/2)}(\beta)$$

$$+ \left\langle 1, \frac{1}{2}; \, 0, \frac{1}{2} \left| \frac{3}{2}, \frac{1}{2} \right\rangle \left\langle 1, \frac{1}{2}; \, 0, \frac{1}{2} \left| \frac{3}{2}, \frac{1}{2} \right\rangle d_{00}^{(1)}(\beta) d_{\frac{1}{2}\frac{1}{2}}^{(1/2)}(\beta)$$

$$+ \left\langle 1, \frac{1}{2}; \, 1, -\frac{1}{2} \left| \frac{3}{2}, \frac{1}{2} \right\rangle \left\langle 1, \frac{1}{2}; \, 0, \frac{1}{2} \left| \frac{3}{2}, \frac{1}{2} \right\rangle d_{10}^{(1)}(\beta) d_{-\frac{1}{2}\frac{1}{2}}^{(1/2)}(\beta)$$

$$= \frac{1}{3} \cos^3\left(\frac{\beta}{2}\right) - \frac{1}{3}\sin(\beta)\sin\left(\frac{\beta}{2}\right) + \frac{2}{3}\cos(\beta)\cos\left(\frac{\beta}{2}\right) - \frac{1}{3}\sin(\beta)\sin\left(\frac{\beta}{2}\right)$$

$$= \left[3\cos^2\left(\frac{\beta}{2}\right) - 2 \right] \cos\left(\frac{\beta}{2}\right)$$

$$= \frac{1}{2}(3\cos\beta - 1)\cos\left(\frac{\beta}{2}\right). \tag{7.428}$$

Similarly, we have

$$d_{\frac{1}{2}-\frac{1}{2}}^{(3/2)}(\beta) = \left\langle 1, \frac{1}{2}; \, 1, -\frac{1}{2} \left| \frac{3}{2}, \frac{1}{2} \right\rangle \left\langle 1, \frac{1}{2}; \, -1, \frac{1}{2} \left| \frac{3}{2}, -\frac{1}{2} \right\rangle d_{1-1}^{(1)}(\beta) d_{-\frac{1}{2}\frac{1}{2}}^{(1/2)}(\beta)$$

$$+ \left\langle 1, \frac{1}{2}; \, 1, -\frac{1}{2} \left| \frac{3}{2}, \frac{1}{2} \right\rangle \left\langle 1, \frac{1}{2}; \, 0, -\frac{1}{2} \left| \frac{3}{2}, -\frac{1}{2} \right\rangle d_{10}^{(1)}(\beta) d_{-\frac{1}{2}-\frac{1}{2}}^{(1/2)}(\beta)$$

$$+ \left\langle 1, \frac{1}{2}; \, 0, \frac{1}{2} \left| \frac{3}{2}, \frac{1}{2} \right\rangle \left\langle 1, \frac{1}{2}; \, -1, \frac{1}{2} \left| \frac{3}{2}, -\frac{1}{2} \right\rangle d_{0-1}^{(1)}(\beta) d_{\frac{1}{2}\frac{1}{2}}^{(1/2)}(\beta)$$

$$+ \left\langle 1, \frac{1}{2}; \, 0, \frac{1}{2} \left| \frac{3}{2}, \frac{1}{2} \right\rangle \left\langle 1, \frac{1}{2}; \, 0, -\frac{1}{2} \left| \frac{3}{2}, -\frac{1}{2} \right\rangle d_{00}^{(1)}(\beta) d_{\frac{1}{2}-\frac{1}{2}}^{(1/2)}(\beta)$$

$$= \frac{1}{3}\sin^3\left(\frac{\beta}{2}\right) - \frac{1}{3}\sin(\beta)\cos\left(\frac{\beta}{2}\right) - \frac{1}{3}\sin(\beta)\cos\left(\frac{\beta}{2}\right) - \frac{2}{3}\cos(\beta)\sin\left(\frac{\beta}{2}\right)$$

$$= -\left[3\cos^2\left(\frac{\beta}{2}\right) - 1 \right] \sin\left(\frac{\beta}{2}\right)$$

$$= -\frac{1}{2}(3\cos\beta + 1)\sin\left(\frac{\beta}{2}\right). \tag{7.429}$$

(b) The remaining ten matrix elements of $d^{(3/2)}(\beta)$ can be inferred from the six elements derived above by making use of the properties of the d-function listed in (7.67). For instance, using $d_{m'm}^{(j)}(\beta) = (-1)^{m'-m}d_{-m'-m}^{(j)}(\beta)$, we can verify that

$$d_{-\frac{3}{2}-\frac{3}{2}}^{(3/2)}(\beta) = d_{\frac{3}{2}\frac{3}{2}}^{(3/2)}(\beta), \qquad d_{-\frac{1}{2}-\frac{1}{2}}^{(3/2)}(\beta) = d_{\frac{1}{2}\frac{1}{2}}^{(3/2)}(\beta), \qquad d_{-\frac{3}{2}-\frac{1}{2}}^{(3/2)}(\beta) = -d_{\frac{3}{2}\frac{1}{2}}^{(3/2)}(\beta),$$

$$\tag{7.430}$$

$$d_{-\frac{3}{2}\frac{1}{2}}^{(3/2)}(\beta) = d_{\frac{3}{2}-\frac{1}{2}}^{(3/2)}(\beta), \qquad d_{-\frac{3}{2}\frac{3}{2}}^{(3/2)}(\beta) = -d_{\frac{3}{2}-\frac{3}{2}}^{(3/2)}(\beta), \qquad d_{-\frac{1}{2}\frac{1}{2}}^{(3/2)}(\beta) = -d_{\frac{1}{2}-\frac{1}{2}}^{(3/2)}(\beta).$$

$$\tag{7.431}$$

Similarly, using $d_{m'm}^{(j)}(\beta) = (-1)^{m'-m}d_{mm'}^{(j)}(\beta)$ we can obtain the remaining four elements:

$$d_{\frac{1}{2}\frac{3}{2}}^{(3/2)}(\beta) = -d_{\frac{3}{2}\frac{1}{2}}^{(3/2)}(\beta), \qquad d_{-\frac{1}{2}\frac{3}{2}}^{(3/2)}(\beta) = d_{\frac{3}{2}-\frac{1}{2}}^{(3/2)}(\beta), \tag{7.432}$$

$$d_{\frac{1}{2}-\frac{3}{2}}^{(3/2)}(\beta) = d_{-\frac{3}{2}\frac{1}{2}}^{(3/2)}(\beta), \qquad d_{-\frac{1}{2}-\frac{3}{2}}^{(3/2)}(\beta) = -d_{-\frac{3}{2}-\frac{1}{2}}^{(3/2)}(\beta). \tag{7.433}$$

Collecting the six matrix elements calculated in (a) along with the ten elements inferred above, we obtain the matrix of $d^{(3/2)}(\beta)$:

$$
\begin{pmatrix}
\cos^3\left(\frac{\beta}{2}\right) & -\sqrt{3}\sin\left(\frac{\beta}{2}\right)\cos^2\left(\frac{\beta}{2}\right) & \sqrt{3}\sin^2\left(\frac{\beta}{2}\right)\cos\left(\frac{\beta}{2}\right) & -\sin^3\left(\frac{\beta}{2}\right) \\
\sqrt{3}\sin\left(\frac{\beta}{2}\right)\cos^2\left(\frac{\beta}{2}\right) & \frac{1}{2}(3\cos\beta-1)\cos\left(\frac{\beta}{2}\right) & -\frac{1}{2}(3\cos\beta+1)\sin\left(\frac{\beta}{2}\right) & \sqrt{3}\sin^2\left(\frac{\beta}{2}\right)\cos\left(\frac{\beta}{2}\right) \\
\sqrt{3}\sin^2\left(\frac{\beta}{2}\right)\cos\left(\frac{\beta}{2}\right) & \frac{1}{2}(3\cos\beta+1)\sin\left(\frac{\beta}{2}\right) & \frac{1}{2}(3\cos\beta-1)\cos\left(\frac{\beta}{2}\right) & -\sqrt{3}\sin\left(\frac{\beta}{2}\right)\cos^2\left(\frac{\beta}{2}\right) \\
\sin^3\left(\frac{\beta}{2}\right) & \sqrt{3}\sin^2\left(\frac{\beta}{2}\right)\cos\left(\frac{\beta}{2}\right) & \sqrt{3}\sin\left(\frac{\beta}{2}\right)\cos^2\left(\frac{\beta}{2}\right) & \cos^3\left(\frac{\beta}{2}\right)
\end{pmatrix},
\tag{7.434}
$$

which can be reduced to

$$
d^{3/2}(\beta) = \frac{\sin\beta}{2}
\begin{pmatrix}
\frac{\cos^2(\beta/2)}{\sin(\beta/2)} & -\sqrt{3}\cos\left(\frac{\beta}{2}\right) & \sqrt{3}\sin\left(\frac{\beta}{2}\right) & -\frac{\sin^2(\beta/2)}{\cos(\beta/2)} \\
\sqrt{3}\cos\left(\frac{\beta}{2}\right) & \frac{3\cos\beta-1}{2\sin(\beta/2)} & -\frac{3\cos\beta+1}{2\cos(\beta/2)} & \sqrt{3}\sin\left(\frac{\beta}{2}\right) \\
\sqrt{3}\sin\left(\frac{\beta}{2}\right) & \frac{3\cos\beta+1}{\cos(\beta/2)} & \frac{3\cos\beta-1}{2\sin(\beta/2)} & -\sqrt{3}\cos\left(\frac{\beta}{2}\right) \\
\frac{\sin^2(\beta/2)}{\cos(\beta/2)} & \sqrt{3}\sin\left(\frac{\beta}{2}\right) & \sqrt{3}\cos\left(\frac{\beta}{2}\right) & \frac{\cos^2(\beta/2)}{\sin(\beta/2)}
\end{pmatrix}.
\tag{7.435}
$$

Following the method outlined in this problem, we can in principle find the matrix of any d-function. For instance, using the matrices of $d^{(1)}$ and $d^{(1/2)}$ along with the Clebsch–Gordan coefficients resulting from the addition of $j_1 = 1$ and $j_2 = 1$, we can find the matrix of $d^{(2)}(\beta)$.

Problem 7.11

Consider two nonidentical particles each with angular momenta 1 and whose Hamiltonian is given by

$$
\hat{H} = \frac{\varepsilon_1}{\hbar^2}(\hat{\vec{L}}_1 + \hat{\vec{L}}_2)\cdot\hat{\vec{L}}_2 + \frac{\varepsilon_2}{\hbar^2}(\hat{L}_{1_z} + \hat{L}_{2_z})^2,
$$

where ε_1 and ε_2 are constants having the dimensions of energy. Find the energy levels and their degeneracies for those states of the system whose total angular momentum is equal to $2\hbar$.

Solution

The total angular momentum of the system is obtained by coupling $l_1 = 1$ and $l_2 = 1$: $\hat{\vec{L}} = \hat{\vec{L}}_1 + \hat{\vec{L}}_2$. This leads to $\hat{\vec{L}}_1 \cdot \hat{\vec{L}}_2 = \frac{1}{2}(\hat{L}^2 - \hat{L}_1^2 - \hat{L}_2^2)$, and when this is inserted into the system's Hamiltonian it yields

$$
\hat{H} = \frac{\varepsilon_1}{\hbar^2}(\hat{\vec{L}}_1 \cdot \hat{\vec{L}}_2 + \hat{L}_z^2) + \frac{\varepsilon_2}{\hbar^2}\hat{L}_z^2 = \frac{\varepsilon_1}{2\hbar^2}(\hat{L}^2 - \hat{L}_1^2 + \hat{L}_2^2) + \frac{\varepsilon_2}{\hbar^2}\hat{L}_z^2.
\tag{7.436}
$$

Notice that the operators $\hat{H}$, $\hat{L}_1^2$, $\hat{L}_2^2$, $\hat{L}^2$, and $\hat{L}_z$ mutually commute; we denote their joint eigenstates by $|\,l,\ m\rangle$. The energy levels of (7.436) are thus given by

$$
E_{lm} = \frac{\varepsilon_1}{2}\left[l(l+1) - l_1(l_1+1) + l_2(l_2+1)\right] + \varepsilon_2 m^2 = \frac{\varepsilon_1}{2}l(l+1) + \varepsilon_2 m^2,
\tag{7.437}
$$

since $l_1 = l_2 = 1$.

The calculation of $|\,l,\ m\rangle$ in terms of the states $|l_1,\ m_1\rangle|l_2,\ m_2\rangle = |l_1, l_2;\ m_1, m_2\rangle$ was carried out in Problem 7.3, page 436; the states corresponding to a total angular momentum of $l = 2$ are given by

$$
|2,\ \pm 2\rangle = |1, 1;\ \pm 1, \pm 1\rangle, \qquad |2,\ \pm 1\rangle = \frac{1}{\sqrt{2}}\left(|1, 1;\ \pm 1, 0\rangle + |1, 1;\ 0, \pm 1\rangle\right),
\tag{7.438}
$$

$$| 2, \, 0\rangle = \frac{1}{\sqrt{6}} \left(|1, 1; \; 1, -1\rangle + 2|1, 1; \; 0, 0\rangle + |1, 1; \; -1, 1\rangle \right). \tag{7.439}$$

From (7.437) we see that the energy corresponding to $l = 2$ and $m = \pm 2$ is doubly degenerate, because the states $| 2, \, \pm 2\rangle$ have the same energy $E_{2,\pm 2} = 3\varepsilon_1 + 4\varepsilon_2$. The two states $| 2, \, \pm 1\rangle$ are also degenerate, for they correspond to the same energy $E_{2,\pm 1} = 3\varepsilon_1 + \varepsilon_2$. The energy corresponding to $| 2, \, 0\rangle$ is not degenerate: $E_{20} = 3\varepsilon_1$.

7.6   Exercises

Exercise 7.1
Show that the linear transformation $y = Rx$ where

$$R = \left(\begin{array}{cc} \cos\phi & \sin\phi \\ -\sin\phi & \cos\phi \end{array} \right), \qquad y = \left(\begin{array}{c} y_1 \\ y_2 \end{array} \right), \qquad x = \left(\begin{array}{c} x_1 \\ x_2 \end{array} \right)$$

is a counterclockwise rotation of the Cartesian $x_1 x_2$ coordinate system in the plane about the origin with an angle ϕ.

Exercise 7.2
Show that the nth power of the rotation matrix

$$R(\phi) = \left(\begin{array}{cc} \cos\phi & -\sin\phi \\ \sin\phi & \cos\phi \end{array} \right).$$

is equal to

$$R^n(\phi) = \left(\begin{array}{cc} \cos(n\phi) & -\sin(n\phi) \\ \sin(n\phi) & \cos(n\phi) \end{array} \right).$$

What is the geometrical meaning of this result?

Exercise 7.3
Using the space displacement operator $U(\hat{\vec{A}}) = e^{-i\hat{\vec{A}}\cdot\hat{\vec{P}}/\hbar}$, where $\hat{\vec{P}}$ is the linear momentum operator, show that $e^{i\hat{\vec{A}}\cdot\hat{\vec{P}}/\hbar} \, \hat{\vec{R}} \, e^{-i\hat{\vec{A}}\cdot\hat{\vec{P}}/\hbar} = \hat{\vec{R}} + \hat{\vec{A}}$.

Exercise 7.4
The components A_j (with $j = x, y, z$) of a vector $\hat{\vec{A}}$ transform under space rotations as $A'_i = R_{ij} A_j$, where R is the rotation matrix.

(a) Using the invariance of the scalar product of any two vectors (e.g., $\hat{\vec{A}} \cdot \hat{\vec{B}}$) under rotations, show that the rows and columns of the rotation matrix R are orthonormal to each other (i.e., show that $R_{lj} R_{lk} = \delta_{j,k}$).

(b) Show that the transpose of R is equal to the inverse of R and that the determinant of R is equal to ± 1.

Exercise 7.5
The operator corresponding to a rotation of angle θ about an axis $\vec{n}$ is given by

$$U_{\vec{n}}(\theta) = e^{-i\theta\vec{n}\cdot\hat{\vec{J}}/\hbar}.$$

Show that the matrix elements of the position operator $\hat{R}$ are rotated through an infinitesimal rotation like $\vec{R}' = \hat{R} + \theta \hat{n} \times \hat{R}$. (i.e., in the case where θ is infinitesimal, show that $U_n^\dagger (\theta) \hat{R}_j U_n(\theta) = \hat{R}_j + \theta(\hat{n} \times \hat{R})_j)$.

Exercise 7.6
Consider the wave function of a particle $\psi(\vec{r}) = (\sqrt{2}x + \sqrt{2}y + z)f(r)$, where $f(r)$ is a spherically symmetric function.

(a) Is $\psi(\vec{r})$ an eigenfunction of $\vec{L}^2$? If so, what is the eigenvalue?

(b) What are the probabilities for the particle to be found in the state $m_l = -1$, $m_l = 0$, and $m_l = 1$?

(c) If $\psi(\vec{r})$ is an energy eigenfunction with eigenvalues E and if $f(r) = 3r^2$, find the expression of the potential $V(r)$ to which this particle is subjected.

Exercise 7.7
Consider a particle whose wave function is given by

$$\psi(\vec{r}) = \left(\frac{1}{\sqrt{5}} Y_{11}(\theta, \varphi) - \frac{1}{5} Y_{1-1}(\theta, \varphi) + \frac{1}{\sqrt{2}} Y_{10}(\theta, \varphi) \right) f(r),$$

where $f(r)$ is a normalized radial function, i.e., $\int_0^\infty r^2 f^2(r)\, dr = 1$.

(a) Calculate the expectation values of $\hat{L}^2$, $\hat{L}_z$, and $\hat{L}_x$ in this state.

(b) Calculate the expectation value of $V(\theta) = 2\cos^2 \theta$ in this state.

(c) Find the probability that the particle will be found in the state $m_l = 0$.

Exercise 7.8
A particle of spin $\frac{1}{2}$ is in a d state of orbital angular momentum (i.e., $l = 2$). Work out the coupling of the spin and orbital angular momenta of this particle, and find all the states and the corresponding Clebsch–Gordan coefficients.

Exercise 7.9
The spin-dependent Hamiltonian of an electron–positron system in the presence of a uniform magnetic field in the z-direction ($\vec{B} = B\vec{k}$) can be written as

$$\hat{H} = \lambda \hat{\vec{S}}_1 \cdot \hat{\vec{S}}_2 + \left(\frac{eB}{mc} \right) \left(\hat{S}_{1_z} - \hat{S}_{2_z} \right),$$

where λ is a real number and $\vec{S}_1$ and $\vec{S}_2$ are the spin operators for the electron and the positron, respectively.

(a) If the spin function of the system is given by $|\frac{1}{2}, -\frac{1}{2}\rangle$, find the energy eigenvalues and their corresponding eigenvectors.

(b) Repeat (a) in the case where $\lambda = 0$, but $B \neq 0$.

(c) Repeat (a) in the case where $B = 0$, but $\lambda \neq 0$.

Exercise 7.10
(a) Show that $e^{-i\pi \hat{J}_z/2} e^{-i\pi \hat{J}_x} e^{i\pi \hat{J}_z/2} = e^{-i\pi \hat{J}_y}$.

(b) Prove $\hat{J}_- e^{-i\pi \hat{J}_x} = e^{-i\pi \hat{J}_x} \hat{J}_+$ and then show that $e^{-i\pi \hat{J}_x} | j, m\rangle = e^{-i\pi j} | j, -m\rangle$.

(c) Using (a) and (b), show that $e^{-i\pi \hat{J}_y} | j, m\rangle = (-1)^{j-m} | j, -m\rangle$.

Exercise 7.11

Using the commutation relations between the Pauli matrices, show that:

(a) $e^{i a \sigma_y} \sigma_x e^{-i a \sigma_y} = \sigma_x \cos(2\alpha) + \sigma_z \sin(2\alpha)$,

(b) $e^{i a \sigma_z} \sigma_x e^{-i a \sigma_z} = \sigma_x \cos(2\alpha) - \sigma_y \sin(2\alpha)$,

(c) $e^{i a \sigma_x} \sigma_y e^{-i a \sigma_x} = \sigma_y \cos(2\alpha) - \sigma_z \sin(2\alpha)$.

Exercise 7.12

(a) Show how $\hat{J}_x$, $\hat{J}_y$, and $\hat{J}_z$ transform under a rotation of (finite) angle α about the x-axis.

(b) Using the results of part (a), determine how the angular momentum operator $\hat{\vec{J}}$ transforms under the rotation.

Exercise 7.13

(a) Show how the operator $\hat{J}_\pm$ transforms under a rotation of angle π about the x-axis.

(b) Use the result of part (a) to show that $\hat{J}_\pm e^{-i\pi \hat{J}_x/\hbar} = e^{-i\pi \hat{J}_x/\hbar} \hat{J}_\mp$.

Exercise 7.14

Consider a rotation of finite angle α about an axis $\vec{n}$ which transforms unit vector $\vec{a}$ into another unit vector $\vec{b}$. Show that $e^{-i\beta \hat{J}_b/\hbar} = e^{i\alpha \hat{J}_n/\hbar} e^{-i\beta \hat{J}_a/\hbar} e^{-i\alpha \hat{J}_n/\hbar}$.

Exercise 7.15

(a) Show that $e^{i\pi \hat{J}_y/2\hbar} \hat{J}_x e^{-i\pi \hat{J}_y/2\hbar} = \hat{J}_z$.

(b) Show also that $e^{i\pi \hat{J}_y/2\hbar} e^{i\alpha \hat{J}_x/\hbar} e^{-i\pi \hat{J}_y/2\hbar} = e^{i\alpha \hat{J}_z/\hbar}$.

(c) For any vector operator $\hat{\vec{A}}$, show that $e^{i\alpha \hat{J}_z/\hbar} \hat{A}_x e^{-i\alpha \hat{J}_z/\hbar} = \hat{A}_x \cos\alpha + \hat{A}_y \sin\alpha$.

Exercise 7.16

Using $\hat{\vec{J}} = \hat{\vec{J}}_1 + \hat{\vec{J}}_2$ show that

$$d^{(j)}_{mm'}(\beta) = \sum_{m_1 m_2} \sum_{m'_1 m'_2} \langle j_1, j_2; m_1, m_2 \mid j, m \rangle \langle j_1, j_2; m'_1 m'_2 \mid j, m' \rangle d^{(j_1)}_{m_1 m'_1}(\beta) d^{(j_2)}_{m_2 m'_2}(\beta).$$

Exercise 7.17

Consider the tensor $A(\theta, \varphi) = \cos\theta \sin\theta \cos\varphi$.

(a) Calculate all the matrix elements $A_{m'm} = \langle l, m' \mid A \mid l, m \rangle$ for $l = 1$.

(b) Express $A(\theta, \varphi)$ in terms of the components of a spherical tensor of rank 2 (i.e., in terms of $Y_{2m}(\theta, \varphi)$).

(c) Calculate again all the matrix elements $A_{m'm}$, but this time using the Wigner–Eckart theorem. Compare these results with those obtained in (a). (The Clebsch–Gordan coefficients may be obtained from tables.)

Exercise 7.18

(a) Express xz/r^2 and $(x^2 - y^2)/r^2$ in terms of the components of a spherical tensor of rank 2.

(b) Using the Wigner–Eckart theorem, calculate the values of $\langle 1, 0 \mid xz/r^2 \mid 1, 1 \rangle$ and $\langle 1, 1 \mid (x^2 - y^2)/r^2 \mid 1, -1 \rangle$.

Exercise 7.19

Show that $\langle j, m' \mid e^{-i\beta \hat{J}_y/\hbar} \hat{J}_z^2 e^{i\beta \hat{J}_y/\hbar} \mid j, m' \rangle = \sum_{m=-j}^{m=j} m^2 \mid d^{(j)}_{m'm}(\beta) \mid^2$.

Exercise 7.20

Calculate the trace of the rotation matrix $D^{(1/2)}(\alpha, \beta, \gamma)$ for (a) $\beta = \pi$ and (b) $\alpha = \gamma = \pi$ and $\beta = 2\pi$.

Exercise 7.21

The quadrupole moment operator of a charge q is given by $\hat{Q}_{20} = q(3\hat{z}^2 - r^2)$. Write $\hat{Q}_{20}$ in terms of an irreducible spherical tensor of rank 2 and then express $\langle j, j | \hat{Q}_{20} | j, j \rangle$ in terms of j and the reduced matrix element $\langle j \parallel r^2 Y^{(2)} \parallel j \rangle$. *Hint:* You may use the coefficient $\langle j, 2; m, 0 | j, m \rangle = (-1)^{j-m}[3m^2 - j(j+1)]/\sqrt{(2j-1)j(j+1)(2j+3)}$.

Exercise 7.22

Prove the following commutation relations:

(a) $\left[J_x, [J_x, T_q^{(k)}] \right] = \sum_{q'} T_{q'}^{(k)} \langle k, q' | J_x^2 | k, q \rangle$,

(b) $\left[J_x, [J_x, T_q^{(k)}] \right] + \left[J_y, [J_y, T_q^{(k)}] \right] + \left[J_z, [J_z, T_q^{(k)}] \right] = k(k+1)\hbar^2 T_q^{(k)}$.

Exercise 7.23

Consider a spin $\frac{1}{2}$ particle which has an orbital angular momentum $l = 1$. Find all the Clebsch–Gordan coefficients involved in the addition of the orbital and spin angular momenta of this particle. *Hint:* The Clebsch–Gordan coefficient $\langle j_1, j_2; j_1, (j_2 - j_1) | j_2, j_2 \rangle$ is real and positive.

Exercise 7.24

This problem deals with another derivation of the matrix elements of $d^{(1)}(\beta)$. Use the relation

$$d_{mm'}^{(j)}(\beta) = \sum_{m_1 m_2} \sum_{m'_1 m'_2} \langle j_1, j_2; m_1, m_2 | j, m \rangle \langle j_1, j_2; m'_1, m'_2 | j, m' \rangle d_{m_1 m'_1}^{(j_1)}(\beta) d_{m_2 m'_2}^{(j_2)}(\beta)$$

for the case where $j_1 = j_2 = \frac{1}{2}$ along with the matrix elements of $d^{(1/2)}(\beta)$, which are given by (7.89), to derive all the matrix elements of $d^{(1)}(\beta)$.

Exercise 7.25

Consider the tensor $A(\theta, \varphi) = \sin^2 \theta \cos(2\varphi)$.

(a) Calculate the reduced matrix element $\langle 2 \parallel Y_2 \parallel 2 \rangle$. *Hint:* You may calculate explicitly $\langle 2, 1 | Y_{20} | 2, 1 \rangle$ and then use the Wigner–Eckart theorem to calculate it again.

(b) Express $A(\theta, \varphi)$ in terms of the components of a spherical tensor of rank 2 (i.e., in terms of $Y_{2m}(\theta, \varphi)$).

(c) Calculate $A_{m'\pm1} = \langle 2, m' | A | 2, \pm1 \rangle$ for $m' = \pm2, \pm1, 0$. You may need this Clebsch–Gordan coefficient: $\langle j, 2; m, 0 | j, m \rangle = [3m^2 - j(j+1)]/\sqrt{(2j-1)j(j+1)(2j+3)}$.

Exercise 7.26

(a) Calculate the reduced matrix element $\langle 1 \parallel Y_1 \parallel 2 \rangle$. *Hint:* For this, you may need to calculate $\langle 1, 0 | Y_{10} | 2, 0 \rangle$ directly and then from the Wigner–Eckart theorem.

(b) Using the Wigner–Eckart theorem and the relevant Clebsch–Gordan coefficients from the table, calculate $\langle 1, m | Y_{1m'} | 2, m'' \rangle$ for all possible values of m, m', and m''. *Hint:* You may find the integral $\int_0^\infty r^3 R_{21}^*(r) R_{32}(r) \, dr = \frac{64 a_0}{15\sqrt{5}} \left(\frac{6}{5} \right)^5$ and the following coefficients useful:

$\langle j, 1; m, 0 | (j-1), m \rangle = -\sqrt{(j-m)(j+m)/[j(2j+1)]}$,

$\langle j, 1; (m-1), 1 | (j-1), m \rangle = \sqrt{(j-m)(j-m+1)/[2j(2j+1)]}$, and

$\langle j, 1; (m+1), -1 | (j-1), m \rangle = \sqrt{(j+m)(j+m+1)/[2j(2j+1)]}$.

Exercise 7.27

A particle of spin $\frac{1}{2}$ is in a d state of orbital angular momentum (i.e., $l = 2$).
  (a) What are its possible states of total angular momentum.

  (b) If its Hamiltonian is given by $H = a + b\hat{\vec{L}} \cdot \hat{\vec{S}} + c\hat{\vec{L}}^2$, where a, b, and c are numbers, find the values of the energy for each of the different states of total angular momentum. Express your answer in terms of a, b, c.

Exercise 7.28

Consider an h-state electron. Calculate the Clebsch–Gordan coefficients involved in the following $\{| j, m\rangle\}$ states of the electron: $| \frac{11}{2}, \frac{9}{2}\rangle, | \frac{11}{2}, \frac{7}{2}\rangle, | \frac{9}{2}, \frac{9}{2}\rangle, | \frac{9}{2}, \frac{7}{2}\rangle$.

Exercise 7.29

Let the Hamiltonian of two nonidentical spin $\frac{1}{2}$ particles be

$$\hat{H} = \frac{\varepsilon_1}{\hbar^2}(\hat{\vec{S}}_1 + \hat{\vec{S}}_2) \cdot \hat{\vec{S}}_1 - \frac{\varepsilon_2}{\hbar}(\hat{S}_{1_z} + \hat{S}_{2_z}),$$

where ε_1 and ε_2 are constants having the dimensions of energy. Find the energy levels and their degeneracies.

Exercise 7.30

Find the energy levels and their degeneracies for a system of two nonidentical spin $\frac{1}{2}$ particles with Hamiltonian

$$\hat{H} = \frac{\varepsilon_0}{\hbar^2}(\hat{S}_1^2 + \hat{S}_2^2) - \frac{\varepsilon_0}{\hbar}(\hat{S}_{1_z} + \hat{S}_{2_z}),$$

where ε_0 is a constant having the dimensions of energy.

Exercise 7.31

Consider two nonidentical spin $s = \frac{1}{2}$ particles with Hamiltonian

$$\hat{H} = \frac{\varepsilon_0}{\hbar^2}(\hat{\vec{S}}_1 + \hat{\vec{S}}_2)^2 - \frac{\varepsilon_0}{\hbar^2}(\hat{S}_{1_z} + \hat{S}_{2_z})^2,$$

where ε_0 is a constant having the dimensions of energy. Find the energy levels and their degeneracies.

Exercise 7.32

Consider a system of three nonidentical particles, each of spin $s = \frac{1}{2}$, whose Hamiltonian is given by

$$\hat{H} = \frac{\varepsilon_1}{\hbar^2}(\hat{\vec{S}}_1 + \hat{\vec{S}}_3) \cdot \hat{\vec{S}}_2 + \frac{\varepsilon_2}{\hbar^2}(\hat{S}_{1_z} + \hat{S}_{2_z} + \hat{S}_{3_z})^2,$$

where ε_1 and ε_2 are constants having the dimensions of energy. Find the system's energy levels and their degeneracies.

Exercise 7.33

Consider a system of three nonidentical particles, each with angular momentum $\frac{3}{2}$. Find the possible values of the total spin S of this system and specify the number of angular momentum eigenstates corresponding to each value of S.

Chapter 8

Identical Particles

Up to this point, we have dealt mainly with the motion of a single particle. Now we want to examine how to describe systems with *many particles*. We shall focus on systems of *identical* particles and examine how to construct their wave functions.

8.1 Many-Particle Systems

Most physical systems—nucleons, nuclei, atoms, molecules, solids, fluids, gases, etc.—involve many particles. They are known as *many-particle* or *many-body* systems. While atomic, nuclear, and subnuclear systems involve intermediate numbers of particles (~ 2 to 300), solids, fluids, and gases are truly many-body systems, since they involve very large numbers of particles ($\sim 10^{23}$).

8.1.1 Schrödinger Equation

How does one describe the dynamics of a system of N particles? This description can be obtained from a generalization of the dynamics of a single particle. The state of a system of N spinless particles (we ignore their spin for the moment) is described by a wave function $\Psi(\vec{r}_1, \vec{r}_2, \ldots \vec{r}_N, t)$, where $|\Psi(\vec{r}_1, \vec{r}_2, \ldots, \vec{r}_N, t)|^2 d^3r_1\, d^3r_2 \cdots d^3r_N$ represents the probability at time t of finding particle 1 in the volume element d^3r_1 centered about $\vec{r}_1$, particle 2 in the volume d^3r_2 about $\vec{r}_2$, ..., and particle N in the volume d^3r_N about $\vec{r}_N$. The normalization condition of the state is given by

$$\int d^3r_1 \int d^3r_2 \cdots \int |\Psi(\vec{r}_1, \vec{r}_2, \ldots, \vec{r}_N, t)|^2 d^3r_N = 1. \tag{8.1}$$

The wave function Ψ evolves in time according to the time-dependent Schrödinger equation

$$i\hbar \frac{\partial}{\partial t} \Psi(\vec{r}_1, \vec{r}_2, \ldots, \vec{r}_N, t) = \hat{H}\Psi(\vec{r}_1, \vec{r}_2, \ldots, \vec{r}_N, t). \tag{8.2}$$

The form of $\hat{H}$ is obtained by generalizing the one-particle Hamiltonian $\vec{P}^2/(2m) + \vec{V}(\vec{r})$ to N

particles:

$$\hat{H} = \sum_{j=1}^{N} \frac{\hat{P}_j^2}{2m_j} + \hat{V}(\vec{r}_1, \vec{r}_2, \ldots, \vec{r}_N, t) = -\sum_{j=1}^{N} \frac{\hbar^2}{2m_j} \nabla_j^2 + \hat{V}(\vec{r}_1, \vec{r}_2, \ldots, \vec{r}_N, t), \qquad (8.3)$$

where m_j and $\hat{P}_j$ are the mass and the momentum of the jth particle and $\hat{V}$ is the operator corresponding to the total potential energy ($\hat{V}$ accounts for all forms of interactions—internal and external—the mutual interactions between the various particles of the system and for the interactions of the particles with the outside world).

The formalism of quantum mechanics for an N-particle system can be, in principle, inferred from that of a single particle. Operators corresponding to different particles commute; for instance, the commutation relations between the position and momentum operators are

$$[\hat{X}_j, \hat{P}_{x_k}] = i\hbar\delta_{j,k}, \qquad [\hat{X}_j, \hat{X}_k] = [\hat{P}_{x_j}, \hat{P}_{x_k}] = 0 \qquad (j, k = 1, 2, 3, \ldots, N), \quad (8.4)$$

where $\hat{X}_j$ is the x-position operator of the jth particle, and $\hat{P}_{x_k}$ the x-momentum operator of the kth particle; similar relations can be obtained for the y and z components.

Stationary states

In the case where the potential $\hat{V}$ is *time independent*, the solutions of (8.2) are given by *stationary* states

$$\Psi(\vec{r}_1, \vec{r}_2, \ldots, \vec{r}_N, t) = \psi(\vec{r}_1, \vec{r}_2, \ldots, \vec{r}_N) \, e^{-iEt/\hbar}, \qquad (8.5)$$

where E is the total energy of the system and ψ is the solution to the time-independent Schrödinger equation $\hat{H}\psi = E\psi$, i.e.,

$$\left[-\sum_{j=1}^{N} \frac{\hbar^2}{2m_j} \vec{\nabla}_j^2 + V(\vec{r}_1, \ldots, \vec{r}_N) \right] \psi(\vec{r}_1, \vec{r}_2, \ldots, \vec{r}_N) = E \, \psi(\vec{r}_1, \vec{r}_2, \ldots, \vec{r}_N). \qquad (8.6)$$

The properties of stationary states for a single particle also apply to N-particle systems. For instance, the probability density $\langle \psi \mid \psi \rangle$, the probability current density $\vec{j}$, and the expectation values of time-independent operators are conserved, since they do not depend on time:

$$\langle \Psi \mid \hat{A} \mid \Psi \rangle = \langle \psi \mid \hat{A} \mid \psi \rangle = \int d^3r_1 \int d^3r_2 \cdots \int \psi^*(\vec{r}_1, \vec{r}_2, \ldots, \vec{r}_N) \hat{A} \psi(\vec{r}_1, \vec{r}_2, \ldots, \vec{r}_N) \, d^3r_N.$$

$$(8.7)$$

In particular, the energy of a stationary state is conserved.

Multielectron atoms

As an illustration, let us consider an atom with Z electrons. If $\vec{R}$ is used to represent the position of the center of mass of the nucleus, the wave function of the atom depends on $3(Z + 1)$ coordinates $\psi(\vec{r}_1, \vec{r}_2, \ldots, \vec{r}_Z, \vec{R})$, where $\vec{r}_1, \vec{r}_2, \ldots, \vec{r}_Z$ are the position vectors of the Z electrons. The time-independent Schrödinger equation for this atom, neglecting contributions from the spin–orbit correction, the relativistic correction, and similar terms, is given by

$$\left[-\frac{\hbar^2}{2m_e} \sum_{i=1}^{Z} \vec{\nabla}_{r_i}^2 - \frac{\hbar^2}{2M} \vec{\nabla}_R^2 - \sum_{i=1}^{Z} \frac{Z e^2}{|\vec{r}_i - \vec{R}|} + \sum_{i>j} \frac{e^2}{|\vec{r}_i - \vec{r}_j|} \right] \psi(\vec{r}_1, \vec{r}_2, \ldots, \vec{r}_Z, \vec{R})$$

$$= E \psi(\vec{r}_1, \vec{r}_2, \ldots, \vec{r}_Z, \vec{R}), \qquad (8.8)$$

where M is the mass of the nucleus and $-\hbar^2 \vec{\nabla}_R^2 / 2M$ is its kinetic energy operator. The term $-\sum_{i=1}^{Z} Ze^2 / |\vec{r}_i - \vec{R}|$ represents the attractive Coulomb interaction of each electron with the nucleus and $\sum_{i>j} e^2 / |\vec{r}_i - \vec{r}_j|$ is the repulsive Coulomb interaction between the ith and the jth electrons; $|\vec{r}_i - \vec{r}_j|$ is the distance separating them. As these (Coulomb) interactions are independent of time, the states of atoms are stationary.

We should note that the Schrödinger equations (8.3), (8.6), and (8.8) are all many-particle differential equations. As these equations cannot be separated into one-body equations, it is difficult, if not impossible, to solve them. For the important case where the N particles of the system do not interact—this is referred to as an *independent particle system*—the Schrödinger equation can be trivially reduced to N one-particle equations (Section 8.1.3); we have seen how to solve these equations exactly (Chapters 4 and 6) and approximately (Chapters 9 and 10).

8.1.2 Interchange Symmetry

Although the exact eigenstates of the many-body Hamiltonian (8.3) are generally impossible to obtain, we can still infer some of their properties by means of symmetry schemes. Let ξ_i represent the coordinates (position $\vec{r}_i$, spin $\vec{s}_i$, and any other internal degrees of freedom such as isospin, color, flavor) of the ith particle and let $\psi(\xi_1, \xi_2, \ldots, \xi_N)$ designate the wave function of the N-particle system.

We define a *permutation operator* (also called *exchange operator*) $\hat{P}_{ij}$ as an operator that, when acting on an N-particle wave function $\psi(\xi_1, \ldots, \xi_i, \ldots, \xi_j, \ldots, \xi_N)$, interchanges the ith and the jth particles

$$\hat{P}_{ij} \psi(\xi_1, \ldots, \xi_i, \ldots, \xi_j, \ldots \xi_N) = \psi(\xi_1, \ldots, \xi_j, \ldots, \xi_i, \ldots, \xi_N); \qquad (8.9)$$

i and j are arbitrary $(i, j = 1, 2, \cdots, N)$. Since

$$\begin{aligned}
\hat{P}_{ji} \psi(\xi_1, \ldots, \xi_i, \ldots, \xi_j, \ldots \xi_N) &= \psi(\xi_1, \ldots, \xi_j, \ldots, \xi_i, \ldots, \xi_N) \\
&= \hat{P}_{ij} \psi(\xi_1, \ldots, \xi_i, \ldots, \xi_j, \ldots \xi_N),
\end{aligned} \qquad (8.10)$$

we have $\hat{P}_{ij} = \hat{P}_{ji}$. In general, permutation operators do not commute:

$$\hat{P}_{ij} \hat{P}_{kl} \neq \hat{P}_{kl} \hat{P}_{ij} \qquad \text{or} \qquad [\hat{P}_{ij}, \hat{P}_{kl}] \neq 0 \qquad (ij \neq kl). \qquad (8.11)$$

For instance, in the case of a four-particle state $\psi(\xi_1, \xi_2, \xi_3, \xi_4) = (3\xi_4 / \xi_2 \xi_3) e^{-i\xi_1}$, we have

$$\hat{P}_{12} \hat{P}_{14} \psi(\xi_1, \xi_2, \xi_3, \xi_4) = \hat{P}_{12} \psi(\xi_4, \xi_2, \xi_3, \xi_1) = \psi(\xi_2, \xi_4, \xi_3, \xi_1) = \frac{3\xi_1}{\xi_4 \xi_3} e^{-i\xi_2}, \qquad (8.12)$$

$$\hat{P}_{14} \hat{P}_{12} \psi(\xi_1, \xi_2, \xi_3, \xi_4) = \hat{P}_{14} \psi(\xi_2, \xi_1, \xi_3, \xi_4) = \psi(\xi_4, \xi_1, \xi_3, \xi_2) = \frac{3\xi_2}{\xi_1 \xi_3} e^{-i\xi_4}. \qquad (8.13)$$

Since two successive applications of $\hat{P}_{ij}$ leave the wave function unchanged,

$$\begin{aligned}
\hat{P}_{ij}^2 \psi(\xi_1, \ldots, \xi_i, \ldots, \xi_j, \ldots, \xi_N) &= \hat{P}_{ij} \psi(\xi_1, \ldots, \xi_j, \ldots, \xi_i, \ldots, \xi_N) \\
&= \psi(\xi_1, \ldots, \xi_i, \ldots, \xi_j, \ldots, \xi_N),
\end{aligned} \qquad (8.14)$$

we have $\hat{P}_{ij}^2 = 1$; hence $\hat{P}_{ij}$ has two eigenvalues ± 1:

$$\hat{P}_{ij} \psi(\xi_1, \ldots, \xi_i, \ldots, \xi_j, \ldots, \xi_N) = \pm \psi(\xi_1, \ldots, \xi_i, \ldots, \xi_j, \ldots \xi_N). \qquad (8.15)$$

The wave functions corresponding to the eigenvalue $+1$ are *symmetric* and those corresponding to -1 are *antisymmetric* with respect to the interchange of the pair (i, j). Denoting these functions by ψ_s and ψ_a, respectively, we have

$$\psi_s(\xi_1, \ldots, \xi_i, \ldots, \xi_j, \ldots, \xi_N) = \psi_s(\xi_1, \ldots, \xi_j, \ldots, \xi_i, \ldots, \xi_N), \qquad (8.16)$$

$$\psi_a(\xi_1, \ldots, \xi_i, \ldots, \xi_j, \ldots, \xi_N) = -\psi_a(\xi_1, \ldots, \xi_j, \ldots, \xi_i, \ldots, \xi_N). \qquad (8.17)$$

Example 8.1

Specify the symmetry of the following functions:

(a) $\psi(x_1, x_2) = 4(x_1 - x_2)^2 + \frac{10}{x_1^2 + x_2^2}$,

(b) $\phi(x_1, x_2) = -\frac{3(x_1 - x_2)}{2(x_1 - x_2)^2 + 7}$,

(c) $\chi(x_1, x_2, x_3) = 6x_1 x_2 x_3 + \frac{x_1^2 + x_2^2 + x_3^2 - 1}{2x_1^3 + 2x_2^3 + 2x_3^3 + 5}$,

(d) $\Phi(x_1, x_2) = \frac{1}{x_2 + 3} e^{-|x_1|}$.

Solution

(a) The function $\psi(x_1, x_2)$ is symmetric, since $\psi(x_2, x_1) = \psi(x_1, x_2)$.

(b) The function $\phi(x_1, x_2)$ is antisymmetric, since $\phi(x_2, x_1) = -\phi(x_1, x_2)$, and ϕ is zero when $x_1 = x_2$: $\phi(x_1, x_1) = 0$.

(c) The function $\chi(x_1, x_2, x_3)$ is symmetric because

$$\begin{aligned} \chi(x_1, x_2, x_3) &= \chi(x_1, x_3, x_2) = \chi(x_2, x_1, x_3) = \chi(x_2, x_3, x_1) \\ &= \chi(x_3, x_1, x_2) = \chi(x_3, x_2, x_1). \end{aligned} \qquad (8.18)$$

(d) The function $\Phi(x_2, x_1)$ is neither symmetric nor antisymmetric, since $\Phi(x_2, x_1) = \frac{1}{x_1 + 3} e^{-|x_2|} \neq \pm\Phi(x_1, x_2)$.

8.1.3   Systems of Distinguishable Noninteracting Particles

For a system of N noninteracting particles that are distinguishable—each particle has a different mass m_i and experiences a different potential $\hat{V}_i(\xi_i)$—the potential $\hat{V}$ is given by

$$\hat{V}(\xi_1, \xi_2, \ldots, \xi_N) = \sum_{i=1}^{N} \hat{V}_i(\xi_i) \qquad (8.19)$$

and the Hamiltonian of this system of N independent particles by

$$\hat{H} = \sum_{i=1}^{N} \hat{H}_i = \sum_{i=1}^{N} \left[-\frac{\hbar^2}{2m_i} \nabla_i^2 + \hat{V}_i(\xi_i) \right], \qquad (8.20)$$

where $\hat{H}_i = -\hbar^2 \nabla_i^2 / 2m_i + \hat{V}_i(\xi_i)$ is the Hamiltonian of the ith particle, known as the single particle Hamiltonian. The Hamiltonians of different particles commute $[\hat{H}_i, \hat{H}_j] = 0$, since $[\hat{X}_i, \hat{X}_j] = [\hat{P}_i, \hat{P}_j] = 0$.

The Schrödinger equation of the N-particle system

$$\hat{H} \psi_{n_1,n_2,\cdots,n_N}(\xi_1, \xi_2, \cdots, \xi_N) = E_{n_1,n_2,\cdots,n_N} \psi_{n_1,n_2,\cdots,n_N}(\xi_1, \xi_2, \cdots, \xi_N) \quad (8.21)$$

separates into N one-particle equations

$$\left[-\frac{\hbar^2}{2m_i} \nabla_i^2 + \hat{V}_i(\xi_i) \right] \psi_{n_i}(\xi_i) = \varepsilon_{n_i} \psi_{n_i}(\xi_i), \quad (8.22)$$

with

$$E_{n_1,n_2,\cdots,n_N} = \varepsilon_{n_1} + \varepsilon_{n_2} + \cdots \varepsilon_{n_N} = \sum_{i=1}^{N} \varepsilon_{n_i} \quad (8.23)$$

and

$$\psi_{n_1,n_2,\cdots,n_N}(\xi_1, \xi_2, \cdots, \xi_N) = \psi_{n_1}(\xi_1) \psi_{n_2}(\xi_2) \cdots \psi_{n_N}(\xi_N) = \prod_{i=1}^{N} \psi_{n_i}(\xi_i). \quad (8.24)$$

We see that, when the interactions are neglected, the N-particle Schrödinger equation separates into N one-particle Schrödinger equations. The solutions of these equations yield the single-particle energies ε_{n_i} and states $\psi_{n_i}(\xi_i)$; the single-particle states are also known as the *orbitals*. The total energy is the sum of the single-particle energies and the total wave function is the product of the orbitals. The number n_i designates the set of all quantum numbers of the ith particle. Obviously, each particle requires one, two, or three quantum numbers for its full description, depending on whether the particles are moving in a one-, two-, or three-dimensional space; if the spin were considered, we would need to add another quantum number. For instance, if the particles moved in a one-dimensional harmonic oscillator, n_i would designate the occupation number of the ith particle. But if the particles were the electrons of an atom, then n_i would stand for four quantum numbers: the radial, orbital, magnetic, and spin quantum numbers $N_i l_i m_{l_i} m_{s_i}$.

Example 8.2
Find the energy levels and wave functions of a system of four distinguishable spinless particles placed in an infinite potential well of size a. Use this result to infer the energy and the wave function of the ground state and the first excited state.

Solution
Each particle moves in a potential which is defined by $\hat{V}_i(x_i) = 0$ for $0 \le x_i \le a$ and $\hat{V}_i(x_i) = \infty$ for the other values of x_i. In this case the Schrödinger equation of the four-particle system:

$$\sum_{i=1}^{4} \left[-\frac{\hbar^2}{2m_i} \frac{d^2}{dx_i^2} \right] \psi_{n_1,n_2,n_3,n_4}(x_1, x_2, x_3, x_4) = E_{n_1,n_2,n_3,n_4} \psi_{n_1,n_2,n_3,n_4}(x_1, x_2, x_3, x_4),$$

$$(8.25)$$

separates into four one-particle equations

$$-\frac{\hbar^2}{2m_i} \frac{d^2 \psi_{n_i}(x_i)}{dx_i^2} = \varepsilon_{n_i} \psi_{n_i}(x_i), \qquad i = 1, 2, 3, 4, \quad (8.26)$$

with

$$\varepsilon_{n_i} = \frac{\hbar^2 \pi^2 n_i^2}{2m_i a^2}, \qquad \psi_{n_i}(x_i) = \sqrt{\frac{2}{a}} \sin\left(\frac{n_i \pi}{a} x_i\right). \qquad (8.27)$$

The total energy and wave function are given by

$$E_{n_1,n_2,n_3,n_4} = \frac{\hbar^2 \pi^2}{2a^2}\left(\frac{n_1^2}{m_1} + \frac{n_2^2}{m_2} + \frac{n_3^2}{m_3} + \frac{n_4^2}{m_4}\right), \qquad (8.28)$$

$$\psi_{n_1,n_2,n_3,n_4}(x_1,x_2,x_3,x_4) = \frac{4}{a^2} \sin\left(\frac{n_1 \pi}{a} x_1\right) \sin\left(\frac{n_2 \pi}{a} x_2\right) \sin\left(\frac{n_3 \pi}{a} x_3\right) \sin\left(\frac{n_4 \pi}{a} x_4\right).$$
$$(8.29)$$

The ground state corresponds to the case where all four particles occupy their respective ground state orbitals, $n_1 = n_2 = n_3 = n_4 = 1$. The ground state energy and wave function are thus given by

$$E_{1,1,1,1} = \frac{\hbar^2 \pi^2}{2a^2}\left(\frac{1}{m_1} + \frac{1}{m_2} + \frac{1}{m_3} + \frac{1}{m_4}\right), \qquad (8.30)$$

$$\psi_{1,1,1,1}(x_1,x_2,x_3,x_4) = \frac{4}{a^2} \sin\left(\frac{\pi}{a} x_1\right) \sin\left(\frac{\pi}{a} x_2\right) \sin\left(\frac{\pi}{a} x_3\right) \sin\left(\frac{\pi}{a} x_4\right). \qquad (8.31)$$

The first excited state is somewhat tricky. Since it corresponds to the next higher energy level of the system, it must correspond to the case where the particle having the largest mass occupies its first excited state while the other three particles remain in their respective ground states. For argument's sake, if the third particle were the most massive, the first excited state would correspond to the configuration $n_1 = n_2 = n_4 = 1$ and $n_3 = 2$; the energy and wave function of the first excited state would then be given by

$$E_{1,1,2,1} = \frac{\hbar^2 \pi^2}{2a^2}\left(\frac{1}{m_1} + \frac{1}{m_2} + \frac{4}{m_3} + \frac{1}{m_4}\right), \qquad (8.32)$$

$$\psi_{1,1,2,1}(x_1,x_2,x_3,x_4) = \frac{4}{a^2} \sin\left(\frac{\pi}{a} x_1\right) \sin\left(\frac{\pi}{a} x_2\right) \sin\left(\frac{2\pi}{a} x_3\right) \sin\left(\frac{\pi}{a} x_4\right). \qquad (8.33)$$

Continuing in this way, we can obtain the entire energy spectrum of this system.

8.2 Systems of Identical Particles

8.2.1 Identical Particles in Classical and Quantum Mechanics

In classical mechanics, when a system is made of identical particles, it is possible to identify and distinguish each particle from the others. That is, although all particles have the same physical properties, we can "tag" each classical particle and follow its motion along a path. For instance, each particle can be colored differently from the rest; hence we can follow the trajectory of each particle separately at each time. *Identical classical particles, therefore, do not lose their identity; they are distinguishable.*

In quantum mechanics, however, identical particles are truly indistinguishable. The underlying basis for this is twofold. First, to describe a particle, we cannot specify more than

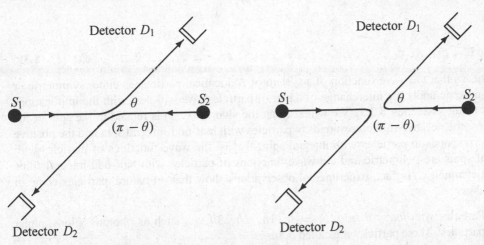

Figure 8.1 When scattering two *identical* particles in the center of mass frame, it is impossible to forcast with certitude whether the particles scatter according to the first process or to the second. For instance, we cannot tell whether the particle fired from source S_1 will make it to detector D_1 or to D_2.

a complete set of commuting observables. In particular, there exists no mechanism to tag the particles as in classical mechanics. Second, due to the uncertainty principle, the concept of the path of a particle becomes meaningless. Even if the position of a particle is exactly determined at a given time, it is not possible to specify its coordinates at the next instant. Thus, *identical particles lose their identity (individuality) in quantum mechanics*.

To illustrate this, consider an experiment in which we scatter two identical particles. As displayed in Figure 8.1, after particles 1 and 2 (fired from the sources S_1 and S_2) have scattered, it is impossible to distinguish between the first and the second outcomes. That is, we cannot determine experimentally the identity of the particles that are collected by each detector. For instance, we can in no way tell whether it is particle 1 or particle 2 that has reached detector D_1. We can only say that a particle has reached detector D_1 and another has made it to D_2, but have no information on their respective identities. There exists no experimental mechanism that allows us to follow the motion of each particle from the time it is fired out of the source till it reaches the detector. This experiment shows that *the individuality of a microscopic particle is lost the moment it is mixed with other similar particles*.

Having discussed the indistinguishability concept on a two-particle system, let us now study this concept on larger systems. For this, consider a system of N *identical* particles whose wave function is $\psi(\xi_1, \xi_2, \ldots, \xi_N)$.

The moment these N particles are mixed together, no experiment can determine which particle has the coordinates ξ_1, or which one has ξ_2, and so on. It is impossible to specify experimentally the identity of the particle which is located at ξ_1, or that located at ξ_2, and so on. The only measurements we can perform are those that specify the probability for a certain particle to be located at ξ_1, another at ξ_2, and so on, but we can never make a distinction as to which particle is which.

As a result, the probability must remain unchanged by an interchange of the particles. For instance, an interchange of particles i and j will leave the probability density unaffected:

$$\left|\psi(\xi_1, \xi_2, \ldots, \xi_i, \ldots, \xi_j, \ldots, \xi_N)\right|^2 = \left|\psi(\xi_1, \xi_2, \ldots, \xi_j, \ldots, \xi_i, \ldots, \xi_N)\right|^2; \quad (8.34)$$

hence we have

$$\psi(\xi_1, \xi_2, \ldots, \xi_i, \ldots, \xi_j, \ldots, \xi_N) = \pm \psi(\xi_1, \xi_2, \ldots, \xi_j, \ldots, \xi_i, \ldots, \xi_N). \qquad (8.35)$$

This means that the wave function of a system of N identical particles is either symmetric or antisymmetric under the interchange of a pair of particles. We will deal with the implications of this result in Section 8.2.3. We will see that the sign in (8.35) is related to the spin of the particles: the negative sign corresponds to particles with half-odd-integral spin and the positive sign corresponds to particles with integral spin; that is, the wave functions of particles with integral spins are symmetric and the wave functions of particles with half-odd-integral spins are antisymmetric. In fact, experimental observations show that, in nature, particles come in two classes:

- Particles with *integral spin*, $S_i = 0$, $1\hbar$, $2\hbar$, $3\hbar$, ..., such as photons, pions, alpha particles. These particles are called *bosons*.

- Particles with *half-odd-integral spin*, $S_i = \hbar/2$, $3\hbar/2$, $5\hbar/2$, $7\hbar/2$, ..., such as quarks, electrons, positrons, protons, neutrons. These particles are called *fermions*.

That is, particles occurring in nature are either bosons or fermions.

Before elaborating more on the properties of bosons and fermions, let us present a brief outline on the interchange (permutation) symmetry.

8.2.2   Exchange Degeneracy

How does the interchange symmetry affect operators such as the Hamiltonian? Since the Coulomb potential, which results from electron–electron and electron–nucleus interactions,

$$V(\vec{r}_1, \vec{r}_2, \vec{r}_3, \ldots, \vec{r}_Z) = -\sum_{i=1}^{Z} \frac{Z e^2}{|\vec{r}_i - \vec{R}|} + \sum_{i>j} \frac{e^2}{|\vec{r}_i - \vec{r}_j|}, \qquad (8.36)$$

is invariant under the permutation of any pair of electrons, the Hamiltonian (8.8) is also invariant under such permutations. This symmetry also applies to the orbital, spin, and angular momenta of an atom. We may thus use this symmetry to introduce another definition of the *identicalness* of particles. *The N particles of a system are said to be identical if the various observables of the system* (such as the Hamiltonian $\hat{H}$, the angular momenta, and so on) *are symmetrical when any two particles are interchanged.* If these operators were not symmetric under particle interchange, the particles would be distinguishable.

The invariance of the Hamiltonian under particle interchanges is not without physical implications: the eigenvalues of $\hat{H}$ are *degenerate*. The wave functions corresponding to all possible electron permutations have the same energy E: $\hat{H}\psi = E\psi$. This is known as the *exchange degeneracy*. For instance, the degeneracy associated with a system of two identical particles is equal to 2, since $\psi(\xi_1, \xi_2)$ and $\psi(\xi_2, \xi_1)$ correspond to the same energy E.

So the Hamiltonian of a system of N identical particles ($m_i = m$) is completely symmetric with respect to the coordinates of the particles:

$$\hat{H}(\xi_1, \ldots, \xi_i, \ldots, \xi_j, \ldots, \xi_N) = \sum_{k=1}^{N} \frac{\vec{P}_k^2}{2m} + \hat{V}(\xi_1, \ldots, \xi_i, \ldots, \xi_j, \ldots, \xi_N)$$

$$= \hat{H}(\xi_1, \ldots, \xi_j, \ldots, \xi_i, \ldots, \xi_N), \qquad (8.37)$$

because $\hat{V}$ is invariant under the permutation of any pair of particles $i \longleftrightarrow j$:

$$\hat{V}(\xi_1, \ldots, \xi_i, \ldots, \xi_j, \ldots, \xi_N) = \hat{V}(\xi_1, \ldots, \xi_j, \ldots, \xi_i, \ldots, \xi_N). \tag{8.38}$$

This property can also be ascertained by showing that $\hat{H}$ commutes with the particle interchange operator $\hat{P}_{ij}$. If ψ is eigenstate to $\hat{H}$ with eigenvalue E, we can write

$$\begin{aligned}
\hat{H}\hat{P}_{ij}\psi(\xi_1, \ldots, \xi_i, \ldots, \xi_j, \ldots, \xi_N) &= \hat{H}\psi(\xi_1, \ldots, \xi_j, \ldots, \xi_i, \ldots, \xi_N) \\
&= E\psi(\xi_1, \ldots, \xi_j, \ldots, \xi_i, \ldots, \xi_N) = E\hat{P}_{ij}\psi(\xi_1, \ldots, \xi_i, \ldots, \xi_j, \ldots, \xi_N) \\
&= \hat{P}_{ij}E\psi(\xi_1, \ldots, \xi_i, \ldots, \xi_j, \ldots, \xi_N) = \hat{P}_{ij}\hat{H}\psi(\xi_1, \ldots, \xi_i, \ldots, \xi_j, \ldots, \xi_N)
\end{aligned} \tag{8.39}$$

or

$$[\hat{H}, \hat{P}_{ij}] = 0. \tag{8.40}$$

Therefore, $\hat{P}_{ij}$ is a *constant of the motion*. That is, if we start with a wave function that is symmetric (antisymmetric), it will remain so for all subsequent times. Moreover, since $\hat{P}_{ij}$ and $\hat{H}$ commute, they possess a complete set of functions that are joint eigenstates of both. As shown in (8.15) to (8.17), these eigenstates have definite parity, either symmetric or antisymmetric.

8.2.3 Symmetrization Postulate

We have shown in (8.35) that the wave function of a system of N identical particles is either symmetric or antisymmetric under the interchange of any pair of particles:

$$\psi(\xi_1, \xi_2, \ldots, \xi_i, \ldots, \xi_j, \ldots, \xi_N) = \pm\psi(\xi_1, \xi_2, \ldots, \xi_j, \ldots, \xi_i, \ldots, \xi_N). \tag{8.41}$$

This result, which turns out to be supported by experimental evidence, is the very essence of the *symmetrization postulate* which stipulates that, in nature, the states of systems containing N identical particles are *either totally symmetric or totally antisymmetric* under the interchange of any pair of particles and that states with mixed symmetry do not exist. Besides that, this postulate states two more things:

- Particles with integral spins, or bosons, have symmetric states.

- Particles with half-odd-integral spins, or fermions, have antisymmetric states.

Fermions are said to obey *Fermi–Dirac statistics* and bosons to obey *Bose–Einstein statistics*. So the wave function of a system of identical bosons is totally symmetric and the wave function of a system of identical fermions is totally antisymmetric.

Composite particles

The foregoing discussion pertains to identical particles that are "simple" or elementary such as quarks, electrons, positrons, muons, and so on. Let us now discuss the symmetry of systems of identical *composite* "particles" where each particle is composed of two or more identical elementary particles. For instance, alpha particles, which consist of nuclei that are composed of two neutrons and two protons each, are a typical example of composite particles. A system of N hydrogen atoms can also be viewed as a system of identical composite particles where each "particle" (atom) consists of a proton and an electron. Protons, neutrons, pions, etc., are themselves composite particles, because protons and neutrons consist of three quarks and pions consist of two. *Quarks* are elementary spin $\frac{1}{2}$ particles.

Composite particles have spin. The spin of each composite particle can be obtained by adding the spins of its constituents. If the total spin of the composite particle is half-odd-integer, this particle behaves like a fermion, and hence it obeys Fermi–Dirac statistics. If, on the other hand, its resultant spin is integer, it behaves like a boson and obeys Bose–Einstein statistics. In general, if the composite particle has an odd number of fermions; it is then a fermion, otherwise it is a boson. For instance, nucleons are fermions because they consist of three quarks; mesons are bosons because they consist of two quarks. For another illustrative example, let us consider the isotopes ^{4}He and ^{3}He of the helium atom: ^{4}He, which is called an alpha particle, is a boson for it consists of four nucleons (two protons and two neutrons), while ^{3}He is a fermion since it consists of three nucleons (one neutron and two protons). The hydrogen atom consists of two fermions (an electron and a proton), so it is a boson.

8.2.4 Constructing Symmetric and Antisymmetric Functions

Since the wave functions of systems of identical particles are either totally symmetric or totally antisymmetric, it is appropriate to study the formalism of how to construct wave functions that are totally symmetric or totally antisymmetric starting from asymmetric functions. For simplicity, consider first a system of two identical particles. Starting from any normalized asymmetric wave function $\psi(\xi_1, \xi_2)$, we can construct symmetric wave functions $\psi_s(\xi_1, \xi_2)$ as

$$\psi_s(\xi_1, \xi_2) = \frac{1}{\sqrt{2}}\left[\psi(\xi_1, \xi_2) + \psi(\xi_2, \xi_1)\right] \tag{8.42}$$

and antisymmetric wave functions $\psi_a(\xi_1, \xi_2)$ as

$$\psi_a(\xi_1, \xi_2) = \frac{1}{\sqrt{2}}\left[\psi(\xi_1, \xi_2) - \psi(\xi_2, \xi_1)\right], \tag{8.43}$$

where $1/\sqrt{2}$ is a normalization factor.

Similarly, for a system of three identical particles, we can construct ψ_s and ψ_a from an asymmetric function ψ as follows:

$$\psi_s(\xi_1, \xi_2, \xi_3) = \frac{1}{\sqrt{6}}\left[\psi(\xi_1, \xi_2, \xi_3) + \psi(\xi_1, \xi_3, \xi_2) + \psi(\xi_2, \xi_3, \xi_1)\right.$$
$$\left. + \psi(\xi_2, \xi_1, \xi_3) + \psi(\xi_3, \xi_1, \xi_2) + \psi(\xi_3, \xi_2, \xi_1)\right], \tag{8.44}$$

$$\psi_a(\xi_1, \xi_2, \xi_3) = \frac{1}{\sqrt{6}}\left[\psi(\xi_1, \xi_2, \xi_3) - \psi(\xi_1, \xi_3, \xi_2) + \psi(\xi_2, \xi_3, \xi_1)\right.$$
$$\left. - \psi(\xi_2, \xi_1, \xi_3) + \psi(\xi_3, \xi_1, \xi_2) - \psi(\xi_3, \xi_2, \xi_1)\right]. \tag{8.45}$$

Continuing in this way, we can in principle construct symmetric and antisymmetric wave functions for any system of N identical particles.

8.2.5 Systems of Identical Noninteracting Particles

In the case of a system of N noninteracting identical particles, where all particles have equal mass $m_i = m$ and experience the same potential $\hat{V}_i(\xi_i) = \hat{V}(\xi_i)$, the Schrödinger equation of the system separates into N identical one-particle equations

$$\left[-\frac{\hbar^2}{2m} \nabla_i^2 + \hat{V}(\xi_i) \right] \psi_{n_i}(\xi_i) = \varepsilon_{n_i} \psi_{n_i}(\xi_i). \tag{8.46}$$

Whereas the energy is given, like the case of a system of N distinguishable particles, by a sum of the single-particle energies $E_{n_1,n_2,...,n_N} = \sum_{i=1}^{N} \varepsilon_{n_i}$, the wave function can no longer be given by a simple product $\psi_{n_1,n_2,...,n_N}(\xi_1,\xi_2,...,\xi_N) = \prod_{i=1}^{N} \psi_{n_i}(\xi_i)$ for at least two reasons. First, if the wave function is given by such a product, it would imply that particle 1 is in the state ψ_{n_1}, particle 2 in the state ψ_{n_2}, ..., and particle N in the state ψ_{n_N}. This, of course, makes no sense since all we know is that one of the particles is in the state ψ_{n_1}, another in ψ_{n_2}, and so on; since the particles are identical, there is no way to tell which particle is in which state. If, however, the particles were distinguishable, then their total wave function would be given by such a product, as shown in (8.24). The second reason why the wave function of a system of identical particles cannot be given by $\prod_{i=1}^{N} \psi_{n_i}(\xi_i)$ has to do with the fact that such a product has, in general, no *definite* symmetry—a mandatory requirement for systems of N identical particles whose wave functions are either symmetric or antisymmetric. We can, however, extend the method of Section 8.2.4 to construct totally symmetric and totally antisymmetric wave functions from the single-particle states $\psi_{n_i}(\xi_i)$. For this, we are going to show how to construct symmetrized and antisymmetrized wave functions for systems of two, three, and N noninteracting identical particles.

8.2.5.1 Wave Function of Two-Particle Systems

By analogy with (8.42) and (8.43), we can construct the symmetric and antisymmetric wave functions for a system of two identical, noninteracting particles in terms of the single-particle wave functions as follows:

$$\psi_s(\xi_1,\xi_2) = \frac{1}{\sqrt{2}} \left[\psi_{n_1}(\xi_1)\psi_{n_2}(\xi_2) + \psi_{n_1}(\xi_2)\psi_{n_2}(\xi_1) \right], \tag{8.47}$$

$$\psi_a(\xi_1,\xi_2) = \frac{1}{\sqrt{2}} \left[\psi_{n_1}(\xi_1)\psi_{n_2}(\xi_2) - \psi_{n_1}(\xi_2)\psi_{n_2}(\xi_1) \right], \tag{8.48}$$

where we have supposed that $n_1 \neq n_2$. When $n_1 = n_2 = n$ the symmetric wave function is given by $\psi_s(\xi_1,\xi_2) = \psi_n(\xi_1)\psi_n(\xi_2)$ and the antisymmetric wave function is zero; we will deal later with the reason why $\psi_a(\xi_1,\xi_2) = 0$ whenever $n_1 = n_2$.

Note that we can rewrite ψ_s as

$$\psi_s(\xi_1,\xi_2) = \frac{1}{\sqrt{2!}} \sum_P \hat{P} \psi_{n_1}(\xi_1)\psi_{n_2}(\xi_2), \tag{8.49}$$

where $\hat{P}$ is the permutation operator and where the sum is over all possible permutations (here we have only two possible ones). Similarly, we can write ψ_a as

$$\psi_a(\xi_1,\xi_2) = \frac{1}{\sqrt{2!}} \sum_P (-1)^P \hat{P} \psi_{n_1}(\xi_1)\psi_{n_2}(\xi_2), \tag{8.50}$$

where $(-1)^P$ is equal to $+1$ for an even permutation (i.e., when we interchange both ξ_1 and ξ_2 and also n_1 and n_2) and equal to -1 for an odd permutation (i.e., when we permute ξ_1 and ξ_2 but not n_1, n_2, and vice versa). Note that we can rewrite ψ_a of (8.48) in the form of a determinant

$$\psi_a(\xi_1, \xi_2) = \frac{1}{\sqrt{2!}} \begin{vmatrix} \psi_{n_1}(\xi_1) & \psi_{n_1}(\xi_2) \\ \psi_{n_2}(\xi_1) & \psi_{n_2}(\xi_2) \end{vmatrix}. \tag{8.51}$$

8.2.5.2 Wave Function of Three-Particle Systems

For a system of three noninteracting identical particles, the symmetric wave function is given by

$$\psi_s(\xi_1, \xi_2, \xi_3) = \frac{1}{\sqrt{3!}} \sum_P \hat{P} \psi_{n_1}(\xi_1) \psi_{n_2}(\xi_2) \psi_{n_3}(\xi_3), \tag{8.52}$$

or by

$$\psi_s(\xi_1, \xi_2, \xi_3) = \frac{1}{\sqrt{3!}} \Big[\psi_{n_1}(\xi_1) \psi_{n_2}(\xi_2) \psi_{n_3}(\xi_3) + \psi_{n_1}(\xi_1) \psi_{n_2}(\xi_3) \psi_{n_3}(\xi_2)$$
$$+ \psi_{n_1}(\xi_2) \psi_{n_2}(\xi_1) \psi_{n_3}(\xi_3) + \psi_{n_1}(\xi_2) \psi_{n_2}(\xi_3) \psi_{n_3}(\xi_1)$$
$$+ \psi_{n_1}(\xi_3) \psi_{n_2}(\xi_1) \psi_{n_3}(\xi_2) + \psi_{n_1}(\xi_3) \psi_{n_2}(\xi_2) \psi_{n_3}(\xi_1) \Big], \tag{8.53}$$

and, when $n_1 \neq n_2 \neq n_3$, the antisymmetric wave function is given by

$$\psi_a(\xi_1, \xi_2, \xi_3) = \frac{1}{\sqrt{3!}} \sum_P (-1)^P \hat{P} \psi_{n_1}(\xi_1) \psi_{n_2}(\xi_2) \psi_{n_3}(\xi_3), \tag{8.54}$$

or, in the form of a determinant, by

$$\psi_a(\xi_1, \xi_2, \xi_3) = \frac{1}{\sqrt{3!}} \begin{vmatrix} \psi_{n_1}(\xi_1) & \psi_{n_1}(\xi_2) & \psi_{n_1}(\xi_3) \\ \psi_{n_2}(\xi_1) & \psi_{n_2}(\xi_2) & \psi_{n_2}(\xi_3) \\ \psi_{n_3}(\xi_1) & \psi_{n_3}(\xi_2) & \psi_{n_3}(\xi_3) \end{vmatrix}. \tag{8.55}$$

If $n_1 = n_2 = n_3 = n$ we have $\psi_s(\xi_1, \xi_2, \xi_3) = \psi_n(\xi_1) \psi_n(\xi_2) \psi_n(\xi_3)$ and $\psi_a(\xi_1, \xi_2, \xi_3) = 0$.

8.2.5.3 Wave Function of Many-Particle Systems

We can generalize (8.52) and (8.55) and write the symmetric and antisymmetric wave functions for a system of N noninteracting identical particles as follows:

$$\psi_s(\xi_1, \xi_2, \ldots, \xi_N) = \frac{1}{\sqrt{N!}} \sum_P \hat{P} \psi_{n_1}(\xi_1) \psi_{n_2}(\xi_2) \cdots \psi_{n_N}(\xi_N), \tag{8.56}$$

$$\psi_a(\xi_1, \xi_2, \ldots, \xi_N) = \frac{1}{\sqrt{N!}} \sum_P (-1)^P \psi_{n_1}(\xi_1) \psi_{n_2}(\xi_2) \cdots \psi_{n_N}(\xi_N), \tag{8.57}$$

or

$$\psi_a(\xi_1, \xi_2, \ldots, \xi_N) = \frac{1}{\sqrt{N!}} \begin{vmatrix} \psi_{n_1}(\xi_1) & \psi_{n_1}(\xi_2) & \cdots & \psi_{n_1}(\xi_N) \\ \psi_{n_2}(\xi_1) & \psi_{n_2}(\xi_2) & \cdots & \psi_{n_2}(\xi_N) \\ \vdots & \vdots & \ddots & \vdots \\ \psi_{n_N}(\xi_1) & \psi_{n_N}(\xi_2) & \cdots & \psi_{n_N}(\xi_N) \end{vmatrix}. \tag{8.58}$$

This $N \times N$ determinant, which involves one-particle states only, is known as the *Slater determinant*. An interchange of any pair of particles corresponds to an interchange of two columns of the determinant; this interchange introduces a change in the sign of the determinant. For even permutations we have $(-1)^P = 1$, and for odd permutations we have $(-1)^P = -1$.

The relations (8.56) and (8.58) are valid for the case where the numbers $n_1, n_2, \ldots, n_N$ are all different from one another. What happens if some, or all, of these numbers are equal? In the symmetric case, if $n_1 = n_2 = \ldots = n_N$ then ψ_s is given by

$$\psi_s(\xi_1, \xi_2, \ldots, \xi_N) = \prod_{i=1}^{N} \psi_n(\xi_i) = \psi_n(\xi_1)\psi_n(\xi_2)\cdots\psi_n(\xi_N). \tag{8.59}$$

When there is a multiplicity in the numbers $n_1, n_2, \ldots, n_N$ (i.e., when some of the numbers n_i occur more than once), we have to be careful and avoid double counting. For instance, if n_1 occurs N_1 times in the sequence $n_1, n_2, \ldots, n_N$, if n_2 occurs N_2 times, and so on, the symmetric wave function will be given by

$$\psi_s(\xi_1, \xi_2, \ldots, \xi_N) = \sqrt{\frac{N_1! N_2! \cdots N_N!}{N!}} \sum_P \hat{P} \psi_{n_1}(\xi_1)\psi_{n_2}(\xi_2)\cdots\psi_{n_N}(\xi_N); \tag{8.60}$$

the summation $\sum_P$ is taken only over permutations which lead to *distinct* terms and includes $N!/N_1! N_2! \cdots N_n!$ different terms. For example, in the case of a system of three independent, identical bosons where $n_1 = n_2 = n$ and $n_3 \neq n$, the multiplicity of n_1 is $N_1 = 2$; hence ψ_s is given by

$$\psi_s(\xi_1, \xi_2, \xi_3) = \sqrt{\frac{2!}{3!}} \sum_P \hat{P} \psi_n(\xi_1)\psi_n(\xi_2)\psi_{n_3}(\xi_3) = \frac{1}{\sqrt{3}} \Big[\psi_n(\xi_1)\psi_n(\xi_2)\psi_{n_3}(\xi_3)$$
$$+ \psi_n(\xi_1)\psi_{n_3}(\xi_2)\psi_n(\xi_3) + \psi_{n_3}(\xi_1)\psi_n(\xi_2)\psi_n(\xi_3) \Big]. \tag{8.61}$$

Unlike the symmetric case, the antisymmetric case is quite straightforward: if, among the numbers $n_1, n_2, \ldots, n_N$, only two are equal, the antisymmetric wave function vanishes. For instance, if $n_i = n_j$, the ith and jth rows of the determinant (8.58) will be identical; hence the determinant vanishes identically. Antisymmetric wave functions, therefore, are nonzero only for those cases where all the numbers $n_1, n_2, \ldots, n_N$ are different.

8.3 The Pauli Exclusion Principle

As mentioned above, if any two particles occupy the same single-particle state, the determinant (8.58), and hence the total wave function, will vanish since two rows of the determinant will be identical. We can thus infer that *in a system of N identical particles, no two fermions can occupy the same single-particle state at a time*; every single-particle state can be occupied by at most one fermion. This is the *Pauli exclusion principle*, which was first postulated in 1925 to explain the periodic table. It states that *no two electrons can occupy simultaneously the same (single-particle) quantum state on the same atom*; there can be only one (or at most one) electron occupying a state of quantum numbers $n_i l_i m_{l_i} m_{s_i}$: $\psi_{n_i l_i m_{l_i} m_{s_i}}(\vec{r}_i, \vec{S}_i)$. The exclusion principle plays an important role in the structure of atoms. It has a direct effect on the *spatial distribution* of fermions.

Boson condensation

What about bosons? Do they have any restriction like fermions? Not at all. There is no restriction on the number of bosons that can occupy a single state. Instead of the exclusion principle of fermions, bosons tend to condense all in the same state, the ground state; this is called *boson condensation*. For instance, all the particles of liquid ^{4}He (a boson system) occupy the same ground state. This phenomenon is known as *Bose–Einstein condensation*. The properties of liquid ^{3}He are, however, completely different from those of liquid ^{4}He, because ^{3}He is a fermion system.

Remark

We have seen that when the Schrödinger equation involves the spin, the wave function of a single particle is equal to the product of the spatial part and the spin part: $\Psi(\vec{r}, \vec{S}) = \psi(\vec{r})\chi(\vec{S})$. The wave function of a system of N particles, which have spins, is the product of the spatial part and the spin part:

$$\Psi(\vec{r}_1, \vec{S}_1; \vec{r}_2, \vec{S}_2; \ldots; \vec{r}_N, \vec{S}_N) = \psi(\vec{r}_1, \vec{r}_2, \ldots, \vec{r}_N)\chi(\vec{S}_1, \vec{S}_2, \ldots, \vec{S}_N). \tag{8.62}$$

This wave function must satisfy the appropriate symmetry requirements when the N particles are identical. In the case of a system of N identical bosons, the wave function must be symmetric; hence the spatial and spin parts must have the same parity:

$$\Psi_s(\vec{r}_1, \vec{S}_1; \vec{r}_2, \vec{S}_2; \ldots; \vec{r}_N, \vec{S}_N) = \begin{cases} \psi_a(\vec{r}_1, \vec{r}_2, \ldots, \vec{r}_N)\chi_a(\vec{S}_1, \vec{S}_2, \ldots, \vec{S}_N), \\ \psi_s(\vec{r}_1, \vec{r}_2, \ldots, \vec{r}_N)\chi_s(\vec{S}_1, \vec{S}_2, \ldots, \vec{S}_N). \end{cases} \tag{8.63}$$

In the case of a system of N identical fermions, however, the space and spin parts must have different parities, leading to an overall wave function that is antisymmetric:

$$\Psi_a(\vec{r}_1, \vec{S}_1; \vec{r}_2, \vec{S}_2; \ldots; \vec{r}_N, \vec{S}_N) = \begin{cases} \psi_a(\vec{r}_1, \vec{r}_2, \ldots, \vec{r}_N)\chi_s(\vec{S}_1, \vec{S}_2, \ldots, \vec{S}_N), \\ \psi_s(\vec{r}_1, \vec{r}_2, \ldots, \vec{r}_N)\chi_a(\vec{S}_1, \vec{S}_2, \ldots, \vec{S}_N). \end{cases} \tag{8.64}$$

Example 8.3 (Wave function of two identical, noninteracting particles)

Find the wave functions of two systems of identical, noninteracting particles: the first consists of two bosons and the second of two spin $\frac{1}{2}$ fermions.

Solution

For a system of two identical, noninteracting bosons, (8.47) and (8.48) yield

$$\Psi_s(\vec{r}_1, \vec{S}_1; \vec{r}_2, \vec{S}_2) = \frac{1}{\sqrt{2}} \begin{cases} \left[\psi_{n_1}(\vec{r}_1)\psi_{n_2}(\vec{r}_2) - \psi_{n_1}(\vec{r}_2)\psi_{n_2}(\vec{r}_1) \right] \chi_a(\vec{S}_1, \vec{S}_2), \\ \left[\psi_{n_1}(\vec{r}_1)\psi_{n_2}(\vec{r}_2) + \psi_{n_1}(\vec{r}_2)\psi_{n_2}(\vec{r}_1) \right] \chi_s(\vec{S}_1, \vec{S}_2), \end{cases} \tag{8.65}$$

and for a system of two spin $\frac{1}{2}$ fermions

$$\Psi_a(\vec{r}_1, \vec{S}_1; \vec{r}_2, \vec{S}_2) = \frac{1}{\sqrt{2}} \begin{cases} \left[\psi_{n_1}(\vec{r}_1)\psi_{n_2}(\vec{r}_2) - \psi_{n_1}(\vec{r}_2)\psi_{n_2}(\vec{r}_1) \right] \chi_s(\vec{S}_1, \vec{S}_2), \\ \left[\psi_{n_1}(\vec{r}_1)\psi_{n_2}(\vec{r}_2) + \psi_{n_1}(\vec{r}_2)\psi_{n_2}(\vec{r}_1) \right] \chi_a(\vec{S}_1, \vec{S}_2), \end{cases} \tag{8.66}$$

where, from the formalism of angular momentum addition, there are three states (a triplet) that are symmetric, $\chi_s(\vec{S}_1, \vec{S}_2)$:

$$\chi_{triplet}(\vec{S}_1, \vec{S}_2) = \begin{cases} \left|\frac{1}{2}\,\frac{1}{2}\right\rangle_1 \left|\frac{1}{2}\,\frac{1}{2}\right\rangle_2, \\ \frac{1}{\sqrt{2}} \left(\left|\frac{1}{2}\,\frac{1}{2}\right\rangle_1 \left|\frac{1}{2}\,-\frac{1}{2}\right\rangle_2 + \left|\frac{1}{2}\,-\frac{1}{2}\right\rangle_1 \left|\frac{1}{2}\,\frac{1}{2}\right\rangle_2 \right), \\ \left|\frac{1}{2}\,-\frac{1}{2}\right\rangle_1 \left|\frac{1}{2}\,-\frac{1}{2}\right\rangle_2, \end{cases} \tag{8.67}$$

and one state (a singlet) that is antisymmetric, $\chi_a(\vec{S}_1, \vec{S}_2)$:

$$\chi_{singlet}(\vec{S}_1, \vec{S}_2) = \frac{1}{\sqrt{2}} \left(\left|\frac{1}{2}\,\frac{1}{2}\right\rangle_1 \left|\frac{1}{2}\,-\frac{1}{2}\right\rangle_2 - \left|\frac{1}{2}\,-\frac{1}{2}\right\rangle_1 \left|\frac{1}{2}\,\frac{1}{2}\right\rangle_2 \right). \tag{8.68}$$

8.4 The Exclusion Principle and the Periodic Table

Explaining the periodic table is one of the most striking successes of the Schrödinger equation. When combined with the Pauli exclusion principle, the equation offers insightful information on the structure of multielectron atoms.

In Chapter 6, we saw that the state of the hydrogen's electron, which moves in the spherically symmetric Coulomb potential of the nucleus, is described by four quantum numbers n, l, m_l, and m_s: $\Psi_{nlm_l m_s}(\vec{r}) = \psi_{nlm_l}(\vec{r})\chi_{m_s}$, where $\psi_{nlm_l}(\vec{r}) = R_{nl}(r)Y_{lm_l}(\theta, \varphi)$ is the electron's wave function when the spin is ignored and $\chi_{m_s} = \left|\frac{1}{2}, \pm\frac{1}{2}\right\rangle$ is the spin's state. This representation turns out to be suitable for any atom as well.

In a multielectron atom, the average potential in which every electron moves is different from the Coulomb potential of the nucleus; yet, to a good approximation, it can be assumed to be spherically symmetric. We can therefore, as in hydrogen, characterize the electronic states by the four quantum numbers n, l, m_l, and m_s, which respectively represent the principal quantum number, the orbital quantum number, the magnetic (or azimuthal) quantum number, and the spin quantum number; m_l represents the z-component of the electron orbital angular momentum and m_s the z-component of its spin.

Atoms have a *shell structure*. Each atom has a number of major shells that are specified by the radial or principal quantum number n. Shells have subshells which are specified by the orbital quantum number l. Subshells in turn have subsubshells, called *orbitals*, specified by m_l; so *an orbital is fully specified by three quantum numbers n, l, m_l*; i.e., it is defined by $|nlm_l\rangle$. Each shell n therefore has n subshells corresponding to $l = 0, 1, 2, 3, \ldots, n-1$, and in turn each subshell has $2l + 1$ orbitals (or subsubshells), since to $m_l = -l, -l+1, -l+2, \ldots, l-2, l-1, l$. As in hydrogen, individual electrons occupy single-particle states or orbitals; the states corresponding to the respective numerical values $l = 0, 1, 2, 3, 4, 5, \ldots$ are called s, p, d, f, g, h, ... states. Hence for a given n an s-state has 1 orbital ($m_l = 0$), a p-state has 3 orbitals ($m_l = -1, 0, 1$), a d-state has 5 orbitals ($m_l = -2, -1, 0, 1, 2$), and so on (Chapter 6). We will label the electronic states by nl where, as before, l refers to s, p, d, f, etc.; for example 1s corresponds to $(n, l) = (1, 0)$, 2s corresponds to $(n, l) = (2, 0)$, 2p corresponds to $(n, l) = (2, 1)$, 3s corresponds to $(n, l) = (3, 0)$, and so on.

How do electrons fill the various shells and subshells in an atom? If electrons were bosons, they would all group in the ground state $|nlm_l\rangle = |100\rangle$; we wouldn't then have the rich diversity of elements that exist in nature. But since electrons are identical fermions, they are

Figure 8.2 Filling orbitals according to the Pauli exclusion principle.

governed by the Pauli exclusion principle, which states that *no two electrons can occupy simultaneously the same quantum state* $|nlm_lm_s\rangle$ *on the same atom*. Hence each orbital state $|nlm_l\rangle$ can be occupied by two electrons *at most*: one having spin-up $m_s = +\frac{1}{2}$, the other spin-down $m_s = -\frac{1}{2}$. Hence, each state nl can accommodate $2(2l+1)$ electrons. So an s-state (i.e., $|n00\rangle$) can at most hold 2 electrons, a p-state (i.e., $|n1m_l\rangle$) at most 6 electrons, a d-state (i.e., $|n2m_l\rangle$) at most 10 electrons, an f-state (i.e., $|n3m_l\rangle$) at most 14 electrons, and so on (Figure 8.2).

For an atom in the ground state, the *electrons fill the orbitals in order of increasing energy*; once a subshell is filled, the next electron goes into the vacant subshell whose energy is just above the previous subshell. When all orbitals in a major electronic shell are filled up, we get a *closed shell*; the next electron goes into the next major shell, and so on. By filling the atomic orbitals one after the other in order of increasing energy, one obtains all the elements of the periodic table (Table 8.1).

Elements $1 \le Z \le 18$

As shown in Table 8.1, the first period (or first horizontal row) of the periodic table has two elements, hydrogen H and helium He; the second period has 8 elements, lithium Li to neon Ne; the third period also has 8 elements, sodium Na to argon Ar; and so on. The orbitals of the 18 lightest elements, $1 \le Z \le 18$, are filled in order of increasing energy according to the sequence: 1s, 2s, 2p, 3s, 3p. The electronic state of an atom is determined by specifying the occupied orbitals or by what is called the *electronic configuration*. For example, hydrogen has one electron, its ground state configuration is $(1s)^1$; helium He has two electrons: $(1s)^2$; lithium Li has three electrons: $(1s)^2(2s)^1$; beryllium Be has four: $(1s)^2(2s)^2$, and so on.

Now let us see how to determine the total angular momentum of an atom. For this, we need to calculate the total orbital angular momentum $\vec{L} = \sum_{i=1}^{Z} \vec{l}_i$, the total spin $\vec{S} = \sum_{i=1}^{Z} \vec{s}_i$, and then obtain total angular momentum by coupling $\vec{L}$ and $\vec{S}$, i.e., $\vec{J} = \vec{L} + \vec{S}$, where $\vec{l}_i$ and $\vec{s}_i$ are the orbital and spin angular momenta of individual electrons. As will be seen in Chapter 9, when the spin–orbit coupling is considered, the degeneracy of the atom's energy levels is partially lifted, introducing a splitting of the levels. The four numbers L, S, J, and M are good quantum numbers, where $|L - S| \le J \le L + S$ and $-J \le M \le J$. So there are $2S + 1$ values of J when $L \ge S$ and $2L + 1$ values when $L < S$. Since the energy depends on J, the levels corresponding to an L and S split into a $(2J + 1)$-multiplet. The issue now is to determine which one of these states has the lowest energy. Before studying this issue, let us introduce the spectroscopic notation according to which the state of an atom is labeled by

$$^{2S+1}L_J, \tag{8.69}$$

where, as before, the numbers $L = 0, 1, 2, 3, \ldots$ are designated by S, P, D, F, $\ldots$, respectively (we should mention here that the capital letters S, P, D, F, $\ldots$ refer to the total orbital angular momentum of an atom, while the small letters s, p, d, f, $\ldots$ refer to individual electrons; that is, s, p, d, f, $\ldots$ describe the angular momentum states of individual electrons). For example, since the total angular momentum of a beryllium atom is $J = 0$, because $L = 0$ (all electrons are in s-states, $l_i = 0$) and $S = 0$ (both electrons in the $(1s)^2$ state are paired and so are the two electrons in the $(2s)^2$ state), the ground state of beryllium can be written as $^1S_0$. This applies actually to all other closed shell atoms such as helium He, neon Ne, argon Ar, and so on; their ground states are all specified by $^1S_0$ (Table 8.1).

Let us now consider boron B: the closed shells 1s and 2s have $L = S = J = 0$. Thus the angular momentum of boron is determined by the 1p electron which has $S = 1/2$ and $L = 1$. A coupling of $S = 1/2$ and $L = 1$ yields $J = 1/2$ or $3/2$, leading therefore to two possible

Table 8.1 Ground state electron configurations, spectroscopic description, and ionization energies for the first four rows of the periodic table. The brackets designate closed-shell elements.

Shell	Z	Element	Ground state configuration	Spectroscopic description	Ionization energy (eV)
1	1	H	$(1s)^1$	$^2S_{1/2}$	13.60
	2	He	$(1s)^2$	$^1S_0$	24.58
2	3	Li	$[He](2s)^1 = (1s)^2(2s)^1$	$^2S_{1/2}$	5.39
	4	Be	$[He](2s)^2$	$^1S_0$	9.32
	5	B	$[He](2s)^2(2p)^1$	$^2P_{1/2}$	8.30
	6	C	$[He](2s)^2(2p)^2$	$^3P_0$	11.26
	7	N	$[He](2s)^2(2p)^3$	$^4S_{3/2}$	14.55
	8	O	$[He](2s)^2(2p)^4$	$^3P_2$	13.61
	9	F	$[He](2s)^2(2p)^5$	$^2P_{3/2}$	17.42
	10	Ne	$[He](2s)^2(2p)^6$	$^1S_0$	21.56
3	11	Na	$[Ne](3s)^1$	$^2S_{1/2}$	5.14
	12	Mg	$[Ne](3s)^2$	$^1S_0$	7.64
	13	Al	$[Ne](3s)^2(3p)^1$	$^2P_{1/2}$	5.94
	14	Si	$[Ne](3s)^2(3p)^2$	$^3P_0$	8.15
	15	P	$[Ne](3s)^2(3p)^3$	$^4S_{3/2}$	10.48
	16	S	$[Ne](3s)^2(3p)^4$	$^3P_2$	10.36
	17	Cl	$[Ne](3s)^2(3p)^5$	$^2P_{3/2}$	13.01
	18	Ar	$[Ne](3s)^2(3p)^6$	$^1S_0$	15.76
4	19	K	$[Ar](4s)^1$	$^2S_{1/2}$	4.34
	20	Ca	$[Ar](4s)^2$	$^1S_0$	6.11
	21	Sc	$[Ar](3d)^1(4s)^2$	$^2D_{3/2}$	6.54
	22	Ti	$[Ar](3d)^2(4s)^2$	$^3F_2$	6.83
	23	V	$[Ar](3d)^3(4s)^2$	$^4F_{3/2}$	6.74
	24	Cr	$[Ar](3d)^4(4s)^2$	$^7S_3$	6.76
	25	Mn	$[Ar](3d)^5(4s)^2$	$^6S_{3/2}$	7.43
	26	Fe	$[Ar](3d)^6(4s)^2$	$^5D_4$	7.87
	27	Co	$[Ar](3d)^7(4s)^2$	$^4F_{9/2}$	7.86
	28	Ni	$[Ar](3d)^8(4s)^2$	$^3F_4$	7.63
	29	Cu	$[Ar](3d)^{10}(4s)^1$	$^2S_{1/2}$	7.72
	30	Zn	$[Ar](3d)^{10}(4s)^2$	$^1S_0$	9.39
	31	Ga	$[Ar](3d)^{10}(4s)^2(4p)^1$	$^2P_{1/2}$	6.00
	32	Ge	$[Ar](3d)^{10}(4s)^2(4p)^2$	$^3P_0$	7.88
	33	As	$[Ar](3d)^{10}(4s)^2(4p)^3$	$^4S_{3/2}$	9.81
	34	Se	$[Ar](3d)^{10}(4s)^2(4p)^4$	$^3P_2$	9.75
	35	Br	$[Ar](3d)^{10}(4s)^2(4p)^5$	$^2P_{3/2}$	11.84
	36	Kr	$[Ar](3d)^{10}(4s)^2(4p)^6$	$^1S_0$	9.81

states:

$$^2P_{1/2} \quad \text{or} \quad ^2P_{3/2}. \tag{8.70}$$

Which one has a lower energy? Before answering this question, let us consider another example, the carbon atom.

The ground state configuration of the carbon atom, as given by $(1s)^2(2s)^2(2p)^2$, implies that its total angular momentum is determined by the two 2p electrons. The coupling of the two spins $s = 1/2$, as shown in equations (7.174) to (7.177), yields two values for their total spin $S = 0$ or $S = 1$; and, as shown in Problem 7.3, page 436, a coupling of two individual orbital angular momenta $l = 1$ yields three values for the total angular momenta $L = 0, 1,$ or 2. But the exclusion principle dictates that the total wave function has to be antisymmetric, i.e., the spin and orbital parts of the wave function must have opposite symmetries. Since the singlet spin state $S = 0$ is antisymmetric, the spin triplet $S = 1$ is symmetric, the orbital triplet $L = 1$ is antisymmetric, the orbital quintuplet $L = 2$ is symmetric, and the orbital singlet $L = 0$ is symmetric, the following states are antisymmetric:

$$^1S_0, \quad ^3P_0, \quad ^3P_1, \quad ^3P_2, \quad \text{or} \quad ^1D_2; \tag{8.71}$$

hence any one of these states can be the ground state of carbon. Again, which one of them has the lowest energy?

To answer this question and the question pertaining to (8.70), we may invoke *Hund's rules*: (a) the lowest energy level corresponds to the state with the largest spin S (i.e., the maximum number of electrons have unpaired spins); (b) among the states with a given value of S, the lowest energy level corresponds to the state with the largest value of L; (c) for a subshell that is less than half full the lowest energy state corresponds to $J = |L - S|$, and for a subshell that is more than half full the lowest energy state corresponds to $J = L + S$.

Hund's third rule answers the question pertaining to (8.70): since the 2p shell of boron is less than half full, the value of J corresponding to the lowest energy is given by $J = |L - S| = 1 - 1/2 = 1/2$; hence $^2P_{1/2}$ is the lower energy state.

To find which one of the states (8.71) has the lowest energy, Hund's first rule dictates that $S = 1$. Since the triplet $S = 1$ is symmetric, we need an antisymmetric spatial wave function; this is given by the spatial triplet $L = 1$. We are thus left with three possible choices: $J = 0, 1,$ or 2. Hund's third rule precludes the values $J = 1$ and 2. Since the 2p shell of carbon is less than half full, the value of J corresponding to the lowest energy is given by $J = |L - S| = 1 - 1 = 0$; hence $^3P_0$ is the lower energy state (Table 8.1). That is, the two electrons are in different spatial states or different orbitals (Figure 8.2). Actually, we could have guessed this result: since the Coulomb repulsion between the two electrons when they are paired together is much larger than when they are unpaired, the lower energy configuration corresponds to the case where the electrons are in different spatial states. The ground state configurations of the remaining elements, oxygen to argon, can be inferred in a similar way (Table 8.1).

Elements $Z \geq 18$
When the 3p shell is filled, one would expect to place the next electron in a 3d shell. But this doesn't take place due to the occurrence of an interesting effect: the 4s states have lower energy than the 3d states. Why? In a hydrogen atom the states 3s, 3p, and 3d have the same energy $(E_3^{(0)} = -\mathcal{R}/3^2 = -1.51 \text{ eV}$, since $\mathcal{R} = 13.6 \text{ eV})$. But in multielectron atoms, these states have different energy values. As l increases, the effective repulsive potential $\hbar^2 l(l+1)/2mr^2$ causes the d-state electrons to be thrown outward and the s-state electrons to remain closer to the nucleus. Being closer to the nucleus, the s-state electrons therefore feel the full attraction of

the nucleus, whereas the d-state electrons experience a much weaker attraction. This is known as the *screening effect*, because the inner electrons, i.e., the s-state electrons, screen the nucleus; hence the outward electrons (the d-state electrons) do not experience the full attraction of the nucleus, but instead feel a weak effective potential. As a result, the energy of the 3d-state is larger than that of the 4s-state. The screening effect also causes the energy of the 5s-state to have a lower energy than the 4d-state, and so on. So for a given n, the energies E_{nl} increase as l increases; in fact, neglecting the spin–orbit interaction and considering relativistic corrections we will show in Chapter 9 (9.90) that the ground state energy depends on the principal and orbital quantum numbers n and l as $E_{nl}^{(0)} = Z^2 E_n^{(0)} \{1 + \alpha^2 Z^2 [2/(2l+1) - 3/4n]/n\}$, where $\alpha = 1/137$ is the fine structure constant and $E_n^{(0)} = -\mathcal{R}/n^2 = -13.6\,\text{eV}/n^2$.

In conclusion, the periodic table can be obtained by filling the orbitals in order of increasing energy E_{nl} as follows (Table 8.1):

$$1s^2,\ 2s^2,\ 2p^6,\ 3s^2,\ 3p^6,\ 4s^2,\ 3d^{10},\ 4p^6,\ 5s^2,\ 4d^{10},\ 5p^6,\ 6s^2,\ 4f^{14},$$
$$5d^{10},\ 6p^6,\ 7s^2,\ 5f^{14},\ 6d^{10},\ 7p^6,\dots .$$

$$(8.72)$$

Remarks

The chemical properties of an element is mostly determined by the outermost shell. Hence elements with similar electron configurations for the outside shell have similar chemical properties. This is the idea behind the structure of the periodic table: it is arranged in a way that all elements in a column have similar chemical properties. For example, the elements in the last column, helium, neon, argon, krypton, and so on, have the outer p-shell completely filled (except for helium whose outside shell is 1s). These atoms, which are formed when a shell or a subshell is filled, are very stable, interact very weakly with one another, and do not combine with other elements to form molecules or new compounds; that is, they are chemically inert. They are very reluctant to give up or to accept an electron. Due to these properties, they are called *noble gases*. They have a very low boiling point (around $-200\,^\circ$C). Note that each row of the periodic table corresponds to filling out a shell or subshell of the atom, up to the next noble gas. Also, there is a significant energy gap before the next level is encountered after each of these elements. As shown in Table 8.1, a large energy is required to ionize these elements; for instance, 24.58 eV is needed to ionize a helium atom.

Atoms consisting of a closed shell (or a noble gas configuration) plus an s-electron (or a valence electron), such as Li, Na, K, and so on, have the lowest binding energy; these elements are known as the alkali metals. In elements consisting of an alkali configuration plus an electron, the second s-electron is more bound than the valence electron of the alkali atom because of the higher nuclear charge. As the p-shell is gradually filled (beyond the noble gas configuration), the binding energy increases initially (as in boron, carbon, and nitrogen) till the fourth electron, then it begins to drop (Table 8.1). This is due to the fact that when the p-shell is less than half full all spins are parallel; hence all three spatial wave functions are antisymmetric. With the fourth electron (as in oxygen), two spins will be antiparallel or paired; hence the spatial wave function is not totally antisymmetric, causing a drop in the energy. Note that elements with one electron more than or one electron less than noble gas configurations are the most active chemically, because they tend to easily give up or easily accept one electron.

Example 8.4

(a) Specify the total angular momenta corresponding to ^4G, ^3H, and ^1D.

(b) Find the spectroscopic notation for the ground state configurations of aluminum Al $(Z = 13)$ and scandium Sc $(Z = 21)$.

Solution

(a) For the term ^4G the orbital angular momentum is $L = 4$ and the spin is $S = 3/2$, since $2S + 1 = 4$. The values of the total angular momentum corresponding to the coupling of $L = 4$ and $S = 3/2$ are given by $|4 - 3/2| \le J \le 4 + 3/2$. Hence we have $J = 5/2, 7/2, 9/2, 11/2$.

Similarly, for ^3H we have $S = 1$ and $L = 5$. Therefore, we have $|5 - 1| \le J \le 5 + 1$, or $J = 4, 5, 6$.

For ^1D we have $S = 0$ and $L = 2$. Therefore, we have $|2 - 0| \le J \le 2 + 0$, or $J = 2$.

(b) The ground state configuration of Al is $[Ne](3s)^2(3p)^1$. The total angular momentum of this element is determined by the 3p electron, because $S = 0$ and $L = 0$ for both [Ne] and $(3s)^2$. Since the 3p electron has $S = 1/2$ and $L = 1$, the total angular momentum is given by $|1 - 1/2| \le J \le 1 + 1/2$. Hence we have $J = 1/2, 3/2$. Which of the values $J = 1/2$ and $J = 3/2$ has a lower energy? According to Hund's third rule, since the 3p shell is less than half full, the state $J = |L - S| = 1/2$ has the lower energy. Hence the ground state configuration of Al corresponds to $^2P_{1/2}$ (Table 8.1), where we have used the spectroscopic notation $^{2S+1}L_J$.

Since the ground state configuration of Sc is $[Ar](4s)^2(3d)^1$, the angular momentum is given by that of the 3d electron. Since $S = 1/2$ and $L = 2$, and since the 3d shell is less than half full, Hund's third rule dictates that the total angular momentum is given by $J = |L - S| = |2 - 1/2| = 3/2$. Hence we have $^2D_{3/2}$.

8.5 Solved Problems

Problem 8.1

Consider a system of three noninteracting particles that are confined to move in a one-dimensional infinite potential well of length a: $V(x) = 0$ for $0 < x < a$ and $V(x) = \infty$ for other values of x. Determine the energy and wave function of the ground state and the first and second excited states when the three particles are (a) spinless and distinguishable with $m_1 < m_2 < m_3$; (b) identical bosons; (c) identical spin $\frac{1}{2}$ particles; and (d) distinguishable spin $\frac{1}{2}$ particles.

Solution

(a) As shown in Example 8.2 on page 459, the total energy and wave function are given by

$$E_{n_1, n_2, n_3} = \frac{\hbar^2 \pi^2}{2a^2} \left(\frac{n_1^2}{m_1} + \frac{n_2^2}{m_2} + \frac{n_3^2}{m_3} \right), \tag{8.73}$$

$$\psi_{n_1, n_2, n_3}(x_1, x_2, x_3) = \sqrt{\frac{8}{a^3}} \sin\left(\frac{n_1 \pi}{a} x_1 \right) \sin\left(\frac{n_2 \pi}{a} x_2 \right) \sin\left(\frac{n_3 \pi}{a} x_3 \right). \tag{8.74}$$

The ground state of the system corresponds to the case where all three particles occupy their respective ground state orbitals, $n_1 = n_2 = n_3 = 1$; hence

$$E^{(0)} = E_{1,1,1} = \frac{\hbar^2 \pi^2}{2a^2} \left(\frac{1}{m_1} + \frac{1}{m_2} + \frac{1}{m_3} \right), \tag{8.75}$$

$$\psi^{(0)}(x_1, x_2, x_3) = \psi_{1,1,1}(x_1, x_2, x_3) = \sqrt{\frac{8}{a^3}} \sin\left(\frac{\pi}{a} x_1 \right) \sin\left(\frac{\pi}{a} x_2 \right) \sin\left(\frac{\pi}{a} x_3 \right). \tag{8.76}$$

Since particle 3 has the highest mass, the first excited state corresponds to the case where particle 3 is in $n_3 = 2$, while particles 1 and 2 remain in $n_1 = n_2 = 1$:

$$E^{(1)} = E_{1,1,2} = \frac{\hbar^2 \pi^2}{2a^2} \left(\frac{1}{m_1} + \frac{1}{m_2} + \frac{4}{m_3} \right), \tag{8.77}$$

$$\psi^{(1)}(x_1, x_2, x_3) = \psi_{1,1,2}(x_1, x_2, x_3) = \sqrt{\frac{8}{a^3}} \sin \left(\frac{\pi}{a} x_1 \right) \sin \left(\frac{\pi}{a} x_2 \right) \sin \left(\frac{2\pi}{a} x_3 \right). \tag{8.78}$$

Similarly, the second excited state corresponds to the case where particles 2 and 3 are in $n_2 = n_3 = 2$, while particle 1 remains in $n_1 = 1$:

$$E^{(2)} = E_{1,2,2} = \frac{\hbar^2 \pi^2}{2a^2} \left(\frac{1}{m_1} + \frac{4}{m_2} + \frac{4}{m_3} \right), \tag{8.79}$$

$$\psi^{(2)}(x_1, x_2, x_3) = \psi_{1,2,2}(x_1, x_2, x_3) = \sqrt{\frac{8}{a^3}} \sin \left(\frac{\pi}{a} x_1 \right) \sin \left(\frac{2\pi}{a} x_2 \right) \sin \left(\frac{2\pi}{a} x_3 \right). \tag{8.80}$$

(b) If all three particles were identical bosons, the ground state will correspond to all particles in the lowest state $n_1 = n_2 = n_3 = 1$ (Figure 8.3):

$$E^{(0)} = E_{1,1,1} = 3\varepsilon_1 = \frac{3\hbar^2 \pi^2}{2ma^2}, \tag{8.81}$$

$$\psi^{(0)} = \psi_1(x_1)\psi_1(x_2)\psi_1(x_3) = \sqrt{\frac{8}{a^3}} \sin \left(\frac{\pi}{a} x_1 \right) \sin \left(\frac{\pi}{a} x_2 \right) \sin \left(\frac{\pi}{a} x_3 \right), \tag{8.82}$$

since $\psi_n(x_i) = \sqrt{2/a} \sin(n\pi x_i / a)$.

In the first excited state we have two particles in ψ_1 (each with energy $\varepsilon_1 = \hbar^2 \pi^2/(2ma^2)$) and one in ψ_2 (with energy $\varepsilon_2 = 4\hbar^2\pi^2/(2ma^2) = 4\varepsilon_1$):

$$E^{(1)} = 2\varepsilon_1 + \varepsilon_2 = 2\varepsilon_1 + 4\varepsilon_1 = 6\varepsilon_1 = \frac{3\pi^2 \hbar^2}{ma^2}. \tag{8.83}$$

The wave function is somewhat tricky again. Since the particles are identical, we can no longer say which particle is in which state; all we can say is that two particles are in ψ_1 and one in ψ_2. Since the value $n = 1$ occurs twice (two particles are in ψ_1), we infer from (8.60) and (8.61) that

$$\psi^{(1)}(x_1, x_2, x_3) = \sqrt{\frac{2!}{3!}} \Big[\psi_1(x_1)\psi_1(x_2)\psi_2(x_3) + \psi_1(x_1)\psi_2(x_2)\psi_1(x_3)$$
$$+ \psi_2(x_1)\psi_1(x_2)\psi_1(x_3) \Big]. \tag{8.84}$$

In the second excited state we have one particle in ψ_1 and two in ψ_2:

$$E^{(2)} = \varepsilon_1 + 2\varepsilon_2 = \varepsilon_1 + 8\varepsilon_2 = 9\varepsilon_1 = \frac{9\pi^2 \hbar^2}{2ma^2}. \tag{8.85}$$

Now, since the value $n = 2$ occurs twice (two particles are in ψ_2) and $n = 1$ only once, (8.60) and (8.61) yield

$$\psi^{(2)}(x_1, x_2, x_3) = \sqrt{\frac{2!}{3!}} \Big[\psi_1(x_1)\psi_2(x_2)\psi_2(x_3) + \psi_2(x_1)\psi_1(x_2)\psi_2(x_3)$$
$$+ \psi_2(x_1)\psi_2(x_2)\psi_1(x_3) \Big]. \tag{8.86}$$

Figure 8.3 Particle distribution among the levels of the ground state (GS) and the first (FES) and second excited states (SES) for a system of three noninteracting identical bosons (left) and fermions (right) moving in an infinite well, with $\varepsilon_1 = \hbar^2\pi^2/(2ma^2)$. Each state of the fermion system is fourfold degenerate due to the various possible orientations of the spins.

(c) If the three particles were identical spin $\frac{1}{2}$ fermions, the ground state corresponds to the case where two particles are in the lowest state ψ_1 (one having a spin-up $| + \rangle = \left|\frac{1}{2}, \frac{1}{2}\right\rangle$, the other with a spin-down $| - \rangle = \left|\frac{1}{2}, -\frac{1}{2}\right\rangle$), while the third particle is in the next state ψ_2 (its spin can be either up or down, $| \pm \rangle = \left|\frac{1}{2}, \pm\frac{1}{2}\right\rangle$); see Figure 8.3. The ground state energy is

$$E^{(0)} = 2\varepsilon_1 + \varepsilon_2 = 2\varepsilon_1 + 4\varepsilon_1 = 6\varepsilon_1 = \frac{3\hbar^2\pi^2}{ma^2}. \tag{8.87}$$

The ground state wave function is antisymmetric and, in accordance with (8.55), it is given by

$$\psi^{(0)}(x_1, x_2, x_3) = \frac{1}{\sqrt{3!}} \begin{vmatrix} \psi_1(x_1)\chi(S_1) & \psi_1(x_2)\chi(S_2) & \psi_1(x_3)\chi(S_3) \\ \psi_1(x_1)\chi(S_1) & \psi_1(x_2)\chi(S_2) & \psi_1(x_3)\chi(S_3) \\ \psi_2(x_1)\chi(S_1) & \psi_2(x_2)\chi(S_2) & \psi_2(x_3)\chi(S_3) \end{vmatrix}. \tag{8.88}$$

This state is fourfold degenerate, since there are four different ways of configuring the spins of the three fermions (the ground state (GS) shown in Figure 8.3 is just one of the four configurations). **Remark:** one should be careful not to erroneously conclude that, since the first and second rows of the determinant in (8.88) are "identical", the determinant is zero. We should keep in mind that the spin states are given by $\chi(S_1) =| \pm \rangle$, $\chi(S_2) =| \pm \rangle$, and $\chi(S_3) =| \pm \rangle$; hence, we need to select these spin states in such a way that no two rows (nor two columns) of the determinant are identical. For instance, one of the possible configurations of the ground state wave function is given by

$$\psi^{(0)}(x_1, x_2, x_3) = \frac{1}{\sqrt{3!}} \begin{vmatrix} \psi_1(x_1) | + \rangle & \psi_1(x_2) | - \rangle & \psi_1(x_3) | + \rangle \\ \psi_1(x_1) | - \rangle & \psi_1(x_2) | + \rangle & \psi_1(x_3) | + \rangle \\ \psi_2(x_1) | + \rangle & \psi_2(x_2) | + \rangle & \psi_2(x_3) | - \rangle \end{vmatrix}. \tag{8.89}$$

This remark applies also to the first and second excited state wave functions (8.90) and (8.92); it also applies to the wave function (8.109).

The first excited state corresponds to one particle in the lowest state ψ_1 (its spin can be either up or down) and the other two particles in the state ψ_2 (the spin of one is up, the other is down). As in the ground state, there are also four different ways of configuring the spins of the three fermions in the first excited state (FES); the FES shown in Figure 8.3 is just one of the four configurations:

$$\psi^{(1)}(x_1, x_2, x_3) = \frac{1}{\sqrt{3!}} \begin{vmatrix} \psi_1(x_1)\chi(S_1) & \psi_1(x_2)\chi(S_2) & \psi_1(x_3)\chi(S_3) \\ \psi_2(x_1)\chi(S_1) & \psi_2(x_2)\chi(S_2) & \psi_2(x_3)\chi(S_3) \\ \psi_2(x_1)\chi(S_1) & \psi_2(x_2)\chi(S_2) & \psi_2(x_3)\chi(S_3) \end{vmatrix}. \tag{8.90}$$

These four different states correspond to the same energy

$$E^{(1)} = \varepsilon_1 + 2\varepsilon_2 = \varepsilon_1 + 8\varepsilon_1 = 9\varepsilon_1 = \frac{9\hbar^2\pi^2}{2ma^2}. \tag{8.91}$$

The excitation energy of the first excited state is $E^{(1)} - E^{(0)} = 9\varepsilon_1 - 6\varepsilon_1 = 3\hbar^2\pi^2/(2ma^2)$.

The second excited state corresponds to two particles in the lowest state ψ_1 (one with spin-up, the other with spin-down) and the third particle in the third state ψ_3 (its spin can be either up or down). This state also has four different spin configurations; hence it is fourfold degenerate:

$$\psi^{(2)}(x_1, x_2, x_3) = \frac{1}{\sqrt{3!}} \begin{vmatrix} \psi_1(x_1)\chi(S_1) & \psi_1(x_2)\chi(S_2) & \psi_1(x_3)\chi(S_3) \\ \psi_1(x_1)\chi(S_1) & \psi_1(x_2)\chi(S_2) & \psi_1(x_3)\chi(S_3) \\ \psi_3(x_1)\chi(S_1) & \psi_3(x_2)\chi(S_2) & \psi_3(x_3)\chi(S_3) \end{vmatrix}. \tag{8.92}$$

The energy of the second excited state is

$$E^{(2)} = 2\varepsilon_1 + \varepsilon_3 = 2\varepsilon_1 + 9\varepsilon_1 = 11\varepsilon_1 = \frac{11\hbar^2\pi^2}{2ma^2}. \tag{8.93}$$

The excitation energy of this state is $E^{(2)} - E^{(0)} = 11\varepsilon_1 - 6\varepsilon_1 = 5\varepsilon_1 = 5\hbar^2\pi^2/(2ma^2)$.

(d) If the particles were distinguishable fermions, there will be no restrictions on the symmetry of the wave function, neither on the space part nor on the spin part. The values of the energy of the ground state, the first excited state, and the second excited state will be similar to those calculated in part (a). However, the wave functions of these states are somewhat different from those found in part (a); while the states derived in (a) are nondegenerate, every state of the current system is eightfold degenerate, since the coupling of three $\frac{1}{2}$ spins yield eight different spin states (Chapter 7). So the wave functions of the system are obtained by multiplying each of the space wave functions $\psi^{(0)}(x_1, x_2, x_3)$, $\psi^{(1)}(x_1, x_2, x_3)$, and $\psi^{(2)}(x_1, x_2, x_3)$, derived in (a), by any of the eight spin states calculated in Chapter 7:

$$\left| 1, \frac{3}{2}, \pm\frac{3}{2} \right\rangle = \left| \frac{1}{2}, \frac{1}{2}, \frac{1}{2}; \pm\frac{1}{2}, \pm\frac{1}{2}, \pm\frac{1}{2} \right\rangle, \tag{8.94}$$

$$\left| 1, \frac{3}{2}, \pm\frac{1}{2} \right\rangle = \frac{1}{\sqrt{3}} \left(| j_1, j_2, j_3; \mp, \pm, \pm\rangle + | j_1, j_2, j_3; \pm, \mp, \pm\rangle + | j_1, j_2, j_3; \pm, \pm, \mp\rangle \right), \tag{8.95}$$

$$\left| 0, \frac{1}{2}, \pm\frac{1}{2} \right\rangle = \frac{1}{\sqrt{2}} \left(| j_1, j_2, j_3; \pm, \pm, \mp\rangle - | j_1, j_2, j_3; \mp, \pm, \pm \rangle \right), \tag{8.96}$$

$$\left| 1, \frac{1}{2}, \pm\frac{1}{2} \right\rangle = \frac{\mp 1}{\sqrt{6}} \left(| j_1, j_2, j_3; \pm, \pm, \mp\rangle - 2 | j_1, j_2, j_3; \pm, \mp, \pm\rangle + | j_1, j_2, j_3; \mp, \pm, \pm\rangle \right). \tag{8.97}$$

Problem 8.2

Consider a system of three noninteracting identical spin $\frac{1}{2}$ particles that are in the same spin state $\left|\frac{1}{2}, \frac{1}{2}\right\rangle$, and confined to move in a one-dimensional infinite potential well of length a: $V(x) = 0$ for $0 < x < a$ and $V(x) = \infty$ for other values of x. Determine the energy and wave function of the ground state, the first excited state, and the second excited state.

Solution

We may mention first that the single-particle energy and wave function of a particle moving in an infinite well are given by $\varepsilon_n = n^2 \hbar^2 \pi^2 / (2ma^2)$ and $\psi_n(x_i) = \sqrt{2/a} \sin(n\pi x_i / 2)$.

The wave function of this system is antisymmetric, since it consists of identical fermions. Moreover, since all the three particles are in the same spin state, no two particles can be in the same state; every energy level is occupied by at most one particle. For instance, the ground state corresponds to the case where the three lowest levels $n = 1, 2, 3$ are occupied by one particle each. The ground state energy and wave function are thus given by

$$E^{(0)} = \varepsilon_1 + \varepsilon_2 + \varepsilon_3 = \varepsilon_1 + 4\varepsilon_1 + 9\varepsilon_3 = 14\varepsilon_1 = \frac{7\hbar^2 \pi^2}{ma^2}, \tag{8.98}$$

$$\psi^{(0)}(x_1, x_2, x_3) = \frac{1}{\sqrt{3!}} \begin{vmatrix} \psi_1(x_1) & \psi_1(x_2) & \psi_1(x_3) \\ \psi_2(x_1) & \psi_2(x_2) & \psi_2(x_3) \\ \psi_3(x_1) & \psi_3(x_2) & \psi_3(x_3) \end{vmatrix} \left|\frac{1}{2}, \frac{1}{2}\right\rangle. \tag{8.99}$$

The first excited state is obtained (from the ground state) by raising the third particle to the fourth level: the levels $n = 1$, 2, and 4 are occupied by one particle each and the third level is empty:

$$E^{(1)} = \varepsilon_1 + \varepsilon_2 + \varepsilon_4 = \varepsilon_1 + 4\varepsilon_1 + 16\varepsilon_3 = 21\varepsilon_1 = \frac{21\hbar^2 \pi^2}{2ma^2}, \tag{8.100}$$

$$\psi^{(1)}(x_1, x_2, x_3) = \frac{1}{\sqrt{3!}} \begin{vmatrix} \psi_1(x_1) & \psi_1(x_2) & \psi_1(x_3) \\ \psi_2(x_1) & \psi_2(x_2) & \psi_2(x_3) \\ \psi_4(x_1) & \psi_4(x_2) & \psi_4(x_3) \end{vmatrix} \left|\frac{1}{2}, \frac{1}{2}\right\rangle. \tag{8.101}$$

In the second excited state, the levels $n = 1, 3, 4$ are occupied by one particle each; the second level is empty:

$$E^{(2)} = \varepsilon_1 + \varepsilon_3 + \varepsilon_4 = \varepsilon_1 + 9\varepsilon_1 + 16\varepsilon_3 = 26\varepsilon_1 = \frac{13\hbar^2 \pi^2}{ma^2}, \tag{8.102}$$

$$\psi^{(2)}(x_1, x_2, x_3) = \frac{1}{\sqrt{3!}} \begin{vmatrix} \psi_1(x_1) & \psi_1(x_2) & \psi_1(x_3) \\ \psi_3(x_1) & \psi_3(x_2) & \psi_3(x_3) \\ \psi_4(x_1) & \psi_4(x_2) & \psi_4(x_3) \end{vmatrix} \left|\frac{1}{2}, \frac{1}{2}\right\rangle. \tag{8.103}$$

Problem 8.3

Find the ground state energy and wave function of a system of N noninteracting identical particles that are confined to a one-dimensional, infinite well when the particles are (a) bosons and (b) spin $\frac{1}{2}$ fermions.

Solution

In the case of a particle moving in an infinite well, its energy and wave function are $\varepsilon_n = n^2\hbar^2\pi^2/(2ma^2)$ and $\psi_n(x_i) = \sqrt{2/a}\sin(n\pi x_i/2)$.

(a) In the case where the N particles are bosons, the ground state is obtained by putting all the particles in the state $n = 1$; the energy and wave function are then given by

$$E^{(0)} = \varepsilon_1 + \varepsilon_1 + \varepsilon_1 + \cdots + \varepsilon_1 = N\varepsilon_1 = \frac{N\hbar^2\pi^2}{2ma^2}, \tag{8.104}$$

$$\psi^{(0)}(x_1, x_2, \ldots, x_N) = \prod_{i=1}^{N}\sqrt{\frac{2}{a}}\sin\left(\frac{\pi}{2}x_i\right) = \sqrt{\frac{2^N}{a^N}}\sin\left(\frac{\pi}{2}x_1\right)\sin\left(\frac{\pi}{2}x_2\right)\cdots\sin\left(\frac{\pi}{2}x_N\right). \tag{8.105}$$

(b) In the case where the N particles are spin $\frac{1}{2}$ fermions, each level can be occupied by at most two particles having different spin states $\left|\frac{1}{2}, \pm\frac{1}{2}\right\rangle$. The ground state is thus obtained by distributing the N particles among the $N/2$ lowest levels at a rate of two particles per level:

$$E^{(0)} = 2\varepsilon_1 + 2\varepsilon_2 + 2\varepsilon_3 + \cdots + 2\varepsilon_{N/2} = 2\sum_{n=1}^{N/2}\frac{n^2\hbar^2\pi^2}{2ma^2} = \frac{\hbar^2\pi^2}{ma^2}\sum_{n=1}^{N/2}n^2. \tag{8.106}$$

If N is large we may calculate $\sum_{n=1}^{N/2}n^2$ by using the approximation

$$\sum_{n=1}^{N/2}n^2 \simeq \int_{1}^{N/2}n^2\,dn \simeq \frac{1}{3}\left(\frac{N}{2}\right)^3; \tag{8.107}$$

hence the ground state energy will be given by

$$E^{(0)} \simeq N^3\frac{\hbar^2\pi^2}{24ma^2}. \tag{8.108}$$

The average energy per particle is $E^{(0)}/N \simeq N^2\hbar^2\pi^2/(24ma^2)$. In the case where N is even, a possible configuration of the ground state wave function $\psi^{(0)}(x_1, x_2, \ldots, x_N)$ is given as follows:

$$\frac{1}{\sqrt{N!}}\begin{vmatrix} \psi_1(x_1)\chi(S_1) & \psi_1(x_2)\chi(S_2) & \cdots & \psi_1(x_N)\chi(S_N) \\ \psi_1(x_1)\chi(S_1) & \psi_1(x_2)\chi(S_2) & \cdots & \psi_1(x_N)\chi(S_N) \\ \psi_2(x_1)\chi(S_1) & \psi_2(x_2)\chi(S_2) & \cdots & \psi_2(x_N)\chi(S_N) \\ \psi_2(x_1)\chi(S_1) & \psi_2(x_2)\chi(S_2) & \cdots & \psi_2(x_N)\chi(S_N) \\ \psi_3(x_1)\chi(S_1) & \psi_3(x_2)\chi(S_2) & \cdots & \psi_3(x_N)\chi(S_N) \\ \psi_3(x_1)\chi(S_1) & \psi_3(x_2)\chi(S_2) & \cdots & \psi_3(x_N)\chi(S_N) \\ \vdots & \vdots & \ddots & \vdots \\ \psi_{N/2}(x_1)\chi(S_1) & \psi_{N/2}(x_2)\chi(S_2) & \cdots & \psi_{N/2}(x_N)\chi(S_N) \\ \psi_{N/2}(x_1)\chi(S_1) & \psi_{N/2}(x_2)\chi(S_2) & \cdots & \psi_{N/2}(x_N)\chi(S_N) \end{vmatrix}, \tag{8.109}$$

where $\chi(S_i) = \left|\frac{1}{2}, \pm\frac{1}{2}\right\rangle$ is the spin state of the ith particle, with $i = 1, 2, 3, \ldots, N$. If N is odd then we need to remove the last row of the determinant.

Problem 8.4

Neglecting the spin–orbit interaction and the interaction between the electrons, find the energy levels and the wave functions of the three lowest states for a two-electron atom.

Solution

Examples of such a system are the helium atom ($Z = 2$), the singly ionized Li$^+$ ion ($Z = 3$), the doubly ionized Be^{2+} ion ($Z = 4$), and so on. Neglecting the spin–orbit interaction and the interaction between the electrons, $V_{12} = e^2/r_{12} = e^2/|\vec{r}_1 - \vec{r}_2|$, we can view each electron as moving in the Coulomb field of the Ze nucleus. The Hamiltonian of this system is therefore equal to the sum of the Hamiltonians of the two electrons:

$$\hat{H} = H_0^{(1)} + H_0^{(2)} = \left(-\frac{\hbar^2}{2\mu}\nabla_1^2 - \frac{Ze^2}{r_1}\right) + \left(-\frac{\hbar^2}{2\mu}\nabla_2^2 - \frac{Ze^2}{r_2}\right), \qquad (8.110)$$

where $\mu = Mm_e/(M + m_e)$, M is the mass of the nucleus, and m_e is the mass of the electron. We have considered here that the nucleus is placed at the origin and that the electrons are located at $\vec{r}_1$ and $\vec{r}_2$. The Schrödinger equation of the system is given by

$$\left[\hat{H}_0^{(1)} + \hat{H}_0^{(2)}\right]\Psi(\vec{r}_1, \vec{S}_1; \vec{r}_2, \vec{S}_2) = E_{n_1 n_2}\Psi(\vec{r}_1, \vec{S}_1; \vec{r}_2, \vec{S}_2), \qquad (8.111)$$

where the energy $E_{n_1 n_2}$ is equal to the sum of the energies of the electrons:

$$E_{n_1 n_2} = E_{n_1}^{(0)} + E_{n_2}^{(0)} = -\frac{Z^2 e^2}{2a_0}\frac{1}{n_1^2} - \frac{Z^2 e^2}{2a_0}\frac{1}{n_2^2}, \qquad (8.112)$$

where $a_0 = \hbar^2/(me^2)$ is the Bohr radius. The wave function is equal to the product of the spatial and spin parts:

$$\Psi(\vec{r}_1, \vec{S}_1; \vec{r}_2, \vec{S}_2) = \psi(\vec{r}_1, \vec{r}_2)\chi(\vec{S}_1, \vec{S}_2); \qquad (8.113)$$

$\vec{S}_1$ and $\vec{S}_2$ are the spin vectors of the electrons.

Since this system consists of two identical fermions (electrons), its wave function has to be antisymmetric. So either the spatial part is antisymmetric and the spin part is symmetric,

$$\Psi(\vec{r}_1, \vec{S}_1; \vec{r}_2, \vec{S}_2) = \frac{1}{\sqrt{2}}\left[\phi_{n_1 l_1 m_1}(\vec{r}_1)\phi_{n_2 l_2 m_2}(\vec{r}_2) - \phi_{n_2 l_2 m_2}(\vec{r}_1)\phi_{n_1 l_1 m_1}(\vec{r}_2)\right]\chi_{triplet}(\vec{S}_1, \vec{S}_2), \qquad (8.114)$$

or the spatial part is symmetric and the spin part is antisymmetric,

$$\Psi(\vec{r}_1, \vec{S}_1; \vec{r}_2, \vec{S}_2) = \phi_{n_1 l_1 m_1}(\vec{r}_1)\phi_{n_2 l_2 m_2}(\vec{r}_2)\chi_{singlet}(\vec{S}_1, \vec{S}_2), \qquad (8.115)$$

where $\chi_{triplet}$ and $\chi_{singlet}$, which result from the coupling of two spins $\frac{1}{2}$, are given by (8.67) and (8.68).

Let us now specify the energy levels and wave functions of the three lowest states. The ground state corresponds to both electrons occupying the lowest state $|nlm> = |100>$ (i.e., $n_1 = n_2 = 1$); its energy and wave function can be inferred from (8.112) and (8.115):

$$E^{(0)} = E_{11} = 2E_1^{(0)} = -2\frac{Z^2 e^2}{2a_0} = -27.2Z^2 \text{ eV}, \qquad (8.116)$$

$$\Psi_0(\vec{r}_1, \vec{S}_1; \vec{r}_2, \vec{S}_2) = \phi_{100}(\vec{r}_1)\phi_{100}(\vec{r}_2)\chi_{singlet}(\vec{S}_1, \vec{S}_2), \qquad (8.117)$$

where $\phi_{100}(\vec{r}) = R_{10}(r)Y_{00}(\Omega) = (1/\sqrt{\pi})(Z/a_0)^{3/2}e^{-Zr/a_0}$.

In the first excited state, one electron occupies the lowest level $|nlm\rangle = |100\rangle$ and the other electron occupies the level $|nlm\rangle = |200\rangle$; this corresponds either to $n_1 = 1$, $n_2 = 2$ or to $n_1 = 2$, $n_2 = 1$. The energy and the wave function can thus be inferred from (8.112) and (8.114):

$$E^{(1)} = E_{12} = E_1^{(0)} + E_2^{(0)} = -\frac{Z^2 e^2}{2a_0} - \frac{1}{4}\frac{Z^2 e^2}{2a_0} = -\frac{5}{4} \times 13.6 Z^2 \text{ eV} = -17.0 Z^2 \text{ eV}, \quad (8.118)$$

$$\Psi_1(\vec{r}_1, \vec{S}_1; \vec{r}_2, \vec{S}_2) = \frac{1}{\sqrt{2}}\left[\phi_{100}(\vec{r}_1)\phi_{200}(\vec{r}_2) - \phi_{200}(\vec{r}_1)\phi_{100}(\vec{r}_2)\right]\chi_{triplet}(\vec{S}_1, \vec{S}_2), \quad (8.119)$$

where $\phi_{200}(\vec{r}) = R_{20}(r)Y_{00}(\Omega) = (1/\sqrt{8\pi})(Z/a_0)^{3/2}(1 - Zr/2a_0)e^{-Zr/2a_0}$.

Finally, the energy and wave function of the second excited state, which correspond to both electrons occupying the second level $|nlm\rangle = |200\rangle$ (i.e., $n_1 = n_2 = 2$), can be inferred from (8.112) and (8.115):

$$E^{(2)} = E_{22} = E_2^{(0)} + E_2^{(0)} = 2E_2^{(0)} = -\frac{1}{2}\frac{Z^2 e^2}{2a_0} = -\frac{1}{2} \times 13.6 Z^2 \text{ eV} = -6.8 Z^2 \text{ eV}, \quad (8.120)$$

$$\Psi_2(\vec{r}_1, \vec{S}_1; \vec{r}_2, \vec{S}_2) = \phi_{200}(\vec{r}_1)\phi_{200}(\vec{r}_2)\chi_{singlet}(\vec{S}_1, \vec{S}_2). \qquad (8.121)$$

These results are obviously not expected to be accurate because, by neglecting the Coulomb interaction between the electrons, we have made a grossly inaccurate approximation. For instance, the numerical value for the ground state energy obtained from (8.112) for the helium atom is $E_{theory}^{(0)} = -108.8\,\text{eV}$ whereas the experimental value is $E_{exp}^{(0)} = -78.975\,\text{eV}$; that is, the theoretical value is 37.8% lower than the experimental value.

In Chapter 9 we will show how to use perturbation theory and the variational method to obtain very accurate theoretical values for the energy levels of two-electron atoms.

Problem 8.5

Find the energy levels and the wave functions of the ground state and the first excited state for a system of two noninteracting identical particles moving in a common external harmonic oscillator potential for (a) two spin 1 particles with no orbital angular momentum and (b) two spin $\frac{1}{2}$ particles.

Solution

Since the particles are noninteracting and identical, their Hamiltonian is $\hat{H} = \hat{H}_1 + \hat{H}_2$, where $\hat{H}_1$ and $\hat{H}_2$ are the Hamiltonians of particles 1 and 2: $\hat{H}_j = -(\hbar^2/2m)d^2/dx_j^2 + m\omega x_j^2/2$ with $j = 1, 2$. The total energy of the system is $E_{n_1 n_2} = \varepsilon_{n_1} + \varepsilon_{n_2}$, where $\varepsilon_{n_j} = \left(n_j + \frac{1}{2}\right)\hbar\omega$.

(a) When the system consists of two identical spin 1 particles, the total wave function of this system must be symmetric. Thus, the space and spin parts must be both symmetric or both antisymmetric:

$$\Psi(x_1, S_1; x_2, S_2) = \frac{1}{\sqrt{2}}\left[\psi_s(x_1, x_2)\chi_s(S_1, S_2) + \psi_a(x_1, x_2)\chi_a(S_1, S_2)\right], \qquad (8.122)$$

where

$$\psi_s(x_1, x_2) = \frac{1}{\sqrt{2}} \left[\psi_{n_1}(x_1)\psi_{n_2}(x_2) + \psi_{n_1}(x_2)\psi_{n_2}(x_1) \right], \tag{8.123}$$

$$\psi_a(x_1, x_2) = \frac{1}{\sqrt{2}} \left[\psi_{n_1}(x_1)\psi_{n_2}(x_2) - \psi_{n_1}(x_2)\psi_{n_2}(x_1) \right], \tag{8.124}$$

where $\psi_n(x)$ is a harmonic oscillator wave function for the state n; for instance, the ground state and first excited state are

$$\psi_0(x) = \frac{1}{\sqrt{\sqrt{\pi} x_0}} \exp\left(-\frac{x^2}{2x_0^2}\right), \qquad \psi_1(x) = \sqrt{\frac{2}{\sqrt{\pi} x_0^3}} x \exp\left(-\frac{x^2}{2x_0^2}\right), \tag{8.125}$$

with $x_0 = \sqrt{\hbar/(m\omega)}$.

The spin states $\chi(S_1, S_2)$ can be obtained by coupling the spins of the two particles, $S_1 = 1$ and $S_2 = 1$: $\vec{S} = \vec{S}_1 + \vec{S}_2$. As shown in Chapter 7, the spin states corresponding to $S = 2$ are given by

$$|2, \pm 2\rangle = |11; \pm 1, \pm 1\rangle, \qquad |2, \pm 1\rangle = \frac{1}{\sqrt{2}} \left(|1, 1; \pm 1, 0\rangle + |1, 1; 0, \pm 1, \rangle \right), \tag{8.126}$$

$$|2, 0\rangle = \frac{1}{\sqrt{6}} \left(|1, 1; 1, -1\rangle + 2|1, 1; 0, 0\rangle + |1, 1; -1, 1\rangle \right), \tag{8.127}$$

those corresponding to $S = 1$ by

$$|1, \pm 1\rangle = \frac{1}{\sqrt{2}} \left(\pm |1, 1; \pm 1, 0\rangle \mp |1, 1; 0, \pm 1\rangle \right), \tag{8.128}$$

$$|1, 0\rangle = \frac{1}{\sqrt{2}} \left(|1, 1; 1, -1\rangle - |1, 1; -1, 1\rangle \right), \tag{8.129}$$

and the one corresponding to $S = 0$ by

$$|0, 0\rangle = \frac{1}{\sqrt{3}} \left(|1, 1; 1, -1\rangle - |1, 1; 0, 0\rangle + |1, 1; -1, 1\rangle \right). \tag{8.130}$$

Obviously, the five states $|2, m_s\rangle$, corresponding to $S = 2$ and $|00\rangle$, are symmetric, whereas the three states $|1, m_s\rangle$ are antisymmetric. Thus, $\chi_s(S_1, S_2)$ is given by any one of the six states $|2, \pm 2\rangle, |2, \pm 1\rangle, |2, 0\rangle$, and $|0, 0\rangle$; as for $\chi_a(S_1, S_2)$, it is given by any one of the three states $|2, \pm 1\rangle$, and $|1, 0\rangle$.

The ground state corresponds to the case where both particles are in their respective ground states $n_1 = n_2 = 0$. The energy is then given by $E^{(0)} = \varepsilon_0 + \varepsilon_0 = \frac{1}{2}\hbar\omega + \frac{1}{2}\hbar\omega = \hbar\omega$. Since $\psi_a(x_1, x_2)$, as given by (8.124), vanishes for $n_1 = n_2 = 0$, the ground state wave function (8.122) reduces to

$$\Psi_0(x_1, S_1; x_2, S_2) = \psi_0(x_1)\psi_0(x_2)\chi_s(S_1, S_2) = \frac{1}{\sqrt{\pi} x_0} \exp\left(-\frac{x_1^2 + x_2^2}{2x_0^2}\right) \chi_s(S_1, S_2), \tag{8.131}$$

where $\psi_0(x)$ is given by (8.125). The ground state is thus sixfold degenerate, since there are six spin states $\chi_s(S_1, S_2)$ that are symmetric.

In the first excited state, one particle occupies the ground state level $n = 0$ and the other in the first excited state $n = 1$; this corresponds to two possible configurations: either $n_1 = 0$ and $n_2 = 1$ or $n_1 = 1$ and $n_2 = 0$. The energy is then given by $E^{(1)} = \varepsilon_0 + \varepsilon_1 = \frac{1}{2}\hbar\omega + \frac{3}{2}\hbar\omega = 2\hbar\omega$. The first excited state can be inferred from (8.122) to (8.124):

$$\Psi_1(x_1, S_1; x_2, S_2) = \frac{1}{2}\left[\psi_0(x_1)\psi_1(x_2) + \psi_0(x_2)\psi_1(x_1)\right]\chi_s(S_1, S_2)$$
$$+ \frac{1}{2}\left[\psi_0(x_1)\psi_1(x_2) - \psi_0(x_2)\psi_1(x_1)\right]\chi_a(S_1, S_2), \qquad (8.132)$$

where $\psi_0(x)$ and $\psi_1(x)$ are listed in (8.125). The first excited state is ninefold degenerate since there are six spin states, $\chi_s(S_1, S_2)$, that are symmetric and three, $\chi_a(S_1, S_2)$, that are antisymmetric.

(b) For a system of two identical fermions, the wave function must be antisymmetric and the space and spin parts must have opposite symmetries:

$$\Psi(x_1, S_1; x_2, S_2) = \frac{1}{2}\left[\psi_s(x_1, x_2)\chi_{singlet}(S_1, S_2) + \psi_a(x_1, x_2)\chi_{triplet}(S_1, S_2)\right], \quad (8.133)$$

where the symmetric spin state, $\chi_{triplet}(S_1, S_2)$, is given by the triplet states listed in (8.67); the antisymmetric spin state, $\chi_{singlet}(S_1, S_2)$, is given by the (singlet) state (8.68).

The ground state for the two spin $\frac{1}{2}$ particles corresponds to the case where both particles occupy the lowest level, $n_1 = n_2 = 0$, and have different spin states. The energy is then given by $E^{(0)} = \varepsilon_0 + \varepsilon_0 = \hbar\omega$ and the wave function by

$$\Psi_0(x_1, S_1; x_2, S_2) = \psi_0(x_1)\psi_0(x_2)\chi_{singlet}(S_1, S_2)$$
$$= \frac{1}{\sqrt{\pi}x_0}\exp\left(-\frac{x_1^2 + x_2^2}{2x_0^2}\right)\chi_{singlet}(S_1, S_2), \qquad (8.134)$$

since $\psi_a(x_1, x_2)$ vanishes for $n_1 = n_2 = 0$. The ground state is not degenerate, since there is only one spin state which is antisymmetric, $\chi_{singlet}(S_1, S_2)$.

The first excited state corresponds also to $n_1 = 0$ and $n_2 = 1$ or $n_1 = 1$ and $n_2 = 0$. The energy is then given by $E^{(1)} = \varepsilon_0 + \varepsilon_1 = 2\hbar\omega$ and the wave function by

$$\Psi_1(x_1, S_1; x_2, S_2) = \frac{1}{2}\left[\psi_0(x_1)\psi_1(x_2) + \psi_0(x_2)\psi_1(x_1)\right]\chi_{singlet}(S_1, S_2)$$
$$+ \frac{1}{2}\left[\psi_0(x_1)\psi_1(x_2) - \psi_0(x_2)\psi_1(x_1)\right]\chi_{triplet}(S_1, S_2). \qquad (8.135)$$

This state is fourfold degenerate since there are three spin states, $\chi_{triplet}(S_1, S_2)$, that are symmetric and one, $\chi_{singlet}(S_1, S_2)$, that is antisymmetric.

8.6   Exercises

Exercise 8.1
Consider a system of three noninteracting identical bosons that move in a common external one-dimensional harmonic oscillator potential. Find the energy levels and wave functions of the ground state, the first excited state, and the second excited state of the system.

Exercise 8.2
Consider two identical particles of spin $\frac{1}{2}$ that are confined in a cubical box of side L. Find the energy and the wave function of this system in the case of no interaction between the particles.

Exercise 8.3
(a) Consider a system of two nonidentical particles, each of spin 1 and having no orbital angular momentum (i.e., both particles are in s states). Write down all possible states for this system.
(b) What restrictions do we get if the two particles are identical? Write down all possible states for this system of two spin 1 identical particles.

Exercise 8.4
Two identical particles of spin $\frac{1}{2}$ are enclosed in a one-dimensional box potential of length L with rigid walls at $x = 0$ and $x = L$. Assuming that the two-particle system is in a *triplet spin state*, find the energy levels, the wave functions, and the degeneracies corresponding to the three lowest states.

Exercise 8.5
Two identical particles of spin $\frac{1}{2}$ are enclosed in a one-dimensional box potential of length L with rigid walls at $x = 0$ and $x = L$. Assuming that the two-particle system is in a *singlet spin state*, find the energy levels, the wave functions, and the degeneracies corresponding to the three lowest states.

Exercise 8.6
Two identical particles of spin $\frac{1}{2}$ are moving under the influence of a one-dimensional harmonic oscillator potential. Assuming that the two-particle system is in a *triplet spin state*, find the energy levels, the wave functions, and the degeneracies corresponding to the three lowest states.

Exercise 8.7
Find the ground state energy, the average ground state energy per particle, and the ground state wave function of a system of N noninteracting, identical bosons moving under the influence of a one-dimensional harmonic oscillator potential.

Exercise 8.8
Find the ground state energy, the average ground state energy per particle, and the ground state wave function of a system of N noninteracting identical spin $\frac{1}{2}$ particles moving under the influence of a one-dimensional harmonic oscillator potential for the following two cases:
   (a) when N is even and
   (b) when N is odd.

Exercise 8.9
Consider a system of four noninteracting particles that are confined to move in a one-dimensional infinite potential well of length a: $V(x) = 0$ for $0 < x < a$ and $V(x) = \infty$ for other values of x. Determine the energies and wave functions of the ground state, the first excited state, and the second excited state when the four particles are
   (a) distinguishable bosons such that their respective masses satisfy this relation: $m_1 < m_2 < m_3 < m_4$, and
   (b) identical bosons (each of mass m).

Exercise 8.10

Consider a system of four noninteracting identical spin 1/2 particles (each of mass m) that are confined to move in a one-dimensional infinite potential well of length a: $V(x) = 0$ for $0 < x < a$ and $V(x) = \infty$ for other values of x. Determine the energies and wave functions of the ground state and the first three excited states. Draw a figure showing how the particles are distributed among the levels.

Exercise 8.11

Consider a system of four noninteracting identical spin $\frac{1}{2}$ particles that are in the same spin state $\left| \frac{1}{2}, \frac{1}{2} \right\rangle$ and confined to move in a one-dimensional infinite potential well of length a: $V(x) = 0$ for $0 < x < a$ and $V(x) = \infty$ for other values of x. Determine the energies and wave functions of the ground state, the first excited state, and the second excited state.

Exercise 8.12

Assuming the electrons in the helium atom to be spinless bosons and neglecting the interactions between them, find the energy and the wave function of the ground state and the first excited state of this (hypothetical) system.

Exercise 8.13

Assuming the electrons in the lithium atom to be spinless bosons and neglecting the interactions between them, find the energy and the wave function of the ground state and the first excited state of this (hypothetical) system.

Exercise 8.14

Consider a system of two noninteracting identical spin 1/2 particles (with mass m) that are confined to move in a one-dimensional infinite potential well of length L: $V(x) = 0$ for $0 < x < L$ and $V(x) = \infty$ for other values of x. Assume that the particles are in a state with the wave function

$$\Psi(x_1, x_2) = \frac{\sqrt{2}}{L} \left[\sin\left(\frac{2\pi x_1}{L}\right) \sin\left(\frac{5\pi x_2}{L}\right) + \sin\left(\frac{5\pi x_1}{L}\right) \sin\left(2\frac{\pi x_2}{L}\right) \right] \chi(s_1, s_2),$$

where x_1 and x_2 are the positions of particles 1 and 2, respectively, and $\chi(s_1, s_2)$ is the spin state of the two particles.

(a) Is $\chi(s_1, s_2)$ going to be a singlet or triplet state?
(b) Find the energy of this system.

Exercise 8.15

Consider a system of two noninteracting identical spin 1/2 particles (with mass m) that are confined to move in a common one-dimensional harmonic oscillator potential. Assume that the particles are in a state with the wave function

$$\Psi(x_1, x_2) = \frac{\sqrt{2}}{\sqrt{\pi} x_0^2} (x_2 - x_1) \exp\left(-\frac{x_1^2 + x_2^2}{2x_0^2}\right) \chi(s_1, s_2),$$

where x_1 and x_2 are the positions of particles 1 and 2, respectively, and $\chi(s_1, s_2)$ is the spin state of the two particles.

(a) Is $\chi(s_1, s_2)$ going to be a singlet or triplet state?
(b) Find the energy of this system.

Exercise 8.16
Consider a system of five noninteracting electrons (in the approximation where the Coulomb interaction between the electrons is neglected) that are confined to move in a common one-dimensional infinite potential well of length $L = 0.5$ nm: $V(x) = 0$ for $0 < x < L$ and $V(x) = \infty$ for other values of x.
(a) Find the ground state energy of the system.
(b) Find the energy of the first state of the system.
(c) Find the excitation energy of the first excited state.

Exercise 8.17
Determine the ground state electron configurations for the atoms having $Z = 40, 53, 70$, and 82 electrons.

Exercise 8.18
Specify the possible J values (i.e., total angular momenta) associated with each of the following states: $^1P, ^4F, ^2G$, and ^1H.

Exercise 8.19
Find the spectroscopic notation $^{2S+1}L_J$ (i.e., find the L, S, and J) for the ground state config-urations of
(a) Sc ($Z = 21$) and
(b) Cu ($Z = 29$).

Chapter 9

Approximation Methods for Stationary States

9.1 Introduction

Most problems encountered in quantum mechanics cannot be solved exactly. *Exact* solutions of the Schrödinger equation exist only for a few *idealized* systems. To solve general problems, one must resort to approximation methods. A variety of such methods have been developed, and each has its own area of applicability. In this chapter we consider approximation methods that deal with *stationary* states corresponding to *time-independent* Hamiltonians. In the following chapter we will deal with approximation methods for explicitly time-dependent Hamiltonians.

To study problems of stationary states, we focus on three approximation methods: *perturbation theory, the variational method*, and *the WKB method*.

Perturbation theory is based on the assumption that the problem we wish to solve is, in some sense, only *slightly* different from a problem that can be solved exactly. In the case where the deviation between the two problems is *small*, perturbation theory is suitable for calculating the contribution associated with this deviation; this contribution is then added as a correction to the energy and the wave function of the exactly solvable Hamiltonian. So perturbation theory builds on the known exact solutions to obtain approximate solutions.

What about those systems whose Hamiltonians cannot be reduced to an exactly solvable part plus a small correction? For these, we may consider the variational method or the WKB approximation. The variational method is particularly useful in estimating the energy eigenvalues of the ground state and the first few excited states of a system for which one has only a qualitative idea about the form of the wave function.

The WKB method is useful for finding the energy eigenvalues and wave functions of systems for which the *classical limit is valid*. Unlike perturbation theory, the variational and WKB methods do not require the existence of a closely related Hamiltonian that can be solved exactly.

The application of the approximation methods to the study of stationary states consists of finding the energy eigenvalues E_n and the eigenfunctions $| \psi_n \rangle$ of a time-independent Hamiltonian $\hat{H}$ that does not have exact solutions:

$$\hat{H} | \psi_n \rangle = E_n | \psi_n \rangle. \tag{9.1}$$

489

Depending on the structure of $\hat{H}$, we can use any of the three methods mentioned above to find the approximate solutions to this eigenvalue problem.

9.2 Time-Independent Perturbation Theory

This method is most suitable when $\hat{H}$ is very close to a Hamiltonian $\hat{H}_0$ that can be solved exactly. In this case, $\hat{H}$ can be split into two time-independent parts

$$\hat{H} = \hat{H}_0 + \hat{H}_p, \tag{9.2}$$

where $\hat{H}_p$ is very small compared to $\hat{H}_0$ ($\hat{H}_0$ is known as the Hamiltonian of the unperturbed system). As a result, $\hat{H}_p$ is called the perturbation, for its effects on the energy spectrum and eigenfunctions will be small; such perturbation is encountered, for instance, in systems subject to *weak* electric or magnetic fields. We can make this idea more explicit by writing $\hat{H}_p$ in terms of a dimensionless real parameter λ which is very small compared to 1:

$$\hat{H}_p = \lambda \hat{W} \qquad (\lambda \ll 1). \tag{9.3}$$

Thus the eigenvalue problem (9.1) becomes

$$(\hat{H}_0 + \lambda \hat{W}) \mid \psi_n \rangle = E_n \mid \psi_n \rangle. \tag{9.4}$$

In what follows we are going to consider two separate cases depending on whether the exact solutions of $\hat{H}_0$ are nondegenerate or degenerate. Each of these two cases requires its own approximation scheme.

9.2.1 Nondegenerate Perturbation Theory

In this section we limit our study to the case where $\hat{H}_0$ has no degenerate eigenvalues; that is, for every energy $E_n^{(0)}$ there corresponds only one eigenstate $\mid \phi_n \rangle$:

$$\hat{H}_0 \mid \phi_n \rangle = E_n^{(0)} \mid \phi_n \rangle, \tag{9.5}$$

where the exact eigenvalues $E_n^{(0)}$ and exact eigenfunctions $\mid \phi_n \rangle$ are known.

The main idea of perturbation theory consists in assuming that the perturbed eigenvalues and eigenstates can both be expanded in power series in the parameter λ:

$$E_n = E_n^{(0)} + \lambda E_n^{(1)} + \lambda^2 E_n^{(2)} + \cdots, \tag{9.6}$$

$$\mid \psi_n \rangle = \mid \phi_n \rangle + \lambda \mid \psi_n^{(1)} \rangle + \lambda^2 \mid \psi_n^{(2)} \rangle + \cdots. \tag{9.7}$$

We need to make two remarks. First, one might think that whenever the perturbation is sufficiently weak, the expansions (9.6) and (9.7) always exist. Unfortunately, this is not always the case. There are cases where the perturbation is small, yet E_n and $\mid \psi_n \rangle$ are not expandable in powers of λ. Second, the series (9.6) and (9.7) are frequently not convergent. However, when λ is small, the first few terms do provide a reliable description of the system. So in practice, we keep only one or two terms in these expansions; hence the problem of nonconvergence of these series is avoided (we will deal later with the problem of convergence). Note that when $\lambda = 0$ the expressions (9.6) and (9.7) yield the unperturbed solutions: $E_n = E_n^{(0)}$ and $\mid \psi_n \rangle = \mid \phi_n \rangle$.

The parameters $E_n^{(k)}$ and the kets $| \psi_n^{(k)} \rangle$ represent the kth corrections to the eigenenergies and eigenvectors, respectively.

The job of perturbation theory reduces then to the calculation of $E_n^{(1)}$, $E_n^{(2)}$, ... and $|\psi_n^{(1)}\rangle$, $|\psi_n^{(2)}\rangle$, ... In this section we shall be concerned only with the determination of $E_n^{(1)}$, $E_n^{(2)}$, and $|\psi_n^{(1)}\rangle$. Assuming that the unperturbed states $| \phi_n \rangle$ are nondegenerate, and substituting (9.6) and (9.7) into (9.4), we obtain

$$\left(\hat{H}_0 + \lambda \hat{W} \right) \left(| \phi_n \rangle + \lambda | \psi_n^{(1)} \rangle + \lambda^2 | \psi_n^{(2)} \rangle + \cdots \right)$$
$$= \left(E_n^{(0)} + \lambda E_n^{(1)} + \lambda^2 E_n^{(2)} + \cdots \right) \left(| \phi_n \rangle + \lambda | \psi_n^{(1)} \rangle + \lambda^2 | \psi_n^{(2)} \rangle + \cdots \right). \tag{9.8}$$

The coefficients of successive powers of λ on both sides of this equation must be equal. Equating the coefficients of the first three powers of λ, we obtain these results:

- Zero order in λ:
$$\hat{H}_0 | \phi_n \rangle = E_n^{(0)} | \phi_n \rangle, \tag{9.9}$$

- First order in λ:
$$\hat{H}_0 | \psi_n^{(1)} \rangle + \hat{W} | \phi_n \rangle = E_n^{(0)} | \psi_n^{(1)} \rangle + E_n^{(1)} | \phi_n \rangle, \tag{9.10}$$

- Second order in λ:
$$\hat{H}_0 | \psi_n^{(2)} \rangle + \hat{W} | \psi_n^{(1)} \rangle = E_n^{(0)} | \psi_n^{(2)} \rangle + E_n^{(1)} | \psi_n^{(1)} \rangle + E_n^{(2)} | \phi_n \rangle. \tag{9.11}$$

We now proceed to determine the eigenvalues $E_n^{(1)}$, $E_n^{(2)}$ and the eigenvector $| \psi_n^{(1)} \rangle$ from (9.9) to (9.11). For this, we need to specify how the states $| \phi_n \rangle$ and $| \psi_n \rangle$ overlap. Since $| \psi_n \rangle$ is considered not to be very different from $| \phi_n \rangle$, we have $\langle \phi_n | \psi_n \rangle \simeq 1$. We can, however, normalize $| \psi_n \rangle$ so that its overlap with $| \phi_n \rangle$ is exactly equal to one:

$$\langle \phi_n | \psi_n \rangle = 1. \tag{9.12}$$

Substituting (9.7) into (9.12) we get

$$\lambda \langle \phi_n | \psi_n^{(1)} \rangle + \lambda^2 \langle \phi_n | \psi_n^{(2)} \rangle + \cdots = 0; \tag{9.13}$$

hence the coefficients of the various powers of λ must vanish separately:

$$\langle \phi_n | \psi_n^{(1)} \rangle = \langle \phi_n | \psi_n^{(2)} \rangle = \cdots = 0. \tag{9.14}$$

First-order correction

To determine the first-order correction, $E_n^{(1)}$, to E_n we need simply to multiply both sides of (9.10) by $\langle \phi_n |$:

$$\boxed{E_n^{(1)} = \langle \phi_n | \hat{W} | \phi_n \rangle,} \tag{9.15}$$

where we have used the facts that $\langle \phi_n | \hat{H}_0 | \psi_n^{(1)} \rangle$ and $\langle \phi_n | \psi_n^{(1)} \rangle$ are both equal to zero and $\langle \phi_n | \phi_n \rangle = 1$. The insertion of (9.15) into (9.6) thus yields the energy to first-order perturbation:

$$\boxed{E_n = E_n^{(0)} + \langle \phi_n | \hat{H}_p | \phi_n \rangle.} \tag{9.16}$$

Note that for some systems, the first-order correction $E_n^{(1)}$ vanishes exactly. In such cases, one needs to consider higher-order terms.

Let us now determine $|\psi_n^{(1)}\rangle$. Since the set of the unperturbed states $|\phi_n\rangle$ form a complete and orthonormal basis, we can expand $|\psi_n^{(1)}\rangle$ in the $\{|\phi_n\rangle\}$ basis:

$$|\psi_n^{(1)}\rangle = \left(\sum_m |\phi_m\rangle\langle\phi_m|\right)|\psi_n^{(1)}\rangle = \sum_{m\neq n}\langle\phi_m|\psi_n^{(1)}\rangle|\phi_m\rangle; \qquad (9.17)$$

the term $m = n$ does not contribute, since $\langle\phi_n|\psi_n^{(1)}\rangle = 0$. The coefficient $\langle\phi_m|\psi_n^{(1)}\rangle$ can be inferred from (9.10) by multiplying both sides by $\langle\phi_m|$:

$$\langle\phi_m|\psi_n^{(1)}\rangle = \frac{\langle\phi_m|\hat{W}|\phi_n\rangle}{E_n^{(0)} - E_m^{(0)}}, \qquad (9.18)$$

which, when substituted into (9.17), leads to

$$|\psi_n^{(1)}\rangle = \sum_{m\neq n}\frac{\langle\phi_m|\hat{W}|\phi_n\rangle}{E_n^{(0)} - E_m^{(0)}}|\phi_m\rangle. \qquad (9.19)$$

The eigenfunction $|\psi_n\rangle$ of $\hat{H}$ to first order in $\lambda\hat{W}$ can then be obtained by substituting (9.19) into (9.7):

$$\boxed{|\psi_n\rangle = |\phi_n\rangle + \sum_{m\neq n}\frac{\langle\phi_m|\hat{H}_p|\phi_n\rangle}{E_n^{(0)} - E_m^{(0)}}|\phi_m\rangle.} \qquad (9.20)$$

Second-order correction

Now, to determine $E_n^{(2)}$ we need to multiply both sides of (9.11) by $\langle\phi_n|$:

$$E_n^{(2)} = \langle\phi_n|\hat{W}|\psi_n^{(1)}\rangle; \qquad (9.21)$$

in obtaining this result we have used the facts that $\langle\phi_n|\psi_n^{(1)}\rangle = \langle\phi_n|\psi_n^{(2)}\rangle = 0$ and $\langle\phi_n|\phi_n\rangle = 1$. Inserting (9.19) into (9.21) we end up with

$$\boxed{E_n^{(2)} = \sum_{m\neq n}\frac{\left|\langle\phi_m|\hat{W}|\phi_n\rangle\right|^2}{E_n^{(0)} - E_m^{(0)}}.} \qquad (9.22)$$

The eigenenergy to second order in $\hat{H}_p$ is obtained by substituting (9.22) and (9.15) into (9.6):

$$\boxed{E_n = E_n^{(0)} + \langle\phi_n|\hat{H}_p|\phi_n\rangle + \sum_{m\neq n}\frac{\left|\langle\phi_m|\hat{H}_p|\phi_n\rangle\right|^2}{E_n^{(0)} - E_m^{(0)}} + \cdots.} \qquad (9.23)$$

In principle one can obtain energy corrections to any order. However, pushing the calculations beyond the second order, besides being mostly intractable, is a futile exercise, since the first two orders are generally sufficiently accurate.

Validity of the time-independent perturbation theory

For perturbation theory to work, the corrections it produces must be small; convergence must be achieved with the first two corrections. Expressions (9.20) and (9.23) show that the expansion parameter is $\langle \phi_m | \hat{H}_p | \phi_n \rangle / (E_n^{(0)} - E_m^{(0)})$. Thus, for the perturbation schemes (9.6) and (9.7) to work (i.e., to converge), the expansion parameter must be small:

$$\left| \frac{\langle \phi_m | \hat{H}_p | \phi_n \rangle}{E_n^{(0)} - E_m^{(0)}} \right| \ll 1 \qquad (n \neq m). \qquad (9.24)$$

If the unperturbed energy levels $E_n^{(0)}$ and $E_m^{(0)}$ were equal (i.e., degenerate) then condition (9.24) would break down. Degenerate energy levels require an approach that is different from the nondegenerate treatment. This question will be taken up in the following section.

Example 9.1 (Charged oscillator in an electric field)

A particle of charge q and mass m, which is moving in a one-dimensional harmonic potential of frequency ω, is subject to a *weak* electric field $\mathcal{E}$ in the x-direction.

(a) Find the exact expression for the energy.

(b) Calculate the energy to first nonzero correction and compare it with the exact result obtained in (a).

Solution

The interaction between the oscillating charge and the external electric field gives rise to a term $\hat{H}_P = q\mathcal{E}\hat{X}$ that needs to be added to the Hamiltonian of the oscillator:

$$\hat{H} = \hat{H}_0 + \hat{H}_p = -\frac{\hbar}{2m}\frac{d^2}{dX^2} + \frac{1}{2}m\omega^2 \hat{X}^2 + q\mathcal{E}\hat{X}. \qquad (9.25)$$

(a) First, note that the eigenenergies of this Hamiltonian can be obtained exactly without resorting to any perturbative treatment. A variable change $\hat{y} = \hat{X} + q\mathcal{E}/(m\omega^2)$ leads to

$$\hat{H} = -\frac{\hbar^2}{2m}\frac{d^2}{dy^2} + \frac{1}{2}m\omega^2 \hat{y}^2 - \frac{q^2\mathcal{E}^2}{2m\omega^2}. \qquad (9.26)$$

This is the Hamiltonian of a harmonic oscillator from which a constant, $q^2\mathcal{E}^2\omega^2/(2m)$, is subtracted. The exact eigenenergies can thus be easily inferred:

$$E_n = \left(n + \frac{1}{2}\right)\hbar\omega - \frac{q^2\mathcal{E}^2}{2m\omega^2}. \qquad (9.27)$$

This simple example allows us to compare the exact and approximate eigenenergies.

(b) Let us now turn to finding the approximate eigenvalues of $\hat{H}$ by means of perturbation theory. Since the electric field is weak, we can treat $\hat{H}_p$ as a perturbation.

Note that the first-order correction to the energy, $E_n^{(1)} = a\langle n | \hat{X} | n \rangle$, is zero (since $\langle n | \hat{X} | n \rangle = 0$), but the second-order correction is not:

$$E_n^{(2)} = q^2\mathcal{E}^2 \sum_{m \neq n} \frac{\left| \langle m | \hat{X} | n \rangle \right|^2}{E_n^{(0)} - E_m^{(0)}}. \qquad (9.28)$$

Since $E_n^{(0)} = \left(n + \frac{1}{2}\right)\hbar\omega$, and using the relations

$$\langle n+1 \mid \hat{X} \mid n \rangle = \sqrt{n+1}\sqrt{\frac{\hbar}{2m\omega}}, \qquad \langle n-1 \mid \hat{X} \mid n \rangle = \sqrt{n}\sqrt{\frac{\hbar}{2m\omega}}, \qquad (9.29)$$

$$E_n^{(0)} - E_{n-1}^{(0)} = \hbar\omega, \qquad\qquad\qquad E_n^{(0)} - E_{n+1}^{(0)} = -\hbar\omega, \qquad (9.30)$$

we can reduce (9.28) to

$$E_n^{(2)} = q^2 \mathcal{E}^2 \left[\frac{|\langle n+1 \mid \hat{X} \mid n \rangle|^2}{E_n^{(0)} - E_{n+1}^{(0)}} + \frac{|\langle n-1 \mid \hat{X} \mid n \rangle|^2}{E_n^{(0)} - E_{n-1}^{(0)}} \right]$$

$$= -\frac{q^2 \mathcal{E}^2}{2m\omega^2}; \qquad (9.31)$$

hence the energy is given to second order by

$$E_n = E_n^{(0)} + E_n^{(1)} + E_n^{(2)} = \left(n + \frac{1}{2}\right)\hbar\omega - \frac{q^2 \mathcal{E}^2}{2m\omega^2}. \qquad (9.32)$$

This agrees fully with the exact energy found in (9.27).

Similarly, using (9.19) along with (9.29) and (9.30), we can easily ascertain that $\mid \psi_n^{(1)} \rangle$ is given by

$$\mid \psi_n^{(1)} \rangle = \frac{q\mathcal{E}}{\hbar\omega}\sqrt{\frac{\hbar}{2m\omega}} \left\{ \sqrt{n} \mid n-1 \rangle - \sqrt{n+1} \mid n+1 \rangle \right\}; \qquad (9.33)$$

hence the state $\mid \psi_n \rangle$ is given to first order by

$$\mid \psi_n \rangle = \mid n \rangle + \frac{q\mathcal{E}}{\hbar\omega}\sqrt{\frac{\hbar}{2m\omega}} \left\{ \sqrt{n} \mid n-1 \rangle - \sqrt{n+1} \mid n+1 \rangle \right\}, \qquad (9.34)$$

where $\mid n \rangle$ is the exact eigenstate of the nth excited state of a one-dimensional harmonic oscillator.

Example 9.2 (The Stark effect)

(a) Study the effect of an external uniform weak electric field, which is directed along the positive z-axis, $\vec{\mathcal{E}} = \mathcal{E}\vec{k}$, on the ground state of a hydrogen atom; ignore the spin degrees of freedom.

(b) Find an approximate value for the polarizability of the hydrogen atom.

Solution

(a) The effect that an external electric field has on the energy levels of an atom is called the Stark effect. In the absence of an electric field, the (unperturbed) Hamiltonian of the hydrogen atom (in CGS units) is:

$$\hat{H}_0 = \frac{\hat{\vec{p}}^2}{2\mu} - \frac{e^2}{r}. \qquad (9.35)$$

The eigenfunctions of this Hamiltonian, $\psi_{nlm}(\vec{r})$, were obtained in Chapter 6; they are given by

$$\langle r\theta\varphi \mid nlm \rangle = \psi_{nlm}(r, \theta, \varphi) = R_{nl}(r)Y_{lm}(\theta, \varphi). \qquad (9.36)$$

When the electric field is turned on, the interaction between the atom and the field generates a term $\hat{H}_p = e\vec{\mathcal{E}} \cdot \vec{r} = e\mathcal{E}\hat{Z}$ that needs to be added to $\hat{H}_0$.

Since the excited states of the hydrogen atom are degenerate while the ground state is not, nondegenerate perturbation theory applies only to the ground state, $\psi_{100}(\vec{r})$. Ignoring the spin degrees of freedom, the energy of this system to second-order perturbation is given as follows (see (9.23)):

$$E_{100} = E_{100}^{(0)} + e\mathcal{E}\langle 100 \mid \hat{Z} \mid 100 \rangle + e^2\mathcal{E}^2 \sum_{nlm \neq 100} \frac{\left| \langle nlm \mid \hat{Z} \mid 100 \rangle \right|^2}{E_{100}^{(0)} - E_{nlm}^{(0)}}. \tag{9.37}$$

The term

$$\langle 100 \mid \hat{Z} \mid 100 \rangle = \int |\psi_{100}(\vec{r})|^2 z \, d^3r \tag{9.38}$$

is zero, since $\hat{Z}$ is odd under parity and $\psi_{100}(\vec{r})$ has a definite parity. This means that there can be no correction term to the energy which is proportional to the electric field and hence there is no linear Stark effect. The underlying physics behind this is that when the hydrogen atom is in its ground state, it has no permanent electric dipole moment. We are left then with only a *quadratic* dependence of the energy (9.37) on the electric field. This is called the quadratic Stark effect. This correction, which is known as the energy shift ΔE, is given by

$$\Delta E = e^2\mathcal{E}^2 \sum_{nlm \neq 100} \frac{\left| \langle nlm \mid \hat{Z} \mid 100 \rangle \right|^2}{E_{100}^{(0)} - E_{nlm}^{(0)}}. \tag{9.39}$$

(b) Let us now estimate the value of the polarizability of the hydrogen atom. The *polarizability* α of an atom which is subjected to an electric field $\vec{\mathcal{E}}$ is given in terms of the energy shift ΔE as

$$\alpha = -2\frac{\Delta E}{\mathcal{E}^2}. \tag{9.40}$$

Substituting (9.39) into (9.40), we obtain the polarizability of the hydrogen atom in its ground state:

$$\alpha = -2e^2 \sum_{nlm \neq 100} \frac{\left| \langle nlm \mid \hat{Z} \mid 100 \rangle \right|^2}{E_{100}^{(0)} - E_{nlm}^{(0)}}. \tag{9.41}$$

To estimate this sum, let us assume that the denominator is constant. Since $n \geq 2$, we can write

$$E_{100}^{(0)} - E_{nlm}^{(0)} \leq E_{100} - E_{200} = \frac{e^2}{2a_0}\left(-1 + \frac{1}{4}\right) = -\frac{3e^2}{8a_0}; \tag{9.42}$$

hence

$$\alpha \leq \frac{16a_0}{3} \sum_{nlm \neq 100} |\langle nlm \mid \hat{Z} \mid 100 \rangle|^2, \tag{9.43}$$

where

$$\sum_{nlm \neq 100} |\langle nlm \mid \hat{Z} \mid 100 \rangle|^2 = \sum_{\text{all } nlm} |\langle nlm \mid \hat{Z} \mid 100 \rangle|^2$$

$$= \langle 100 \mid \hat{Z}\left(\sum_{\text{all } nlm} \mid nlm \rangle\langle nlm \mid\right)\hat{Z} \mid 100 \rangle$$

$$= \langle 100 \mid \hat{Z}^2 \mid 100 \rangle; \tag{9.44}$$

in deriving this relation, we have used the facts that $\langle 100 \mid \hat{Z} \mid 100 \rangle = 0$ and that the set of states $\mid nlm \rangle$ is complete. Now since $z = r\cos\theta$ and $\langle r\theta\varphi \mid 100 \rangle = R_{10}(r)Y_{00}(\theta, \varphi) = R_{10}(r)/\sqrt{4\pi}$, we immediately obtain

$$\langle 100 \mid \hat{Z}^2 \mid 100 \rangle = \frac{1}{4\pi} \int_0^\infty r^4 R_{10}^2(r)\, dr \int_0^\pi \sin\theta \cos^2\theta\, d\theta \int_0^{2\pi} d\varphi = a_0^2. \tag{9.45}$$

Substituting (9.45) and (9.44) into (9.43), we see that the polarizability for hydrogen has an upper limit

$$\alpha \le \frac{16}{3}a_0^3. \tag{9.46}$$

This limit, which is obtained from perturbation theory, agrees with the exact value $\alpha = \frac{9}{2}a_0^3$.

9.2.2 Degenerate Perturbation Theory

In the discussion above, we have considered only systems with nondegenerate $\hat{H}_0$. We now apply perturbation theory to determine the energy spectrum and the states of a system whose unperturbed Hamiltonian $\hat{H}_0$ is degenerate:

$$\hat{H} \mid \psi_n \rangle = (\hat{H}_0 + \hat{H}_p) \mid \psi_n \rangle = E_n \mid \psi_n \rangle. \tag{9.47}$$

If, for instance, the level of energy $E_n^{(0)}$ is f-fold degenerate (i.e., there exists a set of f different eigenstates $\mid \phi_{n_\alpha} \rangle$, where $\alpha = 1, 2, \ldots, f$, that correspond to the same eigenenergy $E_n^{(0)}$), we have

$$\hat{H}_0 \mid \phi_{n_\alpha} \rangle = E_n^{(0)} \mid \phi_{n_\alpha} \rangle \qquad (\alpha = 1, 2, \ldots, f), \tag{9.48}$$

where α stands for one or more quantum numbers; the energy eigenvalues $E_n^{(0)}$ are independent of α.

In the zeroth-order approximation we can write the eigenfunction $\mid \psi_n \rangle$ as a linear combination in terms of $\mid \phi_{n_\alpha} \rangle$:

$$\mid \psi_n \rangle = \sum_{\alpha=1}^f a_\alpha \mid \phi_{n_\alpha} \rangle. \tag{9.49}$$

Considering the states $\mid \phi_{n_\alpha} \rangle$ to be orthonormal with respect to the label α (i.e., $\langle \phi_{n_\alpha} \mid \phi_{n_\beta} \rangle = \delta_{\alpha,\beta}$) and $\mid \psi_n \rangle$ to be normalized, $\langle \psi_n \mid \psi_n \rangle = 1$, we can ascertain that the coefficients a_α obey the relation

$$\langle \psi_n \mid \psi_n \rangle = \sum_{\alpha,\beta} a_\alpha^* a_\beta \delta_{\alpha,\beta} = \sum_{\alpha=1}^f |a_\alpha|^2 = 1. \tag{9.50}$$

In what follows we are going to show how to determine these coefficients and the first-order corrections to the energy. For this, let us substitute (9.48) and (9.49) into (9.47):

$$\sum_\alpha \left[E_n^{(0)} \mid \phi_{n_\alpha} \rangle + \hat{H}_p \mid \phi_{n_\alpha} \rangle \right] a_\alpha = E_n \sum_\alpha a_\alpha \mid \phi_{n_\alpha} \rangle. \tag{9.51}$$

The multiplication of both sides of this equation by $\langle \phi_{n_\beta} \mid$ leads to

$$\sum_\alpha a_\alpha \left[E_n^{(0)} \delta_{\alpha,\beta} + \langle \phi_{n_\beta} \mid \hat{H}_p \mid \phi_{n_\alpha} \rangle \right] = E_n \sum_\alpha a_\alpha \delta_{\alpha,\beta} \tag{9.52}$$

or to

$$a_\beta E_n = a_\beta E_n^{(0)} + \sum_{u=1}^{f} a_\alpha \langle \phi_{n_\beta} \mid \hat{H}_p \mid \phi_{n_\alpha} \rangle, \qquad (9.53)$$

where we have used $\langle \phi_{n_\beta} \mid \phi_{n_\alpha} \rangle = \delta_{\beta, \alpha}$. We can rewrite (9.53) as follows:

$$\sum_{\alpha=1}^{f} \left(\hat{H}_{p\beta\alpha} - E_n^{(1)} \delta_{\alpha, \beta} \right) a_\alpha = 0 \qquad (\beta = 1, 2, \ldots, f), \qquad (9.54)$$

with $\hat{H}_{p\beta\alpha} = \langle \phi_{n_\beta} \mid \hat{H}_p \mid \phi_{n_\alpha} \rangle$ and $E_n^{(1)} = E_n - E_n^{(0)}$. This is a system of f homogeneous linear equations for the coefficients a_α. These coefficients are nonvanishing only when the determinant $|\hat{H}_{p\alpha\beta} - E_n^{(1)} \delta_{\alpha, \beta}|$ is zero:

$$\begin{vmatrix} \hat{H}_{p_{11}} - E_n^{(1)} & \hat{H}_{p_{12}} & \hat{H}_{p_{13}} & \cdots & \hat{H}_{p_{1f}} \\ \hat{H}_{p_{21}} & \hat{H}_{p_{22}} - E_n^{(1)} & \hat{H}_{p_{23}} & \cdots & \hat{H}_{p_{2f}} \\ \vdots & \vdots & \vdots & \ddots & \vdots \\ \hat{H}_{p_{f1}} & \hat{H}_{p_{f2}} & \hat{H}_{p_{f3}} & \cdots & \hat{H}_{p_{ff}} - E_n^{(1)} \end{vmatrix} = 0. \qquad (9.55)$$

This is an fth degree equation in $E_n^{(1)}$ and in general it has f different real roots, $E_{n_\alpha}^{(1)}$. These roots are the first-order correction to the eigenvalues, E_{n_α}, of $\hat{H}$. To find the coefficients a_α, we need simply to substitute these roots into (9.54) and then solve the resulting expression. Knowing these coefficients, we can then determine the eigenfunctions, $\mid \psi_n \rangle$, of $\hat{H}$ in the zeroth approximation from (9.49).

The roots $E_{n_\alpha}^{(1)}$ of (9.55) are in general different. In this case the eigenvalues $\hat{H}$ are not degenerate, hence the f-fold degenerate level $E_n^{(0)}$ of the unperturbed problem is split into f different levels E_{n_α}: $E_{n_\alpha} = E_n^{(0)} + E_{n_\alpha}^{(1)}$, $\alpha = 1, 2, \ldots, f$. In this way, the perturbation lifts the degeneracy. The lifting of the degeneracy may be either total or partial, depending on whether all the roots of (9.55), or only some of them, are different.

In summary, to determine the eigenvalues to first-order and the eigenstates to zeroth order for an f-fold degenerate level from perturbation theory, we proceed as follows:

- First, for each f-fold degenerate level, determine the $f \times f$ matrix of the perturbation $\hat{H}_p$:

$$H_p = \begin{pmatrix} \hat{H}_{p_{11}} & \hat{H}_{p_{12}} & \cdots & \hat{H}_{p_{1f}} \\ \hat{H}_{p_{21}} & \hat{H}_{p_{22}} & \cdots & \hat{H}_{p_{2f}} \\ \vdots & \vdots & \ddots & \vdots \\ \hat{H}_{p_{f1}} & \hat{H}_{p_{f2}} & \cdots & \hat{H}_{p_{ff}} \end{pmatrix}, \qquad (9.56)$$

where $\hat{H}_{p\alpha\beta} = \langle \phi_{n_\alpha} | \hat{H}_p | \phi_{n_\beta} \rangle$.

- Second, diagonalize this matrix and find the f eigenvalues $E_{n_\alpha}^{(1)}$ ($\alpha = 1, 2, \ldots, f$) and their corresponding eigenvectors

$$a_\alpha = \begin{pmatrix} a_{\alpha 1} \\ a_{\alpha 2} \\ \vdots \\ a_{\alpha f} \end{pmatrix} \qquad (\alpha = 1, 2, \ldots, f). \qquad (9.57)$$

- Finally, the energy eigenvalues are given to first order by

$$E_{n_\alpha} = E_n^{(0)} + E_{n_\alpha}^{(1)} \qquad (\alpha = 1, 2, \ldots, f) \tag{9.58}$$

and the corresponding eigenvectors are given to zero order by

$$|\psi_{n_\alpha}\rangle = \sum_{\beta=1}^{f} a_{\alpha\beta} |\phi_{n_\beta}\rangle. \tag{9.59}$$

Example 9.3 (The Stark effect of hydrogen)

Using first-order (degenerate) perturbation theory, calculate the energy levels of the $n = 2$ states of a hydrogen atom placed in an external uniform weak electric field along the positive z-axis.

Solution

In the absence of any external electric field, the first excited state (i.e., $n = 2$) is fourfold degenerate: the states $| nlm \rangle = | 200 \rangle, | 210 \rangle, | 211 \rangle$, and $| 21 - 1 \rangle$ have the same energy $E_2 = -R_y/4$, where $R_y = \mu e^4/(2\hbar^2) = 13.6 \, \text{eV}$ is the Rydberg constant.

When the external electric field is turned on, some energy levels will split. The energy due to the interaction between the dipole moment of the electron ($\vec{d} = -e\vec{r}$) and the external electric field ($\vec{\mathcal{E}} = \mathcal{E}\vec{k}$) is given by

$$\hat{H}_p = -\vec{d} \cdot \vec{\mathcal{E}} = e\vec{r} \cdot \vec{\mathcal{E}} = e\mathcal{E}\hat{Z}. \tag{9.60}$$

To calculate the eigenenergies, we need to determine and then diagonalize the 4×4 matrix elements of $\hat{H}_p$: $\langle 2l'm' | \hat{H}_p | 2lm \rangle = e\mathcal{E}\langle 2l'm' | \hat{Z} | 2lm \rangle$. The matrix elements $\langle 2l'm' | \hat{Z} | 2lm \rangle$ can be calculated more simply by using the relevant selection rules and symmetries. First, since $\hat{Z}$ does not depend on the azimuthal angle φ, $z = r \cos \theta$, the elements $\langle 2l'm' | \hat{Z} | 2lm \rangle$ are nonzero only if $m' = m$. Second, as Z is odd, the states $| 2l'm' \rangle$ and $| 2lm \rangle$ must have opposite parities so that $\langle 2l'm' | \hat{Z} | 2lm \rangle$ does not vanish. Therefore, the only nonvanishing matrix elements are those that couple the 2s and 2p states (with $m = 0$); that is, between $| 200 \rangle$ and $| 210 \rangle$. In this case we have

$$\langle 200 | \hat{Z} | 210 \rangle = \int_0^\infty R_{20}^*(r) R_{21}(r) r^2 dr \int Y_{00}^*(\Omega) z Y_{10}(\Omega) \, d\Omega$$

$$= \sqrt{\frac{4\pi}{3}} \int_0^\infty R_{20}(r) R_{21}(r) r^3 dr \int Y_{00}^*(\Omega) Y_{10}^2(\Omega) \, d\Omega$$

$$= -3a_0, \tag{9.61}$$

since $z = r \cos \theta = \sqrt{4\pi/3}\, r\, Y_{10}(\Omega)$, $\langle \vec{r} | 200 \rangle = R_{20}(r) Y_{00}(\Omega)$, $\langle \vec{r} | 210 \rangle = R_{21}(r) Y_{10}(\Omega)$, and $d\Omega = \sin \theta\, d\theta d\varphi$; $a_0 = \hbar^2/(\mu e^2)$ is the Bohr radius. Using the notations $| 1 \rangle = | 200 \rangle$, $| 2 \rangle = | 211 \rangle$, $| 3 \rangle = | 210 \rangle$, and $| 4 \rangle = | 21 - 1 \rangle$, we can write the matrix of H_p as

$$H_p = \begin{pmatrix} \langle 1 | \hat{H}_p | 1 \rangle & \langle 1 | \hat{H}_p | 2 \rangle & \langle 1 | \hat{H}_p | 3 \rangle & \langle 1 | \hat{H}_p | 4 \rangle \\ \langle 2 | \hat{H}_p | 1 \rangle & \langle 2 | \hat{H}_p | 2 \rangle & \langle 2 | \hat{H}_p | 3 \rangle & \langle 2 | \hat{H}_p | 4 \rangle \\ \langle 3 | \hat{H}_p | 1 \rangle & \langle 3 | \hat{H}_p | 2 \rangle & \langle 3 | \hat{H}_p | 3 \rangle & \langle 3 | \hat{H}_p | 4 \rangle \\ \langle 4 | \hat{H}_p | 1 \rangle & \langle 4 | \hat{H}_p | 2 \rangle & \langle 4 | \hat{H}_p | 3 \rangle & \langle 4 | \hat{H}_p | 4 \rangle \end{pmatrix} \tag{9.62}$$

or as

$$H_p = -3e\mathcal{E}a_0 \begin{pmatrix} 0 & 0 & 1 & 0 \\ 0 & 0 & 0 & 0 \\ 1 & 0 & 0 & 0 \\ 0 & 0 & 0 & 0 \end{pmatrix}. \tag{9.63}$$

The diagonalization of this matrix leads to the following eigenvalues:

$$E^{(1)}_{2\,1} = -3e\mathcal{E}a_0, \qquad E^{(1)}_{2\,2} = E^{(1)}_{2\,3} = 0, \qquad E^{(1)}_{2\,4} = 3e\mathcal{E}a_0. \tag{9.64}$$

Thus, the energy levels of the $n = 2$ states are given to first order by

$$E_{2_1} = -\frac{R_y}{4} - 3e\mathcal{E}a_0, \quad E_{2_2} = E_{2_3} = -\frac{R_y}{4}, \quad E_{2_4} = -\frac{R_y}{4} + 3e\mathcal{E}a_0. \tag{9.65}$$

The corresponding eigenvectors to zeroth order are

$$|\,\psi_2\rangle_1 = \frac{1}{\sqrt{2}}\,(|\,200\rangle + |\,210\rangle), \qquad |\,\psi_2\rangle_2 = |\,211\rangle, \tag{9.66}$$

$$|\,\psi_2\rangle_3 = |\,21-1\rangle, \qquad |\,\psi_2\rangle_4 = \frac{1}{\sqrt{2}}\,(|\,200\rangle - |\,210\rangle). \tag{9.67}$$

This perturbation has only partially removed the degeneracy of the $n = 2$ level; the states $|\,211\rangle$ and $|\,21-1\rangle$ still have the same energy $E_3 = E_4 = -R_y/4$.

9.2.3 Fine Structure and the Anomalous Zeeman Effect

One of the most useful applications of perturbation theory is to calculate the energy corrections for the hydrogen atom, notably the corrections due to the fine structure and the Zeeman effect. The fine structure is in turn due to two effects: spin–orbit coupling and the relativistic correction. Let us look at these corrections separately.

9.2.3.1 Spin–Orbit Coupling

The spin–orbit coupling in hydrogen arises from the interaction between the electron's spin magnetic moment, $\vec{\mu}_S = -e\vec{S}/(m_e c)$, and the proton's orbital magnetic field $\vec{B}$.

The origin of the magnetic field experienced by the electron moving at $\vec{v}$ in a circular orbit around the proton can be explained classically as follows. The electron, within its rest frame, sees the proton moving at $-\vec{v}$ in a circular orbit around it (Figure 9.1). From classical electrodynamics, the magnetic field experienced by the electron is

$$\vec{B} = -\frac{1}{c}\vec{v} \times \vec{E} = -\frac{1}{m_e c}\vec{p} \times \vec{E} = \frac{1}{m_e c}\vec{E} \times \vec{p}, \tag{9.68}$$

where $\vec{p} = m_e \vec{v}$ is the linear momentum of the electron and $\vec{E}$ is the electric field generated by the proton's Coulomb's field: $\vec{E}(\vec{r}) = (e/r^2)(\vec{r}/r) = e\vec{r}/r^3$. For a more general problem of hydrogen-like atoms—atoms with one valence electron outside a closed shell—where an electron moves in the (central) Coulomb potential of a nucleus $V(r) = -e\phi(r)$, the electric field is

$$\vec{E}(\vec{r}) = -\vec{\nabla}\phi(r) = \frac{1}{e}\vec{\nabla}V(r) = \frac{1}{e}\frac{\vec{r}}{r}\frac{dV}{dr}. \tag{9.69}$$

Figure 9.1 (Left) An electron moving in a circular orbit as seen by the nucleus. (Right) The same motion as seen by the electron within its rest frame; the electron sees the nucleus moving in a circular orbit around it.

So the magnetic field of the nucleus calculated in the rest frame of the electron is obtained by inserting (9.69) into (9.68):

$$\vec{B} = \frac{1}{m_e c}\vec{E} \times \vec{p} = \frac{1}{e m_e c}\frac{1}{r}\frac{dV}{dr}\vec{r} \times \vec{p} = \frac{1}{e m_e c}\frac{1}{r}\frac{dV}{dr}\vec{L}, \tag{9.70}$$

where $\vec{L} = \vec{r} \times \vec{p}$ is the orbital angular momentum of the electron.

The interaction of the electron's spin dipole moment $\vec{\mu}_S$ with the orbital magnetic field $\vec{B}$ of the nucleus gives rise to the following interaction energy:

$$\hat{H}_{SO} = -\vec{\mu}_S \cdot \vec{B} = \frac{e}{m_e c}\vec{S} \cdot \vec{B} = \frac{1}{m_e^2 c^2}\frac{1}{r}\frac{dV}{dr}\vec{S} \cdot \vec{L}. \tag{9.71}$$

This energy turns out to be twice the observed spin–orbit interaction. This is due to the fact that (9.71) was calculated within the rest frame of the electron. This frame is not inertial, for the electron accelerates while moving in a circular orbit around the nucleus. For a correct treatment, we must transform to the rest frame of the nucleus (i.e., the lab frame). This transformation, which involves a relativistic transformation of velocities, gives rise to an additional motion resulting from the precession of $\vec{\mu}_S$; this is known as the *Thomas precession*. The precession of the electron's spin moment is a relativistic effect which occurs even in the absence of an external magnetic field. The transformation back to the rest frame of the nucleus leads to a reduction of the interaction energy (9.71) by a factor of 2:

$$\hat{H}_{SO} = \frac{1}{2m_e^2 c^2}\frac{1}{r}\frac{dV}{dr}\vec{S} \cdot \vec{L}. \tag{9.72}$$

As this relation was derived from a classical treatment, we can now obtain the corresponding quantum mechanical expression by replacing the dynamical variables with the corresponding operators:

$$\boxed{\hat{H}_{SO} = \frac{1}{2m_e^2 c^2}\frac{1}{r}\frac{d\hat{V}}{dr}\hat{\vec{S}} \cdot \hat{\vec{L}}.} \tag{9.73}$$

This is the *spin–orbit* energy. For a hydrogen's electron, $V(r) = -e^2/r$ and $dV/dr = e^2/r^2$, equation (9.73) reduces to

$$\hat{H}_{SO} = \frac{e^2}{2m_e^2 c^2}\frac{1}{r^3}\hat{\vec{S}} \cdot \hat{\vec{L}}. \tag{9.74}$$

We can now use perturbation theory to calculate the contribution of the spin–orbit interaction in a hydrogen atom:

$$\hat{H} = \frac{\hat{p}^2}{2m_e} - \frac{e^2}{r} + \frac{e^2}{2m_e^2 c^2 r^3}\hat{\vec{S}}\cdot\hat{\vec{L}} = \hat{H}_0 + \hat{H}_{SO}, \tag{9.75}$$

where $\hat{H}_0$ is the unperturbed Hamiltonian and $\hat{H}_{SO}$ is the perturbation. To apply perturbation theory, we need to specify the unperturbed states—the eigenstates of $\hat{H}_0$. Since the spin of the hydrogen's electron is taken into account, the total wave function of $\hat{H}_0$ consists of a direct product of two parts: a spatial part and a spin part. To specify the eigenstates of $\hat{H}_0$, we have two choices: first, the joint eigenstates $|\, nlm_l m_s\rangle$ of $\hat{L}^2$, $\hat{S}^2$, $\hat{L}_z$, and $\hat{S}_z$ and, second, the joint eigenstates $|\, nljm\rangle$ of $\hat{L}^2$; $\hat{S}^2$, $\hat{J}^2$, and $\hat{J}_z$. While $\hat{H}_0$ is diagonal in both of these representations, $\hat{H}_{SO}$ is diagonal in the second but not in the first, because $\hat{H}_{SO}$ (or $\hat{\vec{S}}\cdot\hat{\vec{L}}$ to be precise) commutes with neither $\hat{L}_z$ nor with $\hat{S}_z$ (Chapter 7). Thus, if $\hat{H}_{SO}$ were included, the first choice would be a bad one, since we would be forced to diagonalize the matrix of $\hat{H}_{SO}$ within the states $|\, nlm_l m_s\rangle$; this exercise is nothing less than tedious and cumbersome. The second choice, however, is ideal for our problem, since the first-order energy correction is given simply by the expectation value of the perturbation, because $\hat{H}_{SO}$ is already diagonal in this representation. We have shown in Chapter 7 that the states $|\, nljm\rangle$,

$$\Psi_{n,l,j=l\pm\frac{1}{2},m} = R_{nl}(r)\left[\sqrt{\frac{l\mp m+\frac{1}{2}}{2l+1}}Y_{l,m+\frac{1}{2}}\left|\frac{1}{2},-\frac{1}{2}\right\rangle \pm \sqrt{\frac{l\pm m+\frac{1}{2}}{2l+1}}Y_{l,m-\frac{1}{2}}\left|\frac{1}{2},\frac{1}{2}\right\rangle\right], \tag{9.76}$$

are eigenstates of $\hat{\vec{S}}\cdot\hat{\vec{L}}$ and that the corresponding eigenvalues are given by

$$\langle nljm\,|\,\hat{\vec{L}}\cdot\hat{\vec{S}}\,|\,nljm\rangle = \frac{\hbar^2}{2}\left[j(j+1) - l(l+1) - \frac{3}{4}\right], \tag{9.77}$$

since $\hat{\vec{S}}\cdot\hat{\vec{L}} = \frac{1}{2}\left[\hat{J}^2 - \hat{L}^2 - \hat{S}^2\right]$.

The eigenvalues of (9.75) are then given to first-order correction by

$$E_{nlj} = E_n^{(0)} + \langle nljm_j\,|\,\hat{H}_{SO}\,|\,nljm_j\rangle = -\frac{e^2}{2a_0}\frac{1}{n^2} + E_{SO}^{(1)}, \tag{9.78}$$

where $E_n^{(0)} = -e^2/(2a_0 n^2) = -(13.6/n^2)$ eV are the energy levels of hydrogen and $E_{SO}^{(1)}$ is the energy due to spin–orbit interaction:

$$E_{SO}^{(1)} = \langle nljm_j\,|\,\hat{H}_{SO}\,|\,nljm_j\rangle = \frac{e^2\hbar^2}{4m_e^2 c^2}\left[j(j+1) - l(l+1) - \frac{3}{4}\right]\left\langle nl\left|\frac{1}{r^3}\right|nl\right\rangle. \tag{9.79}$$

Using the value of $\langle nl\,|r^{-3}|\,nl\rangle$ calculated in Chapter 6,

$$\left\langle nl\left|\frac{1}{r^3}\right|nl\right\rangle = \frac{2}{n^3 l(l+1)(2l+1)a_0^3}, \tag{9.80}$$

we can rewrite (9.79) as

$$E_{SO}^{(1)} = \frac{e^2\hbar^2}{2m_e^2 c^2}\left[\frac{j(j+1) - l(l+1) - \frac{3}{4}}{n^3 l(l+1)(2l+1)a_0^3}\right]$$

$$= \left(\frac{e^2}{2a_0} \frac{1}{n^2} \right) \left(\frac{\hbar}{m_e c a_0} \right)^2 \frac{1}{n} \left[\frac{j(j+1) - l(l+1) - \frac{3}{4}}{l(l+1)(2l+1)} \right] \tag{9.81}$$

or

$$E_{SO}^{(1)} = \frac{|E_n^{(0)}| \alpha^2}{n} \left[\frac{j(j+1) - l(l+1) - \frac{3}{4}}{l(l+1)(2l+1)} \right], \tag{9.82}$$

where α is a dimensionless constant called the *fine structure constant*:

$$\alpha = \frac{\hbar}{m_e c a_0} = \frac{e^2}{\hbar c} \simeq \frac{1}{137}. \tag{9.83}$$

Since $a_0 = \hbar^2/(m_e e^2)$ and hence $E_n^{(0)} = -e^2/(2a_0 n^2) = -\alpha^2 m_e c^2/(2n^2)$, we can express (9.82) in terms of α as

$$E_{SO}^{(1)} = \frac{\alpha^4 m_e c^2}{2n^3} \left[\frac{j(j+1) - l(l+1) - \frac{3}{4}}{l(l+1)(2l+1)} \right]. \tag{9.84}$$

9.2.3.2 Relativistic Correction

Although the relativistic effect in hydrogen due to the motion of the electron is small, it can still be detected by spectroscopic techniques. The relativistic kinetic energy of the electron is given by $\hat{T} = \sqrt{\hat{p}^2 c^2 + m_e^2 c^4} - m_e c^2$, where $m_e c^2$ is the rest mass energy of the electron; an expansion of this relation to $\hat{p}^4$ yields

$$\sqrt{\hat{p}^2 c^2 + m_e^2 c^4} - m_e c^2 \simeq \frac{\hat{p}^2}{2m_e} - \frac{\hat{p}^4}{8m_e^3 c^2} + \cdots. \tag{9.85}$$

When this term is included, the hydrogen's Hamiltonian becomes

$$\hat{H} = \frac{\hat{p}^2}{2m_e} - \frac{e^2}{r} - \frac{\hat{p}^4}{8m_e^3 c^2} = \hat{H}_0 + \hat{H}_R, \tag{9.86}$$

where $\hat{H}_0 = \hat{p}^2/(2m_e) - e^2/r$ is the unperturbed Hamiltonian and $\hat{H}_R = -\hat{p}^4/(8m_e^3 c^2)$ is the relativistic mass correction which can be treated by first-order perturbation theory:

$$E_R^{(1)} = \langle nljm_j \mid \hat{H}_R \mid nljm_j \rangle = -\frac{1}{8m_e^3 c^2} \langle nljm_j \left| \hat{p}^4 \right| nljm_j \rangle. \tag{9.87}$$

The value of $\langle nljm_j \left| \hat{p}^4 \right| nljm_j \rangle$ was calculated in the last solved problem of Chapter 6 (see equation (6.331)):

$$\langle nljm_j \left| \hat{p}^4 \right| nljm_j \rangle = \frac{m_e^4 e^8}{\hbar^4 n^4} \left(\frac{8n}{2l+1} - 3 \right) = \frac{\alpha^4 m_e^4 c^4}{n^4} \left(\frac{8n}{2l+1} - 3 \right). \tag{9.88}$$

An insertion of this value in (9.87) leads to

$$E_R^{(1)} = -\frac{\alpha^4 m_e c^2}{8n^4} \left(\frac{8n}{2l+1} - 3 \right) = -\frac{\alpha^2 |E_n^{(0)}|}{4n^2} \left(\frac{8n}{2l+1} - 3 \right). \tag{9.89}$$

Note that the spin–orbit and relativistic corrections (9.84) and (9.89) have the same order of magnitude, 10^{-3} eV, since $\alpha^2|E_n^{(0)}| \simeq 10^{-3}$ eV.

Remark

For a hydrogenlike atom having Z electrons, and if we neglect the spin–orbit interaction, we may use (9.89) to infer the atom's ground state energy:

$$E_n = Z^2\left(E_n^{(0)} + E_R^{(1)}\right) = Z^2 E_n^{(0)}\left[1 + \frac{\alpha^2}{n}\left(\frac{2}{2l+1} - \frac{3}{4n}\right)\right], \qquad (9.90)$$

where $E_n^{(0)} = -e^4 m_e/(2\hbar^2 n^2) = -\alpha^2 m_e c^2/(2n^2) = -13.6$ eV$/n^2$ is the Bohr energy.

9.2.3.3 The Fine Structure of Hydrogen

The fine structure correction is obtained by adding the expressions for the spin–orbit and relativistic corrections (9.84) and (9.89):

$$E_{FS}^{(1)} = E_{SO}^{(1)} + E_R^{(1)} = \frac{\alpha^4 m_e c^2}{2n^3}\left[\frac{j(j+1) - l(l+1) - \frac{3}{4}}{l(l+1)(2l+1)}\right] - \frac{\alpha^4 m_e c^2}{8n^4}\left[\frac{8n}{2l+1} - 3\right], \qquad (9.91)$$

where $j = l \pm \frac{1}{2}$. If $j = l + \frac{1}{2}$ a substitution of $l = j - \frac{1}{2}$ into (9.91) leads to

$$
\begin{aligned}
E_{FS}^{(1)} &= \frac{\alpha^4 m_e c^2}{8n^4}\left[\frac{4nj(j+1) - 4n\left(j-\frac{1}{2}\right)\left(j+\frac{1}{2}\right) - 3n}{\left(j-\frac{1}{2}\right)\left(j+\frac{1}{2}\right)(2j-1+1)} - \frac{8n}{2j-1+1} + 3\right] \\
&= \frac{\alpha^4 m_e c^2}{8n^4}\left[\frac{4nj - 2n}{2j\left(j-\frac{1}{2}\right)\left(j+\frac{1}{2}\right)} - \frac{4n}{j} + 3\right] = \frac{\alpha^4 m_e c^2}{8n^4}\left[\frac{2n}{j\left(j+\frac{1}{2}\right)} - \frac{4n}{j} + 3\right] \\
&= \frac{\alpha^4 m_e c^2}{8n^4}\left[3 - \frac{4n}{j+\frac{1}{2}}\right].
\end{aligned}
\qquad (9.92)
$$

Similarly, if $j = l - \frac{1}{2}$, and hence $l = j + \frac{1}{2}$, we can reduce (9.91) to

$$
\begin{aligned}
E_{FS}^{(1)} &= \frac{\alpha^4 m_e c^2}{8n^4}\left[\frac{4nj(j+1) - 4n\left(j+\frac{1}{2}\right)\left(j+\frac{3}{2}\right) - 3n}{\left(j+\frac{1}{2}\right)\left(j+\frac{3}{2}\right)(2j+1+1)} - \frac{8n}{2j+1+1} + 3\right] \\
&= \frac{\alpha^4 m_e c^2}{8n^4}\left[\frac{-4nj - 6n}{2\left(j+\frac{1}{2}\right)\left(j+\frac{3}{2}\right)(j+1)} - \frac{4n}{j+1} + 3\right] \\
&= \frac{\alpha^4 m_e c^2}{8n^4}\left[\frac{-2n}{\left(j+\frac{1}{2}\right)(j+1)} - \frac{4n}{j+1} + 3\right] \\
&= \frac{\alpha^4 m_e c^2}{8n^4}\left[3 - \frac{4n}{j+\frac{1}{2}}\right].
\end{aligned}
\qquad (9.93)
$$

As equations (9.92) and (9.93) show, the expressions for the fine structure correction corresponding to $j = l + \frac{1}{2}$ and $j = l - \frac{1}{2}$ are the same:

$$E_{FS}^{(1)} = E_{SO}^{(1)} + E_R^{(1)} = \frac{\alpha^4 m_e c^2}{8n^4}\left(3 - \frac{4n}{j + \frac{1}{2}}\right) = \frac{\alpha^2 E_n^{(0)}}{4n^2}\left(\frac{4n}{j + \frac{1}{2}} - 3\right),\qquad(9.94)$$

where $E_n^{(0)} = -\alpha^2 m_e c^2 / (2n^2)$ and $j = l \pm \frac{1}{2}$.

Since the bracket-terms in (9.82), (9.89), and (9.94) are of the order of unity, the ratios of the spin–orbit, relativistic, and fine structure corrections to the energy of the hydrogen atom are of the order of α^2:

$$\frac{E_{SO}^{(1)}}{|E_n^{(0)}|} \simeq \alpha^2, \qquad \left|\frac{E_R^{(1)}}{E_n^{(0)}}\right| \simeq \alpha^2, \qquad \frac{E_{FS}^{(1)}}{|E_n^{(0)}|} \simeq \alpha^2.\qquad(9.95)$$

All these terms are of the order of 10^{-4} since $\alpha^2 = (1/137)^2 \simeq 10^{-4}$.

In sum, the hydrogen's Hamiltonian, when including the fine structure, is given by

$$\hat{H} = \hat{H}_0 + \hat{H}_{FS} = \hat{H}_0 + \left(\hat{H}_{SO} + \hat{H}_R\right) = \frac{\hat{p}^2}{2m_e} - \frac{e^2}{r} + \left(\frac{e^2}{2m_e^2 c^2 r^3}\hat{\vec{S}} \cdot \hat{\vec{L}} - \frac{\hat{p}^4}{8m_e^3 c^2}\right).\qquad(9.96)$$

A first-order perturbation calculation of the energy levels of hydrogen, when including the fine structure, yields

$$E_{nj} = E_n^{(0)} + E_{FS}^{(1)} = E_n^{(0)}\left[1 + \frac{\alpha^2}{4n^2}\left(\frac{4n}{j + \frac{1}{2}} - 3\right)\right],\qquad(9.97)$$

where $E_n^{(0)} = -13.6\,\text{eV}/n^2$. Unlike $E_n^{(0)}$, which is degenerate in l, each energy level E_{nj} is split into two levels $E_{n\,(l\pm\frac{1}{2})}$, since for a given value of l there are two values of j: $j = l \pm \frac{1}{2}$.

In addition to the fine structure, there is still another (smaller) effect which is known as the *hyperfine structure*. The hydrogen's hyperfine structure results from the interaction of the spin of the electron with the spin of the nucleus. When the hyperfine corrections are included, they would split each of the fine structure levels into a series of hyperfine levels. For instance, when the hyperfine coupling is taken into account in the ground state of hydrogen, it would split the $1S_{1/2}$ level into two hyperfine levels separated by an energy of $5.89 \times 10^{-6}\,\text{eV}$. This corresponds, when the atom makes a spontaneous transition from the higher hyperfine level to the lower one, to a radiation of $1.42 \times 10^9\,\text{Hz}$ frequency and 21 cm wavelength. We should note that most of the information we possess about interstellar hydrogen clouds had its origin in the radioastronomy study of this 21 cm line.

9.2.3.4   The Anomalous Zeeman Effect

We now consider a hydrogen atom that is placed in an external uniform magnetic field $\vec{B}$. The effect of an external magnetic field on the atom is to cause a shift of its energy levels; this is called the *Zeeman effect*. In Chapter 6 we studied the Zeeman effect, but with one major omission: we ignored the spin of the electron. In this section we are going to take it into account. The interaction of the magnetic field with the electron's orbital and spin magnetic dipole moments, $\vec{\mu}_L$ and $\vec{\mu}_S$, gives rise to two energy terms, $-\vec{\mu}_L \cdot \vec{B}$ and $-\vec{\mu}_S \cdot \vec{B}$, whose sum we call the Zeeman energy:

$$\hat{H}_Z = -\vec{\mu}_L \cdot \vec{B} - \vec{\mu}_S \cdot \vec{B} = \frac{e}{2m_ec}\vec{L} \cdot \vec{B} + \frac{e}{m_ec}\vec{S} \cdot \vec{B} = \frac{e}{2m_ec}\left(\vec{L} + 2\vec{S}\right) \cdot \vec{B} = \frac{eB}{2m_ec}\left(\hat{L}_z + 2\hat{S}_z\right),$$
(9.98)

with $\vec{\mu}_L = -e\vec{L}/(2m_ec)$ and $\vec{\mu}_S = -\vec{S}/(m_ec)$; for simplicity, we have taken $\vec{B}$ along the z-axis: $\vec{B} = B\hat{z}$.

When a hydrogen atom is placed in an external magnetic field, its Hamiltonian is given by

$$\hat{H} = \hat{H}_0 + \hat{H}_{FS} + \hat{H}_Z.$$
(9.99)

Like $\hat{H}_{FS}$, the correction due to $\hat{H}_Z$ of (9.99) is expected to be small compared to $\hat{H}_0$; hence it can be treated perturbatively. We may now consider separately the cases where the magnetic field $\vec{B}$ is strong or weak. Strong or weak compared to what? Since $\hat{H}_{SO}$ and $\hat{H}_Z$ can be written as $\hat{H}_{SO} = W\vec{L} \cdot \vec{S}$ (9.74) and since $\hat{H}_Z = B\mu_B(\hat{L}_z + 2\hat{S}_z)/\hbar$, we have $\hat{H}_Z/\hat{H}_{SO} \sim B\mu_B/W$, where μ_B is the Bohr magneton, $\mu_B = e\hbar/(2m_ec)$. Thus, the cases $B \ll W/\mu_B$ and $B \gg W/\mu_B$ would correspond to the weak and strong magnetic fields, respectively.

The strong-field Zeeman effect
The effect of a strong external magnetic field on the hydrogen atom is called the *Paschen–Back effect*. If $\vec{B}$ is strong, $B \gg W/\mu_B$, the term $eB(\hat{L}_z + 2\hat{S}_z)/(2m_ec)$ will be much greater than the fine structure. Neglecting $\hat{H}_{FS}$, we can reduce (9.99) to

$$\hat{H} = \hat{H}_0 + \hat{H}_Z = \hat{H}_0 + \frac{eB}{2m_ec}\left(\hat{L}_z + 2\hat{S}_z\right).$$
(9.100)

Since $\hat{H}$ commutes with $\hat{H}_0$ (because $\hat{H}_0$ commutes with $\hat{L}_z$ and $\hat{S}_z$), they can be diagonalized by a common set of states, $|nlm_lm_s\rangle$:

$$\hat{H} |nlm_lm_s\rangle = \left[\hat{H}_0 + \frac{eB}{2m_ec}\left(\hat{L}_z + 2\hat{S}_z\right)\right] |nlm_lm_s\rangle = E_{nlm_lm_s} |nlm_lm_s\rangle,$$
(9.101)

where

$$E_{nlm_lm_s} = E_n^{(0)} + \frac{eB\hbar}{2m_ec}(m_l + 2m_s) = -\frac{e^2}{2a_0n^2} + \frac{eB\hbar}{2m_ec}(m_l + 2m_s).$$
(9.102)

The energy levels $E_n^{(0)}$ are thus shifted by an amount equal to $\Delta E = B\mu_B(m_l + 2m_s)$ with $\mu_B = e\hbar/(2m_ec)$, known as the Paschen–Back shift (Figure 9.2). When $B = 0$ the degeneracy of each level of hydrogen is given by $g_n = 2\sum_{l=0}^{n-1}(2l + 1) = 2n^2$; when $B \neq 0$ states with the same value of $(m_l + 2m_s)$ are still degenerate.

The weak-field Zeeman effect
If $\vec{B}$ is weak, $B \ll W/\mu_B$, we need to consider all the terms in the Hamiltonian (9.99); the fine structure term $\hat{H}_{FS}$ will be the dominant perturbation. In the case where the Hamiltonian contains several perturbations at once, we should treat them individually starting with the most dominant, then the next, and so on. In this case the eigenstate should be selected to be one that diagonalizes the unperturbed Hamiltonian and the dominant perturbation[1]. In the weak-field

---

[1] When the various perturbations are of approximately equal size, a state that is a joint eigenstate of $\hat{H}_0$ and any perturbation would be an acceptable choice.

Figure 9.2 Splittings of the energy levels $n = 1$ and $n = 2$ of a hydrogen atom when placed in a *strong* external magnetic field; $\mu_B = e\hbar/(2m_e c)$.

Zeeman effect, since $\hat{H}_{FS}$ is the dominant perturbation, the best eigenstates to use are $| nljm_j \rangle$, for they simultaneously diagonalize $\hat{H}_0$ and $\hat{H}_{FS}$. Writing $\hat{L}_z + 2\hat{S}_z$ as $\hat{J}_z + \hat{S}_z$, where $\vec{\hat{J}} = \hat{\vec{L}} + \hat{\vec{S}}$ represents the total angular momentum of the electron, we may rewrite (9.99) as

$$\hat{H} = \hat{H}_0 + \hat{H}_{FS} + \hat{H}_Z = \hat{H}_0 + \hat{H}_{FS} + \frac{eB}{2m_e c}\left(\hat{J}_z + \hat{S}_z\right). \tag{9.103}$$

In a first-order perturbation calculation, the contribution of $\hat{H}_Z$ is given by

$$E_Z^{(1)} = \langle nljm_j | \hat{H}_Z | nljm_j \rangle = \frac{eB}{2m_e c}\langle nljm_j | \hat{J}_z + \hat{S}_z | nljm_j \rangle. \tag{9.104}$$

Since $\langle nljm_j | \hat{J}_z | nljm_j \rangle = \hbar m_j$ and using the expression of $\langle nljm_j | \hat{S}_z | nljm_j \rangle$ that was calculated in Chapter 7,

$$\langle nljm_j | \hat{S}_z | nljm_j \rangle = \frac{\langle nljm_j | \hat{\vec{J}} \cdot \hat{\vec{S}} | nljm_j \rangle}{\hbar^2 j(j+1)}\langle nljm_j | \hat{J}_z | nljm_j \rangle$$

$$= \frac{j(j+1) - l(l+1) + s(s+1)}{2j(j+1)}\hbar m_j, \tag{9.105}$$

we can reduce (9.104) to

$$E_Z^{(1)} = \frac{eB\hbar}{2m_e c}\left[1 + \frac{j(j+1) - l(l+1) + s(s+1)}{2j(j+1)}\right]m_j = \frac{eB\hbar}{2m_e c}g_j m_j = B\mu_B m_j g_j, \tag{9.106}$$

where $\mu_B = e\hbar/(2m_e c)$ is the Bohr magneton for the electron and g_j is the Landé factor or the gyromagnetic ratio:

$$g_j = 1 + \frac{j(j+1) - l(l+1) + s(s+1)}{2j(j+1)}. \tag{9.107}$$

This shows that when $l = 0$ and $j = s$ we have $g_s = 2$ and when $s = 0$ and $j = l$ we have $g_l = 1$. For instance, for an atomic state[2] such as $^2P_{3/2}$, (9.107) shows that its factor is given by $g_{j=3/2} = \frac{4}{3}$, since $j = l + s = 1 + \frac{1}{2} = \frac{3}{2}$; this is how we infer the factor of any state:

$$
\begin{array}{cccccccc}
\text{State} & ^2S_{1/2} & ^2P_{1/2} & ^2P_{3/2} & ^2D_{3/2} & ^2D_{5/2} & ^2F_{5/2} & ^2F_{7/2} \\
g_j & 2 & \frac{2}{3} & \frac{4}{3} & \frac{4}{5} & \frac{6}{5} & \frac{6}{7} & \frac{8}{7}
\end{array}
\tag{9.108}
$$

From (9.107), we see that the Landé factors corresponding to the same l but different values of j (due to spin) are not equal, since for $s = \frac{1}{2}$ and $j = l \pm \frac{1}{2}$ we have

$$
g_{j=l\pm\frac{1}{2}} = 1 \pm \frac{1}{2l+1} = \begin{cases} \frac{2l+2}{2l+1} & \text{for } j = l + \frac{1}{2}, \\ \frac{2l}{2l+1} & \text{for } j = l - \frac{1}{2}. \end{cases}
\tag{9.109}
$$

Combining (9.97), (9.103), and (9.106), we can write the energy of a hydrogen atom in a weak external magnetic field as follows:

$$
\boxed{E_{nj} = E_n^{(0)} + E_{FS}^{(1)} + E_Z^{(1)} = E_n^{(0)} + \frac{\alpha^2 E_n^{(0)}}{4n^2}\left(\frac{4n}{j+\frac{1}{2}} - 3\right) + \frac{eB\hbar}{2m_e c}m_j g_j.}
\tag{9.110}
$$

The effect of the magnetic field on the atom is thus to split the energy levels with a spacing $\Delta E = B\mu_B m_j g_j$. Unlike the energy levels obtained in Chapter 6, where we ignored the electron's spin, the energy levels (9.110) are not degenerate in l. Each energy level j is split into an *even number* of $(2j + 1)$ sublevels corresponding to the $(2j + 1)$ values of m_j: $m_j = -j, -j + 1, \ldots, j - 1, j$. As displayed in Figure 9.3, the splittings between the sublevels corresponding to the same j are constant: the spacings between the sublevels corresponding to $j = l - 1/2$ are all equal to $\Delta\epsilon_1 = B\mu_B(2l)/(2l+1)$, and the spacings between the $j = l + 1/2$ sublevels are equal to $\Delta\epsilon_2 = B\mu_B(2l + 2)/(2l + 1)$. In contrast to the normal Zeeman effect, however, the spacings between the split levels of the same l (and different values of j) are no longer constant, $\Delta\epsilon_1 \neq \Delta\epsilon_2$, since they depend on the Landé factor g_j; for a given value of j, there are two different values of g_j corresponding to $l = j \pm \frac{1}{2}$: $g_{j=l+1/2} = (2l + 2)/(2l + 1)$ and $g_{j=l-1/2} = (2l)/(2l + 1)$; see (9.109). This *unequal* spacing between the split levels is called the *anomalous* Zeeman effect.

9.3 The Variational Method

There exist systems whose Hamiltonians are known, but they cannot be solved exactly or by a perturbative treatment. That is, there is no closely related Hamiltonian that can be solved exactly or approximately by perturbation theory because the first order is not sufficiently accurate. One of the approximation methods that is suitable for solving such problems is the *variational method*, which is also called the Rayleigh–Ritz method. This method does not require knowledge of simpler Hamiltonians that can be solved exactly. The variational method is useful for determining upper bound values for the eigenenergies of a system whose Hamiltonian is known

---

[2] We use here the spectroscopic notation where $^{(2s+1)}L_j$ designates an atomic state whose spin is s, its total angular momentum is j, and whose orbital angular momentum is L where the values $L = 0, 1, 2, 3, 4, 5, \ldots$ are designated, respectively, by the capital letters S, P, D, F, G, H, $\ldots$ (see Chapter 8).

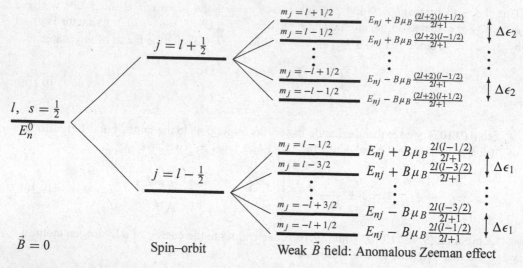

Figure 9.3 Splittings of a level l due to the spin–orbit interaction and to a weak external magnetic field, with $E_{nj} = E_n^0 + E_{SO}^{(1)}$. All the lower sublevels are equally spaced, $\Delta\epsilon_1 = B\mu_B(2l)/(2l + 1)$, and so are the upper sublevels, $\Delta\epsilon_2 = B\mu_B(2l + 2)/(2l + 1)$, with $\mu_B = e\hbar/(2m_e c)$.

whereas its eigenvalues and eigenstates are not known. It is particularly useful for determining the ground state. It becomes quite cumbersome to determine the energy levels of the excited states.

In the context of the variational method, one does not attempt to solve the eigenvalue problem

$$\hat{H} \mid \psi\rangle = E \mid \psi\rangle, \tag{9.111}$$

but rather one uses a variational scheme to find the approximate eigenenergies and eigenfunctions from the variational equation

$$\delta E(\psi) = 0, \tag{9.112}$$

where $E(\psi)$ is the expectation value of the energy in the state $\mid \psi\rangle$:

$$E(\psi) = \frac{\langle \psi \mid \hat{H} \mid \psi\rangle}{\langle \psi \mid \psi\rangle}. \tag{9.113}$$

If $\mid \psi\rangle$ depends on a parameter α, $E(\psi)$ will also depend on α. The variational ansatz (9.112) enables us to vary α so as to minimize $E(\psi)$. The minimum value of $E(\psi)$ provides an upper limit approximation for the true energy of the system.

The variational method is particularly useful for determining the ground state energy and its eigenstate without explicitly solving the Schrödinger equation. Note that for any (arbitrary) trial function $\mid \psi\rangle$ we choose, the energy E as given by (9.113) is always larger than the exact energy E_0:

$$E = \frac{\langle \psi \mid H \mid \psi\rangle}{\langle \psi \mid \psi\rangle} \geq E_0; \tag{9.114}$$

the equality condition occurs only when $|\psi\rangle$ is proportional to the true ground state $|\psi_0\rangle$. To prove this, we simply expand the trial function $|\psi\rangle$ in terms of the exact eigenstates of $\hat{H}$:

$$|\psi\rangle = \sum_n a_n |\phi_n\rangle, \tag{9.115}$$

with

$$H|\phi_n\rangle = E_n|\phi_n\rangle, \tag{9.116}$$

and since $E_0 \geq E_n$ for nondegenerate one-dimensional bound systems, we have

$$E = \frac{\langle \psi|H|\psi\rangle}{\langle \psi|\psi\rangle} = \frac{\sum_n |a_n|^2 E_n}{\sum_n |a_n|^2} \geq \frac{E_0 \sum_n |a_n|^2}{\sum_n |a_n|^2} = E_0, \tag{9.117}$$

which proves (9.114).

To calculate the ground state energy, we need to carry out the following four steps:

- First, based on physical intuition, make an educated guess of a trial function that takes into account all the physical properties of the ground state (symmetries, number of nodes, smoothness, behavior at infinity, etc.). For the properties you are not sure about, include in the trial function *adjustable parameters* $\alpha_1, \alpha_2, \ldots$ (i.e., $|\psi_0\rangle = |\psi_0(\alpha_1, \alpha_2, \ldots)\rangle$) which will account for the various possibilities of these unknown properties.

- Second, using (9.113), calculate the energy; this yields an expression which depends on the parameters $\alpha_1, \alpha_2, \ldots$:

$$E_0(\alpha_1, \alpha_2, \ldots) = \frac{\langle \psi_0(\alpha_1, \alpha_2, \ldots)|\hat{H}|\psi_0(\alpha_1, \alpha_2, \ldots)\rangle}{\langle \psi_0(\alpha_1, \alpha_2, \ldots)|\psi_0(\alpha_1, \alpha_2, \ldots)\rangle}. \tag{9.118}$$

In most cases $|\psi_0(\alpha_1, \alpha_2, \ldots)\rangle$ will be assumed to be normalized; hence the denominator of this expression is equal to 1.

- Third, using (9.118) search for the minimum of $E_0(\alpha_1, \alpha_2, \ldots)$ by varying the adjustable parameters α_i until E_0 is minimized. That is, *minimize* $E(\alpha_1, \alpha_2, \ldots)$ with respect to $\alpha_1, \alpha_2, \ldots$:

$$\frac{\partial E_0(\alpha_1, \alpha_2, \ldots)}{\partial \alpha_i} = \frac{\partial}{\partial \alpha_i} \frac{\langle \psi_0(\alpha_1, \alpha_2, \ldots)|\hat{H}|\psi_0(\alpha_1, \alpha_2, \ldots)\rangle}{\langle \psi_0(\alpha_1, \alpha_2, \ldots)|\psi_0(\alpha_1, \alpha_2, \ldots)\rangle} = 0, \tag{9.119}$$

with $i = 1, 2, \ldots$. This gives the values of $(\alpha_{1_0}, \alpha_{2_0}, \ldots)$ that minimize E_0.

- Fourth, substitute these values of $(\alpha_{1_0}, \alpha_{2_0}, \ldots)$ into (9.118) to obtain the approximate value of the energy. The value $E_0(\alpha_{1_0}, \alpha_{2_0}, \ldots)$ thus obtained provides an upper bound for the exact ground state energy E_0. The exact ground state eigenstate $|\phi_0\rangle$ will then be approximated by the state $|\psi_0(\alpha_{1_0}, \alpha_{2_0}, \ldots)\rangle$.

What about the energies of the excited states? The variational method can also be used to find the approximate values for the energies of the first few excited states. For instance, to find the energy and eigenstate of the first excited state that will approximate E_1 and $|\phi_1\rangle$, we need to choose a trial function $|\psi_1\rangle$ that must be orthogonal to $|\psi_0\rangle$:

$$\langle \psi_1|\phi_0\rangle = 0. \tag{9.120}$$

Then proceed as we did in the case of the ground state. That is, solve the variational equation (9.112) for $|\psi_1\rangle$:

$$\frac{\partial}{\partial\alpha_i}\frac{\langle\psi_1(\alpha_1,\alpha_2,\ldots)|\hat{H}|\psi_1(\alpha_1,\alpha_2,\ldots)\rangle}{\langle\psi_1(\alpha_1,\alpha_2,\ldots)|\psi_1(\alpha_1,\alpha_2,\ldots)\rangle}=0 \qquad (i=1,2,\ldots). \tag{9.121}$$

Similarly, to evaluate the second excited state, we solve (9.112) for $|\psi_2\rangle$ and take into account the following two conditions:

$$\langle\psi_2\mid\psi_0\rangle=0, \qquad \langle\psi_2\mid\psi_1\rangle=0. \tag{9.122}$$

These conditions can be included in the variational problem by means of *Lagrange multipliers*, that is, by means of a constrained variational principle.

In this way, we can in principle evaluate any other excited state. However, the variational procedure becomes increasingly complicated as we deal with higher excited states. As a result, the method is mainly used to determine the ground state.

Remark

In those problems where the first derivative of the wave function is discontinuous at a given value of x, one has to be careful when using the expression

$$-\left\langle\psi\left|\frac{\hbar^2}{2m}\frac{d^2}{dx^2}\right|\psi\right\rangle=-\frac{\hbar^2}{2m}\int_{-\infty}^{+\infty}\psi^*(x)\frac{d^2\psi(x)}{dx^2}\,dx. \tag{9.123}$$

A straightforward, careless use of this expression sometimes leads to a negative kinetic energy term (Problem 9.6 on page 541). One might instead consider using the following form:

$$\boxed{-\left\langle\psi\left|\frac{\hbar^2}{2m}\frac{d^2}{dx^2}\right|\psi\right\rangle=\frac{\hbar^2}{2m}\int_{-\infty}^{+\infty}\left|\frac{d\psi(x)}{dx}\right|^2\,dx.} \tag{9.124}$$

Note that (9.123) and (9.124) are identical; an integration by parts leads to

$$\int_{-\infty}^{+\infty}\left|\frac{d\psi(x)}{dx}\right|^2\,dx=\psi^*(x)\frac{d\psi(x)}{dx}\bigg|_{-\infty}^{+\infty}-\int_{-\infty}^{+\infty}\psi^*(x)\frac{d^2\psi(x)}{dx^2}\,dx=-\int_{-\infty}^{+\infty}\psi^*(x)\frac{d^2\psi(x)}{dx^2}\,dx, \tag{9.125}$$

since $\psi^*(x)d\psi(x)/dx$ goes to zero as $x\longrightarrow\pm\infty$ (this is the case whenever $\psi(x)$ is a bound state, but not so when $\psi(x)$ is a plane wave).

What about the calculation of $\langle\psi\,|-(\hbar^2/(2m))\Delta|\,\psi\rangle$ in three dimensions? We might consider generalizing (9.124). For this, we need simply to invoke Gauss's theorem[3] to show that

$$\int\left(\vec{\nabla}\psi^*(\vec{r})\right)\cdot\left(\vec{\nabla}\psi(\vec{r})\right)\,d^3r=-\int\psi^*(\vec{r})\Delta\psi(\vec{r})\,d^3r. \tag{9.126}$$

To see this, an integration by parts leads to the following relation:

$$\int_S\psi^*(\vec{r})\vec{\nabla}\psi(\vec{r})\cdot d\vec{A}=\int_V\left[\left(\vec{\nabla}\psi^*(\vec{r})\right)\cdot\left(\vec{\nabla}\psi(\vec{r})\right)+\psi^*(\vec{r})\Delta\psi(\vec{r})\right]\,d^3r, \tag{9.127}$$

---

[3] Gauss's theorem states that the surface integral of a vector $\vec{B}$ over a closed surface S is equal to the volume integral of the divergence of that vector integrated over the volume V enclosed by the surface S: $\int_S\vec{B}\cdot d\vec{S}=\int_V\vec{\nabla}\cdot\vec{B}\,dV$.

and since, as $S \longrightarrow \infty$, the surface integral $\int_S \psi^*(\vec{r})\vec{\nabla}\psi(\vec{r}) \cdot d\vec{S}$ vanishes if $\psi(\vec{r})$ is a bound state, we recover (9.126). So the kinetic energy term (9.124) is given in three dimensions by

$$-\left\langle \psi \left| \frac{\hbar^2}{2m}\Delta \right| \psi \right\rangle = \frac{\hbar^2}{2m}\int \left(\vec{\nabla}\psi^*(\vec{r})\right) \cdot \left(\vec{\nabla}\psi(\vec{r})\right) d^3r. \qquad (9.128)$$

Example 9.4

Show that (9.112) is equivalent to the Schrödinger equation (9.111).

Solution

Using (9.113), we can rewrite (9.112) as

$$\delta\left(\langle \psi \mid \hat{H} - E \mid \psi \rangle\right) = 0. \qquad (9.129)$$

Since $\mid \psi \rangle$ is a complex function, we can view $\mid \psi \rangle$ and $\langle \psi \mid$ as two independent functions; hence we can carry out the variations over $|\delta\psi\rangle$ and $\langle\delta\psi|$ independently. Varying first over $\langle\delta\psi|$, equation (9.129) yields

$$\langle\delta\psi|\hat{H} - E \mid \psi\rangle = 0. \qquad (9.130)$$

Since $\mid \psi \rangle$ is arbitrary, then (9.130) is equivalent to $\hat{H} \mid \psi \rangle = E \mid \psi \rangle$. The variation over $|\delta\psi\rangle$ leads to the same result. Namely, varying (9.129) over $|\delta\psi\rangle$, we get

$$\langle\psi|\hat{H} - E|\delta\psi\rangle = 0, \qquad (9.131)$$

from which we obtain the complex conjugate equation $\langle \psi \mid \hat{H} = E\langle \psi \mid$, since $\hat{H}$ is Hermitian.

Example 9.5

Consider a one-dimensional harmonic oscillator. Use the variational method to estimate the energies of (a) the ground state, (b) the first excited state, and (c) the second excited state.

Solution

This simple problem enables us to illustrate the various aspects of the variational method within a predictable setting, because the exact solutions are known: $E_0 = \hbar\omega/2$, $E_1 = 3\hbar\omega/2$, $E_2 = 5\hbar\omega/2$.

(a) The trial function we choose for the ground state has to be even and smooth everywhere, it must vanish as $x \to \pm\infty$, and it must have no nodes. A Gaussian function satisfies these requirements. But what we are not sure about is its width. To account for this, we include in the trial function an adjustable scale parameter α:

$$\psi_0(x, \alpha) = Ae^{-\alpha x^2}; \qquad (9.132)$$

A is a normalization constant. Using $\int_{-\infty}^{+\infty} x^{2n}e^{-\alpha x^2}dx = \sqrt{\pi/a}\, 1 \cdot 3 \cdot 5 \cdots (2n-1)/(2a)^n$, we can show that A is given by $A = (2\alpha/\pi)^{1/4}$. The expression for $E_0(\alpha)$ is thus given by

$$\langle\psi_0|H|\psi_0\rangle = A^2 \int_{-\infty}^{+\infty} e^{-\alpha x^2}\left(\frac{-\hbar^2}{2m}\frac{d^2}{dx^2} + \frac{1}{2}m\omega^2 x^2\right)e^{-\alpha x^2}dx$$

Figure 9.4 Shape of $E_0(\alpha) = \hbar^2\alpha/(2m) + m\omega^2/(8\alpha)$.

$$= A^2\frac{\hbar^2\alpha}{m}\int_{-\infty}^{+\infty}e^{-2\alpha x^2}dx + A^2\left(\frac{1}{2}m\omega^2 - \frac{2\hbar^2\alpha^2}{m}\right)\int_{-\infty}^{+\infty}x^2e^{-2\alpha x^2}dx$$

$$= \frac{\hbar^2\alpha}{m} + \frac{1}{4\alpha}\left(\frac{1}{2}m\omega^2 - \frac{2\hbar^2\alpha^2}{m}\right) = \frac{\hbar^2}{2m}\alpha + \frac{m\omega^2}{8\alpha} \tag{9.133}$$

or

$$E_0(\alpha) = \frac{\hbar^2}{2m}\alpha + \frac{m\omega^2}{8\alpha}. \tag{9.134}$$

Its shape is displayed in Figure 9.4. The value of α_0, corresponding to the lowest point of the curve, can be obtained from the minimization of $E(\alpha)$ with respect to α,

$$\frac{\partial E_0(\alpha)}{\partial\alpha} = \frac{\hbar^2}{2m} - \frac{m\omega^2}{8\alpha^2} = 0, \tag{9.135}$$

yields $\alpha_0 = m\omega/(2\hbar)$ which, when inserted into (9.134) and (9.132), leads to

$$E_0(\alpha_0) = \frac{\hbar\omega}{2} \quad \text{and} \quad \psi_0(x,\alpha_0) = \left(\frac{m\omega}{\pi\hbar}\right)^{1/4}e^{-m\omega x^2/2\hbar}. \tag{9.136}$$

The ground state energy and wave function obtained by the variational method are identical to their exact counterparts.

(b) Let us now find the approximate energy E_1 for the first excited state. The trial function $\psi_1(x,\alpha)$ we need to select must be odd, it must vanish as $x \to \pm\infty$, it must have only one node, and it must be orthogonal to $\psi_0(x,\alpha_0)$ of (9.136). A candidate that satisfies these requirements is

$$\psi_1(x,\alpha) = Bxe^{-\alpha x^2}; \tag{9.137}$$

B is the normalization constant. We can show that $B = (32\alpha^3/\pi)^{1/4}$. Note that $\langle\psi_0\mid\psi_1\rangle$ is zero,

$$\langle\psi_0\mid\psi_1\rangle = B\left(\frac{m\omega}{\pi\hbar}\right)^{1/4}\int_{-\infty}^{+\infty}xe^{-\alpha x^2}e^{-m\omega x^2/2\hbar}dx = 0, \tag{9.138}$$

since the symmetric integration of an odd function is zero; $\psi_0(x)$ is even and $\psi_1(x)$ is odd.

Proceeding as we did for $E_0(\alpha)$, and since $\psi_1(x,\alpha)$ is normalized, we can show that

$$E_1(\alpha) = \langle\psi_1(\alpha)|H|\psi_1(\alpha)\rangle = B^2\int_{-\infty}^{+\infty}xe^{-\alpha x^2}\left[-\frac{\hbar^2}{2m}\frac{d^2}{dx^2} + \frac{1}{2}m\omega^2x^2\right]xe^{-\alpha x^2}dx$$

$$= \frac{3\hbar^2}{2m}\alpha + \frac{3m\omega^2}{8\alpha}. \tag{9.139}$$

The minimization of $E_1(\alpha)$ with respect to α (i.e., $\partial E_1(\alpha)/\partial \alpha = 0$) leads to $\alpha_0 = m\omega/2\hbar$. Hence the energy and the state of the first excited state are given by

$$E_1(\alpha_0) = \frac{3\hbar\omega}{2}, \qquad \psi_1(x, \alpha_0) = \left(\frac{4m^3\omega^3}{\pi\hbar^3}\right)^{1/4} x e^{-m\omega x^2/2\hbar}. \tag{9.140}$$

They are in full agreement with the exact expressions.

(c) The trial function

$$\psi_2(x, \alpha, \beta) = C(\beta x^2 - 1)e^{-\alpha x^2}, \tag{9.141}$$

which includes two adjustable parameters α and β, satisfies all the properties of the second excited state: even under parity, it vanishes as $x \to \pm\infty$ and has two nodes. The term $(\beta x^2 - 1)$ ensures that $\psi_2(x, \alpha, \beta)$ has two nodes $x = \pm 1/\sqrt{\beta}$ and the normalization constant C is given by

$$C = \left(\frac{2\alpha}{\pi}\right)^{1/4} \left[\frac{3\beta^2}{16\alpha^2} - \frac{\beta}{2\alpha} + 1\right]^{-1/2}. \tag{9.142}$$

The trial function $\psi_2(x, \alpha, \beta)$ must be orthogonal to both $\psi_0(x)$ and $\psi_1(x)$. First, notice that it is indeed orthogonal to $\psi_1(x)$, since $\psi_2(x, \alpha, \beta)$ is even while $\psi_1(x)$ is odd:

$$\langle \psi_1 \mid \psi_2 \rangle = C\left(\frac{4m^3\omega^3}{\pi\hbar^3}\right)^{1/4} \int_{-\infty}^{+\infty} x(\beta x^2 - 1)e^{-\alpha x^2}e^{-m\omega x^2/2\hbar}dx = 0. \tag{9.143}$$

As for the orthogonality condition of $\psi_2(x)$ with $\psi_0(x)$, it can be written as

$$\langle \psi_0 \mid \psi_2 \rangle = \int_{-\infty}^{+\infty} \psi_0(x)\psi_2(x, \alpha, \beta)\, dx = \left(\frac{m\omega}{\pi\hbar}\right)^{1/4} C \int_{-\infty}^{+\infty} (\beta x^2 - 1)e^{-(m\omega/2\hbar+\alpha)x^2}dx$$

$$= \left(\frac{m\omega}{\pi\hbar}\right)^{1/4} C \left[\frac{\beta}{2(m\omega/2\hbar + \alpha)} - 1\right]\sqrt{\frac{\pi}{m\omega/2\hbar + \alpha}} = 0. \tag{9.144}$$

This leads to a useful condition between β and α:

$$\beta = \frac{m\omega}{\hbar} + 2\alpha. \tag{9.145}$$

Now let us focus on determining the energy $E_2(\alpha, \beta) = \langle \psi_2 | \hat{H} | \psi_2 \rangle$:

$$E_2(\alpha, \beta) = C^2 \int_{-\infty}^{+\infty} (\beta x^2 - 1)e^{-\alpha x^2}\left[-\frac{\hbar^2}{2m}\frac{d^2}{dx^2} + \frac{1}{2}m\omega^2 x^2\right](\beta x^2 - 1)e^{-\alpha x^2}dx. \tag{9.146}$$

After lengthy but straightforward calculations, we obtain

$$-\frac{\hbar^2}{2m}\left\langle \psi_2 \left| \frac{d^2}{dx^2} \right| \psi_2 \right\rangle = \frac{\hbar^2}{2m}\left(\alpha + \frac{\beta}{2} + \frac{7\beta^2}{16\alpha}\right)C^2\sqrt{\frac{\pi}{2\alpha}}, \tag{9.147}$$

$$\frac{1}{2}m\omega^2\langle \psi_2|x^2 \mid \psi_2\rangle = m\omega^2\left(\frac{15\beta^2}{128\alpha^3} - \frac{3\beta}{16\alpha^2} + \frac{1}{8\alpha}\right)C^2\sqrt{\frac{\pi}{2\alpha}}; \tag{9.148}$$

hence

$$E_2(\alpha, \beta) = C^2\sqrt{\frac{\pi}{2\alpha}}\left(\frac{\hbar^2\alpha}{2m} + \frac{\hbar^2\beta}{4m} + \frac{7\hbar^2\beta^2}{32m\alpha} + \frac{15m\beta^2\omega^2}{128\alpha^3} - \frac{3m\beta\omega^2}{16\alpha^2} + \frac{m\omega^2}{8\alpha}\right). \tag{9.149}$$

To extract the approximate value of E_2, we need to minimize $E_2(\alpha, \beta)$ with respect to α and to β: $\partial E_2(\alpha, \beta)/\partial \alpha = 0$ and $\partial E_2(\alpha, \beta)/\partial \beta = 0$. The two expressions we obtain will enable us to extract (by solving a system of two linear equations with two unknowns) the values of α_0 and β_0 that minimize $E_2(\alpha, \beta)$. This method is lengthy and quite cumbersome; α_0 and β_0 have to satisfy the condition (9.145). We can, however, exploit this condition to come up with a much shorter approach: it consists of replacing the value of β as displayed in (9.145) into the energy relation (9.149), thereby yielding an expression that depends on a single parameter α:

$$E_2(\alpha) = \left(\frac{15\hbar^2\alpha}{18m} + \frac{9\hbar\omega}{8} + \frac{7m\omega^2}{16\alpha} + \frac{15m^3\omega^4}{128\hbar^2\alpha^3} + \frac{9m^2\omega^3}{32\hbar\alpha^2} \right) \left(\frac{3m^2\omega^2}{16\hbar^2\alpha^2} + \frac{m\omega}{4\hbar\alpha} + \frac{3}{4} \right)^{-1} ;$$

(9.150)

in deriving this relation, we have substituted (9.145) into the expression for C as given by (9.142), which in turn is inserted into (9.149). In this way, we need to minimize E_2 with respect to one parameter only, α. This yields $\alpha_0 = m\omega/(2\hbar)$ which, when inserted into (9.145) leads to $\beta_0 = 2m\omega/\hbar$. Thus, the energy and wave function are given by

$$E_2(\alpha_0, \beta_0) = \frac{5}{2}\hbar\omega, \qquad \psi_2(x, \alpha_0, \beta_0) = \left(\frac{m\omega}{4\pi\hbar} \right)^{1/4} \left(\frac{2m\omega}{\hbar}x^2 - 1 \right) e^{-\frac{m\omega}{2\hbar}x^2}. \quad (9.151)$$

These are identical with the exact expressions for the energy and the wave function.

Example 9.6
Use the variational method to estimate the ground state energy of the hydrogen atom.

Solution
The ground state wave function has no nodes and vanishes at infinity. Let us try

$$\psi(r, \theta, \phi) = e^{-r/\alpha}, \tag{9.152}$$

where α is a scale parameter; there is no angular dependence of $\psi(r)$ since the ground state function is spherically symmetric. The energy is given by

$$E(\alpha) = \frac{\langle \psi \mid \hat{H} \mid \psi \rangle}{\langle \psi \mid \psi \rangle} = -\frac{\langle \psi \mid (\hbar^2/2m)\nabla^2 + e^2/r \mid \psi \rangle}{\langle \psi \mid \psi \rangle}, \tag{9.153}$$

where

$$\langle \psi \mid \psi \rangle = \int_0^{+\infty} r^2 e^{-2r/\alpha}\, dr \int_0^\pi \sin\theta\, d\theta \int_0^{2\pi} d\phi = \pi\alpha^3 \tag{9.154}$$

and

$$-\left\langle \psi \left| \frac{e^2}{r} \right| \psi \right\rangle = -4\pi e^2 \int_0^{+\infty} r e^{-2r/\alpha}\, dr = -\pi e^2\alpha^2. \tag{9.155}$$

To calculate the kinetic energy term, we may use (9.128)

$$-\left\langle \psi \left| \frac{\hbar^2}{2m}\nabla^2 \right| \psi \right\rangle = \frac{\hbar^2}{2m} \int \left(\vec{\nabla}\psi^*(r) \right) \cdot \left(\vec{\nabla}\psi(r) \right)\, d^3r, \tag{9.156}$$

where

$$\vec{\nabla}\psi^*(r) = \vec{\nabla}\psi(r) = \frac{d\psi(r)}{dr}\hat{r} = -\frac{1}{\alpha}e^{-r/\alpha}\hat{r}; \tag{9.157}$$

hence

$$-\left\langle \psi \left| \frac{\hbar^2}{2m} \nabla^2 \right| \psi \right\rangle = \frac{4\pi}{\alpha^2} \frac{\hbar^2}{2m} \int_0^{+\infty} r^2 e^{-2r/\alpha} \, dr = \frac{\hbar^2 \pi}{2m} \alpha. \qquad (9.158)$$

Inserting (9.154), (9.155), and (9.158) into (9.153), we obtain

$$E(\alpha) = \frac{\hbar^2}{2m\alpha^2} - \frac{e^2}{\alpha}. \qquad (9.159)$$

Minimizing this relation with respect to α, $dE(\alpha)/d\alpha = -\hbar^2/(m\alpha_0^3) + e^2/\alpha_0^2 = 0$, we obtain $\alpha_0 = \hbar^2/(me^2)$ which, when inserted into (9.159), leads to the ground state energy

$$E(\alpha_0) = -\frac{me^4}{2\hbar^2}. \qquad (9.160)$$

This is the correct ground state energy for the hydrogen atom. The variational method has given back the correct energy because the trial function (9.152) happens to be identical with the exact ground state wave function. Note that the scale parameter $\alpha_0 = \hbar^2/(me^2)$ has the dimensions of length; it is equal to the Bohr radius.

9.4   The Wentzel–Kramers–Brillouin Method

The Wentzel–Kramers–Brillouin (WKB) method is useful for approximate treatments of systems with *slowly* varying potentials; that is, potentials which remain almost *constant* over a region of the order of the de Broglie wavelength. In the case of classical systems, this property is always satisfied since the wavelength of a classical system approaches zero. The WKB method can thus be viewed as a *semiclassical* approximation.

9.4.1   General Formalism

Consider the motion of a particle in a time-independent potential $V(\vec{r})$; the Schrödinger equation for the corresponding stationary state is

$$-\frac{\hbar^2}{2m} \nabla^2 \psi(\vec{r}) + V(\vec{r})\psi(\vec{r}) = E\psi(\vec{r}) \qquad (9.161)$$

or

$$\nabla^2 \psi(\vec{r}) + \frac{1}{\hbar^2} p^2(\vec{r})\psi(\vec{r}) = 0, \qquad (9.162)$$

where $p(\vec{r})$ is the classical momentum at $\vec{r}$: $p(\vec{r}) = \sqrt{2m(E - V(\vec{r}))}$. If the particle is moving in a region where $V(\vec{r})$ is constant, the solution of (9.162) is of the form $\psi(\vec{r}) = Ae^{\pm i\vec{p}\cdot\vec{r}/\hbar}$. But how does one deal with those cases where $V(\vec{r})$ is not constant? The WKB method provides an approximate treatment for systems whose potentials, while not constant, are *slowly* varying functions of $\vec{r}$. That is, $V(\vec{r})$ is almost constant in a region which extends over several de Broglie wavelengths; we may recall that the de Broglie wavelength of a particle of mass m and energy E that is moving in a potential $V(\vec{r})$ is given by $\lambda = h/p = h/\sqrt{2m(E - V(\vec{r}))}$.

In essence, the WKB method consists of trying a solution to (9.162) in the following form:

$$\psi(\vec{r}) = A(\vec{r})e^{iS(\vec{r})/\hbar},$$

(9.163)

where the amplitude $A(\vec{r})$ and the phase $S(\vec{r})$, which are real functions, are yet to be determined. Substituting (9.163) into (9.162) we obtain

$$A\left[\frac{\hbar^2}{A}\nabla^2 A - (\vec{\nabla}S)^2 + p^2(\vec{r})\right] + i\hbar\left[2(\vec{\nabla}A)\cdot(\vec{\nabla}S) + A\nabla^2 S\right] = 0.$$

(9.164)

The real and imaginary parts of this equation must vanish separately:

$$(\vec{\nabla}S)^2 = p^2(\vec{r}) = 2m(E - V(\vec{r})),$$

(9.165)

$$2(\vec{\nabla}A)\cdot(\vec{\nabla}S) + A\nabla^2 S = 0.$$

(9.166)

In deriving (9.165) we have neglected the term that contains $\hbar$ (i.e., $(\hbar^2/A)\nabla^2 A$), since it is small compared to $(\vec{\nabla}S)^2$ and to $p^2(\vec{r})$; $\hbar$ is considered to be very small for classical systems.

To illustrate the various aspects of the WKB method, let us consider the simple case of the *one-dimensional* motion of a single particle. We can thus reduce (9.165) and (9.166), respectively, to

$$\frac{dS}{dx} = \pm\sqrt{2m(E - V)} = \pm p(x),$$

(9.167)

$$2\left(\frac{d}{dx}\ln A\right)p(x) + \frac{d}{dx}p(x) = 0.$$

(9.168)

Let us find the solutions of (9.167) and (9.168). Integration of (9.167) yields

$$S(x) = \pm\int dx\sqrt{2m(E - V(x))} = \pm\int p(x)\,dx.$$

(9.169)

We can reduce (9.168) to

$$\frac{d}{dx}\left[2\ln A + \ln p(x)\right] = 0,$$

(9.170)

which in turn leads to

$$A(x) = \frac{C}{\sqrt{|p(x)|}},$$

(9.171)

where C is an arbitrary constant. So (9.169) and (9.171) give, respectively, the phase $S(x)$ and amplitude $A(x)$ of the WKB wave function (9.163).

Inserting (9.171) and (9.169) into (9.163), we obtain two approximate solutions to equation (9.162):

$$\psi_\pm(x) = \frac{C_\pm}{\sqrt{|p(x)|}}\exp\left[\pm\frac{i}{\hbar}\int^x p(x')dx'\right].$$

(9.172)

The amplitude of this wave function is proportional to $1/\sqrt{p(x)}$; hence the probability of finding the particle between x and $x + dx$ is proportional to $1/p(x)$. This is what we expect for a "classical" particle because the time it will take to travel a distance dx is proportional to the inverse of its speed (or its momentum).

We can now examine two separate cases corresponding to $E > V(x)$ and $E < V(x)$. First, let us consider the case $E > V(x)$, which is called the *classically allowed* region. Here $p(x)$ is a real function; the most general solution of (9.162) is a combination of $\psi_+(x)$ and $\psi_-(x)$:

$$\psi(x) = \frac{C_+}{\sqrt{p(x)}} \exp\left[\frac{i}{\hbar} \int^x p(x')dx'\right] + \frac{C_-}{\sqrt{p(x)}} \exp\left[-\frac{i}{\hbar} \int^x p(x')dx'\right]. \qquad (9.173)$$

Second, in the case where $E < V(x)$, which is known as the *classically forbidden* region, the momentum $p(x)$ is imaginary and the exponents of (9.172) become real:

$$\psi(x) = \frac{C'_-}{\sqrt{|p(x)|}} \exp\left[-\frac{1}{\hbar} \int_x |p(x')|dx'\right] + \frac{C'_+}{\sqrt{|p(x)|}} \exp\left[\frac{1}{\hbar} \int^x |p(x')|dx'\right]. \qquad (9.174)$$

Equations (9.173) and (9.174) give the system's wave function in the allowed and forbidden regions, respectively. But what about the structure of the wave function near the regions $E \simeq V(x)$? At the points x_i, we have $E = V(x_i)$; hence the momentum (9.167) vanishes, $p(x_i) = 0$. These points are called the *classical turning points*, because classically the particle stops at x_i and then turns back to resume its motion in the opposite direction. At these points the wave function (9.172) becomes infinite since $p(x_i) = 0$. One then needs to examine how to find the wave function at the turning points. Before looking into that, let us first study the condition of validity for the WKB approximation.

Validity of the WKB approximation

To obtain the condition of validity for the WKB method, let us examine the size of the various terms in (9.164), notably $A(\vec{\nabla} S)^2$ and $i\hbar A\nabla^2 S$. Since quantities of the order of $\hbar$ are too small in the classical limit, the quasi-classical region is expected to be given by the condition[4]

$$\left|\hbar\nabla^2 S\right| \ll (\vec{\nabla} S)^2, \qquad (9.175)$$

which can be written in one dimension as

$$\hbar \left|\frac{S''}{S'^2}\right| \ll 1 \qquad (9.176)$$

or

$$\left|\frac{d}{dx}\left(\frac{\hbar}{S'}\right)\right| \ll 1, \qquad (9.177)$$

since $\nabla^2 S = d^2 S/dx^2 = S''$ and $|\vec{\nabla} S| = dS(x)/dx = S'$. In what follows we are going to verify that this relation yields the condition of validity for the WKB approximation.

Since $S' = \pm p(x)$ (see (9.167)), we can reduce (9.177) to

$$\left|\frac{d\bar{\lambda}(x)}{dx}\right| \ll 1, \qquad (9.178)$$

where $\bar{\lambda}(x) = \lambda(x)/(2\pi)$ and $\lambda(x)$ is the de Broglie wavelength of the particle:

$$\bar{\lambda}(x) = \frac{\hbar}{p(x)} = \frac{\hbar}{\sqrt{2m(E - V(x))}}. \qquad (9.179)$$

---

[4]The condition (9.175) can be found as follows. Substituting $\psi(\vec{r}) = e^{iS(\vec{r})/\hbar}$ into (9.162) and multiplying by $\hbar^2$, we get $i\hbar\nabla^2 S(\vec{r}) - (\vec{\nabla} S)^2 + p^2(\vec{r}) = 0$. In the classical limit, the term containing $\hbar$, $|i\hbar\nabla^2 S(\vec{r})|$, must be small compared to the terms that do not, $(\vec{\nabla} S)^2$; i.e., $|\hbar\nabla^2 S(\vec{r})| \ll (\vec{\nabla} S)^2$.

The condition (9.178) means that the rate of change of the de Broglie wavelength is small (i.e., the wavelength of the particle must vary only slightly over distances of the order of its size). But this condition is always satisfied for classical systems. So the condition of validity for the WKB method is given by

$$\left| \frac{d\bar{\lambda}(x)}{dx} \right| = \left| \frac{d}{dx} \left(\frac{\hbar}{p(x)} \right) \right| \ll 1. \tag{9.180}$$

This condition clearly breaks down at the classical turning points, $E = V(x_i)$, since $p(x_i) = 0$; classically, the particle stops at $x = x_i$ and then moves in the opposite direction. As $p(x)$ becomes small, the wavelength (9.179) becomes large and hence violates the requirement that it remains small and varies only slightly; when $p(x)$ is too small, the condition (9.180) breaks down. So the WKB approximation is valid in both the allowed and forbidden regions but not at the classical turning points.

How does one specify the particle's wave function at $x = x_i$? Or how does one connect the allowed states (9.173) with their forbidden counterparts (9.174)? As we go through the classical turning point, from the allowed to the forbidden region and vice versa, we need to examine how to determine the particle's wave function everywhere and notably at the turning points. This is the most difficult issue of the WKB method, for it breaks down at the turning points. In the following section we are going to deal with this issue by solving the Schrödinger equation near and at $x = x_i$. We will do so by resorting to an approximation: we consider the potential to be given, near the turning points, by a straight line whose slope is equal to that of the potential at the turning point.

In what follows, we want to apply the WKB approximation to find the energy levels and the wave function of a particle moving in a potential well. We are going to show that the formulas giving the energy levels depend on whether or not the potential well has rigid walls. In fact, it even depends on the number of rigid walls the potential has. For this, we are going to consider three separate cases pertaining to the potential well with: no rigid walls, a single rigid wall, and two rigid walls.

9.4.2 Bound States for Potential Wells with No Rigid Walls

Consider a potential well that has no rigid walls as displayed in Figure 9.5. Here the classically forbidden regions are specified by $x < x_1$ and $x > x_2$, the classically allowed region by $x_1 < x < x_2$; x_1 and x_2 are the classical turning points. This is a suitable and simple example to illustrate the various aspects of the WKB method, notably how to determine the particle's wave function at the turning points. We will see how this method yields the Bohr–Sommerfeld quantization rule from which the bound state energies are to be extracted.

The WKB method applies everywhere in the three regions (1), (2), and (3), except near the two turning points $x = x_1$ and $x = x_2$ at which $E = V(x_1) = V(x_2)$. The WKB approximation to the wave function in regions (1) and (3) can be inferred from (9.174) and the approximation in region (2) from (9.173): the wave function must decay exponentially in regions (1) and (3) as $x \to -\infty$ and $x \to +\infty$, respectively, but must be oscillatory in region (2):

$$\psi_{1WKB}(x) = \frac{C_1}{\sqrt{|p(x)|}} \exp\left[-\frac{1}{\hbar} \int_x^{x_1} |p(x')| \, dx' \right], \qquad x < x_1, \tag{9.181}$$

$$\psi_{2WKB}(x) = \frac{C_2'}{\sqrt{p(x)}} \exp\left[\frac{i}{\hbar} \int_x p(x') dx' \right] + \frac{C_2''}{\sqrt{p(x)}} \exp\left[-\frac{i}{\hbar} \int_x p(x') dx' \right], \qquad x_1 < x < x_2,$$

$$\tag{9.182}$$

Figure 9.5 Potential with no rigid walls: regions (1) and (3) are classically forbidden, while (2) is classically allowed.

$$\psi_{3WKB}(x) = \frac{C_3}{\sqrt{|p(x)|}} \exp\left[-\frac{1}{\hbar} \int_{x_2}^{x} |p(x')| \, dx'\right], \qquad x > x_2; \qquad (9.183)$$

the coefficients C_1, C_2', C_2'', and C_3 have yet to be determined. For this, we must *connect* the solutions $\psi_1(x)$, $\psi_2(x)$, and $\psi_3(x)$ when passing from one region into another through the turning points $x = x_1$ and $x = x_2$ where the quasi-classical approximation ceases to be valid. That is, we need to connect $\psi_3(x)$ to $\psi_2(x)$ as we go from region (3) to (2), and then connect $\psi_1(x)$ to $\psi_2(x)$ as we go from (1) to (2). Since the WKB approximation breaks down at x_1 and x_2, we need to look for the exact solutions of the Schrödinger equation near x_1 and x_2.

9.4.2.1 Connection of $\psi_{3WKB}(x)$ to $\psi_{2WKB}(x)$

The WKB approximation to the wave function in region (2) can be inferred from (9.182):

$$\psi_{2WKB}(x) = \frac{C_2'}{\sqrt{p(x)}} \exp\left[\frac{i}{\hbar} \int_{x}^{x_2} p(x') \, dx'\right] + \frac{C_2''}{\sqrt{p(x)}} \exp\left[-\frac{i}{\hbar} \int_{x}^{x_2} p(x') \, dx'\right], \qquad x_1 < x < x_2;$$

$$\qquad (9.184)$$

this can be written as

$$\psi_{2WKB}(x) = \frac{C_2}{\sqrt{p(x)}} \sin\left(\frac{1}{\hbar} \int_{x}^{x_2} p(x') \, dx' + \alpha\right), \qquad x_1 < x < x_2, \qquad (9.185)$$

where α is a phase to be determined. Since the WKB approximation breaks down near the turning point x_2 (i.e., on both sides of $x = x_2$), we need to find a scheme for determining the wave function near x_2.

For this, let us now look for the exact solution of the Schrödinger equation near $x = x_2$. As mentioned above, if $|x - x_2|$ is small enough, within the region $|x - x_2|$, we can approximately represent the potential by a straight line whose slope is equal to that of the potential at the classical turning point $x = x_2$. That is, expanding $V(x)$ to first order around $x = x_2$, we obtain

$$V(x) \simeq V(x_2) + (x - x_2) \frac{dV(x)}{dx}\bigg|_{x=x_2} = E + (x - x_2)F_0, \qquad (9.186)$$

where we have used the fact that $V(x_2) = E$ and where F_0 is given by $F_0 = \frac{dV(x)}{dx}\big|_{x=x_2}$. Equation (9.186) means that $V(x)$ is approximated by a straight line $(x - x_2)F_0$, where F_0 is the slope of $V(x)$ at $x = x_2$. The Schrödinger equation for the potential (9.186) can be written as

$$\frac{d^2\psi(x)}{dx^2} - \frac{2mF_0}{\hbar^2}(x - x_2)\psi(x) = 0. \tag{9.187}$$

Using the change of variable

$$y = \left(\frac{2mF_0}{\hbar^2}\right)^{1/3}(x - x_2), \tag{9.188}$$

we can transform (9.187) into

$$\left(\frac{2mF_0}{\hbar^2}\right)^{2/3}\left[\frac{d^2\psi(y)}{dy^2} - y\psi(y)\right] = 0 \tag{9.189}$$

or

$$\frac{d^2\psi(y)}{dy^2} - y\psi(y) = 0. \tag{9.190}$$

This is a well-known differential equation whose solutions are usually expressed in terms of the Airy functions[5] $Ai(y)$:

$$\psi(y) = A'Ai(y) = \frac{A'}{\pi}\int_0^\infty \cos\left(\frac{z^3}{3} + yz\right)dz, \tag{9.191}$$

where A' is a normalization constant.

From the properties of the Airy function $Ai(y) = 1/\pi \int_0^\infty \cos(z^3/3 + yz)dz$, the asymptotic behavior of $Ai(y)$ is given for large positive and large negative values of y by

$$Ai(y) \sim \begin{cases} \frac{1}{\sqrt{\pi}|y|^{1/4}}\sin\left[\frac{2}{3}(-y)^{3/2} + \frac{\pi}{4}\right], & y \ll 0, \\ \frac{1}{2\sqrt{\pi}y^{1/4}}\exp\left[-\frac{2}{3}y^{3/2}\right], & y \gg 0. \end{cases} \tag{9.192}$$

The asymptotic expression of (9.191) is therefore given for large positive and large negative values of y by

$$\psi(y) = \begin{cases} \frac{A'}{\sqrt{\pi}|y|^{1/4}}\sin\left[\frac{2}{3}(-y)^{3/2} + \frac{\pi}{4}\right], & y \ll 0, \\ \frac{A'}{2\sqrt{\pi}y^{1/4}}\exp\left[-\frac{2}{3}y^{3/2}\right], & y \gg 0. \end{cases} \tag{9.193}$$

Since $F_0 > 0$ equation (9.188) implies that the cases $y \ll 0$ and $y \gg 0$ correspond to $x \ll x_2$ and $x \gg x_2$, respectively.

Now near the turning point $x = x_2$, (9.186) shows that $E - V(x) = -(x - x_2)F_0$; hence the square of the classical momentum $p^2(x)$ is given by

$$p^2(x) = 2m(E - V(x)) = -2m(x - x_2)F_0, \tag{9.194}$$

which is negative for $x > x_2$ and positive for $x < x_2$. Combining equations (9.188) and (9.194), we obtain

$$p^2(x) = -(2m\hbar F_0)^{2/3}y. \tag{9.195}$$

---

[5]The solution to the differential equation $d^2\phi(y)/dy = y\phi(y)$ is given by the Airy function $\phi(y) = Ai(y) = \frac{1}{\pi}\int_0^\infty \cos(z^3/3 + yz)dz$.

Now since $dx = (\hbar^2/(2m F_0))^{1/3} dy$ (see (9.188)), we use (9.195) to infer the following expression:

$$\frac{1}{\hbar} \int_x^{x_2} p(x') \, dx' = \frac{1}{\hbar} (2m\hbar F_0)^{1/3} \left(\frac{\hbar^2}{2m F_0}\right)^{1/3} \int_y^0 \sqrt{-y'} \, dy' = \int_y^0 \sqrt{-y'} \, dy' = \frac{2}{3}(-y)^{3/2}.$$

(9.196)

Inserting this into (9.193), we obtain

$$\psi(x) = \begin{cases} \frac{A}{\sqrt{p(x)}} \sin\left(\frac{1}{\hbar} \int_x^{x_2} p(x') dx' + \frac{\pi}{4}\right), & x \ll x_2, \\ \frac{A}{2\sqrt{|p(x)|}} \exp\left[-\frac{1}{\hbar} \int_x^{x_2} |p(x')| \, dx'\right], & x \gg x_2, \end{cases}$$

(9.197)

where $A = (2m\hbar F_0)^{1/6} A'/\sqrt{\pi}$. A comparison of (9.197a) with (9.185) and (9.197b) with (9.183) reveals that

$$A = 2C_3, \qquad C_2 = A, \qquad \alpha = \frac{\pi}{4};$$

(9.198)

these expressions are known as the *connection formulas*, for they connect the WKB solutions at either side of a turning point. Since $\alpha = \pi/4$, $\psi_{2WKB}(x)$ of (9.185) becomes

$$\psi_{2WKB}(x) = \frac{C_2}{\sqrt{p(x)}} \sin\left(\frac{1}{\hbar} \int_x^{x_2} p(x') \, dx' + \frac{\pi}{4}\right).$$

(9.199)

9.4.2.2 Connection of $\psi_{1WKB}(x)$ to $\psi_{2WKB}(x)$

The WKB wave function for $x < x_1$ is given by (9.181); the WKB solution for $x > x_1$ can be inferred from (9.182):

$$\psi_{2WKB}(x) = \frac{C_2'}{\sqrt{p(x)}} \exp\left[\frac{i}{\hbar} \int_{x_1}^x p(x') dx'\right] + \frac{C_2''}{\sqrt{p(x)}} \exp\left[-\frac{i}{\hbar} \int_x^{x_1} p(x') dx'\right], \quad x_1 < x < x_2,$$

(9.200)

which can be written as

$$\psi_{2WKB}(x) = \frac{D}{\sqrt{p}} \sin\left(\frac{1}{\hbar} \int_{x_1}^x p(x') \, dx' + \beta\right).$$

(9.201)

Recall that near $x = x_1$ the WKB approximation breaks down.

The shape of the wave function near $x = x_1$ can, however, be found from an exact solution of the Schrödinger equation. For this, we proceed as we did for $x = x_2$. That is, we look for the exact solution of the Schrödinger equation for small values of $|x - x_1|$. Expanding $V(x)$ near $x = x_1$, we obtain a Schrödinger equation similar to (9.190). Its solutions for $x < x_1$ and $x > x_1$ are given by expressions that are similar to (9.197b) and (9.197a) respectively:

$$\psi(x) = \begin{cases} \frac{E}{2\sqrt{|p(x)|}} \exp\left[\frac{1}{\hbar} \int_{x_1}^x |p(x')| \, dx'\right], & x \ll x_1, \\ \frac{E}{\sqrt{p(x)}} \sin\left(\frac{1}{\hbar} \int_{x_1}^x p(x') \, dx' + \frac{\pi}{4}\right), & x \gg x_1. \end{cases}$$

(9.202)

Again, comparing (9.202a) with (9.181) and (9.202b) with (9.201), we obtain the other set of connection formulas:

$$E = 2C_1, \qquad E = D, \qquad \beta = \frac{\pi}{4};$$

(9.203)

hence $\psi_2(x)$ of (9.201) becomes

$$\psi_{2WKB}(x) = \frac{D}{\sqrt{p(x)}} \sin\left(\frac{1}{\hbar}\int_{x_1}^{x} p(x')\,dx' + \frac{\pi}{4}\right).\tag{9.204}$$

9.4.2.3 Quantization of the Energy Levels of the Bound States

Since the two solutions (9.199) and (9.204) represent the same wave function in the same region, they must be equal:

$$\psi_{2WKB}(x) = \frac{D}{\sqrt{p(x)}} \sin\left(\frac{1}{\hbar}\int_{x_1}^{x} p(x')\,dx' + \frac{\pi}{4}\right) = \frac{C_2}{\sqrt{p(x)}} \sin\left(\frac{1}{\hbar}\int_{x}^{x_2} p(x')\,dx' + \frac{\pi}{4}\right).$$

$$\tag{9.205}$$

This is an equation of the form $D\sin\theta_1 = C_2\sin\theta_2$. Its solutions must satisfy the following two relations. The first is $\theta_1 + \theta_2 = (n+1)\pi$, i.e.,

$$\left(\frac{1}{\hbar}\int_{x_1}^{x} p(x')\,dx' + \frac{\pi}{4}\right) + \left(\frac{1}{\hbar}\int_{x}^{x_2} p(x')\,dx' + \frac{\pi}{4}\right) = (n+1)\pi\tag{9.206}$$

or

$$\frac{1}{\hbar}\int_{x_1}^{x_2} p(x)\,dx = \left(n+\frac{1}{2}\right)\pi, \qquad n = 0,\ 1,\ 2,\ 3,\ \ldots;\tag{9.207}$$

and the second is

$$D = (-1)^n C_2.\tag{9.208}$$

Since the integral between the turning points $\int_{x_1}^{x_2} p(x)\,dx$ is equal to half the integral over a complete period of the quasi-classical motion of the particle, i.e., $\int_{x_1}^{x_2} p(x)\,dx = \frac{1}{2}\oint p(x)\,dx$, we can reduce (9.207) to

$$\oint p(x)\,dx = 2\int_{x_1}^{x_2} p(x)\,dx = \left(n+\frac{1}{2}\right)h, \qquad n = 0,\ 1,\ 2,\ 3,\ \ldots.\tag{9.209}$$

This relation determines the *quantized* (WKB) energy levels E_n of the bound states of a semi-classical system. It is similar to the *Bohr–Sommerfeld quantization rule*, which in turn is known to represent an improved version of the Wilson–Sommerfeld rule $\oint p(x)\,dx = nh$, because the Wilson–Sommerfeld rule does not include the zero-point energy term $h/2$ (in the case of large values of n, where the classical approximation becomes reliable, we have $n + 1/2 \simeq n$; hence (9.209) reduces to $\oint p(x)\,dx = nh$). We can interpret this relation as follows: since the integral $\oint p(x)\,dx$ gives the area enclosed by the closed trajectory of the particle in the xp phase space, the condition (9.209) provides the mechanism for selecting, from the continuum of energy values of the semiclassical system, only those energies E_n for which the areas of the contours $p(x, E_n) = \sqrt{2m(E_n - V(x))}$ are equal to $(n + \frac{1}{2})h$:

$$\boxed{\oint p(x, E_n)\,dx = 2\int_{x_1}^{x_2} \sqrt{2m(E_n - V(x))}\,dx = \left(n+\frac{1}{2}\right)h,}\tag{9.210}$$

with $n = 0, 1, 2, 3, \ldots$. So in the xp phase space, the area between two successive bound states is equal to h: $\oint p(x, E_{n+1})\,dx - \oint p(x, E_n)\,dx = h$. Each single state therefore corresponds to an area h in the phase space. Note that the number n present in this relation is equal to the number of bound states; that is, the number of nodes of the wave function $\psi(x)$.

In summary, for a particle moving in a potential well like the one shown in Figure 9.5, the bound state energies can be extracted from the quantization rule (9.210) and the wave function is given in regions (1) and (3) by (9.181) and (9.183), respectively, and in region (2) either by (9.199) or (9.204). Combining the *connection relations* (9.198), (9.203), and (9.208) with the wave functions (9.181), (9.183), (9.199), and (9.204), we get the WKB approximation to the wave function:

$$\psi_{WKB}(x) = \begin{cases} \psi_{1WKB}(x) = \dfrac{(-1)^n C_3}{\sqrt{|p(x)|}} \exp\left[-\dfrac{1}{\hbar}\int_x^{x_1} |p(x')|\, dx'\right], & x < x_1, \\[2mm] \psi_{3WKB}(x) = \dfrac{C_3}{\sqrt{|p(x)|}} \exp\left[-\dfrac{1}{\hbar}\int_{x_2}^x |p(x')|\, dx'\right], & x > x_2. \end{cases} \tag{9.211}$$

In the region $x_1 < x < x_2$, $\psi_{2WKB}(x)$ is given either by (9.199) or by (9.204)

$$\psi_{2WKB}(x) = \begin{cases} \dfrac{2(-1)^n C_3}{\sqrt{p(x)}} \sin\left(\dfrac{1}{\hbar}\int_{x_1}^x p(x')dx' + \dfrac{\pi}{4}\right), & x_1 < x < x_2, \\[2mm] \dfrac{2C_3}{\sqrt{p(x)}} \sin\left(\dfrac{1}{\hbar}\int_x^{x_2} p(x')dx' + \dfrac{\pi}{4}\right), & x_1 < x < x_2. \end{cases} \tag{9.212}$$

The coefficient C_3 has yet to be found from the normalization of $\psi_{WKB}(x)$. This is the wave function of the nth bound state.

Remark

An important application of the WKB method consists of using the quantization rule (9.210) to calculate the energy levels of central potentials. The energy of a particle of mass m bound in a central potential $V(r)$ is given by

$$E = \frac{p_r^2}{2m} + V_{\text{eff}}(r) = \frac{p_r^2}{2m} + V(r) + \frac{\hbar^2}{2m}\frac{l(l+1)}{r^2}. \tag{9.213}$$

The particle is bound to move between the turning points r_1 and r_2 whose values are given by $E = V_{\text{eff}}(r_1) = V_{\text{eff}}(r_2)$ and its bound state energy levels can be obtained from

$$\int_{r_1}^{r_2} dr\, p_r(E, r) = \int_{r_1}^{r_2} dr\, \sqrt{2m\left(E - V(r) - \frac{\hbar^2}{2m}\frac{l(l+1)}{r^2}\right)} = \left(n + \frac{1}{2}\right)\pi\hbar, \tag{9.214}$$

where $n = 0, 1, 2, 3, \ldots$.

Example 9.7

Use the WKB method to estimate the energy levels of a one-dimensional harmonic oscillator.

Solution

The classical energy of a harmonic oscillator

$$E(x, p) = \frac{p^2}{2m} + \frac{1}{2}m\omega^2 x^2 \tag{9.215}$$

leads to $p(E, x) = \pm\sqrt{2mE - m^2\omega^2 x^2}$. At the turning points, x_{min} and x_{max}, the energy is given by $E = V(x) = \frac{1}{2}m\omega^2 x^2$ where $x_{min} = -a$ and $x_{max} = a$ with $a = \sqrt{2E/(m\omega^2)}$. To

Figure 9.6 Potential well with one rigid wall located at $x = x_1$.

obtain the quantized energy expression of the harmonic oscillator, we need to use the Bohr–Sommerfeld quantization rule (9.210):

$$\oint p\,dx = 2 \int_{-a}^{a} \sqrt{2mE - m^2\omega^2 x^2}\,dx = 4m\omega \int_{0}^{a} \sqrt{a^2 - x^2}\,dx. \qquad (9.216)$$

Using the change of variable $x = a \sin \theta$, we have

$$\int_{0}^{a} \sqrt{a^2 - x^2}\,dx = a^2 \int_{0}^{\pi/2} \cos^2 \theta\, d\theta = \frac{a^2}{2} \int_{0}^{\pi/2} (1 + \cos 2\theta)\,d\theta = \frac{\pi a^2}{4} = \frac{\pi E}{2m\omega^2}; \quad (9.217)$$

hence

$$\oint p\,dx = \frac{2\pi E}{\omega}. \qquad (9.218)$$

Since $\oint p\,dq = \left(n + \frac{1}{2}\right)h$ or $2\pi E/\omega = (n + 1/2)h$, we obtain

$$E_n^{WKB} = \left(n + \frac{1}{2}\right)\hbar\omega. \qquad (9.219)$$

This expression is identical with the *exact* energy of the harmonic oscillator.

9.4.3 Bound States for Potential Wells with One Rigid Wall

Consider a particle moving in a potential well that has a rigid wall at $x = x_1$ (Figure 9.6); it is given by $V(x) = +\infty$ for $x < x_1$ and by a certain function $V(x)$ for $x > x_1$. The classically allowed region is specified by $x_1 < x < x_2$; x_1 and x_2 are the turning points.

To obtain the quantization rule which gives the bound state energy levels for this potential, we proceed as we did in obtaining (9.210). The WKB wave function in region $x_1 < x < x_2$ has an oscillatory form; it can be inferred from (9.201):

$$\psi_{WKB}(x) = \frac{A}{\sqrt{p(x)}} \sin\left(\frac{1}{\hbar} \int p(x')\,dx' + \alpha\right), \qquad x_1 \le x \le x_2, \qquad (9.220)$$

where α is a phase factor that needs to be specified. For this, we need to find the WKB wave function near the two turning points x_1 and x_2.

First, near x_2 (i.e., for $x \leq x_2$) we can determine the value of α as we did in obtaining (9.199). That is, expand $V(x)$ around $(x - x_2)$ and then match the WKB solutions at $x = x_2$; this leads to a phase factor $\alpha = \pi/4$ and hence

$$\psi_{WKB}(x) = \frac{B}{\sqrt{p(x)}} \sin\left(\frac{1}{\hbar} \int_x^{x_2} p(x')dx' + \frac{\pi}{4}\right), \qquad x_1 \leq x \leq x_2. \qquad (9.221)$$

Second, since the wave function has to vanish at the rigid wall, $\psi_{WKB}(x_1) = 0$, the phase factor α must be zero; then (9.220) yields

$$\psi_{WKB}(x) = \frac{A}{\sqrt{p(x)}} \sin\left(\frac{1}{\hbar} \int_{x_1}^x p(x')\,dx'\right), \qquad x_1 \leq x \leq x_2. \qquad (9.222)$$

Now, since (9.221) and (9.222) represent the same wave function in the same region, the sum of their arguments must be equal to $(n + 1)\pi$ and $A = (-1)^n B$ (see Eq. (9.208)):

$$\left(\frac{1}{\hbar} \int_{x_1}^x p(x')\,dx'\right) + \left(\frac{1}{\hbar} \int_x^{x_2} p(x')\,dx' + \frac{\pi}{4}\right) = (n + 1)\pi. \qquad (9.223)$$

Thus, the quantization rule which gives the bound state energy levels for potential wells with one single rigid wall is given by

$$\boxed{\int_{x_1}^{x_2} p(x)\,dx = \left(n + \frac{3}{4}\right)\pi\hbar, \qquad n = 0, 1, 2, 3, lcdots.} \qquad (9.224)$$

Remark

From the study carried out above, we may state that the phase factor α of the WKB solution (9.220) is in general equal to

- zero for turning points located at the rigid walls

- $\pi/4$ for turning points that are not located at the rigid walls.

9.4.4   Bound States for Potential Wells with Two Rigid Walls

Consider a potential well that has two rigid walls at $x = x_1$ and $x = x_2$. That is, as shown in Figure 9.7, $V(x)$ is infinite for $x \leq x_1$ and $x \geq x_2$ and given by a certain function $V(x)$ for $x_1 < x < x_2$. The wave function of a particle that is confined to move between the two rigid walls must vanish at the walls: $\psi(x_1) = \psi(x_2) = 0$.

To obtain the quantization rule which gives the bound state energy levels for this potential, we proceed as we did in obtaining (9.224). The WKB wave function has an oscillatory form in $x_1 < x < x_2$ and vanishes at both x_1 and x_2; the phase factor is zero at x_1 and x_2. By analogy with the procedure that led to (9.222), we can show that the WKB wave function in the vicinity of x_1 (i.e., in the region $x > x_1$) is given by

$$\psi_{WKB}(x) = \frac{A}{\sqrt{p(x)}} \sin\left(\frac{1}{\hbar} \int_{x_1}^x p(x')\,dx'\right), \qquad x_1 < x < x_2, \qquad (9.225)$$

and in the vicinity of x_2 (i.e., in the region $x < x_2$) it is given by

$$\psi_{WKB}(x) = \frac{B}{\sqrt{p(x)}} \sin\left(\frac{1}{\hbar} \int_x^{x_2} p(x')dx'\right), \qquad x_1 < x < x_2. \qquad (9.226)$$

Figure 9.7 Potential well with two rigid walls located at x_1 and x_2.

Note that the last two wave functions satisfy the correct boundary conditions at x_1 and x_2:
$\psi_{WKB}(x_1) = \psi_{WKB}(x_2) = 0$.

Since equations (9.225) and (9.226) represent the same wave function in the same region, the sum of the arguments must then be equal to $(n + 1)\pi$ and $A = (-1)^n B$ (see Eq. (9.208)):

$$\left(\frac{1}{\hbar} \int_{x_1}^{x} p(x')\, dx'\right) + \left(\frac{1}{\hbar} \int_{x}^{x_2} p(x')\, dx'\right) = (n + 1)\pi; \tag{9.227}$$

hence the quantization rule for potential wells with two rigid walls is given by

$$\int_{x_1}^{x_2} p(x')\, dx' = (n + 1)\, \pi \hbar, \qquad n = 0, 1, 2, 3, \ldots, \tag{9.228}$$

or by

$$\boxed{\int_{x_1}^{x_2} p(x)\, dx = n\pi \hbar, \qquad n = 1, 2, 3, \ldots.} \qquad \tag{9.229}$$

The only difference between (9.228) and (9.229) is in the minimum value of the quantum number n: the lowest value of n is $n = 0$ in (9.228) and $n = 1$ in (9.229).

Remark

In this section we have derived three quantization rules (9.210), (9.224), and (9.229); they provide the proper prescriptions for specifying the energy levels for potential wells with zero, one, and two rigid walls, respectively. These rules differ only in the numbers $\frac{1}{2}$, $\frac{3}{4}$, and 0 that are added to n. In the cases where n is large, which correspond to the semiclassical domain, these three quantization rules become identical; the semiclassical approximation is most accurate for large values of n.

Example 9.8

Use the WKB approximation to calculate the energy levels of a spinless particle of mass m moving in a one-dimensional box with walls at $x = 0$ and $x = L$.

Solution

This potential has two rigid walls, one at $x = 0$ and the other at $x = L$. To find the energy levels, we make use of the quantization rule (9.229). Since the momentum is constant within the well $p(E, x) = \sqrt{2mE}$, we can easily infer the WKB energy expression of the particle within the well. The integral is quite simple to calculate:

$$\int_0^L p\, dx = \sqrt{2mE} \int_0^L dx = L\sqrt{2mE}. \tag{9.230}$$

Now since $\int_0^L p\, dx = n\pi\hbar$ we obtain

$$L\sqrt{2m E_n^{WKB}} = n\pi\hbar; \tag{9.231}$$

hence

$$E_n^{WKB} = \frac{\pi^2\hbar^2}{2mL^2} n^2. \tag{9.232}$$

This is the exact value of the energy of a particle in an infinite well.

Example 9.9 (WKB method for the Coulomb potential)
Use the WKB approximation to calculate the energy levels of the s states of an electron that is bound to a Ze nucleus.

Solution
The electron moves in the Coulomb field of the Ze nucleus: $V(r) = -Ze^2/r$. Since the electron is bound to the nucleus, it can be viewed as moving between two rigid walls $0 \leq r \leq a$ with $E = V(a), a = -Ze^2/E$; the energy of the electron is negative, $E < 0$.

The energy levels of the s states (i.e., $l = 0$) can thus be obtained from (9.229):

$$\int_0^a \sqrt{2m\left(E + \frac{Ze^2}{r}\right)}\, dr = n\pi\hbar. \tag{9.233}$$

Using the change of variable $x = a/r$, we have

$$\int_0^a \sqrt{2m\left(E + \frac{Ze^2}{r}\right)}\, dr = \sqrt{-2mE} \int_0^a \sqrt{\frac{a}{r} - 1}\, dr = a\sqrt{-2mE} \int_0^1 \sqrt{\frac{1}{x} - 1}\, dx$$

$$= \frac{\pi}{2} a\sqrt{-2mE} = -\pi Ze^2 \sqrt{-\frac{m}{2E}}. \tag{9.234}$$

In deriving this relation, we have used the integral $\int_0^1 \sqrt{1/x - 1}\, dx = \pi/2$; this can be easily obtained by the application of the residue theorem. Combining (9.233) and (9.234) we end up with

$$E_n = -\frac{mZ^2e^4}{2\hbar^2} \frac{1}{n^2} = -\frac{Z^2e^2}{2a_0} \frac{1}{n^2}, \tag{9.235}$$

where $a_0 = \hbar^2/(me^2)$ is the Bohr radius. This is the correct (Bohr) expression for the energy levels.

Figure 9.8 A potential barrier whose classically allowed regions are specified by $x < x_1$ and $x > x_2$ and the forbidden region by $x_1 < x < x_2$.

9.4.5 Tunneling through a Potential Barrier

Consider the motion of a particle of momentum $p_0 = \sqrt{2mE}$ incident from left onto a potential barrier $V(x)$, shown in Figure 9.8, with an energy E that is smaller than the potential's maximum value V_{max}.

Classically, the particle can in no way penetrate inside the barrier; hence it will get reflected backwards. Quantum mechanically, however, the probability corresponding to the particle's tunneling through the barrier and "emerging" to the right of the barrier is not zero. In what follows we want to use the WKB approximation to estimate the particle's probability of passing through the barrier.

In regions (1) and (3) of Figure 9.8 the particle is free:

$$\psi_1(x) = \psi_{incident}(x) + \psi_{reflected}(x) = Ae^{ip_0x/\hbar} + Be^{-ip_0x/\hbar}, \qquad (9.236)$$

$$\psi_3(x) = \psi_{transmitted}(x) = Ee^{ip_0x/\hbar}, \qquad (9.237)$$

where A, B, and E are the amplitudes of the incident, reflected, and transmitted waves, respectively; in region (3) we have outgoing waves only.

What about the wave function in the classically forbidden region (2)? The WKB method provides the answer. Since the particle energy is smaller than V_{max}, i.e., $E < V_{max}$, and if the potential $V(x)$ is a slowly varying function of x, the wave function in region (2) is given by the WKB approximation (see (9.174))

$$\psi_2(x) = \frac{C}{\sqrt{|p(x)|}} \exp\left[-\frac{1}{\hbar}\int_{x_1}^{x}|p(x')|dx'\right] + \frac{D}{\sqrt{|p(x)|}} \exp\left[\frac{1}{\hbar}\int_{x_1}^{x}|p(x')|dx'\right], \quad (9.238)$$

where $p(x) = i\sqrt{2m(V(x) - E)}$. The term $D/\sqrt{|p(x)|}\exp\left\{1/\hbar\int_{x_1}^{x}|p(x')|dx'\right\}$ increases exponentially when the barrier is very wide and is therefore unphysical. We shall be considering the case where the barrier is wide enough so that the approximation $D \simeq 0$ is valid; hence $\psi_2(x)$ becomes

$$\psi_2(x) = \frac{C}{\sqrt{|p(x)|}} \exp\left[-\frac{1}{\hbar}\int_{x_1}^{x}|p(x')|dx'\right]. \qquad (9.239)$$

The probability corresponding to the particle's passage through the barrier is given by the transmission coefficient

$$T = \frac{v_{trans}}{v_{inc}} \frac{|\psi_{trans}(x)|^2}{|\psi_{inc}(x)|^2} = \frac{|E|^2}{|A|^2}, \tag{9.240}$$

since $v_{trans} = v_{inc}$ (the speeds of the incident and transmitted particles are equal). In what follows we are going to calculate the coefficient E in terms of A. For this, we need to use the continuity of the wave function and its derivative at x_1 and x_2. First, using (9.236) and (9.239), the continuity relations $\psi_1(x_1) = \psi_2(x_1)$ and $\psi_1'(x_1) = \psi_2'(x_1)$ lead, respectively, to

$$Ae^{ip_0x_1/\hbar} + Be^{-ip_0x_1/\hbar} = \frac{C}{\sqrt{a_1}}, \tag{9.241}$$

$$\frac{i}{\hbar}p_0(Ae^{ip_0x_1/\hbar} - Be^{-ip_0x_1/\hbar}) = -\frac{a_1}{\hbar\sqrt{a_1}}C, \tag{9.242}$$

where $a_1 = i\sqrt{2m(V(x_1) - E)}$. The continuity of the wave function and its derivative at x_2, $\psi_2(x_2) = \psi_3(x_2)$, and $\psi_2'(x_2) = \psi_3'(x_2)$ lead to

$$\frac{C}{\sqrt{a_2}}\exp\left[-\frac{1}{\hbar}\int_{x_1}^{x_2}|p(x)|dx\right] = Ee^{ip_0x_2/\hbar} \tag{9.243}$$

$$-\frac{a_2}{\hbar\sqrt{a_2}}C\exp\left[-\frac{1}{\hbar}\int_{x_1}^{x_2}|p(x)|dx\right] = \frac{ip_0}{\hbar}Ee^{ip_0x_2/\hbar}, \tag{9.244}$$

where $a_2 = i\sqrt{2m(V(x_2) - E)}$.

Adding (9.241) and (9.242) we get $C = 2A\sqrt{a_1}e^{ip_0x_1/\hbar}/(1 - a_1/ip_0)$ which, when inserted into (9.243), yields

$$\frac{E}{A} = \frac{2}{1 - a_1/ip_0}\sqrt{\frac{a_1}{a_2}}e^{ip_0(x_1-x_2)/\hbar}\exp\left[-\frac{1}{\hbar}\int_{x_1}^{x_2}|p(x)|dx\right], \tag{9.245}$$

which in turn leads to

$$\frac{|E|^2}{|A|^2} = \frac{4}{a_2/a_1 + a_1a_2/p_0^2}\exp\left[-\frac{2}{\hbar}\int_{x_1}^{x_2}|p(x)|dx\right]. \tag{9.246}$$

The substitution of this expression into (9.240) finally yields an approximate value for the transmission coefficient through a potential barrier $V(x)$:

$$\boxed{T \sim e^{-2\gamma}, \qquad \gamma = \frac{1}{\hbar}\int_{x_1}^{x_2}\sqrt{2m(V(x) - E)}\,dx.} \tag{9.247}$$

Tunneling phenomena are common at the microscopic scale; they occur within nuclei, within atoms, and within solids. In nuclear physics, for instance, there are nuclei that decay into an α-particle (helium nucleus with $Z = 2$) and a daughter nucleus. This process can be viewed as the tunneling of an α-particle through the potential (Coulomb) barrier between the α-particle and the daughter nucleus; once formed inside the nucleus, the α-particle cannot escape unless it tunnels through (penetrates) the Coulomb barrier surrounding it. Tunneling also occurs within metals; when a metal is subject to an external electric field, electrons can be emitted from the metal. This is known as *cold emission*; we will study it in Example 9.10.

Example 9.10

Use the WKB approximation to estimate the transmission coefficient of a particle of mass m and energy E moving in the following potential barrier:

$$V(x) = \begin{cases} 0, & x < 0, \\ V_0 - \lambda, x & x > 0. \end{cases}$$

Solution

The transmission coefficient is given by (9.247), where $x_1 = 0$ and the value of x_2, which can be obtained from the relation $V_0 - \lambda x_2 = E$, is given by $x_2 = (V_0 - E)/\lambda$. Setting the values of x_1 and x_2 into (9.247), and since $V(x) - E = (V_0 - E) - \lambda x$, we get

$$\gamma = \frac{1}{\hbar} \int_{x_1}^{x_2} \sqrt{2m(V(x) - E)}\, dx = \frac{\sqrt{2m}}{\hbar} \int_0^{(V_0-E)/\lambda} \sqrt{V_0 - E - \lambda x}\, dx$$

$$= \frac{2\sqrt{2m}}{3\hbar\lambda} (V_0 - E)^{3/2}. \tag{9.248}$$

The transmission coefficient is thus given by

$$T \sim e^{-2\gamma} = \exp\left\{ -\frac{4\sqrt{2m}}{3\hbar\lambda} (V_0 - E)^{3/2} \right\}. \tag{9.249}$$

This problem is useful for the study of *cold emission* of electrons from metals. In the absence of any external electric field, the electrons are bound by a potential of the type $V(x) = V_0$ for $x > 0$, known as the work function of the metal. When we turn on an external electric field $\mathcal{E}$, the potential seen by the electron is no longer V_0 but $V(x) = V_0 - e\mathcal{E}x$. This potential barrier has a width through which the electrons can escape: every electron of energy $E \geq e\mathcal{E}x$ can escape. The quantity $e\mathcal{E}x_2$, where $x_2 = (V_0 - E)/\lambda$, is known as the work function of the metal; the width of the potential barrier of the metal is given by $0 < x < x_2$.

9.5   Concluding Remarks

In this chapter we have studied three approximation methods that apply to stationary Hamiltonians. As we saw, approximation methods offer efficient, short ways for obtaining energy levels that are, at times, identical with the exact results. For instance, in the calculation of the energy levels of the harmonic oscillator and the hydrogen atom, we have seen in a number of solved examples how the variational method and the WKB method lead to the correct energies *without* resorting to solve the Schrödinger equation; the approximation methods deal merely with the solution of a few simple integrals. In Chapters 4 and 7, however, we have seen that, to solve the Schrödinger equation for the harmonic oscillator and for the hydrogen atom, one has to carry out lengthy, laborious calculations.

Approximation methods offer, in general, powerful economical prescriptions for determining reliable results for systems that cannot be solved exactly. In the next chapter we are going to study approximation methods that apply to time-dependent processes such as atomic transitions, decays, and so on.

9.6 Solved Problems

The topic of approximation methods touches on almost all areas of quantum mechanics, ranging from one- to three-dimensional problems, as well as on the various aspects of the formalism of quantum mechanics.

Problem 9.1

Using first-order perturbation theory, calculate the energy of the nth excited state for a spinless particle of mass m moving in an infinite potential well of length $2L$, with walls at $x = 0$ and $x = 2L$:

$$V(x) = \begin{cases} 0, & 0 \le x \le 2L, \\ \infty, & \text{otherwise,} \end{cases}$$

which is modified at the bottom by the following two perturbations:

(a) $V_p(x) = \lambda V_0 \sin(\pi x / 2L)$; (b) $V_p(x) = \lambda V_0 \delta(x - L)$, where $\lambda \ll 1$.

Solution

The exact expressions of the energy levels and of the wave functions for this potential are given by

$$E_n = \frac{\hbar^2 \pi^2}{8mL^2} n^2, \qquad \psi_n(x) = \frac{1}{\sqrt{L}} \sin\left(\frac{n\pi x}{2L}\right). \tag{9.250}$$

According to perturbation theory, the energy of the nth state is given to first order by

$$E_n = \frac{\hbar^2 \pi^2}{8mL^2} n^2 + E_n^{(1)}, \tag{9.251}$$

where

$$E_n^{(1)} = \langle \psi_n | V_p(x) | \psi_n \rangle = \frac{1}{L} \int_0^{2L} \sin^2\left(\frac{n\pi x}{2L}\right) V_p(x)\, dx. \tag{9.252}$$

(a) Using the relation

$$\int \cos nx \sin mx\, dx = -\frac{\cos(m-n)x}{2(m-n)} - \frac{\cos(m+n)x}{2(m+n)}, \qquad m \ne \pm n, \tag{9.253}$$

along with (9.252), we can calculate $E_n^{(1)}$ for $V_p(x) = \lambda V_0 \sin(\pi x / 2L)$ as follows:

$$
\begin{aligned}
E_n^{(1)} &= \frac{\lambda V_0}{L} \int_0^{2L} \sin^2\left(\frac{n\pi x}{2L}\right) \sin\left(\frac{\pi x}{2L}\right) dx \\
&= \frac{\lambda V_0}{2L} \int_0^{2L} \left[1 - \cos\left(\frac{n\pi x}{L}\right) \right] \sin\left(\frac{\pi x}{2L}\right) dx \\
&= \frac{\lambda V_0}{\pi} \left\{ -\cos\left(\frac{\pi x}{2L}\right) + \frac{\cos[(1-2n)\pi x/(2L)]}{2(1-2n)} + \frac{\cos[(1+2n)\pi x/(2L)]}{2(1+2n)} \right\} \Bigg|_0^{2L} \\
&= \frac{2\lambda V_0}{\pi} \frac{4n^2}{4n^2 - 1}. \tag{9.254}
\end{aligned}
$$

Thus, the energy (9.251) would become

$$E_n = \frac{\hbar^2 \pi^2}{8mL^2} n^2 + \frac{2\lambda V_0}{\pi} \frac{4n^2}{4n^2 - 1}. \tag{9.255}$$

(b) In the case of $V_p(x) = \lambda V_0 \delta(x - L)$, (9.252) leads to

$$E_n^{(1)} = \frac{\lambda V_0}{L} \int_0^{2L} \sin^2 \left(\frac{n\pi x}{2L} \right) \delta(x - L) dx = \frac{\lambda V_0}{L} \sin^2 \left(\frac{n\pi}{2} \right); \qquad (9.256)$$

hence, depending on whether the quantum number n is even or odd, we have

$$E_n = \frac{\hbar^2 \pi^2}{8mL^2} n^2 + \begin{cases} 0 & \text{if } n \text{ is even,} \\ \lambda V_0/L, & \text{if } n \text{ is odd.} \end{cases} \qquad (9.257)$$

Problem 9.2

Consider a system whose Hamiltonian is given by $\hat{H} = E_0 \begin{pmatrix} 1+\lambda & 0 & 0 & 0 \\ 0 & 8 & 0 & 0 \\ 0 & 0 & 3 & -2\lambda \\ 0 & 0 & -2\lambda & 7 \end{pmatrix}$,

where $\lambda \ll 1$.

(a) By decomposing this Hamiltonian into $\hat{H} = \hat{H}_0 + \hat{H}_p$, find the eigenvalues and eigenstates of the unperturbed Hamiltonian $\hat{H}_0$.

(b) Diagonalize $\hat{H}$ to find the exact eigenvalues of $\hat{H}$; expand each eigenvalue to the second power of λ.

(c) Using first- and second-order nondegenerate perturbation theory, find the approximate eigenergies of $\hat{H}$ and the eigenstates to first order. Compare these with the exact values obtained in (b).

Solution

(a) The matrix of $\hat{H}$ can be separated as follows:

$$H = \hat{H}_0 + \hat{H}_p = E_0 \begin{pmatrix} 1 & 0 & 0 & 0 \\ 0 & 8 & 0 & 0 \\ 0 & 0 & 3 & 0 \\ 0 & 0 & 0 & 7 \end{pmatrix} + E_0 \begin{pmatrix} \lambda & 0 & 0 & 0 \\ 0 & 0 & 0 & 0 \\ 0 & 0 & 0 & -2\lambda \\ 0 & 0 & -2\lambda & 0 \end{pmatrix}. \qquad (9.258)$$

Notice that $\hat{H}_0$ is already diagonal; hence its eigenvalues are given by

$$E_1^{(0)} = E_0, \quad E_2^{(0)} = 8E_0, \quad E_3^{(0)} = 3E_0, \quad E_4^{(0)} = 7E_0, \qquad (9.259)$$

and its eigenstates by

$$|\phi_1\rangle = \begin{pmatrix} 1 \\ 0 \\ 0 \\ 0 \end{pmatrix}, \quad |\phi_2\rangle = \begin{pmatrix} 0 \\ 1 \\ 0 \\ 0 \end{pmatrix}, \quad |\phi_3\rangle = \begin{pmatrix} 0 \\ 0 \\ 1 \\ 0 \end{pmatrix}, \quad |\phi_4\rangle = \begin{pmatrix} 0 \\ 0 \\ 0 \\ 1 \end{pmatrix}. \qquad (9.260)$$

(b) The diagonalization of $\hat{H}$ leads to the following secular equation:

$$\begin{vmatrix} (1+\lambda)E_0 - E & 0 & 0 & 0 \\ 0 & 8E_0 - E & 0 & 0 \\ 0 & 0 & 3E_0 - E & -2\lambda E_0 \\ 0 & 0 & -2\lambda E_0 & 7E_0 - E \end{vmatrix} = 0 \qquad (9.261)$$

or
$$(E_0 + \lambda E_0 - E)(8E_0 - E)\left[(3E_0 - E)(7E_0 - E) - 4\lambda^2 E_0^2\right] = 0, \tag{9.262}$$

which in turn leads to the following *exact* eigenenergies:

$$E_1 = (1 + \lambda)E_0, \quad E_2 = 8E_0, \quad E_3 = (5 - 2\sqrt{1 + \lambda^2})E_0, \quad E_4 = (5 + 2\sqrt{1 + \lambda^2})E_0. \tag{9.263}$$

Since $\lambda \ll 1$ we can expand $\sqrt{1 + \lambda^2}$ to second order in λ: $\sqrt{1 + \lambda^2} \simeq 1 + \lambda^2/2$. Hence E_3 and E_4 are given to second order in λ by

$$E_3 \simeq (3 - \lambda^2)E_0, \qquad E_4 \simeq (7 + \lambda^2)E_0. \tag{9.264}$$

(c) From nondegenerate perturbation theory, we can write the first-order corrections to the energies as follows:

$$E_1^{(1)} = \langle \phi_1 \mid \hat{H}_p \mid \phi_1 \rangle = E_0(1\ 0\ 0\ 0) \begin{pmatrix} \lambda & 0 & 0 & 0 \\ 0 & 0 & 0 & 0 \\ 0 & 0 & 0 & -2\lambda \\ 0 & 0 & -2\lambda & 0 \end{pmatrix} \begin{pmatrix} 1 \\ 0 \\ 0 \\ 0 \end{pmatrix} = \lambda E_0. \tag{9.265}$$

Similarly, we can verify that the second, third, and fourth eigenvalues have no first-order corrections:

$$E_2^{(1)} = \langle \phi_2 \mid \hat{H}_p \mid \phi_2 \rangle = 0, \quad E_3^{(1)} = \langle \phi_3 \mid \hat{H}_p \mid \phi_3 \rangle = 0, \quad E_4^{(1)} = \langle \phi_4 \mid \hat{H}_p \mid \phi_4 \rangle = 0. \tag{9.266}$$

Let us now consider the second-order corrections to the energy. From nondegenerate perturbation theory, we have

$$E_1^{(2)} = \sum_{m=2,3,4} \frac{\left|\langle \phi_m | \hat{H}_p \mid \phi_1 \rangle\right|^2}{E_1^{(0)} - E_m^{(0)}} = 0, \tag{9.267}$$

since $\langle \phi_2 \mid \hat{H}_p \mid \phi_1 \rangle = \langle \phi_3 \mid \hat{H}_p \mid \phi_1 \rangle = \langle \phi_4 \mid \hat{H}_p \mid \phi_1 \rangle = 0$. Similarly, we can verify that

$$E_2^{(2)} = \sum_{m=1,3,4} \frac{\left|\langle \phi_m | \hat{H}_p \mid \phi_2 \rangle\right|^2}{E_2^{(0)} - E_m^{(0)}} = 0 \tag{9.268}$$

and

$$E_3^{(2)} = \sum_{m=1,2,4} \frac{\left|\langle \phi_m | \hat{H}_p \mid \phi_3 \rangle\right|^2}{E_3^{(0)} - E_m^{(0)}} = \frac{\left|\langle \phi_4 \mid \hat{H}_p \mid \phi_3 \rangle\right|^2}{E_3^{(0)} - E_4^{(0)}} = \frac{(-2\lambda E_0)^2}{(3 - 7)E_0} = -\lambda^2 E_0, \tag{9.269}$$

because

$$\langle \phi_4 \mid \hat{H}_p \mid \phi_3 \rangle = E_0(0\ 0\ 0\ 1) \begin{pmatrix} \lambda & 0 & 0 & 0 \\ 0 & 0 & 0 & 0 \\ 0 & 0 & 0 & -2\lambda \\ 0 & 0 & -2\lambda & 0 \end{pmatrix} \begin{pmatrix} 0 \\ 0 \\ 1 \\ 0 \end{pmatrix} = -2\lambda E_0. \tag{9.270}$$

Similarly, since

$$\langle \phi_3 \mid \hat{H}_p \mid \phi_4 \rangle = E_0 (0\ 0\ 1\ 0) \begin{pmatrix} \lambda & 0 & 0 & 0 \\ 0 & 0 & 0 & 0 \\ 0 & 0 & 0 & -2\lambda \\ 0 & 0 & -2\lambda & 0 \end{pmatrix} \begin{pmatrix} 0 \\ 0 \\ 0 \\ 1 \end{pmatrix} = -2\lambda E_0, \qquad (9.271)$$

we can ascertain that

$$E_4^{(2)} = \sum_{m=1,2,3} \frac{\left| \langle \phi_m \mid \hat{H}_p \mid \phi_4 \rangle \right|^2}{E_4^{(0)} - E_m^{(0)}} = \frac{\left| \langle \phi_3 \mid \hat{H}_p \mid \phi_4 \rangle \right|^2}{E_4^{(0)} - E_3^{(0)}} = \frac{(-2\lambda E_0)^2}{(7-3)E_0} = \lambda^2 E_0. \qquad (9.272)$$

Now, combining (9.265)-(9.272), we infer that the values of the energies to second-order non-degenerate perturbation theory are given by

$$E_1 = E_1^{(0)} + E_1^{(1)} + E_1^{(2)} = (1 + \lambda) E_0, \qquad (9.273)$$

$$E_2 = E_2^{(0)} + E_2^{(1)} + E_2^{(2)} = 8 E_0, \qquad (9.274)$$

$$E_3 = E_3^{(0)} + E_3^{(1)} + E_3^{(2)} = (3 - \lambda^2) E_0, \qquad (9.275)$$

$$E_4 = E_4^{(0)} + E_4^{(1)} + E_4^{(2)} = (7 + \lambda^2) E_0. \qquad (9.276)$$

All these values are identical with their corresponding exact expressions (9.263) and (9.264).
Finally, the first-order corrections to the eigenstates are given by

$$\mid \psi_n^{(1)} \rangle = \sum_{m \neq n} \frac{\langle \phi_m \mid \hat{H}_p \mid \phi_n \rangle}{E_m^{(0)} - E_n^{(0)}} \mid \phi_m \rangle, \qquad (9.277)$$

and hence

$$\mid \psi_1^{(1)} \rangle = \sum_{m=2,3,4} \frac{\langle \phi_m \mid \hat{H}_p \mid \phi_1 \rangle}{E_m^{(0)} - E_1^{(0)}} \mid \phi_m \rangle = \begin{pmatrix} 0 \\ 0 \\ 0 \\ 0 \end{pmatrix}. \qquad (9.278)$$

Similarly, we can show that $\mid \psi_2^{(1)} \rangle$ is also given by a zero column matrix, but $\mid \psi_3^{(1)} \rangle$ and $\mid \psi_4^{(1)} \rangle$ are not:

$$\mid \psi_3^{(1)} \rangle = \sum_{m=1,2,4} \frac{\langle \phi_m \mid \hat{H}_p \mid \phi_3 \rangle}{E_m^{(0)} - E_3^{(0)}} \mid \phi_m \rangle = \frac{\langle \phi_4 \mid \hat{H}_p \mid \phi_3 \rangle}{E_4^{(0)} - E_3^{(0)}} \mid \phi_4 \rangle = \begin{pmatrix} 0 \\ 0 \\ 0 \\ -\lambda/2 \end{pmatrix}, \qquad (9.279)$$

$$\mid \psi_4^{(1)} \rangle = \sum_{m=1,2,3} \frac{\langle \phi_m \mid \hat{H}_p \mid \phi_4 \rangle}{E_m^{(0)} - E_4^{(0)}} \mid \phi_m \rangle = \frac{\langle \phi_3 \mid \hat{H}_p \mid \phi_4 \rangle}{E_3^{(0)} - E_4^{(0)}} \mid \phi_3 \rangle = \begin{pmatrix} 0 \\ 0 \\ 1 \\ \lambda/2 \end{pmatrix}. \qquad (9.280)$$

Finally, the states are given to first order by $\mid \psi_n \rangle = \mid \phi_n \rangle + \mid \psi_n^{(1)} \rangle$:

$$\mid \psi_1 \rangle = \begin{pmatrix} 1 \\ 0 \\ 0 \\ 0 \end{pmatrix}, \quad \mid \psi_2 \rangle = \begin{pmatrix} 0 \\ 1 \\ 0 \\ 0 \end{pmatrix}, \quad \mid \psi_3 \rangle = \begin{pmatrix} 0 \\ 0 \\ 1 \\ -\lambda/2 \end{pmatrix}, \quad \mid \psi_4 \rangle = \begin{pmatrix} 0 \\ 0 \\ \lambda/2 \\ 1 \end{pmatrix}.$$

$$(9.281)$$

Problem 9.3

(a) Find the exact energies and wave functions of the ground and first excited states and specify their degeneracies for the infinite cubic potential well

$$V(x, y, z) = \begin{cases} 0 & \text{if } 0 < x < L, 0 < y < L, 0 < z < L, \\ \infty & \text{otherwise.} \end{cases}$$

Now add the following perturbation to the infinite cubic well:

$$\hat{H}_p = V_0 L^3 \delta\left(x - \frac{L}{4}\right) \delta\left(y - \frac{3L}{4}\right) \delta\left(z - \frac{L}{4}\right).$$

(b) Using first-order perturbation theory, calculate the energy of the ground state.

(c) Using first-order (degenerate) perturbation theory, calculate the energy of the first excited state.

Solution

The energy and wave function for an infinite, cubic potential well of size L are given by

$$E_{n_x,n_y,n_z}^{exact} = \frac{\pi^2 \hbar^2}{2mL^2} \left(n_x^2 + n_y^2 + n_z^2\right), \tag{9.282}$$

$$\phi_{n_x,n_y,n_z}(x, y, z) = \sqrt{\frac{8}{L^3}} \sin\left(\frac{\pi n_x}{L}x\right) \sin\left(\frac{\pi n_y}{L}y\right) \sin\left(\frac{\pi n_z}{L}z\right). \tag{9.283}$$

(a) The ground state is not degenerate; its exact energy and wave function are

$$E_{111}^{exact} = \frac{3\pi^2 \hbar^2}{2mL^2}, \quad \phi_{111}(x, y, z) = \sqrt{\frac{8}{L^3}} \sin\left(\frac{\pi}{L}x\right) \sin\left(\frac{\pi}{L}y\right) \sin\left(\frac{\pi}{L}z\right). \tag{9.284}$$

The first excited state is threefold degenerate: $\phi_{112}(x, y, z)$, $\phi_{121}(x, y, z)$, and $\phi_{211}(x, y, z)$ correspond to the same energy, $E_{112}^{exact} = E_{121}^{exact} = E_{211}^{exact} = 3\pi^2 \hbar^2/(mL^2)$.

(b) The first-order correction to the ground state energy is given by

$$\begin{aligned} E_1^{(1)} &= \langle \phi_{111} \mid \hat{H}_p \mid \phi_{111} \rangle \\ &= 8V_0 \int_0^L \delta\left(x - \frac{L}{4}\right) \sin^2\left(\frac{\pi}{L}x\right) dx \int_0^L \delta\left(y - \frac{3L}{4}\right) \sin^2\left(\frac{\pi}{L}y\right) dy \\ &\quad \times \int_0^L \delta\left(z - \frac{L}{4}\right) \sin^2\left(\frac{\pi}{L}z\right) dz = 8V_0 \sin^2\left(\frac{\pi}{4}\right) \sin^2\left(\frac{3\pi}{4}\right) \sin^2\left(\frac{\pi}{4}\right) \\ &= V_0. \end{aligned} \tag{9.285}$$

Thus, the ground state energy is given to first-order perturbation by

$$E_0 = \frac{3\pi^2 \hbar^2}{2mL^2} + V_0. \tag{9.286}$$

(c) To calculate the energy of the first excited state to first order, we need to use degenerate perturbation theory. The values of this energy are equal to $3\pi^2 \hbar^2/(mL^2)$ plus the eigenvalues of the matrix

$$\begin{pmatrix} V_{11} & V_{12} & V_{13} \\ V_{21} & V_{22} & V_{23} \\ V_{31} & V_{32} & V_{33} \end{pmatrix}, \tag{9.287}$$

with $V_{nm} = \langle n \mid \hat{H}_p \mid m \rangle$, and where the following notations are used:

$$| 1 \rangle = \phi_{211}(x, y, z) = \sqrt{\frac{8}{L^3}} \sin \left(\frac{2\pi}{L} x \right) \sin \left(\frac{\pi}{L} y \right) \sin \left(\frac{\pi}{L} z \right), \qquad (9.288)$$

$$| 2 \rangle = \phi_{121}(x, y, z) = \sqrt{\frac{8}{L^3}} \sin \left(\frac{\pi}{L} x \right) \sin \left(\frac{2\pi}{L} y \right) \sin \left(\frac{\pi}{L} z \right), \qquad (9.289)$$

$$| 3 \rangle = \phi_{112}(x, y, z) = \sqrt{\frac{8}{L^3}} \sin \left(\frac{\pi}{L} x \right) \sin \left(\frac{\pi}{L} y \right) \sin \left(\frac{2\pi}{L} z \right). \qquad (9.290)$$

The calculations of the terms V_{nm} are lengthy but straightforward. Let us show how to calculate two such terms. First, V_{11} can be calculated in analogy to (9.285):

$$V_{11} = 8V_0 \int_0^L \delta \left(x - \frac{L}{4} \right) \sin^2 \left(\frac{2\pi}{L} x \right) dx \int_0^L \delta \left(y - \frac{3L}{4} \right) \sin^2 \left(\frac{\pi}{L} y \right) dy$$

$$\times \int_0^L \delta \left(z - \frac{L}{4} \right) \sin^2 \left(\frac{\pi}{L} z \right) dz = 8V_0 \sin^2 \left(\frac{\pi}{2} \right) \sin^2 \left(\frac{3\pi}{4} \right) \sin^2 \left(\frac{\pi}{4} \right)$$

$$= 2V_0; \qquad (9.291)$$

V_{12} and V_{13} are given by

$$V_{12} = 8V_0 \int_0^L \delta \left(x - \frac{L}{4} \right) \sin \left(\frac{2\pi}{L} x \right) \sin \left(\frac{\pi}{L} x \right) dx \int_0^L \delta \left(y - \frac{3L}{4} \right) \sin \left(\frac{\pi}{L} y \right)$$

$$\times \sin \left(\frac{2\pi}{L} y \right) dy \int_0^L \delta \left(z - \frac{L}{4} \right) \sin^2 \left(\frac{\pi}{L} z \right) dz = -2V_0, \qquad (9.292)$$

$$V_{13} = 8V_0 \int_0^L \delta \left(x - \frac{L}{4} \right) \sin \left(\frac{2\pi}{L} x \right) \sin \left(\frac{\pi}{L} x \right) dx \int_0^L \delta \left(y - \frac{3L}{4} \right) \sin^2 \left(\frac{\pi}{L} y \right) dy$$

$$\times \int_0^L \delta \left(z - \frac{L}{4} \right) \sin \left(\frac{\pi}{L} z \right) \sin \left(\frac{2\pi}{L} z \right) dz = 2V_0. \qquad (9.293)$$

Following this procedure, we can obtain the remaining terms:

$$V = 2V_0 \begin{pmatrix} 1 & -1 & 1 \\ -1 & 1 & -1 \\ 1 & -1 & 1 \end{pmatrix}. \qquad (9.294)$$

The diagonalization of this matrix yields a doubly degenerate eigenvalue and a nondegenerate eigenvalue,

$$E_1^{(1)} = E_2^{(1)} = 0, \qquad E_3^{(1)} = 6V_0, \qquad (9.295)$$

which lead to the energies of the first excited state:

$$E_1 = E_2 = \frac{3\pi^2 \hbar^2}{mL^2}, \qquad E_3 = \frac{3\pi^2 \hbar^2}{mL^2} + 6V_0. \qquad (9.296)$$

So the perturbation has only partially lifted the degeneracy of the first excited state.

Problem 9.4

Consider a hydrogen atom which is subject to two *weak* static fields: an electric field in the xy planes $\vec{\mathcal{E}} = \mathcal{E}(\vec{i} + \vec{j})$ and a magnetic field along the z-axis $\vec{B} = B\vec{k}$, where $\mathcal{E}$ and B are constant. Neglecting the spin–orbit interaction, calculate the energy levels of the $n = 2$ states to first-order perturbation.

Solution

In the absence of any external field, and neglecting spin–orbit interactions, the energy of the $n = 2$ state is fourfold degenerate: four different states $\mid nlm \rangle = \mid 200 \rangle$, $\mid 211 \rangle$, $\mid 210 \rangle$, and $\mid 21 - 1 \rangle$ correspond to the same energy $E_2 = -\mathcal{R}/4$, where $\mathcal{R} = m_e e^4/(2\hbar^2) = 13.6 \, \text{eV}$ is the Rydberg constant.

When the atom is placed in an external electric field $\vec{\mathcal{E}} = \mathcal{E}(\vec{i} + \vec{j})$, the energy of interaction between the electron's dipole moment $(\vec{d} = -e\vec{r})$ and $\vec{\mathcal{E}}$ is given by $-\vec{d} \cdot \vec{\mathcal{E}} = e\mathcal{E}(x + y) = e\mathcal{E}r \sin\theta (\cos\phi + \sin\phi)$. On the other hand, when subjecting the atom to an external magnetic field $\vec{B} = B\vec{k}$, the linear momentum of the electron becomes $\vec{p} \longrightarrow (\vec{p} - e\vec{A}/c)$, where $\vec{A}$ is the vector potential corresponding to $\vec{B}$. So when subjecting a hydrogen atom to both $\vec{\mathcal{E}}$ and $\vec{B}$, its Hamiltonian is given by

$$\hat{H} = \frac{1}{2\mu}\left(\vec{p} - \frac{e}{c}\vec{A}\right)^2 - \frac{e^2}{r} + e\mathcal{E}r \sin\theta(\cos\phi + \sin\phi)$$

$$= \frac{\vec{p}^2}{2\mu} - \frac{e^2}{r} - \frac{e}{2\mu c}\vec{B} \cdot \vec{L} + \frac{e}{2\mu c}A^2 + e\mathcal{E}r \sin\theta(\cos\phi + \sin\phi).$$

$$(9.297)$$

Since the magnetic field is weak, we can ignore the term $eA^2/(2\mu c)$; hence we can write $\hat{H}$ as $\hat{H} = \hat{H}_0 + \hat{H}_p$, where $\hat{H}_0$ is the Hamiltonian of an unperturbed hydrogen atom, while $\hat{H}_p$ can be treated as a perturbation:

$$\hat{H}_0 = \frac{\vec{p}^2}{2\mu} - \frac{e^2}{r}, \qquad \hat{H}_p = -\frac{eB}{2\mu c}\hat{L}_z + e\mathcal{E}r \sin\theta(\cos\phi + \sin\phi). \qquad (9.298)$$

To calculate the energy levels of the $n = 2$ state, we need to use degenerate perturbation theory, since the $n = 2$ state is fourfold degenerate; for this, we need to diagonalize the matrix

$$\begin{pmatrix} \langle 1 \mid \hat{H}_p \mid 1 \rangle & \langle 1 \mid \hat{H}_p \mid 2 \rangle & \langle 1 \mid \hat{H}_p \mid 3 \rangle & \langle 1 \mid \hat{H}_p \mid 4 \rangle \\ \langle 2 \mid \hat{H}_p \mid 1 \rangle & \langle 2 \mid \hat{H}_p \mid 2 \rangle & \langle 2 \mid \hat{H}_p \mid 3 \rangle & \langle 2 \mid \hat{H}_p \mid 4 \rangle \\ \langle 3 \mid \hat{H}_p \mid 1 \rangle & \langle 3 \mid \hat{H}_p \mid 2 \rangle & \langle 3 \mid \hat{H}_p \mid 3 \rangle & \langle 3 \mid \hat{H}_p \mid 4 \rangle \\ \langle 4 \mid \hat{H}_p \mid 1 \rangle & \langle 4 \mid \hat{H}_p \mid 2 \rangle & \langle 4 \mid \hat{H}_p \mid 3 \rangle & \langle 4 \mid \hat{H}_p \mid 4 \rangle \end{pmatrix}, \qquad (9.299)$$

where $\mid 1 \rangle = \mid 200 \rangle$, $\mid 2 \rangle = \mid 211 \rangle$, $\mid 3 \rangle = \mid 210 \rangle$, and $\mid 4 \rangle = \mid 21 - 1 \rangle$. We therefore need to calculate the term

$$\langle 2l'm' \mid \hat{H}_p \mid 2lm \rangle = -\frac{eB}{2\mu c}m\hbar\delta_{l',l}\delta_{m',m} + e\mathcal{E}\langle 2l'm' \mid r \sin\theta(\cos\phi + \sin\phi) \mid 2lm \rangle. \quad (9.300)$$

Since $x = r \sin\theta \cos\phi$ and $y = r \sin\theta \sin\phi$ are both odd, the only terms that survive among $\langle 2l'm' \mid x \mid 2lm \rangle$ and $\langle 2l'm' \mid y \mid 2lm \rangle$ are $\langle 200 \mid x \mid 21 \pm 1 \rangle$, $\langle 200 \mid y \mid 21 \pm 1 \rangle$, and their complex conjugates. That is, x and y can couple only states of different parities $(l' - l = \pm 1)$ and whose azimuthal quantum numbers satisfy this condition: $m' - m = \pm 1$. So we need to calculate only

$$\langle 200 \mid x \mid 21 \pm 1 \rangle = \int_0^{+\infty} R_{20}^*(r) R_{21}(r) r^3 dr \int Y_{00}^*(\Omega) \sin\theta \cos\phi Y_{1\pm1}(\Omega) d\Omega, \quad (9.301)$$

$$\langle 200 \mid y \mid 21 \pm 1 \rangle = \int_0^{+\infty} R_{20}^*(r) R_{21}(r) r^3 dr \int Y_{00}^*(\Omega) \sin\theta \sin\phi Y_{1\pm1}(\Omega) d\Omega, \quad (9.302)$$

where

$$\int_0^{+\infty} R_{20}^*(r) R_{21}(r) r^3 dr = -3\sqrt{3} a_0; \quad (9.303)$$

a_0 is the Bohr radius, $a_0 = \hbar^2/(m_e e^2)$. Using the relations

$$\sin\theta \cos\phi = \sqrt{\frac{2\pi}{3}} \left[Y_{1-1}(\Omega) - Y_{11}(\Omega) \right], \qquad \sin\theta \sin\phi = i\sqrt{\frac{2\pi}{3}} \left[Y_{1-1}(\Omega) + Y_{11}(\Omega) \right],$$
$$(9.304)$$

along with

$$\int Y_{l'm'}^*(\Omega) Y_{lm}(\Omega) d\Omega = \delta_{l',l} \delta_{m',m}, \quad (9.305)$$

we obtain

$$\int Y_{00}^*(\Omega) \sin\theta \cos\phi Y_{11}(\Omega) d\Omega = \frac{1}{\sqrt{4\pi}} \int \sin\theta \cos\phi Y_{11}(\Omega) d\Omega = \frac{1}{\sqrt{6}} \int Y_{1-1}(\Omega) Y_{11}(\Omega) d\Omega$$
$$= -\frac{1}{\sqrt{6}}, \quad (9.306)$$

$$\int Y_{00}^*(\Omega) \sin\theta \sin\phi Y_{11}(\Omega) d\Omega = \frac{i}{\sqrt{6}} \int Y_{1-1}(\Omega) Y_{11}(\Omega) d\Omega = -\frac{i}{\sqrt{6}}. \quad (9.307)$$

Similarly, we have

$$\int Y_{00}^*(\Omega) \sin\theta \cos\phi Y_{1-1}(\Omega) d\Omega = \frac{1}{\sqrt{6}}. \quad (9.308)$$

$$\int Y_{00}^*(\Omega) \sin\theta \sin\phi Y_{1-1}(\Omega) d\Omega = -\frac{i}{\sqrt{6}}. \quad (9.309)$$

Now, substituting (9.303), (9.306), and (9.308) into (9.301), we end up with

$$\langle 200 \mid x \mid 21 \pm 1 \rangle = \pm\frac{3}{\sqrt{2}} a_0, \qquad \langle 200 \mid y \mid 21 \pm 1 \rangle = \frac{3i}{\sqrt{2}} a_0; \quad (9.310)$$

hence

$$\langle 21 \pm 1 \mid x \mid 200 \rangle = \pm\frac{3}{\sqrt{2}} a_0, \qquad \langle 21 \pm 1 \mid y \mid 200 \rangle = -\frac{3i}{\sqrt{2}} a_0. \quad (9.311)$$

The matrix (9.299) thus becomes

$$\begin{pmatrix} 0 & \alpha + i\alpha & 0 & -\alpha + i\alpha \\ \alpha - i\alpha & -\beta & 0 & 0 \\ 0 & 0 & 0 & 0 \\ -\alpha - i\alpha & 0 & 0 & \beta \end{pmatrix}, \quad (9.312)$$

where α and β stand for $\alpha = 3e\mathcal{E}a_0/\sqrt{2}$ and $\beta = e\hbar B/(2\mu c)$.

The diagonalization of (9.312) yields the following eigenvalues:

$$\lambda_1 = -\sqrt{\frac{e^2\hbar^2 B^2}{4\mu^2 c^2} + 18e^2\mathcal{E}^2 a_0^2}, \quad \lambda_2 = \lambda_3 = 0, \quad \lambda_4 = \sqrt{\frac{e^2\hbar^2 B^2}{4\mu^2 c^2} + 18e^2\mathcal{E}^2 a_0^2}. \quad (9.313)$$

Finally, the energy levels of the $n = 2$ states are given to first-order approximation by

$$E_{2_1}^{(1)} = -\frac{R}{4} - \sqrt{\frac{e^2\hbar^2 B^2}{4\mu^2 c^2} + 18e^2\mathcal{E}^2 a_0^2}, \qquad E_{2_2}^{(1)} = -\frac{R}{4}, \quad (9.314)$$

$$E_{2_3}^{(1)} = -\frac{R}{4}, \qquad E_{2_4}^{(1)} = -\frac{R}{4} + \sqrt{\frac{e^2\hbar^2 B^2}{4\mu^2 c^2} + 18e^2\mathcal{E}^2 a_0^2}. \quad (9.315)$$

So the external electric and magnetic fields have lifted the degeneracy of the $n = 2$ level only partially.

Problem 9.5

A system, with an unperturbed Hamiltonian $\hat{H}_0$, is subject to a perturbation $\hat{H}_1$ with

$$\hat{H}_0 = E_0 \begin{pmatrix} 15 & 0 & 0 & 0 \\ 0 & 3 & 0 & 0 \\ 0 & 0 & 3 & 0 \\ 0 & 0 & 0 & 3 \end{pmatrix}, \qquad \hat{H}_1 = \frac{E_0}{100} \begin{pmatrix} 0 & 0 & 0 & 0 \\ 0 & 0 & 1 & 0 \\ 0 & 1 & 0 & 0 \\ 0 & 0 & 0 & 0 \end{pmatrix}.$$

(a) Find the eigenstates of the unperturbed Hamiltonian $\hat{H}_0$ as well as the exact eigenvalues of the total Hamiltonian $\hat{H} = \hat{H}_0 + \hat{H}_p$.

(b) Find the eigenenergies of $\hat{H}$ to first-order perturbation. Compare them with the exact values obtained in (a).

Solution

(a) First, a diagonalization of $\hat{H}_0$ yields the eigenstates

$$|\phi_1\rangle = \begin{pmatrix} 1 \\ 0 \\ 0 \\ 0 \end{pmatrix}, \quad |\phi_2\rangle = \begin{pmatrix} 0 \\ 1 \\ 0 \\ 0 \end{pmatrix}, \quad |\phi_3\rangle = \begin{pmatrix} 0 \\ 0 \\ 1 \\ 0 \end{pmatrix}, \quad |\phi_4\rangle = \begin{pmatrix} 0 \\ 0 \\ 0 \\ 1 \end{pmatrix}. \quad (9.316)$$

The values of the unperturbed energies are given by a nondegenerate value $E_1^{(0)} = 15E_0$ and a *threefold degenerate* value $E_2^{(0)} = E_3^{(0)} = E_4^{(0)} = 3E_0$.

The *exact* eigenvalues of $\hat{H}$ can be obtained by diagonalizing $\hat{H}$. Adopting the notation $\lambda = 1/100$, we can write the secular equation as

$$\begin{vmatrix} 15E_0 - E & 0 & 0 & 0 \\ 0 & 3E_0 - E & \lambda E_0 & 0 \\ 0 & \lambda E_0 & 3E_0 - E & 0 \\ 0 & 0 & 0 & 3E_0 - E \end{vmatrix} = 0 \quad (9.317)$$

or

$$(15E_0 - E)(3E_0 - E)\left[(3E_0 - E)^2 - \lambda^2 E_0^2\right] = 0, \tag{9.318}$$

which in turn leads to the *exact* values of the eigenenergies:

$$E_1 = 15E_0, \quad E_2 = 3E_0, \quad E_3 = (3 - \lambda)E_0, \quad E_4 = (3 + \lambda)E_0. \tag{9.319}$$

(b) To calculate the energy eigenvalues of $\hat{H}$ to first-order degenerate perturbation, and since $\hat{H}_0$ has one nondegenerate eigenvalue, $15E_0$, and a threefold degenerate eigenvalue, $3E_0$, we need to make use of both nondegenerate and degenerate perturbative treatments. First, let us focus on the nondegenerate state; its energy is given by

$$
\begin{aligned}
E_1 &= 15E_0 + \langle \phi_1 \mid \hat{H}_1 \mid \phi_1 \rangle \\
&= 15E_0 + \frac{E_0}{100}(1\ 0\ 0\ 0)\begin{pmatrix} 0 & 0 & 0 & 0 \\ 0 & 0 & \lambda E_0 & 0 \\ 0 & \lambda E_0 & 0 & 0 \\ 0 & 0 & 0 & 0 \end{pmatrix}\begin{pmatrix} 1 \\ 0 \\ 0 \\ 0 \end{pmatrix} \\
&= 15E_0.
\end{aligned} \tag{9.320}
$$

This is identical with the *exact* eigenvalue (9.319) obtained in (a).

Second, to find the degenerate states, we need to diagonalize the matrix

$$V = \begin{pmatrix} V_{11} & V_{12} & V_{13} \\ V_{21} & V_{22} & V_{23} \\ V_{31} & V_{32} & V_{33} \end{pmatrix}, \tag{9.321}$$

where

$$V_{11} = \langle \phi_2 \mid \hat{H}_p \mid \phi_2 \rangle = (0\ 1\ 0\ 0)\begin{pmatrix} 0 & 0 & 0 & 0 \\ 0 & 0 & \lambda E_0 & 0 \\ 0 & \lambda E_0 & 0 & 0 \\ 0 & 0 & 0 & 0 \end{pmatrix}\begin{pmatrix} 0 \\ 1 \\ 0 \\ 0 \end{pmatrix} = 0, \tag{9.322}$$

$$V_{12} = \langle \phi_2 \mid \hat{H}_p \mid \phi_3 \rangle = (0\ 1\ 0\ 0)\begin{pmatrix} 0 & 0 & 0 & 0 \\ 0 & 0 & \lambda E_0 & 0 \\ 0 & \lambda E_0 & 0 & 0 \\ 0 & 0 & 0 & 0 \end{pmatrix}\begin{pmatrix} 0 \\ 0 \\ 1 \\ 0 \end{pmatrix} = \lambda E_0, \tag{9.323}$$

$$V_{13} = \langle \phi_2 \mid \hat{H}_p \mid \phi_4 \rangle = (0\ 1\ 0\ 0)\begin{pmatrix} 0 & 0 & 0 & 0 \\ 0 & 0 & \lambda E_0 & 0 \\ 0 & \lambda E_0 & 0 & 0 \\ 0 & 0 & 0 & 0 \end{pmatrix}\begin{pmatrix} 0 \\ 0 \\ 0 \\ 1 \end{pmatrix} = 0. \tag{9.324}$$

Similarly, we can show that

$$V_{21} = \langle \phi_3 \mid \hat{H}_p \mid \phi_2 \rangle = \lambda E_0, \quad V_{22} = \langle \phi_3 \mid \hat{H}_p \mid \phi_3 \rangle = 0, \quad V_{23} = \langle \phi_3 \mid \hat{H}_p \mid \phi_4 \rangle = 0, \tag{9.325}$$

$$V_{31} = \langle \phi_4 \mid \hat{H}_p \mid \phi_2 \rangle = 0, \quad V_{32} = \langle \phi_4 \mid \hat{H}_p \mid \phi_3 \rangle = 0, \quad V_{33} = \langle \phi_4 \mid \hat{H}_p \mid \phi_4 \rangle = 0. \tag{9.326}$$

Figure 9.9 Shapes of $\psi_0(x) = Ae^{-\alpha|x|}$, $d\psi_0(x)/dx$, and $d^2\psi_0(x)/dx^2$.

So the diagonalization of

$$V = \begin{pmatrix} 0 & \lambda E_0 & 0 \\ \lambda E_0 & 0 & 0 \\ 0 & 0 & 0 \end{pmatrix} \tag{9.327}$$

leads to the corrections $E_2^{(1)} = 0$, $E_3^{(1)} = \lambda E_0$, and $E_4^{(1)} = -\lambda E_0$. Thus, the energy eigenvalues to first-order degenerate perturbation are

$$E_2 = E_2^{(0)} + E_2^{(1)} = 3E_0, \qquad E_3 = E_3^{(0)} + E_3^{(1)} = (3 - \lambda)E_0, \tag{9.328}$$

$$E_4 = E_4^{(0)} + E_4^{(1)} = (3 + \lambda)E_0. \tag{9.329}$$

These are indeed identical with the *exact* eigenenergies (9.319) obtained in (a).

Problem 9.6

Use the variational method to estimate the energy of the ground state of a one-dimensional harmonic oscillator by making use of the following two trial functions:

(a) $\psi_0(x, \alpha) = Ae^{-\alpha|x|}$,   (b) $\psi_0(x, \alpha) = A/(x^2 + \alpha)$,

where α is a positive real number and where A is the normalization constant.

Solution

(a) This wave function, whose shape is displayed in Figure 9.9a, is quite different from a Gaussian: it has a cusp at $x = 0$; hence its first derivative is discontinuous at $x = 0$.

The normalization constant A can be calculated at once:

$$\langle \psi_0 \mid \psi_0 \rangle = A^2 \int_{-\infty}^{0} e^{2\alpha x} \, dx + A^2 \int_{0}^{\infty} e^{-2\alpha x} \, dx = 2A^2 \int_{0}^{\infty} e^{-2\alpha x} \, dx = \frac{A^2}{\alpha}; \tag{9.330}$$

hence $A = \sqrt{\alpha}$. To find $E_0(\alpha)$ we need to calculate the potential and the kinetic terms. Using the integral $\int_{0}^{+\infty} x^n e^{-ax} \, dx = n!/a^{n+1}$ we can easily calculate the potential term:

$$\langle \psi_0 | V(x) | \psi_0 \rangle = \frac{1}{2} m\omega^2 A^2 \int_{-\infty}^{+\infty} x^2 e^{-2\alpha|x|} \, dx = m\omega^2 A^2 \int_{0}^{+\infty} x^2 e^{-2\alpha x} \, dx = \frac{m\omega^2}{4\alpha^2}. \tag{9.331}$$

But the kinetic energy term $-(\hbar^2/2m)\langle\psi_0|d^2/dx^2|\psi_0\rangle$ is quite tricky to calculate. Since the first derivative of $\psi_0(x)$ is discontinuous at $x = 0$, a careless, straightforward calculation of $\langle\psi_0|d^2/dx^2|\psi_0\rangle$, which makes use of (9.123), leads to a negative kinetic energy:

$$
\begin{aligned}
-\frac{\hbar^2}{2m}\left\langle\psi_0\left|\frac{d^2}{dx^2}\right|\psi_0\right\rangle &= -\frac{\hbar^2}{2m}A^2\int_{-\infty}^{+\infty}e^{-\alpha|x|}\frac{d^2e^{-\alpha|x|}}{dx^2}dx \\
&= -\frac{\hbar^2}{m}A^2\int_0^{+\infty}e^{-\alpha x}\frac{d^2e^{-\alpha x}}{dx^2}dx \\
&= -\frac{\hbar^2\alpha^2}{m}A^2\int_0^{+\infty}e^{-2\alpha x}dx = -\frac{\hbar^2\alpha^2}{2m}.
\end{aligned}
$$

(9.332)

So when the first derivative of the wave function is discontinuous, the correct way to calculate the kinetic energy term is by using (9.124):

$$
\begin{aligned}
-\frac{\hbar^2}{2m}\left\langle\psi_0\left|\frac{d^2}{dx^2}\right|\psi_0\right\rangle &= \frac{\hbar^2}{2m}A^2\int_{-\infty}^{+\infty}\left|\frac{de^{-\alpha|x|}}{dx}\right|^2 dx = \frac{\hbar^2\alpha^2}{2m}A^2\int_{-\infty}^{+\infty}e^{-2\alpha|x|}dx \\
&= \frac{\hbar^2\alpha^2}{2m},
\end{aligned}
$$

(9.333)

because $A^2\int_{-\infty}^{+\infty}e^{-2\alpha|x|}dx = 1$.

Why do expressions (9.332) and (9.333) yield different results? The reason is that the correct expression of $d^2e^{-\alpha|x|}/dx^2$ must involve a delta function (Figures 9.9a and 9.9b). That is, the correct form of $d\psi(x)/dx$ is given by

$$
\frac{d\psi_0(x)}{dx} = A\frac{de^{-\alpha|x|}}{dx} = -\alpha\,\psi_0(x)\frac{d|x|}{dx} = -\alpha\,\psi_0(x)\begin{cases} -1, & x < 0, \\ 1, & x > 0. \end{cases}
$$

(9.334)

or

$$
\frac{d\psi_0(x)}{dx} = -\alpha\,[\Theta(x) - \Theta(-x)]\,\psi_0(x),
$$

(9.335)

where $\Theta(x)$ is the Heaviside function

$$
\Theta(x) = \begin{cases} 0, & x < 0, \\ 1, & x > 0. \end{cases}
$$

(9.336)

The second derivative of $\psi_0(x)$ therefore contains a delta function:

$$
\frac{d^2\psi_0(x)}{dx^2} = \frac{d}{dx}\{-\alpha\,[\Theta(x) - \Theta(-x)]\,\psi_0(x)\},
$$

(9.337)

and since

$$
\frac{d\Theta(x)}{dx} = \delta(x), \qquad [\Theta(x) - \Theta(-x)]^2 = 1,
$$

(9.338)

and since $\delta(x) = \delta(-x)$, we have

$$
\begin{aligned}
\frac{d^2\psi_0(x)}{dx^2} &= \alpha^2\,[\Theta(x) - \Theta(-x)]^2\,\psi_0(x) - \alpha\,[\delta(x) + \delta(-x)]\,\psi_0(x) \\
&= \alpha^2\,\psi_0(x) - 2\alpha\,\psi_0(x)\delta(x).
\end{aligned}
$$

(9.339)

So the substitution of (9.339) into (9.332) leads to the same (correct) expression as (9.333):

$$
\begin{aligned}
-\frac{\hbar^2}{2m}\left\langle \psi_0 \left| \frac{d^2}{dx^2} \right| \psi_0 \right\rangle &= -\frac{\hbar^2}{2m}\int_{-\infty}^{+\infty} \psi_0^*(x)\frac{d^2\psi_0(x)}{dx^2}\,dx \\
&= -\frac{\hbar^2}{2m}\int_{-\infty}^{+\infty} \psi_0^*(x)\left[\alpha^2\psi_0(x) - 2\alpha\psi_0(x)\delta(x)\right]dx \\
&= -\frac{\hbar^2}{2m}\alpha^2 + \frac{\hbar^2}{m}\alpha\,|\psi_0(0)|^2 = -\frac{\hbar^2}{2m}\alpha^2 + \frac{\hbar^2}{m}\alpha^2 \\
&= \frac{\hbar^2}{2m}\alpha^2.
\end{aligned}
\tag{9.340}
$$

Now, adding (9.331) and (9.340), we get

$$
E_0(\alpha) = \frac{\hbar^2}{2m}\alpha^2 + \frac{m\omega^2}{4\alpha^2}.
\tag{9.341}
$$

The minimization of $E_0(\alpha)$,

$$
0 = \frac{\partial E_0(\alpha)}{\partial\alpha} = \frac{\hbar^2}{m}\alpha_0 - \frac{m\omega^2}{2\alpha_0^3},
\tag{9.342}
$$

leads to $\alpha_0^2 = m\omega/(\sqrt{2}\hbar)$ which, when inserted into (9.341), leads to

$$
E_0(\alpha_0) = \frac{\hbar^2}{2m}\frac{m\omega}{\sqrt{2}\hbar} + \frac{m\omega^2}{4}\frac{\sqrt{2}\hbar}{m\omega} = \frac{\hbar\omega}{\sqrt{2}} = 0.707\hbar\omega.
\tag{9.343}
$$

This inaccurate result was expected; it is due to the cusp at $x = 0$.

(b) We can show that the normalization constant A is given by $A = (4\alpha^3/\pi^2)^{1/4}$. Unlike $Ae^{-\alpha|x|}$, the first derivative of the trial $A/(1+x^2)$ is continuous; hence we can use (9.123) to calculate the kinetic energy term. The ground state energy is given by

$$
\begin{aligned}
E_0(\alpha) &= \left\langle \psi_0(\alpha)|\hat{H}|\psi_0(\alpha)\right\rangle \\
&= A^2\int_{-\infty}^{+\infty}\frac{1}{x^2+\alpha}\left(-\frac{\hbar^2}{2m}\frac{d^2}{dx^2} + \frac{1}{2}m\omega^2 x^2\right)\frac{1}{x^2+\alpha}\,dx \\
&= -\frac{A^2\hbar^2}{2m}\int_{-\infty}^{+\infty}\frac{6x^2 - 2\alpha}{(x^2+\alpha)^4}\,dx + \frac{1}{2}m\omega^2 A^2\int_{-\infty}^{+\infty}\frac{x^2}{(x^2+\alpha)^2}\,dx \\
&= \frac{\hbar^2}{4m\alpha} + \frac{1}{2}m\omega^2\alpha.
\end{aligned}
\tag{9.344}
$$

The minimization of $E_0(\alpha)$ with respect to α (i.e., $\partial E(\alpha)/\partial\alpha = 0$) yields $\alpha_0 = \hbar/(\sqrt{2}m\omega)$ which, when inserted into (9.344), leads to

$$
E_0(\alpha_0) = \frac{\hbar\omega}{\sqrt{2}}.
\tag{9.345}
$$

This energy, which is larger than the exact value $\hbar\omega/2$ by a factor of $\sqrt{2}$, is similar to that of part (a); this is a pure coincidence. The size of this error is due to the fact that the trial function $A/(x^2+\alpha)$ is not a good approximation to the exact wave function, which has a Gaussian form.

Problem 9.7

For a particle of mass m moving in a one-dimensional box with walls at $x = 0$ and $x = L$, use the variational method to estimate
  (a) its ground state energy and
  (b) its first excited state energy.

Solution

The exact solutions of this problem are known: $E_n^{exact} = \pi^2\hbar^2 n^2/(2mL^2)$.

(a) The trial function for the ground state must vanish at the walls, it must have no nodes, and must be symmetric (i.e., even) with respect to $x = L/2$. These three requirements can be satisfied by the following parabolic trial function:

$$\psi_0(x) = x(L - x); \tag{9.346}$$

no scale parameter is needed here. Since no parameter is involved, we can calculate the energy directly (no variation is required): $E_0 = \langle\psi_0|\hat{H} \mid \psi_0\rangle/\langle\psi_0 \mid \psi_0\rangle$, where

$$\langle\psi_0 \mid \psi_0\rangle = \int_0^L \psi_0^2(x)\, dx = \int_0^L x^2(L^2 - 2Lx + x^2)\, dx = \frac{1}{30}L^5, \tag{9.347}$$

and

$$\langle\psi_0|\hat{H} \mid \psi_0\rangle = \frac{\hbar^2}{2m}\int_0^L \left(\frac{d\psi_0(x)}{dx}\right)^2 dx = \frac{\hbar^2}{2m}\int_0^L (L^2 - 4Lx + 4x^2)\, dx = \frac{\hbar^2 L^3}{6m}. \tag{9.348}$$

Thus, the ground state energy is given by

$$E_0^{VM} = \frac{\langle\psi_0|\hat{H} \mid \psi_0\rangle}{\langle\psi_0 \mid \psi_0\rangle} = 10\frac{\hbar^2}{2mL^2}. \tag{9.349}$$

This is a very accurate result, for it is higher than the exact result by a mere 1%:

$$E_0^{VM} = \frac{10}{\pi^2}E^{exact}. \tag{9.350}$$

(b) The properties of the exact wave function of the first excited state are known: it has one node at $x = L/2$ and must be odd with respect to $x = L/2$; this last property makes it orthogonal to the ground state which is even about $L/2$. Let us try a polynomial function. Since the wave function vanishes at $x = 0$, $L/2$, and L, the trial function must be at least cubic. The following polynomial function satisfies all these conditions:

$$\psi_1(x) = x\left(x - \frac{L}{2}\right)(x - L). \tag{9.351}$$

Again, no scale parameter is needed.
  To calculate E_1^{VM}, we need to find

$$\langle\psi_1 \mid \psi_1\rangle = \int_0^L \psi_1^2(x)\,dx = \int_0^L x^2\left(x - \frac{L}{2}\right)^2(x - L)^2 dx = \frac{1}{840}L^7 \tag{9.352}$$

and

$$\langle \psi_1 | \hat{H} | \psi_1 \rangle = \frac{\hbar^2}{2m} \int_0^L \left(\frac{d\psi_1(x)}{dx} \right)^2 dx - \frac{\hbar^2}{2m} \int_0^L \left(3x^2 - 3Lx + \frac{L^2}{2} \right)^2 dx$$

$$= \frac{\hbar^2 L^5}{40m}. \tag{9.353}$$

Dividing the previous two expressions, we obtain the energy of the first excited state:

$$E_1^{VM} = \frac{\langle \psi_1 | \hat{H} | \psi_1 \rangle}{\langle \psi_1 | \psi_1 \rangle} = 42 \frac{\hbar^2}{2mL^2}. \tag{9.354}$$

This too is a very accurate result; since $E_1^{exact} = (2\pi)^2 \hbar^2/(2mL^2)$ we can write E_1^{VM} as $E_1^{VM} = 42 E_1^{exact}/(2\pi)^2$; hence E_1^{VM} is higher than E_1^{exact} by 6%.

Problem 9.8

Consider an infinite, one-dimensional potential well of length L, with walls at $x = 0$ and $x = L$, that is modified at the bottom by a perturbation $V_p(x)$:

$$V(x) = \begin{cases} 0, & 0 < x < L, \\ \infty, & \text{elsewhere,} \end{cases} \qquad V_p(x) = \begin{cases} V_0, & 0 \le x \le L/2, \\ 0, & \text{elsewhere,} \end{cases}$$

where $V_0 \ll 1$.

(a) Using first-order perturbation theory, calculate the energy E_n.

(b) Calculate the energy E_n in the WKB approximation. Compare this energy with the expression obtained in (a).

Solution

The exact energy E_n^{exact} and wave function $\phi_n(x)$ for a potential well are given by

$$E_n^{exact} = \frac{\pi^2 \hbar^2}{2mL^2} n^2, \qquad \phi_n(x) = \sqrt{\frac{2}{L}} \sin \left(\frac{n\pi x}{L} \right).$$

(a) Since the first-order correction to the energy caused by the perturbation $V_p(x)$ is given by

$$E_n^{(1)} = \langle \phi_n | V_p | \phi_n \rangle = \frac{2}{L} V_0 \int_0^{L/2} \sin^2 \left(\frac{n\pi x}{L} \right) dx$$

$$= \frac{1}{L} V_0 \int_0^{L/2} \left[1 - \cos \left(\frac{2n\pi x}{L} \right) \right] dx = \frac{V_0}{2}; \tag{9.355}$$

hence the energy is given to first-order perturbation by

$$E_n^{PT} = \frac{\pi^2 \hbar^2}{2mL^2} n^2 + \frac{V_0}{2}. \tag{9.356}$$

(b) Since this potential has two rigid walls, the energy within the WKB approximation needs to be extracted from the quantization condition $\int_0^L p(E_n, x) dx = n\pi \hbar$, where

$$\int_0^L p(E_n, x) dx = \sqrt{2m(E_n - V_0)} \int_0^{L/2} dx + \sqrt{2m E_n} \int_{L/2}^L dx$$

$$= \frac{L}{2} \sqrt{2m} \left(\sqrt{E_n - V_0} + \sqrt{E_n} \right); \tag{9.357}$$

hence $L\sqrt{2m}\left(\sqrt{E_n - V_0} + \sqrt{E_n}\right) = 2n\pi\hbar$ or

$$\sqrt{E_n - V_0} + \sqrt{E_n} = \frac{2n\pi\hbar}{L\sqrt{2m}}. \tag{9.358}$$

Squaring both sides of this equation and using the notation $\alpha_n = 2n^2\pi^2\hbar^2/(mL^2)$, we have

$$2\sqrt{E_n(E_n - V_0)} = \alpha_n - 2E_n + V_0. \tag{9.359}$$

Squaring both sides of this equation, we obtain

$$4E_n^2 - 4E_nV_0 = \alpha_n^2 + 4E_n^2 + V_0^2 - 4\alpha_nE_n + 2\alpha_nV_0 - 4E_nV_0, \tag{9.360}$$

which, solving for E_n, leads to

$$E_n = \frac{\alpha_n}{4} + \frac{V_0}{2} + \frac{V_0^2}{4\alpha_n} \tag{9.361}$$

or

$$E_n^{WKB} = \frac{\pi^2\hbar^2}{2mL^2}n^2 + \frac{V_0}{2} + \frac{mL^2V_0^2}{8\pi^2\hbar^2}\frac{1}{n^2}. \tag{9.362}$$

When $n \gg 1$, and since V_0 is very small, the WKB energy relation (9.362) gives back the expression (9.356) that was derived from a first-order perturbative treatment:

$$E_n^{WKB} \simeq E_n^{PT} = \frac{\pi^2\hbar^2}{2mL^2}n^2 + \frac{V_0}{2}. \tag{9.363}$$

Problem 9.9

Consider a particle of mass m that is bouncing vertically and elastically on a reflecting hard floor where $V(z) = \begin{cases} mgz, & z > 0, \\ +\infty, & z \le 0, \end{cases}$ and g is the gravitational constant.

(a) Use the variational method to estimate the ground state energy of this particle.
(b) Use the WKB method to estimate the ground state energy of this particle.
(c) Compare the results of (a) and (b) with the exact ground state energy.

Solution

(a) The ground state wave function of this particle has no nodes and must vanish at $z = 0$ and be finite as $z \rightarrow +\infty$. The following trial function satisfies these conditions:

$$\psi_0(z, \alpha) = Aze^{-\alpha z}, \tag{9.363}$$

where α is a parameter and A is the normalization constant. We can show that $A = 2\alpha^{3/2}$ and hence

$$\psi_0(z, \alpha) = 2\sqrt{\alpha^3}ze^{-\alpha z}. \tag{9.364}$$

The energy is given by

$$\begin{aligned} E_0^{VM}(\alpha) &= 4\alpha^3 \int_0^{+\infty} ze^{-\alpha z}\left[-\frac{\hbar^2}{2m}\frac{d^2}{dz^2} + mgz\right]ze^{-\alpha z}\,dz \\ &= 4\alpha^3\frac{\hbar^2}{2m}\int_0^{+\infty}\left(2\alpha z - \alpha^2 z^2\right)e^{-2\alpha z}\,dz + 4\alpha^3 mg\int_0^{+\infty} z^3e^{-2\alpha z}\,dz \\ &= 2\alpha^3\frac{\hbar^2}{m}\left(\frac{1}{2\alpha} - \frac{1}{4\alpha}\right) + 4mg\alpha^3\frac{3}{8\alpha^4}, \end{aligned} \tag{9.365}$$

or

$$E_0^{VM}(\alpha) = \frac{\hbar^2}{2m}\alpha^2 + \frac{3}{2\alpha}mg. \tag{9.366}$$

The minimization of $E_0(\alpha)$ yields $\alpha_0 = (3m^2g/2\hbar^2)^{1/3}$ and hence

$$E_0^{VM}(\alpha_0) = \frac{3}{2}\left(\frac{9}{2}\right)^{1/3}\left(\frac{1}{2}mg^2\hbar^2\right)^{1/3}. \tag{9.367}$$

(b) Since this potential has one rigid wall at $x = 0$, the correct quantization rule is given by (9.224): $\int_0^{E/mg} p\,dz = (n+\frac{3}{4})\pi\hbar$; the turning point occurs at $E = mgz$ and hence $z = E/mg$. Now, since $E = p^2/2m + mgz$ we have $p(E, z) = \sqrt{2mE}\sqrt{1 - mgz/E}$, and therefore

$$\int_0^{E/mg} p(E_n, z)\,dz = \sqrt{2mE}\int_0^{E/mg}\sqrt{1 - \frac{mg}{E}z}\,dz = \sqrt{2mE}\frac{2E}{3mg} = \sqrt{\frac{8E^3}{9mg^2}}. \tag{9.368}$$

Inserting this relation into the quantization condition $\int_0^{E/mg} p\,dz = (n+\frac{3}{4})\pi\hbar$ gives

$$\sqrt{\frac{8E^3}{9mg^2}} = \left(n + \frac{3}{4}\right)\pi\hbar, \tag{9.369}$$

and we obtain the WKB approximation for the energy:

$$E_n^{WKB} = \left[\frac{9\pi^2}{8}mg^2\pi^2\hbar^2\left(n + \frac{3}{4}\right)^2\right]^{1/3}. \tag{9.370}$$

Hence the ground state energy is given by

$$E_0^{WKB} = \frac{3}{4}\left(3\pi^2\right)^{1/3}\left(\frac{1}{2}mg^2\hbar^2\right)^{1/3}. \tag{9.371}$$

(c) Recall that the *exact* ground state energy, calculated in Problem 4.18, page 275, for a particle of mass m moving in the potential $V(z) = mgz$ is given by

$$E_0^{exact} = 2.338\left(\frac{1}{2}mg^2\hbar^2\right)^{1/3}. \tag{9.372}$$

Combining this relation with (9.367) and (9.371), we see that the variational method overestimates the energy by a 5.9% error, while the WKB method underestimates it by a 0.8% error:

$$E_0^{VM} = \frac{3}{2}\left(\frac{9}{2}\right)^{1/3}\frac{E_0^{exact}}{2.338} \simeq 1.059 E_0^{exact}, \tag{9.373}$$

$$E_0^{WKB} = \frac{3}{4}\left(3\pi^2\right)^{1/3}\frac{E_0^{exact}}{2.338} \simeq 0.992 E_0^{exact}. \tag{9.374}$$

The variational method has given a reasonably accurate result because we succeeded quite well in selecting the trial function. As for the WKB method, it has given a very accurate result because we have used the correct quantization rule (9.224). Had we used the quantization rule (9.210), which contains a factor of $\frac{1}{2}$ instead of $\frac{1}{4}$ in (9.224), the WKB method would have given a very inaccurate result with a 24.3% error, i.e., $E_0^{WKB} \simeq 0.757 E_0^{exact}$.

Problem 9.10

Using first-order perturbation theory, and ignoring the spin of the electron, calculate the energy of the 2p level of a hydrogen atom when placed in a *weak* quadrupole field whose principal axes are along the xyz axes: $\hat{H}_p = \sum_{\mu=-2}^{\mu=2} Q_\mu r^2 Y_{2\mu}(\Omega)$, where Q_μ are real numbers, with $Q_{-1} = Q_1 = 0$ and $Q_{-2} = Q_2$, and $Y_{2\mu}(\Omega)$ are spherical harmonics.

Solution

In the absence of the field, the energy levels of the $| 2, 1, m \rangle$ states are threefold degenerate: $| 2, 1, -1 \rangle, | 2, 1, 0 \rangle$, and $| 2, 1, 1 \rangle$, and hence correspond to the same energy $E_2 = -\mathcal{R}/4$, where $\mathcal{R} = 13.6 \, \text{eV}$ is the Rydberg constant.

When the quadrupole field is turned on, and since Q_{-2}, Q_0, and Q_2 are small, we can treat the quadrupole interaction $\hat{H}_p = Q_{-2} r^2 Y_{2-2}(\Omega) + Q_0 r^2 Y_{20}(\Omega) + Q_2 r^2 Y_{22}(\Omega)$ as a perturbation. To calculate the p split, we need to use degenerate perturbation theory, which, in a first step, requires calculating the matrix

$$
\begin{pmatrix}
\langle 2, 1, -1|\hat{H}_p | 2, 1, -1 \rangle & \langle 2, 1, -1|\hat{H}_p | 2, 1, 0 \rangle & \langle 2, 1, -1|\hat{H}_p | 2, 1, 1 \rangle \\
\langle 2, 1, 0|\hat{H}_p | 2, 1, -1 \rangle & \langle 2, 1, 0|\hat{H}_p | 2, 1, 0 \rangle & \langle 2, 1, 0|\hat{H}_p | 2, 1, 1 \rangle \\
\langle 2, 1, 1|\hat{H}_p | 2, 1, -1 \rangle & \langle 2, 1, 1|\hat{H}_p | 2, 1, 0 \rangle & \langle 2, 1, 1|\hat{H}_p | 2, 1, 1 \rangle
\end{pmatrix},
$$

where

$$
(9.375)
$$

$$
\langle 2, 1, m'|\hat{H}_p | 2, 1, m \rangle = \langle 2, 1|r^2|2, 1\rangle\langle 1, m'|Q_{-2}Y_{2-2}+Q_0 Y_{20}+Q_2 Y_{22}|1, m\rangle. \quad (9.376)
$$

The radial part is easy to obtain (Chapter 6):

$$
\langle n, l|r^2|n, l\rangle = \int_0^{+\infty} r^4 |R_{nl}|^2 \, dr = \frac{1}{2}n^2 \left[5n^2 + 1 - 3l(l+1)\right]a_0^2; \quad (9.377)
$$

hence

$$
\langle 2, 1|r^2|2, 1\rangle = 30a_0^2. \quad (9.378)
$$

As for the angular part, it can be inferred from the Wigner–Eckart theorem:

$$
\langle l', m'|Y_{2\mu}|1, m\rangle = \langle l, 2; m, \mu|l', m'\rangle\langle l' \parallel Y_2 \parallel 1\rangle; \quad (9.379)
$$

the reduced matrix element $\langle l' \parallel Y_2 \parallel 1\rangle$ was calculated in Chapter 7: $\langle l' \parallel Y_2 \parallel 1\rangle = \sqrt{5/4\pi}\sqrt{(2l+1)/(2l'+1)}\langle l, 2; 0, 0|l', 0\rangle$ and hence

$$
\langle l', m'|Y_{2\mu}|1, m\rangle = \sqrt{\frac{5}{4\pi}}\sqrt{\frac{2l+1}{2l'+1}}\langle l, 2; 0, 0|l', 0\rangle\langle l, 2; m, \mu|l', m'\rangle. \quad (9.380)
$$

Using the coefficients $\langle l, 2; m, 0|l, m\rangle = [3m^2 - l(l+1)]/\sqrt{l(2l-1)(l+1)(2l+3)}$ and $\langle l, 2; m, \mp 2, \pm 2|l, m\rangle = \sqrt{\frac{3(l\pm m-1)(l\pm m)(l\mp m+1)(l\mp m+2)}{2l(2l-1)(l+1)(2l+3)}}$, we have

$$
\langle 1, -1 | Y_{2-2} | 1, 1 \rangle = \langle 1, 1 | Y_{22} | 1, -1 \rangle = -\sqrt{\frac{3}{10\pi}}, \quad (9.381)
$$

$$
\langle 1, -1 | Y_{20} | 1, -1 \rangle = \langle 1, 1 | Y_{20} | 1, 1 \rangle = -\frac{1}{\sqrt{20\pi}}, \quad (9.382)
$$

$$
\langle 1, 0 | Y_{20} | 1, 0 \rangle = \frac{1}{\sqrt{5\pi}}. \quad (9.383)
$$

These expressions can also be obtained from the following relations:

$$\int Y_{lm}^*(\Omega) Y_{20}(\Omega) Y_{lm}(\Omega) \, d\Omega = \sqrt{\frac{5}{4\pi}} \frac{l(l+1) - 3m^2}{(2l-1)(2l+3)}, \tag{9.384}$$

$$\int Y_{lm+2}^*(\Omega) Y_{22}(\Omega) Y_{lm}(\Omega) d\Omega = \int Y_{lm}^*(\Omega) Y_{2-2}(\Omega) Y_{lm+2}(\Omega) d\Omega$$

$$= \sqrt{\frac{15}{8\pi}} \frac{\sqrt{(l-m-1)(l-m)(l+m+1)(l+m+2)}}{(2l-1)(2l+3)}. \tag{9.385}$$

Combining (9.376) to (9.383) we can write the matrix (9.375) as

$$30a_0^2 \begin{pmatrix} -\frac{Q_0}{\sqrt{20\pi}} & 0 & -Q_2\sqrt{\frac{3}{10\pi}} \\ 0 & \frac{Q_0}{\sqrt{5\pi}} & 0 \\ -Q_2\sqrt{\frac{3}{10\pi}} & 0 & -\frac{Q_0}{\sqrt{20\pi}} \end{pmatrix}. \tag{9.386}$$

The diagonalization of this matrix leads to the following eigenvalues:

$$E_1^{(1)} = -30\frac{a_0^2}{\sqrt{10\pi}} \left(\frac{Q_0}{\sqrt{2}} + Q_2\sqrt{3} \right), \tag{9.387}$$

$$E_2^{(1)} = 30\frac{Q_0 a_0^2}{\sqrt{5\pi}}, \tag{9.388}$$

$$E_3^{(1)} = 30\frac{a_0^2}{\sqrt{10\pi}} \left(-\frac{Q_0}{\sqrt{2}} + Q_2\sqrt{3} \right). \tag{9.389}$$

Thus, to first-order perturbation theory, the energies of the p level are given by

$$E_{21} = -\frac{R}{4} - 30\frac{a_0^2}{\sqrt{10\pi}} \left(\frac{Q_0}{\sqrt{2}} + Q_2\sqrt{3} \right), \tag{9.390}$$

$$E_{22} = -\frac{R}{4} + 30\frac{Q_0 a_0^2}{\sqrt{5\pi}}, \tag{9.391}$$

$$E_{23} = -\frac{R}{4} + 30\frac{a_0^2}{\sqrt{10\pi}} \left(-\frac{Q_0}{\sqrt{2}} + Q_2\sqrt{3} \right). \tag{9.392}$$

So the quadrupole interaction has lifted all the degeneracies of the p level.

Problem 9.11

Two protons, located on the z-axis and separated by a distance d (i.e., $\vec{r} = d\vec{k}$), are subject to a z-oriented magnetic field $\vec{B} = B\vec{k}$.

(a) Ignoring all interactions between the two protons, find the energy levels and stationary states of this system.

(b) Treating the dipole–dipole magnetic interaction energy between the protons,

$$\hat{H}_p = \frac{1}{r^3} \left[\vec{\mu}_1 \cdot \vec{\mu}_2 - 3\frac{(\vec{\mu}_1 \cdot \vec{r})(\vec{\mu}_2 \cdot \vec{r})}{r^2} \right],$$

as a perturbation, calculate the energy using first-order perturbation theory.

Solution

(a) Since the magnetic moments of the protons are $\vec{\mu}_1 = 2\mu_0 \vec{S}_1/\hbar$ and $\vec{\mu}_2 = 2\mu_0 \vec{S}_2/\hbar$ where $\mu_0 = \hbar e/(2M_p c)$ is the proton magnetic moment, the Hamiltonian of the two-proton system, ignoring all the interactions between the two protons, is due to the interaction of the magnetic moments of the protons with the external magnetic field:

$$\hat{H}_0 = -(\vec{\mu}_1 + \vec{\mu}_2) \cdot \vec{B} = -\frac{2\mu_0}{\hbar}(\vec{S}_1 + \vec{S}_2) \cdot \vec{B} = -\frac{2\mu_0 B}{\hbar}\hat{S}_z. \tag{9.393}$$

As shown in Chapter 7, the eigenstates of a system consisting of two spin-$\frac{1}{2}$ particles are a triplet state and singlet state; the stationary eigenstates of $\hat{H}_0$ are therefore given by

$$|\chi_1\rangle = |1, 1\rangle = \left|\frac{1}{2}, \frac{1}{2}; \frac{1}{2}, \frac{1}{2}\right\rangle, \tag{9.394}$$

$$|\chi_2\rangle = |1, -1\rangle = \left|\frac{1}{2}, \frac{1}{2}; -\frac{1}{2}, -\frac{1}{2}\right\rangle, \tag{9.395}$$

$$|\chi_3\rangle = |1, 0\rangle = \frac{1}{\sqrt{2}}\left[\left|\frac{1}{2}, \frac{1}{2}; \frac{1}{2}, -\frac{1}{2}\right\rangle + \left|\frac{1}{2}, \frac{1}{2}; -\frac{1}{2}, \frac{1}{2}\right\rangle\right], \tag{9.396}$$

$$|\chi_4\rangle = |0, 0\rangle = \frac{1}{\sqrt{2}}\left[\left|\frac{1}{2}, \frac{1}{2}; \frac{1}{2}, -\frac{1}{2}\right\rangle - \left|\frac{1}{2}, \frac{1}{2}; -\frac{1}{2}, \frac{1}{2}\right\rangle\right]. \tag{9.397}$$

The eigenenergies of $|\chi_1\rangle$, $|\chi_2\rangle$, $|\chi_3\rangle$, and $|\chi_4\rangle$ are respectively

$$E_1^{(0)} = -2\mu_0 B, \qquad E_2^{(0)} = 2\mu_0 B, \qquad E_3^{(0)} = E_4^{(0)} = 0. \tag{9.398}$$

So $|\chi_3\rangle$ and $|\chi_4\rangle$ are (doubly) degenerate, whereas $|\chi_1\rangle$ and $|\chi_2\rangle$ are not.

(b) To calculate the energy to first order, we need to calculate the matrix elements of $\hat{H}_p$: $\hat{H}_{P_{ij}} = \langle\chi_i|\hat{H}_p|\chi_j\rangle$, with i, $j = 1, 2, 3, 4$. For this, since $\vec{r} = d\vec{k}$, we have $\vec{\mu}_1 \cdot \vec{r} = 2\mu_0 d\hat{S}_{1_z}/\hbar$ and $\vec{\mu}_2 \cdot \vec{r} = 2\mu_0 d\hat{S}_{2_z}/\hbar$. Thus, we can write $\hat{H}_p$ as

$$\hat{H}_p = \frac{1}{r^3}\left[\vec{\mu}_1 \cdot \vec{\mu}_2 - 3\frac{(\vec{\mu}_1 \cdot \vec{r})(\vec{\mu}_2 \cdot \vec{r})}{r^2}\right] = \frac{4\mu_0^2}{d^3\hbar^2}\left[\hat{\vec{S}}_1 \cdot \hat{\vec{S}}_2 - 3\hat{S}_{1_z}\hat{S}_{2_z}\right]. \tag{9.399}$$

Using the relations

$$2\vec{S}_1 \cdot \vec{S}_2 \,|\, S, \; S_z\rangle = \left[\left(\vec{S}_1 + \vec{S}_2\right)^2 - \vec{S}_1^2 - \vec{S}_1^2\right]|\, S, \; S_z\rangle$$

$$= \hbar^2\left[S(S+1) - S_1(S_1+1) - S_2(S_2+1)\right]|\, S, \; S_z\rangle$$

$$= \hbar^2\left[S(S+1) - \frac{3}{2}\right]|\, S, \; S_z\rangle, \tag{9.400}$$

$$2\hat{S}_{1_z}\hat{S}_{2_z}\,|\, S, \; S_z\rangle = \left[\hat{S}_z^2 - \hat{S}_{1_z}{}^2\hat{S}_{2_z}{}^2\right]|\, S, \; S_z\rangle = \hbar^2\left(S_z^2 - \frac{1}{2}\right)|\, S, \; S_z\rangle, \tag{9.401}$$

along with (9.399), we can rewrite

$$\hat{H}_p\,|\, S, \; S_z\rangle = \frac{2\mu_0^2}{d^3}\left[S(S+1) - \frac{3}{2} - 3\left(S_z^2 - \frac{1}{2}\right)\right]|\, S, \; S_z\rangle$$

$$= \frac{2\mu_0^2}{d^3}\left[S(S+1) - 3S_z^2\right]|\, S, \; S_z\rangle. \tag{9.402}$$

Since the values of S and S_z are given for the triplet state by $S = 1$, $S_z = -1, 0, 1$, and by $S = 0$, $S_z = 0$ for the singlet, the matrix elements of $\hat{H}_p$ are

$$E_1^{(1)} = \langle \chi_1 \mid \hat{H}_p \mid \chi_1 \rangle = -\frac{2\mu_0^2}{d^3}, \qquad E_2^{(1)} = \langle \chi_2 \mid \hat{H}_p \mid \chi_2 \rangle = -\frac{2\mu_0^2}{d^3}, \qquad (9.403)$$

$$E_3^{(1)} = \langle \chi_3 \mid \hat{H}_p \mid \chi_3 \rangle = \frac{4\mu_0^2}{d^3}, \qquad E_4^{(1)} = \langle \chi_4 \mid \hat{H}_p \mid \chi_4 \rangle = 0. \qquad (9.404)$$

All the other matrix elements of $\hat{H}_p$ are zero: $\langle \chi_i | \hat{H}_p | \chi_j \rangle = 0$ for $i \neq j$.

Finally, the energy levels of the two-proton system can be obtained at once from (9.398) along with (9.403) and (9.404):

$$E_1 = E_1^{(0)} + E_1^{(1)} = -2\mu_0 B - \frac{2\mu_0^2}{d^3}, \qquad (9.405)$$

$$E_2 = E_2^{(0)} + E_2^{(1)} = 2\mu_0 B - \frac{2\mu_0^2}{d^3}, \qquad (9.406)$$

$$E_3 = E_3^{(0)} + E_3^{(1)} = \frac{4\mu_0^2}{d^3}, \qquad (9.407)$$

$$E_4 = E_4^{(0)} + E_4^{(1)} = 0. \qquad (9.408)$$

So the dipole–dipole magnetic interaction has lifted the degeneracy of the energy levels in the two-proton system.

Problem 9.12

A spin $\frac{1}{2}$ particle of mass m, which is moving in an infinite, symmetric potential well $V(x)$ of length $2L$, is placed in an external *weak* magnetic field $\vec{B}$ with

$$V(x) = \begin{cases} 0, & -L \leq x \leq L, \\ \infty, & \text{otherwise}, \end{cases} \qquad \vec{B} = \begin{cases} -B\hat{z}, & -L \leq x \leq 0, \\ -B\hat{x}, & 0 \leq x \leq L. \end{cases}$$

Using first-order perturbation theory, calculate the energy of the nth excited state of this particle.

Solution

First, let us discuss the physics of this particle before placing it in a magnetic field. As seen in Chapter 4, the energy and wave function of a *spinless* particle of mass m moving in a symmetric potential well of length $2L$ are

$$E_n = \frac{\hbar^2 \pi^2}{8mL^2} n^2, \qquad \psi_n(x) = \frac{1}{\sqrt{L}} \begin{cases} \cos\left(\frac{n\pi x}{2L}\right), & n = 1, 3, 5, \ldots, \\ \sin\left(\frac{n\pi x}{2L}\right), & n = 2, 4, 6, \ldots. \end{cases} \qquad (9.409)$$

When the spin of the particle is considered, its wave function is the product of a spacial part $\psi_n(x)$ and a spin part $|\chi_{\pm}\rangle$:

$$\psi_n^{\pm}(x) = |\chi_{\pm}\rangle \psi_n(x) = |\chi_{\pm}\rangle \begin{cases} \frac{1}{\sqrt{L}} \cos\left(\frac{n\pi x}{2L}\right), & n = 1, 3, 5, \ldots, \\ \frac{1}{\sqrt{L}} \sin\left(\frac{n\pi x}{2L}\right), & n = 2, 4, 6, \ldots, \end{cases} \qquad (9.410)$$

where $|\chi_\pm\rangle$ represent the spinor fields corresponding to the spin-up and spin-down states, respectively:

$$|\chi_+\rangle = \left|\frac{1}{2}, \frac{1}{2}\right\rangle = \begin{pmatrix} 1 \\ 0 \end{pmatrix}, \qquad |\chi_-\rangle = \left|\frac{1}{2}, -\frac{1}{2}\right\rangle = \begin{pmatrix} 0 \\ 1 \end{pmatrix}. \tag{9.411}$$

Each energy level, $E_n = \hbar^2\pi^2 n^2/(8mL^2)$, of this particle is doubly degenerate, for it corresponds to two different states.

Let us now consider the case where the particle is placed in the magnetic field $\vec{B}$. The interaction between the external magnetic field and the particle's magnetic moment $\vec{\mu}$ is given by

$$\hat{H}_p = -\vec{\mu} \cdot \vec{B} = B\mu_0 \begin{cases} \sigma_z, & -L \leq x \leq 0, \\ \sigma_x, & 0 \leq x \leq L, \end{cases} \tag{9.412}$$

where we have made use of $\vec{\mu} = 2\mu_0\vec{S}/\hbar = \mu_0\vec{\sigma}$; recall that the matrices of σ_x and σ_z are

$$\sigma_x = \begin{pmatrix} 0 & 1 \\ 1 & 0 \end{pmatrix}, \qquad \sigma_z = \begin{pmatrix} 1 & 0 \\ 0 & -1 \end{pmatrix}. \tag{9.413}$$

To estimate the energy of this particle by means of the degenerate perturbation theory, we need to calculate first the matrix

$$\begin{pmatrix} \langle \psi_n^- | \hat{H}_p | \psi_n^- \rangle & \langle \psi_n^- | \hat{H}_p | \psi_n^+ \rangle \\ \langle \psi_n^+ | \hat{H}_p | \psi_n^- \rangle & \langle \psi_n^+ | \hat{H}_p | \psi_n^+ \rangle \end{pmatrix}, . \tag{9.414}$$

where

$$\langle \psi_n^- | \hat{H}_p | \psi_n^- \rangle = \int_{-L}^0 |\psi_n(x)|^2 \langle \chi_-|\hat{H}_p|\chi_-\rangle\, dx + \int_0^L |\psi_n(x)|^2 \langle \chi_-|\hat{H}_p|\chi_-\rangle\, dx$$

$$= \mu_0 B \left[\langle \chi_-|\sigma_z|\chi_-\rangle \int_{-L}^0 |\psi_n(x)|^2\, dx + \langle \chi_-|\sigma_x|\chi_-\rangle \int_0^L |\psi_n(x)|^2\, dx \right]. \tag{9.415}$$

Using $\int_{-L}^0 |\psi_n(x)|^2\, dx = \int_0^L |\psi_n(x)|^2\, dx = \frac{1}{2}$ and since

$$\langle \chi_-|\sigma_z|\chi_-\rangle = (0\ 1) \begin{pmatrix} 1 & 0 \\ 0 & -1 \end{pmatrix} \begin{pmatrix} 0 \\ 1 \end{pmatrix} = -1,$$

$$\langle \chi_-|\sigma_x|\chi_-\rangle = (0\ 1) \begin{pmatrix} 0 & 1 \\ 1 & 0 \end{pmatrix} \begin{pmatrix} 0 \\ 1 \end{pmatrix} = 0, \tag{9.416}$$

we have

$$\langle \psi_n^- | \hat{H}_p | \psi_n^- \rangle = -\frac{\mu_0 B}{2}. \tag{9.417}$$

Following this procedure, we can obtain the remaining matrix elements of (9.414):

$$\frac{\mu_0 B}{2} \begin{pmatrix} -1 & 1 \\ 1 & 1 \end{pmatrix}. \tag{9.418}$$

The diagonalization of this matrix leads to

$$\left(-\frac{\mu_0 B}{2} - E^{(1)}\right)\left(\frac{\mu_0 B}{2} - E^{(1)}\right) - \left(\frac{\mu_0 B}{2}\right)^2 = 0, \tag{9.419}$$

or $E^{(1)} = \pm \mu_0 B / \sqrt{2}$. Thus, the energy of the nth excited state to first-order degenerate perturbation theory is given by

$$E_n = \frac{\hbar^2 \pi^2}{8mL^2} n^2 \pm \frac{\mu_0 B}{\sqrt{2}}. \tag{9.420}$$

The magnetic field has completely removed the degeneracy of the energy spectrum of this particle.

Problem 9.13

Consider a particle of mass m moving in the potential $V(x) = \begin{cases} +\infty, & x \le 0, \\ \frac{1}{2} m\omega^2 x^2, & x > 0. \end{cases}$

Estimate the ground state energy of this particle using
(a) the variational method and (b) the WKB method.

Solution

(a) As seen in Problem 4.9, page 266, the ground state wave function of this potential must be selected from the harmonic oscillator wave functions that vanish at $x = 0$. Only the *odd* wave functions vanish at $x = 0$. So a trial function that, besides being zero at $x = 0$, is finite as $x \longrightarrow +\infty$ is given by

$$\psi_0(x, \alpha) = x e^{-\alpha x^2}. \tag{9.421}$$

Using the results

$$\langle \psi_0 \mid \psi_0 \rangle = \int_0^{+\infty} x^2 e^{-2\alpha x^2} \, dx = \frac{1}{8\alpha} \sqrt{\frac{\pi}{2\alpha}}, \tag{9.422}$$

$$\left\langle \psi_0 \left| \frac{1}{2} m\omega^2 x^2 \right| \psi_0 \right\rangle = \frac{1}{2} m\omega^2 \int_0^{+\infty} x^4 e^{-2\alpha x^2} \, dx = \frac{3m\omega^2}{64\alpha^2} \sqrt{\frac{\pi}{2\alpha}}, \tag{9.423}$$

$$\left\langle \psi_0 \left| -\frac{\hbar^2}{2m} \frac{d^2}{dx^2} \right| \psi_0 \right\rangle = \frac{\hbar^2}{2m} \int_0^{+\infty} \left(3\alpha x^2 - 2\alpha^2 x^4 \right) e^{-2\alpha x^2} \, dx = \frac{3\hbar^2}{16m} \sqrt{\frac{\pi}{2\alpha}}, \tag{9.424}$$

we obtain the ground state energy

$$E_0(\alpha) = \frac{\langle \psi_0(\alpha) | \hat{H} | \psi_0(\alpha) \rangle}{\langle \psi_0 \mid \psi_0 \rangle} = \frac{3\hbar^2}{2m} \alpha + \frac{3m\omega^2}{8\alpha}. \tag{9.425}$$

The minimization of $E_0(\alpha)$ with respect to α yields $\alpha_0 = m\omega / (2\hbar)$ and hence $E_0(\alpha_0) = \frac{3}{2} \hbar\omega$. This energy is identical to the exact value obtained in Chapter 4.

(b) This potential contains a single rigid wall at $x = 0$. Thus the proper quantization rule for this potential is given by (9.224): $\int_0^a p \, dx = (n + \frac{3}{4}) \pi \hbar$; the turning point occurs at $x = a$ with $E = \frac{1}{2} m\omega^2 a^2$ and hence $a = \sqrt{2E/(m\omega^2)}$.

The calculation of $\int_0^a p \, dx$ goes as follows:

$$\int_0^a p \, dx = \int_0^a \sqrt{2mE - m^2\omega^2 x^2} \, dx = m\omega \int_0^a \sqrt{a^2 - x^2} \, dx. \tag{9.426}$$

The change of variable $x = a \sin \theta$ leads to

$$\int_0^a \sqrt{a^2 - x^2} \, dx = a^2 \int_0^{\pi/2} \cos^2 \theta \, d\theta = \frac{a^2}{2} \int_0^{\pi/2} (1 + \cos 2\theta) \, d\theta = \frac{\pi a^2}{4}; \tag{9.427}$$

hence

$$\int_0^a p\,dx = m\omega \frac{\pi a^2}{4} = \frac{\pi E}{2\omega}. \tag{9.428}$$

Since $\int_0^a p\,dx = (n + \frac{3}{4})\pi\hbar$, that is, $\pi E/(2\omega) = (n + \frac{3}{4})\pi\hbar$, we obtain

$$E_n^{WKB} = \left(2n + \frac{3}{2}\right)\hbar\omega, \qquad n = 0, 1, 2, 3, \ldots. \tag{9.429}$$

This relation is identical with the *exact* expression obtained in Chapter 4. The WKB ground state energy is thus given by $E_0^{WKB} = \frac{3}{2}\hbar\omega$.

Problem 9.14

Consider an H_2 molecule where the protons are separated by a wide distance R and both are located on the z-axis. Ignoring the spin degrees of freedom and treating the dipole–dipole interaction as a perturbation, use perturbation theory to estimate an upper limit for the ground state energy of this molecule.

Solution

Assuming the protons are fixed in space and separated by a distance R, we can write the Hamiltonian of this molecule as follows:

$$\hat{H} = \hat{H}_0 + \hat{H}_p = \hat{H}_0^A + \hat{H}_0^B + \hat{H}_p, \tag{9.430}$$

where $\hat{H}_0^A$ and $\hat{H}_0^B$ are the unperturbed Hamiltonians of atoms A and B, and $\hat{H}_p$ is

$$\hat{H}_p = \frac{e^2}{R} + \frac{e^2}{|\vec{R} + \vec{r}_A - \vec{r}_B|} - \frac{e^2}{|\vec{R} + \vec{r}_A|} - \frac{e^2}{|\vec{R} - \vec{r}_B|}, \tag{9.431}$$

where $\vec{r}_A$ and $\vec{r}_B$ are the position vectors of the electrons of atoms A and B as measured from the protons. If $R \gg a_0$, where $a_0 = \hbar^2/(\mu e^2)$ is the Bohr radius, an expansion of (9.431) in powers of $\vec{r}_A/R$ and $\vec{r}_B/R$ yields, to first nonvanishing terms, an expression of the order of $1/R^3$:

$$\hat{H}_p = \frac{e^2}{R^3}\left[\vec{r}_A \cdot \vec{r}_B - 3\frac{(\vec{r}_A \cdot \vec{R})(\vec{r}_B \cdot \vec{R})}{R^2}\right]. \tag{9.432}$$

This is the dipole–dipole interaction energy between the dipole moments of the two atoms.

Since $\vec{R} = R\hat{z}$ we can write (9.432) as

$$\hat{H}_p = \frac{e^2}{R^3}\left(\hat{X}_A\hat{X}_B + \hat{Y}_A\hat{Y}_B - 2\hat{Z}_A\hat{Z}_B\right). \tag{9.433}$$

The ground state energy and wave function of the (unperturbed) molecule are

$$E_0 = E_0^A + E_0^B = 2E_{100} = -\frac{e^2}{a_0}, \qquad |\phi_0\rangle = |\phi_0^A\rangle|\phi_0^B\rangle = |100\rangle_A |100\rangle_B. \tag{9.434}$$

The first-order correction to the molecule's energy, $E^{(1)} = \langle\phi_0|\hat{H}_p|\phi_0\rangle$, is given by

$$E^{(1)} = \frac{e^2}{R^3}\left(\langle\phi_0^A|\hat{X}_A|\phi_0^A\rangle\langle\phi_0^B|\hat{X}_B|\phi_0^B\rangle + \langle\phi_0^A|\hat{Y}_A|\phi_0^A\rangle\langle\phi_0^B|\hat{Y}_B|\phi_0^B\rangle\right.$$
$$\left. - 2\langle\phi_0^A|\hat{Z}_A|\phi_0^A\rangle\langle\phi_0^B|\hat{Z}_B|\phi_0^B\rangle\right). \tag{9.435}$$

Since the operators $\hat{X}$, $\hat{Y}$, and $\hat{Z}$ are odd and the states $|\phi_0^A\rangle$ and $|\phi_0^B\rangle$ are spherically symmetric, then all the terms in (9.435) are zero; hence $E^{(1)} = 0$.

Let us now calculate the second-order correction:

$$E^{(2)} = \sum_{n,l,m;n',l',m'\neq 1,0,0} \frac{\left|\langle n,l,m;n',l',m'|\hat{H}_p|\phi_0\rangle\right|^2}{2E_{100} - E_n - E_{n'}}, \qquad (9.436)$$

where

$$\langle n,l,m;n',l',m'|\hat{H}_p|\phi_0\rangle = \frac{e^2}{R^3} \Big(\langle n,l,m|X_A|1,0,0\rangle_A \langle n',l',m'|X_B|1,0,0\rangle_B$$
$$+ \langle n,l,m|Y_A|1,0,0\rangle_A \langle n',l',m'|Y_B|1,0,0\rangle_B$$
$$+ -2\langle n,l,m|Z_A|1,0,0\rangle_A \langle n',l',m'|Z_B|1,0,0\rangle_B \Big). \qquad (9.437)$$

The terms of this expression are nonzero only if $l = l' = 1$, since the $\hat{X}$, $\hat{Y}$, and $\hat{Z}$ operators are proportional to Y_{1m}. We can evaluate $E^{(2)}$ using a crude approximation where we assume the denominator of (9.436) is constant and we take $E_n \simeq E_{n'}$. Note that, for $n \geq 2$, we have $E_{nlm} \geq E_{200}$. In this case we can rewrite (9.436) as

$$E^{(2)} \leq \frac{1}{2(E_{100} - E_{200})} \sum_{n,l,m;n',l',m'\neq 1,0,0} \left|\langle n,l,m;n',l',m'|\hat{H}_p|\phi_0\rangle\right|^2; \qquad (9.438)$$

since the diagonal term is zero (i.e., $\langle 1,0,0;1,0,0|\hat{H}_p|1,0,0;1,0,0\rangle = 0$), we have

$$\sum_{n,l,m;n',l',m'} \left|\langle n,l,m;n',l',m'|\hat{H}_p|\phi_0\rangle\right|^2$$

$$= \sum_{n,l,m;n',l',m'} \langle 1,0,0;1,0,0|\hat{H}_p|n,l,m;n',l',m'\rangle\langle n,l,m;n',l',m'|\hat{H}_p|1,0,0;1,0,0\rangle$$

$$= \langle 1,0,0;1,0,0|(\hat{H}_p)^2|1,0,0;1,0,0\rangle$$

$$= \frac{e^4}{R^6}\langle 1,0,0;1,0,0|(X_AX_B + Y_AY_B - 2Z_AZ_B)^2|1,0,0;1,0,0\rangle. \qquad (9.439)$$

The calculation of $\langle 1,0,0;1,0,0|(X_AX_B + Y_AY_B - 2Z_AZ_B)^2|1,0,0;1,0,0\rangle$ can be made easier by the use of symmetry. Due to spherical symmetry, the cross terms are zero:

$$\langle X_AY_A\rangle_A = \langle X_AZ_A\rangle_A = \langle Y_AZ_A\rangle_A = \langle X_BY_B\rangle_B = \cdots = \langle Y_BZ_B\rangle_B = 0, \qquad (9.440)$$

while the others are given as follows (see (9.45)):

$$\langle X_A^2\rangle_A = \langle Y_A^2\rangle_A = \langle Z_A^2\rangle_A = \langle X_B^2\rangle_B = \langle Y_B^2\rangle_B = \langle Z_B^2\rangle_B = a_0^2, \qquad (9.441)$$

where $\langle C\rangle_A = \langle\phi_0^A|C|\phi_0^A\rangle$ and $\langle D\rangle_B = \langle\phi_0^B|D|\phi_0^B\rangle$. We can thus obtain

$$\langle 1,0,0;1,0,0|(X_AX_B + Y_AY_B - 2Z_AZ_B)^2|1,0,0;1,0,0\rangle = 6a_0^4. \qquad (9.442)$$

Figure 9.10 Coulomb barrier, $V(r) = Ze^2/r$, seen by a proton of energy E while approaching from the right a nucleus of charge Ze located at the origin.

Inserting (9.442) into (9.439) and then the resultant expression into (9.438), we get

$$E^{(2)} \leq \frac{\langle 1, 0, 0; 1, 0, 0 | (\hat{H}_p)^2 | 1, 0, 0; 1, 0, 0 \rangle}{2(E_{100} - E_{200})} = \frac{3e^4 a_0^4}{R^6} \frac{1}{E_{100} - E_{200}}, \qquad (9.443)$$

or

$$E^{(2)} \leq -\frac{8e^2 a_0^5}{R^6}, \qquad (9.444)$$

because $E_{100} = -e^2/2a_0$ and $E_{200} = -e^2/8a_0$. Finally, the upper limit for the ground state energy of this molecule to second-order perturbation theory is given by

$$E_2 \leq 2E_{100} - \frac{8e^2 a_0^5}{R^6} \qquad \Longrightarrow \qquad E_2 \leq -\frac{e^2}{a_0}\left(1 + 8\frac{a_0^6}{R^6}\right). \qquad (9.445)$$

Problem 9.15
A proton of energy E is incident from the right on a nucleus of charge Ze. Estimate the transmission coefficient associated with the penetration of the proton inside the nucleus.

Solution
To penetrate inside the nucleus (i.e., to the left of the turning point $r = a$ as shown in Figure 9.10), the proton has to overcome the repulsive Coulomb force of the nucleus. That is, it has to tunnel through the Coulomb barrier $V(r) = Ze^2/r$. The transmission coefficient is given in the WKB approximation by (9.247), where $x_1 = a$ and $x_2 = 0$:

$$T = e^{-2\gamma}, \qquad \gamma = \frac{1}{\hbar}\int_a^0 \sqrt{2m(V(r) - E)}\, dr, \qquad (9.446)$$

where a is given by $E = V(a)$: $a = Ze^2/E$. Since $V(r) = Ze^2/r$ we get

$$\gamma = \frac{1}{\hbar}\int_a^0 \sqrt{2m\left(\frac{Ze^2}{r} - E\right)}\, dr = \frac{\sqrt{2mE}}{\hbar}\int_{Ze^2/E}^0 \sqrt{\frac{Ze^2}{Er} - 1}\, dr. \qquad (9.447)$$

The change of variable $x = Er/(Ze^2)$ gives

$$\gamma = \frac{Ze^2}{\hbar}\sqrt{\frac{2m}{E}}\int_0^1 \sqrt{\frac{1}{x} - 1}\, dx = \frac{Ze^2\pi}{\hbar}\sqrt{\frac{m}{2E}}; \tag{9.448}$$

in deriving this relation, we have used the integral $\int_0^1 \sqrt{1/x - 1}\, dx = \pi/2$.

The transmission coefficient is thus given by

$$T = e^{-2\gamma} = \exp\left\{-\frac{Ze^2\pi}{\hbar}\sqrt{\frac{2m}{E}}\right\}. \tag{9.449}$$

The value of this coefficient describes how difficult it is for a positively charged particle, such as a proton, to approach a nucleus.

Problem 9.16

Two identical particles of spin $\frac{1}{2}$ are enclosed in a one-dimensional box potential of length L with walls at $x = 0$ and $x = L$.

(a) Find the energies of the three lowest states.

(b) Then, subjecting the particles to a perturbation

$$\hat{H}_p(x_1, x_2) = -V_0 L^2 \delta\left(x_1 - \frac{L}{2}\right)\delta\left(x_2 - \frac{L}{3}\right),$$

calculate its ground state energy using first-order time-independent perturbation theory.

Solution

Since the two particles have the *same* spin, the spin wave function of the system, $\chi_s(s_1, s_2)$, must be *symmetric*, so χ_s is any one of the triplet states:

$$\chi_s = \begin{cases} |1, 1\rangle &= \left|\frac{1}{2}, \frac{1}{2}\right\rangle_1 \left|\frac{1}{2}, \frac{1}{2}\right\rangle_2, \\ |1, 0\rangle &= \frac{1}{\sqrt{2}}\left[\left|\frac{1}{2}, \frac{1}{2}\right\rangle_1 \left|\frac{1}{2}, -\frac{1}{2}\right\rangle_2 + \left|\frac{1}{2}, -\frac{1}{2}\right\rangle_1 \left|\frac{1}{2}, \frac{1}{2}\right\rangle_2\right], \\ |1, -1\rangle &= \left|\frac{1}{2}, -\frac{1}{2}\right\rangle_1 \left|\frac{1}{2}, -\frac{1}{2}\right\rangle_2. \end{cases} \tag{9.450}$$

In addition, since this two-particle system is a system of *identical fermions*, its wave function must be *antisymmetric*. Since the spin part is symmetric, the spatial part of the wave function has to be antisymmetric:

$$\Psi(x_1, x_2) = \psi_A(x_1, x_2)\chi_s(s_1, s_2); \tag{9.451}$$

that is,

$$\begin{aligned} \psi_A(x_1, x_2) &= \frac{1}{\sqrt{2}}\left[\phi_{n_1}(x_1)\phi_{n_2}(x_2) - \phi_{n_2}(x_1)\phi_{n_1}(x_2)\right] \\ &= \frac{1}{L}\left[\sin\left(\frac{n_1\pi x_1}{L}\right)\sin\left(\frac{n_2\pi x_2}{L}\right) - \sin\left(\frac{n_2\pi x_1}{L}\right)\sin\left(\frac{n_1\pi x_2}{L}\right)\right]. \end{aligned} \tag{9.452}$$

The energy levels of this two-particle system are

$$E = \frac{\pi^2\hbar^2}{2mL^2}(n_1^2 + n_2^2) = E_0(n_1^2 + n_2^2), \tag{9.453}$$

where $E_0 = \pi^2\hbar^2/(2mL^2)$. Note that these energy levels are threefold degenerate because of the spin part of the wave function; that is, there are three different spin states that correspond to the same energy level $\pi^2\hbar^2(n_1^2 + n_2^2)/(2mL^2)$.

(a) Having written the general expressions for the energies and the wave functions, it is now easy to infer the energy levels and wave functions of the three lowest states. First, we should note that the ground state cannot correspond to $n_1 = n_2 = 1$, for the spatial wave function would be *zero*. The ground state corresponds then to $n_1 = 1$, $n_2 = 2$; its energy follows from (9.453),

$$E^{(0)} = E_0(1^2 + 2^2) = 5E_0 = \frac{5\pi^2\hbar^2}{2mL^2}, \tag{9.454}$$

and the wave function $\psi_0(x_1, x_2)$ follows from (9.452).

The first excited state corresponds to $n_1 = 1$, $n_2 = 3$. So the wave function $\psi_1(x_1, x_2)$ can be inferred from (9.452) and the energy from (9.453):

$$E^{(1)} = E_0(1^2 + 3^2) = 10E_0 = \frac{5\pi^2\hbar^2}{mL^2}. \tag{9.455}$$

The second excited state corresponds to $n_1 = 2$, $n_2 = 3$; hence the energy is given by

$$E^{(2)} = 13E_0 = \frac{13\pi^2\hbar^2}{2mL^2}. \tag{9.456}$$

(b) Introducing the perturbation $\hat{H}_p = -V_0L^2\delta(x_1 - L/2)\delta(x_2 - L/3)$, and since $\hat{H}_p$ is diagonal in the spin space, the ground state energy to first-order perturbation theory is given by

$$E = \frac{5\pi^2\hbar^2}{2mL^2} + \langle\psi_0|\hat{H}_p|\psi_0\rangle \tag{9.457}$$

where

$$\langle\psi_0|\hat{H}_p|\psi_0\rangle = \int_0^L dx_1 \int_0^L dx_2 \psi_0^*(x_1, x_2)\hat{H}_p(x_1, x_2)\psi_0(x_1, x_2). \tag{9.458}$$

Since

$$\psi_0^*(x_1, x_2) = \psi_0(x_1, x_2) = \frac{1}{L}\left[\sin\left(\frac{\pi x_1}{L}\right)\sin\left(\frac{2\pi x_2}{L}\right) - \sin\left(\frac{2\pi x_1}{L}\right)\sin\left(\frac{\pi x_2}{L}\right)\right], \tag{9.459}$$

we have

$$\langle\psi_0|\hat{H}_p|\psi_0\rangle = -\frac{V_0L^2}{L^2}\int_0^L dx_1\delta\left(x_1 - \frac{L}{2}\right)\int_0^L dx_2\delta\left(x_2 - \frac{L}{3}\right)$$

$$\times \left[\sin\left(\frac{\pi x_1}{L}\right)\sin\left(\frac{2\pi x_2}{L}\right) - \sin\left(\frac{2\pi x_1}{L}\right)\sin\left(\frac{\pi x_2}{L}\right)\right]^2$$

$$= -V_0\left[\sin\left(\frac{\pi}{2}\right)\sin\left(\frac{2\pi}{3}\right) - \sin(\pi)\sin\left(\frac{\pi}{3}\right)\right]^2$$

$$= -\frac{3}{4}V_0; \tag{9.460}$$

hence

$$E = \frac{5\pi^2 \hbar^2}{2mL^2} - \frac{3}{4} V_0. \qquad (9.461)$$

Problem 9.17

Neglecting the spin–orbit interaction, find the ground state energy of a two-electron atom in these two ways:

(a) Use a first-order perturbation calculation; treat the Coulomb interaction between the two electrons as a perturbation.

(b) Use the variational method.

Compare the results and discuss the merits of the two approximation methods.

Solution

Examples of such a system are the helium atom ($Z = 2$), the singly ionized Li$^+$ ion ($Z = 3$), the doubly ionized Be^{2+} ion ($Z = 4$), and so on. Each electron of these systems feels the effects of two Coulomb fields: one from the Ze nucleus, $V(r) = -Ze^2/r$, and the other from the other electron, $V_{12} = e^2/r_{12} = e^2/|\vec{r}_1 - \vec{r}_2|$; here we consider the nucleus to be located at the origin and the electrons at $\vec{r}_1$ and $\vec{r}_2$. Neglecting the spin–orbit interaction, we can write the Hamiltonian of the two-electron system as

$$\hat{H} = \hat{H}_0 + \hat{V}_{12} = \hat{H}_0 + \frac{e^2}{|\vec{r}_1 - \vec{r}_2|}, \qquad (9.462)$$

where

$$\hat{H}_0 = -\frac{\hbar^2}{2\mu} \left(\nabla_1^2 + \nabla_2^2 \right) - Ze^2 \left(\frac{1}{r_1} + \frac{1}{r_2} \right) \qquad (9.463)$$

is the Hamiltonian of the atom when the interaction between the two electrons is neglected.

We have seen in Chapter 8 that, when the interaction between the two electrons is neglected, the ground state energy and wave function are given by

$$E_0 = -2\frac{Z^2 e^2}{2a} = -27.2 Z^2 \text{ eV}, \qquad (9.464)$$

$$\Psi_0(\vec{r}_1, \vec{S}_1; \vec{r}_2, \vec{S}_2) = \psi_0(\vec{r}_1, \vec{r}_2)\chi_{singlet}(\vec{S}_1, \vec{S}_2), \qquad (9.465)$$

where the spin part is antisymmetric,

$$\chi_{singlet}(\vec{S}_1, \vec{S}_2) = \frac{1}{\sqrt{2}} \left(\left| \frac{1}{2}, \frac{1}{2} \right\rangle_1 \left| \frac{1}{2}, -\frac{1}{2} \right\rangle_2 - \left| \frac{1}{2}, -\frac{1}{2} \right\rangle_1 \left| \frac{1}{2}, \frac{1}{2} \right\rangle_2 \right), \qquad (9.466)$$

and the spatial part is symmetric, $\psi_0(\vec{r}_1, \vec{r}_2) = \phi_{100}(\vec{r}_1)\phi_{100}(\vec{r}_2)$, with

$$\phi_{100}(\vec{r}) = R_{10}(r)Y_{00}(\Omega) = \frac{1}{\sqrt{\pi}} \left(\frac{Z}{a} \right)^{3/2} e^{-Zr_1/a}, \qquad (9.467)$$

that is,

$$\psi_0(\vec{r}_1, \vec{r}_2) = \frac{1}{\pi} \left(\frac{Z}{a} \right)^3 e^{-Z(r_1+r_2)/a}. \qquad (9.468)$$

(a) To calculate the ground state energy using first-order perturbation theory, we have to treat $\hat{V}_{12}$ as a perturbation. A first-order treatment yields

$$E = E_0 + \langle \psi_0 | \hat{V}_{12} | \psi_0 \rangle = -2 \frac{Z^2 e^2}{2a} + \langle \psi_0 | \hat{V}_{12} | \psi_0 \rangle, \tag{9.469}$$

where

$$\begin{aligned}
\langle \psi_0 | \hat{V}_{12} | \psi_0 \rangle &= \int d^3 r_1 \int d^3 r_2 \, \psi_0^*(\vec{r}_1, \vec{r}_2) \hat{V}_{12} \psi_0(\vec{r}_1, \vec{r}_2) \\
&= \int d^3 r_1 \int d^3 r_2 \, |\phi_{100}(\vec{r}_1)|^2 \frac{e^2}{|\vec{r}_1 - \vec{r}_2|} |\phi_{100}(\vec{r}_2)|^2 .
\end{aligned} \tag{9.470}$$

The calculation of this integral is quite involved (I left it as an exercise); the result is

$$\langle \psi_0 | \hat{V}_{12} | \psi_0 \rangle = \frac{5}{8} \frac{Z e^2}{a}, \tag{9.471}$$

which, when combined with (9.469), leads to

$$E = -\frac{Z e^2}{a} \left(Z - \frac{5}{8} \right). \tag{9.472}$$

In the case of helium, $Z = 2$, we have

$$E = -108.8 \, \text{eV} + 34 \, \text{eV} = -74.8 \, \text{eV}; \tag{9.473}$$

this result disagrees with the experimental value, $E_{exp} = -78.975 \, \text{eV}$, by 4 eV or by a 5.3% relative error. Physically, this may be attributed to the fact that, in our calculation, we have not taken into account the "screening" effect: the presence of one electron tends to decrease the net charge "seen" by the other electron. Suppose electron 1 is "between" the nucleus and electron 2; then electron 2 will not "see" Z protons but $(Z-1)$ protons (i.e., electron 2 feels an effective charge $(Z-1)e$ coming from the nucleus).

(b) By analogy with the exact form of the ground state function (9.468), we can choose a trial function that takes into account the screening effect. For this, we need simply to replace Z in (9.468) by a variational parameter α:

$$\psi_0(r_1, r_2) = A e^{-\alpha(r_1 + r_2)/a}, \tag{9.474}$$

where A is a normalization. Using the integral $\int_0^\infty x^n e^{-bx} dx = n!/b^{n+1}$, we can show that $A = (\alpha/a)^3/\pi$; hence

$$\psi_\alpha(r_1, r_2) = \frac{1}{\pi} \left(\frac{\alpha}{a} \right)^3 e^{-\alpha(r_1 + r_2)/a}. \tag{9.475}$$

A combination of this relation with (9.471) leads to

$$E(\alpha) = \langle \psi_\alpha | \hat{H}_0 | \psi_\alpha \rangle + \langle \psi_\alpha | \hat{V}_{12} | \psi_\alpha \rangle = \langle \psi_\alpha | \hat{H}_0 | \psi_\alpha \rangle + \frac{5}{8} \frac{\alpha e^2}{a}. \tag{9.476}$$

The calculation of $\langle\psi_a|\hat{H}_0|\psi_a\rangle$ can be simplified by writing it as

$$\langle\psi_a|\hat{H}_0|\psi_a\rangle = -\frac{\hbar^2}{2\mu}\langle\psi_a|\nabla_1^2 + \nabla_2^2|\psi_a\rangle - Ze^2\langle\psi_a|\frac{1}{r_1} + \frac{1}{r_2}|\psi_a\rangle$$

$$= -\frac{\hbar^2}{2\mu}\langle\psi_a|\nabla_1^2 + \nabla_2^2|\psi_a\rangle - ae^2\langle\psi_a|\frac{1}{r_1} + \frac{1}{r_2}|\psi_a\rangle$$

$$- (Z - a)e^2\langle\psi_a|\frac{1}{r_1} + \frac{1}{r_2}|\psi_a\rangle. \tag{9.477}$$

This form is quite suggestive; since $-\hbar^2\langle\psi_0|\vec{\nabla} \mid \psi_0\rangle/(2\mu) - Ze^2\langle\psi_0|1/r \mid \psi_0\rangle = -Z^2e^2/(2a)$ we can write

$$-\frac{\hbar^2}{2\mu}\langle\psi_a|\nabla_1^2 + \nabla_2^2|\psi_a\rangle - ae^2\langle\psi_a|\frac{1}{r_1} + \frac{1}{r_2}|\psi_a\rangle = -2\frac{a^2e^2}{2a}. \tag{9.478}$$

Now since

$$\langle\psi_a|\frac{1}{r_1}|\psi_a\rangle = \langle\psi_a|\frac{1}{r_2}|\psi_a\rangle = 4\left(\frac{a}{a}\right)^3\int_0^\infty re^{-2ar/a}dr = \frac{a}{a}, \tag{9.479}$$

we can reduce (9.477) to

$$\langle\psi_a|\hat{H}_0|\psi_a\rangle = -2\frac{a^2e^2}{2a} - 2(Z - a)e^2\frac{a}{a}, \tag{9.480}$$

which, when combined with (9.476), leads to

$$E(a) = -2\frac{a^2e^2}{2a} - 2(Z - a)e^2\frac{a}{a} + \frac{5}{8}\frac{ae^2}{a} = \left[a^2 - 2\left(Z - \frac{5}{16}\right)a\right]\frac{e^2}{a}. \tag{9.481}$$

The minimization of $E(a)$, $dE(a)/da = 0$, yields

$$a_0 = Z - \frac{5}{16}; \tag{9.482}$$

hence the ground state energy is

$$E(a_0) = -\left[1 - \frac{5}{8Z} + \left(\frac{5}{16Z}\right)^2\right]\frac{Z^2e^2}{a} \tag{9.483}$$

and

$$\psi(r_1, r_2) = \frac{1}{\pi}\left(\frac{Z}{a} - \frac{5}{16a}\right)^3\exp\left[-\left(\frac{Z}{a} - \frac{5}{16a}\right)(r_1 + r_2)\right]. \tag{9.484}$$

As a numerical illustration, the ground state energy of a helium atom is obtained by substituting $Z = 2$ into (9.483). This yields $E_0 = -77.456\,\text{eV}$, in excellent agreement with the experimental value $E_{exp} = -78.975\,\text{eV}$. The variational method, which overestimates the correct result by a mere 1.9%, is significantly more accurate than first-order perturbation theory. The reason is quite obvious; while the perturbation treatment does not account for the screening effect, the variational method includes it quite accurately. The wave function (9.484) shows that the second electron does not see a charge Ze, but a lower charge $(Z - 5/16)e$.

9.7   Exercises

Exercise 9.1
Calculate the energy of the nth excited state to first-order perturbation for a one-dimensional box potential of length $2L$, with walls at $x = -L$ and $x = L$, which is modified at the bottom by the following perturbations with $V_0 \ll 1$:

(a) $V_p(x) = \begin{cases} -V_0, & -L \le x \le L, \\ 0, & \text{elsewhere}; \end{cases}$

(b) $V_p(x) = \begin{cases} -V_0, & -L/2 \le x \le L/2, \\ 0, & \text{elsewhere}. \end{cases}$

Exercise 9.2
Calculate the energy of the nth excited state to first-order perturbation for a one-dimensional box potential of length $2L$, with walls at $x = -L$ and $x = L$, which is modified at the bottom by the following perturbations with $V_0 \ll 1$:

(a) $V_p(x) = \begin{cases} -V_0, & -L/2 \le x \le 0, \\ 0, & \text{elsewhere}; \end{cases}$

(b) $V_p(x) = \begin{cases} V_0, & 0 \le x \le L/2, \\ 0, & \text{elsewhere}; \end{cases}$

(c) $V_p(x) = \begin{cases} -V_0, & -L/2 \le x \le 0; \\ V_0, & 0 \le x \le L/2, \\ 0, & \text{elsewhere}. \end{cases}$

Exercise 9.3
Calculate the energy of the nth excited state to second-order perturbation and the wave function to first-order perturbation for a one-dimensional box potential of length $2L$, with walls at $x = -L$ and $x = L$, which is modified at the bottom by the following perturbations with $V_0 \ll 1$:

(a) $V_p(x) = \begin{cases} 0, & -L \le x \le 0, \\ V_0, & 0 \le x \le L; \end{cases}$ (b) $V_p(x) = \begin{cases} -V_0(1 - x^2/L^2), & |x| < L, \\ 0, & \text{elsewhere}. \end{cases}$

Exercise 9.4

Consider a system whose Hamiltonian is given by $\hat{H} = E_0 \begin{pmatrix} 3 & 2\lambda & 0 & 0 \\ 2\lambda & -3 & 0 & 0 \\ 0 & 0 & -7 & \sqrt{2}\lambda \\ 0 & 0 & \sqrt{2}\lambda & 7 \end{pmatrix}$,

where $\lambda \ll 1$.

(a) Calculate the *exact* eigenvalues of $\hat{H}$; expand each of these eigenvalues to the second power of λ.

(b) Calculate the energy eigenvalues to second-order perturbation theory and compare them with the exact results obtained in (a).

(c) Calculate the eigenstates of $\hat{H}$ up to the first-order correction.

Exercise 9.5
Consider a particle of mass m that moves in a three-dimensional potential $V(r) = kr$, where k is a constant having the dimensions of a force. Use the variational method to estimate its ground state energy; you may take $R(r) = e^{-r^2/2a^2}$ as the trial radial function where a is an adjustable parameter.

Exercise 9.6
Use the WKB method to estimate the ground state energy of a particle of mass m that moves in a three-dimensional potential $V(r) = kr$, where k is a constant having the dimensions of a force.

Exercise 9.7
Consider a two-dimensional harmonic oscillator Hamiltonian:

$$\hat{H} = \frac{1}{2m}(\hat{P}_x^2 + \hat{P}_y^2) + \frac{1}{2}m\omega^2(\hat{X}^2 + \hat{Y}^2)(1 + \lambda\hat{X}\hat{Y}),$$

where $\lambda \ll 1$.

(a) Give the wave functions for the three lowest energy levels when $\lambda = 0$.

(b) Using perturbation theory, evaluate the first-order corrections of these energy levels when $\lambda \neq 0$.

Exercise 9.8
Consider a particle that has the Hamiltonian $\hat{H} = \hat{H}_0 + \lambda\hbar\omega(\hat{a}^2 + \hat{a}^{\dagger^2})$, where $\hat{H}_0$ is the Hamiltonian of a simple one-dimensional harmonic oscillator, and where $\hat{a}$ and $\hat{a}^\dagger$ are the usual annihilation and creation operators which obey $[\hat{a}, \hat{a}^\dagger] = 1$; λ is a very small real number.

(a) Calculate the ground state energy to second order in λ.

(b) Find the energy of the nth excited state, E_n, to second order in λ and the corresponding eigenstate $\mid \psi_n \rangle$ to first order in λ.

Exercise 9.9
Consider two identical particles of spin $\frac{1}{2}$ that are confined in an isotropic three-dimensional harmonic oscillator potential of frequency ω.

(a) Find the ground state energy and the corresponding wave function of this system when the two particles do not interact.

(b) Consider now that there exists a weakly attractive spin-dependent potential between the two particles, $V(r_1, r_2) = -kr_1r_2 - \lambda\hat{S}_{1z}\hat{S}_{2z}$, where k and λ are two small positive real numbers. Find the ground state to first-order time-independent perturbation theory.

(c) Use the variational method to estimate the ground state energy of this system of two noninteracting spin $\frac{1}{2}$ particles confined to an isotropic three-dimensional harmonic oscillator. How does your result compare with that obtained in (a).

Exercise 9.10
Two identical spin $\frac{1}{2}$ particles are confined to a one-dimensional box potential of size L with walls at $x = 0$ and $x = L$.

(a) Find the ground state energy and the first excited state energy and their respective wave functions for this system when the two particles do not interact.

(b) Consider now that there exists a weakly attractive potential between the two particles:

$$V_p(x) = \begin{cases} -V_0, & 0 \leq x \leq L/2, \\ 0, & L/2 < x \leq L. \end{cases}$$

Find the ground state and first excited state energies to first-order perturbation theory.

(c) Find numerical values for the ground state and first excited state energies calculated in (a) in the case where $L = 10^{-10}$ m, $V_0 = 2$ eV, and the mass of each individual particle is to be taken equal to the electron mass. Compare the sizes of the first order energy corrections with the ground state energy and the first excited state energy (you may simply calculate the ratios between the first-order corrections with the ground state and first excited state energies).

Exercise 9.11
Consider an isotropic three-dimensional harmonic oscillator.

(a) Find the energy of the first excited state and the different states corresponding to this energy.

(b) If we now subject this oscillator to a perturbation $\hat{V}_p(x, y) = -\lambda \hat{x} \hat{y}$, where λ is a small real number, find the energy of the first excited state to first-order degenerate time-independent perturbation theory.   *Hint:* You may use $\hat{x} = \sqrt{\frac{\hbar}{2m\omega}}(\hat{a}_x + \hat{a}_x^\dagger)$, $\hat{y} = \sqrt{\frac{\hbar}{2m\omega}}(\hat{a}_y + \hat{a}_y^\dagger)$, $\hat{z} = \sqrt{\frac{\hbar}{2m\omega}}(\hat{a}_z + \hat{a}_z^\dagger)$, with $\hat{a}_x|n_x\rangle = \sqrt{n_x}|n_x - 1\rangle$, and $\hat{a}_x^\dagger|n_x\rangle = \sqrt{n_x + 1}|n_x + 1\rangle$.

Exercise 9.12
Use the variational method to estimate the ground state energy of the spherical harmonic oscillator by means of the following radial trial functions:

(a) $R(r) = Ae^{-\alpha r^2}$ and

(b) $R(r) = Ae^{-\alpha r}$, where A is a normalization constant that needs to be found in each case and α is an adjustable parameter.

(c) Using the fact that the exact ground state energy is $E_0^{exact} = 3\hbar\omega/2$, find the relative errors corresponding to the energies derived in (a) and (b).

Exercise 9.13
Consider a particle of mass m that is bouncing vertically and elastically on a smooth reflecting floor in the Earth's gravitational field

$$V(z) = \begin{cases} mgz, & z > 0, \\ +\infty, & z \le 0, \end{cases}$$

where g is a constant (the acceleration due to gravity). Use the variational method to estimate the ground state energy of this particle by means of the trial wave function, $\psi(z) = z\exp(-\alpha z^4)$, where α is an adjustable parameter that needs to be determined. Compare your result with the exact value $E_0^{exact} = 2.338 \left(\frac{1}{2}mg^2\hbar^2\right)^{1/3}$ by calculating the relative error.

Exercise 9.14
Calculate the energy of the ground state to first-order perturbation for a particle which is moving in a one-dimensional box potential of length L, with walls at $x = 0$ and $x = L$, when a weak potential $\hat{H}_p = \lambda x^2$ is added, where $\lambda \ll 1$.

Exercise 9.15
Consider a semiclassical system whose energy is given by

$$E = a^2 + \frac{1}{2}\left(\frac{b^2}{4} - a^2\right)p^2 + \frac{1}{2}\left(\frac{4a^2}{b^2/4 - a^2}\right)q^2,$$

where a and b are positive, real constants. Use the Bohr–Sommerfeld quantization rule to extract the expression of the bound state energy E_n for the nth excited state in terms of a.

Exercise 9.16

Use the variational method to estimate the ground state energy of a particle of mass m that is moving in a one-dimensional potential $V(x) = V_0 x^4$; you may use the trial function $\psi_0(x, a) = A e^{-ax^2/2}$, where A is the normalization constant and a is an adjustable parameter that needs to be determined.

Exercise 9.17

Consider a particle of mass m which is moving in a one-dimensional potential $V(x) = V_0 x^4$. Estimate the ground state energy of this particle by means of the WKB method.

Exercise 9.18

Using $\phi_{100}(\vec{r}) = (1/\sqrt{\pi})(Z/a)^{3/2} e^{-Zr/a}$, show that

$$\int \phi_{100}^2(\vec{r}_1) \frac{1}{|\vec{r}_1 - \vec{r}_2|} \phi_{100}^2(\vec{r}_2) d^3r_1 d^3r_2 = \frac{5Z}{8a}.$$

Exercise 9.19

Calculate the ground state energy of the doubly ionized beryllium atom Be^{2+} by means of the following two methods and then compare the two results:
    (a) a first-order perturbation theory treatment,
    (b) the variational method with a trial function $\psi(r_1, r_2) = A \exp[-a(r_1 + r_2)/a]$, where A is the normalization constant, a is an adjusted parameter, and a is the Bohr radius.

Exercise 9.20

Use the variational method to estimate the energy of the second excited state of a particle of mass m moving in a one-dimensional infinite well with walls at $x = 0$ and $x = L$. Calculate the relative error between your result and the exact value (recall that the energy of the second excited state is given by $E_3^{exact} = 9\pi^2\hbar^2/(2mL^2)$).

Exercise 9.21

Consider a spinless particle of orbital angular momentum $l = 1$ whose Hamiltonian is

$$\hat{H}_0 = \frac{\mathcal{E}}{\hbar^2}(\hat{L}_x^2 - \hat{L}_y^2),$$

where $\mathcal{E}$ is a constant having the dimensions of energy.
    (a) Calculate the exact energy levels and the corresponding eigenstates of this particle.
    (b) We now add a perturbation $\hat{H}_p = \alpha \hat{L}_z/\hbar$, where α is a small constant (small compared to $\mathcal{E}$) having the dimensions of energy. Calculate the energy levels of this particle to second-order perturbation theory.
    (c) Diagonalize the matrix of $\hat{H} = \hat{H}_0 + \hat{H}_p$ and find the exact energy eigenvalues. Then expand each eigenvalue to second power in α and compare them with the results derived from perturbation theory in (b).

Exercise 9.22

Consider a system whose Hamiltonian is given by $\hat{H} = E_0 \begin{pmatrix} -5 & 3\lambda & 0 & 0 \\ 3\lambda & 5 & 0 & 0 \\ 0 & 0 & 8 & -\lambda \\ 0 & 0 & -\lambda & -8 \end{pmatrix}$, where

$\lambda \ll 1$.

(a) By decomposing this Hamiltonian into $\hat{H} = \hat{H}_0 + \hat{H}_p$, find the eigenvalues and eigenstates of the unperturbed Hamiltonian $\hat{H}_0$.

(b) Diagonalize $\hat{H}$ to find the exact eigenvalues of $\hat{H}$; expand each eigenvalue to the second power of λ.

(c) Using first- and second-order nondegenerate perturbation theory, find the approximate eigenergies of $\hat{H}$. Compare these with the exact values obtained in (b).

Exercise 9.23

Estimate the ground state energy of the hydrogen atom by means of the variational method using the following two trial functions, find the relative errors, compare the two results, and discuss the merit of each trial function.

(a) $\phi_\alpha(r) = \begin{cases} 1 - r/\alpha, & r \le \alpha, \\ 0, & r > \alpha, \end{cases}$

where α is an adjustable parameter. Find a relation between α_{min} and the Bohr radius.

(b) $\phi_\alpha(r) = Ae^{-ar^2}$.

Exercise 9.24

(a) Calculate to first-order perturbation theory the energy of the nth excited state of a one-dimensional harmonic oscillator which is subject to the following small perturbation: $\hat{H}_p = \lambda(V_0\hat{x}^3 + V_1\hat{x}^4)$, where V_0, V_1, and a are constants and $\lambda \ll 1$.

(b) Use the relation derived in (a) to find the energies of the three lowest states (i.e., $n = 0, 1, 2$) to first-order perturbation theory.

Exercise 9.25

Use the trial function $\psi_0(x, \alpha) = \begin{cases} A(\alpha^2 - x^2)^2, & |x| \le \alpha, \\ 0, & |x| \ge \alpha, \end{cases}$

to estimate the ground state energy of a one-dimensional harmonic oscillator by means of the variational method; α is an adjustable parameter and A is the normalization constant. Calculate the relative error and assess the accuracy of the result.

Exercise 9.26

Use the WKB approximation to estimate the transmission coefficient of a particle of mass m and energy E moving in the following potential barrier:

$$V(x) = \begin{cases} V_0(x/a + 1), & -a < x < 0, \\ V_0(1 - x/a), & 0 < x < a, \\ 0, & \text{elsewhere,} \end{cases}$$

with $0 < E < V_0$; sketch this potential.

Exercise 9.27

Use the variational method to estimate the energy of the ground state of a one-dimensional harmonic oscillator taking the following trial function:

$$\psi_0(x, \alpha) = A(1 + \alpha|x|)e^{-\alpha|x|},$$

where α is an adjustable parameter and A is the normalization constant.

Exercise 9.28
Use the WKB approximation to estimate the transmission coefficient of a particle of mass m and energy E moving in the following potential barrier:

$$V(x) = \begin{cases} V_0(1 - x^2/a^2) & |x| < a \\ 0 & |x| > a, \end{cases}$$

where $0 < E \le V_0$.

Exercise 9.29
Use the WKB approximation to find the energy levels of a particle of mass m moving in the following potential:

$$V(x) = \begin{cases} V_0(x^2/a^2 - 1), & |x| < a, \\ 0, & |x| > a. \end{cases}$$

Exercise 9.30
A particle of mass m is moving in a one-dimensional harmonic oscillator potential, $V(x) = m\omega^2 x^2/2$. Calculate
  (a) the ground state energy, and
  (b) the first excited state energy
to first-order perturbation theory when a small perturbation $\hat{H}_p = \lambda\delta(x)$ is added to the potential, with $\lambda \ll 1$.

Exercise 9.31
A particle of mass m is moving in a one-dimensional harmonic oscillator potential, $V(x) = m\omega^2 x^2/2$. Calculate
  (a) the ground state energy and
  (b) the first excited state energy
to first-order perturbation theory when a small perturbation $\hat{H}_p = \lambda x^6$ is added to the potential, with $\lambda \ll 1$.

Exercise 9.32
A particle of mass m is moving in a three-dimensional harmonic oscillator potential, $V(x) = m\omega^2(x^2 + y^2 + z^2)/2$. Calculate the energy of the nth excited state to first-order perturbation theory when a small perturbation $\hat{H}_p = \lambda\hat{X}^2\hat{Y}^4\hat{Z}^2$ is added to the potential, with $\lambda \ll 1$.

Exercise 9.33
Use the following two trial functions:

  (a) $Ae^{-\alpha|x|}$,   (b) $A(1 + \alpha|x|)e^{-\alpha|x|}$,

to estimate, by means of the variational method, the ground state energy of a particle of mass m moving in a one-dimensional potential $V(x) = \lambda|x|$; α is a scale parameter, λ is a constant, and A is the normalization constant. Compare the results obtained.

Exercise 9.34
Three distinguishable particles of equal mass m are enclosed in a one-dimensional box potential with rigid walls at $x = 0$ and $x = L$. If the three particles are subject to a weak, short-range attractive potential

$$\hat{H}_p = -V_0 \left[\delta(x_1 - x_2) + \delta(x_2 - x_3) + \delta(x_3 - x_1) \right],$$

use first-order perturbation theory to calculate the system's energy levels of
   (a) the ground state, and
   (b) the first excited state.

Exercise 9.35

Three distinguishable particles of equal mass m are in a one-dimensional harmonic oscillator potential $\hat{H}_0 = \sum_{i=1}^{3}(p_i^2/2m + \frac{1}{2}m\omega^2 x_i^2)$. If the three particles are subject to a weak, short-range attractive potential

$$\hat{H}_p = -V_0\left[\delta(x_1 - x_2) + \delta(x_2 - x_3) + \delta(x_3 - x_1)\right],$$

use first-order perturbation theory to calculate the system's energy levels of
   (a) the ground state and
   (b) the first-excited state.

Exercise 9.36

Consider a positronium which is subject to a *weak* static magnetic field in the xz-plane, $\vec{B} = B(\vec{i} + \vec{k})$, where B is a small constant. Neglecting the spin–orbit interaction, calculate the energy levels of the $n = 2$ states to first-order perturbation.

Exercise 9.37

Consider a spherically symmetric top with principal moments of inertia I.
   (a) Find the energy levels of the top.
   (b) Assuming that the top is in the $l = 1$ angular momentum state, find its energy to first-order perturbation theory when a weak perturbation, $\hat{H}_p = \frac{\varepsilon}{I}(\hat{L}_x^2 - \hat{L}_y^2)$, is added where $\varepsilon \ll 1$.

Exercise 9.38

Estimate the approximate values of the ground state energy of a particle of mass m moving in the potential $V(x) = V_0|x|$, where $V_0 > 0$, by means of: (a) the variational method and   (b) the WKB approximation. Compare the two results.

Exercise 9.39

Calculate to first-order perturbation theory the relativistic correction to the ground state of a spinless particle of mass m moving in a one-dimensional harmonic oscillator potential. *Hint:* You need first to show that the Hamiltonian can be written as $\hat{H} = \hat{H}_0 + \hat{H}_p$, where $\hat{H}_0 = \hat{P}^2/(2m) + m\omega^2\hat{X}^2/2$ and $\hat{H}_p = -\hat{P}^4/(8m^3c^2)$ is the leading relativistic correction term which can be treated as a perturbation.

Exercise 9.40

Consider a hydrogen atom which is subject to a small perturbation $\hat{H}_p = \lambda r^2$. Use a first-order perturbation theory to calculate the energy corrections to
   (a) the ground state and
   (b) the 2p state.

Exercise 9.41

   (a) Calculate to first-order perturbation theory the contribution due to the spin–orbit interaction for the nth excited state for a positronium atom.
   (b) Use the result of part (a) to obtain numerical values for the spin–orbit correction terms for the 2p level and compare them to the energy of $n = 2$.

Exercise 9.42

Ignoring the spin of the electron, calculate to first-order perturbation theory the energy of the $n = 2$ level of a hydrogen atom when subject to a *weak* quadrupole field $\hat{H}_p = iQ(y^2 - x^2)$, where Q is a small, real number $Q \ll 1$.

Exercise 9.43

Calculate the energy levels of the $n = 2$ states of positronium in a weak external electrical field $\vec{\mathcal{E}}$ along the z-axis: $\vec{\mathcal{E}} = \mathcal{E}\vec{k}$; positronium consists of an electron and a positron bound by the electric interaction.

Exercise 9.44

   (a) Calculate to first-order perturbation theory the contributions due to the spin–orbit interaction for a hydrogen-like ion having Z protons.

   (b) Use the result of part (a) to find the spin-orbit correction for the 2p state of a C^{5+} carbon ion and compare it with the energy of the $n = 2$ level.

Exercise 9.45

Two identical particles of spin $\frac{1}{2}$ are enclosed in a cubical box of side L.

   (a) Calculate to first-order perturbation theory the ground state energy when the two particles are subject to *weak* short-range, attractive interaction:

$$\hat{V}(\vec{r}_1 - \vec{r}_2) = -\frac{4}{3}\pi a^3 V_0 \delta(\vec{r}_1 - \vec{r}_2).$$

   (b) Find a numerical value for the energy derived in (a) for $L = 10^{-10}$ m, $a = 10^{-12}$ m, $V_0 = 10^{-3}$ eV, and the mass of each individual particle is to be taken to be the mass of the electron.

Chapter 10

Time-Dependent Perturbation Theory

10.1 Introduction

We have dealt so far with Hamiltonians that do not depend explicitly on time. In nature, how-ever, most quantum phenomena are governed by time-dependent Hamiltonians. In this chapter we are going to consider approximation methods treating Hamiltonians that depend explicitly on time.

To study the structure of molecular and atomic systems, we need to know how electro-magnetic radiation interacts with these systems. Molecular and atomic spectroscopy deals in essence with the absorption and emission of electromagnetic radiation by molecules and atoms. As a system absorbs or emits radiation, it undergoes transitions from one state to another.

Time-dependent perturbation theory is most useful for studying processes of absorption and emission of radiation by atoms or, more generally, for treating the transitions of quantum systems from one energy level to another.

10.2 The Pictures of Quantum Mechanics

As seen in Chapter 2, there are many representations of wave functions and operators in quan-tum mechanics. The connection between the various representations is provided by unitary transformations. Each class of representation, also called a *picture*, differs from others in the way it treats the time evolution of the system.

In this section we look at the pictures encountered most frequently in quantum mechanics: the Schrödinger picture, the Heisenberg picture, and the interaction picture. The Schrödinger picture is useful when describing phenomena with time-independent Hamiltonians, whereas the interaction and Heisenberg pictures are useful when describing phenomena with time-dependent Hamiltonians.

10.2.1   The Schrödinger Picture

In describing quantum dynamics, we have been using so far the Schrödinger picture in which *state vectors depend explicitly on time, but operators do not*:

$$i\hbar \frac{d}{dt} \mid \psi(t) \rangle = \hat{H} \mid \psi(t) \rangle, \tag{10.1}$$

where $\mid \psi(t) \rangle$ denotes the state of the system in the Schrödinger picture. We have seen in Chapter 3 that the time evolution of a state $\mid \psi(t) \rangle$ can be expressed by means of the propagator, or time-evolution operator, $\hat{U}(t, t_0)$, as follows:

$$\mid \psi(t) \rangle = \hat{U}(t, t_0) \mid \psi(t_0) \rangle, \tag{10.2}$$

with

$$\hat{U}(t, t_0) = e^{-i(t-t_0)\hat{H}/\hbar}. \tag{10.3}$$

The operator $\hat{U}(t, t_0)$ is unitary,

$$\hat{U}^{\dagger}(t, t_0)\hat{U}(t, t_0) = I, \tag{10.4}$$

and satisfies these properties:

$$\hat{U}(t, t) = I, \tag{10.5}$$

$$\hat{U}^{\dagger}(t, t_0) = \hat{U}^{-1}(t, t_0) = \hat{U}(t_0, t), \tag{10.6}$$

$$\hat{U}(t_1, t_2)\hat{U}(t_2, t_3) = \hat{U}(t_1, t_3). \tag{10.7}$$

10.2.2   The Heisenberg Picture

In this picture the time dependence of the state vectors is completely frozen. The Heisenberg picture is obtained from the Schrödinger picture by applying $\hat{U}$ on $\mid \psi(t) \rangle_H$:

$$\mid \psi(t) \rangle_H = \hat{U}^{\dagger}(t) \mid \psi(t) \rangle = \mid \psi(0) \rangle, \tag{10.8}$$

where $\mid \psi(t) \rangle$ and $\hat{U}^{\dagger}(t)$ can be obtained from (10.2) and (10.3), respectively, by setting $t_0 = 0$: $\hat{U}^{\dagger}(t) = \hat{U}^{\dagger}(t, t_0 = 0) = e^{it\hat{H}/\hbar}$ and $\mid \psi(t) \rangle = \hat{U}(t) \mid \psi(0) \rangle$, with $\hat{U}(t) = e^{-it\hat{H}/\hbar}$. Thus, we can rewrite (10.8) as

$$\mid \psi(t) \rangle_H = e^{it\hat{H}/\hbar} \mid \psi(t) \rangle. \tag{10.9}$$

As $\mid \psi \rangle_H$ is frozen in time we have: $d \mid \psi \rangle_H/dt = 0$. Let us see how the expectation value of an operator $\hat{A}$ in the state $\mid \psi(t) \rangle$ evolves in time:

$$\langle \psi(t) \mid \hat{A} \mid \psi(t) \rangle = \langle \psi(0) \mid e^{it\hat{H}/\hbar} \hat{A} e^{-it\hat{H}/\hbar} \mid \psi(0) \rangle = \langle \psi(0) \mid \hat{A}_H(t) \mid \psi(0) \rangle = {}_H\langle \psi \mid \hat{A}_H(t) \mid \psi \rangle_H, \tag{10.10}$$

where $\hat{A}_H(t)$ is given by

$$\hat{A}_H(t) = \hat{U}^{\dagger}(t)\hat{A}\hat{U}(t) = e^{it\hat{H}/\hbar} \hat{A} e^{-it\hat{H}/\hbar}. \tag{10.11}$$

Equation (10.10) shows that the expectation value of an operator is the same in both the Schrödinger and the Heisenberg pictures. From (10.10) and (10.11) we see that both the Schrödinger and the Heisenberg pictures coincide at $t = 0$, since $\mid \psi(0) \rangle_H = \mid \psi(0) \rangle$ and $\hat{A}_H(0) = \hat{A}$.

10.2.2.1 The Heisenberg Equation of Motion

Let us now derive the equation of motion that regulates the time evolution of operators within the Heisenberg picture. Assuming that $\hat{A}$ does not depend explicitly on time (i.e., $\partial \hat{A}/\partial t = 0$) and since $\hat{U}(t)$ is unitary, we have

$$\frac{d\hat{A}_H(t)}{dt} = \frac{\partial \hat{U}^\dagger(t)}{dt}\hat{A}\hat{U}_{(t)} + \hat{U}^\dagger(t)\hat{A}\frac{\partial \hat{U}(t)}{\partial t} = -\frac{1}{i\hbar}\hat{U}^\dagger\hat{H}\hat{U}\hat{U}^\dagger\hat{A}\hat{U} + \frac{1}{i\hbar}\hat{U}^\dagger\hat{A}\hat{U}\hat{U}^\dagger\hat{H}\hat{U}$$

$$= \frac{1}{i\hbar}\left[\hat{A}_H, \hat{U}^\dagger\hat{H}\hat{U}\right], \tag{10.12}$$

where we have used (10.3) to write $\partial \hat{U}(t)/\partial t = \hat{H}\hat{U}/i\hbar$ and $\partial \hat{U}^\dagger(t)/\partial t = -\hat{U}^\dagger\hat{H}/i\hbar$. Since $\hat{U}(t)$ and $\hat{H}$ commute, we have $\hat{U}^\dagger(t)\hat{H}\hat{U}(t) = H$; hence we can rewrite (10.12) as

$$\boxed{\frac{d\hat{A}_H}{dt} = \frac{1}{i\hbar}\left[\hat{A}_H, \hat{H}\right].} \tag{10.13}$$

This is the *Heisenberg equation of motion*. It plays the role of the Schrödinger equation within the Heisenberg picture. Since the Schrödinger and Heisenberg pictures are equivalent, we can use either picture to describe the quantum system under consideration. The Heisenberg equation (10.13), however, is in general difficult to solve.

Note that the structure of the Heisenberg equation (10.13) is similar to the classical equation of motion of a variable A that does not depend explicitly on time $dA/dt = \{A, H\}$, where $\{A, H\}$ is the Poisson bracket between A and H (see Chapter 3).

10.2.3 The Interaction Picture

The interaction picture, also called the *Dirac picture*, is useful to describe quantum phenomena with Hamiltonians that depend explicitly on time. In this picture *both state vectors and operators evolve in time*. We need, therefore, to find the equation of motion for the state vectors and for the operators.

10.2.3.1 Equation of Motion for the State Vectors

State vectors in the interaction picture are defined in terms of the Schrödinger states $| \psi(t)\rangle$ by

$$\boxed{| \psi(t)\rangle_I = e^{it\hat{H}_0/\hbar} | \psi(t)\rangle.} \tag{10.14}$$

If $t = 0$ we have $| \psi(0)\rangle_I = | \psi(0)\rangle$. The time evolution of $| \psi(t)\rangle$ is governed by the Schrödinger equation (10.1) with $\hat{H} = \hat{H}_0 + \hat{V}$ where $\hat{H}_0$ is time independent, but $\hat{V}$ may depend on time.

To find the time evolution of $| \psi(t)\rangle_I$, we need the time derivative of (10.14):

$$i\hbar\frac{d | \psi(t)\rangle_I}{dt} = -\hat{H}_0 e^{it\hat{H}_0/\hbar} | \psi(t)\rangle + e^{it\hat{H}_0/\hbar}\left(i\hbar\frac{d | \psi(t)\rangle}{dt}\right)$$

$$= -\hat{H}_0 | \psi(t)\rangle_I + e^{it\hat{H}_0/\hbar}\hat{H} | \psi(t)\rangle, \tag{10.15}$$

where we have used (10.1). Since $\hat{H} = \hat{H}_0 + \hat{V}$ and

$$e^{i H_0 t / \hbar} \hat{V} = \left(e^{it \hat{H}_0 / \hbar} \hat{V} e^{-it \hat{H}_0 / \hbar} \right) e^{it \hat{H}_0 / \hbar} = \hat{V}_I(t) e^{it \hat{H}_0 / \hbar}, \tag{10.16}$$

with

$$\hat{V}_I(t) = e^{it \hat{H}_0 / \hbar} \hat{V} e^{-it \hat{H}_0 / \hbar}, \tag{10.17}$$

we can rewrite (10.15) as

$$i\hbar \frac{d \mid \psi(t)\rangle_I}{dt} = -\hat{H}_0 \mid \psi(t)\rangle_I + \hat{H}_0 e^{it \hat{H}_0 / \hbar} \mid \psi(t)\rangle + \hat{V}_I(t) e^{it \hat{H}_0 / \hbar} \mid \psi(t)\rangle, \tag{10.18}$$

or

$$\boxed{i\hbar \frac{d \mid \psi(t)\rangle_I}{dt} = \hat{V}_I(t) \mid \psi(t)\rangle_I.} \tag{10.19}$$

This is the Schrödinger equation in the interaction picture. It shows that the time evolution of the state vector is governed by the interaction $\hat{V}_I(t)$.

10.2.3.2 Equation of Motion for the Operators

The interaction representation of an operator $\hat{A}_I(t)$ is given, as shown in (10.17), in terms of its Schrödinger representation by

$$\boxed{\hat{A}_I(t) = e^{i \hat{H}_0 t / \hbar} \hat{A} e^{-i \hat{H}_0 t / \hbar}.} \tag{10.20}$$

Calculating the time derivative of $\hat{A}_I(t)$ and since $\partial \hat{A} / \partial t = 0$, we can show the time evolution of $\hat{A}_I(t)$ is governed by $\hat{H}_0$:

$$\boxed{\frac{d\hat{A}_I(t)}{dt} = \frac{1}{i\hbar} \left[\hat{A}_I(t),\ \hat{H}_0 \right].} \tag{10.21}$$

This equation is similar to the Heisenberg equation of motion (10.13), except that $\hat{H}$ is replaced by $\hat{H}_0$. The basic difference between the Heisenberg and interaction pictures can be inferred from a comparison of (10.9) with (10.14), and (10.11) with (10.20): in the Heisenberg picture it is $\hat{H}$ that appears in the exponents, whereas in the interaction picture it is $\hat{H}_0$ that appears.

In conclusion, we have seen that, within the Schrödinger picture, the states depend on time but not the operators; in the Heisenberg picture, only operators depend explicitly on time, state vectors are frozen in time. The interaction picture, however, is intermediate between the Schrödinger and the Heisenberg pictures, since both state vectors and operators evolve with time.

10.3 Time-Dependent Perturbation Theory

We consider here only those phenomena that are described by Hamiltonians which can be split into two parts, a time-independent part $\hat{H}_0$ and a time-dependent part $\hat{V}(t)$ that is small compared to $\hat{H}_0$:

$$\hat{H}(t) = \hat{H}_0 + \hat{V}(t), \tag{10.22}$$

where $\hat{H}_0$, which describes the system when unperturbed, is assumed to have exact solutions that are known. Such splitting of the Hamiltonian is encountered in the following typical problem. Consider a system which, when unperturbed, is described by a time-independent Hamiltonian $\hat{H}_0$ whose solutions—the eigenvalues E_n and eigenstates $| \psi_n \rangle$—are known,

$$\hat{H}_0 \, | \, \psi_n \rangle = E_n \, | \, \psi_n \rangle, \tag{10.23}$$

and whose most general state vectors are given by stationary states

$$| \, \Psi_n(t) \rangle = e^{-it\hat{H}_0/\hbar} \, | \, \psi_n \rangle = e^{-iE_nt/\hbar} \, | \, \psi_n \rangle. \tag{10.24}$$

In the time interval $0 \le t \le \tau$ we subject the system to an external time-dependent perturbation, $\hat{V}(t)$, that is small compared to $\hat{H}_0$:

$$\hat{V}(t) = \begin{cases} \hat{V}(t), & 0 \le t \le \tau, \\ 0, & t < 0, \quad t > \tau. \end{cases} \tag{10.25}$$

During the time interval $0 \le t \le \tau$, the Hamiltonian of the system is $\hat{H} = \hat{H}_0 + \hat{V}(t)$ and the corresponding Schrödinger equation is

$$i\hbar\frac{d \, | \, \Psi(t) \rangle}{dt} = (\hat{H}_0 + \hat{V}(t)) \, | \, \Psi(t) \rangle, \tag{10.26}$$

where $\hat{V}(t)$ characterizes the interaction of the system with the external source of perturbation.

How does $\hat{V}(t)$ affect the system? When the system interacts with $\hat{V}(t)$, it either absorbs or emits energy. This process inevitably causes the system to undergo transitions from one unperturbed eigenstate to another. The main task of time-dependent perturbation theory consists of answering this question: If the system is initially in an (unperturbed) eigenstate $| \, \psi_i \rangle$ of $\hat{H}_0$, what is the probability that the system will be found at a later time in another unperturbed eigenstate $| \, \psi_f \rangle$?

To prepare the ground for answering this question, we need to look for the solutions of the Schrödinger equation (10.26). The standard method to solve (10.26) is to expand $| \, \Psi(t) \rangle$ in terms of an expansion coefficient $c_n(t)$:

$$| \, \Psi(t) \rangle = \sum_n c_n(t) e^{-iE_nt} \, | \, \psi_n \rangle, \tag{10.27}$$

and then insert this into (10.26) to find $c_n(t)$ to various orders in the approximation. Instead of following this procedure, and since we are dealing with time-dependent potentials, it is more convenient to solve (10.26) in the interaction picture (10.19):

$$i\hbar\frac{d \, | \, \Psi(t) \rangle_I}{dt} = \hat{V}_I(t) \, | \, \Psi(t) \rangle_I, \tag{10.28}$$

where $| \, \Psi(t) \rangle_I = e^{it\hat{H}_0/\hbar} \, | \, \Psi(t) \rangle$ and $\hat{V}_I(t) = e^{it\hat{H}_0/\hbar}\hat{V}(t)e^{-it\hat{H}_0/\hbar}$. The time evolution equation $| \, \Psi(t) \rangle = \hat{U}(t, t_i) \, | \, \Psi(t_i) \rangle$ may be written in the interaction picture as

$$| \, \Psi(t) \rangle_I = e^{it\hat{H}_0/\hbar} \, | \, \Psi(t) \rangle = e^{it\hat{H}_0/\hbar}\hat{U}(t, t_i) \, | \, \Psi(t_i) \rangle = e^{it\hat{H}_0/\hbar}\hat{U}(t, t_i)e^{-i\hat{H}_0t_i/\hbar} \, | \, \Psi(t_i) \rangle_I, \tag{10.29}$$

or as

$$| \Psi(t) \rangle_I = \hat{U}_I(t, t_i) | \Psi(t_i) \rangle_I, \tag{10.30}$$

where the time evolution operator is given in the interaction picture by

$$\hat{U}_I(t, t_i) = e^{it\hat{H}_0/\hbar} \hat{U}(t, t_i) e^{-i\hat{H}_0 t_i/\hbar}. \tag{10.31}$$

Inserting (10.30) into (10.28) we end up with

$$i\hbar \frac{d\hat{U}_I(t, t_i)}{dt} = \hat{V}_I(t) \hat{U}_I(t, t_i). \tag{10.32}$$

The solutions of this equation, with the initial condition $\hat{U}_I(t_i, t_i) = \hat{I}$, are given by the *integral equation*

$$\hat{U}_I(t, t_i) = 1 - \frac{i}{\hbar} \int_{t_i}^{t} \hat{V}_I(t') \hat{U}_I(t', t_i) \, dt'. \tag{10.33}$$

Time-dependent perturbation theory provides *approximate* solutions to this integral equation. This consists in assuming that $\hat{V}_I(t)$ is *small and then proceeding iteratively*. The first-order approximation is obtained by inserting $\hat{U}_I(t', t_i) = 1$ in the integral sign of (10.33), leading to $\hat{U}_I^{(1)}(t, t_i) = 1 - (i/\hbar) \int_{t_i}^{t} \hat{V}_I(t') \, dt'$. Substituting $\hat{U}_I(t', t_i) = \hat{U}_I^{(1)}(t', t_i)$ in the integral sign of (10.33) we get the second-order approximation:

$$\hat{U}_I^{(2)}(t, t_i) = 1 - \frac{i}{\hbar} \int_{t_i}^{t} \hat{V}_I(t') \, dt' + \left(-\frac{i}{\hbar} \right)^2 \int_{t_i}^{t} \hat{V}_I(t_1) \, dt_1 \int_{t_i}^{t_1} \hat{V}_I(t_2) \, dt_2. \tag{10.34}$$

The third-order approximation is obtained by substituting $\hat{U}_I^{(2)}(t, t_i)$ into (10.33), and so on. A repetition of this iterative process yields

$$\hat{U}_I(t, t_i) = 1 - \frac{i}{\hbar} \int_{t_i}^{t} \hat{V}_I(t') dt' + \left(-\frac{i}{\hbar} \right)^2 \int_{t_i}^{t} \hat{V}_I(t_1) dt_1 \int_{t_i}^{t_1} \hat{V}_I(t_2) dt_2 + \cdots$$
$$+ \left(-\frac{i}{\hbar} \right)^n \int_{t_i}^{t} \hat{V}_I(t_1) dt_1 \int_{t_i}^{t_1} \hat{V}_I(t_2) dt_2 \int_{t_i}^{t_2} \hat{V}_I(t_3) dt_3 \cdots \int_{t_i}^{t_{n-1}} \hat{V}_I(t_n) dt_n + \cdots. \tag{10.35}$$

This series, known as the *Dyson series*, allows for the calculation of the state vector up to the desired order in the perturbation.

We are now equipped to calculate the transition probability. It may be obtained by taking the matrix elements of (10.35) between the eigenstates of $\hat{H}_0$. Time-dependent perturbation theory, where one assumes knowledge of the solutions of the unperturbed eigenvalue problem (10.23), deals in essence with the calculation of the transition probabilities between the unperturbed eigenstates $| \psi_n \rangle$ of the system.

10.3.1 Transition Probability

The transition probability corresponding to a transition from an initial unperturbed state $| \psi_i \rangle$ to another unperturbed state $| \psi_f \rangle$ is obtained from (10.35):

$$P_{if}(t) = \left| \langle \psi_f | \hat{U}_I(t, t_i) | \psi_i \rangle \right|^2 = \left| \langle \psi_f | \psi_i \rangle - \frac{i}{\hbar} \int_0^t e^{i\omega_{fi}t'} \langle \psi_f | \hat{V}(t') | \psi_i \rangle dt' \right.$$

$$+ \left(-\frac{i}{\hbar}\right)^2 \sum_n \int_0^t e^{i\omega_{fn}t_1} \langle \psi_f \mid \hat{V}(t_1) \mid \psi_n \rangle \, dt_1 \int_0^{t_1} e^{i\omega_{ni}t_2} \langle \psi_n \mid \hat{V}(t_2) \mid \psi_i \rangle \, dt_2 + \cdots \Bigg|^2,$$

$$(10.36)$$

where we have used the fact that

$$\langle \psi_f \mid \hat{V}_I(t') \mid \psi_i \rangle = \langle \psi_f \mid e^{iH_0 t'/\hbar} \hat{V}(t') e^{-iH_0 t'/\hbar} \mid \psi_i \rangle = \langle \psi_f \mid V(t') \mid \psi_i \rangle \exp\left(i\omega_{fi}t'\right),$$

$$(10.37)$$

where ω_{fi} is the transition frequency between the initial and final levels i and f:

$$\omega_{fi} = \frac{E_f - E_i}{\hbar} = \frac{1}{\hbar} \left(\langle \psi_f \mid \hat{H}_0 \mid \psi_f \rangle - \langle \psi_i \mid \hat{H}_0 \mid \psi_i \rangle \right). \qquad (10.38)$$

The transition probability (10.36) can be written in terms of the expansion coefficients $c_n(t)$ introduced in (10.27) as

$$P_{if}(t) = \left| c_f^{(0)} + c_f^{(1)}(t) + c_f^{(2)}(t) + \cdots \right|^2, \qquad (10.39)$$

where

$$c_f^{(0)} = \langle \psi_f \mid \psi_i \rangle = \delta_{f,i}, \qquad c_f^{(1)}(t) = -\frac{i}{\hbar} \int_0^t \langle \psi_f \mid \hat{V}(t') \mid \psi_i \rangle e^{i\omega_{fi}t'} \, dt', \ldots \qquad (10.40)$$

The first-order transition probability for $\mid \psi_i \rangle \to \mid \psi_f \rangle$ with $i \neq f$ (and hence $\langle \psi_f \mid \psi_i \rangle = 0$) is obtained by terminating (10.36) at the first order in $V_I(t)$:

$$P_{if}(t) = \left| -\frac{i}{\hbar} \int_0^t \langle \psi_f \mid \hat{V}(t') \mid \psi_i \rangle e^{i\omega_{fi}t'} \, dt' \right|^2. \qquad (10.41)$$

In principle we can use (10.36) to calculate the transition probability to any order in $\hat{V}_I(t)$. However, terms higher than the first order become rapidly intractable. For most problems of atomic and nuclear physics, the first order (10.41) is usually sufficient. In what follows, we are going to apply (10.41) to calculate the transition probability for two cases, which will have later usefulness when we deal with the interaction of atoms with radiation: a *constant* perturbation and a *harmonic* perturbation.

10.3.2 Transition Probability for a Constant Perturbation

In the case where $\hat{V}$ does not depend on time, (10.41) leads to

$$P_{if}(t) = \frac{1}{\hbar^2} \left| \langle \psi_f \mid \hat{V} \mid \psi_i \rangle \int_0^t e^{i\omega_{fi}t'} \, dt' \right|^2 = \frac{1}{\hbar^2} \left| \langle \psi_f \mid \hat{V} \mid \psi_i \rangle \right|^2 \left| \frac{e^{i\omega_{fi}t} - 1}{\omega_{fi}} \right|^2, \qquad (10.42)$$

which, using $|e^{i\theta} - 1|^2 = 4 \sin^2(\theta/2)$, reduces to

$$P_{if}(t) = \frac{4 \left| \langle \psi_f \mid \hat{V} \mid \psi_i \rangle \right|^2}{\hbar^2 \omega_{fi}^2} \sin^2\left(\frac{\omega_{fi}t}{2}\right). \qquad (10.43)$$

Figure 10.1 Plot of $[\sin^2(\omega_{fi}t/2)]/\omega_{fi}^2$ versus ω_{fi} for a fixed value of t; $\omega_{fi} = (E_f - E_i)/2$.

As a function of time, this transition probability is an oscillating sinusoidal function with a period of $2\pi/\omega_{fi}$. As a function of ω_{fi}, however, the transition probability, as shown in Figure 10.1, has an interference pattern: it is appreciable only near $\omega_{fi} \simeq 0$ and decays rapidly as ω_{fi} moves away from zero (here, for a fixed t, we have assumed that ω_{fi} is a continuous variable; that is, we have considered a continuum of final states; we will deal with this in more detail in a moment). This means that the transition probability of finding the system in a state $|\psi_f\rangle$ of energy E_f is greatest only when $E_i \simeq E_f$ or when $\omega_{fi} \simeq 0$. The height and the width of the main peak, centered around $\omega_{fi} = 0$, are proportional to t^2 and $1/t$, respectively, so the area under the curve is proportional to t; since most of the area is under the central peak, the transition probability is proportional to t. The transition probability therefore grows linearly with time. The central peak becomes narrower and higher as time increases; this is exactly the property of a delta function. Thus, in the limit $t \to \infty$ the transition probability takes the shape of a delta function, as we are going to see.

As $t \to \infty$ we can use the asymptotic relation (Appendix A)

$$\lim_{t\to\infty} \frac{\sin^2(yt)}{\pi y^2 t} = \delta(y) \tag{10.44}$$

to write the following expression:

$$\frac{1}{(\frac{1}{2}\omega_{fi})^2} \sin^2\left(\frac{\omega_{fi}t}{2}\right) = 2\pi t\hbar\delta(\hbar\omega_{fi}), \tag{10.45}$$

because $\delta(\omega_{fi}/2) = 2\hbar\delta(\hbar\omega_{fi})$. Now since $\hbar\omega_{fi} = E_f - E_i$ and hence $\delta(\hbar\omega_{fi}) = \delta(E_f - E_i)$, we can reduce (10.43) in the limit of long times to

$$\boxed{P_{if}(t) = \frac{2\pi t}{\hbar} \left| \langle \psi_f | \hat{V} | \psi_i \rangle \right|^2 \delta(E_f - E_i).} \tag{10.46}$$

The *transition rate*, which is defined as a transition probability per unit time, is thus given by

$$\boxed{\Gamma_{if} = \frac{P_{if}(t)}{t} = \frac{2\pi}{\hbar} \left| \langle \psi_f | \hat{V} | \psi_i \rangle \right|^2 \delta(E_f - E_i).} \tag{10.47}$$

The delta term $\delta(E_f - E_i)$ guarantees the conservation of energy: in the limit $t \to \infty$, the transition rate is nonvanishing only between states of equal energy. Hence a constant (time-independent) perturbation neither removes energy from the system nor supplies energy to it. It simply causes energy-conserving transitions.

Transition into a continuum of final states

Let us now calculate the total transition rate associated with a transition from an initial state $| \psi_i \rangle$ into a continuum of final states $| \psi_f \rangle$. If $\rho(E_f)$ is the density of final states—the number of states per unit energy interval—the number of final states within the energy interval E_f and $E_f + dE_f$ is equal to $\rho(E_f)\,dE_f$. The total transition rate W_{if} can then be obtained from (10.47):

$$W_{if} = \int \frac{P_{if}(t)}{t} \rho(E_f)\,dE_f = \frac{2\pi}{\hbar} |\langle \psi_f | \hat{V} | \psi_i \rangle|^2 \int \rho(E_f)\delta(E_f - E_i)\,dE_f, \quad (10.48)$$

or

$$W_{if} = \frac{2\pi}{\hbar} \left| \langle \psi_f | \hat{V} | \psi_i \rangle \right|^2 \rho(E_i). \quad (10.49)$$

This relation is called the *Fermi golden rule*. It implies that, in the case of a constant perturbation, if we wait long enough, the total transition rate becomes constant (time independent).

10.3.3 Transition Probability for a Harmonic Perturbation

Consider now a perturbation which depends harmonically on time (i.e., the time between the moments of turning the perturbation on and off):

$$\hat{V}(t) = \hat{v}e^{i\omega t} + \hat{v}^\dagger e^{-i\omega t}, \quad (10.50)$$

where $\hat{v}$ is a time-independent operator. Such a perturbation is encountered, for instance, when charged particles (e.g., electrons) interact with an electromagnetic field. This perturbation provokes transitions of the system from one stationary state to another.

The transition probability corresponding to this perturbation can be obtained from (10.41):

$$P_{if}(t) = \frac{1}{\hbar^2} \left| \langle \psi_f | \hat{v} | \psi_i \rangle \int_0^t e^{i(\omega_{fi}+\omega)t'}\,dt' + \langle \psi_f | \hat{v}^\dagger | \psi_i \rangle \int_0^t e^{i(\omega_{fi}-\omega)t'}\,dt' \right|^2. \quad (10.51)$$

Neglecting the cross terms, for they are negligible compared with the other two (because they induce no lasting transitions), we can rewrite this expression as

$$P_{if}(t) = \frac{1}{\hbar^2} \left| \langle \psi_f | \hat{v} | \psi_i \rangle \right|^2 \left| \frac{e^{i(\omega_{fi}+\omega)t} - 1}{\omega_{fi} + \omega} \right|^2 + \frac{1}{\hbar^2} \left| \langle \psi_f | \hat{v}^\dagger | \psi_i \rangle \right|^2 \left| \frac{e^{i(\omega_{fi}-\omega)t} - 1}{\omega_{fi} - \omega} \right|^2, \quad (10.52)$$

which, using $|e^{i\theta} - 1|^2 = 4\sin^2(\theta/2)$, reduces to

$$P_{if}(t) = \frac{4}{\hbar^2} \left[\left| \langle \psi_f | \hat{v} | \psi_i \rangle \right|^2 \frac{\sin^2((\omega_{fi} + \omega)t/2)}{(\omega_{fi} + \omega)^2} + \left| \langle \psi_f | \hat{v}^\dagger | \psi_i \rangle \right|^2 \frac{\sin^2((\omega_{fi} - \omega)t/2)}{(\omega_{fi} - \omega)^2} \right]. \quad (10.53)$$

As displayed in Figure 10.2, the transition probability peaks either at $\omega_{fi} = -\omega$, where its maximum value is $P_{if}(t) = (t^2/4\hbar^2)|\langle \psi_f | \hat{v} | \psi_i \rangle|^2$, or at $\omega_{fi} = \omega$, where its maximum

Figure 10.2 Plot of $[\sin^2((\omega_{fi} \pm \omega)t/2)]/(\omega_{fi} \pm \omega)^2$ versus ω_{fi} for a fixed value of t, where $\omega_n = -\omega - n\pi/t$, $\omega'_n = -\omega + n\pi/t$, $\Omega_n = \omega - n\pi/t$, and $\Omega'_n = \omega + n\pi/t$.

value is $P_{if}(t) = (t^2/4\hbar^2)|\langle \psi_f \mid \hat{o}^\dagger \mid \psi_i \rangle|^2$. These are conditions for resonance; this means that the probability of transition is greatest only when the frequency of the perturbing field is close to $\pm \omega_{fi}$. As ω moves away from $\pm \omega_{fi}$, P_{fi} decreases rapidly.

Note that the expression (10.53) is similar to that derived for a constant perturbation, as shown in (10.43). Using (10.45) we can reduce (10.53) in the limit $t \to \infty$ to

$$\Gamma_{if} = \frac{2\pi}{\hbar} \left| \langle \psi_f \mid \hat{o} \mid \psi_i \rangle \right|^2 \delta(E_f - E_i + \hbar\omega) + \frac{2\pi}{\hbar} \left| \langle \psi_f \mid \hat{o}^\dagger \mid \psi_i \rangle \right|^2 \delta(E_f - E_i - \hbar\omega).$$

(10.54)

This transition rate is nonzero only when either of the following two conditions is satisfied:

$$E_f = E_i - \hbar\omega,$$

(10.55)

$$E_f = E_i + \hbar\omega.$$

(10.56)

These two conditions cannot be satisfied simultaneously; their physical meaning can be understood as follows. The first condition $E_f = E_i - \hbar\omega$ implies that the system is initially excited, since its final energy is smaller than the initial energy; when acted upon by the perturbation, the system deexcites by giving up a photon of energy $\hbar\omega$ to the potential $\hat{V}(t)$ as shown in Figure 10.3. This process is called *stimulated emission*, since the system easily emits a photon of energy $\hbar\omega$. The second condition, $E_f = E_i + \hbar\omega$ shows that the final energy of the system is larger than its initial energy. The system then *absorbs* a photon of energy $\hbar\omega$ from $\hat{V}(t)$ and ends up in an excited state of (higher) energy E_f (Figure 10.3). We may thus view the terms $e^{i\omega t}$ and $e^{-i\omega t}$ in $\hat{V}(t)$ as responsible, respectively, for the emission and the absorption of a photon of energy $\hbar\omega$.

In conclusion, the effect of a harmonic perturbation is to transfer to the system, or to receive from it, a photon of energy $\hbar\omega$. In sharp contrast, a constant (time-independent) perturbation neither transfers energy to the system nor removes energy from it.

Figure 10.3 Stimulated emission and absorption of a photon of energy $\hbar\omega$.

Remark

For transitions into a continuum of final states, we can show, by analogy with the derivation of (10.49), that (10.54) leads to the absorption and emission transition rates:

$$W_{if}^{abs} = \frac{2\pi}{\hbar} \left| \langle \psi_f \mid \hat{V}^\dagger \mid \psi_i \rangle \right|^2 \rho(E_f) \Big|_{E_f = E_i + \hbar\omega}, \tag{10.57}$$

$$W_{if}^{emi} = \frac{2\pi}{\hbar} \left| \langle \psi_f \mid \hat{V} \mid \psi_i \rangle \right|^2 \rho(E_f) \Big|_{E_f = E_i - \hbar\omega}. \tag{10.58}$$

Since the perturbation (10.50) is Hermitian, $\langle \psi_f \mid \hat{o} \mid \psi_i \rangle = \langle \psi_i \mid \hat{o}^\dagger \mid \psi_f \rangle^*$, we have $|\langle \psi_f \mid \hat{o} \mid \psi_i \rangle|^2 = |\langle \psi_f \mid \hat{o}^\dagger \mid \psi_i \rangle|^2$; hence

$$\frac{W_{if}^{abs}}{\rho(E_f)\big|_{E_f = E_i + \hbar\omega}} = \frac{W_{if}^{emi}}{\rho(E_f)\big|_{E_f = E_i - \hbar\omega}}. \tag{10.59}$$

This relation is known as the condition of *detailed balancing*.

Example 10.1

A particle, which is initially ($t = 0$) in the ground state of an infinite, one-dimensional potential box with walls at $x = 0$ and $x = a$, is subjected for $0 \le t \le \infty$ to a perturbation $\hat{V}(t) = \hat{x}^2 e^{-t/\tau}$. Calculate to first order the probability of finding the particle in its first excited state for $t \ge 0$.

Solution

For a particle in a box potential, with $E_n = n^2\pi^2\hbar^2/(2ma^2)$ and $\psi_n(x) = \sqrt{2/a}\sin(n\pi x/a)$, the ground state corresponds to $n = 1$ and the first excited state to $n = 2$. We can use (10.41) to obtain

$$P_{12} = \frac{1}{\hbar^2} \left| \int_0^\infty \langle \psi_2 | \hat{V}(t) | \psi_1 \rangle e^{i\omega_{21}t} dt \right|^2 = \frac{1}{\hbar^2} \left| \langle \psi_2 | \hat{x}^2 | \psi_1 \rangle \right|^2 \left| \int_0^\infty e^{-(1/\tau - i\omega_{21})t} dt \right|^2, \tag{10.60}$$

where

$$\langle \psi_2 | \hat{x}^2 | \psi_1 \rangle = \int_0^a x^2 \psi_2^*(x)\psi_1(x)\, dx = \frac{2}{a}\int_0^a x^2 \sin\left(\frac{2\pi x}{a}\right)\sin\left(\frac{\pi x}{a}\right) dx = -\frac{16a^2}{9\pi^2}, \tag{10.61}$$

$$\left| \int_0^t e^{-(1/\tau - i\omega_{21})t} \, dt \right|^2 = \left| \frac{e^{-(1/\tau - i\omega_{21})t} - 1}{1/\tau - i\omega_{21}} \right|^2 = \frac{1 + e^{-2t/\tau} - 2e^{-t/\tau}\cos(\omega_{21}t)}{\omega_{21}^2 + 1/\tau^2}, \quad (10.62)$$

which, in the limit $t \to \infty$, reduces to

$$\left| \int_0^\infty e^{-(1/\tau - i\omega_{21})t} \, dt \right|^2 = \left[\omega_{21}^2 + \frac{1}{\tau^2} \right]^{-1} = \left[\frac{9\pi^4 \hbar^2}{4m^2 a^4} + \frac{1}{\tau^2} \right]^{-1}, \quad (10.63)$$

since $\omega_{21} = (E_2 - E_1)/\hbar = 3\pi^2 \hbar/(2ma^2)$. A substitution of (10.61) and (10.63) into (10.60) leads to

$$P_{12} = \left(\frac{16a^2}{9\pi^2 \hbar} \right)^2 \left[\frac{9\pi^4 \hbar^2}{4m^2 a^4} + \frac{1}{\tau^2} \right]^{-1}. \quad (10.64)$$

10.4 Adiabatic and Sudden Approximations

In discussing the time-dependent perturbation theory, we have dealt with phenomena where the perturbation $\hat{V}(t)$ is small, but we have paid no attention to the rate of change of the perturbation. In this section we want to discuss approximation methods treating phenomena where $\hat{V}(t)$ is not only small but also switched on either *adiabatically* (slowly) or *suddenly* (rapidly). We assume here that $\hat{V}(t)$ is switched on at $t = 0$ and off at a later time t (the turning on and off may be smooth or abrupt).

Since $e^{i\omega_{fi}t} = (1/i\omega_{fi})\partial e^{i\omega_{fi}t}/\partial t$ an integration by parts yields

$$-\frac{i}{\hbar} \int_0^t \langle \psi_f | \hat{V}(t') | \psi_i \rangle e^{i\omega_{fi}t'} \, dt' = -\frac{1}{\hbar \omega_{fi}} \int_0^t \langle \psi_f | \hat{V}(t') | \psi_i \rangle \left(\frac{\partial}{\partial t'} e^{i\omega_{fi}t'} \right) dt'$$

$$= -\frac{1}{\hbar \omega_{fi}} \langle \psi_f | \hat{V}(t) | \psi_i \rangle e^{i\omega_{fi}t} \Big|_{t=0}^t + \frac{1}{\hbar \omega_{fi}} \int_0^t e^{i\omega_{fi}t'} \left(\frac{\partial}{\partial t'} \langle \psi_f | \hat{V}(t') | \psi_i \rangle \right) dt'$$

$$= \frac{1}{\hbar \omega_{fi}} \int_0^t e^{i\omega_{fi}t'} \left(\frac{\partial}{\partial t'} \langle \psi_f | \hat{V}(t') | \psi_i \rangle \right) dt', \quad (10.65)$$

where we have used the fact that $\hat{V}(t)$ vanishes at the limits (when it is switched on at $t = 0$ and off at time t). The calculation of the integral depends on the rate of change of $\hat{V}(t)$. In what follows we are going to consider the cases where the interaction is switched on slowly or rapidly.

10.4.1 Adiabatic Approximation

First, let us discuss briefly the adiabatic approximation without combining it with perturbation theory. This approximation applies to phenomena whose Hamiltonians evolve *slowly* with time; we should highlight the fact that the adiabatic approximation does not require the Hamiltonian to split into an unperturbed part $\hat{H}_0$ and a weak time-dependent perturbation $\hat{V}(t)$. Essentially, it consists in approximating the solutions of the Schrödinger equation at every time by the stationary states (energy E_n and wave functions ψ_n) of the instantaneous Hamiltonian in such a way that the wave function at a given time is continuously and smoothly converted into an eigenstate of the corresponding Hamiltonian at a later time. This result is the basis of an important theorem of quantum mechanics, known as the *adiabatic theorem*, which states that: if a system is initially in the nth state and if its Hamiltonian evolves slowly with time, it will be found at a later time in the nth state of the new (instantaneous) Hamiltonian. That is, *the system will make no transitions*; it simply remains in the nth state of the new Hamiltonian.

Let us now discuss the adiabatic approximation for those cases where the Hamiltonian splits into a time-independent part $\hat{H}_0$ and a time-dependent part $\hat{V}(t)$, which is small enough so that perturbation theory applies and which is turned on and off very slowly. If $\hat{V}(t)$ is turned on at $t = 0$ and off at time t in a slow and smooth way, it will change very little in the time interval $0 \le t' \le t$. The term $\partial \langle \psi_f \mid V(t') \mid \psi_i \rangle / \partial t'$ will be almost constant, so we can take it outside the integral sign in (10.65):

$$P_{if}(t) \simeq \frac{1}{\hbar^2 \omega_{fi}^2} \left| \frac{\partial}{\partial t} \langle \psi_f \mid \hat{V}(t) \mid \psi_i \rangle \right|^2 \left| \int_0^t e^{i\omega_{fi}t'} dt' \right|^2 , \qquad (10.66)$$

or

$$P_{if}(t) \simeq \frac{4}{\hbar^2 \omega_{fi}^4} \left| \frac{\partial}{\partial t} \langle \psi_f \mid \hat{V}(t) \mid \psi_i \rangle \right|^2 \sin^2 \left(\frac{\omega_{fi}t}{2} \right). \qquad (10.67)$$

The adiabatic approximation is valid only when the time change in the energy of the perturbation during one period of oscillation is very small compared with the energy difference $|E_f - E_i|$ between the initial and final states:

$$\left| \frac{1}{\omega_{fi}} \frac{\partial}{\partial t} \langle \psi_f \mid \hat{V}(t) \mid \psi_i \rangle \right| \ll |E_f - E_i|. \qquad (10.68)$$

Since $\sin^2 \alpha \ll 1$ we see from (10.67) that, in the adiabatic approximation, the transition probability is very small, $P_{if} \ll 1$. In fact, if the rate of change of $\hat{V}(t)$, and hence of $\hat{H}(t)$, is very small, we will have $\partial \langle \psi_f \mid \hat{V}(t) \mid \psi_i \rangle / \partial t \to 0$, which in turn implies that the transition probability is practically zero: $P_{if} \to 0$. Once more, we see that *no transition occurs when the perturbation is turned on and off adiabatically*. That is, if a system is initially (at $t = 0$) in the nth state $\mid \psi_n(0) \rangle$ of $\hat{H}_0$ with energy $E_n(0)$, then at the end (at time t) of an *adiabatic perturbation* $\hat{V}(t)$, it will be found in the nth state $\mid \psi_n(t) \rangle$ of the new Hamiltonian ($\hat{H} = \hat{H}_0 + \hat{V}(t)$) with energy $E_n(t)$. As an illustrative example, consider a particle in a harmonic oscillator potential whose constant is being changed *very slowly* from k to, say, $3k$; if the particle is initially in the second excited state, it will remain in the second excited state of the new oscillator.

Note that the transition probability (10.67) was derived by making use of two approximations: the perturbation theory approximation and the adiabatic approximation. It should be stressed, however, that when the perturbation is not weak, but switched on adiabatically, we can still use the adiabatic approximation but no longer in conjunction with perturbation theory.

10.4.2 Sudden Approximation

Again, let us start with a brief discussion of the sudden approximation without invoking perturbation theory. If the Hamiltonian of a system changes abruptly (over a very short time interval) from one form to another, we would expect the wave function not to change much, yet its expansion in terms of the eigenfunctions of the initial and final Hamiltonians may be different. Consider, for instance, a system which is initially ($t < 0$) in an eigenstate $\mid \psi_n \rangle$ of the Hamiltonian $\hat{H}_0$:

$$\hat{H}_0 \mid \psi_n \rangle = E_n^{(0)} \mid \psi_n \rangle, \qquad \mid \psi_n(t) \rangle = e^{i E_n^{(0)} t / \hbar} \mid \psi_n \rangle. \tag{10.69}$$

At time $t = 0$ we assume that the Hamiltonian is suddenly changed from $\hat{H}_0$ to $\hat{H}$ and that it preserves this new form (i.e., $\hat{H}$) for $t > 0$; it should be stressed that the difference between the two Hamiltonians $\hat{H} - \hat{H}_0$ does not need to be small. Let $\mid \phi_n \rangle$ be the eigenfunctions of $\hat{H}$:

$$\hat{H} \mid \phi_n \rangle = E_n \mid \phi_n \rangle, \qquad \mid \phi_n(t) \rangle = e^{i E_n t / \hbar} \mid \phi_n \rangle. \tag{10.70}$$

The state of the system is given for $t > 0$ by

$$\mid \Phi(t) \rangle = \sum_n c_n e^{i E_n t / \hbar} \mid \phi_n \rangle. \tag{10.71}$$

If the system is initially in an eigenstate $\mid \psi_m \rangle$ of $\hat{H}_0$, the continuity condition at $t = 0$ dictates that the system remains in this state just after the change takes place:

$$\mid \Phi(0) \rangle = \sum_n c_n \mid \phi_n \rangle = \mid \psi_m \rangle \implies c_n = \langle \phi_n \mid \psi_m \rangle. \tag{10.72}$$

The probability that a sudden change in the system's Hamiltonian from $\hat{H}_0$ to $\hat{H}$ causes a transition from the mth state of $\hat{H}_0$ to the nth state of $\hat{H}$ is

$$\boxed{P_{mn} = |\langle \phi_n \mid \psi_m \rangle|^2.} \tag{10.73}$$

We should note that the sudden approximation is applicable only for transitions between discrete states.

Let us now look at the sudden approximation within the context of perturbation theory. Consider a system which is subjected to a perturbation that is small and switched on suddenly. When $\hat{V}(t)$ is instantaneously turned on, the term $e^{i \omega_{fi} t'}$ in (10.65) does not change much during the switching-on time. We can therefore take $e^{i \omega_{fi} t'}$ outside the integral sign,

$$P_{if} \simeq \frac{1}{\hbar^2 \omega_{fi}^2} \left| e^{i \omega_{fi} t} \right|^2 \left| \int_0^t \frac{\partial}{\partial t'} \langle \psi_f \mid \hat{V}(t') \mid \psi_i \rangle dt' \right|^2 \tag{10.74}$$

hence the transition probability is given within the sudden approximation by

$$P_{if}(t) \simeq \frac{|\langle \psi_f \mid \hat{V}(t) \mid \psi_i \rangle|^2}{\hbar^2 \omega_{fi}^2}. \tag{10.75}$$

To conclude, notice that both (10.73) and (10.75) give the transition probability within the sudden approximation. Equation (10.73) represents the exact formula, where the change in the Hamiltonians, $\hat{H} - \hat{H}_0$, may be large, but equation (10.75) gives only an approximate result, for it was derived from a first-order perturbative treatment, where we assumed that the change $\hat{H} - \hat{H}_0$ is small, yet sudden.

Example 10.2

A particle is initially ($t < 0$) in the ground state of an infinite, one-dimensional potential well with walls at $x = 0$ and $x = a$.

(a) If the wall at $x = a$ is moved *slowly* to $x = 8a$, find the energy and wave function of the particle in the new well. Calculate the work done in this process.

(b) If the wall at $x = a$ is now *suddenly* moved (at $t = 0$) to $x = 8a$, calculate the probability of finding the particle in (i) the ground state, (ii) the first excited state, and (iii) the second excited state of the new potential well.

Solution

For $t < 0$ the particle was in a potential well with walls at $x = 0$ and $x = a$, and hence

$$E_n = \frac{n^2 \pi^2 \hbar^2}{2ma^2}, \qquad \psi_n(x) = \sqrt{\frac{2}{a}} \sin\left(\frac{n\pi x}{a}\right) \qquad (0 \le x \le a). \qquad (10.76)$$

(a) When the wall is moved slowly, the adiabatic theorem dictates that the particle will make no transitions; it will be found at time t in the ground state of the new potential well (the well with walls at $x = 0$ and $x = 8a$). Thus, we have

$$E_1(t) = \frac{\pi^2 \hbar^2}{2m(8a)^2} = \frac{\pi^2 \hbar^2}{128ma^2}, \qquad \psi_1'(x) = \sqrt{\frac{2}{8a}} \sin\left(\frac{\pi x}{8a}\right) \qquad (0 \le x \le a).$$
$$(10.77)$$

The work needed to move the wall is

$$\Delta W = E_1(t) - E_1 = \frac{\pi^2 \hbar^2}{m(8a)^2} - \frac{\pi^2 \hbar^2}{2ma^2} = -\frac{63\pi^2 \hbar^2}{128ma^2}. \qquad (10.78)$$

(b) When the wall is moved rapidly, the particle will find itself instantly (at $t \ge 0$) in the new potential well; its energy levels and wave function are now given by

$$E_n' = \frac{n^2 \pi^2 \hbar^2}{2m(8a)^2} = \frac{n^2 \pi^2 \hbar^2}{128ma^2}, \qquad \psi_n'(x) = \sqrt{\frac{2}{8a}} \sin\left(\frac{n\pi x}{8a}\right) \qquad (0 \le x \le 8a). \quad (10.79)$$

The probability of finding the particle in the ground state of the new box potential can be obtained from (10.73): $P_{11} = |\langle \psi_1' \mid \psi_1 \rangle|^2$, where

$$\langle \psi_1' \mid \psi_1 \rangle = \int_0^a \psi_1'^*(x) \psi_1(x)\, dx = \frac{2}{\sqrt{8a}} \int_0^a \sin\left(\frac{\pi x}{8a}\right) \sin\left(\frac{\pi x}{a}\right) dx = \frac{16}{63\pi} \sqrt{4 - 2\sqrt{2}};$$
$$(10.80)$$

hence

$$P_{11} = |\langle \psi_1' \mid \psi_1 \rangle|^2 = \left(\frac{16}{63\pi}\right)^2 (4 - 2\sqrt{2}) = 0.0077 \simeq 0.7\%. \qquad (10.81)$$

The probability of finding the particle in the first excited state of the new box potential is given by $P_{12} = |\langle \psi_2' \mid \psi_1 \rangle|^2$, where

$$\langle \psi_2' \mid \psi_1 \rangle = \int_0^a \psi_2'^*(x) \psi_1(x)\, dx = \frac{2}{\sqrt{8a}} \int_0^a \sin\left(\frac{\pi x}{4a}\right) \sin\left(\frac{\pi x}{a}\right) dx = \frac{8}{15\pi}; \qquad (10.82)$$

hence

$$P_{12} = |\langle \psi_2' \mid \psi_1 \rangle|^2 = \left(\frac{8}{15\pi}\right)^2 = 0.1699 \simeq 17\%. \qquad (10.83)$$

A similar calculation leads to

$$P_{13} = |\langle \psi_3' \mid \psi_1 \rangle|^2 = \left| \frac{2}{\sqrt{8a}} \int_0^a \sin\left(\frac{3\pi x}{8a}\right) \sin\left(\frac{\pi x}{a}\right) dx \right|^2 = \left| \frac{16}{55\pi} \sqrt{4 + 2\sqrt{2}} \right|^2 \simeq 24.2\%.$$

(10.84)

These calculations show that the particle is most likely to be found in higher excited states; the probability of finding it in the ground state is very small.

10.5   Interaction of Atoms with Radiation

One of the most important applications of time-dependent perturbation theory is to study the interaction of atomic electrons with an external electromagnetic radiation. Such an application reveals a great deal about the structure of atoms. For simplicity, we assume that only one atomic electron is involved in the interaction and that the electron spin is neglected. We also assume that the nucleus is infinitely heavy.

In the absence of an external perturbation, the Hamiltonian of the atomic electron is $\hat{H}_0 = \vec{P}^2/(2m_e) + V_0(\vec{r})$, where m_e is the mass of the electron and $V_0(\vec{r})$ is the static potential due to the interaction of the electron with the other electrons and with the nucleus.

Now, if electromagnetic radiation of vector potential $\vec{A}(\vec{r}, t)$ and electric potential $\phi(\vec{r}, t)$ is applied on the atom, the Hamiltonian due to the interaction of the electron (of charge $-e$) with the radiation is given by

$$H = \frac{1}{2m_e} \left(\vec{P} + \frac{e}{c}\vec{A}(\vec{r}, t) \right)^2 - e\phi(\vec{r}, t) + V_0(\vec{r})$$

$$= H_0 - e\phi(\vec{r}, t) + \frac{e}{2m_e c} \left[2\vec{A} \cdot \vec{P} - i\hbar\vec{\nabla} \cdot \vec{A} \right] + \frac{e^2\vec{A}^2}{2m_e c^2}, \qquad (10.85)$$

where we have used the relation $\vec{P} \cdot \vec{A} = \vec{A} \cdot \vec{P} - i\hbar\vec{\nabla} \cdot \vec{A}$. Since $\phi(\vec{r}, t) = 0$ for radiation with no electrostatic source and since $\vec{\nabla} \cdot \vec{A} = 0$ (Coulomb gauge), and neglecting the term in $\vec{A}^2$, we may write (10.85) as

$$\hat{H} = \hat{H}_0 + \frac{e}{m_e c}\vec{A} \cdot \vec{P} = \hat{H}_0 + \hat{V}(t), \qquad (10.86)$$

where

$$\hat{V}(t) = \frac{e}{m_e c}\vec{A}(\vec{r}, t) \cdot \vec{P}. \qquad (10.87)$$

This term, which gives the interaction between the electron and the radiation, is small enough (compared to $\hat{H}_0$) to be treated by perturbation theory. We are going to use perturbation theory to study the effect of $\hat{V}(t)$ on the atom. In particular, we will focus on the transitions that are induced as a result of this perturbation.

At this level, we cannot proceed further without calculating $\vec{A}(\vec{r}, t)$. In what follows, we are going to show that, using $\vec{A}(\vec{r}, t)$ for an electromagnetic radiation, we obtain a $\hat{V}(t)$ which has the structure of a harmonic perturbation: $\hat{V}(t) = \hat{v}e^{-i\omega t} + \hat{v}^\dagger e^{i\omega t}$. Therefore, by analogy with a harmonic perturbation, we would expect the atom to emit or absorb photons and then undergo transitions from one state to another. For the sake of completeness, we are going to determine $\vec{A}(\vec{r}, t)$ in two different ways: by treating the radiation *classically* and then *quantum mechanically.* We are going to show that, unlike a quantum treatment, a classical treatment allows only a description of stimulated emission and absorption processes, but not spontaneous emission. Spontaneous emission turns out to be a purely quantum effect.

10.5.1 Classical Treatment of the Incident Radiation

A *classical*[1] treatment of the incident radiation is valid only when *large* numbers of photons contribute to the interaction with the atom (recall that quantum mechanical effects are generally encountered only when a *finite* number of photons are involved).

From classical electrodynamics, if we consider the incident radiation to be a plane wave of polarization $\vec{\varepsilon}$ that is propagating along the direction $\vec{n}$, the vector potential $\vec{A}(\vec{r}, t)$ is given by

$$\vec{A}(\vec{r}, t) = \vec{A}_0(\vec{r}) e^{-i\omega t} + \vec{A}_0^*(\vec{r}) e^{i\omega t} = A_0 \vec{\varepsilon} \left[e^{i(\vec{k}\cdot\vec{r} - \omega t)} + e^{-i(\vec{k}\cdot\vec{r} - \omega t)} \right], \qquad (10.88)$$

with $\vec{k} = k\vec{n}$. Since $\vec{A}(\vec{r}, t)$ satisfies the wave equation $\nabla^2 \vec{A} - (1/c^2)\partial^2 \vec{A}/\partial t^2 = 0$, we have $k = \omega/c$. The Coulomb gauge condition $\vec{\nabla} \cdot \vec{A} = 0$ yields $\vec{k} \cdot \vec{A}_0 = 0$; that is, $\vec{A}(\vec{r}, t)$ lies in a plane perpendicular to the wave's direction of propagation, $\vec{n}$. The electric and magnetic fields associated with the vector potential (10.88) can be obtained at once:

$$\vec{E}(\vec{r}, t) = -\frac{1}{c}\frac{\partial \vec{A}}{\partial t} = \frac{i\omega}{c} A_0 \vec{\varepsilon} \left[-e^{i(\vec{k}\cdot\vec{r} - \omega t)} + e^{-i(\vec{k}\cdot\vec{r} - \omega t)} \right], \qquad (10.89)$$

$$\vec{B}(\vec{r}, t) = \vec{\nabla} \times \vec{A} = i(\vec{k} \times \vec{\varepsilon}) A_0 \left[-e^{i(\vec{k}\cdot\vec{r} - \omega t)} + e^{-i(\vec{k}\cdot\vec{r} - \omega t)} \right] = \vec{n} \times \vec{E}. \qquad (10.90)$$

These two relations show that $\vec{E}$ and $\vec{B}$ have the same magnitude, $|\vec{E}| = |\vec{B}|$.

The energy density (or energy per unit volume) for a single photon of the incident radiation can be obtained from (10.89) and (10.90):

$$u = \frac{1}{8\pi}(|\vec{E}|^2 + |\vec{B}|^2) = \frac{1}{4\pi}|\vec{E}|^2 = \frac{\omega^2}{\pi c^2}|A_0|^2 \sin^2(\vec{k} \cdot \vec{r} - \omega t). \qquad (10.91)$$

Averaging this expression over time, we see that the energy of a single photon per unit volume, $\hbar\omega/V$, is given by $(\omega^2/2\pi c^2)|A_0|^2 = \hbar\omega/V$ and hence $|A_0|^2 = 2\pi\hbar c^2/(\omega V)$, which, when inserted into (10.88), leads to

$$\vec{A}(\vec{r}, t) = \sqrt{\frac{2\pi\hbar c^2}{\omega V}} \left[e^{i(\vec{k}\cdot\vec{r} - \omega t)} + e^{-i(\vec{k}\cdot\vec{r} - \omega t)} \right] \vec{\varepsilon}. \qquad (10.92)$$

Having specified $\vec{A}(\vec{r}, t)$ by means of a classical treatment, we can now rewrite the potential (10.87) as

$$\hat{V}(t) = \frac{e}{m_e c} \left(\frac{2\pi\hbar c^2}{\omega V} \right)^{1/2} \vec{\varepsilon} \cdot \vec{P} \left[e^{i(\vec{k}\cdot\vec{r} - \omega t)} + e^{-i(\vec{k}\cdot\vec{r} - \omega t)} \right] = \hat{v} e^{-i\omega t} + \hat{v}^\dagger e^{i\omega t}, \qquad (10.93)$$

where

$$\hat{v} = \frac{e}{m_e} \left(\frac{2\pi\hbar}{\omega V} \right)^{1/2} \vec{\varepsilon} \cdot \vec{P} e^{i\vec{k}\cdot\vec{r}}, \qquad \hat{v}^\dagger = \frac{e}{m_e} \left(\frac{2\pi\hbar}{\omega V} \right)^{1/2} \vec{\varepsilon} \cdot \vec{P} e^{-i\vec{k}\cdot\vec{r}}. \qquad (10.94)$$

---

[1] A *classical* treatment of the electric and magnetic fields, $\vec{E}(\vec{r}, t)$ and $\vec{B}(\vec{r}, t)$, and their corresponding electric and vector potentials, $\phi(\vec{r}, t)$ and $\vec{A}(\vec{r}, t)$, means that they are described by *continuous* fields.

The structure of (10.93) is identical with (10.50); that is, *the interaction of an atomic electron with radiation has the structure of a harmonic perturbation.* By analogy with (10.50) we can state that the term $e^{-i\omega t}$ in (10.93) gives rise to the *absorption* of the incident photon of energy $\hbar\omega$ by the atom, and $e^{i\omega t}$ to the *stimulated emission* of a photon of energy $\hbar\omega$ by the atom. That is, the absorption process occurs when the atom receives a photon from the radiation, and the stimulated emission when the radiation receives or gains a photon from the decaying atom. At this level, we cannot afford not to mention an important application of stimulated emission. In this process we start with one (incident) photon and end up with two: the incident photon plus the photon given by the atom resulting from its transition to a lower energy level. What would happen if we had a large number of atoms in the same excited state? A single external photon would trigger an avalanche, or chain reaction, of photons released by these atoms in a very short time and all having the same frequency. This would lead to an *amplification* of the electromagnetic field. How does this take place? When the incident photon interacts with the first atom, it will produce two photons, which in turn produce four photons; these four photons then produce eight photons (after they interact with four different atoms), and so on. This process is known as the *amplification by stimulated emission of the* (incident) *radiation.* Two such radiation amplifications have been achieved experimentally and have led to enormous applications: one in the microwave domain, known as *maser* (microwave amplification by stimulated emission of radiation); the other in the domain of light waves, called *laser* (light amplification by stimulated emission of radiation).

Following the approach that led to the transition rates (10.54) from (10.50), we can easily show that the transition rates for the stimulated emission and absorption corresponding to (10.93) are given by

$$\Gamma_{i\to f}^{emi} = \frac{4\pi^2 e^2}{m_e^2 \omega V} \left| \langle \psi_f \mid e^{-i\vec{k}\cdot\vec{r}} \; \vec{\varepsilon} \cdot \vec{P} \mid \psi_i \rangle \right|^2 \delta(E_f - E_i + \hbar\omega), \tag{10.95}$$

$$\Gamma_{i\to f}^{abs} = \frac{4\pi^2 e^2}{m_e^2 \omega V} \left| \langle \psi_f \mid e^{i\vec{k}\cdot\vec{r}} \; \vec{\varepsilon} \cdot \vec{P} \mid \psi_i \rangle \right|^2 \delta(E_f - E_i - \hbar\omega). \tag{10.96}$$

These relations represent the expressions for the transition rates when the radiation is treated classically.

What would happen when there is no radiation? If $\vec{A} = 0$ (i.e., the atom is placed in a vacuum), equations (10.95) and (10.96) imply that no transition will occur since, as equation (10.87) shows, if $\vec{A} = 0$ the perturbation will be zero; hence $\Gamma_{i\to f}^{emi} = 0$ and $\Gamma_{i\to f}^{abs} = 0$. As a result, the classical treatment cannot account for *spontaneous* emission which occurs even in the absence of an external perturbing field. This implies, for instance, that a hydrogen atom in an $n \geq 2$ energy eigenstate remains in this eigenstate unless it is perturbed by an external field. This is in complete disagreement with experimental observations, which show that atoms in the $n \geq 2$ states undergo *spontaneous emissions*; they emit electromagnetic radiation *even when no external perturbation is present.* The spontaneous emission is *a purely quantum effect.*

10.5.2 Quantization of the Electromagnetic Field

We have seen that a classical treatment of radiation leads to transition rates that account only for the processes of absorption and stimulated emission; spontaneous emission of photons by atoms is a typical phenomenon that a classical treatment fails to explain, let alone predict. The classical treatment is valid only when very large numbers of photons contribute to the radiation; that is, when the intensity of the radiation is so high that only its wave aspect is important. At very low intensities, however, the particle nature of the radiation becomes nonnegligible. In this case we have to consider a quantum mechanical treatment of the electromagnetic radiation. To obtain a quantum description of the radiation, we would necessarily need to replace the various fields (such as $\vec{E}(\vec{r}, t)$, $\vec{B}(\vec{r}, t)$, and the potential vector $\vec{A}(\vec{r}, t)$) with operators.

In the absence of charges and currents, the electric and magnetic fields are fully specified by the vector potential $\vec{A}(\vec{r}, t)$. Since $\vec{A}(\vec{r}, t)$ is transverse (perpendicular to the wave vector $\vec{k}$), it has only two nonzero components along the directions of two polarization (unit) vectors, $\vec{\varepsilon}_1$ and $\vec{\varepsilon}_2$, which lie in a plane perpendicular to $\vec{k}$. We can thus expand $\vec{A}(\vec{r}, t)$ in a Fourier series as follows:

$$\vec{A}(\vec{r}, t) = \frac{1}{\sqrt{V}} \sum_{\vec{k}} \sum_{\lambda=1}^{2} \left[A_{\lambda,\vec{k}} \vec{\varepsilon}_\lambda e^{i(\vec{k}\cdot\vec{r}-\omega_k t)} + A_{\lambda,\vec{k}}^* \vec{\varepsilon}_\lambda^* e^{-i(\vec{k}\cdot\vec{r}-\omega_k t)} \right], \tag{10.97}$$

where we have assumed that the electromagnetic field is confined to a large volume V with periodic boundary conditions. We are going to see that, by analogy with the quantization of a classical harmonic oscillator, *the quantization of radiation can be achieved by writing the electromagnetic field in terms of creation and annihilation operators.*

The Hamiltonian of the complete system (atom and the external radiation) is $\hat{H} = \hat{H}_0 + \hat{H}_r + V(t)$, where $\hat{H}_0$ is the Hamiltonian of the unperturbed atom, $\hat{H}_r$ is the Hamiltonian of the electromagnetic field, and $\hat{V}(t)$ is the interaction of the atom with the radiation. To find $\hat{H}_r$ we need to quantize the energy of the electromagnetic field which can be obtained from (10.97):

$$\mathcal{H}_r = \frac{1}{8\pi} \int d^3r \left(\vec{E}^2(\vec{r}, t) + \vec{B}^2(\vec{r}, t) \right) = \frac{V}{8\pi c^2} \sum_{\vec{k}} \sum_{\lambda=1}^{2} (\hbar k)^2 A_{\lambda,\vec{k}}^* A_{\lambda,\vec{k}}, \tag{10.98}$$

with $|\vec{\varepsilon}_\lambda|^2 = 1$, where we have used $\omega_k = ck$, $\vec{E}(\vec{r}, t) = -(1/c)\partial\vec{A}/\partial t$, and $\vec{B}(\vec{r}, t) = \vec{\nabla} \times \vec{A}$. Instead of the two variables $A_{\lambda,\vec{k}}$ and $A_{\lambda,\vec{k}}^*$, we can introduce a new set of two canonically conjugate variables:

$$Q_{\lambda,\vec{k}} = \frac{1}{\sqrt{4\pi c^2}} \left(A_{\lambda,\vec{k}}^* + A_{\lambda,\vec{k}} \right), \qquad P_{\lambda,\vec{k}} = \frac{i\omega_k}{\sqrt{4\pi c^2}} \left(A_{\lambda,\vec{k}}^* - A_{\lambda,\vec{k}} \right). \tag{10.99}$$

Combining (10.98) and (10.99) we can write

$$H_r = \sum_{\vec{k}} \sum_{\lambda=1}^{2} \left(\frac{1}{2} P_{\lambda,\vec{k}}^2 + \frac{\omega_k^2}{2} Q_{\lambda,\vec{k}}^2 \right). \tag{10.100}$$

This expression has the structure of a Hamiltonian of a collection of independent harmonic oscillators. This is compatible with the fact that electromagnetic waves in a vacuum result from the (harmonic) oscillations of the electromagnetic field; hence they can be described by means of a linear superposition of independent vibrational modes. To quantize (10.100) we simply need to find the operators $\hat{Q}_{\lambda,\vec{k}}$ and $\hat{P}_{\lambda\vec{k}}$ that correspond to the variables $Q_{\lambda,\vec{k}}$ and $P_{\lambda\vec{k}}$, respectively, such that they obey the canonical commutation relations:

$$\left[\hat{Q}_{\lambda_1\vec{k}_1}, \hat{P}_{\lambda_2\vec{k}_2} \right] = i\hbar\delta_{\lambda_1,\lambda_2}\delta_{\vec{k}_1,\vec{k}_2}. \tag{10.101}$$

Following the same quantization procedure of a classical harmonic oscillator, and introducing the lowering and raising operators

$$\hat{a}_{\lambda,\vec{k}} = \sqrt{\frac{\omega_k}{2\hbar}}\hat{Q}_{\lambda,\vec{k}} + \frac{i}{\sqrt{2\hbar\omega_k}}\hat{P}_{\lambda,\vec{k}}, \qquad \hat{a}^\dagger_{\lambda,\vec{k}} = \sqrt{\frac{\omega_k}{2\hbar}}\hat{Q}_{\lambda,\vec{k}} - \frac{i}{\sqrt{2\hbar\omega_k}}\hat{P}_{\lambda,\vec{k}}, \qquad (10.102)$$

which lead to $\hat{Q}_{\lambda,\vec{k}} = \sqrt{\hbar/2\omega_k}(\hat{a}^\dagger_{\lambda,\vec{k}} + \hat{a}_{\lambda,\vec{k}})$ and $\hat{P}_{\lambda,\vec{k}} = i\sqrt{\hbar\omega_k/2}(\hat{a}^\dagger_{\lambda,\vec{k}} - \hat{a}_{\lambda,\vec{k}})$, we can show that the Hamiltonian operator corresponding to (10.100) is given by

$$\hat{H}_r = \sum_{\vec{k}}\sum_{\lambda=1}^{2} \hbar\omega_k \left(\hat{N}_{\lambda,\vec{k}} + \frac{1}{2}\right), \qquad (10.103)$$

with $\hat{N}_{\lambda,\vec{k}} = \hat{a}^\dagger_{\lambda,\vec{k}}\hat{a}_{\lambda,\vec{k}}$.

By analogy to the harmonic oscillator, the operators $\hat{a}_{\lambda,\vec{k}}$ and $\hat{a}^\dagger_{\lambda,\vec{k}}$ obey the following commutation relations:

$$\left[\hat{a}_{\lambda_1,\vec{k}_1}, \hat{a}^\dagger_{\lambda_2,\vec{k}_2}\right] = \delta_{\lambda_1,\lambda_2}\delta_{\vec{k}_1,\vec{k}_2}, \qquad \left[\hat{a}_{\lambda_1,\vec{k}_1}, \hat{a}_{\lambda_2,\vec{k}_2}\right] = \left[\hat{a}^\dagger_{\lambda_1,\vec{k}_1}, \hat{a}^\dagger_{\lambda_2,\vec{k}_2}\right] = 0, \qquad (10.104)$$

and serve respectively to *annihilate* and *create* a photon of wave number $\vec{k}$ and polarization λ. The eigenvalues of $\hat{N}_{\lambda,\vec{k}}$ are $n_{\lambda,\vec{k}} = 0, 1, 2, \ldots$; by analogy with the harmonic oscillator, its eigenvectors are

$$|n_{\lambda,\vec{k}}\rangle = \frac{1}{\sqrt{n_{\lambda,\vec{k}}!}}\left(\hat{a}^\dagger_{\lambda\vec{k}}\right)^{n_{\lambda,\vec{k}}} |0\rangle, \qquad (10.105)$$

where $|0\rangle$ is the state with no photons, the vacuum state, and $|n_{\lambda,\vec{k}}\rangle$ is a state of the electromagnetic field with $n_{\lambda,\vec{k}}$ photons with wave vector $\vec{k}$ and polarization λ. The number $n_{\lambda,\vec{k}}$ therefore represents the *occupation number* mode. The actions of $\hat{a}_{\lambda,\vec{k}}$ and $\hat{a}^\dagger_{\lambda,\vec{k}}$ on $|n_{\lambda,\vec{k}}\rangle$ are given by

$$\hat{a}_{\lambda,\vec{k}}|n_{\lambda,\vec{k}}\rangle = \sqrt{n_{\lambda,\vec{k}}}|n_{\lambda,\vec{k}} - 1\rangle, \qquad \hat{a}^\dagger_{\lambda,\vec{k}}|n_{\lambda,\vec{k}}\rangle = \sqrt{n_{\lambda,\vec{k}} + 1}|n_{\lambda,\vec{k}} + 1\rangle. \qquad (10.106)$$

The eigenstates of the Hamiltonian (10.103) can be inferred from (10.105):

$$|n_{\lambda_1\vec{k}_1}, n_{\lambda_2\vec{k}_2}, n_{\lambda_3\vec{k}_3}, \ldots\rangle = \prod_j |n_{\lambda_j\vec{k}_j}\rangle, \qquad (10.107)$$

with the energy eigenvalues (of the radiation)

$$E_r = \sum_{\vec{k}}\sum_{\lambda} \hbar\omega_k \left(n_{\lambda\vec{k}} + \frac{1}{2}\right). \qquad (10.108)$$

The state $|n_{\lambda_1\vec{k}_1}, n_{\lambda_2\vec{k}_2}, n_{\lambda_3\vec{k}_3}, \ldots\rangle$ describes an electromagnetic field with $n_{\lambda_1\vec{k}_1}$ photons in the mode $(\lambda_1, \vec{k}_1)$ (i.e., $n_{\lambda_1\vec{k}_1}$ photons with wave vector $\vec{k}_1$ and polarization λ_1), $n_{\lambda_2\vec{k}_2}$ photons in the mode $(\lambda_2, \vec{k}_2)$, and so on. Substituting (10.99) into (10.102), we get $\hat{a}_{\lambda,\vec{k}} = \sqrt{\omega_k/(2\pi\hbar c^2)}\,\hat{A}_{\lambda,\vec{k}}$ and $\hat{a}^{\dagger}_{\lambda,\vec{k}} = \sqrt{\omega_k/(2\pi\hbar c^2)}\,\hat{A}^{\dagger}_{\lambda,\vec{k}}$; hence

$$\hat{A}_{\lambda,\vec{k}} = \sqrt{\frac{2\pi\hbar c^2}{\omega_k}}\hat{a}_{\lambda,\vec{k}}, \qquad\qquad \hat{A}^{\dagger}_{\lambda,\vec{k}} = \sqrt{\frac{2\pi\hbar c^2}{\omega_k}}\hat{a}^{\dagger}_{\lambda,\vec{k}}. \qquad (10.109)$$

An insertion of these two relations into (10.97) gives the vector potential operator:

$$\hat{\vec{A}}(\vec{r},t) = \sum_{\vec{k}}\sum_{\lambda=1}^{2}\sqrt{\frac{2\pi\hbar c^2}{\omega_k V}}\left[\hat{a}_{\lambda,\vec{k}}e^{i(\vec{k}\cdot\vec{r}-\omega_k t)}\,\vec{\varepsilon}_{\lambda} + \hat{a}^{\dagger}_{\lambda,\vec{k}}e^{-i(\vec{k}\cdot\vec{r}-\omega_k t)}\,\vec{\varepsilon}^{\,*}_{\lambda}\right]. \qquad (10.110)$$

The interaction $\hat{V}(t)$ as given by (10.87) reduces to $\hat{V}(t) = (e/m_e c)\hat{\vec{A}}(\vec{r},t)\cdot\vec{P}$ or

$$\hat{V}(t) = \frac{e}{m_e}\sum_{\vec{k}}\sum_{\lambda}\sqrt{\frac{2\pi\hbar}{\omega_k V}}\left[\hat{a}_{\lambda,\vec{k}}e^{i\vec{k}\cdot\vec{r}}\vec{\varepsilon}_{\lambda}\cdot\vec{P}e^{i\omega_k t} + \hat{a}^{\dagger}_{\lambda,\vec{k}}e^{i\vec{k}\cdot\vec{r}}\vec{\varepsilon}^{\,*}_{\lambda}\cdot\vec{P}e^{-i\omega_k t}\right], \qquad (10.111)$$

or

$$\hat{V}(t) = \sum_{\vec{k}}\sum_{\lambda=1}^{2}\left(\hat{v}_{\lambda,\vec{k}}e^{i\omega_k t} + \hat{v}^{\dagger}_{\lambda,\vec{k}}e^{-i\omega_k t}\right), \qquad (10.112)$$

where

$$\hat{v}_{\lambda,\vec{k}} = \frac{e}{m_e}\sqrt{\frac{2\pi\hbar}{\omega_k V}}\hat{a}_{\lambda,\vec{k}}e^{i\vec{k}\cdot\vec{r}}\vec{\varepsilon}_{\lambda}\cdot\vec{P}, \qquad \hat{v}^{\dagger}_{\lambda,\vec{k}} = \frac{e}{m_e}\sqrt{\frac{2\pi\hbar}{\omega_k V}}\hat{a}^{\dagger}_{\lambda,\vec{k}}e^{-i\vec{k}\cdot\vec{r}}\vec{\varepsilon}^{\,*}_{\lambda}\cdot\vec{P}. \qquad (10.113)$$

The terms $\hat{v}_{\lambda,\vec{k}}$ and $\hat{v}^{\dagger}_{\lambda,\vec{k}}$ correspond to the absorption (annihilation) and emission (creation) of a photon by the atom, respectively. As in the classical case, the interaction (10.112) has the structure of a *harmonic* perturbation.

Remark

The quantization of the radiation is achieved by writing the electromagnetic field in terms of creation and annihilation operators, by analogy with the harmonic oscillator. This process, which is called *second quantization*, leads to the replacement of the various *fields* (such as the vector potential $\vec{A}(\vec{r},t)$, the electric field $\vec{E}(\vec{r},t)$, and the magnetic field $\vec{B}(\vec{r},t)$) by operator quantities, which in turn are expressed in terms of creation and annihilation operators. For instance, the Hamiltonian and the vector potential of the radiation are given in the second quantization representation by equations (10.103) and (10.110), respectively.

10.5.3 Transition Rates for Absorption and Emission of Radiation

Before the atom and the radiation interact, their initial state is given by $\mid\Phi_i\rangle = \mid\psi_i\rangle\mid n_{\lambda,\vec{k}}\rangle$, where $\mid\psi_i\rangle$ is the state of the unperturbed atom and $\mid n_{\lambda,\vec{k}}\rangle$ is the state vector of the radiation. After the interaction takes place, the state of the system is given by $\mid\Phi_f\rangle = \mid\psi_f\rangle\mid n_{\lambda,\vec{k}}\rangle_f$.

 Let us look first at the case of emission of a photon. If after interaction the atom emits a photon, the final state of the system will be given by $\mid\Phi_f\rangle = \mid\psi_f\rangle\mid n_{\lambda,\vec{k}}+1\rangle$, since the electromagnetic field gains a photon; hence its state changes from $\mid n_{\lambda,\vec{k}}\rangle \rightarrow\mid n_{\lambda,\vec{k}}+1\rangle$.

Formally, this process can be achieved by creating a photon, that is, by applying $\hat{v}_{\lambda,\vec{k}}^{\dagger}$ or $\hat{a}_{\lambda,\vec{k}}^{\dagger}$ on the photonic state $\mid n_{\lambda,\vec{k}}\rangle$:

$$\langle \Phi_f \mid \hat{v}_{\lambda,\vec{k}}^{\dagger} \mid \Phi_i \rangle = \frac{e}{m_e}\sqrt{\frac{2\pi\hbar}{\omega_k V}} \langle \psi_f \mid e^{-i\vec{k}\cdot\vec{r}}\vec{\varepsilon}_{\lambda}^{\,*} \cdot \hat{\vec{P}} \mid \psi_i \rangle \langle n_{\lambda,\vec{k}} + 1 \mid \hat{a}_{\lambda,\vec{k}}^{\dagger} \mid n_{\lambda,\vec{k}}\rangle$$

$$= \frac{e}{m_e}\sqrt{\frac{2\pi\hbar}{\omega_k V}}\sqrt{n_{\lambda\vec{k}} + 1}\langle \psi_f \mid e^{-i\vec{k}\cdot\vec{r}}\vec{\varepsilon}_{\lambda}^{\,*} \cdot \hat{\vec{P}} \mid \psi_i \rangle. \qquad (10.114)$$

When $n_{\lambda\vec{k}} = 0$ (i.e., no radiation), equation (10.114) shows that *even in the absence of an external radiation*, the theory can describe events where there is emission of a photon. This is called *spontaneous emission*. This phenomenon cannot be described by means of a classical treatment of radiation. But if $n_{\lambda\vec{k}} \neq 0$, then $n_{\lambda\vec{k}}$ is responsible for *induced* or *stimulated* emissions; the bigger $n_{\lambda\vec{k}}$, the bigger the emission probability.

In the case of a photon absorption, the system undergoes a transition from an initial state $\mid \Phi_i \rangle =\mid \psi_i \rangle \mid n_{\lambda,\vec{k}}\rangle$ to the final state $\mid \Phi_f \rangle =\mid \psi_f \rangle \mid n_{\lambda,\vec{k}} - 1\rangle$. This can be achieved formally by applying the annihilation operator $\hat{a}_{\lambda,\vec{k}}$ on $\mid n_{\lambda,\vec{k}}\rangle$:

$$\langle \Phi_f \mid \hat{v}_{\lambda,\vec{k}} \mid \Phi_i \rangle = \frac{e}{m_e}\sqrt{\frac{2\pi\hbar}{\omega_k V}} \langle \psi_f \mid e^{i\vec{k}\cdot\vec{r}}\vec{\varepsilon}_{\lambda} \cdot \hat{\vec{P}} \mid \psi_i \rangle \langle n_{\lambda,\vec{k}} - 1 \mid \hat{a}_{\lambda,\vec{k}} \mid n_{\lambda,\vec{k}}\rangle$$

$$= \frac{e}{m_e}\sqrt{\frac{2\pi\hbar}{\omega_k V}}\sqrt{n_{\lambda,\vec{k}}}\langle \psi_f \mid e^{i\vec{k}\cdot\vec{r}}\vec{\varepsilon}_{\lambda} \cdot \hat{\vec{P}} \mid \psi_i \rangle. \qquad (10.115)$$

The transition rates corresponding to the emission or absorption of a photon of energy $\hbar\omega_k = \hbar ck$, wave number $\vec{k}$, and polarization λ can be obtained, by analogy with (10.95) and (10.96), from (10.114) and (10.115):

$$\boxed{\Gamma_{i\to f}^{emi} = \frac{4\pi^2 e^2}{m_e^2 \omega_k V}\left(n_{\lambda,\vec{k}} + 1\right)\left| \langle \psi_f \mid e^{-i\vec{k}\cdot\vec{r}}\vec{\varepsilon}_{\lambda}^{\,*} \cdot \vec{P} \mid \psi_i \rangle \right|^2 \delta(E_f - E_i + \hbar\omega_k),} \qquad (10.116)$$

$$\boxed{\Gamma_{i\to f}^{abs} = \frac{4\pi^2 e^2}{m_e^2 \omega_k V}n_{\lambda,\vec{k}}\left| \langle \psi_f \mid e^{i\vec{k}\cdot\vec{r}}\vec{\varepsilon}_{\lambda} \cdot \vec{P} \mid \psi_i \rangle \right|^2 \delta(E_f - E_i - \hbar\omega_k).} \qquad (10.117)$$

10.5.4 Transition Rates within the Dipole Approximation

Approximate expressions of the transition rates (10.116) and (10.117) can be obtained by expanding $e^{\pm i\vec{k}\cdot\vec{r}}$:

$$e^{\pm i\vec{k}\cdot\vec{r}} = 1 \pm i\vec{k}\cdot\vec{r} - \frac{1}{2}(\vec{k}\cdot\vec{r}) \mp \cdots = 1 \pm i\frac{\omega}{c}\vec{n}\cdot\vec{r} - \frac{1}{2}\frac{\omega^2}{c^2}(\vec{n}\cdot\vec{r})^2 \mp \cdots . \qquad (10.118)$$

This expansion finds its justification in the fact that $(\vec{k}\cdot\vec{r})$ is a small quantity, since the wavelength of the radiation (visible or ultraviolet) is very large compared to the atomic size: $kr = 2\pi a_0/\lambda \sim 2\pi \times 10^{-10}\,\text{m}/10^{-6}\,\text{m} \sim 10^{-3}$. In the case of nuclear radiation (such as γ radiation), kr is also in the range of 10^{-3}, with $r_{nucleus} \sim 10^{-15}$ m.

The *electric dipole approximation* corresponds to keeping only the leading term in the expansion (10.118): $e^{\pm i\vec{k}\cdot\vec{r}} \simeq 1$; hence

$$\langle \psi_f \mid e^{\pm i\vec{k}\cdot\vec{r}}\vec{\varepsilon}_\lambda \cdot \vec{P} \mid \psi_i \rangle \simeq \vec{\varepsilon}_\lambda \cdot \langle \psi_f \mid \vec{P} \mid \psi_i \rangle. \tag{10.119}$$

This term gives rise to *electric dipole* or E1 transitions. To calculate this term, we need to use the relation

$$\left[\hat{X}, \hat{H}_0 \right] = \left[\hat{X}, \frac{\hat{\vec{P}}^2}{2m_e} + \hat{V}(\vec{r}) \right] = \left[\hat{X}, \frac{\hat{P}_x^2}{2m_e} \right] = \frac{i\hbar}{m_e} \hat{P}_x, \tag{10.120}$$

which can be generalized to $[\vec{r}, \hat{H}_0] = i\hbar \hat{\vec{P}}/m_e$. Hence, inserting $\hat{\vec{P}} = (m_e/i\hbar)[\vec{r}, \hat{H}_0]$ into (10.119) and using $\hat{H}_0 \mid \psi_i \rangle = E_i \mid \psi_i \rangle$ and $\hat{H}_0 \mid \psi_f \rangle = E_f \mid \psi_f \rangle$, we have

$$\vec{\varepsilon}_\lambda \cdot \langle \psi_f \mid \vec{P} \mid \psi_i \rangle = \frac{m_e}{i\hbar}\vec{\varepsilon}_\lambda \cdot \langle \psi_f \mid [\vec{r}, \hat{H}_0] \mid \psi_i \rangle = \frac{m}{i\hbar}(E_i - E_f)\vec{\varepsilon}_\lambda \cdot \langle \psi_f \mid \vec{r} \mid \psi_i \rangle$$

$$= im_e\omega_{fi}\vec{\varepsilon}_\lambda \cdot \langle \psi_f \mid \vec{r} \mid \psi_i \rangle. \tag{10.121}$$

The substitution of this term into (10.119) leads to

$$\boxed{\langle \psi_f \mid e^{i\vec{k}\cdot\vec{r}}\vec{\varepsilon}_\lambda \cdot \vec{P} \mid \psi_i \rangle = im_e\omega_{fi}\vec{\varepsilon}_\lambda \cdot \langle \psi_f \mid \vec{r} \mid \psi_i \rangle.} \tag{10.122}$$

Inserting (10.122) into (10.116) and (10.117), we obtain the transition rates, within the *dipole approximation*, for the emission and absorption of a photon of energy $\hbar\omega_k$ by the atom:

$$\boxed{\Gamma_{i\to f}^{emi} = \frac{4\pi^2 e^2 \omega_{fi}^2}{\omega_k V}(n_{\lambda,k} + 1) \left| \vec{\varepsilon}_\lambda^* \cdot \langle \psi_f \mid \vec{r} \mid \psi_i \rangle \right|^2 \delta(E_f - E_i + \hbar\omega_k),} \tag{10.123}$$

$$\boxed{\Gamma_{i\to f}^{abs} = \frac{4\pi^2 e^2 \omega_{fi}^2}{\omega_k V}n_{\lambda,k} \left| \vec{\varepsilon}_\lambda \cdot \langle \psi_f \mid \vec{r} \mid \psi_i \rangle \right|^2 \delta(E_f - E_i - \hbar\omega_k).} \tag{10.124}$$

10.5.5 The Electric Dipole Selection Rules

Since $\vec{r}$ is given in spherical coordinates by $\vec{r} = (r \sin\theta \cos\phi)\vec{i} + (r \sin\theta \sin\phi)\vec{j} + (r \cos\theta)\vec{k}$, we can write

$$\vec{\varepsilon}_\lambda \cdot \vec{r} = r(\varepsilon_x \sin\theta \cos\phi + \varepsilon_y \sin\theta \sin\phi + \varepsilon_z \cos\theta). \tag{10.125}$$

Using the relations $\sin\theta \cos\phi = -\sqrt{2\pi/3}(Y_{11} - Y_{1-1})$, $\sin\theta \sin\phi = i\sqrt{2\pi/3}(Y_{11} + Y_{1-1})$, and $\cos\theta = \sqrt{4\pi/3}\,Y_{10}$, we may rewrite (10.125) as

$$\vec{\varepsilon}_\lambda \cdot \vec{r} = \sqrt{\frac{4\pi}{3}}r \left(\frac{-\varepsilon_x + i\varepsilon_y}{\sqrt{2}}Y_{11} + \frac{\varepsilon_x + i\varepsilon_y}{\sqrt{2}}Y_{1-1} + \varepsilon_z Y_{10} \right), \tag{10.126}$$

which in turn leads to

$$\langle \psi_f \mid \vec{\varepsilon}_\lambda \cdot \vec{r} \mid \psi_i \rangle = \sqrt{\frac{4\pi}{3}} \int_0^\infty r^3 R_{n_f l_f}^*(r) R_{n_i l_i}(r)\, dr$$

$$\times \int Y_{l_f m_f}^*(\theta, \varphi) \left(\frac{-\varepsilon_x + i\varepsilon_y}{\sqrt{2}}Y_{11} + \frac{\varepsilon_x + i\varepsilon_y}{\sqrt{2}}Y_{1-1} + \varepsilon_z Y_{10} \right) Y_{l_i m_i}(\theta, \varphi)\, d\Omega, \tag{10.127}$$

where we have used $\langle \vec{r} \mid \psi_i \rangle = R_{n_i l_i}(\vec{r}) Y_{l_i m_i}(\Omega)$ and $\langle \vec{r} \mid \psi_f \rangle = R_{n_f l_f}(\vec{r}) Y_{l_f m_f}(\Omega)$.

The integration over the angular degrees of freedom can be calculated by means of the Wigner–Eckart theorem; we have shown in Chapter 7 that

$$\int d\Omega \, Y_{l_f m_f}^* Y_{1m'} Y_{l_i m_i} = \langle l_f, m_f | Y_{1m'} | l_i, m_i \rangle$$

$$= \sqrt{\frac{3(2l_i + 1)}{4\pi(2l_f + 1)}} \langle l_i, 1; 0, 0 | l_f, 0 \rangle \langle l_i, 1; m_i, m' | l_f, m_f \rangle. \quad (10.128)$$

Inserting (10.128) into (10.123) and (10.124), we obtain $\Gamma_{i\rightarrow f}^{emi} \sim \langle l_i, 1; m_i, m' \mid l_f, m_f \rangle^2$ and $\Gamma_{i\rightarrow f}^{abs} \sim \langle l_i, 1; m_i, m' \mid l_f, m_f \rangle^2$. Thus the *dipole selection rules* are specified by the selection rules of the Clebsch–Gordan coefficient $\langle l_i, 1; m_i, m' \mid l_f, m_f \rangle$:

- The transition rates are zero unless the values of m_f and m_i satisfy the condition $m_i + m' = m_f$ or $m'_f - m_i = m'$. But since m' takes only three values, $m' = -1, 0, 1$, we have

$$m_f - m_i = -1, \ 0, \ 1. \quad (10.129)$$

- The permissible values of l_f must lie between $l_i - 1$ and $l_i + 1$ (i.e., $l_i - 1 \le l_f \le l_i + 1$), so we have $-1 \le l_f - l_i \le 1$ or

$$l_f - l_i = -1, \ 0, \ 1. \quad (10.130)$$

Note that, since the Clebsch–Gordan coefficient $\langle l_i, 1; m_i, m' \mid l_f, m_f \rangle$ vanishes for $l_i = l_f = 0$, no transition between $l_i = 0$ and $l_f = 0$ is allowed.

- Finally, since the coefficient $\langle l_i, 1; 0, 0 | l_f, 0 \rangle$ vanishes unless $(-1)^{l_i + 1 - l_f} = 1$ or $(-1)^{l_i - l_f} = -1$, then $(l_i - l_f)$ must be an odd integer:

$$l_f - l_i = \text{odd integer}. \quad (10.131)$$

This means that, in the case of electric dipole transitions, the final and initial states must have different parities. As a result, electric dipole transitions like $1s \rightarrow 2s$, $2p \rightarrow 3p$, etc., are *forbidden*, while transitions like $1s \rightarrow 2p$, $2p \rightarrow 3s$, etc., are allowed.

10.5.6 Spontaneous Emission

It is clear from (10.123) that the rate of emission of a photon from an atom is not zero even in the absence of an external radiation field ($n_{\lambda, \vec{k}} = 0$). This corresponds to the *spontaneous emission of a photon*. The total transition rate corresponding to spontaneous emission can be inferred from (10.123) by taking $n_{\lambda, \vec{k}} = 0$:

$$\Gamma_{i\rightarrow f}^{emi} = \frac{4\pi^2 \omega_{fi}^2}{\omega V} |\vec{\varepsilon}_\lambda^* \cdot \vec{d}_{fi}|^2 \delta(E_f - E_i + \hbar\omega), \quad (10.132)$$

where $\vec{d}_{fi}$ is the matrix element for the electron's electric dipole moment $\vec{d} = -e\vec{r}$:

$$\vec{d}_{fi} = \langle \psi_f \mid \vec{d} \mid \psi_i \rangle = -e\langle \psi_f \mid \vec{r} \mid \psi_i \rangle. \quad (10.133)$$

The relation (10.132) gives the transition probability per unit time corresponding to the transition of the atom from the initial state $|\psi_i\rangle$ to the final state $|\psi_f\rangle$ as a result of its spontaneous emission of a photon of energy $\hbar\omega$. Thus the final states of the system consist of products of discrete atomic states and a continuum of photonic states. The photon emitted will be detected in general as having a momentum in the momentum interval $(p, p+dp)$ located around $p = \hbar k = \hbar\omega/c$. The transition rate (10.132) needs then to be summed over the continuum of the final photonic states. The number of final photonic states within the unit volume V, whose momenta are within the interval $(p, p + dp)$, is given by

$$d^3 n = \frac{V d^3 p}{(2\pi \hbar)^3} = \frac{V p^2 dp \, d\Omega}{(2\pi \hbar)^3} = \frac{V \hbar^3 \omega^2}{(2\pi \hbar)^3 c^3} d\Omega \, d\omega = \frac{V \omega^2}{(2\pi c)^3} d\Omega \, d\omega. \tag{10.134}$$

Thus, the transition rate corresponding to the emission of a photon in the solid angle $d\Omega$ is obtained by integrating (10.132) over $d\omega$:

$$\begin{aligned}
dW^{emi}_{i \to f} &= \frac{V}{(2\pi)^3 c^3} d\Omega \int \omega^2 \Gamma^{emi}_{i \to f} d\omega = \frac{1}{2\pi c^3} |\vec{\varepsilon}^*_\lambda \cdot \vec{d}_{fi}|^2 d\Omega \int \omega^2_{fi} \omega \delta(E_f - E_i + \hbar\omega) \, d\omega \\
&= \frac{1}{2\pi \hbar c^3} |\vec{\varepsilon}^*_\lambda \cdot \vec{d}_{fi}|^2 d\Omega \int \omega^2_{fi} \omega \delta(\omega_{if} - \omega) \, d\omega,
\end{aligned} \tag{10.135}$$

where we have used the fact $\delta(E_f - E_i + \hbar\omega) = (1/\hbar)\delta(\omega_{if} - \omega)$ with $\omega_{if} = (E_i - E_f)/\hbar$. Carrying out the integration, we can reduce (10.135) to

$$dW^{emi}_{i \to f} = \frac{\omega^3}{2\pi \hbar c^3} |\vec{\varepsilon}^*_\lambda \cdot \vec{d}_{fi}|^2 d\Omega. \tag{10.136}$$

The transition rate (10.136) corresponds to a specific polarization; that is, the photon emitted travels along the direction $\vec{n}$ (since $\vec{k} = k\vec{n}$), which is normal to $\vec{\varepsilon}^*_\lambda$. To find the transition rate corresponding to any polarization, we need to sum over the two polarizations of the photon:

$$\sum_{\lambda=1}^{2} |\vec{\varepsilon}^*_\lambda \cdot \vec{d}_{fi}|^2 = |\varepsilon^*_1 (d_{fi})_1|^2 + |\varepsilon^*_2 (d_{fi})_2|^2 = |\vec{d}_{fi}|^2 - |(d_{fi})_3|^2. \tag{10.137}$$

Since the three directions of $\vec{d}_{fi}$ are equivalent, we have

$$\langle |(d_{fi})_1|^2 \rangle = \langle |(d_{fi})_2|^2 \rangle = \langle |(d_{fi})_3|^2 \rangle = \frac{1}{3} \langle |\vec{d}_{fi}|^2 \rangle. \tag{10.138}$$

Thus, an average over polarization yields

$$\sum_{\lambda=1}^{2} |\vec{\varepsilon}^*_\lambda \cdot \vec{d}_{fi}|^2 = |\vec{d}_{fi}|^2 - \frac{1}{3}|\vec{d}_{fi}|^2 = \frac{2}{3}|\vec{d}_{fi}|^2. \tag{10.139}$$

Substituting (10.139) into (10.136), we obtain the average transition rate corresponding to the emission of the photon into the solid angle $d\Omega$:

$$dW^{emi}_{i \to f} = \frac{\omega^3}{3\pi \hbar c^3} |\vec{d}_{fi}|^2 d\Omega. \tag{10.140}$$

An integration over all possible (photonic) directions ($|\vec{d}_{fi}|^2$ is not included in the integration since we are integrating over the angular part of the photonic degrees of freedom only and not over the electron's) yields $\int d\Omega = 4\pi$. Thus, the transition rate associated with the emission of the photon is

$$W_{i \to f}^{emi} = \frac{4}{3} \frac{\omega^3}{\hbar c^3} \left| \vec{d}_{fi} \right|^2 = \frac{4}{3} \frac{\omega^3 e^2}{\hbar c^3} \left| \langle \psi_f \mid \vec{r} \mid \psi_i \rangle \right|^2, \tag{10.141}$$

where $\omega = (E_f - E_i)/\hbar$.

The total *power (or intensity) radiated* by the electron is obtained by multiplying the total rates (10.141) by $\hbar\omega$:

$$I_{i \to f} = \hbar\omega W_{i \to f}^{emi} = \frac{4}{3} \frac{\omega^4}{c^3} |\vec{d}_{fi}|^2 = \frac{4}{3} \frac{\omega^4 e^2}{c^3} \left| \langle \psi_f \mid \vec{r} \mid \psi_i \rangle \right|^2. \tag{10.142}$$

The transition rates derived above, (10.141) and (10.142), were obtained for single-electron atoms. For atoms that have Z electrons, we must replace the dipole moment $\vec{d} = -e\vec{r}$ with the dipole moment of all Z electrons: $\vec{d} = -e \sum_{j=1}^{Z} \vec{r}_j$.

The mean lifetime τ of an excited state can be obtained by adding together the total transition probabilities per unit time (10.141) for all possible final states:

$$\tau = \frac{1}{W} = \frac{1}{\sum_f W_{i \to f}}. \tag{10.143}$$

Example 10.3

A particle of charge q and mass m is moving in a one-dimensional harmonic oscillator potential of frequency ω_0.

(a) Find the rate of spontaneous emission for a transition from an excited state $\mid n \rangle$ to the ground state.

(b) Obtain an estimate for the rate calculated in (a) and the lifetime of the state $\mid n \rangle$ when the particle is an electron and $\omega_0 = 3 \times 10^{14} \, \text{rad s}^{-1}$.

(c) Find the condition under which the dipole approximation is valid for the particle of (b).

Solution

(a) The spontaneous emission rate for a transition from an excited state $\mid n \rangle$ to $\mid 0 \rangle$ is given by (10.141):

$$W_{n \to 0}^{emi} = \frac{4}{3} \frac{\omega^3 q^2}{\hbar c^3} \left| \langle 0 \mid \hat{X} \mid n \rangle \right|^2, \tag{10.144}$$

where $\omega = (E_n - E_0)/\hbar = (n + \frac{1}{2})\omega_0 - \frac{1}{2}\omega_0 = n\omega_0$. Since $\hat{a} \mid n \rangle = \sqrt{n} \mid n-1 \rangle$ and $\hat{a}^\dagger \mid n \rangle = \sqrt{n+1} \mid n+1 \rangle$, and since $\hat{X} = \sqrt{\hbar/(2m\omega_0)}(\hat{a}^\dagger + \hat{a})$, we have

$$\langle 0 \mid \hat{X} \mid n \rangle = \sqrt{\frac{\hbar}{2m\omega_0}} \langle 0 \mid \hat{a}^\dagger + \hat{a} \mid n \rangle = \sqrt{\frac{\hbar}{2m\omega_0}} \left[\sqrt{n+1}\delta_{0,n+1} + \sqrt{n}\delta_{0,n-1} \right]. \tag{10.145}$$

Thus only a transition from $\mid 1 \rangle$ to $\mid 0 \rangle$ is possible; hence $n = 1$, $\omega = \omega_0$, and $\langle 0 \mid \hat{X} \mid 1 \rangle = \sqrt{\hbar/(2m\omega_0)}$. The emission rate (10.144) then becomes

$$W_{1 \to 0}^{emi} = \frac{4}{3} \frac{\omega^3 q^2}{\hbar c^3} \left| \langle 0 \mid \hat{X} \mid 1 \rangle \right|^2 = \frac{4}{3} \frac{\omega_0^3 q^2}{\hbar c^3} \frac{\hbar}{2m\omega_0} = \frac{2}{3} \frac{\omega_0^2 q^2}{m c^3}. \tag{10.146}$$

(b) If the particle is an electron, we have $q = -e$:

$$W_{1 \to 0}^{emi} = \frac{2}{3} \frac{\omega_0^2 e^2}{m_e c^3} = \frac{2\alpha}{3} \frac{\omega_0^2 \hbar}{m_e c^2} = \frac{2\alpha}{3} \frac{\hbar c}{m_e c^2} \frac{\omega_0^2}{c}. \qquad (10.147)$$

Using $m_e c^2 = 0.511 \, \text{MeV}$, $\hbar c = 197.33 \, \text{MeV fm}$, we have

$$W_{1 \to 0}^{emi} = \frac{2\alpha}{3} \frac{\hbar c}{m_e c^2} \frac{\omega_0^2}{c} = \frac{2}{3 \times 137} \frac{197.33 \, \text{MeV fm}}{0.511 \, \text{MeV}} \frac{9 \times 10^{28} \, \text{s}^{-2}}{3 \times 10^8 \text{m s}^{-1}} = 5.6 \times 10^5 \, \text{s}^{-1}. \quad (10.148)$$

The lifetime of the $| 1 \rangle$ state is

$$\tau = \frac{1}{W_{1 \to 0}^{emi}} = \frac{3}{2} \frac{m_e c^3}{\omega_0^2 e^2} = \frac{1}{5.6 \times 10^5 sec^{-2}} = 0.18 \times 10^{-5} \, \text{s}. \qquad (10.149)$$

(c) For the dipole approximation to be valid, we need $kx \ll 1$, where x was calculated in (10.145) for $n = 1$: $x = \sqrt{\hbar/(2m_e \omega_0)}$. As for k, a crude estimate yields $k = \omega/c = (E_1 - E_0)/(\hbar c) = \omega_0/c$. Thus, we have

$$kx = \frac{\omega_0}{c} \sqrt{\frac{\hbar}{2m_e \omega_0}} = \sqrt{\frac{\hbar \omega_0}{2m_e c^2}} \ll 1 \qquad \Longrightarrow \qquad \hbar \omega_0 \ll 2m_e c^2. \qquad (10.150)$$

This is indeed the case since $2m_e c^2 = 1.022 \, \text{MeV}$ is very large compared to

$$\hbar \omega_0 = \hbar c \frac{\omega_0}{c} = 197.33 \, \text{MeV fm} \times \frac{3 \times 10^{14} \, \text{s}^{-1}}{3 \times 10^8 \text{m s}^{-1}} = 2.0 \times 10^{-7} \, \text{MeV}. \qquad (10.151)$$

10.6 Solved Problems

Problem 10.1

(a) Calculate the position and the momentum operators, $\hat{X}_H(t)$ and $\hat{P}_H(t)$, in the Heisenberg picture for a one-dimensional harmonic oscillator.

(b) Find the Heisenberg equations of motion for $\hat{X}_H(t)$ and $\hat{P}_H(t)$.

Solution

In the Schrödinger picture, where the operators do not depend explicitly on time, the Hamiltonian of a one-dimensional harmonic oscillator is given by

$$\hat{H} = \frac{\hat{P}^2}{2m} + \frac{1}{2} m \omega^2 \hat{X}^2. \qquad (10.152)$$

(a) Using the commutation relations

$$[\hat{H}, \, \hat{X}] = \frac{1}{2m} [\hat{P}^2, \, \hat{X}] = -\frac{i\hbar}{m} \hat{P}, \qquad (10.153)$$

$$[\hat{H}, \, \hat{P}] = \frac{1}{2} m \omega^2 [\hat{X}^2, \, \hat{P}] = i\hbar m \omega^2 \hat{X}, \qquad (10.154)$$

along with

$$e^{\hat{A}}\hat{B}e^{-\hat{A}} = \hat{B} + [\hat{A},\ \hat{B}] + \frac{1}{2!}[\hat{A},\ [\hat{A},\ \hat{B}]] + \frac{1}{3!}[\hat{A},\ [\hat{A},\ [\hat{A},\ \hat{B}]]] + \cdots, \qquad (10.155)$$

we may write (see Eq. (10.11))

$$\hat{X}_H(t) = e^{it\hat{H}/\hbar}\hat{X}e^{-it\hat{H}/\hbar} = \hat{X} + \frac{it}{\hbar}[\hat{H},\ \hat{X}] + \frac{1}{2!}\left(\frac{it}{\hbar}\right)^2[\hat{H},\ [\hat{H},\ \hat{X}]] + \cdots$$

$$= \hat{X} + \frac{t}{m}\hat{P} - \frac{(\omega t)^2}{2!}\hat{X} - \frac{(\omega t)^3}{3!}\frac{1}{m\omega}\hat{P} + \frac{(\omega t)^4}{4!}\hat{X} + \frac{(\omega t)^5}{5!}\frac{1}{m\omega}\hat{P} + \cdots$$

$$= \hat{X}\left[1 - \frac{(\omega t)^2}{2!} + \frac{(\omega t)^4}{4!} + \cdots\right] + \frac{1}{m\omega}\hat{P}\left[(\omega t) - \frac{(\omega t)^3}{3!} + \frac{(\omega t)^5}{5!} + \cdots\right],$$

$$\hfill (10.156)$$

or

$$\boxed{\hat{X}_H(t) = \hat{X}\cos(\omega t) + \frac{1}{m\omega}\hat{P}\sin(\omega t).} \qquad (10.157)$$

A similar calculation yields (see Eq. (10.11))

$$\hat{P}_H(t) = e^{it\hat{H}/\hbar}\hat{P}e^{-it\hat{H}/\hbar} = \hat{P} + \frac{it}{\hbar}[\hat{H},\ \hat{P}] + \frac{1}{2!}\left(\frac{it}{\hbar}\right)^2[\hat{H},\ [\hat{H},\ \hat{P}]] + \cdots$$

$$= \hat{P}\left[1 - \frac{(\omega t)^2}{2!} + \frac{(\omega t)^4}{4!} + \cdots\right] - m\omega\hat{X}\left[(\omega t) - \frac{(\omega t)^3}{3!} + \frac{(\omega t)^5}{5!} + \cdots\right],$$

$$\hfill (10.158)$$

or

$$\boxed{\hat{P}_H(t) = \hat{P}\cos(\omega t) - m\omega\hat{X}\sin(\omega t).} \qquad (10.159)$$

(b) To find the equations of motion of $\hat{X}_H(t)$ and $\hat{P}_H(t)$, we need to use the Heisenberg equation $d\hat{A}_H(t)/dt = (1/i\hbar)[\hat{A}_H(t),\ \hat{H}]$ which, along with (10.153) and (10.154), leads to

$$\frac{d\hat{X}_H(t)}{dt} = \frac{1}{i\hbar}[\hat{X}_H(t),\ \hat{H}] = \frac{1}{i\hbar}e^{it\hat{H}/\hbar}[\hat{X},\ \hat{H}]e^{-it\hat{H}/\hbar} = \frac{1}{i\hbar}\frac{i\hbar}{m}e^{it\hat{H}/\hbar}\hat{P}e^{-it\hat{H}/\hbar},$$

$$\hfill (10.160)$$

$$\frac{d\hat{P}_H(t)}{dt} = \frac{1}{i\hbar}[\hat{P}_H(t),\ \hat{H}] = \frac{1}{i\hbar}e^{it\hat{H}/\hbar}[\hat{P},\ \hat{H}]e^{-it\hat{H}/\hbar} = \frac{(-i\hbar m\omega^2)}{i\hbar}e^{it\hat{H}/\hbar}\hat{X}e^{-it\hat{H}/\hbar},$$

$$\hfill (10.161)$$

or

$$\boxed{\frac{d\hat{X}_H(t)}{dt} = \frac{1}{m}\hat{P}_H(t), \qquad\qquad \frac{d\hat{P}_H(t)}{dt} = -m\omega^2\hat{X}_H(t).} \qquad (10.162)$$

Problem 10.2

Using the expressions derived in Problem 10.1 for $\hat{X}_H(t)$ and $\hat{P}_H(t)$, evaluate the following commutators for a harmonic oscillator:

$$\left[\hat{X}_H(t_1),\ \hat{P}_H(t_2)\right], \quad \left[\hat{X}_H(t_1),\ \hat{X}_H(t_2)\right], \quad \left[\hat{P}_H(t_1),\ \hat{P}_H(t_2)\right].$$

Solution

Using (10.157) and (10.159) along with the commutation relations $[\hat{X}, \hat{P}] = i\hbar$ and $[\hat{X}, \hat{X}] = [\hat{P}, \hat{P}] = 0$, we have

$$[\hat{X}_H(t_1), \hat{P}_H(t_2)] = \left[\hat{X}\cos(\omega t_1) + \frac{1}{m\omega}\hat{P}\sin(\omega t_1), \hat{P}\cos(\omega t_2) - m\omega\hat{X}\sin(\omega t_2)\right]$$

$$= [\hat{X}, \hat{P}]\cos(\omega t_1)\cos(\omega t_2) - [\hat{P}, \hat{X}]\sin(\omega t_1)\sin(\omega t_2)$$

$$= i\hbar\left[\cos(\omega t_1)\cos(\omega t_2) + \sin(\omega t_1)\sin(\omega t_2)\right], \tag{10.163}$$

or

$$\boxed{[\hat{X}_H(t_1), \hat{P}_H(t_2)] = i\hbar\cos\left[\omega(t_1 - t_2)\right].} \tag{10.164}$$

A similar calculation yields

$$[\hat{X}_H(t_1), \hat{X}_H(t_2)] = \left[\hat{X}\cos(\omega t_1) + \frac{1}{m\omega}\hat{P}\sin(\omega t_1), \hat{X}\cos(\omega t_2) + \frac{1}{m\omega}\hat{P}\sin(\omega t_2)\right]$$

$$= \frac{1}{m\omega}[\hat{X}, \hat{P}]\cos(\omega t_1)\sin(\omega t_2) + \frac{1}{m\omega}[\hat{P}, \hat{X}]\sin(\omega t_1)\cos(\omega t_2)$$

$$= \frac{i\hbar}{m\omega}\left[\cos(\omega t_1)\sin(\omega t_2) - \sin(\omega t_1)\cos(\omega t_2)\right], \tag{10.165}$$

or

$$\boxed{[\hat{X}_H(t_1), \hat{X}_H(t_2)] = -\frac{i\hbar}{m\omega}\sin\left[\omega(t_1 - t_2)\right].} \tag{10.166}$$

Similarly, we have

$$[\hat{P}_H(t_1), \hat{P}_H(t_2)] = \left[\hat{P}\cos(\omega t_1) - m\omega\hat{X}\sin(\omega t_1), \hat{P}\cos(\omega t_2) - m\omega\hat{X}\sin(\omega t_2)\right]$$

$$= -m\omega[\hat{P}, \hat{X}]\cos(\omega t_1)\sin(\omega t_2) - m\omega[\hat{X}, \hat{P}]\sin(\omega t_1)\cos(\omega t_2)$$

$$= -i\hbar m\omega\left[\sin(\omega t_1)\cos(\omega t_2) - \cos(\omega t_1)\sin(\omega t_2)\right], \tag{10.167}$$

or

$$\boxed{[\hat{P}_H(t_1), \hat{P}_H(t_2)] = -i\hbar m\omega\sin\left[\omega(t_1 - t_2)\right].} \tag{10.168}$$

Problem 10.3

Evaluate the quantity $\langle n \mid \hat{X}_H(t)\hat{X} \mid n\rangle$ for the nth excited state of a one-dimensional harmonic oscillator, where $\hat{X}_H(t)$ and $\hat{X}$ designate the position operators in the Heisenberg picture and the Schrödinger picture.

Solution

Using the expression of $\hat{X}_H(t)$ calculated in (10.157), we have

$$\langle n \mid \hat{X}_H(t)\hat{X} \mid n\rangle = \langle n \mid \hat{X}^2 \mid n\rangle\cos(\omega t) + \frac{1}{m\omega}\langle n \mid \hat{P}\hat{X} \mid n\rangle\sin(\omega t). \tag{10.169}$$

Since, for a harmonic oscillator, $\hat{X}$ and $\hat{P}$ are given by

$$\hat{X} = \sqrt{\frac{\hbar}{2m\omega}}(\hat{a}^\dagger + \hat{a}), \qquad \hat{P} = i\sqrt{\frac{m\hbar\omega}{2}}(\hat{a}^\dagger - \hat{a}), \tag{10.170}$$

and $\hat{a}^\dagger \mid n\rangle = \sqrt{n+1} \mid n+1\rangle$ and $\hat{a} \mid n\rangle = \sqrt{n} \mid n-1\rangle$, we have

$$\langle n \mid \hat{X}^2 \mid n\rangle = \frac{\hbar}{2m\omega} \langle n \mid \hat{a}^{\dagger\,2} + \hat{a}^2 + \hat{a}\hat{a}^\dagger + \hat{a}^\dagger\hat{a} \mid n\rangle = \frac{\hbar}{2m\omega}(2n+1), \quad (10.171)$$

$$\langle n \mid \hat{P}\hat{X} \mid n\rangle = \frac{i\hbar}{2} \langle n \mid \hat{a}^{\dagger\,2} + \hat{a}^2 - \hat{a}\hat{a}^\dagger + \hat{a}^\dagger\hat{a} \mid n\rangle = -\frac{i\hbar}{2}, \quad (10.172)$$

since $\langle n \mid \hat{a}^{\dagger\,2} \mid n\rangle = \langle n \mid \hat{a}^2 \mid n\rangle = 0$, $\langle n \mid \hat{a}^\dagger\hat{a} \mid n\rangle = n$ and $\langle n \mid \hat{a}\hat{a}^\dagger \mid n\rangle = n+1$. Inserting (10.171) and (10.172) into (10.169), we obtain

$$\langle n \mid \hat{X}_H(t)\hat{X} \mid n\rangle = \frac{\hbar}{2m\omega}\left[(2n+1)\cos(\omega t) - i\sin(\omega t)\right]. \quad (10.173)$$

Problem 10.4

The Hamiltonian due to the interaction of a particle of mass m, charge q, and spin $\vec{S}$ with a magnetic field pointing along the z-axis is $\hat{H} = -(qB/mc)\hat{S}_z$. Write the Heisenberg equations of motion for the time-dependent spin operators $\hat{S}_x(t)$, $\hat{S}_y(t)$, and $\hat{S}_z(t)$, and solve them to obtain the operators as functions of time.

Solution

Let us write $\hat{H}$ in a lighter form $\hat{H} = \omega\hat{S}_z$ where $\omega = -qB/mc$. The commutation of $\hat{H}$ with the components of the spin operator can be inferred at once from $[\hat{S}_x, \hat{S}_z] = -i\hbar\hat{S}_y$ and $[\hat{S}_y, \hat{S}_z] = i\hbar\hat{S}_x$:

$$[\hat{S}_x, \hat{H}] = -i\hbar\omega\hat{S}_y, \qquad [\hat{S}_y, \hat{H}] = i\hbar\omega\hat{S}_x, \qquad [\hat{S}_z, \hat{H}] = 0. \quad (10.174)$$

The Heisenberg equations of motion for $\hat{S}_x(t)$, $\hat{S}_y(t)$, and $\hat{S}_z(t)$ can be obtained from $d\hat{A}_H(t)/dt = (1/i\hbar)[\hat{A}_H(t), \hat{H}] = (1/i\hbar)e^{it\hat{H}/\hbar}[\hat{A}(0), \hat{H}]e^{-it\hat{H}/\hbar}$ which, using (10.174), leads to

$$\frac{d\hat{S}_x(t)}{dt} = \frac{1}{i\hbar}[\hat{S}_x(t), \hat{H}] = \frac{1}{i\hbar}e^{it\hat{H}/\hbar}[\hat{S}_x(0), \hat{H}]e^{-it\hat{H}/\hbar}$$

$$= \frac{-i\hbar\omega}{i\hbar}e^{it\hat{H}/\hbar}\hat{S}_y(0)e^{-it\hat{H}/\hbar} = -\omega\hat{S}_y(t). \quad (10.175)$$

Similarly, we have

$$\frac{d\hat{S}_y(t)}{dt} = \frac{1}{i\hbar}e^{it\hat{H}/\hbar}[\hat{S}_y(0), \hat{H}]e^{-it\hat{H}/\hbar} = \frac{i\hbar\omega}{i\hbar}e^{it\hat{H}/\hbar}\hat{S}_x(0)e^{-it\hat{H}/\hbar} = \omega\hat{S}_x(t), \quad (10.176)$$

$$\frac{d\hat{S}_z(t)}{dt} = \frac{1}{i\hbar}e^{it\hat{H}/\hbar}[\hat{S}_z(0), \hat{H}]e^{-it\hat{H}/\hbar} = 0. \quad (10.177)$$

To solve (10.175) and (10.176), we may combine them into two more conducive equations:

$$\frac{d\hat{S}_\pm(t)}{dt} = \pm i\omega\hat{S}_\pm(t), \quad (10.178)$$

where $\hat{S}_\pm(t) = \hat{S}_x(t) \pm i\hat{S}_y(t)$. The solutions of (10.178) are $\hat{S}_\pm(t) = \hat{S}_\pm(0)e^{\pm i\omega t}$ which, when combined with $\hat{S}_x(t) = \frac{1}{2}[\hat{S}_+(t) + \hat{S}_-(t)]$ and $\hat{S}_y(t) = \frac{1}{2i}[\hat{S}_+(t) - \hat{S}_-(t)]$, lead to

$$\hat{S}_x(t) = \hat{S}_x(0)\cos(\omega t) - \hat{S}_y(0)\sin(\omega t), \tag{10.179}$$

$$\hat{S}_y(t) = \hat{S}_y(0)\cos(\omega t) + \hat{S}_x(0)\sin(\omega t). \tag{10.180}$$

The solution of (10.177) is obvious:

$$\frac{d\hat{S}_z(t)}{dt} = 0 \quad \Longrightarrow \quad \hat{S}_z(t) = \hat{S}_z(0). \tag{10.181}$$

Problem 10.5
Consider a spinless particle of mass m, which is moving in a one-dimensional infinite potential well with walls at $x = 0$ and $x = a$.

(a) Find $\hat{X}_H(t)$ and $\hat{P}_H(t)$ in the Heisenberg picture.

(b) If at $t = 0$ the particle is in the state $\psi(x, 0) = [\phi_1(x) + \phi_2(x)]/\sqrt{2}$, where $\phi_1(x)$ and $\phi_2(x)$ are the ground and first excited states, respectively, with $\phi_n(x) = \sqrt{2/a}\sin(n\pi x/a)$, find the state vector $\psi(x, t)$ for $t > 0$ in the Schrödinger picture.

(c) Evaluate $\langle \psi(x, t) \mid \hat{X} \mid \psi(x, t)\rangle$ and $\langle \psi(x, t) \mid \hat{P} \mid \psi(x, t)\rangle$ as a function of time in the Schrödinger picture.

(d) Evaluate $\langle \psi(x, t) \mid \hat{X}_H(t) \mid \psi(x, t)\rangle$ and $\langle \psi(x, t) \mid \hat{P}_H(t) \mid \psi(x, t)\rangle$ as a function of time in the Schrödinger picture.

Solution
(a) Since the particle's Hamiltonian is purely kinetic, $\hat{H} = \hat{P}^2/2m$, we have $[\hat{H}, \hat{P}] = 0$ and

$$[\hat{H}, \hat{X}] = \frac{1}{2m}[\hat{P}^2, \hat{X}] = -\frac{i\hbar}{m}\hat{P}. \tag{10.182}$$

Using these relations along with (10.155), we obtain

$$\hat{X}_H(t) = e^{it\hat{H}/\hbar}\hat{X}e^{-it\hat{H}/\hbar} = \hat{X} + \frac{it}{\hbar}[\hat{H}, \hat{X}] + \frac{1}{2!}\left(\frac{it}{\hbar}\right)^2 [\hat{H}, [\hat{H}, \hat{X}]] + \cdots, \tag{10.183}$$

and since $[\hat{H}, [\hat{H}, \hat{X}]] = -(i\hbar/m)[\hat{H}, \hat{P}] = 0$, we end up with

$$\hat{X}_H(t) = \hat{X} + \frac{t}{m}\hat{P}. \tag{10.184}$$

On the other hand, since $[\hat{H}, \hat{P}] = 0$, we have

$$\hat{P} = \hat{P}_H(t). \tag{10.185}$$

(b) Since the energy of the nth level is given by $E_n = n^2\pi^2\hbar^2/(2ma^2)$, we have

$$\psi(x, t) = \frac{1}{\sqrt{2}}\left[\phi_1(x)e^{-iE_1t/\hbar} + \phi_2(x)e^{-iE_2t/\hbar}\right]$$

$$= \frac{1}{\sqrt{a}}\left[e^{-iE_1t/\hbar}\sin\left(\frac{\pi x}{a}\right) + e^{-iE_2t/\hbar}\sin\left(\frac{2\pi x}{a}\right)\right]. \tag{10.186}$$

(c) Using (10.186) we can write

$$\langle\psi(x,t)\mid\hat{X}\mid\psi(x,t)\rangle = \frac{1}{2}\Big[\langle\phi_1\mid\hat{X}\mid\phi_1\rangle + \langle\phi_2\mid\hat{X}\mid\phi_2\rangle + \langle\phi_1\mid\hat{X}\mid\phi_2\rangle e^{-i(E_2-E_1)t/\hbar}$$
$$+ \langle\phi_2\mid\hat{X}\mid\phi_1\rangle e^{i(E_2-E_1)t/\hbar}\Big]. \tag{10.187}$$

Since $\langle\phi_n\mid\hat{X}\mid\phi_n\rangle = a/2$ (Chapter 4) and

$$\langle\phi_1\mid\hat{X}\mid\phi_2\rangle = \langle\phi_2\mid\hat{X}\mid\phi_1\rangle = \frac{2}{a}\int_0^a x\sin\left(\frac{\pi x}{a}\right)\sin\left(\frac{2\pi x}{a}\right)dx = -\frac{16a}{9\pi^2}, \tag{10.188}$$

we can rewrite (10.187) as

$$\langle\psi(x,t)\mid\hat{X}\mid\psi(x,t)\rangle = \frac{1}{2}\Big[\frac{a}{2} + \frac{a}{2} - \frac{16a}{9\pi^2}\Big(e^{-i(E_2-E_1)t/\hbar} + e^{i(E_2-E_1)t/\hbar}\Big)\Big]$$
$$= \frac{a}{2} - \frac{16a}{9\pi^2}\cos\left(\frac{3\pi^2\hbar t}{2ma^2}\right), \tag{10.189}$$

since $E_2 - E_1 = 3\pi^2\hbar^2/(2ma^2)$.

A similar calculation which uses $\langle\phi_n\mid\hat{P}\mid\phi_n\rangle = 0$ and

$$\langle\phi_1\mid\hat{P}\mid\phi_2\rangle = -i\hbar\frac{4\pi}{a}\int_0^a \sin\left(\frac{\pi x}{a}\right)\cos\left(\frac{2\pi x}{a}\right)dx = \frac{8i\hbar}{3a} = -\langle\phi_2\mid\hat{P}\mid\phi_1\rangle \tag{10.190}$$

leads to

$$\langle\psi(x,t)\mid\hat{P}\mid\psi(x,t)\rangle = \frac{1}{2}\Big[\langle\phi_1\mid\hat{P}\mid\phi_1\rangle + \langle\phi_2\mid\hat{P}\mid\phi_2\rangle + \langle\phi_1\mid\hat{P}\mid\phi_2\rangle e^{-i(E_2-E_1)t/\hbar}$$
$$+ \langle\phi_2\mid\hat{P}\mid\phi_1\rangle e^{i(E_2-E_1)t/\hbar}\Big], \tag{10.191}$$

or to

$$\langle\psi(x,t)\mid\hat{P}\mid\psi(x,t)\rangle = \frac{1}{2}\Big[\frac{8i\hbar}{3a}e^{-i(E_2-E_1)t/\hbar} - \frac{8i\hbar}{3a}e^{i(E_2-E_1)t/\hbar}\Big] = \frac{8\hbar}{3a}\sin\left(\frac{3\pi^2\hbar t}{2ma^2}\right). \tag{10.192}$$

(d) From (10.184) we have

$$\langle\psi(x,t)\mid\hat{X}_H(t)\mid\psi(x,t)\rangle = \langle\psi(x,t)\mid\hat{X}\mid\psi(x,t)\rangle + \frac{t}{m}\langle\psi(x,t)\mid\hat{P}\mid\psi(x,t)\rangle. \tag{10.193}$$

Inserting the expressions for $\langle\psi(x,t)\mid\hat{X}\mid\psi(x,t)\rangle$ and $\langle\psi(x,t)\mid\hat{P}\mid\psi(x,t)\rangle$ calculated in (10.189) and (10.192), we obtain

$$\langle\psi(x,t)\mid\hat{X}_H(t)\mid\psi(x,t)\rangle = \frac{a}{2} - \frac{16a}{9\pi^2}\cos\left(\frac{3\pi^2\hbar t}{2ma^2}\right) + \frac{8\hbar t}{3ma}\sin\left(\frac{3\pi^2\hbar t}{2ma^2}\right), \tag{10.194}$$

and $\langle\psi(x,t)\mid\hat{P}_H(t)\mid\psi(x,t)\rangle$ is given by (10.192):

$$\langle\psi(x,t)\mid\hat{P}_H(t)\mid\psi(x,t)\rangle = \langle\psi(x,t)\mid\hat{P}\mid\psi(x,t)\rangle = \frac{8\hbar}{3a}\sin\left(\frac{3\pi^2\hbar t}{2ma^2}\right), \tag{10.195}$$

since, as shown in (10.185), we have $\hat{P}_H(t) = \hat{P}$.

Problem 10.6

A particle, initially (i.e., $t \to -\infty$) in its ground state in an infinite potential well whose walls are located at $x = 0$ and $x = a$, is subject at time $t = 0$ to a time-dependent perturbation $\hat{V}(t) = \varepsilon \hat{x} e^{-t^2}$ where ε is a small real number. Calculate the probability that the particle will be found in its first excited state after a sufficiently long time (i.e., $t \to \infty$).

Solution

The transition probability from the ground state $n = 1$ (where $t \to -\infty$) to the first excited state $n = 2$ (where $t \to \infty$) is given by (10.41):

$$P_{1 \to 2} = \frac{1}{\hbar^2} \left| \int_{-\infty}^{+\infty} \langle \psi_2 | \hat{V}(t) | \psi_1 \rangle e^{i\omega_{21}t} dt \right|^2, \qquad (10.196)$$

where

$$\omega_{21} = \frac{E_2 - E_1}{\hbar} = \frac{4\pi^2 \hbar}{2ma^2} - \frac{\pi^2 \hbar}{2ma^2} = \frac{3\pi^2 \hbar}{2ma^2}, \qquad (10.197)$$

$$\langle \psi_2 | \hat{V}(t) | \psi_1 \rangle = \frac{2\varepsilon}{a} e^{-t^2} \int_0^a x \sin\left(\frac{2\pi x}{a}\right) \sin\left(\frac{\pi x}{a}\right) dx = -\frac{16\varepsilon a}{9\pi^2} e^{-t^2}, \qquad (10.198)$$

since $E_n = n^2 \pi^2 \hbar^2/(2ma^2)$ and $\psi_n(x) = \sqrt{2/a} \sin(n\pi x/a)$. Inserting (10.197) and (10.198) into (10.196), we have

$$P_{1 \to 2} = \left(\frac{16\varepsilon a}{9\pi^2 \hbar}\right)^2 \left| \int_{-\infty}^{+\infty} e^{i\omega_{21}t - t^2} dt \right|^2. \qquad (10.199)$$

A variable change $y = t - \frac{i}{2}\omega_{21}$ yields $i\omega_{21}t - t^2 = -\omega_{21}^2/4 - y^2$ and $dt = dy$:

$$P_{1 \to 2} = \left(\frac{16\varepsilon a}{9\pi^2 \hbar}\right)^2 \left| e^{-\omega_{21}^2/4} \int_{-\infty}^{+\infty} e^{-y^2} dy \right|^2 = \pi \left(\frac{16\varepsilon a}{9\pi^2 \hbar}\right)^2 \exp\left(-\frac{9\pi^4 \hbar^2}{8m^2 a^4}\right), \qquad (10.200)$$

since $\omega_{21} = 3\pi^2 \hbar/(2ma^2)$.

Problem 10.7

A particle is initially (i.e., $t = 0$) in its ground state in a one-dimensional harmonic oscillator potential. At $t = 0$ a perturbation $\hat{V}(x, t) = V_0 \hat{x}^3 e^{-t/\tau}$ is turned on. Calculate to first order the probability that, after a sufficiently long time (i.e., $t \to \infty$), the system will have made a transition to a given excited state; consider all final states.

Solution

The transition probability from the ground state $n = 0$ to an excited state n is given by (10.41):

$$P_{0 \to n} = \frac{1}{\hbar^2} \left| \int_0^{+\infty} \langle n | \hat{V}(t) | 0 \rangle e^{i\omega_{n0}t} dt \right|^2 = \frac{V_0^2}{\hbar^2} \left| \langle n | \hat{x}^3 | 0 \rangle \right|^2 \left| \int_0^{+\infty} e^{-(1/\tau - in\omega)t} dt \right|^2, \qquad (10.201)$$

where $\omega_{n0} = \frac{E_n - E_0}{\hbar} = n\omega$ (since $E_n = \hbar\omega(n + \frac{1}{2})$) and the time integration was calculated in (10.63):

$$\left| \int_0^{\infty} e^{-(1/\tau - in\omega)t} dt \right|^2 = \frac{1}{n^2 \omega^2 + 1/\tau^2}. \qquad (10.202)$$

Since $\hat{a} \mid n\rangle = \sqrt{n} \mid n-1\rangle$ and $\hat{a}^\dagger \mid n\rangle = \sqrt{n+1} \mid n+1\rangle$, and since $\hat{X}^3 = (\hbar/2m\omega)^{3/2}(\hat{a}^\dagger + \hat{a})(\hat{a}^2 + \hat{a}^{\dagger 2} + 2\hat{a}^\dagger\hat{a} + 1)$, the only terms that survive in $\langle n \mid \hat{x}^3 | 0\rangle$ are

$$\langle n \mid \hat{X}^3 \mid 0\rangle = \left(\frac{\hbar}{2m\omega}\right)^{3/2} \langle n \mid \hat{a}^{\dagger 3} + \hat{a}\hat{a}^{\dagger 2} + \hat{a}^\dagger \mid 0\rangle = \left(\frac{\hbar}{2m\omega}\right)^{3/2} \left(\sqrt{6}\delta_{n,3} + 3\delta_{n,1}\right).$$

(10.203)

This implies that the particle can be found after a long duration only either in the first or in the third excited state.

Inserting (10.202) and (10.203) into (10.201), we can verify that the probabilities corresponding to the transitions from the ground state to the first, the second and the third excited states are given, respectively, by

$$P_{0\to1} = \frac{V_0^2}{\hbar^2} \left|\langle 1 \mid \hat{x}^3 | 0\rangle\right|^2 \left|\int_0^{+\infty} e^{-(1/\tau - i\omega)t} dt\right|^2 = \left(\frac{\hbar}{2m\omega}\right)^3 \frac{9V_0^2}{(\hbar\omega)^2 + \hbar^2/\tau^2},$$

(10.204)

$$P_{0\to2} = 0$$

(10.205)

$$P_{0\to3} = \frac{V_0^2}{\hbar^2} \left|\langle 3 | \hat{x}^3 | 0\rangle\right|^2 \left|\int_0^{+\infty} e^{-(1/\tau - 3i\omega)t} dt\right|^2 = \left(\frac{\hbar}{2m\omega}\right)^3 \frac{6V_0^2}{(3\hbar\omega)^2 + \hbar^2/\tau^2}.$$

(10.206)

Therefore the system cannot undergo transitions to the second excited state nor to excited states higher than $n = 3$; that is, $P_{0\to2} = 0$, since $\langle 2 \mid \hat{X}^3 \mid 0\rangle = 0$ and $P_{0\to n} = 0$ when $n > 3$, since $\langle n \mid \hat{X}^3 \mid 0\rangle = 0$ for $n > 3$.

Problem 10.8

A hydrogen atom, initially (i.e., $t \to -\infty$) in its ground state, is placed starting at time $t = 0$ in a time-dependent electric field pointing along the z-axis $\vec{E}(t) = E_0 \vec{k}/(\tau^2 + t^2)$, where τ is a constant having the dimension of time. Calculate the probability that the atom will be found in the 2p state after a sufficiently long time (i.e., $t \to \infty$).

Solution

Since the potential resulting from the interaction of the hydrogen's electron with the external field $\vec{E}(t)$ is $V(t) = -e\vec{r} \cdot \vec{E}(t)$, we can use (10.41) to write the transition probability from the 1s state to 2p as

$$P_{1s\to2p} = \frac{1}{\hbar^2} \left|\int_{-\infty}^{+\infty} \langle 210 \mid V(t) \mid 100\rangle e^{i\omega_{fi}t} dt\right|^2,$$

(10.207)

where

$$\langle 210 \mid V(t) \mid 100\rangle = \langle 210 \mid (-e\vec{r} \cdot \vec{E}) \mid 100\rangle = -\frac{eE_0\tau}{\tau^2 + t^2}\langle 210 \mid z \mid 100\rangle.$$

(10.208)

Since $z = r\cos\theta$ and

$$\psi_{1s} = R_{10}(r)Y_{00}(\Omega) = \frac{1}{\sqrt{\pi a_0^3}}e^{-r/a_0}, \qquad \psi_{2p} = R_{21}(r)Y_{10}(\Omega) = \frac{1}{\sqrt{8\pi a_0^3}}\frac{r}{2a_0}e^{-r/2a_0}\cos\theta,$$

(10.209)

and using $\int_0^\pi \sin\theta \cos^2\theta \, d\theta = \int_{-1}^1 x^2 dx = \frac{2}{3}$, we have

$$\langle 210 \mid z \mid 100 \rangle = \int_0^\infty r^3 R_{21}^*(r) R_{10}(r) dr \int_0^\pi \sin\theta \cos^2\theta \, d\theta \int_0^{2\pi} d\phi$$

$$= \frac{4\pi}{3} \frac{1}{4\pi a_0^4 \sqrt{2}} \int_0^\infty r^4 e^{-3r/2a_0} dr = \frac{2^8 a_0}{3^5 \sqrt{2}}. \qquad (10.210)$$

Inserting (10.208) and (10.210) into (10.207) we have

$$P_{1s\to2p} = \frac{2^{15} e^2 E_0^2 \tau^2 a_0^2}{3^{10} \hbar^2} \left| \int_{-\infty}^{+\infty} \frac{e^{i\omega_{fi}t}}{\tau^2 + t^2} dt \right|^2. \qquad (10.211)$$

We may calculate this integral using the method of residues by closing the contour in the upper half of the t-plane. Since the infinite semicircle has no contribution to the integral, the pole at $t = i\tau$ gives

$$\int_{-\infty}^{+\infty} \frac{e^{i\omega_{fi}t}}{\tau^2 + t^2} dt = 2\pi i \operatorname{Res} \left[\frac{e^{i\omega_{fi}t}}{\tau^2 + t^2} \right]_{t=i\tau} = 2\pi i \lim_{t\to i\tau} \left[\frac{e^{i\omega_{fi}t}}{\tau^2 + t^2} \times (t - i\tau) \right]$$

$$= 2\pi i \lim_{t\to i\tau} \left[\frac{e^{i\omega_{fi}t}(t - i\tau)}{(t + i\tau)(t - i\tau)} \right] = \frac{\pi}{\tau} e^{-\omega_{fi}\tau}, \qquad (10.212)$$

where

$$\omega_{fi} = \frac{1}{\hbar}(E_f - E_i) = \frac{1}{\hbar}(E_{2p} - E_{1s}) = \frac{1}{\hbar}\left(\frac{1}{4}E_{1s} - E_{1s}\right) = -\frac{3}{4\hbar}E_{1s} = \frac{3R_y}{4\hbar}, \qquad (10.213)$$

where R_y is the Rydberg constant: $R_y = 13.6\,\text{eV}$. Inserting (10.212) into (10.211), we obtain the transition probability

$$P_{1s\to2p} = \frac{2^{15} e^2 \pi^2 E_0^2 a_0^2}{3^{10} \hbar^2} \exp\left(-2\omega_{fi}\tau\right) = \frac{2^{15} e^2 \pi^2 E_0^2 a_0^2}{3^{10} \hbar^2} \exp\left(-\frac{3R_y}{2\hbar}\tau\right). \qquad (10.214)$$

Problem 10.9

A hydrogen atom is in its excited 2p state. Calculate the transition rate associated with the 2p $\to$ 1s transitions (Lyman) and the lifetime of the 2p state.

Solution

The first expression of the total transition rate is given by (10.141):

$$W_{2p\to1s} = \frac{4\omega_{2p\to1s}^3}{3\hbar c^3} \left| \vec{d}_{2p\to1s} \right|^2, \qquad (10.215)$$

where

$$|\vec{d}_{2p\to1s}|^2 = e^2 |\langle 2p \mid \vec{\varepsilon} \cdot \vec{r} \mid 1s \rangle|^2 = e^2 \left| \int_0^\infty r^3 R_{21}^* R_{10}(r) \, dr \int d\Omega \, Y_{1m}^* \vec{\varepsilon} \cdot \hat{r} Y_{00} \right|^2. \qquad (10.216)$$

First, we need to calculate $\langle 2p \mid \vec{\varepsilon} \cdot \vec{r} \mid 1s \rangle$. The radial integral is given by

$$\int_0^\infty r^3 R_{21}^*(r) R_{10}(r)\, dr = \frac{1}{a_0^4 \sqrt{6}} \int_0^\infty r^4 e^{-3r/2a_0}\, dr = \frac{2^8 a_0}{3^4 \sqrt{6}}. \tag{10.217}$$

The angular part can be calculated from (10.127) as follows:

$$\int d\Omega\, Y_{1m}^*(\Omega)\vec{\varepsilon}.\hat{r} Y_{00}(\Omega) = \sqrt{\frac{4\pi}{3}} \int Y_{1m}^* \left(\frac{-\epsilon_x + i\epsilon_y}{\sqrt{2}} Y_{11} + \frac{\epsilon_x + i\epsilon_y}{\sqrt{2}} Y_{1-1} + \epsilon_z Y_{10} \right) Y_{00} d\Omega$$

$$= \frac{1}{\sqrt{3}} \int Y_{1m}^* \left(\frac{-\epsilon_x + i\epsilon_y}{\sqrt{2}} Y_{11} + \frac{\epsilon_x + i\epsilon_y}{\sqrt{2}} Y_{1-1} + \epsilon_z Y_{10} \right) d\Omega$$

$$= \frac{1}{\sqrt{3}} \left(\frac{-\epsilon_x + i\epsilon_y}{\sqrt{2}} \delta_{m,1} + \frac{\epsilon_x + i\epsilon_y}{\sqrt{2}} \delta_{m,-1} + \epsilon_z \delta_{m,0} \right), \tag{10.218}$$

since $\int Y_{1m}^*(\theta, \phi) Y_{l_i m_i}(\theta, \phi)\, d\Omega = \delta_{l_i,1} \delta_{m,m_i}$. An insertion of (10.217) and (10.218) into (10.216) leads to

$$|\vec{d}_{2p \to 1s}|^2 = 32 \left(\frac{2}{3} \right)^{10} e^2 a_0^2 \left[\frac{1}{2}(\epsilon_x^2 + \epsilon_y^2)(\delta_{m,-1} + \delta_{m,1}) + \epsilon_z^2 \delta_{m,0} \right], \tag{10.219}$$

which, when inserted into (10.215), leads to the total transition rate corresponding to a certain value of the azimuthal quantum number m:

$$W_{2p \to 1s} = \frac{4\omega_{2p \to 1s}^3}{3\hbar c^3} |\vec{d}_{fi}|^2 = \frac{128 e^2 a_0^2 \omega^3}{3\hbar c^3} \left(\frac{2}{3} \right)^{10} \left[\frac{1}{2}(\epsilon_x^2 + \epsilon_y^2)(\delta_{m,-1} + \delta_{m,1}) + \epsilon_z^2 \delta_{m,0} \right]. \tag{10.220}$$

Summing over the three possible m-states, $m = -1, 0, 1$,

$$\sum_{m=-1}^1 \left[\frac{1}{2}(\epsilon_x^2 + \epsilon_y^2)(\delta_{m,-1} + \delta_{m,1}) + \epsilon_z^2 \delta_{m,0} \right] = \epsilon_x^2 + \epsilon_y^2 + \epsilon_z^2 = 1, \tag{10.221}$$

and since, as shown in (10.213), $\omega_{2p \to 1s} = (E_{2p} - E_{1s})/\hbar = 3R_y/(4\hbar) = 3e^2/(8\hbar a_0)$ (because the Rydberg constant R_y is equal to $e^2/(2\hbar a_0)$), we can reduce (10.220) to

$$W_{2p \to 1s} = \frac{128}{3\hbar c^3} \left(\frac{2}{3} \right)^{10} e^2 a_0^2 \omega_{2p \to 1s}^3 = \left(\frac{2}{3} \right)^8 \left(\frac{e^2}{\hbar c} \right)^4 \frac{c}{a_0} = \left(\frac{2}{3} \right)^8 \frac{c \alpha^4}{a_0}, \tag{10.222}$$

where $\alpha = e^2/(\hbar c) = 1/137$ is the fine structure constant and $a_0 = 0.529 \times 10^{-10}$ m is the Bohr radius. The numerical value of the transition rate is

$$W_{2p \to 1s} = \left(\frac{2}{3} \right)^8 \frac{c \alpha^4}{a_0} \simeq \left(\frac{2}{3} \right)^8 \frac{3 \times 10^8 \, \text{m s}^{-1}}{137^4 \times 0.529 \times 10^{-10} \, \text{m}} = 0.628 \times 10^9 \, \text{s}^{-1}. \tag{10.223}$$

The lifetime of the 2p state is then given by

$$\tau = \frac{1}{W_{2p \to 1s}} = \left(\frac{3}{2} \right)^8 \frac{a_0}{c \alpha^4} = \frac{1.5^8 \times 137^4 \times 0.529 \times 10^{-10} \, \text{m}}{3 \times 10^8 \, \text{m s}^{-1}} = 1.6 \times 10^{-9} \, \text{s}. \tag{10.224}$$

This value is in very good agreement with experimental data.

Remark
Another way of obtaining (10.222) is to use the relation

$$W_{2p \to 1s} = \frac{4e^2 \omega_{2p \to 1s}^3}{3\hbar c^3} \frac{1}{3} \sum_{m=-1}^{1} |\langle 21m \mid \vec{r} \mid 100 \rangle|^2$$

$$= \frac{4e^2 \omega_{2p \to 1s}^3}{9\hbar c^3} \sum_{m=-1}^{1} \left[|\langle 21m \mid \hat{x} \mid 100 \rangle|^2 + |\langle 21m \mid \hat{y} \mid 100 \rangle|^2 + |\langle 21m \mid \hat{z} \mid 100 \rangle|^2 \right],$$

$$\tag{10.225}$$

where we have averaged over the various transitions. Using the relations $x = r \sin\theta \cos\phi = -\sqrt{2\pi/3}\, r (Y_{11} - Y_{1-1})$, $y = r \sin\theta \sin\phi = i\sqrt{2\pi/3}\, r(Y_{11} + Y_{1-1})$, and $z = r\cos\theta = \sqrt{4\pi/3}\, r Y_{10}$, we can show that

$$\langle 21m \mid \hat{x} \mid 100 \rangle = -\frac{1}{\sqrt{4\pi}} \sqrt{\frac{2\pi}{3}} \int_0^\infty r^3 R_{21}^*(r) R_{10}(r) dr \int Y_{1m}^*(\Omega)\, (Y_{11} - Y_{1-1})\, d\Omega$$

$$= -\frac{1}{\sqrt{6}} \left[\frac{24}{\sqrt{6}} \left(\frac{2}{3}\right)^5 a_0 \right] (\delta_{m,1} - \delta_{m,-1}), \tag{10.226}$$

$$\langle 21m \mid \hat{y} \mid 100 \rangle = \frac{i}{\sqrt{4\pi}} \sqrt{\frac{2\pi}{3}} \int_0^\infty r^3 R_{21}^*(r) R_{10}(r) dr \int Y_{1m}^*(Y_{11} + Y_{1-1}) d\Omega$$

$$= \frac{i}{\sqrt{6}} \left[\frac{24}{\sqrt{6}} \left(\frac{2}{3}\right)^5 a_0 \right] (\delta_{m,1} + \delta_{m,-1}), \tag{10.227}$$

$$\langle 21m \mid \hat{z} \mid 100 \rangle = \frac{1}{\sqrt{4\pi}} \sqrt{\frac{4\pi}{3}} \int_0^\infty r^3 R_{21}^*(r) R_{10}(r) dr \int Y_{1m}^* Y_{10} d\Omega$$

$$= \frac{1}{\sqrt{3}} \left[\frac{24}{\sqrt{6}} \left(\frac{2}{3}\right)^5 a_0 \right] \delta_{m,0}. \tag{10.228}$$

A combination of the previous three relations leads to

$$\sum_{m=-1}^{1} |\langle 21m \mid \vec{r} \mid 100 \rangle|^2 = 96 a_0^2 \left(\frac{2}{3}\right)^{10} \sum_m \left[\frac{1}{6}(\delta_{m,1} - \delta_{m,-1})^2 + \frac{1}{6}(\delta_{m,1} + \delta_{m,-1})^2 + \frac{1}{3}\delta_{m,0}^2 \right]$$

$$= 96 a_0^2 \left(\frac{2}{3}\right)^{10} \sum_m \left[\frac{1}{6}(\delta_{m,1} + \delta_{m,-1}) + \frac{1}{6}(\delta_{m,1} + \delta_{m,-1}) + \frac{1}{3}\delta_{m,0} \right]$$

$$= \frac{96 a_0^2}{3} \left(\frac{2}{3}\right)^{10} \sum_{m=-1}^{1} (\delta_{m,-1} + \delta_{m,1} + \delta_{m,0}) = 96 \left(\frac{2}{3}\right)^{10} a_0^2. \tag{10.229}$$

Finally, substituting (10.229) into (10.225) and using $\omega_{2p \to 1s} = 3e^2/(8\hbar a_0)$, we obtain

$$W_{2p \to 1s} = \frac{128 e^2 a_0^2}{3\hbar c^3} \omega^3 \left(\frac{2}{3}\right)^{10} = \left(\frac{2}{3}\right)^8 \left(\frac{e^2}{\hbar c}\right)^4 \frac{c}{a_0} = \left(\frac{2}{3}\right)^8 \frac{c\alpha^4}{a_0}. \tag{10.230}$$

Problem 10.10

(a) Calculate the transition rate from the first excited state to the ground state for an isotropic (three-dimensional) harmonic oscillator of charge q.

(b) Find a numerical value for the rate calculated in (a) as well as the lifetime of the first excited state for the case of an electron (i.e., $m_e c^2 = 0.511$ MeV) oscillating with a frequency of an optical radiation $\omega \simeq 10^{15}$ rad s^{-1}.

Solution

As mentioned in Chapter 6, the ground state of an isotropic harmonic oscillator is a 1s state, $(n, l, m) = (0, 0, 0)$, whose energy and wave function are $E_0 = 3\hbar\omega/2$ and

$$\psi_{000}(r, \theta, \phi) = R_{00}(r) Y_{00}(\theta, \phi) = \frac{2}{\sqrt{\sqrt{\pi}}} \left(\frac{m\omega}{\hbar} \right)^{3/4} e^{-m\omega r^2/2\hbar} Y_{00}(\theta, \phi), \qquad (10.231)$$

and the first excited state is a 1p state $(n, l, m) = (1, 1, m)$ whose energy and wave function are $E_1 = 5\hbar\omega/2$ and

$$\psi_{11m}(r, \theta, \phi) = R_{11}(r) Y_{1m}(\theta, \phi) = \sqrt{\frac{8}{3\sqrt{\pi}}} \left(\frac{m\omega}{\hbar} \right)^{5/4} r e^{-m\omega r^2/2\hbar} Y_{1m}(\theta, \phi). \quad (10.232)$$

Using $\int_0^\infty x^4 e^{-x^2} dx = \frac{3}{8}\sqrt{\pi}$ along with a change of variable $x = \sqrt{m\omega/\hbar}\, r$, we have

$$\int_0^\infty r^3 R_{11}^*(r) R_{10}(r)\, dr = 4\sqrt{\frac{2}{3\pi}} \left(\frac{m\omega}{\hbar} \right)^2 \int_0^\infty r^4 e^{-m\omega r^2/\hbar}\, dr = \sqrt{\frac{3\hbar}{2m\omega}}. \qquad (10.233)$$

(a) The transition rate for a 1p → 1s transition is given by

$$W_{1p \to 1s} = \frac{4q^2 \omega_{1p \to 1s}^3}{3\hbar c^3} \frac{1}{3} \sum_{m=-1}^{1} |\langle 11m \mid \vec{r} \mid 000 \rangle|^2$$

$$= \frac{4q^2 \omega_{1p \to 1s}^3}{9\hbar c^3} \sum_{m=-1}^{1} \left[|\langle 11m \mid \hat{x} \mid 000 \rangle|^2 + |\langle 11m \mid \hat{y} \mid 000 \rangle|^2 + |\langle 21m \mid \hat{z} \mid 000 \rangle|^2 \right].$$

$$(10.234)$$

Since $x = r \sin\theta \cos\phi = -\sqrt{2\pi/3}\, r (Y_{11} - Y_{1-1})$, $y = r \sin\theta \sin\phi = i\sqrt{2\pi/3}\, r (Y_{11} + Y_{1-1})$, and $z = r \cos\theta = \sqrt{4\pi/3}\, r Y_{10}$, and using (10.233), we can show by analogy with (10.226) to (10.228) that

$$\langle 11m \mid \hat{x} \mid 000 \rangle = -\frac{1}{\sqrt{4\pi}} \sqrt{\frac{2\pi}{3}} \int_0^\infty r^3 R_{11}^*(r) R_{00}(r)\, dr \int Y_{1m}^*(\Omega) (Y_{11} - Y_{1-1})\, d\Omega$$

$$= -\frac{1}{\sqrt{6}} \sqrt{\frac{3\hbar}{2m\omega}} (\delta_{m,1} - \delta_{m,-1}), \qquad (10.235)$$

$$\langle 11m \mid \hat{y} \mid 000 \rangle = \frac{i}{\sqrt{4\pi}} \sqrt{\frac{2\pi}{3}} \int_0^\infty r^3 R_{11}^*(r) R_{00}(r)\, dr \int Y_{1m}^*(Y_{11} + Y_{1-1})\, d\Omega$$

$$= \frac{i}{\sqrt{6}} \sqrt{\frac{3\hbar}{2m\omega}} (\delta_{m,1} + \delta_{m,-1}), \qquad (10.236)$$

$$\langle 11m \mid \hat{z} \mid 000 \rangle = \frac{1}{\sqrt{4\pi}} \sqrt{\frac{4\pi}{3}} \int_0^\infty r^3 R_{11}^*(r) R_{00}(r)\, dr \int Y_{1m}^* Y_{10}\, d\Omega = \frac{1}{\sqrt{3}} \sqrt{\frac{3\hbar}{2m\omega}} \delta_{m,0}.$$

$$(10.237)$$

A combination of the previous three relations leads to

$$\sum_{m=-1}^{1} |\langle 11m \mid \vec{r} \mid 000 \rangle|^2 = \frac{3\hbar}{2m\omega} \sum_m \left[\frac{1}{6} (\delta_{m,1} - \delta_{m,-1})^2 + \frac{1}{6} (\delta_{m,1} + \delta_{m,-1})^2 + \frac{1}{3} \delta_{m,0}^2 \right]$$

$$= \frac{\hbar}{2m\omega} \sum_{m=-1}^{1} (\delta_{m,-1} + \delta_{m,1} + \delta_{m,0}) = \frac{3\hbar}{2m\omega}. \qquad (10.238)$$

Substituting (10.238) into (10.234), and using $\omega_{1p \to 1s} = (E_1 - E_0)/\hbar = (\frac{5}{2} - \frac{3}{2})\omega = \omega$, we obtain

$$W_{1p \to 1s} = \frac{4q^2 \omega_{1p \to 1s}^3}{9\hbar c^3} \frac{3\hbar}{2m\omega} = \frac{2q^2 \omega^2}{3mc^3}. \qquad (10.239)$$

(b) In the case of an electron ($q = -e$ and $m_e c^2 = 0.511\,\text{MeV}$) which is oscillating with a frequency of $\omega \simeq 10^{15}\,\text{s}^{-1}$, the transition rate is

$$W_{1p \to 1s} = \frac{2e^2 \omega^2}{3m_e c^3} = \frac{2\alpha}{3} \left(\frac{\hbar c}{m_e c^2} \right) \frac{\omega^2}{c}$$

$$= \frac{2}{3} \frac{1}{137} \left(\frac{197\,\text{MeV fm}}{0.511\,\text{MeV}} \right) \frac{10^{30}\,\text{s}^{-2}}{3 \times 10^8\,\text{m s}^{-1}} \simeq 0.64 \times 10^7\,\text{s}^{-1}, \qquad (10.240)$$

where $\alpha = e^2/(\hbar c) = 1/137$ is the fine structure constant. The lifetime of the 1p state for the oscillator is given by

$$\tau = \frac{1}{W_{1p \to 1s}} = \frac{3m_e c^3}{2e^2 \omega^2} \simeq \frac{1}{0.64 \times 10^7\,\text{s}^{-1}} = 1.56 \times 10^{-7}\,\text{s}. \qquad (10.241)$$

Problem 10.11

Show that free electrons can neither emit nor absorb photons.

Solution

If the electron is *free* both before and after it interacts with the photon, its initial and final wave functions are given by *plane* waves: $\psi_i(\vec{r}) = (2\pi)^{-3/2} e^{i\vec{k}_i \cdot \vec{r}}$ and $\psi_f(\vec{r}) = (2\pi)^{-3/2} e^{i\vec{k}_f \cdot \vec{r}}$. Let us assume, for argument sake, that a free electron can absorb and emit a photon; the corresponding absorption and emission transition rates would be given as follows (see (10.95) and (10.96)):

$$\Gamma_{i \to f}^{abs} = \frac{4\pi^2 e^2}{m_e^2 \omega V} \left| (\vec{\varepsilon} \cdot \vec{k}_i) \langle \psi_f \mid e^{i\vec{k} \cdot \vec{r}} \mid \psi_i \rangle \right|^2 \delta(E_f - E_i - \hbar\omega), \qquad (10.242)$$

$$\Gamma_{i \to f}^{emi} = \frac{4\pi^2 e^2}{m_e^2 \omega V} \left| (\vec{\varepsilon} \cdot \vec{k}_i) \langle \psi_f \mid e^{-i\vec{k} \cdot \vec{r}} \mid \psi_i \rangle \right|^2 \delta(E_f - E_i + \hbar\omega), \qquad (10.243)$$

where we have used $\vec{P} \psi_i(\vec{r}) = \vec{k}_i \psi_i(\vec{r})$. Since

$$\langle \psi_f \mid e^{\pm i \vec{k} \cdot \vec{r}} \mid \psi_i \rangle = \frac{1}{(2\pi)^2} \int d^3r \, e^{i(\vec{k}_i - \vec{k}_f \pm \vec{k}) \cdot \vec{r}} = \delta(\vec{k}_i - \vec{k}_f \pm \vec{k}), \qquad (10.244)$$

the delta functions $\delta(\vec{k}_i - \vec{k}_f \pm \vec{k})$ give the conservation laws of the linear momentum for both the absorption and emission processes.

Let us show first that a free electron cannot absorb a photon. For this, we are going to show that the momentum conservation condition $\delta(\vec{k}_i - \vec{k}_f + \vec{k})$ is incompatible with the energy conservation condition $\delta(E_f - E_i - \hbar\omega)$. Combining equations (10.242) and (10.244), we see that the absorption rate is proportional to the product of two delta functions: $\Gamma^{abs}_{i \to f} \sim \delta(\vec{k}_i - \vec{k}_f + \vec{k})\delta(E_f - E_i - \hbar\omega)$, one pertaining to the conservation of momentum

$$\delta(\vec{k}_i - \vec{k}_f + \vec{k}) \quad \Longrightarrow \quad \vec{p}_i - \vec{p}_f + \vec{p}_{photon} = 0, \qquad (10.245)$$

the other dealing with the conservation of energy

$$\delta(E_f - E_i - \hbar\omega) \quad \Longrightarrow \quad E_f - E_i - cp_{photon} = 0, \qquad (10.246)$$

where $\vec{p}_i = \hbar \vec{k}_i$ and E_i are the initial momentum and energy of the electron, $\vec{p}_f = \hbar \vec{k}_f$ and E_f are its final momentum and energy, and $\vec{p}_{photon} = \hbar \vec{k}$ and cp_{photon} are the linear momentum and energy of the absorbed photon. We are now ready to show that the condition (10.245) is incompatible with (10.246). If we work within the rest frame of the initial electron, we have $\vec{p}_i = 0$. Thus, on the one hand, (10.245) leads to $\vec{p}_{photon} = \vec{p}_f$ and, on the other hand, (10.246) leads to $E_f = cp_{photon}$ or $p_f^2/2m_e = cp_{photon}$. Indeed, conditions (10.245) and (10.246) are contradictory since, inserting $\vec{p}_i = 0$ and $\vec{p}_{photon} = \vec{p}_f$ into (10.246), we end up with $p_f^2/2m_e = cp_f$ or $p_f = 2m_ec$. This suggests either that $v_f = 0$ and this is meaningless since, as $\vec{p}_{photon} = \vec{p}_f$, the speed of the photon would also be zero; or that $v_f = 2c$ and this is impossible. So both results are impossible. In summary, having started with the assumption that a free electron can absorb a photon (10.242), we have ended up with a momentum conservation law and an energy conservation law that are contradictory. Thus, a free electron cannot absorb a photon.

Following the same procedure, we can also show that the assumption of a free electron emitting a photon leads to a momentum conservation law and an energy conservation law that are incompatible; thus, a free electron cannot emit a photon.

Problem 10.12

A hydrogen atom in its ground state is placed in an oscillating electric field $\vec{\mathcal{E}}(t) = \vec{\mathcal{E}}_0 \sin(\omega t)$ of angular frequency ω with $\hbar\omega > m_e e^4/(2\hbar^2)$.

(a) Find the transition rate (probability per unit time) that the atom will be ionized.

(b) Use the expression derived in (a) to find the maximum transition rate.

Solution

After ionization we assume the electron to be in free motion: its energy is purely kinetic $E_k = \hbar^2 k^2/2m_e$ and its wave function is a plane wave $\psi_k(\vec{r}) = (2\pi)^{-3/2} e^{i\vec{k} \cdot \vec{r}}$. Since the perturbation resulting from the interaction of the hydrogen's electron with the external field $\vec{\mathcal{E}}(t)$ is *harmonic*,

$$\hat{V}(t) = -e\vec{r} \cdot \vec{\mathcal{E}}(t) = -e\vec{r} \cdot \vec{\mathcal{E}}_0 \sin(\omega t) = \frac{e}{2i} \vec{r} \cdot \vec{\mathcal{E}}_0 e^{-i\omega t} - \frac{e}{2i} \vec{r} \cdot \vec{\mathcal{E}}_0 e^{i\omega t}, \qquad (10.247)$$

we can infer, by analogy with the method that led to (10.54) from (10.50), the transition rate for the ionization of the hydrogen atom:

$$\Gamma_{0k} = \frac{2\pi}{\hbar} \left| \frac{e}{2i} \langle \psi_k | \vec{r} \cdot \vec{\mathcal{E}}_0 \mid 100 \rangle \right|^2 \delta(E_k - E_0 + \hbar\omega)$$

$$+ \frac{2\pi}{\hbar} \left| -\frac{e}{2i} \langle \psi_k | \vec{r} \cdot \vec{\mathcal{E}}_0 \mid 100 \rangle \right|^2 \delta(E_k - E_0 - \hbar\omega), \qquad (10.248)$$

where $E_0 = -m_e e^4/2\hbar^2 = -13.6\,\text{eV}$ is the ground state energy and $E_k = \hbar^2 k/2m_e$ is the final energy of the electron. The first delta term, $\delta(E_k - E_0 + \hbar\omega)$, in (10.248) does not contribute, since if $\hbar\omega = E_0 - E_k$ the ionization could not take place because the electric field would not be strong enough to ionize the atom. The transition rate (10.248) then becomes

$$\Gamma_{0k} = \frac{\pi e^2}{2\hbar} \left| \langle \psi_k | \vec{r} \cdot \vec{\mathcal{E}}_0 \mid 100 \rangle \right|^2 \delta(E_k - E_0 - \hbar\omega). \qquad (10.249)$$

To calculate $\langle \psi_k | \vec{r} \cdot \vec{\mathcal{E}}_0 \mid 100 \rangle$, let us take $\vec{k}$ along the z-axis and hence $\vec{k} \cdot \vec{r} = kr\cos\theta$ and $\psi_k(\vec{r}) = (2\pi)^{-3/2} e^{ikr\cos\theta}$. Taking (θ, ϕ) and (α, β) as the respective polar angles of $\vec{r}$ and $\vec{\mathcal{E}}_0$, we have $\vec{r} = r(\sin\theta\cos\phi\hat{i} + \sin\theta\sin\phi\hat{j} + \cos\theta\hat{k})$ and $\vec{\mathcal{E}}_0 = \mathcal{E}_0(\sin\alpha\cos\beta\hat{i} + \sin\alpha\sin\beta\hat{j} + \cos\alpha\hat{k})$; hence

$$\vec{r} \cdot \vec{\mathcal{E}}_0 = r\mathcal{E}_0(\sin\theta\cos\phi\sin\alpha\cos\beta + \sin\theta\sin\phi\sin\alpha\sin\beta + \cos\theta\cos\alpha)$$

$$= r\mathcal{E}_0 \left[\sin\theta\sin\alpha\cos(\phi - \beta) + \cos\theta\cos\alpha \right]. \qquad (10.250)$$

Since $\psi_{1s} = (\pi a_0^3)^{-1/2} e^{-r/a_0}$ and $d^3r = r^2 dr \sin\theta\cos\phi\, d\theta\, d\phi$, we have

$$\langle \psi_k | \vec{r} \cdot \vec{\mathcal{E}}_0 \mid 100 \rangle = \frac{1}{(2\pi)^{3/2}} \frac{1}{\sqrt{\pi a_0^3}} \int d^3r\, (\vec{r} \cdot \vec{\mathcal{E}}_0) e^{-ikr\cos\theta - r/a_0}$$

$$= \frac{\mathcal{E}_0}{\sqrt{8\pi^4 a_0^3}} \int_0^\infty r^3 e^{-r/a_0} dr \int_0^\pi \sin\theta e^{-ikr\cos\theta} d\theta \int_0^{2\pi} \left[\sin\theta\sin\alpha\cos(\phi - \beta) + \cos\theta\cos\alpha \right] d\phi$$

$$= \frac{2\pi \mathcal{E}_0 \cos\alpha}{\sqrt{8\pi^4 a_0^3}} \int_0^\infty r^3 e^{-r/a_0}\, dr \int_0^\pi \sin\theta\cos\theta e^{-ikr\cos\theta}\, d\theta, \qquad (10.251)$$

where we have used $\int_0^{2\pi} \cos(\phi - \beta) d\phi = 0$, since $\int_0^{2\pi} \cos\phi\, d\phi = 0$ and $\int_0^{2\pi} \sin\phi\, d\phi = 0$. A change of variable $x = \cos\theta$ and an integration by parts leads to

$$\int_0^\pi \sin\theta\cos\theta e^{-ikr\cos\theta} d\theta = \int_{-1}^1 x e^{-ikrx} dx = \frac{1}{-ikr} x e^{-ikrx} \Big|_{-1}^1 - \frac{1}{(-ikr)^2} e^{-ikrx} \Big|_{-1}^1$$

$$= \frac{i}{kr} \left[e^{-ikr} + e^{ikr} \right] + \frac{1}{k^2 r^2} \left[e^{-ikr} - e^{ikr} \right]. \qquad (10.252)$$

When we insert this integral into (10.251), we still need to calculate four radial integrals which can be carried out by parts:

$$\int_0^\infty r e^{\pm ikr - r/a_0} dr = \frac{1}{\pm ik - 1/a_0} r e^{\pm ikr - r/a_0} \Big|_0^\infty - \frac{1}{(\pm ik - 1/a_0)^2} e^{\pm ikr - r/a_0} \Big|_0^\infty = \frac{a_0^2}{(\pm ia_0 k - 1)^2}, \qquad (10.253)$$

$$\int_0^\infty r^2 e^{\pm ikr - r/a_0} dr = \frac{1}{\pm ik - 1/a_0} r^2 e^{\pm ikr - r/a_0} \Big|_0^\infty - \frac{2}{\pm ik - 1/a_0} \int_0^\infty r e^{\pm ikr - r/a_0} dr$$

$$= -\frac{2}{(\pm ik - 1/a_0)^2} r e^{\pm ikr - r/a_0} \Big|_0^\infty + \frac{2}{(\pm ik - 1/a_0)^3} e^{\pm ikr - r/a_0} \Big|_0^\infty$$

$$= -\frac{2a_0^3}{(\pm i a_0 k - 1)^3}. \tag{10.254}$$

Inserting (10.252) to (10.254) into (10.251), we obtain

$$\langle \psi_k | \vec{r} \cdot \vec{\mathcal{E}}_0 | 100 \rangle = \frac{2\pi \mathcal{E}_0 \cos \alpha}{\sqrt{8\pi^4 a_0^3}} \left[\frac{a_0^2}{k^2(-ia_0 k - 1)^2} - \frac{a_0^2}{k^2(ia_0 k - 1)^2} - \frac{2ia_0^3}{k(ia_0 k - 1)^3} \right.$$

$$\left. - \frac{2ia_0^3}{k(-ia_0 k - 1)^3} \right]$$

$$= -\frac{16\mathcal{E}_0 \cos \alpha}{\pi \sqrt{2a_0^5}} \frac{ia_0^6 k}{(a_0^2 k^2 + 1)^3}. \tag{10.255}$$

A substitution of this expression into (10.249) leads to

$$\Gamma_{0k} = \frac{\pi e^2}{2\hbar} \frac{128 \mathcal{E}_0^2 \cos^2 \alpha}{\pi^2 a_0^5} \frac{k^2 a_0^{12}}{(a_0^2 k^2 + 1)^6} \delta(E_k - E_0 + \hbar\omega). \tag{10.256}$$

This relation gives the transition rate for a single final state ψ_k corresponding to a given k. We need to sum over all final states of the electron. These represent a continuum; we must then integrate over all directions of emission and over all possible momenta:

$$\Gamma_0 = \int \Gamma_{0k} d^3 k = \int k^2 dk \int_0^\pi \Gamma_{0k} \sin \alpha \, d\alpha \int_0^{2\pi} d\beta$$

$$= 2\pi \frac{64 e^2 \mathcal{E}_0^2 a_0^7}{\pi \hbar} \int \frac{k^4 \delta(E_k - E_0 - \hbar\omega)}{(a_0^2 k^2 + 1)^6} dk \int_0^\pi \sin \alpha \cos^2 \alpha \, d\alpha$$

$$= \frac{256 e^2 \mathcal{E}_0^2 a_0^7}{3\hbar} \int \frac{k^4 \delta(E_k - E_0 - \hbar\omega)}{(a_0^2 k^2 + 1)^6} dk, \tag{10.257}$$

where we have used $\int_0^\pi \sin \alpha \cos^2 \alpha \, d\alpha = \int_{-1}^1 x^3 dx = \frac{2}{3}$. The integration over k can be converted into an integration over the final energy E_k: since $E_k = \hbar^2 k^2/(2m_e)$, a change of variable $k = \sqrt{2m_e E_k/\hbar^2}$, and hence $k \, dk = (m_e/\hbar^2) dE_k$, reduces (10.257) to

$$\Gamma_0 = \frac{256 e^2 \mathcal{E}_0^2 a_0^7}{3\hbar} \int \frac{k^3 \delta(E_k - E_0 - \hbar\omega)}{(a_0^2 k^2 + 1)^6} k \, dk$$

$$= \frac{m_e}{\hbar^2} \frac{256 e^2 \mathcal{E}_0^2 a_0^7}{3\hbar} \int \frac{(2m_e E_k/\hbar^2)^{3/2} \delta(E_k - E_0 - \hbar\omega)}{\left(2m_e a_0^2 E_k/\hbar^2 + 1\right)^6} dE_k$$

$$= \frac{256 e^2 m_e \mathcal{E}_0^2 a_0^7}{3\hbar^3} \frac{(2m_e/\hbar^2)^{3/2}(E_0 + \hbar\omega)^{3/2}}{\left[2m_e a_0^2 (E_0 + \hbar\omega)/\hbar^2 + 1\right]^6}. \tag{10.258}$$

This relation can be simplified if we use $E_0 = -m_e e^4/(2\hbar^2) = -\hbar\omega_0$, which gives $E_0 + \hbar\omega = \hbar(\omega - \omega_0) = \hbar\omega_0(\omega/\omega_0 - 1)$. Since $a_0 = \hbar^2/(m_e e^2)$, we have $\hbar\omega_0 a_0^2 = m_e e^4 \hbar^4/(2\hbar^2 m_e^2 e^4) = \hbar^2/(2m_e)$ and hence $2m_e a_0^2(E_0 + \hbar\omega)/\hbar^2 = 2m_e \hbar\omega_0 a_0^2(\omega/\omega_0 - 1)/\hbar^2 = \omega/\omega_0 - 1$. Thus, inserting the expressions $E_0 + \hbar\omega = \hbar\omega_0(\omega/\omega_0 - 1)$ and $2m_e a_0^2(E_0 + \hbar\omega)/\hbar^2 + 1 = \omega/\omega_0$ into (10.258), we obtain

$$\Gamma_0 = \frac{256 e^2 m_e \mathcal{E}_0^2 a_0^7}{3\hbar^3} \frac{(2m_e/\hbar^2)^{3/2}(\hbar\omega_0)^{3/2}(\omega/\omega_0 - 1)^{3/2}}{(\omega/\omega_0)^6}. \tag{10.259}$$

Finally, since $(2m_e/\hbar^2)^{3/2}(\hbar\omega_0)^{3/2} = (2m_e/\hbar^2)^{3/2}(m_e e^4/2\hbar^2)^{3/2} = m_e^3 e^6/\hbar^6$ and using $a_0^4 = \hbar^8/(m^4 e^8)$, we can write (10.259) as

$$\Gamma_0 = \frac{256 \mathcal{E}_0^2 a_0^3}{3\hbar} \left(\frac{\omega_0}{\omega}\right)^6 \left(\frac{\omega}{\omega_0} - 1\right)^{3/2}. \tag{10.260}$$

If the frequency of the oscillating electric field is smaller than or equal to ω_0, the atom will not be ionized; at $\omega = \omega_0$ the probability of ionization will be zero.

(b) The maximum transition rate is obtained by taking the derivative of (10.260):

$$\frac{d\Gamma_0}{d\omega} = 0 \implies \frac{2}{\omega}\left(\frac{\omega}{\omega_0} - 1\right) = \frac{1}{2\omega_0} \implies \omega = \frac{4}{3}\omega_0. \tag{10.261}$$

Inserting $\omega = \frac{4}{3}\omega_0$ into (10.260) we obtain the maximum transition rate

$$\Gamma_{0max} = \frac{256 \mathcal{E}_0^2 a_0^3}{3\hbar}\left(\frac{3}{4}\right)^6 \left(\frac{4}{3} - 1\right)^{3/2} = \frac{\mathcal{E}_0^2 a_0^3}{\hbar}\frac{3^{7/2}}{2^4}. \tag{10.262}$$

10.7 Exercises

Exercise 10.1
Consider a spinless particle of mass m in a one-dimensional infinite potential well with walls at $x = 0$ and $x = a$ which is initially (i.e., at $t = 0$) in the state $\psi(x, 0) = [\phi_1(x) + \phi_3(x)]/\sqrt{2}$, where $\phi_1(x)$ and $\phi_3(x)$ are the ground and second excited states, respectively, with $\phi_n(x) = \sqrt{2/a}\sin(n\pi x/a)$.
   (a) What is the state vector $\psi(x, t)$ for $t > 0$ in the Schrödinger picture.
   (b) Evaluate $\langle \hat{X} \rangle$, $\langle \hat{P} \rangle$, $\langle \hat{X}^2 \rangle$, and $\langle \hat{P}^2 \rangle$ as functions of time for $t > 0$ in the Schrödinger picture.
   (c) Repeat part (b) in the Heisenberg picture: i.e., evaluate $\langle \hat{X} \rangle_H$, $\langle \hat{P} \rangle_H$, $\langle \hat{X}^2 \rangle_H$, and $\langle \hat{P}^2 \rangle_H$ as functions of time for $t > 0$.

Exercise 10.2
Evaluate the expectation value $\langle \hat{X}_H(t)\hat{P} \rangle_3$ for the third excited state of a one-dimensional harmonic oscillator.

Exercise 10.3
Evaluate the expectation value $\langle \hat{X}\hat{P}_H(t) \rangle_n$ for the nth excited state of a one-dimensional harmonic oscillator.

Exercise 10.4

Consider a one-dimensional harmonic oscillator which is initially (i.e., at $t = 0$) in the state $| \psi(0) \rangle = (| 0 \rangle + | 1 \rangle)/\sqrt{2}$, where $| 0 \rangle$ and $| 1 \rangle$ are the ground and first excited states, respectively.

(a) What is the state vector $| \psi(t) \rangle$ for $t > 0$ in the Schrödinger picture?

(b) Evaluate $\langle \hat{X} \rangle$, $\langle \hat{P} \rangle$, $\langle \hat{X}^2 \rangle$, and $\langle \hat{P}^2 \rangle$ as functions of time for $t > 0$ in the Schrödinger picture.

(c) Repeat part (b) in the Heisenberg picture.

Exercise 10.5

(a) Calculate the coordinate operator $\hat{X}_H(t)$ for a free particle in one dimension in the Heisenberg picture.

(b) Evaluate the commutator $[\hat{X}_H(t), \hat{X}_H(0)]$.

Exercise 10.6

Consider the Hamiltonian $H = -(eB/mc)\hat{S}_x = \omega \hat{S}_x$.

(a) Write down the Heisenberg equations of motion for the time-dependent operators $\hat{S}_x(t)$, $\hat{S}_y(t)$, and $\hat{S}_z(t)$.

(b) Solve these equations to obtain S_x, S_y, S_z as functions of time.

Exercise 10.7

Evaluate the quantity $\langle n \mid \hat{P}_H(t) \hat{P} \mid n \rangle$ for the nth excited state of a one-dimensional harmonic oscillator, where $\hat{P}_H(t)$ and $\hat{P}$ designate the momentum operators in the Heisenberg picture and the Schrödinger picture, respectively.

Exercise 10.8

The Hamiltonian due to the interaction of a particle of mass m, charge q (the charge is negative), and spin $\vec{S}$ with a magnetic field pointing along the y-axis is $\hat{H} = -(qB/mc)\hat{S}_y$.

(a) Use the Heisenberg equation to calculate $d\hat{S}_x/dt$, $d\hat{S}_y/dt$, and $d\hat{S}_z/dt$.

(b) Solve these equations to obtain the components of the spin operator as functions of time.

Exercise 10.9

A particle is initially (i.e., when $t < 0$) in its ground state in a one-dimensional harmonic oscillator potential. At $t = 0$ a perturbation $\hat{V}(x, t) = V_0 \hat{x}^2 e^{-t/\tau}$ is turned on. Calculate to first order the probability that, after a sufficiently long time (i.e., $t \rightarrow \infty$), the system will have made a transition to a given excited state; consider all final states.

Exercise 10.10

A particle, initially (i.e., when $t < 0$) in its ground state in an infinite potential well whose walls are located at $x = 0$ and $x = a$, is subject, starting at time $t = 0$, to a time-dependent perturbation $\hat{V}(t) = V_0 \hat{x}^2 e^{-t^2}$ where V_0 is a small parameter. Calculate the probability that the particle will be found in its second excited state at $t = +\infty$.

Exercise 10.11

Find the intensity associated with the transition 3s → 2p in the hydrogen atom.

Exercise 10.12

A hydrogen atom in its ground state is placed in a region where, at $t = 0$, a time-dependent electric field is turned on:

$$\vec{E}(t) = E_0(\vec{i} + \vec{j} + \vec{k})e^{-t/\tau},$$

where τ is a positive real number. Using first-order time-dependent perturbation theory, calculate the probability that, after a sufficiently long time (i.e., $t \gg \tau$), the atom is to be found in each of the $n = 2$ states (i.e., consider the transitions to all the states in the $n = 2$ level). *Hint:* You may use: $\int_0^\infty r^3 R_{21}^*(r) R_{10}(r) \, dr = (24a_0/\sqrt{6}) \left(\frac{2}{3}\right)^5$.

Exercise 10.13

(a) Calculate the reduced matrix element $\langle 1 \parallel Y_1 \parallel 2 \rangle$. *Hint:* For this, you may need to calculate $\langle 1, 0|Y_{10}|2, 0\rangle$ directly and then from the Wigner–Eckart theorem.

(b) Using the Wigner–Eckart theorem and the relevant Clebsch–Gordan coefficients from tables, calculate $\langle 1, m|Y_{1m'}|2, m''\rangle$ for all possible values of m, m', and m''.

(c) Using the results of part (b), calculate the 3d $\rightarrow$ 2p transition rate for the hydrogen atom in the dipole approximation; give a numerical value. *Hint:* You may use the integral $\int_0^\infty r^3 R_{21}^*(r) R_{32}(r) \, dr = (64a_0/15\sqrt{5}) \left(\frac{6}{5}\right)^5$ and the following Clebsch–Gordan coefficients:

$$\langle j, 1; m, 0|(j-1), m\rangle = -\sqrt{(j-m)(j+m)/[j(2j+1)]},$$
$$\langle j, 1; (m-1), 1|(j-1), m\rangle = \sqrt{(j-m)(j-m+1)/[2j(2j+1)]}, \text{ and}$$
$$\langle j, 1; (m+1), -1|(j-1), m\rangle = \sqrt{(j+m)(j+m+1)/[2j(2j+1)]}.$$

Exercise 10.14

A particle is initially in its ground state in an infinite one-dimensional potential box with sides at $x = 0$ and $x = a$. If the wall of the box at $x = a$ is *suddenly* moved to $x = 10a$, calculate the probability of finding the particle in

(a) the fourth excited ($n = 5$) state of the new box and

(b) the ninth ($n = 10$) excited state of the new box.

Exercise 10.15

A particle of mass m in the ground state of a one-dimensional harmonic oscillator is placed in a perturbation $\hat{V}(t) = -V_0\hat{x}e^{-t/\tau}$. Calculate to first-order perturbation theory the probability of finding the particle in its first excited state after a long time.

Exercise 10.16

A particle, initially (i.e., when $t < 0$) in its first excited state in an infinite potential well whose walls are located at $x = 0$ and $x = a$, is subject, starting at time $t = 0$, to a time-dependent perturbation $\hat{V}(t) = V_0\tau\hat{x}/(t^2 + \tau^2)$ where V_0 is a small real number. Calculate the probability that the particle will be found in its second excited state at $t = +\infty$.

Exercise 10.17

A one-dimensional harmonic oscillator has its spring constant suddenly reduced by half.

(a) If the oscillator is initially in its ground state, find the probability that the oscillator remains in the ground state.

(b) Find the work associated with this process.

Exercise 10.18

(a) Find the total transition rate associated with the decay of a harmonic oscillator, of charge q and mass m, from the nth excited state to the state just below.

(b) Find the power radiated by this oscillator as a result of its decay.

(c) Find the lifetime of the nth excited state.

(d) Estimate the order of magnitudes for the transition rate, the power, and the lifetime of the fifth excited state ($n = 5$) in the case of a harmonically oscillating electron (i.e., $q = e$) for the case of an optical radiation $\omega \simeq 10^{15}$ rad s^{-1}.

Exercise 10.19

Assuming that $\langle \psi_f \mid \vec{r} \mid \psi_i \rangle$ is roughly equal to the size of the system under study, use a crude calculation to estimate the mean lifetime of

(a) an electric dipole transition in an atom where $\hbar\omega \sim 10$ eV and

(b) an electric dipole transition in a nucleus where $\hbar\omega \sim 1$ MeV.

Exercise 10.20

A particle is initially (i.e., when $t < 0$) in its ground state in the potential $V(x) = -V_0\delta(x)$ with $V_0 > 0$.

(a) If the strength of the potential is changed *slowly* to $3V_0$, find the energy and wave function of the particle in the new potential.

(b) Calculate the work done with this process. Find a numerical value for this work in MeV if this particle were an electron and $V_0 = 200$ MeV fm.

(c) If the strength of the potential is changed *suddenly* to $3V_0$, calculate the probability of finding the particle in the ground state of the new potential.

Exercise 10.21

A hydrogen atom in its ground state is placed at time $t = 0$ in a uniform electric field in the y-direction, $\vec{E}(t) = E_0\vec{j}e^{-t^2/\tau^2}$. Calculate to first-order perturbation theory the probability that the atom will be found in any of the $n = 2$ states after a sufficiently long time ($t = +\infty$).

Exercise 10.22

A particle, initially (i.e., when $t < 0$) in its ground state in an infinite potential well with its walls at $x = 0$ and $x = a$, is subject, starting at time $t = 0$, to a time-dependent perturbation $\hat{V}(t) = V_0\hat{x}\delta(x - 3a/4)e^{-t/\tau}$ where V_0 is a small parameter. Calculate the probability that the particle will be found in its first excited state ($n = 2$) at $t = +\infty$.

Exercise 10.23

Consider an isotropic (three-dimensional) harmonic oscillator which undergoes a transition from the second to the first excited state (i.e., 2s → 1p).

(a) Calculate the transition rate corresponding to 2s → 1p.

(b) Find the intensity associate with the 2s → 1p transition.

Exercise 10.24

Consider a particle which is initially (i.e., when $t < 0$) in its ground state in a three-dimensional box potential

$$V(x, y, z) = \begin{cases} 0, & 0 < x < a, \ 0 < y < 2a, \ 0 < z < 4a, \\ +\infty, & \text{elsewhere.} \end{cases}$$

(a) Find the energies and wave functions of the ground state and first excited state.

(b) This particle is then subject, starting at time $t = 0$, to a time-dependent perturbation $\hat{V}(t) = V_0\hat{x}\hat{z}e^{-t^2}$ where V_0 is a small parameter. Calculate the probability that the particle will be found in the first excited state after a long time $t = +\infty$.

Chapter 11

Scattering Theory

Much of our understanding about the structure of matter is extracted from the scattering of particles. Had it not been for scattering, the structure of the microphysical world would have remained inaccessible to humans. It is through scattering experiments that important building blocks of matter, such as the atomic nucleus, the nucleons, and the various quarks, have been discovered.

11.1 Scattering and Cross Section

In a scattering experiment, one observes the collisions between a beam of incident particles and a target material. The total number of collisions over the duration of the experiment is proportional to the total number of incident particles and to the number of target particles per unit area in the path of the beam. In these experiments, one counts the collision products that come out of the target. After scattering, those particles that do not interact with the target continue their motion (undisturbed) in the forward direction, but those that interact with the target get scattered (deflected) at some angle as depicted in Figure 11.1. The number of particles coming out varies from one direction to the other. The number of particles scattered into an element of solid angle $d\Omega$ ($d\Omega = \sin\theta \, d\theta \, d\varphi$) is proportional to a quantity that plays a central role in the physics of scattering: the *differential cross section*. The differential cross section, which is denoted by $d\sigma(\theta, \varphi)/d\Omega$, is defined as the number of particles scattered into an element of solid angle $d\Omega$ in the direction (θ, φ) per unit time and incident flux:

$$\frac{d\sigma(\theta, \varphi)}{d\Omega} = \frac{1}{J_{inc}} \frac{dN(\theta, \varphi)}{d\Omega}, \tag{11.1}$$

where J_{inc} is the incident flux (or incident current density); it is equal to the number of incident particles per area per unit time. We can verify that $d\sigma/d\Omega$ has the dimensions of an area; hence it is appropriate to call it a differential cross section.

The relationship between $d\sigma/d\Omega$ and the total cross section σ is obvious:

$$\sigma = \int \frac{d\sigma}{d\Omega} d\Omega = \int_0^\pi \sin\theta \, d\theta \int_0^{2\pi} \frac{d\sigma(\theta, \varphi)}{d\Omega} d\varphi. \tag{11.2}$$

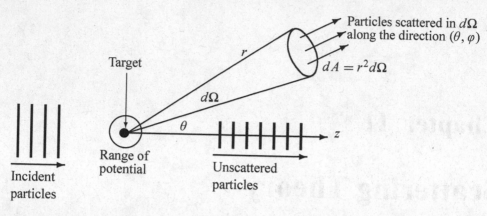

Figure 11.1 Scattering between an incident beam of particles and a fixed target: the scattered particles are detected within a solid angle $d\Omega$ along the direction (θ, φ).

Most scattering experiments are carried out in the laboratory (Lab) frame in which the target is initially at rest while the projectiles are moving. Calculations of the cross sections are generally easier to perform within the center of mass (CM) frame in which the center of mass of the projectiles–target system is at rest (before and after collision). In order to be able to compare the experimental measurements with the theoretical calculations, one has to know how to transform the cross sections from one frame into the other. We should note that the total cross section σ is the same in both frames, since the total number of collisions that take place does not depend on the frame in which the observation is carried out. As for the differential cross sections $d\sigma(\theta, \varphi)/d\Omega$, they are not the same in both frames, since the scattering angles (θ, φ) are frame dependent.

11.1.1 Connecting the Angles in the Lab and CM frames

To find the connection between the Lab and CM cross sections, we need first to find how the scattering angles in one frame are related to their counterparts in the other. Let us consider the scattering of two (structureless, nonrelativistic) particles of masses m_1 and m_2; m_2 represents the target, which is initially at rest, and m_1 the projectile. Figure 11.2 depicts such a scattering in the Lab and CM frames, where θ_1 and θ are the scattering angles of m_1 in the Lab and CM frames, respectively; we are interested in detecting m_1. In what follows we want to find the relation between θ_1 and θ. If $\vec{r}_{1L}$ and $\vec{r}_{1C}$ denote the position of m_1 in the Lab and CM frames, respectively, and if $\vec{R}$ denotes the position of the center of mass with respect to the Lab frame, we have $\vec{r}_{1L} = \vec{r}_{1C} + \vec{R}$. A time derivative of this relation leads to

$$\vec{V}_{1L} = \vec{V}_{1C} + \vec{V}_{CM}, \tag{11.3}$$

where $\vec{V}_{1L}$ and $\vec{V}_{1C}$ are the velocities of m_1 in the Lab and CM frames *before* collision and $\vec{V}_{CM}$ is the velocity of the CM with respect to the Lab frame. Similarly, the velocity of m_1 *after* collision is

$$\vec{V}'_{1L} = \vec{V}'_{1C} + \vec{V}_{CM}. \tag{11.4}$$

From Figure 11.2a we can infer the x and y components of (11.4):

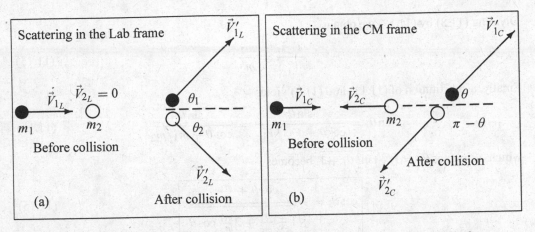

Figure 11.2 Elastic scattering of two structureless particles in the Lab and CM frames: a particle of mass m_1 strikes a particle m_2 initially at rest.

$$V'_{1_L} \cos \theta_1 = V'_{1_C} \cos \theta + V_{CM}, \tag{11.5}$$

$$V'_{1_L} \sin \theta_1 = V'_{1_C} \sin \theta. \tag{11.6}$$

Dividing (11.6) by (11.5), we end up with

$$\tan \theta_1 = \frac{\sin \theta}{\cos \theta + V_{CM}/V'_{1_C}}, \tag{11.7}$$

where V_{CM}/V'_{1_C} can be shown to be equal to m_1/m_2. To see this, since $\vec{V}_{2_L} = 0$, we have

$$\vec{V}_{CM} = \frac{m_1 \vec{V}_{1_L} + m_2 \vec{V}_{2_L}}{m_1 + m_2} = \frac{m_1}{m_1 + m_2} \vec{V}_{1_L}, \tag{11.8}$$

which when inserted into (11.3) leads to $\vec{V}_{1_L} = \vec{V}_{1_C} + m_1 \vec{V}_{1_L}/(m_1 + m_2)$; hence

$$\vec{V}_{1_C} = \left(1 - \frac{m_1}{m_1 + m_2}\right) \vec{V}_{1_L} = \frac{m_2}{m_1 + m_2} \vec{V}_{1_L}. \tag{11.9}$$

On the other hand, since the center of mass is at rest in the CM frame, the total momenta before and after collisions are separately zero:

$$p_C = m_1 V_{1_C} - m_2 V_{2_C} = 0 \quad \Longrightarrow \quad V_{2_C} = \frac{m_1}{m_2} V_{1_C}, \tag{11.10}$$

$$p'_{C_x} = m_1 V'_{1_C} \cos \theta - m_2 V'_{2_C} \cos \theta = 0 \quad \Longrightarrow \quad V'_{2_C} = \frac{m_1}{m_2} V'_{1_C}. \tag{11.11}$$

In the case of *elastic* collision, the speeds of the particles in the CM frame are the same before and after collision; to see this, since the kinetic energy is conserved, a substitution of (11.10) and (11.11) into $\frac{1}{2}m_1 V_{1_C}^2 + \frac{1}{2}m_2 V_{2_C}^2 = \frac{1}{2}m_1 {V'_{1_C}}^2 + \frac{1}{2}m_2 {V'_{2_C}}^2$ yields $V'_{1_C} = V_{1_C}$ and $V'_{2_C} = V_{2_C}$. Thus, we can rewrite (11.9) as

$$\vec{V}'_{1_C} = \vec{V}_{1_C} = \frac{m_2}{m_1 + m_2} \vec{V}_{1_L}. \tag{11.12}$$

Dividing (11.8) by (11.12) we obtain

$$\frac{V_{CM}}{V'_{1C}} = \frac{m_1}{m_2}.$$ (11.13)

Finally, a substitution of (11.13) into (11.7) yields

$$\tan\theta_1 = \frac{\sin\theta}{\cos\theta + V_{2C}/V_{1C}} = \frac{\sin\theta}{\cos\theta + m_1/m_2},$$ (11.14)

which, using $\cos\theta_1 = 1/\sqrt{\tan^2\theta_1 + 1}$, becomes

$$\boxed{\cos\theta_1 = \frac{\cos\theta + \frac{m_1}{m_2}}{\sqrt{1 + \frac{m_1^2}{m_2^2} + 2\frac{m_1}{m_2}\cos\theta}}.}$$ (11.15)

Remark

By analogy with the foregoing analysis, we can establish a connection between θ_2 and θ. From (11.4) we have $\vec{V}'_{2L} = \vec{V}'_{2C} + \vec{V}_{CM}$. The x and y components of this relation are

$$V'_{2L}\cos\theta_2 = -V'_{2C}\cos\theta + V_{CM} = (-\cos\theta + 1)V'_{2C},$$ (11.16)

$$V'_{2L}\sin\theta_2 = -V'_{2C}\sin\theta;$$ (11.17)

in deriving (11.16), we have used $V_{CM} = V'_{2C} = V_{2C}$. A combination of (11.16) and (11.17) leads to

$$\boxed{\tan\theta_2 = \frac{\sin\theta}{-\cos\theta + V_{CM}/V'_{2C}} = \frac{\sin\theta}{1 - \cos\theta} = \cot\left(\frac{\theta}{2}\right) \implies \theta_2 = \frac{\pi - \theta}{2}.}$$ (11.18)

11.1.2   Connecting the Lab and CM Cross Sections

The connection between the differential cross sections in the Lab and CM frames can be obtained from the fact that the number of scattered particles passing through an infinitesimal cross section $d\sigma$ is the same in both frames: $d\sigma(\theta_1, \varphi_1) = d\sigma(\theta, \varphi)$. What differs is the solid angle $d\Omega$, since it is given in the Lab frame by $d\Omega_1 = \sin\theta_1 d\theta_1 d\varphi_1$ and in the CM frame by $d\Omega = \sin\theta \, d\theta d\varphi$. Thus, we have

$$\left(\frac{d\sigma}{d\Omega_1}\right)_{Lab} d\Omega_1 = \left(\frac{d\sigma}{d\Omega}\right)_{CM} d\Omega \implies \left(\frac{d\sigma}{d\Omega_1}\right)_{Lab} = \left(\frac{d\sigma}{d\Omega}\right)_{CM} \frac{\sin\theta}{\sin\theta_1} \frac{d\theta}{d\theta_1} \frac{d\varphi}{d\varphi_1},$$ (11.19)

where (θ_1, φ_1) are the scattering angles of particle m_1 in the Lab frame and (θ, φ) are its angles in the CM frame. Since there is cylindrical symmetry around the direction of the incident beam, we have $\varphi = \varphi_1$ and hence

$$\left(\frac{d\sigma}{d\Omega_1}\right)_{Lab} = \left(\frac{d\sigma}{d\Omega}\right)_{CM} \frac{d(\cos\theta)}{d(\cos\theta_1)}.$$ (11.20)

From (11.15) we have

$$\frac{d\cos\theta_1}{d\cos\theta} = \frac{1 + \frac{m_1}{m_2}\cos\theta}{\left(1 + \frac{m_1^2}{m_2^2} + 2\frac{m_1}{m_2}\cos\theta\right)^{3/2}},$$ (11.21)

which when substituted into (11.20) leads to

$$\boxed{\left(\frac{d\sigma}{d\Omega_1}\right)_{Lab} = \frac{(1 + \frac{m_1^2}{m_2^2} + 2\frac{m_1}{m_2}\cos\theta)^{3/2}}{1 + \frac{m_1}{m_2}\cos\theta}\left(\frac{d\sigma}{d\Omega}\right)_{CM}.} \qquad (11.22)$$

Similarly, we can show that (11.20) and (11.18) yield

$$\boxed{\left(\frac{d\sigma}{d\Omega_2}\right)_{Lab} = 4\cos\theta_2\left(\frac{d\sigma}{d\Omega_2}\right)_{CM} = 4\sin\left(\frac{\theta}{2}\right)\left(\frac{d\sigma}{d\Omega_2}\right)_{CM}.} \qquad (11.23)$$

Limiting cases: (a) If $m_2 \gg m_1$, or when $\frac{m_1}{m_2} \to 0$, the Lab and CM results are the same, since (11.15) leads to $\theta_1 = \theta$ and (11.22) to $\left(\frac{d\sigma}{d\Omega_1}\right)_{Lab} = \left(\frac{d\sigma}{d\Omega}\right)_{CM}$. (b) If $m_2 = m_1$ then (11.15) leads to $\tan\theta_1 = \tan(\theta/2)$ or to $\theta_1 = \theta/2$; in this case (11.22) yields $\left(\frac{d\sigma}{d\Omega_1}\right)_{Lab} = 4\left(\frac{d\sigma}{d\Omega}\right)_{CM}\cos(\theta/2)$.

Example 11.1

In an elastic collision between two particles of equal mass, show that the two particles come out at right angles with respect to each other in the Lab frame.

Solution

In the special case $m_1 = m_2$, equations (11.14) and (11.18) respectively become

$$\tan\theta_1 = \tan\left(\frac{\theta}{2}\right), \qquad \tan\theta_2 = \cot\left(\frac{\theta}{2}\right) = \tan\left(\frac{\pi}{2} - \frac{\theta}{2}\right). \qquad (11.24)$$

These two equations yield

$$\theta_1 = \frac{\theta}{2}, \qquad \theta_2 = \frac{\pi}{2} - \frac{\theta}{2} = \frac{\pi}{2} - \theta_1; \qquad (11.25)$$

hence $\theta_1 + \theta_2 = \pi/2$. In these cases, (11.22) and (11.23) yield

$$\left(\frac{d\sigma}{d\Omega_1}\right)_{Lab} = 4\left(\frac{d\sigma}{d\Omega}\right)_{CM}\cos\theta_1 = 4\left(\frac{d\sigma}{d\Omega}\right)_{CM}\cos\left(\frac{\theta}{2}\right), \qquad (11.26)$$

$$\left(\frac{d\sigma}{d\Omega_2}\right)_{Lab} = 4\left(\frac{d\sigma}{d\Omega_2}\right)_{CM}\cos\theta_2 = 4\left(\frac{d\sigma}{d\Omega}\right)_{CM}\sin\left(\frac{\theta}{2}\right). \qquad (11.27)$$

11.2 Scattering Amplitude of Spinless Particles

The foregoing discussion dealt with definitions of the cross section and how to transform it from the Lab to the CM frame; the conclusions reached apply to classical as well as to quantum mechanics. In this section we deal with the quantum description of scattering. For simplicity, we consider the case of elastic[1] scattering between two spinless, nonrelativistic particles of masses m_1 and m_2. During the scattering process, the particles interact with one another. If the interaction is *time independent*, we can describe the two-particle system with stationary states

$$\Psi(\vec{r}_1, \vec{r}_2, t) = \psi(\vec{r}_1, \vec{r}_2)e^{-iE_T t/\hbar}, \tag{11.28}$$

where E_T is the total energy and $\psi(\vec{r}_1, \vec{r}_2,)$ is a solution of the time-independent Schrödinger equation:

$$\left[-\frac{\hbar^2}{2m_1}\vec{\nabla}_1^2 - \frac{\hbar^2}{2m_2}\vec{\nabla}_2^2 + \hat{V}(\vec{r}_1, \vec{r}_2) \right]\psi(\vec{r}_1, \vec{r}_2) = E_T\psi(\vec{r}_1, \vec{r}_2); \tag{11.29}$$

$\hat{V}(\vec{r}_1, \vec{r}_2)$ is the potential representing the interaction between the two particles.

In the case where the interaction between m_1 and m_2 depends only on their relative distance $r = |\vec{r}_1 - \vec{r}_2|$ (i.e., $\hat{V}(\vec{r}_1, \vec{r}_2) = \hat{V}(r)$), we can, as seen in Chapter 6, reduce the eigenvalue problem (11.29) to two decoupled eigenvalue problems: one for the center of mass (CM), which moves like a free particle of mass $M = m_1 + m_2$ and which is of no concern to us here, and another for a fictitious particle with a reduced mass $\mu = m_1 m_2/(m_1 + m_2)$ which moves in the potential $\hat{V}(r)$:

$$-\frac{\hbar^2}{2\mu}\vec{\nabla}^2\psi(\vec{r}) + \hat{V}(r)\psi(\vec{r}) = E\psi(\vec{r}). \tag{11.30}$$

The problem of scattering between two particles is thus reduced to solving this equation. We are going to show that the differential cross section in the CM frame can be obtained from an asymptotic form of the solution of (11.30). Its solutions can then be used to calculate the probability per unit solid angle per unit time that the particle μ is scattered into a solid angle $d\Omega$ in the direction (θ, φ); this probability is given by the differential cross section $d\sigma/d\Omega$. In quantum mechanics the incident particle is described by means of a wave packet that interacts with the target. The incident wave packet must be spatially large so that spreading during the experiment is not appreciable. It must be large compared to the target's size and yet small compared to the size of the Lab so that it does not overlap simultaneously with the target and detector. After scattering, the wave function consists of an unscattered part propagating in the forward direction and a scattered part that propagates along some direction (θ, φ).

We can view (11.30) as representing the scattering of a particle of mass μ from a fixed scattering center that is described by $V(r)$, where r is the distance from the particle μ to the center of $V(r)$. We assume that $V(r)$ has a *finite* range a. Thus the interaction between the particle and the potential occurs only in a limited region of space $r \le a$, which is called the *range* of $V(r)$, or the scattering region. Outside the range, $r > a$, the potential vanishes, $V(r) = 0$; the eigenvalue problem (11.30) then becomes

$$\left(\nabla^2 + k_0^2\right)\phi_{inc}(\vec{r}) = 0, \tag{11.31}$$

where $k_0^2 = 2\mu E/\hbar^2$. In this case μ behaves as a *free* particle before collision and hence can be described by a *plane* wave

$$\phi_{inc}(\vec{r}) = Ae^{i\vec{k}_0 \cdot \vec{r}}, \tag{11.32}$$

where $\vec{k}_0$ is the wave vector associated with the incident particle and A is a normalization factor. Thus, prior to the interaction with the target, the particles of the incident beam are independent of each other; they move like free particles, each with a momentum $\vec{p} = \hbar\vec{k}_0$.

---

[1] In *elastic* scattering, the internal states and the structure of the colliding particles do not change.

Figure 11.3 (a) Angle between the incident and scattered wave vectors $\vec{k}_0$ and $\vec{k}$. (b) Incident and scattered waves: the incident wave is a plane wave, $\phi_{inc}(\vec{r}) = Ae^{i\vec{k}_0 \cdot \vec{r}}$, and the scattered wave, $\phi_{sc}(\vec{r}) = Af(\theta, \varphi)\frac{e^{i\vec{k} \cdot \vec{r}}}{r}$, is an outgoing wave.

When the incident wave (11.32) collides or interacts with the target, an outgoing wave $\phi_{sc}(\vec{r})$ is scattered out. In the case of an *isotropic* scattering, the scattered wave is *spherically symmetric*, having the form $e^{i\vec{k} \cdot \vec{r}}/r$. In general, however, the scattered wave is not spherically symmetric; its amplitude depends on the direction (θ, φ) along which it is detected and hence

$$\phi_{sc}(\vec{r}) = Af(\theta, \varphi)\frac{e^{i\vec{k} \cdot \vec{r}}}{r}, \tag{11.33}$$

where $f(\theta, \varphi)$ is called the *scattering amplitude*, $\vec{k}$ is the wave vector associated with the scattered particle, and θ is the angle between $\vec{k}_0$ and $\vec{k}$ as displayed in Figure 11.3a. After the scattering has taken place (Figure 11.3b), the total wave consists of a superposition of the incident plane wave (11.32) and the scattered wave (11.33):

$$\psi(\vec{r}) = \phi_{inc}(\vec{r}) + \phi_{sc}(\vec{r}) \simeq A\left[e^{i\vec{k}_0 \cdot \vec{r}} + f(\theta, \varphi)\frac{e^{i\vec{k} \cdot \vec{r}}}{r}\right], \tag{11.34}$$

where A is a normalization factor; since A has no effect on the cross section, as will be shown in (11.40), we will take it equal to one throughout the rest of the chapter. We now need to determine $f(\theta, \varphi)$ and $d\sigma/d\Omega$. In the following section we are going to show that the differential cross section is given in terms of the scattering amplitude by $d\sigma/d\Omega = |f(\theta, \varphi)|^2$.

11.2.1 Scattering Amplitude and Differential Cross Section

The scattering amplitude $f(\theta, \varphi)$ plays a central role in the theory of scattering, since it determines the differential cross section. To see this, let us first introduce the incident and scattered flux densities:

$$\vec{J}_{inc} = \frac{i\hbar}{2\mu}(\phi_{inc}\vec{\nabla}\phi_{inc}^* - \phi_{inc}^*\vec{\nabla}\phi_{inc}), \tag{11.35}$$

$$\vec{J}_{sc} = \frac{i\hbar}{2\mu}(\phi_{sc}\vec{\nabla}\phi_{sc}^* - \phi_{sc}^*\vec{\nabla}\phi_{sc}). \tag{11.36}$$

Inserting (11.32) into (11.35) and (11.33) into (11.36) and taking the magnitudes of the expressions thus obtained, we end up with

$$J_{inc} = |A|^2\frac{\hbar k_0}{\mu}, \qquad J_{sc} = |A|^2\frac{\hbar k}{\mu r^2}\left|f(\theta,\varphi)\right|^2. \tag{11.37}$$

Now, we may recall that the number $dN(\theta,\varphi)$ of particles scattered into an element of solid angle $d\Omega$ in the direction (θ,φ) and passing through a surface element $dA = r^2 d\Omega$ per unit time is given as follows (see (11.1)):

$$dN(\theta,\varphi) = J_{sc}r^2 d\Omega. \tag{11.38}$$

When combined with (11.37) this relation yields

$$\frac{dN}{d\Omega} = J_{sc}r^2 = |A|^2\frac{\hbar k}{\mu}\left|f(\theta,\varphi)\right|^2. \tag{11.39}$$

Now, inserting (11.39) and $J_{inc} = |A|^2\hbar k_0/\mu$ into (11.1), we end up with

$$\boxed{\frac{d\sigma}{d\Omega} = \frac{1}{J_{inc}}\frac{dN}{d\Omega} = \frac{k}{k_0}\left|f(\theta,\varphi)\right|^2.} \tag{11.40}$$

Since the normalization factor A does not contribute to the differential cross section, we will be taking it equal to one. For elastic scattering k_0 is equal to k; hence (11.40) reduces to

$$\boxed{\frac{d\sigma}{d\Omega} = \left|f(\theta,\varphi)\right|^2.} \tag{11.41}$$

The problem of determining the differential cross section $d\sigma/d\Omega$ therefore reduces to that of obtaining the scattering amplitude $f(\theta,\varphi)$.

11.2.2 Scattering Amplitude

We are going to show here that we can obtain the differential cross section in the CM frame from an asymptotic form of the solution of the Schrödinger equation (11.30). Let us first focus on the determination of $f(\theta,\varphi)$; it can be obtained from the solutions of (11.30), which in turn can be rewritten as

$$(\nabla^2 + k^2)\psi(\vec{r}) = \frac{2\mu}{\hbar^2}V(\vec{r})\psi(\vec{r}). \tag{11.42}$$

The general solution to this equation consists of a sum of two components: a general solution to the homogeneous equation:

$$(\nabla^2 + k_0^2)\psi_{homo}(\vec{r}) = 0, \tag{11.43}$$

and a particular solution to (11.42). First, note that $\psi_{homo}(\vec{r})$ is nothing but the incident plane wave (11.32). As for the particular solution to (11.42), we can express it in terms of *Green's function*. Thus, the general solution of (11.42) is given by

$$\psi(\vec{r}) = \phi_{inc}(\vec{r}) + \frac{2\mu}{\hbar^2} \int G(\vec{r} - \vec{r}\,')V(\vec{r}\,')\psi(\vec{r}\,')\,d^3r', \tag{11.44}$$

where $\phi_{inc}(\vec{r}) = e^{i\vec{k}_0\cdot\vec{r}}$ and $G(\vec{r} - \vec{r}\,')$ is Green's function corresponding to the operator on the left-hand side of (11.43). The function $G(\vec{r} - \vec{r}\,')$ is obtained by solving the point source equation

$$(\nabla^2 + k^2)G(\vec{r} - \vec{r}\,') = \delta(\vec{r} - \vec{r}\,'), \tag{11.45}$$

where $G(\vec{r} - \vec{r}\,')$ and $\delta(\vec{r} - \vec{r}\,')$ are given by their Fourier transforms as follows:

$$G(\vec{r} - \vec{r}\,') = \frac{1}{(2\pi)^3} \int e^{i\vec{q}\cdot(\vec{r}-\vec{r}\,')}\tilde{G}(\vec{q})\,d^3q, \tag{11.46}$$

$$\delta(\vec{r} - \vec{r}\,') = \frac{1}{(2\pi)^3} \int e^{i\vec{q}\cdot(\vec{r}-\vec{r}\,')}\,d^3q. \tag{11.47}$$

A substitution of (11.46) and (11.47) into (11.45) leads to

$$\left(-\vec{q}^{\,2} + \vec{k}^{\,2}\right)\tilde{G}(\vec{q}) = 1 \quad \Longrightarrow \quad \tilde{G}(\vec{q}) = \frac{1}{k^2 - q^2}. \tag{11.48}$$

The expression for $G(\vec{r} - \vec{r}\,')$ is obtained by inserting (11.48) into (11.46):

$$G(\vec{r} - \vec{r}\,') = \frac{1}{(2\pi)^3} \int \frac{e^{i\vec{q}\cdot(\vec{r}-\vec{r}\,')}}{k^2 - q^2}d^3q. \tag{11.49}$$

To integrate over the angles in

$$G(\vec{r} - \vec{r}\,') = \frac{1}{(2\pi)^3} \int_0^\infty \frac{q^2\,dq}{k^2 - q^2} \int_0^\pi e^{iq|\vec{r}-\vec{r}\,'|\cos\theta}\sin\theta\,d\theta \int_0^{2\pi} d\varphi, \tag{11.50}$$

we need simply to make the variables change $x = \cos\theta$:

$$\int_0^\pi e^{iq|\vec{r}-\vec{r}\,'|\cos\theta}\sin\theta\,d\theta = \int_{-1}^1 e^{iq|\vec{r}-\vec{r}\,'|x}\,dx = \frac{1}{iq|\vec{r}-\vec{r}\,'|}\left(e^{iq|\vec{r}-\vec{r}\,'|} - e^{-iq|\vec{r}-\vec{r}\,'|}\right). \tag{11.51}$$

Hence (11.50) becomes

$$G(\vec{r} - \vec{r}\,') = \frac{1}{4\pi^2 i|\vec{r}-\vec{r}\,'|} \int_0^\infty \frac{q}{k^2 - q^2}\left(e^{iq|\vec{r}-\vec{r}\,'|} - e^{-iq|\vec{r}-\vec{r}\,'|}\right)dq, \tag{11.52}$$

or

$$G(\vec{r} - \vec{r}\,') = -\frac{1}{4\pi^2 i|\vec{r}-\vec{r}\,'|} \int_{-\infty}^{+\infty} \frac{q e^{iq|\vec{r}-\vec{r}\,'|}}{q^2 - k^2}dq. \tag{11.53}$$

We may evaluate this integral by the method of residues by closing the contour in the upper half of the q-plane: it is equal to $2\pi i$ times the residue of the integrand at the poles. Since there

Figure 11.4 Contours corresponding to outgoing and incoming waves.

are two poles, $q = \pm k$, the integral has two possible values. The value corresponding to the pole at $q = k$, which lies inside the contour of integration in Figure 11.4a, is given by

$$G_+(\vec{r} - \vec{r}\,') = -\frac{1}{4\pi} \frac{e^{ik|\vec{r}-\vec{r}\,'|}}{|\vec{r} - \vec{r}\,'|} \tag{11.54}$$

and the value for the pole at $q = -k$ (Figure 11.4b) is

$$G_-(\vec{r} - \vec{r}\,') = -\frac{1}{4\pi} \frac{e^{-ik|\vec{r}-\vec{r}\,'|}}{|\vec{r} - \vec{r}\,'|}. \tag{11.55}$$

Green's function $G_+(\vec{r} - \vec{r}\,')$ represents an *outgoing spherical wave* emitted from $\vec{r}\,'$ and the function $G_-(\vec{r} - \vec{r}\,')$ corresponds to an *incoming wave* that converges onto $\vec{r}\,'$. Since the scattered waves are outgoing waves, only $G_+(\vec{r} - \vec{r}\,')$ is of interest to us. Inserting (11.54) into (11.44) we obtain the total scattered wave function:

$$\boxed{\psi(\vec{r}) = \phi_{inc}(\vec{r}) - \frac{\mu}{2\pi\hbar^2} \int \frac{e^{ik|\vec{r}-\vec{r}\,'|}}{|\vec{r} - \vec{r}\,'|} V(\vec{r}\,')\psi(\vec{r}\,')\, d^3r'.} \tag{11.56}$$

This is an *integral equation*; it does not yet give the unknown solution $\psi(\vec{r})$ but only contains it in the integrand. All we have done is to rewrite the Schrödinger (differential) equation (11.30) in an integral form (11.56), because the integral form is suitable for use in scattering theory. We are going to show that (11.56) reduces to (11.34) in the asymptotic limit $r \longrightarrow \infty$. But let us first mention that (11.56) can be solved approximately by means of a series of successive or iterative approximations, known as the *Born series*. The zero-order solution is given by $\psi_0(\vec{r}) = \phi_{inc}(\vec{r})$. The first-order solution $\psi_1(\vec{r})$ is obtained by inserting $\psi_0(\vec{r}) = \phi_{inc}(\vec{r})$ into the integral sign of (11.56):

$$\psi_1(\vec{r}) = \phi_{inc}(\vec{r}) - \frac{\mu}{2\pi\hbar^2} \int \frac{e^{ik|\vec{r}-\vec{r}_1|}}{|\vec{r} - \vec{r}_1|} V(\vec{r}_1)\psi_0(\vec{r}_1)\, d^3r_1$$

$$= \phi_{inc}(\vec{r}) - \frac{\mu}{2\pi\hbar^2} \int \frac{e^{ik|\vec{r}-\vec{r}_1|}}{|\vec{r} - \vec{r}_1|} V(\vec{r}_1)\phi_{inc}(\vec{r}_1)\, d^3r_1. \tag{11.57}$$

The second order is obtained by inserting $\psi_1(\vec{r})$ into (11.56):

$$\psi_2(\vec{r}) = \phi_{inc}(\vec{r}) - \frac{\mu}{2\pi\hbar^2} \int \frac{e^{ik|\vec{r}-\vec{r}_2|}}{|\vec{r} - \vec{r}_2|} V(\vec{r}_2)\psi_1(\vec{r}_2)\, d^3r_2$$

Figure 11.5 The distance r from the target to the detector is too large compared to the size r' of the target: $r \gg r'$.

$$= \phi_{inc}(\vec{r}) - \frac{\mu}{2\pi\hbar^2} \int \frac{e^{ik|\vec{r}-\vec{r}_2|}}{|\vec{r}-\vec{r}_2|} V(\vec{r}_2)\phi_{inc}(\vec{r}_2)\, d^3r_2$$

$$+ \left(\frac{\mu}{2\pi\hbar^2}\right)^2 \int \frac{e^{ik|\vec{r}-\vec{r}_2|}}{|\vec{r}-\vec{r}_2|} V(\vec{r}_2)\, d^3r_2 \int \frac{e^{ik|\vec{r}_2-\vec{r}_1|}}{|\vec{r}_2-\vec{r}_1|} V(\vec{r}_1)\phi_{inc}(\vec{r}_1)\, d^3r_1. \tag{11.58}$$

Continuing in this way, we can obtain $\psi(\vec{r})$ to the desired order; the nth order approximation for the wave function is a series which can be obtained by analogy with (11.57) and (11.58).

Asymptotic limit of the wave function
We are now going to show that (11.56) reduces to (11.34) for large values of r. In a scattering experiment, since the detector is located at distances (away from the target) that are much larger than the size of the target (Figure 11.5), we have $r \gg r'$, where r represents the distance from the target to the detector and r' the size of the detector. If $r \gg r'$ we may approximate $k|\vec{r}-\vec{r}'|$ and $|\vec{r}-\vec{r}'|^{-1}$ by

$$k|\vec{r}-\vec{r}'| = k\sqrt{r^2 - 2\vec{r}\cdot\vec{r}' + r'^2} \simeq kr - k\frac{\vec{r}}{r}\cdot\vec{r}' = kr - \vec{k}\cdot\vec{r}', \tag{11.59}$$

$$\frac{1}{|\vec{r}-\vec{r}'|} = \frac{1}{r}\frac{1}{|1 - \vec{r}\cdot\vec{r}'/r^2|} \simeq \frac{1}{r}\left(1 + \frac{\vec{r}\cdot\vec{r}'}{r^2}\right) \simeq \frac{1}{r}, \tag{11.60}$$

where $\vec{k} = k\hat{r}$ is the wave vector associated with the scattered particle. From the previous two approximations, we may write the asymptotic form of (11.56) as follows:

$$\psi(\vec{r}) \longrightarrow e^{i\vec{k}_0\cdot\vec{r}} + \frac{e^{ikr}}{r}f(\theta,\varphi) \qquad (r\to\infty), \tag{11.61}$$

where

$$f(\theta,\varphi) = -\frac{\mu}{2\pi\hbar^2}\int e^{-i\vec{k}\cdot\vec{r}'} V(\vec{r}')\psi(\vec{r}')\, d^3r' = -\frac{\mu}{2\pi\hbar^2}\langle\phi\mid\hat{V}\mid\psi\rangle, \tag{11.62}$$

where $\phi(\vec{r})$ is a plane wave, $\phi(\vec{r}) = e^{i\vec{k}\cdot\vec{r}}$, and $\vec{k}$ is the wave vector of the scattered wave; the integration variable r' extends over the spatial degrees of freedom of the target. The differential cross section is then given by

$$\frac{d\sigma}{d\Omega} = |f(\theta, \varphi)|^2 = \frac{\mu^2}{4\pi^2\hbar^4} \left| \int e^{-i\vec{k}\cdot\vec{r}'} \hat{V}(\vec{r}')\psi(\vec{r}') d^3r' \right|^2 = \frac{\mu^2}{4\pi^2\hbar^4} \left| \langle \phi \mid \hat{V} \mid \psi \rangle \right|^2.$$

(11.63)

11.3 The Born Approximation

11.3.1 The First Born Approximation

If the potential $V(\vec{r})$ is weak enough, it will distort only slightly the incident plane wave. The *first Born approximation* consists then of approximating the scattered wave function $\psi(\vec{r})$ by a *plane wave*. This approximation corresponds to the first iteration of (11.56); that is, $\psi(\vec{r})$ is given by (11.57):

$$\psi(\vec{r}) \simeq \phi_{inc}(\vec{r}) - \frac{\mu}{2\pi\hbar^2} \int \frac{e^{ik|\vec{r}-\vec{r}'|}}{|\vec{r}-\vec{r}'|} V(\vec{r}')\phi_{inc}(\vec{r}') d^3r'.$$

(11.64)

Thus, using (11.62) and (11.63), we can write the scattering amplitude and the differential cross section in the first Born approximation as follows:

$$f(\theta, \varphi) = -\frac{\mu}{2\pi\hbar^2} \int e^{-i\vec{k}\cdot\vec{r}'} V(\vec{r}')\phi_{inc}(\vec{r}') d^3r' = -\frac{\mu}{2\pi\hbar^2} \int e^{i\vec{q}\cdot\vec{r}'} V(\vec{r}') d^3r', \quad (11.65)$$

$$\frac{d\sigma}{d\Omega} = \left| f(\theta, \varphi) \right|^2 = \frac{\mu^2}{4\pi^2\hbar^4} \left| \int e^{i\vec{q}\cdot\vec{r}'} V(\vec{r}') d^3r' \right|^2, \quad (11.66)$$

where $\vec{q} = \vec{k}_0 - \vec{k}$ and $\hbar\vec{q}$ is the momentum transfer; $\hbar\vec{k}_0$ and $\hbar\vec{k}$ are the linear momenta of the incident and scattered particles, respectively.

In *elastic scattering*, the magnitudes of $\vec{k}_0$ and $\vec{k}$ are equal (Figure 11.6); hence

$$q = \left| \vec{k}_0 - \vec{k} \right| = \sqrt{k_0^2 + k^2 - 2kk_0 \cos\theta} = k\sqrt{2(1 - \cos^2\theta)} = 2k \sin\left(\frac{\theta}{2}\right). \quad (11.67)$$

If the potential $V(\vec{r}')$ is *spherically symmetric*, $V(\vec{r}') = V(r')$, and choosing the z-axis along $\vec{q}$ (Figure 11.6), then $\vec{q} \cdot \vec{r}' = qr' \cos\theta'$ and therefore

$$\int e^{i\vec{q}\cdot\vec{r}'} V(\vec{r}') d^3r' = \int_0^\infty r'^2 V(r') dr' \int_0^\pi e^{iqr'\cos\theta'} \sin\theta' d\theta' \int_0^{2\pi} d\varphi'$$

$$= 2\pi \int_0^\infty r'^2 V(\vec{r}') dr' \int_{-1}^1 e^{iqr'x} dx = \frac{4\pi}{q} \int_0^\infty r'V(r') \sin(qr') dr'.$$

(11.68)

Inserting (11.68) into (11.65) and (11.66) we obtain

$$f(\theta) = -\frac{2\mu}{\hbar^2 q} \int_0^\infty r'V(r') \sin(qr') dr',$$

(11.69)

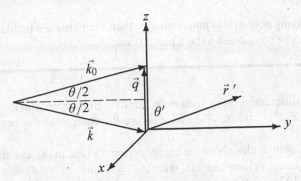

Figure 11.6 Momentum transfer for elastic scattering: $q = |\vec{k}_0 - \vec{k}| = 2k\sin(\theta/2)$, $k_0 = k$.

$$\frac{d\sigma}{d\Omega} = \left| f(\theta) \right|^2 = \frac{4\mu^2}{\hbar^4 q^2} \left| \int_0^\infty r' V(r') \sin(qr')\, dr' \right|^2. \tag{11.70}$$

In summary, we have shown that by solving the Schrödinger equation (11.30) to first-order Born approximation (where the potential $V(\vec{r})$ is weak enough that the scattered wave function is only slightly different from the incident plane wave), the differential cross section is given by equation (11.70) for a spherically symmetric potential.

11.3.2 Validity of the First Born Approximation

The first Born approximation is valid whenever the wave function $\psi(\vec{r})$ is only slightly different from the incident plane wave; that is, whenever the second term in (11.64) is very small compared to the first:

$$\left| \frac{\mu}{2\pi\hbar^2} \int \frac{e^{ik|\vec{r}-\vec{r}'|}}{|\vec{r} - \vec{r}'|} V(r') e^{i\vec{k}_0 \cdot \vec{r}'}\, d^3 r' \right| \ll |\phi_{inc}(\vec{r})|^2. \tag{11.71}$$

Since $\phi_{inc} = e^{i\vec{k}_0 \cdot \vec{r}}$ we have

$$\left| \frac{\mu}{2\pi\hbar^2} \int \frac{e^{ik|\vec{r}-\vec{r}'|}}{|\vec{r} - \vec{r}'|} V(r') e^{i\vec{k}_0 \cdot \vec{r}'}\, d^3 r' \right| \ll 1. \tag{11.72}$$

In elastic scattering $k_0 = k$ and assuming that the scattering potential is largest near $r = 0$, we have

$$\left| \frac{\mu}{\hbar^2} \int_0^\infty r' e^{ik\vec{r}'} V(r')\, dr' \int_0^\pi e^{ikr'\cos\theta'} \sin\theta'\, d\theta' \right| \ll 1 \tag{11.73}$$

or

$$\left| \frac{\mu}{\hbar^2 k} \int_0^\infty V(r') \left(e^{2ikr'} - 1 \right) dr' \right| \ll 1. \tag{11.74}$$

Since the energy of the incident particle is proportional to k (it is purely kinetic, $E_i = \hbar^2 k^2 / 2\mu$), we infer from (11.74) that the Born approximation is valid for large incident energies and weak scattering potentials. That is, when the average interaction energy between the incident particle and the scattering potential is much smaller than the particle's incident kinetic energy, the scattered wave can be considered to be a plane wave.

Example 11.2

(a) Calculate the differential cross section in the first Born approximation for a Coulomb potential $V(r) = Z_1 Z_2 e^2 / r$, where $Z_1 e$ and $Z_1 e$ are the charges of the projectile and target particles, respectively.

(b) To have a quantitative idea about the cross section derived in (a), consider the scattering of an alpha particle (i.e., a helium nucleus with $Z_1 = 2$ and $A_1 = 4$) from a gold nucleus ($Z_2 = 79$ and $A_2 = 197$). (i) If the scattering angle of the alpha particle in the Lab frame is $\theta_1 = 60°$, find its scattering angle θ in the CM frame. (ii) If the incident energy of the alpha particle is 8 MeV, find a numerical estimate for the cross section derived in (a).

Solution

In the case of a Coulomb potential, $V(r) = Z_1 Z_2 e^2 / r$, equation (11.70) becomes

$$\frac{d\sigma}{d\Omega} = \frac{4 Z_1^2 Z_2^2 e^4 \mu^2}{\hbar^4 q^2} \left| \int_0^\infty \sin(qr) \, dr \right|^2, \tag{11.75}$$

where

$$\int_0^\infty \sin(qr) \, dr = \lim_{\lambda \to 0} \int_0^\infty e^{-\lambda r} \sin(qr) \, dr = \frac{1}{2i} \lim_{\lambda \to 0} \left[\int_0^\infty e^{-(\lambda - iq)r} \, dr - \int_0^\infty e^{-(\lambda + iq)r} \, dr \right]$$

$$= \frac{1}{2i} \lim_{\lambda \to 0} \left[\frac{1}{\lambda - iq} - \frac{1}{\lambda + iq} \right] = \frac{1}{q}. \tag{11.76}$$

Now, since $q = 2k \sin(\theta/2)$, an insertion of (11.76) into (11.75) leads to

$$\frac{d\sigma}{d\Omega} = \left(\frac{2 Z_1 \mu Z_2 e^2}{\hbar^2 q^2} \right)^2 = \left(\frac{Z_1 Z_2 \mu e^2}{2\hbar^2 k^2} \right)^2 \sin^{-4} \left(\frac{\theta}{2} \right) = \frac{Z_1^2 Z_2^2 e^4}{16 E^2} \sin^{-4} \left(\frac{\theta}{2} \right), \tag{11.77}$$

where $E = \hbar^2 k^2 / 2\mu$ is the kinetic energy of the incident particle. This relation is known as the *Rutherford formula* or the Coulomb cross section.

(b) (i) Since the mass ratio of the alpha particle to the gold nucleus is roughly equal to the ratio of their atomic masses, $m_1/m_2 = A_1/A_2 = \frac{4}{197} = 0.0203$, and since $\theta_1 = 60°$, equation (11.14) yields the value of the scattering angle in the CM frame:

$$\tan 60° = \frac{\sin \theta}{\cos \theta + 0.0203} \qquad \Longrightarrow \qquad \theta = 61°. \tag{11.78}$$

(ii) The numerical estimate of the cross section can be made easier by rewriting (11.77) in terms of the fine structure constant $\alpha = e^2/\hbar c = \frac{1}{137}$ and $\hbar c = 197.33$ MeV fm:

$$\frac{d\sigma}{d\Omega} = \frac{Z_1^2 Z_2^2}{16 E^2} \left(\frac{e^2}{\hbar c} \right)^2 (\hbar c)^2 \sin^{-4} \left(\frac{\theta}{2} \right) = \left(\frac{Z_1 Z_2 \alpha}{4} \right)^2 \left(\frac{\hbar c}{E} \right)^2 \sin^{-4} \left(\frac{\theta}{2} \right). \tag{11.79}$$

Since $Z_1 = 2$, $Z_2 = 79$, $\theta = 61°$, $\alpha = \frac{1}{137}$, $\hbar c = 197.33$ MeV fm, and $E = 8$ MeV, we have

$$\frac{d\sigma}{d\Omega} = \left(\frac{2 \times 79}{4 \times 137} \right)^2 \left(\frac{197.33 \text{ MeV fm}}{8 \text{ MeV}} \right)^2 \sin^{-4}(30.5°)$$

$$= 30.87 \text{ fm}^2 = 0.31 \times 10^{-28} \text{ m}^2 = 0.31 \text{ barn}, \tag{11.80}$$

where 1 barn $= 10^{-28}$ m^2.

11.4 Partial Wave Analysis

So far we have considered only an approximate calculation of the differential cross section where the interaction between the projectile particle and the scattering potential $V(\vec{r})$ is considered small compared with the energy of the incident particle. In this section we are going to calculate the cross section without placing any limitation on the strength of $V(\vec{r})$.

11.4.1 Partial Wave Analysis for Elastic Scattering

We assume here the potential to be *spherically symmetric*. The angular momentum of the incident particle will therefore be conserved; a particle scattering from a central potential will have the same angular momentum before and after collision. Assuming that the incident plane wave is in the z-direction and hence $\phi_{inc}(\vec{r}) = \exp(ikr\cos\theta)$, we may express it in terms of a superposition of angular momentum eigenstates, each with a definite angular momentum number l (Chapter 6):

$$e^{i\vec{k}\cdot\vec{r}} = e^{ikr\cos\theta} = \sum_{l=0}^{\infty} i^l (2l+1) j_l(kr) P_l(\cos\theta). \tag{11.81}$$

We can then examine how each of the partial waves is distorted by $V(r)$ after the particle scatters from the potential. The most general solution of the Schrödinger equation (11.30) is

$$\psi(\vec{r}) = \sum_{lm} C_{lm} R_{kl}(r) Y_{lm}(\theta, \varphi). \tag{11.82}$$

Since $V(r)$ is central, the system is symmetrical (rotationally invariant) about the z-axis. The scattered wave function must not then depend on the azimuthal angle φ; hence $m = 0$. Thus, as $Y_{l0}(\theta, \varphi) \sim P_l(\cos\theta)$, the scattered wave function (11.82) becomes

$$\psi(r, \theta) = \sum_{l=0}^{\infty} a_l R_{kl}(r) P_l(\cos\theta), \tag{11.83}$$

where $R_{kl}(r)$ obeys the following radial equation (Chapter 6):

$$\left[\frac{d^2}{dr^2} + k^2 - \frac{l(l+1)}{r^2}\right] (r R_{kl}(r)) = \frac{2m}{\hbar^2} V(r) (r R_{kl}(r)). \tag{11.84}$$

Each term of (11.83), which is known as a *partial wave*, is a joint eigenfunction of $\vec{L}^2$ and $\hat{L}_z$. A substitution of (11.81) into (11.34) with $\varphi = 0$ gives

$$\psi(r, \theta) \simeq \sum_{l=0}^{\infty} i^l (2l+1) j_l(kr) P_l(\cos\theta) + f(\theta)\frac{e^{ikr}}{r}. \tag{11.85}$$

The scattered wave function is given, on the one hand, by (11.83) and, on the other hand, by (11.85).

In almost all scattering experiments, detectors are located at distances from the target that are much larger than the size of the target itself; thus, the measurements taken by detectors pertain to scattered wave functions at large values of r. In what follows we are going to show that, by establishing a connection between the asymptotic forms of (11.83) and (11.85), we can determine the scattering amplitude and hence the differential cross section.

First, since the limit of the Bessel function $j_l(kr)$ for large values of r (Chapter 6) is given by

$$j_l(kr) \longrightarrow \frac{\sin(kr - l\pi/2)}{kr} \qquad (r \longrightarrow \infty), \tag{11.86}$$

the asymptotic form of (11.85) is

$$\psi(r, \theta) \longrightarrow \sum_{l=0}^{\infty} i^l (2l+1) P_l(\cos\theta) \frac{\sin(kr - l\pi/2)}{kr} + f(\theta)\frac{e^{ikr}}{r}, \tag{11.87}$$

and since $\sin(kr - l\pi/2) = [(-i)^l e^{ikr} - i^l e^{-ikr}]/2i$, because $e^{\pm il\pi/2} = \left(e^{\pm i\pi/2}\right)^l = (\pm i)^l$, we can write (11.87) as

$$\psi(r, \theta) \longrightarrow -\frac{e^{-ikr}}{2ikr}\sum_{l=0}^{\infty} i^{2l}(2l+1) P_l(\cos\theta) + \frac{e^{ikr}}{r}\left[f(\theta) + \frac{1}{2ik}\sum_{l=0}^{\infty} i^l(-i)^l (2l+1) P_l(\cos\theta) \right]. \tag{11.88}$$

Second, to find the asymptotic form of (11.83), we need first to determine the asymptotic form of the radial function $R_{kl}(r)$. At large values of r, the scattering potential is effectively zero, for it is short range. In this case (11.84) becomes

$$\left(\frac{d^2}{dr^2} + k^2\right)(r R_{kl}(r)) = 0. \tag{11.89}$$

As seen in Chapter 6, the general solution of this equation is given by a linear combination of the spherical Bessel and Neumann functions

$$R_{kl}(r) = A_l j_l(kr) + B_l n_l(kr), \tag{11.90}$$

where the asymptotic form of the Neumann function is

$$n_l(kr) \longrightarrow -\frac{\cos(kr - l\pi/2)}{kr} \qquad (r \longrightarrow \infty). \tag{11.91}$$

Inserting of (11.86) and (11.91) into (11.90), we obtain the asymptotic form of the radial function:

$$R_{kl}(r) \longrightarrow A_l \frac{\sin(kr - l\pi/2)}{kr} - B_l \frac{\cos(kr - l\pi/2)}{kr} \qquad (r \longrightarrow \infty). \tag{11.92}$$

If $V(r) = 0$ for all r (free particle), the solution of (11.84), $r R_{kl}(r)$, must vanish at $r = 0$; thus $R_{kl}(r)$ must be finite at the origin. Since the Neumann function diverges at $r = 0$, the cosine term in (11.92) does not represent a physically acceptable solution; hence, it needs to be discarded near the origin. By rewriting (11.92) in the form

$$R_{kl}(r) \longrightarrow C_l \frac{\sin(kr - l\pi/2 + \delta_l)}{kr} \qquad (r \longrightarrow \infty), \tag{11.93}$$

we have $A_l = C_l \cos \delta_l$ and $B_l = -C_l \sin \delta_l$, hence $C_l = \sqrt{A_l^2 + B_l^2}$ and

$$\tan \delta_l = -\frac{B_l}{A_l} \qquad \Longrightarrow \qquad \delta_l = -\tan^{-1}\left(\frac{B_l}{A_l}\right). \tag{11.94}$$

We see that, with $\delta_l = 0$, the radial function $R_{kl}(r)$ of (11.93) is finite at $r = 0$, since (11.93) reduces to $j_l(kr)$. So δ_l is a real angle which vanishes for all values of l in the absence of the scattering potential (i.e., $V = 0$); δ_l is called the *phase shift*. It measures, at large values of r, the degree to which $R_{kl}(r)$ differs from $j_l(kr)$ (recall that $j_l(kr)$ is the radial function when there is no scattering). Since this "distortion," or the difference between $R_{kl}(r)$ and $j_l(kr)$, is due to the potential $V(r)$, we would expect the cross section to depend on δ_l. Using (11.93) we can write the asymptotic limit of (11.83) as

$$\psi(r, \theta) \longrightarrow \sum_{l=0}^{\infty} a_l P_l(\cos\theta) \frac{\sin(kr - l\pi/2 + \delta_l)}{kr} \qquad (r \longrightarrow \infty). \tag{11.95}$$

This wave function is known as a *distorted plane wave*, for it differs from a plane wave by having phase shifts δ_l. Since $\sin(kr - l\pi/2 + \delta_l) = [(-i)^l e^{ikr} e^{i\delta_l} - i^l e^{-ikr} e^{-i\delta_l}]/2i$, we can rewrite (11.95) as

$$\psi(r, \theta) \longrightarrow -\frac{e^{-ikr}}{2ikr} \sum_{l=0}^{\infty} a_l i^l e^{-i\delta_l} P_l(\cos\theta) + \frac{e^{ikr}}{2ikr} \sum_{l=0}^{\infty} a_l (-i)^l e^{i\delta_l} P_l(\cos\theta). \tag{11.96}$$

Up to now we have shown that the asymptotic forms of (11.83) and (11.85) are given by (11.96) and (11.88), respectively. Equating the coefficients of e^{-ikr}/r in (11.88) and (11.96), we obtain $(2l + 1)i^{2l} = a_l i^l e^{-i\delta_l}$ and hence

$$a_l = (2l + 1)i^l e^{i\delta_l}. \tag{11.97}$$

Substituting (11.97) into (11.96) and this time equating the coefficient of e^{ikr}/r in the resulting expression with that of (11.88), we have

$$f(\theta) + \frac{1}{2ik} \sum_{l=0}^{\infty} i^l (-i)^l (2l + 1) P_l(\cos\theta) = \frac{1}{2ik} \sum_{l=0}^{\infty} (2l + 1) i^l (-i)^l e^{2i\delta_l} P_l(\cos\theta), \tag{11.98}$$

which, when combined with $(e^{2i\delta_l} - 1)/2i = e^{i\delta_l} \sin \delta_l$ and $i^l(-i)^l = 1$, leads to

$$f(\theta) = \sum_{l=0}^{\infty} f_l(\theta) = \frac{1}{2ik} \sum_{l=0}^{\infty} (2l + 1) P_l(\cos\theta)(e^{2i\delta_l} - 1) = \frac{1}{k} \sum_{l=0}^{\infty} (2l + 1) e^{i\delta_l} \sin \delta_l P_l(\cos\theta),$$
$$\tag{11.99}$$

where $f_l(\theta)$ is known as the partial wave amplitude.

From (11.99) we can obtain the differential and the total cross sections

$$\frac{d\sigma}{d\Omega} = |f(\theta)|^2 = \frac{1}{k^2} \sum_{l=0}^{\infty} \sum_{l'=0}^{\infty} (2l + 1)(2l' + 1) e^{i(\delta_l - \delta_{l'})} \sin \delta_l \sin \delta_{l'} P_l(\cos\theta) P_{l'}(\cos\theta),$$
$$\tag{11.100}$$

$$\sigma = \int \frac{d\sigma}{d\Omega} d\Omega = \int_0^\pi |f(\theta)|^2 \sin\theta \, d\theta \int_0^{2\pi} d\varphi = 2\pi \int_0^\pi |f(\theta)|^2 \sin\theta \, d\theta$$

$$= \frac{2\pi}{k^2} \sum_{l=0}^{\infty} \sum_{l'=0}^{\infty} (2l+1)(2l'+1) e^{i(\delta_l - \delta_{l'})} \sin\delta_l \sin\delta_{l'} \int_0^\pi P_l(\cos\theta) P_{l'}(\cos\theta) \sin\theta \, d\theta.$$

(11.101)

Using the relation $\int_0^\pi P_l(\cos\theta) P_{l'}(\cos\theta) \sin\theta \, d\theta = [2/(2l+1)]\delta_{ll'}$, we can reduce (11.101) to

$$\boxed{\sigma = \sum_{l=0}^{\infty} \sigma_l = \frac{4\pi}{k^2} \sum_{l=0}^{\infty} (2l+1) \sin^2 \delta_l,}$$

(11.102)

where σ_l are called the *partial cross sections* corresponding to the scattering of particles in various angular momentum states. The differential cross section (11.100) consists of a superposition of terms with different angular momenta; this gives rise to *interference* patterns between different partial waves corresponding to different values of l. The interference terms go away in the total cross section when the integral over θ is carried out. Note that when $V = 0$ everywhere, all the phase shifts δ_l vanish, and hence the partial and total cross sections, as indicated by (11.100) and (11.102), are zero. Note that, as shown in equations (11.99) and (11.102), $f(\theta)$ and σ are given as infinite series over the angular momentum l. We may recall that, for cases of practical importance with the exception of the Coulomb potential, these series converge after a finite number of terms.

We should note that in the case where we have a scattering between particles that are in their respective s states, $l = 0$, the scattering amplitude (11.99) becomes

$$f_0 = \frac{1}{k} e^{i\delta_0} \sin\delta_0 \qquad (l=0),$$

(11.103)

where we have used $P_0(\cos\theta) = 1$. Since f_0 does not depend on θ, the differential and total cross sections are given by the following simple relations:

$$\frac{d\sigma}{d\Omega} = |f_0|^2 = \frac{1}{k^2} \sin^2 \delta_0, \qquad \sigma = 4\pi |f_0|^2 = \frac{4\pi}{k^2} \sin^2 \delta_0 \qquad (l=0).$$

(11.104)

An important issue here is the fact that the total cross section can be related to the *forward scattering amplitude* $f(0)$. Since $P_l(\cos\theta) = P_l(1) = 1$ when $\theta = 0$, equation (11.99) leads to

$$f(0) = \frac{1}{k} \sum_{l=0}^{\infty} (2l+1) \left(\sin\delta_l \cos\delta_l + i \sin^2 \delta_l \right),$$

(11.105)

which when combined with (11.102) yields the connection between $f(0)$ and σ:

$$\boxed{\frac{4\pi}{k} \operatorname{Im} f(0) = \sigma = \frac{4\pi}{k} \sum_{l=0}^{\infty} (2l+1) \sin^2 \delta_l.}$$

(11.106)

This is known as the *optical theorem* (it is reminiscent of a similar theorem in optics which deals with the scattering of light). The physical origin of this theorem is the conservation of particles (or probability). The beam emerging (after scattering) along the incidence direction ($\theta = 0$)

contains fewer particles than the incident beam, since a number of particles have scattered in various directions. This decrease in the number of particles is measured by the total cross section σ; that is, the number of particles removed from the incident beam along the incidence direction is proportional to σ or, equivalently, to the imaginary part of $f(0)$. We should note that, although (11.106) was derived for elastic scattering, the optical theorem, as will be shown later, is also valid for inelastic scattering.

11.4.2   Partial Wave Analysis for Inelastic Scattering

The scattering amplitude (11.99) can be rewritten as

$$f(\theta) = \sum_{l=0}^{\infty} (2l + 1) f_l(k) P_l(\cos\theta), \tag{11.107}$$

where

$$f_l(k) = \frac{1}{k} e^{i\delta_l} \sin\delta_l = \frac{1}{2ik}\left(e^{2i\delta_l} - 1\right) = \frac{1}{2ik}\left(S_l(k) - 1\right), \tag{11.108}$$

with

$$S_l(k) = e^{2i\delta_l}. \tag{11.109}$$

In the case where there is no flux loss, we must have $|S_l(k)| = 1$. However, this requirement is not valid whenever there is *absorption* of the incident beam. In this case of flux loss, $S_l(k)$ is redefined by

$$S_l(k) = \eta_l(k)e^{2i\delta_l}, \tag{11.110}$$

with $0 < \eta_l(k) \le 1$; hence (11.108) and (11.107) become

$$f_l(k) = \frac{\eta_l e^{2i\delta_l} - 1}{2ik} = \frac{1}{2k}\left[\eta_l \sin 2\delta_l + i(1 - \eta_l \cos 2\delta_l)\right], \tag{11.111}$$

$$f(\theta) = \frac{1}{2k} \sum_{l=0}^{\infty} (2l + 1)\left[\eta_l \sin 2\delta_l + i(1 - \eta_l \cos 2\delta_l)\right] P_l(\cos\theta). \tag{11.112}$$

The *total elastic* scattering cross section is given by

$$\sigma_{el} = 4\pi \sum_{l=0}^{\infty} (2l + 1)|f_l|^2 = \frac{\pi}{k^2} \sum_l (2l + 1)(1 + \eta_l^2 - 2\eta_l \cos 2\delta_l). \tag{11.113}$$

The *total inelastic* scattering cross section, which describes the loss of flux, is given by

$$\sigma_{inel} = \frac{\pi}{k^2} \sum_{l=0}^{\infty} (2l + 1)\left(1 - \eta_l^2(k)\right). \tag{11.114}$$

Thus, if $\eta_l(k) = 1$ there is no inelastic scattering, but if $\eta_l = 0$ we have total absorption, although there is still elastic scattering in this partial wave. The sum of (11.113) and (11.114) gives the total cross section:

$$\sigma_{tot} = \sigma_{el} + \sigma_{inel} = \frac{2\pi}{k^2} \sum_{l=0}^{\infty} (2l + 1)\left(1 - \eta_l \cos(2\delta_l)\right). \tag{11.115}$$

Next, using (11.107) and (11.111), we infer

$$\text{Im } f(0) = \sum_{l=0}^{\infty} (2l+1) \text{ Im } f_l = \frac{1}{2k} \sum_{l=0}^{\infty} (2l+1)(1 - \eta_l \cos(2\delta_l)). \tag{11.116}$$

A comparison of (11.115) and (11.116) gives the optical theorem relation, $\text{Im } f(0) = k\sigma_{tot}/4\pi$; hence the optical theorem is also valid for inelastic scattering.

Example 11.3 (High-energy scattering from a black disk)
Discuss the scattering from a black disk at high energies.

Solution
A black disk is totally absorbing (i.e., $\eta_l(k) = 0$). Assuming the values of l do not exceed a maximum value l_{max} ($l \le l_{max}$) and that k is large (high-energy scattering), we have $l_{max} = ka$ where a is the radius of the disk. Since $\eta_l = 0$, equations (11.113) and (11.114) lead to

$$\sigma_{inel} = \sigma_{el} = \frac{\pi}{k^2} \sum_{l=0}^{ka} (2l+1) = \frac{\pi}{k^2}(ka+1)^2 \simeq \pi a^2; \tag{11.117}$$

hence the total cross section is given by

$$\sigma_{inel} = \sigma_{el} + \sigma_{inel} = 2\pi a^2. \tag{11.118}$$

Classically, the total cross section of a disk is equal to πa^2. The factor 2 in (11.118) is due to purely quantum effects, since in the high-energy limit there are two kinds of scattering: one corresponding to waves that hit the disk, where the cross section is equal to the classical cross section πa^2, and the other to waves that are diffracted. According to Babinet's principle, the cross section for the waves diffracted by a disk is also equal to πa^2.

11.5   Scattering of Identical Particles

First, let us consider the scattering of two identical *bosons* in their center of mass frame (we will consider the scattering of two identical fermions in a moment). *Classically*, the cross section for the scattering of two identical particles whose interaction potential is central is given by

$$\sigma_{cl}(\theta) = \sigma(\theta) + \sigma(\pi - \theta). \tag{11.119}$$

In quantum mechanics there is no way of distinguishing, as indicated in Figure 11.7, between the particle that scatters at an angle θ from the one that scatters at $(\pi - \theta)$. Thus, the scattered wave function must be symmetric:

$$\psi_{sym}(\vec{r}) \longrightarrow e^{i\vec{k}_0 \cdot \vec{r}} + e^{-i\vec{k}_0 \cdot \vec{r}} + f_{sym}(\theta)\frac{e^{ikr}}{r}, \tag{11.120}$$

and so must also be the scattering amplitude:

$$f_{boson}(\theta) = f(\theta) + f(\pi - \theta). \tag{11.121}$$

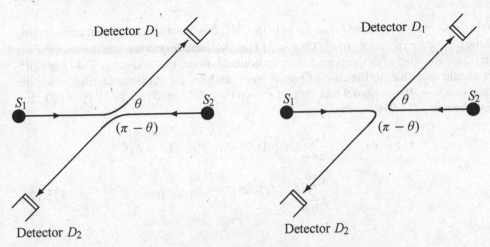

Figure 11.7 When scattering two *identical* particles in the center of mass frame, it is impossible to distinguish between the particle that scatters at angle θ from the one that scatters at $(\pi - \theta)$

Therefore, the differential cross section is

$$\frac{d\sigma}{d\Omega}\bigg|_{boson} = \left| f(\theta) + f(\pi - \theta) \right|^2 = \left| f(\theta) \right|^2 + \left| f(\pi - \theta) \right|^2 + f(\theta)^* f(\pi - \theta) + f(\theta) f^*(\pi - \theta)$$

$$= \left| f(\theta) \right|^2 + \left| f(\pi - \theta) \right|^2 + 2\,\mathrm{Re}\left[f^*(\theta) f(\pi - \theta) \right]. \tag{11.122}$$

In sharp contrast to its classical counterpart, equation (11.122) contains an interference term $2\,\mathrm{Re}\left[f^*(\theta) f(\pi - \theta) \right]$. Note that when $\theta = \pi/2$, we have $(d\sigma/d\Omega)_{boson} = 4\,|f(\pi/2)|^2$; this is twice as large as the classical expression (which has no interference term): $(d\sigma/d\Omega)_{cl} = 2\,|f(\pi/2)|^2$. If the particles were distinguishable, the differential cross section will be four times smaller, $(d\sigma/d\Omega)_{distinguishable} = |f(\pi/2)|^2$.

Consider now the scattering of two identical spin $\frac{1}{2}$ particles. This is the case, for example, of electron–electron or proton–proton scattering. The wave function of a two spin $\frac{1}{2}$ particle system is known to be either symmetric or antisymmetric. When the spatial wave function is symmetric, that is the two particles are in a spin singlet state, the differential cross section is given by

$$\frac{d\sigma_S}{d\Omega} = |f(\theta) + f(\pi - \theta)|^2, \tag{11.123}$$

but when the two particles are in a spin triplet state, the spatial wave function is antisymmetric, and hence

$$\frac{d\sigma_A}{d\Omega} = |f(\theta) - f(\pi - \theta)|^2. \tag{11.124}$$

If the incident particles are unpolarized, the various spin states will be equally likely, so the triplet state will be three times as likely as the singlet:

$$\frac{d\sigma}{d\Omega}\bigg|_{fermion} = \frac{3}{4}\frac{d\sigma_a}{d\Omega} + \frac{1}{4}\frac{d\sigma_s}{d\Omega} = \frac{3}{4}|f(\theta) - f(\pi - \theta)|^2 + \frac{1}{4}|f(\theta) + f(\pi - \theta)|^2$$

$$= |f(\theta)|^2 + |f(\pi - \theta)|^2 - \mathrm{Re}\left[f^*(\theta) f(\pi - \theta) \right]. \tag{11.125}$$

When $\theta = \pi/2$, we have $(d\sigma/d\Omega)_{fermion} = |f(\pi/2)|^2$; this quantum differential cross section is half the classical expression, $(d\sigma/d\Omega)_{cl} = 2|f(\pi/2)|^2$, and four times smaller than the expression corresponding to the scattering of two identical bosons, $(d\sigma/d\Omega)_{boson} = 4|f(\pi/2)|^2$.

We should note that, in the case of partial wave analysis for elastic scattering, using the relations $\cos(\pi - \theta) = -\cos\theta$ and $P_l(\cos(\pi - \theta)) = P_l(-\cos\theta) = (-1)^l P_l(\cos\theta)$ and inserting them into (11.99), we can write

$$
\begin{aligned}
f(\pi - \theta) &= \frac{1}{k}\sum_{l=0}^{\infty}(2l+1)e^{i\delta_l}\sin\delta_l P_l(\cos(\pi - \theta)) \\
&= \frac{1}{k}\sum_{l=0}^{\infty}(-1)^l(2l+1)e^{i\delta_l}\sin\delta_l P_l(\cos\theta),
\end{aligned}
\tag{11.126}
$$

and hence

$$
f(\theta) \pm f(\pi - \theta) = \frac{1}{k}\sum_{l=0}^{\infty}\left[1 \pm (-1)^l\right](2l+1)e^{i\delta_l}\sin\delta_l P_l(\cos\theta).
\tag{11.127}
$$

Example 11.4

Calculate the differential cross section in the first Born approximation for the scattering between two identical particles having spin 1, mass m, and interacting through a potential $V(r) = V_0 e^{-ar}$.

Solution

As seen in Chapter 7, the spin states of two identical particles with spin $s_1 = s_2 = 1$ consist of a total of nine states: a quintuplet $|2, m\rangle$ (i.e., $|2, \pm2\rangle, |2, \pm1\rangle, |2, 0\rangle$) and a singlet $|0, 0\rangle$, which are symmetric, and a triplet $|1, m\rangle$ (i.e., $|1, \pm1\rangle, |1, 0\rangle$), which are antisymmetric under particle permutation. That is, while the six spin states corresponding to $S = 2$ and $S = 0$ are symmetric, the three $S = 1$ states are antisymmetric. Thus, if the scattering particles are unpolarized, the differential cross section is

$$
\frac{d\sigma}{d\Omega} = \frac{5}{9}\frac{d\sigma_S}{d\Omega} + \frac{1}{9}\frac{d\sigma_S}{d\Omega} + \frac{3}{9}\frac{d\sigma_A}{d\Omega} = \frac{2}{3}\frac{d\sigma_S}{d\Omega} + \frac{1}{3}\frac{d\sigma_A}{d\Omega},
\tag{11.128}
$$

where

$$
\frac{d\sigma_S}{d\Omega} = \left|f(\theta) + f(\pi - \theta)\right|^2, \qquad \frac{d\sigma_A}{d\Omega} = \left|f(\theta) - f(\pi - \theta)\right|^2.
\tag{11.129}
$$

The scattering amplitude is given in the Born approximation by (11.69):

$$
\begin{aligned}
f(\theta) &= -\frac{2V_0\mu}{\hbar^2 q}\int_0^{\infty} re^{-ar}\sin(qr)\,dr = -\frac{V_0\mu}{i\hbar^2 q}\int_0^{\infty} re^{-(a-iq)r}\,dr + \frac{V_0\mu}{i\hbar^2 q}\int_0^{\infty} re^{-(a+iq)r}\,dr \\
&= \frac{V_0\mu}{\hbar^2 q}\frac{\partial}{\partial q}\int_0^{\infty} e^{-(a-iq)r}\,dr + \frac{V_0\mu}{\hbar^2 q}\frac{\partial}{\partial q}\int_0^{\infty} e^{-(a+iq)r}\,dr \\
&= \frac{V_0\mu}{\hbar^2 q}\frac{\partial}{\partial q}\left(\frac{1}{a-iq}\right) + \frac{V_0\mu}{\hbar^2 q}\frac{\partial}{\partial q}\left(\frac{1}{a+iq}\right) = \frac{V_0\mu}{\hbar^2 q}\left[\frac{i}{(a-iq)^2} + \frac{-i}{(a+iq)^2}\right] \\
&= -\frac{4V_0\mu a}{\hbar^2}\frac{1}{(a^2+q^2)^2} = -\frac{4V_0\mu a}{\hbar^2}\frac{1}{\left(a^2 + 4k^2\sin^2(\theta/2)\right)^2},
\end{aligned}
\tag{11.130}
$$

where we have used $q = 2k \sin(\theta/2)$, with $\mu = m/2$. Since $\sin[(\pi - \theta)/2] = \cos(\theta/2)$, we have

$$\frac{d\sigma_S}{d\Omega} = \frac{16V_0^2\mu^2 a^2}{\hbar^4} \left[\frac{1}{\left(a^2 + 4k^2 \sin^2(\theta/2)\right)^2} + \frac{1}{\left(a^2 + 4k^2 \cos^2(\theta/2)\right)^2} \right]^2, \quad (11.131)$$

$$\frac{d\sigma_A}{d\Omega} = \frac{16V_0^2\mu^2 a^2}{\hbar^4} \left[\frac{1}{\left(a^2 + 4k^2 \sin^2(\theta/2)\right)^2} - \frac{1}{\left(a^2 + 4k^2 \cos^2(\theta/2)\right)^2} \right]^2. \quad (11.132)$$

11.6 Solved Problems

Problem 11.1
(a) Calculate the differential cross section in the Born approximation for the potential $V(r) = V_0 e^{-r/R}/r$, known as the Yukawa potential.
(b) Calculate the total cross section.
(c) Find the relation between V_0 and R so that the Born approximation is valid.

Solution
(a) Inserting $V(r) = V_0 e^{-r/R}/r$ into (11.70), we obtain

$$\frac{d\sigma}{d\Omega} = \frac{4\mu^2 V_0^2}{\hbar^4 q^2} \left| \int_0^\infty e^{-r/R} \sin(qr)\, dr \right|^2, \quad (11.133)$$

where

$$\int_0^\infty e^{-r/R} \sin(qr)\, dr = \frac{1}{2i} \int_0^\infty e^{-(1/R - iq)r}\, dr - \frac{1}{2i} \int_0^\infty e^{-(1/R + iq)r}\, dr$$

$$= \frac{1}{2i} \left[\frac{1}{1/R - iq} - \frac{1}{1/R + iq} \right] = \frac{q}{1/R^2 + q^2}; \quad (11.134)$$

hence

$$\frac{d\sigma}{d\Omega} = \frac{4\mu^2 V_0^2}{\hbar^4} \frac{1}{(1/R^2 + q)^2} = \frac{4\mu^2 V_0^2}{\hbar^4} \frac{1}{\left[1/R^2 + 4k^2 \sin^2(\theta/2)\right]^2}. \quad (11.135)$$

Note that a connection can be established between this relation and the differential cross section for a Coulomb potential $V(r) = Z_1 Z_2 e^2/r$. For this, we need only to insert $V_0 = -Z_1 Z_2 e^2$ into (11.135) and then take the limit $R \to \infty$; this leads to (11.77):

$$\left(\frac{d\sigma}{d\Omega} \right)_{Rutherford} = \lim_{R \to \infty} \left(\frac{d\sigma}{d\Omega} \right)_{Yukawa}. \quad (11.136)$$

(b) The total cross section can be obtained at once from (11.135):

$$\sigma = \int \frac{d\sigma}{d\Omega} \sin\theta\, d\theta d\varphi = 2\pi \int_0^\pi \frac{d\sigma}{d\Omega} \sin\theta\, d\theta = 2\pi \frac{4\mu^2 V_0^2 R^4}{\hbar^4} \int_0^\pi \frac{\sin\theta\, d\theta}{\left(1 + 4k^2 R^2 \sin^2(\theta/2)\right)^2}. \quad (11.137)$$

The change of variable $x = 2kR \sin(\theta/2)$ leads to $\sin\theta \, d\theta = x \, dx/(k^2 R^2)$; hence

$$\sigma = \frac{8\pi \mu^2 V_0^2 R^4}{\hbar^4} \frac{1}{k^2 R^2} \int_0^{2kR} \frac{x \, dx}{(1 + x^2)^2} = \frac{16\pi \mu^2 V_0^2 R^4}{\hbar^4} \frac{1}{1 + 4k^2 R^2}$$

$$= \frac{16\pi \mu^2 V_0^2 R^4}{\hbar^4} \frac{1}{1 + 8\mu E R^2/\hbar^2}, \qquad (11.138)$$

where we have used $k^2 = 2\mu E/\hbar^2$; E is the energy of the scattered particle.

(c) The validity condition of the Born approximation is

$$\frac{\mu V_0}{\hbar^2 k^2} \left| \int_0^\infty \frac{e^{-ar}}{r} (e^{2ikr} - 1) \, dr \right| \ll 1, \qquad (11.139)$$

where $a = 1/R$. To evaluate the integral

$$I = \int_0^\infty \frac{e^{-ar}}{r} (e^{2ikr} - 1) \, dr \qquad (11.140)$$

let us differentiate it with respect to the parameter a:

$$\frac{\partial I}{\partial a} = -\int_0^\infty e^{-ar} (e^{2ikr} - 1) \, dr = -\frac{1}{a - 2ik} + \frac{1}{a}. \qquad (11.141)$$

Now, integrating over the parameter a such that $I(a = +\infty) = 0$, we obtain

$$I = \ln a - \ln(a - 2ik) = -\ln\left(1 - 2i\frac{k}{a}\right) = -\frac{1}{2}\ln(1 + \frac{4k^2}{a^2}) + i \tan^{-1}\left(\frac{2k}{a}\right). \qquad (11.142)$$

Thus, the validity condition (11.139) becomes

$$\frac{\mu V_0}{\hbar^2 k^2} \left\{ \frac{1}{4}\left[\ln(1 + 4k^2 R^2)\right]^2 + \left(\tan^{-1}(2kR)\right)^2 \right\}^{1/2} \ll 1. \qquad (11.143)$$

Problem 11.2

Find the differential and total cross sections for the scattering of slow (small velocity) particles from a spherical delta potential $V(r) = V_0 \delta(r - a)$ (you may use a partial wave analysis). Discuss what happens if there is no scattering potential.

Solution

In the case where the incident particles have small velocities, only the s-waves, $l = 0$, contribute to the scattering. The differential and total cross sections are given for $l = 0$ by (11.104):

$$\frac{d\sigma}{d\Omega} = |f_0|^2 = \frac{1}{k^2} \sin^2 \delta_0, \qquad \sigma = 4\pi |f_0|^2 = \frac{4\pi}{k^2} \sin^2 \delta_0 \qquad (l = 0). \qquad (11.144)$$

We need now to find the phase shift δ_0. For this, we need to consider the Schrödinger equation for the radial function:

$$-\frac{\hbar^2}{2m} \frac{d^2 u(r)}{dr^2} + \left[V_0 \delta(r - a) + \frac{l(l + 1)\hbar^2}{2mr^2} \right] u(r) = Eu(r), \qquad (11.145)$$

where $u(r) = r R(r)$. In the case of s states and $r \neq a$, this equation yields

$$\frac{d^2 u(r)}{dr^2} = -k^2 u(r),$$

(11.146)

where $k^2 = 2mE/\hbar^2$. The acceptable solutions of this equation must vanish at $r = 0$ and be finite at $r \to \infty$:

$$u(r) = \begin{cases} u_1(r) = A \sin(kr), & 0 < r < a, \\ u_2(r) = B \sin(kr + \delta_0), & r > a. \end{cases}$$

(11.147)

The continuity of $u(r)$ at $r = a$, $u_2(a) = u_1(a)$, leads to

$$B \sin(ka + \delta_0) = A \sin(ka).$$

(11.148)

On the other hand, integrating (11.145) (with $l = 0$) from $r = a - \varepsilon$ to $r = a + \varepsilon$, we obtain

$$-\frac{\hbar^2}{2m} \int_{a-\varepsilon}^{a+\varepsilon} \frac{d^2 u(r)}{dr^2} dr + V_0 \int_{a-\varepsilon}^{a+\varepsilon} \delta(r - a) u(r) dr = E \int_{a-\varepsilon}^{a+\varepsilon} u(r) dr,$$

(11.149)

and taking the limit $\varepsilon \to 0$, we end up with

$$\frac{du_2(r)}{dr}\bigg|_{r=a} - \frac{du_1(r)}{dr}\bigg|_{r=a} - \frac{2m V_0}{\hbar^2} u_2(a) = 0.$$

(11.150)

An insertion of $u_1(r)$ and $u_2(r)$ as given by (11.147) into (11.150) leads to

$$B\left[k \cos(ka + \delta_0) - \frac{2m V_0}{\hbar^2} \sin(ka + \delta_0)\right] = Ak \cos(ka).$$

(11.151)

Dividing (11.151) by (11.148), we obtain

$$k \cot(ka + \delta_0) - \frac{2m V_0}{\hbar^2} = k \cot(ka) \implies \tan(ka + \delta_0) = \left[\frac{1}{\tan(ka)} + \frac{2m V_0}{k\hbar^2}\right]^{-1}.$$

(11.152)

This equation shows that, when there is no scattering potential, $V_0 = 0$, the phase shift is zero, since $\tan(ka + \delta_0) = \tan(ka)$. In this case, equations (11.103) and (11.104) imply that the scattering amplitude and the cross sections all vanish.

If the incident particles have small velocities, $ka \ll 1$, we have $\tan(ka) \simeq ka$ and $\tan(ka + \delta_0) \simeq \tan(\delta_0)$. In this case, equation (11.152) yields

$$\tan \delta_0 \simeq \frac{ka}{1 + 2m V_0 a/\hbar^2} \implies \sin^2 \delta_0 \simeq \frac{k^2 a^2}{k^2 a^2 + (1 + 2m V_0 a/\hbar^2)^2}.$$

(11.153)

Inserting this relation into (11.144), we obtain

$$\frac{d\sigma}{d\Omega_0} \simeq \frac{a^2}{k^2 a^2 + (1 + 2m V_0 a/\hbar^2)^2}, \qquad \sigma_0 \simeq \frac{4\pi a^2}{k^2 a^2 + (1 + 2m V_0 a/\hbar^2)^2}.$$

(11.154)

Problem 11.3

Consider the scattering of a particle of mass m from a hard sphere potential: $V(r) = \infty$ for $r < a$ and $V(r) = 0$ for $r > a$.

(a) Calculate the total cross section in the low-energy limit. Find a numerical estimate for the cross section for the case of scattering $5\,\text{keV}$ protons from a hard sphere of radius $a = 6\,\text{fm}$.

(b) Calculate the total cross section in the high-energy limit. Find a numerical estimate for the cross section for the case of $700\,\text{MeV}$ protons with $a = 6\,\text{fm}$.

Solution

(a) As the scattering is dominated at low energies by s-waves, $l = 0$, the radial Schrödinger equation is

$$-\frac{\hbar^2}{2m}\frac{d^2u(r)}{dr^2} = Eu(r) \qquad (r > a), \tag{11.155}$$

where $u(r) = r R(r)$. The solutions of this equation are

$$u(r) = \begin{cases} u_1(r) = 0, & r < a, \\ u_2(r) = A\sin(kr + \delta_0), & r > a, \end{cases} \tag{11.156}$$

where $k^2 = 2mE/\hbar^2$. The continuity of $u(r)$ at $r = a$ leads to

$$\sin(ka + \delta_0) = 0 \implies \tan\delta_0 = -\tan(ka) \implies \sin^2\delta_0 = \sin^2(ka), \tag{11.157}$$

since $\sin^2\alpha = 1/(1 + \cot^2\alpha)$. The lowest value of the phase shift is $\delta_0 = -ka$; it is negative, as it should be for a repulsive potential. An insertion of $\sin^2\delta_0 = \sin^2(ka)$ into (11.104) yields

$$\sigma_0 = \frac{4\pi}{k^2}\sin^2\delta_0 = \frac{4\pi}{k^2}\sin^2(ka). \tag{11.158}$$

For low energies, $ka \ll 1$, we have $\sin(ka) \simeq ka$ and hence $\sigma_0 \simeq 4\pi a^2$, which is four times the classical value πa^2.

To obtain a numerical estimate of (11.158), we need first to calculate k^2. For this, we need simply to use the relation $E = \hbar^2 k^2/(2m_p) = 5\,\text{keV}$, since the proton moves as a free particle before scattering. Using $m_p c^2 = 938.27\,\text{MeV}$ and $\hbar c = 197.33\,\text{MeV fm}$, we have

$$k^2 = \frac{2m_p E}{\hbar^2} = \frac{2(m_p c^2)E}{(\hbar c)^2} = \frac{2(939.57\,\text{MeV})(5 \times 10^{-3}\,\text{MeV})}{(197.33\,\text{MeV fm})^2} = 0.24 \times 10^{-3}\,\text{fm}^{-2}. \tag{11.159}$$

Thus $k = 0.0155\,\text{fm}^{-1}$; the wave shift is given by $\delta_0 = -ka = -0.093\,\text{rad} = -5.33°$. Inserting these values into (11.158), we obtain

$$\sigma = \frac{4\pi}{0.24 \times 10^{-3}\,\text{fm}^{-2}}\sin^2(5.33) = 449.89\,\text{fm}^2 = 4.5\,\text{barn}. \tag{11.160}$$

(b) In the high-energy limit, $ka \gg 1$, the number of partial waves contributing to the scattering is large. Assuming that $l_{max} \simeq ka$, we may rewrite (11.102) as

$$\sigma = \frac{4\pi}{k^2}\sum_{l=0}^{l_{max}}(2l + 1)\sin^2\delta_l. \tag{11.161}$$

Since so many values of l contribute in this relation, we may replace $\sin^2 \delta_l$ by its average value, $\frac{1}{2}$; hence

$$\sigma \simeq \frac{4\pi}{k^2} \frac{1}{2} \sum_{l=0}^{l_{max}} (2l + 1) = \frac{2\pi}{k^2} (l_{max} + 1)^2, \qquad (11.162)$$

where we have used $\sum_{l=0}^{n}(2l + 1) = (n + 1)^2$. Since $l_{max} \gg 1$ we have

$$\sigma \simeq \frac{2\pi}{k^2} l_{max}^2 = \frac{2\pi}{k^2} (ka)^2 = 2\pi a^2. \qquad (11.163)$$

Since $a = 6$ fm, we have $\sigma \simeq 2\pi (6 \text{ fm})^2 = 226.1 \text{ fm}^2 = 2.26$ barn. This is almost half the value obtained in (11.160).

In conclusion, the cross section from a hard sphere potential is (a) four times the classical value, πa^2, for low-energy scattering and (b) twice the classical value for high-energy scattering.

Problem 11.4
Calculate the total cross section for the low-energy scattering of a particle of mass m from an attractive square well potential $V(r) = -V_0$ for $r < a$ and $V(r) = 0$ for $r > a$, with $V_0 > 0$.

Solution
Since the scattering is dominated at low energies by the s partial waves, $l = 0$, the Schrödinger equation for the radial function is given by

$$-\frac{\hbar^2}{2m} \frac{d^2 u(r)}{dr^2} - V_0 u(r) = E u(r) \qquad (r < a), \qquad (11.164)$$

$$-\frac{\hbar^2}{2m} \frac{d^2 u(r)}{dr^2} = E u(r) \qquad (r > a), \qquad (11.165)$$

where $u(r) = r R(r)$. The solutions of these equations for positive energy states are

$$u(r) = \begin{cases} u_1(r) = A \sin(k_1 r), & r < a, \\ u_2(r) = B \sin(k_2 r + \delta_0), & r > a, \end{cases} \qquad (11.166)$$

where $k_1^2 = 2m(E + V_0)/\hbar^2$ and $k_2^2 = 2mE/\hbar^2$. The continuity of $u(r)$ and its first derivative, $u'(r) = du(r)/dr$, at $r = a$ yield

$$\left. \frac{u_2(r)}{u_2'(r)} \right|_{r=a} = \left. \frac{u_1(r)}{u_1'(r)} \right|_{r=a} \implies \frac{1}{k_2} \tan(k_2 a + \delta_0) = \frac{1}{k_1} \tan(k_1 a), \qquad (11.167)$$

which yields

$$\delta_0 = -k_2 a + \tan^{-1} \left[\frac{k_2}{k_1} \tan(k_1 a) \right]. \qquad (11.168)$$

Since

$$\tan(k_2 a + \delta_0) = \frac{\sin(k_2 a) \cos \delta_0 + \cos(k_2 a) \sin \delta_0}{\cos(k_2 a) \cos \delta_0 - \sin(k_2 a) \sin \delta_0} = \frac{\tan(k_2 a) + \tan \delta_0}{1 - \tan(k_2 a) \tan \delta_0}, \qquad (11.169)$$

we can reduce Eq. (11.167) to

$$\tan \delta_0 = \frac{k_2 \tan(k_1 a) - k_1 \tan(k_2 a)}{k_1 + k_2 \tan(k_1 a) \tan(k_2 a)}. \tag{11.170}$$

Using the relation $\sin^2 \delta_0 = 1/(1 + 1/\tan^2 \delta_0)$, we can write

$$\sin^2 \delta_0 = \left[1 + \left(\frac{k_1 + k_2 \tan(k_1 a) \tan(k_2 a)}{k_2 \tan(k_1 a) - k_1 \tan(k_2 a)} \right)^2 \right]^{-1}, \tag{11.171}$$

which, when inserted into (11.104), leads to

$$\sigma_0 = \frac{4\pi}{k_1^2} \sin^2 \delta_0 = \frac{4\pi}{k_1^2} \left[1 + \left(\frac{k_1 + k_2 \tan(k_1 a) \tan(k_2 a)}{k_2 \tan(k_1 a) - k_1 \tan(k_2 a)} \right)^2 \right]^{-1}. \tag{11.172}$$

If $k_2 a \ll 1$ then (11.170) becomes $\tan \delta_0 \simeq \frac{\tan(k_1 a) - k_1 a}{k_1/k_2 + k_2 a \tan(k_1 a)}$, since $\tan(k_2 a) \simeq k_2 a$. Thus, if $k_2 a \ll 1$ and if E (the scattering energy) is such that $\tan(k_1 a) \simeq k_1 a$, we have $\tan \delta_0 = 0$; hence there will be no s-wave scattering and the cross section vanishes. Note that if the square well potential is extended to a hard sphere potential, i.e., $E \to 0$ and $V_0 \to \infty$, equation (11.168) yields the phase shift of scattering from a hard sphere $\delta_0 = -ka$, since $(k_2/k_1) \tan(k_1 a) \to 0$.

Problem 11.5

Find the differential and total cross sections in the first Born approximation for the elastic scattering of a particle of mass m, which is initially traveling along the z-axis, from a nonspherical, double-delta potential $V(\vec{r}) = V_0 \delta(\vec{r} - a\vec{k}) + V_0 \delta(\vec{r} + a\vec{k})$, where $\vec{k}$ is the unit vector along the z-axis.

Solution

Since $V(\vec{r})$ is not spherically symmetric, the differential cross section can be obtained from (11.66):

$$\frac{d\sigma}{d\Omega} = \frac{m^2}{4\pi^2 \hbar^4} \left| \int V_0 \left[\delta(\vec{r} - a\vec{k}) + \delta(\vec{r} + a\vec{k}) \right] e^{i\vec{q}\cdot\vec{r}} \, d^3 r \right|^2 = \frac{m^2 V_0}{4\pi^2 \hbar^4} |I|^2. \tag{11.173}$$

Since $\delta(\vec{r} \pm a\vec{k}) = \delta(x)\delta(y)\delta(z \pm a)$ we can write the integral I as

$$I = \int \delta(x) e^{ixq_x} \, dx \int \delta(y) e^{iyq_y} \, dy \int \left[\delta(z - a) + \delta(z + a) \right] e^{izq_z} \, dz$$

$$= e^{iaq_z} + e^{-iaq_z} = 2\cos(aq_z). \tag{11.174}$$

The calculation of q_z is somewhat different from that shown in (11.67). Since the incident particle is initially traveling along the z-axis, and since it scatters elastically from the potential $V(\vec{r})$, the magnitudes of its momenta before and after collision are equal. So, as shown in Figure 11.8, we have $q_z = q \sin(\theta/2) = 2k \sin^2(\theta/2)$, since $q = |\vec{k}_0 - \vec{k}| = 2k \sin(\theta/2)$. Thus, inserting $I = 2\cos(aq_z) = 2\cos\left[2ak \sin^2(\theta/2) \right]$ into (11.173), we obtain

$$\frac{d\sigma}{d\Omega} = \frac{m^2 V_0}{\pi^2 \hbar^4} \cos^2 \left(2ak \sin^2 \frac{\theta}{2} \right). \tag{11.175}$$

Figure 11.8 Particle traveling initially along the z-axis (taken here horizontally) scatters at an angle θ, with $q = |\vec{k}_0 - \vec{k}| = 2k \sin(\theta/2)$, since $k_0 = k$ and $q_z = q \sin(\theta/2)$.

The total cross section can be obtained at once from (11.175):

$$
\begin{aligned}
\sigma &= \int \frac{d\sigma}{d\Omega} \sin\theta \, d\theta d\varphi = 2\pi \int_0^\pi \frac{d\sigma}{d\Omega} \sin\theta \, d\theta \\
&= 2\pi \frac{m^2 V_0}{\pi^2 \hbar^4} \int_0^\pi \sin\theta \cos^2\left(2ak \sin^2 \frac{\theta}{2}\right) d\theta,
\end{aligned}
\tag{11.176}
$$

which, when using the change of variable $x = 2ak \sin^2(\theta/2)$ with $dx = 2ak \sin(\theta/2) \cos(\theta/2) \, d\theta$, leads to

$$
\begin{aligned}
\sigma &= \frac{2m^2 V_0}{\pi \hbar^4} \int_0^\pi 2\sin\left(\frac{\theta}{2}\right) \cos\left(\frac{\theta}{2}\right) \cos^2\left(2ak \sin^2 \frac{\theta}{2}\right) d\theta \\
&= \frac{2m^2 V_0}{\pi ak \hbar^4} \int_0^1 \cos^2(x) \, dx \\
&= \frac{m^2 V_0}{\pi ak \hbar^4} \int_0^1 [1 + \cos(2x)] \, dx \\
&= \frac{m^2 V_0}{\pi ak \hbar^4}.
\end{aligned}
\tag{11.177}
$$

Problem 11.6

Consider the elastic scattering of 50 MeV neutrons from a nucleus. The phase shifts measured in this experiment are $\delta_0 = 95°$, $\delta_1 = 72°$, $\delta_2 = 60°$, $\delta_3 = 35°$, $\delta_4 = 18°$, $\delta_5 = 5°$; all other phase shifts are negligible (i.e., $\delta_l \simeq 0$ for $l \geq 6$).
  (a) Find the total cross section.
  (b) Estimate the radius of the nucleus.

Solution

(a) As $\delta_l \simeq 0$ for $l \geq 6$, equation (11.102) yields

$$
\begin{aligned}
\sigma &= \frac{4\pi}{k^2} \sum_{l=0}^{6} (2l + 1) \sin^2 \delta_l \\
&= \frac{4\pi}{k^2} \left(\sin^2 \delta_0 + 3\sin^2 \delta_1 + 5\sin^2 \delta_2 + 7\sin^2 \delta_3 + 9\sin^2 \delta_4 + 11\sin^2 \delta_5\right) = \frac{4\pi}{k^2} \times 10.702.
\end{aligned}
\tag{11.178}
$$

To calculate k^2, we need simply to use the relation $E = \hbar^2 k^2/(2m_n) = 50\,\text{MeV}$, since the neutrons move as free particles before scattering. Using $m_n c^2 = 939.57\,\text{MeV}$ and $\hbar c = 197.33\,\text{MeV fm}$, we have

$$k^2 = \frac{2m_n E}{\hbar^2} = \frac{2(m_n c^2) E}{(\hbar c)^2} = \frac{2(939.57\,\text{MeV})(50\,\text{MeV})}{(197.33\,\text{MeV fm})^2} = 2.41\,\text{fm}^{-2}. \tag{11.179}$$

An insertion of (11.179) into (11.178) leads to

$$\sigma = \frac{4\pi}{2.41\,\text{fm}^{-2}} \times 10.702 = 55.78\,\text{fm}^2 = 0.558\,\text{barn}. \tag{11.180}$$

(b) At large values of l, when the neutron is at its closest approach to the nucleus, it feels mainly the effect of the centrifugal potential $l(l+1)\hbar^2/(2m_n r^2)$; the effect of the nuclear potential is negligible. We may thus use the approximations $E \simeq l(l+1)\hbar^2/(2m_n r_c^2) \simeq 42\hbar^2/(2m_n r_c^2)$ where we have taken $l \simeq 6$, since $\delta_l \simeq 0$ for $l \geq 6$. A crude value of the radius of the nucleus is then given by

$$r_c \simeq \sqrt{\frac{21\hbar^2}{m_n E}} = \sqrt{\frac{21(\hbar c)^2}{(m_n c^2) E}} = \sqrt{\frac{21(197.33\,\text{MeV fm})^2}{(939.57\,\text{MeV})(50\,\text{MeV})}} = 4.17\,\text{fm}. \tag{11.181}$$

Problem 11.7

Consider the elastic scattering of an electron from a hydrogen atom in its ground state. If the atom is assumed to remain in its ground state after scattering, calculate the differential cross section in the case where the effects resulting from the identical nature of the electrons (a) are ignored and (b) are taken into account (in part (b), discuss the three cases when the electrons are in (i) a spin singlet state, (ii) a spin triplet state, or (iii) an unpolarized state).

Solution

(a) By analogy with (11.63) we may write the differential cross section for this process as

$$\frac{d\sigma}{d\Omega} = |f(\theta)|^2 = \left| -\frac{\mu}{2\pi\hbar^2}\langle \Psi_f \mid \hat{V} \mid \Psi_i \rangle \right|^2, \tag{11.182}$$

where $\mu \simeq m_e/2$, since this problem can be viewed as the scattering of a particle whose reduced mass is half that of the electron. Assuming the atom to be very massive and that it remains in its ground state after scattering, the initial and final states of the system (incident electron plus the atom) are given by $\Psi_i(\vec{r}, \vec{k}_0, \vec{r}\,') = e^{i\vec{k}_0 \cdot \vec{r}} \psi_0(\vec{r}\,')$ and $\Psi_f(\vec{r}, \vec{k}, \vec{r}\,') = e^{i\vec{k} \cdot \vec{r}} \psi_0(\vec{r}\,')$, where $e^{i\vec{k}_0 \cdot \vec{r}}$ and $e^{i\vec{k} \cdot \vec{r}}$ are the states of the incident electron before and after scattering, and $\psi_0(\vec{r}\,') = (\pi a_0^3)^{-1/2} e^{-r'/a_0}$ is the atom's wave function. We have assumed here that the nucleus is located at the origin and that the position vectors of the incident electron and the atom's electron are given by $\vec{r}$ and $\vec{r}\,'$, respectively. Since the incident electron experiences an attractive Coulomb interaction $-e^2/r$ with the nucleus and a repulsive interaction $e^2/|\vec{r} - \vec{r}\,'|$ with the hydrogen's electron, we have

$$f(\theta) = -\frac{\mu}{2\pi\hbar^2} \int d^3r\, e^{i\vec{q}\cdot\vec{r}} \int d^3r'\, \psi_0^*(\vec{r}\,') \left[-\frac{e^2}{r} + \frac{e^2}{|\vec{r} - \vec{r}\,'|} \right] \psi_0(\vec{r}\,'), \tag{11.183}$$

with $q = |\vec{k}_0 - \vec{k}| = 2k \sin(\theta/2)$, since $k = k_0$ (elastic scattering). Using $\int_0^\infty \sin(qr)\, dr = 1/q$ (see (11.76)), and since $\int_0^\pi e^{iqr\cos\theta} \sin\theta\, d\theta = \int_{-1}^1 e^{iqrx}\, dx = (2/qr)\sin(qr)$, we obtain the following relation:

$$\int d^3r \frac{e^{i\vec{q}\cdot\vec{r}}}{r} = \int_0^\infty r\, dr \int_0^\pi e^{iqr\cos\theta} \sin\theta\, d\theta \int_0^{2\pi} d\varphi = \frac{4\pi}{q} \int_0^\infty dr \sin(qr) = \frac{4\pi}{q^2},$$
(11.184)

which, when inserted into (11.183) and since $\int d^3r'\, \psi_0^*(\vec{r}')\psi_0(\vec{r}') = 1$, leads to

$$f(\theta) = \frac{\mu}{2\pi\hbar^2} \left[\frac{4\pi e^2}{q^2} - \int d^3r\, e^{i\vec{q}\cdot\vec{r}} \int d^3r'\, \psi_0^*(\vec{r}') \frac{e^2}{|\vec{r}-\vec{r}'|} \psi_0(\vec{r}') \right].$$
(11.185)

By analogy with (11.184), we have $\int d^3r\, e^{i\vec{q}\cdot|\vec{r}-\vec{r}'|}/|\vec{r}-\vec{r}'| = 4\pi/q^2$; hence we can reduce the integral in (11.185) to

$$\int d^3r\, e^{i\vec{q}\cdot\vec{r}} \int d^3r'\, \psi_0^*(\vec{r}') \frac{e^2}{|\vec{r}-\vec{r}'|} \psi_0(\vec{r}') = e^2 \int d^3r'\, \psi_0^*(\vec{r}') e^{i\vec{q}\cdot\vec{r}'} \psi_0(\vec{r}') \int d^3r\, \frac{e^{i\vec{q}\cdot|\vec{r}-\vec{r}'|}}{|\vec{r}-\vec{r}'|}$$

$$= \frac{4\pi e^2}{q^2} \int d^3r'\, \psi_0^*(\vec{r}') e^{i\vec{q}\cdot\vec{r}'} \psi_0(\vec{r}').$$
(11.186)

The remaining integral of (11.186) can, in turn, be written as

$$\int d^3r'\, \psi_0^*(\vec{r}') e^{i\vec{q}\cdot\vec{r}'} \psi_0(\vec{r}') = \frac{1}{\pi a_0^3} \int_0^\infty r'^2 e^{-2r'/a_0}\, dr' \int_0^\pi e^{iqr'\cos\theta'} \sin\theta'\, d\theta' \int_0^{2\pi} d\varphi'$$

$$= \frac{4}{qa_0^3} \int_0^\infty r' e^{-2r'/a_0} \sin(qr')\, dr' = \left(1 + \frac{a_0^2 q^2}{4} \right)^{-2}, \quad (11.187)$$

where we have used the expression for $\int_0^\infty r e^{-ar} \sin(qr)\, dr$ calculated in (11.130). Inserting (11.187) into (11.186), and the resulting expression into (11.185), we obtain

$$f(\theta) = \frac{2\mu e^2}{\hbar^2 q^2} \left[1 - \left(1 + \frac{a_0^2 q^2}{4} \right)^{-2} \right] = \frac{\mu e^2}{2k^2\hbar^2 \sin^2(\theta/2)} \left[1 - \left(1 + a_0^2 k^2 \sin^2 \frac{\theta}{2} \right)^{-2} \right].$$
(11.188)

We can thus reduce (11.182) to

$$\frac{d\sigma}{d\Omega} = \frac{4\mu^2 e^4}{\hbar^4 q^4} \left[1 - \left(1 + \frac{a_0^2 q^2}{4} \right)^{-2} \right]^2 = \frac{\mu^2 e^4}{4k^4\hbar^4 \sin^4 \frac{\theta}{2}} \left[1 - \left(1 + a_0^2 k^2 \sin^2(\theta/2) \right)^{-2} \right]^2,$$
(11.189)

with $q = 2k \sin(\theta/2)$.

   (b) (i) If the electrons are in their spin singlet state (antisymmetric), the spatial wave function must be symmetric; hence the differential cross section is

$$\frac{d\sigma_S}{d\Omega} = \left| f(\theta) + f(\pi - \theta) \right|^2,$$
(11.190)

where $f(\theta)$ is given by (11.188) and

$$f(\pi - \theta) = \frac{2\mu e^2}{\hbar^2 q^2}\left[1 - \left(1 + \frac{a_0^2 q^2}{4}\right)^{-2}\right] = \frac{\mu e^2}{2k^2\hbar^2 \cos^2\frac{\theta}{2}}\left[1 - \left(1 + a_0^2 k^2 \cos^2(\theta/2)\right)^{-2}\right],$$
(11.191)

since $\sin(\pi - \theta/2) = \cos(\theta/2)$.

(ii) If, however, the electrons are in their spin triplet state, the spatial wave function must be antisymmetric; hence

$$\frac{d\sigma_A}{d\Omega} = \left| f(\theta) - f(\pi - \theta)\right|^2.$$
(11.192)

(iii) Finally, if the electrons are unpolarized, the differential cross section must be a mixture of (11.191) and (11.192):

$$\frac{d\sigma}{d\Omega} = \frac{1}{4}\frac{d\sigma_S}{d\Omega} + \frac{3}{4}\frac{d\sigma_A}{d\Omega} = \frac{1}{4}|f(\theta) + f(\pi - \theta)|^2 + \frac{3}{4}|f(\theta) - f(\pi - \theta)|^2.$$
(11.193)

Problem 11.8

In an experiment, 650 MeV π^0 pions are scattered from a heavy, totally absorbing nucleus of radius 1.4 fm.

(a) Estimate the total elastic and total inelastic cross sections.

(b) Calculate the scattering amplitude and check the validity of the optical theorem.

(c) Using the scattering amplitude found in (b), calculate and plot the differential cross section for elastic scattering. Calculate the total elastic cross section and verify that it agrees with the expression found in (a).

Solution

(a) In the case of a totally absorbing nucleus, $\eta_l(k) = 0$, the total elastic and inelastic cross sections, which are given by (11.113) and (11.114), become equal:

$$\sigma_{el} = \frac{\pi}{k^2}\sum_{l=0}^{l_{max}}(2l + 1) = \sigma_{inel}.$$
(11.194)

This experiment can be viewed as a scattering of high-energy pions, $E = 650$ MeV, from a black "disk" of radius $a = 1.4$ fm; thus, the number of partial waves involved in this scattering can be obtained from $l_{max} \simeq ka$, where $k = \sqrt{2m_{\pi^0}E/\hbar^2}$. Since the rest mass energy of a π^0 pion is $m_{\pi^0}c^2 \simeq 135$ MeV and since $\hbar c = 197.33$ MeV fm, we have

$$k \simeq \sqrt{\frac{2m_{\pi^0}E}{\hbar^2}} = \sqrt{\frac{2(m_{\pi^0}c^2)E}{(\hbar c)^2}} = \sqrt{\frac{2(135\,\text{MeV})(650\,\text{MeV})}{(197.33\,\text{MeV fm})^2}} = 2.12\,\text{fm}^{-1};$$
(11.195)

hence $l_{max} = ka \simeq (2.12\,\text{fm}^{-1})(1.4\,\text{fm}) = 2.97 \simeq 3$. We can thus reduce (11.194) to

$$\sigma_{el} = \sigma_{inel} = \frac{\pi}{k^2}\sum_{l=0}^{3}(2l + 1) = \frac{16\pi}{k^2} \simeq \frac{16\pi}{(2.12\,\text{fm}^{-1})^2} = 40.1\,\text{fm}^2 = 0.40\,\text{barn}.$$
(11.196)

Figure 11.9 Plot of $\frac{d\sigma}{d\Omega} = \frac{1}{4k^2}\left[1 + 3\cos\theta + \frac{5}{2}(3\cos^2\theta - 1) + \frac{7}{2}(5\cos^3\theta - 3\cos\theta)\right]^2$.

The total cross section

$$\sigma_{tot} = \sigma_{el} + \sigma_{inel} = \frac{32\pi}{k^2} = 0.80 \text{ barn.} \qquad (11.197)$$

(b) The scattering amplitude can be obtained from (11.112) with $\eta_l(k) = 0$:

$$f(\theta) = \frac{i}{2k}\sum_{l=0}^{3}(2l+1)P_l(\cos\theta)$$

$$= \frac{i}{2k}\left[1 + 3\cos\theta + \frac{5}{2}(3\cos^2\theta - 1) + \frac{7}{2}(5\cos^3\theta - 3\cos\theta)\right], \qquad (11.198)$$

where we have used the following Legendre polynomials: $P_0(u) = 1$, $P_1(u) = u$, $P_2(u) = \frac{1}{2}(3u^2 - 1)$, $P_3(u) = \frac{1}{2}(5u^3 - 3u)$. The forward scattering amplitude ($\theta = 0$) is

$$f(0) = \frac{i}{2k}\left[1 + 3 + \frac{5}{2}(3 - 1) + \frac{7}{2}(5 - 3)\right] = \frac{8i}{k}. \qquad (11.199)$$

Combining (11.197) and (11.199), we get the optical theorem: $\text{Im} f(0) = (k/4\pi)\sigma_{tot} = 8/k$.

(c) From (11.198) the differential elastic cross section is

$$\frac{d\sigma}{d\Omega} = |f(\theta)|^2 = \frac{1}{4k^2}\left[1 + 3\cos\theta + \frac{5}{2}(3\cos^2\theta - 1) + \frac{7}{2}(5\cos^3\theta - 3\cos\theta)\right]^2. \qquad (11.200)$$

As shown in Figure 11.9, the differential cross section displays an interference pattern due to the superposition of incoming and outgoing waves. The total elastic cross section is given by $\sigma_{el} = \int_0^\pi |f(\theta)|^2 \sin\theta\, d\theta \int_0^{2\pi} d\varphi$ which, combined with (11.200), leads to

$$\sigma_{el} = \frac{2\pi}{4k^2}\int_0^\pi\left[1 + 3\cos\theta + \frac{5}{2}(3\cos^2\theta - 1) + \frac{7}{2}(5\cos^3\theta - 3\cos\theta)\right]^2 \sin\theta\, d\theta = \frac{16\pi}{k^2}. \qquad (11.201)$$

This is the same expression we obtained in (11.196). Unlike the differential cross section, the total cross section displays no interference pattern because its final expression does not depend on any angle, since the angles were integrated over.

11.7 Exercises

Exercise 11.1
Consider the scattering of a 5 MeV alpha particle (i.e., a helium nucleus with $Z_1 = 2$ and $A_1 = 4$) from an aluminum nucleus ($Z_2 = 13$ and $A_2 = 27$). If the scattering angle of the alpha particle in the Lab frame is $\theta_1 = 30°$,
  (a) find its scattering angle θ in the CM frame and
  (b) give a numerical estimate of the Rutherford cross section.

Exercise 11.2
  (a) Find the differential and total cross sections for the classical collision of two hard spheres of radius r and R, where R is the radius of the larger sphere; the larger sphere is considered to be stationary.
  (b) From the results of (a) find the differential and total cross sections for the scattering of pointlike particles from a hard stationary sphere of radius R. *Hint:* You may use the classical relation $d\sigma/d\Omega = -[b(\theta)/\sin\theta]db/d\theta$, where $b(\theta)$ is the impact parameter.

Exercise 11.3
Consider the scattering from the potential $V(r) = V_0 e^{-r^2/a^2}$. Find
  (a) the differential cross section in the first Born approximation and
  (b) the total cross section.

Exercise 11.4
Calculate the differential cross section in the first Born approximation for the scattering of a particle by an attractive square well potential: $V(r) = -V_0$ for $r < a$ and $V(r) = 0$ for $r > a$, with $V_0 > 0$.

Exercise 11.5
Consider the elastic scattering from the delta potential $V(r) = V_0 \delta(r - a)$.
  (a) Calculate the differential cross section in the first Born approximation.
  (b) Find an expression between V_0, a, μ, and k so the Born approximation is valid.

Exercise 11.6
Consider the elastic scattering from the potential $V(r) = V_0 e^{-r/a}$, where V_0 and a are constant.
  (a) Calculate the differential cross section in the first Born approximation.
  (b) Find an expression between V_0, a, μ, and k so the Born approximation is valid.
  (c) Find the total cross section using the Born approximation.

Exercise 11.7
Find the differential cross section in the first Born approximation for the elastic scattering of a particle of mass m, which is initially traveling along the z-axis, from a nonspherical, double-delta potential:
$$V(\vec{r}) = V_0 \delta(\vec{r} - a\vec{k}) - V_0 \delta(\vec{r} + a\vec{k}),$$
where $\vec{k}$ is the unit vector along the z-axis.

Exercise 11.8
Find the differential cross section in the first Born approximation for neutron–neutron scattering in the case where the potential is approximated by $V(r) = V_0 e^{-r/a}$.

Exercise 11.9
Consider the elastic scattering of a particle of mass m and initial momentum $\hbar k$ off a delta potential $V(\vec{r}) = V_0 \delta(x)\delta(y)\delta(z-a)$, where V_0 is a constant.
   (a) What is the physical dimensions of the constant V_0?
   (b) Calculate the differential cross sections in the first Born approximation.
   (c) Repeat (b) for the case where the potential is now given by

$$V(\vec{r}) = V_0 \delta(x) \left[\delta(y-b)\delta(z) + \delta(y)\delta(z-a) \right].$$

Exercise 11.10
Consider the S-wave ($l = 0$) scattering of a particle of mass m from a repulsive spherical potential $V(r) = V_0$ for $r < a$ and $V(r) = 0$ for $r > a$, with $V_0 > 0$.
   (a) Calculate S–wave ($l = 0$) phase shift and the total cross section.
   (b) Show that in the limit $V_0 \to \infty$, the phase shift is given by $\delta_0 = -ka$. Find the total cross section.

Exercise 11.11
Consider the S-wave neutron–neutron scattering where the interaction potential is approximated by $V(r) = V_0 \vec{S}_1 \cdot \vec{S}_2 e^{-r/a}$, where $\vec{S}_1$ and $\vec{S}_2$ are the spin vector operators of the two neutrons, and $V_0 > 0$. Find the differential cross section in the first Born approximation.

Exercise 11.12
Consider the S-partial wave scattering ($l = 0$) between two identical spin $1/2$ particles where the interaction potential is given approximately by

$$\hat{V}(r) = V_0 \vec{S}_1 \cdot \vec{S}_2 \delta(r-a),$$

where $\vec{S}_1$ and $\vec{S}_2$ are the spin vector operators of the two particles, and $V_0 > 0$. Assuming that the incident and target particles are unpolarized, find the differential and total cross sections.

Exercise 11.13
Consider the elastic scattering of 170 MeV neutrons from a nucleus or radius $a = 1.05$ fm. Consider the hypothetical case where the phase shifts measured in this experiment are given by $\delta_l = \frac{180°}{l+2}$.
   (a) Estimate the maximum angular momentum l_{max}.
   (b) Find the total cross section.

Appendix A

The Delta Function

A.1 One-Dimensional Delta Function

A.1.1 Various Definitions of the Delta Function

The delta function can be defined as the limit of $\delta^{(\varepsilon)}(x)$ when $\varepsilon \to 0$ (Figure A.1):

$$\delta(x) = \lim_{\varepsilon \to 0} \delta^{(\varepsilon)}(x), \tag{A.1}$$

where

$$\delta^{(\varepsilon)}(x) = \begin{cases} 1/\varepsilon, & -\varepsilon/2 < x < \varepsilon/2, \\ 0, & |x| > \varepsilon/2. \end{cases} \tag{A.2}$$

The delta function can be defined also by means of the following integral equations:

$$\int_{-\infty}^{+\infty} f(x)\delta(x)\,dx = f(0), \tag{A.3}$$

$$\int_{-\infty}^{+\infty} f(x)\delta(x-a)\,dx = f(a). \tag{A.4}$$

We should mention that the δ-function is not a function in the usual mathematical sense. It can be expressed as the limit of analytical functions such as

$$\delta(x) = \lim_{\varepsilon \to 0} \frac{\sin(x/\varepsilon)}{\pi x}, \qquad \delta(x) = \lim_{a \to \infty} \frac{\sin^2(ax)}{\pi a x^2}, \tag{A.5}$$

or

$$\delta(x) = \lim_{\varepsilon \to 0} \frac{1}{\pi} \frac{\varepsilon}{x^2 + \varepsilon^2}. \tag{A.6}$$

The Fourier transform of $\delta(x)$, which can be obtained from the limit of $\frac{\sin(x/\varepsilon)}{\pi x}$, is

$$\delta(x) = \frac{1}{2\pi} \int_{-\infty}^{+\infty} e^{ikx}\,dk, \tag{A.7}$$

which in turn is equivalent to

$$\frac{1}{2\pi} \int_{-\infty}^{+\infty} e^{ikx}\,dk = \frac{1}{2\pi} \lim_{\varepsilon \to 0} \int_{-1/\varepsilon}^{+1/\varepsilon} e^{ikx}\,dk = \lim_{\varepsilon \to 0} \frac{\sin(x/\varepsilon)}{\pi x} = \delta(x). \tag{A.8}$$

653

Figure A.1 The delta function $\delta(x)$ as defined by $\delta(x) = \lim_{\varepsilon \to 0} \delta^{(\varepsilon)}(x)$.

A.1.2   Properties of the Delta Function

The delta function is even:

$$\delta(-x) = \delta(x) \qquad \text{and} \qquad \delta(x - a) = \delta(a - x). \qquad (A.9)$$

Here are some of the most useful properties of the delta function:

$$\int_a^b f(x)\delta(x - x_0)\, dx = \begin{cases} f(x_0), & \text{if } a < x_0 < b, \\ 0, & \text{elsewhere,} \end{cases} \qquad (A.10)$$

$$\delta(x) = 0 \qquad \text{for } x \neq 0, \qquad (A.11)$$

$$x\delta(x) = 0, \qquad (A.12)$$

$$\delta(ax) = \frac{1}{|a|}\delta(x) \qquad (a \neq 0), \qquad (A.13)$$

$$f(x)\delta(x - a) = f(a)\delta(x - a), \qquad (A.14)$$

$$\int_c^d \delta(a - x)\delta(x - b)\, dx = \delta(a - b) \qquad \text{for } c \leq a \leq d, \quad c \leq b \leq d, \qquad (A.15)$$

$$\int_a^b \delta(x)\, dx = 1 \qquad \text{for } a \leq 0 \leq b \qquad (A.16)$$

$$\delta[g(x)] = \sum_i \frac{1}{|g'(x_i)|}\delta(x - x_i), \qquad (A.17)$$

where x_i is a zero of $g(x)$ and $g'(x_i) \neq 0$. Using (A.17), we can verify that

$$\delta[(x - a)(x - b)] = \frac{1}{|a - b|}[\delta(x - a) + \delta(x - b)] \qquad (a \neq b), \qquad (A.18)$$

$$\delta(x^2 - a^2) = \frac{1}{2|a|}[\delta(x - a) + \delta(x + a)] \qquad (a \neq 0). \qquad (A.19)$$

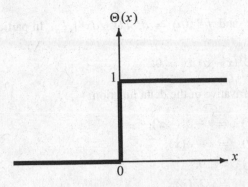

Figure A.2 The Heaviside function $\Theta(x)$.

A.1.3 Derivative of the Delta Function

The Heaviside function, or step function is defined as follows; see Figure A.2:

$$\Theta(x) = \begin{cases} 1, & x > 0, \\ 0, & x < 0. \end{cases} \tag{A.20}$$

The derivative of the Heaviside function gives back the delta function:

$$\frac{d}{dx}\Theta(x) = \delta(x). \tag{A.21}$$

Using the Fourier transform of the delta function, we can write

$$\frac{d\delta(x)}{dx} = \delta'(x) = \frac{i}{2\pi}\int_{-\infty}^{+\infty} ke^{ikx}\,dk. \tag{A.22}$$

Another way of looking at the derivative of the delta function is by means of the following integration by parts of $\delta'(x - a)$:

$$\int_{-\infty}^{\infty} f(x)\delta'(x-a)\,dx = f(x)\delta(x-a)|_{-\infty}^{\infty} - \int_{-\infty}^{\infty} f'(x)\delta(x-a)\,dx = -f'(a), \tag{A.23}$$

or

$$\int_{-\infty}^{\infty} f(x)\delta'(x-a)\,dx = -f'(a), \tag{A.24}$$

where we have used the fact that $f(x)\delta(x - a)$ is zero at $\pm\infty$. Following the same procedure, we can show that

$$\int_{-\infty}^{\infty} f(x)\delta''(x-a)\,dx = (-1)^2 f''(a) = f''(a). \tag{A.25}$$

Similar repeated integrations by parts lead to the following general relation:

$$\int_{-\infty}^{\infty} f(x)\delta^{(n)}(x-a)\,dx = (-1)^n f^{(n)}(a), \tag{A.26}$$

where $\delta^{(n)}(x - a) = d^n[\delta(x - a)]/dx^n$ and $f^{(n)}(a) = d^n f(x)/dx^n|_{x=a}$. In particular, if $f(x) = 1$ and $n = 1$, we have

$$\int_{-\infty}^{\infty} \delta'(x - a)\,dx = 0. \tag{A.27}$$

Here is a list of useful properties of the derivative of the delta function:

$$\delta'(x) = -\delta'(-x), \tag{A.28}$$
$$x\delta'(x) = -\delta(x), \tag{A.29}$$
$$x^2\delta'(x) = 0, \tag{A.30}$$
$$x^2\delta''(x) = 2\delta(x). \tag{A.31}$$

A.2   Three-Dimensional Delta Function

The three-dimensional form of the delta function is given in Cartesian coordinates by

$$\delta(\vec{r} - \vec{r}') = \delta(x - x')\delta(y - y')\delta(z - z') \tag{A.32}$$

and in spherical coordinates by

$$\begin{aligned}
\delta(\vec{r} - \vec{r}') &= \frac{1}{r^2}\delta(r - r')\delta(\cos\theta - \cos\theta')\delta(\varphi - \varphi') \\
&= \frac{1}{r^2\sin\theta}\delta(r - r')\delta(\theta - \theta')\delta(\varphi - \varphi'), \tag{A.33}
\end{aligned}$$

since, according to (A.17), we have $\delta(\cos\theta - \cos\theta') = \delta(\theta - \theta')/\sin\theta$.

The Fourier transform of the three-dimensional delta function is

$$\delta(\vec{r} - \vec{r}') = \frac{1}{(2\pi)^3}\int d^3k\, e^{i\vec{k}\cdot(\vec{r} - \vec{r}')}, \tag{A.34}$$

and

$$\int d^3r\, f(\vec{r})\delta(\vec{r}) = f(0), \qquad \int d^3r\, f(\vec{r})\delta(\vec{r} - \vec{r}_0) = f(\vec{r}_0). \tag{A.35}$$

The following relations are often encountered:

$$\vec{\nabla}\cdot\left(\frac{\hat{r}}{r^2}\right) = 4\pi\,\delta(\vec{r}), \qquad \nabla^2\left(\frac{1}{r}\right) = -4\pi\,\delta(\vec{r}), \tag{A.36}$$

where $\hat{r}$ the unit vector along $\vec{r}$.

We should mention that the physical dimension of the delta function is one over the dimensions of its argument. Thus, if x is a distance, the physical dimension of $\delta(x)$ is given by $[\delta(x)] = 1/[x] = 1/L$, where L is a length. Similarly, the physical dimensions of $\delta(\vec{r})$ is $1/L^3$, since

$$[\delta(\vec{r})] = [\delta(x)\delta(y)\delta(z)] = \frac{1}{[x]}\frac{1}{[y]}\frac{1}{[z]} = \frac{1}{L^3}. \tag{A.37}$$

Appendix B

Angular Momentum in Spherical Coordinates

In this appendix, we will show how to derive the expressions of the gradient $\vec{\nabla}$, the Laplacian ∇^2, and the components of the orbital angular momentum in spherical coordinates.

B.1 Derivation of Some General Relations

The Cartesian coordinates (x, y, z) of a vector $\vec{r}$ are related to its spherical polar coordinates (r, θ, φ) by

$$x = r \sin\theta \cos\varphi, \qquad y = r \sin\theta \sin\varphi, \qquad z = r \cos\theta. \tag{B.1}$$

The orthonormal Cartesian basis $(\hat{x}, \hat{y}, \hat{z})$ is related to its spherical counterpart $(\hat{r}, \hat{\theta}, \hat{\varphi})$ by

$$\hat{x} = \hat{r} \sin\theta \cos\varphi + \hat{\theta} \cos\theta \cos\varphi - \hat{\varphi} \sin\varphi, \tag{B.2}$$

$$\hat{y} = \hat{r} \sin\theta \sin\varphi + \hat{\theta} \cos\theta \sin\varphi + \hat{\varphi} \cos\varphi, \tag{B.3}$$

$$\hat{z} = \hat{r} \cos\theta - \hat{\theta} \sin\theta. \tag{B.4}$$

Differentiating (B.1), we obtain

$$dx = \sin\theta \cos\varphi \, dr + r \cos\theta \cos\varphi \, d\theta - r \sin\theta \sin\varphi \, d\varphi, \tag{B.5}$$

$$dy = \sin\theta \sin\varphi \, dr + r \cos\theta \sin\varphi \, d\theta + r \sin\theta \cos\varphi \, d\varphi, \tag{B.6}$$

$$dz = \cos\theta \, dr - r \sin\theta \, d\theta. \tag{B.7}$$

Solving these equations for dr, $d\theta$, and $d\varphi$, we obtain

$$dr = \sin\theta \cos\varphi \, dx + \sin\theta \sin\varphi \, dy + \cos\theta \, dz, \tag{B.8}$$

$$d\theta = \frac{1}{r} \cos\theta \cos\varphi \, dx + \frac{1}{r} \cos\theta \sin\varphi \, dy - \frac{1}{r} \sin\theta \, dz, \tag{B.9}$$

$$d\varphi = -\frac{\sin\varphi}{r \sin\theta} \, dx + \frac{\cos\varphi}{r \sin\theta} \, dy. \tag{B.10}$$

We can verify that (B.5) to (B.10) lead to

$$\frac{\partial r}{\partial x} = \sin\theta\cos\varphi, \qquad \frac{\partial \theta}{\partial x} = \frac{1}{r}\cos\varphi\cos\theta, \qquad \frac{\partial \varphi}{\partial x} = -\frac{\sin\varphi}{r\sin\theta}, \qquad (B.11)$$

$$\frac{\partial r}{\partial y} = \sin\theta\sin\varphi, \qquad \frac{\partial \theta}{\partial y} = \frac{1}{r}\sin\varphi\cos\theta, \qquad \frac{\partial \varphi}{\partial y} = \frac{\cos\varphi}{r\sin\theta}, \qquad (B.12)$$

$$\frac{\partial r}{\partial z} = \cos\theta, \qquad \frac{\partial \theta}{\partial z} = -\frac{1}{r}\sin\theta, \qquad \frac{\partial \varphi}{\partial z} = 0, \qquad (B.13)$$

which, in turn, yield

$$\frac{\partial}{\partial x} = \frac{\partial}{\partial r}\frac{\partial r}{\partial x} + \frac{\partial}{\partial \theta}\frac{\partial \theta}{\partial x} + \frac{\partial}{\partial \varphi}\frac{\partial \varphi}{\partial x}$$

$$= \sin\theta\cos\varphi\frac{\partial}{\partial r} + \frac{1}{r}\cos\theta\cos\varphi\frac{\partial}{\partial \theta} - \frac{\sin\varphi}{r\sin\theta}\frac{\partial}{\partial \varphi}, \qquad (B.14)$$

$$\frac{\partial}{\partial y} = \frac{\partial}{\partial r}\frac{\partial r}{\partial y} + \frac{\partial}{\partial \theta}\frac{\partial \theta}{\partial y} + \frac{\partial}{\partial \varphi}\frac{\partial \varphi}{\partial y}$$

$$= \sin\theta\sin\varphi\frac{\partial}{\partial r} + \frac{1}{r}\cos\theta\sin\varphi\frac{\partial}{\partial \theta} + \frac{\cos\varphi}{r\sin\theta}\frac{\partial}{\partial \varphi}, \qquad (B.15)$$

$$\frac{\partial}{\partial z} = \frac{\partial}{\partial r}\frac{\partial r}{\partial z} + \frac{\partial}{\partial \theta}\frac{\partial \theta}{\partial z} + \frac{\partial}{\partial \varphi}\frac{\partial \varphi}{\partial z} = \cos\theta\frac{\partial}{\partial r} - \frac{\sin\theta}{r}\frac{\partial}{\partial \theta}. \qquad (B.16)$$

B.2   Gradient and Laplacian in Spherical Coordinates

We can show that a combination of (B.14) to (B.16) allows us to express the operator $\vec{\nabla}$ in spherical coordinates:

$$\vec{\nabla} = \hat{x}\frac{\partial}{\partial x} + \hat{y}\frac{\partial}{\partial y} + \hat{z}\frac{\partial}{\partial z} = \hat{r}\frac{\partial}{\partial r} + \hat{\theta}\frac{1}{r}\frac{\partial}{\partial \theta} + \hat{\varphi}\frac{1}{r\sin\theta}\frac{\partial}{\partial \varphi}, \qquad (B.17)$$

and also the Laplacian operator ∇^2:

$$\nabla^2 = \vec{\nabla}.\vec{\nabla} = \left(\hat{r}\frac{\partial}{\partial r} + \frac{\hat{\theta}}{r}\frac{\partial}{\partial \theta} + \frac{\hat{\varphi}}{r\sin\varphi}\frac{\partial}{\partial \varphi}\right).\left(\hat{r}\frac{\partial}{\partial r} + \frac{\hat{\theta}}{r}\frac{\partial}{\partial \theta} + \frac{\hat{\varphi}}{r\sin\theta}\frac{\partial}{\partial \varphi}\right). \qquad (B.18)$$

Now, using the relations

$$\frac{\partial \hat{r}}{\partial r} = 0, \qquad \frac{\partial \hat{\theta}}{\partial r} = 0, \qquad \frac{\partial \hat{\varphi}}{\partial r} = 0, \qquad (B.19)$$

$$\frac{\partial \hat{r}}{\partial \theta} = \hat{\theta}, \qquad \frac{\partial \hat{\theta}}{\partial \theta} = -\hat{r}, \qquad \frac{\partial \hat{\varphi}}{\partial \theta} = 0, \qquad (B.20)$$

$$\frac{\partial \hat{r}}{\partial \varphi} = \hat{\varphi}\sin\theta, \qquad \frac{\partial \hat{\theta}}{\partial \varphi} = \hat{\varphi}\cos\theta, \qquad \frac{\partial \hat{\varphi}}{\partial \varphi} = -\hat{r}\sin\theta - \hat{\theta}\cos\theta, \qquad (B.21)$$

we can show that the Laplacian operator reduces to

$$\nabla^2 = \frac{1}{r^2}\left[\frac{\partial}{\partial r}\left(r^2\frac{\partial}{\partial r}\right) + \frac{1}{\sin\theta}\frac{\partial}{\partial \theta}\left(\sin\theta\frac{\partial}{\partial \theta}\right) + \frac{1}{\sin^2\theta}\frac{\partial^2}{\partial \varphi^2}\right]. \qquad (B.22)$$

B.3 Angular Momentum in Spherical Coordinates

The orbital angular momentum operator $\vec{L}$ can be expressed in spherical coordinates as

$$\hat{\vec{L}} = \hat{\vec{R}} \times \hat{\vec{P}} = (-i\hbar r)\hat{r} \times \vec{\nabla} = (-i\hbar r)\hat{r} \times \left[\hat{r}\frac{\partial}{\partial r} + \frac{\hat{\theta}}{r}\frac{\partial}{\partial \theta} + \frac{\hat{\varphi}}{r\sin\theta}\frac{\partial}{\partial \varphi}\right], \tag{B.23}$$

or as

$$\hat{\vec{L}} = -i\hbar\left(\hat{\varphi}\frac{\partial}{\partial \theta} - \frac{\hat{\theta}}{\sin\theta}\frac{\partial}{\partial \varphi}\right). \tag{B.24}$$

Using (B.24) along with (B.2) to (B.4), we express the components $\hat{L}_x, \hat{L}_y, \hat{L}_z$ within the context of the spherical coordinates. For instance, the expression for $\hat{L}_x$ can be written as follows:

$$\hat{L}_x = \hat{x}.\vec{L} = -i\hbar\left(\hat{r}\sin\theta\cos\varphi + \hat{\theta}\cos\theta\cos\varphi - \hat{\varphi}\sin\varphi\right)\cdot\left(\hat{\varphi}\frac{\partial}{\partial\theta} - \frac{\hat{\theta}}{\sin\theta}\frac{\partial}{\partial\varphi}\right)$$

$$= i\hbar\left(\sin\varphi\frac{\partial}{\partial\theta} + \cot\theta\cos\varphi\frac{\partial}{\partial\varphi}\right). \tag{B.25}$$

Similarly, we can easily obtain

$$\hat{L}_y = i\hbar\left(-\cos\varphi\frac{\partial}{\partial\theta} + \cot\theta\sin\varphi\frac{\partial}{\partial\varphi}\right), \tag{B.26}$$

$$\hat{L}_z = -i\hbar\frac{\partial}{\partial\varphi}. \tag{B.27}$$

From the expressions (B.25) and (B.26) for $\hat{L}_x$ and $\hat{L}_y$, we infer that

$$\hat{L}_\pm = \hat{L}_x \pm i\hat{L}_y = \pm \hbar e^{\pm i\varphi}\left(\frac{\partial}{\partial\theta} \pm i\cot\theta\frac{\partial}{\partial\varphi}\right). \tag{B.28}$$

The expression for $\vec{L}^2$ is

$$\vec{L}^2 = -\hbar^2 r^2(\hat{r}\times\vec{\nabla})\cdot(\hat{r}\times\vec{\nabla}) = -\hbar^2 r^2\left[\nabla^2 - \frac{1}{r^2}\frac{\partial}{\partial r}\left(r^2\frac{\partial}{\partial r}\right)\right]; \tag{B.29}$$

it can be easily written in terms of the spherical coordinates as

$$\vec{L}^2 = -\hbar^2\left[\frac{1}{\sin\theta}\frac{\partial}{\partial\theta}\left(\sin\theta\frac{\partial}{\partial\theta}\right) + \frac{1}{\sin^2\theta}\frac{\partial^2}{\partial\varphi^2}\right]. \tag{B.30}$$

This expression was derived by substituting (B.22) into (B.29).

Note that, using the expression (B.29) for $\vec{L}^2$, we can rewrite ∇^2 as

$$\nabla^2 = \frac{1}{r^2}\frac{\partial}{\partial r}\left(r^2\frac{\partial}{\partial r}\right) - \frac{1}{\hbar^2 r^2}\vec{L}^2 = \frac{1}{r}\frac{\partial^2}{\partial r^2}r - \frac{1}{\hbar^2 r^2}\vec{L}^2. \tag{B.31}$$

Appendix C

C++ Code for Solving the Schrödinger Equation

This C++ code is designed to solve the one-dimensional Schrödinger equation for a harmonic oscillator (HO) potential as well as for an infinite square well (ISW) potential as outlined in Chapter 4. My special thanks are due to Dr. M. Bulut and to Prof. Dr. H. Mueller-Krumbhaar and his Ph.D. student C. Gugenberger who have worked selflessly hard to write and test the code listed below. Dr. Mevlut wrote an early code for the ISW, while Prof. Mueller-Krumbhaar and Gugenberger not only wrote a new code (see the version listed below) for the HO but also designed it in a way that it applies to the ISW potential as well (they have also added effective didactic comments so that our readers can effortlessly understand the code and make use of it).

Note: to shift from the harmonic oscillator code to the infinite square well code, one needs simply to erase the first double forward-slash (i.e., "//") from the oscillator's program line below:

```
E_pot[i] = 0.5*dist*dist; // E_pot[i]=0;//E_pot=0:Infinite Well!
```

Of course, one still needs to rescale the energy and the value of 'xRange' in order to agree with the algorithm outlined at the end of Chapter 4.

──────────────── **The C++ Code: osci.cpp** ────────────────

```
/* osci.cpp: Solution of the one-dimensional Schrodinger equation fo
   a particle in a harmonic potential, using the shooting method.
   To compile and link with gnu compiler, type: g++ -o osci osci.cpp
   To run the current C++ program, simply type: osci
   Plot by gnuplot: /GNUPLOT> set terminal windows
                    /GNUPLOT> plot "psi-osc.dat" with lines   */
#include <cstdio>
#include <cstdlib>
#include <cmath>
#define MAX(a, b) (((a) > (b)) ? (a) : (b))
int main(int argc, char*argv[])
{// Runtime constants
const static double Epsilon = 1e-10; // Defines the precision of
                //... energy calculations
```

```
const static int N_of_Divisions = 1000;
const static int N_max = 5; //Number of calculated Eigenstates

FILE *Wavefunction_file, *Energy_file, *Potential_file;
Wavefunction_file = fopen("psi-osc.dat","w");
Energy_file = fopen("E_n_Oszillator.dat","w");
Potential_file = fopen("HarmonicPotentialNoDim.dat", "w");
if (!(Wavefunction_file && Energy_file && Potential_file))
{ printf("Problems to create files output.\n"); exit(2); }

/* Physical parameters using dimensionless quantities.
 ATTENTION: We set initially:  hbar = m = omega = a = 1, and
 reintroduce physical values at the end.  According to Eq.(4.117),
 the ground state energy then is  E_n = 0.5. Since the wave function
 vanishes only at -infinity and  +infinity, we have to cut off the
 calculation somewhere, as given by  'xRange'. If xRange is chosen
 too large, the open (positive) end of  the  wave function can
 diverge numerically in this simple shooting  approach. */

const static double xRange = 12; // xRange=11.834 corresponds to a
  //... physical range of -20fm < x < +20fm, see after Eq.(4.199).
const static double h_0 = xRange / N_of_Divisions;
double* E_pot = new double[N_of_Divisions+1];
double dist;

for (int i = 0; i <= N_of_Divisions; ++i)
{ // Harmonic potential, as given in Eq. (4.115), but dimensionless
dist = i*h_0 - 0.5*xRange;
E_pot[i] = 0.5*dist*dist; // E_pot[i]=0;//E_pot=0:Infinite Well!
fprintf(Potential_file, "%16.12e \t\t %16.12e\n", dist, E_pot[i]);
}
fclose(Potential_file);

/* Since the Schrodinger equation is linear, the amplitude of the
 wavefunction will be fixed by normalization.
At left we set it small but nonzero. */
const static double Psi_left = 1.0e-3; // left boundary condition
const static double Psi_right = 0.0; // right boundary condition

double *Psi, *EigenEnergies;// Arrays to hold the results
Psi = new double[N_of_Divisions+1]; //N_of_Points = N_of_Divisions+
EigenEnergies = new double[N_max+1];
Psi[0] = Psi_left;
Psi[1] = Psi_left + 1.0e-3; // Add arbitrary small value

int N_quantum;//N_quantum is Energy Quantum Number
int Nodes_plus; // Number of nodes (+1) in wavefunction
```

```
double K_square;// Square of wave vector
// Initial Eigen-energy search limits
double E_lowerLimit = 0.0;// Eigen-energy must be positive
double E_upperLimit = 10.0;
int End_sign = -1;
bool Limits_are_defined = false;
double Normalization_coefficient;
double E_trial;


// MAIN LOOP begins:---------------------------------
for(N_quantum=1; N_quantum <= N_max; ++N_quantum)
{
// Find the eigen-values for energy. See theorems (4.1) and (4.2).
Limits_are_defined = false;
while (Limits_are_defined == false)
{ /* First, determine an upper limit for energy, so that the wave-
 function Psi[i] has one node more than physically needed.    */
Nodes_plus = 0;
E_trial = E_upperLimit;
for (int i=2; i <= N_of_Divisions; ++i)
{ K_square = 2.0*(E_trial - E_pot[i]);
// Now use the NUMEROV-equation (4.197) to calculate wavefunction
Psi[i] = 2.0*Psi[i-1]*(1.0 - (5.0*h_0*h_0*K_square / 12.0))
/(1.0 + (h_0*h_0*K_square/12.0))-Psi[i-2];
if (Psi[i]*Psi[i-1] < 0) ++Nodes_plus;
}
/* If one runs into the following condition, the modification
 of the upper limit was too aggressive. */
if (E_upperLimit < E_lowerLimit)
E_upperLimit = MAX(2*E_upperLimit, -2*E_upperLimit);
if (Nodes_plus > N_quantum) E_upperLimit *= 0.7;
else if (Nodes_plus < N_quantum) E_upperLimit *= 2.0;
else Limits_are_defined = true; // At least one node should appear.
} // End of the loop: while (Limits_are_defined == false)
// Refine the energy by satisfying the right boundary condition.
End_sign = -End_sign;
while ((E_upperLimit - E_lowerLimit) > Epsilon)
{ E_trial = (E_upperLimit + E_lowerLimit) / 2.0;
for (int i=2; i <= N_of_Divisions; ++i)
{ // Again eq.(4.197) of the Numerov-algorithm:
K_square = 2.0*(E_trial - E_pot[i]);
Psi[i] = 2.0*Psi[i-1] * (1.0 - (5.0*h_0*h_0*K_square / 12.0))
/(1.0 + (h_0*h_0*K_square/12.0))-Psi[i-2];
}
if (End_sign*Psi[N_of_Divisions] > Psi_right) E_lowerLimit = E_trial
else E_upperLimit = E_trial;
} // End of loop: while ((E_upperLimit - E_lowerLimit) > Epsilon)
```

```cpp
// Initialization for the next iteration in main loop
E_trial = (E_upperLimit+E_lowerLimit)/2;
EigenEnergies[N_quantum] = E_trial;
E_upperLimit = E_trial;
E_lowerLimit = E_trial;

// Now find the normalization coefficient
double Integral = 0.0;
for (int i=1; i <= N_of_Divisions; ++i)
{ // Simple integration
Integral += 0.5*h_0*(Psi[i-1]*Psi[i-1]+Psi[i]*Psi[i]);
}
Normalization_coefficient = sqrt(1.0/Integral);
// Output of normalized dimensionless wave function
for (int i=0; i <=N_of_Divisions; ++i)
{ fprintf(Wavefunction_file, "%16.12e \t\t %16.12e\n",
 i*h_0 - 0.5*xRange, Normalization_coefficient*Psi[i]);
}
fprintf(Wavefunction_file,"\n");
} // End of MAIN LOOP. --------------------------------
fclose(Wavefunction_file);

/*Finally convert dimensionless units in real units. Note that
  energy does not depend explicitly on the particle's mass anymore:
    hbar = 1.05457e-34;// Planck constant/2pi
    omega = 5.34e21; // Frequency in 1/s
    MeV = 1.602176487e-13; // in J
    The correct normalization would be hbar*omega/MeV = 3.514846114
    but we use the approximation 3.5 for energy-scale as in chap. 4
const static double Energyscale = 3.5;// in MeV
// Output with rescaled dimensions; assign Energy_file
printf("Quantum Harmonic Oscillator, program osci.cpp\n");
printf("Energies in MeV:\n");
printf("n \t\t E_n\n");
for (N_quantum=1; N_quantum <= N_max; ++N_quantum)
{ fprintf(Energy_file,"%d \t\t %16.12e\n", N_quantum-1,
  Energyscale*EigenEnergies[N_quantum]);
  printf("%d \t\t %16.12e\n", N_quantum-1,
  Energyscale*EigenEnergies[N_quantum]);
}
fprintf(Energy_file,"\n");
fclose(Energy_file);
printf("Wave-Functions in File: psi_osc.dat \n");
printf("\n");
return 0;
}
```

Index

Physical Constants

Quantity	Symbol, equation	Value
Speed of light	c	$2.9979 \times 10^8\,\mathrm{m\,s^{-1}}$
Electron charge	e	$1.602 \times 10^{-19}\,\mathrm{C}$
Planck constant	h	$6.626 \times 10^{-34}\,\mathrm{J\,s}$
Planck constant, reduced	$\hbar = h/2\pi$	$1.055 \times 10^{-34}\,\mathrm{J\,s}$
Conversion constant	$\hbar c$	$197.327\,\mathrm{MeV\,fm} = 197.327\,\mathrm{eV\,nm}$
Electron mass	m_e	$9.109 \times 10^{-31}\,\mathrm{kg} = 0.511\,\mathrm{MeV/c^2}$
Proton mass	m_p	$1.673 \times 10^{-27}\,\mathrm{kg} = 938.272\,\mathrm{MeV/c^2}$
Neutron mass	m_n	$1.675 \times 10^{-27}\,\mathrm{kg} = 939.566\,\mathrm{MeV/c^2}$
Fine structure constant	$\alpha = e^2/\hbar c$	$1/137.036$
Classical electron radius	$r_e = e^2/m_e c^2$	$2.818 \times 10^{-15}\,\mathrm{m}$
Electron Compton wavelength	$\lambda = h/m_e c = r_e/\alpha$	$2.426 \times 10^{-12}\,\mathrm{m}$
Proton Compton wavelength	$\lambda = h/m_p c$	$1.321 \times 10^{-15}\,\mathrm{m}$
Bohr radius	$a_0 = r_e/\alpha^2$	$0.529 \times 10^{-10}\,\mathrm{m}$
Rydberg energy	$\mathcal{R} = m_e c^2 \alpha^2/2$	$13.606\,\mathrm{eV}$
Bohr magneton	$\mu_B = e\hbar/2m_e$	$5.788 \times 10^{-11}\,\mathrm{MeV\,T^{-1}}$
Nuclear magneton	$\mu_N = e\hbar/2m_p$	$3.152 \times 10^{-14}\,\mathrm{MeV\,T^{-1}}$
Avogadro number	N_A	$6.022 \times 10^{23}\,\mathrm{mol^{-1}}$
Boltzmann constant	k	$1.381 \times 10^{-23}\,\mathrm{J\,K^{-1}}$
		$= 8.617 \times 10^{-5}\,\mathrm{eV\,K^{-1}}$
Gas constant	$R = N_A k$	$8.31\,\mathrm{J\,mol^{-1}\,K^{-1}}$
Gravitational constant	G	$6.673 \times 10^{-11}\,\mathrm{m^3\,kg^{-1}\,s^{-2}}$
Permittivity of free space	$\epsilon_0 = 1/\mu_0 c^2$	$8.854 \times 10^{-12}\,\mathrm{F\,m^{-1}}$
Permeability of free space	μ_0	$4\pi \times 10^{-7}\,\mathrm{N\,A^{-2}}$

Conversion of units

$1\,\mathrm{fm} = 10^{-15}\,\mathrm{m}$, $1\,\mathrm{barn} = 10^{-28}\,\mathrm{m^2} = 100\,\mathrm{fm^2}$, $1\,\mathrm{G} = 10^{-4}\,\mathrm{T}$
$1\,\mathrm{atmosphere} = 101\,325\,\mathrm{Pa}$, Thermal energy at $T = 300\,\mathrm{K}$: $kT = [38.682]^{-1}\,\mathrm{eV}$
$0\,^\circ\mathrm{C} = 273.15\,\mathrm{K}$, $1\,\mathrm{eV} = 1.602 \times 10^{-19}\,\mathrm{J}$, $1\,\mathrm{eV/c^2} = 1.783 \times 10^{-36}\,\mathrm{kg}$

Essential Relations

Bohr model: $a_0 = \dfrac{\hbar^2}{m_e e^2}, \quad r_n = n^2 a_0, \quad v_n = \dfrac{\alpha}{n}c = \dfrac{1}{137n}c, \quad E_n = -\dfrac{e^2}{2a_0 n^2}$

General relations:

$$e^{\hat{A}}e^{\hat{B}} = e^{\hat{A}+\hat{B}}e^{[\hat{A},\ \hat{B}]/2}$$

$$e^{\hat{A}}\hat{B}e^{-\hat{A}} = \hat{B} + [\hat{A},\ \hat{B}] + \frac{1}{2!}[\hat{A},\ [\hat{A},\ \hat{B}]] + \frac{1}{3!}[\hat{A},\ [\hat{A},\ [\hat{A},\ \hat{B}]]] + \cdots$$

Generalized uncertainty principle: $\Delta A \Delta B \geq \dfrac{1}{2}\left|\langle[\hat{A},\ \hat{B}]\rangle\right|$, where $\Delta A = \sqrt{\langle\hat{A}^2\rangle - \langle\hat{A}\rangle^2}$

Canonical commutator: $[\hat{x},\ \hat{p}] = i\hbar$

Heisenberg uncertainty principle: $\Delta x \Delta p \geq \dfrac{\hbar}{2}, \qquad \Delta E \Delta t \geq \dfrac{\hbar}{2}$

Measurement probability: $\hat{A}|\psi_n\rangle = a_n|\psi_n\rangle, \quad P_n(a_n) = \dfrac{|\langle\psi_n|\psi\rangle|^2}{\langle\psi|\psi\rangle}$

Expectation value: $\langle\hat{A}\rangle = \dfrac{\langle\psi|\hat{A}|\psi\rangle}{\langle\psi|\psi\rangle} = \displaystyle\sum_n a_n P_n(a_n)$

Time evolution of expectation values: $\dfrac{d}{dt}\langle\hat{A}\rangle = \dfrac{1}{i\hbar}\langle[\hat{A},\ \hat{H}]\rangle + \langle\dfrac{\partial\hat{A}}{\partial t}\rangle$

Commutators and Poisson brackets: $\dfrac{1}{i\hbar}[\hat{A},\ \hat{B}] \longrightarrow \{A,\ B\}_{classical}$

Time-dependent Schrödinger equation: $i\hbar\dfrac{\partial|\Psi(t)\rangle}{\partial t} = \hat{H}|\Psi(t)\rangle$

Probability density: $\rho(\vec{r}, t) = \Psi^*(\vec{r}, t)\Psi(\vec{r}, t)$

Probability current density: $\vec{J}(\vec{r}, t) = \dfrac{i\hbar}{2m}(\Psi\vec{\nabla}\Psi^* - \Psi^*\vec{\nabla}\Psi)$

Conservation of probability: $\dfrac{\partial\rho(\vec{r}, t)}{\partial t} + \vec{\nabla}\cdot\vec{J} = 0$

Angular momentum:

$$[\hat{J}_x, \hat{J}_y] = i\hbar\hat{J}_z, \qquad [\hat{J}_y, \hat{J}_z] = i\hbar\hat{J}_x, \qquad [\hat{J}_z, \hat{J}_x] = i\hbar\hat{J}_y$$

$$\hat{J}^2\,|\,j,\ m\rangle = \hbar^2 j(j+1)\,|\,j,\ m\rangle, \qquad \hat{J}_z\,|\,j,\ m\rangle = \hbar m\,|\,j,\ m\rangle$$

$$\hat{J}_\pm\,|\,j,\ m\rangle = \hbar\sqrt{j(j+1) - m(m\pm 1)}\,|\,j,\ m\pm 1\rangle$$

$$\langle j,\ m\,|\,\hat{J}_x^2\,|\,j,\ m\rangle = \langle j,\ m\,|\,\hat{J}_y^2\,|\,j,\ m\rangle = \frac{\hbar^2}{2}\left[j(j+1) - m^2\right]$$

For $j = \frac{1}{2}$: $J_k = \frac{\hbar}{2}\sigma_k$ $(k = x, y, z)$, where σ_x, σ_y, and σ_z are the Pauli matrices:

$$\sigma_x = \begin{pmatrix} 0 & 1 \\ 1 & 0 \end{pmatrix}, \qquad \sigma_y = \begin{pmatrix} 0 & -i \\ i & 0 \end{pmatrix}, \qquad \sigma_z = \begin{pmatrix} 1 & 0 \\ 0 & -1 \end{pmatrix}$$

For $j = 1$: the matrices of J_x, J_y, and J_z are

$$\hat{J}_x = \frac{\hbar}{\sqrt{2}}\begin{pmatrix} 0 & 1 & 0 \\ 1 & 0 & 1 \\ 0 & 1 & 0 \end{pmatrix}, \quad \hat{J}_y = \frac{\hbar}{\sqrt{2}}\begin{pmatrix} 0 & -i & 0 \\ i & 0 & -i \\ 0 & i & 0 \end{pmatrix}, \quad \hat{J}_z = \hbar\begin{pmatrix} 1 & 0 & 0 \\ 0 & 0 & 0 \\ 0 & 0 & -1 \end{pmatrix}$$

Time-independent potentials: $|\Psi(t)\rangle = |\psi(x)\rangle \exp(-iEt/\hbar)$

Time-independent Schrödinger equation: $-\dfrac{\hbar^2}{2m}\nabla^2\psi(x) + V(x)\psi(x) = E\psi(x)$

Infinite square well: $E_n = -\dfrac{\hbar^2\pi^2}{2ma^2}n^2$, $\qquad \psi_n(x) = \sqrt{\dfrac{2}{a}}\sin\left(\dfrac{n\pi}{a}x\right) \qquad (n = 1, 2, 3, \ldots)$

Harmonic oscillator:

$$\hat{H} = \frac{\hat{P}^2}{2m} + \frac{1}{2}m\omega^2\hat{X}^2 = \hbar\omega\left(\hat{a}^\dagger\hat{a} + \frac{1}{2}\right), \qquad E_n = \hbar\omega\left(n + \frac{1}{2}\right)$$

$$\hat{a}|n\rangle = \sqrt{n}|n-1\rangle, \qquad \hat{a}^\dagger|n\rangle = \sqrt{n+1}|n+1\rangle, \qquad [\hat{a}, \hat{a}^\dagger] = 1$$

$$\langle n \mid \hat{X}^2 \mid n\rangle = \frac{\hbar}{2m\omega}(2n+1), \qquad \langle n \mid \hat{P}^2 \mid n\rangle = \frac{m\hbar\omega}{2}(2n+1)$$

Hydrogen atom: radial equation and averages: $-\dfrac{\hbar^2}{2\mu}\dfrac{d^2U(r)}{dr^2} + \left[\dfrac{l(l+1)\hbar^2}{2\mu r^2} - \dfrac{e^2}{r}\right]U(r) = EU(r)$

$$\langle nl| r |nl\rangle = \frac{1}{2}[3n^2 - l(l+1)]a_0, \qquad \langle nl| r^2 \mid nl\rangle = \frac{1}{2}n^2[5n^2 + 1 - 3l(l+1)]a_0^2$$

$$\langle nl \mid r^{-1} \mid nl\rangle = \frac{1}{n^2 a_0}, \qquad\qquad \langle nl|r^{-2}|nl\rangle = \frac{2}{n^3(2l+1)a_0^2}$$

Time-independent perturbation theory:

$$(\hat{H}_0 + \hat{H}_p)\mid\psi_n\rangle = E_n\mid\psi_n\rangle, \qquad \hat{H}_0\mid\phi_n\rangle = E_n^{(0)}\mid\phi_n\rangle \qquad (\hat{H}_p \ll \hat{H}_0)$$

$$E_n = E_0 + \langle\phi_n\mid\hat{H}_p\mid\phi_n\rangle + \sum_{m\neq n}\frac{\left|\langle\phi_m|\hat{H}_p\mid\phi_n\rangle\right|^2}{E_n^{(0)} - E_m^{(0)}} + \cdots$$

$$\mid\psi_n\rangle = \mid\phi_n\rangle + \sum_{m\neq n}\frac{\langle\phi_m\mid\hat{H}_p\mid\phi_n\rangle}{E_n^{(0)} - E_m^{(0)}}\mid\phi_m\rangle + \cdots$$

Quantization condition: $\oint p(x, E_n)\,dx = 2\int_{x_1}^{x_2}\sqrt{2m(E_n - V(x))}\,dx = (n + \frac{1}{2})h$

Time-dependent potentials:

Heisenberg and interaction pictures:

$$\mid\psi(t)\rangle_H = e^{it\hat{H}/\hbar}\mid\psi(t)\rangle, \quad \hat{A}_H(t) = e^{it\hat{H}/\hbar}\hat{A}e^{-it\hat{H}/\hbar}, \quad \frac{d\hat{A}_H}{dt} = \frac{1}{i\hbar}[\hat{A}_H, \hat{H}]$$

$$\mid\psi(t)\rangle_I = e^{i\hat{H}_0 t/\hbar}\mid\psi(t)\rangle, \quad \hat{V}_I(t) = e^{i\hat{H}_0 t/\hbar}\hat{V}e^{-i\hat{H}_0 t/\hbar}, \quad i\hbar\frac{d\mid\psi(t)\rangle_I}{dt} = \hat{V}_I(t)\mid\psi(t)\rangle_I$$

Time-dependent perturbation theory: $P_{if}(t) = \left|-\dfrac{i}{\hbar}\displaystyle\int_0^t \langle\psi_f\mid\hat{V}(t')\mid\psi_i\rangle e^{i\omega_{fi}t'}\,dt'\right|^2$

Intensity of radiation emitted: $I_{i\rightarrow f} = \hbar\omega W_{i\rightarrow f}^{emi} = \dfrac{4}{3}\dfrac{\omega^4}{c^3}|\vec{d}_{fi}|^2 = \dfrac{4}{3}\dfrac{\omega^4 e^2}{c^3}\left|\langle\psi_f\mid\vec{r}\mid\psi_i\rangle\right|^2$

Scattering:

Differential cross section (Born approximation): $\dfrac{d\sigma}{d\Omega} = |f(\theta,\varphi)|^2 = \dfrac{\mu^2}{4\pi^2\hbar^4}\left|\int e^{i\vec{q}\cdot\vec{r}'}V(\vec{r}')\,d^3r'\right|^2$

Partial wave analysis: $f(\theta) = \displaystyle\sum_{l=0}^{\infty} f_l(\theta) = \frac{1}{k}\sum_{l=0}^{\infty}(2l+1)e^{i\delta_l}\sin\delta_l P_l(\cos\theta)$

Printed in the United States
By Bookmasters